Veterinary Assisting

Fundamentals & Applications

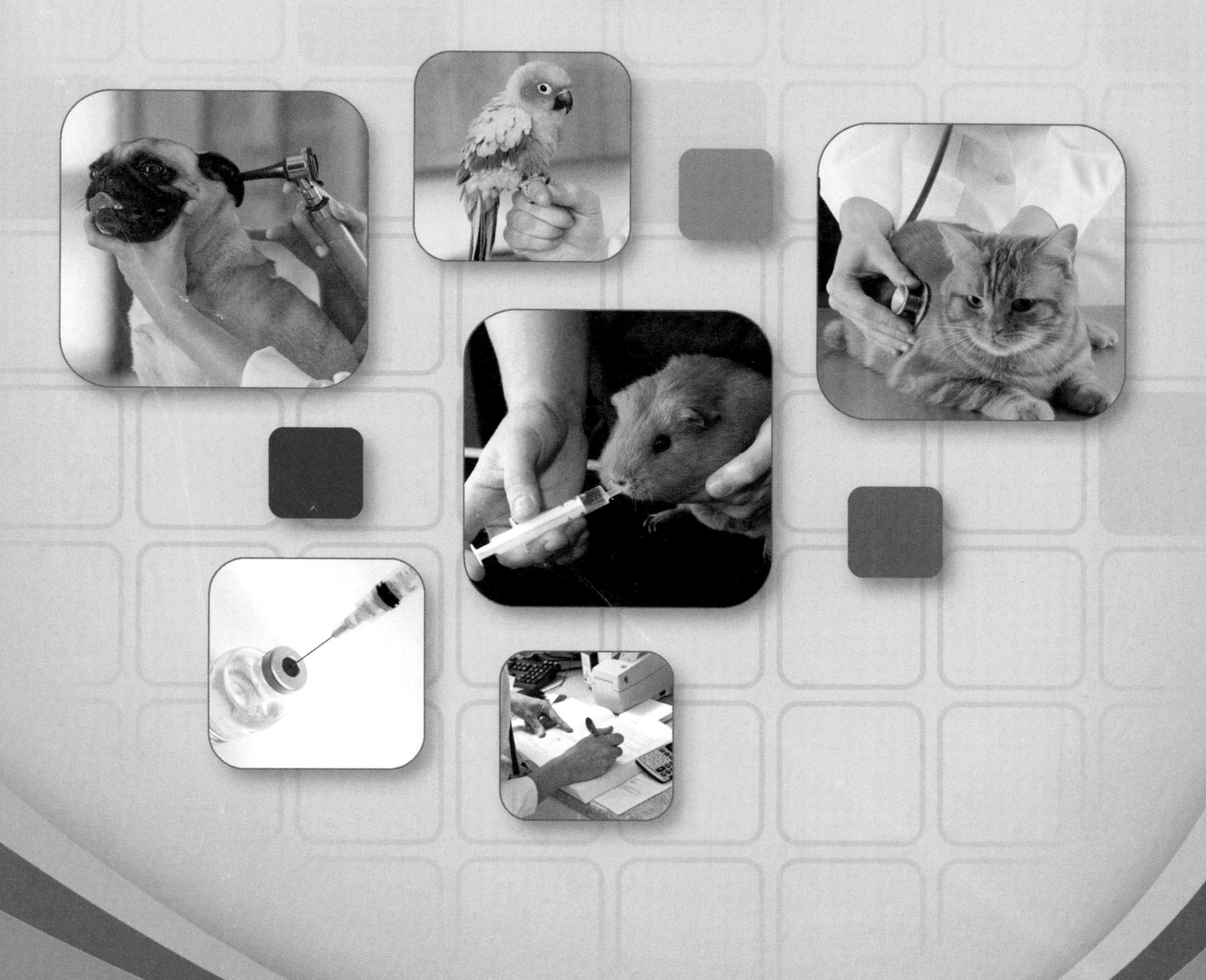

Veterinary Assisting

Fundamentals & Applications

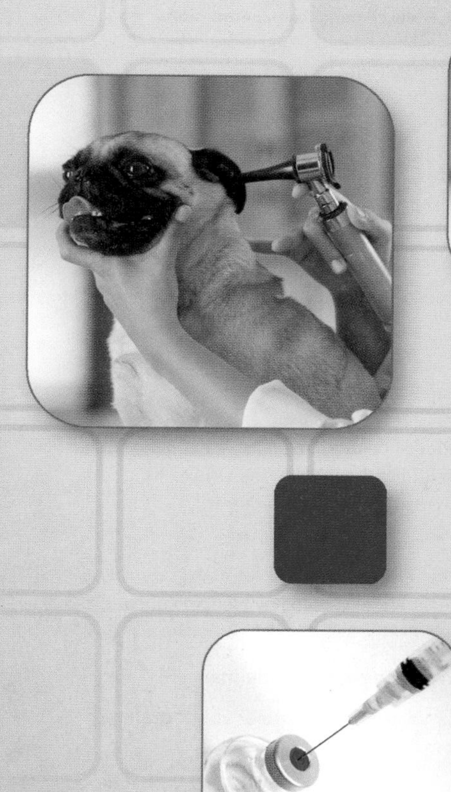

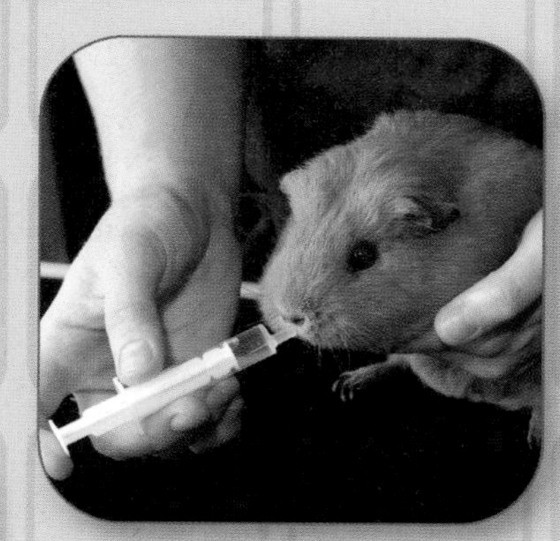

Beth Vanhorn, CVT, AS

Robert W. Clark, PhD

DELMAR
CENGAGE Learning™

Australia • Brazil • Japan • Korea • Mexico • Singapore • Spain • United Kingdom • United States

Veterinary Assisting: Fundamentals & Applications
Beth Vanhorn and Robert W. Clark

Vice President, Career and Professional Editorial: Dave Garza

Director of Learning Solutions: Matthew Kane

Acquisitions Editor: Benjamin Penner

Managing Editor: Marah Bellegarde

Senior Product Manager: Darcy M. Scelsi

Editorial Assistant: Scott Royael

Vice President, Career and Professional Marketing: Jennifer Ann Baker

Marketing Manager: Erin Brennan

Marketing Coordinator: Erin DeAngelo

Production Director: Carolyn Miller

Production Manager: Andrew Crouth

Content Project Manager: Allyson Bozeth

Senior Art Director: David Arsenault

Design Image Credits

Cover and Half-Title and Title Pages:
Microscope: Rodolfo Clix/stock.xchng
Syringe: Brian Hoskins/stock.xchng
Bird: © Corbis
Cat: © Comstock Images/Getty Images
Dog: © Ariel Skelley/Blend Images/Getty Images
Guinea pig: © iStockphoto/Mark Penny
RefBook: © iStockphoto/DenGuy

Library of Congress Control Number: 2009941808

ISBN-13: 978-1-4354-5387-6

ISBN-10: 1-4354-5387-5

Delmar
5 Maxwell Drive
Clifton Park, NY 12065-2919
USA

Cengage Learning is a leading provider of customized learning solutions with office locations around the globe, including Singapore, the United Kingdom, Australia, Mexico, Brazil, and Japan. Locate your local office at:
international.cengage.com/region

Cengage Learning products are represented in Canada by Nelson Education, Ltd.

To learn more about Delmar, visit **www.cengage.com/delmar**

Purchase any of our products at your local college store or at our preferred online store **www.cengagebrain.com**

Notice to the Reader

Publisher does not warrant or guarantee any of the products described herein or perform any independent analysis in connection with any of the product information contained herein. Publisher does not assume, and expressly disclaims, any obligation to obtain and include information other than that provided to it by the manufacturer. The reader is expressly warned to consider and adopt all safety precautions that might be indicated by the activities described herein and to avoid all potential hazards. By following the instructions contained herein, the reader willingly assumes all risks in connection with such instructions. The publisher makes no representations or warranties of any kind, including but not limited to, the warranties of fitness for particular purpose or merchantability, nor are any such representations implied with respect to the material set forth herein, and the publisher takes no responsibility with respect to such material. The publisher shall not be liable for any special, consequential, or exemplary damages resulting, in whole or part, from the readers' use of, or reliance upon, this material.

Printed in the United States of America
5 14

This book is dedicated to
all the veterinarians and staff members
who made working in and learning about
the veterinary industry worthwhile.

Contents

Section I ■ Practice Management and Client Relations

Preface

The popularity and demand nationwide of veterinarians and trained staff have caused the industry to reach out to the high school career and technical schools to allow students the experience and understanding of the veterinary field. Thus, Ms. Vanhorn began creating a high school–based veterinary assistant program to allow students the experience of and a basic introduction to the world of veterinary medicine. On developing the program, Vanhorn found that the majority of veterinary-related educational materials and textbooks were written at a college-to-postsecondary level. In an effort to allow secondary students to understand and learn about veterinary assisting as a career opportunity, this textbook was created for veterinary assistant students and as a resource for practicing veterinary assistants, veterinary receptionists, hospital managers, and technicians.

The authors believe that the basic layout of the units, as well as the clinical experiences, offers students a way to see veterinary medicine. This approach provides the learner with a preview of the veterinary industry and the necessary education needed to achieve success in the profession, as well as allowing the student to critically think and problem-solve realistic scenarios.

Supplements

To help you learn and teach from ***Veterinary Assisting: Fundamentals & Applications***, a variety of additional materials have been prepared for you.

Anticipation/reaction guide

This assignment is completed at the beginning of a lesson and helps gauge student prior knowledge and assumptions. It can then aid the instructor by identifying areas that may need more class instruction time. Students are then asked to complete the assignment again after they have learned the content in class. This approach allows students to assess what they have learned and where they can benefit by honing in on weak areas that may need more study and review.

Content map

This is a tool to help students focus their attention and outline a study plan. It highlights the key concepts in each chapter, uses questions to reinforce the learning of the key concepts, and lists vocabulary terms relevant to the key concepts. This tool may also help instructors organize and plan their classroom discussions and activities.

Study guide

This is a tool for students to help them study chapter contents. Objectives, Key Terms, and a summary of the chapter are provided for review. Additional activities, checklists, resources, and Web links also help students learn the content.

Assignment sheet

This is a tool that lists assignments that can be completed outside of class. Instructors may assign these projects for grades or as extra credit.

Presentations on PowerPoint

These presentations can be used to study and review material. Presentations have been developed for each chapter and correlate to the chapter content.

Acquisition lesson planning form

This is a tool for instructors to help in planning classroom discussion, activities, and assignments. It provides suggested activities and assignments.

Computerized test banks

These computerized test banks contain a wealth of questions related to each chapter. A variety of question styles are provided: multiple choice, true/false, completion, matching, and short answer. Instructors can customize and create quizzes or final examinations. Online quizzing can also be implemented.

Evaluation sheet

This is a tool for instructors to aid in grading skills and competencies in each chapter.

All supplemental material can be accessed online.

Student Instructions for accessing online materials:

1. Go to: **http://www.cengagebrain.com**
2. type author, title, or ISBN in the **Search** window
3. Locate the desired product and click on the title
4. When you arrive at the Product Page, click on the **Free Stuff** tab
5. Use the **Click Here** link to be brought to the Companion site
   - Note: You will only see the Click Here link if there is a companion product available
6. click on the **Student Resources** link in the left navigation pane to access the resources

Instructor Instructions for accessing online materials:

1. Go to: **http://www.cengagebrain.com**
2. Type author, title, or ISBN in the **Search** window
3. Locate the desired product and click on the title
4. When you arrive at the Product Page, click on the **Free Stuff** tab
5. Use the **Click Here** link to be brought to the Companion site
   - Note: You will only see the Click Here link if there is a companion product available
6. At the Companion Product Page, click on the **Instructor Resources** link
7. Type the username and password, and then click the **Login** button

About the Authors

Beth Vanhorn, CVT, AS, currently works as a certified veterinary technician in educational programs training both high school students and adults as veterinary assistants. In addition, Ms. Vanhorn works in the veterinary industry as a CVT in small, large, and exotic animal practices. She earned her AS degree in veterinary medical technology from Wilson College. She is a past member of the Pennsylvania State Board of Veterinary Medicine and was instrumental in developing veterinary assistant programs within high school programs, working with both the Pennsylvania Veterinary Medical Association (PVMA) and the Pennsylvania Department of Agriculture. Vanhorn shows and breeds Quarter Horses.

Robert W. Clark, PhD, currently works as an Associate Professor of Workforce Education and Development in the College of Education at Penn State University. Dr. Clark is responsible for the Vocational Director certification program at Penn State. He teaches courses in career and technical education at the graduate and undergraduate levels.

Dr. Clark holds a BS and MEd in Agricultural Education from North Carolina State University in Raleigh, North Carolina. Dr. Clark received his PhD from Penn State in 1993 in Agricultural Education with an emphasis in career and technical education and educational leadership. Dr. Clark's background includes teaching agricultural education at the secondary level and serving as an administrator in a comprehensive career and technical center.

Acknowledgments

I would like to thank all of my students and staff members at Dauphin County Technical School for their help and support in providing photographs and case material. The hands-on approach to career and technical school training programs made for an invaluable number of photographs.

I would like to also thank the Delmar Cengage Learning Team for all of their support and help through the long process of learning to write a textbook and for making all of this work possible.

Reviewers

Gail A. Chesnut, MAg
Leto High School
Tampa, Florida

Anne Duffy, RVT
Kirkwood Community College
Cedar Rapids, Iowa

Joseph A. Durso, BS, RVT
Norwalk Community College
Norwalk, Connecticut

Shannon Ramsey, AAS, RVT
Veterinary Staffing Solutions
Woodlands, Texas

Karen G. Stanford, RVT
The University of Texas at Arlington
Arlington, Texas

Michelle Stephens, BS
Hart County Board of Education
Hartwell, Georgia

Frances Turner, RVT
McLennan Community College
Waco, Texas

Section I

Practice Management and Client Relations

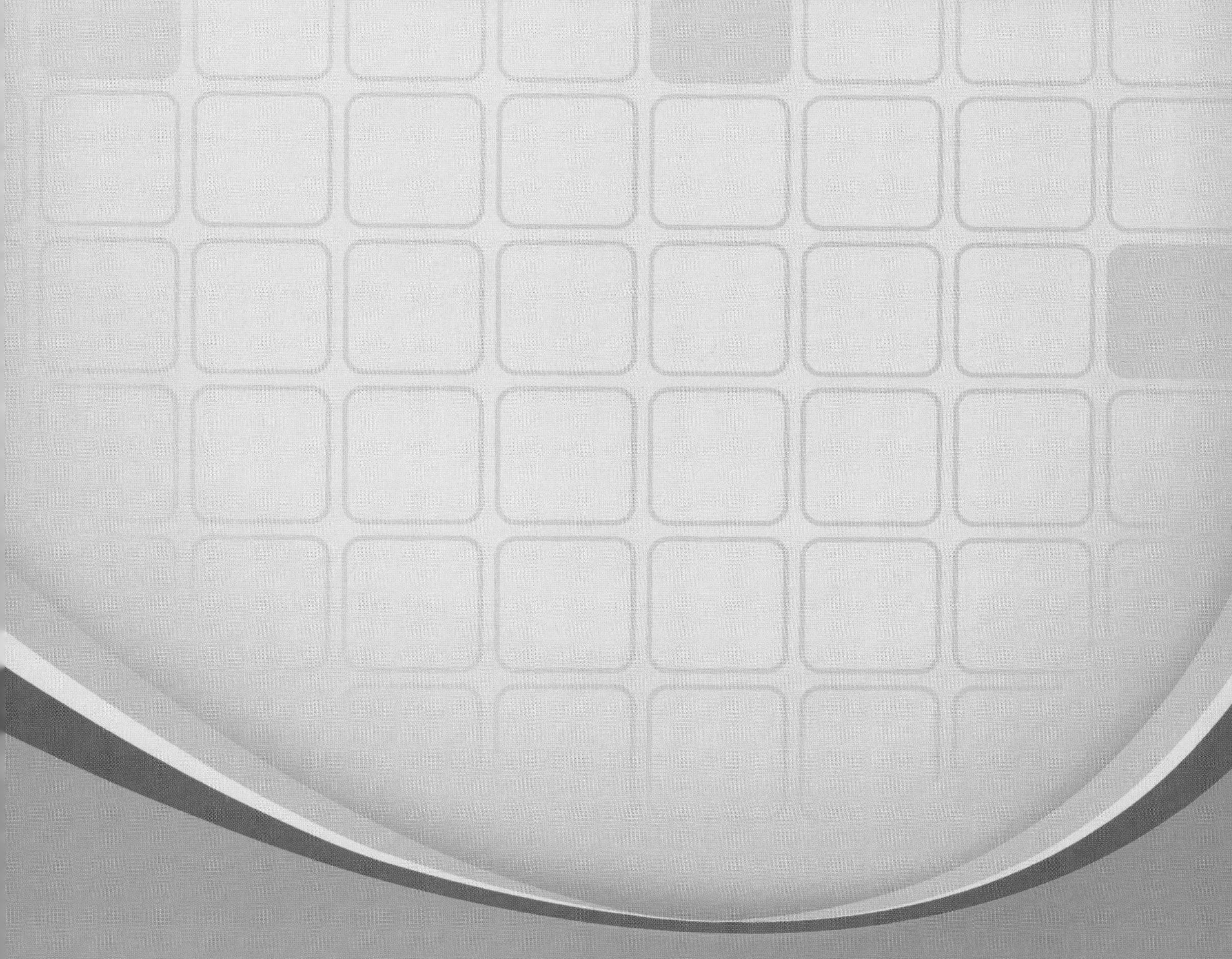

1

Delmar/Cengage Learning

Veterinary Medical Terminology and Abbreviations

Objectives

Upon completion of this chapter, the reader should be able to:

- Analyze veterinary medical terms to determine the prefix, suffix, and root word
- Identify the meaning of common prefixes
- Identify the meaning of common suffixes
- Identify the meaning of common root words
- Define common veterinary medical terms related to direction, species, patient history, and pharmacy
- Identify and use common veterinary medical abbreviations

Introduction

The field of veterinary medicine has adopted a common language used by all veterinary professionals working at all levels of the industry. A basic knowledge and foundation of veterinary medical terminology will help the veterinary assistant communicate with other veterinary professionals, understand treatments and procedures within the veterinary facility, and allow the veterinary assistant to properly enter and record medical records in patient charts. This chapter will focus on beginning to learn the basics of how to pronounce a word, how to break down and interpret word parts, and how these words apply to animals and areas of veterinary medical practice.

Veterinary Medical Terminology

Learning new vocabulary terms is a part of learning any new language, and veterinary medical terminology is the language used in the field of veterinary medicine. Some of the terms may be familiar; however, many may be foreign. Veterinary terms are often based on Latin and Greek words. Learning how to break down a term to define the meaning is the key to learning this new language. Some words may seem long and complex, but working with each part of the word will create a structure for learning terminology. Veterinary assistants should also have a veterinary dictionary to verify spelling and definitions of new words that may be encountered.

FIGURE 1-1 These veterinary assistants are using flashcards together to practice terminology.

Learning to Dissect a Veterinary Term

Words are made of prefixes, root words, and suffixes. Learning each of these parts and learning to **dissect** or separate them to define their meanings will make learning veterinary terminology much easier. The **prefix** is the word part at the beginning of a term. The **root word** is the part of the word that gives the term its essential meaning. A term may consist of more than one root word. The **suffix** is the part of the word at the end of the term. Some words have a **combining vowel** that is attached to a root word and allows for certain terms to be pronounced more easily. The combining vowel is commonly the letter *o*.

> **TERMINOLOGY TIP**
>
> Create flash cards to learn new terms. Use 3" × 5" or larger index cards and write the term, root word, prefix, or suffix on one side and a definition on the other side. Practice new terms 5–10 minutes each day (see Figure 1-1).

Common Root Words

A root word gives a term its essential meaning. Terms may have one or more root words. Some root words are formed with a combining vowel, usually the letter *o* that is placed at the end of the root word to allow the term to be more easily pronounced. Table 1-1 contains commonly used root words that veterinary assistants should be familiar with. Again, it is important to reference a veterinary dictionary when encountering new root words.

TABLE 1-1

Common Root Words

ROOT	MEANING
ARTHR/O-	joint
CARDI/O-	heart
CHEM/O-	chemical
COL/O-	colon
CUTANE/O-	skin
CYST/O-	urinary bladder
DENT/O-	teeth
ELECTR/O-	electricity
ENTER/O-	intestines
GASTR/O-	stomach
HEM/O-	blood
HEPAT/O-	liver
HYSTER/O-	uterus
LAPAR/O-	abdomen
MAST-	mammary gland
NAS/O-	nose
NEPHR/O-	kidneys
OSTE/O-	bone
OVARI/O-	ovary
RADI/O-	radiation
RECT/O-	rectum
RHIN/O-	nose
URIN/O-	urine
UTER/O-	uterus

Common Prefixes

The meaning of a prefix never changes; however, changing the prefix on a root word will change the meaning of the word. Prefixes are added to the beginning of words to qualify as follows:

- Numbers: bi- = two, tri- = three, quadra- = four
- Measurements: hyper- = excessive, hypo- = less than normal
- Position and/or direction: sub- = under, supra- = above, peri- = around
- Negatives: an- = without, anti- = against
- Color: cyan/o = blue, erythr/o = red, jaund/o = yellow

Example

Cardi/o is the root word meaning heart.

The suffix *–ia* means condition.

The prefix *brady* means abnormally slow; therefore, *bradycardia* means the condition of an abnormally slow heartbeat.

The prefix *tachy* means abnormally fast; therefore, *tachycardia* means the condition of an abnormally fast heartbeat.

Many prefixes have opposites that can be used to learn their meanings.

Example

ab- = away from, as in *abduction*

ad- = toward, as in *adduction*

pre- = before, as in *preoperative*

post- = after, as in *postoperative*

Table 1-2 lists prefixes commonly used in veterinary medicine. When encountering an unfamiliar prefix, it is important to use a dictionary to determine the meaning. Using the dictionary to look up new vocabulary is a good habit to acquire.

Common Suffixes

Suffixes are located at the end of the word and commonly define surgical and medical procedures and conditions. Several suffixes have the same meaning and are called the "pertaining to" suffixes (see Table 1-3). To define these terms, look at the root word for the essential meaning. Table 1-4 lists suffixes commonly used in veterinary medicine. Locate the meaning of new suffixes in a dictionary.

Combining Forms

The combining form includes the combining vowel, which is usually the letter *o*, but may also be such vowels as *a*, *e*, *i*, and *u*. A combining vowel is placed between the prefix and suffix in certain words to allow the word to

TABLE 1-2

Common Prefixes

PREFIX	MEANING	EXAMPLE	DEFINITION
A- or AN-	without; no	anemia	without or no blood cell production
AB-	away from	abduction	away from the center of the body
AD-	toward	adduction	toward the center of the body
ANTI-	against; to stop	anticoagulant	medicine used to stop bleeding
DYS-	difficult; painful	dysuria	difficult or painful urination
ECTO-	outside	ectoskeleton	bones located outside the body
ENDO-	within, inside	endothermic	body temperature controlled within the body
EXO-	outside	exothermic	body temperature controlled outside of the body
HYPER-	above normal	hyperglycemia	high blood sugar
HYPO-	below normal	hypoglycemia	low blood sugar
INTER-	between	interdigital	between the toes
INTRA-	within	intramuscular	within the muscle
OLIGO-	very little	oliguria	very little urine production
PERI-	around	perioperative	around or during the surgery
POLY-	many, excessive	polyuria	excessive urine production
POST-	after	postoperative	after the surgery
PRE-	before	preanesthetic	before anesthesia
SUB-	below	subcutaneous	below the skin
SUPER-	above	superimposed	above the surface

TABLE 1-3

Suffixes Meaning "Pertaining To"

SUFFIX	TERM	MEANING
-AC	cardiac	pertaining to the heart
-AL	renal	pertaining to the kidney
-AN	ovarian	pertaining to the ovary
-AR	lumbar	pertaining to the loin or lower back
-ARY	alimentary	pertaining to the gastrointestinal tract
-EAL	laryngeal	pertaining to the larynx
-IC	enteric	pertaining to the intestines
-INE	uterine	pertaining to the uterus
-OUS	cutaneous	pertaining to the skin
-TIC	nephrotic	pertaining to the kidneys

be more easily pronounced. Some words have a poor flow between the prefix, suffix, and root word combination; placing a vowel within the word allows the newly combined form to be easy to pronounce.

Putting It All Together

With a basic understanding of prefixes, suffixes, and root words, it becomes easy to break words down into parts to define new terminology. When dissecting a term, use a slash mark to isolate each part of the word. Then focus on the meaning of each dissected part of the term. Once the meaning of each word part is determined, place the meaning into a logically formed definition. Below are some examples of dissecting veterinary terms.

Example

ARTHR/IT IS	inflammation of the joint
CARDIO/LOGY	the study of the heart
CARDIO/MEGALY	enlargement of the heart
CYSTO/CENTESIS	to puncture into the urinary bladder
HEMAT/URIA	blood in the urine
HEPATITIS	inflammation of the liver
GASTR/IC	pertaining to the stomach

TABLE 1-4

Common Suffixes

SUFFIX	MEANING	EXAMPLE	DEFINITION
-CENTESIS	surgical puncture into	cystocentesis	surgical puncture with a needle into the urinary bladder
-ECTOMY	surgical removal of	ovariohysterectomy	surgical removal of the ovaries and uterus
-EMIA	blood	hypocalcemia	low blood calcium
-GRAM	a record of	electrocardiogram	a record of the electrical activity of the heart
-GRAPH	to record with an instrument	radiograph	a record made using radiation; x-ray
-GRAPHY	the act of recording using an instrument	radiography	the act of taking a picture using radiation
-ITIS	inflammation	colitis	inflammation of the colon
-LOGY	the study of	histology	the study of tissues
-LYSIS	to break down	urinalysis	the breakdown of urine into parts
-MEGALY	enlargement of	cardiomegaly	enlargement of the heart
-OSIS	condition	osteoporosis	condition of bone loss
-PATHY	disease	cardiopathy	heart disease
-PEXY	to suture to	gastropexy	to suture to the stomach
-PLASTY	to surgically repair	rhinoplasty	surgical repair of the nose
-RRHAGE-	to burst	hemorrhage	the bursting of blood; bleeding
-RRHEA-	to flow	diarrhea	the flow of feces
-SCOPE	instrument used to view	microscope	instrument used to view small items
-SCOPY	the act of using an instrument for viewing	endoscopy	the act of using a scope to view the inside of the body
-STOMY	to create a new surgical opening	cystostomy	to create a new surgical opening in the urinary bladder
-THERAPY	treatment	chemotherapy	treatment of chemicals
-TOMY	to cut into surgically; to make an incision	cystotomy	to make an incision into the urinary bladder

LAPARO/TOMY	surgical incision into the abdomen
MAST/ECTOMY	the surgical removal of the mammary gland
OVARIO/HYSTER/ ECTOMY	surgical removal of ovaries and uterus
Abbreviated OHE	
RADIO/GRAPH	to record using radiation
RHINO/PLASTY	the surgical repair of the nose
URINA/LYSIS	to break down urine

Remember, when defining, pronouncing, or spelling unfamiliar veterinary medical terms, it is best to have a veterinary medical dictionary to reference prior to recording in medical records or communicating with others. Veterinary medical terminology is a foreign language for some that is necessary in the veterinary facility, and over time and with consistent use it will a comfortable second language.

TERMINOLOGY TIP

Practice makes perfect. Complete practice assignments to study and review veterinary terms. Additional assignments can be found on the online companion for this book.

Common Directional Terms

Directional terms relate to the body position and are useful in recording locations relating to the body; they are also useful in surgical or radiographic positioning. These terms relate to a specific area on the body of an animal and allow for better communication when referring to an animal's anatomy (see Table 1-5, Figure 1-2, and Figure 1-3).

TABLE 1-5

Common Directional Terms

TERM	MEANING
ASPECT	area
CAUDAL	toward the tail
CRANIAL	toward the head
DISTAL	away from the center of the body
DORSAL	toward the back area
LATERAL	side of the body; toward the outside
MEDIAL	inside of an area; toward the inside
PALMAR	the bottom of the front feet
PLANTAR	the bottom of the rear feet
PROXIMAL	closer to the center of the body
RECUMBENCY	lying in position
RECUMBENT	lying
ROSTRAL	toward the nose
TRANSVERSE	across an area dividing it into cranial and caudal sections
VENTRAL	toward the abdomen or belly area

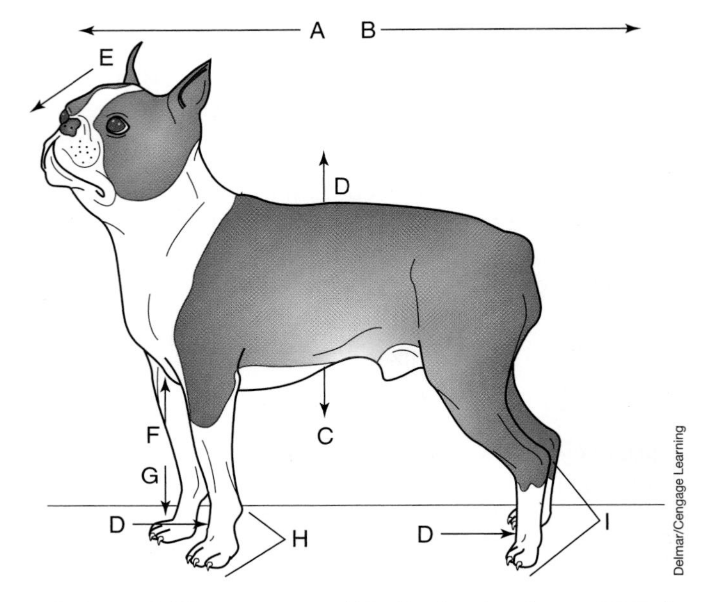

FIGURE 1-2 The arrows on this Boston terrier represent the following directional terms: A = cranial, B = caudal, C = ventral, D = dorsal, E = rostral, F = proximal, G = distal, H = palmer, I = plantar.

FIGURE 1-3 Medial versus lateral. The lines on these cats represent the directional terms *medial* and *lateral*.

Common Terms and Abbreviations Used in Veterinary Practice

The veterinary field has developed common universal terms that are used in the industry to relate to patient history, species information, the physical exam, and pharmacy terms. Many of these terms also have common abbreviations used in medical recording. This section will outline commonly used terms and abbreviations used in recording medical records and treatment boards (see Table 1-6 through Table 1-15, and Figure 1-4).

TABLE 1-6

Species Terms

AMPHIBIAN	frog or toad
AVIAN	bird
BOVINE	cow
CANINE	dog
CAPRINE	goat
CAVY	guinea pig
EQUINE	horse
FELINE	cat
LAGOMORPH	rabbit
MURINE	rat or mouse
OVINE	sheep
PORCINE	pig or swine
POULTRY	chicken and turkey
PRIMATE	monkey and ape
REPTILE	snake and lizard
TERRAPIN	turtle

TABLE 1-7

Common Veterinary Abbreviations

C or cast	castrated
C-sect	C-section or caesarian section
d	day
d/c	discharge
DLH	domestic long hair (cat)
DSH	domestic short hair (cat)
d/o	drop off
EX	exotic
F	female
K-9	dog or canine
M	male
Mo	month
NM	neutered male
o	owner
p/u	pick up
rec	recommend
S or SF	spayed or spayed female
S/R	suture removal
Sx	surgery
Wk	week
y or yr	year

TABLE 1-8

Terms Related to Patient History

ANOREXIA	not eating or decreased appetite
BM	bowel movement
D	diarrhea
DYSURIA	difficulty or trouble with urination
Dz	disease
HBC	hit by car
HEMATURIA	blood in the urine
Hx	history
LETHARGIC	tired or inactive
PD	polydipsia (increased thirst)
PU	polyuria (increased urination)
U	urine
V	vomiting
V/D	vomiting and diarrhea

TABLE 1-9

Terms Related to Physical Examination

ACUTE	short term
ANALS or AG	anal glands
BAR	bright, alert, responsive
CHRONIC	long term
CRT	capillary refill time
Dx	diagnosis
FeLV	feline leukemia virus
FIP	feline infectious peritonitis
FIV	feline immunodeficiency virus
HR	heart rate
L	left
LN	lymph node
mm	mucous membranes
N or (−)	negative
NR	nothing reported
NSF	no signs found
PE	physical exam
Px	prognosis

(*Continues*)

TABLE 1-9 (*Continued*)

QAR	quiet, alert, responsive
R	right
RR	respiratory rate
Rx	prescription
SOAP	Subjective, Objective, Assessment, Plan
TPR	temperature, pulse, respiration
Tx	treatment
URI	upper respiratory infection
UTI	urinary tract infection
WNL	within normal limits
Wt	weight
+	positive

TABLE 1-10

Laboratory Terms and Abbreviations

Bx	biopsy
CBC	complete blood count
CHEM	blood chemistry panel
C/S or C & S	culture and sensitivity
Cysto	cystocentesis
Fecal	fecal or stool sample
HW	heart worm
PCV	packed cell volume
T4	thyroid test
UA	urinalysis

TABLE 1-11

Pharmacy Terms and Abbreviations

BID	twice a day	prn	refill as needed
Cap	capsule	q	every
cc	cubic centimeter	qd	every day
d	day	QID	four times a day
EOD	every other day	Rx	prescription
h	hour	SID	once a day
kg	kilogram	Tab	tablet
mg	milligram	TID	three times a day
ml	milliliter	w	week
NPO	nothing by mouth	/	per
oz	ounces	# or lb	pound
PO	by mouth	#	number of tablets to dispense

TABLE 1-12

Eyes and Ears

AD	right ear
AS	left ear
AU	both ears
OD	right eye
OS	left eye
OU	both eyes

TERMINOLOGY TIP

When reading a pharmacy prescription label, write out all the information on the label including the abbreviations and verify it is correct before typing and labeling the container.

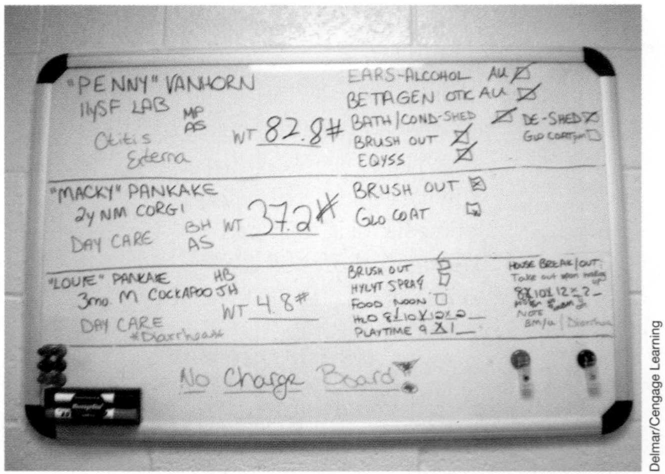

Delmar/Cengage Learning

FIGURE 1-4 It is important to understand veterinary abbreviations as used on this treatment board to ensure proper care and handling of the animals in the clinic.

TABLE 1-13

Routes of Medicinal Administration

Adm.	administer
IC	intracardiac (into the heart)
ID	intradermal (within the layers of skin)
IM	intramuscular (into the muscle)
IN	intransal (into the nasal cavity)
IO	intraoseous (into the bone)
IP	intraperitoneal (into the peritoneum or lining of the abdomen)
IT	intratracheal (into the trachea or windpipe)
IV	intravenous (into the vein)
PO	by mouth or orally
SQ	subcutaneous (under the skin)
SUB-Q	subcutaneous (under the skin)

TABLE 1-14

Association Abbreviations

AAHA	American Animal Hospital Association
AVMA	American Veterinary Medical Association
NAVTA	National Association of Veterinary Technicians of America
OFA	Orthopedic Foundation of America

TABLE 1-15

Common Animal Terms

Bitch	intact female dog	Chick	young parrot; young chicken
Litter	group of newborn dogs	Cock	male parrot; male chicken
Puppy	young dog	Flock	group of birds; group of chickens, turkeys, or ducks
Stud dog	intact male dog	Hen	female parrot; female chicken; female turkey
Whelping	the labor process of dogs	Clutch	group of eggs
Kitten	young cat	Poult	young turkey; young chicken
Tom	intact male cat; male turkey	Capon	young castrated male chicken
Queen	intact female cat	Cockerel	immature male chicken
Queening	the labor process of cats	Pullet	immature female chicken
Buck	male rabbit; male goat; male deer	Rooster	male chicken
Doe	female rabbit; female goat; female deer	Drake	male duck
Kindling	labor process of rabbits and ferrets	Duck	female duck
Kit	young rabbit; young ferret	Duckling	young duck
Lapin	neutered male rabbit	Barrow	young castrated male pig
Gib	neutered male ferret	Farrowing	labor process of swine
Hob	intact male ferret	Gilt	young female pig that has not farrowed
Jill	intact female ferret	Piglet	young pig
Sprite	neutered female ferret	Stag	male pig castrated after maturity
Boar	male guinea pig; male pig	Colt	young male horse
Pup	young guinea pig; young mouse; young rat; young dog	Filly	young female horse
Sow	female guinea pig; female pig	Foal	young male or female horse
Dam	female rat; female mouse; term for a female parent that is breeding	Gelding	castrated male horse; castrated male llama
Sire	male rat; male mouse; term for a male parent that is breeding	Hand	the measurement of a horse equal to 4 inches

(*Continues*)

TABLE 1-15 (*Continued*)

Herd	group of horses	Lambing	labor process of sheep
Horse	horse over 14.2 hands in height	Ram	intact male sheep
Mare	intact adult female horse	Wether	castrated male sheep; castrated male goat
Pony	horse under 14.2 hands in height	Freshening	labor process of dairy-producing animals
Stallion	intact adult male horse	Kid	young goat
Weanling	young horse under a year of age	Kidding	labor process of goats
Yearling	young horse between 1 and 2 years of age	Bull	intact male cow; intact male llama
Donkey	donkey crossed with a donkey	Cow	intact female cow; intact female llama
Hinny	cross of a male horse and female donkey	Cria	young llama
Jack	intact male donkey	Calf	young cow
Jenny	intact female donkey	Calving	labor process of cows
Mule	cross of a male donkey and female horse	Heifer	young female cow that has not been bred
Ewe	intact female sheep	Stag	mature castrated male cow
Lamb	young sheep	Steer	young castrated male cow

SUMMARY

Veterinary medical terminology and veterinary abbreviations are a vital part of the veterinary industry. Each member of the veterinary health care team must have a basic knowledge of terminology and abbreviations to complete his or her job. This may mean reading a hospital treatment board, reading a patient's medical record, reading a cage card, or reading directions on a medication label. For the veterinary assistant to perform his or her duties properly, it is essential to have a working knowledge and understanding of the veterinary language.

Key Terms

dissect to separate into pieces, or break down into parts, to identify the meaning of a word

prefix the word part at the beginning of a term

suffix the word part at the end of a term

root word origin or main part of the word that gives the term its essential meaning

combining vowel a letter, usually a vowel, that is placed between the prefix and suffix that makes the word easier to pronounce

REVIEW QUESTIONS

Matching

Match the prefix, suffix, or root word to its definition.

1. __ -itis
2. __ cardiac
3. __ -ectomy
4. __ hepatic
5. __ hyper-
6. __ post-
7. __ hemo-
8. __ therapy
9. __ cysto-
10. __ gastric

a. pertaining to the stomach
b. blood
c. treatment
d. urinary bladder
e. inflammation
f. above normal, increased, excessive
g. after
h. pertaining to the heart
i. surgical removal of
j. pertaining to the liver

Interpretation of Abbreviations

11. Write a description of the meaning of the following situation involving an animal:

    The veterinarian has given you "Sassy" Smith's dx and tx plan and you are to complete the directions and d/c "Sassy." You read the following in the medical record: Dx: gastritis; NPO × 24 h, soft diet × 7 d and then return to normal diet; adm. 100 ml Lactated ringers SQ; d/c; recheck 3 d. o to monitor v/d.

12. You read the following notation in a medical record as asked by the veterinarian:

    1/7/08 OHE sx. ; NR; S/R 8-10 d; limit exercise and monitor sx site

    What do you tell the owner? Please write your complete response.

Problem Solving

Provide a solution for the following three pharmaceutical problems.

13. The veterinarian instructs you to call Mrs. Smith and have her change "Sweetie's" dose from (1) 100 mg capsule PO SID to (1) 100 mg capsule PO TID × 5 d. What will you tell Mrs. Smith?
14. You are asked to label the following prescription for Dr. Daniels: Place (3) drops AU BID × 10 d. What will you place on the label?
15. You read the medical record for "Phoebe" Jones. It instructs the veterinary assistant to "Give 3 ml of Amoxicillin EOD × 1 w prn. Keep medication refrigerated." What will you do for "Phoebe's" treatment?

Identification

Identify the common abbreviations that you will need to know when recording in medical records.

16. F ____________
17. SID ____________
18. R ____________
19. DSH ____________
20. wt ____________
21. BAR ____________
22. Hx ____________
23. BM ____________
24. TPR ____________
25. HBC ____________
26. L ____________
27. NM ____________
28. BID ____________
29. V-D ____________
30. q ____________
31. TID ____________
32. DLH ____________
33. UA ____________
34. Dx ____________
35. S ____________

Bonus Question

What is meant by the term *SOAP*.

Clinical Situation

© iStock/Mohamed Sadath

Complete the following activity by providing the common abbreviation for the underlined medical terms.

A 6 year old spayed female domestic long hair cat was seen at the veterinary clinic for a physical examination due to vomiting and diarrhea. The cat has a history of being within normal limits. A temperature, pulse, and respiration rate was evaluated with normal signs found. The cat's weight is recorded as 11 pounds. The vet evaluated a complete blood count and radiographs to rule out infection and disease. The diagnosis is found to be acute colitis. The treatment is prepared. The cat's owner is to give 2 tablets of Metronidazole by mouth once a day. The owner is to monitor the cat's bowel movements. A recheck visit was scheduled.

Competency Skills Task

Learning Veterinary Medical Terminology and Abbreviations

Objective:

Have a basic knowledge and understanding of common veterinary terminology and abbreviations.

Preparation:

- Obtain 3" × 5" or larger index cards.
- Obtain colored markers, pens, or pencils.

Procedure:

1. Obtain a vocabulary terminology word list. Use the terms presented in the tables in this chapter.
2. Write each vocabulary term on one side of each index card.
3. Write the definition of each word on the opposite side of the index card.
4. Study each word term and then try reciting the definition.
5. Study each definition and then try reciting the word.

Practice daily by studying each term and definition.

2

Medical Records

Objectives

Upon the completion of this chapter, the reader should be able to:

- ☐ Identify the various forms included in the medical record
- ☐ Properly create and label veterinary medical records
- ☐ Demonstrate how to record the information in the medical record
- ☐ Properly locate, file, and refile medical records
- ☐ Discuss the importance of the legal requirements of the veterinary medical record

Introduction

The veterinary assistant is responsible for many office procedures and administrative duties, including the critically important creation, maintenance, and organization of medical records. The medical records furnish documentary evidence of the patient's illness, hospital care, and treatment, and they serve as a basis for review, study, and evaluation of the care and treatment rendered by the veterinarian.

The Veterinary Medical Record

The primary purpose of the veterinary **medical record** is to record detailed information for each veterinary **patient** (the animal seen by the facility; see Figure 2-1). This information should include patient and **client** (the animal's owner) information, patient history, medical and surgical records, progress notes, and laboratory information. The medical record serves as a diary of the animal's health. This is especially important in multiple-doctor facilities and when medical records are transferred from location to location. The veterinary medical record is owned by the veterinary facility and is the property of the facility that originally activated the record. It is a legal document that is private and confidential, even within the veterinary team. This is also true for any radiographs, ultrasound images, and other diagnostic tests that are produced by the facility.

This legal document allows for the **Veterinary-Client-Patient Relationship** or **VCPR** to be established. The VCPR legally allows the veterinarian to treat the patient and dispense medications. The VCPR must be maintained on a yearly basis. It serves as documentation and protects the veterinarian and staff through written recordings. The VCPR relationship satisfies the following criteria: (1) the veterinarian has assumed responsibility for the patient in making judgment regarding the health of an animal and the need for veterinary treatment, and the owner has agreed to follow the instructions of the veterinarian; (2) the veterinarian has sufficient knowledge of the animal species to initiate a general, preliminary, or tentative diagnosis of the medical condition of the animal; (3) the veterinarian is competent with keeping and caring for the animal through examination and timely visits to the facility where the animal is housed; (4) the veterinarian is available for consultation in adverse reactions or failure of the deemed therapy; and (5) the veterinarian maintains medical records of the animal's treatments.

Delmar/Cengage Learning

FIGURE 2-1 Accurate record keeping is an important duty of the veterinary assistant.

The original medical record must remain in the facility for at least 1–3 years from any patient's last visit, depending on state law. Many facilities keep records for at least 7 years depending on the patient's status. This may be done in a storage area that allows easy access to the files. The original record must not be removed from the facility, but copies may be made and sent with the owner or placed in the mail for continued animal health treatments. A facility may not withhold the release of the veterinary medical record contents.

Creating a Medical Record

Each veterinary medical record must contain certain information and paperwork. Owner information and patient information should be obtained and recorded (see Table 2-1). This information may change and should be maintained and updated on a regular basis. The owner and patient information should be typed on a label and placed on the folder in the location established by the veterinary facility.

The medical file should include the following sections or forms:

- The client and patient information sheet should be at the beginning for easy retrieval (see Figure 2-2).
- The master problem list details a patient's history and previous medical problems, vaccines, or surgeries (see Figure 2-3).

TABLE 2-1

Owner and Patient Information

OWNER INFORMATION	PATIENT INFORMATION
Name	Name
Address	Species
Phone	Breed
Work Phone	Color
Emergency Contact Numbers	Gender/Sex
Client ID or Chart Number	Age/Birth date
	Vaccine history
	Allergies
	Surgical History
	Chief Complaint

DATE ____________ CASE NUMBER ____________

CLIENT/PATIENT INFORMATION FORM

Please provide the following information for our records: **PLEASE PRINT!**

OWNER INFORMATION

Owner's Name			Social Security Number
Street Address			
City/State		Zip Code	Country
Telephone (Include Area Code)	Home		Business
Driver's License Number	Place of Employment		How Long Employed?

ANIMAL INFORMATION

Animal Species (Dog, Cat, Other)			Breed
Animal's Name		Sex	Has the animal been altered? ☐ YES ☐ NO
Color	Birth Date	THE UNDERSIGNED OWNER OR AGENT CERTIFIES THAT THE HEREIN DESCRIBED ANIMAL HAS A MAXIMUM VALUE OF APPROXIMATELY $.______	

REFERRAL INFORMATION

Were you referred by a veterinarian?

	☐ YES	☐ NO	If so, complete the following information.
Veterinarian's Name			Phone
Street Address			
City/State			Zip Code

You will be advised of estimated cost and anticipated procedures. Please feel free to discuss the proposed treatment and any costs with the veterinarian. A minimum deposit of 50% of the initial estimated charges will be required for hospitalization of the patient.

STATEMENT OF OWNERSHIP AND CONSENT: I am the owner of the above–described animal, or have authorization from the owner to consent to its treatment.

I hereby authorize the performance of professionally accepted diagnostic, therapeutic, anesthetic, and surgical procedures necessary for its treatment.

I accept financial responsibility for these services.

I have read the above consent and understand why these procedures may be necessary. I have also been told of the possible complications and alternatives to the anticipated procedures.

PAYMENT CHOICE: ☐ Cash ☐ Check ☐ Bank Card

SIGNATURE (Owner/Agent)	DATE

FIGURE 2-2 Sample client/patient information form.

CITY ANIMAL HOSPITAL
Master Problem List

OWNER INFORMATION

Owner Name ☐ Mr. ☐ Miss ☐ Ms. ☐ Mrs. Patient/Pet's Name

Address City/State/Zip

Home Phone Business Phone

PATIENT/PET INFORMATION

Chart # ____________
Patient ____________ Species ____________ Breed ____________
Color ____________ Sex ☐ F ☐ M ☐ N Birth Date ____________
Vax History ____________ Weight ____________

IMMUNIZATION/PREVENTATIVE RECORD

DATE											
RABIES											
DA2PL											
FVR-CP											
FELV											
FECAL											

PROBLEM LIST

PROBLEM	DATE ENTERED	DATE RESOLVED
1.		
2.		
3.		
4.		
5.		
6.		
7.		
8.		
9.		
10.		
11.		
12.		
13.		

FIGURE 2-3 Sample master problem list.

- **Progress notes** that allow for chronological log entries to be recorded each time a patient is seen and treatment is completed (see Figure 2-4).
- Laboratory reports, which include veterinary test results.
- Radiology reports that detail X-rays, including proper identification of hospital name, date, name of client, name of patient, and positional marker.
- Pharmacy records note all medications prescribed. If the medications prescribed are controlled substances, a controlled substance log must also be maintained.
- Surgical and anesthesia reports that identify surgical procedures.
- **Consent forms** and other forms that may be recorded and documented (see Figure 2-5).

PROGRESS NOTES

PROGRESS NOTES	
Client Name: ______________________________	Address: Phone:

Pet Name:	Species:	Breed:	Color:	Gender: Age:

Date/Initials	
	S
	O
	A
	P

Delmar/Cengage Learning

FIGURE 2-4 Sample progress note.

CITY HOSPITAL
ANY STREET
ANY TOWN, SS 00000

CONSENT FORM

Owner's Name: ______________________ Animal's Name: ______________________
Address: ______________________ Species: ______________________
______________________ Breed: ______________________
Case Number: ______________________ Sex: ______________________

I am the owner or agent for the owner of the above-described animal and have the authority to execute this consent.

I hereby consent to the hospitalization of the above-described animal and authorize the veterinarian and staff to administer any tests, medications, anesthesia, or surgical procedures that the veterinarian deems necessary for the health, safety, and well-being of the animal.

I specifically request the following procedure(s) or operation(s):

____

I understand that during the course of the above-mentioned procedure(s) or operation(s), unforeseen conditions may be discovered that necessitate an extension of the above-mentioned procedure(s) or operation(s) or additional procedure(s) or operation(s). I hereby consent to and authorize the performance of such procedure(s) or operation(s) as are necessary according to the veterinarian's professional judgment.

I also authorize the use of appropriate anesthetics and other medications. I understand that the veterinary support personnel will be employed as necessary according to the veterinarian's professional judgment.

I have been advised as to the nature of the procedure(s) or operation(s) to be performed and the risks involved. I realize that results cannot be guaranteed.

I understand that all fees for professional services are due at the time of discharge.

I have read and understand this authorization and consent.

Additional Comments/Information:

____

____

____

Date ______________ Signature of Owner or Agent ______________________

Signature of Witness ______________________

FIGURE 2-5 Sample consent form.

When assembling the medical record, it may also be helpful to make a **cage card** for the animal. The cage card is used to identify and locate each patient within the facility (see Figure 2-6).

Each veterinary facility will have a preference for medical forms that are used and the sequence in which they will be placed in the medical record. It is important that each medical file be kept in the same format with information in the same location for ease of use.

Once the veterinarian recommends a specific treatment, an estimate sheet will be prepared for the client. This outlines the recommended procedures or treatments and the costs associated with them. This will be reviewed with the client. The client will be required to sign

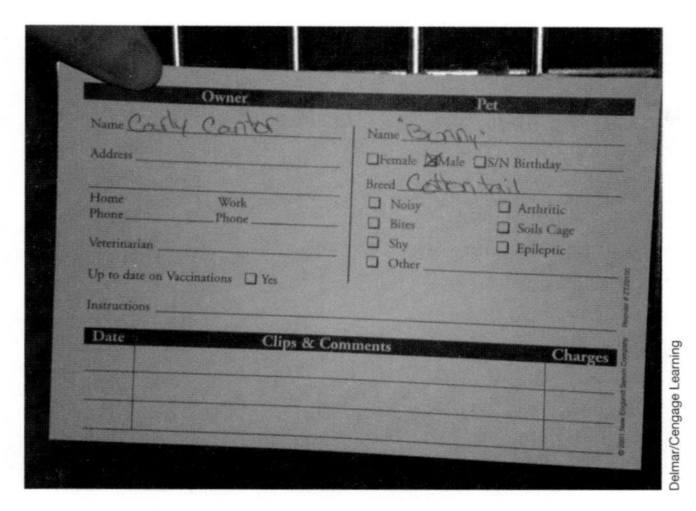
Owner
Name Carly Cantor
Address
Home Phone
Work Phone
Veterinarian
Up to date on Vaccinations ☐ Yes
Instructions

Pet
Name "Bunny"
☐ Female ☒ Male ☐ S/N Birthday
Breed Cottontail
☐ Noisy ☐ Arthritic
☐ Bites ☐ Soils Cage
☐ Shy ☐ Epileptic
☐ Other

Date	Clips & Comments	Charges

FIGURE 2-6 A cage card helps to identify and keep track of the animal while it is in the facility.

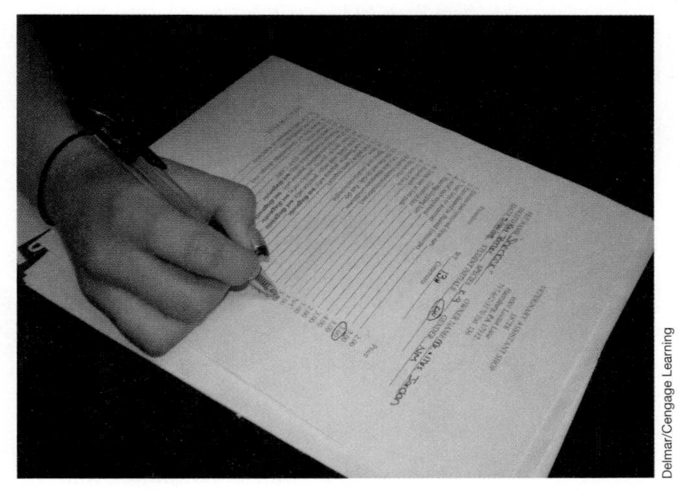

FIGURE 2-7 An invoice will be completed and presented to the client upon discharge.

authorization or consent forms before the veterinarian can begin to provide the recommended procedures or treatments. The authorization form serves as a contract between the client and the veterinarian.

After treatments or procedures are completed, the client will receive a discharge sheet. This contains instructions for the client on care of the animal while recuperating, such as how to give prescribed medications or care for surgical incision sites.

The invoice should be placed on the top of the chart for easy access for the veterinarian and all staff members to keep a working list of itemized charges. The invoice can easily be removed during the patient's discharge and totaled out for the client (see Figure 2-7).

Recording Information in the Medical Record

Due to the legal requirements for veterinary medical records, there are several rules when recording on medical forms. First, *all* information must be recorded in black or blue ink—never in pencil or other ink colors. Second, all information should be accurate and legible. If a mistake is made, one should place a single line through the error and initial the error and then place the corrected statement after the entry (see Figure 2-8). This indicates an error in writing occurred rather than suggesting information was changed. Never erase, use a correction fluid such as White-Out, or scribble out information! Table 2-2 lists the Dos and Don'ts of recording in medical records.

Record all communications and phone conversations held with clients and detail the conversation in the medical record following the date entry and the initial of the team member who was involved in the discussion. This allows others to properly identify and communicate with the patient's caregiver as necessary.

Each patient record should contain one medical record for that patient only. In the case of a veterinary facility that treats large animals and farm production systems, one chart may be used for the entire herd as long as each animal treated is identified by a tag number or registration number to determine which animals have been examined and treated by the veterinarian. Laboratory research facilities should maintain their records in a similar fashion and may have regulations outlined by other veterinary laboratory animal associations.

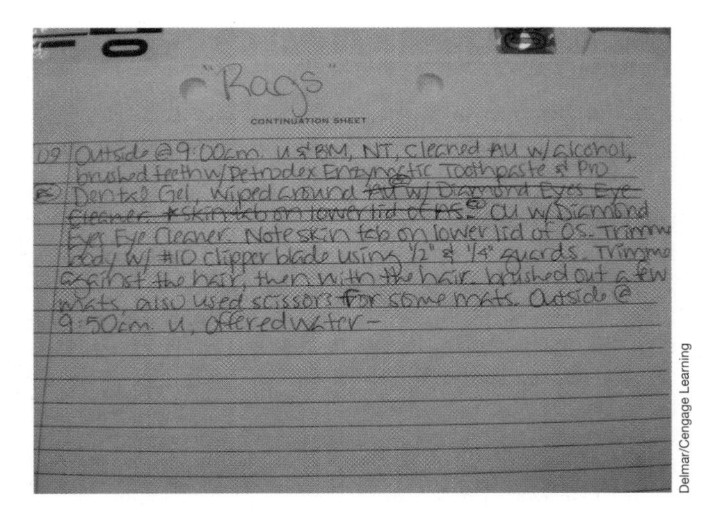
"Rags"
CONTINUATION SHEET

09 Outside @ 9:00am. U & BM, NT, cleaned AU w/ alcohol, brushed teeth w/ Petrodex Enzymatic Toothpaste & Pro Dental Gel, wiped around ~~AU w/ Diamond Eyes Eye Cleaner. *Skin tab on lower lid of AS.~~ OU w/ Diamond Eyes Eye Cleaner. Note skin tab on lower lid of OS. Trimmed body w/ #10 clipper blade using 1/2" & 1/4" guards. Trimmed against the hair, then with the hair. brushed out a few mats. also used scissors for some mats. Outside @ 9:50am. U, offered water –

FIGURE 2-8 To properly correct a documentation error, place a single line through the incorrect content and initial it. Write the correct content following this entry.

TABLE 2-2

Recording in Medical Records

DOS	DON'TS
Use blue or black ink	Use pencil
Write legibly and neat	Use red or light colors of ink
Place one single line through written errors	Use a correction fluid (e.g., Wite-Out)
Initial all mistakes	Erase mistakes
Check spelling and grammar	Cross or scratch out errors
Use proper veterinary medical terminology	Write illegibly
Record dates and times	Use common language
Identify the initials of the person treating the animal	Use bad grammar or spelling
Maintain one record per patient	Place multiple animals within one record without properly identifying the animals
Record all client communications	

When recording information in the medical record, some facilities use a format called **SOAP**. This stands for **Subjective, Objective, Assessment, and Plan**. The **subjective information** is information that is based on the animal's overall appearance and the health care team's descriptions of the animal. This may include the animal's attitude and the veterinary assistant's feelings on the appearance of the pet. The **objective information** is measured facts that can be recorded based on the patient. This includes the vital signs, such as temperature, heart rate, and weight. The **assessment** of the patient is what the veterinarian determines to be the diagnosis or the patient's problem. The **plan** is the treatment or procedure that is to be given to the patient. The SOAP is commonly used in the progress notes and the physical examination section of the medical records.

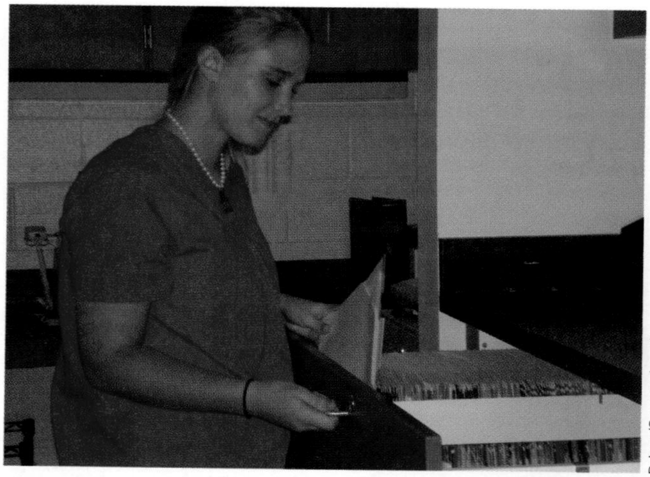

Delmar/Cengage Learning

FIGURE 2-9 The veterinary assistant must understand the filing system in use by the facility.

Filing Medical Records

The veterinary medical record must be kept in a file system such as a paper file or as a computer file. Many facilities will use a combination of both paper and computerized records. The paper-based filing system must meet the legal requirements outlined in the state's veterinary practice act. Paper forms and information are usually stored in a manila file that may have pockets or clasps to contain each form.

Medical records can be filed and stored in the veterinary office by using several methods. The most common filing method is the use of an **alphabetical filing system**. The purpose of alphabetical filing is for locating files by letter, usually by the client's last name. Each letter of the alphabet is assigned a color code and the first two to three letters of the owner's last name are attached to the outside of the medical file. The concept is for charts to be easy to identify according to color and letter sequencing on a shelf or in a file cabinet (see Figure 2-9).

Another filing method used in veterinary facilities is the **numerical filing system**. The numerical filing system can be used in two ways: the client may be assigned a number, or each individual patient is assigned a number. Each number digit is assigned a **color code**. The color-code method assigns a color for each label so that when filed together, it is easy to note a misfiled chart by the color of the labels. The numbers are placed on labels and then placed on the file, aiming for ease of use and location of file sequencing. Each file should also contain a label for the year the patient was last seen. This allows the staff to determine which patients are overdue for health exams. Most files are kept for 7 years from the patient's last visit. This may be due to animal's passing away or clients moving or transferring to another hospital. When the medical record is no longer in use, it may be deleted or inactivated. After 7 years it may be discarded. Remember, some clients may move back to the area or decide to return to the clinic.

Consent Forms and Certificates

The veterinary **consent form** exists to identify to each client the procedures and prices that will be incurred while a patient is under the care of the veterinary facility. The client should read and sign the consent form to show agreement to the medical care necessary and understanding of any potential risks that may be involved. The consent form also outlines the costs of the procedures and thus enters the client into a legal agreement and authorization for the veterinary facility to provide care to the patient. The owner should receive a copy of the consent form and the original should be placed in the medical record.

Patients that have undergone surgical alteration to remove the reproductive organs should be issued a **neuter certificate** (see Figure 2-10). This provides the owner with proof that the pet has been spayed or castrated and is no longer sexually intact or capable of reproducing. This certificate may allow pet owners to purchase licenses at a reduced fee or may be required if a pet has been adopted from a shelter.

A **rabies certificate** is issued after a pet has received its rabies vaccine (see Figure 2-11). Many animals are legally required to have rabies vaccines, and thus owners must show proof if an animal bites or injures another person or pet. State laws for pet ownership vary from state to state and each pet owner should review their state's dog and animal laws.

A **health certificate** must be issued if an animal is being transported out of state or out of the country (see Figure 2-12). This requirement must include a physical examination from a licensed veterinarian that states the animal is healthy and free of disease and that the required vaccines are up to date. A health certificate usually includes vaccine records, a statement of health, and any required test results that may be necessary for travel or show purposes.

VETERINARY HOSPITAL
ANY STREET, ANY TOWN, XX 00000
NEUTER CERTIFICATE

Patient name: __________ Breed: __________

Gender: _______ Age: _______ Color: _______

Client Name: ____________________

Address: ____________________

This neuter certificate certifies that the above name animal has been surgically altered and is no longer capable of reproduction. The surgical procedure was performed by the veterinarian signing this certificate.

(*Signature of Veterinarian*) Date: __________

FIGURE 2-10 Sample neuter certificate.

Medical Records as Legal Documents

Veterinarians and their staff shall protect the personal privacy of clients by maintaining **confidentiality** of all information of the client and patient records. Legally, all information within the veterinary medical record is private and not to be discussed unless the client gives approval for the information to be released. This may be necessary when veterinarians release records for clients in cases of moving to a new location, changing veterinary facilities, or needing a referral to a veterinary specialist. When a record is released from the facility, it may only be done as a summary or copy of the record. Veterinary assistants should be familiar with using a copier. When requested by law or by the client, a veterinarian may not withhold the release of a copy of the medical record. The client should sign a waiver of confidentiality form to allow medical records to be released (see Figure 2-13).

Veterinary medical records serve as legal documents and as communication among members of the veterinary staff. The records provide documented evidence of an animal's illness, hospital care, treatment, and surgical history. The records must be maintained in a way that any veterinary staff member may read the notes and be able to proceed with the proper care and treatment of the animal. The records must be accurate and properly maintained should they ever be necessary to be reviewed in a legal case.

CITY HOSPITAL
ANY STREET
ANY TOWN, SS 00000

RABIES CERTIFICATE

Patient Information

Client Name: ____________________ Patient Name: ____________________

Address: ____________________ Species: ____________________

____________________ Breed: ____________________

Telephone (H): ____________________ Sex: ____________________

(W): ____________________ License No.: ____________________

Vaccination Information

Vaccination Date: ____________________ Expiration Date: ____________________

Vaccine Serial No.: ____________________ Producer Killed: ____________________

Comments:

____

____

____

____

Signature of Veterinarian

Delmar/Cengage Learning

FIGURE 2-11 Sample rabies certificate.

Invoicing

The veterinary assistant should be familiar with the invoicing procedures of the facility. This may be completed using a paper invoice or a computerized invoice. A written invoice is often kept with the medical record to note billable activities that are completed while the animal is in the facility. Computerized invoices are kept in the computer system and items are entered as they are completed, and the invoice is then printed out at the time the patient is discharged.

VETERINARY HOSPITAL
ANY STREET, ANY TOWN, XX 00000

HEALTH CERTIFICATE

Date: ______________

Client Name: ______________ Patient Name: ______________
Address: ______________ Species/Breed: ______________
Telephone: ______________ Color: ________ Gender: ________ Age: ________
Vaccination: ________ Date: ________ Vaccination: ________ Date: ________
Vaccination: ________ Date: ________ Vaccination: ________ Date: ________
Vaccination: ________ Date: ________ Vaccination: ________ Date: ________
Test: ________ Result: ________ Date: ________
Test: ________ Result: ________ Date: ________
Test: ________ Result: ________ Date: ________

This health certificate certifies that the animal listed is in adequate health and has been vaccinated according to the above information. The animal has been examined by a licensed veterinarian with 30 days of issue of this health certificate.

(*Signature of Veterinarian*)

(*License Number*)

FIGURE 2-12 Sample health certificate.

CONSENT TO DISCLOSURE OF MEDICAL RECORD
Waiver of Confidentiality
By Authorized Patient Representative

I, ______(name of owner or agent)______, the ______(owner or agent of owner)______
of ______(name of animal)______, a ______(breed)______,
do understand that the information contained in ______(name of animal)______'s medical record
is confidential. However, I specifically give my consent for ______(name of veterinary practice)______
to release the following information concerning ______(name of animal)______
to ______(name of party to whom information is being released)______.

The above-listed information is to be disclosed for the specific purpose of

____

____

It is further understood that the information released is for professional purposes only.
This information may not be given in whole or part to any person other than that stated above.

____
Signature of Authorized Representative

____
Date

FIGURE 2-13 Sample consent of disclosure certificate.

SUMMARY

All veterinary assistants are required to have a basic knowledge and understanding of veterinary medical record keeping. Medical records are used to summarize a patient's history. Medical records are legal documents that allow veterinarians and the veterinary health care team to properly care for each patient seen at the facility and as a means of communicating information to each other. A patient's medical records tell the history of the patient in a diary-like script each time an animal visits the facility. It is important that every member of the veterinary health care team learn to accurately record in the medical record. Medical records are a vital tool for every veterinarian and must be properly assembled, recorded, and maintained. Medical records are legal documents and valuable tools in the veterinary facility. The veterinarian and health care team rely on medical records to provide a history of each patient. This includes proper labeling, information, forms, and accurate notations. It is important to record everything and be accurate and neat.

Key Terms

alphabetical filing system organization of medical records by patient or client identification number and color coding

assessment veterinary diagnosis of problem

cage card identification and location of patient while in care of facility

client the owner of the animal

color code method used to file medical records for visual ease of use

confidentiality maintenance of privacy of medical information

consent form permission given by owner to treat animal

health certificate documentation that an animal is in good health and can be transported out of state or out of the country

medical record written and recorded information on the care and treatment of the animal

neuter certificate documentation the animal has been spayed or castrated and is no longer able to reproduce

numerical filing system organization of medical records by client or patient last name

objective information measured facts, e.g., vital signs

patient the animal being treated

plan treatments or procedures provided

progress notes documentation chronologically detailing the patient's history, treatments, and outcomes

rabies certificate documentation the animal has received proper vaccination for rabies

SOAP subjective information, objective information, assessment, plan; one method used for documentation in the medical record

subjective information observations of the animal's appearance and behavior made by the veterinary staff

Veterinary-client-patient relationship (VCPR) relationship established between patient, client, and veterinary staff based on trust, expertise, and duty to care for animal

REVIEW QUESTIONS

True or False

Read each statement and determine if it is true or false. Circle T for true and F for false. If a statement is false, rewrite or modify the statement to make it a true statement.

1. The subjective information in a veterinary medical record includes factual information, such as the animal's respiratory rate and temperature. T F
2. When taking a patient's information over the phone, you may record the information in pencil, but use only blue or black ink when recording in the medical record. T F
3. Each medical file should have a format or sequence of where each form is placed in the file. T F
4. The SOAP format is used as a system for filing medical records. T F
5. Upon signing a consent form, the original should be given to the owner and a copy placed in the medical record. T F
6. All errors in a medical chart should be erased, or place correction fluid over the mistakes and correct them immediately. T F

Short Answer

The following items were recorded incorrectly in a medical chart. Describe the proper way to correct the mistakes.

7. Wt. 75.6 lbs—The weight should have been recorded as 57.6 lbs. How will you correct the mistake?
8. T-102 degrees F—The temperature should have been recorded as 101 degrees F. How will you correct the mistake?
9. Why is it a standard practice to keep medical records at least 7 years after the patient's last visit?
10. What two systems are commonly used when filing medical records? Explain each system.

Clinical Situation

Delmar/Cengage Learning

Mrs. Smith scheduled an examination for her dog, "Gretchen," with Dr. Jones for a strange swelling on its abdomen. The veterinary assistant greeted the client and escorted her to an exam room. The veterinary assistant quickly reviewed "Gretchen's" medical file and noted the size, location, and appearance of the swelling. When Dr. Jones arrived to examine "Gretchen," he reviewed the chart and asked Mrs. Smith some questions about the dog's recent health. The vet noticed in the chart that "Gretchen" had been seen in the clinic several years ago for a similar problem, which had been an abscess or build up of fluid from a minor infection.

On examination of the dog, Dr. Jones quickly determined that "Gretchen" had another abscess. He explained the situation to Mrs. Smith, who had forgotten about the previous problem. The vet drained the fluid and prescribed the same antibiotic used for the last infection. A recheck exam appointment was scheduled for 5 days.

1. How did the use of the medical records play out in this case?
2. What might have happened if the medical record had not been reviewed?
3. How did proper medical record filing help the veterinary assistant in this situation?

Competency Skill

Medical Record Assembly

Objective:

To be able to create a medical record for a new patient.

Preparation:

- Familiarize yourself with the proper procedures used in the facility for completion of the medical record. Note the forms used and the sequence or order within the record they are to be placed.
- Manila file folder
- Alphabetical or numerical labels
- Blank label
- Year labels
- Client/patient forms, patient history forms, progress notes, laboratory forms, surgical forms, radiology forms, miscellaneous forms, cage card
- Blue or black pen
- Select an adequate space to assemble the medical record.

Procedure:

1. Place proper labels on the manila file folder according to the filing system used in the facility. This includes the owner and patient information label and filing labels.
2. Place the year label on the manila file folder.
3. Record the owner and patient information on each form using a blue or black ink pen.
4. Secure the forms and file them in proper sequence and location within the medical record.
5. Correctly file the medical record in the facility filing system.
6. Put all materials in their proper storage area.

Competency Skills

Filing Medical Records

Objective:

To properly file records as per office protocol.

Preparation:

- Determine the type of filing system used by the facility.
- Locate completed files and filing area.

Procedure: Alphabetical Method:

1. Locate the medical record file and the owner's last name.
2. The medical record is filed alphabetically by the owner's last name in the filing cabinet.
3. If there are multiple last names, then the record is filed alphabetically according to the owner's first name.
4. If there are multiple files for the same name, the record is filed alphabetically according to the patient name.
5. Check each record filed to make sure no misfiling errors occur.

Procedure: Numeric Method:

1. Each client is assigned a number according to the facility filing system.
2. A separate recording system, either index cards listed alphabetically by name or a computer system, should record each client number.
3. Each record is then filed by number in the filing cabinet.

Procedure: Files for Appointments:

1. Pull the record from the shelf or cabinet according to the filing system method. Pull each file for the daily scheduled appointments.
2. Place the files in a holding area for the daily appointments. File each medical record in order of scheduled appointments.
3. After the appointment or patient discharge, place files in a bin for refiling.
4. Complete filing according to the facilities filing method. Recheck.
5. Before leaving the veterinary office area, make sure all files have been returned to the proper location.

3

Scheduling and Appointments

Delmar/Cengage Learning

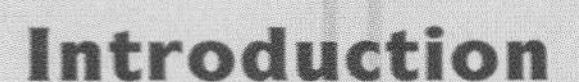

Objectives

Upon completion of this chapter, the reader should be able to:

- Properly schedule appointments in a veterinary clinic
- Properly discuss the necessary client and patient information needed to set up an appointment
- Identify the importance of hospital policies related to scheduling appointments
- Implement hospital policies when scheduling appointments
- Manage the appointment schedule efficiently and accurately

Introduction

Appointment scheduling is an important part of the administrative duties in a veterinary medical facility. It is important to follow the facility's scheduling procedures to pace each veterinarian's day appropriately. This area of veterinary office procedures requires efficiency and knowledge of common medical problems so that the veterinary health care team is not overworked or stressed during office hours.

The Veterinary Appointment Book

The appointment book is a valuable tool for recording notes and managing the daily workings of the veterinary facility (see Figure 3-1). Many styles and formats of appointment books or schedules can be used in a veterinary facility. The type of appointment book used is determined by the type of schedule that the hospital has established.

Some facilities may use veterinary software that has a **computerized appointment book**. A computerized appointment book is scheduled and maintained on the veterinary computer system. The procedures that accompany the software should be used when recording information in the computer. The computerized program works much like the traditional written appointment book schedule, but it allows a more organized scheduling system that saves all information for future use. Many veterinary practices are becoming completely computerized and moving away from using paper in facilities.

Types of Appointment Schedules

There are three basic types of schedules used in veterinary facilities: wave schedule, flow schedule, and fixed office hours. It is important to have a general understanding of each type of schedule.

Wave Schedule

The **wave schedule** is maintained by a veterinarian seeing a total number of patients within a set amount of time. This is usually determined by the veterinarian who will see a certain number of patients within an hour.

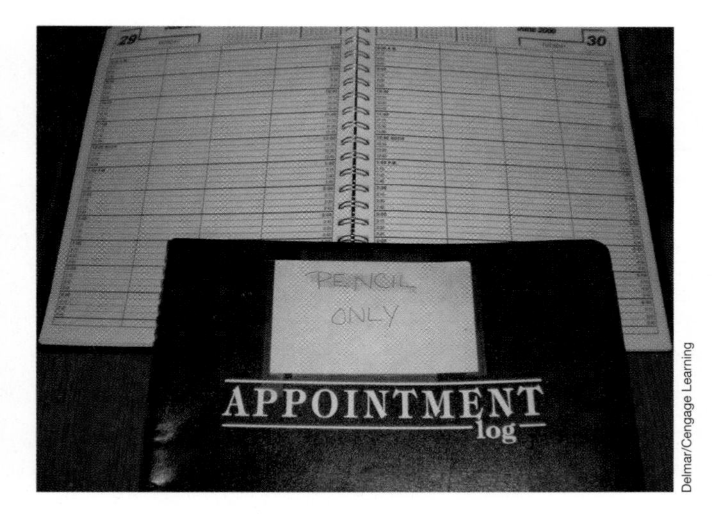

FIGURE 3-1 The appointment book.

They are usually seen in the order they arrive within the time frame. For example, a wave schedule may be set in the morning for the veterinarian to see 4 patients between 9 a.m. and 10 a.m. or over a one-hour time frame. This allows 15 minutes for each patient. The wave schedule is helpful in situations where an appointment may not show up or is late.

Flow Schedule

The **flow schedule** is based on each patient being seen at a specific time, or **interval**, such as every 15 or 30 minutes. An interval is a set amount of time that is required to see a patient. The most common interval is 15 minutes (see Figure 3-2); thus, appointments are scheduled for every 15 minutes in the appointment book. Some medical problems may require more or less time than the scheduled interval.

The flow schedule is the most common type of veterinary schedule. It allows for monitoring time and keeping the staff on schedule. However, if not managed appropriately, it is easy to get behind schedule or overbook patients.

Fixed Office Hours

A facility with **fixed office hours** may have office hours listed throughout the day when clients arrive and the patients are seen in the order they enter the facility. This is a schedule set for walk-in appointments. Some clients will sign in or check in at the front desk and be seen by the veterinarian on a first-come basis. A disadvantage to this schedule type is a loss of control over the flow of patients.

Types of Appointments

It is important for employees who schedule appointments to be trained and knowledgeable in common medical conditions so they are able to determine approximately how long the veterinarian will need to be with the patient. If it is believed the appointment may go over the scheduled interval, it is good practice to confirm this with the attending veterinarian to gain approval to schedule the appointment and allow more or less time accordingly.

Routine Appointments

Routine appointments are scheduled in the appointment book as physical exams during the office appointment schedule. Routine appointment examples would be yearly physical exams, vaccinations, and sick patient exams. Each routine appointment may be spaced out according to how long the expected procedure or exam will take. Each veterinarian may have a different amount of time needed for specific appointments.

Appointment Schedule

8	00	
	15	
	30	
	45	
9	00	
	15	
	30	
	45	
10	00	
	15	
	30	
	45	
11	00	
	15	
	30	
	45	
12	00	
	15	
	30	
	45	
1	00	
	15	
	30	
	45	
2	00	
	15	
	30	
	45	
3	00	
	15	
	30	
	45	
4	00	
	15	
	30	
	45	
5	00	
	15	
	30	
	45	

FIGURE 3-2 Sample appointment book style.

Surgical Appointments

Surgical appointments are usually kept in a separate appointment book and schedule. Surgeries are commonly done at specific times in the day and set by the veterinarian scheduled for surgical duty on a particular day. Surgeries involve anesthesia and a sterile or clean, specialized environment to perform specific procedures. There may be specific surgery days or specific surgery times designated by the hospital. The appointment schedule for surgical cases should include the veterinarian's name, the type of procedure, and the surgery times. Many veterinary hospitals have a set **drop-off time** for surgical patients, which is a specific time an animal arrives prior to a surgery or another procedure.

When scheduling a surgery appointment, the following information should be provided on the schedule: owner name, phone number, patient name, breed or species, and surgical procedure. Veterinarians usually have presurgical instructions that are given to clients, and these instructions should be provided to the owner at the time the surgery appointment is set up. Common presurgical instructions that should be communicated to the client at the time the surgical appointment is scheduled include the following:

- Do not feed 12 hours prior to surgery.
- Do not give water 8 hours prior to surgery.
- Bring in your pet at the specified drop-off time.
- Sign the surgical release form.
- Sign presurgical blood work consent forms.
- Sign any estimates or form disclosing payment for services.
- Bring any medications currently prescribed to the pet.

Many hospitals will hand out or mail these instructions to their clients.

Emergency and Walk-in Appointments

Many hospitals offer **emergency services** and **walk-in appointments**. An emergency is an appointment that must be seen immediately and is a life or death situation. Common emergencies in veterinary medicine include hit by cars, bloats, and severe bleeding from wounds. Emergencies should be clearly identified as needing immediate attention by the veterinarian. Walk-ins are clients that do not have an appointment and arrive at the hospital wishing to see the veterinarian. Even if the facility doesn't allow walk-in clients, they will still occasionally arrive.

The best way to serve emergency situations and walk-ins is to have several slots on the daily appointment schedule reserved for such cases. This will allow clients the service of knowing that if there is an emergency, the clinic will possibly be able to see the animal. It will serve the veterinary health care team in keeping on schedule and not delaying the services of the clients. If the emergency or walk-in slots are not used, it is usually a benefit to have some catch-up time for the staff. When emergencies or walk-ins call or arrive at the hospital, the veterinary technician or veterinarian should be informed immediately so the situation can be handled appropriately. Once the technician or veterinarian has control over the case, the patient's chart is pulled or assembled if it is a new patient.

Drop-off Appointments

Another service that many veterinary facilities will offer is **drop-off appointments**. Clients may set a drop-off time according to their schedule and the veterinarian's schedule for routine appointments, surgical procedures, or grooming services. This allows flexibility for both the owner and the staff to work on the patient throughout the day. There is a set drop-off time and a set pick-up time. Charts for these appointments are usually kept with surgical appointments and pulled when the patient arrives. It is important to note that drop-off patients have owner consent forms for the services that will be given and contact numbers for the owner if the health care team has any questions that may arise.

House Calls

Many clinics offer **house calls** and **mobile services**. These services are provided at the client's home. Mobile services usually include a vehicle that has equipment and an area to do procedures within the vehicle. The appointments for these services should be noted on the appointment schedule or confirmed with the veterinarian before setting the appointment. The facility should have a policy on scheduling house calls. It may be necessary to take a message and call the client back with a date and time of the house call. Once a date and time are scheduled, this should be noted in the appointment book. The staff will need to take into account that the veterinarian will be off the premises and may need additional staff members. The medical file for house calls and mobile services should be assembled and completed prior to the date and time of the appointment. It is beneficial to include extra forms that may be necessary. Also, include the reason for the appointment so the veterinary health care team has adequate information on the equipment and supplies that will be needed for the appointment.

Organization of the Appointment Book

It is important to make sure the appointment book allows enough space for the time frame that the office is open for appointments. It is also important to make sure the space is adequate to record the information necessary for the appointment and the scheduled time intervals (see Figure 3-3A and Figure 3-3B). Most facilities record the following information in the appointment book:

- the client's name
- phone number
- patient name
- breed or species
- a brief note stating the reason for the visit

This information is critical to the veterinarian and the health care team in being prepared and efficient for each scheduled appointment. Some clinics or veterinarians may request other items of information in the appointment space. It is important to know the facility's expectations and rules for scheduling appointments.

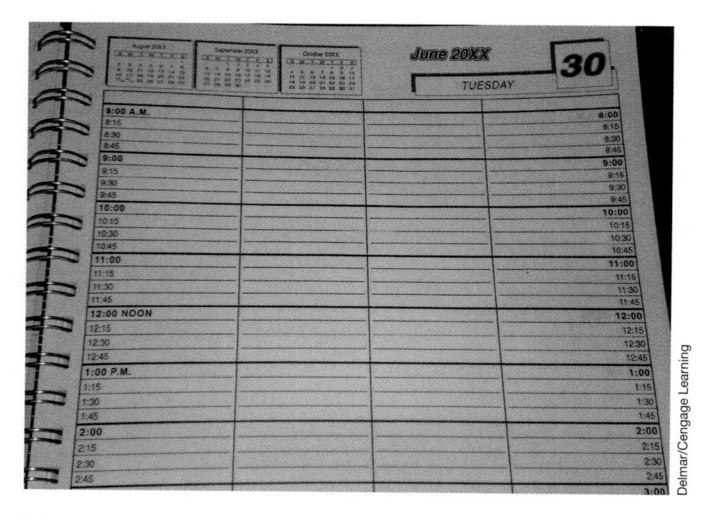

	DAY DATE	MON 6/23
Julie Simon, 555-0001; Persian; Amos; Eye Infection	8:	00
		(15)
		30
		45

Delmar/Cengage Learning

FIGURE 3-3 There should be enough room allowed in the appointment book for the veterinary assistant to write in all appropriate information.

Scheduling Appointments

The top priority when scheduling appointments should be efficiency. Many facilities will have designated time frames for many common and routine procedures and appointments (see Figure 3-4). Information in the appointment book should always be recorded in pencil so that changes can be easily made should clients need to reschedule their appointments (see Figure 3-5).

Common Veterinary Procedures and Time Estimates	
• Yearly exam and vaccines	20 min
• New puppy or kitten exam	30 min
• Repeat puppy and kitten vaccines	15 min
• Recheck exam	10 min
• Suture removal	5 min
• Hospital discharge appointment	15 min
• Sick pet exam	30 min
• Consultation exam/second opinion	30 min
• Behavioral consult	45 min
• Bird, reptile or exotic exam	45 min
• Reproductive exam	30 min
• Artificial insemination	30 min
• Chronic problem exams (ears, skin)	15 min
• Nail trims	10 min
• Pre-surgical blood-work collection	10 min
• X-Rays or Radiographs	30 min
• Heartworm check	10 min
• FeLV test	10 min
• Fecal check	10 min

Delmar/Cengage Learning

FIGURE 3-4 The veterinary assistant should be knowledgeable about the times in the schedule that are allotted to routine types of appointments and procedures.

FIGURE 3-5 Pencil should be used when entering information into the appointment book.

This allows for erasing appointments that may be cancelled or rescheduled. It is not appropriate to scratch out or use correction fluid to cover the entry in the appointment book, as this makes the book difficult to read, and it may confuse staff members if the book is not kept neat and legible. The appointment is always recorded using the same method as the filing system. If the file system method is alphabetical, then record the client name by last name first, then first name in the appointment book. If the file system method is numeric, then record the client number and the client name in the appointment space. Also, when scheduling appointments, check to make sure the information is accurate, neat, and complete on a daily basis. The more time spent managing the appointment book, the more the day will be on time and less stressful for the entire veterinary health care team.

Working with the Client

When an appointment is scheduled, it is recommended that you read the date and time of the appointment back to the client and make sure all client and patient information is accurate. This will allow any necessary changes to be updated in the charts. When at all possible, give **appointment cards** to remind clients of their appointment dates. It is important to make sure new clients have directions to the facility so they do not get lost or arrive late. It is also a benefit to clients and the veterinary staff to confirm all appointments a day in advance. After reaching a client and confirming the appointment, note this in the schedule or medical chart (see Figure 3-6). Many hospitals have a policy on clients that do not show up for an appointment, and this can serve as a history in the medical chart.

FIGURE 3-6 The veterinary assistant should call the client in advance to confirm the appointment and note any changes in the appointment book.

Cancellations

It is important to ask clients to call if they cannot keep their scheduled appointment, and many hospital policies require cancellations 24 hours in advance. Make a note of any cancelled appointment or when an owner doesn't show for his or her scheduled visit.

Blocking Times

When scheduling the day it is necessary to consider the times during the day when surgeries are being performed; when the office is closed; when employees have lunches, breaks, or days off; and when emergencies are scheduled. Marking the daily schedule with times blocked off for emergency appointments will allow for daily emergencies to be seen without disrupting the flow of the scheduled appointments. This may be done by highlighting areas or drawing an "X" through areas that the veterinarian would prefer to use as a catch-up time or for emergencies that may occur during the day (see Figure 3-7). An example would be to block off a 30-minute time frame in the morning and another in the afternoon to account for emergencies that are brought into the clinic; this prevents the staff from getting behind schedule and keeping scheduled clients waiting.

Scheduling for Multiple Veterinarians

If the facility has multiple veterinarians on staff, each doctor is scheduled according to his or her available appointment times. The appointment book should accommodate the number of veterinarians that work on any given day and may work during the same office hours. Some veterinary facilities offer appointments with the veterinary technicians so there may need to be an elaborate or customized appointment book according to each facility. Often a color-coded system will be used to denote multiple doctor's schedules and times when appointments should *not* be scheduled. Highlighters and colored pencils may be useful (see Figure 3-8).

Scheduling Mishaps

A carelessly created schedule will cause chaos, angry clients, and stress for the veterinary health care team and patients. No one likes to wait, and this tends to lead to unhappy and dissatisfied clients. An unhappy client will not return to the facility. An employer will not be tolerant of poor scheduling, either. Figure 3-9 lists some of the dos and don'ts for scheduling appointments. It is important to note that certain days or times of the day may require scheduling issues, such as when staff members and veterinarians will not be available, including lunch

Monday, May 9

8	00			
	15		*Out*	
	30			
	45		*Accountant*	
9	00			
	15			
	30			
	45			
10	00			
	15			
	30			
	45			
11	00			
	15			
	30			
	45			
12	00			
	15		*Lunch*	
	30			
	45			

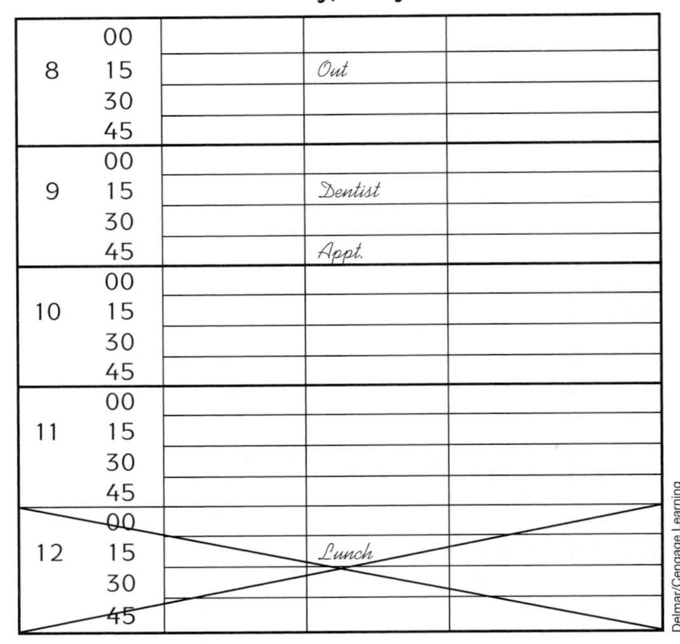

Tuesday, May 10

8	00			
	15		*Out*	
	30			
	45			
9	00			
	15		*Dentist*	
	30			
	45		*Appt.*	
10	00			
	15			
	30			
	45			
11	00			
	15			
	30			
	45			
12	00			
	15		*Lunch*	
	30			
	45			

FIGURE 3-7 Times that are not available for appointments should be blocked off in the appointment book.

Multiple Doctor Schedule

Dr. Jones	Dr. Smyth
Date:	Date:
8:00	8:00
8:15	8:15
8:30	8:30
8:45	8:45
9:00	9:00
9:15	9:15
9:30	9:30
9:45	9:45
10:00	10:00
10:15	10:15
10:30	10:30
10:45	10:45
11:00	11:00
11:15	11:15
11:30	11:30
11:45	11:45
12:00	12:00
12:15	12:15
12:30	12:30
12:45	12:45

FIGURE 3-8 Sample of an appointment book for multiple veterinarians. (*Continues*)

Dr. Jones	Dr. Smyth
1:00	1:00
1:15	1:15
1:30	1:30
1:45	1:45
2:00	2:00
2:15	2:15
2:30	2:30
2:45	2:45
3:00	3:00
3:15	3:15
3:30	3:30
3:45	3:45
4:00	4:00
4:15	4:15
4:30	4:30
4:45	4:45
5:00	5:00
5:15	5:15
5:30	5:30

FIGURE 3-8 *(Continued)*

Appointment Book Dos and Don'ts

DOS

- Use pencil to schedule appointments
- Confirm all appointments
- Give reminder cards with the date and time of the appointment
- Give hospital directions to new clients
- Allow open slots for emergencies and walk-ins
- Make sure every appointment is accurate and complete
- Record cancellations and no-shows
- Keep the appointment schedule neat and easy to read
- Identify multiple doctors schedules clearly
- Identify when there are NO office hours by an X over the closed times

DON'TS

- Use pen to schedule appointments
- Assume clients remember everything they're told
- Overbook the veterinarian and staff
- Neglect to follow hospital scheduling policies
- Schedule house calls without consulting the veterinarian
- Scratch out or white out the appointment book
- Forget to get a phone number for all appointments

FIGURE 3-9 Appointment book Dos and Don'ts.

time, surgery time, and treatment times. These scheduling issues may need to be addressed in the appointment schedule to prevent conflicts and avoid scheduling an appointment when a doctor is not available. Times when staff or veterinarians are not available should be noted in the appointment book. This may be best done by drawing an "X" through a time frame when scheduling should *not* be done.

Policies and Procedures

The veterinary clinic should have a detailed policy and procedures manual that is provided to each staff member so everyone will understand the rules and regulations of the facility (see Figure 3-10). The **policy manual** provides information regarding what the staff expectations

Policy and Procedures Manual

POLICY MANUAL ITEMS	PROCEDURES MANUAL ITEMS
Dress Code/Attire	Payment Policy
Personal Appearance	Appointment Policy
Vacation time	Surgery Policy
Sick time	Scheduling Policy
Personal time	Opening/closing procedures
Attendance Policy	Admission/Discharge Procedures
Continuing Education	Emergency procedures (fires, robbery, injuries, etc.)
Benefits (Employee discounts, uniform allowance, reimbursements)	Answering phones
Codes of conduct	Invoicing and billing
	Computer system procedures
	Client visitation hours

FIGURE 3-10 Information found in the policy and procedures manual.

are, such as the dress code, policies on vacation time, personal time, and sick leave. The policy manual is provided so that all staff members have an understanding of the employee rules. This will apply to all staff employees, and some regulations may be based on the length of time that an individual has been employed by the facility. The **procedures manual** provides the rules regarding the workings of the veterinary facility itself, such as client rules, scheduling methods, payment procedures, and client visitation times. This information is based on the rules of the clinic and is part of a training method for all new employees to follow so that all staff members are completing their duties and responsibilities in the same manner. These manuals will outline the employer's expectations of the employees and provide a basis of consistency and how practice issues are handled.

SUMMARY

The appointment book and art of scheduling must be an important part of the clinic's policy and procedures manual. Each employee should know how to schedule appointments and know the policies affecting each type of appointment. It is important to be efficient and manage the schedule closely. Learning to work ahead and have medical charts and equipment prepared ahead of time will make a day in the veterinary facility less stressful. Efficient and accurate scheduling will also keep clients happy about the facility's services and will keep them coming back. The veterinary health care team must be able to work together to make an efficient and safe environment.

Key Terms

appointment cards a written reminder for clients that includes a date and time to remember the appointment scheduled for their pet

computerized appointment book software item used to schedule appointments and maintain medical information on the veterinary computer system; allows future use of information in an organized manner

drop-off appointment a set time for the client to bring in an animal and leave the patient for exams, surgeries, and other procedures according to the veterinarian's schedule

drop-off time set time when an animal is expected at the veterinary facility for a later exam or surgery

emergency services an appointment that must be seen immediately and is a life or death situation

fixed office hours appointments seen throughout the day when clients arrive and the patients are seen in the order they enter the facility during set times

flow schedules appointments scheduled based on each patient being seen at a specific time by the veterinarian during the day, such as an appointment scheduled for 1:00 p.m.

house calls appointment scheduled for the veterinarian to examine a patient in the owner's home

intervals a set amount of time that is required to see a patient, such as 15 minutes

mobile services a vehicle equipped with veterinary equipment, tools, and supplies that travels to a client's home to examine a patient

policy manual information provided to all staff members as to what the staff expectations are, such as the dress code, policies on vacation time, personal time, and sick leave

procedures manual information provided to all staff members on the workings of the veterinary facility itself, such as client rules, scheduling methods, payment procedures, and client visitation times

routine appointment appointment scheduled in the appointment book as physical exams during the office appointment schedule; examples would be vaccinations, yearly exams, and rechecks

surgical appointment appointment scheduled during the veterinarian's set time for doing surgeries and anesthesia, usually separate from routine office hours

walk-in appointment a client that does not have an appointment and arrives at the hospital wishing to see the veterinarian by walking in the door

wave schedule appointments maintained by a veterinarian seeing a total number of patients within a set amount of time during the day and is based on each hour the facility is open

REVIEW QUESTIONS

Matching

Match the term to its definition.

1. appointment cards	a. set time when an animal is expected at the veterinary facility for a later exam or surgery
2. computerized appointment book	b. an appointment that must be seen immediately and is a life or death situation
3. drop-off time	c. written reminder for clients that includes a date and time to remember the appointment scheduled for their pet
4. emergency services	d. information provided to all staff members as to what the staff expectations are, such as the dress code, policies on vacation time, personal time, and sick leave
5. fixed office hours	e. appointment scheduled in the appointment book as physical exams during the office appointment schedule; examples would be vaccinations, yearly exams, and rechecks
6. flow schedules	f. appointment scheduled during the veterinarian's set time for doing surgeries and anesthesia, usually separate from routine office hours
7. intervals	g. appointments maintained by a veterinarian seeing a total number of patients within a set amount of time during the day and is based on each hour the facility is open
8. mobile services	h. appointments seen throughout the day when clients arrive and the patients are seen in the order they enter the facility during set times

9. surgical appointment
10. wave schedule

i. software item used to schedule appointments and maintain medical information on the veterinary computer system that allows future use of information in an organized manner

j. a set time for the client to bring in an animal and leave the patient for exams, surgeries, and other procedures according to the veterinarian's schedule

k. appointments scheduled based on each patient being seen at a specific time by the veterinarian during the day, such as an appointment scheduled for 1:00 p.m.

l. appointment scheduled for the veterinarian to examine a patient in the owner's home

m. a set amount of time that is required to see a patient, such as 15 minutes

n. a vehicle equipped with veterinary equipment, tools, and supplies that travels to a client's home to examine a patient

o. a client that does not have an appointment and arrives at the hospital wishing to see the veterinarian by walking in the door

p. information provided to all staff members on the workings of the veterinary facility itself, such as client rules, scheduling methods, payment procedures, and client visitation times

Short Answer

1. List the information you should obtain when setting up an appointment.
2. As a veterinary assistant, you work in a multiple veterinarian practice and need to set up the appointment book. List the factors you should consider in doing the scheduling and ways to be accurate in scheduling.
3. What should be done to the schedule to ensure no appointments are scheduled during lunch and when the office is closed?
4. What types of policies regarding appointment scheduling should you be aware of when you work at a veterinary facility?
5. What type of appointments should you consult with the veterinarian before scheduling in the appointment book?

Clinical Situation

© iStock/Ruth Ann Johnston

The veterinary assistant, Cathy, is working at a small animal clinic, answering phones and handling the appointment schedule. Cathy answers the phone: "Good afternoon, Green Valley Veterinary Hospital, this is Cathy, how may I help you?"

"Yes, this is Mrs. Haines. I need to make an appointment for my dog, 'Timmy.' He has not been feeling well."

"What has been the problem with 'Timmy,' Mrs. Haines?"

"He has been vomiting for a few days and not eating. He has been very lazy and he has me concerned. Can Dr. Bloom see him today?"

"I think that 'Timmy' should be seen as soon as possible. Can you bring him in at 2:00?"

"That is fine. Thank you so much."

"We will see you and 'Timmy' today at 2:00," Cathy said, placing Mrs. Haines's name and phone number in the appointment book. She also wrote "Timmy's" name and information including his history of vomiting in the schedule. When she pulled the medical record, she noticed that "Timmy" is overdue for his yearly exam and vaccines. Cathy noted this in the medical record and on the appointment schedule. She also noted that "Timmy" is a 10 yr old NM Collie.

- Which form of scheduling does this clinic most likely use?
- How does placing the overdue vaccine and yearly exam information in the appointment book help the veterinary staff?
- What information can be determined by the age and breed of the animal when listed on the appointment schedule?

Competency Skill

Appointment Scheduling

Objective:

To accurately and efficiently schedule appointments in the appointment book.

Preparation:

- Have knowledge of any computer-based veterinary software
- Sharpened pencils
- Erasers
- Appointment book

Procedure:

1. Have the appointment book open to the current date when appointments begin.
2. Block off unavailable times for appointments using a single line or X through the time frames.
3. Determine the type of appointment as routine, surgical, emergency, drop-off, or house call.
4. Determine the amount of time necessary for the appointment according to the schedule intervals.
5. Ask the client for a preferred date or time for the appointment.
6. Check the availability of the date and time. Schedule the appointment based on the client and the clinic suitability.
7. Enter the data into the appointment book as follows:
   - Client's name (confirm)
   - List as alphabetical filing or numeric filing according to the clinic medical record system
   - Patient's name, age, breed or species
   - Client's phone number (confirm)
   - Reason for appointment or chief complaint
   - Record veterinarian to see the patient
8. If the client is present when the appointment is scheduled, complete a reminder appointment card with the date and time and give it to the owner.
9. Call the client a day in advance to remind him or her of the date and time of the scheduled appointment.
10. Update and check for accurate information with each client and patient when he or she arrives for the appointment.

© iStock/Ilda Masa

Computer and Keyboarding Skills

Objectives

Upon completion of this chapter, the reader should be able to:

- Identify the basic parts of the computer and computer equipment
- Explain the importance of veterinary software programs
- Discuss the uses of a computer system in a veterinary facility
- Implement basic keyboarding skills within the veterinary clinic

Introduction

Many veterinary offices use computers and veterinary software programs that maintain records, schedule appointments, record client and patient information, and handle invoicing and accounting for the veterinary facility. Everyone has a different level of computer experience and knowledge. Many computerized facilities will have expectations of their employees to have a general understanding of computers and basic keyboarding skills. Therefore, it is important to develop an understanding of the computer system and software programs within the veterinary clinic.

Basic Computer Equipment

Most computer systems in a veterinary facility will use a basic and simple set up that provides easy-to-use **hardware**. The hardware refers to the physical parts of the computer and its equipment. Hardware consists of input devices, output devices, and the CPU. The primary input devices are the **keyboard**, which allows for typing; a **mouse**, which allows the user to move the **cursor**, or blinking line, on the computer screen; and the **scanner**, which allows documents and images to be copied and stored electronically. The primary output devices of the computer are the **monitor**, which is the screen; the **modem**, which allows Internet access; and the **printer**, which allows for printing forms and invoices from the computer system (see Figure 4-1). The **central processing unit (CPU)** is the brains of the computer. It carries out the instructions designed by particular **software** (a program or set of instructions the computer follows). It takes the data input into the system, configures it based on the instruction received from the software, and sends it to the output device. In addition to a general knowledge of operating the computer system, the veterinary assistant should also know how to maintain the printer, such as how to load the paper or change the **toner**, which is either black or colored ink that allows printing.

Many clinics will purchase veterinary software that simplifies tasks pertaining to the office and hospital procedures. These software programs include **word-processing programs** that allow typing of forms and notes similar to using a typewriter. Commonly used word-processing programs include Microsoft Word and WordPerfect. Other components to a veterinary software program include patient and client information; medical recording; surgical reports; laboratory reports; anesthesia reports; radiology reports; scheduling appointments; invoicing and billing; inventory management; employee records and information; and consent forms used by the clinic. There are many veterinary software programs available, and it is important to complete specific training on the program used in the clinic.

Veterinary clinics also use such equipment as a copier machine and a fax machine. The **copier machine**

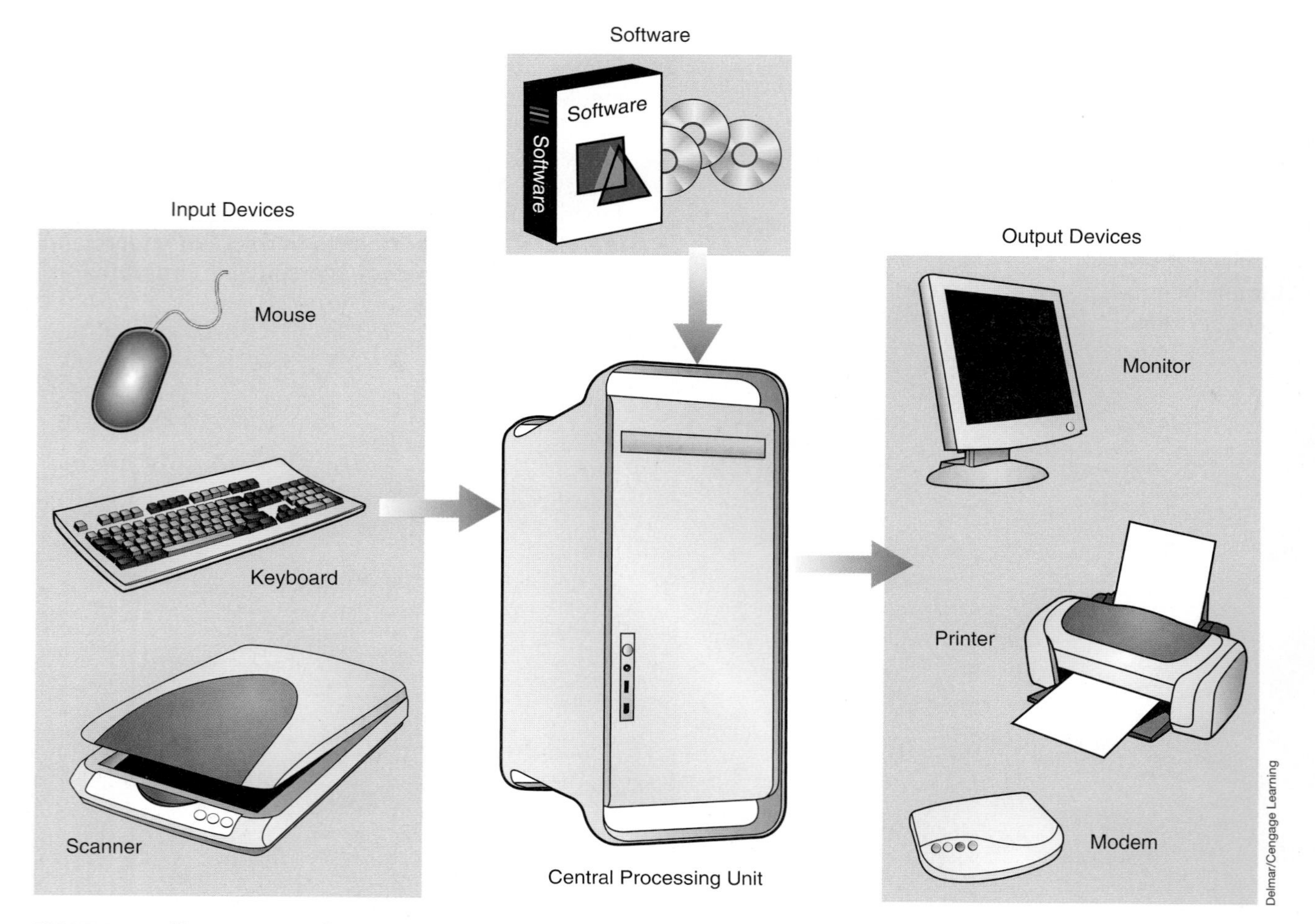

FIGURE 4-1 Components of a computer system.

is used to make paper copies of records and other documents (see Figure 4-2). Copiers must be maintained with toner and paper. They may need additional maintenance and service from the manufacturer. A **fax machine** is used to copy and send documents via a telephone line (see Figure 4-3). Fax machines copy paper documents and transmit them to other fax machines via a telephone line. A fax machine requires toner and regular maintenance and service, too. Some veterinary offices may use a cash register to keep money sorted and contained (see Figure 4-4).

Keyboarding Skills

Keyboarding skills include development of basic understanding and use of functions on the computer keyboard and overall typing skill. The veterinary assistant should know basic keyboarding techniques and habits and have an understanding of keyboard functions (see Figure 4-5). Keyboarding skills will also come in handy when working with cash registers, typewriters, fax machines, and other electronic equipment with keyboards. It usually is not important how many **words per minute** the veterinary assistant can type, but rather the typing and use of the keyboarding functions is accurate. The importance of keyboarding skills is learning the positioning of the keys and the functions that can be performed from the keyboard (see Figure 4-6). There are programs and online tutorials

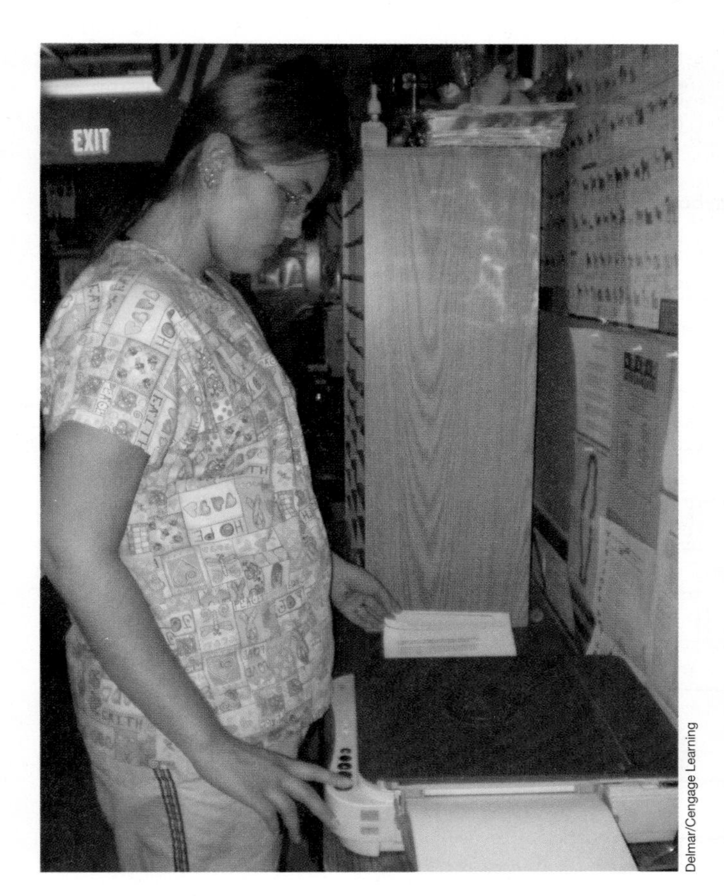

Delmar/Cengage Learning

FIGURE 4-2 Veterinary assistants will need to understand the use and maintenance of a copier.

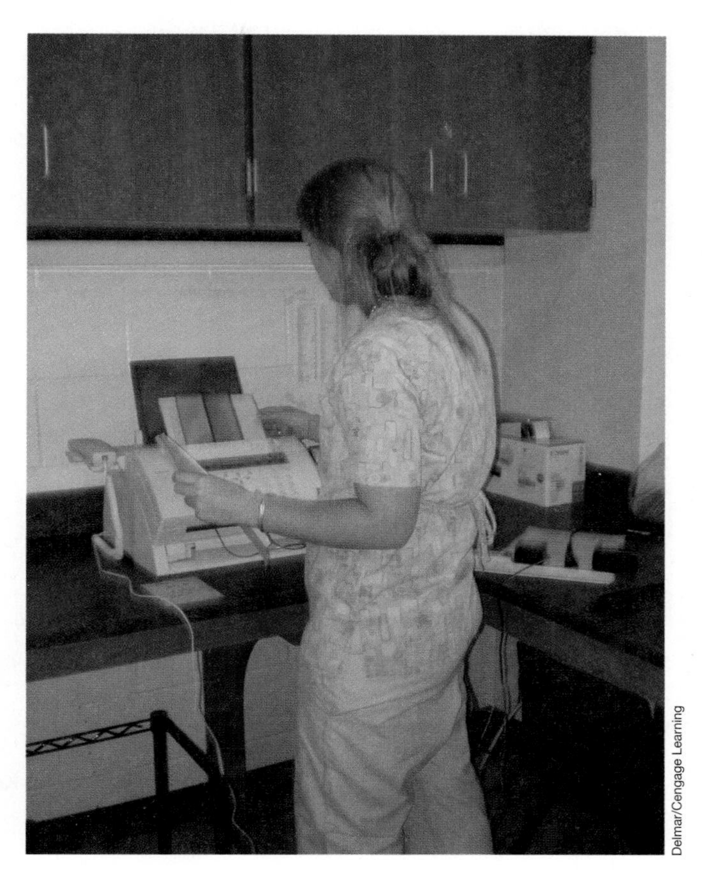

Delmar/Cengage Learning

FIGURE 4-3 Veterinary assistants may often be required to fax information to other offices or to clients.

Delmar/Cengage Learning

FIGURE 4-4 Veterinary assistants may be required to operate a cash register in the veterinary facility.

Delmar/Cengage Learning

FIGURE 4-5 A basic understanding of keyboarding will assist the veterinary assistant in her ability to efficiently perform her duties.

that allow practice and development of keyboarding skills. A common program used to make **spreadsheets** is called Excel. A spreadsheet is used to track large amounts of items in a single document. The program is used by veterinary facilities to track large numbers of items, such as inventory supplies, clients, patients, and invoices. Some common sites and resources to learn keyboarding and develop keyboarding skills include the following:

- www.qwerty.com
- Type to Learn software
- Mavis Beacon keyboarding software
- Typing Tutor
- UltraKey

SUMMARY

Computers are set standards in today's veterinary profession. Employees should have a basic understanding of computers, equipment, and software programs. Accuracy of keyboarding skills and functions can be maximized by employee training and use of the veterinary software. The veterinary assistant should make sure all information entered is accurate and spelled properly. Training may be provided in the clinic or through use of a self-training program and printed manuals. Practice makes perfect!

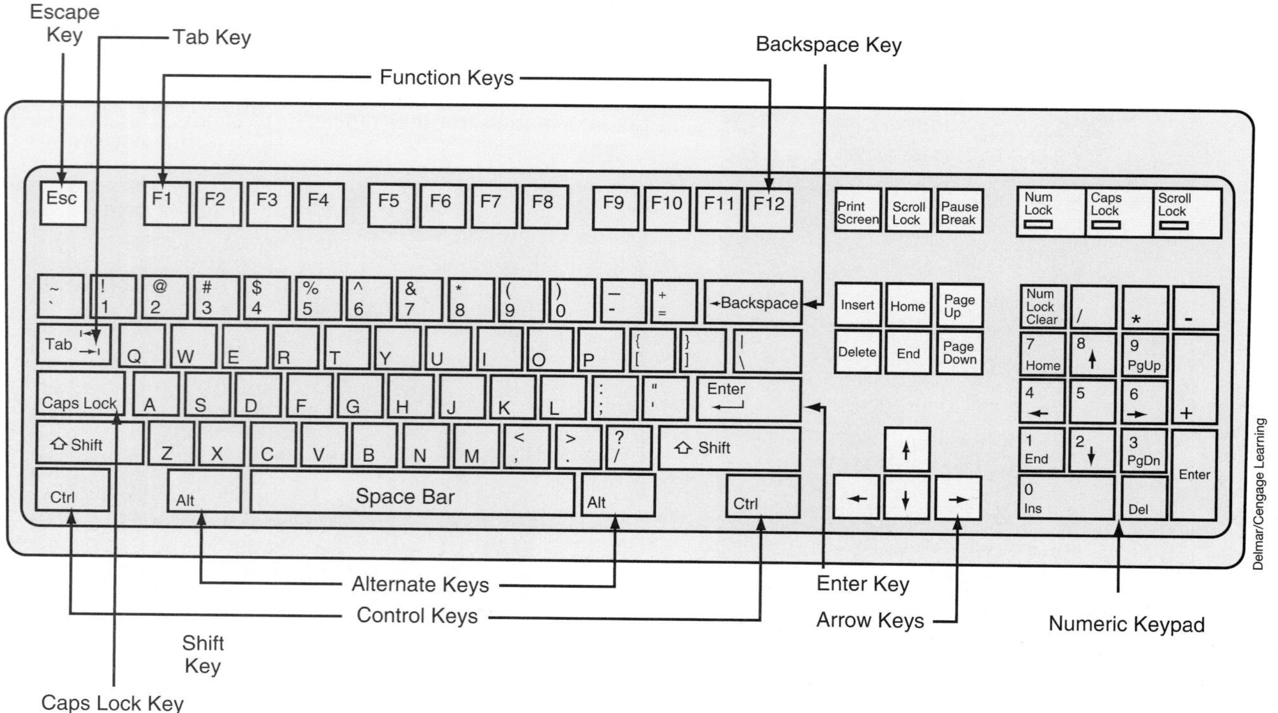

FIGURE 4-6 Standard keyboard.

Key Terms

central processing unit (CPU) the brain of the computer that carries out instructions asked by the software

copier machine equipment used to make exact copies of paper or other documents

cursor blinking line on the computer screen that determines where typing is occurring

fax machine equipment used to copy and send documents via a telephone line

hardware the physical parts of the computer and its equipment

keyboard buttons that have letters that allow typing and word formation on a computer

keyboarding act of typing and development of skills used in computer functions and keys

modem box-like equipment that allows access to the Internet

monitor the computer screen that looks similar to a TV screen

mouse movable instrument attached to a computer that allows the user to move the cursor

printer allows printing forms and invoices from the computer onto paper

scanner equipment that allows documents and images to be copied and stored electronically

software a program or set of instructions the computer follows

spreadsheet program used to track large quantities of items in one file

toner black or colored ink that allows printing in a printer, fax machine, or copier

words per minute the number of words that can be typed within one minute or 60 seconds

word-processing program software that allows typing of a document

REVIEW QUESTIONS

Matching

Match the term to the definition.

1. central processing unit
2. fax machine
3. hardware
4. keyboard
5. modem
6. monitor
7. mouse
8. scanner
9. software
10. toner

a. box-like equipment that allows access to the Internet
b. equipment used to make exact copies of paper or other documents
c. movable instrument attached to a computer that allows the user to move the cursor
d. equipment that allows documents and images to be copied and stored electronically
e. the brain of the computer that carries out instructions asked by the software
f. the computer screen that looks similar to a TV screen
g. allows printing forms and invoices from the computer onto paper
h. program used to track large quantities of items in one file
i. act of typing and development of skills used in computer functions and keys
j. black or colored ink that allows printing in a printer, fax machine, or copier
k. a program or set of instructions the computer follows
l. buttons that have letters that allow typing and word formation on a computer
m. blinking line on the computer screen that determines where typing is occurring
n. equipment use to copy and send documents via a telephone line
o. the physical parts of the computer and its equipment

Short Answer

1. List the components of a computer that are considered hardware.
2. What veterinary tasks can be accomplished using a veterinary software program?
3. Name some commonly used word-processing programs that may be used in a veterinary facility.
4. Medical records may be computerized. T F
5. Why is it important to develop good keyboarding skills?

Clinical Situation

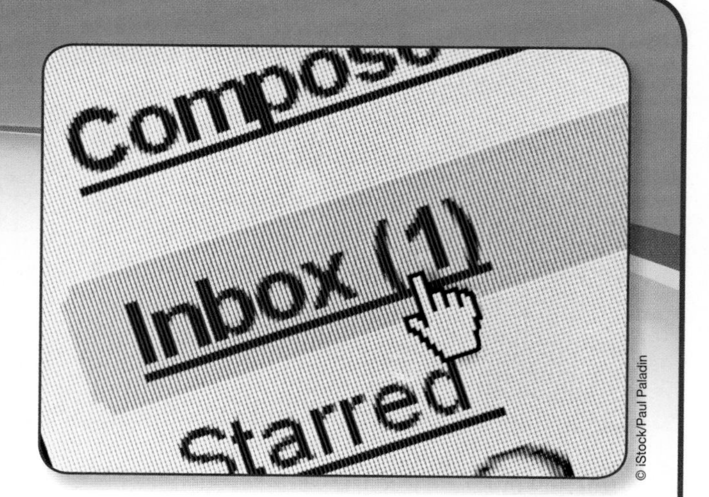

© iStock/Paul Paladin

Kim is a veterinary assistant at the Smithville Vet Clinic. She works the morning shift and opens up the clinic. One of her daily responsibilities is to boot up the computers in the clinic and check the hospital email account each morning to see if any clients have contacted the hospital. Kim notes that the computers are working slowly this morning, and eventually she is able to access the email account. Two clients have sent emails requesting medication refills. She notes the information and client names and writes a memo to Dr. Calvin, who is the veterinarian on duty this morning. One client has also sent an email about changing an appointment time to 9:00 this morning. "That is not enough time to contact the client," she thought, as it was already 8 a.m. "If only she had emailed or called earlier, I would have been able to accommodate her." Unfortunately, the client emailed too late to reschedule the appointment. Kim emailed the client back with other available times and then logged off.

- How did Kim use the computer to aid her work?
- How could Kim have better served the client with the scheduling issue?
- What should Kim do about the computer problems she encountered?
- How does the use of a computer make the veterinary facility more efficient?

Competency Skill

Computer Skills

Objective:

To perform basic computer operations and maintenance.

Preparation:

- Establish basic keyboarding skills (30–45 words per minute typed correctly is the average).
- Review basic computer terminology as well as components of hardware and software.
- Familiarize yourself with basic word processing.

Procedure:

1. Connect or turn on the computer.
2. Learn the veterinary software program and how it is used in the veterinary facility. It is often helpful to train with someone who is competent and knows the following:
   - The responsibilities and duties that are computer based
   - The veterinary program components you are expected to use
   - How to navigate through the program
   - How to use the printers
3. Practice with the program and review your work for accuracy.
4. Double check all information and that it is properly recorded.

5

Veterinary Office Management

Objectives

Upon completion of this chapter, the reader should be able to:

- Explain the roles and duties of each veterinary health care team member
- Explain the fees and prices of the veterinary facility
- Discuss invoicing procedures
- Properly receive and account for cash, credit, and check methods of payment
- Prepare an invoice and collect payments
- Identify the role of payroll services
- Describe billing and collection procedures
- Discuss and practice appropriate inventory management and control

Introduction

Many veterinary facilities have a person in charge of the management and financial matters of the clinic. This person may be trained as an office manager or may have other duties and responsibilities in the hospital. It is important to have knowledge in handling management duties and some basic training in handling finances. The management of the facility is a critical factor of success or failure.

The Veterinary Team

The members of the veterinary health care team are all involved in the facility's daily workings. These duties range from practice management to client communication to treating the patient.

The Veterinarian

The **veterinarian** is a doctor and holds the ultimate responsibility within the facility. Legally the veterinarian is held responsible for the safety of the staff and the actions of each employee within the facility. Veterinarians are the only team members that may legally diagnose patients, prescribe medications, discuss prognosis of patients, and perform surgery (see Figure 5-1). The education of a veterinarian includes four years of a pre-vet program resulting in a **Bachelor of Science (BS)** degree in a science-related course of study. Veterinary medical colleges then offer a four-year doctorate program. In the United States, veterinary schools must be **American Veterinary Medical Association (AVMA)** accredited for state licensure (see Appendix A for a list of accredited schools). The AVMA sets the standards of veterinary medicine and the quality of care veterinarians provide to animal owners. The **American Animal Hospital Association (AAHA)** sets the standards for small companion animal hospitals. Upon graduation, veterinarians are required to take a **State Board examination**. Once the State Board exam has been passed, the veterinarian will be licensed to practice in the state he or she resides in. Veterinarians are required to update their licenses on a regular basis and complete a set number of hours of **continuing education**. Continuing education (CE) refers to a required amount of professional development that a licensed veterinarian and veterinary technician must complete in a set amount of time. This education is meant to keep the staff up to date on changes in the industry and allow the facility to keep learning new information within the industry. It is important to note that *all* staff members should receive continuing education credits.

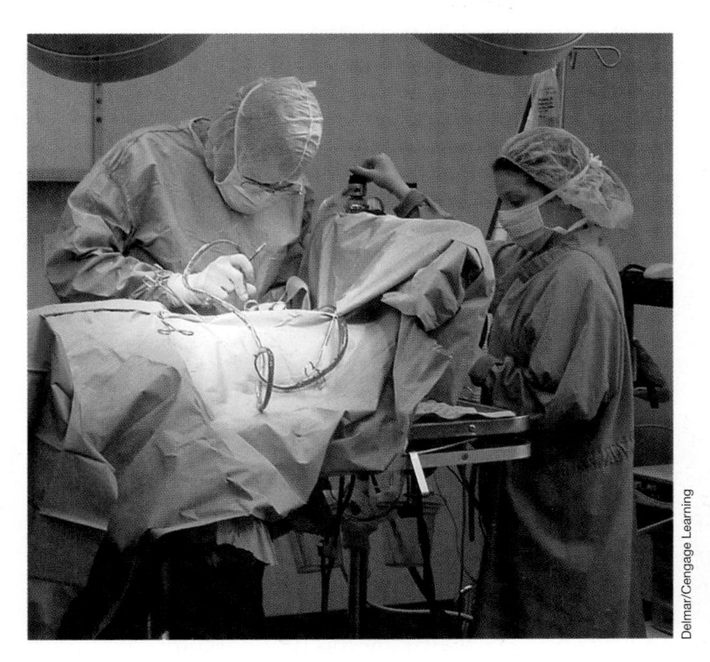

FIGURE 5-1 The veterinarian is the only staff member qualified to perform surgery.

The Veterinary Technician and Veterinary Technologist

Veterinary technicians and **veterinary technologists** have specialized training in areas of veterinary health and surgical care of many types of animal species (see Figure 5-2). The specialized training is acquired at an AVMA-accredited institution (see Appendix B for a list of schools). The veterinary technician graduates with a two-year **associate degree** and the veterinary technologist graduates with a four-year bachelor degree. The associate degree results in an Associate of Science (AS) diploma. The bachelor degree results in a Bachelor of Science (BS) diploma.

The **National Association of Veterinary Technicians in America (NAVTA)** sets the standards of veterinary care for veterinary technicians and technologists. Upon completing the program, graduates will then take a State Board examination as required by the state in which they intend to work. Successfully passing this exam will result in state licensure. Depending on the state, the technicians will be recognized as **Certified Veterinary Technicians (CVT)**, **Registered Veterinary Technicians (RVT)**, or **Licensed Veterinary Technicians (LVT)**.They

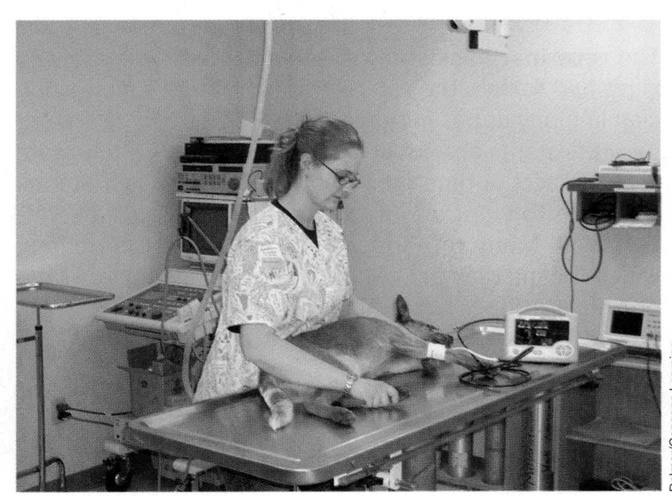

FIGURE 5-2 The veterinary technician performs various procedures within the veterinary facility.

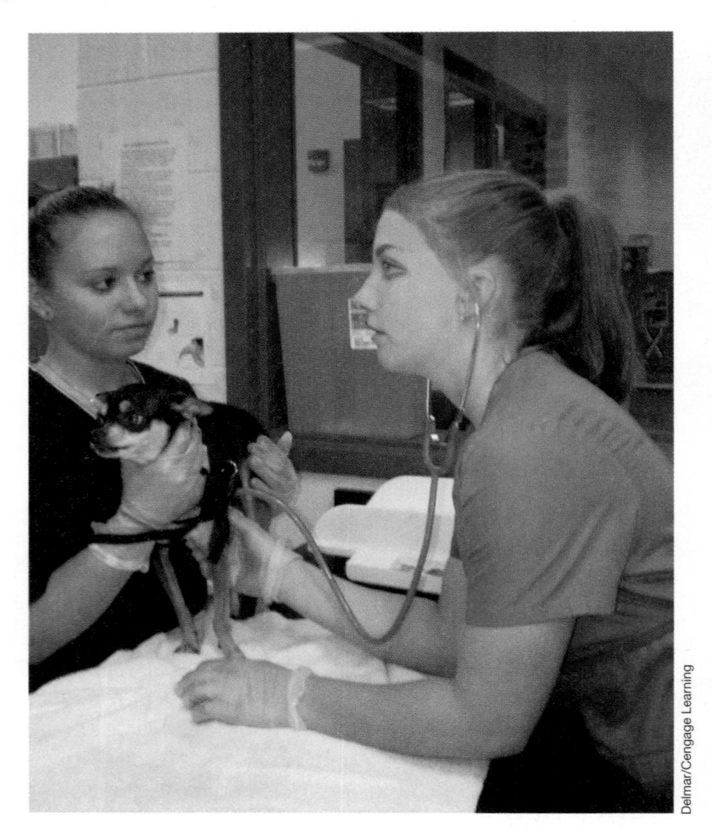

Delmar/Cengage Learning

FIGURE 5-3 The veterinary assistant is a critical part of the veterinary health care team.

work under indirect supervision of the veterinarian. The responsibilities of the veterinarian and veterinary technician are outlined by each state's **veterinary practice act**. The practice act is a set of rules and regulations that govern veterinary medicine.

The Veterinary Assistant

The **veterinary assistant** is a member of the team who is trained to help the veterinarian and veterinary technician in many duties in the facility (see Figure 5-3). Veterinary assistants are not licensed by the state. Many state practice acts outline the rules for veterinary assistants. They are to work under direct supervision of the veterinarian. Many veterinary assistants are cross-trained, which means they are capable of working in every area of the veterinary hospital and of completing many duties and responsibilities.

The Veterinary Hospital Manager

The **veterinary hospital manager** is the person in charge of the internal workings of the facility (see Figure 5-4). This person needs to be trained in business management. The hospital or practice manager is

Photo courtesy of iStock

FIGURE 5-4 The veterinary hospital manager runs the business aspects of the veterinary facility.

responsible for scheduling the staff, paying bills, maintaining inventory, and overseeing the day-to-day business within the hospital. This is usually the person who is interviewing and hiring employees. The manager is the employee who maintains information in the **personnel manual** and the **procedures manual**. The personnel manual contains staff job descriptions, the attendance code, vacation and sick days, the dress code, and employee expectations and conduct. The procedures manual outlines the events of the facility, hospital rules, and staff and client rules. Some hospital managers may wish to become certified veterinary hospital managers.

The Kennel Attendant

The **kennel attendant** is the person in charge of the kennels and animal wards within the facility (see Figure 5-5). Duties of the kennel attendant include cleaning cages, feeding and watering patients, monitoring and exercising patients, and cleaning and sanitizing the facility. Kennel attendants must be trained to handle animals and have a basic knowledge of animal care. They are responsible for noting any abnormal signs or changes in the patients.

The Receptionist

The veterinary **receptionist** is the person responsible for maintaining the front office of the facility (see Figure 5-6). Receptionists have similar duties to that of an administrative assistant. Their roles include scheduling appointments, answering telephones, invoicing and collecting payments, and patient admissions. The receptionist is vital to the daily flow of the hospital. The receptionist is usually the first person clients have contact with and must have excellent communication and office skills.

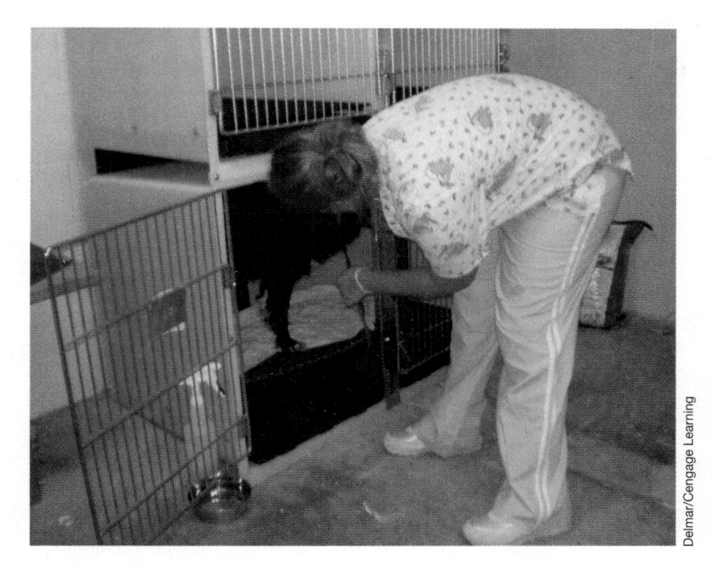
Delmar/Cengage Learning

FIGURE 5-5 The kennel attendant maintains the kennel area and the animals housed in the facility.

Delmar/Cengage Learning

FIGURE 5-6 The receptionist meets and greets clients, answers phones, and fields questions in the veterinary facility.

Professional Dress and Appearance

Professional dress and appearance will vary from veterinary facility to facility. Most commonly, the veterinary health care team including veterinarians, veterinary technicians, and veterinary assistants will dress in professional scrub uniforms, including tops and pants (see Figure 5-7). It is also common for these veterinary members to wear lab coats. Scrub uniforms and lab coats are worn to prevent dress clothing from becoming contaminated, soiled, or torn. These items are easily washed and replaced when necessary.

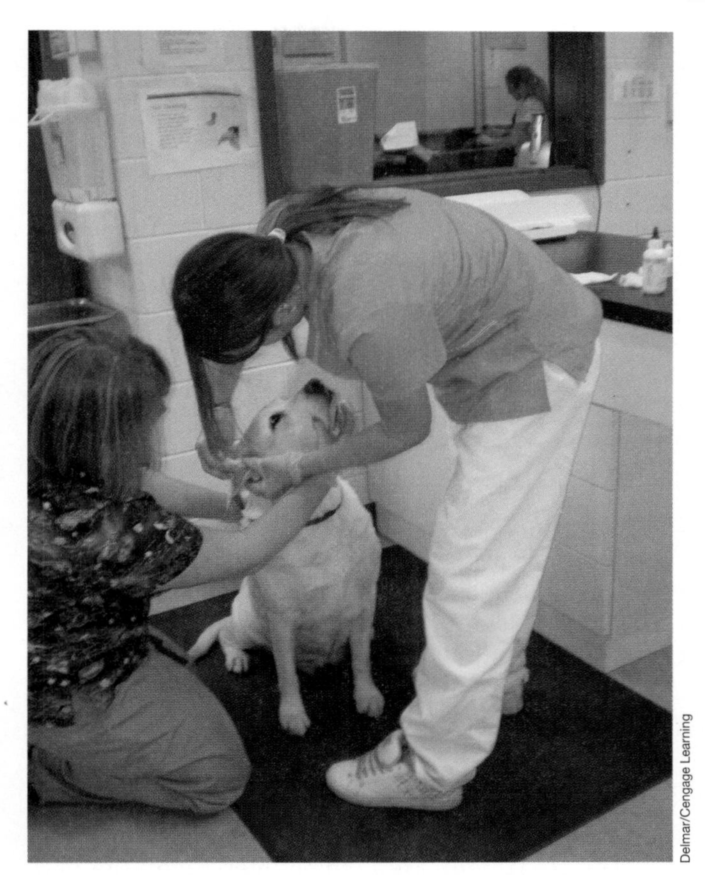
Delmar/Cengage Learning

FIGURE 5-7 The veterinary assistant should dress in a professional manner and be well-groomed.

Some facilities have a set dress code or color to be worn by all staff members. The kennel staff and receptionists may have a similar dress code or may be asked to wear special hospital shirts, such as polo shirts, and dress pants such as khakis. All employees should wear fully enclosed shoes that are nonslip, preferably sneaker-type shoes. This is for safety as well as for ease of cleaning and controlling the spread of contaminants. When working in a veterinary facility, hazards such as animal bites, needles, and glass can cause a foot injury. In large-animal facilities, employees may wear heavier boot-type footwear for protection. Items that should be carried in the pockets or easily available include a blue or black pen, watch or timer, calculator, gauze roll, scissors, ½-inch tape, a nylon leash, and a stethoscope. Long hair should be pulled back out of the face. Jewelry should ideally not be worn or be kept at a minimum. No long bracelets or necklaces or other dangling items, such as hoop earrings, should be worn, as they are safety issues to the employee and to animals. Fingernails should be kept short, neat, and clean and not extend beyond the end of the finger. Artificial nails are not to be worn by staff members handling animals due to an increased potential for bacterial contamination and injury to animals or other staff members. Tattoos and body piercings should be kept concealed, and some facilities may have regulations against these items. The dress code and appearance of all hospital employees should be listed in the hospital manual and outline the expectations of the employer. The professional image the veterinary assistant portrays is the image that the entire hospital portrays. Remember, first impressions are lasting impressions. Clients will judge the staff based on their appearance.

The Veterinary Facility

The veterinary hospital is composed of multiple sections and locations where different veterinary medical procedures and functions are completed. The front entrance typically leads into the veterinary **waiting room** and **reception area** (see Figure 5-8). This is the area where clients wait to see the vet and are greeted and signed in by the front office staff.

The **exam room** is where patients receive routine exams. There may be one or more exam rooms depending on the size of the facility and the number of veterinarians on staff (see Figure 5-9).

The **treatment area**, sometimes called the prep area, is a location in the hospital where patients are treated or prepared for surgical procedures. This is typically an area for veterinary staff only (see Figure 5-10).

The **pharmacy** is the area where medications are stocked and prepared (see Figure 5-11).

The **laboratory** is where patient tests and samples are prepared for sampling. Some examples of laboratory procedures may include fecal samples, urine samples, and blood tests (see Figure 5-12).

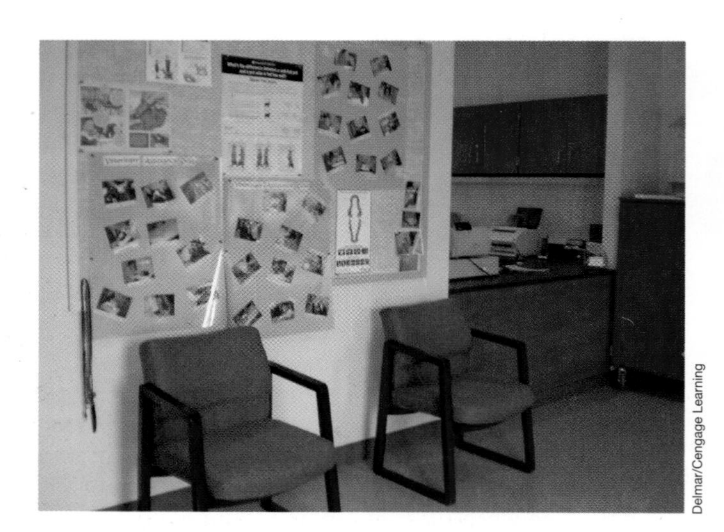

Delmar/Cengage Learning

FIGURE 5-8 The waiting room and reception area.

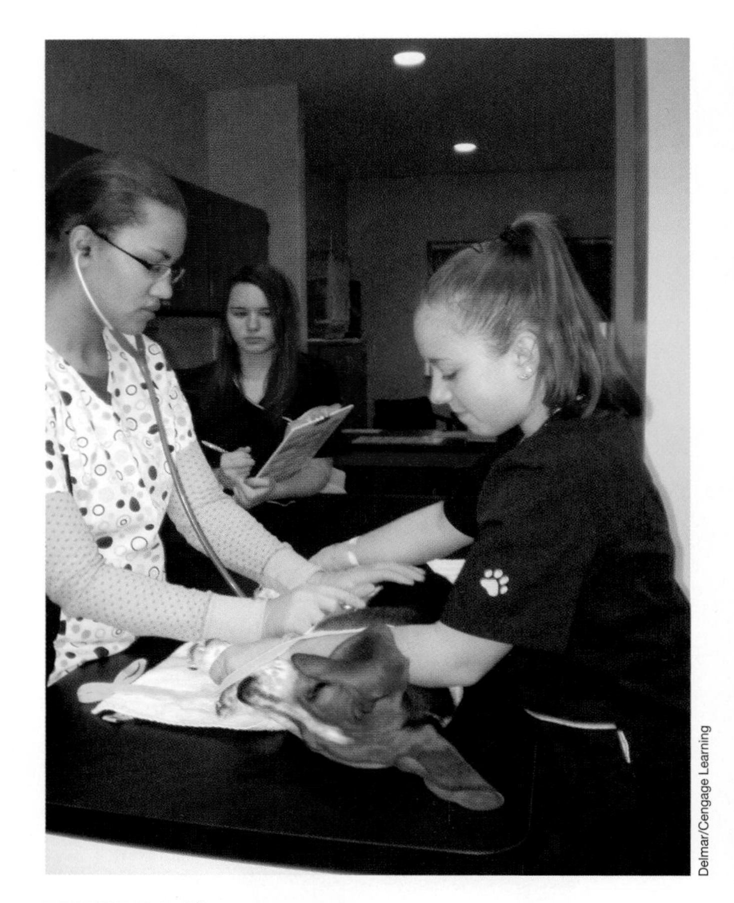

Delmar/Cengage Learning

FIGURE 5-9 The exam room.

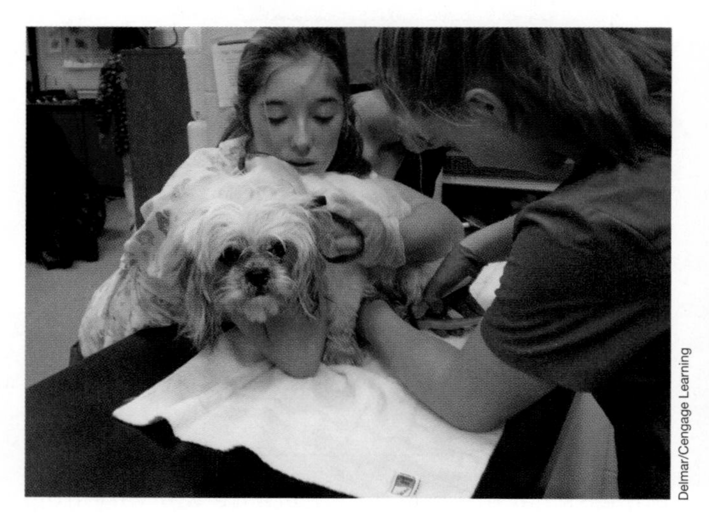

Delmar/Cengage Learning

FIGURE 5-10 The treatment room.

Radiology is a specific area that meets veterinary regulations for use of radiation and special chemicals to develop X-rays (see Figure 5-13).

The **intensive care unit** or ICU is where critical patients needing immediate care are treated and housed. This area has emergency equipment to care for injured or sick animals.

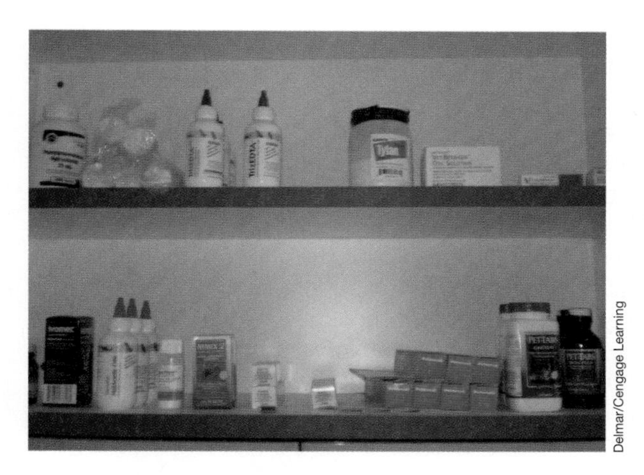

FIGURE 5-11 The pharmacy area.

FIGURE 5-12 The laboratory area.

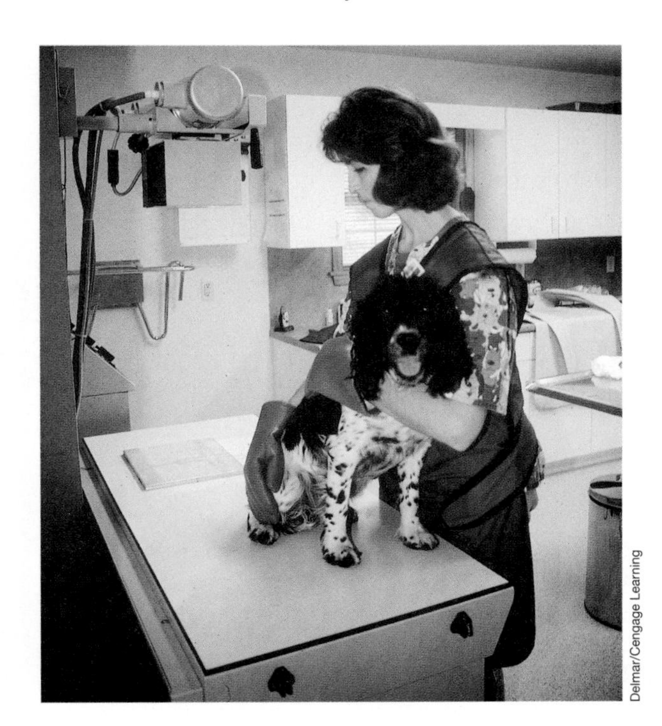

FIGURE 5-13 The radiology area.

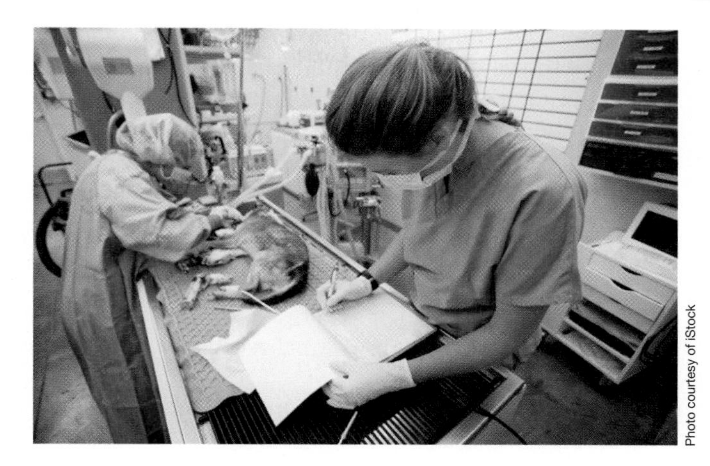

FIGURE 5-14 The surgical suite.

FIGURE 5-15 The kennel ward.

The **surgical suite** is another specialized area that houses the surgery table, anesthesia machine, and instruments and equipment used in surgical procedures (see Figure 5-14).

The **isolation ward** is a room that contains cages and equipment to treat animals that may be contagious and pass diseases to other patients.

The **kennel ward** is the area that houses patients waiting for surgery or treatment, or they may be boarding in the facility (see Figure 5-15).

Managing the Veterinary Facility

Managing a veterinary facility includes many duties and responsibilities from office management to staff management. The staff members must be scheduled on a daily basis to cover morning, afternoon, and evening shift hours. It is necessary to have enough staff coverage to keep the facility functioning properly. The management control of the veterinary hospital is usually completed by the owner or owners and the hospital manager. These staff members typically have experience in management and business. Some other management duties include maintaining employee records, ordering supplies and materials, paying invoices and bills for the facility, and completing tax and insurance information. The manager must have good business and communication skills.

Invoicing and Fees

It is important for every member of the veterinary health care team to have a basic knowledge of the invoicing and billing procedures used in the practice. Generally, the veterinary receptionist collects client payments and handles invoicing and receipts. However, some circumstances may occur when other members of the team may be in the situation of handling these tasks. This requires training in handling money, basic mathematical skills, and communication skills.

The **invoice** is commonly called the bill and outlines the fees that the client must pay (see Figure 5-16). The receipt states the bill has been paid and what type of payment was made. An invoice and receipt may be the same item that notes the bill has been paid. The fee structure in clinics is based on the type of services and procedures that are provided. The fee structure lists the prices or costs of the service. Some facilities will use a **travel sheet** that lists codes according to the procedure or service (see Figure 5-17). The travel sheet is used by the veterinary team to circle or note the services that have been provided to the patient. The veterinary receptionist or other veterinary team member will then enter the fees into a computer or onto an invoice. This is how a client is charged for the services the veterinary hospital has provided. It is important for all members of the team to understand the costs involved with the services. Many procedures involve the use of equipment, supplies, staff assistance, and inventory fees. Customers may ask why certain services cost the amount they do. The veterinary assistant must have a basic knowledge of these services to explain to the client all that is involved with the fees behind the procedures.

Payment plans should begin with the client receiving an estimate. The **estimate** details the approximate cost of procedures that are required by the patient. This will be helpful to the client to determine a type of payment plan that will be necessary at the time of discharge. Payment options can then be reviewed by the receptionist. Many clinics accept payments in cash, check, or credit card form. Some facilities also offer **credit applications**, which allow clients to obtain a line of credit specifically to pay the costs owed to the veterinary clinic. Occasionally, veterinarians may allow clients to set up a **specialized payment plan**. This usually

ANIMAL HOSPITAL
1823 Main St
Philadelphia, Pa 15555

Client Name:	Patient Name:	Date:	
Doctor:	Breed:	Age:	Gender:
DHLPPC Due:	FVRCP Due:	Rabies due:	

INVOICE

Physical Exam/Office Visit	$35.00
Ear Flush	$14.00
Betagen Otic Ear Drops	$12.00
Amoxicillin Tablets	$15.00
Nail Trim	$ 8.00
BALANCE DUE:	$84.00

Paid—Check #: Cash:

Thank you for your business!

FIGURE 5-16 A sample bill or invoice.

INVOICE SHEET

Owner Name: ______________________ **Phone:** ____________

Address: ______________________ **Date:** ____________

Patient: ______________________ **Age:** ____________ **Breed:** ____________

Office Call	OC Routine	OCM Multiple Pet	OCE Office Call extended	OCC Office Call Complimentary
Exam	PE Comprehensive PE	PEA PE Annual Visit	PER PE Recheck	PEC PE Complimentary
Treatment	HOSP Hospital care	# DAYS ________	IV IV Fluids	# DAYS ________
	SQ SQ Fluids	# DAYS ________	CBC Complete Blood Count	CHEM Chemistry Panel
	FEC Fecal Sample	URIN Urine Sample	HWT Heartworm Test	FeLV Feline Leukemia test
Medication	HW Heartworm meds	FLEA Flea/tick meds	WORM Wormer medicine	TYPE ________ Qty ________
	MEDS:	NAME ________	QTY ________	
	MEDS:	NAME ________	QTY ________	
	MEDS:	NAME ________	QTY ________	
Diet	CD Hills c/d diet QTY ________	AD Hills a/d diet QTY ________	RD Hills r/d diet QTY ________	UD Hills u/d diet QTY ________
	FEED Daily hospital food	QTY ________		

FIGURE 5-17 A sample travel sheet.

includes a down payment with a set amount of money and smaller payments collected on set dates until the total balance is paid.

The invoice should itemize each fee so that the client has an understanding of how much each part of the procedure cost. Through an **account history** the customer's previous invoices can be reviewed at any time. This is also a way for the veterinarians to see what procedures were done and what charges incurred.

Payroll and Accounting

Payroll includes staff salaries and writing checks for each employee on a regular basis. Employees will receive their wages and compensation as gross pay (see Figure 5-18). **Gross pay** is the total amount earned by the employee before tax deductions. The **salary** is the amount the employee earns each year based on the hourly or set wage. Hourly wages are calculated by the number of hours the employee works each week. The law states that a work week includes 40 hours. Any employee working overtime, which is the amount of time in excess of 40 hours, is entitled to one and one-half times the regular set wage. The **net pay** is the amount received after tax deductions are made. **Payroll deductions** include insurance costs, union dues, and investment plans. Payroll deductions are also incurred from federal income taxes that are withheld from the gross wages based on what the employee claims each year. These claims are called **exemptions** and represent additional funds that are included in net pay. For example, an employee can claim an exemption for each dependent family member such as a child or spouse. Exemptions are determined when the employee claims a certain amount of yearly wages that is exempt from income tax. This is done when each employee files an Employee Withholding Allowance Certificate, which is better known as a **Form W-4** (see Figure 5-19). This form is filled out according to

EMPLOYEE'S EARNINGS RECORD

NAME Mary Crane
ADDRESS 810 Columbia Street
Newport Beach, CA 92663

SOC. SEC. No. 000-00-0000
EMPLOYEE NO. 3

DATE EMPLOYED 8-16-2000 DATE OF TERMINATION
DATE OF BIRTH 7-2-72
MARRIED X SINGLE NUMBER PAY
MALE FEMALE X ALLOWANCES 1 RATE 6.00/hr.
OCCUPATION Clerk
CLOCK NO. 6123

Pay Period Ended	Total Hours	Regular Pay	Overtime	Gross Pay	Cumulative Pay	FICA Taxes	Fed. Income Taxes	Hosp. Ins.	Other Deductions	Total Deductions	Net Pay	Check. No.
1/6	40	240.00		240.00	240.00	18.36	26.40	6.25		51.01	188.99	46
1/13	40	240.00		240.00	480.00	18.36	26.40	6.25		51.01	188.99	59
1/20	40	240.00		240.00	720.00	18.36	26.40	6.25		51.01	188.99	72
1/27	48	240.00	54.00	294.00	1,014.00	22.49	36.90	6.25		65.64	228.36	85
2/4	40	240.00		240.00	1,254.00	18.36	26.40	6.25		51.01	188.99	98
2/11	40	240.00		240.00	1,494.00	18.36	26.40	6.25		51.01	188.99	111
2/18	40	240.00		240.00	1,734.00	18.36	26.40	6.25		51.01	188.99	124
2/25	40	240.00		240.00	1,974.00	18.36	26.40	6.25		51.01	188.99	137
4/25	40	248.00*		248.00*	3,902.00	18.97	26.40	6.25		51.01	196.38	372

*Rate Increased to $6.20/hr.

FIGURE 5-18 Sample employment payroll record.

Form **W-4** | Department of the Treasury | Internal Revenue Service

Employee's Withholding Allowance Certificate

▶ **For Privacy Act and Paperwork Reduction Act Notice, see page 2.**

OMB No. 1545-0010

1 Type or print your first name and middle initial	Last name	2 Your social security number
Home address (number and street or rural route)	3 ☐ Single ☐ Married ☐ Married, but withhold at higher Single rate. *Note: If married, but legally separated, or spouse is a nonresident alien, check the "Single" box.*	
City or town, state, and ZIP code	4 **If your last name differs from that shown on your social security card, check here. You must call 1-800-772-1213 for a new card.** ▶ ☐	

5 Total number of allowances you are claiming (from line **H** above **or** from the applicable worksheet on page 2) . . . 5

6 Additional amount, if any, you want withheld from each paycheck . . . . 6 $

7 I claim exemption from withholding for 2003, and I certify that I meet **both** of the following conditions for exemption:
- Last year I had a right to a refund of **all** Federal income tax withheld because I had **no** tax liability **and**
- This year I expect a refund of **all** Federal income tax withheld because I expect to have **no** tax liability.

If you meet both conditions, write "Exempt" here . . . . ▶ 7

Under penalties of perjury, I certify that I am entitled to the number of withholding allowances claimed on this certificate, or I am entitled to claim exempt status.

Employee's signature
(Form is not valid
unless you sign it.) ▶ **Date** ▶

8 Employer's name and address (Employer: Complete lines 8 and 10 only if sending to the IRS.)	9 Office code (optional)	10 Employer identification number

Cat. No. 10220Q

FIGURE 5-19 Employer's withholding allowance certificate, Form W-4.

the employee's wage bracket and marital status. Each employee also has taxes withheld from state and local **income tax** earnings (see Figure 5-20). The **FICA tax** is a federal government tax and is withheld for social security taxes. This is a legal tax under the Federal Insurance Contribution Act.

Inventory Management and Control

Most veterinary facilities have systems to maintain and manage inventory supplies. One or more people may be in charge of **inventory management**. This involves knowing how much of each drug or supply is currently in the hospital. It also includes knowing the price, manufacturer, pharmaceutical company that handles the product, and expiration dates. Routinely, each facility will do an **inventory** count. This means that a physical count of all medications and supplies within the clinic is completed to make sure accurate records are kept and maintained. Expiration dates are checked and expired products are removed from the facility. A **want list** is helpful to determine when a product is becoming low and needs to be reordered. This is a list posted somewhere in the facility in a book, on a clipboard, or on a whiteboard where staff members can note when supplies are becoming low. Many facilities have a minimum amount of commonly used items that must be reordered before they are completely out. The hospital manager is typically in charge of ordering these items.

Many veterinary facilities use **computer inventory systems** to monitor the inventory and assist in purchasing items. The inventory of each product and supply item is entered into the computer system and relates to the invoicing system so that any item sold or used is tied to the client invoice and removed from the inventory total. The hospital manager is typically in charge of maintaining the inventory and ordering supplies. Every staff member, including the veterinary assistant, plays a vital role in inventory management. Each person must record when items are running low. They must also track inventory on **consumable products**, such as paper towels, gauze sponges, and cotton balls. These products are used on a daily basis but not kept itemized in an invoice. They may be items used in treatments or surgical procedures. Another important part of staff involvement in inventory control is unpacking order shipments. When orders arrive, the first step is to open the box and located the **packing slip**. The packing slip itemizes each item that was ordered and has arrived in the package. It is important to make sure each item is accounted for and not damaged. Expiration dates of the new items should be noted in the computer and the new quantities ordered should also be placed in the computer inventory system. Items should be stored on shelves or in cabinets following the veterinary facility's procedure.

Some facilities may not be computerized or have a computer inventory system and use an **automated inventory system**. This means that items are counted and controlled by hand and numbers are kept logged in a notebook. This involves a physical inventory, which means counting every item in the practice and noting the quantity on hand—office supplies, medical supplies, and pharmacy supplies. This system must be controlled

MARRIED Persons—**WEEKLY** Payroll Period

If the wages are—		And the number of withholding allowances claimed is—										
At least	But less than	0	1	2	3	4	5	6	7	8	9	10
		The amount of income tax to be withheld is—										
$740	$750	$93	$86	$79	$72	$65	$58	$51	$44	$37	$30	$23
750	760	95	88	81	74	67	60	53	45	38	31	24
760	770	96	89	82	75	68	61	54	47	40	33	26
770	780	98	91	84	77	70	63	56	48	41	34	27
780	790	99	92	85	78	71	64	57	50	43	36	29
790	800	101	94	87	80	73	66	50	51	44	37	30
800	810	102	95	88	81	74	67	60	53	46	39	32
810	820	105	97	90	83	76	69	62	54	47	40	33
820	830	108	98	01	84	77	70	63	56	49	42	35
830	840	111	100	93	86	79	72	65	57	50	43	36
840	850	114	101	94	87	80	73	66	59	52	45	38
850	860	116	103	96	89	82	75	68	60	53	46	30
860	870	119	106	97	90	83	76	69	62	55	48	41
870	880	122	109	99	92	85	78	71	63	56	49	42
880	890	125	112	100	93	86	79	72	65	58	51	44
890	900	128	114	102	95	88	81	74	66	59	52	45
900	910	130	117	104	96	89	82	75	68	61	54	47
910	920	133	120	107	98	91	84	77	69	62	55	48
920	930	136	123	110	99	92	85	78	71	64	57	50
930	940	139	126	112	101	94	87	80	72	65	58	51
940	950	142	128	115	102	95	88	81	74	67	60	53
950	960	144	131	118	105	97	90	83	75	68	61	54
960	970	147	134	121	108	98	01	84	77	70	63	56
970	980	150	137	124	110	100	93	86	78	71	64	57
980	990	153	140	126	113	101	04	87	80	73	66	59
990	1,000	156	142	129	116	103	06	80	81	74	67	60
1,000	1,010	158	145	132	119	106	97	00	83	76	69	62
1,010	1,020	161	148	135	122	108	09	92	84	77	70	63
1,020	1,030	164	151	138	124	111	100	93	86	79	72	65
1,030	1,040	167	154	140	127	114	102	05	87	80	73	66
1,040	1,050	170	156	143	130	117	104	06	89	82	75	68
1,050	1,060	172	159	146	133	120	106	98	90	83	76	69
1,060	1,070	175	162	149	136	122	109	99	92	85	78	71
1,070	1,080	178	165	152	138	125	112	101	93	86	79	72
1,080	1,090	181	168	154	141	128	115	102	95	88	81	74
1,090	1,100	184	170	157	144	131	118	104	96	89	82	75
1,100	1,110	186	173	160	147	134	120	107	98	91	84	77
1,110	1,120	189	176	163	130	136	123	110	99	92	85	78
1,120	1,130	192	179	166	152	139	126	113	101	94	87	80
1,130	1,140	195	182	168	155	142	129	116	102	95	88	81
1,140	1,150	198	184	171	158	145	132	118	105	97	90	83
1,150	1,160	200	187	174	161	148	134	121	108	98	91	84
1,160	1,170	203	190	177	164	150	137	124	111	100	93	86
1,170	1,180	206	193	180	166	153	140	127	114	101	94	87
1,180	1,190	209	196	182	160	156	143	130	116	103	96	89
1,190	1,200	212	198	185	172	159	146	132	119	106	97	00
1,200	1,210	214	201	188	175	162	148	135	122	109	99	92
1,210	1,220	217	204	191	178	164	151	138	125	112	100	93
1,220	1,230	220	207	194	130	167	154	141	128	114	102	95
1,230	1,240	223	210	196	183	170	157	144	130	117	104	96
1,240	1,250	226	212	199	186	173	160	146	133	120	107	98
1,250	1,260	228	215	202	189	176	162	149	136	123	110	99
1,260	1,270	231	218	205	192	178	165	152	139	126	112	101
1,270	1,280	234	221	208	194	181	168	155	142	128	115	102
1,280	1,290	237	224	210	197	184	171	158	144	131	118	105
1,290	1,300	240	226	213	200	187	174	160	147	134	121	108
1,300	1,310	242	229	216	203	100	1 76	163	150	137	124	110
1,310	1,320	245	232	219	206	192	179	166	153	140	126	113
1,320	1,330	248	235	222	208	195	182	169	156	142	129	116
1,330	1,340	251	238	224	21 1	198	185	172	158	145	137	110
1,340	1,350	254	240	227	214	201	188	174	161	148	135	122
1,350	1,360	259	243	203	217	204	190	177	164	151	138	124
1,360	1,370	259	246	233	220	206	103	180	167	154	140	127
1,370	1,380	262	249	236	222	209	196	183	170	156	143	130
1,380	1,390	265	252	238	225	212	199	186	172	159	146	133

$1,390 and over

FIGURE 5-20 Federal weekly wage bracket withholding table.

carefully to avoid running out of stock of supplies. Many facilities perform a yearly physical inventory to make certain all records are accurate.

SUMMARY

Veterinary management and practice methods are essential for keeping a business successful. Collecting payments and handling invoicing can be a stressful part of the job for every veterinary health care team member. It requires good knowledge of veterinary medicine and excellent communication skills. Understanding the facility's methods and policies is important in keeping a facility from failing. Inventory management is an important part of managing the daily needs of the clinic. Lack of supplies and medicines will mean a lack of fees collected for the staff payroll. The veterinary clinic relies on clients to pay the fees of the hospital and staff and ensure the success of the entire veterinary team.

Key Terms

account history client's previous invoices that can be viewed at any time to see past payments and charges

American Animal Hospital Association (AAHA) sets the standards of small animal and companion animal hospitals

American Veterinary Medical Association (AVMA) accredits college programs for veterinarian and veterinary technician state licensure and sets the standards of the veterinary medical industry

associate degree a two-year degree program

automated inventory system system of counting all supplies and products within the facility by hand and keeping a list of how many items are in stock

Bachelor of Science (BS) a four-year college degree program

certified veterinary technician (CVT) a veterinary technician that has passed a state licensing exam

computer inventory system method of monitoring supplies and products within the facility and knowing when items need to be reordered

consumable products supplies used on a daily basis that are not considered in the client invoice, such as paper towels, gauze sponges, and cotton balls

credit application paperwork completed by clients to allow them to obtain a line of credit to pay for veterinary services

estimate the approximate cost of services

exam room area where patients receive routine exams

exemptions fees determined when an employee claims a set of yearly wages from their income taxes

Form W-4 Employee Withholding Allowance Certificate; form filled out according to the employee wage bracket and marital status for tax purposes

gross pay total amount of wages earned by an employee before tax deductions

intensive care unit (ICU) area of the facility where critical patients needing immediate and constant care are treated and housed

inventory physical count of every medicine and supply item within the facility to record accurate records

inventory management knowing the price, manufacturer, pharmaceutical company, amount, and expiration date of each item in the facility

invoice bill given to a client that lists the costs of services and procedures

isolation ward area that contains cages and equipment to care for patients that are contagious and may spread disease

kennel attendant person working in the kennel area and animals wards, providing food, exercise, and clean bedding

kennel ward area where animals are housed in cages for surgery, boarding, and care

laboratory location in the facility where tests and samples are prepared or completed for analysis

licensed veterinary technician (LVT) a veterinary technician that has passed a state licensing exam

(Continues)

National Association of Veterinary Technicians in America (NAVTA) sets the standard of veterinary technology and the care of animals

net pay amount of money received after tax deductions

packing slip paper enclosed in a package that details the items in the shipment

payroll salary paid to each employee by writing checks on a regular basis

payroll deductions money withheld from a paycheck for tax purposes or payment of medical insurance

personnel manual book that describes the staff job descriptions, attendance policy, dress code, and employer expectations

pharmacy area where medications are kept and prepared

procedures manual book that describes the rules and regulations of the facility, such as visiting hours and opening and closing procedures

radiology area of the facility where X-rays are taken and developed; this area is regulation due to chemical and radiation exposure

reception area area where clients arrive and check in with the receptionist

receptionist personnel greeting and communicating with patients as they come in and leave the veterinary facility

registered veterinary technician (RVT) a veterinary technician that has passed a state licensing exam

salary amount an employee earns each year based on the hourly wage

specialized payment plan a plan approved by the veterinarian that allows a down payment with additional regular payments until the balance is paid

State Board examination exam that veterinarians and veterinary technicians are required to pass for state licensure in order to practice

surgical suite area where surgery is performed and harmful anesthesia gases are used; the area must be kept sterile to prevent contamination

travel sheet paperwork list that codes services and procedures that allows the staff to track fees to charge the client

treatment area area where animals are treated or prepared for surgery

veterinarian the doctor that holds a state license to practice veterinary medicine and holds the ultimate responsibility within the facility

veterinary assistant staff member trained to assist veterinarians and technicians in various duties

veterinary hospital manager staff member in charge of the business workings of the entire hospital, including staff scheduling, paying bills, and payroll

Veterinary Practice Act set of rules and regulations that govern veterinary medicine and the state

veterinary technician two-year associate degree in specialized training to assist veterinarians

veterinary technologist four-year bachelor degree in specialized training to assist veterinarians

waiting room area where clients and patients wait for their appointments

want list items that are needed or low in stock within the facility and need to be reordered

REVIEW QUESTIONS

Matching

Match the term with the definition.

1. account history
2. consumable products
3. estimate
4. intensive care unit (ICU)
5. inventory

a. client's previous invoices that can be viewed at any time to see past payments and charges

b. physical count of every medicine and supply item within the facility to retain accurate records

c. area that contains cages and equipment to care for patients that are contagious and may spread disease

d. supplies used on a daily basis that are not considered in the client invoice, such as paper towels, gauze sponges, and cotton balls

e. items that are needed or low in stock within the facility and need to be reordered

6. isolation ward
7. kennel attendant
8. laboratory
9. travel sheet
10. want list

f. area of the facility where X-rays are taken and developed; this area is regulation due to chemical and radiation exposure
g. person working in the kennel area and animal wards, providing food, exercise, and clean bedding
h. bill given to a client that lists the costs of services and procedures
i. area of the facility where critical patients needing immediate and constant care are treated and housed
j. the approximate cost of services
k. location in the facility where tests and samples are prepared or completed for analysis
l. area where patients receive routine exams
m. treatment area—area where animals are treated or prepared for surgery
n. paperwork list that codes services and procedures that allows the staff to track fees to charge the client
o. area where surgery is performed and harmful anesthesia gases are used; the area must be kept sterile to prevent contamination
p. paper enclosed in a package that details the items in the shipment
q. paperwork completed by clients to allow them to obtain a line of credit to pay for veterinary services

Short Answer

Please respond to the questions below:

1. List the essential members of the veterinary health care team and their duties and responsibilities within the hospital.
2. Who determines the fees and costs of procedures within the facility?
3. What are some differences between the AS or Associate of Science degree and a BS or Bachelor of Science degree?
4. What associations are important to veterinary medicine?
5. What are the sections of the veterinary facility and what is each area's importance?
6. What items should each staff member receive when beginning a veterinary job position?
7. List essential parts of the inventory management.
8. Explain the employee payroll and how wages are determined.

Clinical Situation

Kate, the hospital manager at All Creatures Animal Hospital, was performing an end-of-the-month inventory control and maintaining client accounts. One client account was still not paid and the remaining balance of $200 that was owed for 2 months following a surgery procedure was still outstanding. Kate decided to mail the client another invoice with a friendly, but firm, reminder that the balance was due.

Kate also noticed when reviewing the inventory list that an order placed 3 weeks ago was still not received. She looked up the pharmaceutical company information on the computer and called the company. There was no answer with the pharmaceutical representative, so she left a message concerning the order not arriving and asked that the agent, Melissa, give her a call today.

- How did Kate handle the overdue payment situation?
- How did Kate handle the missing order situation?
- How did the computer system help in these situations?

Competency Skill

Professional Appearance and Dress

Objective:

To conduct oneself in both manner and appearance befitting a professional.

Preparation:

- Scrub outfit
- Non-slip enclosed shoes
- Gauze roll

- ½″ to 1″ adhesive tape roll
- Watch or timer
- Stethoscope
- Nylon leash
- Memo pad
- Scissors
- Pen
- Calculator

Procedure:

1. Review your personnel and procedures manual that outlines the staff rules and dress code for the facility.
2. The appropriate dress code should be followed. That includes the following:
   - A clean and durable scrub outfit, including top and pants that fit well and are comfortable. The neckline should be modest and pants should fit without falling down during bending and movement. Pockets are desired.
   - Nonslip shoes that are enclosed over the entire foot. White sneakers are preferable. Shoes should fit well and be comfortable and supportive to the feet.
   - The above listed preparation items should be carried in pockets and be readily available at all times.
   - Jewelry should not be worn or be kept minimal. Items may become damaged, cause patient or staff injury, or may become contaminated. Loose items such as earrings, necklaces, and bracelets may cause injury to patients or the staff member.
   - Hair should be pulled back from the face to allow safe patient handling and a clear view of the work area.
   - Fingernails should be kept trimmed and neat. Long and fake nails tend to harbor bacterial and infectious materials under them and are a higher risk of contamination to patients.
   - Tattoos and piercings should be kept concealed. Many facilities have rules that should be noted in the personnel manual.

Competency Skill

Invoicing and Payment Collection

Objective:

To properly process and invoice and collect payment on services rendered.

Preparation:

- Maintain excellent communication skills.
- Familiarize yourself with the process of invoicing and collecting payments.
- Act efficiently and accurately.

Procedure:

1. Acknowledge the client and patient.
2. Make sure the medical record and invoice belong to the correct client and patient.
3. Make sure all medications, supplies, and patient items have been received.
4. Ask if the client has any questions.
5. Total the fees for services.
6. Print an invoice and present the total balance in an itemized structure.
7. Inquire about the method of payment.
8. Collect the fee and handle the processing methods as indicated by the facility.
9. Prepare the receipt of payment and give it to the client.
10. Ask the client if there are any additional questions.
11. Thank the owner and compliment the patient.
12. Let the owner know you look forward to seeing him or her again.

Competency Skills

Inventory Management Procedure

Objective:

To maintain an accurate accounting of the products and goods in the veterinary office.

Preparation:

- Locate a list of all products in the veterinary facility.
- Have a pen and paper available.

Procedure:

1. Locate items on inventory list.
2. Count all items in each category and list the amount of each item located in the facility.
3. Check the inventory list for amount accuracy.
4. Check each item for expiration dates.
5. Remove any outdated or expired products from the shelves.
6. Place items on shelves according to expiration dates so that items to be outdated soon will be used first.
7. Store items in correct location.
8. Update inventory lists to be accurate according to product amount, manufacturer, pharmaceutical company ordering information, and date of expiration.

6

Communication and Client Relations

Objectives

Upon completion of this chapter, the reader should be able to:

- Discuss the communication process and basic components of communication
- Distinguish between appropriate and inappropriate communication and professional interactions with people
- Demonstrate ways to better improve speech and listening skills
- Describe how to communicate with people on the phone
- Explain how interpersonal communication skills apply to clients' perception
- Discuss how to handle difficult clients and situations
- Discuss strategies for people on how to handle the grief process

Introduction

Veterinary assistants, as well as the entire health care team, rely on **interpersonal communication** skills when working closely with people. Interpersonal communication is any form of message and response between two or more people. It may be in the form of the written word, verbal exchanges, or body language cues. The people the veterinary assistant will regularly interact with include clients, staff members, veterinary professionals from other facilities, and businesspeople the practice may work with such as a salesperson the facility makes purchases from.

The Communication Process

The **communication process** has five essential components: the sender, the message, the channel, the receiver, and the feedback (see Figure 6-1). The process begins with the **sender**, the person trying to relay an idea. This idea is called the **message**. The message is sent by the sender to another person called the **receiver**. The receiver is the person intended to understand the message. The message is sent through a **channel**, which is the chosen route of communication. Examples of routes of communication include verbal, nonverbal, and written. Once this process of communication is complete the receiver may return a message called **feedback**. This process is continued until the conversation is complete. Communication is a way of passing along information from one person to another or to a group of people.

FIGURE 6-2 Verbal communication plays a large role in the duties of the veterinary assistant.

Verbal Communication

Verbal communication is spoken words used between two or more people to form an understanding. Verbal communication is the most common form of communication between people. This type of communication is needed between the veterinary assistant and other staff members, as well as with clients (see Figure 6-2). It is important to be able to speak well to other people. Verbal communication in medicine can now also mean recording notes on tape recorders to be written at a later date, allowing for more accurate notes.

FIGURE 6-3 Body language can communicate more than spoken words.

Nonverbal Communication

Nonverbal communication is interaction between people without the use of spoken words. This may be through body language and physical expressions.

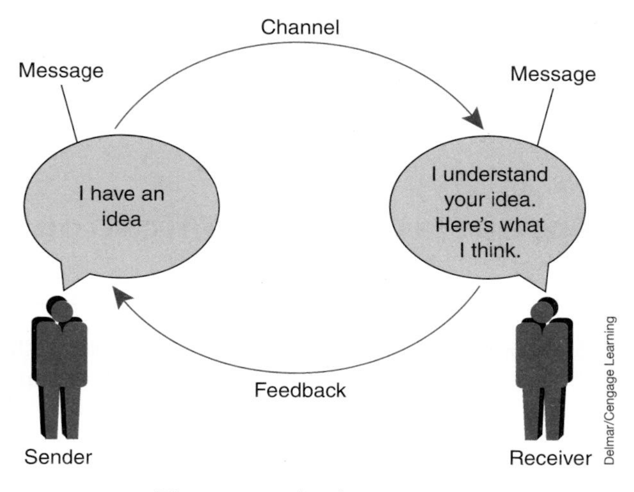

FIGURE 6-1 The communication process.

Body language is the use of mannerisms and gestures that tell how a person feels. Nonverbal communication is used commonly with staff members and may be helpful when dealing with upset, angry, or grieving clients. Examples of nonverbal communication include smiling, frowning, glaring, or shrugging the shoulders (see Figure 6-3). It is easy to confuse the meaning of the message when only nonverbal communication is used. To effectively communicate, positive body language can smooth the process along. Some positive body language gestures are identified below (see Figure 6-4):

- Hold arms at side or gently folded in front of body.
- Stand up straight with a relaxed position.

FIGURE 6-4 Positive body language can enhance communication between the veterinary assistant and the client.

- Face the person and keep eye contact.
- Speak with a moderate voice and clear tone.
- Stay at least 1 arm's length from people to form a comfort zone.
- Dress professionally and wear a name tag.
- Smile and keep a relaxed facial expression.
- Practice good hygiene.

Written Communication

Written communication is used daily in written hospital notes for progress records. Written communication may also be used in writing letters, memos, emailing, or texting someone. Email and text messaging have become popular methods of communication. Written communication must be clear, accurate, and understandable. In a veterinary facility, it is important to have a memo pad and a pen or pencil located at every telephone so that accurate messages and details can be written. Note that one should record all telephone messages and give the memo to the recipient immediately. It is important to write a date, time of call, caller's name and phone number, message, and your name on the memo. Make certain that all information recorded is neat and accurate. Medical records and progress notes must have excellent and accurate information to allow proper communication between staff members (see Figure 6-5).

FIGURE 6-5 Written communication is an important part of the communication among the veterinary health care team.

Appropriate Communication Skills

When people are communicating with each other, they should consider several qualities that will enhance positive routes of communication. These qualities include courtesy, kindness, patience, tactfulness, sympathy, and empathy. **Courtesy** means putting someone else's needs and concerns before your own. It includes such characteristics as sharing, giving, and cooperation. All people should be treated with respect and professionalism. **Kindness** is a characteristic that exemplifies being helpful, understanding, and working in a friendly manner. Each client should be treated in the manner in which you wish to be treated by others. **Patience** is a trait that demonstrates calm demeanor in all situations without any negative complaints. Patience may need to be practiced. Some people and animals will require more patience than others. **Tactfulness** is a characteristic that means doing and saying the appropriate things at the correct time. This will allow maintenance of good relations with others. This trait involves thinking and planning before speaking and saying something that may be offensive to another person. **Sympathy** involves sharing feelings with each other in a time of sadness. **Empathy** is being able to understand another's feelings at a particular time. Table 6-1 provides examples of these traits in common situations.

TABLE 6-1

Examples of Appropriate Communication Skills

Courtesy	• Holding a door • Helping a person to his or her vehicle • Carrying a bag of dog food • Walking someone's dog to an exam room
Empathy	• To tell someone you understand how they feel when putting an animal to sleep
Kindness	• Asking if you can help someone • Making a phone call for someone • Drawing a diagram of how to do something
Patience	• Walking an elderly person to his or her car • Repeating information for a person hard of hearing • Offering children coloring books
Sympathy	• Saying you are sorry to a client who has lost a pet • Writing a sympathy note to a client • Sending flowers to a staff member who is sick
Tactfulness	• Speaking politely • Not swearing • Not yelling • Discussing a difficult situation professionally

Interacting with People

Interpersonal communication includes speaking, listening, and observing others to allow for good communication to occur.

Speaking

Speech communication requires practice and experience. This can be a frustrating part of any career. Some people will begin conversations that have no bearing on the reason they are in the veterinary clinic. They may also ask many questions, some of which may be of little importance. When interacting with people, show respect to everyone encountered. Remember that people are different; therefore, treat each person as an individual. Speaking with other team members will allow for safety of working with animals and allow everyone's job to become a little easier. Interaction with other veterinary health care team members may be necessary to complete the treatment of an animal or complete a progress report (see Figure 6-6). There are several methods to use to show respect during communication with individuals:

- Always greet people upon first coming in contact with them.
- Treat everyone equally, with respect and honesty.
- Address people with professionalism—Dr. Jones, Mrs. Smith, Professor Lee.
- Talk slowly and patiently when explaining information.
- Keep all areas of the facility neat and clean for the comfort of all who work in and enter the facility.
- Offer assistance to everyone that appears could use help.
- Smile even when you don't feel like smiling.

FIGURE 6-6 It is often necessary to get help with completing a written report from other staff members.

TERMINOLOGY TIP

Students should reference a veterinary medical dictionary or a Webster's dictionary to determine proper speech and usages if they are not certain if a word is correct.

Listening

Another way of interacting with clients involves listening and reflection with clients. **Listening skills** are essential in the veterinary field. Listening involves training oneself to focus on what is being said. Good listeners hear exactly what is being said and think about the words to clearly understand the main point. Good listening skills also include looking at a person's body language to determine his or her mood. Good listeners know when not to speak and allow people to finish their thoughts. It is necessary to take the time to understand what is being said and think about what an appropriate response will be. Remember that silence is a way to fully understand what a person has communicated. It may be a good practice to acknowledge the silence as taking the time to think about what has been said to not upset the speaker. After reaching an understanding of what was said, it is always a good idea to repeat how you understand it back to the speaker. This allows both the sender and receiver to have a common understanding and make sure the information is accurate. In a veterinary situation, the veterinary assistant may need to communicate to another staff member how to properly restrain an animal or ask for help when necessary (see Figure 6-7).

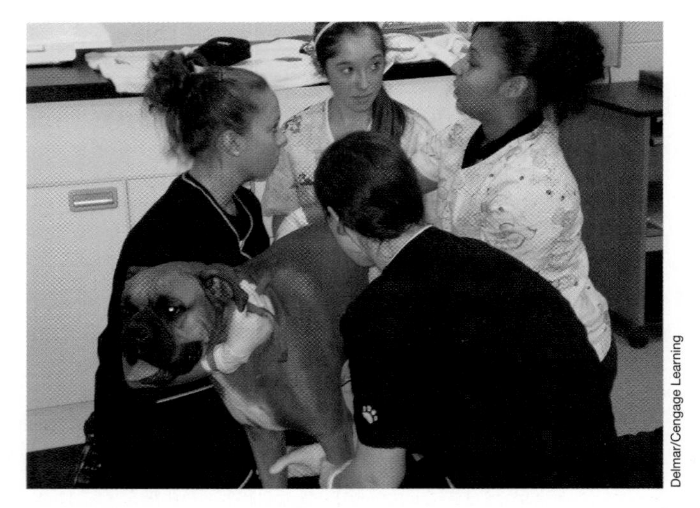

Delmar/Cengage Learning

FIGURE 6-7 Stop and listen to what your coworker is saying to ensure your safety, the safety of the others you are working with, and the animal's safety.

Observation

Observation skills are used to pay attention to a person's body language and speech. A person's body language can reveal nonverbal communication. Nonverbal communication is behavior that is recognized without speaking. These behaviors can tell us how a person feels through gestures or expressions. Many people may not be aware they are expressing these behaviors. Others may use the expressions to allow a person to understand their point. Some examples of client nonverbal communication an employee in a veterinary practice may observe include the following:

- Smiling
- Heavy sighing
- Shaking the head
- Slouching
- Nervous tapping of the fingers
- Arms folded
- Pacing
- Frowning
- Glaring

Communicating with Difficult Clients

The unfortunate occasion will occur when the veterinary assistant must handle an angry or difficult person. It is best to read a person's body language and note when he or she is becoming upset. This may help to avoid a difficult situation. However, if a difficult situation arises, it may cause shock and frustration, but it is best to remain calm. There is no exact method for dealing with an angry person. It is best to begin by asking the person what the problem or situation may be. Ask for an explanation of what happened. Many times, if an angry person has a chance to discuss or describe a bad situation, that helps relieve some of the tension. This will also provide some time for the assistant to collect his or her thoughts and think of a plan of action. The following list shares some pointers to consider when handling difficult people.

- Keep a positive and assertive attitude.
- Treat each client with courtesy and respect.
- Communicate invoice and procedure costs to clients before they pick up a pet.
- Explain all procedures and ask if there are any questions.

- Obtain client permission with signed consent forms at all times.
- Be helpful at all times.
- Stay calm and relaxed.
- Minimize interruptions.
- Do not argue or speak negatively to the person.
- Do not place blame on anyone for the situation.
- Apologize for any misunderstanding that has occurred.
- Talk to the person in a private place with a third-party witness.
- Remain objective and listen attentively.
- Assure the person the matter will be resolved.
- Find a neutral party to help resolve the matter.

Angry people may say things that they may later regret. Do not take personal offense if this situation arises. If a client becomes enraged, it may be best to alert a veterinarian to the situation. Treat every person that becomes difficult with tact and respect. These matters are not easy for anyone to handle but sometimes are unavoidable.

Telephone Skills

Speaking on the telephone is usually the main line of communication in a veterinary facility (see Figure 6-8). This is one area where good or bad impressions can be made. Anyone answering a telephone should be comfortable and knowledgeable in telephone communication skills. There are several rules that must be followed when speaking on a telephone. First, always answer the phone promptly; try to answer by the third ring. Second, speak pleasantly and greet the caller. Saying "good morning" or "thank you for calling" can brighten someone's day. Also, identify who you are by telling the caller who they are speaking to and where they are calling. Next, be courteous and polite. Give the caller your full attention and use polite and simple language. Speak clearly and loud enough to be understood. Some callers may be hard of hearing so it may be necessary to adjust the tone of your voice. For this reason, it is good practice to repeat important information for accuracy. Finally, thank the caller and end the conversation with a pleasant good-bye.

When another line rings while speaking on a telephone, politely ask the first caller if you may place him or her on hold. Then wait for a response. This is an example of courtesy and should be done to make sure the client doesn't have an emergency or need immediate assistance. When placing a line on hold, note that if the hold time is longer than 1 minute, ask the caller for a name and phone number to return the call promptly. Remember that 1 minute on hold on a telephone feels like 10 minutes.

Delmar/Cengage Learning

FIGURE 6-8 Courtesy and pleasant telephone communication skills will make a good impression on clients.

Grief and Communication

Grief is the emotion that people feel after they have lost a pet or loved one. Members of the veterinary health care team may also feel a certain amount of grief, either feeling the loss of a pet or becoming close to the client and patient through hospital visits. This close relationship developed between people and animals is known as the human–animal bond. Many people think of their pets as children or a part of the family. The company of an animal tends to make people happy. Due to this bond, it can be difficult for a pet owner to deal with an old or severely ill animal. When owning a pet, grief is inevitable. A common procedure in the veterinary facility is **euthanasia**. Euthanasia is the process of humanely ending an animal's life. The term **humane** is used to describe what is considered acceptable by people in regards to an animal's physical, mental, and emotional well-being. Euthanasia is considered a humane way to end an animal's suffering. This can be a difficult situation for both the owner and the veterinary staff. Making the decision to euthanize a pet is a difficult choice for everyone. The process should be explained to the client. A person going through the process of grieving may exhibit some of the following emotions. **Shock** is the emotion of the sudden death of a pet and that may lead to denial. **Denial** is when a person will not accept the pet's death right away. **Bargaining** allows the person an attempt to resolve the pet's problem by any means possible. That leads to **acceptance**, which is when the owner understands and accepts the pet has passed away. This may lead to the owner becoming angry, which is a natural emotion in the event of the trauma. The **anger** may be at themselves, a family member, or a veterinary staff member. Some people will also face several other emotions after acceptance. They include sorrow, depression, and guilt. **Sorrow** is an emotion that leads to sadness over the loss of the pet. This is usually a positive emotion when the person will cry and talk about the pet. It is a huge part of the healing process. **Depression** is a state of sadness where a person becomes so sad he or she can't handle the normal functions of daily life. This state requires a doctor or therapist to aid in the recovery of the person. **Guilt** is a stage when a person feels he or she is to blame for the pet's death and that he or she should have been able to do something to save the pet. These emotions and stages of the grief process are something the veterinary assistant should be prepared to deal with. Clients and staff can deal with grief by offering some helpful strategies, such as talking to a family member or friend, talking with other people who have lost pets, joining a pet loss support group, or seeking a professional grief counselor.

SUMMARY

Positive and effective interpersonal communication is a necessity within the veterinary facility. The veterinary assistant must know how to get a message delivered clearly and with understanding and also be able to listen effectively. Proper communication skills including speaking, listening, observing, and use of body language are critical for the veterinary assistant to achieve. This is as important with the veterinary staff as it is with clients. The use of respect, courtesy, understanding, and caring coupled with the use of communication skills will make for a much more successful and enjoyable veterinary environment.

Key Terms

acceptance emotion when a person understands and accepts that a pet has passed away

anger an emotional stage when confronting the death of a pet

bargaining stage of grief that allows the person an attempt to resolve the pet's problem by any means possible

body language the use of mannerisms and gestures that are observed and tell how a person feels

channel a chosen route of communication

communication process the four essential steps to relaying information to others including the sender, the receiver, the message, the channel and feedback

courtesy the emotion of placing one's needs and concerns before your own

denial the emotion that occurs when a person will not accept a pet's death

depression the emotion of sadness where a person becomes so sad that he or she can't handle the normal functions of daily life

empathy the emotion of being able to understand another's feelings

euthanasia the process of humanely putting an animal to sleep

feedback the return message that is sent by the receiver

grief the emotion of sadness that people feel after the loss of a pet or loved one

guilt stage of grief when a person feels he or she is to blame for the pet's death and that he or she should have been able to save the pet's life

humane what is considered acceptable by people in regards to an animal's physical, mental, and emotional well-being

interpersonal communication allows people to discuss and understand information with other people

kindness characteristic that means you are helpful, understanding, and work in a friendly manner

listening skills involves hearing what is being said by someone and understanding what they mean

message an idea being passed along through a route of communication

nonverbal communication behaviors recognized and passed along without the need for speaking

observation skills used to pay attention to a person's body language and speech

patience trait that means you behave calmly in all situations without any negative complaints

receiver person who gets a message in the route of communication

sender person trying to relay an idea or message in the route of communication

shock emotion of a lack of feelings caused by the sudden death of an animal or loved one

sorrow positive emotion that leads to sadness over the loss of a pet and causes crying and talking about the pet.

speech communication talking in front of others to pass along information verbally

sympathy sharing feelings with another in the time of sadness

tactfulness characteristic that means doing and saying the appropriate things at the correct time

verbal communication the use of speech to pass along information in the communication process

written communication the use of writing, emailing, or texting information to pass along in the communication process

REVIEW QUESTIONS

Matching

Match the term to its definition.

1. bargaining
2. body language
3. communication process
4. empathy
5. grief
6. humane
7. receiver
8. sender
9. sympathy
10. tactfulness

a. stage of grief that allows the person an attempt to resolve the pet's problem by any means possible
b. what is considered acceptable by people in regards to an animal's physical, mental, and emotional well-being
c. person trying to relay an idea or message in the route of communication
d. characteristic that means doing and saying the appropriate things at the correct time
e. the emotion of sadness that people feel after the loss of a pet or loved one
f. an idea being passed along through a route of communication
g. sharing feelings with another in the time of sadness
h. the use of mannerisms and gestures that are observed and tell how a person feels
i. person that gets a message in the route of communication
j. the emotion of being able to understand another's feelings
k. the return message that is sent by the receiver
l. the four essential steps to relaying information to others including the sender, the receiver, the message, the channel and feedback

Please respond to the following questions:

1. Give three examples of channels of communication.
2. Which of the following responses is the best example of effective communication?
   a. I am sorry to hear about "Kitty" being hit by a car, Mrs. Campbell. You must be taking it pretty hard. Cats get themselves into so much trouble. That is why I like dogs better.
   b. Of course there is no ashtray; didn't you see the no-smoking sign?
   c. I know you were in a hurry, Mrs. Smith, but payment is still expected when services are provided.
   d. Hello, Mr. Tate, how are you and "Fluffy" today? Please have a seat and we will be with you shortly.
3. List seven ways that a person may display nonverbal communication.
4. The gestures and mannerisms that a person uses to communicate are called ________ ____________.
5. Explain how you would deal with a difficult client who is upset with his bill.
6. Fill in the correct communication that you feel should be said in the following telephone conversations:
   a. "Hello, please hold."
   b. "Can I help you?" (after 3 minutes being on hold)
   c. "What did you say your name was?"
   d. "See you later." (Hangs up phone laughing)
   e. "Come at 6." (Sighs and hangs up phone)

Clinical Situation

© iStock/Steve Luker

The veterinary assistant, Emily, is being shouted at by a client because his appointment was 35 minutes ago and he has a sick dog waiting to see the vet. Emily is stunned by the client's behavior and responds, "It is not my fault. Stop screaming at me!"

a. What did Emily do wrong?
b. Write a script of how you would have responded to the client.
c. What are the stages of the grief process?

Competency Skill

Interpersonal Communication Skills

Objective:

To effectively communicate with clients and coworkers.

Preparation:

- Determine what needs to be stated.
- Use references as needed, such as a dictionary or veterinary medical terminology text.
- If communication is written, use a memo form and dark pen.
- Use a computer with a word-processing program when possible.

Procedure:

Verbal Communication:

1. Listen carefully by focusing attention on the speaker.
2. Use good eye contact when listening.
3. Record notes as necessary for accuracy.
4. Speak with a good, clear tone that is audible and easy to understand.
5. Speak using eye contact.
6. Speak slowly.
7. Use correct grammar and be polite.

Written Communication:

1. Check all spelling, punctuation, and grammar for accuracy.
2. Proofread for meaning and understanding.
3. Use reference materials as necessary.
4. Be neat and legible and record in blue or black ink.
5. Place all written items in the proper location.

Competency Skill

Client Communication Skills

Objective:

To effectively communicate with clients.

Preparation:

- Determine the type of client communications used in the facility.
- Determine which type of forms are appropriate for the situation.

Procedure:

Written Communications:

1. Check all written information for accuracy and clarity.
2. Read all information to the client for understanding.
3. Ask the client if he or she has any questions regarding the information.
4. Place all original paperwork in the client's file. Make sure the client receives a copy of the paperwork.

Verbal Communication:

1. Speak clearly and slowly with an audible tone.
2. Maintain eye contact with the client.
3. Read the client's body language for understanding of information.
4. Ask the client if he or she has any questions regarding the information.
5. Note all verbal communication details within the patient's medical record.

Competency Skills

Telephone Communication Skills

Objective:

To foster a clear and concise demeanor when communicating via telephone.

Preparation:

- Have a memo pad or paper and functional pen at every telephone location.
- Have a good understanding and working knowledge of the telephone system.
- Have client and patient information forms near each telephone.
- Understand the proper scheduling technique of the facility.
- Understand how to properly schedule an appointment and maintain the appointment book.
- Have a price list available by each telephone.
- Keep a smile on your face to reflect a smile in your voice.
- Answer by the second to third ring.

Procedure:

1. Greet the caller and state the name of the facility. Example: "Good morning, this is Homestead Animal Hospital."
2. State your name for the caller.
3. Follow with, "How may I help you?"
4. Avoid placing a caller on hold if possible. If a caller must be placed on hold, ask the caller if you may do so first. Then wait for a response.
5. If the hold time is longer than one minute, ask the caller for a name and phone number to return the call promptly.
6. When answering a line that has been put on hold, tell the caller, "Sorry for the hold, this is Amanda, how may I help you?" and direct the call as necessary.
7. Record all telephone messages and give the memo to the recipient immediately. Write a date, time of call, caller's name and phone number, message, and your name on the memo. Be neat and accurate.
8. Direct all personal phone calls as needed. Ask the caller his or her name and the reason for calling so that you may direct the call appropriately.
9. Be polite and patient at all times. Speak slowly and clearly with an audible tone.

Competency Skills

Handling Difficult Clients

Objective:

To create a strategy for handling difficult client situations when they arise in the veterinary facility.

Preparation:

- Try to avoid creating a difficult situation.
- Treat each client with courtesy and respect.
- Use a warm, caring, and positive attitude at all times.
- Minimize interruptions during client communication.
- Minimize client waiting time.
- Communicate costs with estimates before the services are provided. Each client should be given a copy of an estimate.
- Obtain client permission with signed consent forms.
- Explain all procedure details and ask the client if he or she has any questions.
- Be helpful at all times.

Procedure:

1. Remain calm.
2. Ask the client to explain the problem. Allow the client to talk before asking questions. Listen to the client's needs, problems, and perspectives.
3. Determine who can best resolve the situation.
4. Determine if the situation should be dealt with in private and with a third-party witness.
5. Do not be defensive or place blame on someone else.
6. Be understanding.
7. Ask client what he or she would like done to help resolve the problem. Offer an alternative as necessary.
8. If the client becomes disrespectful or verbally abusive, locate a veterinarian immediately.

Competency Skill

Grief Counseling

Objective:

To create a strategy for handling the death of a client's animal and the client's grieving process.

Preparation:

- Grief counseling phone numbers
- Handouts on euthanasia
- Coloring books on euthanasia for children

Procedure:

1. Limit discussions of grief if at all possible. Grief counseling should be done by professionals.
2. Offer grief counseling hotlines or phone numbers.
3. All clients should receive handouts of the grief process and on euthanasia.
4. Children should be given a euthanasia coloring book that discusses the situation in a manner that helps children understand what is occurring.
5. A sympathy card should be prepared and signed by the veterinarian and staff members.

7

Veterinary Ethics and Legal Issues

Objectives

Upon completion of this chapter, the reader should be able to:

- ☐ Discuss the differences between ethical and unethical practices
- ☐ Explain emergency care to patients
- ☐ Describe the Veterinary Practice Act
- ☐ Explain the Rules and Regulations of the Veterinary Practice Act
- ☐ Demonstrate professional responsibility in veterinary medicine
- ☐ Describe the difference between common law, state law, and federal law

Introduction

Veterinarians and all veterinary staff members are required to uphold a standard of ethical conduct. There are laws that are in place in each state that put the needs of the patient first. It has become a responsibility to the profession to set a high standard of patient care and outline ethics and laws within the profession. Each state's veterinary practice act outlines these laws. Many practices go beyond the laws and determine rules and regulations for the facility.

Veterinary Ethics

Ethics are rules that govern proper conduct. Veterinary ethics dictate the **moral** conduct of veterinarians and their professional staff members. These morals indicate what a person believes as right and wrong. The AVMA has set standards in veterinary medical ethics. The most important responsibility of the veterinary staff is to put the patient's needs first and relieve any suffering and pain in animals. When these obligations are fulfilled, it is considered to be **ethical**. When these obligations are not completed, then it is considered to be **unethical**. Listed below are some examples of unethical practices in veterinary medicine:

- Not following the laws pertaining to the Practice Act
- **Misrepresentation** as a veterinarian (not licensed but working as a vet)
- **Slander**—speaking negatively of other veterinary professionals
- Violating confidentiality
- Practicing below the standards of patient care
- Substance abuse problems
- Animal abuse or neglect
- Prescribing medicine without a VCPR

The veterinarian makes all the decisions based on patient care. These decisions are established through the veterinarian-client-patient relationship, commonly called the VCPR. This means that legally the veterinarian must examine an animal before making any diagnosis, providing treatment, surgery, prescription medicines, or prognosis of diseases (see Figure 7-1). This relationship must be maintained within a year's time frame to provide the listed services. Each year a reexamination must take place to continue the legal responsibility to the client and patient. Once the veterinarian has begun this relationship, he or she is then responsible for continuing the standard of care for the patient. This includes emergency services that are to be available as specified by the veterinary facility. The VCPR may be ended by the client at any time. The VCPR may be ended by the veterinarian only after an animal is provided treatment.

Medical records are ethically considered **confidential**. This means that patient information is private and not to be shared by anyone except during the care of the patient by the necessary staff members. Personal client information is not to be disclosed. The medical records are the property of the veterinary facility and are not owned by the client or veterinarian. Copies may be made and provided to the client. The client may also request copies of the record to be transferred with written permission. The physical medical record shall legally remain in the veterinary facility at all times.

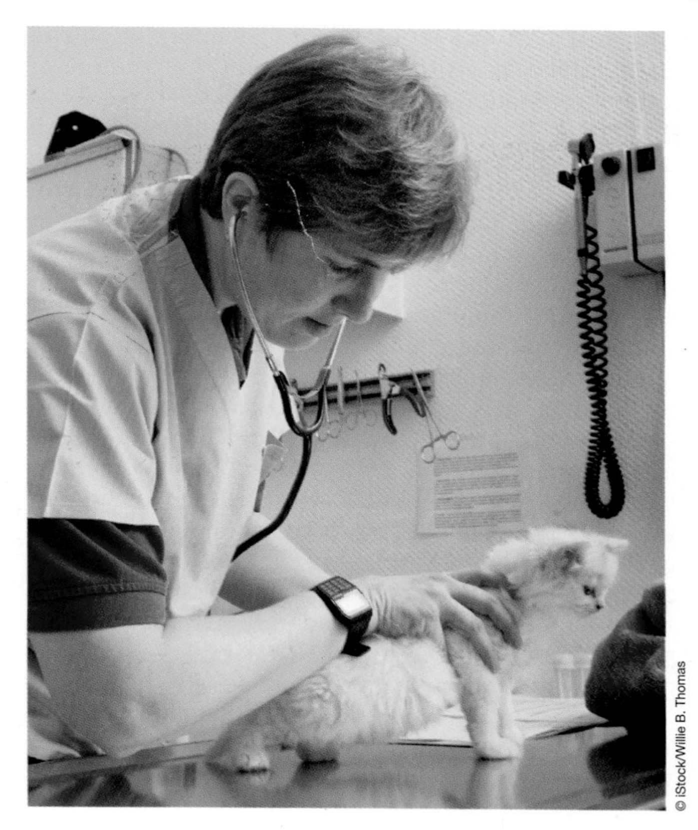

FIGURE 7-1 The veterinarian is required to conduct an examination of the patient before administering treatments or rendering a diagnosis.

Veterinary Laws and Veterinary Practice Acts

Veterinary laws are based on each state's **Veterinary Practice Act**. The practice act is a legal document that outlines rules and regulations of veterinary professionals. These laws will vary from state to state. It is important that all staff members obtain and understand their state's practice act and follow the rules. The practice act is regulated and overseen by the **State Board of Veterinary Medicine**. Each state has veterinary professionals and public members, along with legal counsel that oversees the veterinary laws within the state. The board serves the public by overseeing complaints of veterinary professionals and taking disciplinary action as deemed necessary for the veterinary profession. The veterinary assistant should research his or her state Veterinary Practice Act and become familiar with the rules and regulations. This information can be obtained by visiting the American Association of Veterinary State Boards (AAVSB) Website: http://www.aavsb.org/. It contains licensing information and each state's laws according to the Practice Act. Each state's Practice Act outlines that all veterinary licensed professionals are required to keep abreast of changes in veterinary technology and medical information updates on a regular basis by attending **continuing education** or **CE** seminars. This allows each staff member to obtain updated education within his or her area of specialty and learn more about the profession (see Figure 7-2). It is considered ethical for all members of the profession, whether licensed or not, to update their training and learn as much as they can within the industry. Veterinary staff members are expected to be monitored by the veterinarian through either direct or indirect supervision. Direct supervision states the supervisor needs to be on the premises and within the same room in case of the need of assistance. Indirect supervision states the supervisor needs to be on the premises and near enough to be of assistance.

© iStock/Ben Blankenburg

FIGURE 7-2 Continuing education (CE) is a requirement in the field of veterinary medicine and may be completed by taking online courses.

Veterinary Common Law

Common laws are based on legal violations. **Statutory laws** are based on written government laws. Veterinary medicine is regulated through both types of laws. Knowledge of veterinary laws is essential to all people in the veterinary industry. It is not an acceptable excuse or defense for the veterinary assistant to state that he or she was not aware of the laws within the profession. Common law is based on a court and judge's decision. Statutory law is based on the legislature and government laws. Animals are considered private property, the same as a car or house. The owners have rights based on damage, loss, or injury to their pets. This governing action applies to the veterinary facility, as well. This is known as **liability** or a legal responsibility. This may extend to the owner, as well as the pet, when on the premises of the facility. An example of liability is if a client falls on a slippery floor while in the veterinary hospital. **Malpractice** is working below the standard of practice, knowingly or unknowingly. **Negligence** is a failure to do what should be done. In these cases, the client may have a right to file legal actions in these situations.

The following are examples of ways to prevent malpractice and negligence in a veterinary facility:

- Document everything in the medical record.
- Have every client sign a consent form with listed procedures and give estimates to all clients.
- Lock all windows, doors, and drug cabinets.
- Provide leashes and carriers for all small pets.
- Animal restraint should legally be performed by only trained staff members.
- Keep animals separate.
- Use ID cards and neck bands on all hospitalized patients.
- Record all controlled substances in a log that is correct and maintained daily.
- Check all areas of the hospital for safety hazards on a daily basis.
- Post signs that warn clients of possible hazards.
- Suggest that any complaints be made to the State Board of Veterinary Medicine.

Federal Veterinary Laws

Federal laws are laws that are passed by the U.S. Congress and upheld through the federal court system. They are broader and hold greater authority than state laws. Federal veterinary laws pertain to both people and pets. Federal laws affecting people pertain mostly to veterinary staff members. There are laws regarding safety within the workplace that are governed by the **Occupational Safety and Health Administration (OSHA)**. The **Americans with Disabilities Act (ADA)** governs accommodations for employees who have physical and mental needs. Another law is the **Fair Labor Standards Act**, which governs children and **child labor laws**. There are guidelines for hiring children at certain ages and the duties that they may perform. Federal laws can pertain to animals, too. The **Animal Welfare Act** affects how animals are handled and cared for and takes a large interest in monitoring research facilities. Laws may also pertain to **animal rights**. This is the notion that animals have feelings and emotions that entitle them to the same rights as people. Many of these rights relate to **animal abuse** laws. The neglect and abuse of animals is typically handled by humane societies and state dog wardens.

Veterinary professionals are also governed by federal agencies that monitor medications used within the industry. The **Food and Drug Administration (FDA)** sets manufacturing standards of food additives and medications used in animals, specifically those animals and products used for human consumption. The **Drug Enforcement Agency (DEA)** is responsible for monitoring drugs and issuing veterinary DEA licenses to those professionals prescribing medicines that have the potential for addiction and abuse. Many of these drugs are referred to as **controlled substances**. Controlled substances must legally be kept locked at all times and are to be logged each time they are prescribed to a patient. These records must be kept updated on a daily basis.

SUMMARY

Veterinary ethics and laws are important to everyone within the veterinary field. Knowledge of these legal responsibilities is essential to each employee, but the veterinarian has ultimate responsibility. Many general rules apply to safety and the use of common sense. It is best to know the law and follow it. Each person is expected to behave in a professional manner and pass along respect and a high standard of quality care to both the patient and the client.

Key Terms

Americans with Disabilities Act (ADA) agency that governs the accommodations of people with physical and mental problems to allow them to work more efficiently

animal abuse laws rules that govern the neglect and abuse of animals; typically regulated by a humane society or dog warden

animal rights rules that govern how animals are handled and cared for, especially in research facilities

Animal Welfare Act rules governing how animals are handled and cared for; takes a large interest in monitoring research facilities

child labor laws laws and guidelines that regulate children working and the conditions in which they work

common laws regulations and rules based on legal violations

confidential information that is private and not to be shared by anyone outside of the facility without permission

continuing education (CE) seminars and classes for employees that train and educate veterinary health care members on topics in the profession

controlled substance drugs that have a potential for addition and abuse

Drug Enforcement Agency (DEA) agency responsible for enforcing and monitoring drugs that may be addicted and issuing DEA licenses to veterinarians

ethical the act of doing what is right

ethics rules and regulations that govern proper conduct

Fair Labor Standards Act agency that governs the age of children working and the duties they can perform

Food and Drug Administration (FDA) agency that sets manufacturing standards of food additives and medications used in animals, especially for human consumption

liability a legal responsibility

(Continues)

malpractice the act of working below the standards of practice

misrepresentation working or acting as someone that you are not qualified to be

moral what a person believes is right and wrong

negligence failure to do what is necessary or proper

Occupational Safety and Health Administration (OSHA) agency that governs laws for safety in the workplace

slander talking negatively about someone in an improper manner

State Board of Veterinary Medicine each state's agency of veterinary members and public members that take disciplinary action in the profession as deemed necessary and monitor the rights of the public

statutory laws rules and regulations based on written government law

unethical the act of doing something wrong or improper and knowingly doing it

Veterinary Practice Act legal document that outlines rules and regulations of veterinary professionals

REVIEW QUESTIONS

Matching

Match the term to the proper definition.

1. common laws — a. the act of working below the standards of practice
2. confidential — b. failure to do what is necessary or proper
3. controlled substance — c. talking negatively about someone in an improper manner
4. direct supervision — d. drugs that have a potential for addition and abuse
5. ethics e. regulations and rules based on legal violations
6. indirect supervision — f. what a person believes is right and wrong
7. liability — g. rules and regulations based on written government law
8. malpractice — h. rules and regulations that govern proper conduct
9. misrepresentation — i. working or acting as someone that you are not qualified to be
10. moral — j. a legal responsibility
11. negligence — k. information that is private and not to be shared by anyone outside of the facility without permission
12. slander — l. individual is working while veterinarian is within the facility or area to assist as necessary
13. statutory laws m. individual is working while veterinarian is not within the facility or property

— n. individual is working while veterinarian is immediately within the same room if assistance is necessary

True/False

Read statements 1 through 10 and determine if they are TRUE or FALSE. If the statement is FALSE, correct it to be a true statement.

1. Veterinarians are required to provide emergency care to all patients. T F
2. Veterinary assistants should never allow clients to think they are the veterinarian. T F
3. The VCPR may be ended at any time. T F
4. Only veterinarians should set a high standard of patient care. T F
5. It is acceptable to question the ability of a veterinarian. T F
6. Veterinarians are morally expected to care for each patient that needs medical care. T F
7. All veterinary health care members are required to take CE courses to update their training. T F
8. Common laws and state laws are the same. T F
9. The VCPR is legally required within a 2-year time frame. T F
10. The FDA regulates controlled substances within the veterinary industry. T F

Short Answer

1. Describe five situations that would be considered unethical in a veterinary hospital.
2. List four areas in veterinary medicine that are governed by federal laws.
3. What government agencies regulate areas of veterinary ethics?
4. What is the difference between direct and indirect supervision?

Clinical Situation

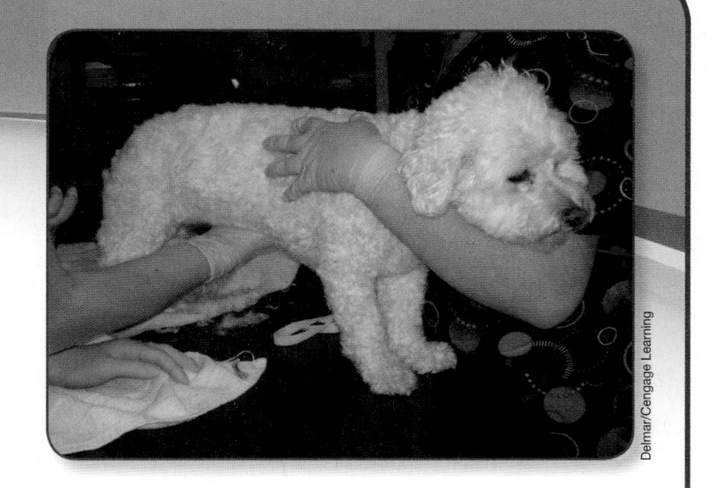
Delmar/Cengage Learning

Kelly, a veterinary assistant at All Creatures Great and Small Veterinary Hospital, was working in the reception area when a client came through the door carrying a small dog that was bleeding.

"Please help my dog, she is hurt really bad."

Rushing to the waiting area, Kelly said, "Please try to calm down and tell me what happened."

"I was walking 'Patches' in the park this morning when she got away from me and ran into the street. A car hit her and I thought she was dead," she replied. "She needs to see the vet now!"

Kelly looked at the dog and said, "She is bleeding very badly. Let me get you a towel so you can apply pressure to the wound. She needs X-rays and will probably need stitches. She also may be going into shock."

"Please get the vet to look at her."

"I'm sorry, but Dr. Hughes is not in yet and Dr. Carr is out on a farm call this morning. I am writing down directions to the emergency clinic that is about 15 minutes away. I will call and let them know you are bringing a dog in that is in need of immediate attention. They are not the best doctors, but they should be able to help 'Patches.'"

"Please, can't you do anything to help 'Patches'?"

"No, I'm sorry, there is nothing I can do."

"Please, she may die!"

"I can't help her. Your dog needs to see a veterinarian and the only one available is at the emergency clinic. I recommend you apply pressure to the area that is bleeding and take her there immediately."

"Why won't you help me?" she replied, walking to the door. "Fine, I'll take her somewhere that cares and will help her."

- What did Kelly do that was ethical in this situation?
- What did Kelly do that was unethical in this situation?
- What could Kelly have done to better handle this situation?

Competency Skills

Professional Veterinary Ethics

Objective:

To reach an understanding regarding what is appropriate and inappropriate practice in the veterinary facility.

Preparation:

- Know the rules and regulations of the veterinary facility.
- Know the difference between ethical and unethical practices.
- Use common sense and respect.

Procedure:

1. The patient needs come before all other needs by reducing pain, suffering, and disease.
2. Obey all laws.
3. Be fair and honest with everyone.
4. The veterinarian makes all decisions based on patient care.
5. Every patient must develop a VCPR within a year, which includes a physical exam in order to diagnose, prescribe medication, perform surgery, or supply a prognosis. This relationship must legally be updated every year.
6. The veterinarian is responsible for all aspects of patient care including emergency services.
7. Medical records are confidential, even between staff members.

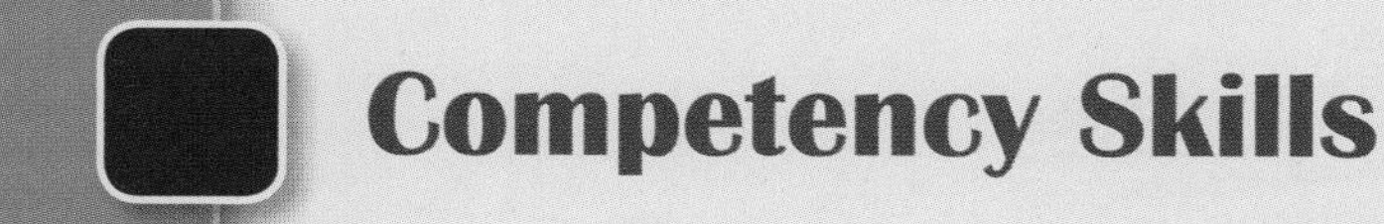

Competency Skills

Professional Veterinary Laws

Objective:

To acquire an understanding of the legal obligations of the veterinary profession.

Preparation:

- Obtain a copy of the state's Veterinary Practice Act and Rules and Regulations.
- Know common and federal laws relating to the veterinary profession.
- Use common sense at all times.

Procedure:

1. Know all laws and obey them.
2. Always document everything within the medical record.
3. Have every client sign a consent form.
4. Give estimates to every client.
5. Keep all doors and windows closed and secure.
6. Provide leashes and carriers for all small pets.
7. Animal restraint should legally be performed by only trained staff members.
8. Keep animals separate.
9. Use ID cards and neck bands on all hospitalized patients.
10. Record all controlled substances in a log that is correct and maintained daily.
11. Check all areas of the hospital for safety hazards on a daily basis.
12. Post signs that warn clients of possible hazards.
13. Report any misconduct to the State Board of Veterinary Medicine.

Section II

Veterinary Animal Production

8

Dog Breed Identification and Production Management

Objectives

Upon completion of this chapter, the reader should be able to:

- Define common veterinary terms relating to the dog
- Describe the biology and development of the dog
- Identify common breeds of dogs recognized by the American Kennel Association (AKC)
- Discuss proper dog selection methods
- Discuss nutritional factors for dogs
- Describe normal and abnormal dog behaviors
- Explain basic training methods in dogs
- Properly and safely restrain dogs for various procedures
- Discuss proper grooming needs of dogs
- Discuss basic health care and maintenance of dogs
- Discuss basic vaccine health programs for dogs
- Explain dog reproduction and breeding methods
- Identify common dog diseases and prevention
- Identify common internal and external parasites of dogs
- Discuss common surgical procedures in dogs

Introduction

Dogs have become one of the most popular companion animals and one of the most important species of animals in society by providing some type of service to people. This is one reason that dogs have become known as "man's best friend" (see Figure 8-1). Research has shown that owning a dog relieves stress, and this in turn results in a healthier immune system. Touching and sharing time with a dog has a positive effect on people and their health. For this reason, veterinary medicine is largely dedicated to the health and well-being of the dog.

FIGURE 8-1 Dogs are the most popular companion animals.

Veterinary Terminology

In veterinary medicine, it is essential to know species-specific terms and when to properly use them in discussions. Table 8-1 contains some common canine terms that are used in the veterinary industry. The veterinary assistant should also be familiar with the external (outside) parts of the canine body structure. Figure 8-2 shows the location of each body part that the assistant should be familiar with and the term that is used for the canine.

TABLE 8-1

Canine-Specific Terminology

Canine (K-9)	Dog (derived from Latin)
Bitch	Female dog
Stud	Male dog of breeding age
Intact	Capable of breeding; still has reproductive organs
Neuter	Surgical removal of the reproductive organs
Spay	Surgical removal of the reproductive organs in the female
Castration	Surgical removal of the reproductive organs in the male
Litter	Group of young dogs from the same parents
Puppy	Young dog under the age of one year
Pack	Group of dogs in a household
Gestation	Length of pregnancy
Whelping	Labor process of a female dog

Biology

Dogs were bred from wolves and were domesticated over 10,000 years ago. **Domestication** means that the animal has become tame and is bred to be near humans for companionship. The canine can be described by its genus and species known as **Canis familiaris.** This identifies it as a relative of the wolf. Dogs are mammals that are known as **carnivores**, or meat eaters. They are able to regulate their body temperature internally in the same manner as humans and are thus called **endothermic** animals. Dogs have a simple stomach digestive system called **monogastric**. Dogs are built for specific uses and purposes and are categorized by size, weight, age, and coat type. They also have a unique **anatomy**, or body structure. The body structure has similar functions to the human body. The body functions are known as **physiology.** Dogs live different lengths of time or **life span** based on their size and health status. Smaller dogs typically live longer than larger-sized dogs. Life span will vary by weight range, breed (some breeds may be at greater risk of genetic diseases), and how well the dogs are cared for by their owners.

Breeds

The American Kennel Club (AKC) recognizes 155 breeds of dogs categorized into seven groups. The seven groups are the sporting dogs, hounds, working dogs, terriers, toys, herding dogs, and nonsporting dogs (see Table 8-2). A miscellaneous group has been established to classify popular breeds that have not been admitted to one of the seven groups. **Hybrid** or **designer breeds** are also increasing in popularity. An example of a designer dog breed is the labradoodle, or a mixture of a Labrador retriever and a poodle. These are crosses between two common **purebred** breeds. A purebred dog is a dog that is from parents that are registered and have known parentage (see Figure 8-3). Some dogs are **mixed breeds**, or a mixture of two or more breeds of dogs that has no known parentage (see Figure 8-4).

Breed Selection

It is common in the veterinary field for clients to seek out veterinary assistance and knowledge in selecting the proper dog for a household. There are many factors for potential dog owners to consider. When offering

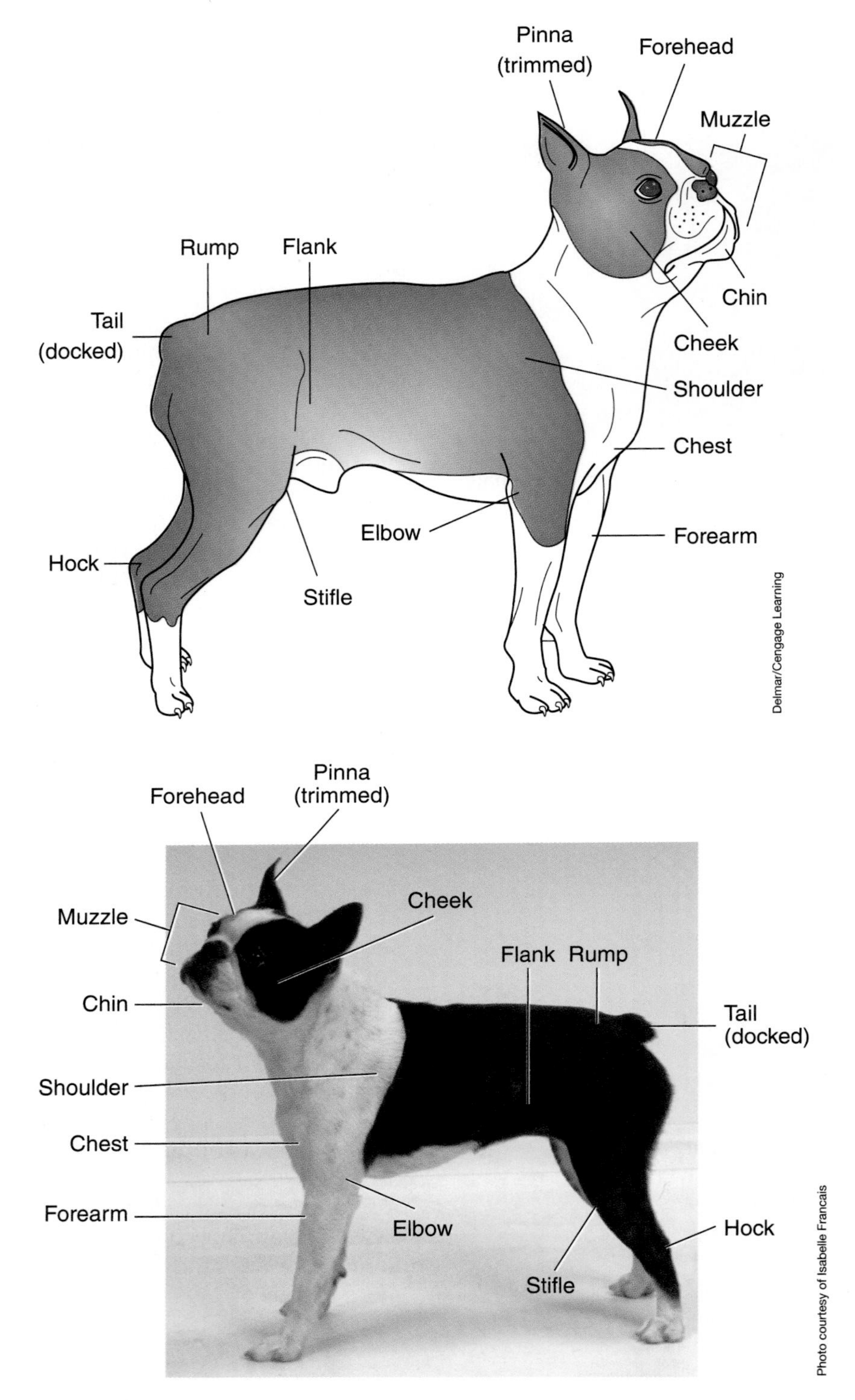

FIGURE 8-2 Anatomical parts of a dog.

dog selection information, guide the client to the proper factors for consideration when choosing a breed. For instance, the client should consider if he or she wants the responsibility of a puppy or an adult dog. Puppies require a lot of time and care with housebreaking, training, and health care. Some factors that must be discussed include the following:

- Size of the dog as an adult
- Indoor or outdoor dog

TABLE 8-2

Dog Groups and Breeds

GROUP	NUMBER OF BREEDS	EXAMPLES OF BREEDS	CHARACTERISTICS
Sporting	27	Pointers, Retrievers, Setters, Spaniels	Bred for hunting, active, need regular exercise
Hound	23	Bassett, Beagle, Bloodhound, Greyhound	Bred for hunting, strong scent abilities, good stamina, vocal
Terrier	27	Airedale, Bull, Fox	Bred to hunt vermin, feisty, energetic, not good with other animals
Working	26	Boxer, Great Dane, Siberian Husky, Saint Bernard	Bred for jobs such as pulling sled, large, intelligent, quick learners
Herding	22	Collie, German Shepherd, Sheepdog, Corgi	Bred to control movement of other animals, intelligent, good for training exercises
Toy	21	Chihuahua, Pekingese, Pug, Shih Tzu	Small in size, don't require a lot of space
Nonsporting	17	Boston Terrier, Bulldog, Dalmatian, Poodle	Sturdy, very varied
Miscellaneous	11	Icelandic Sheepdog, Bluetick Coonhound	Dog breeds not yet placed in one of the seven dog groups recognized by the AKC
Hybrid		Puggle, Labradoodle, Schnoodle	Breed mixtures that are not yet recognized by AKC. Examples: Pug and Beagle; Schnauzer and Poodle; Labrador and Poodle.

Delmar/Cengage Learning

FIGURE 8-3 A purebred dog has characteristics and features unique to one specific breed.

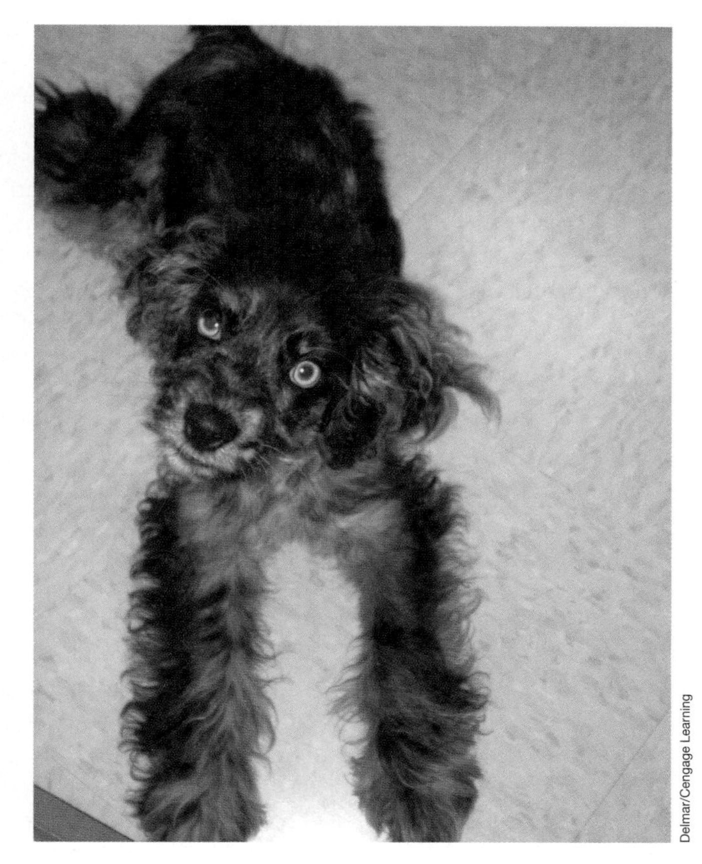

Delmar/Cengage Learning

FIGURE 8-4 A mixed-breed dog will exhibit characteristics and features of two or more different breeds.

- Space needed for dog
- Yard size or fencing
- Shelter type
- Short hair or long hair
- Grooming needs
- Health care needs
- Breed health problems
- Used for breeding or show purposes
- Training needs
- Nutritional needs
- Restrictions in the area

The best advice to offer clients is to research several breeds to determine which type of dog will best suit their lifestyle and family needs. Veterinary clinics may keep a list of reputable breeders that are located in the area to provide to interested clients. Refer people to local humane shelters to seek information in adopting a dog. Keep reference materials that will allow future dog owners to become educated and responsible pet owners. Some Websites that offer information on various dog breeds include the following:

- American Kennel Club—http://www.akc.org
- Dog Breed Info Center—http://dogbreedinfo.com
- Continental Kennel Club—http://continentalkennelclub.com

Nutrition

A dog's nutritional needs will change as it grows from a puppy to an adult. These changes are known as **life stages.** Each life stage requires different nutrients for the age, health, and lifestyle of the dog (see Table 8-3).

Feeding Puppies

Puppies should be **weaned** from their mother around 4–5 weeks of age. Weaning is the process of stopping nursing and beginning to eat a solid food diet. Some puppies may need to nurse longer depending on size and health conditions. During the weaning process,

TABLE 8-3
Feeding Requirements for Dogs

LIFE STAGE	REQUIREMENTS
Puppy (orphaned)	• High-quality puppy formula, such as Esbilac or KMRFeed 60 mL/day • 4 daily feedings until 2 weeks of age • 3 daily feedings until weaned • High-protein, high-calcium, high-fat diet
Puppy Weaning	• Wean at 4–5 weeks • Begin on high-quality dry puppy food that is size appropriate • Add warm water to soften as necessary • High-protein, high-calcium diet • Large- or giant-breed dogs should be fed moderate to low protein
Puppy to 1 year	• High-quality puppy diet • Large-breed puppy diet • Moderate protein, calcium, and phosphorus diet
Adult (1 to 7 years old)—active	• Maintenance diet • Moderate protein, calcium, and phosphorus diet with low fat
Adult—overweight, inactive	• Reduced calorie diet • Healthy weight diet • Low-fat, low-calorie diet • Low-protein, moderate calcium and phosphorus diet
Senior (over 7 years old)	• Senior diet • Reduced-calorie diet • Low-fat, low-calorie diet • High-phosphorus, high-calcium diet with moderate protein

pups should be fed either canned food or dry puppy food that has been softened with warm water. The dry food should be size appropriate. The water can be decreased and dry food offered to allow the puppy to be completely weaned by 6–8 weeks of age. This transition food should be fed several times a day and the puppy allowed food for 15 minutes. This will teach the puppy to learn to eat at scheduled times. As the puppy grows, it should be fed twice a day. Sometimes breeders will deal with **orphaned** puppies due to the mother refusing to nurse, illness, or death. In the case of a puppy that is orphaned, a specialized formula diet is necessary. Examples of commercial formulas include Esbilac and KMR (see Figure 8-5). A homemade formula can be prepared as well, but a specific commercial high-quality puppy formula is best for the health of the orphan. When preparing or opening a can of formula, always date it and then keep it refrigerated. It will need to be warmed for each use and should be used within 24–48 hours.

Homemade Puppy Formula
Three egg yolks
One cup homogenized milk
One tablespoon corn oil
One dropper full of liquid vitamins such as Pet Tinic

When preparing a homemade formula care must be taken to maintain the purity of the ingredients.

The formula may be fed with a pet nurser, an eye dropper, a syringe, or a rubber feeding tube (see Figure 8-6). A pet nurser is a bottle used to feed a puppy similar to how a human baby is fed. It is important to weigh each puppy daily prior to feeding. This will allow the veterinary assistant to monitor weight gains. The typical formula rate for puppies is 60 mL per pound per day. The total amount should be divided into 4 feedings until 2 weeks of age and then decreased to 3 feedings until weaned.

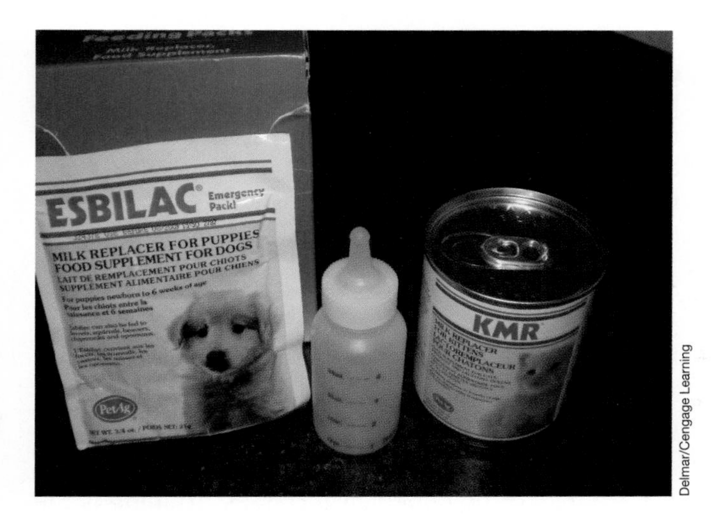

FIGURE 8-5 A commercial puppy formula can be bottle fed to puppies that are orphaned before weaning age.

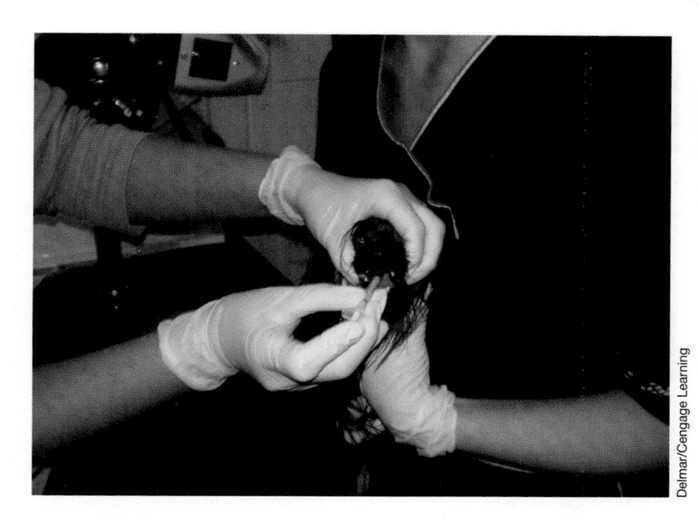

FIGURE 8-6 Syringe feeding of a puppy.

Calculating Puppy Formula Rates
Formula Rate = 60 mL/pound/day
Feed puppies under 2 weeks of age 4 times daily
Feed puppies over 2 weeks of age, and until weaned, 3 times daily

Example

To calculate the amount per feeding for a 1.6-pound puppy that is 1 week old, the following process should be used:

Number of pounds x 60 mL = total amount feed in one day

1.6# x 60 mL = 96 mL

Total number of mL for day divided by number of feedings per day = amount to be fed per feeding

96 mL / 4 daily feedings = 24 mL per feeding.

It's important to remember that the term *pound* can be written with the number sign (#) to mean pounds of body weight.

Feeding Adult Dogs

Adult dogs should ideally be fed a dry food diet to help prevent tartar buildup on teeth. The food size should be appropriate for the size of the dog. It is best to feed a high-quality pet food that is easily digestible. Examples of high-quality dog foods are Eukanuba, Iams, Purina, Purina One, and Nestle.

Adult dogs that have regular exercise and are between the average age of 1 and 7 years should be fed a **maintenance diet**. This will allow the dog to maintain the same weight for the activity it receives on a daily

FIGURE 8-7 Young active dogs should be fed a maintenance diet.

basis (see Figure 8-7). Dogs over 7 years, especially medium to giant breeds, are considered senior pets and should be fed a **senior diet** (see Figure 8-8). These diets usually consist of fewer calories for less-active dogs. Dogs that are overweight may be placed on a **reduced-calorie diet**. These diets are much lower in calories to help dogs lose weight and meet the daily energy needs of an overweight or less-active dog. A dog should be fed for the ideal weight of the dog based on age, breed, and activity level. The pet-food label should list feeding requirements. Make sure to feed any overweight dogs for their ideal size rather than their actual weight. The label will state the necessary **ration**, or the amount of food that should be fed.

MAKING THE CONNECTION

For a general discussion of nutritional needs of animals, please refer to Chapter 30.

FIGURE 8-8 Older dogs, particularly large breeds, should be fed a senior diet.

Behavior

Dogs may be happy, scared, or **aggressive** (ready to attack). The happy dog is alert, ears up and forward, mouth open and panting, tail up and wagging, with a relaxed stance and appearance. A scared dog will appear with the tail tucked under its body, the body lowered to the ground, ears flat and back on the head, avoiding eye contact, jumping when noises occur, raising a paw when approached, and shaking from nervousness. An angry or aggressive dog may appear growling, showing teeth, hackles up over the back and shoulders, head lowered to the ground, staring with direct eye contact, the tail raised and bristled, and a stiff stance.

Dogs are naturally pack animals. They will bark to warn others of approaching danger or strangers. Barking is a natural behavior that must be taken into consideration when owning and housing a dog. Barking is a way of communication among dogs. Another common dog behavior is marking, or eliminating with urine on areas to mark its territory. This creates a scent alerting other animals to where it lives. Marking may become a problem if a dog does this behavior inside a home. Marking is decreased by neutering dogs at an early age.

Basic Training

Veterinary assistants should discuss basic training information of puppies and dogs and offer referrals of local dog trainers or dog obedience classes. **Training** is the process of helping an animal understand and acquire specific desired habits. These habits should include housebreaking, obedience training, and socialization. **Housebreaking** includes teaching the dog to eliminate outside at specific times. Inappropriate elimination is when a dog has an accident inside the house, usually from urinating. **Obedience** includes teaching a dog to walk on a leash, sit, stay, and behave with proper manners. **Socialization** includes teaching a dog to act properly with other dogs and people. Some facilities will offer puppy socialization classes to discuss what protocols new puppy owners should address while their pets are young. It is not uncommon for clients to ask questions related to behavior and training issues for their dogs. Some common topics dog owners inquire about are excessive barking, marking territory, digging, and inappropriate elimination. Some pet owners may inquire about specialized training for their dogs such as hunting behaviors, tricks, or sports-related behaviors.

Equipment and Housing Needs

Dogs will require some special facilities and equipment. All dogs should have a collar and leash, a crate or kennel area, a food and water bowl, and appropriate bedding (see Figure 8-9, A–D). Toys are optional but recommended to keep dogs from becoming bored.

Restraint and Handling

Animals have more developed senses than humans. They use these senses to perceive changes within the environment. Dogs, therefore, should be handled with

FIGURE 8-9 Essential equipment needed to care for a dog: (A) collar and leash; (B) food and water bowls; (C) crate and bedding; (D) toys.

care. Dogs see prolonged staring or direct eye contact as threatening. Quick movements from people can also cause a negative reaction. Large numbers of people can be intimidating to a dog. Restrainers should move slowly, calmly, talk quietly, and avoid eye contact with any nervous animal. Knowledge of dog behavior and body language is essential to a veterinary assistant for use in safe and adequate restraint skills. As a restrainer, the veterinary assistant has the ability to influence the behavior of the animals he or she is working with. Each situation and patient should be monitored carefully to prevent injury to the handler, the patient, and others involved in working with the animal. Sometimes less restraint is more, but the veterinary assistant must monitor patients at all times for changes in behavior. Restraint may include **physical, verbal, chemical restraint,** or a combination of these methods. Physical restraint is a person controlling the position of an animal through the use of the body and/or equipment. Verbal restraint is the use of the voice to control an animal, such as the words NO, SIT, or DOWN. Chemical restraint is the use of sedatives or tranquilizers to calm the animal to allow procedures to be done.

TERMINOLOGY TIP

A **sedative** or **tranquilizer** is a type of medication used to calm an animal and prevent it from moving. It is important to note, however, that these medications may not relieve pain sensations.

Physical Restraint

There are several common physical restraints that may be used for dogs. These include sitting and standing restraint techniques as well as placing the dog in **recumbency** positions. Standing restraint may be used on the floor or on a table. One arm is placed around the neck to control the head. This is called the *bear-hug*. The other arm is placed under the stomach to prevent sitting and to control the body (see Figure 8-10). If the dog appears to become fearful or aggressive, the hand holding the neck can be moved to the muzzle to prevent biting.

Sitting restraint is done in the same manner, except the dog is sitting (see Figure 8-11). If a large dog resists the restraint, it is helpful to use a wall to prevent the dog from backing up. The restrainer can place the back side of his or her body against the wall and place the dog in a sitting position with the rear side between the legs. The head is then controlled with the arms (see Figure 8-12). The same method can be applied for small dogs and the restrainer uses his or her own body as the wall.

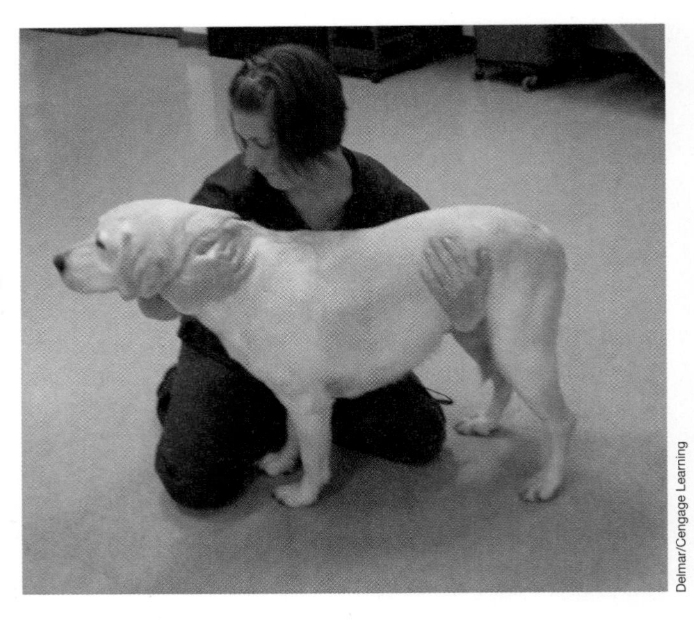

FIGURE 8-10 Standing restraint.

FIGURE 8-11 Sitting restraint.

Lateral recumbency, placing the animal on its side for restraint, is commonly used for X-ray positioning and other procedures. The dog should be restrained in a sitting position and carefully placed in a lying position on its chest. It is then carefully placed on the necessary side. For small dogs, the restrainer must hold the legs that are on the down side and nearest the floor or table. This will prevent the dog from standing. The arms and upper body of the restrainer are used to gently hold the dog in

FIGURE 8-12 Sitting restraint of large dog with use of wall and handler's body.

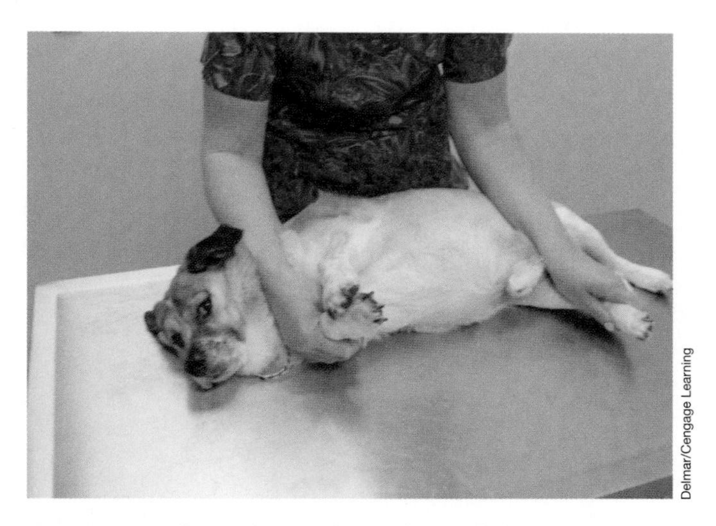

FIGURE 8-13 Lateral recumbency in small dog.

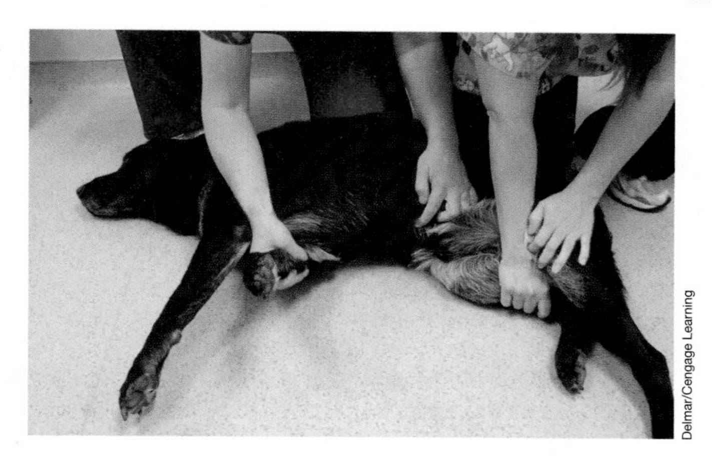

FIGURE 8-14 Lateral recumbency in large dog.

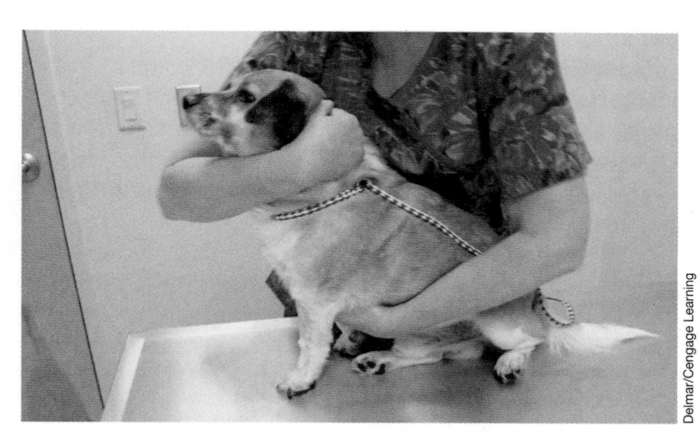

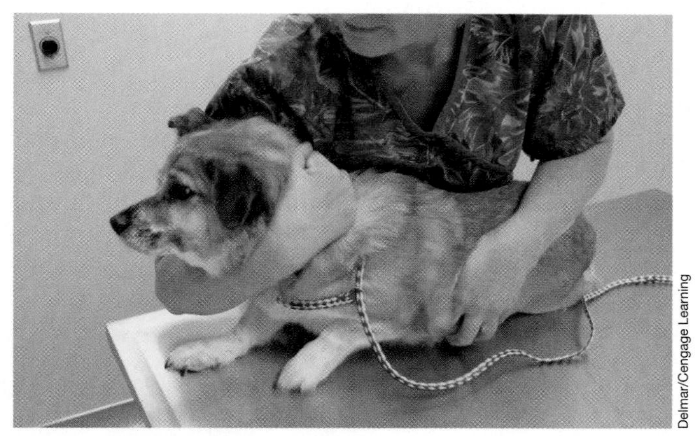

FIGURE 8-15 Sternal recumbency.

position (see Figure 8-13). For large-sized dogs, this restraint will require two to three people. One person will restrain the front of the body and one will restrain the rear of the dog (see Figure 8-14). It is important to continue restraint of the legs on the downside of the dog, as this is what will keep the dog from rising. The care of the head must also be considered so that the dog doesn't hit its head against the floor during restraint.

Sternal recumbency is a restraint position with the dog lying on its chest or sternum. This requires placing the dog in a sitting position and applying pressure to gently place it in a lying position on its chest (see Figure 8-15). For small dogs, the restrainer can use one arm to control the head and the other arm and body to gently hold the dog in position. For large dogs, one person can bear-hug the head of the dog and gently apply body pressure to the shoulders and back and another person can restrain the rear of the dog.

Dorsal recumbency is when the dog is positioned on its back, commonly used in X-ray positioning. The dog is placed in lateral recumbency and then carefully rolled onto its back. Two or more people should be restraining the head, middle of the body, and the rear of the body. The front legs are pulled forward toward the head of the dog and the rear limbs are pulled backward toward the tail (see Figure 8-16).

Cephalic venipuncture is blood collection from the cephalic vein located on the inside of the front legs. The dog is typically placed in sitting or sternal recumbency and a front leg is extended for the blood collection (see Figure 8-17). Some vets and technicians will ask the restrainer to hold off the vein by placing pressure around the elbow of the dog.

Jugular venipuncture is blood collection from the jugular vein located on either side of the neck. The common restraints are sitting or sternal recumbency (see Figure 8-18). For smaller dogs, the front limbs can be extended over the edge of a table to allow more space to collect the blood sample.

Saphenous venipuncture is blood collection from the inside of the thigh in the rear limbs. Dogs are commonly placed in lateral recumbency and the restrainer may be asked to hold off the pressure around the upper thigh area of the rear limb (see Figure 8-19).

Lifting Techniques

Lifting dogs requires safety for both the restrainers and the animal. When lifting any sized dog, a restrainer should always keep his or her back straight and bend at the knees. This will help prevent any back injuries. Small dogs may be lifted by supporting the front and

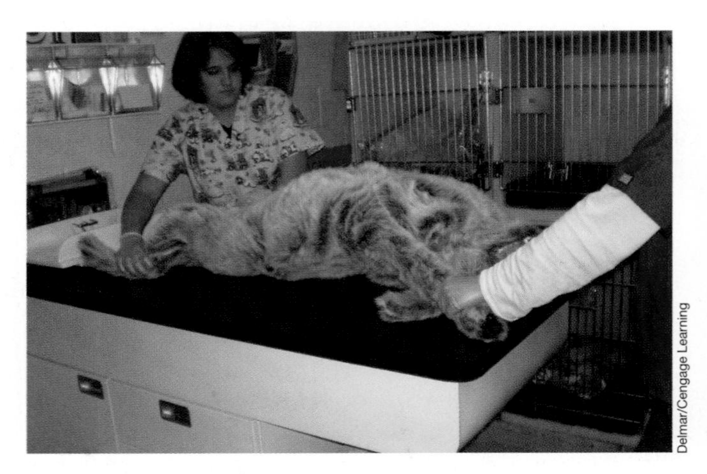

FIGURE 8-16 Dorsal recumbency.

FIGURE 8-17 Restraint for cephalic venipuncture.

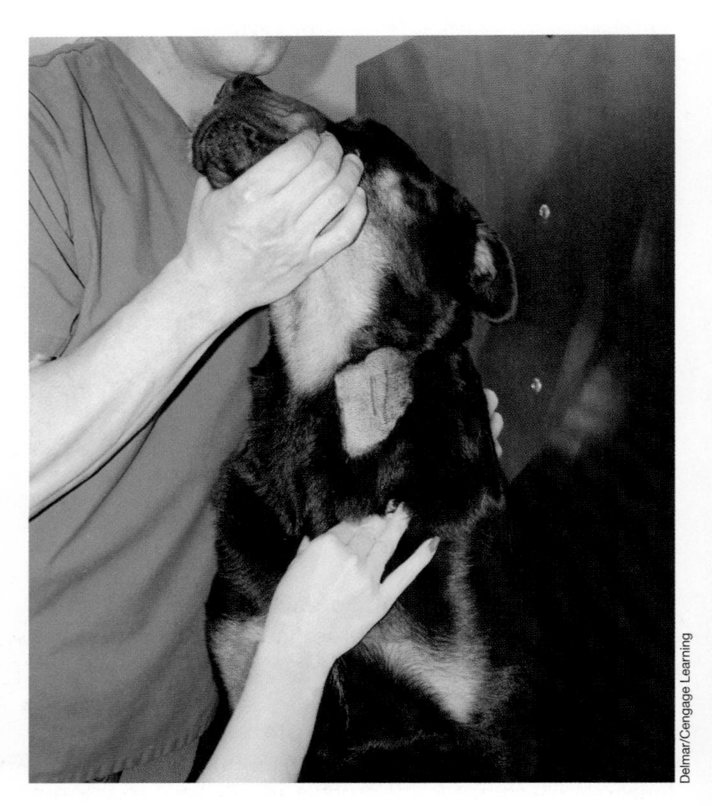

FIGURE 8-18 Restraint for jugular venipuncture.

FIGURE 8-19 Restraint for saphenous venipuncture.

back ends similar to standing restraint. Large dogs will require two or more people. One person should control the head and one or more will need to control the rear of the body. Lifting should occur with pressure on the abdomen and chest area. It is important for two or more people to always work on the same side of the dog. Teamwork involves communication to indicate to each other when to lift.

Use and Application of a Muzzle

Dogs will sometimes need restraint equipment to help keep them from biting or to keep them calm. A common piece of equipment used to prevent biting is the **muzzle**. The muzzle may be made of commercial nylon, a plastic or wire basket, tape, a leash, or gauze material. A commercial muzzle is usually adjustable with a snap that locks in place (see Figure 8-20). They are come in various sizes to fit the size of the dog. The muzzle should provide a snug fit. The short side of the muzzle should be on top of the dog's nose and the wide side should be placed under the jaw. The snap goes behind the ear flaps and snaps securely in place (see Figure 8-21, A–B). The dog's legs should be

FIGURE 8-20 Various sizes of commercial muzzles.

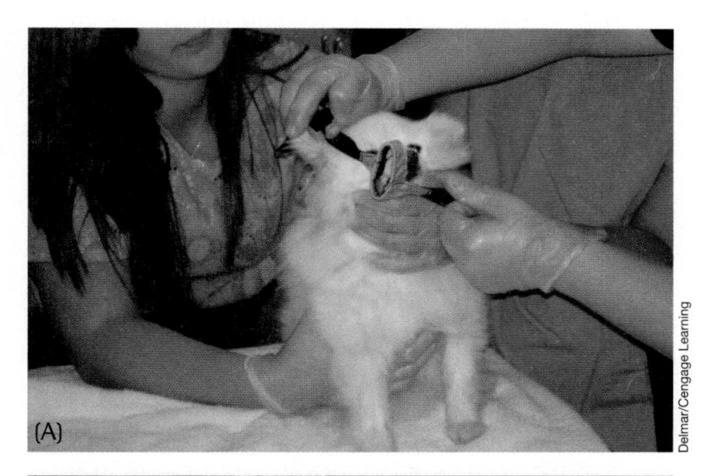

FIGURE 8-21 Use of a commercial muzzle: (A) applying a muzzle; (B) snapping the muzzle into place.

properly restrained to prevent removing the muzzle. When applying the muzzle, note that for safety, the restrainer should stand at the side or the rear of the animal. Never work directly in front of the dog. A tape or gauze muzzle can be used if a commercial muzzle is not available. Make sure to use an adequate amount of material that will fit the dog. **Gauze** is thin material that may come on a roll and is made of a mesh-like cotton. Gauze material should be heavy to prevent tearing. Tape should be folded upon itself to prevent glue from attaching to the dog's hair or skin. Begin by making a large loop that is placed over the muzzle and secured as in tying a shoelace. Pull tight and cross ends snugly under the jaw and tie in a knot under the jaw. Take each end on either side of the ears and tie in a bow behind the ear flaps (see Figure 8-22, A–C). When removing the muzzle, release the bow and gently pull off the muzzle (see Figure 8-23). If the dog is likely to try biting, untie or unsnap the muzzle and allow the dog to remove it. Dogs that have a flat face lacking a long muzzle, such as pugs or Bulldogs, are called **brachycephalic**. These breeds are prone to difficulty breathing. Care must be taken when muzzling these breeds. A tape or gauze muzzle can be placed over the nose and one end looped under the material and pulled tight. Both ends are pulled between the eyes and over the head and tied securely in a bow around the nape of the neck.

Use of a Rabies or Snare Pole

A restraint device used to remove aggressive dogs from a cage is called the **rabies pole** or **snare pole**. The pole is a long extension with a heavy loop on the end that is adjusted by sliding the pole handle (see Figure 8-24). It acts as a leash to walk the dog or control the dog to be properly restrained. Once the loop is over the head, the dog should be guided forward with the restrainer behind the dog. Do not choke the animal by pulling it. Care must be used in this type of restraint to avoid injury to the dog.

Use of Sedation

Some dogs that are too aggressive to handle safely may be sedated for the safety of the veterinary staff. Sedation is used to calm down a dog and allow it to safely be restrained. The use of sedation must be prescribed by the veterinarian and should be used as a final resort when all other options have been used. Sedation is an example of chemical restraint.

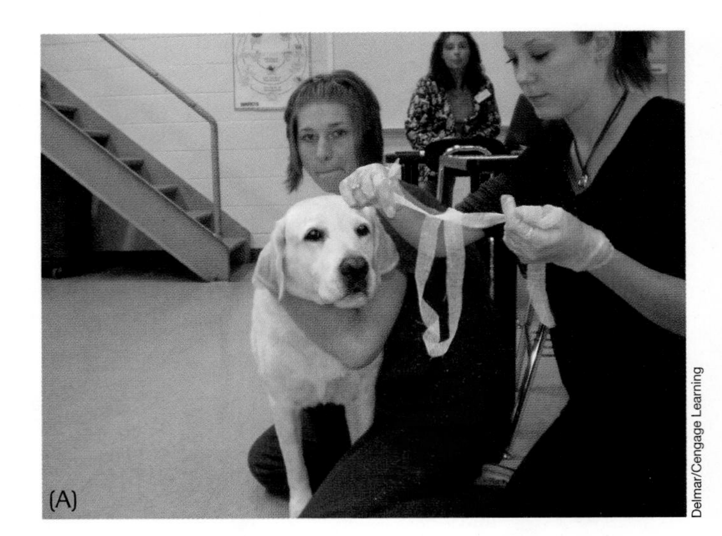
(A)

Delmar/Cengage Learning

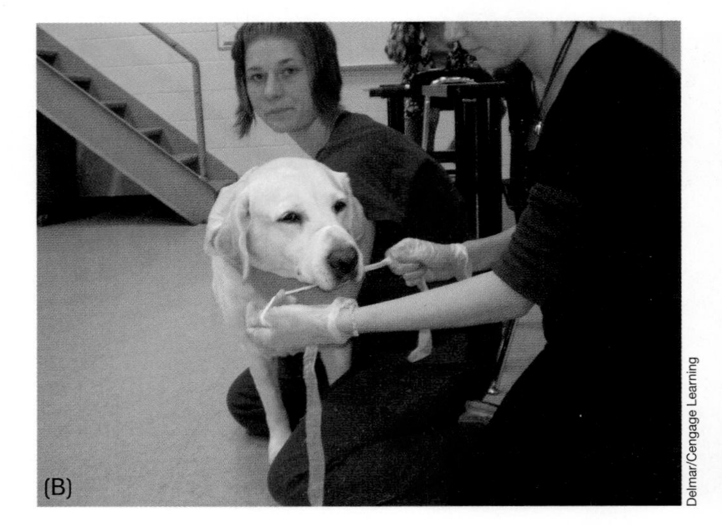
(B)

Delmar/Cengage Learning

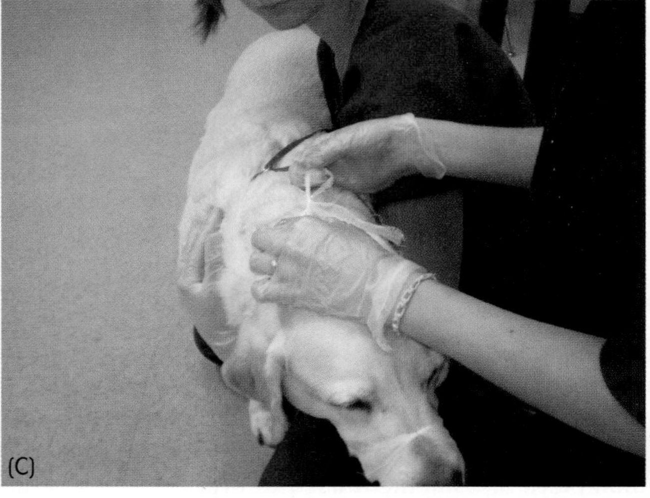
(C)

Delmar/Cengage Learning

FIGURE 8-22 Use of gauze for a muzzle: (A) make a loop with the gauze; (B) secure the loop over the dog's nose and mouth; (C) tie the ends of the loop behind the dog's ears.

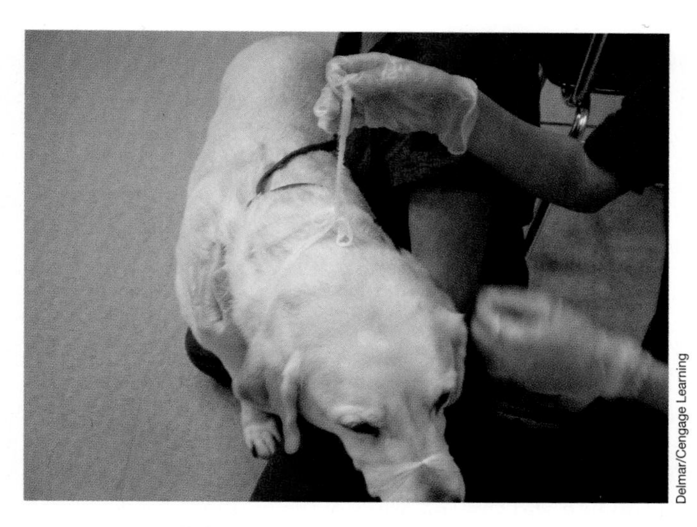

FIGURE 8-23 Release the muzzle by pulling the loose ends of the bowtie.

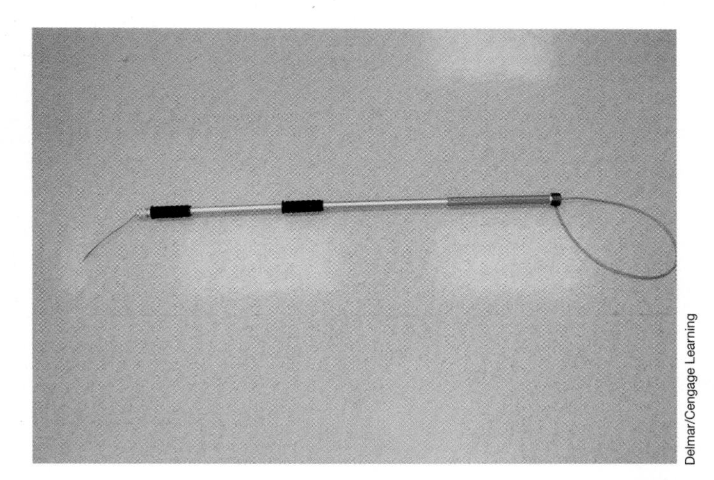

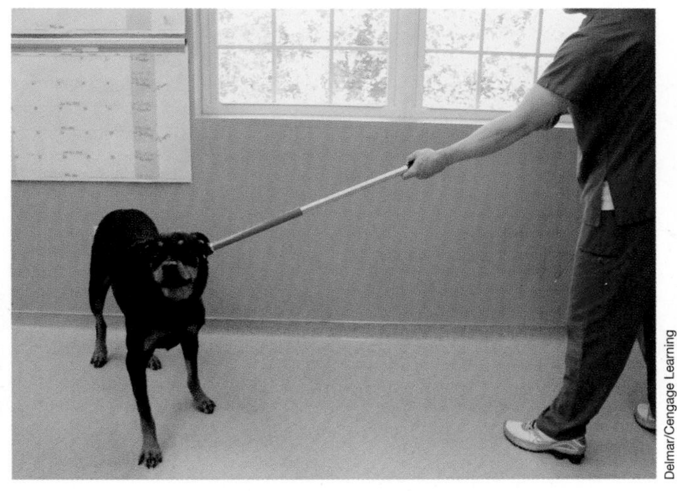

FIGURE 8-24 Restraint using a rabies pole.

MAKING THE CONNECTION

For a more general discussion of restraint techniques used for animals, please see Chapter 27.

Grooming

Grooming services may be provided by the veterinary facility or veterinary assistants may be asked to provide grooming care to hospitalized patients. **Grooming** is the process of trimming and bathing in an appropriate manner for the best care of the patient (see Figure 8-25). Other grooming services may include ear cleaning, brushing the hair coat, brushing teeth, trimming nails, trimming hair, shaving, and ear plucking. Brushing is done first to allow the hair to be lying in the proper position and to remove any mats or items that may have collected within the coat. This allows the veterinary

FIGURE 8-25 Grooming services are provided in some veterinary facilities.

assistant to see any issues with the hair coat and have a neatly organized coat to work with.

Tools

Clippers should be maintained with oil to keep clipper blades lubricated, cool lube to prevent the clippers from overheating, and blade wash to keep the clippers and blades sanitary and properly disinfected. Clippers and blades are commonly cleaned with a toothbrush or wire brush to remove all hair and debris. Special chemicals are used to clean and disinfect the clippers and blades to prevent rusting and damage. Clipper blades can be sharpened if they become dull. The clipper blades come in many sizes (see Table 8-4). The blade size to use is determined by the type of hair coat and the length of hair being clipped or shaved. Clipper blades can be removed from the clippers for cleaning and for changing to different sizes depending on the area to be trimmed or hair coat length of a dog.

TABLE 8-4

Clipper Blade Selection

BLADE SIZE	CUT LENGTH	USAGE
#50	1/125"	Surgical prep; poodle feet and face
#40	1/100"	Surgical prep; poodle feet; under snap-on combs
#30	1/50"	Poodle feet; between feet pads
#15	3/64"	Sensitive feet and faces; Poodles, Cockers, Terriers
#10	1/16"	Sanitary trim on genitals and rectum; cat clipping
#9	5/64"	Sporting breeds such as cockers and schnauzers
#8.5	3/16"	Head, face, neck, and back
#7	1/8"	Matted dogs; body clipping
#5	1/4"	Body clipping
#4	3/8"	Body clipping and trimming wire-haired breeds; short trims
#3 ¾	1/2"	Medium to short cuts; puppy cuts
#5/8	5/8"	Longer cuts and puppy cuts
#3/4	3/4"	Body work on longer breeds

It is important for veterinary assistants to know how to properly clean and care for grooming equipment and instruments. Daily maintenance is required to keep grooming supplies sanitary and working properly. Sanitation of tools and grooming supplies should be routine with the use of disinfectants and cleaners that are labeled for cleaning veterinary equipment. The spread of parasites and diseases is common in grooming items if they are not properly sanitized and sterilized between uses. Some veterinary assistants may be interested in pursuing a career in grooming and learning how to trim and clip animals by meeting the breed standards.

Trimming or Shaving

Trimming and shaving areas of the hair may be done with scissors, thinning shears, or clippers. Scissors should be used with caution as they can easily injure the animal. It is best to use a blunt tip, which is rounded and not sharp. Clippers are used to shave down areas of the hair coat. For example, some long-haired dog breeds may need to be shaved around the rectum or urogenital area to prevent feces and urine from collecting on the hair coat (see Figure 8-26). Another common area

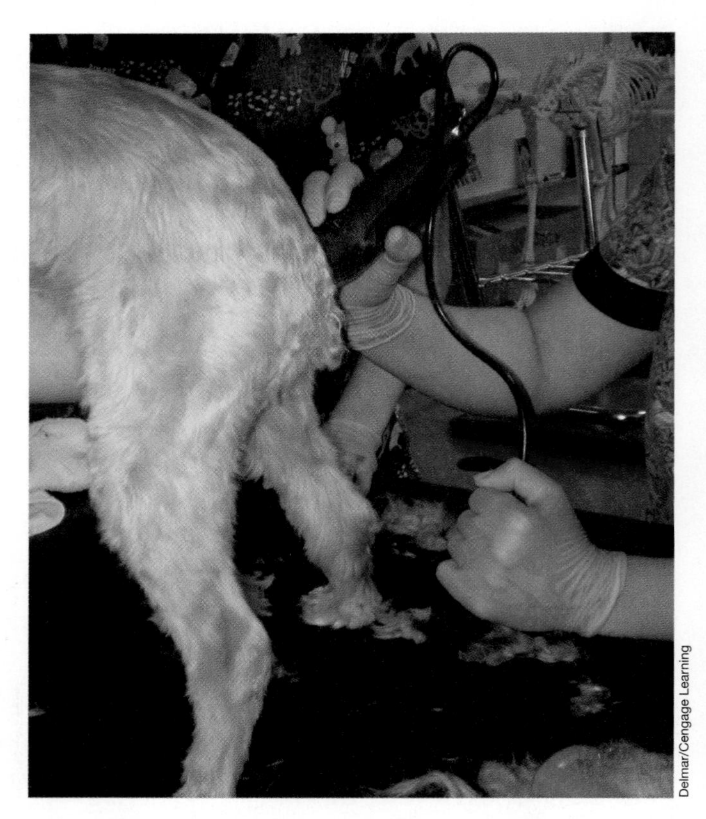

FIGURE 8-26 Some dog breeds may require clipping around the urogenital area to prevent infection and maintain the dog's cleanliness.

to shave is between the digital pads of the feet where long hair may collect and become irritating to dogs (see Figure 8-27). Ear flaps may also be a common area to shave or trim to allow better air circulation for breeds prone to infections (see Figure 8-28).

> **SAFETY ALERT**
>
> *Care must be used when trimming or shaving animals. Be wary of cutting or injuring the animal, especially when using scissors.*

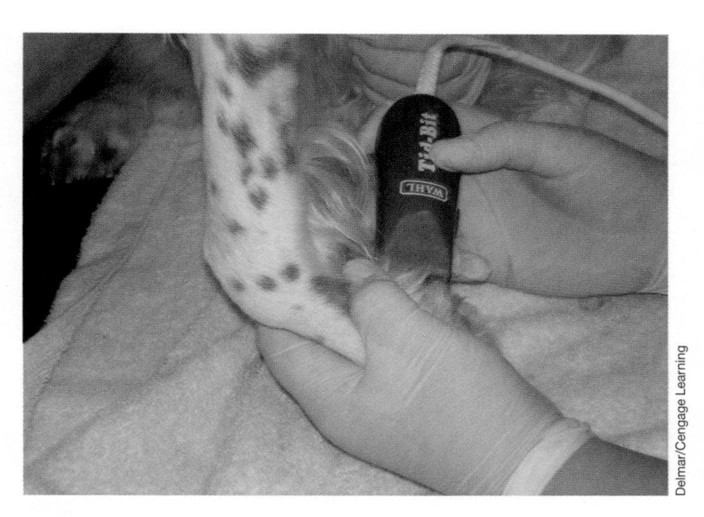

FIGURE 8-27 Dogs may grow hair between the digital pads of the feet that may require clipping for the comfort of the animal.

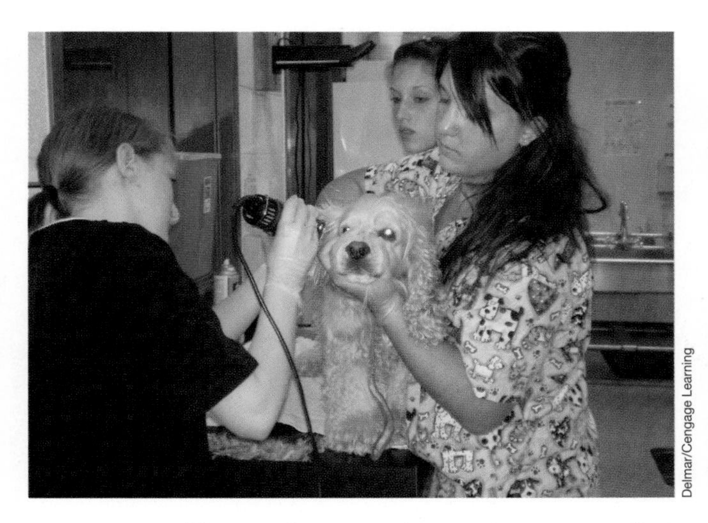

FIGURE 8-28 The ear flaps are also a common area that may require clipping to prevent infection.

Bathing

Bathing should be done with warm water and gentle shampoo appropriate for the dog's skin. Most dog breeds need bathing every few months. Caution should be taken to avoid overbathing, which will cause the skin to become dry and flaky. This causes natural oils in the body to be removed and the skin to dry out and develop dandruff, which can lead to other skin conditions. Bathing requires applying a protective eye ointment that lubricates the eyes to prevent injury from shampoo or water (see Figure 8-29). Shampoo and conditioner should be applied to the hair coat and worked into the hair and skin (see Figure 8-30). The shampoo

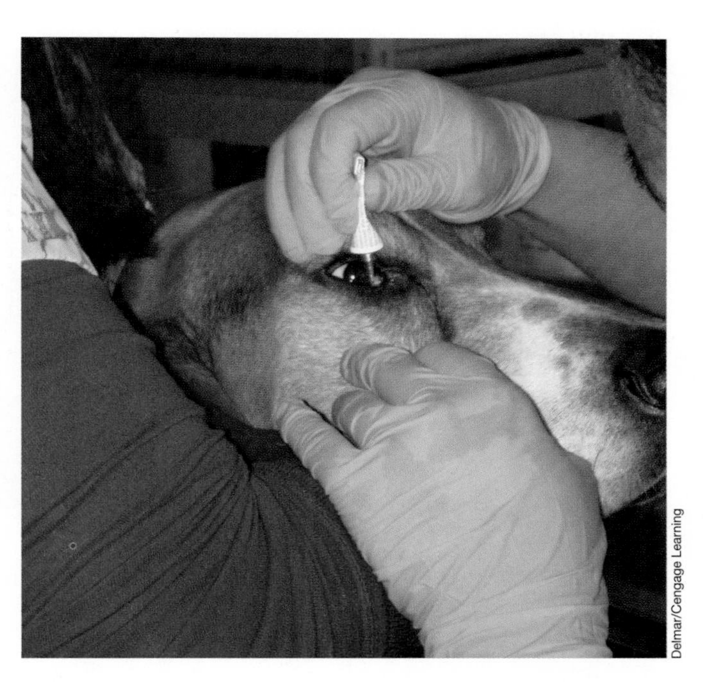

FIGURE 8-29 Application of eye ointment before bathing.

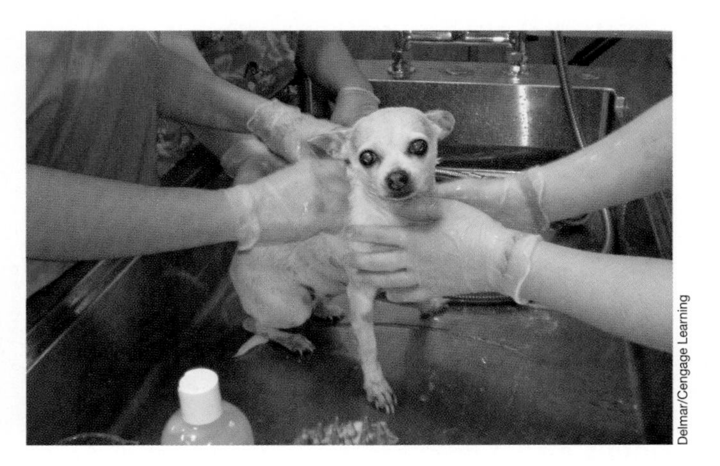

FIGURE 8-30 The dog should be lathered with shampoo and conditioner.

and conditioner should be kept on the hair coat for 5–10 minutes before rinsing. Rinse with warm water and make sure all chemicals have been removed from the coat (see Figure 8-31). The dog should be dried with a towel and placed under a dryer until the hair coat is dry. A blow dryer is often used to dry small dogs. A dryer can be placed in front of the cage to completely dry a dog (see Figure 8-32). Make sure the dryer is placed on the appropriate temperature and doesn't cause burns to the animal. When the dog is dry, a **slicker brush** can be used to brush and detangle the hair (see Figure 8-33). A spray-on **coat conditioner** should be applied to prevent the hair and skin from drying out (see Figure 8-34).

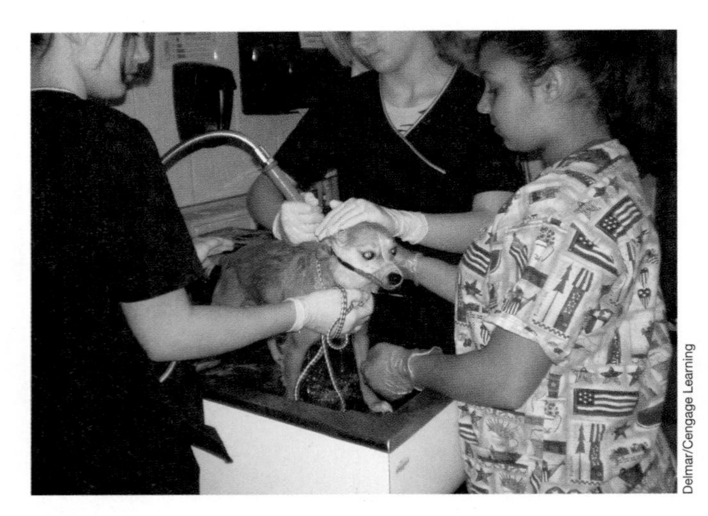
Delmar/Cengage Learning

FIGURE 8-31 Rinse with warm water.

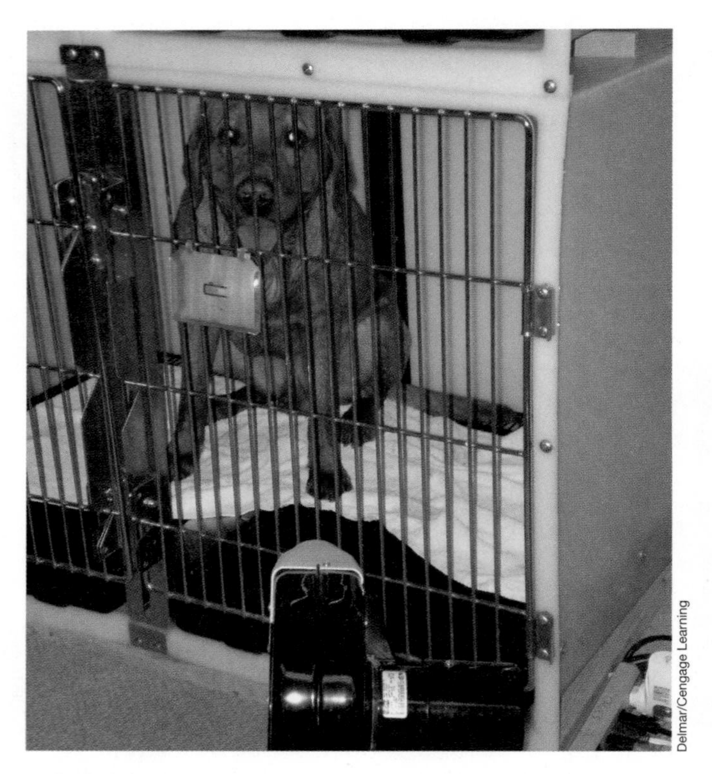
Delmar/Cengage Learning

FIGURE 8-32 A dryer can be placed in front of a cage to completely dry the dog. It is critical to make sure the temperature is not too high.

SAFETY ALERT

Dryers should be used with caution. If the heat settings are too high, the patient can be burned. There have been instances in which the improper use of a drying device has resulted in the death of the patient. Never use a drying device without monitoring the patient while the device is in use.

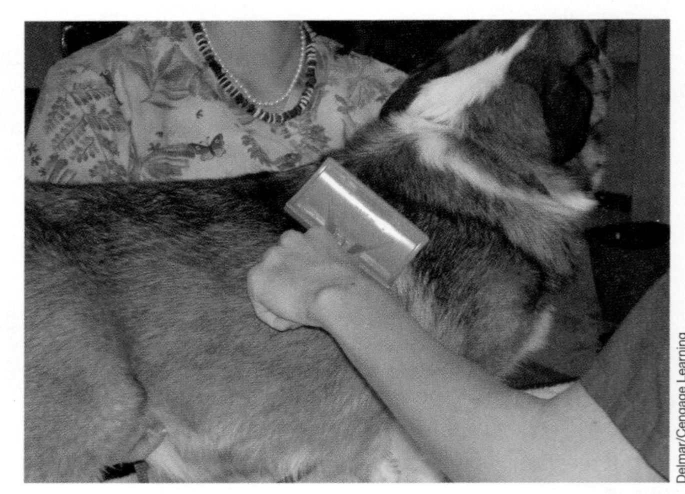
Delmar/Cengage Learning

FIGURE 8-33 A slicker brush is used to brush the dog's coat once it is dry.

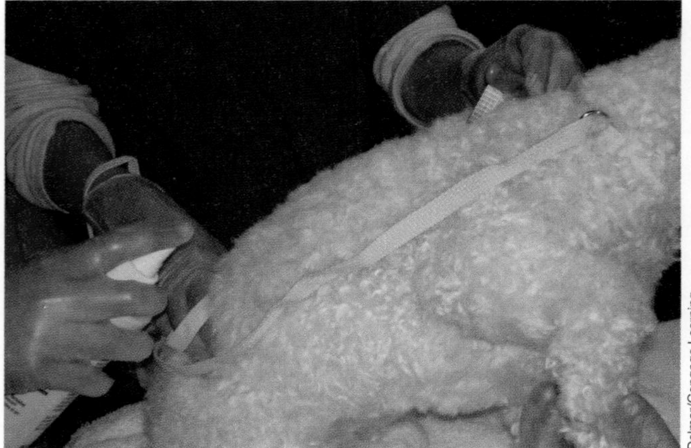
Delmar/Cengage Learning

FIGURE 8-34 A conditioner is sprayed on the dog's coat to prevent the hair and skin from drying out.

Brushing

It is best to routinely brush the hair coat of a dog on a daily to weekly basis. This will keep the hair coat healthy. Long-haired breeds may develop **mats**, which are tangled pieces of hair. Mats may need to be trimmed or shaved. Some long-haired breeds will require daily brushing. It is important to know how to properly care for the hair coat based on the breed of the dog. Some breeds will require regular bathing, while bathing others too often may cause damage to the hair coat.

Basic Health Care and Maintenance

Each year a dog should be seen by the veterinarian for a **physical examination (PE)**. This exam will help determine the health status of the dog and if any illnesses or conditions are developing. Dogs require daily exercise, and some breeds will require more activity than others. Regular maintenance should include nail trimming (NT), ear cleaning, anal gland expression, and teeth brushing.

Ear Cleaning

Dogs should have their ears checked and cleaned as necessary. Some breeds will need more cleaning than others. This will help maintain breeds that are prone to ear infections. Ears can be cleaned with cotton balls, cotton-tipped applicators, or gauze pads (see Figure 8-35). Caution should be used with applicators since the wooden sticks can break off and become trapped in the ear canal. The veterinary assistant should clean only as far as can be seen in the ear canal (see Figure 8-36). The ear is L-shaped and can be damaged if cleaning is not done properly (see Figure 8-37). It is best to use an ear cleaner that is not water based. If only a small amount of debris or wax is noted, an ear cleaner may not be needed. Some breeds have large amounts of hair that grows within the ear canal that may need to be plucked (pulled) or shaved. The ear canal should be examined for any dirt, wax, or debris that may build up within the inner ear. The opening of the ear is under the ear flap, with the ear canal lying internally within the ear in an L-shape. The end of the L-shape is the eardrum. This area can be damaged or punctured during ear cleaning. It is also easy to pack debris into the inner ear, causing severe irritation and possible infection. When cleaning ears, it is best to use gauze sponges or cotton balls rather than a cotton-tipped applicator, which can cause ear damage if used improperly. Begin cleaning the external ear flap by wiping the outer area with alcohol or an ear cleaner to remove any dirt or debris. Wipe away from the internal

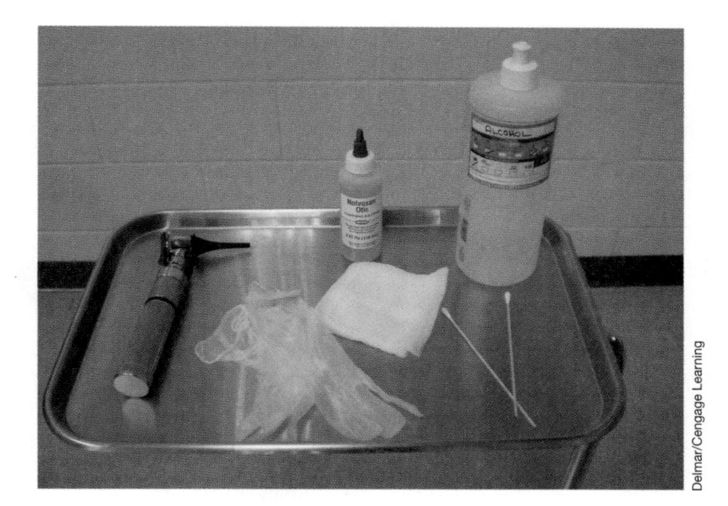

Delmar/Cengage Learning

FIGURE 8-35 Tools required to clean a dog's ears.

Delmar/Cengage Learning

FIGURE 8-36 Cleaning the dog's ears.

canal area. Next, it is best to clean the internal ear by placing ear cleaning solution on a gauze pad or cotton ball and wiping the ear backward and toward the outer ear to avoid pushing debris into the eardrum. Cleaning should continue until no debris is noted. Severely dirty ears may need to be flushed with ear cleaner based on the veterinarian's assessment upon physical examination. When ear cleaning has been completed, use dry gauze sponges or cotton balls to dry out the ear canal. Ears may be protected from water during a bath by placing a large cotton ball into each ear canal. Once the ear is clean, some animals may require medication administration by a veterinarian.

Nail Trimming

Nail trims are done based on the length of the nail and the type of surface the dog spends time on. Long nails can get caught on items and torn or may scratch

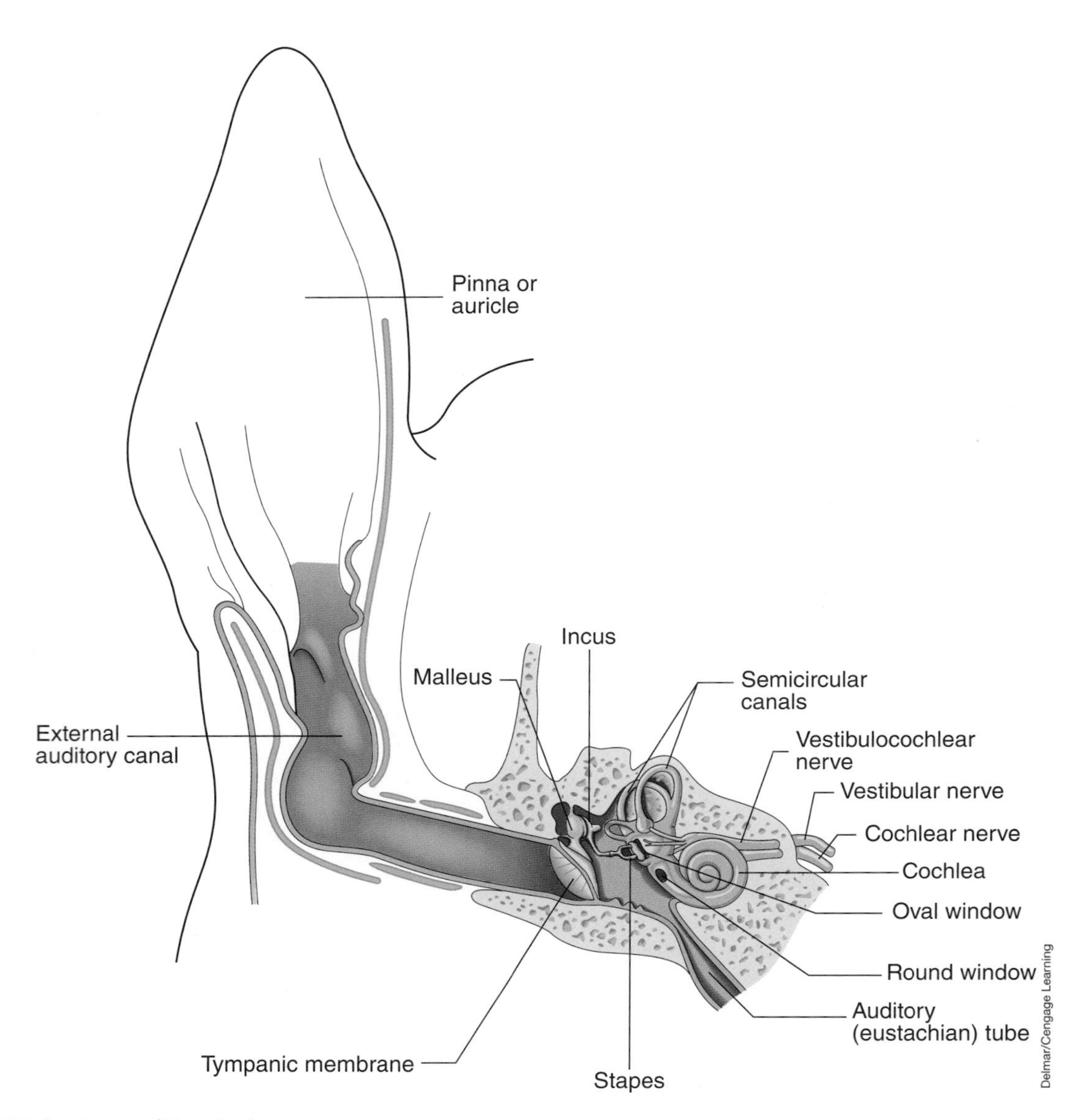

FIGURE 8-37 Anatomy of the dog's ear.

people or damage furniture. Trimming nails requires some knowledge of the nail anatomy and practice at clipping. Veterinary assistants should be able to use handheld nail trimmers as well as a **dremel** tool that grinds down nails (see Figure 8-38). The nail bed has a blood supply known as the **quick**. If a nail is trimmed too short, it will start to bleed. There are several ways to **clot** or stop the bleeding. A veterinary assistant may use a special powder that clots the blood called **styptic powder** (see Figure 8-39). If the bleeding is difficult to control, a **silver nitrate stick** may be used. This is a chemical that burns the blood vessel and causes the blood to clot. Clients who trim their dog's nails at home may use corn starch, flour, or a bar of soap packed into the bleeding nail. If the nail is cut or torn very short a bandage may be necessary.

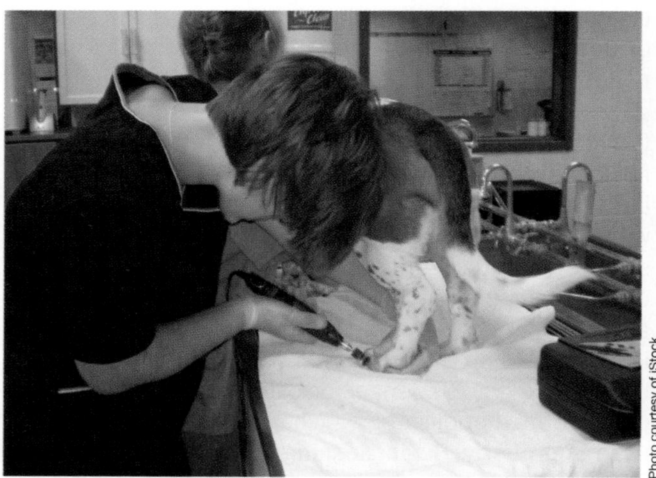

FIGURE 8-38 Use of a dremel tool to trim a dog's nails.

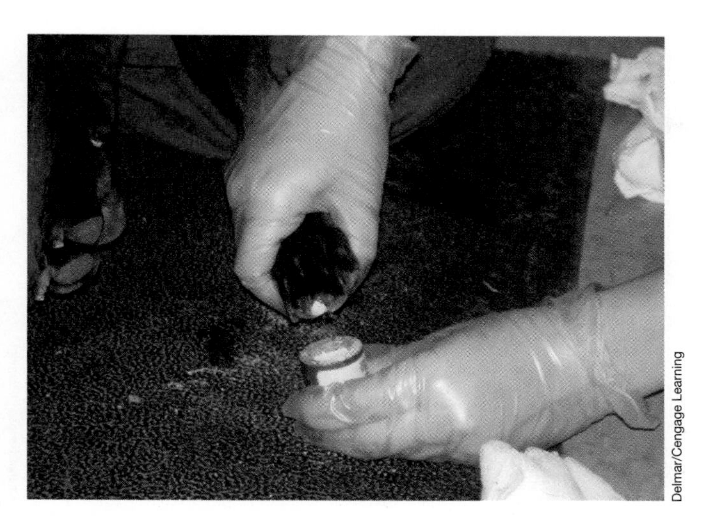

FIGURE 8-39 Styptic powder is used to stop bleeding that may occur when nails are cut.

Ear Plucking

Ear plucking is the removal of hair visible to the eye from the inside of the ear canal. The hair may be removed with **forceps**, which grasp the hair and allow it to be pulled out. This is done in certain dog breeds that have excessive amounts of hair within the ear canal. The poodle is one such breed. The growth of hair within the ear canal can cause irritation and hold in dirt and debris. Plucking of the hair can prevent dirt and wax from collecting in the ear canal.

Expression of Anal Glands

Anal glands are the scent glands located on either side of the rectum (see Figure 8-40). The **scent glands** allow dogs to determine who is who in the dog world. Smelling each other's scent glands is the way dogs greet each other. These sacs hold small amounts of fluid from the dog's bowel movements. Over time, the sacs fill with the fluid and cause pressure on the rectum. Some dogs are capable of releasing the pressure on their own during a bowel movement. Other dogs are not able to and will show signs of needing the glands expressed. Excessive licking at the rectum, discomfort in the rectal area, and scooting the rear over the floor are signs of anal glands needing expression. Anal glands can be expressed internally and externally. Veterinary assistants should be

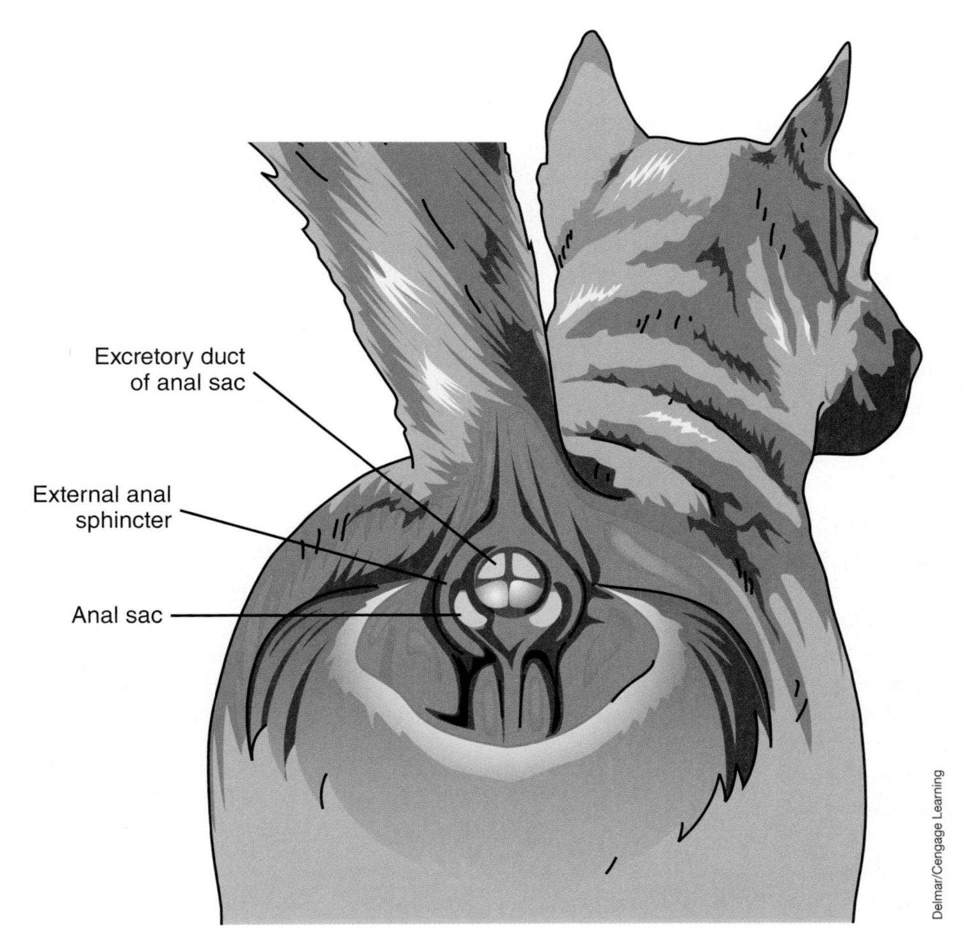

FIGURE 8-40 Location of the anal glands.

trained to express anal glands externally. This procedure is done with the dog standing. Exam gloves should be worn and paper towels and a bowl of warm soapy water should be available for cleanup. The external method of expressing anal glands is to locate the sacs on either side of the rectum; as if on a face of a clock, the glands are located at the 3 o'clock and 9 o'clock positions. Using the thumb and pointer finger gently apply pressure to the sac area by massaging the site (see Figure 8-41). The sacs should press against each other and release the fluid. Be warned not to stand directly behind the dog when performing this procedure. The glands tend to release and spray an odorous fluid that can be projected several feet. If the glands become **impacted** or difficult to express due to thickening, a veterinarian or veterinary technician will need to express the glands internally. When describing anal gland expression procedures the term *external* means on the outside of the rectal area and the term *internal* means on the inside of the rectum.

Teeth Brushing

Brushing the teeth of dogs is ideal for preventing tartar buildup and to reduce bad breath odor. Veterinary assistants should be able to brush a dog's teeth and note any dental problems (see Figure 8-42). Assistants should also be able to show clients how to brush their dog's teeth at home. Dogs require special toothpaste that has specific digestive enzymes that are absorbed in the body. Human toothpaste is not digestible and can cause toxic effects in dogs. There are several types of toothbrushes that may be used. One is a long-handled toothbrush similar to the ones humans use. Another is a **finger brush** that fits on the end of the pointer finger and allows for ease of brushing. There are also dental sprays that help reduce bacteria that cause bad breath. Typically the outer surfaces of the teeth are brushed

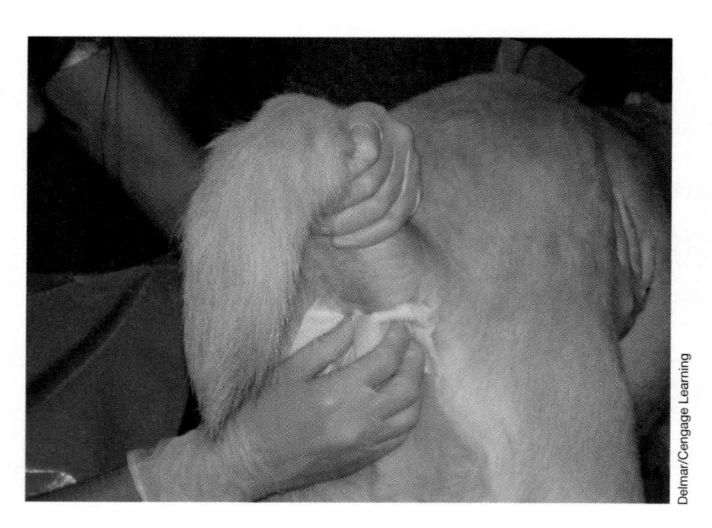

FIGURE 8-41 Expression of the anal glands.

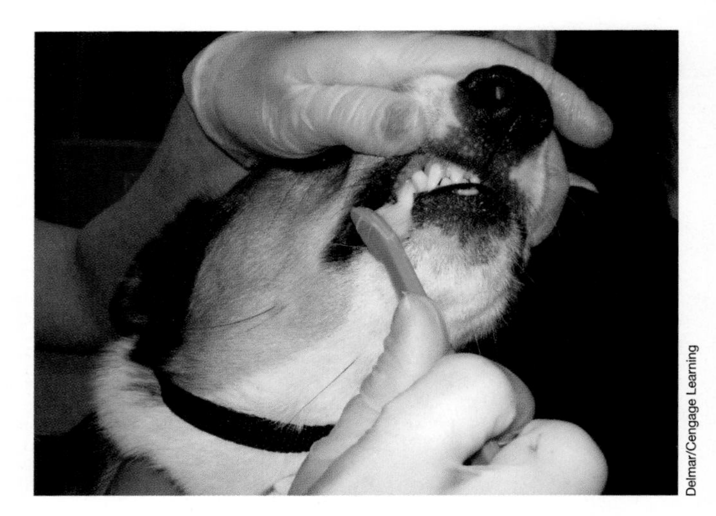

FIGURE 8-42 Tooth brushing is an important part of the dental care needs of a dog.

using a downward angle that brushes the gums and the teeth. The gums are sensitive and should be brushed gently. The lips should be lifted for ease of viewing the teeth. During the brushing procedure, any broken, missing, or deciduous (baby) teeth should be noted.

Vaccinations

A **vaccine program** should begin for the puppy during 6–8 weeks of age (see Table 8-5). The vaccines will provide the puppy with protection from common canine diseases. Most puppies receive a vaccine series, or multiple vaccines, in one dose. The most common canine vaccine series is the DHLPPC combination, also known as the distemper series. Each letter represents the disease from which the vaccination protects the puppy:

- D—Distemper
- H—Hepatitis
- L—Leptospirosis
- P—Parainfluenza
- P—Parvo virus
- C—Corona virus

TABLE 8-5

Vaccination Schedule

AGE	VACCINE
6–8 weeks old	DHLPPC
10–12 weeks old	DHLPPC
14–16 weeks	DHLPPC
16 weeks	RV
Yearly or more	DHLPPC, RV

The DHLPPC is given in **boosters**, or multiple times, to build up the immune system's protection. The boosters are typically done a month apart for 3 to 4 times ending around 16 weeks of age. A rabies vaccine (RV) is given between 12 and 16 weeks of age and is valid for 1 year; there are some rabies vaccines that are good for 1 year and some that will protect for 3 years. The rabies vaccine must legally be given to any owned dog. A rabies tag and a proof of certificate is issued at the time the vaccine is given. These vaccines are commonly given into a muscle or **intramuscular (IM)** or under the skin known as **subcutaneous (SQ)**. Some veterinarians may suggest other vaccines against diseases common to the area, such as Lyme disease or kennel cough.

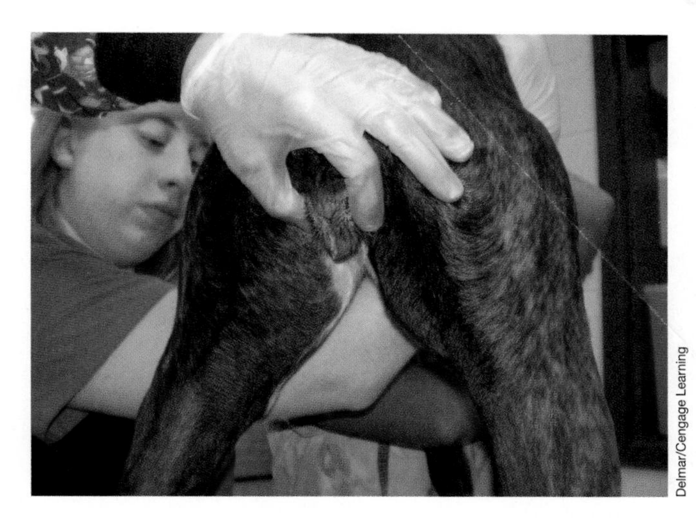

FIGURE 8-43 Swollen vulva indicating a female dog in estrus.

Reproduction and Breeding

Canine reproduction and breeding techniques require a general knowledge of a dog's reproductive system and the **estrus cycle**, commonly known as being in **heat**. This is the time the female is receptive to the male and will allow breeding to occur. The term **receptive** means that the female dog will allow the male dog to breed with her. Estrus normally begins between 6 and 24 months of age depending on the size and breed of the dog. This time frame is known as **puberty**. This is when the dog nears adulthood and reproductive organs are completely developed. Small-sized dogs typically begin puberty around 6–12 months of age and large-sized dogs begin around 8–24 months of age. Estrus typically lasts 10–14 days. Signs of a female dog in estrus include the following:

- Swollen vulva (see Figure 8-43)
- Vaginal discharge—usually yellowish to bloody in appearance
- Licking at vulva
- Swollen mammary glands
- Tail held high and to the side
- Behavior changes

Breeding usually begins 10–11 days after seeing signs of estrus. The female will allow contact with the male and breeding should occur every other day until the female will no longer accept the male. Once a dog is bred, the gestation, or length of pregnancy, is an average of 63 days. The labor process of dogs is called whelping. A whelping box may be provided to the female 2–3 weeks before the due date. The box should be placed in a quiet, warm, and darkened area to relieve stress on the dog. It is a good idea to monitor the dog for signs of labor and when she begins the whelping process to allow for privacy. Signs of whelping include the following:

- Restless and anxious behavior
- Stop eating 24 hours prior to labor
- Nesting habits (collect items and make a nest)
- Licking vulva
- Circling and uncomfortable
- Milk develops at mammary glands 12–24 hours prior to labor
- Panting

Normal labor occurs within 8–12 hours depending on litter size. On occasion, some dogs may be in labor up to 24 hours. Normal pup presentation is the head first. When a dog is having a difficult labor it is called **dystocia**. Dystocia may occur for several reasons, such as the female tires during labor, a puppy becomes stuck within the birth canal, or the puppy is too large to enter the birth canal. When a female dog shows signs of having a difficult labor, this becomes an emergency and a veterinarian should be contacted immediately. The mother will lick at the puppies to stimulate them to breath and to help dry them. Once the puppies are born it is important that they begin nursing as soon as possible to obtain the mother's **colostrum**, or first 24 hours of milk that contains **antibodies** to protect the immune system (see Figure 8-44). The antibodies will protect the puppies from disease until a vaccine program is started. The temperature should be kept at 85 degrees for the first 7 days, and then it can be gradually lowered to around 70–75 degrees. The pups will nurse every 2 hours and the mother will lick each

FIGURE 8-44 Chihuahua giving birth, one baby is about to be born, while two are nursing.

puppy to stimulate urine and waste production. The eyes and ears of the pups will open at 10–14 days of age. They begin walking around 2 weeks of age and tail wagging and barking will begin around 3 weeks of age. At this time, socialization should begin to help form behaviors and personalities.

Common Diseases

Dogs may develop many types of infectious diseases. There are five classes of diseases in animals: bacterial, viral, fungal, protozoan, and rickettsial.

Bacterial Disease

Bacterial diseases are caused by bacteria that invade the body. Some examples of bacterial diseases in dogs include **tetanus**, commonly known as lockjaw, and **Lyme disease**, transmitted by the bite of a tick. Lyme disease is transmitted by a tick, commonly the deer tick, but is not a rickettsial disease as the tick bite transmits bacteria into the bloodstream causing this disease.

Leptospirosis is a bacterial disease that affects dogs and is transmitted in the urine. Signs of leptospirosis include fever, **lethargy** (inactivity), blood in urine, skin bleeding, and liver and kidney damage. Vaccination and treatment are available. This disease is zoonotic and care should be taken to wear gloves when handling or cleaning urine.

Viral Disease

Viral diseases are caused by a virus that invades the body and is not treatable, but must run its course. Examples of viruses in dogs are **rabies**, caused by saliva transmitted in bites and scratches, and **hepatitis**, a virus that causes an infection in the liver.

Canine distemper is a virus that can be fatal, especially in puppies. There are different levels of severity and it takes 9–14 days for **incubation**, or the time that the dog is infected, to occur. Sanitation is very important to control this virus. A vaccine is available but treatment is limited to controlling symptoms. Signs of distemper include fever, lethargy, **anorexia** (not eating), eye discharge, nasal discharge, vomiting, diarrhea, coughing, and in some cases seizures may develop. Dogs that recover from distemper can have **neurological** problems that affect the way they move.

Parvovirus is another viral infection that occurs in dogs. It is usually fatal in puppies. Dogs that transmit this disease are highly contagious and should be kept in isolation. Sanitation is the best control method. Signs of parvovirus include vomiting, bloody and odorous diarrhea, anorexia, lethargy, and fever. A vaccine is available for protection.

Rabies is a fatal virus that affects all mammals and humans. It is transmitted through saliva into a wound, such as a bite or scratch. Each state has **rabies laws** that should be reviewed and understood by all veterinary staff members. All dogs are legally required to be vaccinated for rabies. Signs include a change in behavior, aggression, foaming at the mouth, drooling, fever, **paralysis** (loss of leg and body movement), and death. Any dog that is suspected of having rabies should be immediately quarantined in a veterinary facility. Veterinary professionals should consider a pre-exposure vaccination against rabies. A **pre-exposure vaccine** is one that is given before an animal is in contact with a diseased animal or contaminated area. This will hopefully prevent the animal from getting the disease.

Fungal Disease

Fungal diseases are caused by a fungus that lives on the outside of the body. An example of a fungal disease is **ringworm**. All mammals are susceptible to the fungus and it is contagious. It has the potential to be highly **zoonotic**, or transmitted to humans. Signs of ringworm include skin irritation, scratching, hair loss, and circular crusty skin lesions. Testing of the site is necessary to determine the fungus. All bedding and grooming items should be disinfected to prevent spreading. Proper sanitation is necessary for control.

Protozoan Disease

Protozoan diseases are caused by single-celled organisms that invade the body. Examples of protozoan diseases include **giardiasis**, caused by water contamination, and **coccidiosis**, caused by contaminated bird droppings in water or soil.

Rickettsial Disease

The final class is **rickettsial** diseases caused by parasites, such as fleas and ticks. An example of a rickettsial disease is **Rocky Mountain spotted fever**, which is caused by a tick bite.

Common Parasites and Prevention

Parasites that are inside the body are commonly called "worms." Parasites may live inside the body, usually in the intestinal tract, or on the outside of the body, such as the skin of an animal. Canines are prone to several types of internal and external parasites.

External Parasites

The veterinary assistant should be able to identify common external parasites, such as fleas and ticks. It is also important to know the signs that may suggest a dog has external parasites. Signs of external parasites include scratching, chewing at skin, skin sores, visible parasites, and flea dirt on hair coat.

Fleas

Fleas are wingless insects that jump and seek heat to survive. Fleas feed on the blood of animals. They are brown to black in color. The bite of a flea may cause an allergic reaction in dogs due to sensitivity of the flea's saliva, and fleas will also bite humans. They are known to transmit certain diseases. A female flea can lay up to 50 eggs a day; the eggs will hatch in 48 hours and the young become adults within 15 days of hatching. This life cycle continues and can become a large problem within a household. The *life cycle* of a parasite is how the parasite begins and the stages it goes through when an animal becomes infected. Dogs that have fleas will scratch, bite at the fur, and may develop sores in the area of the bites. There are several ways to determine if a dog has fleas. Fleas produce fecal droppings that appear to look like dirt on the hair coat. This is a warning sign of a flea problem.

To check for fleas, use a flea comb and brush the hair coat both with the growth of hair and against it. If any dirt particles are removed from the animal's body, apply water to the dirt particles and watch for a red or rust color; if red color appears, that is the blood meal within the feces from a flea. A warm white towel can also be placed on the floor of the house. Turn off any lights and wait several minutes and then turn the lights on and check the towel for fleas. Fleas seek heat.

Fleas can be prevented using several methods. There are **topical** medications that are applied to the

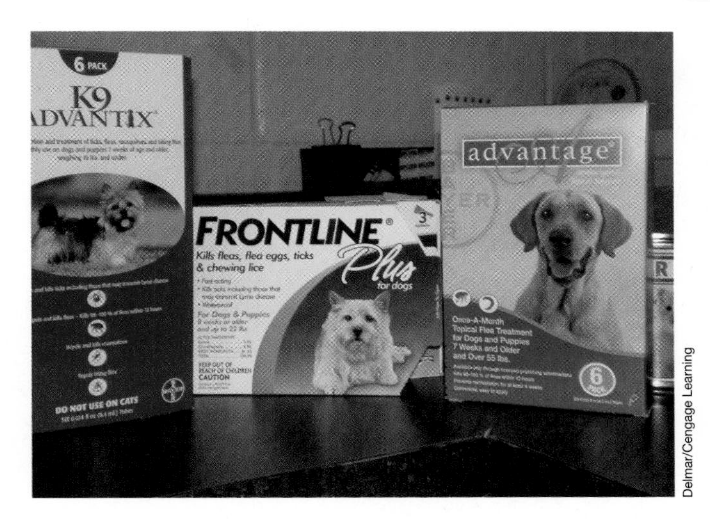

Delmar/Cengage Learning

FIGURE 8-45 Various flea control products.

skin and hair coat, **oral** medications that are given by mouth, flea collars, and flea shampoos and dips that kill or repel fleas (see Figure 8-45 and Table 8-6). There are also sprays and foggers that are used in a household for controlling fleas.

Ticks

Ticks are wingless insects that seek movement and attach to animals and people by embedding their mouth into skin and feeding on blood (see Figure 8-46). During feeding, bacteria may be transmitted that cause disease. As the tick feeds on the blood meal, the body of the tick will swell up in size. It is important for veterinary assistants to know removal of a tick must be done carefully and the head must not be detached from the body. This may cause a skin infection. Ticks lay their eggs on the ground and take 2 years to reach the adult stage. Signs of a tick bite include finding a live tick, itching at site, skin irritation, and redness. A skin infection causes redness, heat, pain, and pus at the site of the problem. Control is through routine brushing and checking skin for ticks and applying topical products that kill or repel ticks.

Mites

Mites are insects that can live on the skin, hair coat, or within the ears. Some mites are barely visible to the naked eye and others are **microscopic**, or only visible with the use of a microscope. **Ear mites** live within the ear canal and cause a thick, dark, crusty wax-like material that builds up within the ear. They are not visible to the eye. They cause itchiness within the ear canal and dogs will scratch their ears and sometimes cause open sores within the ear flap due to scratching. Ear mites are **contagious** and are transmitted to all animals. The term *transmission* or *transmitted* means passing a disease by some route or method. A disease may be transmitted

TABLE 8-6

Common Flea Product Comparison Chart

PRODUCT NAME	APPLICATION TYPE	PARASITES CONTROLLED	AGE OF USE	WATERPROOF OR RESISTANT	LENGTH OF APPLICATION
Advantage	Topical	Fleas	7 weeks	Water resistant	Monthly
Advantix	Topical	Fleas, ticks, flies, mosquitoes	7 weeks	Water resistant	Monthly
Capstar	Oral	Fleas	4 weeks	Waterproof	Daily or as needed
Frontline	Topical	Fleas	8 weeks	Waterproof	Monthly
Frontline Plus	Topical	Fleas, ticks, lice	8 weeks	Waterproof	Monthly
Program	Oral/ Injectable	Fleas	4 weeks	Waterproof	Monthly-oral 6 months-inj
Promeris	Topical	Fleas, ticks	8 weeks	Waterproof	Every 4–6 weeks
Revolution	Topical	Fleas, ticks, mites, heartworm	6 weeks	Waterproof	Monthly
Sentinel	Oral	Fleas, hookworms, roundworms, whipworms, heartworm	4 weeks	Waterproof	Monthly

FIGURE 8-46 Tick.

by direct contact, in the air, or by being bitten by an infected animal. Ears should be cleaned and evaluated under a microscope for mites. Ear medications are used to treat and kill the mites.

Skin mites occur within the hair and skin. There are two types of skin mites that cause a condition called mange. The sarcoptes mite causes **sarcoptic mange**. This mite is contagious and can spread to humans, in which the condition is known as **scabies**. The demodex mite causes **demodectic mange**. The proper medical term for hair loss is **alopecia**. Demodectic mange causes patches or areas of hair that are bald or thinning on the hair coat. Demodex is naturally occurring in small amounts on all dogs' skin, but the condition can become problematic and cause a skin irritation and infection. Both types of mange will cause severe itchiness that leads to scratching, hair loss, skin lesions and sores, skin infection, and discoloration of the skin. Both types of mites are not visible to the eye and the skin must be scraped by a veterinarian or technician and evaluated under a microscope. It is important for veterinary assistants to disinfect all grooming items and equipment since this is a common route of transmission for mange mites. All skin lesions should be evaluated for mange mites.

Lice

Lice are tiny insects that live on the hair. Lice are species specific and are only contagious within a species. Signs of lice include scratching, hair coat damage, and visible spots of white on the hair. The small spots of white seen on the hair coat are the lice larvae called **nits**. They appear to stick to the hair and can be removed with a piece of tape for microscopic evaluation. Lice must be treated with topical insecticides to kill the insects. All bedding and areas of the house where the dog has access should be treated.

Internal Parasites

The veterinary assistant should also be aware of signs of internal parasites. There are some common internal parasites that occur within the body, usually within the intestinal tract. Internal parasites feed on the blood of animals by surviving inside the body. Some internal parasites are microscopic and some may be visible after passing through the dog's intestinal system.

Heartworms

Heartworm disease is a serious issue among dog owners and veterinary professionals. It is caused by an internal parasite that lives in the heart, lungs, and bloodstream. It is transmitted by the mosquito, which is known as the **vector**, or route of transmission. The young heartworms, called **microfilaria**, are a serious stage in heartworm disease, as they can cause a blockage in the heart or bloodstream that results in possible death. The heartworms reach the adult stage in 6 months. They can be as long as 12 inches and look like a long piece of white spaghetti. Prevention is the best method. Several products are available to prevent heartworm and other parasites (see Table 8-7). Veterinarians require a negative blood test before placing a patient on a product. This should be done routinely to monitor for control. Dogs that test positive for heartworm disease can be treated with caution. Signs of heartworm include shortness of breath, coughing, exercise intolerance, tire easily, and possible death.

Roundworms

Roundworms or **ascarids** are the most common intestinal parasites of dogs and puppies. They live and feed off of the small intestine. Common causes are dogs ingesting rodents, contaminated soil, and through passing eggs from the placenta or nursing in puppies. When passed in waste material, roundworms appear as long, thin, white pieces of spaghetti. Signs of roundworm infection include a swollen abdomen, diarrhea, vomiting, weight loss, and poor hair coat appearance. Roundworms are zoonotic to humans and can cause a condition known as **visceral larval migrans**, which is caused commonly through contaminated soil that is ingested. The eggs can invade the lungs and intestines.

Hookworms

Hookworms are another internal parasite found in the intestinal tract that feed on blood. Transmission may occur through contaminated soil that is ingested, eggs may penetrate through the pads and skin, and puppies may acquire infection through nursing. Signs of infection include **anemia** (low red blood cell count), diarrhea, vomiting, weight loss, and poor coat appearance. Hookworms are zoonotic to humans and cause a condition known as **cutaneous larval migrans**, in which hookworm eggs penetrate the skin and enter the intestinal tract.

Whipworms

Whipworms are internal parasites that attach to the colon and feed on intestinal blood and tissues. They are transmitted from contaminated soil that is ingested. The eggs survive in soil for several years and are very hard to control. A **fecal egg analysis** is done to determine if parasites are infecting an animal (see Figure 8-47). This procedure may be completed in a variety of ways, as discussed in Chapter 38, Laboratory Procedures. Signs of whipworm infection include anemia, weight loss, diarrhea, and poor coat appearance.

TABLE 8-7

Canine Heartworm Product Comparison Chart

PRODUCT NAME	APPLICATION TYPE	PARASITES CONTROLLED	AGE OF USE	LENGTH OF PROTECTION
Advantage Multi	Topical	Heartworms, roundworms, hookworms, whipworms, fleas	7 weeks	Monthly
Heartgard	Oral	Heartworm	6 weeks	Monthly
HeartgardPlus	Oral	Heartworm, roundworms, hookworms	6 weeks	Monthly
Interceptor	Oral	Heartworms, roundworms, hookworms, whipworms	4 weeks	Monthly
Iverhart Max	Oral	Heartworms, roundworms, hookworms, tapeworms	6 weeks	Monthly
Revolution	Topical	Heartworms, fleas, ticks, ear mites, sarcoptic mange mites	6 weeks	Monthly
Sentinel	Oral	Heartworms, roundworms, hookworms, whipworms, fleas	4 weeks	Monthly
Tri-Heart	Oral	Heartworms, roundworms, hookworms	6 weeks	Monthly

Delmar/Cengage Learning

FIGURE 8-47 Conducting a fecal egg analysis test.

Tapeworms

Tapeworms are intestinal parasites that live in the small intestine. They are commonly transmitted by ingesting rodents or by ingesting a flea. The tapeworms form segments called **proglottids** that shed as the tapeworms grow in size (see Figure 8-48). These segments may be seen in waste material or around the anus and appear as flat, white pieces similar to rice. Signs of tapeworm infection include flea problems, diarrhea, weight loss, and scooting on the floor. Diagnosis of internal parasites is through fecal egg analysis and treatment is with **anthelmintics**, commonly called de-wormers. The best way to prevent parasites is through product control and removing all waste materials from the yard area.

Common Canine Surgical Procedures

Veterinary assistants will routinely discuss canine surgical procedures with dog owners. Several common canine surgical procedures should be understood by the veterinary assistant, such as the **ovariohysterectomy (OHE)**, commonly called a spay; the castration; the dewclaw removal; and the tail docking. These surgical procedures are common in dogs.

The spay surgery is the removal of a female dog's uterus and ovaries to prevent reproduction. The castration surgery is the removal of a male dog's testicles to prevent reproduction. Both male and female dogs that are altered surgically to prevent breeding may be called neutered. This procedure is highly recommended in dogs that are not breeding quality and are not intended

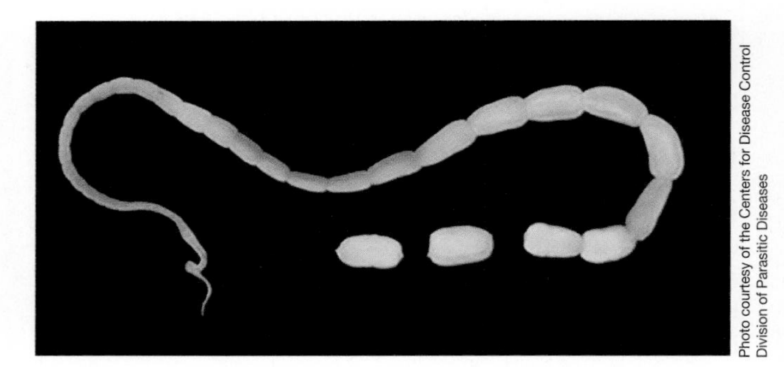

Photo courtesy of the Centers for Disease Control Division of Parasitic Diseases

FIGURE 8-48 Tapeworm.

for breeding or show situations, as neutering is healthier for the pet and prevents unwanted animal populations.

Dewclaw surgery is the removal of the dewclaw or first digit on the foot pads that doesn't have contact with the ground. These nails often become caught on objects and can be torn, causing a painful injury.

Tail docking may be offered by the veterinarian and is a cosmetic surgery to shorten the length of the tail. This is best done within 3–5 days of birth. It is important that the veterinary assistant has a basic understanding of the procedures to discuss the benefits of the surgery to the client.

Spay and Castration

Dogs are typically neutered around 6 months of age. Some veterinarians will neuter dogs at a much earlier age. The OHE is performed in the female dog with an incision made over the abdominal midline, or the center of the lower abdomen. Female dogs have two seasonal estrus cycles a year (with the exception in the Basenji, which has one seasonal estrus cycle). They are capable of having two litters of puppies a year. The OHE will decrease the dog population, reduce the attraction of male dogs, decrease mammary gland tumors, and decrease the incidence of spraying and marking territory. The spay is performed under general anesthesia. The uterus and ovaries are removed and the incision is sutured. Sutures are removed in 7–10 days with a limit on exercise.

The **orchidectomy** (castration) in the male dog is much less complex than in the female dog. The testicles are exposed, **ligated** (tied off), and excised. There are typically a few sutures in a dog castration and sutures are removed in 7–10 days with a limit on exercise. The castration procedure in dogs decreases the likelihood of marking, wandering for females, and chances of testicular tumors. The castration surgery is usually done under general anesthesia.

SUMMARY

Veterinary assistants will work with dogs on a regular basis. It is important for them to have a basic knowledge of canine restraint, grooming, care and maintenance, and the necessary information on health care needed to educate clients on proper dog care. Veterinary assistants will also need to use this basic information to educate clients within the veterinary facility. They may need to show dog owners how to do a variety of procedures at home for the well-being of their pet. This is an important part of hands-on skills as many other animals may need these same types of basic health care needs and maintenance.

Key Terms

aggressive behavior of an animal exhibiting an attempt to do harm such as biting

alopecia hair loss

anal glands scent glands or sacs located on each side of the rectum

anatomy the study of body structures

anemia low red blood cell count due to inability of body to make or replicate blood cells

anorexia not eating

anthelmintic medication also known as a dewormer; used to treat and prevent parasitic infection

antibody substance that acts to protect the immune system

ascarid intestinal parasite, long white worm, roundworm

bacterial disease caused by bacterium

booster series of vaccines given multiple times to build up the immune system

brachycephalic short, nonexistent nasal area, giving a flat face appearance

canine distemper viral infection may be fatal to puppies; dogs who recover may have long-term neurological affects

Canis familiaris genus and species of domesticated dog

carnivore meat eating

cephalic venipuncture blood collection from vein inside the front of the legs

chemical restraint use of sedatives or tranquilizers to calm and control an animal

clot stop or control bleeding

coat conditioner spray placed on hair coat to eliminate drying of skin and hair

coccidiosis protozoan disease cause by contaminated bird droppings in water and soil sources

colostrum milk produced my female within 24 hours of giving birth; provides antibodies that protect the immune system

contagious capable of spreading disease

cutaneous larval migrans hookworm infection in humans caused by penetrating the skin

demodectic mange disease caused by demodex mange mite on the skin and hair of dogs

designer breed the cross breeding of two breeds to form a new breed not yet recognized by the AKC

dewclaw first digit on the paw; doesn't make contact with the ground

domestication taming of an animal to coexist peaceably with humans

dorsal recumbency restraint position with animal placed on its back

dremel tool used to grind nails

dystocia difficult labor process

ear mite insects living in ear canal

ear plucking removal of hair from the ear canal

endothermic ability to regulate body temperature within the body

estrus cycle period of time when a female dog is receptive to a male dog for mating; also known as the heat cycle

fecal egg analysis test to determine the type of parasite found in fecal matter of animal

(*Continues*)

finger brush tool used to clean a dog's teeth

flea external parasite; wingless, feeds off blood of animal

forcep tool used to remove hair or parasite from body

fungal disease caused by a fungus that lives or grows on the body

gauze thin cotton like material used for bandages and for making quick restraint muzzles

giardiasis protozoan disease cause by water contamination

grooming process of trimming and bathing in an appropriate manner

heartworm disease caused by internal parasite that lives in the hear and bloodstream; transmitted by mosquito

heat period of time when a female dog is receptive to a male dog for mating; also known as the estrus cycle

hepatitis inflammation of the liver

hookworm internal parasite; may pass through skin and pads from contaminated soil

housebreaking training a dog to eliminate outside

hybrid the cross breeding of two breeds to form a new breed not yet recognized by the AKC

impacted difficult or unable to express

incubation time an animal becomes infected with disease

intramuscular (IM) injection given into a muscle

jugular venipuncture restraint position with an animal placed in sitting recumbency with the head and neck extended upward to pull blood from the vein in the neck

lateral recumbency restraint with animal lying on its side

leptospirosis bacterial disease transmitted through contact with contaminated urine

lethargy inactive; tired

lice tiny insects that live on hair, are species specific, and contagious

lifespan length of time an animal will live; based on body size and health

life stage particular periods in one's life characterized by specific needs during that time

ligated to tie off

Lyme disease bacterial disease transmitted through bite of tick

maintenance diet nutrition that allows animal to stay at the same weight

mat tangled pieces of hair

microfilaria young heartworm stage

microscopic very small; unable to view with the naked eye

mite insects that live on skin, hair, or in ears

mixed breed animal of unknown parentage that is a combination of two or more breeds

monogastric one single, simple stomach to digest food

muzzle equipment placed over the nose to prevent biting

nail trim cutting the length of the nail to prevent injury and promote comfort

neurological pertaining to the brain and spinal cord

neuter surgically removal of the reproductive organs or either gender

nit larvae of lice

obedience training a dog to obey commands

oral pertaining to the mouth

orchidectomy castration; surgical removal of the testes

orphaned young animal that is rejected by or loses its mother

ovariohysterectomy (OHE) surgical removal of the uterus and ovaries

paralysis unable to move one or more body parts

parasite internal or external insects that live off an animal

parvovirus viral infection causing severe and bloody diarrhea; can be fatal

physical examination (PE) assessment of the animal to determine overall health status

physical restraint use of a person's body to control the position of an animal

physiology study of body functions

pre-exposure vaccine

proglottid tapeworm segment that is shed as the tapeworm grows

protozoan disease caused by a single-celled organism

puberty sexual maturity

purebred animal with known parentage of one breed

quick blood supply within the nail bed

rabies viral infection spread through saliva; is fatal

rabies laws legal standards in each state to prevent the spread of rabies infections

rabies pole tool used to restrain and control an animal

ration specified amount of food

receptive accepting of

recumbency lying position

reduced-calorie diet meals low in calories to promote weight loss

rickettsial disease caused by parasites that invade the body; spread by biting insects

ringworm fungus that grows on the skin and is contagious

Rocky Mountain spotted fever rickettsial disease caused by tick bite

roundworms intestinal parasite that appears long and white

saphenous venipuncture restraint position with an animal placed in lateral recumbency to draw blood from the outer thigh area

sarcoptic mange disease caused by a mite that lives on hair and skin; is contagious; in humans known as scabies

scabies disease caused by a mite that lives on hair and skin; is contagious; in animals known as sarcoptic mange

scent glands sacs located on either side of the rectum also called anal glands

sedative medication used to calm an animal

senior diet nutritional requirements for an older, less-active animal

silver nitrate stick applicator used to apply a chemical that will stop bleeding

skin mite insects that live within hair and skin; not visible to the naked eye

slicker brush tool used to brush the hair coat

snare pole tool used to restrain and control an animal

socialization interaction with other animals and people to become used to them

sternal recumbency restraint position with the animal lying on its chest and abdomen

styptic powder chemical used to stop bleeding

subcutaneous (SQ) injection given under the skin

tail docking surgical removal of all or part of the tail

tapeworm internal parasite caused by ingesting fleas; long segmented bodies

tetanus bacterial disease caused by a wound; commonly called lockjaw

tick wingless insect; feeds off blood of animal

topical application of medication to skin or outside of body

training teaching an animal to understand specific desired habits

tranquilizer medication used to calm an animal

vaccine program injections given to protect animals from common diseases and build up the immune system

vector an insect or organism that transmits a disease

verbal restraint voice commands used to control an animal

viral disease caused by a virus

visceral larval migrans roundworm infection in humans

weaned process of young animal to stop nursing and eat solid foods

whipworm internal parasites ingested from soil

zoonotic a disease that is transmitted from animals to humans

REVIEW QUESTIONS

1. What is the difference between a purebred dog and mixed-breed dog?
2. List the seven groups of dogs and give three examples of breeds in each group.
3. What is the difference between lateral recumbency and sternal recumbency?
4. What are factors to consider when selecting a dog?
5. What are signs of a female dog in estrus?
6. When should a dog be fed a maintenance diet?
7. What is the ideal vaccine program for a puppy?
8. What are some general health practices that should be maintained for a dog?
9. What is the difference between internal and external parasites?
10. What are signs that a dog may have parasites?
11. What is the term for a disease passed from an animal to a human?
12. What are the five classes of diseases? Give an example of each type.
13. What is the importance of training a puppy or dog?
14. What grooming services should veterinary assistants be familiar with?
15. What is the importance of educating clients about zoonotic diseases and parasites?

Clinical Situation

Courtesy of Isabelle Francis

A 5-year-old male Basset Hound is presented to the veterinary facility for a castration surgery. The dog has a history of odorous bad breath and difficulty eating hard food. The veterinarian discussed a dental procedure with the clients on the visit last week where a physical exam and yearly vaccines were completed. The owners are dropping the dog off in the morning and ask if they can pick the dog up that afternoon. The veterinarian noted in the medical chart the dog will require general anesthesia and need to stay overnight. Food and water is to be withheld the night before the surgery and the day of the surgery. The owners appear very distracted and in a hurry.

- What are some questions the veterinary assistant should ask in this situation?
- How would you answer the client's question regarding pick up of the dog in the afternoon?
- What will you discuss as far as the castration surgery and the dental procedure?

9

Cat Breed Identification and Production Management

Objectives

Upon completion of this chapter, the reader should be able to:

- ☐ Identify common breeds of cats recognized by the Cat Fanciers' Association (CFA)
- ☐ Describe the biology and development of the cat
- ☐ Define common veterinary terms relating to the cat
- ☐ Describe normal and abnormal behaviors in the cat
- ☐ Properly and safely restrain cats for various procedures
- ☐ Discuss proper cat selection methods
- ☐ Describe health management practices for cats
- ☐ Describe nutritional factors of cats
- ☐ Explain cat breeding and reproduction methods
- ☐ Discuss basic vaccine programs in cats
- ☐ Recognize common feline diseases and prevention
- ☐ Recognize common internal and external parasites of cats
- ☐ Discuss common feline surgical procedures
- ☐ Discuss proper grooming needs of cats

Introduction

Cats are the most popular companion animal in today's society (see Figure 9-1). Cats are relatively easy to care for and tend to be self-sufficient. Cats have evolved from their wild counterpart, the lion. There are over 500 million domestic cats throughout the world and more than 30 different recognized breeds. The United States has over 70 million cats as pets. Cats are known to be playful, active, and alert. They have not been domesticated as long as the dog, but they have become an important part of the pet industry.

FIGURE 9-1 More cats are owned as pets than dogs.

Veterinary Terminology

In veterinary medicine, it is essential to know species-specific terms and when to properly use them in a facility. Table 9-1 contains some common feline terms and abbreviations that are used frequently in veterinary medicine. Figure 9-2 shows the location of each body part and the term used to describe it on the feline.

Biology

Cats descended from the wild cats, called **Felis catus**, commonly known as the lion. There are over 75 breeds of cats, and over 30 of those are recognized as pure breeds. The average lifespan of a cat ranges from 12 to 18 years old. The oldest known living cats have been in their late 20s to early 30s. Cats are mammals that are carnivores, or meat

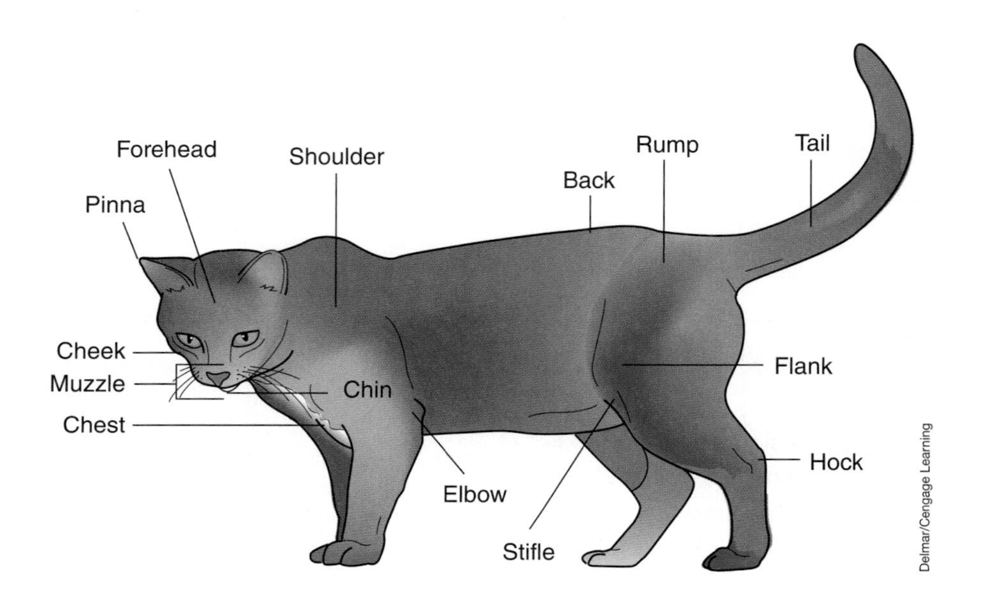

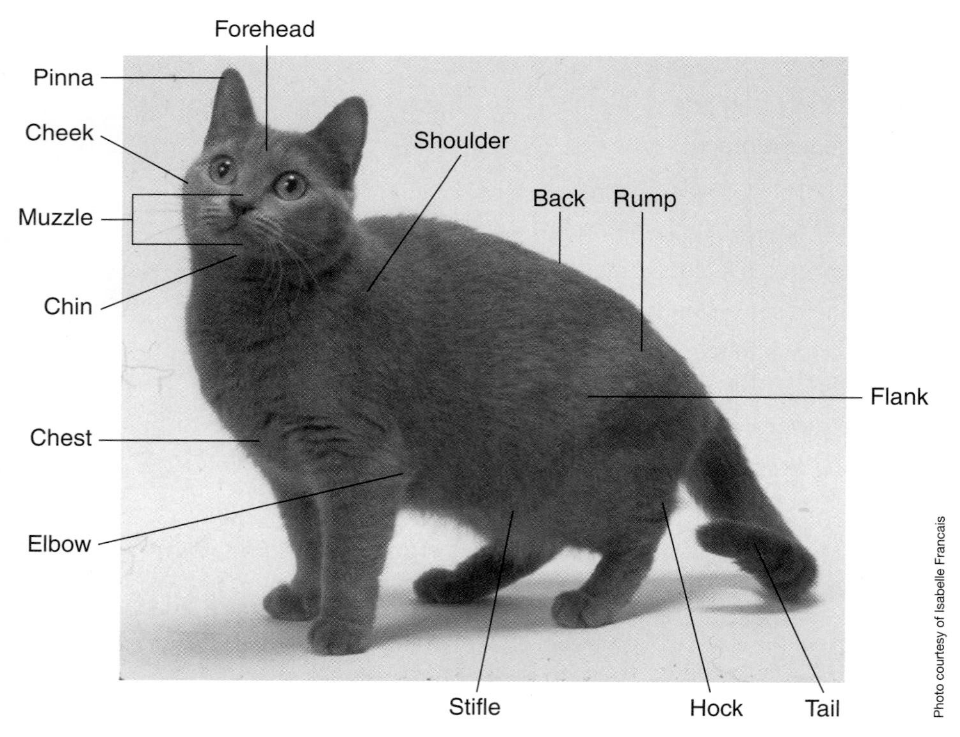

FIGURE 9-2 Anatomical parts of the cat.

TABLE 9-1

Feline-Specific Terminology

FELINE	CAT
Queen	Female cat of breeding age
Tom	Male cat of breeding age
Kitten	Young cat under 1 year of age
Gib	A male cat that is no longer able to reproduce
Spay	A female cat that is no longer able to reproduce
Bevy	Group of cats in household
Queening or kittening	The labor process of the female cat
DLH	Domestic long hair
DSH	Domestic short hair

eaters. Cats are similar to dogs in that they are endothermic, or capable of regulating their own body temperature internally, and they have a monogastric digestive system. *Monogastric* means a simple and single stomach within the digestive system. Much of the similarities end there. Cats are known for their well-developed sensory systems. They have a sensitive nose, sensitive paws, extra smell and taste organs, sensitive ears that detect the slightest of sounds, 3-D and night vision, and specialized whiskers that serve as feelers and as a balance mechanism. Cats feet differ from any other species of domestic animal. They have five toes on each paw and each toe has a **retractable** claw. This means the cat has control of placing the

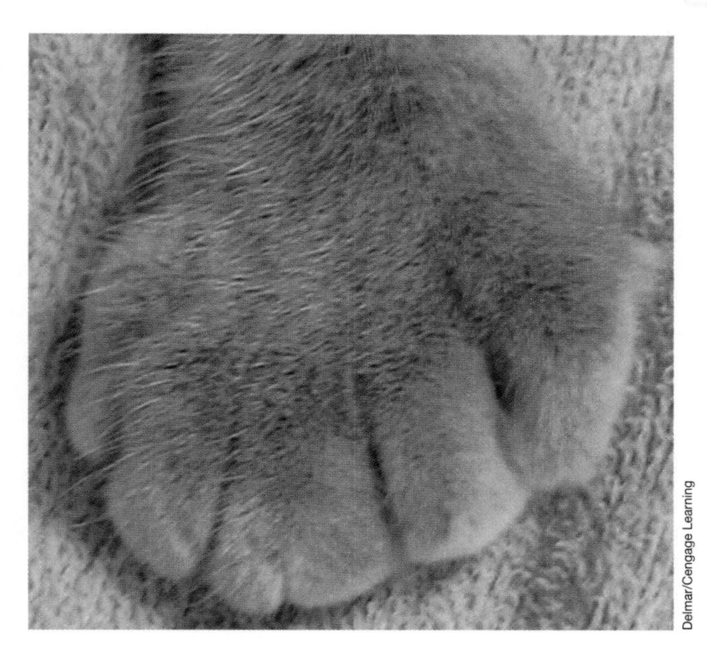

FIGURE 9-3 A polydactyl cat has extra toes.

nail outside or inside the toe. There are four long toes and one dewclaw. Many cats have extra toes called **polydactyl** (see Figure 9-3). The pads of the feet are soft layers of flesh that serve as cushions on the sole of the foot, which reduces the noise during the stalking of prey.

The cat's body and structure is designed for speed and quickness. Cats have an average of 250 bones in the body (see Figure 9-4). A cat's body consists of 500 muscles, the largest of which are in the rear legs to aid in climbing and jumping.

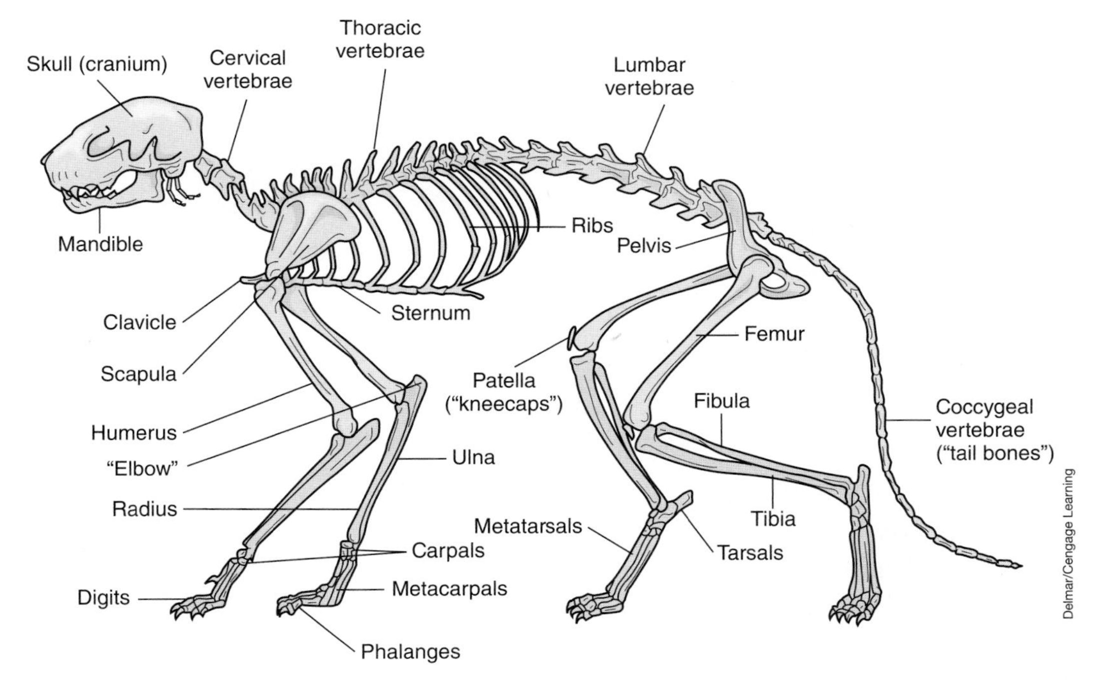

FIGURE 9-4 Skeletal structure of the cat.

Breeds

Over 30 breeds of cats are recognized by the **Cat Fanciers' Association (CFA)** (Table 9-2). The CFA is a breed registry that promotes the health and responsible breeding of cats. Cats that are not registered and have unknown parentage are usually called **domestics** (see Figure 9-5). There are a variety of different hair coat types but most cats are distinguished by either **long hair** or **short hair** appearances. Shorthair breeds tend to be low maintenance. Longhair breeds are high maintenance and require daily brushing and routine grooming. Longhair breeds are more susceptible to **hairballs**, which are clumps of hair ingested from grooming that build up and collect in the digestive tract. Hairballs can cause obstructions and can become a severe health problem. Cats come in a variety of sizes ranging from 5 pounds to 25 pounds as an adult weight.

TABLE 9-2

Cat Breeds

SHORTHAIR BREEDS	
Abyssinian	Japanese Bobtail
American Shorthair	Korat
American Wirehair	Malayan
Bombay	Manx
Burmese	Ocicat
British Shorthair	Oriental Shorthair
Charteux	Russian blue
Colorpoint	Scottish Fold
Cornish Rex	Showshoe Breed
Devon Rex	Siamese
Egyptian Mau	Sinapura
Exotic Shorthair	Sphynx
Havana Brown	Tonkinese
LONGHAIR BREEDS	
Balinese and Javanese	Norwegian Forest
Birman	Persian
Cymric	Ragdoll
Himalayan and Kashmir	Somali
Maine Coon	Turkish Angora
MISCELLANEOUS BREEDS	
American Curl	Siberian
LaPerm	Turkish Van
Selkirk Rex	

FIGURE 9-5 Domestic cat.

Breed Selection

Clients may ask veterinary assistants information on selecting the proper cat. There are many factors that people will need to consider when offering cat selection information. The client should consider if he or she wants a kitten or an adult cat. Kittens will require more training time than an adult cat (see Figure 9-6). Also, clients should consider if they want a purebred cat or if they would be interested in adopting a cat from a local shelter. Some factors that must be discussed when selecting a cat include the following:

- Indoor or outdoor cat
- Longhair or shorthair cat
- Grooming needs
- Health care needs
- Breed health problems
- Nutritional needs
- Breeding or show purposes

The best advice to offer clients is to research several breeds and visit local animal shelters to determine what type of cat will best suit their lifestyle. Keep a list of reputable cat breeders in the area to refer clients. Also, keep reference materials to help future cat owners become educated and responsible pet owners. The CFA provides breed information on their Website (http://cfa.org).

Nutrition

Cats require specialized nutrients in their diet. They should be fed a high-quality dry food that helps in tartar buildup. Cats should never be fed dog food since

FIGURE 9-6 Kittens require more training than an adult cat.

the nutrient needs of both species are very different. Cats require a high-protein and high-fat diet with the added nutrients of taurine and thiamine. High-protein diets are necessary since cats break down protein very slowly. **Taurine** is necessary for proper development and body system functions. **Thiamine** is a compound of Vitamin B that promotes a healthy coat. Cat foods are available in dry, semi-moist, and moist forms. Cats should not be given milk as it upsets the digestive system and causes diarrhea.

Feeding Kittens

Kittens require colostrum from the mother in the first 24 hours after birth. This will protect them from diseases until vaccines begin. They will nurse from the queen for 3 weeks. Around 3 to 4 weeks of age kittens should be offered canned or moist food. Over the next 2 weeks they should begin dry food that is moistened with warm water. By 6–8 weeks of age the kittens should be weaned and on solid kitten food. The feedings should be divided into several times a day. It is important when feeding a dry kitten food to feed a size-appropriate diet. Kitten dietary requirements include the following:

- Protein needs—35–50%
- Fat needs—17–35%

Kitten energy needs are identified as follows:

- 2 months of age—175 kcal/day
- 3 months of age—260 kcal/day
- 6 months of age—280 kcal/day

Kittens that may be orphaned should be fed a commercial formula, such as KMR or goat's milk. Kittens require 30 mL per pound body weight daily and can be fed three to four feedings a day until 4 weeks of age. Kittens may be fed with a nurser bottle or a syringe (see Figure 9-7). A daily weight should be recorded to determine that the kitten is growing. Kittens should be placed on heating pads for increased body temperatures for the first 2 weeks of life (see Figure 9-8). Between 4–5 weeks of age, kittens can begin eating a soft solid kitten food diet. The diet can be softened using warm water, if needed.

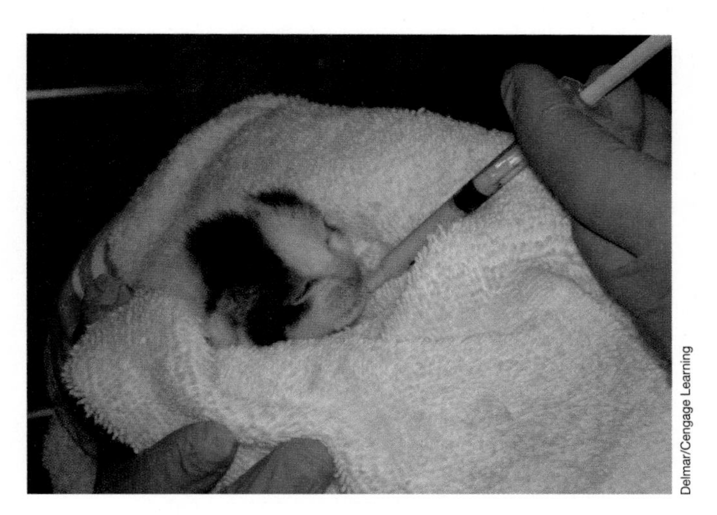

FIGURE 9-7 Orphaned kittens can be fed formula using a syringe.

FIGURE 9-8 Kittens should be kept warm by placing them on a heating pad.

Feeding Adult Cats

Adult cats should be fed a dry diet and a ration according to their ideal weight range. Cats between 1 and 7 years of age that are healthy should be fed a maintenance diet. Senior cats should be fed a senior diet that is reduced in calories. Overweight cats should be fed a weight management diet. Adult cats should be fed twice daily. Some specialized cat foods to consider are tartar control to prevent buildup on the teeth and hairball formula diets to aid in digesting hairballs in the digestive tract. Adult cats on a maintenance diet require around 280 kcal/day. Pregnant queens should be fed **ad lib** or as much as they want with food available at all times during the last 20 days of gestation. This can continue post-queening until the kittens begin weaning. During weaning, the queen should have her food reduced by ¼ rations until the ideal amount is reached. This will allow lactation to stop.

FIGURE 9-9 The happy cat.

Behavior

Cats have many natural instincts that they inherited from wild cats. Cats have developed hunting skills, marking behaviors, purring traits, and social skills from their ancestors. The domestication of the cat has added in the companionship factor. Similar to the dog, cats have three distinct behaviors. Each of these behaviors is displayed in the cat's body language, and most cats will display their behaviors relatively quickly. There is the happy cat, the angry cat, and the fearful cat.

FIGURE 9-10 The angry cat.

The Happy Cat

Signs of a happy cat include rubbing against objects or people, ears up and forward, tail held high, purring and meowing, and a relaxed appearance (see Figure 9-9).

The Angry Cat

The angry cat is likely to become aggressive. Signs of an angry cat include a puffed hair appearance, tail flicking or moving rapidly, hissing, growling, ears flattened back on head, eyes wide, and pupils **dilated** (widened) (see Figure 9-10). Cats have several reasons that they may become aggressive. There is territorial aggression, which is similar to that of dogs, but much more so in cats. Cats are very protective of their territory, especially in multiple cat households. **Intermale aggression** is common in adult male cats. These behaviors may be due to sexual dominance or as a social status. The occurrence is higher in reproductively intact males than those that are neutered. **Defensive aggression** is a common occurrence in the veterinary facility. This behavior is a way for the cat to protect itself. Cats will display this behavior toward people or other animals. **Redirected aggression** is a behavior toward another animal that didn't provoke the cat in any way. This is most likely due to the social status of the cat.

The Fearful Cat

The fearful cat is another common occurrence in the veterinary facility. Cats that are scared have a tendency to turn aggressive. Signs of a fearful cat include a lowered body position, tail tucked under the body, ears flattened on the side of the head, hair up over the body, hiding, and wide eyes (see Figure 9-11). When cats become upset they will try to scratch and bite to protect themselves.

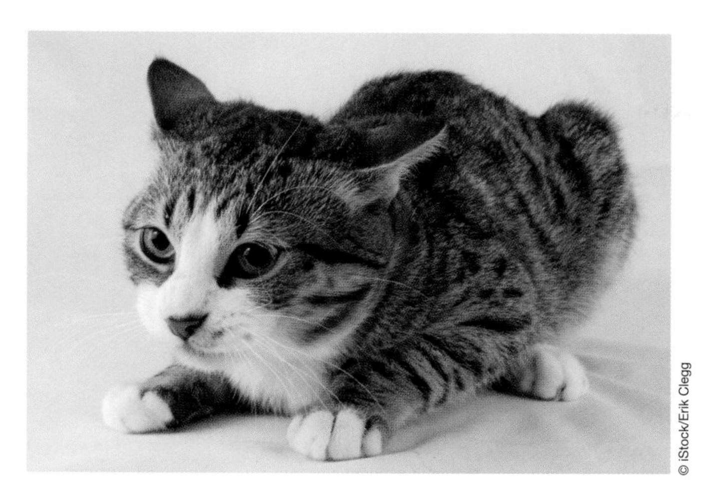

FIGURE 9-11 The fearful cat.

Basic Training

Cats should be **litter trained** to use a **litter box**. This is done by placing the cat or kitten inside the litter box frequently and helping them make a scratching motion. It is recommended to place several litter boxes throughout the house and keep them in a quiet and private location. Cats require an isolated area when eliminating. The litter boxes should be cleaned daily and fresh litter provided on a routine basis.

Equipment and Housing Needs

Cats should have a food and water bowl located in the household. Other equipment that is recommended for cat owners includes the following:

- Carrier
- Litter pans
- Litter
- Scratching post
- Toys

Restraint and Handling

The most important factor to consider when handling a cat is to avoid being bitten or scratched. Cats can cause severely infected bites and scratches; therefore, care must be taken when restraining a cat. Veterinary assistants will need to be properly trained in cat restraint. Cats are quick and agile and a cat can bite or scratch in an instant. Cats may be restrained using physical, verbal, chemical, or a combination of methods. One item that is helpful when handling a cat is the use of a towel. Cats feel less threatened when they are hiding under a towel and the towel can serve as a barrier if the cat becomes difficult to handle (see Figure 9-12). A towel can be used to remove a cat from a cage by gently placing the towel over the cat or in certain circumstances throwing the towel over the cat. The towel can also be used to wrap the cat in a "**kitty taco**." The kitty taco is made by placing a towel on a table and placing the cat on the towel. The sides of the towel are wrapped around the cat to contain the legs and body (see Figure 9-13). The head is left outside the towel and gives the appearance of a taco shell surrounding the cat. This allows better control of the cat to access and control the head.

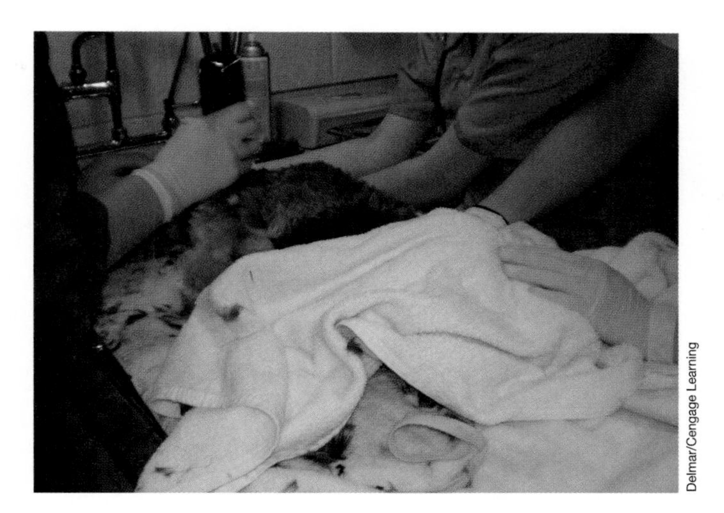

FIGURE 9-12 A towel can be used to help calm and control a cat.

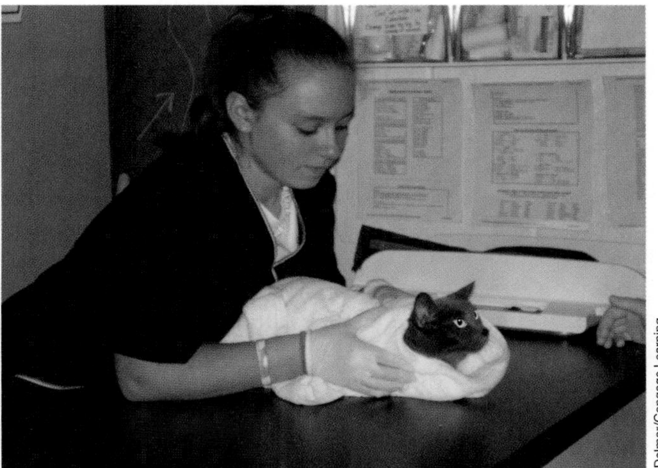

FIGURE 9-13 A towel can be used to create a "kitty taco" to restrain the cat.

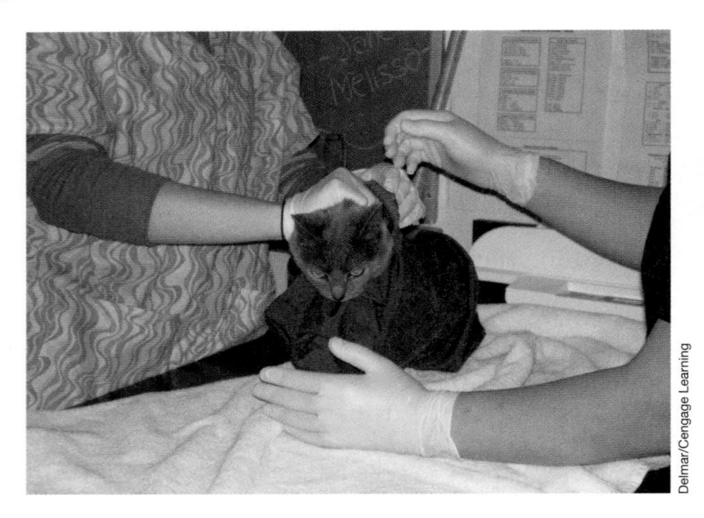

FIGURE 9-14 A cat bag is a device that can be used to restrain the cat while still allowing access to limbs when necessary.

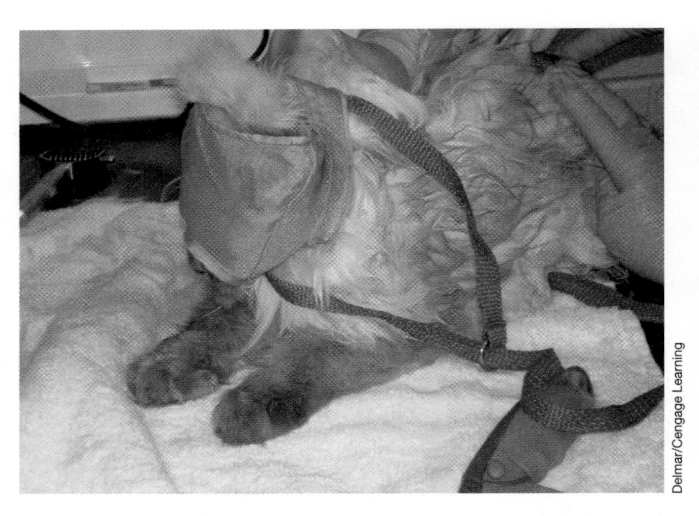

FIGURE 9-15 Use of a cat muzzle will aid in calming the cat and keep the cat from biting.

Another way to control a cat is with the use of a **cat bag.** The cat bag is similar to the kitty taco, but also allows access to the limbs or parts of the body. The cat is placed in a bag that has zippers to allow access to parts of the cat's body to perform certain procedures (see Figure 9-14). This allows better control of the head and legs to prevent injury.

Cat muzzles are similar to dog muzzles. They are placed over the face of the cat and typically cover the eyes, nose, and mouth. A small hole at the end of the nose area allows for breathing. Cats relax better when they are not able to see anything. They feel a sense of security. The cat muzzle is fashioned similar to a dog muzzle in which the top of the muzzle is shorter than the bottom. Cat muzzles are closed with a Velcro strip. They are best applied while standing to the back or the side of the cat (see Figure 9-15). Cats will try to scratch off a muzzle so care should be taken to control the legs. Care should be used when muzzling brachycephalic cat breeds to ensure the cat is capable of breathing. *Brachycephalic* means a flat face that may prevent the animal from breathing properly.

Cats can be restrained using a technique called the "**scruff**." The scruff is the loose skin over the back of the neck. The scruff is grasped firmly with the fist of one hand (see Figure 9-16). Many times this will cause the cat to become trance-like and allow for better control of the cat's head and body. When "scruffing" the cat, one should hold the cat against a table to prevent injury. One should use the free hand to hold the front legs of the cat to prevent scratches, if necessary.

Cats may also be placed in a **stretch technique**. This restraint is done by stretching the body of the cat. First, one should scruff the neck of the cat and then

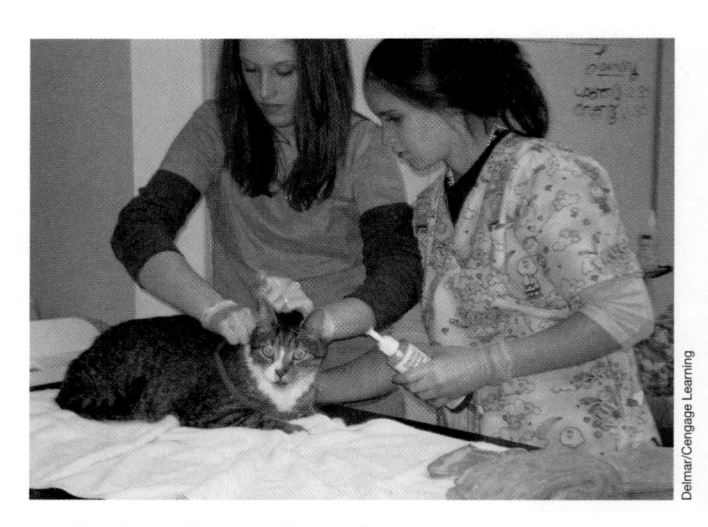

FIGURE 9-16 The scruff technique.

place the cat on its side in lateral recumbency, keeping a firm grasp on the scruff. The rear limbs should be held to avoid being scratched and to control the body (see Figure 9-17). This position often immobilizes the cat and keeps it quiet and still.

Cephalic venipuncture restraint is done by scruffing the cat with one hand to control the head and use the free hand to reach around the opposite side to extend the front leg (see Figure 9-18). Pressure can be applied around the limb to aid in blood collection. Sometimes the cat will sit quietly with less restraint by restraining the cat under the jaw.

Jugular venipuncture restraint is accomplished similar to that of a small dog. The head is grasped under the jaw and the head is extended upward (see Figure 9-19). The free hand and arm are used to hold the front limbs to avoid scratching. The front limbs may be extended over the edge of a table.

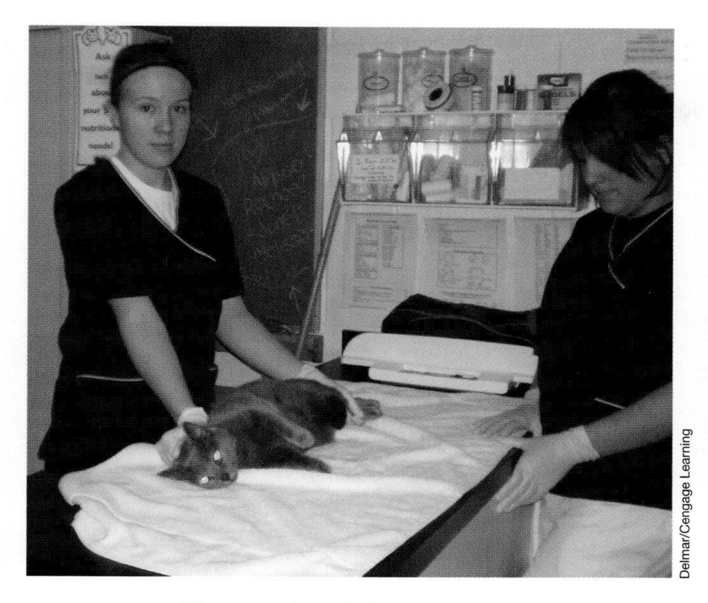

FIGURE 9-17 The stretch technique.

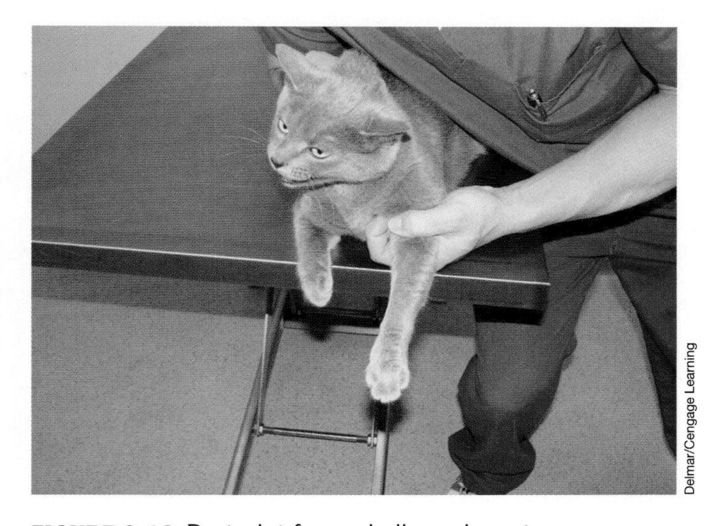

FIGURE 9-18 Restraint for cephalic venipuncture.

The saphenous venipuncture restraint is done using the scruff technique with one hand in combination with the stretch technique. A karate chop pressure hold is placed over the limb to allow the blood collector to visualize the vein.

Cats may also be restrained using a squeeze chute or an anesthesia chamber. The **squeeze chute** is a wire cage that is used to restrain **feral** (wild) cats, rabies suspect cases, and very aggressive cats that are too difficult to handle. The box-like cage is used to hold a cat in place for examinations and procedures. The sides of the cage are wire openings that have been made to slide. This allows the cat to be pulled inward against the side of the cage. Injections may be given to vaccinate cats or administer sedatives for further restraint. The **anesthesia chamber** is used in a similar fashion. A cat may be placed in the glass enclosure for examination or further sedation. The chamber is an aquarium-like box that has openings in the top to attach to an anesthesia machine. This allows anesthesia to be filtered in for further restraint. **Welding gloves** are also a common piece of restraint equipment in a veterinary facility. The gloves are layers of leather that act as protection to the hands and arms when handling aggressive cats. Caution must be used when wearing the gloves as they are bulky and can be used too harshly on a cat, possibly causing injury, or the cat may be able to escape if the restrainer loses his or her grip.

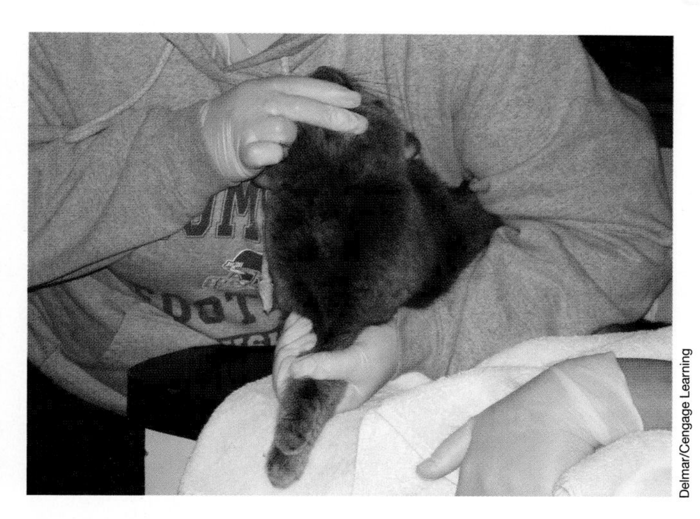

FIGURE 9-19 Restraint for jugular venipuncture.

Grooming

Cats require the same grooming needs as dogs. Longhair and shorthair cats should be brushed on a daily basis using a medium to soft bristle brush, especially if they have medium to long hair (see Figure 9-20). A flea comb should be used on a monthly basis to watch for fleas and flea dirt. Bathing should be done on a routine basis (see Figure 9-21). Cats may also be groomed similar to that in dogs. Longhaired cats may need trimming or shaving, especially if the hair coat develops mats (see Figure 9-22). Cats may be given a traditional bath with shampoo and water, or if the cat tends to be difficult, a waterless shampoo product may be used (see Figure 9-23). **Waterless shampoo** is applied to the hair coat and worked into the skin and allowed to dry. It is not rinsed with water. The product is a self-cleaner and naturally disinfects the hair coat. Many cats are not cooperative for a bath and caution

FIGURE 9-20 Cats should be brushed daily.

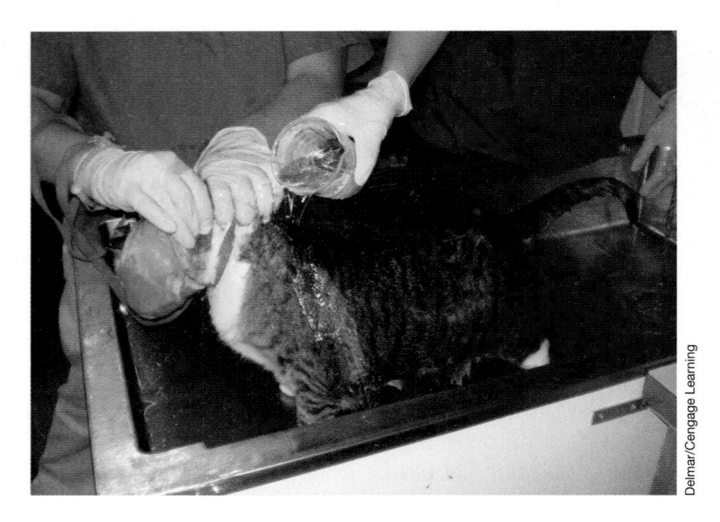

FIGURE 9-21 Cats should be bathed routinely.

should be used to not be bitten or scratched. It is best to trim a cat's nails prior to bathing to reduce scratches.

Brushing, nail trimming, ear cleaning, teeth brushing, and anal gland expression should be done on a regular basis as outlined in "Canine Production Management." Nails may be trimmed with cat-sized nail

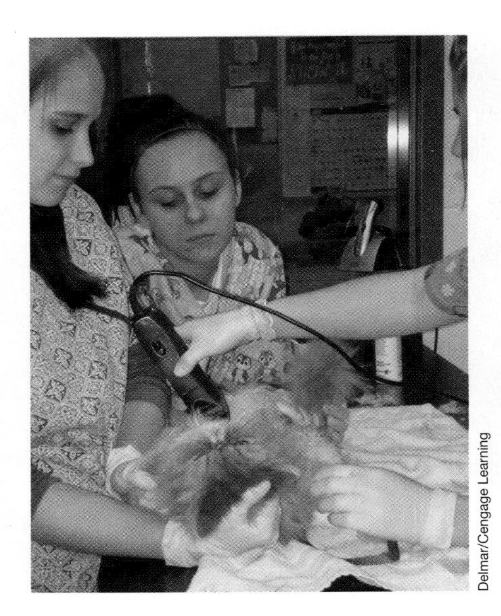

FIGURE 9-22 Longhair cats may need to be shaved if they develop mats.

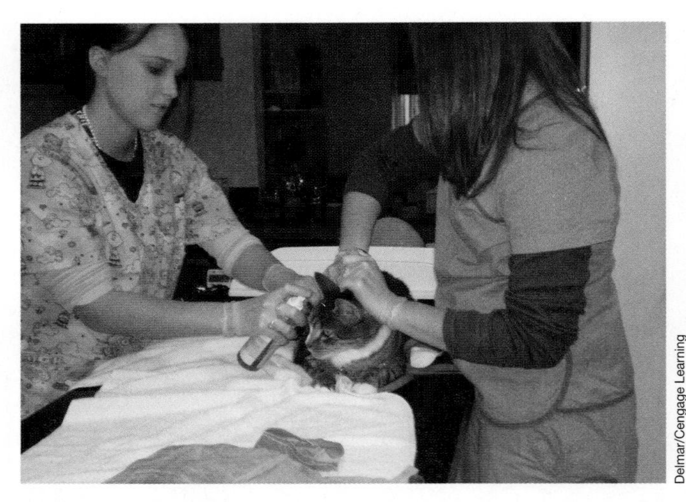

FIGURE 9-23 Waterless shampoo can be used on a cat that will not tolerate a bath with traditional shampoo and water.

clippers or human fingernail trimmers (see Figure 9-24). It is recommended to trim a cat's nails first before any other procedure to reduce the likelihood of scratching injuries.

Basic Health Care and Maintenance

Each year a cat should be seen by the veterinarian for a physical examination. This examination will evaluate the health status of the cat and any illnesses or

FIGURE 9-24 Trimming a cat's nails.

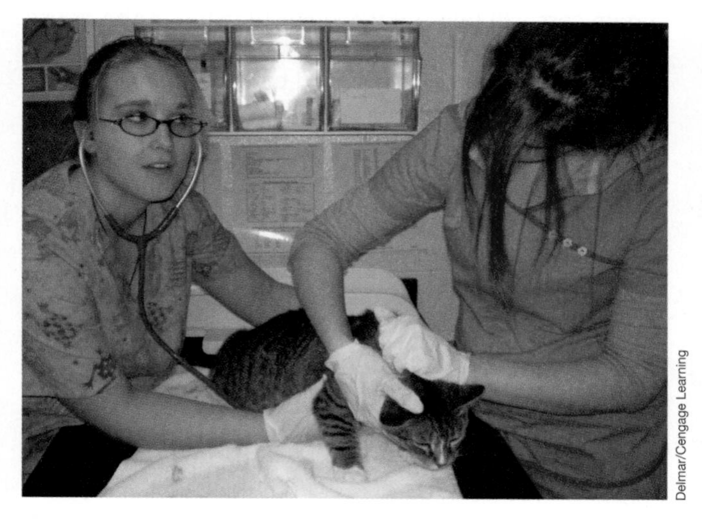

FIGURE 9-25 Cats should be seen annually by the veterinarian for a physical examination.

conditions that may be developing (see Figure 9-25). It is recommended that every cat be tested for feline leukemia, feline AIDS, and feline infectious peritonitis. A simple blood test may be done in the veterinary facility, called a **SNAP test** (see Figure 9-26). The test results will be available in 10 minutes and test for the antigens and antibodies of the diseases within the patient's bloodstream. It is important to know the health status of every cat that enters the veterinary facility.

The ears, eyes, and teeth should be monitored on a regular basis (see Figure 9-27). Like dogs, cats also have anal glands that may need to be expressed. Weights should be monitored each year (see Figure 9-28). Veterinary assistants should be properly trained to perform basic grooming skills on cats, such as shaving hair, nail trims, cleaning ears, brushing teeth, and anal gland expression.

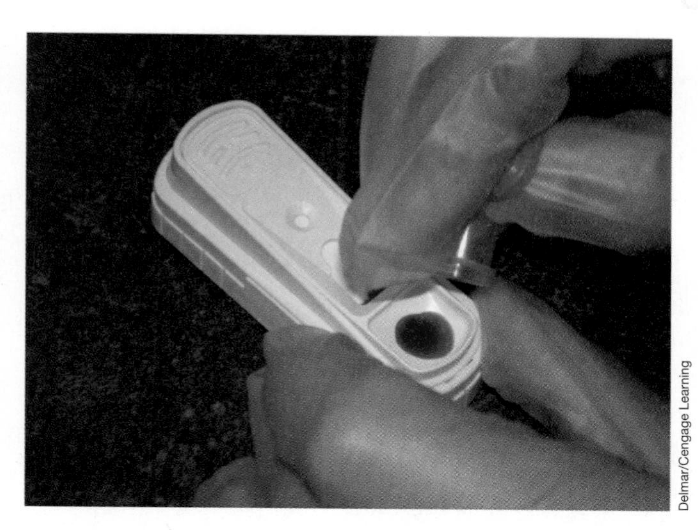

FIGURE 9-26 SNAP test.

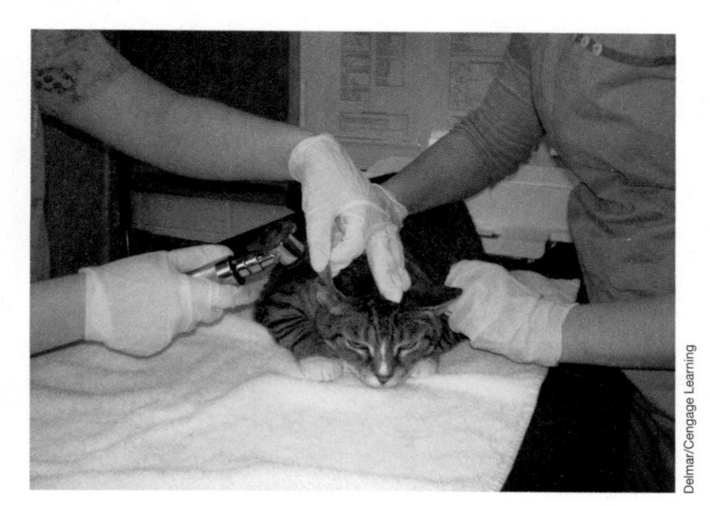

FIGURE 9-27 Checking the cat's ears.

Vaccinations

Kittens should begin a vaccine program between 6 and 8 weeks of age (see Table 9-3). The most common feline vaccine series is the **FVRCP** combination, also known as

TABLE 9-3

Vaccination Program for Cats

AGE	VACCINE
6–8 weeks old	FVRCP
10–12 weeks old	FVRCP
12 weeks old	FeLV
14–16 weeks old	FVRCP
16 weeks old	FeLV
16 weeks old	RV
Yearly or more	FVRCP, FeLV, RV

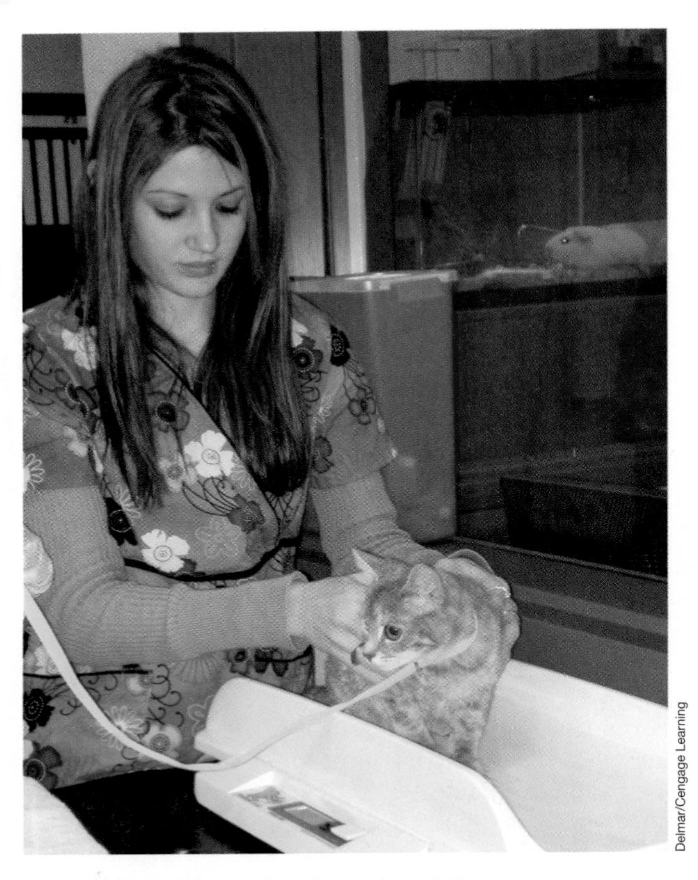

FIGURE 9-28 Monitoring the cat's weight.

the feline distemper series. Each letter is an abbreviation for the disease it protects the kitten against.

- FVR—Feline Viral Rhinotracheitis
- C—Calicivirus
- P—Panleukopenia

The FVRCP is given in boosters once a month for 3–4 times until the kitten reaches 16 weeks of age. A rabies vaccine is given at 16 weeks of age. All vaccines are then good for one year. Some veterinarians may suggest other vaccines against feline diseases, such as the Feline Leukemia or Feline Infectious Peritonitis. The feline leukemia vaccine, or **FeLV**, is given at 12 weeks of age and requires one booster in 3–4 weeks and is then given annually.

Reproduction and Breeding

Female cats are known as **polyestrus**, which means that they have multiple heat cycles during a season. Typically, their heat cycle occurs every 15 to 21 days and will last for 5 to 7 days. Felines reach puberty between 5 and 8 months of age. Signs of a female cat in estrus include the following:

- Growling
- Behavior changes
- Vocal
- Very friendly and rubbing on objects
- Rolling and laying upside down (see Figure 9-29)
- Chewing on objects
- Assuming the **lordosis** position—prayer position of lowering the front of the body and raising the rear

Cats are **induced ovulators** and will become pregnant only after breeding has occurred. Gestation in cats is between 60–65 days with an average of 63 days. The labor process is known as queening or kittening. Some breeders will provide a maternity box, but more female cats will find a dark, quiet location in the house to give birth. A bedroom or a closet are typical locations. Cats are private and noise should be kept at a minimum.

Signs of queening:

- Restless and anxious
- Nesting behaviors
- Panting
- Stop eating 12–24 hours prior to labor
- Licking at vulva

Cats should be monitored for dystocia, which is an abnormal or difficult labor. A normal labor process is front feet first followed by the head and body. The female will lick the kitten to stimulate breathing and to help dry the newborn. The kittens should not be handled for 3–4 days. A typical litter size is between 3 and 4 kittens.

FIGURE 9-29 Rolling around or laying upside down is a sign of a cat in heat.

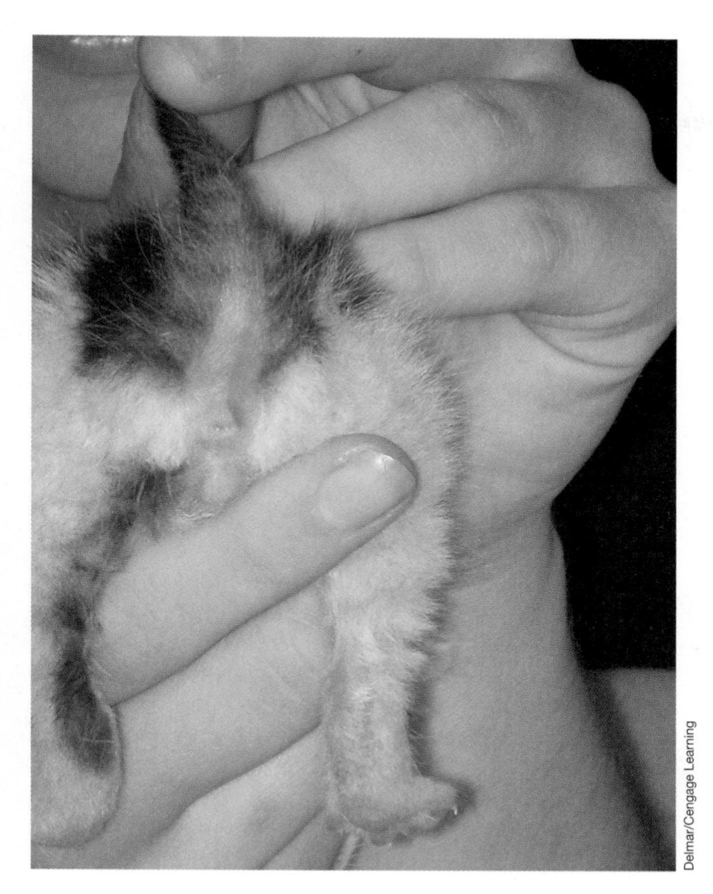

FIGURE 9-30 It is possible to determine the sex of a kitten within weeks of birth.

Kittens should be kept at a room temperature of 85 degrees for several weeks. Kittens' eyes and ears open in 10–14 days. Kittens can be sexed within weeks of birth (see Figure 9-30). Kitten gender can be determined by the distance of the anogenital area from the anus to the genital opening. Females have a shorter distance than do males. The female genital opening forms a slit and the male genital opening is round. Males that have not yet been castrated may have already developed testicles.

Common Diseases

Cats may develop many types of infectious diseases. Some of the more common feline diseases seen in the veterinary facility include **panleukopenia**, commonly called **feline distemper**; **rhinotracheitis**, **calicivirus**, and the rabies virus. Several fatal diseases are of a concern for cats as well. These include **feline leukemia**, **feline immunodeficiency virus (FIV)**, and **feline infectious peritonitis (FIP).** Clients of any cats diagnosed with one of these fatal diseases that are extremely ill should be counseled about treatment versus euthanasia.

Panleukopenia

Panleukopenia is also known as "feline distemper." This is a systemic virus that affects the cat by causing a decreased white blood cell count. It is transmitted through urine, feces, and direct contact. In kittens it is usually fatal. There is a vaccine available to prevent the disease. Signs of feline panleukopenia include the following:

- Vomiting
- Diarrhea
- Depression
- Dehydration
- Anorexia
- Seizures
- Death

Rhinotracheitis

Rhinotracheitis is also a virus that affects the upper respiratory tract of cats. It is spread through direct contact of saliva and nasal discharges. The disease typically runs a 2–4 week course until signs begin to improve. A vaccine is available for prevention. Signs of feline rhinotracheitis include the following:

- Nasal discharge
- Ocular Discharge
- Sneezing
- Drooling/salivation
- Anorexia

Feline Calicivirus

Feline calicivirus is a highly contagious virus that also affects the upper respiratory tract. Cats that survive the disease usually have a permanent head tilt. The disease is spread through direct contact and body secretions. A vaccine is available for prevention. Signs of feline calicivirus include the following:

- Nasal discharge
- Ocular discharge
- Oral ulcers
- Pneumonia
- Head tilt

Feline Leukemia

Several fatal diseases occur in cats. These conditions are highly contagious and once a cat is diagnosed with the disease, there is no treatment. Feline leukemia is the most common of these diseases. It is a virus that is passed through direct contact and body secretions. A vaccine and test are available. Cats that test positive for the disease should be isolated from all other cats to prevent transmission. Cats may show signs of illness or they may be a **carrier** of the disease, which means they show no signs of disease but are still capable of passing the disease.

Feline Immunodeficiency Virus

Feline immunodeficiency virus (FIV) is commonly called feline AIDS. It is species specific to cats but acts in the same manner as the human AIDS virus. It is highly contagious in cats and is typically spread through direct contacts and bite wounds. The immune system is greatly affected and many cats die due to other illnesses. There is a test available but no vaccine has been developed. Cats that test positive should be placed in strict isolation. As with feline leukemia, cats may show any sign of sickness or be in a carrier state. A simple SNAP test may also be done in a clinic to determine a cat's FIV status (see Figure 9-31).

Feline Infectious Peritonitis

Feline infectious peritonitis (FIP) is a highly contagious disease that may occur in a wet or dry form. This disease affects the lungs and chest cavity, causing severe respiratory problems that tend to lead to fatal pneumonia. **Pneumonia** is an abnormal condition that affects the lungs. There is a test and vaccine available.

Rabies

Rabies is also a fatal disease that may occur in cats. As in dogs, it is spread through saliva from bite wounds or scratches. It has a zoonotic potential and state laws should be reviewed for cat owners. Feral cats are at a greater risk for rabies. Veterinary professionals should consider a pre-exposure vaccination against rabies.

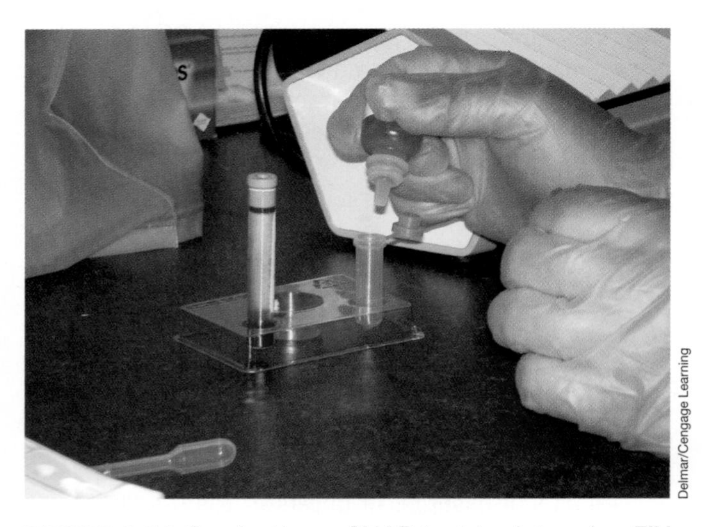

Delmar/Cengage Learning

FIGURE 9-31 Conducting a SNAP test to determine FIV status.

Common Parasites and Prevention

Cats may become infected with internal and external parasites similar to dogs. Cats may become infested with fleas, ticks, lice, mange mites, and ear mites as listed in the "Canine Production Management" chapter; review Chapter 8 for more information on external parasites. The veterinary assistant should be able to identify these external parasites. He or she should also be familiar with products that are available to prevent and treat external parasites (see Table 9-4).

Cats may also become infected with internal parasites. Cats are susceptible to roundworms, hookworms, and tapeworms. Cats rarely develop whipworm infections. Cats also become infected with heartworms. Internal parasites of felines occur in the same manner as those in dogs. Refer to the "Canine Production Management" chapter to review the information on internal parasites. Veterinary assistants should be knowledgeable in products that are used to prevent and control internal parasites in cats (see Table 9-5).

Another parasite that affects cats and is zoonotic to humans and a large concern especially for pregnant women is called **toxoplasmosis**. Toxoplasmosis is a protozoan parasite that is passed in infected cat feces. An infected cat will shed the parasite in a 12–24 hour time frame. It is important for female clients who are cat owners to be educated about this parasite especially if they are pregnant. Pregnant women should not handle cat feces or clean litter boxes due to the possibility that the infection may cause spontaneous abortion or miscarriage in women or cause birth defects in the baby. Cats typically become infected with the organism through soil contamination.

Common Surgical Procedures

Veterinary assistants will routinely discuss feline surgical procedures with cat owners. Several canine surgical procedures were discussed in "Canine Production Management," such as the ovariohysterectomy (spay), castration, and declaw surgery. These surgical procedures are common in cats, as well. The procedures are similar to that of the dog. **Declaw** surgeries in cats are the removal of the retractable claw or digit on the feet.

TABLE 9-4

Feline Flea Product Comparison Chart

PRODUCT NAME	APPLICATION TYPE	PARASITES CONTROLLED	AGE OF USE	WATERPROOF OR RESISTANT	LENGTH OF APPLICATION
Advantage	Topical	Fleas	8 weeks	Water resistant	Monthly
Advantage Multi	Topical	Fleas, hookworms, roundworms, ear mites, heartworm	8 weeks	Water resistant	Monthly
Capstar	Oral	Fleas	4 weeks	Waterproof	Daily or as needed
Frontline	Topical	Fleas	8 weeks	Waterproof	Monthly
Frontline Plus	Topical	Fleas, ticks, lice	8 weeks	Waterproof	Monthly
Program	Oral/ Injectable	Fleas	4 weeks	Waterproof	Monthly-oral 6 months-Inj
Promeris	Topical	Fleas	8 weeks	Waterproof	Monthly to every 6 weeks
Revolution	Topical	Fleas, hookworms, roundworms, ear mites, heartworm	8 weeks	Waterproof	Monthly

TABLE 9-5

Feline Heartworm Product Comparison Chart

PRODUCT NAME	APPLICATION TYPE	PARASITES CONTROLLED	AGE OF USE	LENGTH OF APPLICATION
Advantage Multi	Topical	Heartworms, fleas, ear mites, hookworms, roundworms	9 weeks	Monthly
Heartgard	Oral	Heartworm, hookworms	6 weeks	Monthly
Interceptor	Oral	Heartworm, hookworms, roundworms	6 weeks	Monthly
Revolution	Topical	Heartworms, fleas, ear mites, hookworms, roundworms	8 weeks	Monthly

It is important that the veterinary assistant has a basic understanding of the procedures in order to discuss the benefits of the surgery to the client.

Spay

Cats are typically neutered around 6 months of age. Some veterinarians will neuter cats at a much earlier age. The ovariohysterectomy (OHE) is performed in the female cat with the same procedure discussed in the dog. Female cats go in and out of estrus seasonally until they are bred. They are capable of having two to three litters of kittens a year. The OHE will decrease the cat population, reduce the attraction of male cats, and decrease the incidence of spraying and marking territory. The spay is performed under general anesthesia with a midline incision on the abdomen. The uterus and ovaries are removed and the incision is sutured. Sutures are removed in 7–10 days with a limit on exercise.

The **orchidectomy** (castration) in the male cat is much less complex than in the male dog. The testicles are exposed, ligated, and excised. There are typically no sutures in a cat castration so suture removal is not necessary. The castration procedure in cats decreases the likelihood of spraying, wandering for females, contracting diseases or wounds, and strong urine odor.

Onychectomy

Cats are commonly seen in the veterinary facility for a procedure known as **onychectomy**, commonly called a declaw surgery. This surgery results in the surgical removal of the digits or claws. The procedure involves the amputation of the distal digit where the toenail begins. The procedure may be done in the front paws, rear paws, or both. After the digits are removed, the toes are either sutured closed or a glue-like **tissue adhesive** is placed in the area. The paws are then bandaged to control bleeding. The bandages are usually removed in 48 to 72 hours depending on the age of the cat. Owners are instructed to not use litter as it will enter into the declaw sites and cause infection. Many owners will purchase litter pellets made specifically for declaw surgeries or will shred newspaper to place in the litter box. Cats are usually healed in 2 weeks. Declawed cats must be kept indoors. They will have a less likely chance of defending themselves outside. Some veterinarians will not perform this procedure as it may be a painful procedure and requires increased amounts of pain management. Cat owners should be counseled on the advantages and disadvantages of the declaw procedure. An alternative to declawing is placing plastic tips over the nail to reduce scratching and damage in the household.

SUMMARY

Cats have become popular companion animals. Veterinary assistants will need to be knowledgeable in feline health care and maintenance needs. Cats require an experienced restrainer and the veterinary assistant should have a good working knowledge of normal and abnormal cat behavior. Veterinary medicine is becoming more specialized each day with feline-only practices.

Key Terms

ad lib giving something, such as food or water, as much as necessary or so that it is available at all times

anesthesia chamber glass or plastic aquarium-like equipment with a lid and opening that attach to the anesthesia machine to sedate cats and small animals, such as rodents

calicivirus highly contagious virus that affects the upper respiratory system of cats and may cause a head tilt

carrier box-like object that safely holds a cat for transport; also a cat with a disease that spreads the disease but shows no signs of having the disease

cat bag restraint equipment that holds the body and limbs of the cat and controls for certain procedures, such as a nail trim

Cat Fanciers' Association (CFA) cat breed registry association that promotes the health and responsible breeding of purebred cats

cat muzzles restraint equipment placed over the face and mouth of a cat to prevent biting

declaw the surgical removal of the digits or claws on the feet of a cat

defensive aggression natural behavior that occurs when a cat is attempting to protect itself from people or other animals

dilated the occurrence of widening or opening, such as the pupils

domestics cats with unknown parentage that are of two or more breeds and not registered

feline distemper virus called panleukopenia that causes a decreased white blood cell count in cats

feline immunodeficiency virus (FIV) commonly called feline AIDS virus; fatal virus that attacks the immune system of cats

feline infectious peritonitis (FIP) fatal virus of cats that has a dry and wet form that affects the lungs and chest cavity

feline leukemia fatal virus in cats that affects the white blood cells and immune system

Felis catus genus and species of cats

FeLV abbreviation for feline leukemia virus

feral cat that is wild with a domesticated appearance

FVRCP feline vaccine combination series also known as the feline distemper series; abbreviation for Feline Viral Rhinotracheitis, Calicivirus, and Panleukopenia

hairballs large amounts of hair that collects in the digestive system of cats due to excessive grooming

induced ovulators female cats that become pregnant after breeding with a male as the release of an egg occurs at the time of mating

intermale aggression behavior tendency that is common in adult male cats due to sexual dominance or social ranking

kitty taco restraint procedure of wrapping a cat in a towel to restrain it

litter box equipment filled with pebble-like material and used by a cat to eliminate in

litter trained process of teaching cats to eliminate in a specific area of the house

longhair hair coat that is long and requires regular brushing and grooming

lordosis prayer position of female cats that are in heat; lower the front of their body and raise the rear of the body while in estrus

onychectomy veterinary term for the declaw surgical procedure of a cat

orchidectomy veterinary medical term for the removal of the testicles, also known as castration

panleukopenia commonly called feline distemper; systemic viral disease that causes decreased white blood cells in cats

pneumonia lung infection in cats

polydactyl having many or multiple toes or digits

polyestrus having multiple heat or estrus cycles during a season

redirected aggression aggressive behavior toward another animal that did not provoke it but caused it to become upset

retractable capable of projecting inward and outward, as in the nail of a cat

rhinotracheitis viral that affects the upper respiratory system of cats

scruff the loose skin over the base of the neck of a cat that is sometimes used to restrain a cat

shorthair hair coat that is smooth in cats

SNAP test simple blood test completed in the clinic that tests for certain disease antigens in the bloodstream

squeeze chute wire cage used to control a cat by sliding the sides closed for such procedures as exams and injection

stretch technique restraint procedure by holding the loose skin of the neck with one hand and pulling the rear limbs backward with the other hand

taurine nutrient required by cats for allowing proper development

thiamine nutrient compound of Vitamin B that promotes a healthy coat

tissue adhesive glue-like substance that is used to close incisions

toxoplasmosis protozoan parasite passed in cat feces that may cause abortion in pregnant women

waterless shampoo product applied to the hair coat that doesn't require rinsing with water and cleanses the coat

welding gloves heavy layers of leather placed over the hands to restrain cats and small animals

REVIEW QUESTIONS

1. Explain the differences between the cat and dog.
2. List the signs of the happy cat, angry cat, and scared cat.
3. Why is it important to be experienced and knowledgeable in cat restraint methods?
4. What selection factors should be considered when considering a cat?
5. What are the signs of a female cat in estrus?
6. What are the nutritional requirements of the cat?
7. What should an ideal feline vaccine program include?
8. What fatal diseases of cats should clients be educated about?
9. What are some common internal parasites that occur in the cat?
10. What common surgical procedures are performed in cats?

Clinical Situation

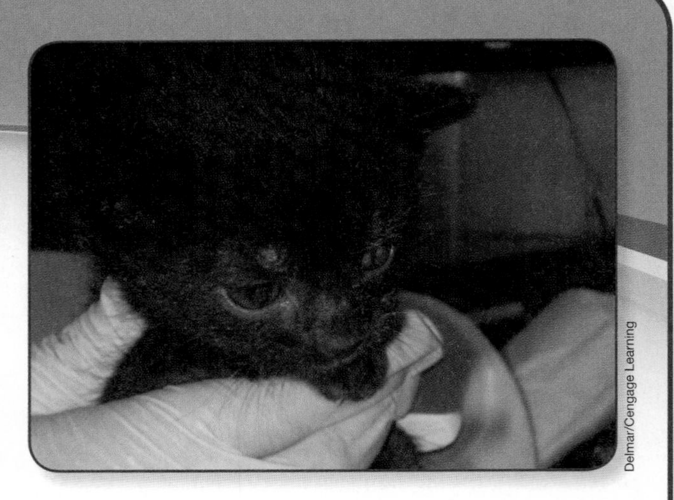
Delmar/Cengage Learning

A 3-week-old kitten is presented to the veterinary facility. The people who brought the kitten to the clinic found it this morning and were concerned because it seems so young. They attempted to feed it some milk from their refrigerator, but it didn't seem to want to eat. It appears bright and alert and vocal. The veterinarian examines the kitten and feels it should be placed on a commercial formula and fed 3–4 times a day. The veterinarian leaves the room and the owners begin questioning the veterinary assistant, Anne.

"We were not expecting this today. We don't have animals and really do not want to care for a kitten," says the husband.

Anne responds with, "Dr. Smith went to get you some formula and we will show you how to feed the kitten. It really is easy and shouldn't be much work."

The wife answers with, "I don't think we have any money to spare for a kitten. Plus, we are planning on going out of town for vacation next weekend. I don't think we can care for the little guy."

- What would you do in this situation?
- What are some options to discuss in this clinical situation?

10

© Photo Disc

Avian Breed Identification and Production Management

Objectives

Upon completion of this chapter, the reader should be able to:

- Identify and explain common veterinary terms relating to avians
- Describe the biology and development of avians
- Describe the classes of birds
- Identify common breeds of exotic avian species
- Discuss the nutritional requirements of birds
- Understand normal and abnormal behaviors of birds
- Properly and safely restrain small and large birds for various procedures
- Discuss breeding and reproduction in birds
- Discuss the health care and maintenance of avians

Introduction

Birds are one of the animal species that have only been domesticated as companion animals in the last 50 years. They are also the third most popular pet in the United States (see Figure 10-1). Birds are considered exotic and make excellent pets in homes or apartments. Each breed of bird has different needs and requirements. Special equipment and facilities are necessary for the proper care and health of a bird. For this reason, the veterinary assistant should have a basic understanding and knowledge in avian species.

© iStock/Jill Lang

FIGURE 10-1 Birds are a popular pet in the United States.

Veterinary Terminology

There are a few terms that relate specifically to the avian. In veterinary medicine, it is essential to be familiar with these terms and know when to properly use them in the facility (see Table 10-1).

TABLE 10-1

Avian Veterinary Terms

AVIAN	BIRD
Hen	Female bird of breeding age
Cock	Male bird of breeding age
Brood	Young birds in a nest at one time
Clutch	Total number of eggs in the nest
Chick	Newborn bird
Juvenile	Young bird that has not yet developed feathers
Fledgling	Young bird that has left the nest but is not eating on its own
Weanling	Young bird that has left the nest and is feeding on its own
Flock or company	A group of birds

Avian Biology

A bird (avian) is an animal that has feathers, two legs, and two wings. Many birds are capable of flying but some breeds are not. Birds have a **beak** that serves as a mouth. Birds stand upright on two limbs, and each limb has four toes for grasping and climbing. Some birds are very small, such as the hummingbird, and others are very large, such as the ostrich. This chapter mainly discusses veterinary medicine related to pet caged birds.

The avian external body parts are very different from other companion animals (see Figure 10-2). The structure and terms vary greatly. The veterinary assistant should be familiar with the following terms and locations shown in the diagram of the bird's external body parts (see Table 10-2). Bird feathers have anatomy parts, such as the barb, barbule, and calamus (see Figure 10-3). The anatomy of the feather is important when performing restraint procedures and wing clipping procedures. Bird health can be determined by examining the external parts of the bird for signs of disease.

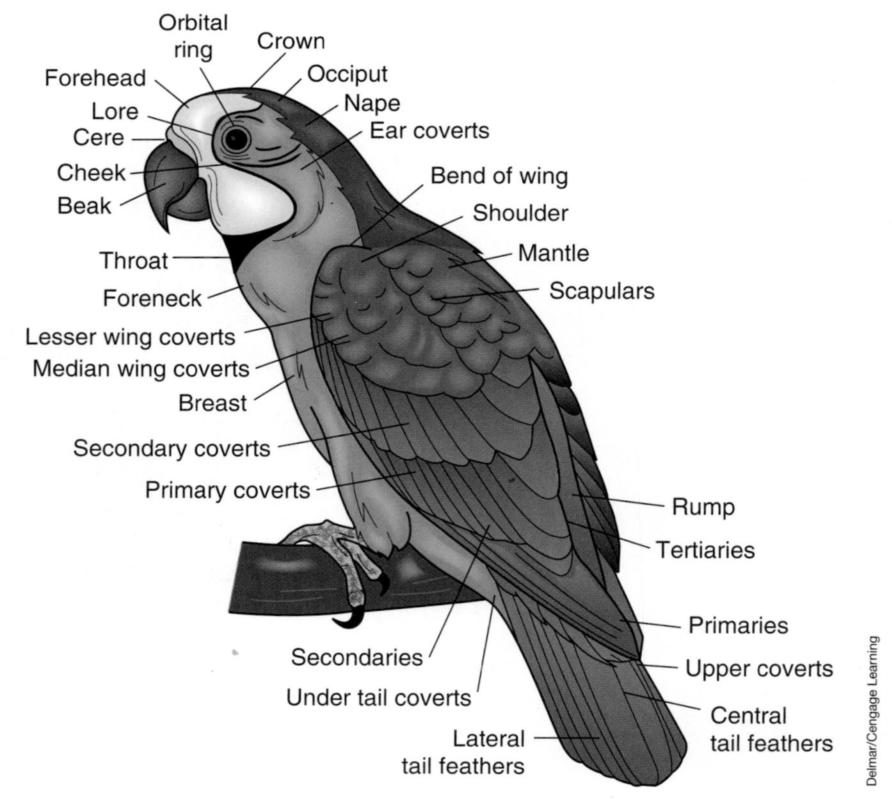

FIGURE 10-2 External anatomical structure of a bird.

TABLE 10-2

Common Avian External Anatomical Terms

Barb	Individual section that projects from the wing and forms the feather
Barbule	Small projection on the edge of the barb of a feather
Beak	Hard structure that forms the mouth of a bird
Breast	Front area of the chest
Calamus	Quill of the feather that attaches to the skin
Cere	Thick skin at the base of the beak that holds the nostrils; may be in different colors according to the gender
Crown	The top of the head
Feather	Similar to hair parts that form the wings and allow flight in certain species
Keel	Breast bone
Nape	Area at the back of the head
Orbital ring	Area that encircles the eye
Primary feathers	Lower feathers of the wing that form the first row
Secondary feathers	Upper feathers of the wing that form the second row
Shaft	Part of the feather where the barbs attach
Tail feathers	Feathers that form at the lower back of the bird
Talon	Claws or nails of the feet
Throat	Area below the beak
Wing	Formed by feathers on each side of a bird that act as arms

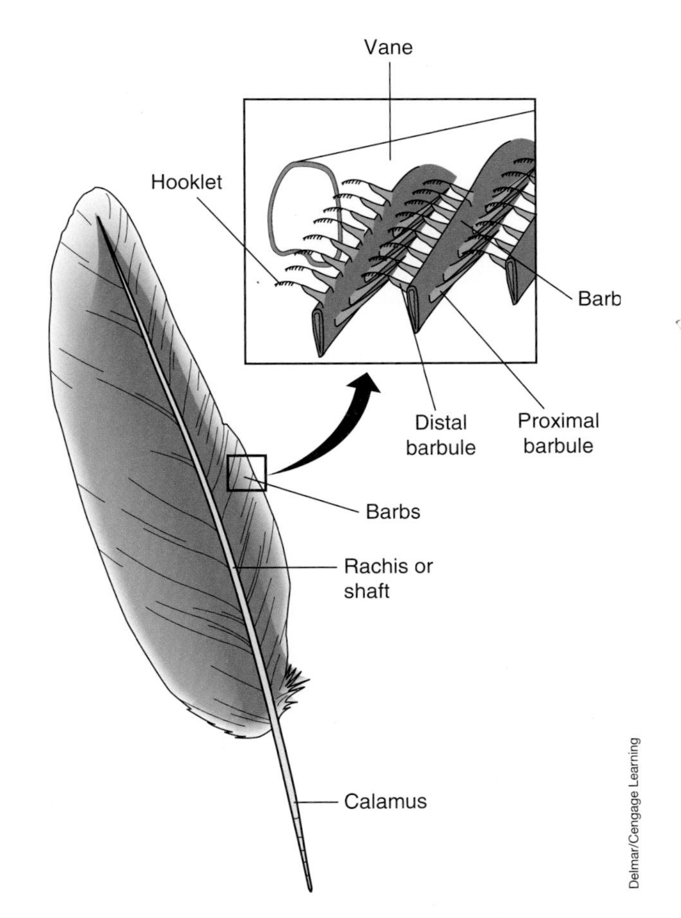

FIGURE 10-3 Anatomical structure of a feather.

Birds are **omnivores** in that they eat both meat and plant sources. Some breeds are considered **herbivores** in the wild in that they only eat plant sources. In **captivity** (living in a cage and cared for by humans), birds learn to eat many types of foods. Birds have a specialized digestive system known as an avian digestive tract (see Figure 10-4). They have specialized organs used to hold food and then break it down for ease of digestion. These parts include the **crop**, which acts as a holding tank for food and is located below the esophagus. The **gizzard**, which serves as a filter system that breaks down hard foods, such as seed shells or bones, is located below the crop. The **proventriculus** is the tube that connects the crop to the gizzard. Then the food moves through the intestines and passes waste materials via the **cloaca**, which serves as the rectum. The cloaca is also called the **vent**, which is the external area of the digestive system.

Birds are capable of living long life spans. Some larger species have been known to outlive humans. Birds reproduce by laying eggs.

The skeletal system of birds is somewhat similar to other companion animals. The two legs of birds act as the rear limbs in animals such as cats and dogs. The two feathers act as the front limbs of other animals. Birds only have vertebrae bones in the neck and in the tail bone. They have a keel bone that lies over the lower chest and abdomen area that serves as a breastbone to protect the internal organs (see Figure 10-5).

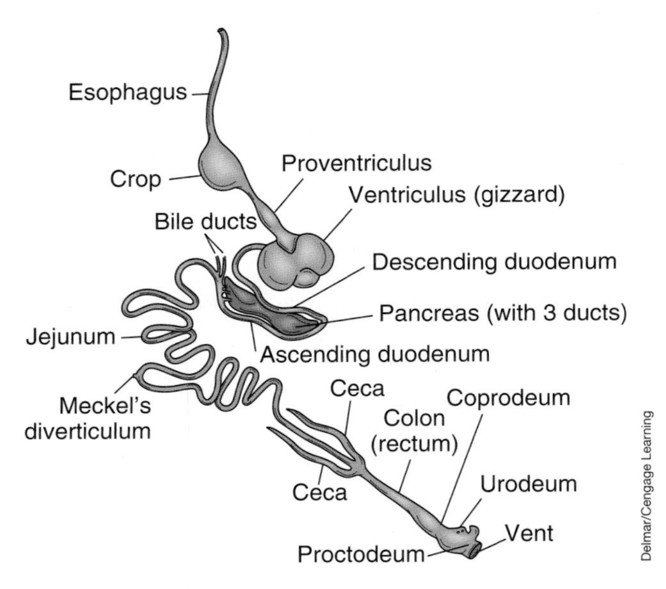

FIGURE 10-4 Avian digestive tract.

Breeds

Birds are classified into 28 different orders. There are thousands of breeds of birds. This chapter will discuss only those bird breeds that are most popular among breeders and bird owners. Each group has different characteristics, behaviors, attitudes, and a uniqueness that makes them popular with people.

Cockatiels

Cockatiels are a common breed of bird small in size and relatively easy to care for and train. They are great beginner birds for someone who is just learning how to care for avians. Adult birds have a variety of feather colors including yellow, gray, white, and orange (see Figure 10-6). Cockatiels originated in Australia. They are friendly and lovable and tend to enjoy singing and

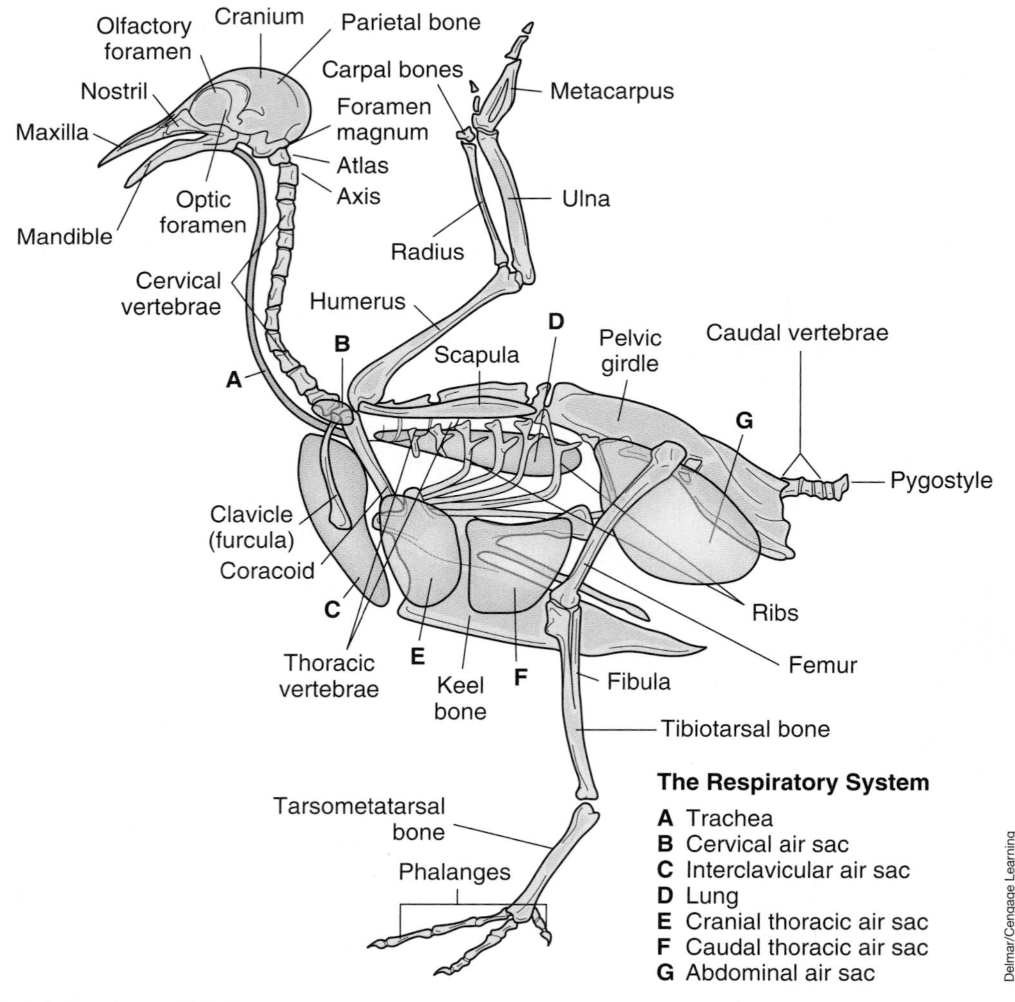

FIGURE 10-5 Skeletal system of birds.

Photo courtesy of Isabelle Francais

FIGURE 10-6 Cockatiel.

Photo courtesy of Isabelle Francais

FIGURE 10-7 Parakeet.

Photo courtesy of Isabelle Francais

FIGURE 10-8 Society finches.

whistling. Cockatiel eggs hatch in 25–26 days after being laid. The babies begin to mature around 55 days of age. Adults eat cereal grains, fruits, seeds, and green foods. They do well on a high-quality commercial bird food. They range in size from 10 to 18 inches in height.

Parakeets

Parakeets are also known as **budgies** or **budgerigars**. They come in a variety of body colors, such as yellow, white, and blue (see Figure 10-7). They have black bars (lines) over the wing and tail feathers. They originated in Australia. They are a small breed that is easy to care for and are excellent for beginners. They are very active, will develop some extensive talking habits, but tend to talk very rapidly. The eggs hatch in 25–26 days. They eat cereal grains, seeds, and fruits. They do well on a high-quality commercial diet. Their sizes range from 8 to 10 inches in height. They do tend to have some health problems.

Finches

Finches come in a variety of sizes, appearance, and colors (see Figure 10-8). Some species originated in Australia and others in Africa. They are a very small breed that is easy to care for and raise. Eggs hatch in 12 days and they begin to mature in 18–21 days. The males tend to be more vocal and enjoy singing. Finches are great beginner birds for breeders. They eat seeds and small cereal grains. Adults range in size from 4 to 5 inches in height.

Lovebirds

Lovebirds are a small-sized bird that comes in a variety of colors (see Figure 10-9). They are excellent cage birds. Lovebirds originated in Ethiopia. They are easy to tame when they are young. Adults are much more difficult to tame and handle. Lovebirds may sing but do not talk. They eat cereal grains, seeds, and fruits. Adults are about 5–6 inches in height.

Photo courtesy of Isabelle Francais

FIGURE 10-9 Lovebird.

Photo courtesy of Isabelle Francais

FIGURE 10-10 Canary.

Canaries

Canaries originated in the Canary Islands. They are small breeds known for their solid bright colors (see Figure 10-10). They are popular pets and love to sing. The males tend to be better singers. They require larger cages due to an increased amount of activity and energy. Their eggs hatch in 14 days and begin to mature in 28 days. They prefer a seed diet with small amounts of green foods. Adults range in size from 5 to 7 inches in height.

Conures

Conures are medium-sized birds and come in a variety of colors (see Figure 10-11). They originated in South America. They are very active and playful. Eggs hatch in 23 days and the birds mature in about a year. They eat cereal grains, fruits, vegetables, and green foods. Adults range in size from 12 to 14 inches in height.

Parrots

Parrots come in a variety of colors, appearances, and sizes (see Figure 10-12). They are considered medium breeds. Many originated in Central and South America. They are very active, playful, and tend to bond to one

Photo courtesy of Isabelle Francais

FIGURE 10-11 Conure.

Photo courtesy of Isabelle Francais

FIGURE 10-12 Parrots.

specific family member. They are easy to train and love performing tricks and have the ability for an extensive vocabulary. They are considered highly intelligent. Eggs hatch in 28–30 days and the birds reach maturity in about a year. They eat cereal grains, nuts, fruit, vegetables, and pellets. They are typically several pounds and reach a variety of heights.

Macaws

Macaws are a large breed of bird. They come in a variety of bright colors (see Figure 10-13). They originated in Central and South America. They are playful and very trainable. They are smart and tend to get in trouble easily. They need large amounts of room due to their sizes. They require more health care than smaller breeds. They are hardy but do require protection from cold weather. Eggs hatch in 25 days and adults mature in 1–2 years. They eat seeds, fruits, and green foods. Occasionally some breeds will eat small amounts of meat. They range in size from 15 to 36 inches in height.

Amazons

Amazons are also a large breed of bird. They also originated in Central and South America. They are quite talkative and entertaining. They have cuddly personalities and love human company. They come in a variety of bright colors (see Figure 10-14). They eat commercial pellets, fruits, and green foods. They need large cages and range in size from 15 to 20 inches.

FIGURE 10-13 Macaw.

FIGURE 10-14 Amazon.

FIGURE 10-15 Cockatoo.

Cockatoos

Cockatoos are a large breed of bird. They originated in Australia. They are typically a white body color (see Figure 10-15). They have the ability to have large vocabularies and tend to be loud, often screaming for attention. They are easy to train and love human companionship. Eggs hatch in 25–30 days and adults reach maturity in 1–2 years. They eat a variety of seeds, nuts, fruits, vegetables, and green foods. They need a large cage and range in size from 15 to 24 inches in height.

Table 10-3 presents an overview of a variety of bird breeds and the characteristics specific to the breed.

TABLE 10-3

Characteristics of Birds by Breed

BREED OF BIRD	WEIGHT (GRAMS)	AVERAGE LIFE SPAN (YEARS)	CHARACTERISTICS
Finches	10–15	5–10	Easy to care for and breed
Canaries	15–30	10–20	Easy to care for
Parakeets	30–35	10–20	Easy to care for; great beginner birds
Cockatiels	80–90	10–20	Easy to care for and train
Lovebirds	40–50	40–50	Taming and training more difficult
Parrots	400–500	60–70	Easy to train to talk and perform tricks
Macaws	1,000–1,400	60–70	Entertaining, can train to talk
Amazons	900–1,200	60–70	Entertaining, enjoy talking
Conures	100–200	40–50	Easy to train and care for
Cockatoos	300–800	50–70	Very vocal; easy to train

Breed Selection

When selecting a breed of bird to house as a pet, it is important to research the following before purchasing:

- Breed behaviors and characteristics
- Lifespan
- Cage needs
- Supplies and environmental needs
- Nutritional needs and diets
- Health care and veterinary needs
- Cost of bird
- Cost of care of the bird
- Common health problems or conditions

Cage birds need a lot of attention, specific health care, specific cage needs, and a variety of supplies. These items can be costly depending on the breed and size of the bird. Many new bird owners want to purchase a young bird as older birds may have developed behavioral problems, such as biting or screaming. Young birds are easier to train and bond with new people. Adult birds will have much more difficulty adapting to change. Selecting and acquiring a pet bird is a lifetime commitment and should be taken seriously. Some bird species can outlive humans and this must be taken into consideration when selecting a bird as a pet. The following Websites may be helpful in researching various breeds of birds:

- Bird Breeds—http://www.bird-breeds.com
- BirdBreeders.com—http://www.birdbreeders.com

Nutrition

Birds eat a variety of different foods based on their breed and size. Many commercial diets are available for pet birds. The best source of a commercial bird diet is a **pelleted diet**. A pellet diet is very similar to that of a rabbit. The pellets are small pressed cubes that vary with the size of the bird. Each pellet contains the necessary amounts of protein, vitamins, and minerals to meet the daily nutritional requirements of avians. Another type of commercial diet is a **seed diet**. Seed diets are not the best nutritional food. Seed diets have a variety of seed types, colors, and sizes in them. Many birds will select the best-tasting seeds, which tend to be high in fat. Thus, they are not receiving the necessary nutrition. Many bird species require supplements of fruit and **green foods**. Green foods are non-dried vegetables with some juice substances still present. Examples include cabbage, carrots, fresh corn, peas, and bean sprouts. **Grit** is sometimes provided to a bird to help break down foods, such as seeds and nuts, in the digestive system. The grit substance is a finely ground hard material similar to sand that rubs against food particles to better allow digestion to occur. Fresh food and water should be provided daily (see Figure 10-16). Do not feed stale or spoiled foods. Use caution to not feed fresh vegetables or fruits that may have been sprayed for insects or weeds. When introducing a new food or diet to the bird, it is best to eat the food for the bird and then offer the item as birds learn by example. Some birds will prefer foods that are a certain color, substance, or temperature. Like dogs and cats, birds can easily become overweight. Some of the items included in a suggested avian commercial diet include the following:

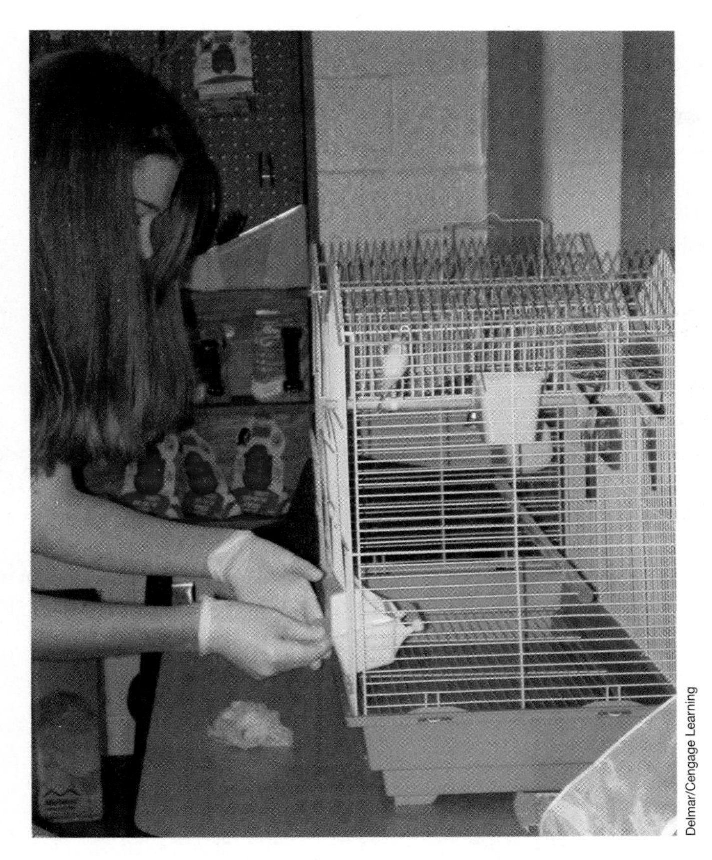
Delmar/Cengage Learning

FIGURE 10-16 It is important to provide fresh food and water daily to birds.

- Lafeber Diets
- Harrison's Diets
- Kaytee Diets
- Rowdybush Diets
- ZuPreem Diets

Table 10-4 lists safe fruits and vegetables to feed pet birds.

The following are unsafe foods that should not be fed to pet birds:

- High-fat junk food (potato chips, doughnuts, etc.)
- Avocado (guacamole)
- Chocolate
- Alcohol or caffeine
- Fruit pits
- Persimmons
- Table salt
- Onions
- Apple seeds
- Mushrooms

TABLE 10-4

Fruits and Vegetables That Are Safe for Birds

FRUITS	
	Carrots
Apples	
Apricots	
Banana	
Berries	
Cantaloupe	
Cherries	
Cranberries	Broccoli
Grapefruit	
Grapes	Cauliflower
Kiwi	Cooked sweet potatoes
	Collard greens
	Cooled red potatoes
	Corn
	Cucumbers
	Eggplant
	Green beans
	Lettuce
Mango	Kale
	Peas
Oranges	Radishes
Papaya	
Plums	
Peaches	
Pears	
Pineapple	Tomatoes

Behavior

Although birds have been domesticated, they have many wild behaviors that influence how they interact with humans. In fact, many bird breeds are only one to three generations removed from the wild. It is important for the veterinary assistant to understand normal avian behaviors and to know how to educate clients about abnormal behaviors. For this reason, the veterinary assistant should have some knowledge about behavior basics and training of avian species. Table 10-5 lists factors that affect a bird's behavior.

Avian behaviors are designed for survival. In the wild they are on the food chain of predators. Therefore, birds are natural prey and adapt behaviors for protection. This includes biting, flapping wings, and screaming. These are typical negative behaviors as pets. In the wild, each bird in a flock has a **pecking order**; each

TABLE 10-5

Factors Affecting Behavior in Birds

FACTOR	EXAMPLES
Breed and genetics	Some birds are very vocal; some have gender appearances in color or characteristics
Socialization as a juvenile	Ease of training; some adults are less adaptable to change and acceptance of new owners
Rearing and environmental conditions	Daily handling is important; some breeds will become territorial and difficult to handle
Post-weaning experiences	Learning to eat the proper diet, such as seeds, pellets, or green foods
Training methods	Learning to be removed from the cage; learning to be carried; learning not to bite
State of health	Learning abnormal signs of bird health; birds mask signs of illness
Environment	Cage too small; not providing enough toys; birds become bored easily

member has a social ranking. In a household, the bird will try to be the head of the household. Proper training methods are essential for appropriate behaviors to be learned. Some factors that lead to behavioral problems include the following:

- over-bonding with one person
- prolonged periods alone, absolute confinement
- no toys for mental stimulation
- solitude
- frequent environmental changes

Therefore, it is important that people have a basic understanding of the behaviors of the bird species they own. They must be handled properly to avoid developing bad habits.

Need for a Routine and Stable Environment

Birds do not adapt well to change. In an unstable environment, the slightest change will stress a bird and cause a behavioral problem. Examples that are common in birds not adapting to change may be moving the cage to a new location and the bird stops eating; changing to a new diet and the bird begins to bite; or having house guests that the bird is not accustomed to and the bird begins screaming for attention.

Socialization and Stimulation

Birds have an intellect at the level of a 5- or 6-year-old child. Emotionally they are at the level of a 2-year-old child. As with children, they need boundaries, discipline, and guidance to learn appropriate behaviors.

One training method to use with avians includes teaching a bird to step up and step down on command when inside or outside of a cage (see Figure 10-17). This teaches the bird respect and commands and that the owner is the leader. The step-up process is done by taking one hand held vertically like a perch and sweeping the hand under the feet while saying "step up" and forcing the bird to perch (see Figure 10-18). The bird is then placed back in the cage with the command "step down." Birds should not be allowed to climb over the body of the handler, as this can potentially become a problem if the bird begins to bite. A thumb can be placed over the digits when holding a bird by the hands to prevent them from climbing up the arm (see Figure 10-19). Birds learn with repetition or repeating. Birds also need privacy. It is recommended that they be placed in their cage, which is their territory. During the night they may be covered with a sheet to feel secure and allow for sleep and privacy. Some birds that are confined to the cage for long periods may become territorial and protective of their cage.

FIGURE 10-17 Teaching a bird to step up and down off finger upon command.

FIGURE 10-18 To teach a bird to step up, sweep the hand back toward the bird and force it to perch.

FIGURE 10-19 Holding the bird's talons by placing a thumb over the digits will prevent the bird from moving up the arm.

Vocalization

Screaming and vocalization are normal behaviors for birds in the wild. In a wild environment, screaming is a way to alert the flock to danger. It is also a social ranking. In captivity, it is developed as an attention-seeking behavior. Birds need frequent attention and must adapt to a set schedule that rarely varies from day to day.

Screaming is best handled by placing a sheet over the cage when vocalization becomes too loud. Do not attempt to scream or yell back at the bird. They enjoy a good argument and this reinforces the behavior. Instead, walk away and ignore the bad behaviors. Approach the cage when the bird is quiet and reinforce the good behavior with attention or a toy.

FIGURE 10-20 Biting is an instinctual behavior in wild birds but can become problematic in captive birds.

Biting

Biting is also a defensive behavior. It is a way in the wild to test trees for perching and capturing food. In captivity, biting is a way for a bird to get attention (see Figure 10-20). This can lead to further aggression if not handled properly.

Birds that bite should be handled cautiously. When they are perched on a hand and try to bite, one method to teach them that biting is not acceptable is a technique called the "**earthquake**." This is done by holding the thumb over the toes to prevent the bird from moving up the arm or having too much freedom during handling. When the bird tries to bite, the handler quickly and carefully shakes the hand while holding on to the feet. The loss of balance causes the bird to be distracted. At the same time the handler should say a firm "NO" each time the bird attempts to bite.

When handling birds that bite, it is recommended that the cage and perches be kept lowered to the floor and below the level of the head and shoulders of humans. Birds that are kept high in a cage or capable of climbing up high have a feeling of dominance over humans, and this will increase any behavioral issues. Perches can also be used to handle birds that bite to avoid biting injuries (see Figure 10-21).

Feather Picking

Birds commonly develop the behavior of feather picking. A bird that picks at and bites its feathers may also pull the feathers out of the body. Some birds will pick feathers at a particular spot of the body, where others will pick feathers out over the entire body. Feather picking should be evaluated by a veterinarian to rule out a physical problem, such as mites or other parasites.

FIGURE 10-21 Using a perch can be helpful when working with birds that have a tendency to bite.

Birds that have a behavior problem may begin picking feathers out of boredom. Others may pick feathers for attention. Sometimes stress is a factor that leads to feather picking, and the bad habit becomes difficult to control or stop. Determining the factor that has caused the feather picking may not be possible. E-collars, or Elizabethan collars, similar to those used in dogs and cats, may be placed on the neck area of birds to control feather picking. Some birds may become so destructive in picking their feathers that they may begin traumatizing their own skin, causing open sores that could become infected.

TABLE 10-6

Minimum Cage Size for Birds

SPECIES	CAGED PAIR LENGTH × WIDTH × HEIGHT	SINGLE BIRD LENGTH × WIDTH × HEIGHT
African gray	4' × 3' × 4'	3' × 2' × 2'
Amazon parrots	4' × 3' × 4'	3' × 2' × 2'
Budgerigars	24" × 14" × 8"	*
Canaries	18" × 10" × 10"	*
Cockatiels	4' × 2' × 3'	26" × 20" × 20"
Cockatoos	4' × 4' × 3.5'	4' × 3' × 4'
Conures	4' × 4' × 4'	4' × 3' × 4'
Finches	2' × 2' × 2'	12" × 12" × 12"
Lovebirds	4' × 4' × 4'	*
Macaws	6' × 6' × 6'	3' × 2' × 3.5'
Mynah birds	†6' × 3' × 3'	6' × 3' × 3'

*These birds prefer the company of other birds and should not be caged singly.

†The cage should be at least this size for a pair.

Equipment and Housing Needs

All birds will require a cage. The cage provided should be as large as possible (see Table 10-6). Cages should be wider than they are tall to accommodate the bird's flying and natural movements. Perches should be provided in the cage and should be of various heights and widths. Newspaper is an appropriate floor material for the cage. Nothing that will be toxic or harmful to the bird should be used as flooring. The cage should be placed in an area of the house where the bird will be around its owner and free of drafts. Toys can be provided to birds to help keep them from becoming bored (see Figure 10-22).

FIGURE 10-22 Offer birds toys to keep them occupied and alleviate boredom.

Restraint and Handling

Veterinary assistants must learn the essentials of proper avian restraint. Birds are fragile animals and therefore must be handled with extreme care and with as little force as possible. Birds can stress due to the shock of handling and die of heart failure if not accustomed to being handled. Small birds are more at risk of this occurrence. Before handling a bird, it is best to talk to it in a soft voice and place its cage in a quiet and low-lighted area for comfort. When handling any bird, it is important not to apply large amounts of pressure over the

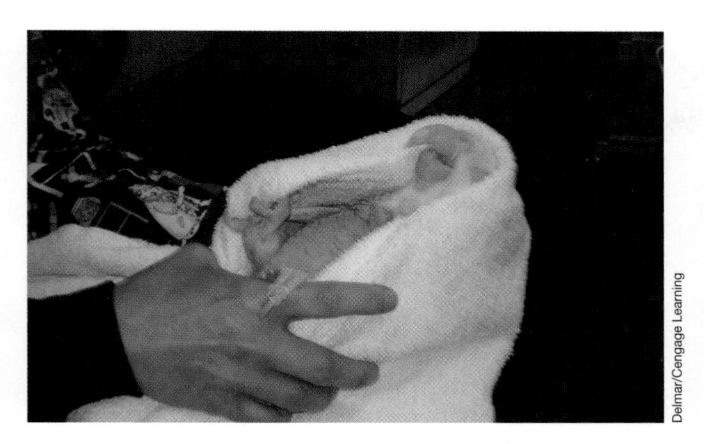

FIGURE 10-23 Do not place too much pressure over the bird's sternum during restraint as the bird requires the ability to expand and contract its chest to breathe.

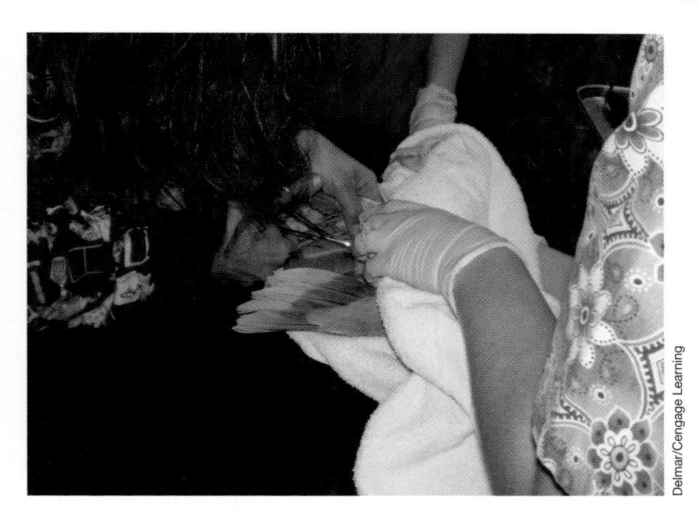

FIGURE 10-24 A towel can be used to control or restrain a large-breed bird.

chest or **diaphragm** of the bird (see Figure 10-23). The lungs and heart lie within the diaphragm that encases the chest cavity. The chest area must be able to expand (open) during **inspiration** and restrict (close) during **expiration** for the bird to breathe. This is one of the hardest restraints to remember since all other animals are capable of breathing while restraining the chest area.

It is important to note that when handling medium- to large-breed birds, they may become aggressive and bite if they feel threatened and are not used to people. Care must be taken to avoid severe injuries when handling certain birds. Large breeds, such as macaws and amazons, are capable of breaking a finger or even amputating a finger. They have very strong beaks that can cause severe injuries. It is always a good idea to have a heavy towel available when working with larger-breed birds (see Figure 10-24). Just like with cat restraint, it may be wise to use a pair of welding gloves or heavy leather gloves when working with aggressive birds. Care must be taken to not over-restrain when wearing bulky gloves.

Small pet birds, such as parakeets and cockatiels, can be picked up using one or two hands. This is done by grasping the bird from behind and securing both wings against the side of the body. Using the thumb and forefinger to restrain the head, the other hand can cradle the bird's body (see Figure 10-25). The area of the chest should be avoided. It is helpful to show clients how to train their bird to hop onto a finger. Small birds can also be restrained using a small hand towel or paper towel to prevent injury to the bird or restrainer. Large birds can be held by perching them on the hand if they are trained to do so. Always hold a thumb over the toes to prevent the bird from moving up the arm or flying. If a bird is unfamiliar with perching on a hand, it is a good idea to use a heavy towel. This will help protect the hands and arms and help in protecting the bird

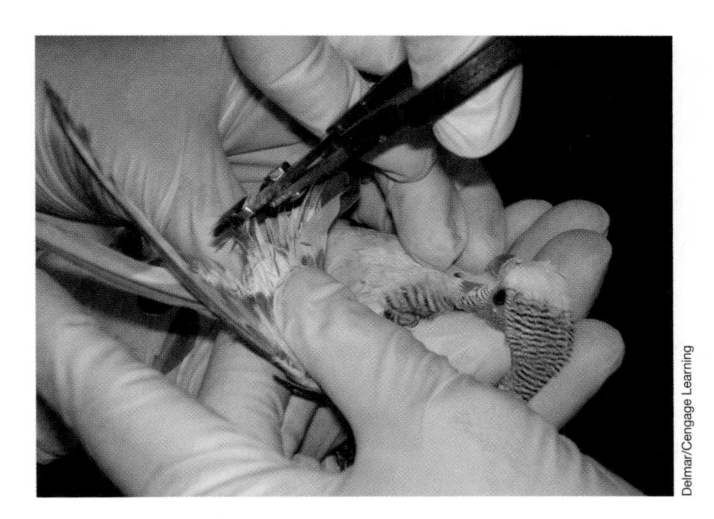

FIGURE 10-25 Restraint of a small-breed bird.

from injury. Never hold a large bird close to the face. Large birds can be restrained by holding the thumb and fingers on either side of the head to prevent biting. The pointer finger can be placed on the top of the head for added control. The free hand can be used to hold the wings at the side of the body. The use of a towel makes this restraint much safer and easier. It is always wise to have a second person helping restrain a bird, especially for hospital procedures. In some circumstances, the veterinarian may opt to place the bird in an anesthesia chamber and sedate the bird for safer handling.

Removing a bird from a cage may be done with a towel or by the hands, depending on how well trained the bird is and the behavior of the bird. Toys and items attached to the cage may need to be removed before attempting to remove the bird. This will make it easier to capture the bird during the restraint procedure.

When returning a bird to a cage, it is important to place it on the bottom of the cage. This will keep it from falling from a perch or injuring itself by flying.

Grooming

Birds should have a daily bath provided by a spray of water or a shallow bowl for bathing. This allows birds to keep their feathers clean so they can then **preen**, or groom themselves.

Birds should be maintained with regular grooming practices including nail trims, beak trims, and wing trims. Nail trims should be done with caution as birds have the same type of quick, or blood supply in their claws, as dogs and cats. The blood supply continues to grow and the blood vessel is not visible in the nail bed as in dogs. It is best to use a handheld **cautery unit** that burns through the nail bed and clots the blood. If this is not available, a cat nail trimmer or human nail clipper may be used, but styptic powder or silver nitrate sticks should always be available to control bleeding.

Birds' beaks continue to grow in the same fashion as the nails. Birds will require a **beak trim** based on their size and type of foods in their diet. It is important to make sure the beak tips do not overlap or grow into each other as this will prevent birds from eating properly. Beaks should be maintained using a dremel tool to grind down the edges of the beak (see Figure 10-26). The dremel tool is an instrument that is used to grind down beaks and smooth the tip from becoming sharp and overgrown. This allows for smooth edges that can be shaped into the natural lines of the beak. The sharp point of the tip of the beak should be rounded to prevent injuries to the bird, reduce damage in the house, and reduce injuries to people handling the bird.

Wing trims should be done to prevent birds from taking flight (see Figure 10-27). This is important in households where pet birds are given freedom to roam in the house. Birds that take flight may fly into windows, fans, or walls causing serious injury. They may also escape the house and die due to a lack of proper care in the wild. **Flight feathers** should be trimmed according to the size of the bird and how well it has adapted to flying (see Figure 10-28). The flight feathers are the adult feathers located over the edges of the wings. As these feathers begin to grow, they are called **blood feathers** since they have a blood vessel visible at the end of the feather that allows it to grow. Caution should be used to not trim a blood feather. Small breeds of birds can suffer severe blood loss from trimming a blood feather. Seasonally, adult birds will shed their flight feathers in a process called **molting**. Molting occurs similar to dogs shedding their winter coats. During this time, blood feathers will be present and wing trims should be avoided.

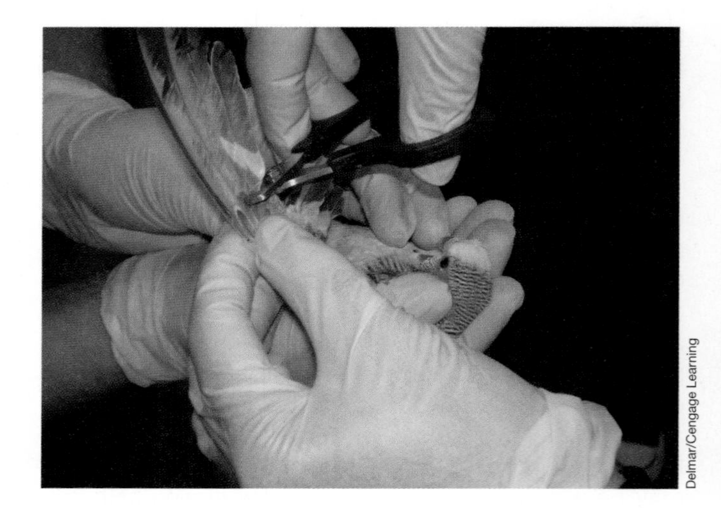
Delmar/Cengage Learning

FIGURE 10-27 Wing trim.

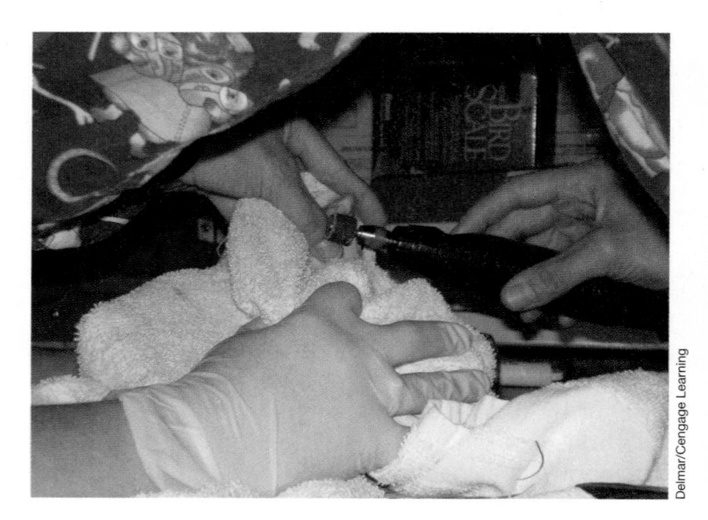
Delmar/Cengage Learning

FIGURE 10-26 Beak trim.

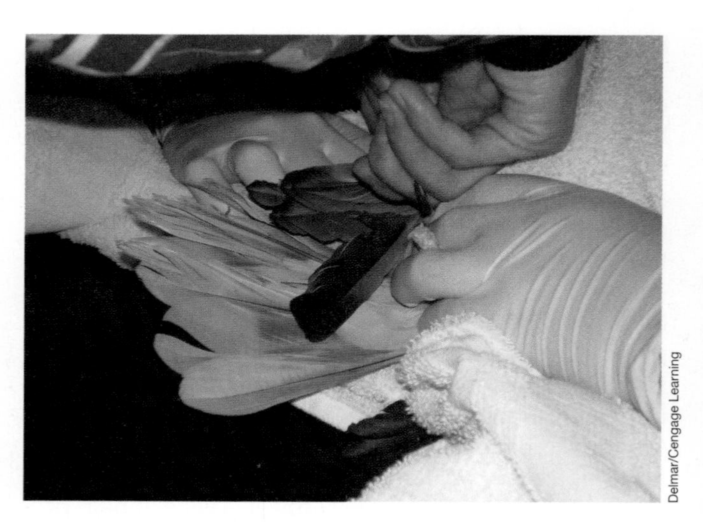
Delmar/Cengage Learning

FIGURE 10-28 Trimming the flight feathers.

Basic Health Care and Maintenance

Birds require several necessities to be healthy and stress free. Birds do not adapt well to stress or change, and they should have a constant environment. They should have a cage that is at least 1½ times the wingspan of the bird. Several perches should be included in the cage in areas where droppings will not contaminate the food and water sources. Birds should be given toys with bright colors to keep them stimulated and from becoming bored. Birds that are bored tend to develop behavioral problems. Cages should be maintained and monitored especially when several perches or large amounts of toys are provided. It is not uncommon for birds to injure themselves on items within the cage. Wings can be damaged and toenails can be torn off (see Figure 10-29).

Birds should have a yearly physical examination by a veterinarian, just as other companion animals. The physical exam helps to evaluate each body system and any underlying changes or diseases that may be affecting the bird (see Figure 10-30). Birds have adapted in the wild by masking signs of illness or disease, so a proper physical examination will reveal any hidden diseases or health problems. Birds often appear normal even when they are not feeling well. It is important to educate bird owners to understand that when a bird appears to not be feeling well it has been sick for a long time. As a result, many birds that show signs of illness do not recover well. Many veterinarians who specialize in aviary medicine suggest yearly blood work to determine the bird's internal health and changes in the bird's physical health due to this occurrence.

FIGURE 10-29 Birds can injure themselves if there are a lot of materials in the cage.

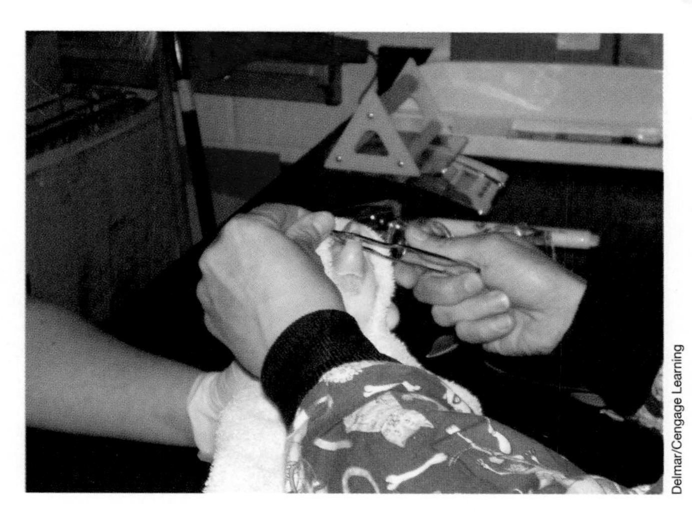

FIGURE 10-30 Birds should have a yearly physical examination by a veterinarian.

Veterinary assistants play a vital role in avian restraint and handling for hospital procedures and in educating bird owners on the needs and required health care of avians. It is important to educate pet bird owners on dangers to pet birds. Many items in the household can cause injury or death to birds. The same goes for items placed in the bird's cage.

The following items are dangerous to pet birds:

- Ceiling fans
- Cigarette smoke
- Teflon cooking pans
- Cleaning products
- Perfume
- Flowers and houseplants (poinsettias, holly, mistletoe, azalea, calla lily, philodendron)
- Lead and zinc
- Pressure-treated wood materials
- Cedar, wild cherry, and some oak woods

The following wood is safe for pet birds:

- Apple
- Maple
- Elm
- Ash
- Peach

Cages should be cleaned on a regular basis; all areas of the cage must be cleaned as bird droppings can

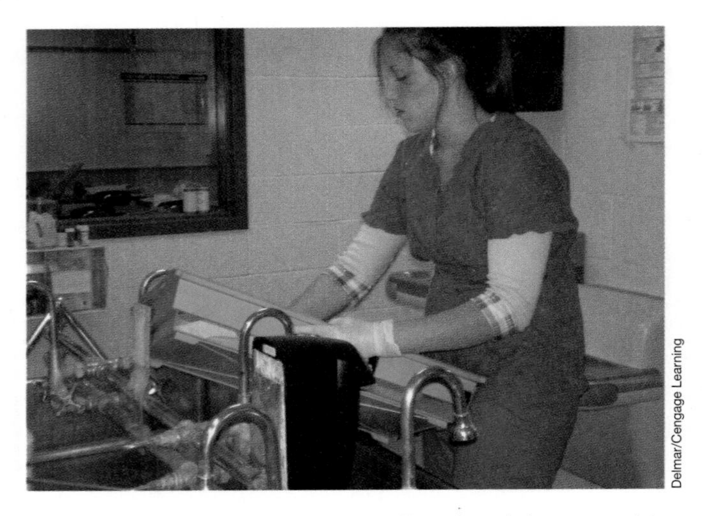

Delmar/Cengage Learning

FIGURE 10-31 Thoroughly clean all areas of the cage with a pet-safe cleaning product.

accumulate anywhere in the cage (see Figure 10-31). The cage should be disinfected using a pet-safe cleaner.

Vaccinations

Vaccines for bird species are very rare. There is no specific vaccine program for birds as there is for animals such as dogs or cats. Veterinarians will provide information to bird owners if they have a bird that is a high risk for specific diseases and suggest if the bird should be vaccinated.

Reproduction and Breeding

In general, birds reproduce by the females laying eggs. All female birds are capable of laying eggs, but a male must be present with the female to **fertilize** the eggs. If a male is not housed with the female and she lays an egg it is called **infertile**. In some species, the male and female have the same characteristics and coloring. It is advisable to **DNA test** any birds that will be used for breeding purposes to determine their gender. The DNA test will determine if the bird is male or female. This is done with a simple blood sample or feather sample submitted to an outside lab.

When a male and female are breeding, the male will mate with the female several days before the female lays a fertilized egg. Some breeders will allow the eggs to remain in the nest and the birds to **incubate**, or keep the eggs warm until hatching, naturally. Other times the eggs may be removed for **artificial incubation**. In this case, the eggs are placed in an **incubator**, a machine that keeps the eggs at a constant temperature until hatching. The best method is **natural incubation** although some birds do not take care of the eggs and they may not hatch. Nests should be placed within the cage.

The incubation times of each bird breed vary from 2 to 4 weeks. It is recommended that bird owners who wish to breed research the type of avian they own and learn as much as possible about the reproduction needs and cycle of the bird. Some female birds will require diet supplements that increase the calcium intake to allow for egg production.

The following are some basic and essential facts about breeding avians:

- The breeding pair must be mature, typically between 1 and 3 years old depending on species.
- Use caution when housing a male and female together; not all breeding pairs will bond.
- House the breeding pair in a large enough cage with a nest or nesting box.
- Provide adequate food and water and necessary supplements for the female, such as a mineral block or calcium supplement.
- Provide a quiet, warm area with slight humidity.
- Provide a light source during the day and cover the cage at night.
- Clean the cage daily.
- Do not handle the eggs.
- Monitor the chicks as they hatch, making sure none are being pushed away.
- Monitor the chicks' weights daily until weaned.

One possible condition that can result in breeding females is a problem called egg binding or **egg bound**. This condition occurs when an egg doesn't pass through the reproductive system at a normal rate. This condition can occur in any female bird regardless if a male is present. It occurs most commonly in small bird breeds. It can be treated in the early stages. Some signs of the egg-bound female include the following:

- Abdominal straining
- Wide stance with legs
- Depression
- Loss of appetite
- Swollen abdomen
- Feathers fluffed
- Droppings stuck on vent
- Egg noted in vent

Common Diseases

Avian species contract a lot of different contagious and fatal diseases, as well as internal and external parasites. The most common avian diseases of importance are psittacosis, psittacine beak and feather disease (PBFD), fatty liver disease, and knemidokoptes mite infestation.

Psittacosis

Psittacosis is commonly called "parrot fever" or chlamydia. It is a disease that has zoonotic potential especially in young children and anyone with a poor immune system. In humans it causes severe flulike symptoms. The disease is transmitted in the air or by direct contact. Birds shed the disease during times of stress and may be carriers of the disease. There are no specific signs of psittacosis and any abnormal appearance could be a sign of the disease. Diagnosis is made through blood testing, and prevention occurs through quarantine of all new birds and avoiding keeping birds in close quarters. This disease can be fatal to birds.

Psittacine Beak and Feather Disease

Psittacine beak and feather disease or **PBFD** is a contagious and fatal viral disease that affects the beak, feathers, and immune system of avians. It is transmitted in the air and through direct contact or contamination. There are two forms, an **acute** form, which occurs short term, and a **chronic** form that occurs long term. The following are identified signs of PBFD:

- Depression
- Diarrhea
- Weight loss
- Anorexia
- Abnormal feather development
- Growths and deformities of the beak
- Oral lesions
- Death

Fatty Liver Disease

Fatty liver disease, also called **hepatic lipidosis**, is a condition in which large amounts of fat are deposited in the liver. It is a serious condition that can result in death. Causes of the condition include high-fat diet, overfeeding, nutritional deficiencies, hereditary factors, and toxic substances. A low-fat diet and medications are helpful if the condition is diagnosed early. Signs of fatty liver disease include the following:

- Sudden loss of appetite
- Lethargy
- Swollen abdomen
- Green droppings
- Obesity

Common Parasites

Bird species are also prone to both internal and external parasites, like cats and dogs. Certain parasites are more common in pet cage birds. Knowing the signs of parasites is helpful to the veterinary assistant.

Knemidokoptes

Knemidokoptes mites are commonly called scaly leg and face mites. They are external parasites that occur on the skin, beak, and feathers of birds. The mites spend the entire life cycle on a bird, burrowing into the top layer of skin, and are transmitted through direct contact and contamination of cages and equipment. White to gray lesions occur over areas of the skin, legs, face, and beak. Birds may show signs of itching and feather loss. Treatment is with ivermectin wormers. The following are signs of a sick bird:

- Fluffed feathers
- Sitting on bottom of cage
- Depression
- Anorexia
- Lethargic
- Head tucked under wings
- Eyes closed

Common Surgical Procedures

Many bird owners do not typically have cage birds neutered. However, it is possible to spay a bird that may be a chronic egg layer or that becomes easily egg bound. Many avian veterinarians will have specialized skills in avian surgical procedures. General anesthesia is typically used with bird species. Sometimes sedation with general anesthesia is required for blood collection. Blood collection is commonly done at the site of the jugular vein on the neck, which in birds is located easily on the right side of the neck on a featherless tract, which is an area where feathers do not grow naturally.

SUMMARY

Birds have become popular companion animals. There are over 40 million pet birds in the United States. Veterinary assistants play an essential role in restraining and handling birds, educating bird owners on the care and needs of birds, and in basic hospital care of avians. Birds can offer the veterinary facility a challenging experience. It is important to have a basic knowledge and understanding of the behaviors and requirements of avian breeds.

Key Terms

acute condition that occurs short term

amazons large breed of cage bird from South or Central America that is talkative, entertaining, and very trainable

artificial incubation providing a controlled temperature in which eggs hatch, such as an incubator

beak the mouth of an avian or bird species used for chewing and grasping food

beak trim procedure to shorten or round the sharp edges of the beak

blood feathers growing feathers that have a blood vessel in the center

budgerigar commonly called a parakeet; small bird breed that comes in a variety of bright colors with bars located over the wings and back

budgie abbreviated name for a budgerigar or parakeet

canaries small breed of bird in solid bright colors that originated in the Canary Islands and is known for its singing ability

captivity an animal that is housed in an enclosure or with people.

cautery unit handheld equipment that heats to high temperatures to burn through nails and clot them at the same time

chronic condition occurring long term

cloaca the rectum area of a bird where waste materials pass

cockatiels small bird breed that is a combination of yellow, gray, white, and orange colors and is very easy to train

cockatoos large bird breed that is commonly white and has a large speaking ability

conures medium bird breed that is brightly colored from Central America and is active and playful

crop sac in the digestive system that acts as a holding tank for food

diaphragm the chest area of a bird that must expand during the breathing process

DNA test test using blood or a feather that helps determine a bird's gender

"earthquake" technique procedure used in training a caged bird to stand on the hand and be held that teaches a bird to not bite; when a bird attempts to bite, the hand is shaken slightly to cause a balance check

egg bound a female bird that is attempting to lay an egg that has become stuck in the cloaca

expiration to breathe out and expand the chest

fatty liver disease condition of birds in which large amounts of fat are deposited in the liver

fertilize the ability of a male to mate with a female to create an egg with an embryo

finches a small songbird that comes in a variety of colors and patterns and is easy to raise and breed

flight feathers adult feathers located over the edges of the wings

gizzard digestive organ of birds that serves as a filter system that breaks down hard foods, such as seed shells or bones; located below the crop

green foods foods such as fruits and vegetables that are green in color and have additional calcium

grit substance provided to birds to help break down hard substances, such as seed shells

hepatic lipidosis condition of birds in which large amounts of fat are deposited in the liver

herbivores animals that eat a plant-based diet

incubate to keep at a constant warm temperature

incubator equipment used to keep eggs at a constant warm temperature for hatching

infertile not able to reproduce; also known as sterile

inspiration to breathe in air and allow the chest to depress

knemidokoptes mites external parasites that occur on the skin, beak, and feathers of birds; commonly called scaly leg mites

lovebirds small songbirds of bright colors that originated in Ethiopia and are popular cage birds

macaws large bird breed that is brightly colored and originated in South and Central America; playful and easily trained

molting the natural seasonal dropping of feathers

natural incubation the process of a female or male bird nesting on eggs to hatch them

omnivores animals that eat both plant- and meat-based diets

parakeet small songbird from Australia also known as a budgie that is brightly colored with bars located over the wings

parrots variety of medium-sized bird breeds that come from Central America and are bright in color, intelligent, playful, and easy to train

pecking order the dominance order of a group of birds

pelleted diet food fed to a bird in a size-appropriate pellet that is composed of the necessary ingredients

preen to clean the feathers

proventriculus the tube that connects the crop to the gizzard

psittacine beak and feather disease (PBFD) contagious and fatal viral disease that affects the beak, feathers, and immune system of birds

psittacosis commonly called "parrot fever" or Chlamydia; disease that has zoonotic potential especially in young children and anyone with a poor immune system and causes severe flulike symptoms

seed diet food composed of natural seeds eaten by birds

vent the area where birds pass their waste materials; also called the cloaca

wing trim the procedure of clipping the wings to prevent a bird from flying

REVIEW QUESTIONS

1. List the major classes of birds that are the most common pets.
2. List the factors that affect a bird's behavior.
3. What are the most common bird behavior problems?
4. What are the signs of egg-bound birds?
5. What are some foods that should not be fed to birds?
6. What grooming practices are necessary with pet birds?
7. What is psittacosis?
8. What are the causes of fatty liver disease?

Clinical Situation

© iStock/Sven Klaschik

Mrs. James scheduled an appointment for her Amazon, "Goldie." "Goldie" has been known to be difficult to handle, both biting and screaming when handled by the veterinary staff members. It has been recommended by the veterinarian, Dr. Black, that "Goldie" be brought in as a drop-off appointment and sedated for a physical exam.

Mrs. James has been uncooperative about scheduling the appointment. She feels she can safely handle "Goldie" and refuses to have her sedated. The veterinary assistant, John, is attempting to discuss the situation with Mrs. James.

"Mrs. James, we really feel that 'Goldie' will be less stressed by a drop-off appointment. Dr. Black has suggested sedation for the exam. This will be helpful as we may need to collect blood, as well."

"I do not want 'Goldie' sedated! She will be fine if I hold her. Please, let's schedule her for tomorrow morning and I will gladly hold her for the exam."

- What would you suggest John do in this situation?
- What potential safety issues could be harmful in this situation?
- How would you handle this situation?

11

Pocket Pet Health and Production Management

Objectives

Upon completion of this chapter, the reader should be able to:

- ☐ Define common veterinary terms relating to rodent species
- ☐ Describe the biology and development of rodent species
- ☐ Identify common breeds of rodent or pocket pet species
- ☐ Explain the nutritional requirements of rodents
- ☐ Predict and describe normal and abnormal behaviors of rodents
- ☐ Properly and safely restrain rodents for various procedures
- ☐ Discuss breeding and reproduction in rodent species
- ☐ Develop a health care and maintenance plan for rodents
- ☐ Identify and describe common diseases affecting rodents

Introduction

A **rodent** is a mammal that has large front teeth designed for chewing or gnawing. Common rodents are mice, rats, hamsters, and guinea pigs. Rodents, also called pocket pets since they are a size that can fit into one's pockets, have increased in popularity as companion animals (see Figure 11-1). Several species are more commonly kept. However, there are rodent species that are illegal to house as pets in California and some other locations. Several states are currently considering changing their laws regarding ownership of species of hamsters, guinea pigs, and gerbils. States such as Hawaii and California have laws against owning certain species of animals, such as hedgehogs and ferrets. The veterinary assistant should have a basic knowledge in proper restraint and handling of rodent species, as many rodents will bite to protect themselves.

FIGURE 11-1 Rodents, such as this hamster, have increased in popularity as companion animals.

Mice and Rats

Mice and rats are curious and friendly pocket pets. They are usually tame animals, especially when acquired at a young age. The younger the rodent, the easier it is to tame. The best way to tame a rodent is by hand-feeding it. Both rats and mice can be trained to do tricks. Many times these animals are used in research and nutritional studies. Unfortunately, both have received a negative reputation from their wild ancestors, which cause damage and spread disease.

Veterinary Terminology

murine = mouse or rat

sire = intact male mouse or rat

dam = intact female mouse or rat

pup = young mouse or rat

Biology

Rats and mice typically live an average lifespan of 1–3 years. Rats and mice are in the family *Rodentia* and the veterinary term that refers to mice and rats is **murine** (see Figure 11-2). Rats and mice have poor vision but a well-developed sense of smell. Species that have red- or pink-colored eyes have reduced vision. Many species of rodents are **nocturnal**, which means they sleep during the day and are active at night. Many rodent species share common anatomical features. Common rat and mouse external body terms include the following:

- Claws—nails on feet
- Snout—nose and mouth area
- Tail—long part that extends off the back and is hairless and scaly
- Whiskers—located on nose and used as sensory devices

Some types of mice and rats are bred specifically for research, while others are bred as pets.

Breeds

Mice and rats come in a variety of colors and types, more so than breeds. Rats are usually white, brown, black, or a variety called **hooded** (see Figure 11-2A). Hooded rats are white in the body color with a black or brown coloring that is located on the head and shoulders, giving it a hooded appearance. White rats are commonly bred as research and laboratory species. Mice also come in a variety of colors including black, white, tan, and spotted.

(A)

Photo courtesy of iStock Photo

(B)

Photo courtesy of iStock Photo

FIGURE 11-2 (A) Rat with hooded markings. (B) Albino mouse.

There are species of fancy mice, as well as white mice, much like rat species.

Breed Selection

Rats and mice that are being selected as pets should be purchased young and handled often to allow them to be socialized. Most rats and mice that are handled frequently and with care seldom bite. They are timid and social pets that can be entertaining. They make excellent pets for children if cared for and handled properly.

Nutrition

Mice and rats have the same basic nutritional requirements. They will eat almost any type of food and do well on commercial rodent pellets. They enjoy human foods such as seeds, nuts, bread, cereals and grains, and raw vegetables. They can be fed **ad lib** or as much as they want, within reason. It is best to provide food in small dishes, which should be kept clean and free of bedding. Typical maintenance diets contain about 14% protein and 4% to 5% fat, while diets for growth and reproduction contain 17% to 19% protein and 7% to 11% fat. Seed diets are also formulated for mice and rats, but these diets should only supplement the basic rodent pellet as a treat item. Rodents prefer sunflower-based diets to pellets, but these seeds are low in calcium and high in fat and cholesterol. When fed exclusively, seed diets can lead to obesity and nutritional deficiencies.

Water should be provided in bottles that are secured to the side of the cage. Fresh water should be provided daily.

Behavior

Rats and mice are timid animals that must learn to trust people (see Figure 11-3). When they become accustomed to people they become very curious pets. They are always looking to escape and should be housed in well-confined cages. They are chewers and will chew through wood or plastic. Both species can be easily frightened and should be handled quietly and calmly. They should be housed in well-lit areas, as they can become stressed when kept in dark or unlit rooms. A stressed or upset rat will show anger by whipping its tail. A happy rat will chatter or click its teeth and occasionally make a vocal squeaking noise. Some rats and mice may become territorial of their cage and should be handled frequently to prevent this occurrence.

FIGURE 11-3 It is important to handle rats and mice at an early age to get them accustomed to people.

Basic Training Methods

Mice and rats are intelligent animals that can be easily trained. They will do almost anything for food and respond willingly to noises. The more they are handled and trained with consistency, the easier the animal will be to handle.

Equipment and Housing Needs

Cages should be made of heavy wire or metal because rodents will chew through wooden or plastic cages. Items placed in the cage should be small enough that mice and rats can't escape or climb out of the cage. They are both curious animals and need toys that help keep them from boredom. Perches, tunnels, and exercise wheels are common items. Bedding should be free of dust and could be sawdust, cedar shavings, shredded paper, pellets, or cat litter (see Figure 11-4). At least one inch of bedding should be provided. The cages should be kept clean and dry. Adult mice require a minimum floor area of 15 square inches and a cage height of 5 inches. Rats need at least 40 square inches of floor space and a minimum of 7 inches in height. Breeder mice and rats require much larger areas.

FIGURE 11-4 Rats and mice will require clean bedding materials.

Restraint and Handling

When handled frequently, a mouse or rat will typically sit in the hand and be easily held. If they are not handled a lot or used to people, they do have a tendency to bite. Rats and mice should never be touched or picked up when they are sleeping. Rats should be handled by the body and never picked up by the tail. The tail can be "stripped" of its skin or broken if used to pick up the rat. When the body is secure, grasp the rat by the tail base and place on a firm surface. Then, with the opposite hand, encircle the body over the shoulder and neck region. It is necessary to perform this maneuver expeditiously, or the rat may turn and bite. Do not grip the chest so tightly that respiration is hindered. Once the head and body are secured, the rat may be lifted (see Figure 11-5).

Mice can be picked up and handled by their tails, making sure to secure the base of the tail and not the end of the tail. Always handle rats and mice with quiet, calm, and confident manners. Most rats and mice will bite when a person shows fear or handles the animal incorrectly. Grasp the mouse by the tail base and place on a surface the animal is able to grip with its feet. Then, pull gently on the tail as the mouse holds onto the surface with its forelimbs. While maintaining traction on the tail, grasp the loose skin over the neck, immediately behind the ears, with the thumb and forefinger of the opposite hand. It is necessary to perform this maneuver quickly to avoid being bitten. If the skin over the shoulders or an insufficient amount of loose skin is grasped, the mouse will be able to turn and bite. If the skin is grasped too tightly, the animal's breathing can be compromised. Once the loose skin over the neck is appropriately secured, the mouse may be lifted and the tail secured by the fourth and fifth fingers of the same hand (see Figure 11-6).

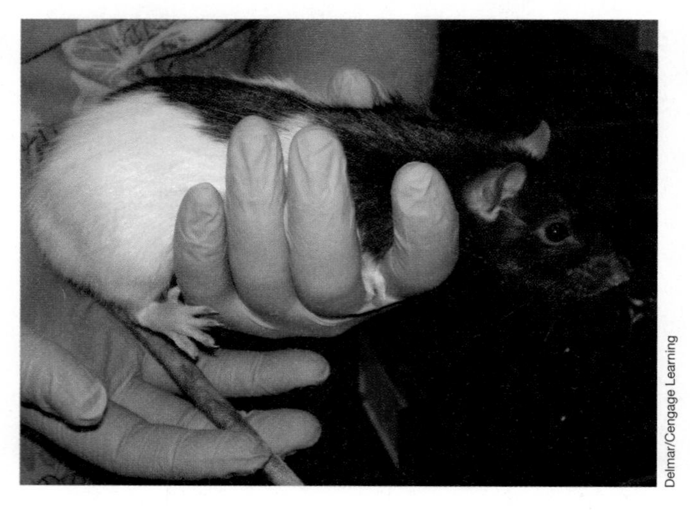

FIGURE 11-5 A tame rat or mouse can be picked up and will sit in the palm of the hand.

FIGURE 11-6 Mice can carefully be picked up by the base of the tail.

Grooming

Mice and rats, like other members of the rodent family, do not require bathing or any additional grooming. They are fastidious groomers and will keep their coats shiny and clean if they are kept healthy. If the coat does appear to be untidy, this may be a sign of disease.

Basic Health Care Maintenance

Mice and rats should have items in their cages to chew on to keep their teeth healthy and in good condition. Some teeth will get long and overgrown, which could cause eating problems or sore mouths. On occasion, they may need their teeth trimmed.

Their coats should be monitored for external parasites such as fleas and mites. They may be treated with topical parasite medications labeled for rodents. Do not use dog or cat products as they may be toxic.

Vaccinations

Rats and mice do not require vaccinations. There are no currently labeled vaccines for rats and mice.

Reproduction and Breeding

Mice reach puberty at an early age, usually by 8–10 weeks old. Rats usually reach puberty at 3 months of age. They can be bred at this time and both species have a gestation length of 21 days. They will have an average of 6–8 babies in a litter. Males and females may be housed together until labor occurs; the male should be removed until the babies are weaned, around

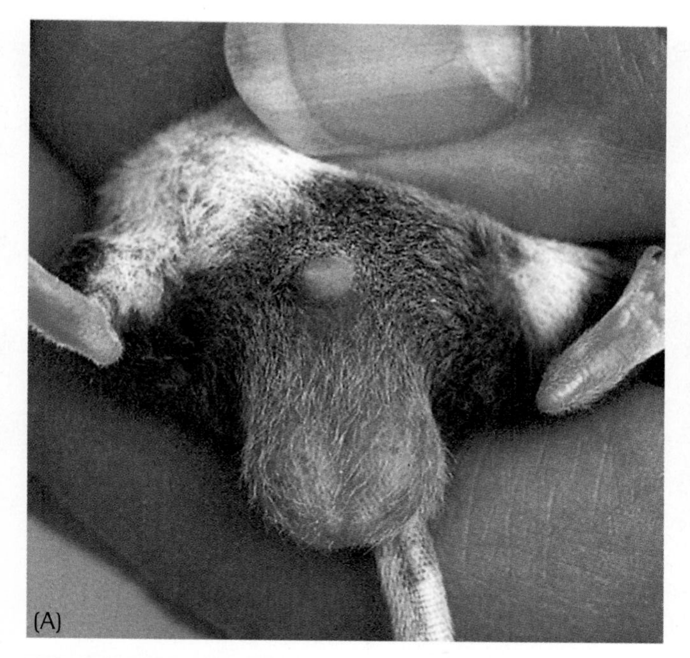

Photos courtesy of Dean Warren

FIGURE 11-7 (A) anogenital area of a male mouse; (B) anogenital area of a female mouse.

3 weeks of age. Baby mice are active and learning to climb and jump. Housing should be done with caution so that they do not escape from their cage. It is important to separate the babies at weaning and determine their gender as they will continue to breed. Sexing mice and rats is done through viewing their **anogenital** area (see Figure 11-7 and Figure 11-8). This is the area located around the far stomach between the rear legs and the base of the tail. The distance is measured from the anus

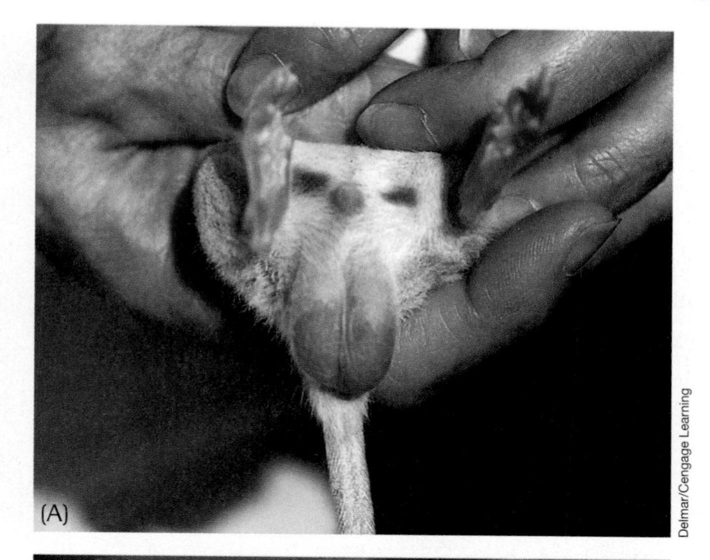

Delmar/Cengage Learning

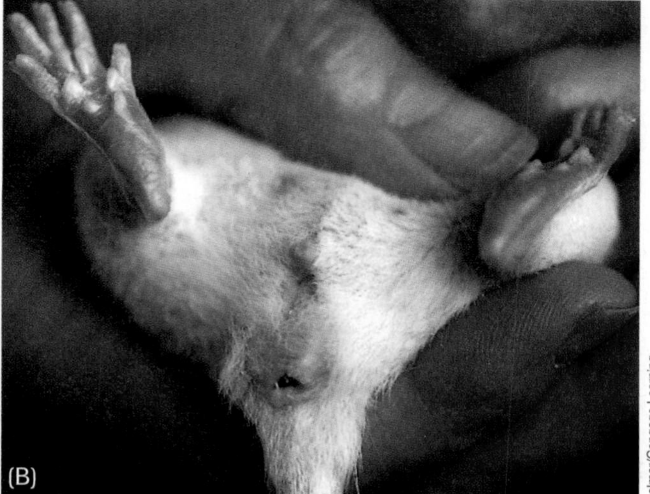

Delmar/Cengage Learning

FIGURE 11-8 (A) anogential area of a male rat; (B) anogenital area of a female rat.

to the genital area. The anogenital distance in a male is much greater than that of a female. In a female the distance is measured from the **urethra** (area where females urinate) to the anus; in the male the distance is measured from the **penile area** (area where males urinate) to the anus. A male will also have a **scrotum**, which holds the **testicles** where sperm is produced. Determining gender can be done at several days of age.

Common Diseases

Rats and mice are relatively healthy animals. They rarely get sick or acquire diseases. They will occasionally develop respiratory infections with such signs as nasal discharge, sneezing, coughing, loss of appetite, and weight loss. They are also prone to tumors that may be malignant or cancer causing.

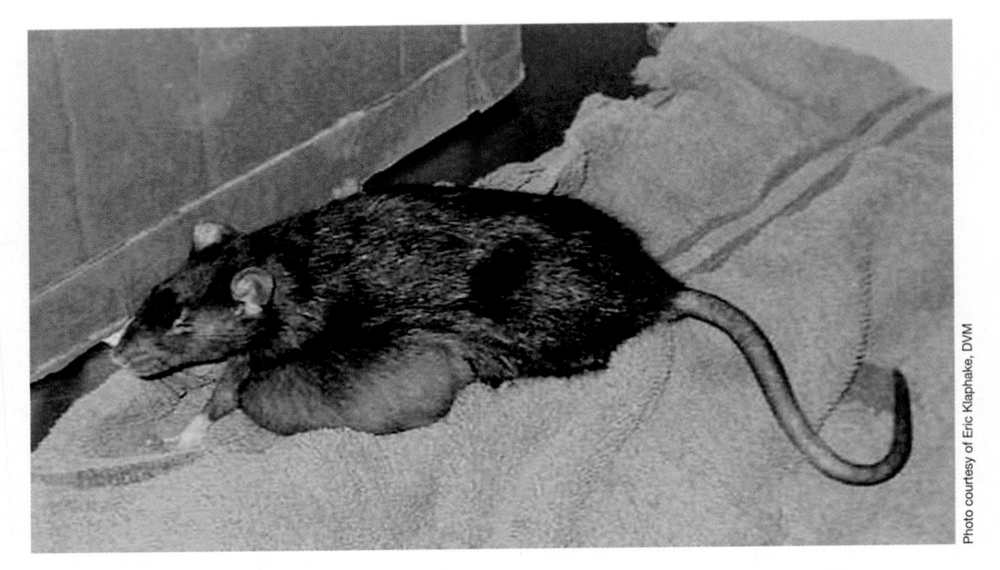
Photo courtesy of Eric Klaphake, DVM

FIGURE 11-9 Tumors can be a problem in mice and rats. They can become large enough to restrict movement. This is an example of a vascular tumor in a rat.

Common Parasites

Mice and rats may commonly develop external parasites such as fleas, fur mites, and lice. The signs of external parasites include hair loss, itching, and occasional scratches on the skin from the claws. Rats and mice may also be affected by internal parasites, such as tapeworms and protozoans, similar to those parasites that affect dogs and cats. Fecal analysis and treatment with deworming medications is necessary. It is important that products used to treat internal and external parasites are labeled for use in mice and rats.

Common Surgical Procedures

Surgical procedures of rats and mice are uncommon, mostly due to the short lifespan of the rodents. Rats and mice tend to develop tumors that may be surgically removed. Many tumors are **benign**, or noncancerous, but occasionally the tumors may be **malignant**, which means they are cancerous tumors. Many tumors grow so large that they cause difficulty moving. Most rats and mice are euthanized when tumors develop (see Figure 11-9).

Hamsters

Hamsters are very active and playful companions (see Figure 11-10). They are popular pets and are also a commonly used species in laboratory research. Hamsters are native to warm, humid desert climates and temperatures. The species was developed in areas of Asia and Europe.

Delmar/Cengage Learning

FIGURE 11-10 Hamsters are very active and playful.

Biology

Hamsters have small round bodies with lots of skin, short legs, and a short tail. Like mice and rats, they are nocturnal and active at night. Common hamster external body terms include the following:

- Cheek pouch—open areas located inside the cheeks used to store food
- Ears—small flaps on either side of the head that are hairless
- Incisors—long teeth located on upper and lower jaw
- Whiskers—long hair located on the face next to the nose

Hamsters' average lifespan is 18 to 24 months. They are best kept alone and are naturally solitary animals.

Breeds

Hamsters come in a variety of coat types and colors. The Syrian hamster, also called the Golden hamster or teddy bear hamster, is one of the most popular breeds of pet hamsters (see Figure 11-11). They may reach sizes as large as 5 inches in body length. The Chinese hamster is smaller in size and most often used as a research species. Several variations and mixtures of the breeds have been developed for different color characteristics and size variations. Dwarf hamsters are small in size and may be an inch in size or smaller (see Figure 11-12).

Delmar/Cengage Learning

FIGURE 11-12 Note the small size of the Chinese hamster.

Breed Selection

When selecting a pet hamster, it's important to observe the animal to ensure that it's alert, active, and healthy. Some hamsters are susceptible to disease if they spend their first few weeks of life at a pet store. A hamster that is not active may be ill. Finding a healthy hamster with a winning personality will lead to years of enjoyment and companionship.

It is also important to select a breed or variation that will fit your cage space. Dwarf hamsters and some mixtures are more susceptible to disease. Selecting a younger hamster will allow it to better bond with a new owner. It is best to select a hamster that enjoys being picked up and handled. Golden hamsters generally should be housed alone.

© iStock/Syagci

FIGURE 11-11 A teddy bear hamster.

Nutrition

Hamsters eat about ½ ounce of food a day. They should be fed in the evenings when they are most active. Most commercial feeds include a variety of seeds, grains, and cereals with common ingredients being corn, oats, and wheat. They enjoy treats and will often store food and treats into their **cheek pouches**, as a storage area to then **burrow** or tunnel the food items for later use (see Figure 11-13). It is common for hamsters to burrow food in an area of their cage.

Behavior

Hamsters should be housed in areas where the temperature is between 60 and 80 degrees Fahrenheit. This is due to the hamsters' instinct to **hibernate** if the temperature goes below 45 degrees. Hibernation is a natural condition where the animal goes into a long sleep and the body systems will slow down until temperatures rise. If this occurs, it is best to slowly warm the hamster and when awake offer it warm milk until it revives.

Basic Training

Training a hamster is similar to training rats and mice. They can be trained with consistency, proper handling, and using food as treats. Training a young hamster between 4 and 7 weeks of age will be easier than an older animal. Hamsters enjoy exercising on wheels and in balls for additional activities.

FIGURE 11-13 (A) Hamsters store food in their cheek pouches to burrow it later. (B) Hamsters check pouches are large and may expand to several inches as items are stored.

Equipment and Housing Needs

Hamsters can be housed in aquariums or rodent cages. The housing method should be monitored as hamsters have the ability to squeeze through very small openings. They should not be housed in wooden or plastic cages as they love to chew and can easily escape. They are great climbers as well. They enjoy toys such as exercise wheels and tunnels, just like rats and mice. Hamsters should have their cages cleaned routinely and bedding that is soft and dust free (see Figure 11-14).

Restraint and Handling

Hamsters are easy to handle if they are accustomed to people. They are an animal species that must be handled frequently to stay tame. It is best to begin by petting a hamster's back to allow it to adjust to touching. Never attempt to pet or pick up a sleeping hamster; they *will* bite out of a defensive reaction. A **defensive reaction** is when an animal protects itself from danger, and an example of protection is biting or scratching. They can

FIGURE 11-14 Hamsters can be housed in an aquarium that must be kept clean with clean bedding materials.

be picked up by placing a hand underneath them for support. Many hamsters enjoy climbing in shirt or coat pockets. Older hamsters are considered by some to be difficult to handle. However, if simple precautions are taken, hamsters can be routinely handled with minimal stress to the animal and handler. Nonetheless, care must be taken to avoid being bitten. Factors that affect hamster restraint are the large amount of loose skin, which must be gathered to control the animal, and the tendency to bite when startled, such as when touched during sleep. Hamsters can be removed from their cage using a small can or cup. To do so, place a can or cup into the animal's cage and gently encourage the hamster to enter. After entering, the opening can be covered and the hamster removed from its cage. Alternatively, hamsters may be picked up by the loose skin over their back. With the hamster on a flat surface, place the palm of the hand over the hamster and press gently downward to control the animal while gathering the large amount of loose skin over its neck and back (see Figure 11-15). When an adequate amount of skin is secured, the hamster will be restrained and unable to bite.

Grooming

Hamsters are easy pets to maintain, requiring little in the way of grooming aside from the occasional fur brushing and checking of teeth and nails. Like rats and mice, hamsters groom themselves to keep clean. Animals that are not grooming themselves, appear unkempt, or appear to have areas of wetness over the coat may be ill.

Basic Health Care and Maintenance

Hamsters can be given regular exams like dogs and cats (see Figure 11-16). They should be monitored for any sign of illness, such as breathing problems, skin

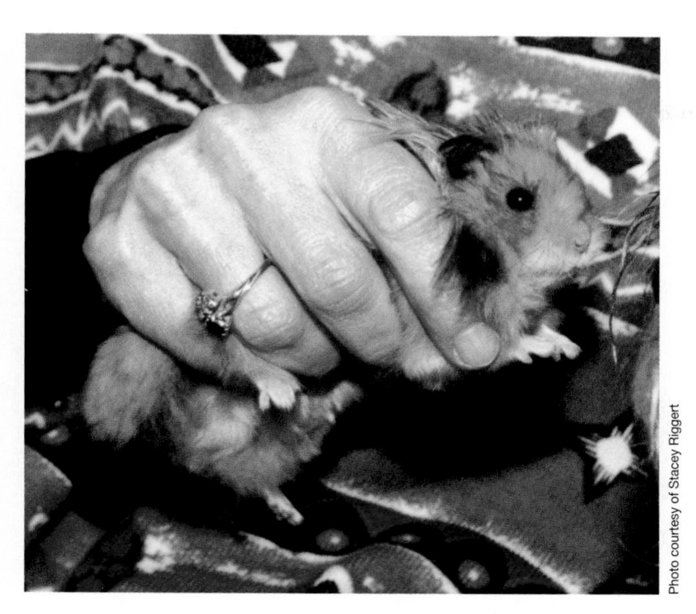

Photo courtesy of Stacey Riggert

FIGURE 11-15 A hamster should be picked up firmly between the thumb and forefinger.

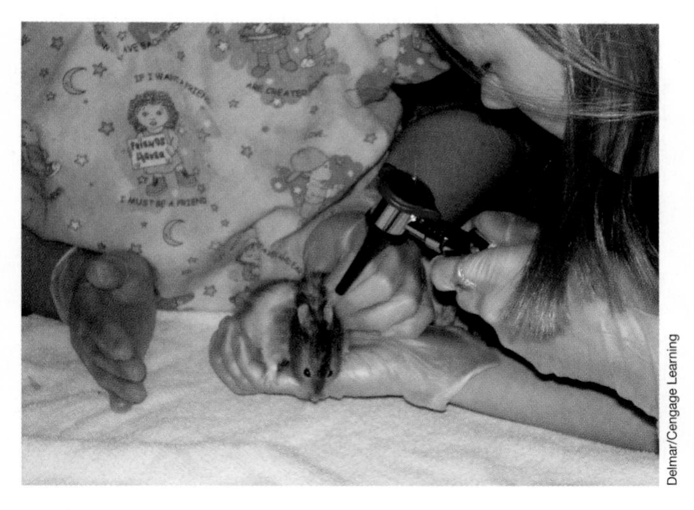

Delmar/Cengage Learning

FIGURE 11-16 Physical examination of a hamster.

infections, or decreased activity. They may require regular nail trimming and teeth trimming, as both continuously grow and if not being worn down can become overgrown. The teeth should be monitored to make sure the hamster can eat appropriately.

Vaccinations

Hamsters do not require vaccinations and no vaccines are currently labeled for hamsters.

Reproduction and Breeding

Hamsters reach maturity around 2 months of age. The males and females can be housed together for breeding purposes by placing the female in the male's cage. If the male is placed in the female's cage it is likely she will kill him due to feeling threatened and being territorial. If the female is not in estrus, they may fight. It is recommended to wear gloves to protect the hands from being bitten. The female should be placed with the male at night when breeding typically takes place. Otherwise, they should not be housed together. The estrus cycle lasts 4 days. Female hamsters should not be bred before 3 months of age. Many females do not produce **offspring**, or young, after a year of age. The gestation length is 16 days. A typical litter size is 6–8 babies. The babies should not be touched for about 7–10 days after birth. The cage should not be cleaned and all contact with the female and young should be avoided. Female hamsters are very protective and feel threatened, and if disturbed they may kill their young. This is known as **cannibalism**. Hamsters can be sexed in the same manner as rats and mice (see Figure 11-17). The young should be separated and weaned within 2–3 weeks.

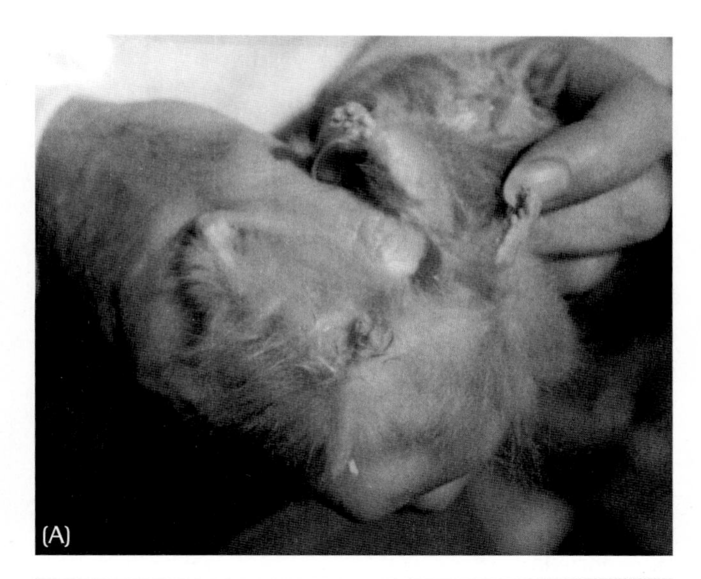

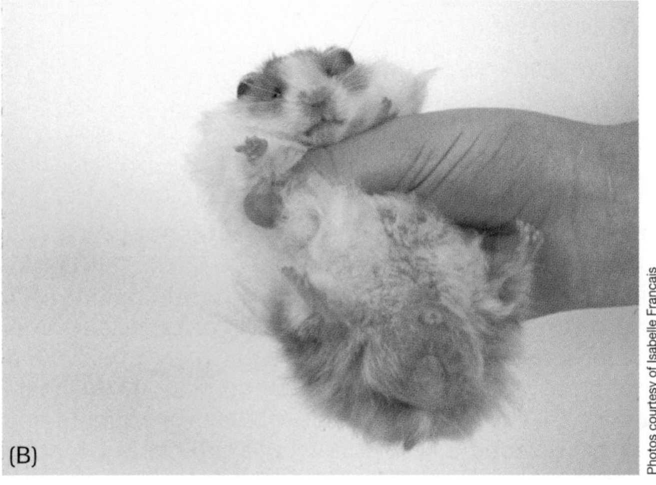

Photos courtesy of Isabelle Francais

FIGURE 11-17 (A) the genital area of a female hamster; (B) the genital area of a male hamster.

Common Diseases

Hamsters are prone to several diseases and conditions. One common condition is called "**wet tail**." This is caused by a bacterial disease that spreads rapidly through direct contact or bacterial spores. Causes include overcrowded cages, poor nutrition, poor sanitation methods, and stress. The hamsters can be treated with **antibiotics** or medications and fluid therapy. They should be isolated from other hamsters and cages thoroughly sanitized. They are most susceptible to this condition between the ages of 3 and 8 weeks old. Signs of wet tail include the following:

- Watery diarrhea
- Dehydration
- Weight loss
- Eye discharge
- Nasal discharge
- Lethargy
- Anorexia
- Wet tail appearance
- Irritable attitude

Another common condition is respiratory infections and diseases caused by bacteria. This may come from poor sanitation or excessive dust from bedding. If diagnosed in time they may respond to antibiotics. The cage conditions must be improved. Signs of respiratory disease in hamsters include the following:

- Nasal discharge
- Eye discharge
- Sneezing
- Labored or difficult breathing
- Anorexia
- Depression
- Weight loss
- Dehydration

Common Parasites

Hamsters may also become infested with mange mites, similar to those in dogs (see Figure 11-18). This is usually passed in poor bedding or other animals within the household. Signs include hair loss, especially large clumps, and scratching. They may also get other external parasites such as fleas, lice, and ticks. They should only be treated with topicals labeled for rodents and never with dog or cat products. A **topical** is a type of medicine or chemical applied to the outside on the skin or hair coat.

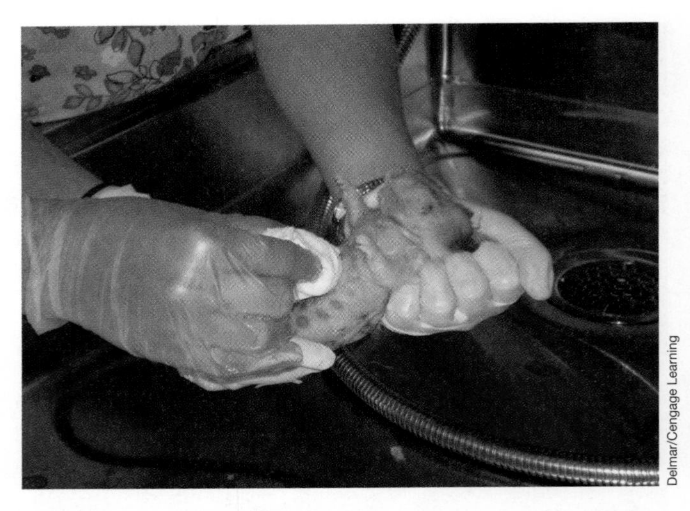

Delmar/Cengage Learning

FIGURE 11-18 Hamsters can become infected with parasites.

Guinea Pigs

Guinea pigs are excellent pets with lots of personality. They are known for greeting people with whistles. Guinea pigs are excellent starter pets for older children. They have entertaining personalities and are docile animals. Larger than hamsters, but smaller than rabbits, guinea pigs can weigh a couple of pounds and generally live for 5 to 7 years.

Veterinary Terminology

cavy = guinea pig

boar = intact male guinea pig

sow = intact female guinea pig

kit = young guinea pig

Biology

The guinea pig is a rodent with a short, heavy body, short legs, and no tail. Common guinea pig external body terms include the following:

- Cheek pouches—area inside mouth for storing food
- Fore feet—4 claws
- Hind feet—3 claws

Guinea pigs live an average lifespan of 5–7 years but have been known to live beyond 10 years. Guinea pigs are social animals that prefer to live in small groups. Two or more females will usually get along well in the same living quarters. If housing two males together, it's smart to choose two babies from the same litter. Since guinea pigs, like all rodents, multiply rapidly, keeping males and females together is not recommended.

Breeds

There are several breeds of guinea pigs that come in a variety of colors and hair coats: some short and smooth, some long and amusing, and there is also a hairless breed. The three most common breeds of guinea pig are the Smooth or American, with short, glossy fur; the Abyssinian, whose hair grows in fluffy tufts all over the body; and the Peruvian, with long, silky hair that flows to the ground (see Figure 11-19).

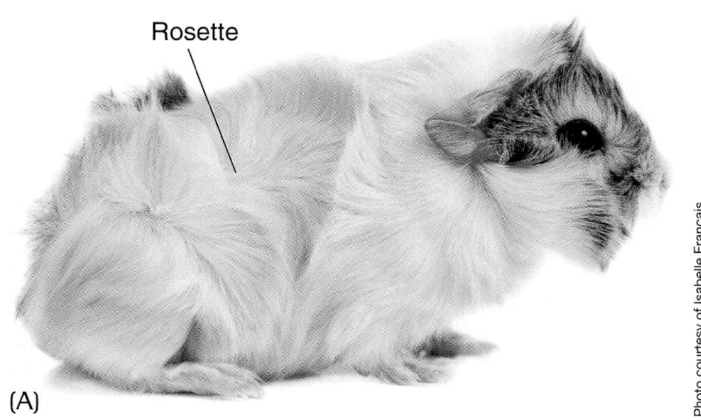

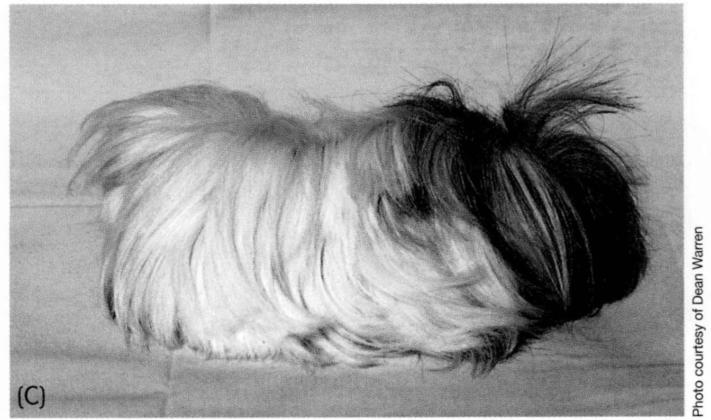

FIGURE 11-19 Examples of guinea pigs: (A) Abyssinian; (B) American short-hair; and (C) Peruvian.

Breed Selection

Guinea pig selection should be based on coat type and grooming needs. Some longer-haired breeds will need more maintenance and brushing (see Figure 11-20). Otherwise, they are relatively docile pets and should be handled to determine their behavior.

Nutrition

Guinea pig nutrition should include a high-quality commercial food that is typically a mix of pellets that are formulated for the needs of the species. They require a special need for **vitamin C** in their diets. They can't make their own vitamin C naturally as other animals so it is essential that they receive foods high in vitamin C. These foods include fruits, vegetables, and greens, such as alfalfa hay, apples, carrots, lettuce, celery, and spinach. Food should be available ad lib. Water should also be available, but many guinea pigs drink less water than other rodents due to receiving large amounts in their food sources.

FIGURE 11-20 Long-haired guinea pigs will need to be brushed to maintain their hair coat.

Behavior

Guinea pigs are very social and can be housed with other pigs. They are not known to fight, jump, or climb like other rodent species. Guinea pigs are usually mild and timid animals but will bite if excited, scared, or in pain. Guinea pigs love to hide when they play, so be sure to place cardboard tubes and/or empty coffee cans with smoothed edges in the enclosure for this purpose. Plastic pipes and rocks for climbing will be much appreciated. All guinea pigs need an area for sleeping and resting. When happy and content, guinea pigs will make a whistle noise.

Basic Training

Like mice, rats, and hamsters, guinea pigs can be trained. Time and correct handling are important.

Equipment and Housing Needs

Guinea pigs can be housed in wire cages or aquariums and should have at least one square foot of floor space per pig. They should have dust-free bedding and cages should be cleaned on a regular basis (see Figure 11-21). Room temperatures should be kept between 70 and 80 degrees. The young will stop growing if the temperature goes below 65 degrees or becomes excessively hot.

Restraint and Handling

Most guinea pigs are easy to handle and rarely bite, but they can scratch with their nails. Guinea pigs should always be restrained using both hands (see Figure 11-22). One hand should gently secure the animal around the torso while the second hand supports the hindquarters (see Figure 11-23). This is particularly important with large or pregnant animals to avoid internal injuries. If animals become scared or fractious, subdued lighting and covering the eyes may have a calming effect.

Delmar/Cengage Learning

FIGURE 11-21 Guinea pig enclosures must be cleaned on a regular basis.

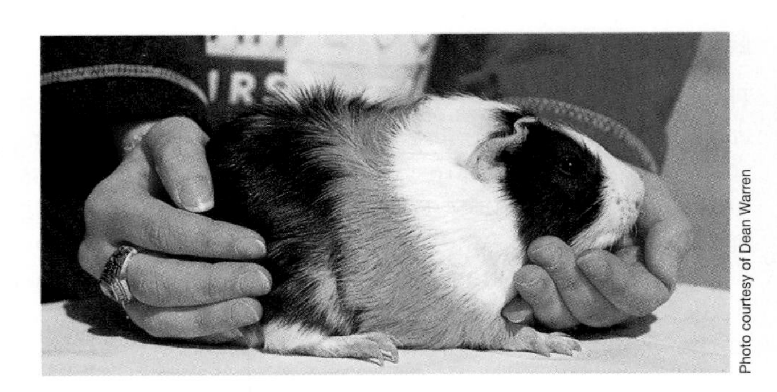

Photo courtesy of Dean Warren

FIGURE 11-22 Use both hands to secure the guinea pig.

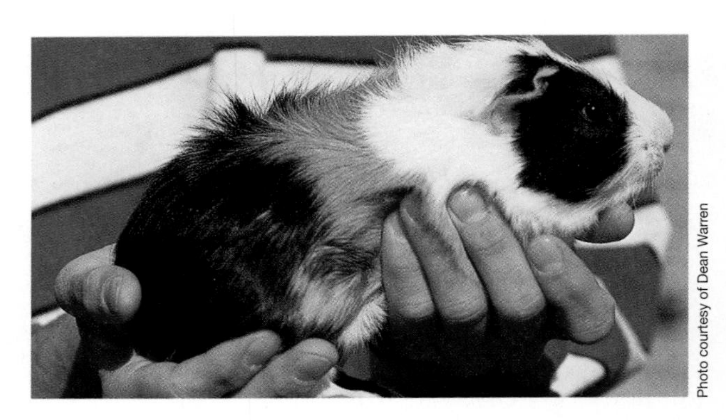

Photo courtesy of Dean Warren

FIGURE 11-23 When picking up a guinea pig, one hand should support the torso and one hand should support the hind end.

Grooming

Grooming practices in rodents should include trimming the teeth if they become overgrown, nail trims, and bathing as necessary. Guinea pig's teeth, like other rodents, grow continuously. Providing hard items for them to chew on will keep their teeth healthy. Routine evaluation of the teeth is important to make sure the teeth are not becoming overgrown (see Figure 11-24). Rodents can be bathed using a safe waterless shampoo. Some breeds will require regular brushing, especially if a long-haired breed.

Basic Health Care and Maintenance

Guinea pigs should be examined on a regular basis (see Figure 11-25). Common signs of illness in guinea pigs include sneezing, coughing, diarrhea, and lethargy. Guinea pigs are also susceptible to external parasites such as mites and lice.

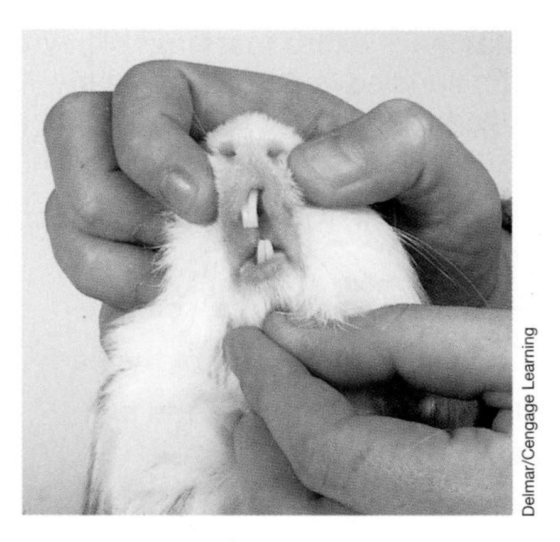

FIGURE 11-24 A guinea pig's teeth should be checked regularly to make sure they are aligned and wearing down properly.

FIGURE 11-25 Routine exams are encouraged for guinea pigs.

Vaccinations

Guinea pigs do not require vaccinations and no vaccines are currently labeled for guinea pigs.

Reproduction and Breeding

Guinea pigs reach maturity between 3 and 4 months of age. They should not be bred before this age. The male should be placed in the female's cage for the same reasons as in breeding hamsters. They may be left together for 3 weeks during which the estrus cycle will occur. Gestation occurs between 63 and 72 days. Sows will cycle and breed again within a few hours of the labor process. If they are not bred during this time then the female should not be bred until the litter is weaned. Babies are born with hair, eyes open, and the ability to eat solid food within the first day of birth. The sow will

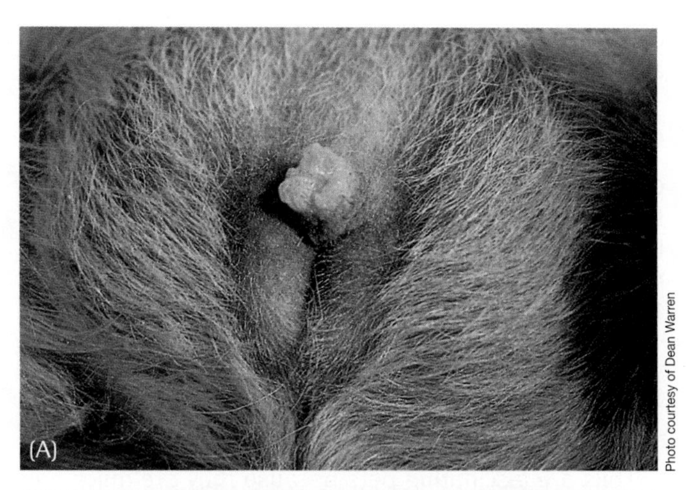

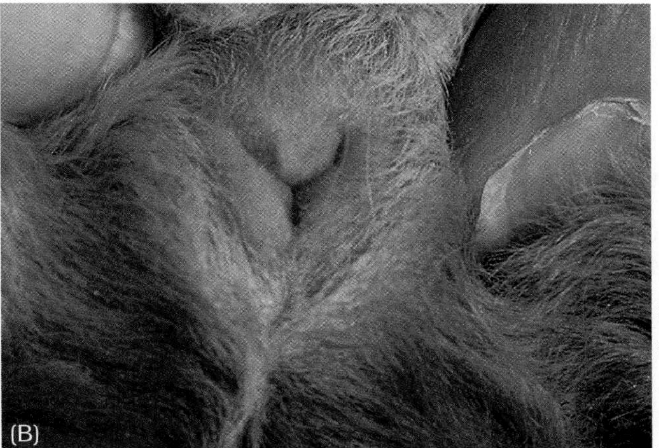

FIGURE 11-26 (A) genital area of the male guinea pig; (B) genital area of the female guinea pig.

nurse the young. Weaning should occur around three weeks of age and males and females separated to prevent breeding. Sexing guinea pigs is done by examining the genital area (see Figure 11-26). The female guinea pig has a Y-shaped anogenital region. The **anogenital** region is the area of the anus where bowel movements are produced and the genital area is where urine is produced. The male has a straight slit and gentle pressure will expose the penis. Male guinea pigs will develop larger testicles than the other rodents.

Common Diseases

Guinea pigs tend to be healthy pets. Some signs that a pig is not feeling well include sitting very still, sitting hunched up, a ruffled or messy coat, anorexia, weight loss, and watery droppings. If any of these symptoms appear, the pig should be seen by a veterinarian. Guinea pigs are commonly known for developing respiratory problems, usually from dusty bedding. Signs of respiratory problems include wheezing, sneezing, eye and nasal discharge, and a decreased appetite.

Common Parasites

Occasionally, guinea pigs may become infected with internal and external parasites, such as fleas, mites, or lice. They should be treated for external parasites with medications labeled for use in guinea pigs. Internal parasites are not a common problem in guinea pigs.

Gerbils

Gerbils are fascinating pets because they are quick and curious animals (see Figure 11-27). Gerbils are clean and easy to maintain and handle. Most gerbils are **agouti** in color, which is a mixture of two or more colors. Other popular colors include white and black.

Veterinary Terminology

Gerbils are known as **jirds** or burrowing rodents with short bodies and a hunched appearance.

- sire = male gerbil
- dam = female gerbil
- kit = young gerbil

Biology

Gerbils are different than other rodents since they are **diurnal**, or they sleep during the night and are active during the day. This makes them great pets. The gerbil is about 4 inches in length and resembles a hamster but has a long tail that is typically as long as the body. Adults weigh about 3 ounces. They are usually a bark or light-brown color and may have white markings. They can be shy and should be handled gently and regularly. Many times scratching a gerbil behind the ears will help it relax. Common gerbil external terms include the following:

- Eyes—large, black, round areas on either side of the face
- Snout—nose and mouth area on face
- Tail—very long extension off of the back
- Tuft—dark hair at the end of the tail

FIGURE 11-27 Gerbil.

Breeds

Mongolian gerbils are native to areas of Mongolia and China. They live in a variety of arid terrains, including deserts, low plains, grasslands, and mountain valleys.

Breed Selection

Gerbils should be housed in pairs and selection is similar to that of other rodents. Gerbils are social animals and should usually be purchased in bonded, same-sex pairs. While personalities vary, and not every gerbil will run gladly into your hand right away, an emotionally healthy gerbil will be curious and display a friendly demeanor. They should be easy and willing to be handled. They should be free of any signs of illness.

Nutrition

Gerbils may be fed a commercial food that contains a variety of seeds, grains, corn, sunflower seeds, and oats. They also enjoy fruit seeds, apples, lettuce, and fresh grass. Gerbils will usually eat about a tablespoon of food once a day. Water should always be available.

Behavior

Gerbils will use one area of their cage to eliminate. It is important to clean their cages on a regular basis. Gerbils should be housed in temperatures between 65 and 80 degrees. They do not do well in cool temperatures. Gerbils are very social animals, and it is not a good idea to keep them singly. Pair-bonded or family units of gerbils are usually quite affectionate with each other. They will play, chasing each other around, wrestling and boxing. They will also groom one another, sleep in piles, and cuddle together.

Basic Training

Gerbils are intelligent and may be easily trained. The more they are handled and trained with consistency, the easier the animal will be to handle.

Equipment and Housing Needs

Gerbils are best housed in a 10-gallon aquarium. A glass fish tank is better than a wire hamster cage because gerbils like to dig in deep substrate and will kick the litter out of a wire enclosure. With a pair of gerbils, it can be helpful to put a wire cage topper above the aquarium to give them more room to run around. They are curious and will attempt to escape if given the opportunity. They do enjoy climbing, jumping, and chewing. A block of wood can help keep their teeth healthy. Gerbils like to dig and build tunnels, so put at least 3 inches of substrate at the bottom of their habitat. A common bedding is cedar chips or shavings that are dust free.

Restraint and Handling

Gerbils can be restrained by grasping the base of the tail with one hand and the skin on the back of the neck with the opposite hand. Extreme care must be taken not to pull on the tail because gerbils are especially prone to tail injuries from improper handling. Gerbils should never be restrained by the tail (see Figure 11-28). If additional restraint is desired, gerbils may be grasped from above, securing the animal's head between the index and middle fingers and the body with the thumb and remaining fingers.

Grooming

Gerbils do not need large amounts of grooming care. They will groom each other and keep each other clean.

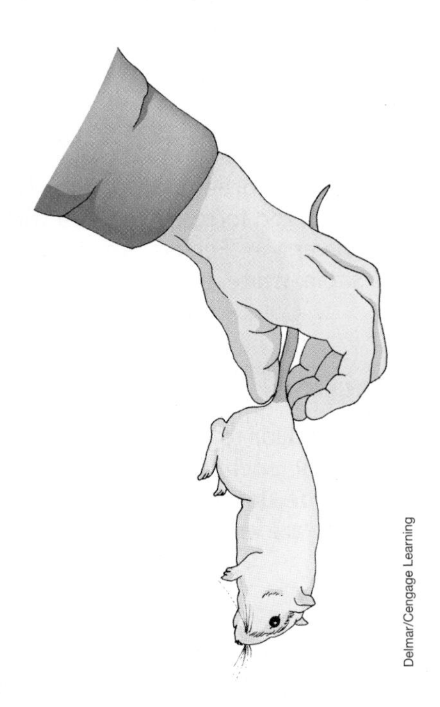

FIGURE 11-28 Never restrain or pick up a gerbil by the tail.

The nails and teeth should be monitored for overgrowth. Signs of a sick gerbil include unkempt hair coats and decreased amounts of self-grooming.

Basic Health Care and Maintenance

Gerbils should be examined regularly as with any other pets. They come from a dry natural habitat; therefore, they are designed to conserve water. As a result they produce scant amounts of urine and dry droppings, making it fairly easy to keep their cage fresh and clean. They should be provided toys and items to chew to make sure their teeth do not become overgrown.

Vaccinations

Gerbils do not require vaccinations and no vaccines are currently labeled for gerbils.

Reproduction and Breeding

Gerbils mature around 3 months of age. They will select one mate for life. They can be housed together but should allow time to grow acquainted. The gestation length is 24–25 days. Most females have a litter of 5 babies. Both the male and female can be housed with the babies, which are typically weaned around 6 weeks of age. People should not handle the young until their eyes are open. Sexing is done as in all other rodents with the male anogenital distance being greater than that of the female gerbil (see Figure 11-29).

Common Diseases

Gerbils are relatively disease-free. They should be monitored for any signs of respiratory diseases with signs that include sneezing, nasal discharge, increased breathing, and wheezing.

Common Parasites

Common external parasites of gerbils include fleas and mites. Internal parasites are rarely seen, but occasionally tapeworms may develop. Gerbils should be medicated with label-approved medications.

Ferrets

Ferrets are a member of the weasel family and are considered rodents. Domestic ferrets are small, furry mammals whose average size ranges from 1 to 5 pounds at maturity (see Figure 11-30). Ferrets are not wild animals. They are very curious and entertaining pets. Ferrets originated from a larger species of weasel known as the polecat. They developed in Europe and were originally used for hunting rabbits.

Veterinary Terminology

Hob = intact male ferret

Jill = intact female ferret

Gib = neutered male ferret

Sprite = spayed female ferret

Kit = young ferret

Kindling = giving birth to ferrets

(A)

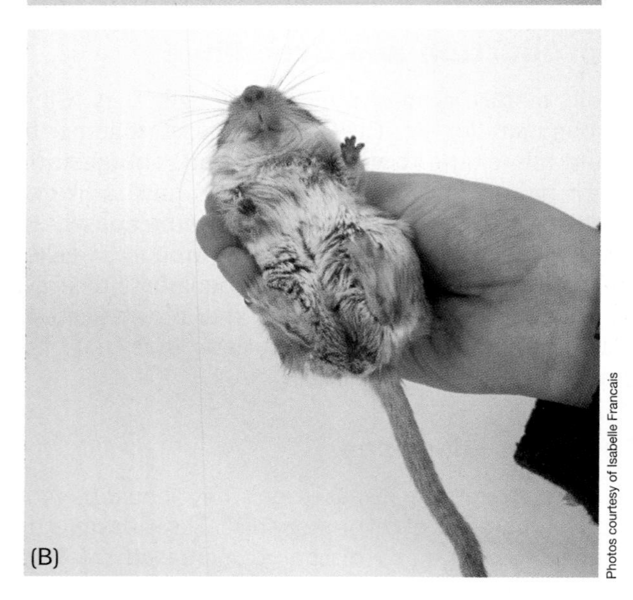

(B)

Photos courtesy of Isabelle Francais

FIGURE 11-29 (A) the genital area of the male gerbil; (B) the genital area of the female gerbil.

Photo courtesy of Isabelle Francais

FIGURE 11-30 Sable ferret.

Biology

Ferrets are interesting animals that have a long, slim body with long tails. They come in several colors, with sable being the most common. Common ferret external body terms include the following:

- Body—long and thin central area
- Mask—black coloring around eyes on the face
- Tail—bushy area extending off the back

Ferrets have scent glands that give them a strong musky odor, which is much more notable in a male. They are primarily nocturnal. They have an average lifespan of 5–8 years. Ferrets combine the best features of dogs and cats with some unique features of their own. Like cats, ferrets are small and quiet. Like dogs, they are affectionate, playful, and enjoy human interaction. They are independent yet enjoy being with people. Their mischievous and playful nature, retained well into old age, makes them entertaining companions.

Breeds

There are no essential breeds of ferrets. They are related to the black-footed ferret, which is a wild cousin of the domesticated version. Ferrets are classified by their colors, such as sable, white, silver, and white-footed (see Figure 11-31).

Breed Selection

Ferret selection is similar to that of other rodents. A ferret should be playful, social, and enjoy being handled. Some ferrets may be descented and neutered, which makes their health care and maintenance more ideal.

Nutrition

Ferrets' nutritional needs include a commercial ferret food or high-quality cat food. Ferrets are true carnivores. They will eat both dry and canned types of food.

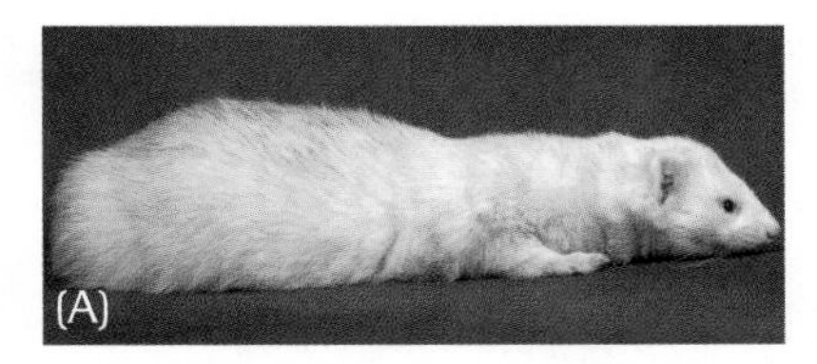

FIGURE 11-31 (A) white ferret; (B) silver mitt ferret; (C) sterling silver ferret.

They need a high-protein diet and fresh water daily. Table scraps should be limited. An appropriate commercial diet for ferrets might include the following:

- Totally Ferret (Performance Foods, Inc.)
- Ferret Forti Diet (Kaytee Products)
- Ferret Chow (Purina)
- Marshall Ferret Diet (Marshall Products)

Behavior

Ferrets are intelligent animals and are amazing with what they can do and learn. They recognize their name, respond to verbal and visual commands, and can even learn to do tricks. Ferrets can also be litter-box trained. A healthy, well-trained ferret should not bite. Like all pets, ferrets need to be taught what acceptable behavior is. Ferrets have a lower bite rate than other household pets, such as a hamster. Ferrets are very curious and love to explore. They will easily get into trouble and may chew on items that can harm them, so care must be taken to ensure their safety. Ferrets are "pack rats" and they will take anything that interests them and drag it off to their hiding place, usually under a couch or chair.

FIGURE 11-32 Ferrets can be trained to walk on a leash.

Basic Training

Ferrets can be litter trained in a similar manner as that of cats and kittens. They can learn to perform tricks with consistent training and rewards. Like dogs, ferrets can be trained to come when their name is called. They can be trained to sit, beg for treats, and ride on the owner's shoulder. They may also be walked on a leash and harness (see Figure 11-32).

Equipment and Housing Needs

Ferrets should be kept in secure cages. Wire or metal cages are best. Make sure they can't squeeze through any openings. When they are outside of a cage they should be watched very closely as they are curious animals and get into trouble easily.

Restraint and Handling

Ferrets vary greatly in temperament. Although some animals may be nonaggressive, others may be aggressive, especially when restrained. Ferrets should be initially grasped around the neck and shoulders, similar to the scruff method in cats (see Figure 11-33). The handler should hold the ferret with one hand under the shoulders with a thumb under the jaw and the other hand supporting the animal's hindquarters (see Figure 11-34). Ferrets can move quickly and with their long and thin body they can be a challenge to handle and control. Ferrets that are handled often become easier to handle. Ferrets can be allowed outdoors if they are on a leash and under close supervision. For walks, a leash can be attached to an H-type harness worn by the ferret.

FIGURE 11-33 Ferrets can be picked up by the nape of the neck.

FIGURE 11-34 To hold a ferret, place one hand under the front of the body supporting the head and one hand under the hind end of the body.

Grooming

Ferrets have a natural musky odor, even when descented. It is not recommended to give frequent baths as the naturally oily skin will continue to produce a more musky smell. Ferrets' health care programs should also include regular nail trims. Regular brushing may be necessary, but most ferrets will groom themselves and keep their hair coat clean (see Figure 11-35).

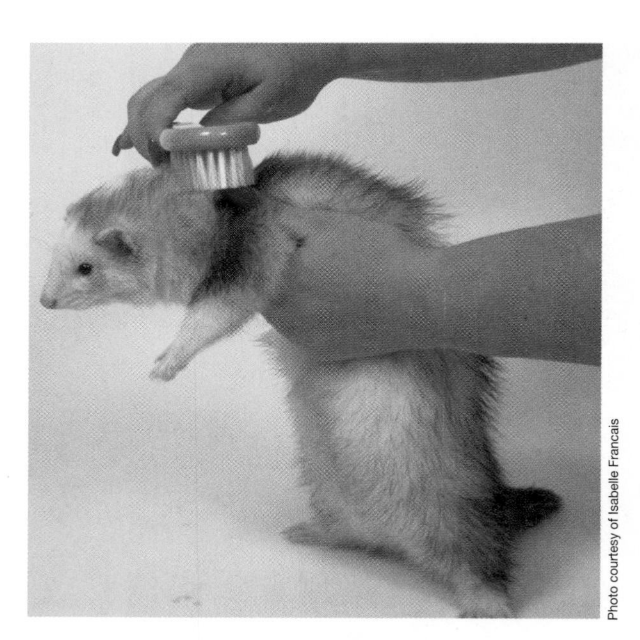

FIGURE 11-35 Ferrets can benefit from brushing.

Basic Health Care and Maintenance

Ferrets should have an annual visit to a ferret-knowledgeable veterinarian, which helps to identify potential problems early. This yearly visit should include a careful physical exam, inspection of the ears for mites, and inspection of the teeth (see Figure 11-36). Dental cleanings should be performed as necessary. A test for Aleutian Disease Virus should be done at least once a year to ensure that the ferret is not carrying this highly contagious and potentially fatal disease.

Vaccinations

Ferrets are susceptible to canine distemper and are typically vaccinated to prevent the disease. The distemper vaccine (Fervac-D) is generally given between 6–8 weeks old and a booster is repeated monthly until 14 weeks of age. The vaccine is then given yearly. Some states will also allow and require a veterinarian to vaccinate a ferret for rabies.

Reproduction and Breeding

Ferrets reach maturity around 10 months of age. The females usually are bred in the spring and gestation lasts an average of 42 days. Ferrets usually have 6–8 young in a litter. The eyes open in 3–4 weeks. The young should start weaning at this time and should be weaned by 8 weeks old. They reach their adult weight by 14 weeks. Females should be spayed if not being bred. If a female goes into the estrus cycle and is not

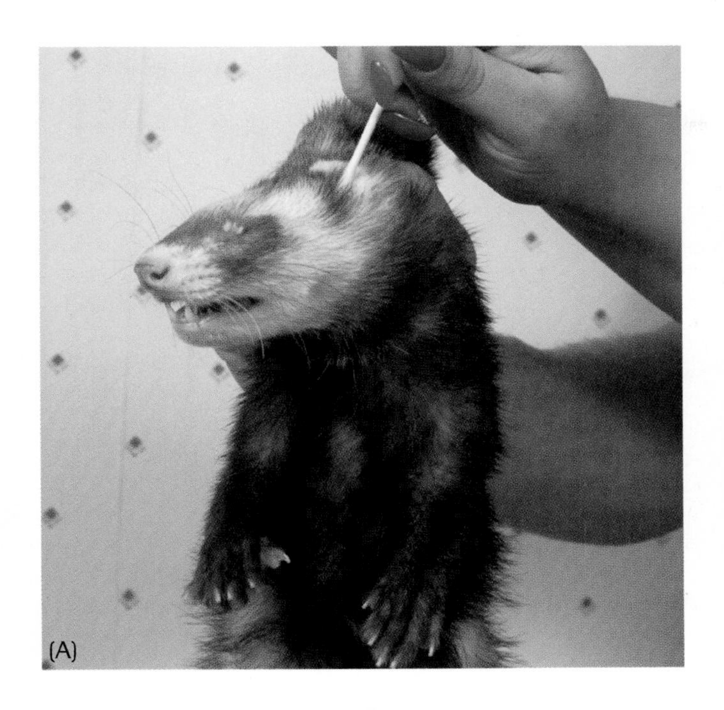

FIGURE 11-36 (A) cleaning the ears of a ferret; (B) trimming a ferret's nails.

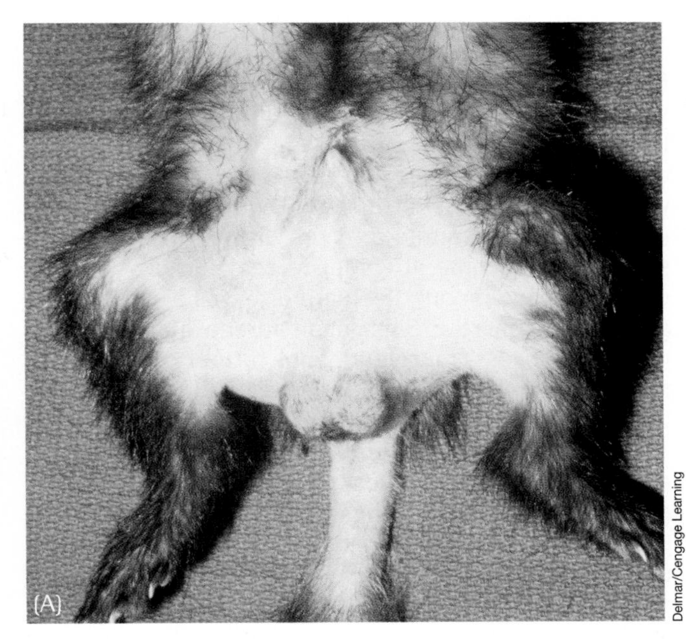

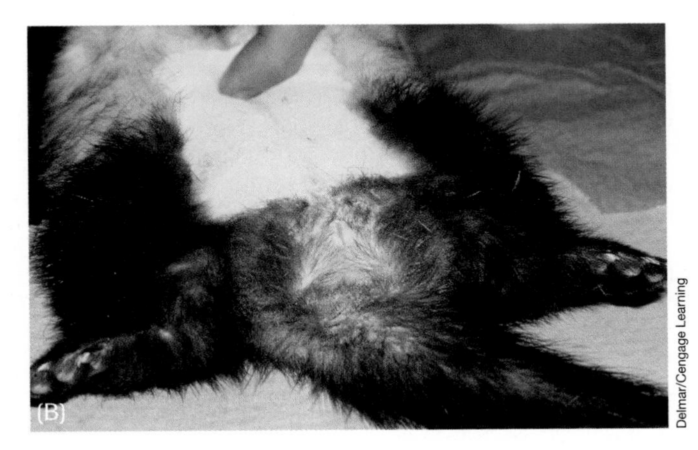

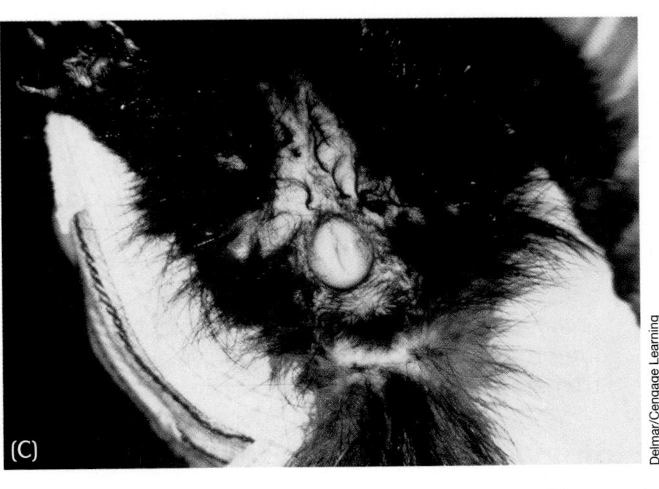

FIGURE 11-37 (A) genital area of the male ferret; (B) genital area of the nonestrus female ferret; (C) genital area of an estrous female ferret.

bred, she will continue to cycle and bleeding may cause death. Spaying or castrating a ferret will decrease the musky body odor to some extent. Ferrets have scent glands located at either side of the rectum, similar to dogs and cats. These glands create a musky odor and are usually removed during neutering. The ferret will still have some odor. Ferrets should be neutered around 6 months of age. Sex determination of the ferret can be done as in other rodent species by examination of the anogenital area (see Figure 11-37).

Common Diseases

Ferrets are relatively healthy animals, but ferrets are susceptible to canine distemper and Aleutian disease. In addition, they can come down with flulike symptoms or respiratory illnesses, similar to the "common cold," which can be transmitted by human companions. It is important to handle a ferret with extreme caution should you become ill. Older ferrets can develop diseases. Most commonly seen are diseases of the adrenal glands and pancreas. Signs of an adrenal gland disorder include hair loss, muscle atrophy, urinary blockage in males, and enlarged vulva in females. Signs of pancreatic disease include lethargy, nausea, and seizures.

Any digestive problem (changes in bowel routine, extreme weight gain or loss, vomiting) a ferret experiences is potentially serious. The best way to prevent these problems is to keep the ferret in an environment that is clean and free of dangerous objects. Foam packaging peanuts, rubber chew toys, erasers, rubber bands, latex, or plastic items should be kept away from ferrets.

Common Parasites

Ferrets are susceptible to external parasites common to those in cats, such as fleas, ticks, and skin and ear mites. They can be treated with medications safely labeled for use in ferrets. Ferrets are also capable of getting internal parasites, such as tapeworms. A fecal sample can be evaluated for internal parasites. Ferrets are also susceptible to heartworm disease.

Hedgehogs

Hedgehogs are interesting pets that can be unpredictable and tend to curl into a ball when handled. They have short spines that protect their body and can cause injury when handling. They are as common in Europe and Britain as skunks are in the United States.

Veterinary Terminology

- Hoglet or pup—young hedgehogs
- Herd—group of hedgehogs
- Self-anointed—the act of frothing at the mouth, arching the head back over the shoulders, depositing the frothy saliva onto the quills, an action triggered by pungent smells and new tastes

Biology

Hedgehogs are interesting animals that have **quills** or spines over their back and body that are between 1 and 1½ inches in length (see Figure 11-38). The muscles over their back cause the quills to stick out and protect them from predators. They do not release or throw out their quills. They are a "salt and pepper" color. They are diurnal and live an average of 4–7 years. When they are frightened or stressed, they will curl up into a ball with the quills serving as their protection. They will also roll into a ball to sleep. Occasionally, hedgehogs make quick movements or dart to get away from danger. Common hedgehog external terms include the following:

Photo courtesy of Isabelle Francais

FIGURE 11-38 The hedgehog is covered in spines.

- Muzzle—mouth area located on face
- Neck—located behind ears
- Nose—long area located at the point of the face
- Quills—spines located over the back and body
- Underside—belly area that has hair and no spines

Hedgehogs are about the size of a guinea pig and weigh between one and three pounds.

Breeds

There are no essential breeds of hedgehogs, but they originally developed in England, Europe, and Africa. Common colors are a mixture of brown, black, and white.

Breed Selection

Hedgehogs are naturally shy and tend to curl into a ball when handled. They require early socialization and should be handled often when they are young.

Nutrition

Hedgehogs are meat eaters and enjoy a diet of meal worms and insects with a small amount of vegetables and fruits in their diet. They will eat a commercial hedgehog diet that is high in protein. They will also eat high-quality cat foods. They are prone to developing obesity problems and care must be taken to not overfeed them.

Behavior

They are relatively **solitary** animals and do best housed alone. Hedgehogs are nocturnal. They usually start their nightly activities such as foraging as the sun goes down. In the wild, hedgehogs may hibernate depending on the weather. During hibernation the heart rate of the hedgehog will drop almost 90% to save energy. Hibernation periods may last from a few weeks to six months depending on the severity of the winter. The primary defense mechanism of the hedgehog is rolling into a tight ball and protruding its quills (see Figure 11-39). The hedgehog is able to form a ball by means of muscles located around the body. These muscles pull the skin around the sides and down over the feet and head. The hedgehog's spines can be moved individually by a complex layer of muscles beneath the skin. The hedgehog erects its spines for protection from potential predators. Also when the hedgehog feels threatened it may hiss or make clicking sounds. Many hedgehogs are shy and independent.

FIGURE 11-39 Hedgehogs will roll up in a ball as a defense mechanism when frightened.

FIGURE 11-40 Aquariums make good enclosures for hedgehogs.

Basic Training

Hedgehogs are not typically trained due to their naturally shy behaviors. Hedgehogs can be trained to use a litter box.

Equipment and Housing Needs

An aquarium is ideal housing for a hedgehog (see Figure 11-40). They require dust-free bedding on the bottom of the cage. This may be shredded newspaper, pellets, or dust-free shavings. Ideal room temperatures are between 75 and 85 degrees. Lower temperatures may make hedgehogs sluggish as they naturally are hibernating animals. Hedgehogs need places to hide, especially during the day when they sleep. They can be provided a plastic container with a hole in its side, a wide-diameter section of PVC pipe, or a cardboard tube. They do like exercise and will run in wheels similar to hamsters.

Restraint and Handling

The most important thing to remember when beginning an examination on a hedgehog is that if the hedgehog rolls up it will be nearly impossible to examine. The more you fight with the pet, the fewer your chances of getting it unrolled. However, most pets are fairly docile and can be handled with latex exam gloves with relative ease and a minimum of discomfort. An open hedgehog should be scooped up from underneath and lifted off the table. Supported on two hands off the table, they will seldom attempt to roil up. Less cooperative hedgehogs may be scruffed by grasping the skin between the ears and lifting the hedgehog off the table. A latex glove will add a greater measure of control than a bare hand. A third technique is a leg hold. One can grasp the rear legs firmly and gently lift the rear end of the hedgehog slightly off the table. Some animals will squirm and try to pull away from the hold, but many pets will simply remain still when held in this manner.

Another method of unrolling a hedgehog is to wait a few minutes for it to uncurl on the examination table and then place the animal abdomen side down. Heavy stroking of the spines toward the rump may help in this method. Sometimes none of these methods will work and as a last resort the best examination technique becomes the use of isoflurane. A large cone may be placed over the entire animal until it starts to relax, then a small cone may be placed over the face.

Grooming

Hedgehogs may become dirty and need an occasional bath. This can be done using a toothbrush and warm water. Some hedgehogs will allow light spraying of water to remove dirt and debris. They should be towel dried well in a warm location. They may need regular nail trims, as well.

Basic Health Care and Maintenance

A yearly physical exam is ideal but consideration must be taken that the hedgehog may need to be sedated for an adequate exam.

Vaccinations

Hedgehogs do not require vaccinations and no vaccines are currently labeled for hedgehogs.

Reproduction and Breeding

Breeding hedgehogs should be done after maturity around 7 weeks of age. They have a short gestation period of 35 days and typically a small litter. The male and female should be housed together only for breeding. The female is an induced ovulator and estrus lasts about 2–5 days. Males are known to kill and eat the babies if left with the female. The female will nurse the litter for about 5 weeks, after which they can be weaned (see Figure 11-41).

Common Diseases

Hedgehogs are prone to obesity as discussed in the nutritional section. They also are prone to bacterial skin infections and injuries from their spines. Dental disease may also be a problem and hedgehogs that are having difficulty eating should be examined for dental problems. They may also develop respiratory conditions due to dusty bedding materials, so care should be taken to monitor breathing patterns.

Photo courtesy of Isabelle Francais

FIGURE 11-41 Young hedgehogs can be weaned by 5 weeks of age.

Common Parasites

Hedgehogs are prone to mange mites, fleas, and ticks. Signs of mange mites include heavy dandruff, quill and/or hair loss, crusty thickened skin, thickening of ear margins, easy plugs in ear canals, and overall itchiness. Hedgehogs are not prone to internal parasites.

Chinchillas

A chinchilla is a small silky-haired rodent (see Figure 11-42). They are bred as pets and for their fine hair. Chinchillas make excellent pets but require more care and maintenance than other small rodents.

Veterinary Terminology

- Kit—baby chinchilla
- Litter—group of newborn chinchillas

Biology

Chinchillas have a round body with a squirrel-like tail. They come in a variety of colors with slate gray being

FIGURE 11-42 Chinchilla.

the most common color. Common chinchilla external body terms include the following:

- Ears—large flaps on either side of the head
- Forelimbs—short front legs used for grasping
- Hindlimbs—long back legs
- Pads—bottom surface of the feet
- Tail—large bushy area extending off of the back

When stressed, chinchillas will throw out fine clumps of their hair. They are nocturnal and love to be active while jumping and bouncing around in a very quick manner. They are unpredictable in their movements and can easily escape. They live between 9 and 17 years of age. They weigh about a pound at adult size.

Breeds

There are no essential breeds in chinchillas. There are several color variations, such as gray, white, beige, and black (see Figure 11-43). The chinchilla developed in South America.

Breed Selection

Chinchillas are social animals that may be selected similar to other rodent species. A chinchilla usually is housed alone as they may not tolerate other chinchillas. They should tolerate being handled and not bite. The more they are handled, the tamer they become. They do demand a lot of attention.

Nutrition

Chinchillas are usually fed a commercial diet that consists of pellets similar to rabbits. They enjoy snacks of fruits, leafy vegetables, and raisins. They should be fed high-quality grass or alfalfa hay.

Behavior

Chinchillas are excellent jumpers and very quick. They are social but don't tend to like to be held and cuddled.

FIGURE 11-43 Common chinchilla colors (A) white chinchilla; (B) beige chinchilla; (C) black chinchilla.

Sudden moves and loud noises will startle and frighten them. Chinchillas won't bite unless they smell food on the fingers. They will sometimes nibble on clothing and belts. They are quite curious and should have as many places to climb into and on top of as possible. When stressed they will lose a large amount of their fine hair as in a dog or cat that is shedding.

Basic Training

Chinchillas may be trained but it may take several weeks to months for chinchillas to accept a person's trust. Consistency and frequent handling are the best training methods of chinchillas.

Equipment and Housing Needs

Chinchillas are acrobatic animals and therefore require a lot of space to house. Since they like to climb and jump in both horizontal and vertical directions, a large multilevel cage is recommended. Ideally, the larger the cage, the better for the chinchilla. The cage should be constructed of a small welded wire mesh to prevent leg or foot injury. An area of solid flooring should also be available. Drop pans below the cage are ideal for facilitating cleaning. Since chinchillas are shy animals, they need a place to hide. Nondestructive boxes work well, such as wooden or metal boxes.

Environmental temperature at or above 80 degrees Fahrenheit coupled with humidity will lead to heat stroke. A recommended temperature range is 50 to 68 F. Chinchillas need to chew hard items, so it is best to provide wooden chew items to allow them to keep their teeth in ideal shape.

Restraint and Handling

Chinchillas should never be captured or handled by their tails. This may cause severe injury or a broken tail. It is best to carry and lift a chinchilla by having a firm grip on the body with one hand under the abdomen or around the scruff and one hand holding the base of the tail (see Figure 11-44). Holding the base of the tail will help prevent the chinchilla from jumping. Sometimes the use of a towel is best to allow the chinchilla to feel more secure as though it is in hiding.

Grooming

Chinchillas require a **dust bath** every day to every other day. The dust bath is a special dust substance that chinchillas role in to cleanse their delicate fur (see Figure 11-45). This hydrates and cleans their hair coat and protects it from damage. They will need about one inch of dust placed on the bottom of the cage. Commercial dust baths are available.

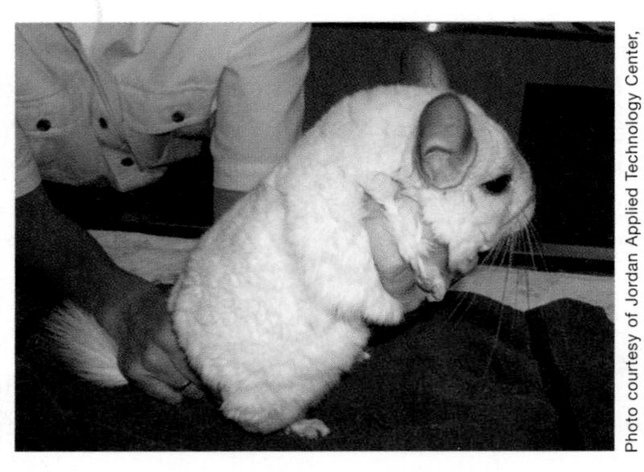

Photo courtesy of Jordan Applied Technology Center, West Jordan, Utah

FIGURE 11-44 To restrain a chinchilla, place one hand under the torso while the other hand grasps the base of the tail.

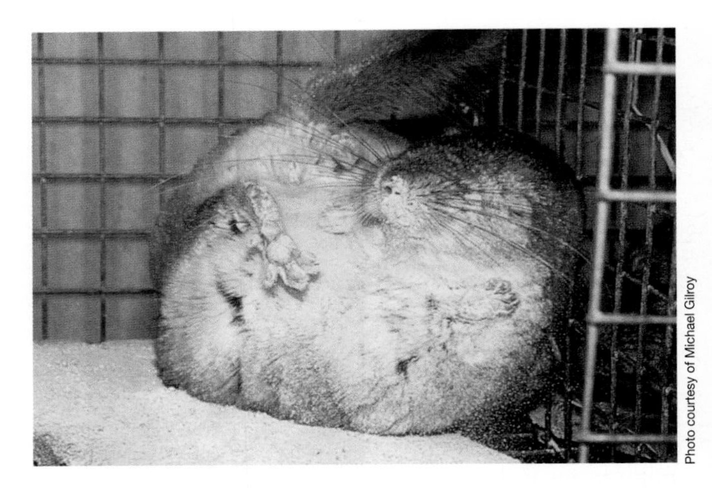

Photo courtesy of Michael Gilroy

FIGURE 11-45 A chinchilla taking a dust bath.

Basic Health Care and Maintenance

Yearly physical exams are recommended for chinchillas. They should have frequent dental exams as their teeth continuously grow and they can develop difficulty with eating. They should also be monitored for hair balls called **trichobezoars** as they do groom themselves similar to cats. Respiratory problems may develop, so signs of difficulty breathing, wheezing, and sneezing should be monitored.

Vaccinations

Chinchillas do not require vaccinations and no vaccines are currently labeled for chinchillas.

Reproduction and Breeding

Chinchillas have a spontaneous estrus cycle that ranges from 30–50 days. The gestation length is 111 days. The babies should be weaned between 3–8 weeks of age. Chinchillas are sexed in the same manner as in other rodents (see Figure 11-46).

Common Diseases

Chinchillas are healthy pocket pets. Most commonly chinchillas should be monitored for dental disease, signs of heat stroke, and respiratory problems.

Common Parasites

Chinchillas may have flea issues, but both external and internal parasites are rare in chinchillas.

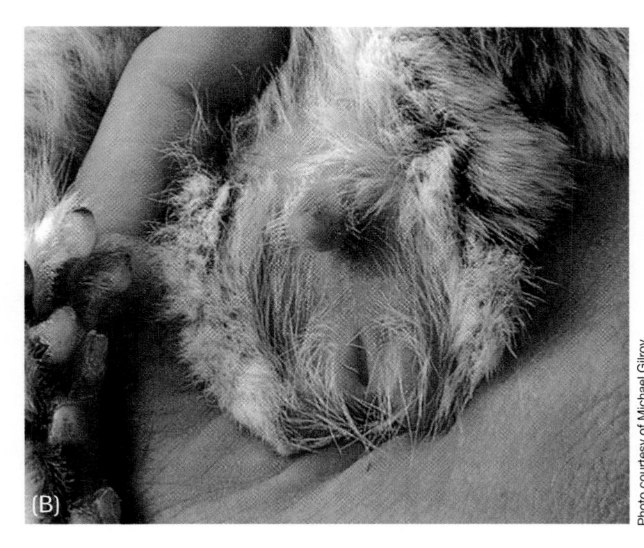

FIGURE 11-46 (A) genital are of the female chinchilla; (B) genital area of the male chinchilla.

SUMMARY

Rodents include some of the most interesting and most kept companion animals. Gerbils, rats, guinea pigs, mice, and hamsters are the most widely kept of the pocket pets. It is important for the veterinary assistant to have knowledge in the proper handling, health care, housing needs, nutrition, and reproductive care of the common rodent species that are kept as pets. This area of veterinary medicine is becoming more popular in the animal industry, as well as in specialty medicine.

Key Terms

ad lib to feed an amount of food that is available all day or when needed

agouti a mixture of two or more colors

anogenital the area located around the far stomach between the rear legs and the base of the tail

antibiotics medications used to treat infection and disease

benign noncancerous tumor

burrow to dig a tunnel into an area

cannibalism the act of rodents eating their young

cheek pouches area within the mouth where food is stored

defensive reaction behavior when an animal protects itself from danger

diurnal the term for animals that are active during the day and sleep at night

dust bath method of grooming of the chinchilla in which the animal rolls in a dusty medium to condition its skin and hair coat

hibernate the process of an animal sleeping when its body temperature decreases

hooded variety of rats that are white in the body color with a black or brown coloring that is located on the head and shoulders, giving it a hooded appearance

jird burrowing rodent

malignant cancerous tumor

murine veterinary term for mice or rats

nocturnal term for animals that sleep during the day and are active at night

offspring young newborn animals

penile area area in males where urine is passed through the outside of the body

quills spines that project and protect from the body of a hedgehog

rodents mammals that have large front teeth designed for chewing or gnawing, such as rats, mice, and hamsters

scrotum sac of skin that holds the testicles in males

(Continues)

solitary alone and independent behavior

testicles two glands where sperm is produced in males

topical substance applied to the skin

trichobezoar a hairball

urethra body area in females where urine is passed to the outside of the body

vitamin C essential nutrient and supplement needed by guinea pigs as they don't produce their own in their body

wet tail condition in hamsters caused by a bacterial disease that spreads rapidly through direct contact or bacterial spores

REVIEW QUESTIONS

1. Explain the differences in the estrus cycles of the hamster and the guinea pig.
2. Why should ferrets be spayed if they are not going to be used for breeding?
3. What is the difference between nocturnal and diurnal?
4. Why do rodents develop respiratory problems?
5. What nutritional needs do guinea pigs have and why?
6. Why are rodents kept as pets commonly called "pocket pets"?
7. What common bacterial disease in hamsters causes diarrhea?
8. Which rodent species may be given vaccines as a health program?

Clinical Situation

Courtesy of Isabelle Francais

Mr. and Mrs. Kline bring in their daughter's ferret, "Mikey," a 6-year-old NM that has been showing some signs of not eating well and being very lethargic. The Kline's are concerned that the ferret may not survive. Nicole, the veterinary assistant, is getting a weight on the animal and noting its listless appearance.

"How long has 'Mikey' not been feeling well?" she asks the Kline's.

"He has been sick for 2–3 days," they answer.

"Well, Dr. Dawson will be in shortly to examine him. I am certain he will be fine. Ferrets rarely get sick so I am sure he just has a cold." As Nicole leaves the exam room, the Kline's start discussing how they were under the impression that ferrets did have several health problems to be concerned about. They are now confused about the information they have previously been given.

- What was done incorrectly in this situation?
- How would you have handled the situation?
- What is an ideal way to handle a situation that you may not be familiar with?

12

Rabbit Identification and Production Management

Objectives

Upon completion of this chapter, the reader should be able to:

- ☐ Define common veterinary terms relating to rabbits
- ☐ Explain the biology and development of the rabbit
- ☐ Identify common breeds of rabbits
- ☐ Discuss the nutritional requirements of rabbits
- ☐ Define normal and abnormal behaviors of rabbits
- ☐ Properly and safely restrain rabbits for various procedures
- ☐ Discuss the health care and maintenance of rabbits
- ☐ Detail breeding and reproduction in rabbits
- ☐ Identify and describe common diseases affecting rabbits

Introduction

Rabbits have become very popular indoor pets and companion animals (see Figure 12-1). They are also commonly bred for meat and pelt production. Rabbits require special handling and health care needs compared to rodent species.

FIGURE 12-1 Rabbits are popular pets and companion animals.

Veterinary Terminology

lagomorph = rabbit
buck = intact male rabbit
doe = intact female rabbit
lapin = neutered male rabbit
kit = young (blind, deaf) rabbit
kindling = giving birth to rabbits
herd = group of rabbits
junior = a rabbit under 6 months of age
senior = a rabbit over 6 months of age

Biology

Rabbits are mammals that are herbivores. They have long ears, a round body, very short tails, and four large teeth. They are related to hares, which are larger in body size and have longer ears with black tips on the end. Rabbits are endothermic but have a different digestive system compared to dogs and cats. Rabbits are **non-ruminants**, which means they have one simple stomach that is made to filter and break down grasses and roughages, like horses. They are not able to vomit and lack a gall bladder. The digestive tract is hardy and must move constantly to break down heavy foods such as hay and grass. Common rabbit external body part terms are identified below (see Figure 12-2):

- Ears—long flaps that are floppy and located on top of the head
- Flank—lower thigh area of the rear legs
- Hock—joint or point of the rear leg
- Loin—the middle back area
- Neck—area behind the head
- Rump—lower back area just above the tail
- Tail—small cotton ball–like structure extending off of the back
- Toe—individual digits located on the feet

Figure 12-3 illustrates the skeletal structure of the rabbit.

Rabbits were developed for many uses and purposes. They include meat production, fur production, research, show, and as pets. Rabbits live an average lifespan of 5–6 years.

Breeds

Rabbits come in a variety of breeds, colors, and sizes. There are around 45 breeds of rabbits that are classified by their body shapes and sizes. Sizes range from miniature to giant. **Miniature** rabbits are fewer than

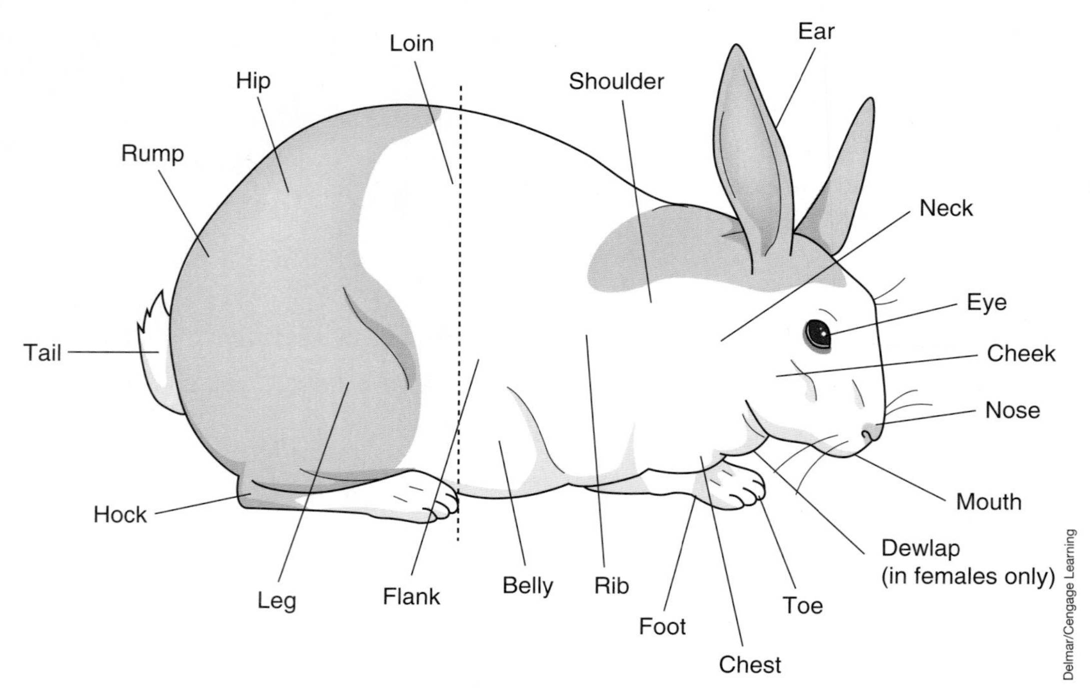

FIGURE 12-2 External anatomy of the rabbit.

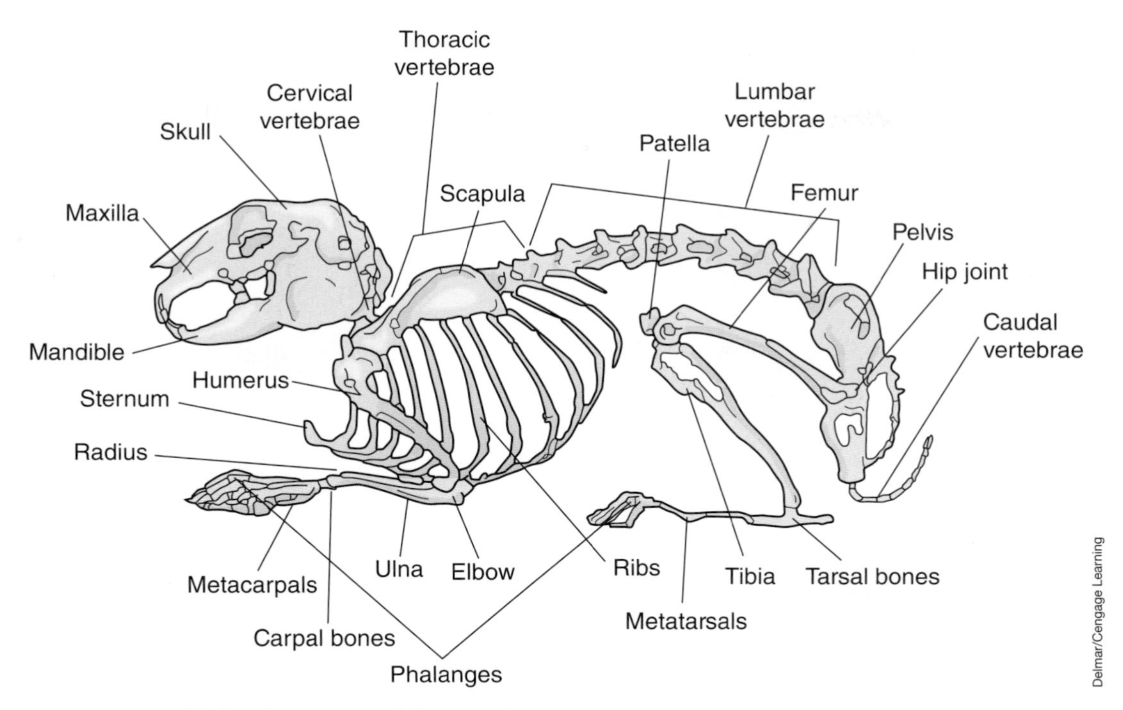

FIGURE 12-3 Skeletal anatomy of the rabbit.

2 pounds as adults and are sometimes also called **dwarf**. Examples are the Mini Lop and Netherland Dwarf (see Figure 12-4). The small-breed rabbits are between 2 and 7 pounds as adults and include Dutch and Polish breeds (see Figure 12-5). Medium rabbits are between 8 and 12 pounds in size and are sometimes referred to as **commercial** breeds. Examples are the New Zealand and Rex rabbits (see Figure 12-6). The large-breed rabbits are between 12 and 14 pounds and are also known as **Semi Arch** breeds, which include the Californian and Chinchilla breeds (see Figure 12-7). The largest group is the **Giant** or **Full Arch** breeds that are over 14 pounds in size. An example is the Flemish Giant (see Figure 12-8).

Breed Selection

Rabbit selection is dependent on the size, hair coat, and health and maintenance needs of the breed. Rabbits come in a variety of sizes and hair coat types. Potential owners should research breeds to determine the amount of care and maintenance they will require prior to purchasing them as a pet. Each animal should be evaluated for temperament and be docile and easy to handle. Rabbits are active and should be monitored for good health.

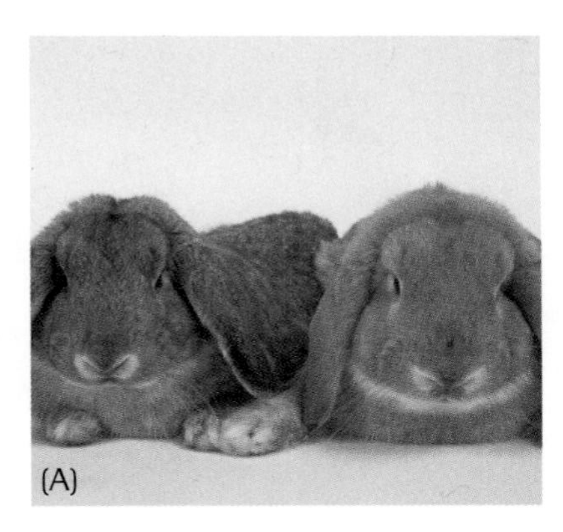

FIGURE 12-4 Miniature rabbits: (A) Mini Lop; (B) Netherland dwarf.

Photos courtesy of Isabelle Francais

FIGURE 12-5 Small-breed rabbits: (A) Dutch; (B) Polish.

Photos courtesy of Isabelle Francais

FIGURE 12-6 Medium or commercial breed rabbits: (A) New Zealand; (B) Short-haired Rex.

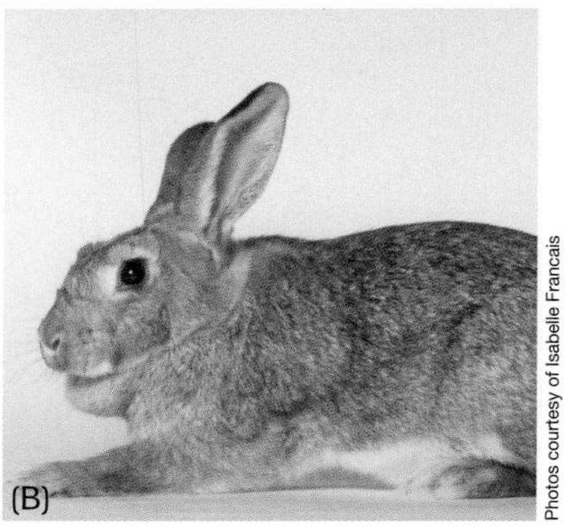

Photos courtesy of Isabelle Francais

FIGURE 12-7 Large or semi arch breeds: (A) Californian; (B) Chinchilla.

Photo courtesy of Isabelle Francais

FIGURE 12-8 Giant or full arch breed: Flemish Giant.

Nutrition

Rabbits are herbivores and eat a plant-based diet. They are usually fed a commercial rabbit food that is in pellet form (see Figure 12-9). They should be fed high-quality hay such as timothy or orchard grass that should be available **free choice**, or available at all times. Changes in food will cause an upset in the digestive system. Rabbits enjoy fruits, vegetables, and green foods that may

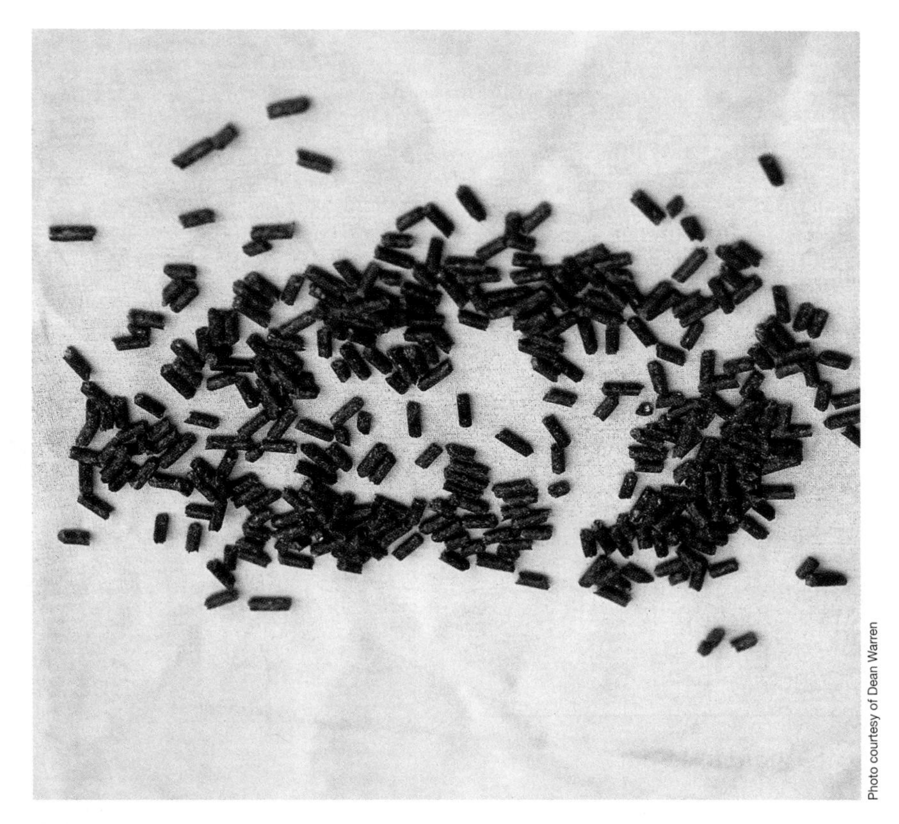

Photo courtesy of Dean Warren

FIGURE 12-9 Commercial rabbit feed.

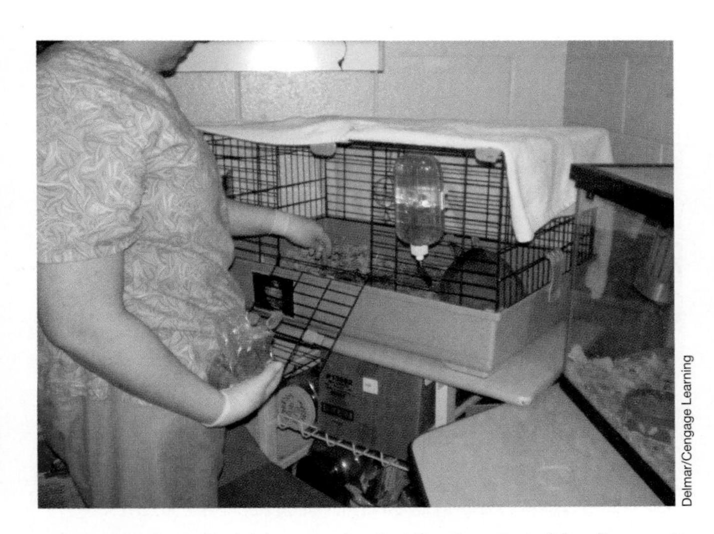

Delmar/Cengage Learning

FIGURE 12-10 Rabbits can be fed fresh vegetables in small amounts.

Delmar/Cengage Learning

FIGURE 12-11 Orphaned bunnies should be weighed daily to monitor weight gain.

be fed in small amounts (see Figure 12-10). Too much of these items will cause digestive problems. Rabbits should not be overfed as they will develop obesity problems. Rabbits should eat about 4% of their body weight daily as a maintenance diet.

Rabbits will continue to feed during the night by eating their feces, which become slightly softer at night and provide nutrition for the digestive system by allowing for an ease of breaking down plant materials. These feces are known as the **night feces**.

Sometimes young rabbits will become orphaned and require formula feeding. If this occurs, bunnies should be weighed daily to monitor weight gain (see Figure 12-11). A similar formula for kittens and puppies may be given, or a commercial formula, such as KMR formula, may be given. If a commercial formula is used, an egg yolk should

TABLE 12-1

Formula Requirements for Young Rabbits

Newborn	5 ml	Divide into two daily feedings
One week old	12–17 ml	Divide into two daily feedings
Two weeks old	25–27 ml	Divide into two daily feedings
Three weeks old	30 ml	Divide into two daily feedings
Three weeks to weaning	30 + ml	Divide into two daily feedings

be added for protein. Newborns should receive 5 ml daily divided into 2 feedings. Table 12-1 outlines the formula requirements for young rabbits.

Behavior

Rabbits make nice, quiet, and gentle pets and most have docile attitudes. On occasion, a rabbit may produce a warning scream or growl to alert that it is unhappy. They do have the instinct to protect themselves by biting, kicking, and striking with the front limbs. They will give a warning sign that they are upset by thumping a foot. Rabbits can be hypnotized for short periods by rubbing their abdomen when upside down (see Figure 12-12). The term **hypnotized** means putting in a trance-like position where the rabbit is still.

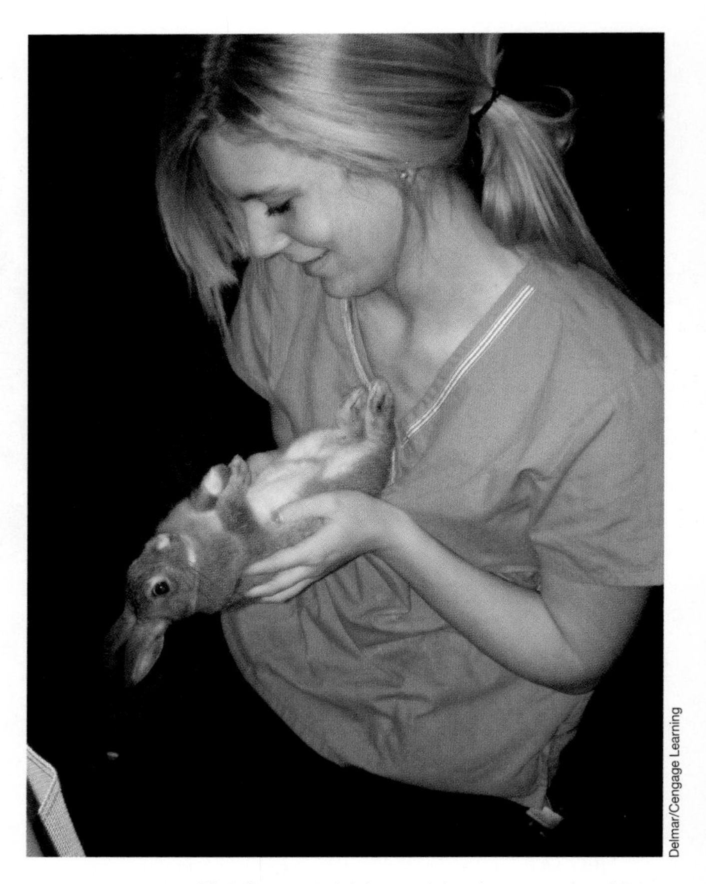

FIGURE 12-12 Holding a rabbit upside down and rubbing its abdomen will cause hypnosis for a short period of time.

Basic Training

Rabbits are trainable and can be litter-box trained similar to cats (see Figure 12-13). Many people that house rabbits as pets allow them to have some free range in the household. This includes the need for training to prevent chewing and destructive habits from occurring. Consistency and time will allow a rabbit to be trained fairly easily. It is important to note that rabbits will chew on anything, so an area where they are allowed free range should be monitored for any safety issues, such as telephone cords or electrical wires.

FIGURE 12-13 Rabbits can be litter-box trained.

Equipment and Housing Needs

Rabbits are often housed in cages called **hutches**. A hutch is a special cage designed for a rabbit that has a covered area to allow the rabbit to escape the weather and an open area that allows proper air and ventilation. These pens typically are made of wire and have a solid wood bottom. They should be free of drafts and provide one square foot of space per rabbit. Many people are keeping rabbits as pets in the household. They must be

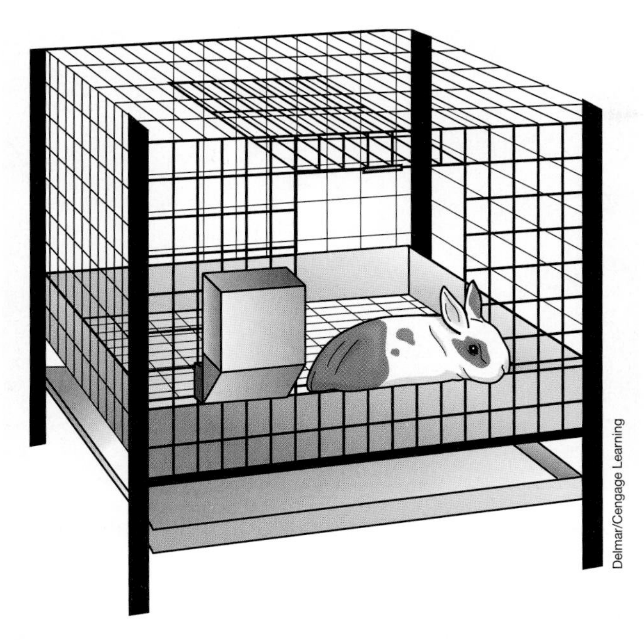

FIGURE 12-14 Rabbit housing.

FIGURE 12-15 Carriers should be used when transporting a rabbit.

monitored closely as they are curious and love to chew and may be injured if they bite into electrical wires or items they could possible ingest. Rabbit breeders will typically provide nest boxes, feeders, and water bottles within the cage or hutch.

Rabbits may also be housed in commercial rabbit pens made of metal, wire, plastic, or wood. These cages should be made of solid materials and provide enough space for the rabbit (see Figure 12-14). Rabbits chew, and this should be taken into consideration when selecting a cage. The bottom of the cage may be solid or wire with a tray that collects waste materials. Routine cleaning using shredded paper, pelleted litter, or dust-free shavings should be done on a regular basis.

Restraint and Handling

Improper handling and restraint of rabbits can result in the injury of the rabbit or an injury to the restrainer. Rabbits should never be lifted or restrained by the ears. Cat carriers should be used when traveling with a rabbit (see Figure 12-15). Rabbits can be "hypnotized" by gently rubbing the belly for short periods of time. When carrying or lifting a rabbit, one should always support the rear legs and back (see Figure 12-16). If a rabbit kicks its rear limbs, it may break its back or injure its spine. Wrapping a rabbit in a towel is helpful in keeping it calm and properly restrained (see Figure 12-17).

FIGURE 12-16 When handling a rabbit, always support the rear limbs.

Grooming

Rabbits require regular brushing, which may include bathing and grooming depending on the hair type of the rabbit (see Figure 12-18). They should also be evaluated

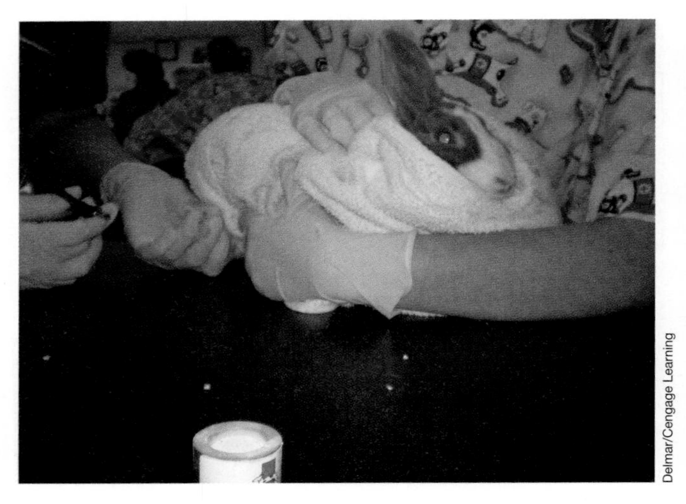

FIGURE 12-17 A towel can be used to help restrain the rabbit and may help keep the rabbit calm.

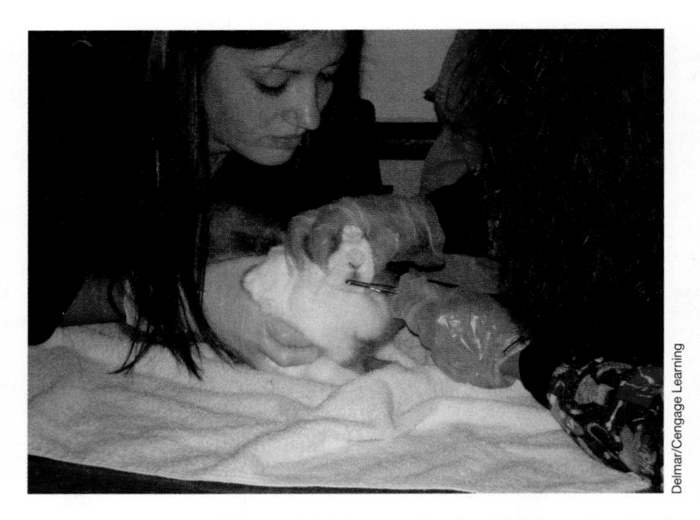

FIGURE 12-19 The rabbit's teeth should be checked regularly and trimmed if necessary.

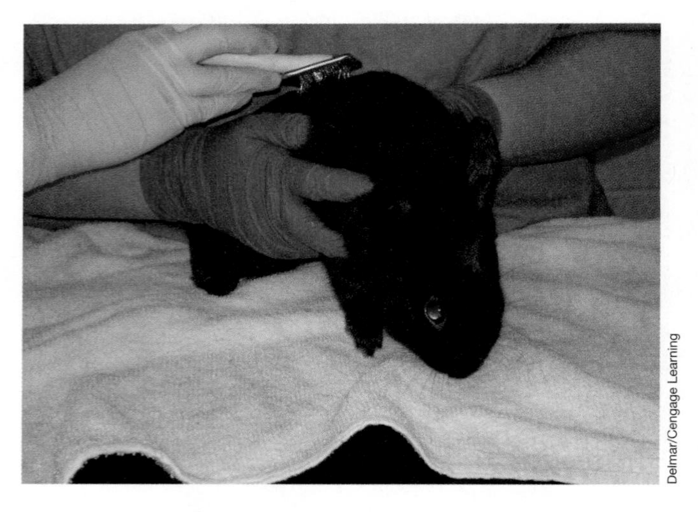

FIGURE 12-18 Rabbits require regular brushing.

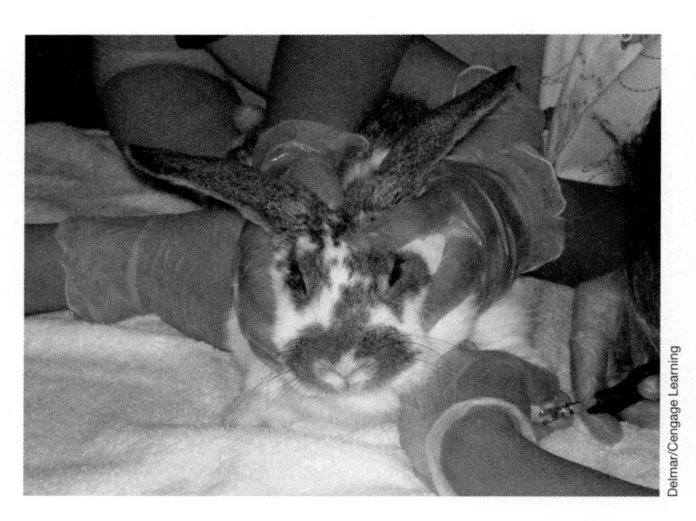

FIGURE 12-20 It is necessary to periodically trim a rabbit's nails.

for the need for nail trims, teeth trimming, and dental problems (see Figure 12-19). The nails and teeth continue to grow over the life of the rabbit and may require regular trimming and care (see Figure 12-20). Small handheld nail trimmers, styptic powder, and an oral speculum should be provided to complete these grooming duties. Rabbits are self-groomers and will clean their coats on a daily basis.

Basic Health Care and Maintenance

Rabbits are relatively healthy animals but can easily develop respiratory problems if not properly housed. Rabbit cages must be cleaned on a regular basis. Dust-free bedding should be used to keep cages clean and well bedded to prevent sores on feet. Avoid cool and drafty cage locations as rabbits can get colds. Rabbits are prone to diarrhea from food changes and stress. They can also become **constipated** and not be able to produce a bowel movement. Therefore, diet changes should be done slowly over a few weeks to allow the digestive system time to adjust. Long-haired breeds are prone to hairballs, or trichobezoars, similar to cats. Rabbits can be given hairball medications or pineapple or papaya juice to help break down and prevent hairballs. Feline laxative medicines are commonly used. A **laxative** is commonly called a stool softener. They are used to soften stool to allow animals to have a bowel movement.

Rabbits should be monitored for sores over the legs and feet, especially in the area of the **hocks**, or the back of the feet. They have long rear feet that are used

FIGURE 12-21 Rabbits should be weighed as part of their annual examination.

FIGURE 12-22 Baby rabbits grow quickly. This one is being examined in the veterinary facility.

to hop around, and pushing around on a wire or hard surface can cause sores. A sign of an illness in a rabbit is a change in behavior. Rabbits should have a physical exam on a yearly basis and be taken to a veterinarian that specializes in rabbits if any signs of illness occur. It is important to note to the veterinary assistant that for any rabbits that are seen in a veterinary facility that require blood work or surgery, the hair should be **plucked**, not shaved, from the area of the body. Plucking is pulling hair to remove it from the skin. This is done because shaving will tear the very delicate skin of rabbits. A regular weight check should also be noted (see Figure 12-21).

Vaccinations

Rabbits are not commonly given vaccines and this topic should be discussed with the veterinarian in cases where certain diseases are more common or show regulations require certain vaccinations in rabbits.

Reproduction and Breeding

Rabbits reach maturity around 6 months of age. The female rabbit should be taken to the male's cage for breeding, as the female is more territorial of her environment and will fight and possibly injure the male. The mating should be monitored and the male and female separated after breeding has occurred. The gestation length of rabbits is 30–32 days. Females will begin building nests just prior to the labor process. Nest building is typically done with items such as straw, paper, or shavings. The female will pluck hair from her abdomen to line the nest. Most rabbit breeds have an average litter size of 6 kits. Baby rabbits are born hairless and the eyes and ears are closed similar to puppies and kittens. The young are known as **precocial**, which means they grow very quickly (see Figure 12-22). Baby rabbits are able to move away from danger in a short amount of time. This is an instinct they adapt from nature. The kits will nurse only one to two times daily. The eyes open in 10 days. When the eyes open they can begin eating pellets, hay, and leafy vegetables. They are usually weaned between 6 and 8 weeks of age.

Common Diseases

Rabbits are prone to several types of diseases and health conditions. A healthy rabbit will be active, eating well and showing signs of a bright and alert appearance. It is important for the veterinary assistant to be aware of common diseases that affect rabbits.

Respiratory Disease

Rabbits are prone to respiratory disease. Rabbits tend to develop excessive nasal discharge often called "runny nose." The causes are poor cage ventilation, dusty bedding, high temperature and humidity, dental disease, and foreign objects in the nasal passage. The respiratory cause must be determined to treat the condition. Signs of respiratory disease in rabbits include the following:

- Nasal discharge
- Sneezing
- Clumping of hair on the front paws

Pasteurella

Respiratory disease can sometimes be wrongly diagnosed, when in fact the true cause is the bacteria **Pasteurella**, a naturally occurring bacteria in the respiratory tract of rabbits. Pasteurella is commonly called "snuffles" and is similar to a cold. This may be an acute or chronic inflammation of the mucous membranes in the air passages and lungs. Mucus is discharged from the nose and eyes. Affected rabbits will rub their eyes and noses. The fur on the face and paws may become matted and caked with dried mucus. The infected animals usually show signs of sneezing and coughing. Rabbits that are affected have a depressed immune system or are under increased amounts of stress. This condition is treatable with antibiotics. Rabbits that survive this disease often remain carriers. It is recommended that any rabbits showing signs of this condition be isolated and a strict sanitation program be followed.

Tyzzer's Disease

Another common bacterial disease is called **Tyzzer's disease**. This infection is spread through spores in the air. The disease affects the intestinal tract of rabbits. There is no treatment and no vaccine, so prevention is the key. This includes frequent sanitation of the cage, eliminating dust in the bedding, and avoiding overcrowding in rabbits. Signs of Tyzzer's disease in rabbits include the following:

- Watery diarrhea
- Feces or staining on rectum
- Anorexia
- Lethargic
- Dehydration
- Weight loss
- Death

Common Parasites

Rabbits commonly develop external parasites, such as fleas, ticks, and mites. Ear mites are a common condition that must be considered. Regular grooming practices will decrease the occurrence of these parasites. Fleas and ticks may be seen on the hair or skin coat and cause itching and skin irritation. Ear mites produce a thick, dark, and crusty wax-like appearance in the ear. Rabbits will scratch excessively at the ears and may cause blood vessels in the ear to break, thus causing a **hematoma**, or blood-filled mass, within an area. Ear mites are treatable with topical ear medications. The

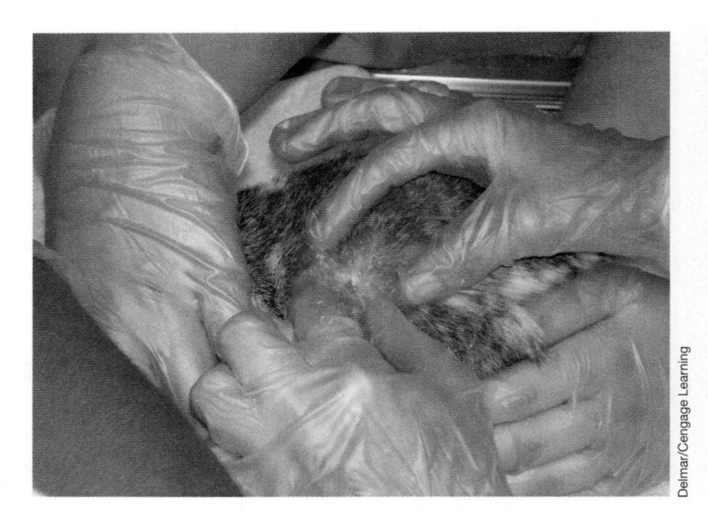

FIGURE 12-23 The rabbit's fur should be checked for signs of parasites.

skin and hair should be evaluated for signs of parasites (see Figure 12-23).

Internal parasites may also occur in rabbits. A fecal sample should be evaluated yearly to determine if any internal parasites are occurring. A common intestinal protozoan in rabbits is coccidia. It is difficult to rid rabbits of coccidian parasites. The organism may infect the stomach or the liver. Signs of the parasite include a pot-belly appearance, diarrhea, weight loss, and a loss of appetite. Mild cases may show no signs.

Common Surgical Procedures

Common surgical procedures in the rabbit include castration, spay, and dental extraction procedures. Rabbits are frequently neutered to prevent reproduction and to decrease aggression and behavioral problems in pets. Most rabbits can be castrated at three to four months of age, although others have recommended even earlier castration because rabbits may become sexually mature before four months of age. Castration decreases urine spraying as well. Most rabbit castrations are done similar to cat castration surgeries.

Rabbit ovariohysterectomy or spay surgery is helpful in decreasing aggression and breeding, as well as false pregnancies and reproductive tumors. The surgical procedure is done similar to a dog or cat spay. The spay surgery is typically recommended between 5 and 6 months of age but may be done earlier due to sexual maturity.

Dental problems and extractions are another common occurrence in rabbits as their teeth continuously grow. Rabbits have 6 incisor teeth in the front and several cheek teeth that sometimes overlap each other. This can cause trauma to the mouth and make it difficult for the rabbit to eat properly. Signs of dental disease in rabbits include the following:

- Dropping food
- Tooth grinding
- Excessive salivation
- Anorexia
- Nasal discharge

The teeth may need to be trimmed or extracted depending on the extent of the disease.

SUMMARY

Rabbits make excellent pets that can be housebroken and litter trained and allowed to roam inside the house. Rabbits are also used in production programs for meat and fur. They are commonly called rodents but are actually an entirely different species. Rabbits require special handling, health care needs, and nutritional needs. Veterinary assistants should understand the basic care involved with rabbits and be able to properly retrain and handle rabbits in a clinical setting.

Key Terms

commercial medium rabbits are between 8 and 12 pounds in size

constipated unable to pass waste materials and feces

dwarf rabbits are fewer than 2 pounds as adults

free choice food given so that it is available at all times or as necessary

Full Arch giant breeds of rabbits that are over 14 pounds in size

Giant breeds of rabbits that are over 14 pounds in size

hematoma collection of blood under the skin

hocks the tarsal joints at the point of the rear leg

hutches special cage designed for a rabbit that has a covered area to allow the rabbit to escape the weather and an open area that allows proper air and ventilation

hypnotize to place in a trance-like state

claxative stool softener

Miniature small rabbit breed under 2 pounds in size as an adult

night feces slightly softer feces that occur at night and provide nutrition for the digestive system by allowing for an ease of breaking down plant materials

non-ruminants animals with a specialized digestive system with a simple stomach built for slow digestion

Pasteurella a naturally occurring bacteria in the respiratory tract of rabbits

precocial growth that occurs very quickly in bunnies so that they are able to move away from danger in a short amount of time

plucked to pull out hair by hand

Semi Arch large-breed rabbits that are between 12 and 14 pounds

Tyzzer's disease bacterial infection in rabbits that affects the digestive system that spreads through spores in the air

REVIEW QUESTIONS

1. Explain the non-ruminant digestive system.
2. What are the night feces and their importance in rabbit nutrition?
3. What is the gestation length of rabbits?
4. What is a trichobezoar and how is it prevented in rabbits?
5. What are the nutritional needs of an adult rabbit?
6. How are rabbits housed?
7. What grooming and health needs do rabbits require?
8. Why is it important to "pluck" the fur on a rabbit?
9. What injuries can occur to a rabbit's hocks?
10. What are signs of a respiratory disease in rabbits?

Clinical Situation

Photo courtesy of Isabelle Francais

Miss Crawford, a new rabbit owner, called the veterinary clinic asking several questions about her new bunny. She recently purchased an Angora rabbit. She is asking what health program the rabbit needs, what care she will need to provide the rabbit, and if the rabbit should be examined like her kitten "Louie" was after he was purchased.

Sally, the veterinary assistant, takes the phone call and responds with "Hello, Miss Crawford. It would be ideal to have the bunny examined. Are you noticing any problems or concerns?"

"No, he seems very healthy. Actually, I don't even know if it is a boy. I have never owned a rabbit before and have very little information about them. Can we set up an exam for sometime next week?"

"That sounds good; how is Friday at 1 p.m.?" asks Sally.

"Great, in the meantime, can you give me some basic information of what I should be doing or what supplies I may need?"

- How would you answer Miss Crawford's questions?
- What information should Miss Crawford be given immediately?
- Was this situation handled effectively?

13

Courtesy of PhotoDisc

Reptile and Amphibian Breed Identification and Production Management

Objectives

Upon completion of this chapter, the reader should be able to:

- Understand common veterinary terms relating to reptiles and amphibians
- Describe the biology and development of a reptile and amphibian
- Identify common breeds of reptiles
- Identify common breeds of amphibians
- Identify the nutritional requirements of reptiles and amphibians
- Describe normal and abnormal behaviors of reptiles and amphibians
- Properly and safely restrain reptiles for various procedures
- Properly and safely restrain amphibians for various procedures
- Describe the health care maintenance needs of reptiles and amphibians
- Explain the process of breeding and reproduction in reptiles and amphibians
- Explain common diseases affecting reptiles and amphibians

Introduction

Reptiles and amphibians are becoming more popular as pets. With the increase in popularity, there is now a need for knowledge and understanding in the health care of these companion animals. Reptiles and amphibians are alike in several ways. Caution must be used to have the proper environment and needs for these delicate pets. Reptiles include species of snakes, lizards, turtles, crocodiles, and alligators. The most commonly kept reptile species are lizards and snakes (see Figure 13-1). There are over 6,500 species of reptiles. Common examples of amphibians are frogs, toads, and salamanders.

FIGURE 13-1 Lizards are a popular reptile to keep as a pet.

FIGURE 13-2 Some species of reptiles and amphibians live in aquatic conditions.

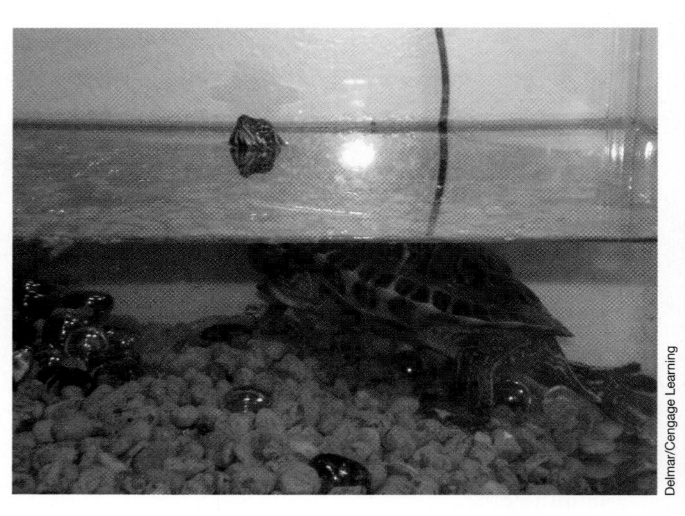

FIGURE 13-3 Some species of reptiles and amphibians live both in water and on land.

Veterinary Terminology

Herpetology is the study of reptiles and amphibians. Some species are **aquatic** or live in the water (see Figure 13-2). Others are **terrestrial** or live on the land. Others are **semi-aquatic** or live both in the water and on land (see Figure 13-3). Snakes are commonly called **serpents**. Turtles are commonly called **chelonians** or **terrapins.** A group of reptiles or amphibians is called a **brood**. When an animal is carrying live young or eggs the term is **gravid**, or pregnant. When eggs are laid the group of eggs is called a **clutch.** Reptiles and amphibians are known to hibernate when their body temperatures decrease. In these species, this process is called **brumation**. This is a semi-hibernation state when the animal sleeps a majority of the time and eats very little. Many of these species have specific terms for their anatomy. The **cloaca**, or rectum, is also commonly known as the **vent**. This is the area that contains the genital organs and excretes wastes. The **crest** is the top of the animal's head. Reptiles and amphibians are housed in **terrariums** or the enclosure that houses the habitat. The **substrate** of the environment is the type of floor material that is provided in the cage.

Both reptiles and amphibians may be **venomous** (poisonous) or **nonvenomous** (nonpoisonous). They may also lay eggs or bear live young depending on the species. They need proper environmental conditions and temperatures to survive. Reptiles and amphibians may be herbivores, carnivores, or **omnivores** (eat both meat and plant sources).

Biology of Reptiles

A **reptile** is an animal with dry, scaly skin that uses the outside temperature and environment to adjust body temperature. This type of process is called **ectothermic**. They are commonly called *cold-blooded*. This term was developed because reptiles feel cool to the touch.

Snakes

Snakes have long, thin bodies and tails, but no legs, ears, or eyelids (see Figure 13-4). They have a smooth skin texture but are not slimy as many people think. Some common snake external body part terms include the following:

- Body—long, thin central area of the snake
- Fang—long teeth located in the front of the mouth
- Forked tongue—thin area in mouth that has a slit in the center and is used to smell food
- Head—front part of the snake, shaped like a triangle
- Neck—area behind the head
- Rattles—located on the tail of certain species and are used to sound a warning of danger
- Scales—outer skin covering the body
- Tail—end of the body

Figure 13-5 illustrates the anatomical structure of the snake.

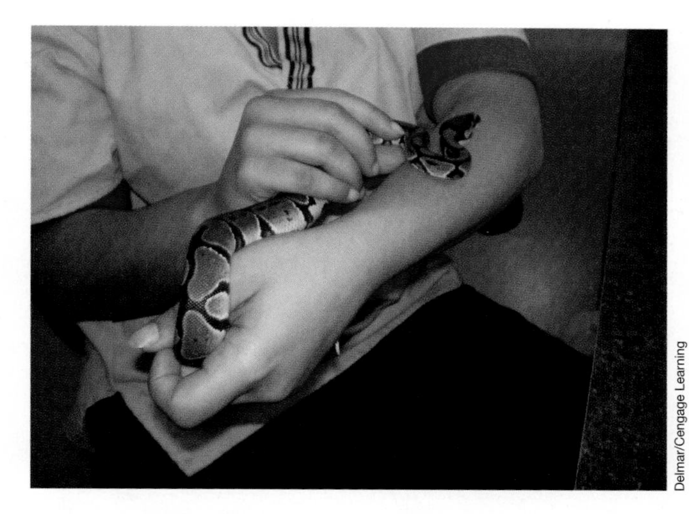

FIGURE 13-4 Snakes have long, thin bodies and tails, but no legs, ears, or eyelids.

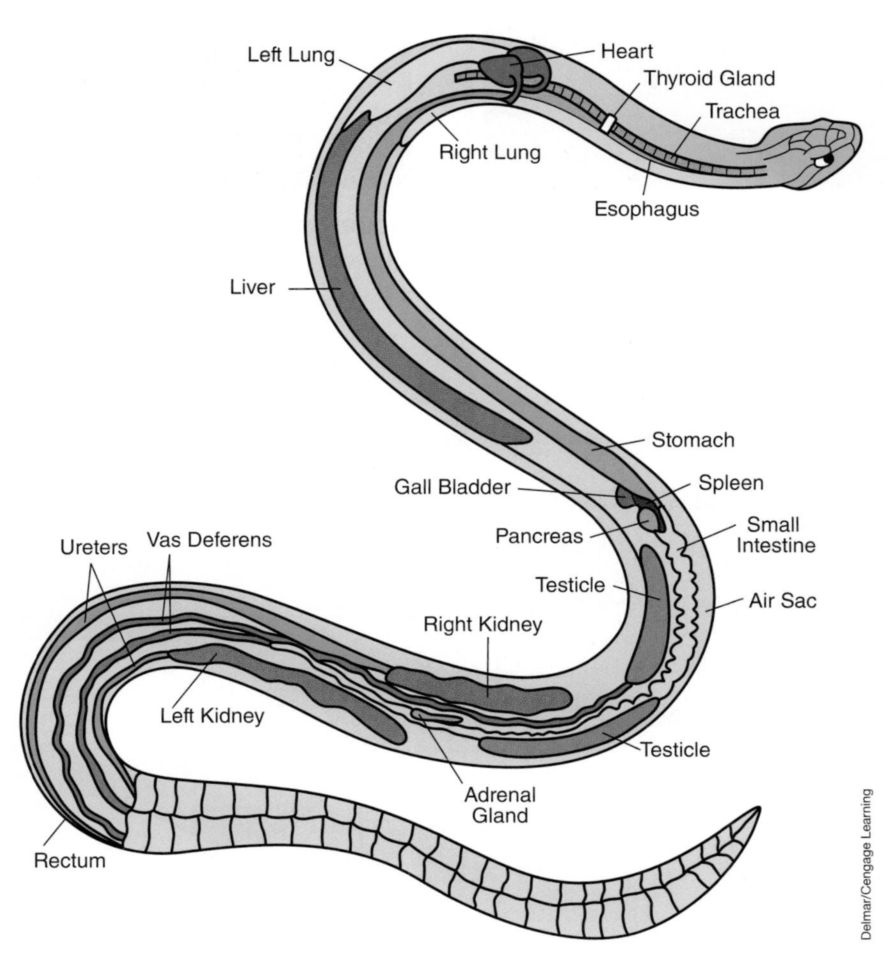

FIGURE 13-5 External anatomy of the snake.

Lizards

Most lizards have four legs, long tails, movable eyelids, and ear openings. Iguanas tend to be the most popular lizard species (see Figure 13-6). Common lizard external body part terms include the following:

- Caudal crest—spines over the lower back
- Crest—neck area located behind the head
- Dewlap—loose skin fold located under the chin
- Dorsal crest—spines over the middle of the back
- Ear—hole in either side of the head located behind the eyes
- Jowel—area of the head located below the ear holes
- Spines—long projections located over the back
- Tail—long extension off of the back
- Vent—opening located on the underside of the body behind the rear legs

Figure 13-7 illustrates the anatomical structure of the lizard.

Turtles

Turtles have shells and four legs and can pull their head, limbs, and tail into the shell for protection (see Figure 13-8).

FIGURE 13-6 Iguanas are a popular lizard species often kept as pets.

FIGURE 13-8 Some species of turtles can be kept as pets.

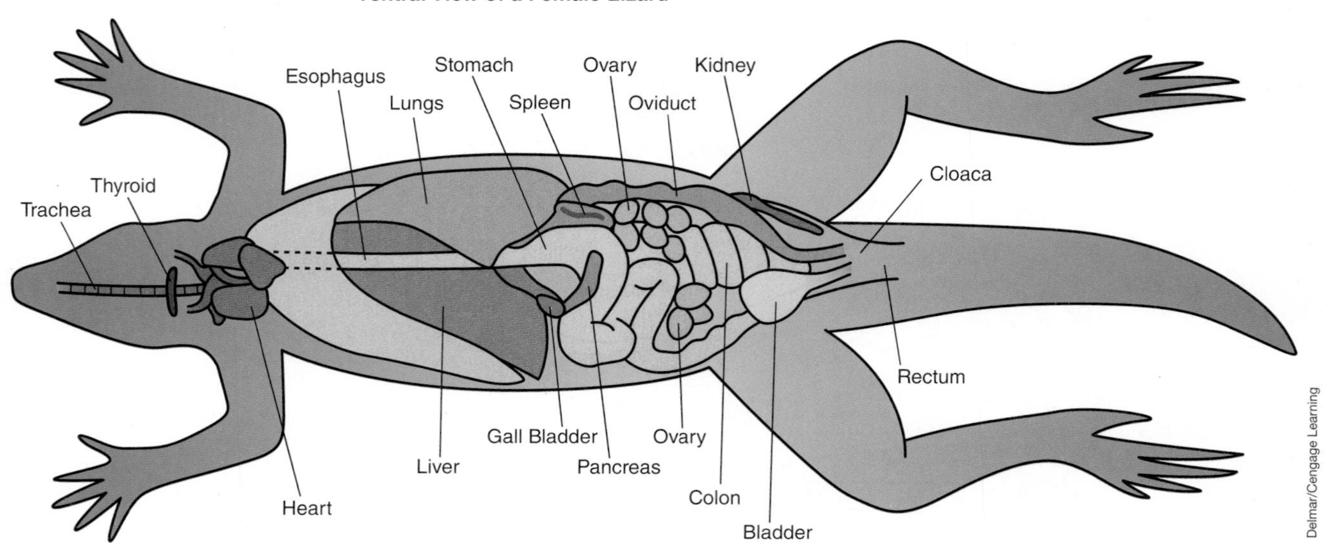

FIGURE 13-7 External anatomy of the lizard.

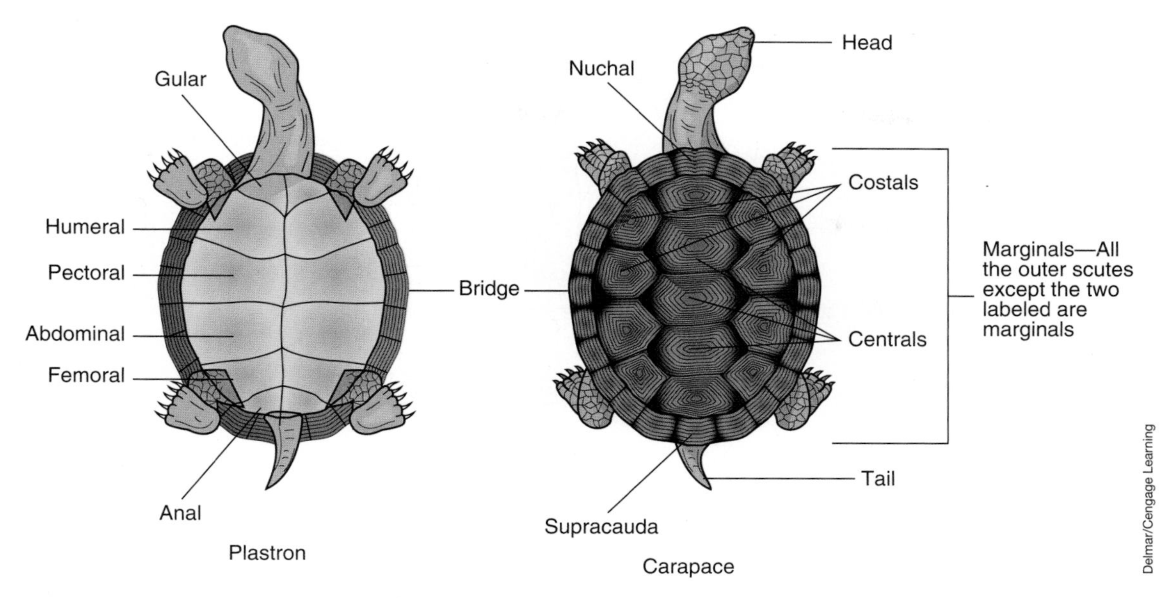

FIGURE 13-9 External anatomy of the turtle.

Turtles can live on land or in the water. Common turtle external anatomy terms include the following (see Figure 13-9):

- Carapace—hard covering over the upper body; upper shell
- Head—located at the front of the body and shell
- Hinge—area where the upper and lower shells meet on either side of the body
- Plastron—hard covering over the bottom of the body; lower shell
- Scutes—individual plates located over the shell
- Shell—hard covering located over the body
- Tail—located at the back of the body and shell

Biology of Amphibians

An **amphibian** is an animal with smooth skin and spends part of its life on land and part in the water.

Frogs and Toads

Frogs and toads are very similar. They spend their **juvenile**, or young, stages in the water. As adults, they have four legs and no tail (see Figure 13-10). Common frog external body terms include the following (see Figure 13-11):

- Dorsal surface—back area of the body
- Nares—nostril openings located on the face

FIGURE 13-10 American bullfrog.

- Tympanum—ear opening located behind the eyes on either side of the head
- Ventral surface—belly area of the body

Salamanders

Salamanders have a long body and tail and two or four legs depending on the species (Figure 13-12). Common salamander external body parts terms include the following:

- Body—midsection of the salamander
- Crest—neck area located behind the head
- Tail—long extension off of the back
- Vent—opening located on the underside of the body behind the rear legs

Figure 13-13 illustrates the anatomical features of the salamander.

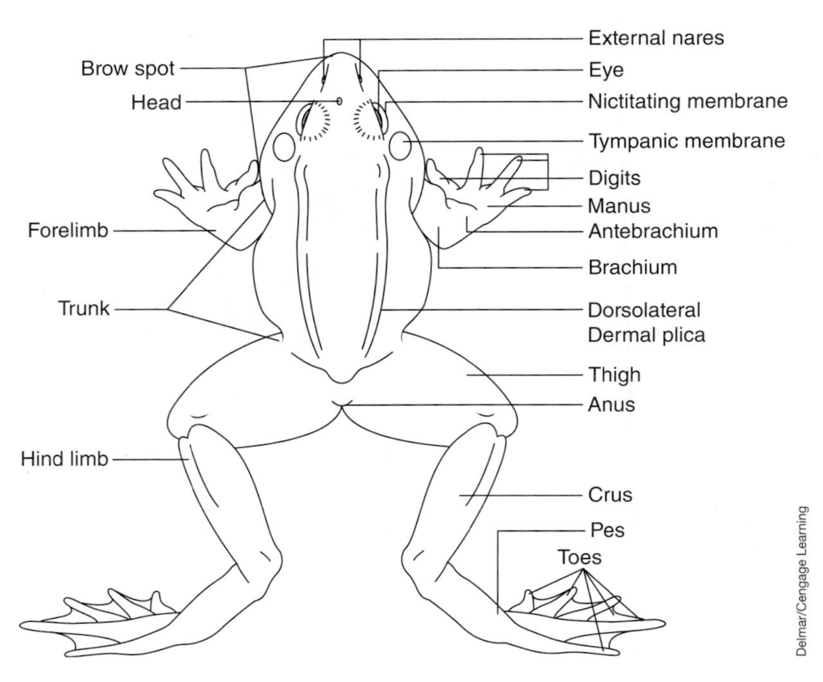

FIGURE 13-11 External anatomy of a frog.

FIGURE 13-12 Salamander.

Breeds

Reptile species come in a variety of appearances, sizes, and colors. Reptiles can live 10–20 years depending on the species and health care provided. Amphibian species also have a variety of appearances, sizes, and colors and have an average lifespan of 5–20 years or longer depending on the species.

Snakes

Snakes come in a variety of breeds. The most common breeds of snakes that are housed as pets include the boa constrictor, python, and green snakes. These snakes grow to a variety of sizes, some of which can be over 12 feet in length. Many larger species of snakes eat live prey, which they constrict to eat. They enjoy heat sources and climbing. Cage temperatures should range from 85 to 90 degrees. Most of these breeds of snakes make good pets, but the larger species are not recommended for children. A juvenile ball python is seen in Figure 13-14.

Lizards

Lizards come in a variety of sizes and appearances. Some commonly kept pet lizard breeds include iguanas, geckos, chameleons, and anoles. The green iguana is the most commonly kept iguana breed (see Figure 13-15). Lizards tend to be high maintenance because they require strict diets and environmental care. They may have nutritional and health problems if their needs are

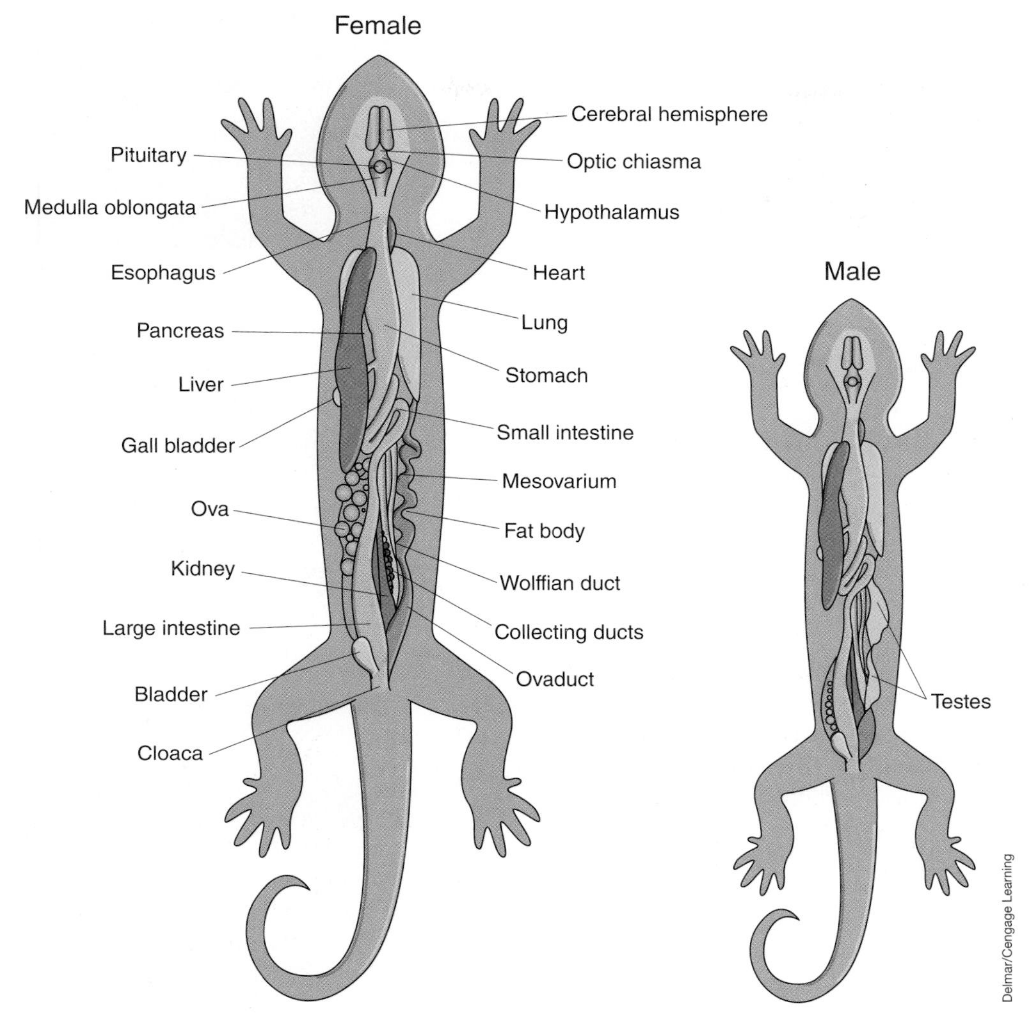

FIGURE 13-13 External anatomy of a salamander.

FIGURE 13-14 Juvenile ball python.

not met. The best starter lizards are the geckos and anoles. The most commonly kept gecko breeds are the leopard gecko (see Figure 13-16), fat-tailed gecko, and crested gecko. The green anole is a common small lizard breed that is readily available, requires little tank space, and doesn't need special lighting. Bearded dragons and

FIGURE 13-15 Green iguana.

FIGURE 13-16 Leopard gecko.

some varieties of skinks are also commonly found lizard breeds. Chameleons tend to be slightly higher maintenance, with the veiled chameleon, Jackson's chameleon, and the Panther chameleon being the most common.

Turtles

Turtles are ideal pets that require little maintenance and are relatively easy to care for. Some commonly kept turtle breeds include the Red Eared Slider, map turtles, and box turtles (see Figure 13-17). Any turtle or hatchling under 4 inches in size is illegal to purchase and house in the United States.

Frogs

Frogs are readily available as pets and are very entertaining and relatively easy to house and care for. Common breeds of frogs kept as pets include the bullfrog, green frog, and tree frogs (see Figure 13-18).

Salamanders

Salamanders are less commonly kept as pets but are relatively easy to care for. The most common breeds of salamanders are the tiger salamander (see Figure 13-19) and the spotted salamander.

FIGURE 13-17 (A) red-eared turtle; (B) map turtle; (C) box turtle.

FIGURE 13-18 Green tree frog.

Breed Selection

People who have reptiles as pets must research the species and have knowledge of its environmental and health needs. Many of these species require specialized diets, proper housing, warm temperatures, and adequate cage space. They must be handled carefully and properly housed to prevent escape. Nutritional needs are important with reptiles. Some reptiles become very large in size and may outgrow their cages and can become aggressive. Some species will require more care and time than others. Also, it must be considered that all reptiles and amphibians are natural hosts for salmonella bacteria, which may be a potential health problem to some people.

© iStock/Al Braunworth

FIGURE 13-19 Tiger salamander.

Nutrition

The nutritional requirements of reptiles and amphibians are essential and important to the well-being and survival of the animal. This is an area that should be researched according to the species that is being housed.

Snakes

Snakes are relatively carnivores. Many people try to train snakes to eat their prey dead, especially when feeding rodents. Rodents can bite the snake and cause trauma or wounds. Some snakes will not eat if the prey is not alive. It is best to observe the feeding until the prey is killed so that injuries do not occur. Food should be fed on a smooth surface so that the substrate particles are not ingested. Substrate examples are sand, pebbles, pellets, water, or cedar chips. The amount and frequency of feedings is based on the size and species of the animal.

Lizards

Lizards that are herbivores include the iguana species. They will eat leafy green vegetables that are high in calcium. All plant materials should be washed, chopped, and served at room temperature or slightly warmer. Newborn lizards should be fed twice daily. Juveniles up to 2.5 years of age are typically fed once a day. Adult species will vary from once-a-day feeding to every 48 hours. Most lizards eat in the early morning or late afternoon. Some common iguana foods include the following:

- Dark-green leafy vegetables
- Grapes
- Kale
- Collard greens
- Dandelions
- Parsley
- Spinach
- Alfalfa pellets
- Turnip greens

Lizards that are carnivores should be fed pre-killed prey. Live rodents will bite and cause trauma to lizards. Most carnivorous lizards are fed once a day. Ground beef sources do not provide adequate amounts of calcium. Many lizards that are fed a meat-based diet will require calcium supplements. Common meat-based foods include the following:

- Meal worms
- Crickets
- Grasshoppers
- Katydids
- Ants
- Pinkie mice
- Fish
- Snails

Turtles

Many turtles are mostly herbivores as adults. In the juvenile stages, most species are more carnivorous than adults. As a turtle ages, the need for a meat-based diet decreases and the need for a calcium-based diet increases. Many species remain omnivores and eat a variety of foods in their diets. Most tortoise species are herbivores, eating grasses, clover, leaves, and flowers. Snapping turtles are carnivores, eating insects, fish, worms, and snails. Box turtles are omnivores, eating a variety of fruits, vegetables, flowers, insects, worms, and small fish. Red-eared sliders, map turtles, and painted turtles eat water plants, worms, insects, and small fish.

Frogs and Toads

Amphibians in general need a variety of food sources. In the winter months, there may be a need to increase the amount of food fed on a daily basis. Frogs and toads that are terrestrial require insects, flies, crickets, and meal worms. Large toads may eat pinkie mice and adult rodents. Aquatic species eat a variety of insects, worms, fish, and commercial pellets. They should be fed in the water. Tadpoles should have access to **algae**. This is a plant-based source of food that grows spontaneously in water when left in the sunlight. Boiled dark-green leafy vegetables may also be fed. Vitamin C is essential for tadpoles to grow. They should be fed several times a day or they will cannibalize and eat each other. They should stop feeding as their legs begin to form. When adults, they should gradually be introduced to their new diet needs.

Salamanders

Salamanders eat insects, worms, flies, shrimp, and insect larvae. Many commercial diets have been formulated into pellets for reptiles and amphibians.

Behavior

Most reptiles and amphibians are not relatively social animals. They may become accustomed to humans, but excessive handling may cause stress and result in their refusing to eat.

Snakes

Snakes are relatively docile when handled as young and accustomed to people. When upset or scared, many species will bite. The constrictors will strike and apply pressure by curling around parts of the body. Each species should be researched for additional normal behaviors. Snakes will go through a skin shed cycle known as **ecdysis** (see Figure 13-20). When a snake begins the shedding cycle, the eyes will appear cloudy and the skin may develop a milky coloration. At mid-cycle the eyes and skin turn clear. The skin will shed about 1–4 days after this occurs. Snakes should not be handled during the shed cycle as their skin is very fragile and the new skin may be damaged. Snakes and some lizards use their tongues as sensory functions. They can smell and hear vibrations through the air and ground using the tongue. The tongue will flick as they sense their surroundings.

Lizards

Lizards are highly territorial or protective of their cages. Males are more so than females. Males will be more aggressive with other males than with females. Women who are menstruating should use caution when handling lizards. Male lizards are capable of detecting hormonal changes and odors and their pheromones can lead to behavioral changes, such as aggression.

Turtles

Most species of turtles and tortoises are docile and easy to handle. When scared, they will pull their head, legs, and tail into the shell for protection. Contrary to belief, several species of turtles are fast and can move readily when necessary.

Delmar/Cengage Learning

FIGURE 13-20 Snake shedding its skin.

Frogs and Toads

Frogs and toads are docile and easy to handle. Some species have the ability to emit a bad odor when frightened or scared as a protective behavior. They tend to be more vocal during the evening and during breeding season.

Salamanders

Salamanders are relatively docile and easy to handle. Salamanders are lively, excellent climbers, and very curious. Some salamanders are nocturnal, while others will be active during the day.

Basic Training

In general, reptiles' and amphibians' basic training includes being handled and accepting of humans. This is achieved by frequent and proper handling techniques. Socializing young reptiles and amphibians will allow them to become more accepting of handling.

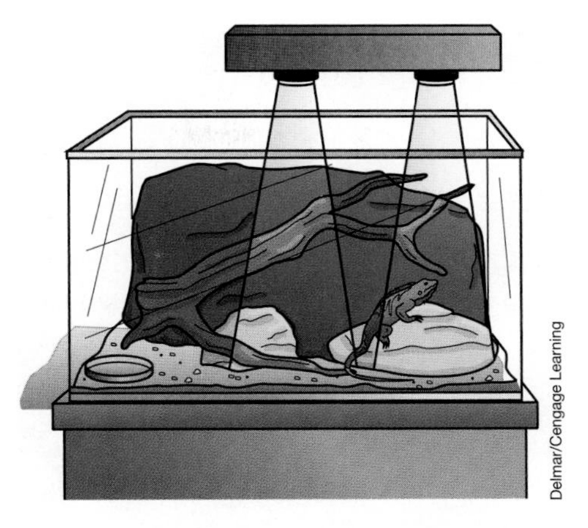

FIGURE 13-21 A glass aquarium with a heat source is ideal for housing some reptiles or amphibians.

Equipment and Housing Needs

All reptiles and amphibians should be kept in secure enclosures and provided adequate environments. Substrates should be selected carefully. Bark and woodchip products are not safe for most species. They may ingest these items and the digestive tract may become obstructed. Substrates for cages include the following:

- Newspaper
- Wood or pine shavings
- Synthetic bedding
- Branches
- Sand
- Lab animal bedding pellets

Glass aquariums and homemade Plexiglas cages are ideal for most species. Snakes are escape artists and should have a well-sealed cage. Branches of various heights and sizes should be provided. Many species are **arboreal** and require items to climb on. Some containment areas for hiding, such as logs, rocks, or boxes, may be included for privacy. Food and water bowls should be able to be easily accessed. Cages should be sanitized on a regular basis and food and water bowls cleaned daily. Temperatures are important to reptiles and amphibians. Most species of reptiles require a cage temperature of 75–85 degrees Fahrenheit with a relative humidity of 60–80%. This can be provided through the use of heat rocks, heating pads, and incandescent light bulbs (see Figure 13-21). A thermometer should be placed in the cage to monitor daily temperatures. Some heat items can cause burns.

Daytime temperatures may need to be increased and nighttime temperatures decreased to provide a natural environment. Sunlight or a light source should be provided for 12–14 hours on a daily basis. Systems that require water for semi-aquatic animals will require a **filtration system** to clean the water and provide oxygen. The optimum pH level of the water should be 6.5–8.5 and clear. The pH level of water is the necessary acidic balance of water that is required for an animal to survive.

Amphibians are commonly kept in aquariums with a part covered in water and a part above the waterline that is dry, often covered in pebbles, stone, soil, or wood. Many amphibians that are housed have been captured from the wild.

Restraint and Handling

It is important to be familiar with the safe manners in which to handle and restrain reptiles and amphibian species.

Snakes

The most important factor to consider with snakes in a veterinary setting is to determine if they are poisonous. Many veterinary facilities are not equipped to deal with venomous species. Snakes may be handled by hand, with restraint poles, or within anesthesia chambers. Snakes that are accustomed to human handling should be restrained by holding the head with one hand and the body with the other hand (see Figure 13-22). Larger species of snakes may require more restrainers. It is best to avoid allowing larger species of snakes to wrap around any body parts of the restrainer. Some snakes may need to be placed in a bag or pillow case for capture and carrying. Small species should be handled with care as to not hold them too tight and cause injury (see Figure 13-23).

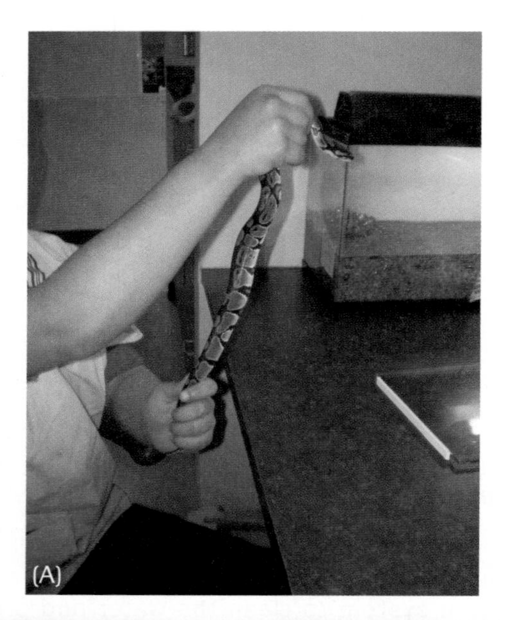

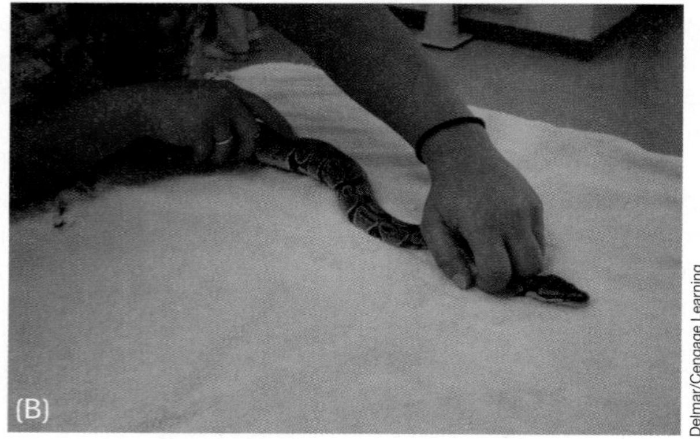

FIGURE 13-22 (A) Proper method for handling a snake. (B) Proper handling includes secure control of the head and body.

FIGURE 13-23 Care should be taken not to hold the snake too tightly.

Lizards

Lizards should be restrained well at both the front end of the body, including the head, and the tail end of the body. Many species of lizards have a tail that may be whipped and cause injury to the handler. It is best to control the front limbs and head with one hand and the rear limbs and tail with the other hand (see Figure 13-24). Larger species of lizards may require two or more restrainers. Lizards are capable of biting, and some iguanas and larger species can cause a serious bite wound. Lizards that are not accustomed to restraint may be difficult to handle. Lizards should never be picked up or handled by the tail. Many species can strip their tail or break off the tail stump as a defense mechanism for protection. Small lizards should be held carefully as holding them too tightly can cause injury.

Turtles

Turtles are relatively docile to handle and the most irritating instance in restraint is their ability to retract into their shell (see Figure 13-25). When lifting and restraining a turtle, the body of the shell should be held firmly

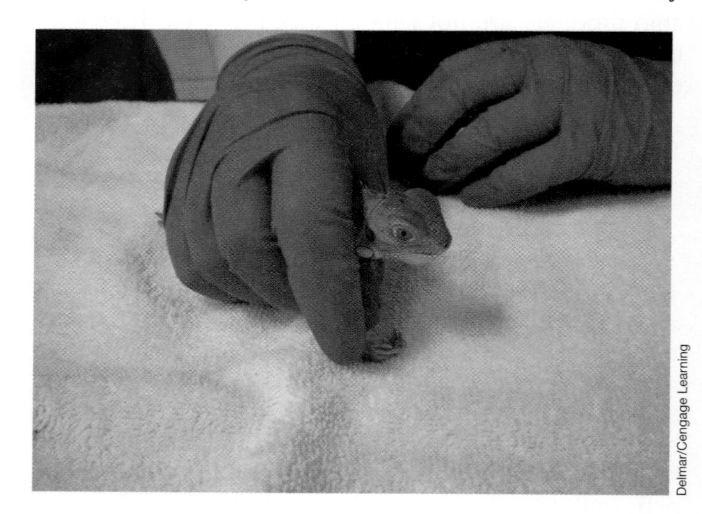

FIGURE 13-24 Proper handling technique for a lizard.

FIGURE 13-25 A turtle's ability to retract into its shell can make handling and restraint a challenge.

between one or both hands. To prevent a turtle from retracting into the shell, the index finger may be placed under the shell and in front of the hind limbs. The shell will not close but may cause a pinch depending on the size of the animal. Some turtle species may bite, causing injury.

Frogs and Toads

Frogs and toads may be picked up and restrained using one or both hands. One hand holds and restrains the head and the other hand restrains and holds the body. Larger species may be suspended in the air with the rear limbs hanging. Small species should be handled with care as to not cause injury.

Salamanders

Salamanders can be quick and if in a water source or tank may need to be captured using a net. When picking up and restraining a salamander, one hand should control the head and front limbs and the other hand should control the body and rear limbs. Aquatic species may need a water source to soak in. Larger species may cause injury with their tail. Salamanders should never be lifted or handled by the tail, as they will be stripped and break off.

Grooming

Some reptile species will need regular bathing, which may be done by providing a water source in the cage, placing in a water bath, or using a spray bottle to cleanse the animal (see Figure 13-26). Some species with nails may require routine nail trims.

Delmar/Cengage Learning

FIGURE 13-26 Some reptile or amphibian species will require a water source for bathing.

Basic Health Care and Maintenance

Reptiles and amphibians should be observed daily for changes in behavior or appearance. Teeth should be checked for dental problems. The vent should be examined for waste material buildup. The **urates** or urine and waste output should be noted. A fecal sample should be obtained with the yearly physical examination. The weight should be monitored for changes. All new animals should be **quarantined** or isolated from others for 2–3 months to make sure they are disease free. All reptiles and amphibians should be handled with caution as they have natural amounts of **salmonella** bacteria on their skin. This can cause a potential zoonotic concern. Exam gloves should be worn and hands washed after handling and cleaning the cage (see Figure 13-27).

Delmar/Cengage Learning

FIGURE 13-27 Care should be taken by wearing gloves and washing the hands after handling reptiles or cleaning their enclosures.

Vaccinations

Reptiles and amphibians do not require vaccinations.

Reproduction and Breeding

Reproduction and breeding in reptiles and amphibians is a popular hobby. Many species are easy to breed. Some animals may become aggressive and should be monitored for aggression and injuries. Some reptile and amphibian species are easy to determine the gender of the species, whereas others are not easily identified. It is important to know that a male and female are paired together when breeding.

Snakes

Before breeding snakes, it is important to determine the gender as many species of snakes look alike. Snakes are sexed by the **probe method**. A small metal probe is inserted into the **hemipenis** (male reproductive organ) area on the edge of the cloaca, or rectum (see Figure 13-28). In males, the hemipenis opening is much larger than in females, which only have scent glands in this area. The depth of the probe method is based on the length of the snake's scales. Males will probe between 6–12 scales depending on size and species. Females will probe only 2–4 scales. Males and females can be housed together for breeding purposes. Many species are aggressive and should be separated after breeding. Some species of snakes bear live young and others lay eggs that must be **incubated**, or warmed until hatched, similar to birds. Incubation in snake eggs may be done in a nest or the eggs are laid in the ground and naturally incubated. Garter snakes and boa constrictors bear live young. King snakes and pythons lay eggs. Gestation or incubation lengths vary from species to species. Garter snakes range from 90–100 days gestation and boa constrictors range from 120–240 days. Burmese pythons have an incubation period of 56–65 days. Ball pythons' incubation is 90 days.

Photo courtesy of Jordan Applied Technology Center, West Jordan, Utah

FIGURE 13-28 Determining the sex of a snake.

Lizards

Lizards can be sexed by visual exam. Males have taller dorsal spines over the back and tail, a larger **dewlap** (loose flap of skin under the chin), and bulges at the base of the tail. In males, the hemipenis can be everted with light pressure over the base of the tail. The term **everted** means that the area can be expressed or exposed to the outside of the body. Some lizard species have horns or plates that distinguish male and female. Many lizard species undergo color changes during breeding season (see Figure 13-29).

Males and females should be housed together for breeding purposes only, as they will also become aggressive with each other. Iguanas, geckos, and anoles are egg layers and chameleons bear live young.

Photo courtesy of Erika Klaphake, DVM

FIGURE 13-29 Color change occurs in males when ready for breeding.

TABLE 13-1

Species and Incubation and Gestation Period

SPECIES	INCUBATION/GESTATION
Green Anole	60–90 days; eggs
Green Iguana	73–93 days; eggs
Jackson's Chameleon	90–180 days; live young
Leopard Gecko	55–60 days; eggs

Table 13-1 outlines the incubation and gestation of common lizard species.

Turtles

Turtles and tortoises can be sexed using a visual exam as well. The males have a longer tail and a wider tail base than the females. The cloaca is more caudal and the shell area around the cloaca is **concave** or goes inward (see Figure 13-30). In some species, the males have longer claws on the front limbs. Females are usually larger in size. The male hemipenis can be expressed with gentle pressure and the use of warm water over the vent area. Most male and female turtles can be housed together without fighting. Most turtle species lay eggs and have a variety of incubation periods. The box turtle has an incubation length of 50–90 days and the red-eared slider turtle 53–93 days. Most tortoise species are around 60 days of incubation. Amphibians typically have some **dimorphism** characteristics, which means that the male and female are different. In most species the female is larger in size. During the breeding season, many male species have a skin-color change that is evident.

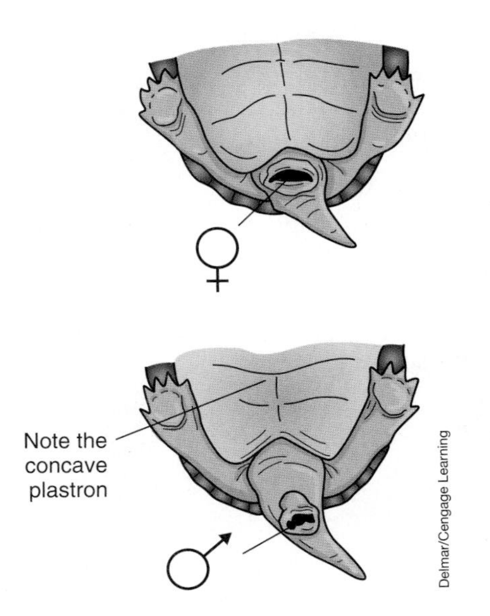

FIGURE 13-30 Determining sex of turtle.

Frogs and Toads

Frogs and toads begin as **tadpoles**. This is the larval stage of the newborn. Each species has a different time frame when it reaches the adult stage.

Salamanders

Salamanders are less commonly kept as pets but are relatively easy to care for. The most common breeds of salamanders are the tiger salamander and the spotted salamander. Salamanders have a head, body, limbs and tail but spend part of their time on land and in water. They resemble lizards but have smooth skin versus scales.

Common Diseases

The most common concern with reptiles and amphibians is the possibility of **salmonella infection**. An estimated 3% of the population owns a reptile or amphibian. An estimated 70,000 people contract salmonella from contact with a reptile. Snakes, lizards, and turtles are the most common animals that pass salmonella to people. **Salmonellosis** is the bacterial infection that affects people. All reptile and amphibian owners should be educated on the possible risks of salmonella and how to prevent such an outbreak. Children and people with compromised immune systems are at increased risk and should avoid contact with these animals. Anyone handling an animal or a cage should wash his or her hands thoroughly using soap and water. Reptiles and amphibians should be contained to a cage and not be allowed to freely roam the house. Any area that the animal comes in contact with should be properly sanitized.

Infections from wounds and burns are common due to poor husbandry and inappropriate housing. Housing multiple reptiles and amphibians can cause fighting and aggression, and bite wounds become easily infected. Bacterial infections may also occur from injuries within the cage. Rubbing against sharp objects can be avoided by choosing to safely set up a cage. Thermal burns occur due to direct contact with hot surfaces, usually hot rocks, light bulbs, or direct sunlight. A physical barrier should be provided for such objects within a cage. When such injuries occur, treatments with antibiotics and topical medications are necessary to prevent further health issues.

Common Parasites

Reptiles and amphibians can become infected with internal and external parasites. Hookworms, roundworms, tapeworms, and pinworms are possible. Coccidia is also of importance. **Coccidia** is a one-celled parasite that affects all animals and is passed in contaminated water sources, commonly from bird feces. Care should be taken when handling these pets and cleaning their cages. Mites and ticks occur and are visible to the eye. Mites may come from substrate or items placed in the cage.

Common Surgical Procedures

Most reptiles and amphibians do not require routine elective surgical procedures. However, many species may be curious and possibly ingest foreign objects that are not able to pass through the digestive system. These situations may require surgical repair. It is recommended that only veterinarians specializing in reptiles and amphibians attempt these procedures.

SUMMARY

Reptiles and amphibians are becoming more popular as companion animals. The health and environmental needs become the responsibility of the pet owner and many of these people rely on veterinary staff members to educate them on the proper care of these animals. Many of these species are fragile and delicate and easily become injured or sick due to the negligence of the owners. Therefore, it is important that the veterinary assistant be knowledgeable in the husbandry, nutritional needs, reproduction, health care, and disease control of these creatures.

Key Terms

algae plant-based source of food that grows spontaneously in water when left in the sunlight

amphibian animal with smooth skin and spends part of its life on land and part of its life in the water

aquatic animal that lives in the water

arboreal animal that lives in trees

brood a group of reptiles or amphibians

brumation process in reptiles similar to hibernation when their body temperature decreases

chelonians veterinary term for turtles

cloaca rectum in reptiles; also called the vent

clutch group of eggs

coccidia fungal disease

concave inward appearance

crest top of an animal's head

dewlap flap of skin under the chin

dimorphism the occurrence of having both male and female body parts and appearances

ecdysis process of snakes shedding their skin

ectothermic the body temperature of an animal, controlled externally by the environment

evert turn inside out

filtration system water system that cleans the water source

gravid term in reptiles meaning pregnant

hemipenis the male reproductive organ in reptiles and amphibians

herpetology the study of reptiles and amphibians

incubated the process of keeping eggs warm for hatching

juvenile young stage of growth

nonvenomous not poisonous

omnivores animals that eat a meat- and plant-based diet source

probe method process of passing an instrument into the cloaca to determine the sex of a reptile or amphibian

quarantined to isolate

reptile animal with dry, scaly skin that uses the outside temperature and environment to adjust body temperature

salmonella infection bacterial infection that causes severe diarrhea

salmonellosis bacterial infection that occurs from touching reptiles or amphibians

semi-aquatic living part of the time in the water and part of the time on land

serpents the veterinary term for snakes

substrate the type of material on the bottom of the cage

tadpoles young newborn frogs or toads

terrapins veterinary term for turtles

terrariums plant-sourced housing for reptiles and amphibians

terrestrial living on land

urates the urine waste materials of reptiles and amphibians

venomous poisonous

vent the rectum in reptiles; also called the cloaca

REVIEW QUESTIONS

1. Describe the characteristics of the snake, lizard, and turtle.
2. What are the differences between a reptile and an amphibian?
3. Explain the probe method of sexing a reptile.
4. What is the difference between gestation and incubation?
5. What are the signs of a reptile going through a shed cycle?
6. Why should reptiles and amphibians not be fed live food?
7. What are some good examples of iguana foods?
8. List some ideal substrates for reptiles or amphibians.
9. What is the ideal temperature for reptiles?
10. How long should quarantines occur for reptiles or amphibians?

Clinical Situation

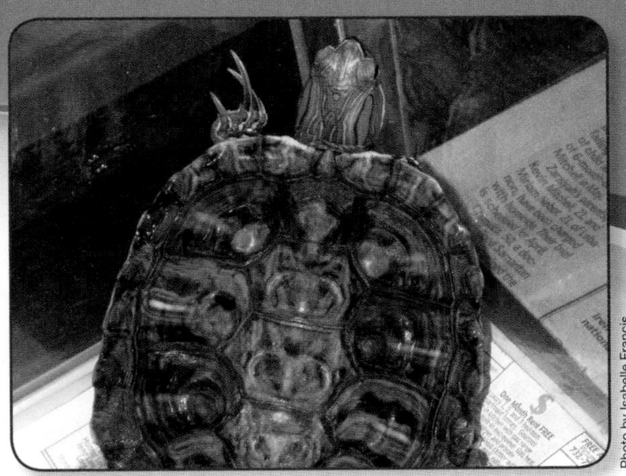

Photo by Isabelle Francis

Mr. Charles, an elderly client, calls the veterinary clinic about a turtle that he has just purchased as a gift for his grandson. He has some questions and concerns about his grandson handling and playing with the turtle.

"Hello, I was calling about a Red Eared Slider turtle I just got for my grandson, Billy. I was wondering if there are any health concerns with the turtle. Also, how should Billy hold it?"

The veterinary assistant, Katie, has had several turtles and says, "Mr. Charles, I have owned many turtles and nothing has ever happened to me. Turtles make excellent pets. Billy will enjoy his new pet. As far as picking the turtle up, he should hold it by its shell. He can also let it walk around the house if he likes."

"Oh, I am so glad to hear that. Someone at the pet shop told me about some handling concerns and after I left the store I was worried. Thank you so much for answering my questions."

- Did Katie provide proper information?
- How would have you answered Mr. Charles's questions?
- What could potentially happen in this situation?

14

Courtesy of PhotoDisc

Ornamental Fish Identification and Production Management

Objectives

Upon completion of this chapter, the reader should be able to:

- Explain the common veterinary terms relating to ornamental fish
- Describe the biology and development of ornamental fish
- Identify common species of ornamental fish
- Develop a nutritional plan for ornamental fish
- Demonstrate the safety methods of capturing and handling ornamental fish
- Explain the water sources and housing requirements of ornamental fish
- Establish an appropriate water environment for ornamental fish
- Create a health management plan for ornamental fish
- Describe breeding and reproduction in ornamental fish
- Discuss common diseases and health conditions of ornamental fish

Introduction

Ornamental fish are a large and popular industry in the United States. Many homes have a variety of sizes of aquariums that house all types of fish. These fish are more valuable as pets than food fish due to their small size and low weight. Ornamental fish require specialized housing and supplies. The popularity of ornamental fish production has evolved from people discovering that watching fish swim and interact in their aquatic environments provides stress relief.

Veterinary Terminology

Ornamental fish are primarily **omnivores**. Omnivores are animals that eat both plant and meat sources of food. There are no specific terms relating to the male and female fish. However, newborn fish are called **fry**. When fish mature, they are referred to as **yearlings** when they become a year old and **fingerlings** in their second year of life. A group of fish is called a **run** or **school**.

Biology

Ornamental fish are fish species that are kept for their color and appearance and due to a special appeal to people. Many species and breeds of ornamental fish have been developed based on color, fancy fins, interesting characteristics, and overall enjoyment. These fish are not raised as a food source. **Companion fish** are species kept in people's homes as pets or companions. They are kept for entertainment, amusement, and leisure. **Tropical fish** form another group of ornamental fish. They are small, brightly colored fish that require warm water and are naturally found in warm, tropical locations. Most ornamental fish are not native to the United States or North America. Many species are found in locations of South America, Asia, and the Pacific Islands. Hawaii is the largest U.S. state that has the largest amount of native fish species found in a tropical environment. Several states have the temperature and environment for producing quality fish species that have been imported from other locations. Florida is the leading U.S. state in ornamental fish production. Releasing nonnative fish species into the wild is illegal. Many fish species are captured in their natural locations and raised in confined breeding programs. There is a large **diversity** (difference) of ornamental fish. The overall body shape has adapted fish to their specific habitat. **Surface dwelling fish** live and feed near the surface or top of a body of water. **Bottom dwelling fish** live and feed on the bottom or floor of a body of water. Surface dwelling species have an upturned mouth and a flat back. Bottom dwelling fish have flat bodies and small mouths. Tall, laterally compressed species, such as the discus or angel fish, are made for slow moving waters. Slender, streamlined fish are made for fast moving waters.

Fins have been adapted for movement, floating, and breeding purposes (see Table 14-1). They may be single or paired. Many aquarium fish have long fins that are ideal for tanks. They have been developed through selective breeding. The **caudal fin** is also the tail fin. It is used to move the fish through the water. The **dorsal fin** is located on the back of fish and is used to steer the fish through the water.

TABLE 14-1

Tail Fin Shapes

Forked tail fin	Fast swimmers
Round tail fin	Quick movements
Long tail fin	Attract mates

Most fish species are covered by **scales**. These body coverings serve as protection. The color of fish is determined by the pigment and light reflection. **Pigment** represents the color of the skin. Some species are dark in color to blend into their environment. Others are brightly colored to attract mates. **Gills** are the organs that allow fish to breathe. Through the gills, fish absorb oxygen to exchange gases within the water (see Figure 14-1). Fish have a **lateral line** that is an organ located just under the scales that picks up vibrations in the water. This is what allows fish to detect danger, food, and navigate the water. Fish also have an organ called a **swim bladder** that allows them to float. The swim bladder is an

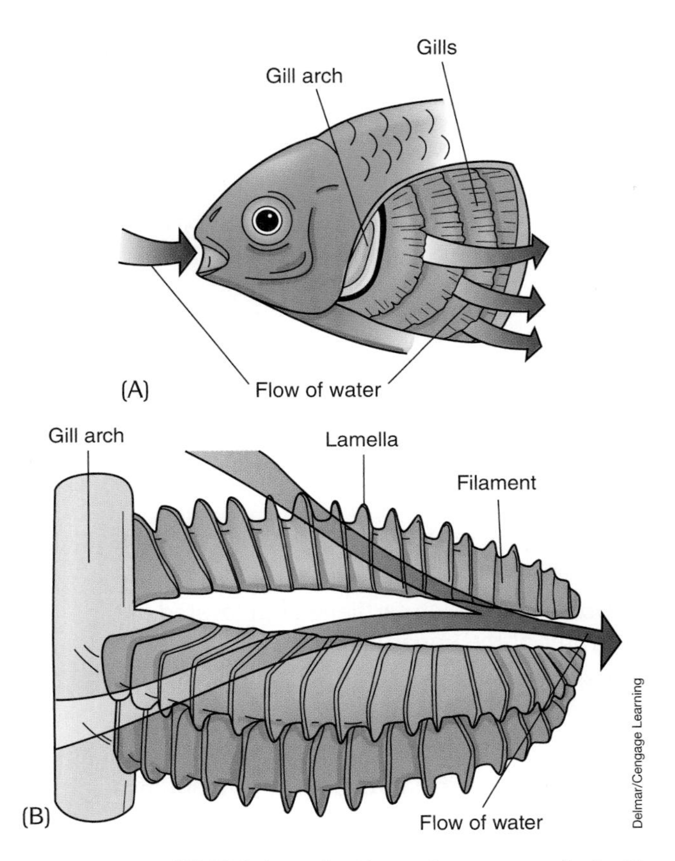

FIGURE 14-1 (A) Fish breathe through organs called gills; (B) each gill consists of many filaments.

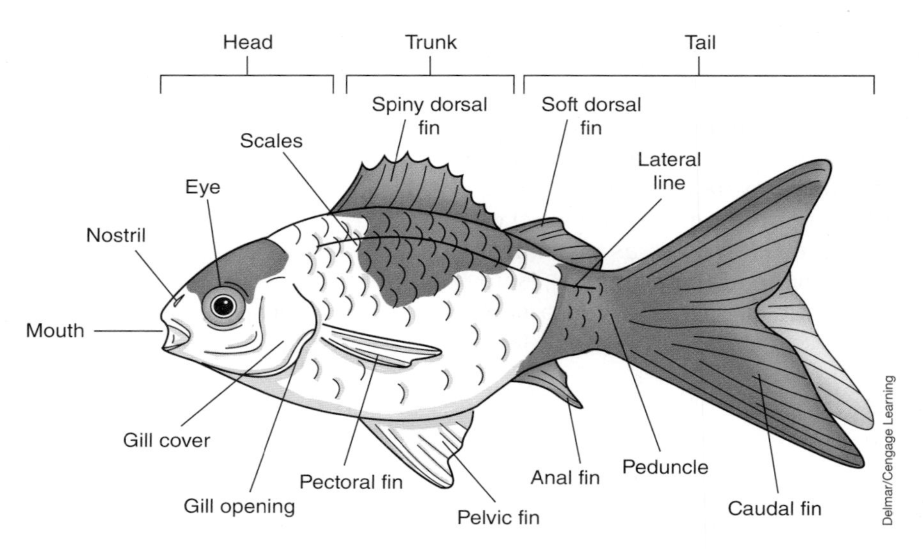

FIGURE 14-2 External anatomy of the fish.

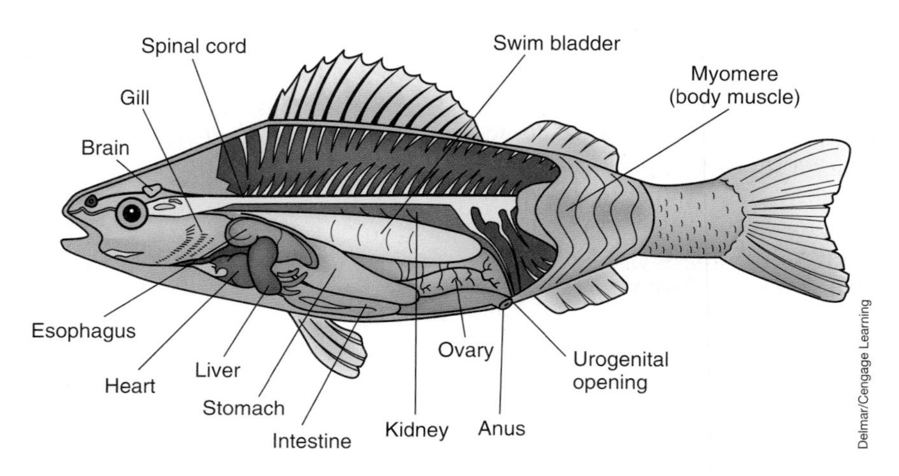

FIGURE 14-3 Internal anatomy of the fish.

air-filled sac that prevents fish from sinking. Common external fish body terms include the following:

- Anal fin—cartilage area located on the underside of the back of the body
- Caudal fin—cartilage area located at the base of the body, acting as a tail
- Dorsal fin—spiny cartilage area located in the center of the upper back
- Ear flap—opening located on either side of the head behind the eyes
- Gills—openings holding the lungs located on either side of the body in front of the pectoral fins and behind the head
- Lateral line—located on either side of the body, running along the entire body lying just under the scales
- Pectoral fin—cartilage areas located on either side of the body acting as arms
- Pelvic fin—cartilage area located under the pectoral fins on the underside of the front part of the body

Figure 14-2 and Figure 14-3 illustrate the external and internal anatomy of the fish.

Ornamental Fish Breeds

Fish may live in freshwater or saltwater. **Freshwater fish** are species that live in freshwater that has no or very little amount of salt. They are the most popular ornamental fish. **Saltwater fish** live in and require salt in their water source. This water is a complex mixture of salt and various minerals. They require a synthetic seawater environment commonly called a marine aquarium.

Goldfish

Goldfish are one of the oldest breeds of ornamental fish (see Figure 14-4). They have been raised and bred for over 2,000 years. Goldfish are small and commonly a bright orange color, but many new and fancy colors and interesting characteristics have been developed. Fancy fins and large eyes, as well as a variety of colors, have been bred. They are hardy and adapt well to a wide range of environments. Their size depends on the size of the tank they are housed in and their nutrition. Some are 1 to 2 inches in size and others can grow as large as 2 feet in length. Goldfish require about two gallons of fresh water to grow 2 inches in size. They are great beginner fish. They require regular cleaning as they are a messy species compared to other fish. They have been known to live as long as 15 years. They can lay up to 10,000 eggs that hatch in three to six days.

Koi

Koi are a variety of carp that look similar to large goldfish (see Figure 14-5). They were developed by selective breeding for certain colors. Colors range from bright red, black, white, and gold to combinations of colors. They are freshwater fish that may reach sizes in length of 3–4 feet and weigh up to 10 pounds or more. Most koi are kept in ponds or large tanks that may be housed indoors or outdoors.

Labyrinth Fish

Labyrinth fish are nest builders, and the males maintain the nest and incubate the eggs. The nests are built from bubbles the male builds within plants. Eggs hatch in 24–30 hours. Labyrinth fish come in a variety of sizes.

Photo courtesy of Aaron Norman

FIGURE 14-4 Bubble-eye variety of goldfish.

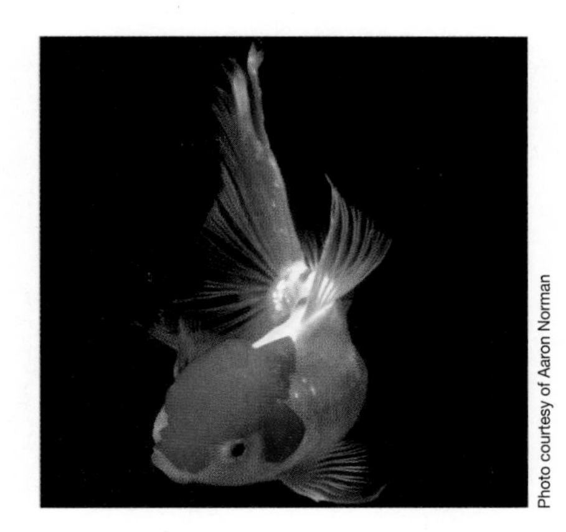

Photo courtesy of Aaron Norman

FIGURE 14-5 Koi.

Photo courtesy of Aaron Norman

FIGURE 14-6 Kissing gourami.

Gourami

Gourami come in a variety of breeds (see Figure 14-6). They are large freshwater fish with interesting characteristics. They grow to lengths of 12 inches or more. Some examples of gourami include the kissing gourami, blue gourami, and three-spot gourami. Gouramis lay eggs.

Barbs

Barb fish are a popular species (see Figure 14-7). They are easy to raise and are very hardy. They are related to goldfish and require fresh water. They typically range in sizes of 2 to 4 inches in length. They need plenty of light. Caution should be used with breeding as some barbs eat their eggs after spawning.

Tetras

Tetras tend to require more care than some other species of fish. The most common breed is the neon tetra, known for its bright electric-blue stripe that runs the entire length of the body (see Figure 14-8). Many grow to lengths of 1 to 3 inches. They lay eggs that hatch within 30 hours.

Catfish

Catfish have been developed into ornamental fish in small sizes and varieties. Most species prefer dark areas

FIGURE 14-7 Tiger barb.

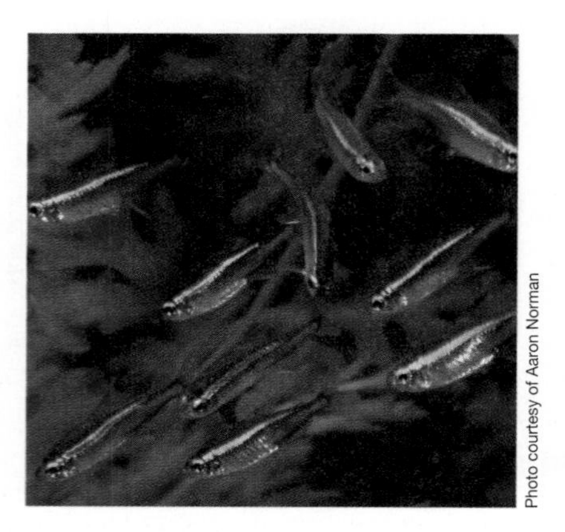

FIGURE 14-8 Neon tetra.

and like to hide in objects within the freshwater tank. They feed off of the bottom of the tank and can be used to keep the tank clean. Some favorite breeds include the upside down catfish, which swims upside down, and the glass catfish, which is clear and may have a colored appearance (see Figure 14-9). Catfish lay about 100 eggs that hatch in four to five days.

Guppy

Guppies are the most popular live-bearing fish. They come in a variety of colors and fancy-shaped tails and fins. They need larger amounts of food than other species. They usually reach lengths of 2–3 inches. Females tend to give birth to 200 babies at a time; only about 50 babies will survive. They give birth every four to six weeks. Many adults will eat the young if not separated. They are easy to raise and tend to be a hardy type of fish.

Swordtails

Swordtails are named for their long caudal fin. They come in a variety of colors and range in lengths of 3–5 inches. They bear live young and live in fresh water. The adults tend to eat the babies and should be separated from young. They eat a variety of food.

FIGURE 14-9 (A) Glass catfish; (B) upside down catfish.

FIGURE 14-10 Black molly.

Molly

Mollies have two types, the small fin and large fin breeds. Some large fin types get so large that they have difficulty swimming. Most mollies are black, but breeders breed for selective colors (see Figure 14-10). They live in large freshwater groups and bear live young.

Platy

Platys are a popular aquarium fish that come in a variety of bright colors. They grow to about 3 inches in length and live in fresh water. They are easy to keep and are very hardy. They also bear live young.

Photo courtesy of Aaron Norman

FIGURE 14-11 Angelfish.

Angelfish

Angelfish come in a variety of colors, shapes, and sizes (see Figure 14-11). Most live in salt water, but a few species can live in fresh water. They appear delicate but are hardy and capable of living a long time in a well-managed tank. They average in sizes of 6–10 inches in height. They lay eggs that hatch in about 30 hours. They are mouth brooders and will place the eggs on plants or sand sources prior to hatching.

Basslets

Basslets are small colorful fish that are popular in saltwater tanks. They are hardy fish and excellent for beginners just starting with saltwater fish. They tend to do best with other species but become aggressive with their own kind.

Butterfly Fish

Butterfly fish are beautiful and colorful saltwater fish. They require a large aquarium space. They range in sizes from 6–10 inches. They are territorial and are best isolated from other species and do best when housed alone.

Breed Selection

When beginning an aquarium, it is important to think about the types of fish that will be housed and determine if they will be compatible. Many pet shops and ornamental fish breeders will be able to guide owners about passive and aggressive fish breeds and which will not be compatible in the same tank. Another factor to consider is the size the fish will grow to as an adult and the number of fish that will be housed in the tank. Water type and plant sources will also need to be considered and are discussed in this chapter.

Nutrition

Fish can easily be overfed. The amount a fish is to be fed each day is called the ration. The ration should be provided in the correct diet that meets the needs of the fish species. The amount of food a fish eats in a day is directly related to the water temperature as well as the type of fish and the life stage. As water temperatures rise, the need for food increases. The food should be fed and eaten within a few minutes. Thus, careful observation will allow an owner to know how much to feed. Overfeeding creates more waste products in the water. Most people feed a commercial fish food made from flakes or pellets. Most of these diets include protein and grain sources. It is important to read all food labels for the ingredient list and what amounts are recommended for certain species. Table 14-2 summarizes typical nutritional needs for some common breeds of fish.

Equipment Needs

Ornamental fish require a specialized environment to survive and maintain a healthy life. These environments must be artificially created to act as their natural habitat. The needs will vary for each species. For example, a goldfish can survive in a basic system, whereas an angelfish or gourami requires a more complex system. Basic equipment begins with a water container. These items can be decorated with gravel or pebbles, plants, and other creative structures to add eye appeal to the system. Many types of fish water containers are available commercially.

TABLE 14-2

Fish Nutritional Needs

SPECIES	DIET NEEDS	SUPPLEMENTS
Cichlids	Commercial flakes	Seaweed
Angelfish	Commercial flakes; frozen bloodworms	Brine shrimp
Barbs	Commercial flakes	Brine shrimp
Bettas	Betta flakes; freeze-dried bloodworms	Brine shrimp
Catfish	Commercial flakes	Brine shrimp; sinking wafers
Discus	Discus flakes; frozen discus formula	Freeze-dried bloodworms
Goldfish	Goldfish flakes	Commercial flakes
Gouramis	Commercial flakes	Brine shrimp
Loaches	Commercial flakes	Freeze-dried brine shrimp
Tetras	Commercial flakes	Brine shrimp

The most common type for ornamental fish is the **tank** or **aquarium**. These containers hold water and come in a variety of shapes and sizes. The most basic tank or aquarium is a small fish bowl. This is a rounded glass bowl that holds about a gallon of water. It may be used for a single fish that is hardy and can survive in water that doesn't need added oxygen. The next size of a basic tank is the rectangular or square glass aquarium. They range in size from 10 gallons to large 100 gallons or more in size. All aquariums should be watertight and sealed to prevent leaking. Glass materials are preferred, especially with saltwater tanks. This allows for ease of cleaning and a smaller chance of leaks. The glass should be between 1/4 and 3/8 inches thick. Figure 14-12 illustrates a typical aquarium setup. The most important terms involved in learning the operation of the aquarium are the following:

- Air hose—allows flow of oxygen into the tank
- Air pump—controls the flow of oxygen into the tank
- Cover—the lid that lies on top of the tank
- Filtration system—allows removal of gases and solid materials from the water
- Gravel—pebbles located on the bottom of the tank that hold plants and other objects
- Heater—electric source that warms the water
- Light—electric source that lies within the cover to provide light into the tank
- Thermostat—area that provides the water temperature in a number scale

Basic and complex tank systems require proper equipment and supplies to maintain the tanks and the water quality. It is also necessary to maintain the proper health of the fish.

A **pool** or **fountain** may be used to house ornamental fish. These may be indoors or outdoors. They are usually decorative and may contain several species of fish, such as koi and goldfish. The species housed in these types of water containers must be hardy. The structures are usually made of plastic, fiberglass, or concrete materials. They must be watertight and easy to clean.

A **vat** is a large tank of water commonly used for reproduction and breeding systems. They are made of concrete or fiberglass that is heavy enough to hold hundreds to thousands of gallons of water. They must have proper water circulation (movement), aeration (oxygen or air), and drainage to maintain healthy fish.

Ponds are commonly arranged outside for larger species of hardy fish. They must be constructed for the climate and usually require proper heating sources and a way to keep the fish safe from predators.

Oxygenation is the process of keeping air in the water by using dissolved oxygen (DO). Oxygen that has been dissolved in the water is removed by the fish's gills and used for breathing. Water that has little to no movement has very little oxygen. If the fish do not have adequate amounts of oxygen, they will die. Several methods are available to provide oxygen to the tank. The first is **aeration**. This is the process of making bubbles within the water to create movement. This allows gases to transfer from the water and air. Oxygen is thus created in the water. Another way to allow oxygen into the water system is through the use of an air pump. The pump is used to force air into the tank through a plastic tube located in the bottom of the tank. In some ponds or vats, an air pump may

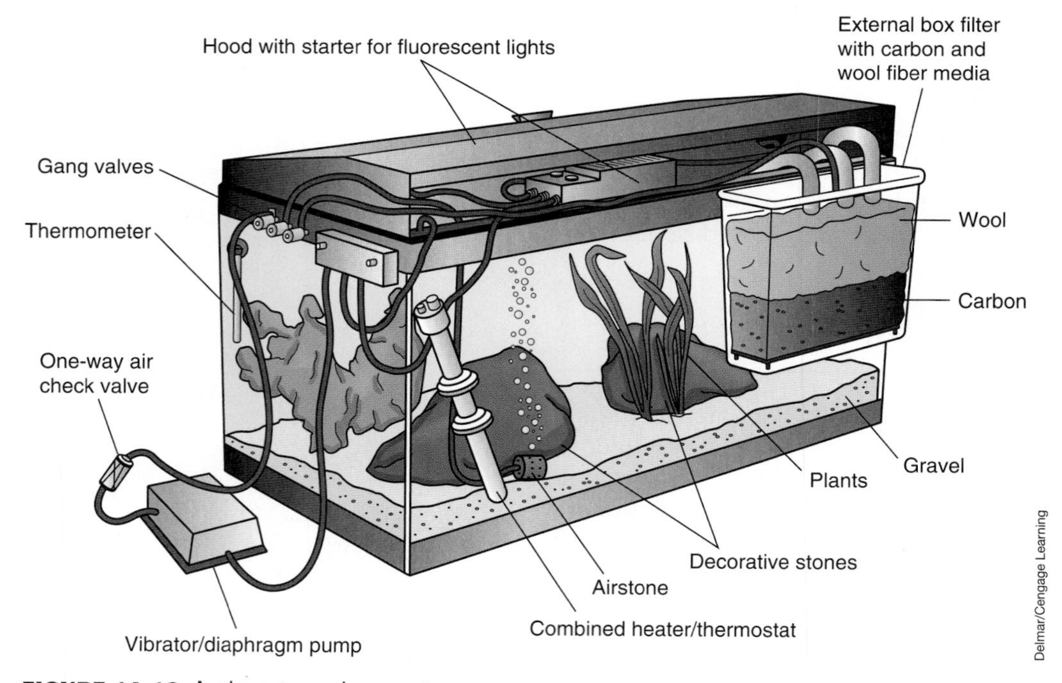

FIGURE 14-12 A glass aquarium system.

lie on the surface of the water. Finally, some tanks can be oxygenated through the use of live plants. This is known as **biological oxygenation**. Tiny plants called **plankton** and other type of live plants that are grown within the tank provide a natural source of oxygen to the water. This growth is called **photosynthesis** and allows for the release of oxygen in a natural process.

Filtration is a way to rid the tank water of solid waste materials and gases to provide a good environment for the fish. There are several filtration methods that may be used. First, **biological filtration** is used by providing bacteria and other living organisms in the water to change harmful materials into forms that are safe. Bacteria will feed on fish wastes, uneaten food, and gases that are created in the water. Snails, crawfish, loaches, and catfish may also be used to scavenge or remove foreign matter from the water. Through this method, the bacteria change nitrogen and ammonia gas forms into dissolved oxygen. **Mechanical filtration** is used with various types of equipment that filters the water to remove harmful particles and keep the water clear. Water flows over a filter placed in the water. Filters may be made of gravel, charcoal, and fiber materials called **floss**. Filters must be cleaned often and may become clogged with waste matters and stop functioning. Most of these systems have combination oxygen systems. Figure 14-13 shows various filtration systems. Table 14-3 summarizes various types of filters.

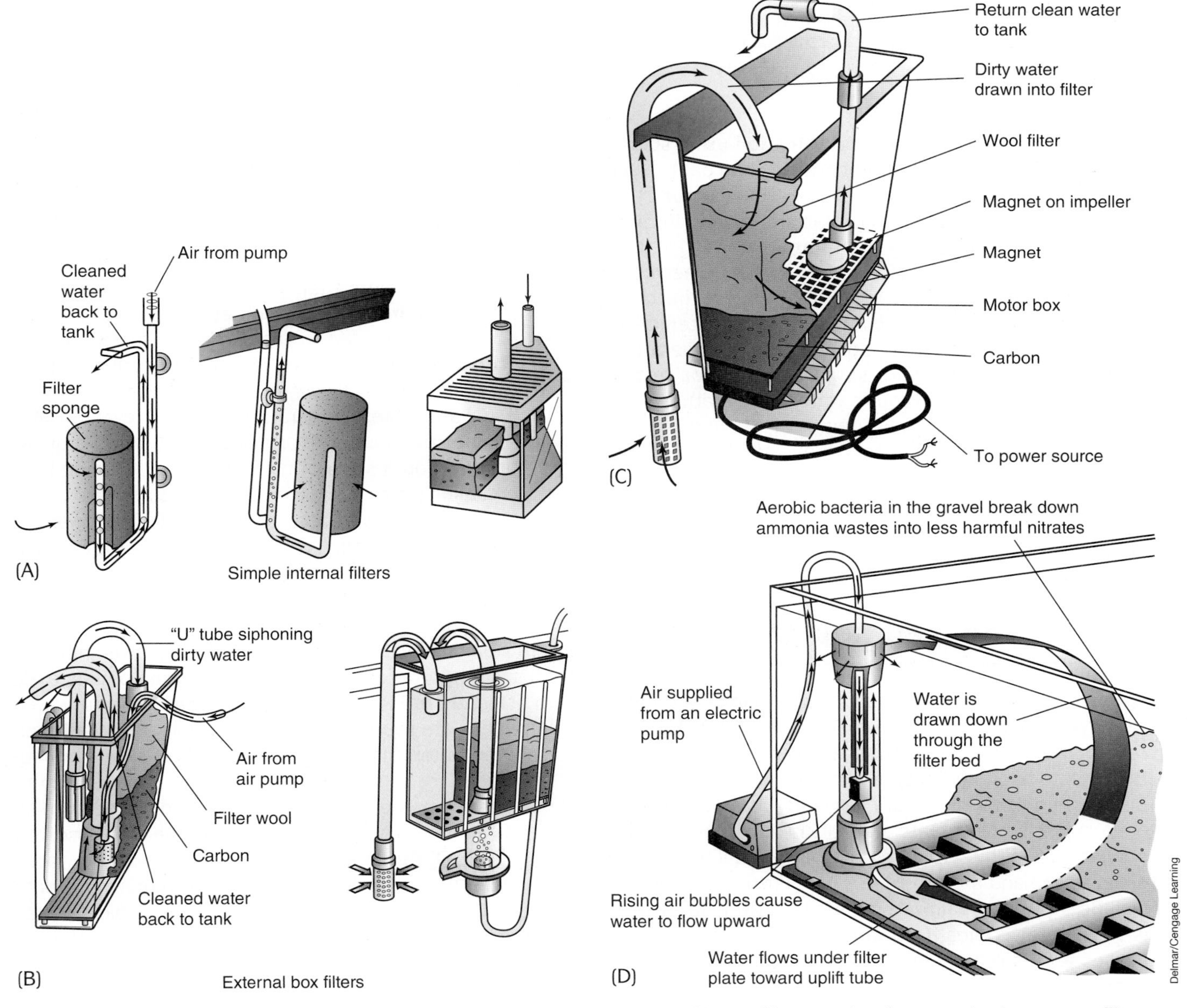

FIGURE 14-13 (A) Types of internal filters; (B) examples of external box filters; (C) example of a motorized or power filter; (D) example of an undergravel filter system.

TABLE 14-3

Types of Filters

Undergravel filter	Placed on the bottom of the tank; water pulled through the bed of gravel to screen out materials
Canister filter	Located outside of the tank; includes pumps and tubes that move water through a cup-like tube to remove materials; must be cleaned or replaced every other week
Outside filter	Hang on the back of the tank; water is moved through a tube holding filter materials; common for small tanks
Nitrifier/Dentrifier filter	Supplement the action of bacteria; expensive and only used in large, specialized tanks

Another method is **chemical filtration** using special chemicals such as ozone and activated charcoal to keep the water clear and from becoming a yellowish color.

Other types of equipment that are needed in more complex systems include a **thermometer**, which is used to measure the temperature of the water. Thermometers are often attached to or float within the tank. A **thermostat** is used to control the water temperature when the water goes below a certain temperature. Heaters keep tropical waters warm enough for certain fish species. Many fish require water temperatures that range from 70 degrees to 85 degrees Fahrenheit. Lights are also used within the tank to allow visibility of the tank and to provide a light source for fish and plant sources for growth. A light source should be placed on the tank for 10–12 hours a day. Caution should be used with electricity around water sources.

Water Sources and Quality

Water sources require a **habitat** that mimics the natural environment that an animal lives in. This requires research on the species that are to be housed within the tank to allow the proper climate, plants, rocks, fish breeds, and other necessary items. By using this knowledge, a person can create an artificial environment that provides a healthy setting for fish. Water quality depends on its source and the substances in it. Freshwater sources include tap water, rainwater, and well water. **Tap water** is water that comes from faucets in homes and businesses and is readily available but many times is not adequate for most tanks. Many public water systems have excessive amounts of chlorine, sulfur, and other chemicals in the water. These substances are safe for humans but are **toxic** (poisonous) to fish. Tap water can be prepared for tanks by allowing it to age. **Aging** is done by collecting water and allowing it to stand in an open container for at least 5 days. Some larger tanks will require large amounts of water. The containers that hold this water must be clean. Special tablets that clean the water can also be purchased to put in tap water sources. **Rain water** can be used as long as it is not polluted or from a contaminated source. **Acid rain** and other substances will be toxic for fish. **Well water** is used by breeders and people who have large vats or ponds. The water source should be tested to determine its contents since some well water may be unsafe for fish. Well water can be aged in the same manner as tap water. Saltwater can be provided from actual saltwater sources or **synthetic seawater** sources, which are manmade. Natural seawater can be collected from oceans and lakes. Synthetic seawater is made by mixing freshwater with a salt-based mix. Water quality is determined by the temperature, the pH levels, the nitrogen cycle, the amount of dissolved oxygen, and the amount of salt in the water. Water temperature should be within the appropriate range for the species that are housed in the tank. In some circumstances, the water will need to be warmed. Fish should be placed in water that varies only by one to two degrees from their previous water source. Fish do not do well in sudden temperature changes and will stress and likely die. Most ornamental fish species require water temperatures between 70 and 85 degrees. Saltwater varieties are usually slightly warmer. Heaters may be used to warm water temperatures. The **pH level** of water determines the **alkalinity** or **acidity** of water. Water pH is measured on a scale of 0 to 14, with 7.0 being neutral. Alkaline water is measured at levels above 7.0 and acid water is measured at levels below 7.0. The lower the number of the pH rating, the more acidic the water. Most ornamental fish require pH levels of 7.8 to 8.3. Changes in the pH can be due to materials in the water.

The **nitrogen cycle** is a process that converts wastes into ammonia and then into nitrites and then into nitrates. This is a natural occurrence in water. Ammonia is toxic to fish and produces an odor that comes from the water. When fish are overfed or waste materials are not filtered from the water, there may be a buildup of excess ammonia. There are commercial test kits that may be used to determine these levels. They must be used in a quick manner as ammonia quickly evaporates in the air. **Dissolved oxygen** (DO) forms a gas state of oxygen in the water. Fish use the DO to breathe. Most fish require 5.0 ppm of DO. Below 3.0 ppm DO, the fish will likely die. Oxygen meters can be used to measure the DO levels in the water. **Salinity** is the amount of salt in the water. Freshwater species of fish cannot survive in water with salt contents. Salt content is measured as **specific gravity** (SG) using a hydrometer. Water has a specific gravity or weight of 1.0. Saltwater has a higher SG and these species require water values between 1.020 and 1.024. The salinity can be increased by adding prepared salts to the water. Water sources should be changed periodically. This can be a huge job in cases of large tanks, vats, or ponds. Water management should be monitored to keep water clean and healthy. Fish will need to be temporarily relocated during the water changing process.

Catching and Handling Ornamental Fish

When the need arises to catch a fish from an aquarium, it is best to first empty some of the water from the tank. This will allow less space for the fish to move around, which means less time chasing it in the tank. The first and most common method of capturing fish is with the use of a net. Nets come in a variety of sizes and colors. The best color for catching fish is black or dark green. A white net may be used to direct fish away toward another net for capture. Nets should be soft and movable. If a net becomes coarse and stiff it should be replaced. The net may be placed in the water a few seconds before use to allow it to soften. It will also allow fish to grow accustomed to the net. The net must be moved slowly in the water. When the fish swims into the net, it should be removed immediately from the tank and placed in the necessary area. Sometimes fish may become tangled in the net. If this occurs, simply put the net and fish in a large water source and turn the net inside out gently to free the fish. If the fins are caught and not able to be freed, the use of scissors to cut the net may be necessary. Tips for using a net to catch fish are as follows:

- Guppies can be caught best by bringing a net up from beneath them. If they can't see the net coming, they will not try to escape.
- Use two nets to catch fast-moving fish or bottom dwellers. One net will herd them and one net will catch them.
- Stand a large net in one corner of the tank. Use a stick or straw to herd the fish into the net.
- Use a plastic baggie to catch angelfish and other delicate tropical fish to avoid damaging their fins.

Another method of capturing a fish is with the use of a Styrofoam cup, plastic baggie, or bottle. The objects used to catch fish should be dark in color. These objects should be clean and chemical free. Heavy items will sink easier and be less difficult to keep from floating to the top of the tank. The fish can be herded into the object and carefully removed from the tank. Catching a fish by using the hands may be necessary with larger species of fish. This must be done carefully as some fish can bite or have spines or whiskers used for defense. Use one hand to grasp the fish gently in front of the tail and the other hand to support the fish from underneath the body. This requires patience and care. The fish can then be removed and placed in the necessary area. The hands should be washed thoroughly after handling fish and equipment.

Basic Health Care and Maintenance

A regular **aquarium maintenance schedule** or AMS should be followed to ensure an ideal tank system is being kept (see Table 14-4). This is a list of important duties that should be completed and the frequency necessary for an adequate fish tank. Some duties are necessary daily while others may be completed weekly, monthly, or annually.

Breeding and Reproduction in Fish

Reproduction and breeding in fish has become a popular hobby, as well as a profitable industry. Male and female fish are housed together for spawning or mating

TABLE 14-4

AMS Fish Checklist

Daily Items	Provide food Check water temperature Monitor aeration Monitor filtration Remove any dead fish Observe fish behavior Note any unusual behavior
Weekly Items	Check water level Check pH levels Add water as needed Add chemicals as needed
Monthly Items	Change and add water Clean material on the tank bottom Remove algae Observe plants
Quarterly Items	Clean filter Check electrical connections Check pump Check hoses
Annual Items	Clean tank thoroughly Rinse bottom gravel or pebbles Change bottom every other year Replace light bulbs

to occur. Many fish will breed spontaneously and others will pair up. Sometimes fish will need to be separated from each other prior to mating to develop an interest in each other. Some fish will bear live young and others are egg layers. The **egg-laying fish** reproduce when the female fish releases eggs that are fertilized by the male's sperm. Incubation of the eggs lasts only a few days, depending on species. Some examples of egg-laying fish include goldfish, koi, gouramis, tetras, catfish, and barbs. The majority of ornamental fish species are egg-laying fish. Even with egg-laying species there are still many variations of how the eggs are incubated. Some breeds are **egg scatterers**. This means the fish spontaneously scatter their eggs around the tank as they swim. When housed in a tank with numerous fish, these eggs commonly are eaten by the other adult fish. Eggs being eaten in a tank can be avoided by placing pebbles, marbles, or plants within the tank. The eggs will rest within the bottom or stick on plants for safety. This is a common occurrence with goldfish.

Some species of fish deposit their eggs on plants, rocks, or other objects. Examples of fish that are **egg depositors** include cichlids and tetras. This allows eggs to be protected from predators. Other species will build a nest for egg protection. Examples of nest builders include bettas and gouramis. A few species of fish will carry their eggs in their mouths and are known as mouth brooders. A few fish species are **live bearing fish**. They will give birth to live young. These fish usually live in groups of five or more breeders and are called a **shoal**. Live young bearers mate and the male deposits the sperm inside the female. The fertilized eggs will develop into fully formed baby fish. Some females will store the sperm of the male until it is needed for reproduction. Examples of live bearing fish include mollies, guppies, swordtails, and platys. These fish are easy to breed and make great beginner fish for new breeders.

Fish Disease and Health

Fish need positive bacteria within their environment to remain healthy. The bacteria are necessary within the filtration system to keep the water quality proper and clean. Many tanks do not have enough good bacteria within the water source. In those cases, products that contain live bacterial sources can be added to the water. Natural bacteria growth occurs over several days to several weeks. Maintaining adequate amounts of bacteria within a fish's environment also helps to keep them healthy. This reduces the chances of disease outbreaks. Stress occurs when fish are handled frequently or when there is a sudden change in their environment. Fish that show a change in behavior may have a disease or be ill. Signs of diseases in fish include the following:

- Not eating
- Increase gill movement or respiration
- Gasping at the water surface
- Scratching on objects
- Folded fins
- Frayed fins
- Cloudy eyes
- Blood spots on body
- Fuzzy patches over body
- White spots on body
- Color changes
- Protruding eyes
- Large, swollen abdomen
- Material hanging from anus

Sick fish should be removed from the tank and held in isolation. A veterinarian who has fish knowledge should be contacted. It is recommended to treat the sick fish as well as the healthy tank where the fish was housed. When purchasing or receiving new fish, a quarantine tank should be used. All new fish should be observed for signs of disease before being placed with healthy fish. Some common diseases occur in ornamental fish. Veterinary assistants should be familiar with some basic information relating to these diseases. The following are some common fish diseases and conditions.

Fin Rot

Fin rot is a bacterial disease that begins when fin tissues start to erode and the tissue becomes inflamed. This may be caused by poor water quality, stress, or aggressive fish that have bitten or that attacked the fish. Treatment is done with a bath immersion in salt water if the fish species is salt tolerant, topical antibiotics, and antibiotics applied to the water.

Lymphocytosis

Lymphocytosis is commonly called "cauliflower disease." This is a viral disease that affects certain species of fish. Signs include white to gray growths over the fins and body. Causes may be stress or that a carrier fish is spreading the disease within the tank. There is no treatment and death is rare. The only means of control and prevention of the disease is through water quality.

Ich

Ich is a parasitic protozoan disease caused by a one-celled organism. It is commonly called "whitespot" due to white spots that develop over the fins and body. Ich may be caused from stress, poor nutrition, and poor water quality from contamination. All new fish and plants should be quarantined. The treatment is through treating the water for 10–14 days.

Fish Fungus

Fish fungus is also known as "cotton-wool disease" due to the appearance of tufts of growth that occur on the skin of the face, gills, or eyes, which resemble cotton or wool. The cause is fungus spores in the water. Treatment is to isolate infected fish and treat with antifungal medications. Salt can be placed in the water of salt-tolerant species by using 2 grams of salt per liter of water.

Cloudy Eye

Cloudy eye is a bacterial infection that may affect one eye or both eyes of tropical fish. It can cause damage to the nervous system. The cause is poor water quality. Prevention is through monitoring the water and changing it frequently while allowing proper filtration methods. Treatment is done with the use of antibacterial medications and adding salt to the water.

Ascites

Ascites may be caused by a bacterial or viral infection and is also called "dropsy" due to the appearance of the fish with this condition. The body will swell due to a fluid buildup and this causes the fish to lie on its side or have difficulty swimming. The body will develop a pinecone-like appearance to the scales. The causes are poor water quality and a buildup of ammonia in the water. Treatments include the use of antibacterial medications and salt in the water.

Swim Bladder Disease

Swim bladder disease is caused by a bacterial infection, poor diet, and from genetic deformities. The disease causes a fish to have difficulty swimming to the water's surface and in depths of water. It is common in goldfish. Some fish will not recover from the disease, but treatments may include a diet change, improved water quality, and feeding flakes instead of pellets.

SUMMARY

Ornamental fish are kept for many reasons, from breeding programs to entertainment to personal appeal. Most of these species are bred in fish tanks and not in their natural habitats. Some of these fish bear live young and others lay eggs. They require research as to nutrition and health care. Fish are also high maintenance when it comes to water sources, water quality, and housing conditions. Fish should be monitored closely and observed daily for any changes in behavior that may be a sign of disease. Veterinary assistants may work in a facility that tends to fish care or may become an ornamental fish breeder.

Key Terms

acidity water pH levels below 7.0

acid rain polluted rainwater

aeration oxygen or air created in the water by making bubbles

aging allowing a water source to stand in a container for at least 5 days

alkalinity water pH levels above 7.0

aquarium basic container that holds water and fish

aquarium maintenance schedule (ams) list of important duties that should be completed and the frequency necessary for an adequate fish tank

ascites bacterial or viral infection, also called "dropsy," that causes the fish to swell with fluid

biological filtration process of providing bacteria and other living organisms in the water to change harmful materials into forms that are safe

biological oxygenation the use of plants to create oxygen in the water

bottom dwelling fish fish that live and feed on the bottom or floor of a body of water

chemical filtration using special chemicals such as ozone and activated charcoal to keep the water clear and from becoming a yellowish color

(Continues)

cloudy eye bacterial infection that may affect one eye or both eyes of tropical fish

companion fish fish kept in people's homes as stress relievers and as entertainment

dissolved oxygen (do) forms a gas state of oxygen in the water

diversity difference

egg depositers fish lay eggs on areas such as plants, rocks, or the bottom of a water source for protection

egg scatterers the fish spontaneously scatter their eggs around the tank as they swim

egg-laying fish reproduce when the female fish releases eggs that are fertilized by the male's sperm

filtration method that allows removal of gases and solid materials from the water

fin rot bacterial disease that begins when fin tissues start to erode and the tissue becomes inflamed

fingerlings fish that are two years of age

fish fungus fungal disease known as "cotton-wool disease" due to the appearance of tufts of growth that occur on the skin of the face, gills, or eyes, which resemble cotton or wool

floss material within a filter made of gravel, charcoal, and fiber materials

fountain structures made of plastic, fiberglass, or concrete that hold fish species and must be watertight

freshwater fish species of fish that live in freshwater with very little salt content

fry young newborn fish

gills organs that act as lungs and allow fish to breath by filtering oxygen in the water

habitat area where an animal lives

ich parasitic protozoan disease caused by a one-celled organism

lateral line organ located just under the scales that picks up vibrations in the water

live-bearing fish fish that bear or give birth to live young

lymphocytosis viral disease commonly called "cauliflower disease"

mechanical filtration various types of equipment that filters the water to remove harmful particles and keep the water clear

nitrogen cycle a process that converts wastes into ammonia and then into nitrites and then into nitrates

ornamental fish popular pet kept in aquariums, which are small in size and vary in color and appearance

oxygenation process of keeping air in the water by using dissolved oxygen

pH level the quality of water on a scale of 0 to 14, with 7.0 being neutral

photosynthesis the process of sunlight allowing plants to grow in a water source

pigment the color of the skin

plankton tiny plants that grow in water sources

ponds outside structure that varies in size as to how much water and space it contains

pool structure that holds a large amount of water and is made of plastic, fiberglass, or concrete and holds large amounts of fish

rain water source of water from the sky that can only be used if not contaminated or polluted

run group of fish

salinity the amount of salt in a water source

saltwater fish species of fish that require salt in their water source

scales body covering that serves as skin and protects the outside of the fish

school group of fish

shoal fish usually live in groups of five or more breeders

specific gravity (sg) the weight of a liquid measurement

surface-dwelling fish fish that live and feed near the surface or top of a body of water

swim bladder air-filled sac that prevents fish from sinking and allows them to float and move through depths of water

swim bladder disease bacterial disease that causes a fish to have difficulty swimming to the water's surface and in depths of water

synthetic seawater water that is manmade

tank basic container that holds water and fish

tap water water that comes from faucet

thermometer used to measure the temperature of the water

thermostat used to control the water temperature when the water goes below a certain temperature

toxic poisonous

tropical fish small, brightly colored fish that require warm water and are naturally found in warm, tropical locations

vat large structure that is made of metal or concrete that holds large amounts of water and is commonly used for breeding and reproduction of fish

well water water source from a well

yearlings fish that are a year of age

REVIEW QUESTIONS

1. Where do ornamental fish occur naturally?
2. Give six examples of ornamental fish and note if they are freshwater or saltwater fish and if they bear live young or lay eggs.
3. List equipment that is needed for an ornamental fish tank.
4. What are oxygenation and filtration? Why are they important to fish?
5. What are the types of filters used in aquariums?
6. What are some sources of water for tanks?
7. What are the areas of importance in water quality?
8. What are some signs of disease in fish?
9. Describe a regular maintenance schedule for aquariums.
10. What are some guidelines for feeding fish?

Clinical Situation

Photo by Aaron Norman

Allie, a veterinary assistant at The Feline Hospital, greets a client who walks in the front door. She asks how she can help the young man: "Hi, how may I help you?"

"I am not a client here but I have some tropical fish that have been sick and some are now dying. They were very expensive and I would like to talk to someone about what can be done." The man seems very upset about the situation; however, the veterinarian at the hospital doesn't work with tropical fish. Allie explains this and feels bad that she can't help the man.

"Dr. Wills doesn't treat fish, he sees cats and some small animal species. I am not even sure if he has any experience or knowledge with fish. He is currently in surgery and not available. I am sorry we can't help you."

"Thanks anyway," he says and leaves the hospital.

- What could have been done in this situation to better serve the client?
- How would you have handled this situation?
- What was done correctly in this situation?

15

Courtesy of Photo Disc

Wildlife Management and Rehabilitation

Objectives

Upon completion of this chapter, the reader should be able to:

- Identify common wildlife species
- Explain and understand the importance of wildlife in veterinary medicine
- Explain the classes of wildlife
- Discuss important health management practices in wildlife
- Implement wildlife management practices
- Discuss and describe the career of wildlife rehabilitation

Introduction

Wildlife describes wild animals that occur in their natural environments. These animals are enjoyed by people in many ways such as animal watching, photography, and hunting. Wild animals may become sick or injured and need veterinary attention. Some careers focus on the care of wild animals. Some facilities receive wild animals that have been found sick or injured. Wildlife management has become increasingly an area for veterinary medicine as the natural environments of many species are destroyed.

FIGURE 15-1 The American crocodile in its natural habitat.

The Importance of Wildlife

Animal wildlife is any living animal that has not been domesticated. This includes mammals, birds, insects, fish, rodents, reptiles, and many other types of animals (see Figure 15-1). These animals may benefit humans in several ways. They may provide food, products (such as fur), or entertainment.

Wildlife as a Food Source

A **game animal** is one that is hunted for food. Common game animals in the United States include deer, elk, bear, ducks, and fish (see Figure 15-2). Game animals may be raised in captivity for release into the wild where populations are decreased or sparse. Some areas have increased and overpopulated areas of game animals, which causes environmental damage and spread of disease. Wildlife is hunted as food sources and for use of byproducts, such as skin pelts, antlers, or hooves. Many areas of the country have specific hunting seasons, depending on the animal species.

FIGURE 15-2 The white-tailed deer is often hunted in the United States.

Wildlife as a Resource

Animals may also be hunted or trapped for fur or hides. **Trapping** is typically done with smaller mammals that have furs and pelts that are of economic value. Trapping involves the capture of small animals within wire cages or traps. Some of these animals may be used as food sources as well. Many areas of the country have trapping seasons based on the animal species. Examples of animals that may be trapped include beaver, muskrats, and weasels.

Wildlife as Entertainment

Many people enjoy animal watching, photography, and drawing wildlife. People enjoy the beauty of wild animals. Many areas of the world have locations that are visited to view animals that occur naturally in the environment, serving as a vacation and recreational purpose to people. Many parks and conservation areas dedicate their land and services to protecting the animals that live there. Zoos and wildlife parks have increased in popularity. This allows wild animals that may not be able to survive in the wild a safe place to live and be observed (see Figure 15-3). Many of these facilities have veterinary staff who care for the well-being of the

FIGURE 15-3 Zoos provide people with the opportunity to see wild animals in a manner that is safe for both the animal and the spectator.

FIGURE 15-4 The American alligator is endangered.

animals. They may also have breeding and reproduction programs that continue to increase populations in species that are endangered or extinct in their natural habitats. **Endangered animals** are a species that are decreasing in numbers in the wild and are close to becoming extinct (see Figure 15-4). **Extinct** means that a species of animal no longer exists in the wild and may only be found in zoos or wildlife parks.

LIST OF TOP U.S. ENDANGERED WILDLIFE

- Polar Bear
- Humpback Whale
- Grizzly Bear
- Gray Wolf
- Florida Panther
- Jaguar
- Otter
- Chinook Salmon
- Bighorn Sheep
- Manatee
- Puma (cougar)
- Green Sea Turtle
- American Alligator

Control of Wildlife

Hunting is also a means of wildlife population and disease control. Some species become overpopulated in the wild. This causes any disease breakouts to pass quickly throughout the species. Fishing also serves the same means for certain fish species. Hunting and fishing are delegated by each state with seasons that are controlled by when a person may hunt or fish a specific species. These control methods are a humane way of controlling large populations of animals.

The Role of Veterinary Medicine in Wildlife Management

As land begins to disappear, the habitats and food sources of wildlife species will disappear. This occurrence is known as **urbanization**. Building cities and homes for the benefit of people takes away the environments of animals that can't survive in a human world. This continues to be a threat to many wild animals. Wildlife veterinarians may work in research, zoos, and wildlife parks or in government areas to serve wild animals that may become sick or injured. Their role is in serving the state or country in discovering disease outbreaks, the handling and care of wildlife, investigating the decrease in numbers of species, and working with public concerns regarding wildlife.

Classes of Wildlife

There are several classes of wildlife. The game animals are the wildlife species that are used for food and other products produced by wild animals. These species may be large or small. Many locations have designated hunting seasons for a variety of animals. The State Department and **Game Commission** are agencies that monitor these seasons by providing rules, regulations, and licenses for hunting. Every state within the United States has its own Game Commission and State Department. The State Departments govern the laws of each state. The Game Commission agencies also serve educational and conservation purposes for wildlife and people. Popular species of game animals include deer, bear, rabbit, and squirrel (see Figure 15-5). **Non-game animals** are wild species that are not hunted for food. They provide beauty and entertainment only. Some of these animals are food sources for predators or act as scavengers in the environment. Many of these animal species are found in urban and residential areas. Examples of non-game animals include skunks, opossum, mice, raccoons, and porcupine (see Figure 15-6).

FIGURE 15-5 The bear is an example of a game animal often hunted for food or fur.

FIGURE 15-6 The raccoon is an example of a non-game animal often found in residential locations.

Game fish are fish species that are in the wild and are used for food and sport purposes. Game fish exist in both fresh and salt waters. Many states have fishing seasons also monitored and licensed by the Game Commission. Examples of freshwater game fish include bass, trout, and catfish (see Figure 15-7). Saltwater examples of game fish are tuna, flounder, and snapper (see Figure 15-8). **Non-game fish** are species of fish that are too small or too large to be caught on a line. They may be found in any type of water, both fresh and salt. Examples of non-game fish include sunfish, bluegills, and sturgeon.

Game birds are wild bird species hunted for food and sport. They are divided further into three categories. **Migratory birds** are birds that move from location to location, sometimes out of the local area or state. Examples of migratory game birds are the crow and woodcock. **Waterfowl** are birds that swim and spend a large amount of time on the water. They include a variety of species of ducks and geese (see Figure 15-9). **Upland game birds** are wild species of birds that spend most of their time in the woods and do not move out of the area. They include the pheasant, quail, and turkey (see Figure 15-10). Game birds are regulated by the state and federal government and by the Game Commission.

Birds of prey or **raptors** are large species of birds that are protected by law and are not legal to hunt or shoot. They are hunters that prey on other wildlife such as rabbits and squirrels. They have large **talons**, or claws used for capturing prey and sitting trees.

FIGURE 15-7 The brown trout is an example of a freshwater game fish.

FIGURE 15-8 Tuna is an example of a saltwater game fish.

Examples of birds of prey include eagles, owls, and hawks (see Figure 15-11). **Non-game birds** include all other species of birds and songbirds. Many are found in areas near homes and cities. They live in a variety of environments from woodlands to waterways. Examples of non-game birds include robins, pelicans, blue jays, doves, cardinals, and wrens (see Figure 15-12).

FIGURE 15-9 Geese are an example of migratory birds.

FIGURE 15-10 Turkeys are an example of upland game birds.

FIGURE 15-11 An eagle is an example of a bird of prey.

FIGURE 15-12 The cardinal is an example of a non-game bird.

Wildlife Health and Management Practices

The habitat of wildlife is the area where animals live and provides food, water, space, and cover. All four of these components must be met to support the life of wild animals. If one is missing or lacking, it is known as a **limiting factor**. Many animals eat browse, which are wood-like, broad-leafed plants used as forage. When homes are built in wooded areas and browse is removed, it becomes a limiting factor. Even when an area meets the needs of the wildlife in its habitat, there are limits to how much of a population can survive and be sustained. This is known as the habitat's **carrying capacity** or the number of animals that the area can support. This number will increase and decrease over the year. When an animal claims an area as its own, it becomes its **territory.** Wildlife will protect this area and keep other wildlife from entering. Space can easily become a limiting factor.

Wildlife management is the practice of researching the needs of wildlife, providing them the essentials of life, and monitoring their survival. Many people begin their career in wildlife management by taking courses in **mammalogy**, or the study of mammals. There are many college programs that offer **wildlife biologist** degrees. Wildlife biologists research and study data on all types of wild animals to determine how wildlife can be saved and protected (see Figure 15-13). Many biologists study a particular species of wildlife for a number of years to determine the needs of the animal. Wildlife health practices include monitoring wild species for diseases and focusing on the control of those diseases. That may mean determining a treatment for the disease and how to treat the population of the species as a whole. It may also mean how to humanely control the populations by increasing hunting or capturing and destroying sick animals affected by the problem. Other practices include manufacturing vaccines to control the spread of disease and capturing

Photo courtesy of U.S. Fish and Wildlife Service. Photo by Pedro Ramirez, Jr

FIGURE 15-13 Wildlife biologists study animal species to find ways to control and protect the animals.

and tagging wildlife to monitor their health, provide treatment, and monitor the range the animal travels in the area.

Capture and Restraint of Wildlife

Capturing and handling wildlife can be a risk for both the animal and the wildlife team working with the animal. Most wildlife is captured by trapping in humane cages that contain the animal safely or by tranquilizers that are used to sedate the animal for safe handling. Tranquilizers are administered with a dart gun so that people do not have to get close to wildlife and become injured. Both of these restraint methods are safe for the animal and for the people working with them. However, stress can be caused in any wild animal that can make it a risk for capture. This is why it is important that anyone working with wildlife be properly trained. The term *sedate* means to put to sleep with medicine that allows an animal to be handled without it moving. Sedated animals must be monitored for complications and for the possibility of awaking before procedures are complete. Proper sedatives, doses, and medications should be determined by a veterinarian. It is also important to keep accurate medical records of these situations.

Wildlife Rehabilitation

Wildlife and people can live together in certain areas. However, animals and people living in the same habitats means more injuries, more disease, and more death to the wildlife. Areas that have populations of wildlife usually have wildlife rehabilitation centers or wildlife rehabilitators. **Wildlife rehabilitation centers** are areas that house and care for wildlife that are sick or injured (see Figure 15-14). These centers offer veterinary staff who are knowledgeable and experienced in handling and caring for wild animals. The staff usually consists of veterinarians, veterinary technicians, and veterinary assistants who have been trained to properly handle wildlife. Most centers also rely on volunteers to help with some basic duties and raise money for these programs, as finding profits to purchase supplies and pay employees is difficult. These centers are not businesses run like a veterinary practice where owners bring a pet for services and pay for the care of them. Rather, a rehab center is where wildlife are cared for and either returned to the wild or placed in a zoo or wildlife park

Delmar/Cengage Learning

FIGURE 15-14 Wildlife rehabilitation centers will take in and care for sick, injured, or young abandoned animals.

because they will no longer be able to survive on their own in a natural setting. **Wildlife rehabilitators** are individual people who care for wild animals that are sick or injured. They may offer the care in their homes, in private veterinary facilities, or at local nonprofit centers that they have developed specifically for wildlife. Many states require that wildlife rehabilitators take a course and exam to become a licensed rehabilitator. Many wild animals are protected within the states and housing them is illegal, and government agencies have been created to oversee wildlife rehabilitators and require them to be licensed. Handling wildlife can be dangerous to the handlers and the people in the local area. For this reason, most states also require a permit to house any wild animal on private property. Most rehabilitators work with a veterinarian who has experience in wildlife and who can offer a diagnosis of illness and injury and recommend the proper treatment for the animal. Wildlife rehabilitators should be able to administer basic first aid care and basic physical therapy treatments to wildlife species. State agencies for wildlife rehabilitation include the following:

- State Game Commission
- U.S. Fish and Wildlife Service
- U.S. Department of Agriculture

Wildlife rehabilitators' education should begin with studies in biology and ecology. **Biology** is the study of life. **Ecology** is the study of the environments that animals live in. It is important that animal behavior courses are taken to learn normal and abnormal behaviors in animals. Other areas of knowledge include wildlife nutrition, natural history, environmental needs, and proper caging requirements. A college degree is not required to become a wildlife rehabilitator, but it is essential that a person become educated and learn the proper handling and health care techniques required with wild animals. Many veterinary technology schools and veterinary schools are offering courses in wildlife management and rehabilitation.

The goal of wildlife rehabilitation is to provide professional care for diseased, injured, and orphaned wildlife so they can ultimately be returned to their natural habitats. Wildlife rehabbing is not an attempt to turn wild animals into pets. Fear of humans is a necessary trait that these animals must have to survive.

SUMMARY

Wildlife is an important part of the environment and means a lot to many people. They serve as beauty, entertainment, food, and other products to our society. Wildlife management has become an important part of today's world, as necessary resources for wildlife are being lost. This includes food, water, space, and cover. Wildlife is being threatened as more people move into their environments. For this reason, wild animals face more disease and injuries than ever before. Wildlife rehabilitation and veterinary care has become an important part of the veterinary field. For additional information on the field of wildlife rehabilitation or wildlife services, see Appendix C.

Key Terms

biology the study of life

birds of prey large birds that are hunters and eat small mammals and rodents; also protected by law and are not legal to hunt

carrying capacity the number of animals living in an area that can be supported and sustained by that area

ecology the study of the environment that animals live in

endangered animal species that are decreasing in numbers in the wild and are close to becoming extinct

extinct animal species that are no longer found in their natural habitats

game species of wildlife that are hunted for food or other products

game animals wildlife species that are used for food and other byproducts produced by wild animals

game birds wild bird species hunted for food and sport

game commission government agency that monitors the seasons by providing rules, regulations, and licenses for hunting and serves the public in education and conservation purposes

game fish fish species that are caught for food or sport purposes

(*Continues*)

hunting the tracking and killing of animals for sport, food, or other resources

limiting factor missing or lacking one of the essential components to support the life of a wild animal

mammalogy the study of animals

migratory birds that move from location to location throughout the year

non-game animals animals that are not hunted for food or other products

non-game birds species of birds and songbirds that are not hunted for food

non-game fish fish species that are either too small or too large to catch and eat as a food source

permit a form that states a person may legally house and care for a wild animal

raptors large birds of prey

talons claws on a bird of prey

territory an area of land that has been claimed by an animal

tranquilizers used to sedate animals for safe handling

trapping the capture and restraint of wild animals

upland game birds wild species of birds that spend most of their time in the woods and do not move out of an area

urbanization disappearance of wildlife due to loss of land, food, and other habitat resources as people take over their land

waterfowl birds that swim in water and spend a large amount of time on the water

wildlife any living animal that has not been domesticated

wildlife biologist person who researches and studies data on all types of wild animals to determine how wildlife can be saved and protected

wildlife management the practice of researching the needs of wildlife, providing them the essentials of life, and monitoring their survival

wildlife rehabilitation centers areas that house and care for sick or injured wild animals, often run by a veterinary staff

wildlife rehabilitators individual people who care for sick, injured, or orphaned wild animals

REVIEW QUESTIONS

1. What animal species are considered wildlife?
2. What is the difference between endangered and extinct animals?
3. How do animals become endangered or extinct?
4. What are the classes of wildlife? Give some examples of each class.
5. What are the four requirements of a wild animal's habitat?
6. What is a wildlife biologist?
7. What is a wildlife rehabilitator?
8. What are the requirements of a wildlife rehabilitator?

Clinical Situation

iStock/Lisa DeMore

Alison, a veterinary assistant and wildlife rehabilitator, works at Davis Animal Hospital. The veterinarians at the hospital are also wildlife specialists. Alison has just received a call about a young fawn that was hit by a car along a local road. The people who hit the young deer are bringing it to the facility immediately. They stated it had some bleeding injuries and seemed dazed and scared. They would be arriving within 10 minutes. Alison alerts the staff of the emergency and they prepare the treatment area.

When the clients arrive with the young injured deer, they say the fawn has been trying to stand.

"The deer is very scared and is fighting us to keep it lying down," says one client as they carry the deer into the treatment area.

The staff begins treatment and Alison shows the clients to the front door and thanks them for bringing in the injured animal. They ask, "What will happen to the animal now?"

- How would you answer the client's question?
- Who assumes the responsibility for the animal?
- What would you have done if the facility was not equipped to care for wildlife?

16

Courtesy of PhotoDisc

Zoo and Exotic Animal Production Management

Objectives

Upon completion of this chapter, the reader should be able to:

- Identify common exotic and zoo animal species
- Explain the definition of an exotic animal
- Explain the purposes and uses of exotic animals
- Describe the major classes of exotic animals
- Discuss the health practices and needs of exotic animal production and management

Introduction

Exotic animals are popular in zoos and, depending on the species of animals, may be an exotic pet for some people. Many people enjoy watching and learning about the numerous types of different animals that they would not normally see in a natural or wild setting. These exotic animals require specialized health care, diets, handling and restraint, and environmental resources that allow them to remain healthy and stress free.

New species of animals are developed from several of these exotic animals. Many times veterinary professionals do not have information on how to properly care for these new species. These new species depend on humans to survive and often learning about their needs is by trial and error. They also require more care and attention than the average domestic animal.

The Exotic Animal

An **exotic animal** is an animal that is not native to the area where it is raised and may be rarely found in its natural habitat. Exotic animals are now being bred for new characteristics, such as color or hair coats, and as exotic pets. Many of these species require specialized care and additional resources to survive and be healthy. Most exotic animals are kept as companion animals or breeding and production animals. Most are newly domesticated animals that still have some wild instincts. Some may be used as food sources or for products they provide to humans. They may also be used for entertainment purposes, such as zoo and circus performances. Others may be used for investment purposes, such as sales or breeding.

Photos by Bill Tarpenning. Courtesy of U.S. Department of Agriculture

FIGURE 16-1 (A) Llamas are bred and raised for meat, hair, and as pack animals; (B) young llama.

Purposes and Uses of Exotics

A variety of exotic animals are raised as food sources or other products used by people. These animals may be considered agriculture in respect to food sources. In the United States, llamas, alpacas, bison, deer, elk, alligator, and ostrich are all raised for food and other **byproducts**. These animals are raised, fed, and cared for by humans in a similar environment to that of a beef or dairy farm. They are produced for the benefit of humans and must be raised in a controlled environment. Some species are valuable and costly. They can also be challenging to care for as there are limited veterinary resources for certain species of exotic animals.

Llamas

Llamas are mammals that originated in South America and are used for their meat, hair, and as pack animals. Many are now becoming alternative pets, meaning they are being used in a way that they were not bred for. Llama meat is low in fat and has a desirable taste. Few llamas are eaten in the United States. Breeding llamas can sell from several hundred dollars to several thousand dollars. Another alternative use for llamas is as a guard animal. They can be aggressive and will protect property and other animals. They are now being used for guarding flocks of sheep and goats. They range in sizes of 250–450 pounds and grow between 5 and 6 feet tall (see Figure 16-1). Their average lifespan is 15–25 years. Llamas are ruminants and can spit regurgitated food when upset or scared. A *ruminant* is an animal that has a stomach with four sections for breaking down food as it passes in the digestive tract. *Regurgitation* is the process in which food that has been chewed and swallowed is brought back into the mouth to be re-chewed for further break down. This is a natural process in ruminants.

Alpacas

Alpacas are mammals that are similar to llamas and they also originated in South America. Alpacas are smaller in size by about half and have much finer and silkier hair coats than those of llamas (see Figure 16-2). Alpacas are raised for their hair, which is used to make clothing and other linen products. The hair coat of the alpaca is very soft and provides insulation and warmth for winter clothing and other products. Alpacas are profitable animals. They range in weight from 100–175 pounds. They have similar lifespans to that of a llama. **Breeding stock** alpacas are very expensive because it costs a lot to achieve high hair quality. The term *breeding stock*

FIGURE 16-2 Alpacas are bred mainly for their hair.

FIGURE 16-3 Bison are raised for beef production.

means that the male and female animals are raised specifically for reproductive purposes. Females may sell for $20,000–$30,000 and a male breeding alpaca can range from $50,000–$60,000 depending on the bloodlines and the quality of the animal.

Bison

Bison are mammals that are ruminants, related closely to cattle. They are also known as the American Buffalo. Bison are raised for meat and byproducts, such as their skin pelts and horns (see Figure 16-3). Bison are large animals that may become aggressive and dangerous. They require large amounts of land, secure types of fencing, and specialized handling and restraint equipment. Bison average around 1,000–1,500 pounds, stand between 6 and 7 feet tall as adults, and live to an average age of 30 years. They are normally quiet animals but do have the ability to jump over 6 feet high and run at speeds faster than many horses. The average price for breeding quality bison is several thousand dollars. Many states have regulations on bison farming and reproduction. Bison ranches have become increasingly popular over the last several years due to the fact that bison meat is low in cholesterol and lower in fat content than beef. Thus, the industry has become an alternative profitable business method to that of the beef industry. Several producers have begun **cross-breeding** bison and domesticated beef cattle breeds to produce beef animals with healthier and tastier meat products. The cross of these two animals is called the **beefalo**. *Cross-breeding* is taking two different breeds of related animals and producing an animal with a combination of characteristics from each animal.

Elk

Elk breeding and marketing has become a rewarding and profitable business. Elk are relatively easy to breed and maintain. They need adequate amounts of land but are adaptable to a variety of habitats. Elk are bred for their meat and byproducts, including the **velvet** that is shed from the antlers and used in medications and as nutritional supplements (see Figure 16-4). The byproducts of an animal are the parts that are not the main profit or production source, such as the antlers, hooves, hair coat, or internal organs. Elk antlers have the ability to produce ½ pound of new growth in a single day. Elk are also raised for hunting and release programs, which breed and release elk back into the wild where populations are decreasing. They are ruminants that are related to deer and cattle. In fact, elk are raised very similar to cows. Elk are considered the livestock of the future.

FIGURE 16-4 Elk are bred for meat and byproducts.

Deer

Deer are raised for their meat, which is called **venison** (see Figure 16-5). They are also raised for breeding stock, scents for hunting, and their antlers. Deer are ruminants that graze on pasture. They require specialized fencing due to their ability to jump great heights. Many different species of deer are raised, such as red deer, mule deer, and white-tailed deer. Deer are relatively easy to raise but a concern for the spread of diseases has been an issue within the industry. Production management must be thorough to keep all deer healthy within the farm site.

Ostrich

Ostrich are large, flightless birds that are raised for their meat, feathers, oils, and other byproducts. Ostrich stand about 7 feet tall and may live up to 70 years (see Figure 16-6). They can run at speeds of 40 mph and can kick causing severe injury. Ostrich do not require large amounts of land, but they do need proper fencing and handling training. They can produce between 30 and 50 eggs per year. They are adaptable to a variety of habitats. Ostrich production has considerably reduced as the popularity of ostrich farming in the last several years caused an overabundance of ostrich, and the needs in the industry have decreased. However, rearing ostrich still remain a marketable business.

Alligators

Alligators are large reptiles that are raised for their meat, skin, byproducts, and for release into the wild (see Figure 16-7). There are over 1.5 million alligators in the wild in Florida alone. In some locations, alligators are still an endangered species. Alligator production became popular in the 1980s in the southern United States and thus the numbers in the wild have been increasing. In fact, several southern states have organized hunts for alligators. The meat is a delicacy in

Photo courtesy of PhotoDisc

FIGURE 16-5 Deer are raised for meat called venison.

Photo courtesy of PhotoDisc

FIGURE 16-6 Ostrich are raised for meat, feathers, and other byproducts.

Photo courtesy of PhotoDisc

FIGURE 16-7 Alligators are raised for meat, skin, and byproducts.

parts of the United States and other countries. It has a firm texture and distinct flavor. The skin is used to make leather products such as boots, belts, and purses. Alligators vary in their needs regarding nutrition, housing, and climate. They grow and reproduce well in captivity. Alligators are dangerous wild animals and handling and restraint are critical for safety purposes, both to the handler and the animal.

Rare Breeds

Other **rare breeds** of animals are bred and raised for a variety of purposes. Some are considered alternative livestock and others are exotic species. These include mammals, reptiles, birds, and aquatics. Examples are caiman, flamingos, zebu cattle, and zebra. Rare breeds are animals that are newly developed and not common or old breeds that were once popular and are now decreased in numbers. Producers and breeders take pride in having a rare breed thrive and exist. Many of these animals are domesticated and live well in the care of humans. Efforts are made to continue quality breeding and keep these animals rare. The popularity of breeds usually leads to poor breeding practices and genetic health problems that reduce the quality of the species. Genetic health problems occur in animals that are not carefully bred; these animals inherit unwanted characteristics or diseases from their parents.

Other species that are slowly becoming popular as exotic alternative species for reproduction, food, and byproducts include eels, mink, turtles, antelope, snakes, sharks, and several species of game birds.

FIGURE 16-8 Bear playing with a ball in the zoo.

Exotic Animal Businesses

Exotic animals are obtained from a variety of sources. Some are captured in the wild. Others are purchased from existing breeding programs. Others are zoo or circus performing animals used as entertainment (see Figure 16-8). Each type of business requires that animals be cared for by an experienced veterinary staff. Many small and large game animals that are marketed from wild animals, such as deer, elk, and alligators, require specialized health care. Exotic animals that are purchased from breeders as future investments include llamas and alpacas. These animals are raised for profit. The general care, housing, purchase, nutrition, and health needs are usually much more costly than those of a domestic animal. Pot-bellied pigs were once considered exotics and due to a lack of interest in the product, profits were not plentiful for breeders. Some exotic animal species become pets, as well. Monkeys, tigers, deer, bear, and other species have been raised in captivity as pets. These animals are not always safe around humans. The law requires a permit to keep any exotic species of animal. This is usually regulated by the state's Department of Agriculture or the U.S. Department of Agriculture.

Exhibition (Zoo) and Performance Animals

A **zoo** is also called a *zoological garden* that houses wild and exotic animals and plants in a park-like setting. The animals are seen by the public in an area that closely resembles the animal's native home (see Figure 16-9). Many of the species are from other countries or locations and kept in enclosures that are much like their native lands. Many of these animals are housed to breed in captivity to allow the species to continue to thrive. Zoos serve several purposes such as entertainment, educational resources, research, and conservation. Many of the employees of a zoo have a background in **zoology**, which is the study of animal life. A **zookeeper** is a person who manages the

Delmar/Cengage Learning

FIGURE 16-9 Wild animals are housed and raised in zoos for study and entertainment.

TABLE 16-1

Famous Zoos of the United States

Baltimore Aquarium	Baltimore, Maryland
Bronx Zoo	Bronx, New York
Busch Gardens	Williamsburg, Virginia Tampa Bay, Florida
Cincinnati Zoo	Cincinnati, Ohio
Henry Doorly Zoo	Omaha, Nebraska
Philadelphia Zoo	Philadelphia, Pennsylvania
Pittsburgh Zoo	Pittsburgh, Pennsylvania
San Diego Zoo and Wild Animal Park	San Diego, California
Sea World	San Diego, California Orlando, Florida San Antonio, Texas
Smithsonian Institute and National Zoo	Washington, D.C.
St. Louis Zoo	St. Louis, Missouri

zoo and the animals that are kept there. Zookeepers train the staff, oversee the health care needs of the animals, decide on proper nutritional needs of the animals, clean the animal enclosures, and give educational tours to people.

Animals are acquired by zoos in several different ways. Some animals are donated or given to zoos by people who can't properly care for them or are not licensed to house them. Animals that are retired from circuses or other performance areas will be donated to zoos. Many zoos exchange animals with other zoos to improve the breeding programs throughout the country. Zoos also purchase animals from **exotic animal dealers**. These are professional people who buy and sell exotic species of animals and help zoos meet the needs of certain species. Exotic animal dealers must have a license to purchase, sell, house, and handle a majority of these species. Table 16-1 outlines large zoos found in the United States.

Zoo animals require specialized health care through zoo veterinarians. Zoo vets are trained in the same way as other veterinarians but specialize in exotic and zoo animal species. They typically graduate and receive training and additional education in zoo species and how to treat them, provide proper health management programs, and how to provide anesthesia and surgical services and proper restraint and handling for all types of zoo animals. Veterinary technicians and assistants are also trained in zoo and exotic animal care.

An **animal refuge** is a large protected area where exotic animals are housed in a constructed setting that resembles their natural habitat. These areas house all types of exotic species, from mammals, reptiles, birds, and amphibians to species that are naturally found in the surrounding location. Animal refuges may have a variety of habitats in the area, from swamplands to plains, to desert-like areas. The United States has over 475 National Wildlife and exotic animal refuges. These animals are **feral**, meaning they are wild and not domesticated and live in the area where they are provided the land that is maintained by humans. Typically, the staff and people maintaining the refuge do not feed or provide routine health care for these animals. They only provide the environment. If an animal becomes sick or injured, then the staff will contact a wildlife or zoo veterinarian or wildlife rehabilitator to care for the animal.

Some animal refuges have biologists on staff who have backgrounds and knowledge in animal behavior and a degree in biology. Animal refuges are protected by law, and hunting and fishing are prohibited. Many refuges are open for public visitation and people may hike, boat, and visit the area to take photographs.

Performing animals are exotic animals that are raised and trained to perform an unusual activity that is not natural for the animal. Many people feel that performing animals are mistreated or forced to do a job that they do not want to do. This has become a controversial topic over the years. The circus is a variety show that houses many species of animals that perform tricks and acts that entertain crowds of people. These animals are raised and trained and cared for closely by humans. Some have been domesticated and others remain wild. The people who work with these animals are usually trained in animal behavior, are animal trainers, and have degrees in an area of animal science or biology.

One circus animal that has been adapted from being trained in circus performing is the elephant. The elephant is a species of animal that can be trained to work alongside humans. Elephants have been used in many countries where they are native to the land. They have been trained by people to remove trees and pull large loads very much like a draft animal. Elephants are over 12,000 pounds and may reach 12 feet in height. They are strong enough to push over 30-foot trees. They may live up to 60 years in captivity. Due to these qualities, the elephant must always be treated as a wild animal that has the potential to be dangerous.

Exotic Animal Care and Management

The use of exotic animals in breeding production programs, zoos, circuses, or as exotic pets and alternative livestock means that they must be profitable. **Profitability** is making money in a business after all other costs have been paid for in purchasing and caring for the product. The product in this case is an exotic animal. Exotic animals require research in how to house them, feed them, care for them, and market them. The startup costs for an exotic business can be immense. It is wise that the people involved in this business venture have a background in animal science or livestock production to better understand the industry. Exotic animals require specialized equipment and housing, as well as fencing needs, that can cost several thousand dollars. Some items to consider in the exotic animal business are listed in Table 16-2.

It is important that people owning or operating an exotic animal–type of business research the animal species they are considering housing. Many locations have limitations, such as environment, climate, state and local regulations, and **marketability** of the species. Many exotic species of animals have various environmental and climate limitations. They may need a tropical habitat, wooded habitat, or arctic habitat. It is important to determine if the exotic animal will be able to live and thrive in the surrounding environment and various climates that occur there seasonally. If the species will not do well in the natural area, it is necessary to determine if it is possible to create its needed environment. Fortunately, there are several exotic species that are not sensitive to unnatural environments. Some animals will need to be kept indoors year round or at specific times of the year. These limitations will need to be determined before purchasing the exotic animal. It is also important to research the federal, state, and local regulations in the state of occupancy. Many laws are enacted that pertain to exotic animals. Some locations require permits or licenses for housing or owning exotic species. Regulations on certain species are very strict on owning and housing exotic animals that may be dangerous to others or domestic pets. States such as Florida, Texas, and Louisiana have experienced problems with wild and exotic animals and have, therefore, had to place regulations and restrictions on the owning and housing of exotics (see Table 16-3). Exotic animals that are placed in the wrong hands, such as tigers, lions, bears, and wolves, do not mix well in the general public. Regulations are also placed on exotics to prevent the outbreak and spread of disease. Some exotic species are closely related to canine, felines,

TABLE 16-2

Operating Expenses of Exotic Animals

Cost of the exotic animals and their availability	Varies between locations	$1,000–$100,000
Equipment	Sometimes difficult to locate	$5,000–$50,000
Housing and fencing	Extensive supplies, upkeep, and may require permits	$10,000–$50,000
Nutrition and food supplies	May be difficult to locate	$1,000–$20,000
Veterinary medical care	May be difficult to locate veterinary specialists	$10,000–$80,000
Staff and labor	Must locate trained and knowledgeable employees	$15,000–$50,000
Marketing and advertising	May need to educate the public	$10,000–$25,000
Utilities—lighting, electric, water, telephone, etc.	Varies with species needs	$5,000–$20,000
Permits and licenses	Depend on state and federal laws	$1,000–$10,000
Insurance and liability	Depend on state and federal laws and locating insurance companies with experience	$20,000–$100,000 +

TABLE 16-3

States with Regulations on Exotics

Alaska	Arizona
Arkansas	California
Colorado	Connecticut
Delaware	Florida
Georgia	Hawaii
Illinois	Indiana
Iowa	Kansas
Kentucky	Louisiana
Maine	Maryland
Massachusetts	Michigan
Minnesota	Mississippi
Nebraska	New Hampshire
New Jersey	New Mexico
New York	North Dakota
Oklahoma	Oregon
Pennsylvania	Rhode Island
South Dakota	Tennessee
Texas	Utah
Vermont	Virginia
Washington	Wyoming

equine, and bovines and can easily spread a disease to domesticated animals, some of which are intended for human consumption. Research should be done on the marketability or how well the product will sell in the local area. It is important to know that there is a market for the type of exotic investment.

There are many similarities between exotics and livestock or other agricultural animals; however, there are even more differences. Raising an animal that requires specialized care and housing is the ultimate responsibility of the owner. Breeders, dealers, and private owners must know how to properly care for and undertake the responsibility of the animal's needs. That means the animal should have the required housing, fencing, diet, sanitation, health care, and environmental needs. Exotic animals' nutritional needs will vary among species. The rations of domestic livestock and companion animals have been thoroughly studied and researched. Exotic animals may not have commercial diets available to sustain them. In those cases, it is the responsibility of the owner to purchase and provide the necessary nutrients the animal would eat in the wild. Many animal diets have been studied by biologists and zoos and with research should be available from the proper sources. Housing and space requirements will also vary among exotics. Research will need to determine the needs of the animal according to the following:

- Indoor, outdoor, or a combination?
- Temperature needs?
- Housed alone or in groups?
- Amount of space per animal?
- Housing materials?
- Fencing materials?
- Restraint equipment?

Finally, the most important requirements involve the health and veterinary care of exotics. A limited number of veterinarians and other health care members are familiar with exotics. Even then they may not know the particular needs of the exotic species. A majority of exotic veterinary professionals are located at universities, zoos, aquariums, and animal rehabilitation centers. It is important that these staff members know how to sedate and anesthetize exotics, perform various surgical procedures, and handle diagnosis and treatments pertaining to exotic animals. When handling and restraining exotic animals, veterinary staff should treat these animals as wildlife and handle them accordingly. The stress of restraint may cause aggression and injury may occur even in the most gentle species of exotic animals. This means that special restraint equipment and sedation may be necessary.

SUMMARY

Exotic animals are expensive and costly investments that may be used for a variety of purposes. The major types of exotics include food and fiber animals, production and breeding animals, zoo and performing animals, and investment animals. No matter what the type of business or private use, exotic animals have specialized needs. Many differences occur between exotics and domesticated species of animals. It is necessary to research the nutritional, environmental, health, and housing needs of the animal. It is also important to determine any regulations that may be in place on owning exotic animals.

Key Terms

alligators large reptiles raised for meat, skin, byproducts, and for release into the wild

alpacas ruminant mammals similar to llamas but smaller and that originated in South America; they have silky fine coats and are raised for their hair

animal refuge large protected area of land where exotic animals are housed in a manmade setting that resembles their own habitat

beefalo cross breed of bison and domesticated beef cattle to produce beef animals with healthier and tastier meat products

bison ruminant mammals closely related to cattle

breeding stock the male and female animals of a species raised specifically for reproductive purposes

byproducts the parts of animals that are not the main profit or production source, such as the hooves, antlers, hair coat, or internal organs

cross-breeding mating of two different breeds of related animal species and producing an animal with a combination of characteristics from each animal

deer ruminant animals that graze on pasture and have antlers

elk ruminant animal, much like a deer, grazes on pasture and has antlers

exotic animal animal not native to the area where it is raised and may be rarely found in its natural habitat

exotic animal dealers people who buy and sell exotic animals to private collectors and zoos

feral animal that is wild and not domesticated to live in an area maintained by humans

llamas ruminant animals native to South America and raised for their hair, meat, and used as pack animals

marketability a measure of how well a product will sell in an area

ostrich large, flightless bird raised for meat, feathers, oils, and other byproducts

profitability making money in a business after all other costs have been paid for in purchasing and caring for the product

rare breeds animals that are bred and raised for a variety of purposes but not found locally

velvet material shed from antlers of deer and used in medications and nutritional supplements

venison deer meat

zoo zoological garden that houses wild and exotic animals and plants for people to visit and observe their behaviors and beauty

zookeeper person who manages and cares for animals in a zoo

zoology the study of animal life

REVIEW QUESTIONS

1. What is an exotic animal?
2. List the similarities and differences between exotics and domestic animals.
3. What are some purposes and uses of exotic animals?
4. Discuss several species of exotic animals and how they are used.
5. What are the purposes of zoos?
6. How do zoos get their animals?
7. What items need to be considered before selecting an exotic animal?
8. What are the housing needs of exotic animals?
9. What is the difference between profitability and marketability?
10. What are feral animals?

Clinical Situation

Dr. Andrews is a vet who works with a local zoo. He routinely has his veterinary assistants, Mark and Amanda, and his veterinary technician, Bethany, accompany him for any medical and surgical cases. Today, they are going to the zoo to look at a mountain lion that may have a tooth abscess. The entire staff is excited to work at the zoo as it makes for an interesting and challenging time.

When they arrive, the zookeeper, Anthony, shows them to the veterinary medical center so they can prepare for the day's work. The vet and Anthony prepare to tranquilize the mountain lion with a dart gun, which Anthony has been trained in using. The assistants and technician set up the surgical area for a possible dental extraction and exam. The anesthesia machine is ready for use. The team goes together to sedate the lion.

"Okay, here is 'Romeo.' He is a 6-year-old male mountain lion that has been here for about 3 years. He can be aggressive and difficult, which is probably territorial and protective of his mate, 'Juliet.' I have put 'Juliet' outside so we don't have to worry about her during the capture."

"I will dart him and when he is down, everyone move quickly to lift him on the stretcher and move him to the surgery area."

The lion is darted and down and the team is moving him quickly to the surgery area. When they arrive in the center, "Romeo" is beginning to wake up.

- What is the problem with this situation?
- What could have caused this issue?
- How should the veterinary staff handle this situation?
- How could this problem has been avoided?

17

Delmar/Cengage Learning

Laboratory Animal and Research Animal Management

Objectives

Upon completion of this chapter, the reader should be able to:

- ☐ Identify common breeds and species of animals used in laboratory medicine
- ☐ Describe methods and types of laboratory animal research
- ☐ Discuss the use of animal models and computers in laboratory medicine
- ☐ Describe the career opportunities in veterinary laboratory animal medicine
- ☐ Discuss the history of animal research in relationship to disease discoveries
- ☐ Explain the health management practices of laboratory animals
- ☐ Implement a management plan for laboratory animals

Introduction

Laboratory Animal Science is a field of veterinary medicine devoted to the production, care, and study of laboratory animals used in biomedical research and education. It is a large field comprised of professionals from many scientific, educational, and veterinary disciplines. Professionals within this field have a deep sense of purpose and are dedicated to helping society and improving the lives of people and animals alike. Much controversy has evolved over the years with a difference of opinion regarding the use of animals in research. Animal well-being and high standards of care have been the goal of animal research over the last 10 years.

Animal Research

Scientists sometimes use animals in **research** to answer questions about medical conditions that affect humans and animals. From rats to dogs, pigs to fruit flies, a variety of animal species contribute to medical breakthroughs that save millions of human and animal lives each year. Through research on these animals, scientists have discovered cures and preventions for a number of human and animal ailments, as well as medications to treat these conditions. The **American Medical Association (AMA)** represents the human side of medicine and conducts many studies that benefit the advancement of human medicine and surgical procedures. In 1950, the **American Association for Laboratory Animal Science** or **AALAS** was developed for the exchange of information and expertise in the care and use of laboratory animals, to promote the humane care and treatment of laboratory animals, and to support the quality research that leads to scientific gains that benefit people and animals. Benefits of AALAS include the following:

- Promotion of the humane care and use of laboratory animals (animal welfare)
- Encouragement of responsible research
- Education of research animal personnel
- Establishment of standards and licensure of laboratory animal personnel

Lab animals are raised for use in research laboratories. Table 17-1 outlines some of the uses and discoveries that have come from animal research. They are bred, raised, and cared for in controlled environments to ensure accurate results are obtained and that they are genetically suitable for the work. Rats, mice, and other rodents compose 90% of the lab animals that are used in today's research studies (see Figure 17-1). Less than 1% of animals are dog or cat species. Primate use in labs is also on the decrease. One of the main goals of laboratory animal scientists is to ensure that research animals are not exposed to any unnecessary pain or stress. While the **Animal Welfare Act** strictly regulates experiments on animals, statistics show that most experiments are not painful to animals. The act lists regulations on the unnecessary use of animals and recognizes that animals have rights and feelings and should not suffer in any way. According to a 2001 report by the U.S. Department of Agriculture, over 90% of research was not painful to animals. In a majority of cases, animals are not exposed to or involved in any painful procedures. Most animals receive anesthesia or pain-relieving medications during procedures that could possibly involve some pain or distress. The protocols that are used in animal labs many times exceed the standards of veterinary medicine.

TABLE 17-1

Animal Use Research Facts

Animal	Use
Armadillo	Used in the development of a leprosy vaccine; current research in a cure for leprosy
Cat	Treatments for ocular conditions such as lazy eye, cross eye, glaucoma, and cataracts
Dog	Used in discovery of diabetes; used to develop insulin; used in the development of open-heart surgeries; used in development of organ transplants; used in the development of anesthesia
Horse	Used in developing an antitoxin and vaccine for diphtheria; used to develop treatment for tetanus
Ferret	Used in menstrual cycle and reproductive research; used in developing a bird flu vaccine; used to develop methods to control ovulation
Guinea Pig	Used to discover vitamin C; used to develop a vaccine for tuberculosis (TB); used to develop anticoagulants
Hamster	Used to study brain disorders
Drosophila (fruit fly)	Used to develop drugs to treat skin infections, pneumonia, and meningitis; current use in researching breast cancer
Jellyfish	Used to study chemicals for treating cancer and Huntingdon's disease
Kangaroo	Used in researching obesity and high blood pressure
Rabbit	Used to develop treatments for cystic fibrosis and asthma; used to develop polio vaccine; used to develop treatments of arthrosclerosis (hardening of the arteries)
Mouse	Used to develop the human genome
Pig	Used to transplant organs; used to develop treatments for strokes; used to grow heart valves for humans
Rat	Used to develop medicines used in cancer, stroke, and bone repair; used in discovery of DNA
Monkey	Used to develop measles vaccine
Tamarin	Used to develop treatments for colon cancer

Animals are also used for their **by-products tissue** or tissue that is removed from an animal when it is used for other purposes in research or for human consumption. Many tissues are collected from meat animals that are produced for human meat sources. Other animal products that are commonly used include chicken eggs, which are used in manufacturing vaccines and medications.

FIGURE 17-1 Rats are commonly used for laboratory and research purposes.

Types of Research

Animals are used in three types of research (see Table 17-2). The first is **basic research**, which is studied in a lab setting to determine knowledge and understanding in life processes and diseases. This type of research uses observation, measurements, and experiments to determine data. An example of basic research was the discovery of anthrax, which was discovered to be bacteria while spores were isolated in a research experiment. **Applied research** is done for a specific purpose, such as developing a new vaccine or medication. This type of research is completed by using existing information. Finally, **clinical research** is research that is conducted in a lab setting and focuses on human or veterinary related issues. An example is using a dog or guinea pig to study specific diseases on canines or rodents.

TABLE 17-2

Types of Research

Basic Research	■ Laboratory setting ■ Conducting wide variety of experiments ■ Increase knowledge in life processes and diseases ■ No predetermined facts ■ Use of observation, measurements, descriptions ■ Provides data as answers ■ Studies a species of animal
Applied Research	■ Any type of controlled setting ■ Specific topics ■ Builds on existing knowledge ■ Uses all types of animal models
Clinical Research	■ Medical or veterinary research lab setting ■ Focus on human or veterinary medicine ■ Uses specific species based on the topic of research ■ Supports facts that have been proven

Animal Models and Laboratory Medicine

The use of animals in a research setting must be completely controlled and monitored for accuracy. Laboratory medicine has developed several acceptable **animal models** for animal research. These models are used to assist in gaining knowledge and information on human and veterinary medicine. Four basic models exist for animal research:

- living animals
- animal tissues
- nonliving systems
- computer or simulated programs

These models are used to study and better understand the human and animal bodies, behavior, diseases, surgical procedures, treatments, and the effects of medications on the body. **Living animals** are any species that is alive and in some way responds to a stimulus (see Figure 17-2). Another model is the use of **animal tissue**,

FIGURE 17-2 Guniea pigs can be used as living animal models.

Delmar/Cengage Learning

FIGURE 17-3 Animal tissue can be cultured and used in experiments for cancer and chemotherapy treatments.

which is a living part of an animal that is cultured in a lab and grown for specific needs (see Figure 17-3). This usually is research dealing with cancer and chemotherapy treatments. **Nonliving systems** are models that are mechanical in nature and mimic the species of animal that is being studied. They are typically used in studies that require the monitoring of movement or injuries on living animals. An example is the artificial hip in humans or the artificial ball-and-socket joint in canine hip dysphasia. These models help develop a pattern or design without the use of a living animal. The last model type is the **computer system** or the **simulated system**. These models mimic an animal's behavior and internal and external structures. They require significant amounts of research on animal and human subjects to develop programs that work in the same manner as a live animal. Conducting experiments through the use of a computer or simulator may not be perfect, but it is a step in the right direction for someday not needing a live animal in research. An example of a computer-simulated model is using a simulation of frog dissection for biology classes.

Common Species of Animals Used in Research

Many types of animals are used in laboratory research. The species vary greatly and are commonly companion animals and livestock so that research conducted for human and veterinary medicine is true to the need of the experiment. Animals that are used in experiments should be as closely related to the desired outcome as possible.

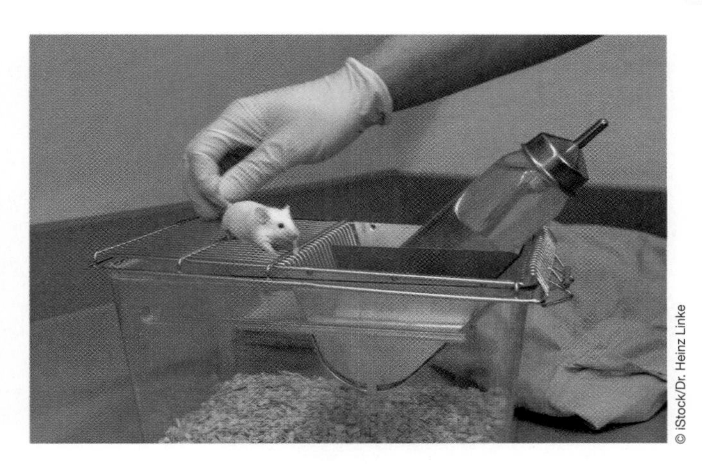

© iStock/Dr. Heinz Linke

FIGURE 17-4 Mice are commonly used in labs for research purposes.

Mice

Mice are known as **Mus musculus** in terms of laboratory species. Along with rats and other rodents, mice make up the majority of animals used in medical research. Their small size, ease of handling, and low cost make them ideal for laboratory experiments (see Figure 17-4). In addition, with their efficiency in breeding, scientists can breed different strains of mice with natural genetic problems to achieve specific models of human and animal diseases. Researchers have developed mice with leukemia, breast cancer, and many other types of cancer through breeding and genetics. This allows new treatments to be tested on animal models, instead of humans. Scientists are searching continuously for the best animal model for the study of human AIDS. Much that we know about this disease has come from research using mice, which has also helped develop vaccines for influenza, polio, yellow fever, and rabies. Mice are popular for studies that involve large numbers of animals. There have been more than 100 bloodlines developed by researchers.

Rats

The larger cousin of the mouse is the rat, known as **Rattus norvegicus** (see Figure 17-5). Rats are commonly selected for use in studies concerning behavioral and nutritional problems. Rats are known for naturally developing tumors; thus, cells from human cancer tumors can be placed in rats and studied for cures (see Figure 17-6). Five bloodlines of rats have been developed in research from the wild brown Norwegian rat.

Rabbits

Rabbits are known in lab settings as **Oryctolagus cuniculus**. There are several similarities between the physiology of rabbits and humans, which makes the rabbit a good model for research into human

FIGURE 17-5 Rats are often studied for behavioral and nutritional problems.

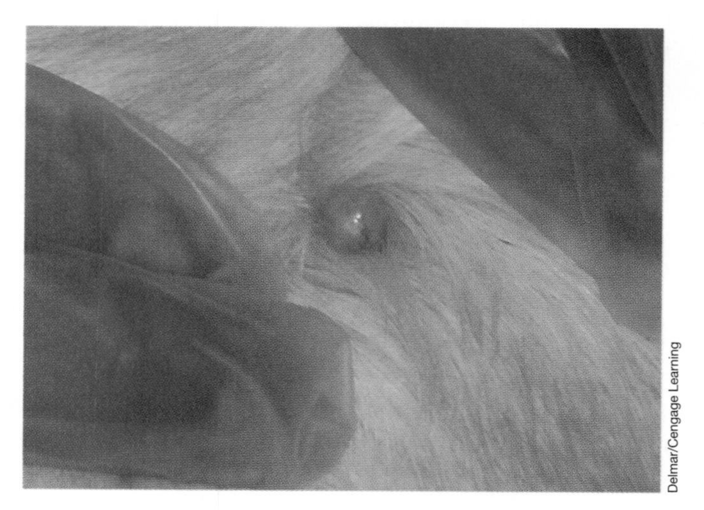

FIGURE 17-6 Tumor on a rat.

disease (see Figure 17-7). The term physiology refers to the functions of the body and how the systems within an animal work. Rabbits have the ability to produce tumors, similar to rats, and this makes them useful models to study for chemotherapy treatments, as well as the prevention of certain cancers. Rabbits are also used to study ear infections, which affect millions of infants and children each year. In cases of veterinary medicine,

FIGURE 17-7 Rabbits have similarities to humans that make them good candidates for living research models.

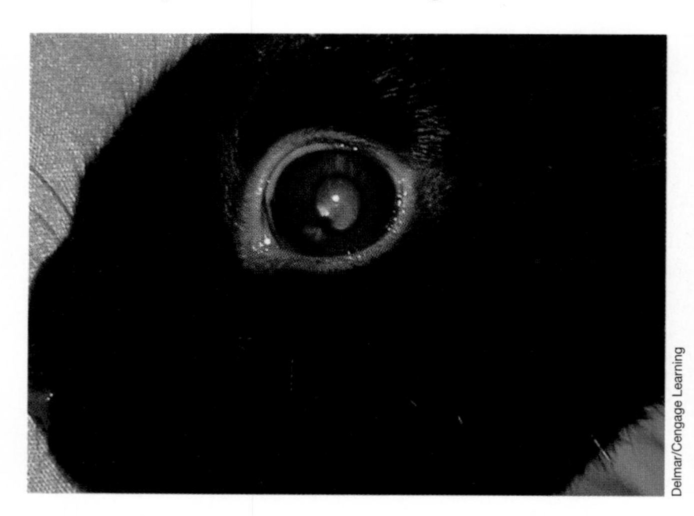

FIGURE 17-8 Rabbits are used to study disorders of the eye.

the rabbit has proven useful in finding treatments for diseases and problems involving the eyes, such as finding treatments for cataracts and glaucoma (see Figure 17-8).

Primates

Primates are animals, such as monkeys and apes, that have movable thumbs and in many ways resemble humans; due to this resemblance, primates are commonly called nonhuman primates in lab settings. Several species of monkeys are used in research. Monkeys can be divided into two classes, New World monkeys and Old World monkeys (see Table 17-3). New World monkeys originated in South America and Old World monkeys originated in Africa and Asia. They have some differing features that distinguish them from each other. Primates are studied based on their wild behaviors and ability to communicate through verbal and nonverbal communication (Figure 17-9).

TABLE 17-3

Characteristics of Monkey Types

NEW WORLD MONKEYS	OLD WORLD MONKEYS
■ Nostrils open to the side	■ Nostrils are close together and open downward
■ No cheek pouches	■ Check pouches to carry food
■ No pads on buttocks	■ Pads on buttocks
■ Long tails for grasping	

© iStock/syagci

FIGURE 17-9 Primates are used in research studies due to similarities between them and humans.

Research Regulations and Guidelines

Animal research studies are closely monitored and regulated by several government agencies and humane associations. Guidelines have been put in place to protect the rights and uses of animals in research because even though many benefits have been achieved from animal research, a great debate has ensued regarding the necessity of using animals in research and testing. One agency that monitors and approves studies for research is known as the Institutional Animal Care and Use Committee (IACUC). All research studies involving animals must offer a proposal to the IACUC based on the type of the study, how animals will be used and cared for, an outline of the personnel and veterinary staff and their credentials, and a list of any drugs or anesthetics that may be used. The IACUC then reviews the proposal and accepts or denies the study (see Figure 17-10). In 1985, the Animal Welfare Act (AWA) was developed to further

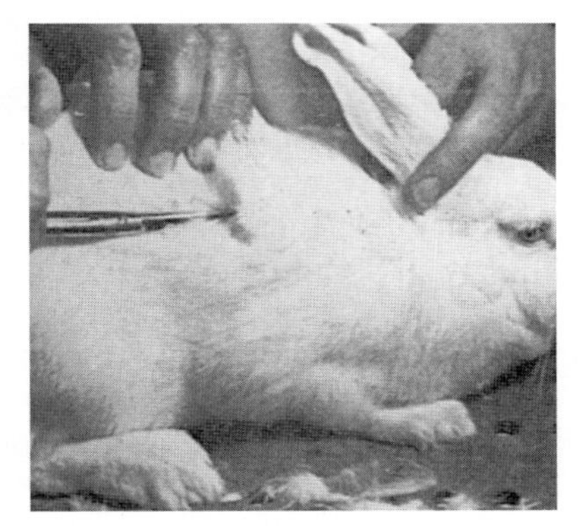

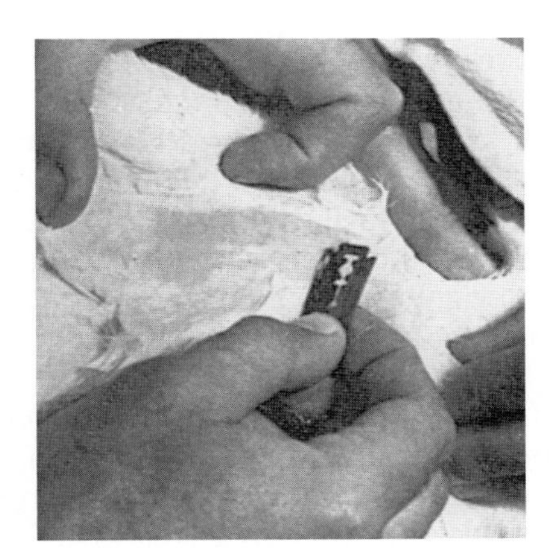

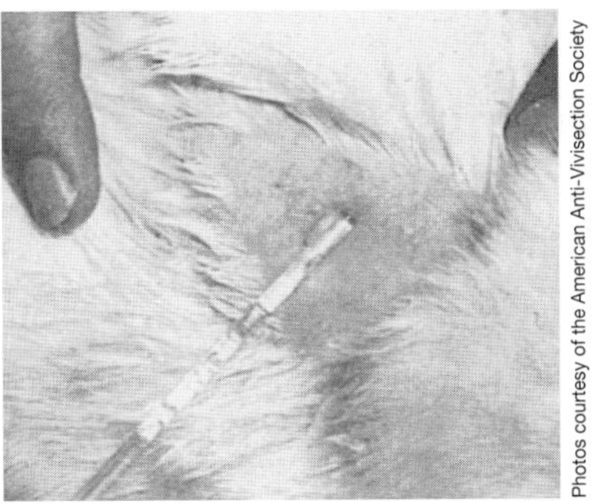
Photos courtesy of the American Anti-Vivisection Society

FIGURE 17-10 All studies, such as this skin irritancy testing, must be approved by the IACUC.

regulate animal use in research on a federal level. This legislation was brought about due to the debate on the need for animal testing and the need for proper use of animals in research. The AWA is enforced by the U.S. Department of Agriculture, which regulates and inspects all facilities that conduct animal-related studies.

Lab Animal Management and Health Practices

Lab animals require controlled conditions in which they are housed and cared for by proper hygiene and asepsis, or a sterile environment. These facilities must use excellent sanitation methods. This is due to the types of studies that may increase the chances for disease outbreaks or the contamination of a study source. Cages and equipment must be properly sterilized. Blood and tissue samples must be handled with extreme care. All bedding, food, and waste materials must be disposed of properly. Lab animal nutrition should include high-quality, balanced food. Each animal must be fed a strict diet that is approved by the study proposal. It is important to know that research can be compromised if a proper diet is not followed. Diets will vary according to animal species, age, size, maturity level, and environment. Lab animal facilities require a specialized environment that must meet specific conditions. The biggest concern is regarding the space. All rodent species are required to be housed in at least 6 square feet of space. Livestock require at least 144 square feet of space. Every animal species has a specific space requirement available from the National Institutes of Health (NIH). The ideal temperature should be between 65 and 85 degrees with humidity between 60 and 70%. Light is required on a set schedule by providing an amount of light as well as an amount of dark and must be uniform throughout. Proper ventilation allows for fresh air to circulate, which will enable eliminating odors, eliminating vapor, and reducing airborne disease chances. Noise barriers must be in place for both the animals and the outside area.

Lab Animal Careers

There are several career opportunities in the animal research and lab animal areas. The first career option is to become a laboratory animal technician. Laboratory animal technicians are specially trained to care for the animals used in laboratory research (see Figure 17-11). They are responsible for feeding the animals, cleaning their housing, monitoring each animal's overall health, and providing basic care, such as giving medicine, feeding, and assisting other employees. Lab animal technicians have a basic knowledge of the needs and behaviors of animals. They play an important role before and after a scientific experiment. The technician will be involved with the planning stages of an experiment to help determine the care an animal will need before, during, and after a procedure. The technician is also involved in the analysis or study of the research data following an experiment.

Photo courtesy of PhotoDisc

FIGURE 17-11 Laboratory technicians work closely in labs assisting with the research studies and care of the animals in the study.

Another possible career field is as a **laboratory veterinarian**. Some veterinarians provide medical care to animals involved in research; others provide their skill to improve surgical techniques for humans and animals. A number of veterinarians work for government agencies, universities, or corporations. Laboratory vets perform a variety of jobs, ranging from management to product development and research (see Figure 17-12).

Another extremely important job is concerned with the humane care and treatment of laboratory animals. Through the U.S. Department of Agriculture, these doctors visit research laboratories and ensure the animals are being treated according to federal law.

Another area of the profession is animal behavior. **Animal behaviorists** study animals to collect data on their behavior in captivity. Some may help rehabilitate animals from zoos or research facilities. Behaviorists also provide an important role in research by discovering health programs and housing methods to promote a high quality of laboratory animal medicine.

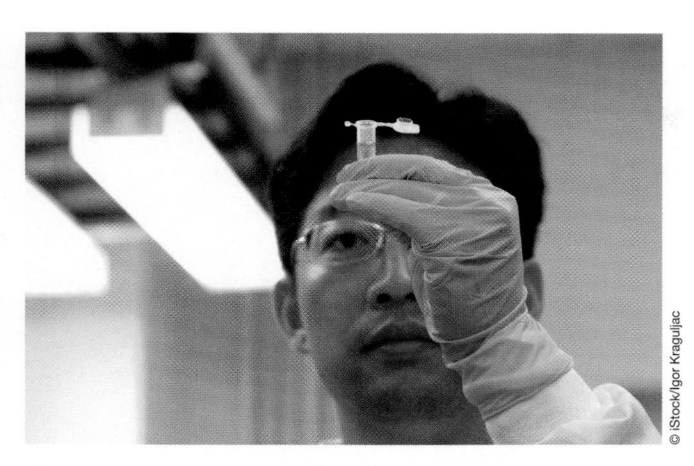

© iStock/Igor Kraguljac

FIGURE 17-12 Laboratory veterinarians provide medical care to animals involved in research.

Finally, there is the career of **research scientist**. Researchers are committed to human and animal health. They are responsible for planning the experiments to be done on animals and ensuring the animals are not in pain. Researchers are responsible for obeying the protocol set up by the federal government.

Veterinary medicine and animal research have led to cures, treatments, vaccines, or advancements in diseases and conditions such as the following:

- allergies
- arthritis
- asthma
- birth defects
- bronchitis
- cancer
- deafness
- diabetes
- epilepsy and seizures
- glaucoma
- heart disease
- hypertension
- influenza
- kidney disease
- leukemia vaccine
- Lyme disease vaccine
- measles
- tetanus vaccine
- tuberculosis
- ulcers

SUMMARY

Animals and people live longer and better lives thanks to animal research. This research is a systematic and controlled attempt to answer questions on a specific topic that is based on a carefully planned procedure. Laboratory animals are bred and raised in controlled environments and used in a variety of research studies. Many types of animal species are used in these studies. Ninety percent of the animals used today are from the rodent family. Advancements in animal research have contributed to finding the causes of disease, treatment and prevention of disease, and cures for former fatal diseases both in human and veterinary medicine. Due to the controversy that society has created on the topic of animal research, the standards in health care and management practices of lab animals many times exceed those of doctors and veterinarians.

Key Terms

American Association for Laboratory Animal Science (AALAS) association developed for the exchange of information and humane care of laboratory animals used in research

American Medical Association (AMA) agency that represents the human side of medicine and conducts many studies that benefit the advancement of human medicine and surgical procedures

animal behaviorists people who study animals to collect data on their behavior in captivity

animal models live animals used to assist in gaining knowledge and information on human and veterinary medicine by studying behavior, diseases, procedures, treatments, and the effects of medications on the body

animal tissue living part of an animal that is cultured in a lab and grown for specific needs

Animal Welfare Act strictly regulates experiments on animals; statistics show that most experiments are not painful to animals

(Continues)

applied research research done for a specific purpose, such as developing a vaccine or medication

asepsis technique of keeping a sterile environment

basic research research studied in a lab setting in order to determine knowledge and understanding in life processes and diseases

by-products tissue tissue that is removed from an animal when it is used for other purposes in research or for human consumption

clinical research research that is conducted in a lab setting and focuses on human and veterinary related issues

computer systems computerized models that mimic animal behaviors and internal or external structures; also known as a simulated system

Institutional Animal Care and Use Committee (IACUC) agency that monitors and approves studies for research prior to being conducted

lab animals animals bred, raised, and cared for specifically in controlled environments to ensure accurate results are obtained and that the animals are genetically suitable for work

laboratory animal technician specially trained professionals that care for animals used in research by feeding, cleaning, monitoring, and caring for their overall health

laboratory veterinarian doctors who provide medical care to animals involved in research and perform surgical procedures used in improving techniques for humans and animals

living animals any species of animal that is alive and in some way responds to a stimulus

Mus musculus genus and species of the mouse

National Institutes of Health (NIH) agency that specifies animal space requirements of a species housed for research purposes

New World monkeys monkey species that originated in South America

nonhuman primates animals that resemble humans and have movable thumbs

nonliving systems models that are mechanical in nature and mimic the species of animal that is being studied

Old World monkeys monkeys that originated in Africa and Asia

Oryctolagus cuniculus the genus and species of the rabbit

physiology study of living organisms

primates animals, such as monkeys and apes, that have movable thumbs and in many ways resemble humans

Rattus norvegicus genus and species of the rat

research studies used to answer questions about medical conditions that affect humans and animals

research scientists staff responsible for planning the experiments to be done on animals and ensuring the animals are not in pain

simulated systems models that mimic an animal's behavior and internal or external structures to conduct research

REVIEW QUESTIONS

1. What organizations are involved in setting standards and regulations on animal research?
2. What species of animals are commonly used in animal research studies?
3. What are the three kinds of research types? Explain each type.
4. What are the four animal models used in animal research? Briefly explain each type.
5. What are the two types of monkeys? What are the differences between each type?
6. What are the career options in laboratory animal medicine?
7. What are some conditions or disease advancements that have been made in veterinary medicine due to animal research?
8. How is animal research regulated?

Clinical Situation

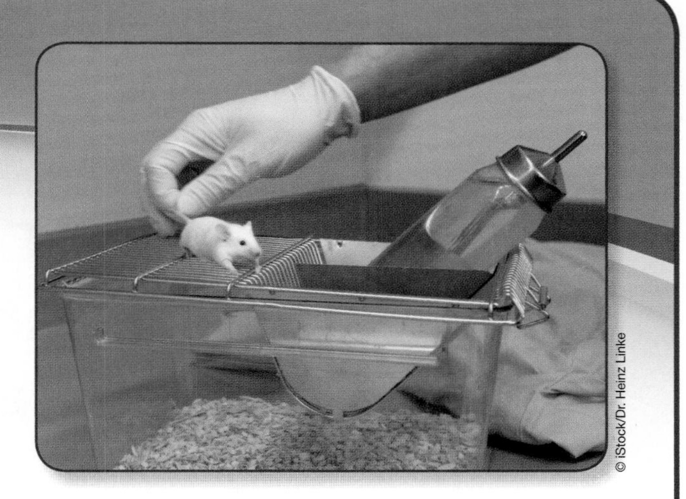

Mary and James, both laboratory animal technicians, are working in a research facility that is currently housing rats, mice, and rabbits for both veterinary and human medicine research. When the techs enter the lab housing area, they note the health status of each animal in the medical record. Today, the techs notice that several animals are showing signs of respiratory disease. Each animal is being used for several different studies. Some are housed separately and others are housed with 2–3 animals per cage.

- What signs might be seen with respiratory disease in these animal species?
- What should the techs do as far as noting respiratory conditions?
- What protocols should the techs take in this situation?

18

Beef and Dairy Cattle Breed Identification and Production Management

Objectives

Upon completion of this chapter, the reader should be able to:

- Understand and discuss common veterinary terms relating to cattle
- Identify common breeds of beef cattle
- Identify common breeds of dairy cattle
- Discuss the nutritional requirements of cattle
- Understand normal and abnormal behaviors of cattle
- Properly and safely restrain cattle for various procedures
- Discuss the health care and maintenance of cattle
- Discuss breeding and reproduction in cattle
- Discuss common diseases affecting cattle
- Explain the importance of beef cattle production
- Explain the importance of dairy cattle production
- Discuss the types of beef production systems
- Discuss dairy production and milking
- Discuss management practices with cattle

Introduction

Cattle are among the most widely raised livestock in the United States (see Figure 18-1). Americans enjoy both meat and milk, which make raising both beef and dairy cattle a large, profitable business. The cattle industry provides many opportunities in the veterinary industry, from herd health management to reproduction and products for human consumption.

FIGURE 18-1 The cattle industry provides many opportunities for the practice of veterinary care.

Veterinary Terminology

Cattle are known by the term **bovine** in veterinary medicine. The adult female is called the **cow** and the adult male is called the **bull**. A castrated male is called a **steer**. A young female cow that has not yet been bred is called a **heifer**. A young female cow that is pregnant with her first calf is called a **springing heifer**. A young cow is known as a **calf**. The labor process of cows is called **calving**; the term **freshening** also means the labor process in dairy producing animals. A group of cattle is known as a **herd**. Occasionally, an adult cow is sterile and not able to produce and is called a **freemartin**. Additional important cattle-related terms are the following:

- **Polled**—no horn growth
- **Forager**—animal that eats grass and pasture
- **Marbling**—the appearance of intramuscular fat within the meat as striations or lines through the meat layers
- **Cutability**—quality and quantity of meat from a beef animal
- **Dual purpose breed**—a breed that serves more than one purpose, such as both a meat and dairy animal

Biology

Beef and dairy cattle are similar in appearance but are raised for very different purposes. **Beef cattle** are produced for meat and the expectations are to have increased body size to produce more quality meat per cow (see Figure 18-2). **Dairy cattle** are raised for quality milk and are raised with the expectations of producing

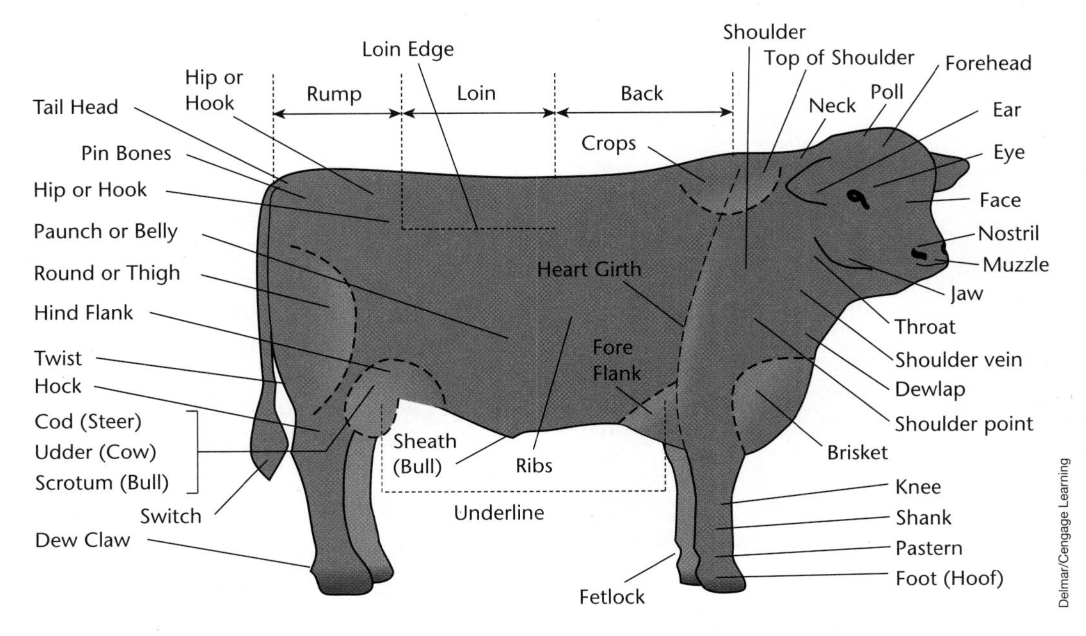

FIGURE 18-2 Body parts of the beef cow.

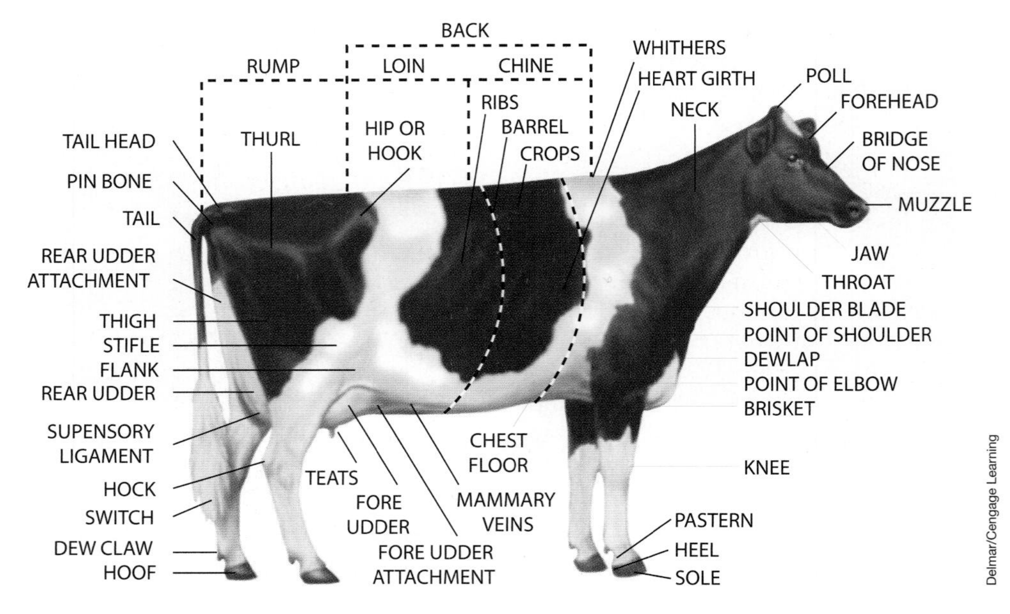

FIGURE 18-3 Body parts of the dairy cow.

large amounts of milk per cow (see Figure 18-3). Cows were one of the first domesticated livestock species. Today there are over 1.5 billion cattle in the world. In many parts of the world, cattle are considered sacred animals and worshipped and valued greatly for their milk and meat or as a living animal.

Breeds

Cattle breeds vary greatly in size, structure, and color. Some cattle are miniature in size, such as the zebu cattle, and others are very large in size, such as the Limousine and Brahman cattle. Beef breeds are bred to be heavier in appearance and body weight (see Figure 18-4). Beef

FIGURE 18-4 Beef cattle are raised for a large body size to maximize meat production.

FIGURE 18-5 Dairy cattle are slender in body composition.

cattle should have large amounts of muscling over the hindquarters, body, and pelvis. Dairy cattle are more angular in shape and, due to smaller amounts of body weight and muscle, have a bony appearance over the pelvis and hind end (see Figure 18-5). Some breeds of cattle are dual purpose, meaning they serve more than one purpose in their breed. For instance, some cattle breeds are used for both meat and milk production, such as the South Devon and Salers.

Beef Cattle

Beef cattle are raised specifically for meat that is called *beef*. The goal of beef cattle production is to select a beef animal that produces a large amount of meat. This goal must also include a quality product, which is the result of a high standard of health care and a nutritional program. There are over 95 million beef cattle raised on farms in the United States. These farms bring in a profit of $74 billion a year, making the industry an important part of today's agriculture. Beef is consumed over 75 billion times a day in America alone, and the market for beef is increasing daily worldwide. The demand on the beef industry has increased over time; however, the number of cattle producers continues to decrease. Beef is one of the most frequently eaten foods due to its nutritional value of protein, vitamins, iron, and other essential nutrients. There has been an improvement in the genetics of beef cattle, the overall heath care, and updated research to make beef quality even better.

Dairy Cattle

Dairy cattle are raised specifically for milk. There are over 90,000 dairy farms located throughout the United States. Cows produce over 175 billion pounds of milk each year (see Table 18-1). A single cow can produce around 18,000 pounds of milk a year. In a similar fashion to the beef industry, the dairy cattle industry demand is on the rise, with the number of dairy cattle decreasing. Milk is one of the most wholesome foods found today and has more nutrients than calories. The goal of dairy production is to produce the highest quality and quantity of milk per cow. Dairy cattle production is one of the oldest agriculture programs, dating back almost 7,000 years.

TABLE 18-1

Milk Production by State

STATE	AMOUNT IN MILLION POUNDS
1. California	36,000
2. Wisconsin	22,000
3. New York	11,000
4. Pennsylvania	10,000
5. Idaho	9,000
6. Minnesota	8,000
7. New Mexico	6,700
8. Michigan	6,300
9. Texas	6,000
10. Washington	5,000
11. Ohio	4,500
12. Iowa	3,800
13. Arizona	3,600
14. Indiana	2,900
15. Vermont	2,500

Source: United States Department of Agriculture

Breed Selection

Selecting a quality cow should be based on the purpose or use of the cow, the breed, and the industry standards that are placed on cattle. The cow should bring the best market price based on the product.

Beef Cattle

Beef cattle are rated on cattle **selection guidelines** (see Figure 18-6). These guidelines state the standards of the beef type cow and how it is chosen for a production program. The beef breed's standards should be researched and selected based on showing quality traits for the breed. The **pedigree**, or parents' breeding lines, should be evaluated for any known **genetic flaws**. Genetic flaws are undesired traits or characteristics that

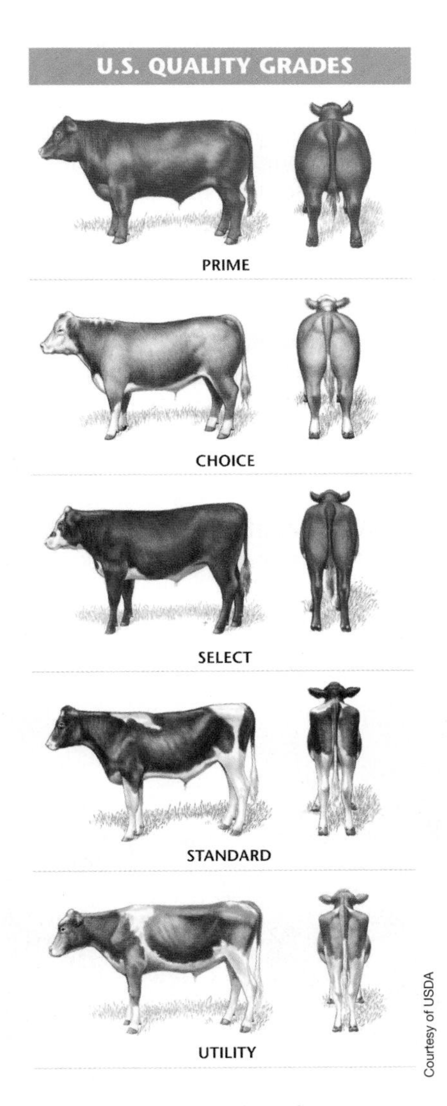

FIGURE 18-6 Quality grades of beef.

are passed from one or both parents to the **offspring**, or the young. Quality beef cattle should have no genetic flaws. Superior animals should be selected that show ideal weights and health status. Cows should be **sound**, which means they show no signs of lameness or injury to the legs and hooves. Beef cattle should have excellent **conformation**, which is their body shape and form. The breed should be able to survive well in the type of environment they will be raised in. The cow should be maintained easily with the proper diet. The marketability of the cow and the product should be at the high end of the production scale. The cow should be healthy and disease- and parasite-free.

Dairy Cattle

Dairy cattle standards are based on specific conformation requirements. Dairy cows should have an angular body shape that is rectangular in appearance (see Figure 18-7). They lack muscle mass in comparison to beef cattle. Dairy cows should have well-attached **udders**, or mammary glands. Depending on the breed, the udders should hold a total capacity of 50–70 pounds of milk. Each udder should contain four **teats** that are sections of the mammary gland used to produce milk, called **lactation**. All dairy cattle are bred to produce quality milk (see Figure 18-8).

The dairy industry has developed the **DHI program** or Dairy Herd Improvement program. This program is a production testing and record-keeping system of dairy herds. The U.S. Department of Agriculture (USDA) works with individual dairy farms to compare information on local, state, and national herds. All cattle within the program have permanent identification numbers and the milk is tested every 15–45 days to determine the quality and quantity of the product. The DHI program also has a registry that records the information of dairy

Delmar/Cengage Learning

FIGURE 18-7 Dairy cows should have an angular body shape that is rectangular in appearance.

Photo courtesy of Getty Images/B. Drake

FIGURE 18-8 A cow should produce up to 70 pounds of milk.

herd's milk production rates and the breeding records of bulls. The DHI program and registry have rules that dairy producers must follow to be part of the program. These rules include the following:

- All cattle must be registered breeds.
- All cattle must have copies of a pedigree on record.
- All cattle within the entire herd must be tested.
- Testing is required every month.
- All cattle must have permanent IDs.
- Testing must be completed within 24 hours.
- Milk or fat samples above normal levels must be retested.
- The USDA must supervise testing.
- Inaccurate files are liable for fraud charges.
- Surprise testing may be done at any time.

DHI records also monitor the **progeny** or offspring of the cattle. The progeny records help to identify any genetic flaws in breeding lines of cattle. The genetic flaws can then be traced to the **sire** (father) or the **dam** (mother). Each cow has a permanent ID number, and when a calf is entered into the system it records the progeny data under this identification number. This can help provide information to dairy breeders to research which bulls to breed to their cows to increase milk production rates in the dairy herd. This information can be used to determine the estimates of traits a bull or cow will pass along to offspring to determine the genetic superiors and inferiors; the estimation record is called **predicted transmission ability**.

Culling

When beef or dairy cattle are no longer producing a quality product, they may be **culled** from the herd. This means selecting cattle that should be removed from the herd. Culling cattle depends on the production rates of the cow, genetic flaws that are being produced, and the quality and quantity of beef or milk products being produced.

Nutrition

Cattle require nutrients that work to convert their food into the product that they are being bred for, such as meat or milk. Cattle are fed between three and four times a day. Cattle producers feed their herds a **total mix ration** or TMR that consists of all the nutrients that they need in a day in a mixture of high-quality products (see Figure 18-9).

Lactating Dairy Cows

Lactating cattle require the most nutrients. Lactating dairy cattle require increased amounts of food on a daily basis. They also require large amounts of water. Both food and water nutrients are converted to milk. All lactating cattle are bred so their bodies produce milk. The first four months of the gestation period are the most critical for nutrition. Over this time frame, cows need increased calcium and fat for energy to produce increased amounts of milk. Lactation feeding can be the most challenging nutritional stage, and several factors must be considered when feeding the lactating cow. These factors include the size of the cow, the stage of lactation, the quality of milk, the labor capability, the food quality, the costs of food, and the food **palatability** (how well the food tastes and is eaten). Cows are usually fed 1 pound of high-quality grain for every 2.5–3 pounds of milk produced a day. High-quality hay is fed at 1.5–2% of the cow's body weight. High quality grain products include the following:

- 20% protein
- 2% calcium
- 2% phosphorus

Photo by Michael Dzaman

FIGURE 18-9 The daily feed and rations of cattle are important to the outcomes of milk and meat production.

Cows produce milk for a 10-month period and after calving are given time for their reproductive systems to relax. This period of two months of not producing milk is called the *dry period*.

Feeding Dairy Calves

After the cows freshen, or give birth to the calf, the calves are allowed to nurse for the first 24 hours to receive the **colostrum** that provides natural antibodies to keep the calves free of disease until vaccines are given. During this time the calf should gain 12% of its body weight from the time of birth. After 24 hours, the calves are removed from the mother and fed by hand. The first three days, milk is collected from the cow and fed to the calf by hand-feeding in a nursing bottle (see Figure 18-10). This allows the calf to learn to nurse from a bottle. Gradually, the calf will be placed on milk replacer for a 2–3 week period. Milk replacer has increased amounts of fat and protein for growth. Then the calf will be weaned to begin a solid food source once

Photo by Peggy Greb. Courtesy of U.S. Department of Agriculture

FIGURE 18-10 Calves are hand-fed by bottle following the first 24 hours of nursing from the cow.

the digestive system has developed enough to support solid materials. This is usually a calf **starter food** that is easily digestible. The starter food diet is fed when the calf is between four and six weeks of age and is typically fed at a pound and a half per day. At five weeks of age, calves will begin eating **roughage**, or a hay or grass source. Most producers place calves on a high-protein diet to increase the size of the calf and will feed an **alfalfa** diet, which is hay high in protein. Calves are placed on **free choice** hay that is available at all times (see Figure 18-11). During this time, the calf starter food is increased to around six pounds a day per calf. By 8–12 weeks of age, the calf should begin eating a grain source with concentrated supplements that will promote continued growth.

Body condition scoring (BCS) is used from the time a calf is born through its entire adult purpose, whether that is as a lactating dairy cow or a beef cow used for meat production. The BCS is rated on a scale of numbers from 1 to 9 (see Table 18-2). This information is used by producers to estimate how much food an animal should be fed for optimal growth and maintenance.

Feeding the Dry Cow

The dry cow period nutrition will vary from the lactation nutrition. The cow is placed only on good-quality grass hay that is fed for 2–3 weeks. Around days 25–30 of the dry period, the cow is placed on grain at 4–7 pounds per day. During the last week of the dry period, this amount is increased to 7–14 pounds per day. Then the cow will resume the lactation feeding once the cow has been bred.

Photo by Sarah Kane. Courtesy of Cabe Brothers Farm

FIGURE 18-11 By five weeks of age, calves are fed free choice hay diets.

TABLE 18-2

Cattle Body Condition Scoring

Score 1	Emaciated appearance with little muscle left
Score 2	Very thin appearance; no fat noted; bones visible
Score 3	Thin appearance; ribs visible
Score 4	Borderline—most ribs not visible
Score 5	Moderate—neither fat nor thin
Score 6	Good—smooth body appearance
Score 7	Very Good—smooth appearance with fat noted over back and tail head
Score 8	Fat—blocky appearance; bone over back not visible
Score 9	Very fat—tail base buried in fat
Score 3.5–4	Dairy cattle ideal at end of lactation (DRY PERIOD)
Score 2–2.5	Dairy cattle peak lactation @ 60–90 days

Beef Cattle

Beef cattle convert their nutrients to meat and muscle mass. The adults are typically fed high-quality grains or pellet-based diets that are high in protein and fat for energy and growth. Beef cattle should be fed 2–3 pounds of grain per day, which is 2–3% of their body weight daily. They also need to have free access to hay or a pasture source and water. The calves are typically kept with their mothers to nurse until weaning. Calves typically begin a starter food and then begin a grain supplement in much the same manner as dairy calves. When the weaning process begins, the adult cattle will eat the grain if they are given the opportunity. Calves are typically fed in **creep feeders** to prevent the adults from consuming all of the food. Creep feeders are small food troughs that have bars with spaces in them that allow only the calves to eat. Calves are fed 2–3% of their body weight until they reach market weight.

Behavior

Cows scare easily and become difficult to work with once frightened. It may be difficult to settle down the cow once spooked. Cattle are capable of kicking, stomping, and trampling people. When cattle are around people or things that they are not used to they will become nervous. This can lead to them running, jumping, or kicking, and their size and actions can cause severe injury to people. Cattle are very different from other types of livestock and should not be handled according to other animals, such as horses. Cattle are herd animals and considered prey, so when placed in a threatening or frightening situation, they will go into a defensive reaction and rely on their speed, body size, head, and hooves to defend themselves. Cattle often head butt and kick as a first means of protection. Cattle often kick out to the side rather than straight back, so standing near their rear legs is discouraged. Most dairy cattle are used to being handled by humans as they are often milked several times a day. Beef cattle may not be as accustomed to people. Regardless, care must be taken when working around cattle. Signs of aggression in cattle include the following:

- Direct staring with head lowered
- Pawing at the ground with the front feet
- Lowering and shaking the head
- Snorting
- Short charging motions with the body
- Tail swishing quickly

Basic Training

Cattle that are shown or handled regularly for milking procedures are trained to follow routines and may be led with a halter and lead rope (see Figure 18-12). Cattle are often trained to herd to move them from locations. Training of cattle must be done in a calm manner. Cattle have a pressure point at their shoulder, so moving past the shoulder and going toward the rear of the cow will prompt it to move forward. A person doesn't need to be close to the cow to allow this to happen. Moving from the head to the point of the shoulder will make a cow stop and turn and move away from you. This herding must be done slowly and calmly.

Equipment and Housing Needs

Cattle equipment may include the need for chutes or stanchions, which are used to restrain cattle to administer medications or vaccines or to collect blood. These large pieces of equipment safely contain and control the cow and prevent it from kicking and injuring the handler (see Figure 18-13). There are many different types of metal chutes and stanchions, most of which have a head gate that locks in and controls the head of the cow. They also have horizontal bars located on the sides to prevent kicking. Sometimes cows will need to be haltered and led or tied in certain locations, requiring a cow halter and lead rope. Sometimes anti-kick bars may be placed over the hips or pelvis area of a cow to prevent it from kicking. These are metal devices in the shape of a V that clamp around the animal's flank.

Housing Needs

Cattle may be housed in barns or pastures depending on the type of weather and environment (see Figure 18-14). There are two types of housing for cattle: **warm housing** and **cold housing**. Warm housing is a heated barn or building facility during the winter that includes an insulated area with dairy stalls for each cow. Cold housing is a barn or building with no heat where the natural air circulates out moisture. Cold housing is usually a large open area where the herd is kept together. This is a more common housing method for calves and dry period cows.

FIGURE 18-12 Cattle can be trained to walk on a lead rope.

Photo by Kieth Weller. Courtesy of the U.S. Department of Agriculture

FIGURE 18-13 A chute helps to safely restrain the cow for performing medical procedures or exams.

Photo courtesy of Getty Images

FIGURE 18-14 It is common for cows to be housed in barns.

Milking Needs

Dairy cattle are milked in either stalls or milking parlors. Stall milking entails milking each cow in the area it is housed. This requires a pipeline that transports the milk to an outside bulk tank where it is then collected (see Figure 18-15). Milking parlors are specialized areas where cows are milked for improved efficiency and sanitation methods. The parlors typically include an elevated platform where the cow is restrained for milking. The floor area, known as the pit, is where the staff prepares the cows for milking.

Pasture

Feeding management requires rotating cattle from pastures to allow the grass to grow and to provide sanitation methods to the fields to decrease diseases and parasites. Pastures are usually rotated every 48 hours. This practice improves nutrition and forces good grazing

Delmar/Cengage Learning

FIGURE 18-15 Bulk tank where milk is collected.

Delmar/Cengage Learning

FIGURE 18-16 Pastures should have structures or areas where the cows can be protected from the weather.

methods. It will help maximize the number of cows in the herd and maintain good health practices. Many pastures should have access to shelter for cattle to get out of the weather (see Figure 18-16).

Feed Storage Needs

Proper storage of cattle feed products is important for good health practices. Grains are typically stored in **silos**, which are tall storage buildings that keep the food free of moisture. These buildings can also be controlled to feed specific amounts of food to the cattle during the day.

Waste Material Systems

Manure management is a concern for the health of the cattle herd and the health and sanitation of humans. Each cow produces about 8% of its body weight in waste materials on a daily basis. In one year a cow can produce

15–20 tons of waste. The waste is called **manure**, which is high in nitrogen, phosphorus, and potassium. Waste material systems must be in place on all cattle production farms to practice quality management and sanitation. The systems should provide efficient collection and removal methods. A **solid manure system** is used on some farms, where the costs are low and the manure is collected and hauled away on a daily basis. The wastes are stored on piles away from the cattle and then removed periodically. A **liquid manure system** is used in larger production systems and must meet environmental standards. This system is costly and requires large storage tanks that may be below or above the ground. Water is typically added to the system, as it requires pumping to remove the wastes on a regular basis of every 4–5 months.

Restraint and Handling

Beef cattle are more difficult to restrain than dairy cattle, mainly because they tend to be handled by people less than dairy cows. Dairy cattle grow accustomed to humans during the milking process. Cattle are sometimes shown and must learn how to be handled with a **halter** and **lead**. The halter fits over the head and face of the cow and the lead acts as a leash that is held to walk the animal. Cattle should be taught to tie for both showing and milking purposes (see Figure 18-17). Tying cattle is typically done using a **square knot**. The square knot is easy to make and will contain the cattle, yet it is easy to untie when necessary.

Cattle can be handled by moving them from place to place without the use of hands. This is called "**pushing**" cattle by walking toward them so they move away from a person's movement. This can be done by waving the arms and moving behind cattle so they move away from you and to the area in which you are trying to contain them. When working around cattle, a person should always stand at their shoulder area and never to the side of their rear legs. Cattle can kick to the side, known as **cow kicking**. Cattle can also be restrained using a **squeeze chute** or **stanchion**. The squeeze chute is a cage-like structure made of metal pipes that hold the cow and prevent it from kicking when being restrained. They can be tied in the chute or left free. Squeeze chutes are useful for veterinary staff to complete physical exams, give medication, collect blood, or conduct reproductive procedures. The stanchion is commonly called a head gate. This equipment may restrain the head of the cow or the entire body. The cow is walked into the stanchion and a head gate is closed to contain the head of the cow and restrain it from moving (see Figure 18-18). Some stanchions can squeeze the cow's body in place to restrain the entire cow.

Cattle may also be restrained using some hands-on restraint methods. The **tail switch restraint** prevents cattle from moving and kicking while being handled. The tail is twisted at the base, and this prevents the cow from moving and kicking (see Figure 18-19).

Delmar/Cengage Learning

FIGURE 18-17 Tying of cattle.

Delmar/Cengage Learning

FIGURE 18-18 A head gate is used to hold cattle for treatment.

FIGURE 18-19 Tail switch restraint.

Cattle can also be restrained using a rope placed around a front leg and lifting the leg from the ground. This prevents the cow from moving and provides minimal restraint. Cattle can also be reefed using the rope. **Reefing** is a restraint procedure using a heavy rope tied around the cow and pulling it to the ground for restraint. The rope is placed in a manner that when pulled, it forces the cow gently to the ground and a handler can kneel on the neck to prevent the cow from standing (see Figure 18-20).

FIGURE 18-20 Reefing restraint of cow.

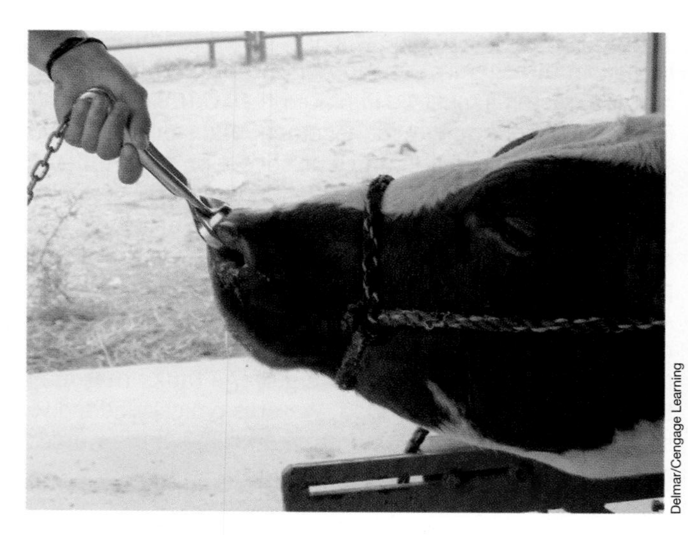

FIGURE 18-21 Use of nasal tongs for restraint of the cow.

Cattle are sometimes restrained and led by **nose tongs** that place pressure on the **nasal septum**, which is the cartilage located in the nostrils. Occasionally, some cattle will have a **nose ring** placed through the septum to be led. This pressure allows for cattle to be handled similar to a halter and lead (see Figure 18-21).

Grooming

Cattle that are shown need to be groomed on a regular basis. Show cattle are given baths, dried using high-powered vacuums, and brushed on a regular basis. This requires close contact with humans. Show cattle are usually handled at a young age and learn to be groomed and handled by people. The grooming practices in cattle are similar to those of a dog or horse.

Basic Health Care and Maintenance

Cattle production requires specific management and health practices of the herds because both dairy and beef cattle products are primarily used for human consumption. Thus, strict guidelines must be followed when caring for cattle. First, it is important that the producers and staff have knowledge and experience in cattle production and availability of veterinary staff who have a background in bovine medicine. Cattle require some human interaction and should be used to people so that health care and maintenance are relatively stress-free for the herds. There should be a **herd health manager** who records all health program and production information. The herd health manager is a person who has a background in bovine veterinary care and reproduction and should be able to maintain herd health records, vaccine programs, deworming schedules, reproductive care and records, and work closely with the veterinarian to establish a high standard of care for the entire herd of cattle. The goal of the herd health manager is to provide a parasite- and disease-free herd. Cattle should be placed on a regular daily schedule to provide a stress-free environment. Cattle should be fed at the same times each day, dairy cattle should be milked at the same times each day, all cattle should be provided adequate space and pasture, and a regular health care program should be established. Cattle should have a controlled climate, whether housed indoors or outdoors. This includes appropriate temperatures, proper shelter, appropriate humidity, and ventilation. Cattle should be provided shelter free from wind, rain, and cold weather. Water and feed troughs should be kept clean. Fencing should be monitored for the need of repair and maintenance. The herd health manager will monitor the following:

- Vaccine program
- Deworming program
- Heat cycle maintenance
- Body condition scoring
- Breeding log
- Veterinary maintenance
- Nutritional program
- Sanitation program
- Fly control system
- Pasture control
- Proper housing
- Proper fencing

All individual cattle should have a type of permanent identification. Cattle ID methods include **ear tag identification, neck tag identification, microchips, tattoos,** and **branding**. Ear tags are applied to the ears in a similar fashion as ear piercing. The ear tags have a number system that identifies each cow within the herd (see Figure 18-22). Neck tags are worn around the neck in a similar manner as a necklace. Each cow is given a number that is recorded on the neck tag. Microchips are electronic identification computer chips placed under the skin. Each chip has an individual number that is used for each cow within the herd. This identification method requires a scanner that reads the microchip number to record data information. The tattoo method is used by applying a number to the skin of the cow through the use of an electronic needle and ink that is placed on an area of the body that is easy to locate, such as an ear or the gums. Branding is a method that uses extreme

FIGURE 18-22 Calf identified with an ear tag.

heat or cold to burn or freeze a number on the hair coat of each cow. The branding methods cause the hair to change color and record the information on the hair. Occasionally, **ear marks** may be used where notches are made on the ear edges to identify cattle. Some producers are using new technology identification methods such as **DNA testing** and **nose printing**. DNA testing is done by using a blood or hair sample to identify each cow and its breeding lines. Nose printing is similar to fingerprinting, where the nose print is recorded since each cow nose has its own skin lines that differ from cow to cow.

Dehorning

Dehorning is the removal of horns to prevent injury to humans and other cattle. Dehorning methods include using **caustic** chemicals that burn the horn buds before the horns begin to grow. This is the most common method of dehorning and causes less pain, less stress, and easier after-care. Cattle that have developed horns must have them surgically removed with the use of sedation and dehorning equipment. Dehorners are specialized tools that have blades that remove the **horn bud**, or area of horn growth. The most important part of the after-care is treating for flies and monitoring the site for infection.

Castration

Castration of cattle is done to prevent reproduction and is the surgical removal of the testicles. Castration is done either by the **banding** method, which is the use of a tight rubber band that is placed over the testicle area and with time will cut off the circulation to the area and cause the tissue to **slough** or fall off. The band is placed using an **elastrator**, which is a tool used to stretch the band over the testicle area. This method is usually done in young calves. A surgical procedure is necessary in older, developed bulls. This is done with the use of an **emasculator**, which is a surgical tool that cuts the spermatic cord and allows the veterinarian to remove the testicles surgically.

Vaccinations

Cattle vaccination programs should be discussed with the herd veterinarian, as in some locations cattle may be at higher risk of certain diseases than others. Young beef calves are typically vaccinated 14 to 21 days prior to weaning. Many calves are castrated and dehorned at this time. Dairy calves are typically vaccinated with boosters at 3–4 months of age and 5–6 months of age and then yearly as the veterinarian feels necessary. Adult breeding cattle are usually vaccinated prior to breeding and prior to calving. Table 18-3 lists the common vaccines used in cattle.

Common Diseases

Cattle diseases are important in today's society. Veterinary staff and cattle producers must work together to provide quality health care to establish a cattle herd that is both disease- and parasite-free. Cattle diseases that are common or of a public concern must be monitored within the herd and a well-developed vaccine program must be developed. Veterinarians and herd health managers should work together to determine which diseases and parasites are prevalent in their area and what health program should be in place. Some diseases affect both dairy and beef cattle equally, and others are more prevalent in one cattle type.

Brucellosis

Brucellosis, commonly called **Bang's disease**, is a reproductive disease of cattle. The disease is passed through breeding and causes **abortion** (loss of the baby) in cows and **infertility** (not able to reproduce) in both bulls and cows. There is a vaccine available to prevent this disease, but no treatment has been discovered. All breeding animals should be tested prior to reproduction. This disease is zoonotic to humans and all bodily fluids should be handled with caution. Bodily fluids can include saliva, urine, feces, milk, semen, or vaginal discharge.

Bovine Viral Diarrhea

Bovine viral diarrhea (BVD) is a virus in cattle that causes many signs. There is a vaccine available but no treatment and the virus must run its course, with supportive care as the therapy. Signs of BVD include severe diarrhea, fever, coughing and other respiratory problems, nasal discharge, possible lameness, and a poor hair coat. Pregnant cows that develop this viral infection may abort.

Infectious Bovine Rhinotracheitis

Infectious bovine rhinotracheitis (IBR) is a virus that affects cattle and is commonly called "red nose." IBR is prevented through vaccination, but like BVD there is no treatment and the virus must run its course along with supportive therapy based on the signs. Clinical signs include respiratory problems, coughing, nasal and ocular discharge, fever, and weight loss. The term *nasal* means pertaining to the nose and *ocular* means pertaining to the eye. Severe cases may develop staggering and seizures and eventually death.

TABLE 18-3

Common Vaccines in Cattle

IBR (INFECTIOUS BOVINE RHINOTRACHEITIS)	BVD (BOVINE VIRAL DIARRHEA)	PARA INFLUENZA (FLU)
Bovine Respiratory Virus	Clostridium	Leptospirosis
Pasteurella	Bordetella	E-coli
Rotavirus	Coronavirus	Rabies

Leptospirosis

Leptospirosis is a bacterial disease of cattle that is transmitted in the urine. There is a vaccine to prevent the disease and treatments are available. There is a zoonotic potential with urine contamination. Signs of this disease include bloody urine, fever, rapid respiratory rate, and occasional lameness or stiffness.

Campylobacter

Campylobacter is a bacterial infection that affects either the intestinal tract or the reproductive system of cattle. The intestinal form causes diarrhea and digestive problems in cattle; this form is zoonotic in people. The reproductive form is more serious as infertility may occur. Signs of the reproductive form include possible abortion, decreased conception rates, and erratic heat cycles. Breeding animals should be vaccinated yearly.

Mastitis

Mastitis is the inflammation of the mammary gland and can occur in any female cow, but most commonly the condition affects dairy cattle. Bacteria typically enter the teat and cause an infection within the mammary glands. This can be prevented through good sanitation practices of the barn, pasture, and milking equipment. There is no vaccine for mastitis but it can be easily treated. It is estimated that the average cost per cow to treat for mastitis in a herd is $200 per year. The **California Mastitis Test** or **CMT** is used to diagnose and measure the level of cell infection within the glands. Most cows will need to be placed in a dry period for treatment of mastitis.

Bloat

Another condition that may affect all cattle is **bloat.** This is air that is ingested into the stomach that causes the stomach to swell and possibly rotate, which is called a **displaced abomasum**. This may be caused by overeating or by feeding a very rich pasture. Signs of the condition include restlessness, not eating, unable to have a bowel movement, and not able to regurgitate. Most stomachs rotate and are displaced to the left.

Retained Placenta

A **retained placenta** is a reproductive condition that can affect any female cow within 8 hours post-labor. The placenta is the afterbirth that has not been passed and may cause a severe infection within the female's body. This condition can lead to **sepsis**, which is caused by toxins entering the bloodstream. In the case of a retained placenta, the toxins are the waste materials of the newborn. Signs of a retained placenta include fever, anorexia, low milk production, foul odor, discharge from vulva, and membranes hanging from vulva. This condition should be treated as an emergency and veterinary attention is necessary.

Grass Tetany

Grass tetany is a condition of cattle that are placed on rich green pasture that is high in nitrogen that they are not used to eating. This leads to the cow developing low magnesium and calcium levels in the bloodstream. It is more common in older cows during the lactation phase and will occur in the springtime. It can be prevented by feeding forage or placing on new pastures for limited amounts of time. Signs of grass tetany include abdominal pain, seizures, and occasionally death.

Metabolic Disorders in Dairy Cattle

Dairy cattle are affected by other **metabolic diseases** that cause a chemical change in the body, usually from stress. Stress may occur with calving, milking, and changes in the nutrition or environment.

Hypocalcemia

Hypocalcemia or **milk fever** occurs in lactating cows and is due to low blood calcium. This condition may occur from improper calcium supplements in the diet or too much calcium in the lactating animal. It commonly occurs near calving time but may occur at any time. The condition can be avoided by providing phosphorus in the diet and adding in a supplement of vitamin D.

Ketosis

Ketosis is a condition of dairy cattle that causes low blood sugar, similar to diabetes. It may occur early in the lactation phase or may be related to a nutritional imbalance. Signs include a poor appetite, a dull coat, depression, incoordination, increased licking motion, and a fruity odor to the breath.

Blackleg

Blackleg is a fatal disease of cattle caused by bacteria that can live in soil for many years. The bacteria are ingested through grazing and enter the digestive system or invade the body through wounds that become infected with the bacteria. The bacteria that causes this disease are a type called *Clostridium*. Cattle between 6 months and 2 years old that are in pastures and eating high-protein diets are most susceptible to the disease. Signs of the disease include lameness, loss of appetite, difficult or increased breathing, depression, fever, swelling, and pain. Cows usually die within 12 to 48 hours. As the disease progresses the swelling causes the limbs to become infected, causing the tissue to become black in color. This disease is not contagious from cow to cow. The disease is preventable through vaccination.

Anaplasmosis

Anaplasmosis affects dogs, cats, and ruminants. The most common animal affected by this parasitic protozoan disease is the cow. The disease is not contagious but is transmitted most commonly by ticks, but it is also transmitted via flies and mosquitoes. The disease invades the red blood cells. Signs include anemia, fever, weight loss, difficulty breathing, uncoordinated movements, abortion, and death. This disease can cause cattle to become carriers. This can be a huge concern to dairy and beef producers. Anaplasmosis is a major problem in many southern states but is frequently diagnosed in the Midwest. Several states require that cattle be tested for anaplasmosis before importation is allowed. Infected herds are not eligible to export cattle to those states. A strict interpretation of disease control regulations for both interstate and intrastate movement of cattle precludes sale of cattle from herds that are infected with anaplasmosis. A vaccine is available and appropriate insect control methods are recommended.

Common Parasites

All parasites affect all cattle and a dewormer and sanitation program should be clearly developed. External parasites that are the most common problem in cattle include flies, ticks, mites, and lice. Fly control is an important part of cattle production. This includes the need for **insecticides**, sanitation control, and the use of **back rubbers**. Insecticides are sprays or pour-on chemicals that repel flies and other insects and prevent them from biting the cattle. Back rubbers are equipment placed in the barn or pasture that allow cows to rub their backs on a rolling bar that spreads on fly control chemicals. Sanitation control is the removal of body wastes to decrease the amount of flying insects. Internal parasites in cattle typically cause diarrhea, weight loss, poor hair quality, and a dull hair coat. Many types of external parasites may occur, the most common of which are strongyles, flatworms, and roundworms. Cattle are treated with **anthelmintics** or dewormers that are given 1–2 times a year in oral, injectible, pour-on, or topical methods. A different product should be rotated each time to prevent a buildup of chemical resistance in the body system.

Common Surgical Procedures

Male cattle are often castrated if they are not going to be used as breeding bulls. Castration allows beef cattle to put on weight much faster for beef production systems. Young calves are commonly castrated using an Elastrator band, which applies thick rubber bands to the testicle area and remain on until the tissue sloughs off, allowing for a nonsurgical procedure. Older cattle that are castrated must be sedated, and standing castration is done as a surgical procedure, similar to that in dogs or horses.

Reproduction and Breeding Beef Cattle

There are three primary types of beef cattle production systems that determine how beef cattle are raised and used for profit. These systems are the cow-calf system, the back-grounding system, and the finishing system.

The Cow-Calf System

The first type of production is called the **cow-calf system**. The goal of this system is to raise cattle and breed them as adults. Therefore, this type of production is based on beef cattle reproduction. One or more cattle breeds may be raised in this type of production system, but only high-quality cows of a specific breed are used. The cows begin breeding around 2 years of age and the calves are born in late winter or early spring and are then sold as a profit. The cows are bred each season as long as they are producing a quality calf. The cows are maintained on pasture and the cow-calf system requires a large amount of land to hold the number of cattle in the production system.

The Back-Grounding System

Another type of beef cattle production is the **back-grounding system**. The back-grounding system goal is to raise a calf to market size for a profit. Cows are bred to produce calves that are weaned and continued to be fed to an ideal weight that increases their size and profit. This ideal calf size is called **feedlot size**. The calves are fed quality pasture, hay, and grain sources that increase their body mass. They may also be fed supplements that allow their body to convert more food to muscle. When they reach feedlot size, they are sold for meat.

The Finishing System

The last type of beef cattle production is the **finishing system**. The finishing system produces a calf that grows through the entire adult stage. The adult cattle are sold for profit as either a breeding animal or for the meat. Most adult cattle weigh between 1,100 and 1,300 pounds as adults at **market weight**. These cattle are fed **concentrates**, or a mixture of foods high in energy and fat, as well as minerals that help promote growth. This system is more costly and takes a longer time for a profit, but the profit is much larger due to the size of the animal.

Reproduction and Breeding of Dairy Cattle

Dairy production is based on the cow producing a quality milk product. That requires the cow to be bred for lactation to occur. Most milk production systems have large numbers of cows that are producing milk for around 305 days. A cow's gestation cycle, or the length of pregnancy, is 283 days. The cow produces milk through this entire phase. Milk production is typically stopped for 50–60 days and this is called the **dry cow period**. This allows the cow's reproductive system time for rest and then the cow is bred and the milking process begins again. The milk content is about 87% water and a mixture of fat, protein, and sugar nutrients. Milk is produced in the udder, or the bag-like structure that contains four mammary glands or **quarters** that are sections of the glands that store the milk. Each quarter has a teat that is used to remove the milk from the cow. Milk nutrients are converted to milk by tiny structures within the udder called **alveoli**. Dairy production is a 24-hour-a-day job with a long-term financial growth possibility. However, dairy cattle are a costly investment with many risks. The size and type of the dairy herd may vary between farms. A **purebred business** consists of a breed of dairy cattle that is registered with a known pedigree. A **commercial business** consists of a grade type of dairy cow that is a mixed breed. Sizes range from a **family size herd** that is fewer than 100 head of cattle or a **large size herd** that is more than 100 head of cattle. Many dairy producers have both a **milking herd** and a **total herd**. The milking herd is the group of cattle that are used to produce a high quality and quantity of milk. Total herds are cows that have replacement heifers when the adult cattle become too old to continue to reproduce. The heifers then replace the adult aged cattle in the milk production system. Each farm herd also needs bulls to reproduce with the cows and continue the profitability of the herd. Small herds need between five and eight bulls for breeding purposes. Large herds need between 10 and 15 bulls. **Dairies** are buildings where cattle are milked two to three times a day. The milking process must be thorough and sanitary due to the product being for human consumption. Milk is made mostly of water, which means that dairy cattle must drink large quantities of water to convert it to milk. The milking process consists of disinfecting the udder of each cow within the herd (see Figure 18-23). The udder is cleansed and dried and then **primed**, which means applying the milking machine devices to the udders to begin collecting milk. The milking machine is attached to tubes that connect to pipes that transport the milk through a system and into a cooling tank. When all of the milk has been removed from the udder, iodine solution is applied to the udder to prevent disease and bacteria from entering the mammary glands and to help close the muscle of the udders to prevent milk from continuing to drain.

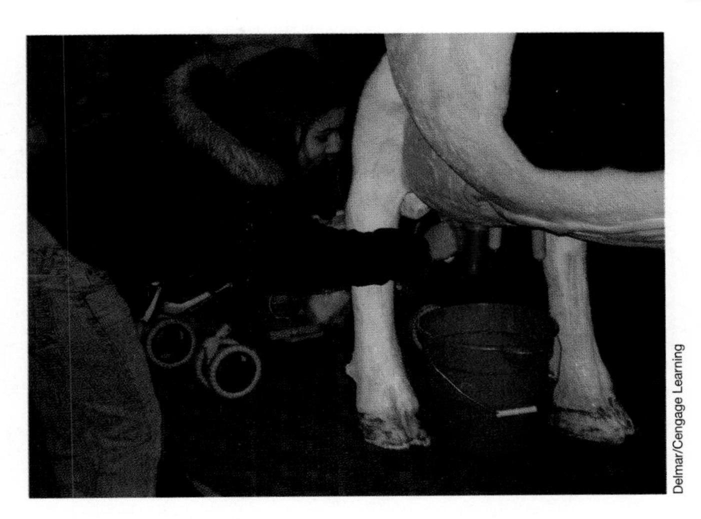

FIGURE 18-23 Disinfecting and priming the udder.

SUMMARY

Beef and dairy cattle production systems are different in their types and practices; however, both require strict sanitary practices and have regulations that must be followed due to both products being used for human consumption. Several types of production systems exist as to the product type and the market profit that each system will bring the producer. All cattle must have proper nutrition, health care, and management practices to ensure a healthy outcome. It is important for veterinary assistants to have some experience in cattle behavior, restraint, and handling when working with these large animals.

Key Terms

abortion condition that causes a pregnant female to lose her baby

alfalfa hay high in protein

alveoli tiny structures within the udder where nutrients are converted to milk

anthelmintics dewormer medications used to control parasites

back rubbers equipment used to allow cows to rub their back on a rolling bar that applies fly control chemicals

back-grounding system production system that raises calves to market size for profit

banding method of castrating male livestock by using a tight rubber band placed over the testicle area to cut off circulation and cause the tissue to fall off

Bang's disease common name for brucellosis

beef cattle cows specifically raised for meat

bloat air ingested into the stomach causing the stomach to swell

bovine veterinary term for a cow

Bovine viral diarrhea (BVD) virus in cattle causing diarrhea and respiratory signs

branding the use of extreme heat or cold temperatures to mark the skin with a number or symbol

brucellosis reproductive disease of mammals passed through breeding practices

bull adult male cow of breeding age

calf young newborn cow of either gender

California Mastitis Test (CMT) test used to diagnose and measure the level of cell infection within the mammary glands

calving the labor process of cows

campylobacter bacterial infection that affects the intestinal tract or reproductive tract in cattle

castration surgical removal of the testicles to prevent reproduction

caustic chemicals used to burn horn buds to prevent horn growth

cold housing building with no heat where air circulates out moisture and usually houses the entire herd as a group

colostrum antibodies produced in the first 24 hours of the mother's milk to protect the immune system of the calf

commercial business grade type of cow that is a mixed breed

concentrates mixtures of foods high in energy and fat

conformation the body's shape and form

cow adult female cow of breeding age

cow kicking the direction of cows kicking to the side with their rear legs

cow-calf system production system that raises cattle to breed them as adults

creep feeders equipment used to feed young calves that prevents adult cattle from eating the food

culled to remove from the herd

cutability quality and quantity of meat from a beef animal

dairies buildings where cows are milked

dairy cattle cows raised specifically for milk

dam mother or female parent

dehorning process of removing the horns to prevent injury to people and other animals

DHI program Dairy Herd Improvement program that maintains records of dairy herd information

displaced abomasums condition in cattle that causes the stomach to rotate out of place

DNA testing blood or hair samples used to identify an animal's parents

dry cow period time that milk production is stopped to allow the reproductive system to rest

dual purpose breed breed that serves more than one purpose; in the case of cattle, both milk- and meat-producing breed

ear marking notches made on the edges of the ear flaps to identify cattle

ear tag identification tags with numbers applied to the ears by piecing the ear flap

elastrator instrument used to stretch bands over the testicles for castration

emasculator surgical tool used to cut the spermatic cord during a castration procedure

family size herd a group of fewer than 100 head of cattle

feedlot size the ideal weight of calves that increases size and profit

finishing system production system that produces a calf through the entire adult stage and then sold as meat or for breeding

forager animal that eats grass and pasture

free choice food that is given in large quantities and available at all times

freemartin adult cow that is sterile and not able to reproduce

freshening the labor process of dairy animals

genetic flaws undesired traits or characteristics passed from one or both parents to the offspring

gestation cycle length of pregnancy

grass tetany condition in cattle due to eating rich pasture high in nitrogen gases; causes abdominal pain

halter equipment that fits over the head and is used to control the animal

heifer young female cow that has not yet been bred

herd a group of cattle

herd health manager person that maintains the records and health care of a herd of cattle

horn bud area of horn growth on top of the head

hypocalcemia milk fever; condition causing low blood calcium

infectious bovine rhinotracheitis (IBR) respiratory virus affecting cattle; commonly called "red nose'

infertility not capable of reproducing

insecticides sprays or pour-on chemicals used to control flies and other insects

ketosis condition in dairy cattle that causes low blood sugar

lactation the process of milk production

large size herd a group of more than 100 head of cattle

lead rope that attaches to a halter and is used to walk an animal

leptospirosis bacterial disease transmitted in urine of infected animals

liquid manure system costly waste removal system used on large farms and requires above- or below-ground storage tanks with water added for ease of pumping

manure waste material high in nitrogen, phosphorus, and potassium

marbling appearance of intramuscular fat within the meat

market weight adult weight of cattle when they are sold for beef

mastitis inflammation of the mammary gland

metabolic diseases conditions that cause a chemical change within the body, usually from stress

microchips electronic identification in a computer chip placed under the skin

milk fever low blood calcium condition called hypocalcemia

milking herd group of cattle used to produce high-quality amounts of milk

nasal septum cartilage between the nostrils

neck tag identification tag with numbers worn around the neck of cows to identify them

nose printing methods similar to finger printing in which the lines of the nose are imprinted for identification

nose ring metal ring placed through the septum of the nose to allow a cow to be led

nose tongs equipment placed in the nasal septum to apply pressure and allow leading

offspring young animals produced by a male and female animal

palatability how well food tastes and is eaten

pedigree the parents and breeding lines of animals' offspring

polled no horn growth

predicted transmission ability estimation record to determine the possible traits that a bull or cow will pass along to offspring

progeny offspring

purebred business registered breed of cattle that has pedigrees

(*Continues*)

pushing moving cattle by walking toward them quietly and calmly so they move away into another area

quarters sections of the mammary glands that store milk

reefing restraint procedure using rope to place a cow on the ground for restraint

retained placenta reproductive condition in female animals where the afterbirth materials have not passed within eight hours after labor

roughage hay or grass source fed to livestock and ruminants

selection guidelines set of rules that state the standards of the type of cow and how it is chosen for a production program

sepsis conditions when toxins enter the bloodstream causing a severe infection

silos tall storage buildings that keep food free of moisture

sire male parent

slough tissue that dies and falls off

solid manure system waste material system low in cost where manure is collected and removed on a daily basis and stored on a pile away from cattle

sound showing no signs of lameness or injury

springing heifer young female cow that is pregnant with the first calf

square knot knot tie looped into a square and used to tie and untie cattle easily

squeeze chute cage-like structure made of metal pipes that holds a cow and prevents the cow from kicking during restraint

stanchion head gate that holds the head of a cow in place during restraint

starter food easily digestible food for calves as they begin eating

steer castrated male cow

tail switch restraint restraint method of twisting the tail at the base to prevent a cow from kicking and moving

tattoos number applied to the skin using an electronic needle with ink

teats sections of the mammary gland used to produce milk

total herd cows that have replacement heifers when the adult cows become too old to continue milk production and breeding

total mix ration nutrients needed daily within a mixture of high-quality products

udders mammary glands in dairy animals that produce milk

warm housing heated building that holds cattle in the winter and is insulated with individual stalls for each cow

REVIEW QUESTIONS

1. What are the differences between beef and dairy cattle?
2. What are the types of beef cattle production systems? Explain each type.
3. What is the basic concept of dairy cattle production?
4. What are the types of dairy production systems? Explain each type.
5. What is the DHI program?
6. What is the importance of the DHI program?
7. What are some methods of restraining cattle?
8. What is the role of the herd health manager?
9. What are some management practices in cattle?
10. What are some common diseases that may affect cattle?
11. What are some methods of cattle identification?
12. How much should adult dairy cattle be fed on a daily basis?
13. How much should adult beef cattle be fed on a daily basis?
14. What is the difference between warm and cold housing for cattle?
15. What are some management practices for waste materials of cattle?

Clinical Situation

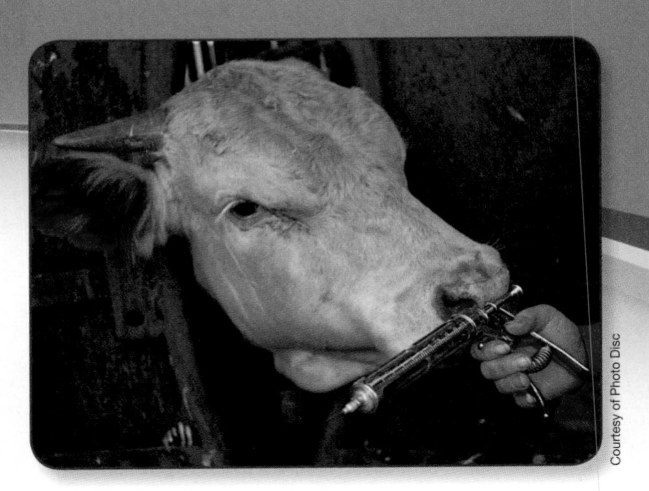
Courtesy of Photo Disc

Brittany, a veterinary assistant at the Delmont Bovine Practice, answered a phone call from Mrs. Shore, the wife of Mr. Shore, a local dairy producer. Brittany remembered that Mr. Shore's Holstein dairy farm was one of the largest in the area.

"Hello, Mrs. Shore, how can I help you?"

"Hi, my husband asked me to call and set up a farm call appointment for his cattle herd to be vaccinated and examined. This is the yearly visit. He also said he needs about 100 head of cattle checked for pregnancy."

"Okay, Mrs. Shore, Dr. Evans is not in the office at the moment, so I will need to check with him to see what day works best for his schedule. How many head of cattle are in the herd?"

"I really don't know; it's my husband's business and I was just asked to call and set up an appointment."

"Well, let's tentatively set it up for next Wednesday. It looks like Dr. Evans has a light schedule. I will call you back if that day doesn't work."

- How did Brittany handle the situation?
- What would you have done in this situation?
- What ways could Brittany have obtained a history on this client and the dairy herd?

19

Courtesy of PhotoDisc

Equine Breed Identification and Production Management

Objectives

Upon completion of this chapter, the reader should be able to:

- Understand and discuss common veterinary terms relating to the horse
- Describe the biology of the horse
- Identify common breeds of horses
- Distinguish the different classes of horses
- Discuss the nutritional needs of horses
- Describe normal and abnormal behaviors of horses
- Describe housing, facility, and equipment requirements for horses
- Properly and safely restrain horses for veterinary procedures
- Discuss the necessary health care and veterinary practices of horses
- Describe breeding practices with horses
- Discuss common diseases and parasites of horses

Introduction

Horses have become one of the most popular types of livestock for both breeders and animal lovers alike. Horses have a grace and beauty that are unlike any other animal (see Figure 19-1). They also have a great ability to bond to humans and serve many purposes to humans. Most horses are owned for recreation and entertainment, but some are also used for working and in some countries as a food source. Today, there are over 9.5 million horses in the United States.

Delmar/Cengage Learning

FIGURE 19-1 Horses are owned and raised for recreation, entertainment, and work.

Veterinary Terminology

Horses are classified in terms based on their age, gender, size, and type of use. Several different terms are used to describe horses. An adult female horse is called a **mare** and an adult male horse that is capable of reproduction is called a **stallion**. A female horse that has not been bred is called a **maiden** horse. The male parent is known as a **sire** and the female parent is called a **dam**. The male horse used in a breeding program is called a **stud** horse. When a male horse is castrated it is called a **gelding**. A young newborn horse is called a **foal**; the female is specifically called a **filly** and the male is known as a **colt** (see Figure 19-2). The labor or birth process of horses is known as **foaling**. In veterinary medicine, horses are called **equines**. They are also classified by their sizes and purposes or uses. There are **light horses**, **draft horses**, and **ponies**. Equine species are measured in hands. A **hand** is equal to 4 inches and is used to measure the height of a horse at the highest point of the shoulder. Light horses are typically riding horses that stand between 14.3 and 17 hands in height and weigh between 900 and 1,400 pounds (see Figure 19-3). Draft horses are known as large and muscular animals that weigh over 1,400 pounds and stand over 15 hands in height (see Figure 19-4). Ponies are small in size, standing under 14.2 hands in height and weighing less than 900 pounds. They are usually favored by small children. Other types of equines are donkeys and mules. A **donkey** is a member of the equine family and sometimes called a *burro*. The female donkey is called a **jenny** and the male is called a **jack**. A **mule** is the cross of a female horse and a male donkey. A male horse crossed with a female donkey will produce a **hinny**.

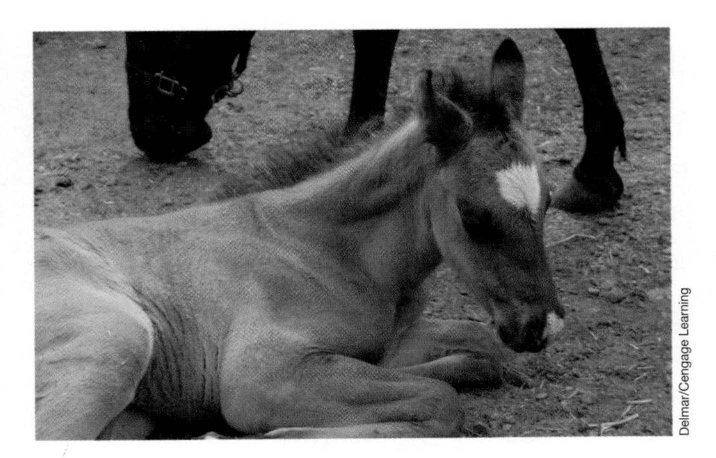

Delmar/Cengage Learning

FIGURE 19-2 A young horse is known as a foal.

Delmar/Cengage Learning

FIGURE 19-3 Example of a light horse.

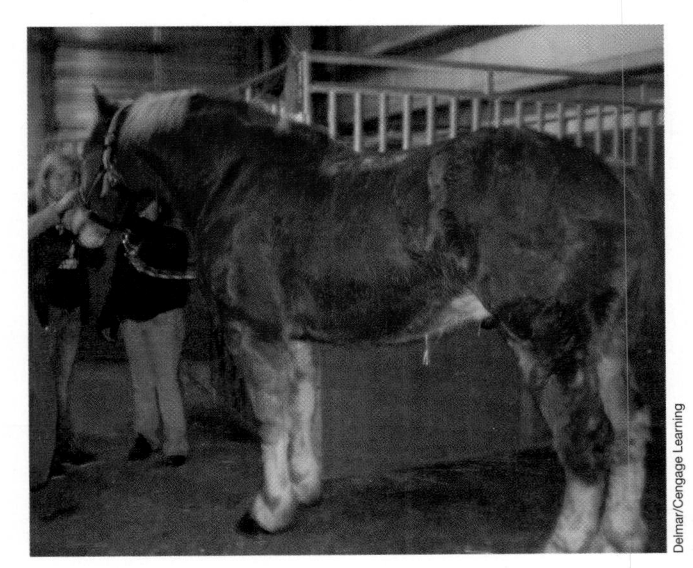

Delmar/Cengage Learning

FIGURE 19-4 Example of a draft horse.

Biology

Horses are large animals that are classified as livestock, similar to cattle. Horses are from the genus and species ***Equus caballus*** and developed from prehistoric types of horses. The purposes and use of horses have changed over the last several hundred years. Many breeds and varieties of horses have been developed.

Horses are herbivores and feed on a plant-based diet (see Figure 19-5). They are mammals that have a specialized digestive system called **nonruminant**. A nonruminant digestive system is similar in appearance to humans but requires small amounts of food to be eaten and digested throughout the day, similar to cattle and other grazing animals. They have a relatively small stomach with a long intestinal tract that allows food to flow all day. The horse's digestive system

FIGURE 19-5 Horses primarily eat plants and are known as herbivores.

FIGURE 19-6 Horses are hoofed animals.

also has an enlarged **cecum**, which is an area within the intestinal tract that breaks down grasses and roughage. Horses do not have a gall bladder and are not capable of vomiting.

Horses are hoofed animals with the hoof wall being made of hard material similar to that of a fingernail (see Figure 19-6). Their legs and hooves are critical parts of the horse and may easily become injured. Hooves continually grow and must be trimmed and cared for on a regular basis. The legs are very fragile and can suffer severe damage.

Horses are endothermic or **warm-blooded**, meaning they control their body temperature internally. Horses' bodies and structures are based on the type of work they are used for (see Figure 19-7). They have around 205 bones in their skeletal system (see Figure 19-8). Horses can typically live into their 20s with proper health care.

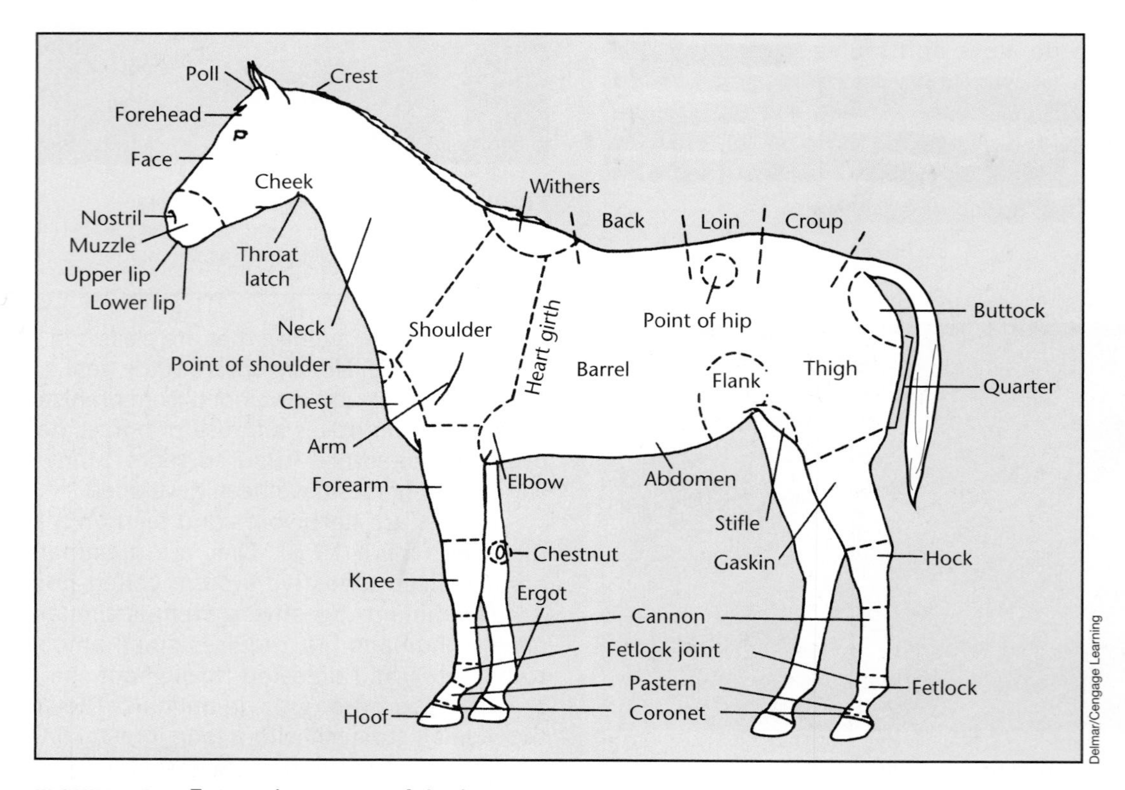

FIGURE 19-7 External anatomy of the horse.

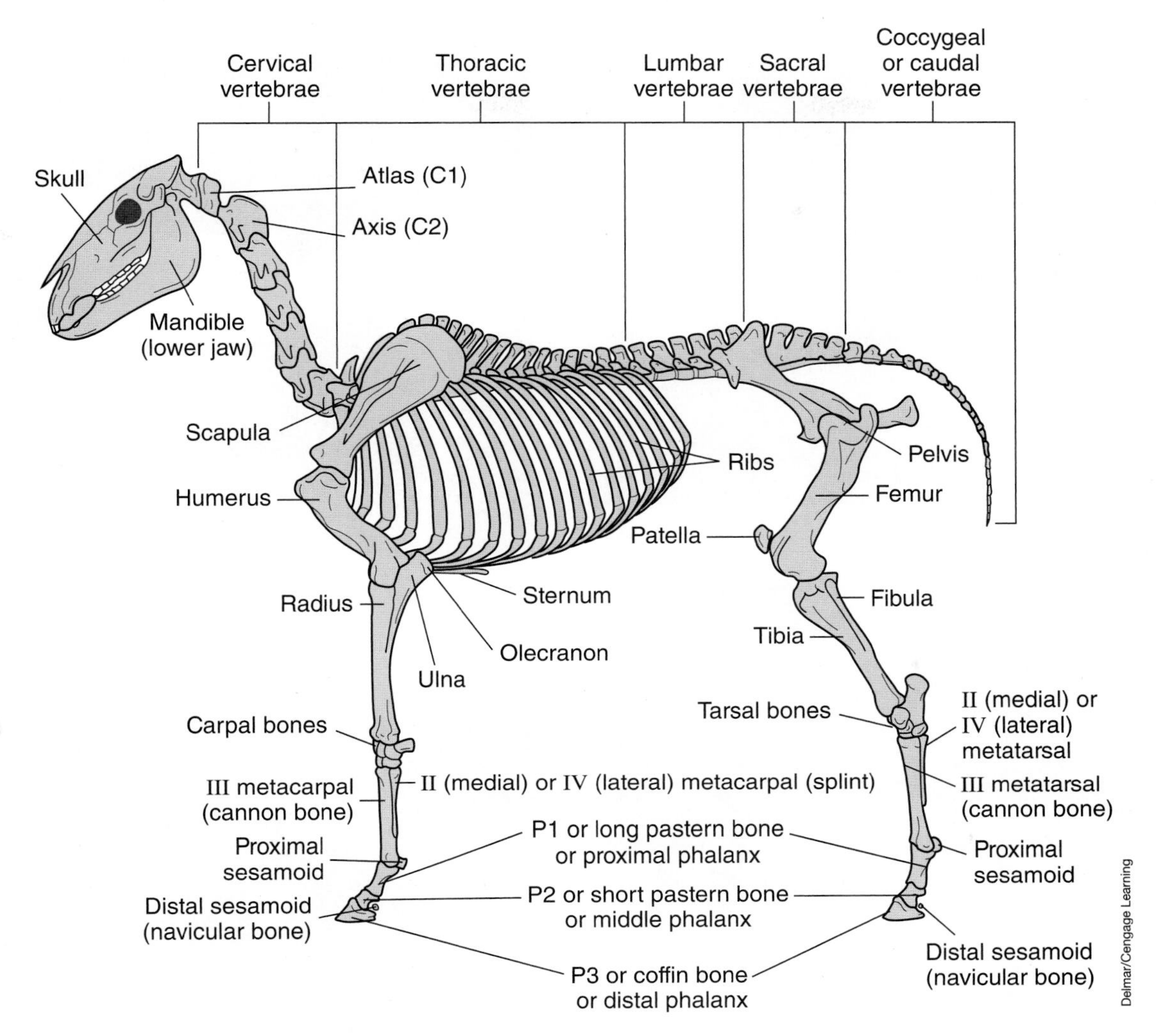

FIGURE 19-8 Skeletal anatomy of the horse.

Breeds of Horses

Horses are classed according to their body type, size, and purpose or use. Horse breeds come in a variety of sizes and colors, some of which are riding horses, which include the Quarter Horse, Paint, Thoroughbred, and Appaloosa (see Figure 19-9); draft horses, which include the Clydesdale and Percheron (see Figure 19-10); and ponies, which include the Shetland and Welsh (see Figure 19-11).

Riding Horses

Horses that are used as **riding horses** are ridden for pleasure and showing purposes. Riding horses are further classed by their type of gaits and the uses they provide to people. Some of the most popular riding horses are the Quarter Horse, Paint, Appaloosa, and Tennessee Walking Horse (see Figure 19-12).

Gaited Horses

Gaited horses are classified by the way they move in their **gaits**, or the way they walk and run. Gaited horses have a rhythm to their movement that is easy to sit and provides the rider with little movement or fewer rough rides. These horses are popular with older riders and those that may have back problems. There are different speeds at each gait that were developed by people that used to spend a lot of time in the saddle and chose these horses due to the easy gait and disposition of the animal. Popular gaited horse breeds include the Tennessee Walking Horse, American Saddlebred, and the Rocky Mountain Horse (see Figure 19-13).

Stock Horses

A **stock horse** is a type of horse that was once used on ranches for herding and working cattle. They are hardy, athletic, able to serve many purposes, agile and surefooted, fast and durable. They are also known for their

Courtesy of the American Quarter Horse Journal

Courtesy of the American Paint Horse Association

Delmar/Cengage Learning

Courtesy of the Appaloosa Horse Club

FIGURE 19-9 (A) American Quarter Horse (B) American Paint horse (C) Thoroughbred (D) Appaloosa.

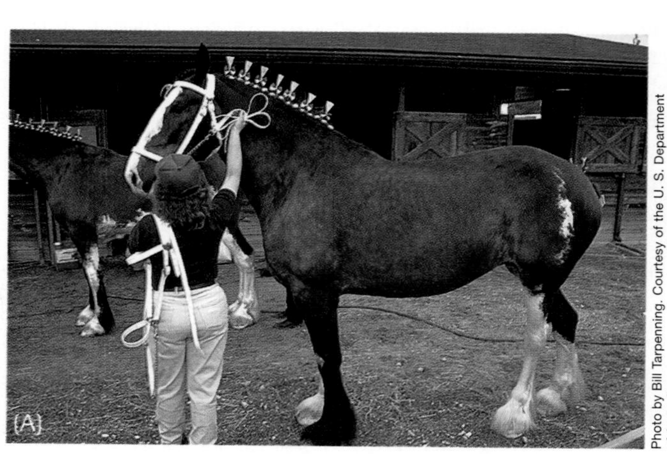

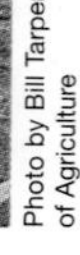

Photo by Bill Tarpenning. Courtesy of the U. S. Department of Agriculture

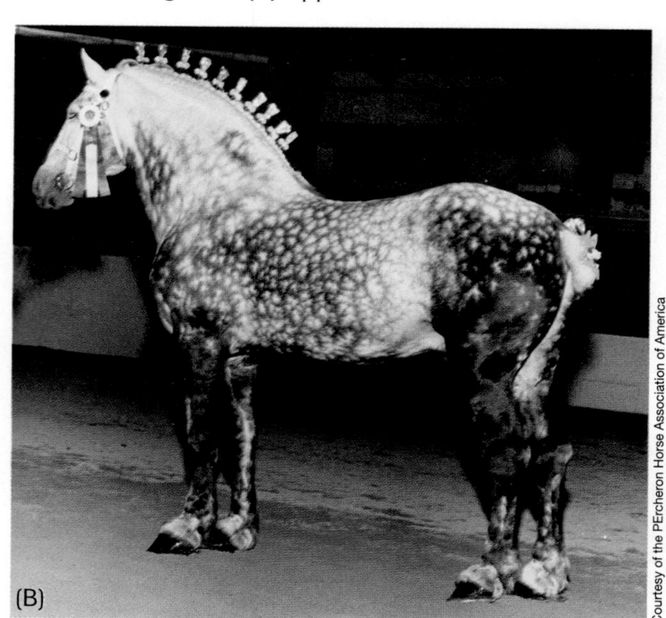

Courtesy of the PErcheron Horse Association of America

FIGURE 19-10 (A) Clydesdale (B) Percheron.

Courtesy of the American Shetland Pny Club

FIGURE 19-11 American Shetland pony.

Delmar/Cengage Learning

FIGURE 19-12 Riding horses are used in shows and competitions.

muscular and athletic body structure. Popular stock horse breeds include the Quarter Horse, Appaloosa, and Paint horse (see Figure 19-14).

Hunters and Jumpers

Hunters and **jumpers** are horses that were bred for fox hunting and cross country riding. They tend to be tall, athletic, and muscular and have speed and endurance for jumping large fences. Hunter horses usually are used for cross country fox hunting and jumping smaller fences, whereas jumpers are used for jumping large fences on a set course (see Figure 19-15). Popular horse breeds used for hunters and jumpers include Thoroughbreds and a variety of warm bloods.

Courtesy of the Tennessee Walking Horse Breeders Association

FIGURE 19-13 The Tennessee Walking horse is a breed of gaited horse.

Delmar/Cengage Learning

FIGURE 19-14 Stock horses are sturdy horses often used for working.

Racehorses

Racehorses are a type of horse that were bred for speed and are used for racing (see Figure 19-16). This includes horses raced under saddle and in harnesses. They often have the ability to be used for other sports-related activities. Horse racing has become a popular sport, with some of the famous races being the Kentucky Derby and Belmont Stakes. Popular race horses are the Quarter Horse and Thoroughbred. Popular harness racing breeds include the Standardbred and Morgan horse.

Courtesy of Botts/Watson Photography

FIGURE 19-15 Jumpers are used on courses in competitive riding events.

Courtesy of USDA

FIGURE 19-16 Race horses are bred for speed.

Ponies

A pony is a small type of horse that is commonly used by children. They stand under 14.2 hands in height. They tend to be gentle and intelligent. The most popular pony breeds include the Shetland, Welsh, and Pony of the Americas (see Figure 19-17).

Draft Horses

Draft horses are large and muscular and bred to pull heavy loads. They are also known as work horses, as many are used today in helping work fields and pull large wagons. Some popular draft breeds include the Clydesdale, Belgian, and Percheron.

FIGURE 19-17 Pony of the Americas.

Breed Selection

Horse selection is based on several factors. Some areas that prospective horse owners need to take into consideration are the cost involved in purchasing, caring for, and maintaining a horse. Horse breed selection is based on the size, disposition, breed, gait, purpose, gender, and color.

Horse price ranges vary greatly depending on the breed and purpose. Show horses or specially trained horses may be more costly than another type of horse. It is important to know the cost factors in both purchasing and owning a horse. Some other factors to consider with the cost are boarding bills, hay and feed cost, veterinary care, shoeing costs, fencing and building costs, and supplies. The disposition of the horse must be evaluated, as it is important that anyone working with the horse is able to get along and handle the animal. Beginners will need an equine that is calm, quiet, and gentle as well as well broke for the purpose or use.

The size of the horse should be optimal for the size of the rider as far as the rider's height and weight and ability to control the size of the horse. Breed, color, and gender are personal preferences that should be considered in the purchase.

Nutrition

Horses have a nonruminant digestive system that requires them to continually digest food to keep their system from binding up. The digestive system has a cecum, or a sac located between the small and large intestine that helps break down roughage and protein. A horse's stomach is always digesting food and should have a notable gurgling sound at all times. Horses should be fed grains that include oats, corn, wheat, or barley mixtures. This supplies the horse with adequate amounts of energy and protein. Some horses require additional supplements or concentrates depending on the health and activity levels of the horse. Common supplements include soybean meal or linseed oil for skin and coat conditioning. Horses can be fed in grain or pellet formulas. Pellet feed is easier to digest and break down in the horse's intestinal system. Grains can be ground down, cracked, or shelled for better digestion. All foods should be free of dust and mold. The usual maintenance diet for a horse requires one pound of protein per 1,000 pounds body weight. Horses are typically fed twice daily with adequate amounts of pasture or hay (see Figure 19-18). **Forages**, or hay and pasture, may include timothy, orchard grass, or alfalfa sources. Hay should be kept dry and monitored for signs of mold. Hay should be fed at one to two pounds per 100 pounds of body weight and may be provided several times daily. Horses on pasture should have at least three acres of space for grazing. Some horses and foals may require minerals and vitamins for necessary growth and development of teeth, bones, and tissues.

The most common nutritional deficiencies include calcium and phosphorus. Feed supplements should be fed based on individual horse basis. Horses require additional sources of salt that can be supplied in **mineral blocks** or **salt blocks**. Vitamin A may need to be given to horses on poor or inadequate pasture or hay sources. Vitamin B may be needed for horses stalled indoors on a regular basis. Horses that are underweight may be fed higher amounts of protein or flax seed oil for additional fat sources. All horses require daily fresh water sources (see Figure 19-19). The average horse requires 10–12 gallons of water a day. Increased amounts of water will be necessary during high amounts of activity, during

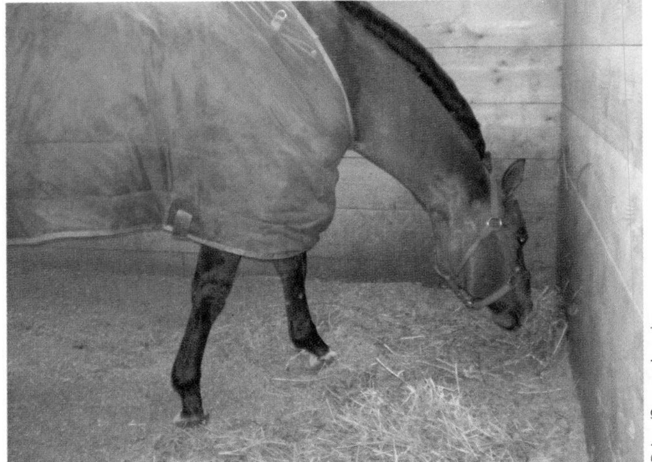

FIGURE 19-18 Horses eat hay and grains.

FIGURE 19-19 Horses should be supplied with water regularly.

TABLE 19-1

Recommended Maintenance Diet Rations	
Protein	10–12%
Fat	6–8%
Fiber	12–15%
Calcium	1–3%
Phosphorus	0.5–1%
Sodium	0.2–0.7%

lactation, and in the winter time. Water intake should be monitored as horses can easily dehydrate. **Electrolytes** or powdered supplements can be placed in the water or food source to encourage horses to drink more and prevent dehydration. Horses should not be allowed to drink immediately after exercise or when sweated. This can lead to severe stomach pain. Any changes in diet should be done gradually over a 7–10 day time frame to avoid digestive problems. Table 19-1 outlines the recommended rations for horses.

Behavior

Horses are herd animals and prefer to live in groups. Within a group or herd, each horse has its place, known as a **hierarchy**. This is also called the "pecking order" in which animals have behaviors that direct others and allow for more control over the group. Horses are also comforted by being near or with other horses. This can determine their type of behavior depending on the situation. Horses are suspicious by nature and can quickly learn to detect a nervous person working or handling them. This can decrease the amount of control a person has over the horse. Due to being domesticated thousands of years ago, horses have learned to become trusting of people. The horse is an animal that is not always accustomed to loud noises and will react in various ways, such as **spooking** or startling. Horses are known to display many types of behaviors that are normal for the horse but can be dangerous for humans handling or working near them. Such normal behaviors for a startled horse may include kicking, **bucking** (lifting and kicking with the rear limbs), **rearing** (lifting the front limbs off the ground and raising the body into the air), striking with the front legs, or biting. A horse's body language will show if a horse is happy, angry, or scared. The happy horse will appear relaxed, ears forward and alert, with a relaxed tail and a relaxed eye expression. The angry horse will appear with its ears pinned back on its head, bared teeth, flicking tail, stomping feet, and kicking or bucking (see Figure 19-20). The scared horse will appear wide-eyed with the whites around the eyes visible, the head will be held high in the air, ears will flick back and forth, the tail may be held high in the air or clamped down, the horse may shake or move around nervously, and snorting sounds may occur. Some normal signs of horse behavior include the following:

- Behavior consistent with other herd animals
- Displaying of fight or flight instinct
- Prey animals—feel threatened by animals that look at them as a food source
- Vocal communication (neighing, snorting, squealing)
- Grooming on other horses
- Nuzzling
- Ears forward and alert
- Tail down and quiet

FIGURE 19-20 This horse is displaying signs of anger.

Some abnormal signs of horse behavior include the following:

- Biting
- Kicking
- Pawing
- Chewing
- Striking
- Head tossing
- Ears moving quickly
- Ears flattened on head
- Tail swishing

Basic Training

Horses are trainable in many disciplines. Horses may be ridden, driven, used as pack animals, and kept as companions. There are many areas of training involved with horses depending on their purpose and use. Professional trainers should work in training young or inexperienced horses as many problems, such as kicking, rearing, or bucking, may occur with someone who is not used to training a horse. All horses should be taught to **lead** on a rope, walking calmly and quietly with a person (see Figure 19-21). They should be taught to be tied on a post or chain and also on cross ties, which are two ropes placed on either side of the halter and then secured to a wall or post (see Figure 19-22). Horses should also be trained to work on a longe line, which is a long rope that allows a horse to work at various gaits while circling the handler. Riding and driving horses should be taught basic commands such as the walk, jog or trot, and lope. They should also be trained to stop and back up. A large amount of basic training in the horse is gaining the horse's respect and trust.

Equipment and Housing Needs

Horse facilities should consist of a barn or shelter area. Shelter may be a run-in shed with at least three sides that will protect the horse from the weather. Barns may consist of a large open area or single stalls for individual horses. Stall space should consist of at least 10 feet by 10 feet areas with a height of at least 8 feet. Cross tie areas or walkways should be around 5 feet wide. The lower areas of the barn should be reinforced to prevent injury and damage should a horse kick. Paddocks and pastures should be built of high-quality fencing safe for horses and at least one acre of space per horse. Barns should have the availability of water and electricity.

FIGURE 19-21 Horse can be trained to walk on a lead with a person.

FIGURE 19-22 Horse can be trained to be tied or secured to a wall or post by cross ties.

Restraint and Handling

Restraining horses requires experience, knowledge, and a good eye for reading a horse's body language. Horses can be calmed using a low, even tone of voice. It is best for the restrainer to allow a horse to evaluate the person by smelling the restrainer's hands and allowing the restrainer to scratch or pat the neck to show the horse it is meant no harm. Horses should have a **halter** applied to control the head; this will allow a handler to properly lead them (see Figure 19-23). Horses should be approached from the front and slightly near the left or **near side**. Horses are accustomed to being handled and trained from the left side. It is important to stay to the front of the horse and within its range of vision. A horse doesn't see well directly in front and directly behind it. Approaching the horse should be done quietly and calmly while watching its body language. If the horse tries to move away, stop and talk calmly to it and resume moving with slow movements toward its head. Once close enough, pat the horse on the neck before applying the halter. A lead rope can be slipped over the horse's neck with a knot tied to prevent the lead from slipping off. Stand on the left side to the side of the head and neck and gently place the halter over the muzzle and slip the strap behind the ears. Then properly secure the halter and apply a lead rope to the center ring on the bottom of the halter. When handling and restraining a horse, one should never wrap the lead rope around the hand or wrist. Hold the lead in neat loops with the left hand and hold the lead rope close to the halter with the right. This allows the restrainer or handler to move freely from one side to another to control the horse. **Leading** a horse is done by walking with the horse near its left shoulder about one foot from the horse. Caution must be taken to avoid being stepped on by the horse. Tying a horse should be done with a **slip knot** that is used in case of a quick release. Tie the horse to allow 2–3 feet of lead rope to allow some movement and allow it to hold its head even with its withers. Horses can also be placed on **cross ties**, which are short ropes placed on either side of the horse's halter to allow safe movement around all areas of the horse. The ties are placed on the lateral cheek ring of the halter. When handling a horse, some distractions can be made while the horse is examined, given an injection, or having blood collected. Distraction techniques include grasping the ear at the base and gently bending it back and forth, the eye cover by slowly sliding a hand up from the cheek to cover the eye, the skin roll by rubbing or grasping the skin at the shoulder (see Figure 19-24), or the leg lift to prevent a horse from moving. Another technique to control a nervous or difficult horse is the use of the **twitch**. A twitch is a chain or rope loop attached to a long handle that is placed over the lip of the horse (see Figure 19-25). This instrument is used to control a horse for certain procedures that may cause discomfort to the horse. It allows the horse to stand under better control. A **chain shank** can be used for better control as well. The chain is attached to a lead rope and placed on

FIGURE 19-23 Haltering a horse will allow the handler to properly lead the horse and control it's movement.

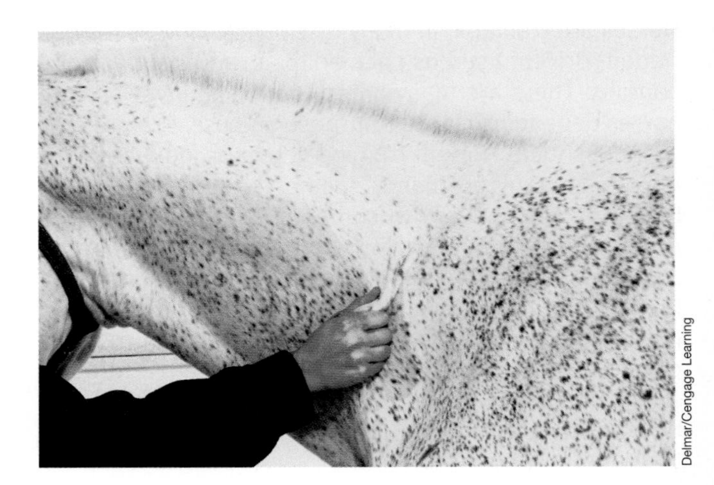

FIGURE 19-24 Pinching of the skin roll to provide distraction when the horse is having procedures done.

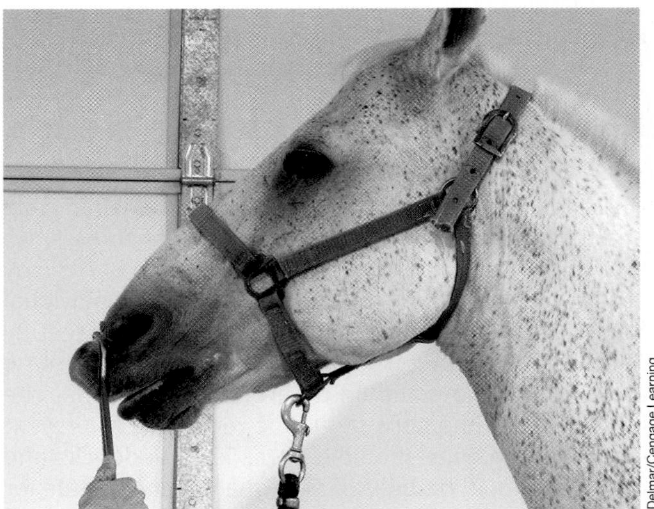

FIGURE 19-25 Application of a mechanical twitch.

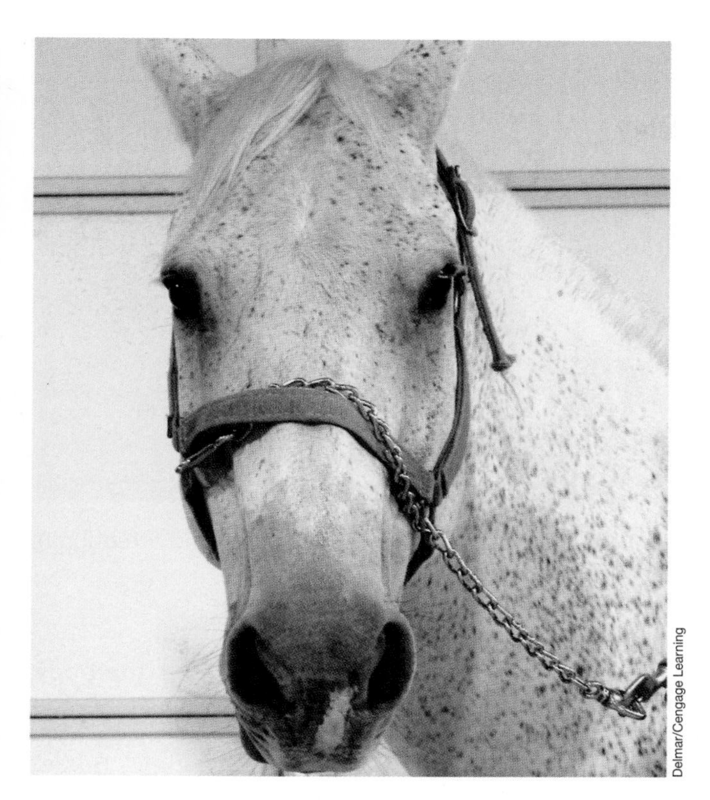

FIGURE 19-26 Restraint by using a chain over the nose.

the halter to apply pressure over the muzzle, under the chin, under the lip, or in the mouth (see Figure 19-26). This is used as a control method as well as an additional distraction when working with a horse. Lifting a horse's feet should be done with the restrainer standing with the body facing the rear of the horse. When working with the front feet and legs, one should stand at the shoulder of the horse and run a hand down the rear aspect of the limb and gently squeeze the fetlock joint, just above the hoof. When the horse lifts its foot, hold it by the hoof and balance it on the knee or firmly in the hand. When lifting the rear feet, stand facing the rear of the horse and fairly close to the side of the hindquarters. Run the hand down the leg in the same manner as the front limb. When the horse lifts the foot, grasp it at the hoof and move slightly forward to stretch the leg out and allow the hoof to rest on the handler's knee.

Some other methods of equine restraint include **tail tying** to adjust a horse's weight and movement or using **hobbles** to prevent a horse from kicking. Tail tying is generally used with a sedated horse to move it or to keep it out of the way for certain procedures. The tail should be tied to the horse's body by securing a rope below the base of the tail bone (see Figure 19-27). The rope can then be secured to the horse's front leg or neck. Hobbles are leg restraints placed around the limbs to prevent horses from kicking. The safest method of restraint on a scared, aggressive, or difficult

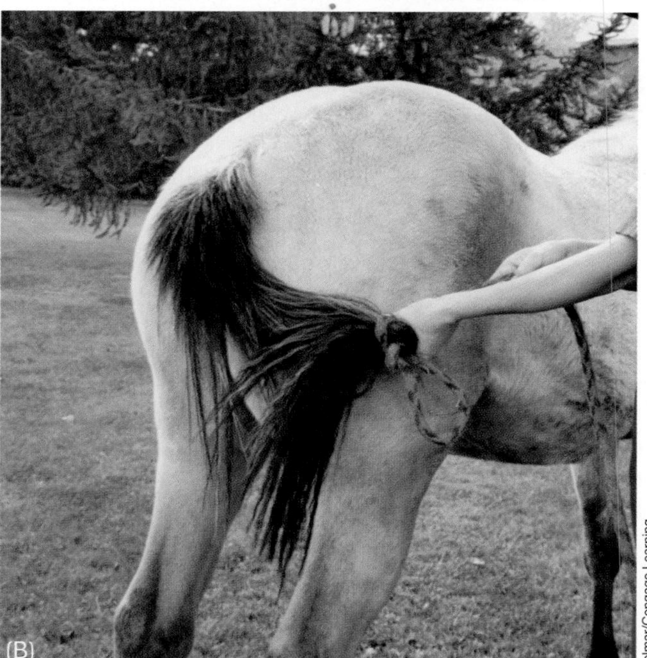

FIGURE 19-27 (A) Pass the bight through the tail loop (B) finished tail tie.

to handle horse is through sedation. Many sedatives can be given that allow the horse to remain in a standing position. Care must be taken to monitor the horse and body language as the anesthesia can easily wear off. Foals can be restrained by grasping one arm around the front of the chest and the other arm around the rump and keeping it against a stall wall

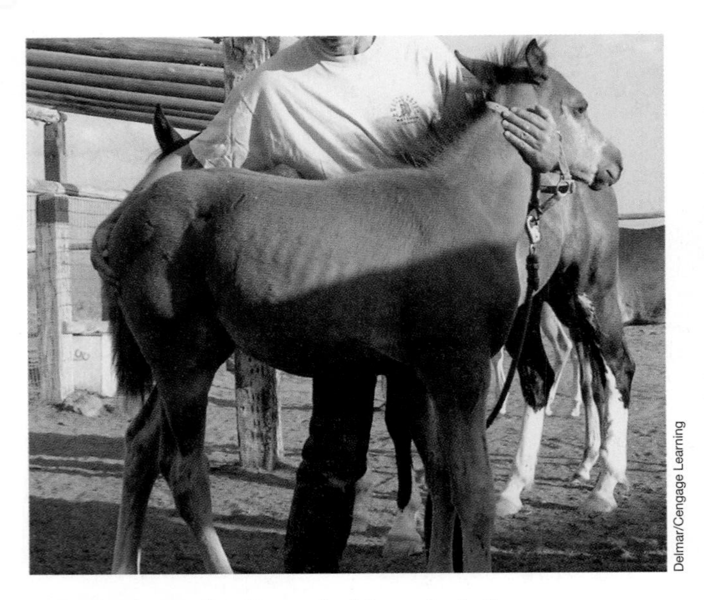

Delmar/Cengage Learning

FIGURE 19-28 Cradling a foal to restrain it.

(see Figure 19-28). The tail can be grasped and pulled upward and forward over its back to prevent movement. When working with any horse, remember to talk and comfort the horse throughout the restraint and handling. Use caution when walking or standing anywhere near the back end of the horse, as kicking could easily injure the restrainer. A handler should never turn his or her back on a restrained horse.

Grooming

Horses need regular and consistent grooming, such as brushing, bathing, and hoof maintenance. Grooming equipment tools and supplies are essential for all horse owners. Some of the basic tools and equipment needed are the following:

- Body brush
- Mane/tail brush
- Curry comb
- Hoof pick
- Body sponge
- Shedding blade
- Sweat scraper
- Conditioning spray
- Fly spray
- Hoof oil
- Clippers and blades
- Scissors

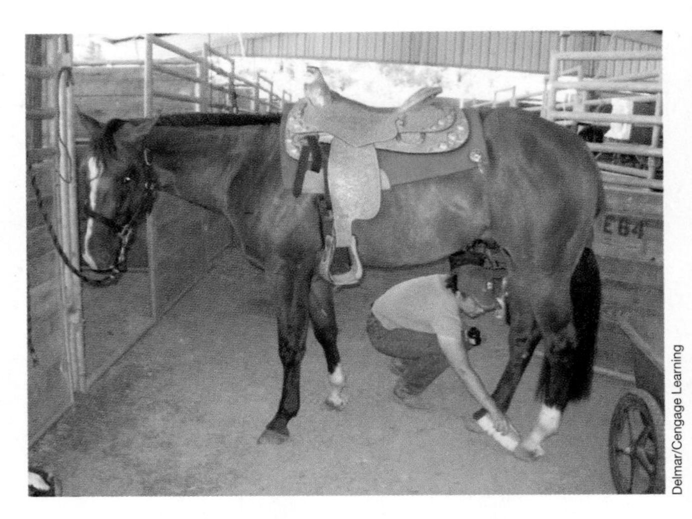

Delmar/Cengage Learning

FIGURE 19-29 Horses require daily care and attention to their hooves.

Horses require regular brushing over the body, mane, and tail. Each horse should have its own halter and lead rope. The halter should be well fitting over the face and throatlatch and not be too tight. Horses should have their hooves cleaned and picked out daily (see Figure 19-29).

Basic Health Care and Maintenance

The key to quality health and management of horses is prevention. Horses should be on a regular health program that includes a knowledgeable and experienced equine veterinarian. Horses should be monitored on a daily basis for any health problems or lameness. They should be on a quality exam, vaccine, and dewormer schedule. A horse's stall must be cleaned on a daily basis. Pastures should be rotated on a regular basis and have a regular sanitation and disinfectant program to control parasites. Every horse should be on a high-quality nutritional program that meets its energy, health, and age needs. A horse should have a complete physical exam at least once a year. New horses should have a prepurchase or postpurchase exam before entering the barn area. It is important to determine the soundness of a horse, as the legs and hooves are the most important part of the horse.

Health Testing

Horses are highly susceptible to contagious diseases and organisms, so a health exam should be performed before allowing new horses on a farm. Exams may contain a complete physical, blood work, and radiographs. A **Coggins test** should be completed every six months to a year depending on state requirements. The Coggins is a blood test to determine the horse's status for **equine infectious anemia** or EIA. This deadly virus has no vaccine and no cure and spreads rapidly in horses by transmission of biting insects. All horses that are transported to shows, events, and travel throughout the country are required to have a negative test.

Dental Care

Horses also require regular dental care. Horse teeth continue to grow as the horse ages. The **points** of the teeth or the corners and edges become sharp and can make a horse's mouth sore and prevent it from eating properly. Horses can live well into their 30s with proper health care. A horse is in its prime years from 3–12 years of age. The teeth can help determine the age of the horse through their appearance and wear (see Figure 19-30). Young horses will have temporary teeth that are smaller in size and length and whiter in appearance than the adult teeth. Young horses begin to lose their temporary teeth between 2 and 3 years of age and should have all of their adult teeth in place by 5 years of age. The shape and wear of the adult teeth will change as the horse ages. A **bit** is a piece of equipment placed in the horse's mouth for riding purposes. It is a metal bar that lies over the tongue and provides pressure and control while riding. The **cusp** or the center of the tooth will wear. Until the age of 12, horses have oval-shaped teeth. After 12, the teeth become triangular in shape and begin to slant forward. Adult horses have between 36 and 40 teeth. Females have 36 and do not usually have canine teeth. Males usually have 40 teeth with canines being present. Some horses will develop **wolf teeth**. These are small teeth about the size of an end of a pencil and located in front of the first molar. They most commonly occur on the upper jaw and occur in about 15% of horses. They can cause many problems and pain in the horse's mouth and are easy to extract, only having one root. Horse teeth should have dental care and may require routine floating procedures. **Floating** is the filing of the teeth as they tend to wear down unevenly due to the upper jaw being wider than the lower jaw. A **float** or **rasp** is used to file the teeth and sharp edges (see Figure 19-31). Sharp edges will cause discomfort in eating and having a bit placed in the mouth. The teeth should be checked each year with the physical exam.

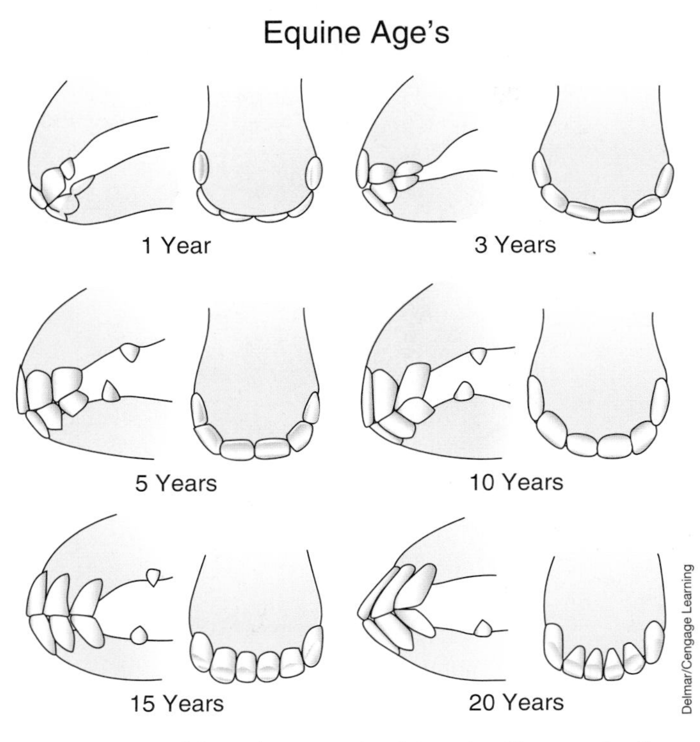

FIGURE 19-30 A horse's age can be determined by examination of its teeth.

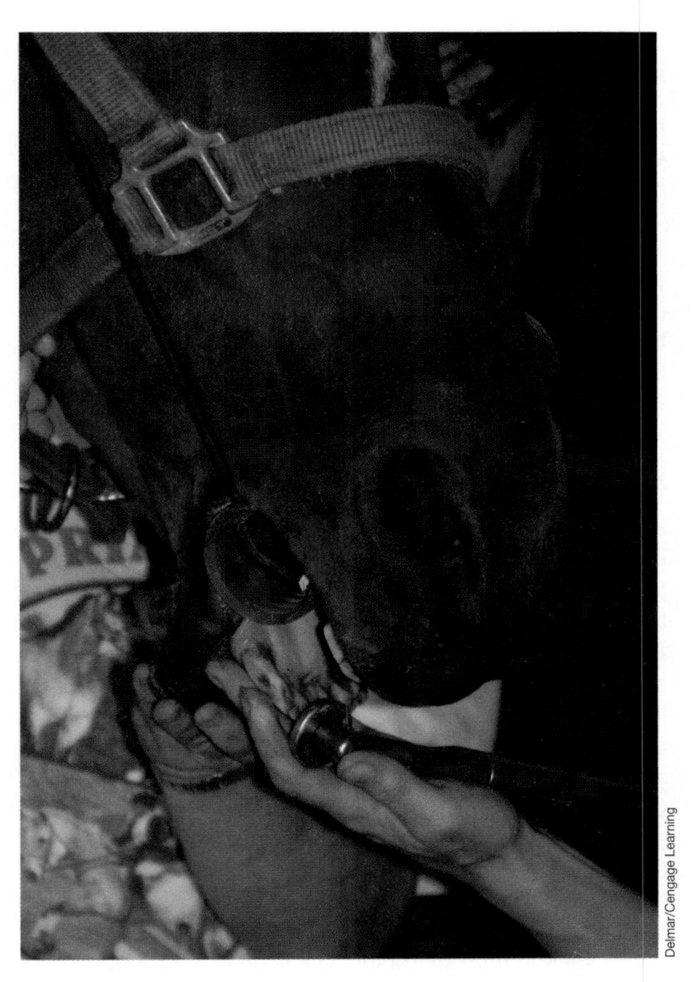

FIGURE 19-31 A float or rasp is used to file a horse's teeth.

Some horses will require more dental care than others. The following diagram details dental characteristics for horses of different ages.

Hoof Care

A horse will also require routine hoof care. This is typically provided by a **farrier**. A farrier is a professional horseshoer, sometimes called a blacksmith. Hooves need to be trimmed every 4–6 weeks, and some horses will require shoes depending on their purpose and use. Hooves grow continuously similar to fingernails. They need to be a proper length and shape for the balance and comfort of the horse. The hooves should also be cleaned out or picked on a daily basis. A **hoof pick** is used to remove dirt and debris from the hoof. Rocks and other objects can become wedged into the hoof and sole and can cause lameness and infection. The **frog** is the soft padded V-shaped area at the causal aspect of the hoof (see Figure 19-32). It should not get too wet or too dry; otherwise, it will lead to possible soreness. The **hoof wall** or outer covering of the hoof may become dry and begin to crack. These injuries will cause a horse to become unsound.

Castration

Male horses commonly are castrated to prevent reproduction and allow for ease of handling. Castration in horses is usually done around 6 months of age. Most castration procedures are done with the horse standing. The incision usually remains open due to the location and size of the testicles. This will decrease swelling and allow the surgical site to drain. However, care must be taken to prevent bacteria from entering the incision. Castrations done in late fall or early spring will help prevent flies from irritating the incision.

Assessment of Lameness

Horses should be evaluated for lameness by a veterinarian. The veterinary assistant may be asked to walk or jog the horse in hand with a lead rope to allow the vet to evaluate the gaits. It is important to know the horse's gaits and natural movements. Many vets will also ask to see the horse on a **longe line** and under saddle. A longe line is a long lead rope that allows the horse to exercise in circles while under control. The movements of the horse include the walk, the jog, the trot, the lope, and the canter (see Figure 19-33). The **walk** is a relaxed slow four-beat gait. A Western horse is ridden in ranch or rodeo events using a Western saddle or heavy saddle. The **jog** is a slow two-beat gait of Western horses. The **trot** is also a two-beat gait with a longer stride or distance between legs and is seen in English horses. The **lope** is a three-beat gait of Western horses and the **canter** is a three-beat gait in

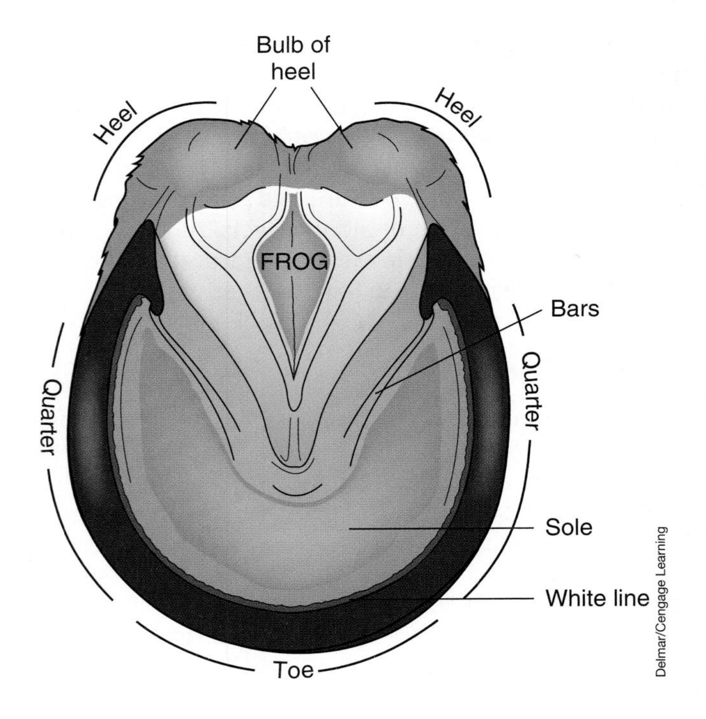

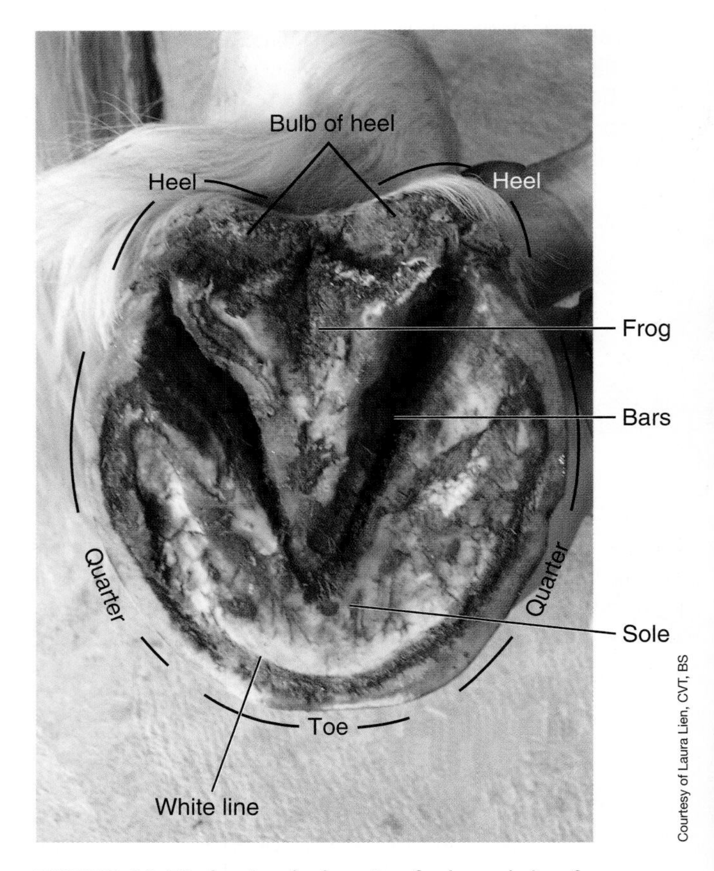

FIGURE 19-32 Anatomical parts of a horse's hoof.

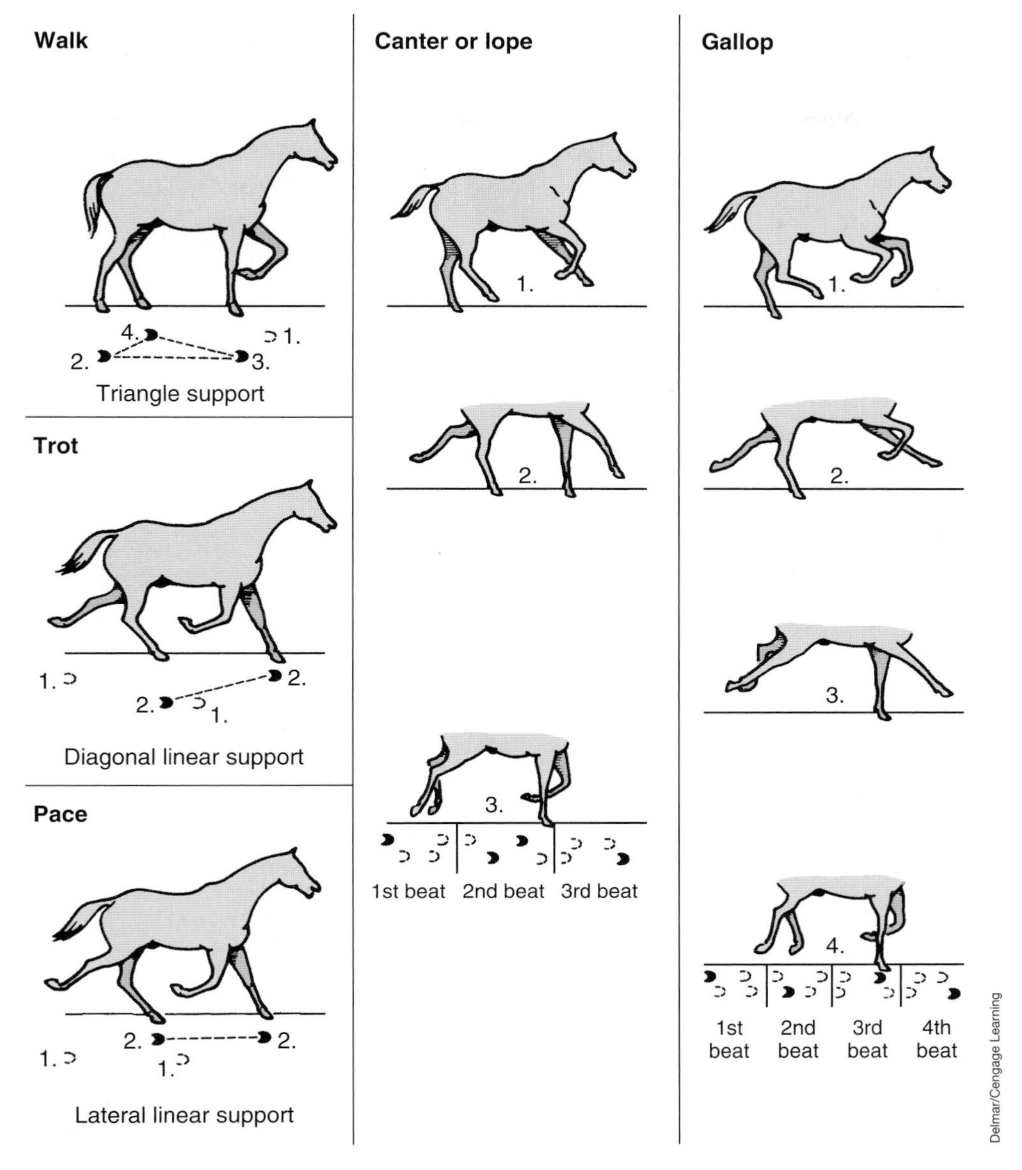

FIGURE 19-33 Basic gaits of horses: walk, trot, pace, cantor, and gallop.

English horses. An English horse is ridden in light horse events using an English or light saddle. The **gallop** is a four-beat fast gait or run where each foot hits the ground at a different time. There is a brief time between each stride where all four feet are off the ground.

Vaccinations

The equine vaccine program should begin in foals around 4–5 months of age and boosters given once a month for 2–3 series and then yearly as recommended by the veterinarian. The location and weather of the environment may dictate the type of vaccines and schedule a horse should be on. A regular dewormer program should begin in foals shortly after birth and continue monthly until 6 months of age. Adult horses should be on a regular dewormer every 6–8 weeks. It is important to rotate wormers so that a horse's immune system does not become immune to the product. This is done by changing the class of wormer with each worming. Table 19-2 and Table 19-3 highlight the necessary vaccines and dewormers for horses.

Reproduction and Breeding

Reproduction in horses can be difficult if a person is not experienced and knowledgeable in the horse's reproductive system. Horses have the lowest conception rate of

TABLE 19-2

Common Equine Vaccines

VACCINE NAME	FREQUENCY	DISEASE/CONDITION
EEE/WEE/VEE	yearly	Encephalomyelitis
Tetanus	Yearly	Tetanus toxoid
Flu	Every six months	Influenza
Rhino or EHV	Every six months	Rhinopneumonitis Herpesvirus
Rabies	Yearly	Rabies virus
Strangles	Every six months	Strangles intra nasal virus
West Nile	Every six months	West Nile virus
Potomac	Yearly to every six months	Potomac horse fever

TABLE 19-3

Common Equine Dewormers

CLASS	TRADE NAME	PARASITE CONTROL
Ivermectin	Zimecterin, Eqvalan, Rotation	Stongyles, stomach worms, pinworms, ascarids, bots, threadworms
Moxidectin	Quest	Stongyles, hair worms, pinworms, ascarids, bots
Fenbendazole	Safe-Guard	Strongyles, pinworms, ascarids
Pyrantel Pamoate	Strongid	Strongyles, pinworms, ascarids
Praziquantel	EquiMax, Zimectrin Gold, Quest Plus	Stongyles, hair worms, pinworms, ascarids, bots, stomach worms, threadworms, tapeworms

all livestock species. Typically, 50% of all horses bred will become pregnant during a breeding season. The breeding season in horses is usually mid-January through early summer. A mare's estrus cycle begins on average between 12 and 15 months of age but can begin as early as 6 months old. Breeding should not begin before 3–4 years of age. Horses in quality health can be bred into their late teens or early 20s. The estrus cycle of a mare is every 21 days and occurs from 4–10 days in length. Signs of estrus include the following:

- Relaxed vulva
- Increased urination
- "Winking" of the vulva
- Increased vocalization
- Slight clear mucous vaginal discharge
- Tail held high and to the side

A mare that is used for breeding purposes is called a **broodmare**. Broodmares should be kept in shape with adequate amounts of pasture and exercise (see Figure 19-34). Many broodmares are housed together. Broodmares should be examined prior to each breeding for health issues and cultured for possible bacterial infections or diseases that could be transmitted to the stallion. Exams include a **palpation** or feeling of the ovaries for **follicle** development, which determines when eggs will be released for fertilization to occur. This timing is critical for the mare to be bred. Some veterinarians will use **ultrasound** or the use of radio and sounds waves to view the ovaries for follicle

FIGURE 19-34 Broodmare.

development. Stallions should be evaluated for breeding purposes based on breed quality, ease of handling, and semen quality. Semen should be evaluated for **fertility** to ensure breeding will be successful. The semen should have a high sperm count, good **motility** (movement), and proper **morphology** (correct appearance). Stallions should be kept from other horses due to aggressive tendencies and the potential for accidental breeding.

Breeding can be achieved by several methods. **Live cover** breeding is when the stallion actually breeds with the mare. This may be done through **in-hand breeding** where the stallion and mare are both restrained to allow natural breeding. This is typically done when the mare is showing signs of estrus and occurs every other day until the mare ovulates and signs of estrus are no longer present. **Pasture breeding** is when the mare and stallion are turned out together to allow natural breeding to occur. This is typically over a specific period of time to ensure breeding occurs. **Ovulation** is the release of the egg to unite with the sperm. Breeding can also occur by using **artificial insemination**. This is the process of collecting semen from the stallion and inseminating or placing the semen into the mare at the appropriate time of estrus during the ovulation phase. Artificial insemination must be timed correctly through an experienced reproductive equine veterinarian.

Semen can be shipped through several methods, usually cooled or frozen. **Cooled semen** is shipped in special containers within a 24-hour period and kept cool with ice packs. **Frozen semen** is shipped in special tanks with the use of liquid nitrogen to keep the sperm frozen. The semen is evaluated and prepared for shipping with the use of specific medications that help preserve the sperm during transport. Typically, mares are inseminated one to two times prior to ovulation, so timing must be accurate. The gestation length of mares is on average 340 days or 11 months. The process of **parturition** or labor should occur with a 15–30 minute time frame. Broodmares may be kept in a large stall or in a large pasture to foal. Signs of a mare going into labor will begin 12–24 hours prior to the parturition process. It is important that mares be kept in a dry and clean area during labor. Some signs will be noted 7–10 days prior to the labor process, especially in experienced broodmares. Signs of labor include the following:

- Large distended mammary glands
- Milk present at udder (lactation)
- Restless and anxious
- Stop eating 12 hours prior to labor
- Laying down
- Sweating
- Vulva distention and loosening
- Muscles over back and hind end drop

The labor process begins when the mare's water breaks, releasing the fetal membranes, and a large amount of fluid will be expelled. **Expelled** means to release or push out, usually within the body. Foaling should occur shortly after the water breaks. The foal is born with normal presentation being front legs with the feet pointed down followed by the head and rest of the body (see Figure 19-35). The foal should be monitored to see that breathing begins shortly after birth. The mare will begin to lick and clean the foal by nudging and working at the sac or membrane that encases the foal during development. The membranes are removed and this should stimulate the foal to begin breathing (see Figure 19-36). The mare will continue to lick and clean the foal to dry it (see Figure 19-37). The **placenta** may be attached to the foal or be expelled shortly after birth. The placenta is the waste materials of the foal during the gestation time (see Figure 19-38). Some mares will attempt to eat the placenta and afterbirth membranes that contain the foal's waste products. This is a natural instinct to remove the membranes, which may attract predators to the foal. The placenta and membranes should be removed and examined to make sure all tissues have been expelled from the

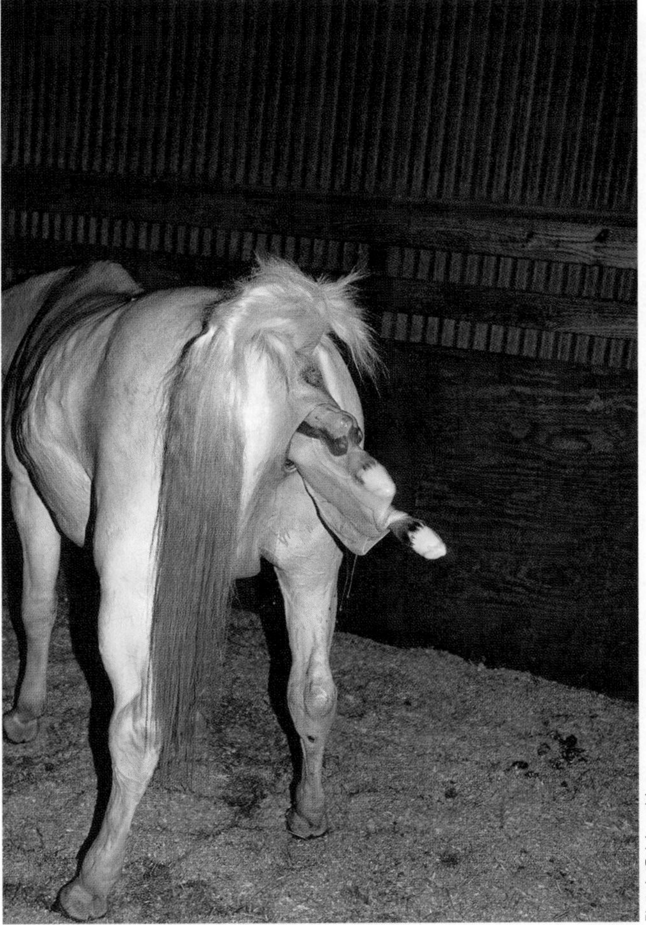

Photo by Bob Langrish

FIGURE 19-35 Normal presentation of the foal at birth.

FIGURE 19-36 The amniotic membranes may be pulled away from the foal. The mare will sometimes do this on her own.

FIGURE 19-37 The mare will lick the foal after birth to clean it.

mare. A retained placenta will cause severe infection and should be expelled within hours of the labor process. The foal should rise within 30 minutes to two hours after birth. Their movements are wobbly and take several days to become coordinated. Foals should begin nursing shortly after they stand (see Figure 19-39). The foal's umbilical cord should be treated with **iodine** to prevent bacteria from entering the body and causing an infection; this should be done shortly after birth. When the foal begins to stand, an **enema** consisting of warm soapy water should be given rectally to begin the foal's bowel movements. The mare can be given small amounts of hay and water.

FIGURE 19-38 The placenta should be examined to be sure it has been completely expelled.

FIGURE 19-39 The foal should begin nursing shortly after birth.

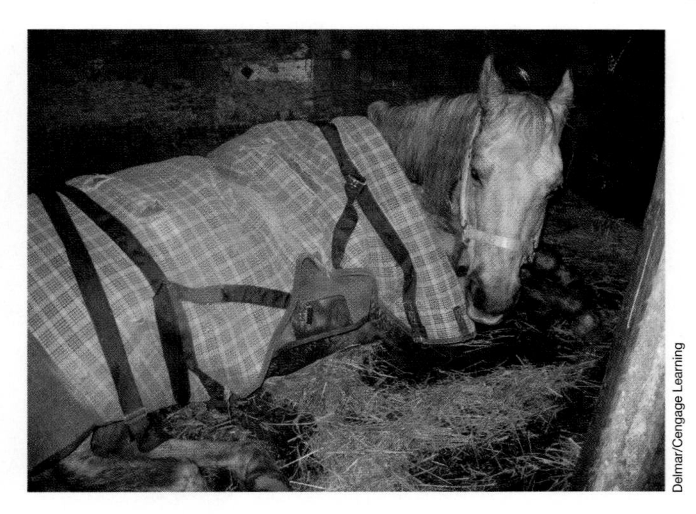

FIGURE 19-40 Colic.

Common Diseases

Horses are known to be fragile animals when it comes to diseases and injuries. They have delicate digestive systems and fragile bones that can be easily damaged. It is important for anyone working with horses to know the signs of pain, lameness, and illness. Horses that become sick can become gravely ill very quickly.

Colic

A common equine medical condition is known as **colic**. Colic is a severe stomach pain that can lead to the death of the horse if not cared for properly. A horse that colics tends to lay down and roll due to the abdominal pain and discomfort (see Figure 19-40). This can cause the stomach and intestines to twist and cut off circulation, leading to the death of the horse. Colic can also cause the horse to become **constipated** or not able to have a bowel movement. This will slow down or stop digestion. A horse may colic due to many reasons, such as internal parasites, overfeeding or overeating, mold or poor food sources, ingesting too much water, eating or drinking shortly after exercise or being overheated, high fever, too rich hay or protein sources, and from a sudden change in food. Horses that show signs of colic should be walked to prevent them from lying down and rolling. This is an immediate emergency and a veterinarian should be contacted. Signs of colic include the following:

- Flanking or looking at the abdomen
- Bloated or distended abdomen
- Frequent lying down and standing
- Rolling
- Kicking at abdomen
- Biting at abdomen
- Sweating
- Restless
- Constipation or no bowel movements
- Anorexia (food and water)

Laminitis

Another common condition in horses is **laminitis,** commonly called "**founder.**" Laminitis is an inflammation of the lamina within the hoof bone and tissue connecting to the hoof wall. Laminitis can be caused by many factors, including overfeeding, rich grain or hay, a rapid change in diet, excessive amounts of cold water, fever or high body temperatures, inflammation of the uterus following labor, and overworking or stress. Laminitis can cause severe lameness and a rotation in the **coffin bone** located within the hoof joint. This can cripple a horse and causes severe pain. A horse that is foundering will show signs of not wanting to move, standing with the front and rear feet close together, pain noted in the hoof area, lameness, fever, and a parked out appearance (see Figure 19-41). A **parked out** horse stands with the front legs and back legs spread wide apart and causes the back to be lowered or sunken in.

Equine Infectious Anemia

Another condition in horses that will lead to death is equine infectious anemia (EIA) or **swamp fever**. This is a virus caused by biting insects, such as flies or mosquitoes, and can also be spread by contaminated needles. It is highly contagious and there is no vaccine and no cure. All horses should be tested yearly by a licensed veterinarian with a blood sample being submitted to the state veterinary lab and the Department of Agriculture. This test

FIGURE 19-41 Laminitis.

is called a Coggins test. Many states have legal requirements of a negative Coggins test for show and travel purposes. Some horses show no clinical signs and are only carriers of the disease. Horses that test positive for EIA are usually euthanized and the property placed under quarantine. Signs of EIA include the following:

- High fever
- Stiffness
- Weakness
- Weight loss
- Icterus (yellow color of tissues)
- Anemia
- Swelling of limbs

Equine Encephalomyelitis

Equine encephalomyelitis, also called equine sleeping sickness, is a virus caused and transmitted by mosquitoes. It can be zoonotic. There are several strains known as Eastern (EEE), Western (WEE), and Venezuelan (VEE). There is a vaccine available but no treatment. Insecticides and pesticides are recommended for insect control. Signs of EEE/WEE/VEE include the following:

- Lethargy
- Sleepy appearance
- Anorexia
- Fever
- Depression
- Local paralysis of lips and bladder
- Unable to swallow

Equine Influenza

Equine **influenza** or flu is another highly contagious virus. Outbreaks can cause statewide emergencies and a vaccine is available as part of the yearly horse health program. Signs include typical flu symptoms such as fever, anorexia, depression, rapid breathing, coughing, weakness, and nasal and ocular discharge. Horses tend to injure themselves easily through wounds, punctures, and cuts. The possibility of infection from bacteria is likely. Bacteria, known as ***Clostridium***, are naturally found in horse manure and can survive long periods in the ground. Bacterial infection causes the disease known as **tetanus** or lockjaw. Prevention and sanitation are key and a vaccination is recommended as part of the yearly horse health program. Booster vaccines are usually given after a severe wound or injury to a horse, especially from wounds involving metal or nails. A horse that develops tetanus will eventually die. Signs of tetanus include stiffness, difficulty chewing, inability to swallow, and reluctance to eat. These signs worsen over time.

Common Parasites

Common internal parasites include roundworms, bots, strongyles, and pinworms. Many of these parasites feed off of the blood in the intestinal tract. The best control method of internal parasites is a regular dewormer, daily stall sanitation and cleaning, rotating pastures, cleaning and disinfecting buckets, and routine yearly fecal exams. External parasites that should be controlled include flies, mosquitoes, mites, ticks, lice, and ringworm fungus. External parasites carry many diseases, some of which are zoonotic. The best control methods for external parasites include insecticide sprays to control biting insects, regular stall cleaning, and regular manure removal.

SUMMARY

Horses are mainly used today as recreational and income sources. Horses have become a popular livestock to own, show, and breed. There are many types of horses that have been developed. They require exceptional health care, an experienced handler, and a knowledgeable veterinary staff. Routine maintenance and preventative care is the key to overall good equine health. Veterinary assistants can play a vital role in the care and handling of horses.

Key Terms

artificial insemination process of collecting semen from a stallion to place in a mare for breeding

bit equipment placed in a horse's mouth when riding a horse

broodmare female horse used for reproduction and breeding purposes

bucking lifting and kicking action of the hind limbs

canter three-beat running gait of English horses

cecum area within the intestinal tract located between the small and large intestine that helps break down roughage and protein

chain shank chain placed around the head and halter of a horse for added control

Clostridium bacteria that causes tetanus and is found naturally in manure and soil

coffin bone bone located within the hoof joint that rotates when laminitis occurs

Coggins test blood test for equine infectious anemia

colic severe stomach pain in the horse

colt young male horse

constipation inability to have a bowel movement

cooled semen semen shipped in a special contained kept in ice

cross ties short ropes placed on either side of the horse to allow safe movement while working around the animal

cusp center of the tooth that wears with age

dam female parent of the horse

donkey member of the equine family with large ears; also known as a burro

draft horse large and muscular horse standing over 17 hands in height

electrolytes powdered supplements placed in food or water to encourage horses to drink more water

enema warm soapy water used to produce a bowel movement

equine veterinary term for the horse

equine encephalomyelitis viral disease of horses found in different strains transmitted by mosquitoes; also called sleeping sickness

equine infectious anemia (EIA) deadly virus in horses caused and transmitted by mosquitoes

Equus caballus genus and species of the horse

expel to remove

farrier professional who trims and shoes horses hooves; also known as a blacksmith

fertility ability to reproduce

filly young female horse

float instrument used to file down teeth; also called a rasp

floating filing process of horse teeth

foal young newborn horse of either gender

foaling the labor process of horses

follicle development in the ovary that releases the egg

forage hay or pasture source

founder common term for laminitis

frog V-shaped soft pad in the bottom of the hoof

frozen semen semen shipped in a special container and kept frozen with liquid nitrogen

gallop fast four-beat running gait where each foot hits the ground at a different time

gait movement or way a horse moves

gaited horse horse that moves with a specific rhythm and motion

gelding castrated male horse

halter equipment applied to the head of the horse for control and handling

hand the measurement method of determining the height of a horse; a hand equals 4 inches

hierarchy the place a horse has in the herd

hinny cross of a male horse and female donkey

hobbles leg restraints to prevent a horse from kicking

hoof pick equipment used to remove rocks and debris from horse hooves

hoof wall outer covering of the hoof

hunter horse ridden over small fences

influenza equine flu virus

in-hand breeding method of holding the mare and stallion for breeding purposes

iodine chemical disinfectant used to destroy bacteria

(*Continues*)

jack male donkey

jenny female donkey

jog slow two-beat gait of Western horses

jumper horse ridden over large fences on a set course

laminitis inflammation of the lamina within the hoof bone and surrounding tissue

lead rope applied to a halter to handle, control, and move a horse

leading the act or motion of moving or walking a horse

light horse riding horse between 14.3 and 17 hands in height

live cover natural breeding method of the mare and stallion

longe line long lead line used to exercise a horse in circles

lope three-beat running gait of Western horses

maiden female horse that has not been bred

mare the adult intact female horse capable of reproduction

mineral block minerals supplied in a block for horses to lick

morphology correct appearance

motility movement

mule cross with a female horse and male donkey

near side left side of the horse

nonruminant specialized digestive system of the horse that requires small amounts of food to be digested throughout the day

ovary female reproductive sex organ

ovulation the time the egg is released from the mare's ovary

palpation to feel

parked out standing with the front and rear legs wide apart

parturition veterinary term for the labor process

pasture breeding method of breeding when the mare and stallion are turned out in pasture

placenta membranes that contain waste materials of a foal

points sharp edges or corners of teeth

pony small equine under 14.2 hands in height, commonly used by children

racehorse horse bred for speed and racing purposes

rasp instrument used to file down horse teeth; also called a float

rearing lifting action of the front limbs off the ground and raising the body into the air

riding horse horse ridden for show or recreational purposes

salt block salt provided in a block for horses to lick

sire male parent of a horse

slip knot type of tie used as a quick release

spooking startling action

stallion adult intact male horse capable of reproduction

stock horse horse built to work on a ranch that has athletic ability

stud adult male intact horse of breeding age

swamp fever common name for equine infectious anemia

tail tying method of tying the tail to keep it out of the way or to move a sedated horse

trot slow two-beat gait of English horses

tetanus bacterial condition known as lockjaw

twitch chain or rope loop placed over the lip of a horse as a restraint

ultrasound use of sound and radio waves as a diagnostic procedure

walk relaxed four-beat slow gait

warm-blooded ability for controlling the body temperature internally

wolf teeth small teeth that may develop in front of the first molars

REVIEW QUESTIONS

1. What is a hand in relationship to a horse?
2. What are the classes of horses? Explain each class type.
3. What are some normal behaviors of horses?
4. Explain the horse digestive system.
5. What are pieces of equipment that can be used to restrain a horse?
6. What methods can be used to breed a horse?
7. How can a horse's age be determined by its teeth?
8. What are the signs of colic in a horse?
9. What are the movements of a horse? Explain each type of movement.
10. What is EIA? What is the testing procedure for EIA?

Clinical Situation

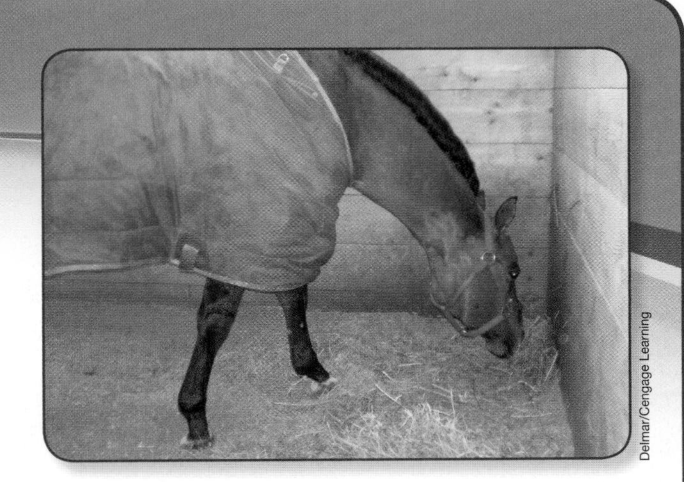
Delmar/Cengage Learning

Dr. Reese and Emily, a veterinary assistant at Lineville Equine Center, are on a farm call visit for a horse that is not eating, has been showing signs of not having bowel movements, and is in notable pain by lying down and rolling in its stall. These signs have been occurring several hours today and the owner was resistant to having the horse examined. Finally, the owner agreed to an exam but was not sure if he would be able to pay for any treatment.

"Sir, the horse is in a lot of pain and really should be seen at the clinic for possible surgery. At least, allow me to give the horse some pain medication to make him more comfortable." Dr. Reese is trying to explain the importance of care needed by the horse.

"Well, I don't really have any money to spend on the horse. If the horse is really that bad, then I guess it should be put down."

Dr. Reese and Emily are discussing the situation and what may be wrong with the horse. The owner allowed only a minimal exam of the horse and was not willing to give much background history on the animal.

- How would you handle the situation as a veterinary assistant?
- What are some likely issues with the horse's health?
- What are some options to give the horse owner?

20

Courtesy of Photo Disc

Draft Animal Breed Identification and Production Management

Objectives

Upon completion of this chapter, the reader should be able to:

- Understand and discuss common veterinary terms relating to draft animals
- Identify common breeds of draft animals
- Perform draft animal selection
- Discuss the purpose and uses of draft animals in society
- Demonstrate draft animal production methods
- Discuss draft animal health care and management practices

Introduction

Draft animals are types of livestock that are muscular, strong, and powerful. They are animals that were originally developed to help people with their work. People still use some draft animals to help them pull wagons, plow fields, and for farming activities. They are also kept as pets and recreational entertainment. Animals that were originally used in farming, travel, and the military still have a major role in today's society.

Veterinary Terminology

- Equine—veterinary term for horses, donkeys, and mules
- **Donkey—**equine species with large head, large ears, and a tail with a switch on the end; may be miniature to mammoth in size
- Mammoth—large-sized draft donkey
- **Mule—**result of crossing a male donkey and female horse; equine species that has long ears, narrow body, horse-like tail and smooth hair coat and may be miniature or draft in size

Biology

Draft animals include several species of large animals that were developed to work and pull heavy loads. Many of these animals were developed for specific purposes and were domesticated to help serve humans in today's society. Draft animals include mules, oxen, and draft horses.

Breeds

Several species of animals are used for draft purposes. Horses, oxen, mules, donkeys, camels, elephants, and bison are the most common draft animals in use today. Draft animals have been bred and trained for pulling heavy loads and providing strong power for farming purposes. People in the United States still use horses, mules, donkeys, and oxen as draft animals. Camels, buffalo, and elephants are used in some countries in similar ways.

Draft Horse

Draft horses are the most common draft animal in the United States. Draft horses are larger and heavier than other equine breeds. They are powerful animals built to pull heavy loads. They were developed to weigh over 1,500 pounds and several breeds are well over 2,500 pounds. Their body structure is wide, deep, and large-boned with a low center of gravity, which holds a heavier muscle build. There are five recognized draft horse breeds still used as work animals in the United States. These breeds include the Belgian, the Clydesdale, the Percheron, the Shire, and the Suffolk.

The **Belgian** is a Flemish breed that developed in the Middle Ages as a farm horse. They are light brown to sorrel in color with a lighter **flaxen** (blonde) mane and tail. They are popular draft animals due to their large size, tremendous power, and excellent muscle build. They are the most popular and numerous draft horse breed today.

The **Percheron** is the second-most-numerous draft horse. They originated in France and have a more refined build than the Belgian. They were used in the Middle Ages as battle mounts and then became popular carriage horses. Percheron's colors range from black to dapple grey to white (see Figure 20-1).

The **Clydesdale** is the third-most-numerous draft horse breed. They have become notable for pulling heavy wagons and their fame as mascots for a national beverage corporation. The Clydesdale is a Scottish breed that is usually brown or black in color with white faces and legs. They have long amounts of hair on their legs known as **feathers** (see Figure 20-2).

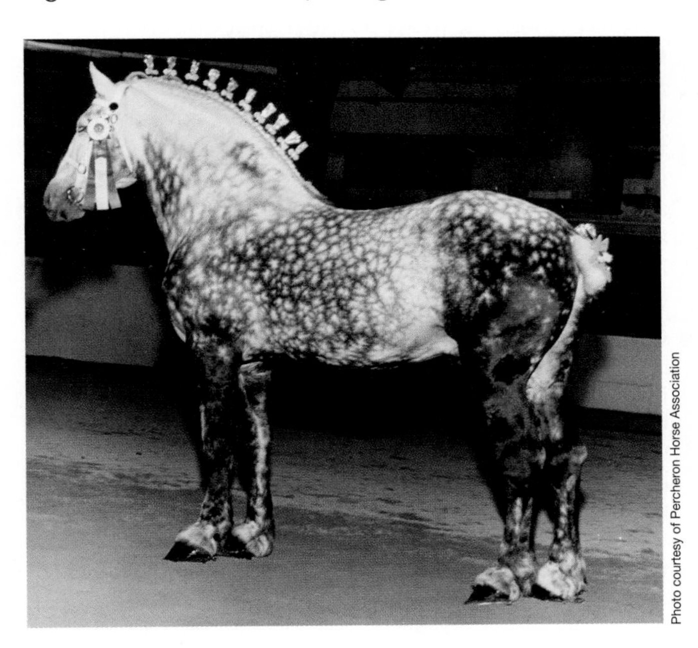

Photo courtesy of Percheron Horse Association

FIGURE 20-1 Percheron.

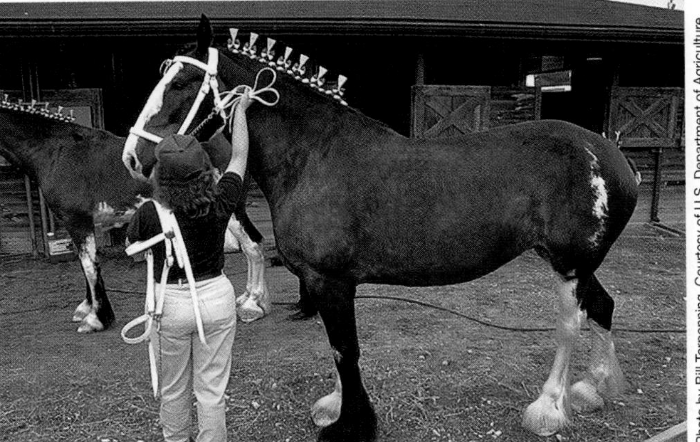

Photo by Bill Tarpenning. Courtesy of U.S. Department of Agriculture

FIGURE 20-2 Clydesdale.

The **Shire** is a draft horse breed that developed in England and is the tallest of the draft horse breeds. They are typically black or bay in color and also have feathers on their legs.

The **Suffolk** was also developed in England and was developed solely for agricultural purposes. They are chestnut in color and tend to have very little white. They are easier keepers than other draft breeds and require less feed than other heavy breeds. They are smaller is size than the other draft breeds.

Donkeys and Mules

Donkeys and mules are also used as draft animals and are members of the equine family. A donkey is also called a burro or ass and was originally domesticated from a wild version of the African wild ass. They are not hybrids of horses. Most donkeys that are bred to each other tend to be sterile, some are capable of being bred. A donkey is pony-sized, with long ears, a shade of gray, with a bristle-like coarse coat. Donkeys have a vocalization sound known as a "**bray**" that can be heard from a long distance. Donkeys tend to be tough and versatile animals, which makes them hardy and handy as draft animals. They do also tend to have a stubborn attitude and move at a slower pace than a horse. The female donkey is called a jenny and the male is called a jack (see Figure 20-3).

The mule is a hybrid cross of a female horse and a male donkey. Mules are almost always sterile. A mule is larger in size than a donkey and more the size of a horse with the appearance of a donkey. They have long ears and move at a moderate pace between that of a donkey and horse. Mules come in a variety of colors, most ranging in bay, brown, or black. They are powerful and tend to be docile to work around. They can be trained to ride like a horse and are tough draft animals. A female donkey crossed with a male horse is called a hinny. A hinny is a rare occurrence compared to that of a mule. Mules and donkeys both endure heat well and are much less sensitive to digestive problems and leg injuries.

Photo courtesy of American Donkey and Mule Society, Lewisville, Texas

FIGURE 20-3 Two wild jacks registered as standard donkeys/wild burros, owned by Elmer Zeiss of Valley, Nebraska.

© iStock/Andy Gehrig

FIGURE 20-4 A team of oxen.

Oxen

Oxen are members of the bovine family and largely resemble large-sized cattle. *Oxen* refers to two or more animals and represents the plural form of the word. The singular form is *ox* and refers to one animal. They have long horns that sit vertically on their heads. Males are used more often than females as draft animals. Oxen are ruminants and use their feed better than horses and other equines. However, they are much slower paced animals. Oxen are multipurpose animals as they can be used for work, beef, and milk (see Figure 20-4).

Breed Selection

Draft animals are mostly used for work purposes but also are bred as show animals. Horses, donkeys, mules, and oxen can all be shown as their respectable breeds or species. They may be judged based on their conformation or body structure, power, or athletic ability. Some events they may compete in are horse pulls, draft hitch classes, farming demonstrations, or log pulling. **Horse pulls** are contests of power to see which animal can pull the heaviest load with the least damage to the land. **Draft hitch classes** are based on the number of animals used to pull a wagon and how the team works together to complete the job. **Farming demonstrations** include draft animals pulling plows to work the land in farming

FIGURE 20-5 Horse pulling load of hay on a sled.

activities (see Figure 20-5). **Log pulling** is a contest to see which animal or team can pull the largest amount of logs the greatest distance. The term ***horsepower*** was developed in England based on the work done by large horses and other draft animals. It now refers to the unit or power of an engine or motor. It is defined as the amount of work done by lifting or moving 550 pounds at a distance of one mile per second. One horsepower is also equal to 746 watts of electricity. Many of the draft animal contests measure this horsepower.

The term **draft capacity** is the force an animal must exert to move or pull at a constant rate of 1/10 of its body weight. Power is the combination of the draft capacity and speed. Draft horses are the most popular animal for use in these contests of power as they have better skill, behavior, and coordination for team working events. Draft animals are selected based on the type of job or work they are intended to complete. Other factors that affect selection are the needs of the animal, gender, age, cost and upkeep, and health care. Several draft animals kept together will establish a social rank. This is their place or order within the group. This depends highly on the animal's size, strength, and age. Some factors to consider when selecting a draft animal are outlined in Table 20-1.

TABLE 20-1

Selection Factors of Draft Horses and Oxen

DRAFT HORSES	OXEN
–High purchase price	–Low purchase cost
–Require large amounts of food	–Less food intake
–Work fast	–Work slow
–Require large amounts of space	–Require moderate amounts of space

Nutrition

Draft animal nutrition is also similar to the animal species discussed in their respective chapters. The nutritional needs are based on the individual animal according to species, size, purpose or use, and age. Increased amounts of food are necessary when lactation or increased work is occurring. Increased amounts of protein and fiber are needed for muscling and strength. Draft animals require at least 8–10 gallons of water on a daily basis. Draft horses have a nonruminant digestive system as other equines. They require increased amounts of protein, salt, calcium, and phosphorus in their diets. They may also need additional vitamins and minerals based on the quality of food and their overall health. The feeding schedule should be based on allowing one to two hours of rest before food or water after periods of work. They should not be worked for at least 45 minutes to an hour after being fed.

Oxen are ruminants as are all bovine species. They typically graze for 4–9 hours a day. They require increased amounts of protein and fat when working. All draft animals should be placed on a routine feeding schedule consisting of either hand feeding, self-feeding, or automated feeding. Hand feeding requires a person to feed the animals every day at specific times, usually at least twice daily. The needs will depend on the type of animal and the owner's schedule and type of work. Self-feeding is done by allowing food to be available at all times. These animals are usually kept in a pasture and have a grass or hay source available, as well as a water source. Automated feeding is done mechanically at a set time by the use of an auger or silo that releases a set amount of food per animal at specific times.

Behavior

Draft animals are typically **gregarious** in behavior, which means they have a natural instinct to join with the herd or group. Some draft animals that work alone may stress while away from the herd, which makes the animal difficult to handle and decreases work and production. Draft animals that work in teams must be able to get along to be safe to handle and prevent injuring each other. It is best to make sure that when working with a team or group of draft animals, they have established a pecking order in which their ranking in the group is how they are placed in a team. Some animals will kick or bite at each other, possibly causing injury to the team or the handler.

Basic Training

Draft animals' health status is based on **conditioning**. This is the practice of working an animal to build up its body and strength for its work needs. Conditioning begins slowly over time and the amount of work increases at regular intervals. This is similar to training an athlete for a race or a competition. If an animal is not conditioned properly, it may pull a muscle or otherwise physically injure itself.

Equipment and Housing Needs

Draft animals require specialized equipment while working. Common types of equipment include a harness for equines, which are leather straps that fit the horse and attach to a wagon or plow used for pulling. The harness attaches to a collar that lies around the neck and shoulders and holds the harness in place. **Hame straps** attach to the harness and collar through pieces of leather. A bridle is placed on the head and attaches to reins that are used to guide and control the animal (see Figure 20-6). Oxen require a whipple tree and a yoke to guide and control them. The **yoke** is a wooden bar that fits over the neck and shoulders and connects a team together. The **whipple tree** is a harness-like attachment that hooks onto the yoke and connects to the wagon or plow. Draft animals will need to be kept in stalls and fenced areas that are strong and will contain large-sized animals that may be powerful enough to easily break out of areas.

FIGURE 20-6 A Clydesdale in full tack and harness.

Restraint and Handling

Restraint and handling of these large animal draft species is similar to that of horses and cows. Many of these animals have been trained for a specific purpose and have a respect for the work they do and the humans they work alongside. As with any large animal, safety should be practiced at all times.

Grooming

Draft animals require similar grooming techniques as horses and cattle. Some animals have hooves and others are cloven-hoofed and need trimming and cleaning of the hooves on a regular basis.

Basic Health Care and Maintenance

Regular physical exams should be completed at least yearly and all vaccine programs should be followed according to the species recommendations. Regular deworming schedules should also be in place. Equines and oxen will require regular hoof trimming and farrier services to maintain their hooves and feet.

Vaccinations

Vaccinations will be dependent on the animal species. Equine and bovine species should follow their regular health care vaccine programs as deemed necessary by the veterinarian.

Reproduction and Breeding

Draft animal reproduction is similar to other livestock species. Draft horses have the same estrus cycle and breeding methods as all other horse breeds as discussed in Chapter 19. The only difference is that draft horses tend to reach puberty around 12–15 months of age. Oxen have similar estrus cycles and breeding methods as cattle as discussed in Chapter 18. Oxen reach puberty between 8 and 15 months of age. Estrus cycles occur for 12–20 hours every 21 days. The gestation period is around 270–295 days.

Common Diseases

Common diseases will be related to the species of animals, such as equine- and bovine-related diseases. These animals are as easily susceptible to these diseases as their domesticated counterparts.

Common Parasites

Common parasites will be related to the species of animals, such as equine- and bovine-related internal and external parasites. These animals are as easily susceptible to these parasites as their domesticated counterparts.

Common Surgical Procedures

The most common surgical procedure with draft animals is the castration procedure in male animals. This surgery would be completed as described with horses or cattle.

SUMMARY

Draft animals have been developed to pull heavy loads and work with people for such tasks as farming, pulling wagons, and competitive competitions. Draft horses, donkeys, mules, and oxen make up some of the more common draft animals used today. Many considerations and management practices are needed for draft animals. It is important to have experience and knowledge in the needs of these animals.

Key Terms

Belgian Flemish draft horse breed that is light brown in color with a lighter mane and tail

bray loud vocal sound of a mule or donkey

Clydesdale Scottish draft horse breed that is dark in color with white face and legs and is famous as the Budweiser mascot

conditioning process of working an animal to build up its body and strengthen the animal for work

donkey member of the equine family known as a burro or ass

draft animals type of animal that is muscular, powerful, and strong and used for pulling heavy loads

draft capacity force an animal must exert to move or pull at a constant rate of 1/10 of its body weight

draft hitch class contest based on the number of animals used to pull a wagon and how the team works together

farming demonstration draft animals used in plowing and farming activities

feathers long hair over the lower legs of horses located just above the hoof

flaxen color similar to blonde

gregarious instinct or behavior to join a herd or group of like animals

hame strap leather parts that attach the harness and collar

horse pulls contests of power to see which horse can pull the heaviest load

horsepower unit of power on an engine or motor that equals 746 watts of electricity or the amount of work done by lifting or moving 550 pounds at a distance of one mile per second

log pulling draft animal contest where logs are pulled to see which team can pull the heaviest amount

mule hybrid cross of a female horse and a male donkey

oxen members of the bovine family that resemble cattle with long vertical horns on the head

Percheron French horse breed that has a refined build and is a popular carriage horse and is black to grey in color

Shire English horse breed that is black or bay in color with feathers on the legs

Suffolk English horse breed that is chestnut, smaller in size, and an easy keeper

whipple tree harness-like attachment that connects the yoke to the wagon by hooks

yoke wooden bar that fits over the neck and shoulders of team of oxen

REVIEW QUESTIONS

1. What is a draft animal? List some examples of draft animal species.
2. What are the major differences between draft horses and oxen?
3. What equipment is needed for draft animals?
4. What feeding methods may be used with draft animals? Explain each method.
5. How do mules and donkeys differ? How are they similar?
6. What is "horsepower"?

Clinical Situation

Courtesy of Photo Disc

"Hello, Edgemont Large Animal Services, this is Betsy, how may I help you?"

"This is Mr. Greene, I have two mules that Dr. Simms sees each year. They do not seem right. They have been lying around all day and are not eating well. I use them to plow my fields, so I really need them seen soon."

"Okay, Mr. Greene, I will have Dr. Simms come out to see them this afternoon. I will also have him call you when he is on the way. Is there anything else I can do for you?"

"Nope, just get the doc out here soon. Thanks a lot."

- Was this situation handled properly?
- How would you have handled this situation?
- What other questions could Betsy have asked for a history on the mules?

21

Courtesy of Image Source

Swine Breed Identification and Production Management

Objectives

Upon completion of this chapter, the reader should be able to:

- ☐ Discuss and understand common veterinary terms pertaining to swine
- ☐ Understand swine biology
- ☐ Identify common breeds of swine
- ☐ Discuss the nutritional needs of swine
- ☐ Understand normal swine behavior
- ☐ Practice proper swine restraint methods and handling safety
- ☐ Discuss the vaccine program of swine
- ☐ Discuss swine reproduction factors and methods
- ☐ Describe common health problems and diseases of swine
- ☐ Discuss the swine production industry
- ☐ Discuss swine selection factors
- ☐ Discuss swine body types
- ☐ Describe swine production methods
- ☐ Discuss common health practices used in the swine industry

Introduction

Pigs, *hogs*, and *swine* are all terms for the popular animal that provides humans with the meat known as pork. Swine have become a popular production animal due to their ability to reproduce at a high rate and increase their body weight rapidly. Swine production requires low amounts of labor and has a high investment and profit rate. The swine industry is governed with strict regulations in how pigs are housed and cared for due to the use of the meat for human consumption. This means care of swine plays an important role in the veterinary industry.

Veterinary Terminology

Swine or **porcine** is the veterinary term for pigs and hogs. The female pig is called the **sow** and the male is called the **boar**. The labor process in swine is known as **farrowing**. Young pigs are called **piglets**. A **barrow** is a young castrated male pig. A male pig castrated after maturity is called a **stag**. A female pig that has not yet been bred is called a **gilt**. The amount of meat produced by one pig is known as the **dressing**.

Biology

Swine have organ systems similar to other animals and a digestive system that is monogastric, similar to dogs, cats, and humans. Their food must be more concentrated. They have **cloven shaped** hooves, which means they have a split in the toe of the hoof that separates the hoof into two parts. Most swine breeds weigh between 220 and 240 pounds, but some select bacon breeds can top 400 pounds. Swine used for food products are usually young pigs as the older hogs, especially the boars, will develop a strong flavor in the meat. Most hogs in meat production systems will yield a dressing of 65–80% of their body weight. Pigs should farrow around 7 to 12 piglets a year and have a gestation length of 114 days. The anatomical features of the swine are illustrated in Figure 21-1.

Breeds

Several swine breeds are found throughout the United States. Swine provide pork and other meat products for human consumption. There are many different types of

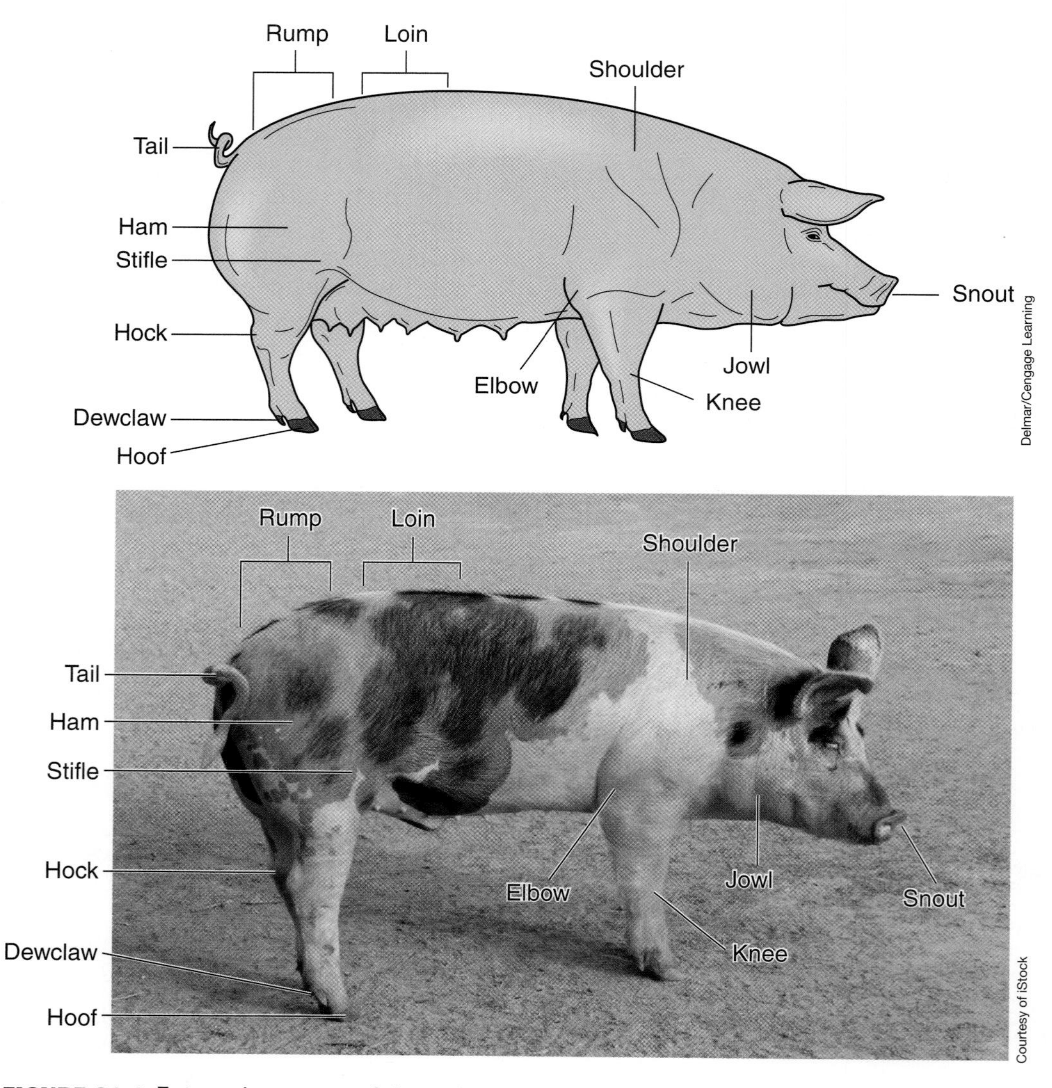

FIGURE 21-1 External anatomy of the swine.

swine that are classified by meat type or bacon type producers. Many breeds are now being developed through hybrid breeding or cross-breeding two or more purebreds based on specific traits. The conformation of the breed is the most critical factor. Swine breeds vary in size from 50 pounds, such as the Vietnamese Potbellied Pig, to several hundred pounds, such as the Yorkshire or Landrace.

Breed Selection

Swine have several requirements for ideal production systems that are based on the overall appearance and profitability. There are two types of swine—the **meat type hog** and the **bacon type hog**. Meat type hogs have a large amount of meat, known as ham, on the body. Bacon type hogs have a large amount of fat, known as bacon, on the body. Both types grow fast. The meat type hog is muscular and the producers use the cuts of meat as ham, loins, and roasts that are high in cost. These meat cuts have low amounts of fat. Bacon type hogs are fatter than muscle-based meat, such as bacon and sausage. Producers want to select a hog that meets the ideal type of swine for the product produced. Most producers will select a hog that has a verified pedigree and will measure the amount of meatiness on the pig's body. There are several measurement methods for measuring the body fat versus the body muscle. The probe method is a tool used to measure the thickness of backfat, which is found over the mid-back and loin area. A lean meter is used to measure the difference in fat versus muscle thickness over areas of the body. The ultrasound method is also used to measure both fat and muscle through the use of sound waves that reflect the pulses of the hog. Registered breeds must meet meat certification program guidelines for a production program. Sow selection factors include selecting gilts from large litters that have no genetic flaws and have backfat measurements of less than 1.2 inches of fat. Figure 21-2 illustrates the grading of hogs.

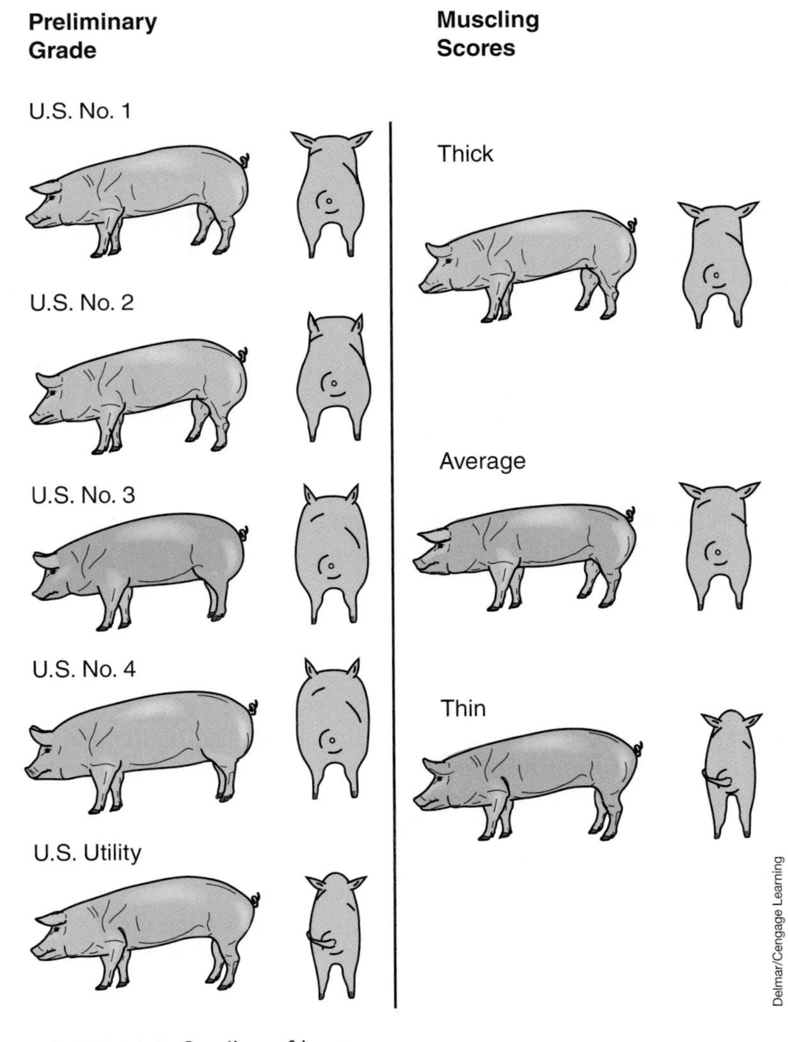

FIGURE 21-2 Grading of hogs.

Sows should be reevaluated between 150 and 200 pounds of growth. This is to ensure that sows are growing adequately and will be excellent producers for breeding or meat production programs. How well the sow gains weight will determine if the sow may be replaced or put into a reproduction program based on the data.

Boar selection is based on the temperament and body structure. Boars should be selected from large litters, have a medium to large body structure, should reach the adult size by 155 days of age or less, and have less than one inch of backfat. Ideal weight gain for breeding boars is 2 pounds of body weight gain per day. It is necessary for reproduction programs to have one boar for every 15–20 sows. The boar should enter the breeding program between 6 and 7 months of age and begin breeding sows around 8 months of age.

Nutrition

Swine nutrition is based on the life stage and production type system. Mature boars are usually fed similar to pregnant sows. That means they should be fed around 4–5 pounds of feed per day. Young boars under a year of age may need more nutrients than adults. Piglets that have been weaned can be placed on a starter food. Growing pigs will generally need more nutrients based on the type of production they will be used for. Breeding sows should be fed increased amounts in the last two weeks of the gestation phase, which is typically 4–5 pounds of feed per day. Many producers feed ad lib as pigs will not overeat. The term *ad lib* means as much as the animal will eat or having food available at all times, or free choice. The increased nutrients will help stimulate milk production. Food should slowly be decreased for the sow around three days after farrowing for a short period to prevent digestive problems. Sows can be managed on a lactation diet that is started two weeks after farrowing and consists of 10–12 pounds of feed per day to maximize milk production. Many commercial feeds are balanced for the life stage of the pig.

General maintenance swine nutrition is classified by the production method as well as the size and life stage of the pig. Pigs that have been weaned and are between 40 and 125 pounds require a 16–18% protein diet. Pigs over 125 pounds require a 14–16% protein food source. Most producers feed a manufactured food source that meets the needs of their production system. Many are corn and soybean meal based. Supplements may be added as necessary. Swine that are used as meat sources for human consumption as well as breeding systems are typically fed a high-energy diet of fats and carbohydrates. The biggest source is cereal grains, such as corn, barley, and wheat (see Figure 21-3). Pigs do not eat roughage; it doesn't provide adequate sources of energy. Protein is an important part of a swine's diet and is necessary for muscle development, fat development, and tissue quality, and it also builds and repairs the body's cells. Protein is broken down in the stomach into amino acids that further help in muscle development. The most common protein is soybean meal. A corn-based diet will require protein supplements of 35–40%, as corn is primarily fat (see Table 21-1).

(A)

(B)

(C)

Courtesy of U.S. Department of Agriculture

FIGURE 21-3 Hogs are fed grains such as (A) corn, (B) wheat, (C) oats.

Swine are the most common livestock species to suffer a mineral deficiency. Salt, calcium, and phosphorous minerals are necessary in large amounts. Copper, iodine, iron, zinc, and selenium are needed in small amounts.

TABLE 21-1

Protein and Diets

PROTEIN DIET SOURCES	PERCENTAGE OF PROTEIN
Growth diets	12–16% protein
Finishing diet	12–16% protein
Boar breeding diet	13–16% protein
Lactation diet	13–16% protein

TABLE 21-2

Water Requirements

CLASS OF PIG	WATER INTAKE (GALLONS/PIG/DAY)
Sow and litter	8
Nursery pig	1
Growing pig	3
Finishing pig	4
Gestating sow	5
Lactating sow	7
Boar	5

Vitamins A, D, E, and B12 are usually supplemented in the food source as well. Water is essential for the health of the pig, but the requirements are less than most other types of livestock. The amount of water is 1/4 to 1/3 gallon per pound of dry feed (see Table 21-2). A 60-pound pig needs 3/4 gallon of water daily. A 60–100 pound pig should have around 2.5 gallons a day. A 100–250 pound pig needs 4 gallons a day. A mature boar and a pregnant sow will typically drink around 5 gallons a day. A lactating sow will require 7 gallons a day as milk is predominantly formed by water.

Behavior

Pigs are generally stubborn, contrary, and vocal but are very intelligent. They are also known for being aggressive, especially if a sow has piglets or a boar is present. These various characteristics make pigs one of the most difficult animals to handle and restrain. Pigs have poor eyesight and are easily scared, and their legs are susceptible to injury if handled improperly. It is important to move slowly and calmly when working around swine.

A pig's defense is its teeth. Swine have fine, needle-sharp teeth, even in piglets (see Figure 21-4). The neck and shoulder muscles are very strong and pigs will use them to push and pin people against objects. Larger pigs can be dangerous due to their body weight and size. Pigs will chase people if they feel threatened or cornered.

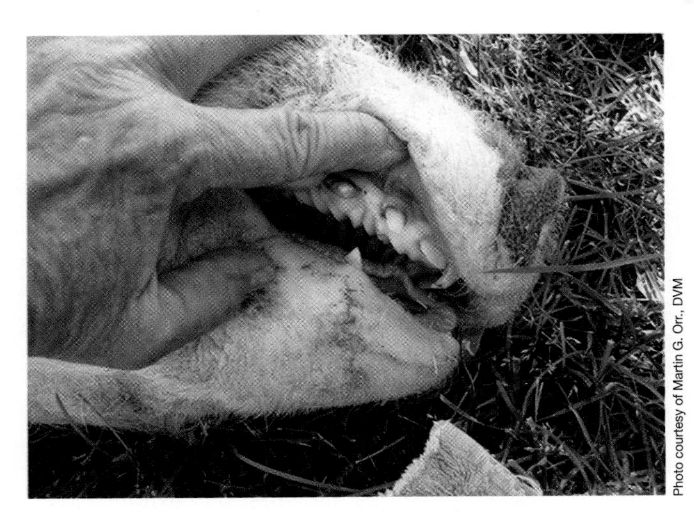
Photo courtesy of Martin G. Orr., DVM

FIGURE 21-4 Swine have very sharp teeth.

Basic Training

Pigs can be trained to move from location to location by using specialized equipment designed to protect the handler's lower body. Pigs are then herded or pushed to the areas the handlers want them to move to. The more swine are handled by humans, especially at a young age, the more acceptable pigs will be to handling.

Equipment and Housing Needs

Swine are usually kept in large groups. Housing needs of swine are constructed based on the production system. Gestation facilities are used to house sows for breeding. Many producers are using free-access gestation housing, which allows the sows to move about and socialize and access an individual stall when they want to eat or rest (see Figure 21-5). Sows are moved to farrowing facilities shortly before the birth of the piglets. Farrowing is usually done in pens or crates (see Figure 21-6). During the pigs' growing period (weaning to 100 pounds), they may be housed on pastures or in permanent facilities. Pigs over 100 pounds will be housed in finishing facilities. This is often done in confinement housing or can be done on pasture. Confinement housing is temperature and moisture controlled (see Figure 21-7).

Photo courtesy of Chore-Time Hog Production Systems, Milford, Indiana

FIGURE 21-5 Free-access gestation stalls allow sows to gather in a socialization area and enter and exit individual stalls as desired.

Photo courtesy of Chore-Time Hog Production Systems, Milford, Indiana

FIGURE 21-6 Farrowing crates are used when sows give birth.

Restraint and Handling

The principles of swine restraint are not different from those of other large animal species. The method used to handle pigs is determined by the size and age of the animal and the nature of the procedure being performed. Swine are intelligent animals that may be stubborn to handle. They do not have a strong herd instinct but prefer to be with other pigs to that of being alone. They become very vocal when angry, stressed, or scared. Some pigs also have large amounts of fat that when stressed or handled in high temperatures can cause overheating. Small boned legs can easily be injured if grasped or tied in the wrong manner. People must learn to move slowly and deliberately when handling pigs; gentle handling and talking in a soft voice is necessary. Pigs will bite and cause serious injury with their teeth, even in young piglets. They are also very strong, with muscular shoulders and necks, and can squeeze easily through openings and escape. Moving pigs can best be done with the use of a

Photo courtesy of Chore-Time Hog Production Systems, Milford, Indiana

FIGURE 21-7 A modern swine finishing facility.

hurdle, or a panel board that is flat and kept between the handler and the pig to act as a wall (see Figure 21-8). **Hurdles** are commonly solid pieces of wood, plastic, or metal. They are large enough to cover the legs of the handler and used to isolate a pig against a wall or to move it from one area to another. Paddles can also be used in a similar manner. **Paddles** are similar to hurdles with a long handle that acts as an extension of the arm.

Pigs less than 50 pounds of weight can be restrained for exams, vaccines, or medication administration by lifting them by the rear feet and placing the head between the legs for control. The front legs can touch the ground for support of the body weight (see Figure 21-9). To catch a pig, it is best to grasp one rear leg. Piglets less than 5 pounds can be handled similar to carrying a dog. Quickly place one hand and arm under the piglet's chest and the other over its shoulders. Lifting may be done by supporting and controlling the head and back end. Holding pigs firmly but gently is necessary. Pigs will likely squeal and become vocal. This can be dangerous if working near a sow with piglets, as they may become aggressive. In these cases, piglets and sows should be separated when handling. Restraint equipment includes a **hog snare**, **snubbing rope**, or **trough**. The hog snare is the restraint tool of choice and is a long pipe with a cable loop attached to the end. The loop is used to rope the upper jaw of the pig and then tightened around the snout (see Figure 21-10). The snare is kept elevated above the pig to control the animal for 20–30 minute periods. This can be used to restrain a pig for blood collection, exams, and injections. A snubbing rope can be used to tie a pig for restraint. The rope is used in the same manner as the hog snare over the snout. The rope can be tied to a post for similar restraint. Smaller pigs can be restrained using a trough. This is a V-shaped table that is used to hold a pig on its back for surgery and other procedures. A rope may be used to keep the pig still.

Delmar/Cengage Learning

FIGURE 21-9 Manual pig restraint technique.

Photo courtesy of Ron Fabrizius, DVM, Diplomate ACT

FIGURE 21-8 Use of a hog hurdle.

Photo courtesy of Ron Fabrizius, DVM, Diplomate ACT

FIGURE 21-10 Use of a hog snare.

Grooming

Pigs do not routinely need grooming. They tend to bathe themselves in mud and water sources to keep their skin moist. Pigs that are show animals may require bathing. The hooves of swine may need routine trimming and care. Teeth may need occasional trimming if sharp edges appear.

Basic Health Care and Maintenance

Swine producers typically raise pigs in **aseptic** and controlled environments to be disease free. Herds that are disease free and in excellent health are known as **Specific Pathogen Free** or **SPF**. A *pathogen* is the veterinary term for a disease. Each pig should be disease free at birth, raised in an aseptic (sterile) environment, and have no physical exam findings or internal or external parasites, swine diseases, or snout distortion (poorly shaped).

Tail Docking

Another common practice done shortly after piglets are born is **tail docking**. This is done to control feces from collecting on the tail and helps to keep proper sanitation. This practice is also done to prevent the piglets from biting each other's tails and causing serious injury. Tail biting commonly occurs when pigs are kept in confinement. The tail is cut about one inch or slightly shorter from the tail bone (see Figure 21-11).

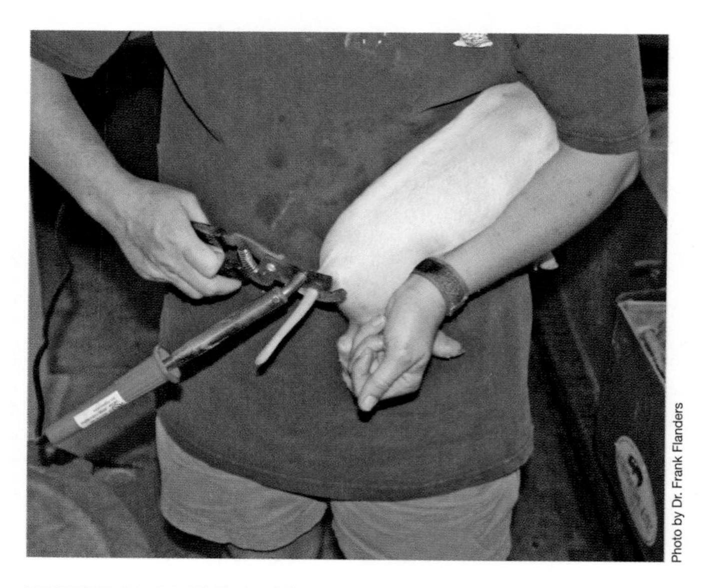

FIGURE 21-11 Tail docking.

Tail docking is done shortly after birth at the same time as the teeth are trimmed and the umbilical cord is treated.

Ear Notching

A common method used for swine identification is **ear notching**, which is the practice of making a small cut into the ear in different locations to identify the pig and litter number (see Figure 21-12). A **V-notcher** is the tool used to make cuts in the ears. The practice of identification is especially necessary in large herds. Breeding stock and replacement pigs are easily identified through this method. The right ear identifies the litter number and the left ear identifies the pig number. Each ear notch position has a specific value. Other methods of identification that may be used include ear tags, tattoos, and branding. Ear tags tear easily and tattoos and brands are difficult to read on many breeds of pigs.

Castration

Male swine that are not used for breeding purposes are typically castrated. Those raised for meat are usually castrated before weaning. When the procedure is done the pigs should be kept in a clean and sanitary environment. This practice is commonly done in the late fall or winter to prevent fly contamination.

Additives

Swine management includes feed and medication additives to meet the needs of a pig's health status. This is a standard practice for ideal health that is not related to nutrition. This may include antibiotics, anthelmintics, or hormones. Antibiotics may be used to increase weight by 10% and help to decrease the feed consumption by 15%. A common growth hormone used in swine production is called **porcine somatotropin (pST)**. This hormone is used to increase protein synthesis that helps the body increase food use by 20–30%. The average daily weight gain is 15–20% of the body weight. This increases the muscle mass by 10–15%. No additives are fed for a specific time prior to slaughter. The term *slaughter* means to kill humanely for the meat source. The removal of additives is called *withdrawal time*.

Each medication or hormone has a specific time they are not to be used in the body that assures that the meat will be drug free. Each additive's label will state the necessary **withdrawal time**, which is regulated by law. The withdrawal times of products range from 2–70 days depending on the product. Injections are commonly given by intramuscular (IM) and subcutaneous (SQ) routes. The IM injection should be given into the front shoulder

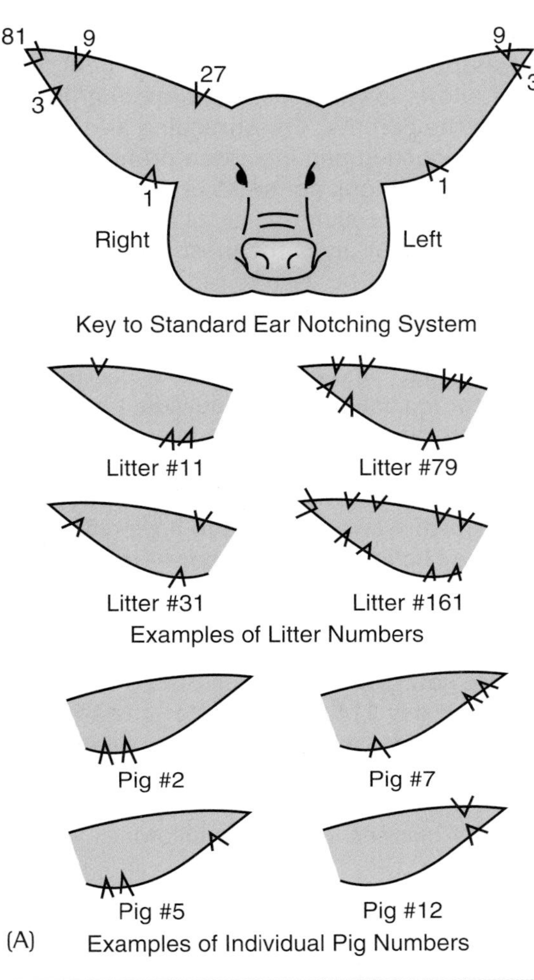

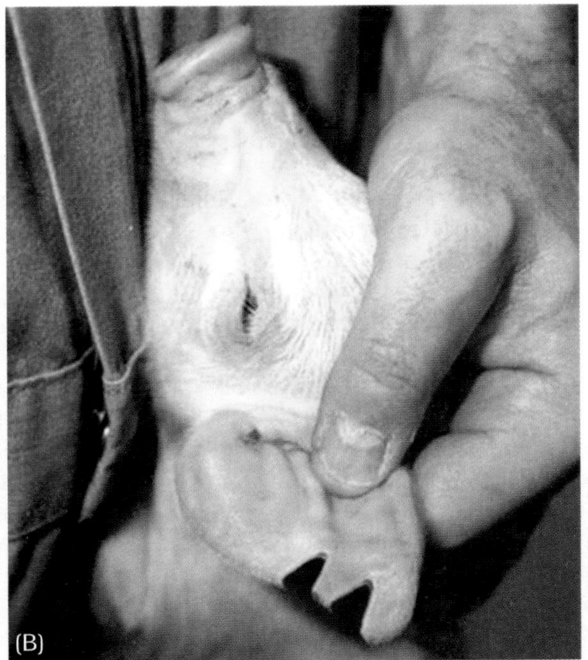

FIGURE 21-12 (A) Ear notching identification in swine; (B) pig with notched ears.

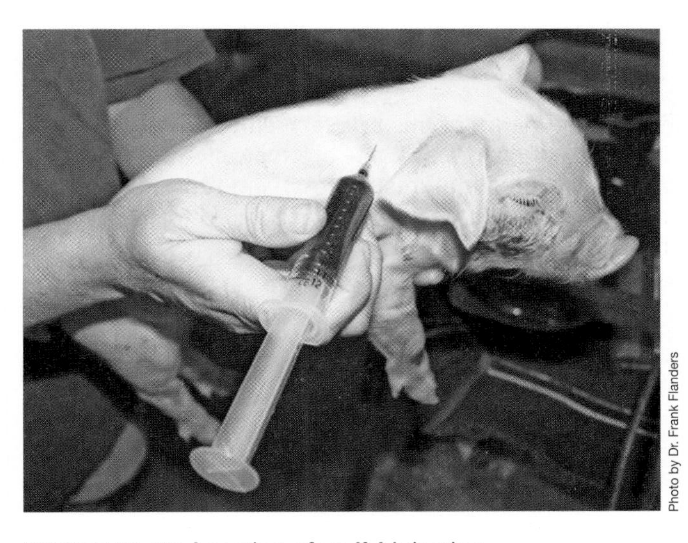

FIGURE 21-13 Location of an IM injection.

just behind and below the ear (see Figure 21-13). The ham and loin areas should *never* be used for injections. The SQ injection can be given into loose flaps of skin over the flank or behind the elbow. Injections should be given only into clean and dry locations.

Trimming Needle Teeth

Pigs are born with eight **needle teeth**, very long, sharp teeth-like needles. There are two teeth located on each side of the upper and lower jaw. The teeth serve no purpose and can cause severe injury to the sow during nursing. The teeth are trimmed with forceps or nippers shortly after birth at the same time the umbilical cord is treated with iodine. Forceps or nippers are instruments that are used to grasp teeth and trim them short or completely remove them from the mouth (see Figure 21-14).

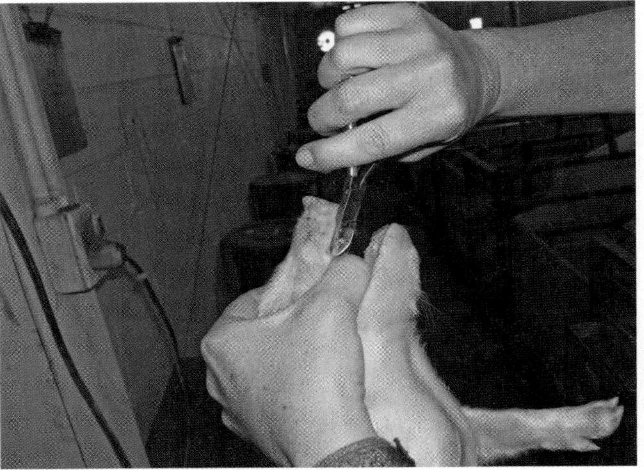

FIGURE 21-14 Trimming of the needle teeth.

Vaccinations

Swine vaccine programs will vary with location and high incidence of diseases. Swine are typically vaccinated for necessary intestinal, respiratory, and reproductive diseases. Swine are commonly vaccinated for protection from the following conditions:

- Pneumonia
- Rhinitis (inflammation of the nose)
- Flu
- E. coli
- Salmonella
- Leptospirosis
- Pseudo rabies
- Brucellosis
- Parvo virus

Swine vaccine programs begin with piglets receiving an E. coli vaccine within 12 hours after birth. They should also have an iron injection for nutrition at this time. At 2 to 3 weeks of age they should receive a vaccine for pneumonia, salmonella, and flu. Piglets begin to produce natural antibodies between 5 and 6 weeks of age. At 7 to 8 weeks of age all necessary vaccines should be given depending on the veterinarian's recommendation and booster vaccines given within 30 days, then re-vaccinate yearly. All breeding pigs should be vaccinated four to six weeks prior to breeding and repeat in three to four weeks. Sows should be vaccinated two weeks prior to farrowing and again two weeks after farrowing.

Reproduction and Breeding

Swine reach puberty between 4 and 8 months of age. Gilts usually begin breeding around 11 or 12 months of age and boars around 8 months old. Gilts and sows are in estrus from 1–5 days with 2–3 days being the average. Signs of estrus in swine include the following:

- Restlessness
- Mounting other pigs
- Vaginal swelling
- Vaginal discharge
- Increased urination
- Increased vocalization

Gestation occurs for approximately 114 days or 3 months, 3 weeks, and 3 days. Swine are either pure breeds or cross breeds. Cross breeding is a favorable method as producers can develop an ideal type of hog. This allows for heterosis or breeding lines that outperform the parents'. Crossbreeding also allows for increased production and increased profits. Over 90% of breeding operations are cross breeds. Pigs may be bred naturally or through the use of artificial insemination; breeding naturally is the most common method. Breeding crates may be used to reduce fighting and injuries. Sows are checked for pregnancy between 30 and 45 days after mating. Ultrasounds may be done to determine pregnancy and are around 85% accurate. The sows are kept in confinement in individual pens, usually with hands-on care and making them easier to manage, especially after labor. Sows should be monitored closely during the labor process as they tend to have difficulties during birth. Sows that tend to have difficulty with farrowing may be induced into labor at days 111–113. Farrowing should then occur within 18–36 hours. Only about 70% of piglets will reach weaning, due to disease, difficulty during farrowing, poor nutrition, or being injured by the sow. Sows should be placed in an area to farrow around day 110. Farrowing or gestation crates may be used to decrease injuries to the piglets. The sow and piglets may be kept in crates for several weeks until the piglets are large enough to no longer be injured. Signs of labor in swine include the following:

- Nervousness
- Anxiousness
- Restlessness
- Swollen vulva
- Vulva discharge
- Nesting
- Lactation

Sows that are kept in a nonsterile environment may be bathed prior to farrowing. Nesting material should include wheat, barley, rye, or oat straw. Chopped hay or corn cobs are also ideal for gestation crates. The area should be disinfected daily. After the sow has farrowed and the piglets have been removed, the area should be sanitized and left empty for 5–7 days. Sows and piglets are sensitive to weather and should be housed in areas with adequate protection and ventilation; ideal temperatures are 60–70 degrees. In winter or cold weather, heat lamps or hot pads may be necessary to keep piglets warm (see Figure 21-15). Boars should be kept in individual areas and have plenty of exercise. They should be de-tusked to prevent injury to each other and humans, and they should be handled regularly and easy to manage. It is important to keep reproductive records to determine if breeding problems may exist.

FIGURE 21-15 Heat lamps may be used to keep piglets warm.

Common Diseases

Swine are highly susceptible to diseases and parasites and thus typically are raised in a controlled environment when produced for human consumption. Pigs are also easily stressed. Stress can be brought on by transport, overcrowding, disease, or as a genetic factor.

Porcine Stress Syndrome

Pigs are prone to a condition known as **porcine stress syndrome** or **PSS**. This is a non-pathological disorder, meaning it is not a disease that is spread but rather a condition that occurs due to the conditions and environment of an animal. This condition occurs most often in animals that are heavily muscled and is usually genetic or spread through breeding lines. The stress condition can cause sudden death.

Brucellosis

Brucellosis or **Bang's disease** is a concern with swine breeders. This reproductive disease causes a virus known as "**Hog Cholera**," which is highly contagious and has no treatment. Pigs that are affected by this disease must be destroyed as they are no longer capable of breeding and their meat will cause contamination that is zoonotic to humans. Signs of this virus include fever, weakness, loss of appetite (anorexia), and increased thirst.

Leptospirosis

Another disease of concern to the swine industry is **leptospirosis**, a bacterial condition spread by urine contamination. Signs of the disease include fever, poor appetite, and blood in the urine, or **hematuria**. *Hemato-* is the root word for blood and *–uria* means pertaining to the urine. A vaccine is available for this disease and antibiotics may be helpful to treat this condition. Breeders and producers will be affected if this disease is found in their program. Female sows tend to abort if pregnant.

Pneumonia

Swine are prone to a bacterial condition known as **pneumonia**. This condition affects the lungs and causes a secondary condition to the respiratory tract and lungs through lesions, or areas of the tissue that are damaged. The lesions cause a chronic cough. The term *chronic* means occurring long term; *acute* means occurring suddenly or for a short time. There are no control methods to this bacteria spreading in a swine program, but antibiotics may be helpful to control or prevent a secondary infection. Lungworms and lungworm larvae may also cause lungs lesions to form.

Pseudo Rabies

Another condition that is a concern for swine producers is called **pseudo rabies**. This condition is caused by a virus. Swine may show clinical signs or be carriers of the disease. A **carrier** may not show symptoms of the disease but is still contagious and capable of passing along the disease. Young nursing pigs may show signs of fever, nasal discharge, paralysis, or coma. Adults may show similar signs and pregnant sows may also abort or have stillbirths. Nasal and oral secretions spread the disease. A vaccine is available.

Swine Dysentery

Another common condition in pigs is **swine dysentery**, also known as "bloody scours." This is spread in fecal matter and is typically ingested by young pigs. Late summer and early fall are the most common times that pigs are infected. This condition is hard to control and requires quality contamination methods as well as proper antibiotic control. Signs of the disease include soft and watery feces, bloody diarrhea, loss of appetite, and a slight fever.

Common Parasites

Parasites are of concern in swine programs, including both internal and external parasites. Parasites can be caused by soil contamination, poor sanitation practices, food contamination, and poor health practices. Parasites are more likely in a pasture situation versus an aseptic controlled environment. Roundworms and tapeworms are the most common internal swine parasites. Regular deworming and sanitation methods are required. Signs of internal parasites include the following:

- Weight loss
- Poor skin and body condition
- Anemia
- Diarrhea

Common external parasites include lice and mites. Lice are common and spread rapidly in closely confined pig herds. Lice should be suspected in pigs that commonly rub on trees, posts, or fences. Most cases occur in winter and can cause severe itching, skin lesions, and anemia. **Mange dermatitis** is another possible condition caused by mites. Mange mites cause itching and hair loss over the head and neck and spread rapidly over the body and from pig to pig. If not treated, severe skin infections can cause death. Back rubbers can be provided with the use of insecticides. Dipping and high-pressure sprays can also be used to treat and prevent external parasites. Pigs should be housed in a controlled environment at 60–75 degree temperatures. Small herds that are housed outside are best kept at a ratio of 10 pigs per acre. Sows with piglets should be kept at 7 pigs per acre. Boars should be isolated on 1/4 acre per pig. Fencing should be used to keep pigs contained safely and should be at least 3 feet high and made of strong wire. Shelter must be provided for pigs to escape from the weather. Ventilation is important to control temperatures, reduce disease, control odor, and remove moisture. Manure disposal is important as government regulations are maintained to avoid pollution and maintain good health practices. Pigs housed in controlled environments need 2.25 square feet of space per 30 pounds of body weight.

Swine Production Industry

Swine domestication began thousands of years ago and was introduced to the United States by Columbus in 1493. Wild hogs, known as wild boars, were already in North America and were hunted as a meat source. The domestic swine were bred with each other as well as the wild boars to create new breeds. Many of today's swine breeds are the result of cross-breeding two or more pure breeds of swine. Swine are used as meat producers for either pork or bacon sources. Pork has become known as "the other white meat" and is a tasty and tender type of meat. Today there are around 100 million pigs in the United States. Swine production is the second-largest livestock production industry in North America. Pigs are known for rapidly converting the food they eat to meat and muscle. Compared to other livestock species, pigs can eat smaller amounts of food and convert and produce larger amounts of meat. Less than 5 pounds of food is necessary to produce 1 pound of pork. A lot of technology and research has been put into the swine industry, thus making the importance of swine production in veterinary medicine necessary.

Corn is an industry that parallels that of swine production. The "Corn Belt" is the area in the Midwest where corn yield is high, the price is low, and the swine industry is growing. States in the Corn Belt are the top-producing swine states in the country. These states include, Illinois, Indiana, Iowa, Minnesota, Missouri, Nebraska, and North Carolina. More than 75% of the swine produced in the United States are raised in confinement. Many producers are contracted by large companies or corporations that process their own meat. There are many factors in the swine industry that make it a favorable business: Swine are efficient in converting their food to meat. Swine are prolific, which means they are capable of producing a large amount of young in a short time frame. Labor requirements are low due to swine being self-sufficient. The startup costs for swine programs are low because they require less land and space depending on the type of production method. There is a relatively short amount of time in which a profit return is made, usually less than 10 months. Although many factors make this a favorable business, other unfavorable factors should also be considered:

- Susceptibility to disease and parasites
- The need for large amounts of concentrates to be added to food
- The need for close monitoring during labor
- Economic issues and conditions
- Waste management laws and regulations
- Production permit requirements
- High cost of swine feed; swine do not eat pasture or forage sources

Swine Production Methods and Systems

Most swine producers have become corporations, from small farms to large factory farms. They are contracted by companies, usually markets or grocery stores, for their meat sources. A *contract* is an agreement by two or more people in writing between the producer and the buyer. The contract will list the breed of swine, type of production system, and price of each hog. This is typically arranged with feeder pigs or market hogs. Due to the susceptibility of disease and parasites, most swine are raised in isolation with limited human contact. All employees must decontaminate and be sterile before entering the swine's environment. This includes changing clothes and sanitizing footwear.

Feeder Pig Production Systems

Feeder pig production systems are typically used for breeding purposes as well as meat market sources. A herd of brood sows is bred and farrowed. The litter is raised to 40 pounds at weaning and sold to continue feeding until market size. There are boars used for breeding usually through artificial insemination to continue an aseptic environment.

Finishing System

The **finishing system** includes feeder pigs that are fed until market weight. Market size is typically 200–240 pounds based on breed and hog type. The ideal outcome of this system is the best weight gain at a minimal cost. This is the most common type of swine production method.

Farrow to Finish Systems

Farrow to finish systems are less specialized production systems. The sows are bred, farrow out, and the piglets are raised to market weight on the same farm. This type of production system requires multiple housing areas for breeding and feeding.

SUMMARY

Swine production is important to the veterinary industry as well as to consumers and producers. Meat production is a large part of the swine industry and is regulated by the government; therefore, the health care practices and a controlled environment are essential for proper swine management. More than 60% of all swine production systems are in controlled and aseptic environments. Veterinary assistants play a large role in the care and sanitation for large herd operations and programs. Knowledge and experience in restraint and handling, management practices, and sanitation are essential for proper swine production.

Key Terms

aseptic sterile environment condition

bacon type hog hogs raised for the large amount of meat, known as bacon, on their body

barrow young castrated male pig

boar adult male pig of reproductive age

brucellosis reproductive disease that is highly contagious and has no treatment; also known as Bang's disease

carrier animals that are infected by a disease but do not show signs

cloven shaped split in the toe of the hoof that separates the hoof into two parts

dressing the amount of meat produced on one pig

ear notching practice of making a small cut on the ears to identify the pig number and litter number

farrowing the labor process of pigs

farrow to finish systems production system where sows farrow out and the piglets are raised to market weight

finishing system production system where pigs are fed to market size

gilt young female pig that has not yet been bred

hematuria blood in the urine

hog cholera name of the virus that causes brucellosis

hog snare restraint tool with loop on the end of a metal pole used to restrain the pig by pressure over the snout

hurdle wood, plastic, or metal board used to direct and move pigs

leptospirosis bacterial condition spread by urine contamination

mange dermatitis condition caused by mange mites that cause itching and hair loss over the head and neck and can lead to severe skin infection

meat type hog hogs raised for the large amount of meat, known as ham, on their body

needle teeth eight long, sharp teeth that are like needles located on each side of the upper and lower jaw

(Continues)

paddle long-handled board similar to a hurdle that act as an extension of the hand to move pigs

pathogen Free or **SPF** a disease or disease source

piglets newborn pigs

pneumonia condition of the lungs and respiratory tract

porcine veterinary term for pigs, hogs, and swine

porcine somatotropin pST; hormone used to increase protein synthesis and increases food use and produces weight gain

porcine stress syndrome PSS; condition of heavily muscled pigs that can die from stress, such as overcrowding

pseudo rabies condition caused by a virus that shows similar signs as rabies and is spread through oral and nasal secretions

snubbing rope rope placed over snout and tied to a post for restraint

sow adult female pig of reproductive age

specific pathogen free (SPF) disease free at birth

stag a male pig castrated after maturity

swine pigs or hogs

swine dysentery condition in pigs that causes diarrhea and known as bloody scours and spreads rapidly through contamination

tail docking cutting the tail short to prevent pigs from chewing and injuring the tail and keep the tail free of feces

V-notcher tool used to make cuts in a pig's ear

withdrawal time time when no additives are fed prior to slaughter

REVIEW QUESTIONS

1. What are the different types of swine?
2. What are some common restraint methods used with swine?
3. What are some factors used in selecting swine?
4. How should a sow be managed during farrowing?
5. Describe the nutritional requirements of swine.
6. What are the production system types in swine programs?
7. What is the importance of swine production?
8. What management practices are necessary with young pigs?
9. What is the difference between a barrow and a stag?
10. What are the requirements of an SPF herd?

Clinical Situation

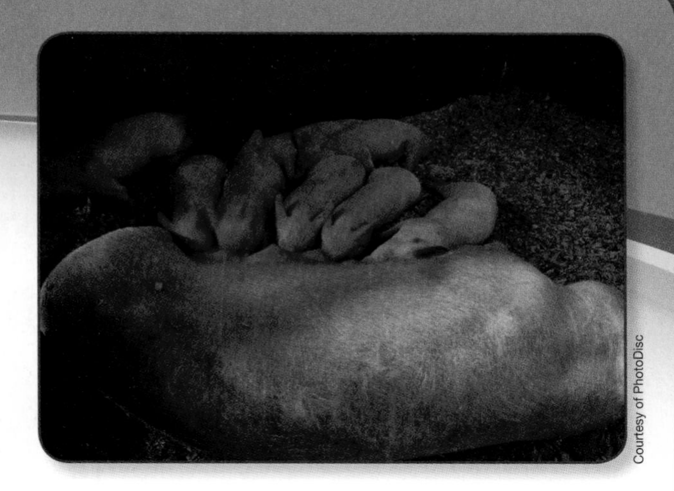

Mr. Townson, a local dairy farmer, has decided to purchase several Yorkshire piglets. He has phoned the Davidson Large Animal Clinic and is talking to Devin, the veterinary assistant.

"I have just acquired 10 Yorkshire piglets and the sow. I have never raised pigs before and really don't know a lot about them. Normally, Dr. Richards comes out to check my cattle herd every 2–3 months. I guess I need to know what care is needed for the pigs."

"Mr. Townson, first, we should schedule an exam for the sow and the piglets. They may need vaccines and some routine health care. Dr. Richards is available this Friday afternoon. How is that day for you?"

"That is fine. Thank you."

- What other information could Devin have provided Mr. Townson?
- What production factors should be considered in this situation?
- What are some basic areas that Dr. Richards will probably have to address with the sow and piglets?

22

Courtesy of Image Source

Sheep Breed Identification and Production Management

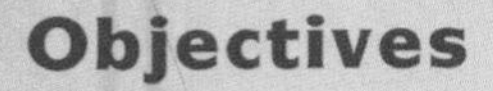

Objectives

Upon completion of this chapter, the reader should be able to:

- ☐ Understand and discuss veterinary terms relating to sheep
- ☐ Discuss sheep biology
- ☐ Identify common breeds of sheep
- ☐ Discuss the nutritional needs of sheep
- ☐ Understand normal sheep behavior
- ☐ Understand and discuss the restraint and handling of sheep
- ☐ Understand and describe common health practices of sheep
- ☐ Describe the reproduction of sheep
- ☐ Discuss diseases and conditions that affect sheep
- ☐ Discuss the purposes and uses of sheep
- ☐ Discuss the production types of sheep

Introduction

Sheep have been raised for food and clothing purposes for thousands of years. Sheep were domesticated more than 8,000 years ago in Europe and Asia. They were brought to America by settlers over 400 years ago. Sheep can be used as meat animals, for wool and fiber, and as pets. There are six major breeds that have been instrumental in developing the different breeds of today. The United States is seeing a decline in the number of sheep producers. Australia and New Zealand are the leading sheep producers in the world.

Veterinary Terminology

Sheep are known as **ovine** in veterinary medicine. The **ewe** is the adult female and the **ram** is the adult male. A **lamb** is a young sheep under a year of age. The labor process of sheep is called **lambing**. A **wether** is a castrated male sheep. Breeding types include the ram, the ewe, and the dual purpose sheep. A dual purpose sheep is bred for both meat and wool production, and occasionally milk. Sheep meat from an animal over a year of age is known as **mutton.** The meat from a sheep under a year of age is referred to as lamb.

Biology

Sheep are ruminants, have cloven hooves, and are members of the *Bovine* family. Sheep breeds range in size from 100–225 pounds. Sheep live an average life span of 7–13 years. Some breeds' wool may weigh 15 pounds or more. Sheep have a gestation length of 148–150 days. Sheep herds are best moved from behind and are a commonly herded livestock. Figure 22-1 illustrates the anatomical features of sheep.

Breeds

There are six major breeds of sheep that have been the basis for the development of over 200 total breeds of sheep seen today. Those breeds are the Dorset, Suffolk, Hampshire, Rambouillet, Border Leicester, and the Columbia (see Figure 22-2). These breeds have been cross-bred over time with over ¾ of all sheep breeds resulting from the six major breeds. Sheep are classified by their purpose and by their wool quality (see Table 22-1). Many of these breeds of sheep have become popular in North America.

TABLE 22-1

Sheep Classified by Wool Types

Fine-wool ■ Jacob	Medium-wool ■ Suffolk, Dorset, Hampshire, Oxford, Columbia
Long-wool ■ Rambouillet, Border Leicester, Cheviot, Merino	Crossbred wool ■ Southdown
Carpet-wool ■ Corriedale	Fur bred ■ Katahdin

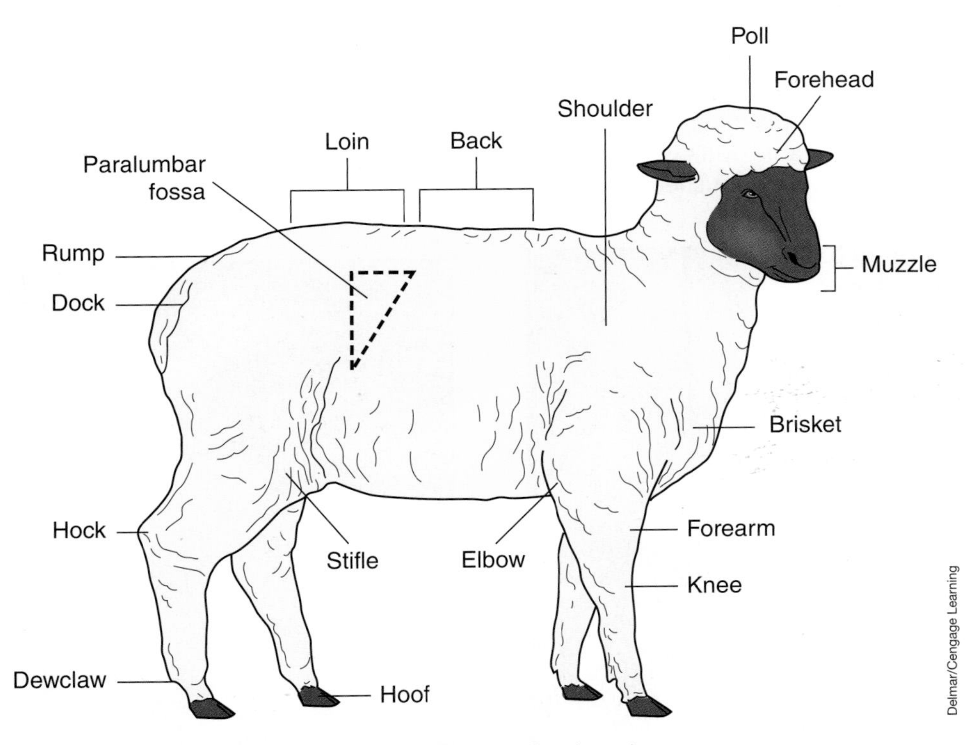

FIGURE 22-1 External anatomy of the sheep. (*Continues*)

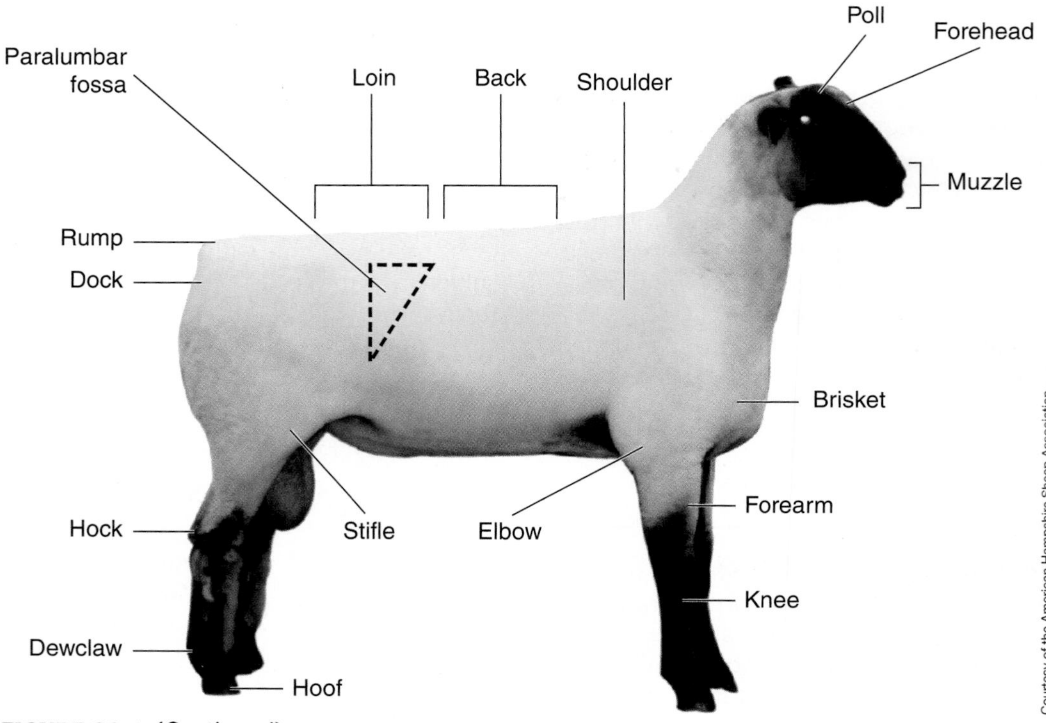

FIGURE 22-1 (*Continued*)

Breed Selection

Sheep are used for several purposes, mostly as meat and wool products (see Figure 22-3). However, sheep can also be used for milk products. Lamb is popular since it is very delicate meat, whether from a male or female lamb. Mutton tends to be tougher and have a stronger taste and is less desirable. Milk is not popular but is available from sheep. Milk from sheep tends not to be as digestible as other milk sources. Another popular product of sheep is their hair or **wool**, a soft fiber that acts as a coat. A by product is the skin of sheep, which is called a **chammy** and is a leather-like material that is commonly used in polishing cars.

Sheep selection should include researching the characteristics of the breeds, the size of the flock, the time of the year, the overall health status, age, cost, a crossbred or a purebred, the purpose, and the space and facilities.

FIGURE 22-2 (A) Dorset ram (B) Hampshire ram (*Continues*)

Photo courtesy of the Sheep Breeder Magazine

(D)

Photo by Ken Hammond. Courtesy of U. S. Department of Agriculture

Photo courtesy of the Columbia Sheep Breeders

FIGURE 22-2 (C) Suffolk ran (D) Leicester sheep (E) Columbia ram (*Continued*)

Photo by Keith Hammond. Courtesy of the U.S. Department of Agriculture

FIGURE 22-3 Wool samples. Long hair wool on left and short hair wool on right.

Nutrition

Sheep are ruminants that have a roughage-based diet and require quality hay and pasture sources (see Figure 22-4). Typically, most sheep will not need a grain source unless during gestation or other health issues require additional supplements. Sheep use carbohydrates and fats for energy, usually provided through pasture, hay, silage, or grains. Protein is acquired through alfalfa hay, linseed oil, or peanut oil supplements. Vitamins A, D, E, and K are needed and provided in quality hay and pasture sources. Vitamin B is naturally produced in the rumen of the digestive system. The rumen is the filter system of the ruminant animal digestive system. This helps break down fiber sources, such as hay and grass. Minerals that are necessary in small amounts are salt, phosphorus, calcium, and potassium. Water is required at about one gallon per sheep per day.

Photo courtesy of the U.S. Department of Agriculture

FIGURE 22-4 Pasture or other roughage is the basic feed for sheep.

Behavior

Sheep have an instinct to remain with a flock and when separated they can be easily stressed, so the importance of handling and knowledge of their behavior is important to veterinary assistants. Many veterinary professionals work with sheep still in the flock, if capable (see Figure 22-5). Sheep behavior and body language are based on a defense mechanism. Sheep show they are upset or angry by stomping their front feet or **butting** with their heads. Head butting is the action of using the head to hit an object, usually a person or other sheep or animal. Rams will use this action when angered and also during the breeding season. This along with moving into the flock is their defense mechanism. A *defense mechanism* is how an animal uses its body or an action to protect itself, usually from prey or a stressful situation, such as handling and restraint.

Equipment and Housing Needs

Sheep usually are maintained in pastures or barns with adequate ventilation. Sheep flocks require shelter sources to protect sheep from the weather, especially those raised for wool. Stalls should be kept well-bedded and dry. Muddy and moist surfaces are not ideal for sheep. There is a need for electricity and water sources especially in wintertime. Water troughs should be kept from freezing and fresh water provided daily. Indoor housing for ewes should include free stalls and large open areas for young lambs to exercise. Sheep breeds with horns or rams that develop horns should be trimmed to prevent injury to other sheep or people.

Photo courtesy of Getty Images

FIGURE 22-5 Sheep tend to remain in flocks and can be best handled if allowed to remain with the flock.

Fencing requirements are wire types built 60 inches or higher with 4–5 inches between strands.

Restraint and Handling

When working with sheep, it is important to stay calm, quiet, and gentle and work with assurance. Sheep can be easily injured by a careless handler as they have fragile bones that may be broken. This is important when attempting to grasp and catch a sheep. Sheep that have a full wool coat can easily become overheated due to the stress of handling. Wool can also be damaged in handling and will reduce the quality and profit of the product. Sheep are otherwise easy to work with; they rarely bite or kick during restraint. Capturing a sheep is best done within the herd, by approaching the individual sheep and swinging an arm around the neck and front shoulders and the other arm around the **dock** (tail area). Grasping the dock and then steering the animal to the necessary location is relatively easy. Halters can be placed on sheep, but due to the short noses and odd-shaped head, they can slip off and occlude the nostrils, causing breathing problems. There are several methods of restraint depending on the procedures being completed. One of the most common and easiest restraint methods is the "rump" method. The **rump method** is sitting the sheep up on its hind end. Sheep rarely struggle and this allows for examination, hoof trimming, shearing, or injections. When holding the sheep, stand on the left side and place one hand on the chin and pull the sheep close to your legs. Using head control and a quick twist one person can usually sit a sheep on its rump. Larger sheep can be held with the restrainer's back against a wall for support. A restraint for medicating is typically holding the sheep against a wall or using the limbs to restrain and hold the head in place. Newborn lambs can be held similar to small dogs.

Grooming

Many breeds are raised for wool. Few breeds shed their own wool and require regular **shearing**, or shaving of the wool. Shearing is done carefully with clippers or shears that easily remove and prevent damage to the wool quality (see Figure 22-6). Most producers will shear several times a year, but care must be taken to not allow sheep to overheat in the summer or become sick in the winter due to the length of the wool coat. Some equipment that is necessary for shearing includes electric shearers, shearing blades, trimmer disinfectant, shearing tables or chutes, and blankets and hoods.

Photo courtesy of iStock

FIGURE 22-6 Shearing sheep.

Basic Health Care and Maintenance

Sheep are highly susceptible to diseases and parasites and require a quality health care program. There are many signs of illness that should be of concern to veterinary assistants and sheep producers. The signs of an unhealthy sheep include poor wool or coat quality, loss of wool, discharge from eyes and nose, pale mucous membranes, anorexia, no signs of chewing cud, diarrhea, and signs of standing alone from the flock.

When new sheep are added to a program, it is important to quarantine and isolate all sheep for 30 days to determine the health status. All sheep flocks should have yearly physical exams, updated yearly vaccines, dewormers administered every three months, isolation of all sick animals, and daily cleaning and sanitation of stalls.

Lambs tails are usually docked by the banding method, which involves placing an elastic band tightly over the tail area, causing it to slough off. This is done within three to five days of age and is not painful. The tail docking prevents a buildup of feces on the wool. Adult sheep may be docked using a tail docking instrument that cuts the tail (see Figure 22-7). Young male sheep can be castrated through the banding method while they are still in the juvenile stage. Adult sheep should be castrated through the use of an elastrator, similar to cattle and goats.

Vaccinations

Vaccine programs will vary from location and veterinarian recommendations. There are numerous sheep vaccines available and the proper vaccines should be discussed with the veterinarian. Many lambs are vaccinated with a combination vaccine of the most common diseases for that particular area. Lambs should begin the vaccine program between six and eight weeks of age with booster vaccines in two to four weeks. Vaccines for breeding programs may require boosters every six months to a year; other sheep production programs will require yearly vaccine boosters.

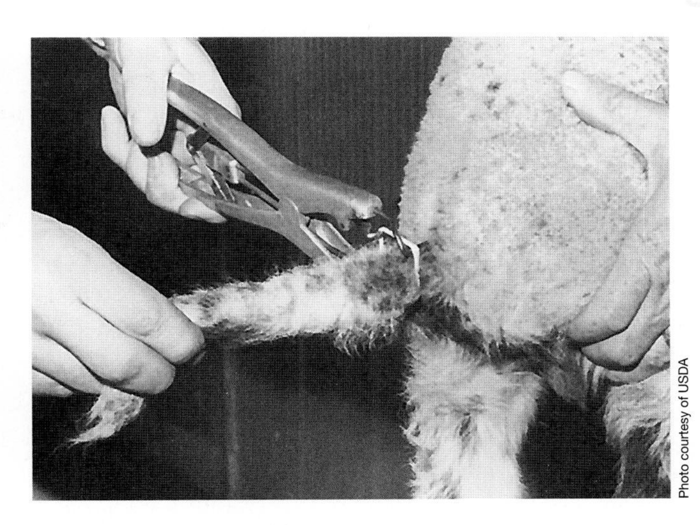

Photo courtesy of USDA

FIGURE 22-7 Tail docking.

Breeding and Reproduction

Sheep producers strive to have a high number of lambs born each season (see Figure 22-8). The average lamb crop is two lambs per ewe per breeding. Ewes reach puberty around 8 to 10 months of age, while rams reach puberty around 5 to 7 months old. Ewes typically are bred for the first time around 2 years of age. Ewes can

Photo by Ken Hammond. Courtesy of U.S. Department of Agriculture

FIGURE 22-8 Ewe with lamb.

sometimes be difficult to breed as they do not normally show any signs of heat. The estrus cycle lasts an average of 30 hours, sometimes making it difficult to determine a breeding time. The estrus cycle occurs every 16 to 17 days. The gestation length is between 148 and 150 days. The labor process in sheep is known as lambing. It is important to ensure that each ewe has been bred, because a sheep that doesn't lamb is not profitable. Rams are normally kept isolated from ewes. One may be placed in a flock of ewes for short time frames to allow breeding to occur, but timing must be correct. When not breeding, the rams should be kept isolated to prevent fighting and aggression. They require large amounts of exercise and are best kept in a pasture. They do not require large amounts of grain, but they should have quality grass or hay sources. The pregnant ewe should have quality pasture and hay sources and adequate exercise. Grains should be given until close to labor. When the lambing process nears, the ewe should be placed in a dry and sanitary area. The udders should be sheared for proper sanitation and to allow the lambs the ease of finding a milk source. There are four phases of the lambing process: dilation, accelerated contractions, expulsion of the lamb, and the expulsion of the placenta. *Expulsion* refers to removal from the body by internal muscular contractions.

Sheep typically have a quick and easy labor but should be monitored for difficulty, especially when twins are expected. Many ewes will have twins and this is the ideal goal of a reproductive program. When giving birth to twins, the ewe may need some help in reviving and drying the lambs. During the **dilation phase**, the **cervix** (birth canal) will **dilate** or expand. This prepares the birth canal for the lamb, which should be born within the next 12–24 hours. The dilation phase is completed when a clear to whitish color discharge appears at the vulva. This leads to the **contraction phase**. The contractions or wave-like movements of the uterus allow the lamb to move into the birth canal. This phase occurs over a 6 to 12 hour period and is completed when the water breaks and a clear, water-like fluid is passed. This indicates the lamb is beginning to be born. The labor phase should be monitored for **dystocia** or difficult birth signs. The normal birth of a lamb is the tip of the nose and front feet delivered first. When the lamb is born, the ewe will clean the head and nasal passage first to stimulate breathing. If a twin is then born, the ewe will lose interest in the first lamb as contractions begin. The placenta or afterbirth is expelled within 30 minutes to an hour after the labor process is completed. It is important to note the passing of the placenta, as sheep will consume the membranes as a natural instinct to not attract predators. A **retained placenta** that is not expelled within 24 hours after birth will lead to a serious and possibly fatal infection. A veterinarian should be contacted immediately. Occasionally, a ewe may die or reject a lamb and the lamb becomes known as an **orphaned lamb**. Some sheep will become a foster mother if they have a lamb at their side. Otherwise, a milk replacer or cow's milk may need to be supplemented through bottle feeding.

Common Diseases

Sheep are easily susceptible to diseases. This is important to note in production programs where sheep are raised for meat products and human consumption.

Foot Rot

Sheep tend to be affected by a common condition that affects their hooves and is caused by a bacterial infection. Bacteria can enter a wound or the hoof wall and cause a condition known as **foot rot** or an **abscess**. Foot rot is the decay and damage of the soft tissues within the cloven hoof. An abscess is a buildup of pus caused by an infection and is typically localized to one area. The infection causes the sheep to become lame and limping will occur. The joints and tendons may also be affected. Treatment and prevention should be quality hoof care by trimming away any areas of damage, applying a topical disinfectant and antibiotic, and keeping the sheep in a clean and dry location until healed. Sheep kept in moist, muddy, or wet areas are more prone to bacterial infections. Abscesses may need to be opened and drained to relieve pressure buildup from the pus and infection. A vaccine is available for bacteria prevention from foot rot. Signs of foot rote or abscesses include the following:

- Limping and lameness
- Standing without putting weight on hoof
- Weight loss
- Anorexia
- Fever
- Foul odor from the hoof area
- Pus drainage from the hoof
- Swollen joints

Mastitis

Another common condition that affects ewes is **mastitis**, or inflammation of the mammary glands. The udder infection occurs during lactation when bacteria enters the teat glands and causes an infection. The udder may feel hard, hot, and tender to the touch. Antibiotics should be prescribed by a veterinarian and the sheep may need to stop nursing to allow the udder to dry up.

Actinobacillosis

Another infection that must be considered is called **actinobacillosis,** commonly called "lumpy jaw." This disease is of importance to producers as there is an economic loss when this occurs, as the meat is condemned and cannot be used for human consumption. This disease affects the head and jaw and causes tumor-like lumps of yellow pus to build up within the jaw. The infection spreads to the muscles, organs, and tissues of the body. The affected areas are not legal to slaughter for human use. This condition does not cause death but can spread to other herd animals.

Actinomycosis

Another disease that affects sheep is called **actinomycosis** or "wooden tongue." This bacterial infection causes lesions on the head that spread to the lymph nodes located in the neck, causing swelling. Hard lesions will also form on the tongue, causing difficulty eating and drinking. Weight loss and death will occur as the disease progresses. This is caused from contamination of food and water sources and all infected animals should be isolated from the herd to decrease spread of the disease.

Blue Tongue

Blue tongue is a virus that is spread by gnats, which are tiny flying insects that bite and feed off of blood meal of animals and humans. Blue tongue has no treatment and the virus rapidly weakens the sheep's immune system. A secondary infection typically causes death of the animal. A vaccine is available and should be administered at the time of shearing to boost the immune system. Signs of blue tongue include the following:

- Fever
- Weight loss
- Anorexia
- Difficulty eating
- Lethargy/sluggish behavior
- Swelling on the mouth and ears
- Ulcers on the mouth and tongue

Enterotoxemia

Enterotoxemia is called the "overeating disease." This bacterial disease affects lambs and show animals due to feeding too much grain. Bacteria within the rumen develop and produce toxins that can't be absorbed by the body. This leads to seizure activity with death usually occurring. There is no treatment of this condition but vaccination is recommended and good health practices reduce the occurrence. Vaccines should be administered two and six weeks prior to lambing.

Johne's Disease

Johne's disease is a digestive and intestinal condition that causes the intestinal wall to thicken. Signs include diarrhea, weight loss, and possible death. No treatment has been developed but research continues to determine the causes of this disease. A prepurchase exam and past medical history is important prior to purchasing new sheep.

Soremouth

Soremouth is a viral disease that is highly contagious and zoonotic. This disease causes sores that form around the mouth and lips of lambs. Healing can be done by scrubbing the sores with iodine. Scabs form and should be removed, and scrubbing should continue until the sores have cleared. Isolation and sanitation are keys to treatment and prevention. Exam gloves should be worn when treating young lambs. A vaccination is available and should be given shortly after birth.

Lamb Dysentery

Lamb dysentery is caused by bacteria that affect newborn lambs around 1–5 days of age. This disease spreads rapidly through newborn lambs and requires good sanitation and control management. Signs include loss of appetite, diarrhea, depression, and sudden death.

Polyarthritis

Young lambs may also be affected by a condition known as **polyarthritis**. This is a condition that affects lambs around three to five weeks of age and affects multiple joints causing pain and swelling. Lambs affected by this condition will show signs of reluctance to move, inactivity, weight loss, or no weight gain. Lambs can be treated with tetracycline antibiotics.

Common Parasites

Parasite prevention programs are also a vital part of a sheep production program, as many sheep are housed outdoors and feed off of pasture. This makes them more susceptible for internal and external parasites. External parasites that are common in sheep include lice, mange mites, ear mites, and ringworm. All external parasites will severely affect the wool quality of the coat. Ringworm is highly contagious to other sheep and humans. Mites can be treated with common parasitic medicine and disinfectants, such as Fulvicin, nolvasan, or bleach solution. Common internal parasites of sheep include lungworms, whipworms, tapeworms, coccidian, liver flukes, and stomach worms. Parasite preventions and treatments include topical dips, insecticide sprays, anthelmintics (dewormers), waste control and removal, and pasture rotations. Complete eradication of parasites is impossible. Dewormers should be given every three months through a process known as **drenching**. Drenching is the method of giving medicine by mouth with a dose syringe, helpful in giving large amounts of medicine at once for large numbers of animals. Signs of parasites in sheep include the following:

- Diarrhea
- Weight loss
- Rough coat
- Wool loss
- Anemia
- Respiratory problems
- Anorexia
- Bloated appearance
- Depression

Sheep Production

Sheep production has its advantages and disadvantages in relation to other types of livestock programs. Sheep are the best-suited livestock for grazing, are excellent scavengers, are very efficient at converting food to meat, serve dual purposes, produce a low amount of waste material, are easy to raise and handle, and require little space. Disadvantages of sheep include industry competition, low wool prices, low meat production, susceptibility to disease, and their existence as easy prey. There are several types of production systems used in sheep production.

Farm Flock Method

The most popular is the **farm flock method**. This system's primary goal is to produce market lambs that will be used as a lamb meat source as the primary product and wool as the secondary product (see Figure 22-9). Typically, this system has producers that have other livestock and may raise the sheep with other types of livestock in a pasture setting.

Purebred Flock Method

The **purebred flock method** is usually a small production system of purebred sheep. The goal of this system is to produce quality rams and ewes for sheep breeding programs. This requires a high level of management

Photo courtesy of Dr. Joseph Daniel, Berry College

FIGURE 22-9 Example of a farm flock sheep production system.

and a high standard of the breed. Producers are raising ideal types of sheep that meet the standards of that particular breed. This is a full-time business where more labor is required compared to other production systems. Pasture space is also necessary.

Range Band Method

The **range band method** is seen more in the Midwest where larger amounts of land are used to raise large sheep herds. The herd is known as a "**band**" or large group that ranges free over a vast amount of land and is moved and cared for by people known as **herders**. Herders move the band of sheep over the land to different pasture sources. It is important that the land is in an area where a large amount of vegetation and rainfall are prominent. The goal of the range band method is either a focus on wool or lamb production.

Confinement Method

The **confinement method** is becoming a more popular method as an aseptic system in raising sheep for human consumption. Sheep are raised indoors in a sterile and controlled environment to prevent little to no health or parasite problems. There is no need for land. Young are produced due to the health status and care of the herd. This system requires more skill and care in the production management of the sheep.

Lamb Feeding

Lamb feeding is a specialized production system in which lambs are raised to weaning and then sold to feedlots. The knowledge and care of proper breeding management is necessary. This type of system is typically seen on a large farm with quality pasture space (see Figure 22-10).

FIGURE 22-10 Lamb feeding systems raise lambs to age of weaning before selling them.

SUMMARY

Sheep production is an important part of the United States and essential for economic success. However, a decrease in the number of producers and sheep flocks has been occurring over the last several years. Sheep are used for meat, wool, and milk sources. They tend to be an easier and more efficient livestock source to raise and manage. Due to the high susceptibility to disease and parasites, a quality health care program must be followed to ensure quality products.

Key Terms

abscess bacterial infection that causes pus to build up in a localized area

actinobacillosis disease of the head and jaw that causes tumor-like lumps of yellow pus to build up within the jaw; also called "lumpy jaw"

actinomycosis bacterial infection of the head and neck lymph nodes and causes hard lesions to form over the tongue; also called "wooden tongue"

band large group

blue tongue virus spread by gnats that weakens the immune system and causes secondary infections to occur

butting the action of using the head to hit an object or a person

cervix birth canal

chammy by product of sheep skin used as a leather product for polishing

confinement method system of raising sheep in an aseptic environment for human consumption purposes

contraction phase second phase of the labor process that causes wave-like motions in the uterus to move the fetus into the birth canal

dilate to open, expand, or widen

(*Continues*)

dilation phase first phase of the labor process causing the birth canal to open and expand

dock tail area

drenching method of giving medicine by mouth with a dose syringe to large numbers of animals

dystocia difficult or abnormal birth

enterotoxemia bacterial disease from overeating that causes toxins to build up within the rumen that are not absorbed by the body

ewe the adult female sheep capable of reproduction

farm flock method most popular sheep system that involves market lamb production for meat sources as the primary system and wool production as the secondary system

foot rot bacterial condition of the hoof that decays or damages the hoof tissue

herder person in charge of a group of sheep

Johne's disease digestive and intestinal condition that causes the intestinal wall to thicken

lamb a young sheep under a year of age; also the meat from a young lamb under a year of age

lamb dysentery bacterial diarrhea condition in newborn lambs

lamb feeding specialized production system of raising lambs to weaning and then selling to feed lots

lambing the labor process of sheep

mastitis inflammation of the mammary glands

mutton sheep meat from an animal over a year of age

orphaned lamb young sheep that has lost its mother or has been rejected by the mother and must be bottle fed

ovine the veterinary term for sheep

polyarthritis condition of young lambs that causes multiple joints to become inflamed and painful

purebred flock method small production system that raises purebred sheep for reproductive purposes

ram the adult male sheep capable of reproduction

range band method raising large sheep herds on large amounts of land

retained placenta fetal membranes of waste materials not passed from the uterus after the labor process

rump method restraint method of placing a sheep on its hind end for handling

shearing method of shaving wool from sheep

soremouth highly contagious viral disease that is zoonotic and causes lesions over the mouth and lips of sheep

wether male castrated sheep

wool soft fiber of sheep that acts as a coat

REVIEW QUESTIONS

1. How are sheep classified by type?
2. What are the advantages and disadvantages of sheep production?
3. What are the types of sheep production systems? Describe each type.
4. What is necessary for ewe management in regard to reproductive needs?
5. What nutritional requirements are needed for sheep?
6. What health needs are required by sheep?
7. What are signs of parasites in sheep?
8. What treatments and preventions are necessary for foot rot and abscesses?

Clinical Situation

© iStock/Lee Pettet

Jim and Chris Smith, clients at Millers Animal Clinic, have an appointment for their dog to have its yearly exam and vaccines. During the visit, the couple asks Amy, the veterinary assistant, who is taking a patient history, about their newest pets, twin lambs.

"We just got twin lambs that are bottle babies. We got them for our kids. Would Dr. Anderson see them like he sees our dog?"

"Dr. Anderson does see both small and large animals. He would schedule a farm call visit to examine the lambs. We usually only see large animals in the clinic if they need to be hospitalized."

"What would the lambs need as far as veterinary care?" Mrs. Smith asks.

- What would be the best answer that Amy could give Mrs. Smith?
- How would you go about scheduling the lambs for a farm call appointment?
- What educational information should you give to the Smiths about the lambs?
- What information will Amy need to schedule the farm call visit?

23

Goat Breed Identification and Production Management

Courtesy of Photo Disc

Objectives

Upon completion of this chapter, the reader should be able to:

- ☐ Define and discuss veterinary terms relating to goats
- ☐ Discuss goat biology
- ☐ Identify common breeds of goats
- ☐ Discuss the nutritional needs of goats
- ☐ Explain normal goat behavior
- ☐ Demonstrate the restraint and handling of goats
- ☐ Identify and describe common health practices of goats
- ☐ Describe the reproduction of goats
- ☐ Discuss diseases and conditions that affect goats
- ☐ Discuss the purposes and uses of goats

Introduction

Goats are similar to sheep in several ways, but they have some different characteristics to set them apart. Goats have a hair coat, rather than wool. Goats and sheep have similar shaped heads and ears, but goats tend to be slightly larger. Both are cloven-hoofed. Some breeds have horns and goats have beards. Goats tend to be more intelligent, independent, and have a better ability to protect themselves.

Goats were one of the first types of livestock to be domesticated and developed over 8,000 years ago. They have been used as grazers to keep grass trimmed, to clean up plants and weeds, to provide company to other types of livestock, as a source of meat, as by-products and fiber sources, and as pets. Goats are grouped by their purpose and use. Today, China and India are the leading countries in goat production. Goats are now used as food sources in products such as milk and meat.

Veterinary Terminology

Goats are known as **caprine** in veterinary medicine. A female goat is known as a **doe** or **nanny**; the male goat is called the **buck** or **billy**. A young goat under a year of age is called a **kid**. The labor process of goats is known as **kidding**. A group of goats is known as a **herd**.

Biology

Goats, like sheep, are ruminants. Goats range in sizes from 20 pounds to well over 150 pounds and range in height between 1.5 and 4 feet. The life span of goats averages from 8–10 years. Male goats are known for having a strong odor that is due to hormones that are present on the top of the head, usually behind the horns. The **scent glands** create an unpleasant odor that is typically used to rub objects or areas to mark their territory. Young male goats' scent glands are usually removed during dehorning. Figure 23-1 illustrates the anatomical features of the goat.

Breeds

There are over 300 breeds of domesticated goats. Wild goats are still found in the United States and various other countries. There are five groups of goat breeds: Angora, Dairy, meat, **cashmere**, and pygmy. Goats are kept as pets, for milk production, for meat production, and for breeding purposes.

Angora

Angora goats are produced for their hair coat, which is used to make clothing and fiber products (see Figure 23-2). The fiber produced from Angora goats is called **mohair**. Angora goats are also used for meat.

FIGURE 23-2 Angora goat.

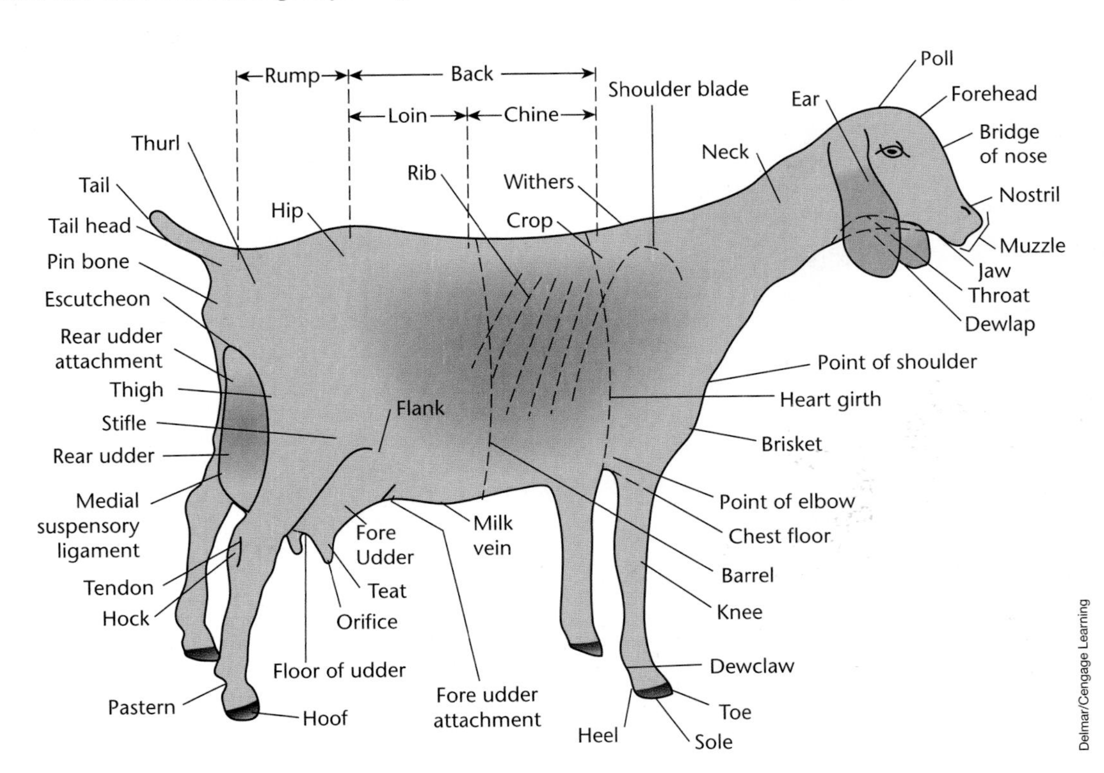

FIGURE 23-1 External anatomy of the goat.

Dairy

Milk from **dairy goats** is a desirable product to people who have digestion problems and sensitivities to dairy products. Goat milk is easy to digest, has more vitamin A than other animals, and has less fat due to the particles known as **globules**. There are six common breeds of dairy goat in the United States: French alpine, Oberhasli, LaMancha, Nubian (see Figure 23-3), Saanen, and Toggenburg.

Meat

The **meat goats** are raised for meat sources and are bred for quality muscles. Most meat goats are from Spanish breeds. A popular meat breed goat is the Boer goat (see Figure 23-4).

Cashmere

Cashmere refers to the soft down or undercoat produced by most breeds of goat. Angora goats do not produce cashmere. There is no registered breed for cashmere goats, although goats have been bred to improve on this characteristic.

Pygmy

The **Pygmy goats** are miniature-sized goats commonly used as pets and in research (see Figure 23-5). They are great entertainers and very playful. Pygmy goats can be produced for milk and meat. Due to their small size, these goats are relatively easy to handle.

Photo courtesy of iStock

FIGURE 23-3 Nubian goat.

Photo by Dr. Frank Flanders

FIGURE 23-4 Boer goat.

Photo courtesy of Linda Henwood, National Pygmy Goat Association

FIGURE 23-5 Pygmy goat.

Breed Selection

Goat selection should be based on the type of production or purpose, such as raising dairy or meat goats. Hair and coat characteristics are important considerations for fiber goat production. The primary concern in dairy goats is milk production.

Goat Nutrition

Goats will eat anything, especially weeds and **browse**, which is a woody type of plant (see Figure 23-6). The major components of a goat's diet include carbohydrates,

Photo by Scott Bauer. Courtesy of U.S. Department of Agriculture

FIGURE 23-6 Goats will eat browse or woody plants.

fats, protein, vitamins, minerals, and water. Like sheep, goats have a ruminant digestive system. Their diet is mainly roughage and plants. Goats should be fed based on their use and purpose. Goat diets will depend on the needs of the purpose they serve, such as dairy goats, meat goats, and show goats. Goats chew cud to break down the plants and roughage they eat to allow it to filter through their digestive system. Browse is the primary diet, consisting of shrubs, weeds, woody plants, and briars. Many goats do not get adequate nutrients from this type of diet, especially does during pregnancy and lactation. It's important to note that some weeds and plants are toxic to goats such as azalea, cherry, hemlock, and laurel.

The best maintenance diet for goats includes a 14–16% protein diet with added salt, phosphorous, and calcium. Grains or pellet-based diets may be fed. Growing kids and lactating females will require added calcium supplements to their diet. Minerals required in small amounts include cobalt, copper, iodine, iron, manganese, selenium, and zinc. Copper is required more in goats than other ruminant species. Most commercial grain or pellet diets are fed at 1 to 1.5 pounds per day with high-quality roughage. Hay, grass, and browse needs are between 3.5 and 4.5 pounds per day. Kids will nurse for about 60 days and then weaning should begin with a starter diet that consists of soft pellet formulated food. The young goats should begin a regular diet as soon as possible.

Behavior

Goats do not have the same instincts as sheep to join a group and tend to scatter when herded. In a herd of goats, there will be a lead goat that is usually a doe. Most other goats will follow the lead goat. Normal goat behavior is typically docile and friendly, which makes them easy to handle. They can become agitated and will butt similar to sheep, usually rearing up on their hind legs. Signs of anger and aggression include hair rising over the back similar to hackles on a dog, holding the tail close to the back, sneezing, stomping, and snorting. Goats have a naturally playful nature and are commonly kept as pets. They become less playful with age. Nanny goats remain calmer as adults than billy goats. Billy goats also have a stronger-smelling odor due to the scent glands on their heads. Male goats will rub their scents over objects to mark their territory.

Basic Training

Goats may be trained to halter and lead similar to sheep and cattle. When shown, goats must be able to lead in hand and walk with a handler. They should be taught to stand quietly to be judged. Goats can be brushed and groomed as cattle with regular brushing, bathing, and hoof trimming.

Equipment and Housing Needs

Housing of goats may be indoor or outdoor facilities. Indoor barns or sheds should serve as shelter from the weather and any open sides should open to the south. Stalls should be kept well-bedded and dry. Proper ventilation is necessary to allow air to circulate through the open area. The barn should include running water, electricity, and food and water troughs. If several goats and kids are housed together, an open area should be large enough to contain the herd comfortably. Goats should be dehorned to prevent injury to people and other animals. Goats kept outdoors in a fenced area should be contained in an area at least 60 inches in height, as goats are excellent jumpers. Fence panels should be at least 4 to 5 inches between strands or boards. Goats that are shown or regularly transported should have loading chutes and weight crates to determine the weight of the animal. Other equipment that may be necessary for goats includes blankets, hoof trimmers, clippers, shearing table, blankets, halters, leads, and fans.

Restraint and Handling

Goats will struggle if handled roughly or improperly. The minimum amount of restraint used the better for the goat and the handler. Goats shouldn't be handled or treated like sheep.

FIGURE 23-7 Proper restraint technique for jugular access.

The best method of catching a goat is to grasp one leg and lift it. Most goats will stand still and allow many types of procedures to be completed in this manner. The best way to do this restraint is by reaching over the back and grasping one front leg. This is a common restraint for hoof trimming. Goats that have not been handled or are scared may need to be caught by grasping a hind leg, using caution as goats may kick and struggle. Goats can be placed against walls, similar to sheep, holding it against a solid structure. Goats should never be restrained by their rumps. Lateral recumbency as in dogs is another method of restraining a goat. This can be done by grasping the head and muzzle with one hand and reaching over the back and grasping one front leg, and then move the other hand to a front leg. Gently raise the rear legs and lay the goat on its side holding down the limbs to prevent it from standing. Goats can be trained to be haltered and will lead. Halters should be fitted so they don't fall off the nose or occlude the nostrils. Head restraint can be done similar to holding a dog. Venipuncture restraint may also be done as in the dog or cat with the front limb extended for cephalic blood collection and extending the head upward for jugular blood collection (see Figure 23-7).

Grooming

Goats can be washed and kept clean through the use of water and mild soap. They may be brushed on a regular basis with a stiff brush that helps remove dirt and shedding hair. Goats can be sheared like sheep. The hooves require routine trimming and maintenance. Goats that are shown commonly are bathed, brushed, trimmed, or sheared before a show. Blankets are placed over the body to keep the animal clean and dry. Some goats have hair and others have mohair, which may need to be sheared rather than clipped.

TABLE 23-1

Normal Goat Vital Signs

Temperature (T)	103–103.6 degrees F
Heart Rate or Pulse (HR)	80–85 bpm
Respiratory Rate (RR)	25–30 br/min

Basic Health Care and Maintenance

Goats are relatively healthy, and good care and management allow them to lead long lives. Goats that are healthy will graze on pasture and plants throughout the day, chew on cud, have a shiny and healthy hair coat, strong legs and hooves, appear social, and have bright and clear eyes. Table 23-1 outlines normal vital signs for a goat.

Goats require the same health practices and programs as that of sheep. The scent glands are usually removed when they are kids with a caustic ointment or electric dehorner. Males are also banded as kids within 3–5 days of age.

Vaccinations

Vaccines should begin between 6 and 8 weeks of age and follow similar recommendations to that of sheep and that of the veterinarian. Most goats are routinely vaccinated for overeating disease, tetanus, soremouth, and goat pox, but the vaccine programs will vary with location and occurrence of disease. Kids should begin a deworming program 4–8 weeks after they begin to graze.

Reproduction and Breeding

Goats are similar to sheep in that the ideal number of young kids produced each year is two to three per breeding season. This means that the ideal breeding doe will have twins or triplets each season. The breeding seasons are late summer and late fall. Does have a gestation length of 148–151 days and the breeding is timed for kids to be born in the early spring or late fall. Does are typically bred around 2 years of age. The estrus cycles are regulated by the climate and weather and occur every 16–17 days. They are usually receptive to the ram for 24–48 hours. When the kids are born, they usually average around 5 pounds in weight. Goats are best kept in open areas with shelter provided from the weather. Rarely do the does have reproductive problems or dystocia issues. The normal labor presentation is similar to the sheep with the head and front limbs presented first.

Common Diseases

Goats are relatively healthy but can acquire many of the diseases that sheep are susceptible to, so a review of Chapter 22 is valuable in learning common conditions and possible diseases of goats. Signs of an unhealthy or diseased goat include a poor or rough hair coat, hair loss, abnormal temperature, swelling on joints and body, pale mucous membranes of the mouth and eye mucosa, and thickened nasal and ocular discharge.

TERMINOLOGY TIP

The term *mucosa* refers to the mucous membranes located throughout the body. The mucous membranes include the gums, eye tissues, nasal tissues, and rectal tissues. The mucous membranes allow for color determination of blood circulation and oxygen flow through the body.

Joint Ill

Young kids may be affected by **joint ill**, a bacterial condition where the umbilical cord site becomes contaminated and bacteria enters the blood stream, causing severe inflammation and pain in the growing joints. Breeding goats should be tested for brucellosis prior to reproduction.

Bloat

Bloat is possible when goats ingest foreign objects that cause severe stomach pains and digestive upset. Bloat in goats is handled in a similar method as that in cattle. Indigestion may be a common issue in goats that tend to have access to poor nutrition or objects that are ingested, such as garbage.

Goat Pox

A viral disease of goats that affects mostly young or immune-compromised goats is **goat pox**. This virus causes flulike symptoms and spreads rapidly through herds. Infected goats should be isolated. Affected animals will develop lesions over the head and face that eventually open up and drain. The lesions should be cleaned daily with hydrogen peroxide and topical antibiotics should be applied.

Hypocalcemia

Hypocalcemia or milk fever often occurs in does that have been lactating. Calcium supplements are necessary due to the body's need to use up the calcium to produce milk. Signs of milk fever in goats include the following:

- Inability to stand
- Decreased respiratory rate
- Gasping for breath
- Coma
- Possible death

MAKING THE CONNECTION

Other common conditions are mastitis, foot rot, and hoof abscesses. These conditions are further discussed in Chapter 22.

Common Parasites

The parasites that affect the sheep also are of importance in the goat. The most common external condition in goats is the fungus **ringworm**. Goats that are housed outside and have access to wet and muddy areas are more susceptible to the fungus. Ringworm causes hair loss in patches on areas of the coat. This condition is highly contagious to all animals and humans.

SUMMARY

Sheep and goats are closely related and require similar needs and practice. Goats have relatively minor differences in physical characteristics. Goats may be used for milk, meat, fiber, and by-products. They have become more popular as pets. The importance of their health care as both livestock and products for human consumption has made them a vital part of veterinary medicine.

Key Terms

Angora goat group of goats that produce hair

billy common name for the adult male goat

bloat air in stomach causing a swollen abdomen

buck adult male goat

browse woody type of plant

caprine veterinary term for goats

cashmere fine down undercoat of goats that is warm and used in clothing

dairy goat group of goats that produce milk

doe adult female goat

globules particles in milk that reduce the fat content

goat first type of livestock domesticated that are used for meat, dairy, and fiber products

goat pox viral disease of goats that affects the immune system and causes flulike symptoms

herd a group of goats

hypocalcemia low blood calcium

joint ill bacterial infection caused by contamination of the umbilical cord that causes pain and inflammation within a joint

kid young newborn goat

kidding labor process of goats

meat goat group of goats that produce meat

mohair long silky locks of hair from the Angora goat used to make clothing and bedding products

nanny common name for the adult female goat

Pygmy goat group of small-sized goats used as pets

ringworm common contagious external fungus located over the skin and hair coat

scent glands area on the head located behind the horns that produces a strong hormonal odor in male goats

REVIEW QUESTIONS

1. How are goats and sheep similar in appearance? How are they different?
2. Describe how goats are grouped.
3. What is the normal behavior of a goat?
4. What is the best restraint method for catching a goat?
5. What is the estrus cycle of a female goat?
6. How long is the length of gestation in goats?
7. What are the normal vital signs of goats?
8. What are some restraint methods used on goats?

Clinical Situation

Courtesy of Photo Disc

Mr. Jones, a local goat farmer, calls to schedule a farm call appointment. The veterinary assistant, Greg, tells Mr. Jones that the veterinarian, Dr. Johnson, is out of town on vacation. Greg states that he can schedule Mr. Jones's goat herd for exams in two weeks.

"I really need my meat goat herd seen as soon as possible. Some of the herd is showing signs of nose and eye discharge. Can you set up an appointment for tomorrow?"

"I am sorry Mr. Jones, but Dr. Johnson is on vacation until next week. I would recommend calling another veterinarian to see your goats." Greg then hangs up the phone.

- What was done wrong in this situation?
- What should have been done in this situation?
- What is the importance of having Mr. Jones's meat goat herd seen immediately?

24

Courtesy of Image Source

Poultry Breed Identification and Production Management

Objectives

Upon completion of this chapter, the reader should be able to:

- ☐ Define common veterinary terms used in the poultry industry
- ☐ Describe the biology of poultry
- ☐ Explain egg anatomy, color, and quality
- ☐ Identify common breeds of poultry species
- ☐ Describe the different species and classes of poultry
- ☐ Discuss the nutritional needs of poultry
- ☐ Discuss common health practices and management of poultry
- ☐ Describe common poultry diseases and parasites
- ☐ State the importance of poultry production
- ☐ Explain poultry production systems
- ☐ Discuss poultry reproduction and chick development

Introduction

Poultry species include chickens, ducks, geese, turkeys, and other bird species that may be seen in the wild or have been domesticated for a particular use or product to serve humans (see Figure 24-1). Poultry production needs have increased greatly over the last 10 years, with chicken being the most popular poultry. The average person eats over 75 pounds of chicken per year. Poultry production is both economical and a large part of the veterinary industry.

Delmar/Cengage Learning

FIGURE 24-1 Chicken is the most popular poultry production species.

Veterinary Terminology

Poultry are also called birds or **fowl**. Many terms are used in veterinary medicine that relate specifically to chickens. Many times these terms are used interchangeably with other poultry species. The adult female is known as a **hen** and the adult male is called a **rooster** or **cock**. A young chicken is called a **chick**. A male chicken under a year of age is known as a **cockerel**. A young chicken raised to lay eggs is called a **pullet**. The adult female hen that lays eggs is known as a **layer**. As the layer hens age and begin to no longer produce eggs they become known as **spent hens**. Spent hens are typically used as a meat source in processed foods. Chickens that are used for food purposes have different terms. A young chick used for meat production between 6 and 8 weeks of age is called a **broiler**. They weigh 4 pounds or less and may be male or female. Their meat is tender and easy to cook and may also be known as **fryers**. A larger chick over 4 pounds and usually over 8 weeks of age is a **roaster**. A male chicken that has been castrated and thus will gain more weight, usually over 6 pounds and 6 months of age, is called a **capon**. A group of similar poultry is called a **flock**. Additional terms related to poultry can be found in Table 24-1.

Biology

Poultry have an avian digestive tract. Thus, they are much like caged pet birds. Modern poultry do not have a need to fly and many breeds grow too large to fly. They use their beaks to break food particles into small pieces for ease of digestion. Like other avians, poultry have a **gizzard**, which is a muscular organ used to break down food particles. They also ingest **grit** to help break down hard substances, such as seed shells. The beak can grow sharp and may become overgrown, causing difficulty with eating. Most producers will **de-beak** young chicks to prevent them from pecking and injuring each other. The de-beak process is trimming of the sharp end of the beak with small clippers. The external features of most poultry are similar to other bird species (see Figure 24-2). The heads of poultry have certain head and neck features that help identify the species and breed. Chickens have **combs** on the tops of their heads, which are flesh-like projections (see Figure 24-3). Some combs are complete and some have **serrations.** The serrations are spaces between finger-like sections called points. Some poultry species also have a flesh-like projection under the chin known as a **wattle**. Turkeys have a small amount of hair beneath the wattle known as a **beard**. Some species of geese and ducks have a projection located on the top of the beak called a **knob**.

Males tend to have larger bodies and head features, such as the comb and wattle. The eyes, skin, wattle, and comb are useful in determining the overall health and condition of the bird. Most breeds should have a large, red comb as a sign of good health (see Figure 24-4).

TABLE 24-1

Additional Terms Related to Poultry

Poult	Young turkey
Drake	Adult male duck
Duckling	Young duck
Goose	Adult female goose
Gander	Adult male goose
Gosling	Young goose
Peacock	Adult male peafowl
Peahen	Adult female peafowl

Poultry Species and Classes

Poultry are considered domesticated birds with feathers, have two legs, two wings, and a beak. Poultry species are raised primarily for eggs, meat, feathers, and by-products.

Chickens are the most important and the most popular type of poultry. They are raised for eggs and meat and some breeds are better producers in their respective products. There are four classes of chickens: the American class, English class, Asiatic class, and Mediterranean class.

The **American class** developed in North America and was created based on the need for top egg and meat production requirements. Examples of breeds in the American class include the Rhode Island Red and Plymouth Rock (see Figure 24-5).

The **English class** developed in England as a meat group that is commonly cross bred with other breeds for improved meat quality. An example of the English class is the Cornish.

The **Asiatic class** developed in Asia primarily as a show quality group bred for size and appearance. An example of the Asiatic class is the Cochin (see Figure 24-6).

The **Mediterranean class** developed in the Mediterranean as an egg production group bred for large size and quality eggs.

Chicken production is the most economical program in the poultry industry. Chicken provides humans with a healthy food source in eggs and meat and provides feathers and by-products for numerous other markets. Chicken eggs are useful in research in developing medications and

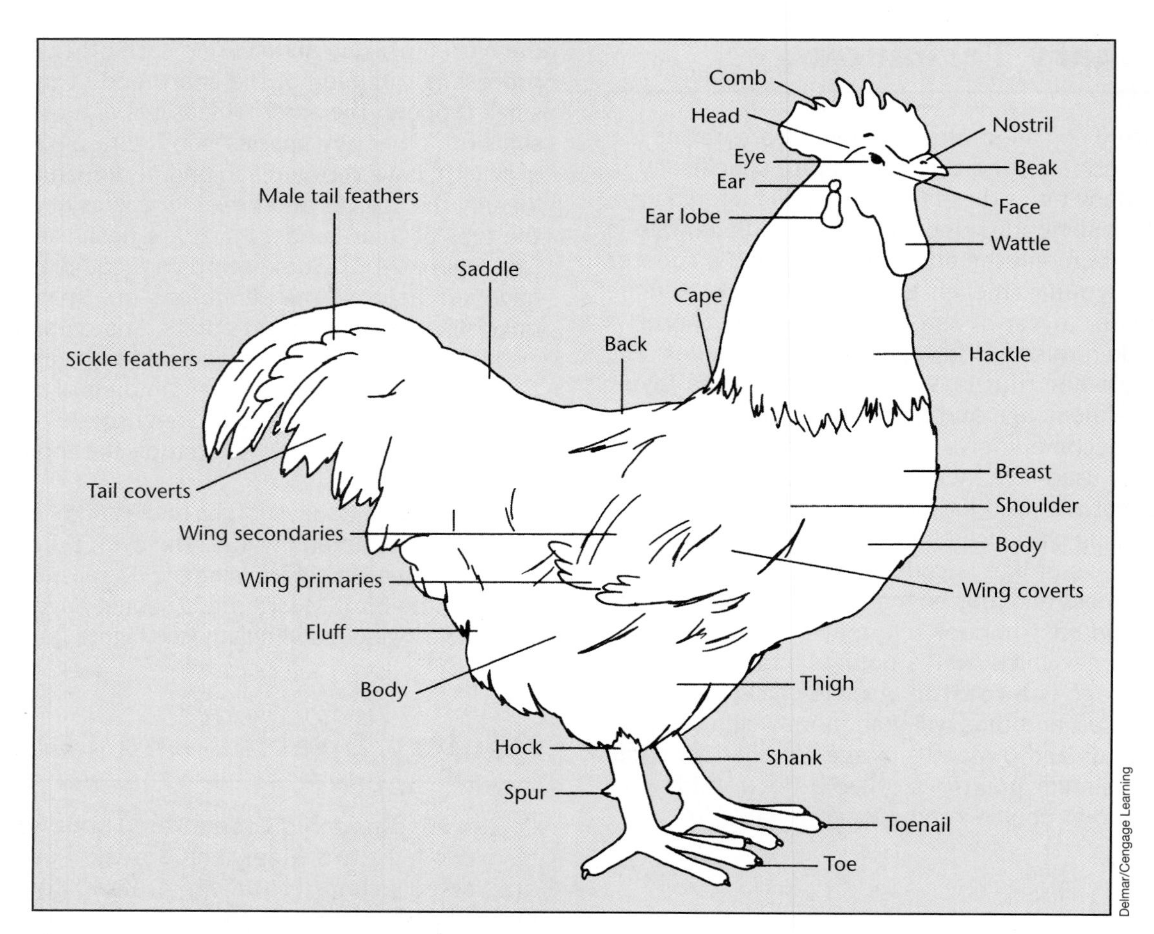

FIGURE 24-2 External anatomy of the chicken.

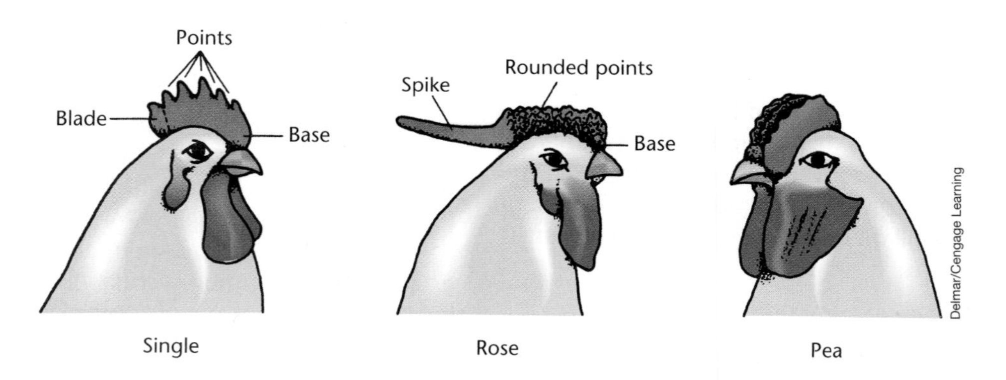

FIGURE 24-3 Comb types found on poultry.

vaccines. Chickens are typically raised in confinement and are under strict regulations and guidelines from the U.S. Department of Agriculture. Raising chickens has also become a popular hobby. Over eight billion chickens are raised each year in the United States for production purposes.

Turkeys

Turkeys are a large poultry breed raised specifically for meat (see Figure 24-7). Producers and consumers want to raise turkeys with a large amount of white meat. Only about 200 million turkeys are raised each year in the United States, which is a decrease in production over

FIGURE 24-4 A large, red comb is a sign of a healthy bird.

FIGURE 24-6 Cochin hen.

(A)
Photo courtesy of Andrea Glover, Glover Eggs, Edgewood, New Mexico

(B)

Photo by Charles Wabeck

FIGURE 24-5 (A) Rhode Island Red (B) White Plymouth Rock.

Photo courtesy of the U.S. Department of Agriculture

FIGURE 24-7 Beltsville Small White Turkey.

the past several years. Most turkey producers are from commercial farms raising over 100,000 birds a year on a single farm. Turkeys are usually raised and marketed around 20 weeks of age, with males being more popular due to an increased size.

Ducks

Ducks are water birds raised for meat, eggs, feathers, and soft down (see Figure 24-8). **Down** is the soft feather covering that grows under the primary feathers of young ducks. Ducks are not raised as commonly as chickens and turkeys. A little over 15 million ducks are raised in the United States each year. Ducks are economical in the fact that they grow faster and heavier than chickens.

FIGURE 24-8 Ducks are raised for eggs, meat, down, and feathers.

FIGURE 24-9 Geese.

Geese

Geese are water birds that are larger in size than ducks and are usually raised for meat, eggs, feathers, and down (see Figure 24-9). A little over one million geese are raised in the United States each year. Geese enjoy eating grass and weeds and are also kept as ornamental pets to keep weeds limited, and they are also known to be good watch animals. They are vocal, can be aggressive, and are resistant to many common poultry diseases. They can also become destructive to crops in large numbers.

Peafowl

Peafowl are poultry raised for their beautiful colors and large feathers. The male peafowl is known for the spread of their tail feathers, called a train (see Figure 24-10). It is not uncommon for the train of the male to grow to five times the length of the body. Most peafowl are raised for ornamental pets and eggs.

FIGURE 24-10 Peacock.

FIGURE 24-11 Swan.

Swans

Swans are water birds similar to ducks and geese. They range in a variety of colors and are known for having a long, thin, and graceful neck (see Figure 24-11). They are raised for ornamental purposes.

Other Poultry Species

Several species of wild birds have been domesticated for poultry production purposes. **Guinea fowl** are raised for meat, eggs, and hunting purposes. They are a small bird that is very hardy. They produce eggs that have a thicker shell than chickens and are often used for dying and decorations as they are not easily broken. **Pigeons** are small birds that are common in the wild and have been raised in captivity for meat and competition purposes. Many people raised carrier pigeons, which are bred and trained to carry small messages or identification bands from one location to another, common in competitive events. **Quail** are another common wild species raised in captivity. They are small, round-bodied birds known for their "bobwhite" whistle sound. They are raised for both meat and eggs. **Pheasant** is also a wild species raised in captivity. They are larger in size and are raised primarily for meat and for hunting purposes (see Figure 24-12).

Ratites

Ratites are a group of large flightless birds that have become popular in commercial poultry production programs. The largest and best recognized ratite is the ostrich. **Ostrich** can weigh over 350 pounds as adults and stand over 10 feet in height (see Figure 24-13). They are known to have a long life span, capable of living over 70 years. Ostrich are raised for meat, eggs, feathers, and by-products, such as oil and skin. Their feathers are called plumes and used in decorations. **Emus** are native to Australia and are smaller than the ostrich, but they have a similar body structure. Another common ratite is the **rhea**, which is native to South America and is the smallest of the three.

© iStock/Stephen Muskie

FIGURE 24-12 Pheasant.

© iStock

FIGURE 24-13 Ostrich.

Selection of Species or Breed

Poultry selection is based on the ability to select a bird that produces the largest number of eggs or meat as quickly as possible. For layers, this means producing one egg every day. Chicks should be in excellent condition and continue to grow to a large size based on breed standards. Meat breeds should reach market weight in 6 weeks, an efficient time to reduce the cost of production. Other factors that should be considered when selecting poultry include how well the bird eats and uses the food efficiently, a fast return on the investment, a high profit compared to the cost of feed, the amount of space needed, and the amount of labor involved.

Nutrition

Adult poultry should receive a commercial feed that is formulated for the species of fowl. The maintenance diet of adult poultry should be a 12–15% protein-based grain diet. Young chicks begin eating starter food that is between 18 and 20% protein and should be placed on a commercial diet as quickly as possible. Poultry will eat off of the ground or in pans and dishes. Most producers use an automated system to prevent contamination or disease in aseptic environments. Automated feeder systems consist of a trough or tray that releases specific amounts of food at certain times (see Figure 24-14). This is done through a conveyer belt system that weighs the feed and distributes a set amount. Water may be provided in a similar manner but must be monitored for appropriate daily consumption. Water troughs should be cleaned on a daily basis to prevent contamination.

Behavior

Poultry are known for their adaptable behavior to any type of environment and changing conditions. The domestic fowl by nature is a wary, shy animal. Poultry have excellent vision and hearing, but other senses tend to be poorly developed. Many poultry species are gregarious in nature and have a social status within the flock. Fowl have specific body postures associated with the head position that allow communication with other birds. They maintain contact with flock mates by sight and vocal communication. They often become

FIGURE 24-14 Automated feeder system.

nervous and more vocal when flock mates get out of sight. A male rooster will establish hens as his territory and will become aggressive and fight with other roosters that may attempt to enter the flock. Poultry will become aggressive with humans and may become territorial and attempt to chase and flop using the wings as weapons.

Basic Training

Poultry species that may be shown can be trained to be handled and restrained rather easily. This should begin when the fowl are young and continue with routine handling. The wings should be extended for observation and examination (see Figure 24-15). The feet and legs should be handled, as well as feathers checked for proper appearance. Poultry will easily adjust to humans with regular handling.

Equipment and Housing Needs

Housing poultry will depend on the type of production and the number of birds within the flock. Housing may be as basic as a cage or shed or an elaborate building that houses 40,000 birds (see Figure 24-16). Housing must meet the needs of the birds as far as temperature

FIGURE 24-15 Poultry can become accustomed to being handled with care.

FIGURE 24-16 Enclosure for a small flock of chickens.

and humidity. Day-old chicks require 1/4 to 1/3 of a square foot of space per bird. Space requirements increase with the growth of the chick. Growers will commonly use wire portable fencing to contain chicks and remove the fence as the birds grow. Floor and cage coverings must be gentle and dust-free. **Litter** is commonly used as bedding and is made of fine wood shavings that absorb the droppings. Proper lighting is necessary for growth and ideal health. Artificial lights are commonly placed on timers to increase growth and egg production. Temperatures should be controlled at 85–95 degrees for young growers and after 6 weeks of age can be reduced to 70–75 degrees. Proper ventilation is required and a relative humidity of 50–75% is ideal. Emergency generators should be available in large commercial programs in case of power failures.

Restraint and Handling

Poultry should be restrained and handled by holding the wings against the bird's body to keep them from flapping, which could injure the handler or the animal. The body should be restrained securely with one arm and the head held around the neck with the other hand (see Figure 24-17). Poultry are like caged birds in that they must breathe using their diaphragm and chest and should never be held tightly around this area. Young chicks often lie quietly in the hand and can be easily examined (see Figure 24-18).

Grooming

Poultry do not typically require grooming but species that are shown should be clean and free of external parasites, dirt, and debris. Feathers can be gently sprayed with water to remove dust and debris. Fowl will preen and clean their wings and feathers many times throughout the day. Some owners may want the nails trimmed, wings clipped to prevent flying, and beaks dremeled due to overgrowth (see Figure 24-19).

Basic Health Care and Maintenance

All birds go through a process of shedding their feathers, known as **molting**. This process allows for new feather growth. During this time, hens will decrease production of eggs or may stop lying all together. The body is depleted of calcium as new feather development and

FIGURE 24-17 Proper methods to restrain poultry.

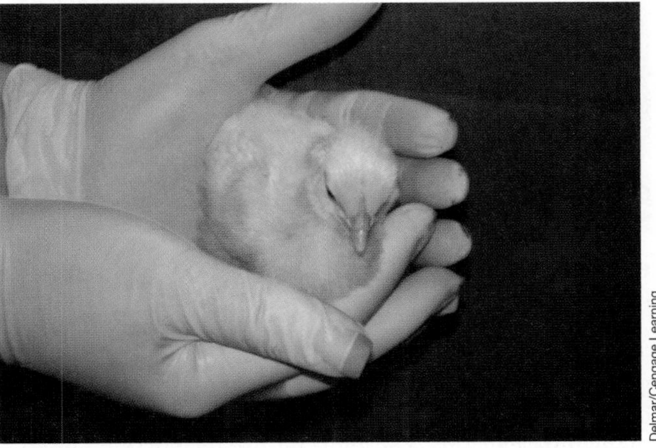

FIGURE 24-18 A young bird will sit quietly in the hand.

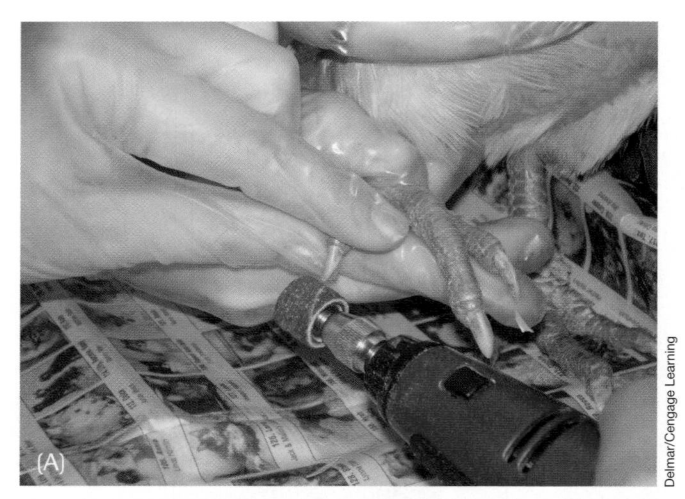

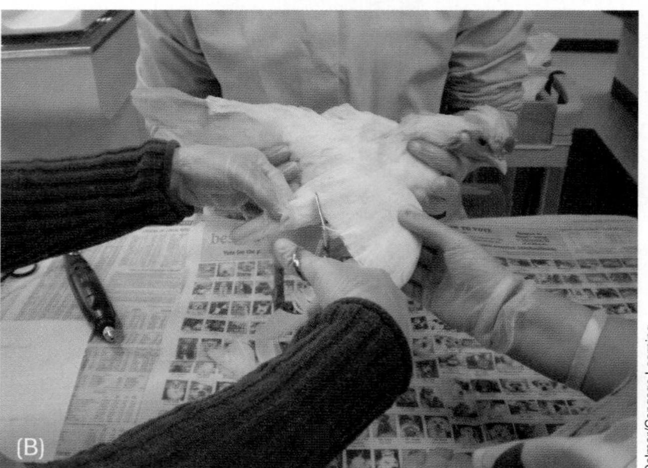

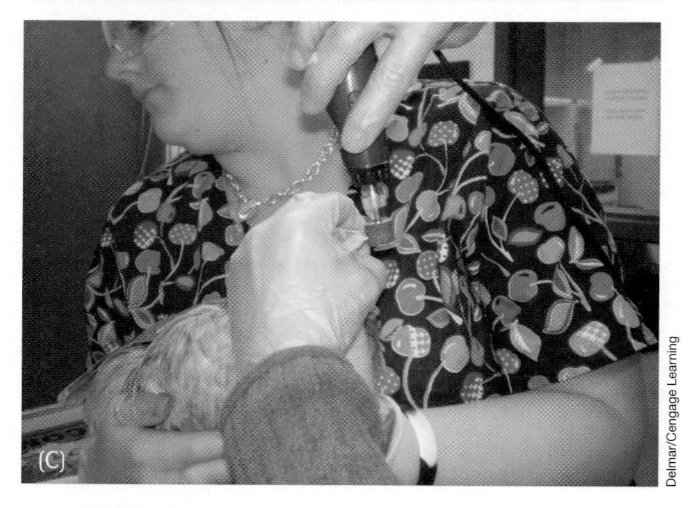

FIGURE 24-19 (A) Trimming nails; (B) clipping wings; (C) filing beak.

growth occurs. Most producers will rest hens during this phase to allow time for the body to build up and replenish its calcium. The molting process usually begins around a year of age and occurs in the fall just prior to winter. Molting occurs over a four-month period until the new feathers develop. Producers may do a forced molting of some hens so that egg production profits do not stop for all layers. The **forced molting** is the process of using lights to trigger the molting phase to begin at a time of the year when it doesn't normally occur. This is done by decreasing the light during the day through the use of artificial lights. Allowing eight hours of light and a longer dark period will trigger the molting phase to begin. After the molting occurs, the light should be returned to 14–16 hours of light.

Sanitation is essential for all poultry systems as one sick bird can kill an entire flock quickly. Good sanitation control methods are required to ensure disease-free flocks. Insect control and removal of waste materials is important. Sanitation control methods include the following:

- Disinfection of cages, equipment, and clothing
- Use of insecticides
- Waste control
- Disposal of dead birds
- Removal of litter
- Limitations on access of people and vehicles

Dead bird disposal follows guidelines and regulations due to the potential for human contact and the spread of disease in wild bird species. Any dead birds should be removed immediately. The best control method of disease is through incineration. This is the process of burning the **carcass** (dead animal's body) to ashes. Other methods of disposing of a dead bird include **composting**, which is the process of layering the carcass with waste materials and bedding to allow decomposition or wearing away. **Burial** is a method of placing the carcass in the ground and covering.

Vaccinations

Poultry should be on a regular vaccination program that helps to develop immunity to specific diseases (see Figure 24-20). **Fertile** eggs can be vaccinated using the in vivo method where the embryo is injected with a vaccine during the incubation phase. This is usually done on the 18th day of the incubation. Chicks will hatch with immunity, and this decreases the stress of regular vaccination and allows for a healthier chick that is ready for market two days earlier than conventional vaccination. A specialized egg injection system is used to vaccinate eggs at a high rate of speed through the use of a needle that punctures the egg shell and delivers a set dose of vaccine. This machine is capable of

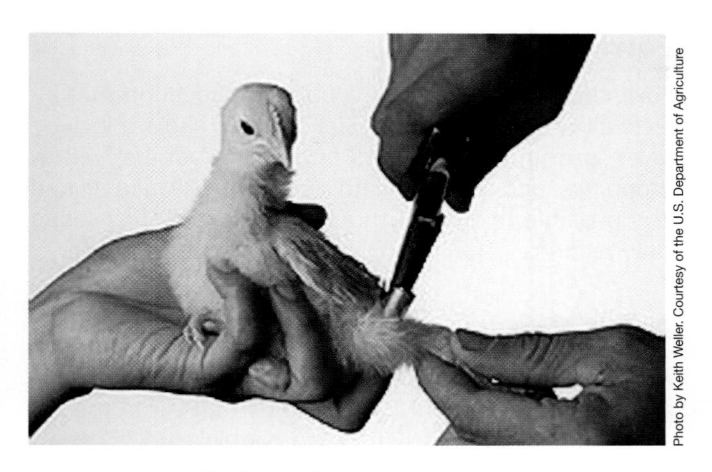

Photo by Keith Weller. Courtesy of the U.S. Department of Agriculture

FIGURE 24-20 Chick receiving vaccine.

Delmar/Cengage Learning

FIGURE 24-21 Eggs in an incubator.

injecting 20,000–30,000 eggs per hour. If the in vivo method is not used, vaccines should be administered on day one of the hatch at the time of the de-beak procedure. The vaccine can be administered into a wing, within the eye membrane, or via a nostril. The program should vaccinate for common diseases and should have all vaccines and boosters completed by 12 weeks of age. Breeders typically vaccinate every three months or as necessary.

Reproduction and Breeding

Female hens lay eggs that are sterile if no male is present and fertile eggs if a male rooster is available within the flock. The roosters and hens must mate for **fertilization** of eggs to occur. The eggs may be incubated or left on the nests to be hatched out. The **incubation period** is different from species to species and occasionally by breed and egg size (see Table 24-2). The hatching also depends on the temperature and humidity of the environment. Most producers will hatch out eggs in an **incubator**, which is a piece of equipment with controlled temperature and humidity settings (see Figure 24-21).

TABLE 24-2

Incubation Lengths of Poultry

Chickens	Average of 21 days
Ducks	28–35 days
Geese	29–31 days
Turkeys	27–28 days

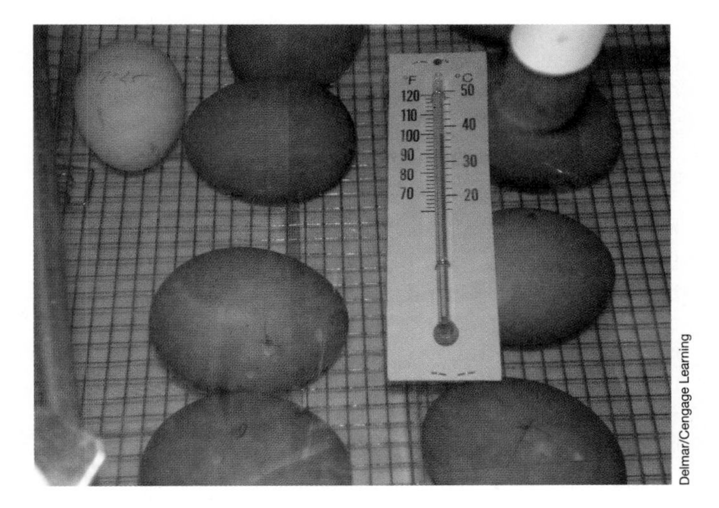

Delmar/Cengage Learning

FIGURE 24-22 Temperature must be monitored so that the eggs will hatch.

Hens do not need to mate for eggs to be produced. There is no difference in taste or quality to the consumer between a fertile and sterile egg. Fertile eggs must be in a controlled temperature environment for eggs to hatch. Fertile eggs placed in a refrigerator will not produce an embryo. Artificial insemination is not used in chicken production, but it is commonly used in turkey production systems. Turkeys commonly have difficulty breeding due to their size.

The incubation of chicks varies between species. Most incubation temperatures are between 99 and 100 degrees Fahrenheit (see Figure 24-22). The humidity should remain between 82 and 84%. Ideally, eggs should be rotated daily for uniform temperature and development.

Common Diseases

Poultry are susceptible to 33 different disease pathogens. They are highly susceptible to viruses and bacterial infections and the most common problems occur from nutritional problems. Poultry are costly to treat, especially when large flocks are affected. When fowl become sick, many will die. When fowl become sick they become less efficient and decrease or stop egg production or are not able to be used for human consumption. This causes a loss in profits.

Marek's Disease

One disease that affects poultry is called **Marek's disease**, also known as **range paralysis**. This condition is caused by a **herpes virus** that causes diarrhea, weight loss, leg or wing paralysis, and death. There is currently no treatment, but a vaccine is available as a prevention of the disease. Some poultry species are genetically predisposed and have a poor resistance to the virus.

Newcastle Disease

Newcastle disease is caused by a virus that affects the respiratory system causing wheezing, gasping for breath, open-mouthed breathing, possible paralysis, soft-shelled eggs, and possibly no egg production. One notable sign with the disease is birds will attempt to twist their necks around due to respiratory distress. A vaccine is available for prevention, but no treatment has been found.

Infectious Bronchitis

Infectious bronchitis is a respiratory virus that affects only chickens; no other species of poultry have been affected. Mostly young birds are affected but older birds may become infected and decrease or stop egg production. A vaccine is available and it can be prevented through proper sanitation and isolation of sick birds. Signs of infectious bronchitis include the following:

- Bronchitis
- Wheezing
- Nasal discharge
- Gasping for air
- Poor appetite
- Ruffled feathers
- Depression

Fowl Cholera

Fowl cholera is caused by a bacterial infection that infects all species of poultry. Signs of fowl cholera include fever, purple coloration to heads and combs, yellow droppings, and sudden death. A vaccine is available and it is possible to treat with a class of antibiotics called sulfonamides.

Avian Pox

Avian pox is a virus spread by mosquitoes and is a slow-spreading disease that has several strains; it is species specific. This virus is difficult to control and no treatment is available. Vaccination is the key to prevention. Signs of the virus include respiratory distress and wart-like growths on the skin and beak.

Avian Influenza

The most recent and worrisome disease affecting the poultry industry is **avian influenza**, commonly called **bird flu**. This virus affects all breeds of poultry and occurs from a naturally occurring virus within the intestinal tract. It is transmitted directly through body fluids, such as nasal discharge or droppings. It is zoonotic to humans, but no human to human contact transmission has been recorded. There are several strains, including a highly fatal strain. So far the United States has not had any fatal outbreaks. There is huge concern with this virus as studies show that it is capable or mutating or changing forms to survive. There has been a human vaccine developed, but currently there are no avian vaccines or treatments. Isolation of all sick birds is key.

Common Parasites

Poultry are susceptible to 10 parasites. Internal and external parasites are common in poultry. Isolated cages, confinement methods, and specialized production systems help to reduce the possibility of infection. Proper sanitation controls are the key to prevention. Control of insects and wild birds is helpful.

Coccidiosis

Coccidiosis or coccidia is the most common internal parasite that affects all birds. Coccidiosis is caused by a **protozoan** (single-celled) parasite. This is commonly seen in the droppings of wild birds and is spread by feces in the soil, food, or water sources. Companion animals, livestock, and people may easily be infected from contaminated sources. Treatments are through antibiotics and dewormers given by mouth or in a water source. Signs of coccidiosis include the following:

- Lethargy or depressed appearance
- Bloody droppings
- Pale skin
- Weight loss
- Loss of appetite
- Ruffled feathers

Large Roundworms

Large roundworms may occur in the intestinal tract and can reach up to 3 inches in length. Signs of roundworm infestation include diarrhea, weight loss, **emaciated** (severely dehydrated and appearing very thin) appearance, and droopy wings. Prevention and treatment is through sanitation and dewormers.

Tapeworms

Tapeworms are more common in poultry with access to the outdoors as most birds are infected through ingesting hosts including snails, earthworms, beetles, and flies. Prevention is through diet control, dewormers, and proper sanitation. Signs of tapeworms include pale skin over the body, head, and legs.

Mites

Mites are the most common external parasite affecting poultry. Lice, ticks, and chiggers may also occur. The key to prevention is through the use of approved insecticides and limited access to the outdoors and wild birds. Signs of external parasites include visible parasites, droopy wings, listless appearance, itching, feather loss or damage, and pale skin color.

Poultry Production

Poultry production began over 5,000 years ago in Asia. The practice began when poultry was introduced to America in the 1600s. Families began raising small flocks of chickens for both eggs and meat sources. Turkeys were later introduced by the Aztec Indians and settlers began raising and domesticating wild turkeys. In the 1950s poultry production became a large industry in the United States. The confinement method of production began with standards set by the government. Currently, California is the largest state poultry egg producer. Due to the popularity of poultry

FIGURE 24-23 Poultry science courses are offered focusing on the study of breeding, incubation, raising, housing, and marketing.

and egg production, many agricultural and animal science college programs offer a course in **poultry science**. This is the study of poultry and includes breeding, incubation, raising, housing, and marketing (see Figure 24-23). The goal of poultry science is to learn to produce a quality product at a reasonable price. Most poultry houses hold 40,000 birds. One hen can produce over 300 eggs a year. Over the last 50 years, poultry science has developed methods for increasing production rates of commercial fowl.

Poultry uses have gone beyond the conventional food and fiber sources. Eggs are currently being used in the research and development of vaccines for veterinary and human medicine. Many types of human and animal medications are now egg based.

Poultry Production Systems

There are three main production systems in poultry production, mostly relating to chickens. Poultry production systems are used in raising poultry, eggs, meat, feathers, or other products for human use.

Broiler Production

The first system type is called **broiler production.** The goal of this system is to produce the largest amount of meat as quickly as possible. Chicks are raised to 6 weeks of age and should weigh around 4 to 4.5 pounds in size. They are raised in large poultry houses in confinement to prevent disease and contamination (see Figure 24-24). They must stay healthy to continue adequate growth. The chicks should grow fast in an efficient amount of time to reduce costs. The mortality (death rate) should be kept at less than 5%. The larger the number of livestock and production animals that are produced, the larger the death rate.

Broilers should grow at a rate of 1 pound weight gain for every 2 pounds of food. Young newborn chicks are fed starter food, which is finely ground grain meal. Starter food should be between 18 and 20% protein for ideal growth. As the chicks age, the protein amount should decrease to 12–15%. Most starter food is soybean meal with fish or a meat-meal base. As they age, they can begin eating a corn and grain-based diet with added vitamins and minerals for continued growth.

Egg Production System

The second type of production method is the **egg production system.** The goal for egg production is to produce a high-quality egg for human consumption. Each hen should produce one egg per day. The eggs are not fertile and are not hatched or incubated. They are graded by their size and quality. The grades range from small to jumbo. They are placed in egg cartons with the small end down and the large end up to protect the air cell of the egg, which helps to protect the quality of the egg. Ninety percent of all egg production systems use layers kept in specialized cages with nests that have slatted trays that collect the eggs as they are laid. The eggs may be collected at regular times or through the use of an automatic conveyer belt (see Figure 24-25). The facilities are kept sanitary to protect the eggs from contamination and damage.

Photo by Joe Valbuena. Courtesy of USDA

FIGURE 24-24 Broiler chicks in a commercial broiler operation.

Photo courtesy of Chore-Time Poultry Production Systems, Milford, Indiana

FIGURE 24-25 Eggs roll to the front of the layer cages and are moved on an automated system to the packing house.

Egg Anatomy

The parts of an egg are important to egg production and in determining the quality of an egg (see Figure 24-26). The **eggshell** is the outer layer of the egg that is made of calcium. The inner yellow membrane is the **yolk**, which is a cell created by the hen's ovary. The **albumin** is the egg white layer that surrounds the yolk. The **air cell** is the empty area located at the large end of an egg where oxygen is stored. Within the egg yolk is a **germinal disc**, which is a white spot within the egg where the sperm enters the cell to allow fertilization to occur. The egg also has a **chalaza**, which is the part of the egg that anchors the yolk into the egg white. The egg parts are held together by two membranes; one is the **inner shell membrane**, or the thin area located outside of the albumin, and the other is the **outer shell membrane**, or the thin area just inside of the shell.

Egg Quality

The **egg quality** is determined both by the internal and external appearance of the egg (see Figure 24-27). The size, shape, color, and condition are important factors to egg production. Cracks, blemishes, and dirt are signs of a poor-quality egg. The shell should be of good quality and consistency, as it is made of calcium and what protects the egg. Ninety-five percent of eggs sold in markets are white; however, the demand for brown eggs is

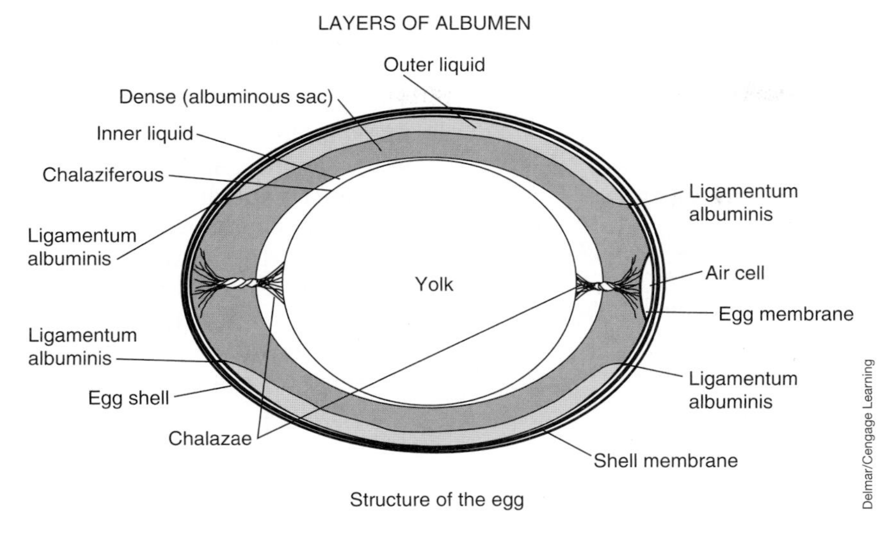

FIGURE 24-26 Parts of an egg.

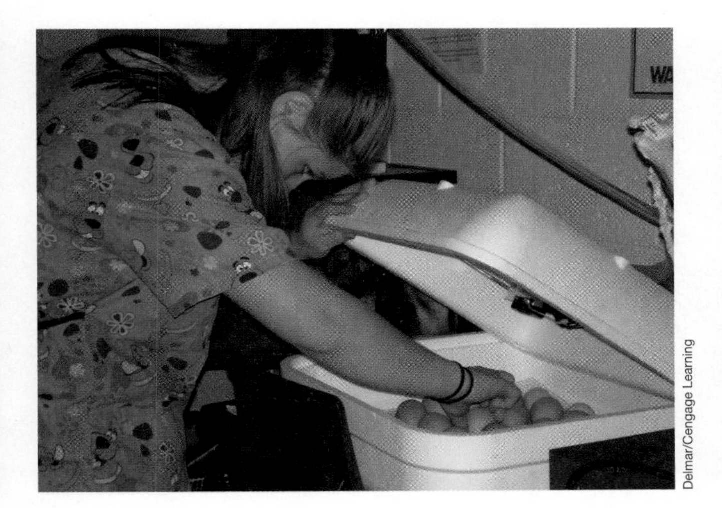

FIGURE 24-27 Eggs should be examined for cracks and blemishes.

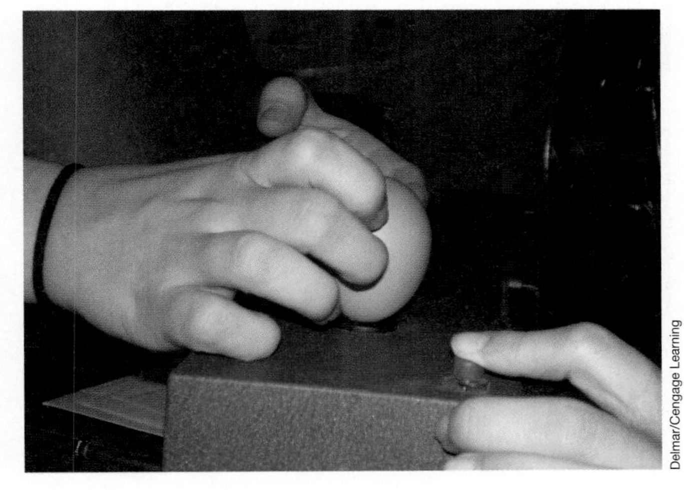

FIGURE 24-28 Device used for candling eggs to inspect structure.

increasing as more chicken breeds produce brown eggs than those producing white eggs. There is no difference in taste or anatomy.

The internal structures of the egg can be viewed through a process called **candling**. Candling is done by shining a light source through the air cell to visualize the inside egg parts (see Figure 24-28). Fresh eggs will have small air cells. Older eggs will have a larger air cell. The yolk should appear nice and yellow due to high amounts of **carotene**, a protein that promotes the health of organs, such as the human eye. The albumin should cling near the egg yolk when rotated. If the albumin moves or spreads out with movement, that is an indication of age and deterioration. Occasionally, fertilized eggs may show a blood cell in the yolk that is a sign of embryo development. Eggs that are transparent or show signs of dark spots are damaged eggs.

Pullet Production

The next type of production system is the **pullet production**. Pullet hens are specially raised for fertile egg production. They are selected as day-old chicks and raised similar to broilers. Only the females are selected. They are selected for maximum egg production. The pullets are raised for 20 weeks and then placed in breeding systems. The hens usually begin laying eggs around 24 weeks of age and reach maximum production rates between 30 and 35 weeks of age. Pullets are also called **replacement hens**. This is usually a contracted commercial industry that raises breeding poultry. One cockerel is placed per 8–10 pullets and raised together for future breeding. As they mature they may be placed in a larger flock.

Broiler Egg Production System

The last type of production is the **broiler egg production system**. Pullets and cockerels are raised together specifically to produce broilers. The females should have quality eggs and large production numbers and the males should be large in size. The chicks are hatched out and selected for broiler production systems.

SUMMARY

The poultry industry has become one of the most marketable and economical animal production systems in the United States. Chicken and turkey production are the most popular species of poultry. Poultry science programs have developed due to the popularity and need for further development into the industry. Meat, eggs, feathers, and numerous by-products of poultry have become popular due to the health benefits that these products have for humans. Poultry species include chickens, turkeys, ducks, geese, and ratites.

Key Terms

air cell empty area located at the large end of an egg that provides oxygen

albumin egg white that surrounds the yolk

American class chicken class developed in North America for egg and meat production

Asiatic class chicken class developed in Asia for show and ornamental breeds

avian influenza commonly called bird flu and occurs as a naturally occurring virus within the intestinal tract; affects poultry and humans

avian pox virus of poultry spread by mosquitoes; lowers the immune system slowly

beard hairs located under the wattle in turkeys

bird flu influenza virus found in avian species that may be zoonotic to humans

broiler young chick between 6 and 8 weeks of age and under 4 pounds used for meat

broiler egg production system system where chicks are raised for large production numbers and selected for broiler programs

broiler production system used to produce the most amount of meat as possible

Burial process of placing a carcass in the ground and covering to allow decomposition

candling technique of using a light to view the internal structures of an egg

capon male castrated chicken over 6 pounds and 6 months of age

carcass dead body of an animal

carotene protein that promotes the health of organs

chalaza part of the egg that anchors the yolk to the egg white

chick young chicken

chicken most important type of poultry and largest industry with four different classes

coccidiosis internal protozoan parasite of poultry and wild birds that is spread through water, food, and soil contamination

cock adult male chicken

cockerel male chicken under a year of age

comb flesh-like projection on the top of the head

composting process of disposing of dead birds by layering the carcass with waste materials and bedding to allow decomposition or wearing away

de-beak procedure to remove the tip of a chicks beak to prevent pecking damage to other chicks

down soft feather covering that grows under primary feathers

drake adult male duck

duckling young ducks

ducks small water birds used for meat, feathers, and down

egg production system used to produce a quality egg for human consumption

eggshell outer calcium layer that protects an egg

egg quality the size, shape, and color of the egg determined by the internal and external appearance of the egg

emaciated severe weight loss and dehydration

emu large flightless bird native to Australia; smaller than the ostrich

English class chicken class developed in England for better meat production

fertile egg with an embryo present due to a male fertilizing the egg

fertilization uniting of the egg and sperm to create an embryo

flock a group of poultry

forced molting using a light source to cause feathers to shed and re grow at specific times of the year

fowl another term for poultry

fowl cholera bacterial infection of poultry that causes a purple coloration to the head

fryer another term for broiler since they have tender meat

gander adult male goose

geese large water birds used for meat, eggs, feathers, and down

germinal disc white spot in the egg yolk where the sperm enters the egg cell to allow fertilization

gizzard muscular organ used to break down food to allow digestion

goose adult female goose

gosling young goose

grit substance used to break down hard materials, such as seed shells, to allow food to digest

guinea fowl small wild birds domesticated for meat, eggs, and hunting

hen adult female chicken

herpes virus virus that causes diarrhea, weight loss, wing or leg paralysis, and death in bird species

incubation period length of time until eggs hatch

incubator equipment used to house eggs at a controlled temperature and humidity

infectious bronchitis respiratory virus that affects only chickens and affects egg production

inner shell membrane thin area located outside of the albumin

knob projection on the top of the beak

layer adult female hen that lays eggs

litter wood shavings used as bedding to absorb droppings

Marek's disease poultry disease caused by a herpes virus also known as range paralysis due to causing wing or leg paralysis

Mediterranean class chicken class developed in the Mediterranean for large quantities of egg producers

molting process of shedding feathers to allow new feather growth

Newcastle disease respiratory virus that causes poultry to twist their necks

ostrich largest and most common ratite that grows over 10 feet tall

outer shell membrane thin area just inside of the shell

peacock adult male pea fowl

peafowl poultry rose for colorful, large feathers

peahen adult female pea fowl

Pheasant wild species of game bird raised for meat and hunting purposes

pigeons small wild birds raised in captivity for meat and competition purposes

poult young turkey

poultry domesticated birds with feather, two legs, two wings, and used as a source of eggs, meat, feathers, and by-products

poultry science study of poultry breeding, incubation, production, and management to learn to produce a quality product at a reasonable price

protozoan single celled organism or parasite.

pullet young chicken raised to lay eggs

pullet production system where hens are raised for fertile egg production

quail small wild bird that makes the whistle sound "bobwhite" and is raised for eggs

range paralysis common name for Marek's disease caused by a herpes virus

ratite large flightless birds raised for poultry production purposes

replacement hens pullets used to replace older hens that are no longer capable of egg production

rhea moderate-sized flightless bird native to South America

(*Continues*)

roaster chick over 8 weeks of age and 4 pounds used as meat

rooster adult male chicken

serration spaces on the top of a comb

spent hen adult female hen that has aged and decreased or stops egg production and is used as meat sources in processed foods

swan large water birds with long, thin necks raised as ornamental poultry

Turkeys large size poultry breed with large body and long, thin neck, specifically raised for meat

wattle flesh-like area located under the chin

yolk inner yellow cell membrane created by the hen's ovary

REVIEW QUESTIONS

1. What species of birds are considered poultry?
2. What are the classes of chickens and where did they develop? What were they developed for?
3. What are some purpose and uses of poultry?
4. What are the four types of poultry production systems? Explain each type.
5. What is poultry science? What is its importance to the poultry industry?
6. What are the parts of an egg?
7. Explain the difference between normal molting and forced molting.
8. What methods can be used to dispose of dead birds?
9. What is candling? In what ways can it be useful in determining egg quality?
10. What are the needs of eggs during incubation?

Clinical Situation

Dr. Ross, a poultry veterinarian, and Harry, a veterinary assistant, have just arrived at a poultry egg production facility. Today is a monthly scheduled health inspection visit for the several thousand chickens housed on the property.

Dr. Ross and Harry begin applying sterile uniforms and face masks to enter the facility. They are in route to entering the first poultry barn when Harry realizes he has not placed his sterile booties over his shoes. He tells Dr. Ross of this mistake and returns to the veterinary truck to obtain sterile booties.

Courtesy of Chore-Time Poultry Production Systems, Milford, Indiana

- What was done correctly in this situation?
- What was done incorrectly in this situation?
- What should have been done at the beginning of the arrival to the facility?

25

Aquaculture Identification and Production Management

Objectives

Upon completion of this chapter, the reader should be able to:

- Define veterinary terms relating to aquatics
- Discuss aquatic biology
- Identify common species of aquatic animals
- Discuss the nutritional needs of aquatics
- Explain the importance of aquaculture
- Discuss the reproductive methods of aquatics
- Describe common diseases and conditions of aquatics
- Discuss and define aquaculture production systems
- Discuss and detail aquaculture management
- Describe water quality

Introduction

Aquatic animals are decreasing in number in their natural environments as water sources are drying up or becoming polluted. Aquaculture has become an industry that provides the resources for such animals and also allows a market for the species of aquatics used for human consumption. Aquaculture is a large industry for major food sources in the United States. Aquatic animals include such species as fish, shrimp, alligators, and lobsters.

Aquaculture Industry

Aquaculture is the production of aquatic plants, animals, and other species of living organisms that grow and live in or around a water source. Most aquaculture in the United States involves fish and **crustacean** (animals with a bony outer shell) species. Each species of aquatic organism is different and requires specific needs and management. Water is necessary for proper living conditions. The products are thought of as crops that are being harvested and marketed; thus, they are referred to as **aquacrops**. An aquacrop is a commercially produced plant or animal species marketed for income and profit. They are produced as food, sport, recreation, and ornamental pets. The climate, water, and environmental conditions are important factors in aquaculture, and several species are more adapted to grow and live in certain areas. The market is also an important consideration when raising an aquatic species, as the demand will control the profitability. Species in high demand in the United States include shrimp, oysters, lobster, salmon, trout, and catfish.

TERMINOLOGY TIP

The term *aquatic* means pertaining to water.

General requirements must be met when considering an aquaculture program. A market must be available and a method to transport products to consumers must be developed. A suitable species must be selected and be adaptable to the environment. The water quality and type must be appropriate for the species. The nutritional needs of the species must be available. Sanitation and disease control must be developed. The water and environmental temperature must be managed for proper growth and health.

Aquaculture requires proper **production intensity**. Production intensity is the assurance that the size of the body of water will hold the appropriate amount of food sources and be large enough for a high production level of products. This will also factor into the **intensity biomass** or the number of aquatic species in a volume of water. Too many or too few fish in a body of water will affect production. Some aquatic species may overpopulate a body of water or land and not be able to survive properly. For example, in fish production, every 6,000 to 8,000 fish should have one acre of water for proper survival. Reproduction will be occurring and this must factor into the size of the water source. In this example, the fish would be capable of producing 8,000 pounds of fish per acre of water per year.

Veterinary Terminology

- Broodstock—adult fish retained for spawning
- Clutch—number of eggs incubated at one time
- Cultured fish—farm-raised fish or shellfish
- Fingerling—stage of life characterized by 1 inch in length to length at one year of age
- Fry—stage of life from hatching to 1 inch in length
- Roe—eggs of fish
- Spawn—mass of eggs deposited in water

Biology

Fish are **vertebrates**, which means they have a backbone. They must live in water to survive and have such characteristics as a fragile bony skeleton, a body covered in **scales**, **gills** that act as lungs for breathing, and **fins** that aid in swimming. The scales are individual segments that act as the skin, but are waterproof. Some fish have thin skin rather than scales. The gills act as lungs and as water passes over them, oxygen is filtered from the water to allow breathing, and then dissolves into the water (see Figure 25-1).

The fins act as arms and legs and allow movement and balance in the water; some fins have sharp ends that act as protection from predators. Fish have a streamlined body structure that helps propel them through water. Fish have a variety of colors and markings, mostly to help them blend into their surroundings. Fish have three segments that make up their body: **head**, **trunk,** and **tail**. The head contains the eyes, mouth, and gills. An **operculum**, or gill cover, divides the head from the trunk. The trunk contains the body, fins, and anus. The tail contains the **caudal fin**, which acts as a propeller to move the fish through the water. Figure 25-2 illustrates the external anatomy of the fish.

The internal anatomy of fish is somewhat like other animals, but some organ systems have more developed features for aquatic life (see Figure 25-3). A fish's body is made of soft bones and cartilage that

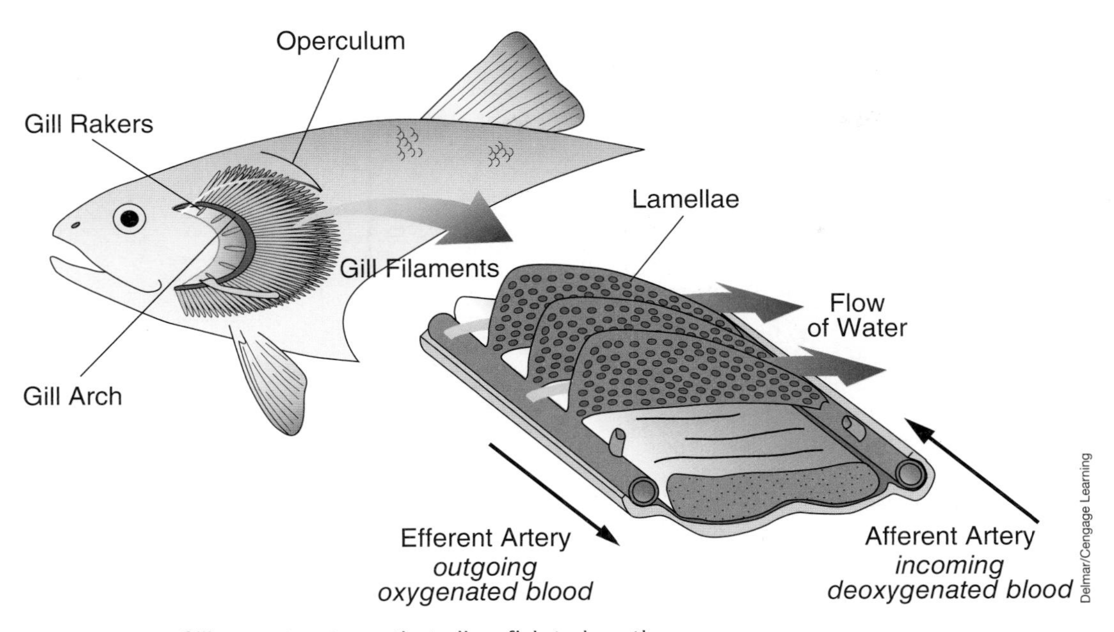

FIGURE 25-1 Gills are structures that allow fish to breathe.

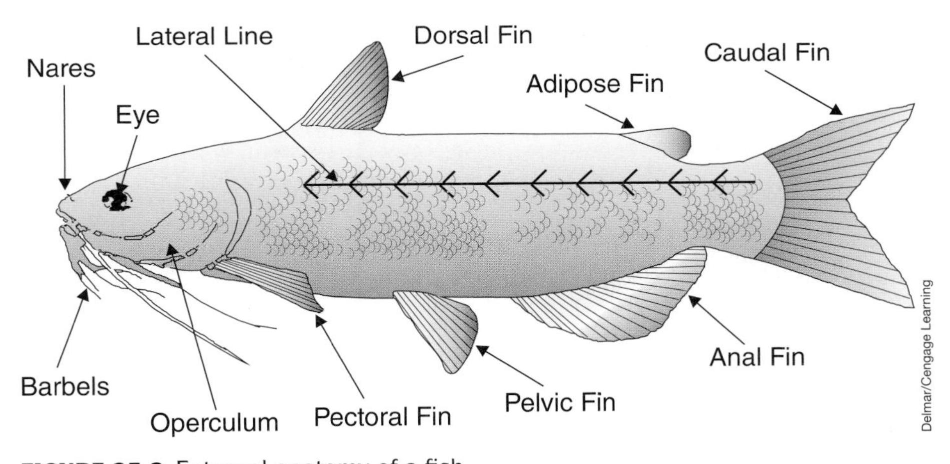

FIGURE 25-2 External anatomy of a fish.

serve as protection in the water and maintain the body structure. The muscular system is the largest body system in fish, with the largest muscles located in the tail. Fish are adapted with **barbels** or **barbs**, which act as sensory devices to detect food, danger, and movement. Fish have a monogastric digestive system. Female fish may lay eggs or give birth to live young depending on species. The most unusual and important feature of fish is the **swim bladder**. This structure acts like a small balloon that inflates and gives the fish buoyancy to float and balance in all levels of water. Without a swim bladder, a fish would not be capable of moving through deep water.

Crustaceans are an aquatic species that have a shelled body with legs. Crustaceans are considered **arthropods** as they have an **exoskeleton** made of cartilage. Arthropods also have a segmented body and jointed legs. Insects are also in this class of organisms. **Decapods** are crustaceans with five pairs of legs. Both arthropods and decapods have a body structure consisting of a head, **carapace** or thorax, an abdomen, and limbs. They also have a pair of **pincers** located on the

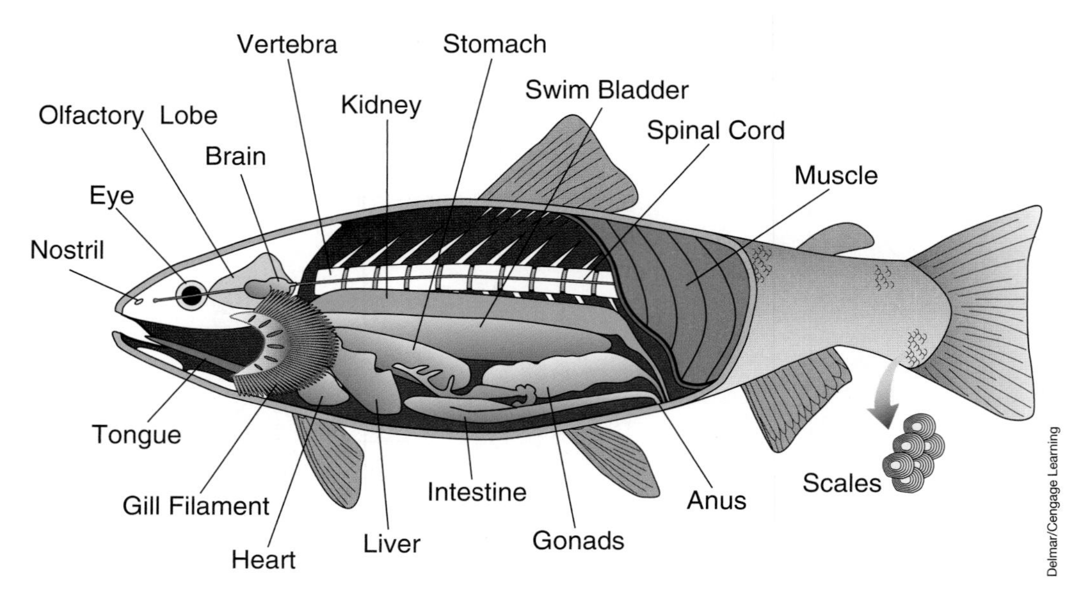

FIGURE 25-3 Internal anatomy of a fish.

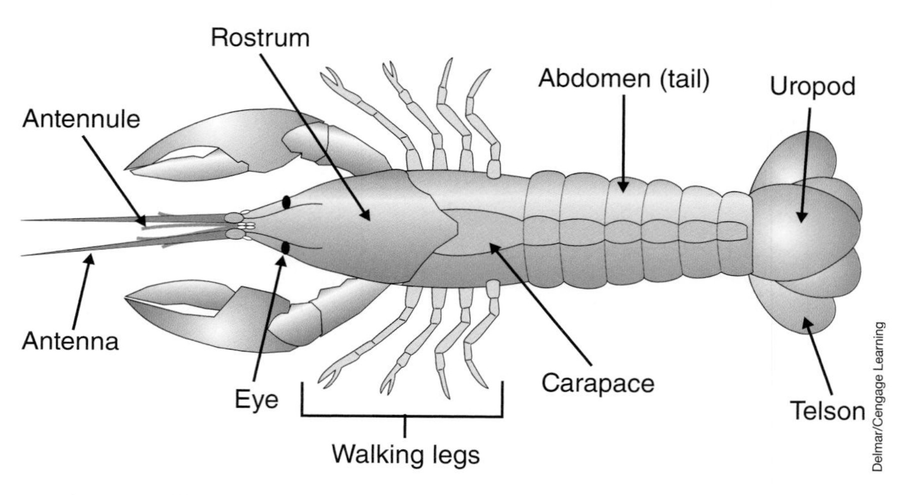

FIGURE 25-4 External anatomy of a crustacean.

head for protection, smell, and feeling. Figure 25-4 and Figure 25-5 illustrate the internal and external anatomy of crustaceans.

Mollusks are aquatic species that have only a thick, hard shell. They have no limbs but instead have **valves** like gills for breathing. Their shells are hinged to open and feed and breathe. Some types of mollusks are **bivalves**, meaning they have two shells. Bivalves have a strong **abductor muscle** that allows the shell to open and close. They are filter feeders and remove particles from the water to feed. Other mollusks are **univalves**, meaning they only have a partial shell. One example of a mollusk is an oyster; oysters are farmed for their popularity in pearl production, which develop naturally from a single grain of sand. The internal anatomy of the mollusk is illustrated in Figure 25-6. Table 25-1 lists examples of aquatic species.

TABLE 25-1

Examples of Aquatics

Crab	Crustacean-arthropod
Shrimp	Crustacean-decapod
Lobster	Crustacean-decapod
Oysters	Mollusk-bivalve
Clam	Mollusk-bivalve
Mussels	Mollusk-bivalve
Abalone	Mollusk-univalve
Snail	Mollusk-univalve
Bass	Fish
Tilapia	Fish
Salmon	Fish

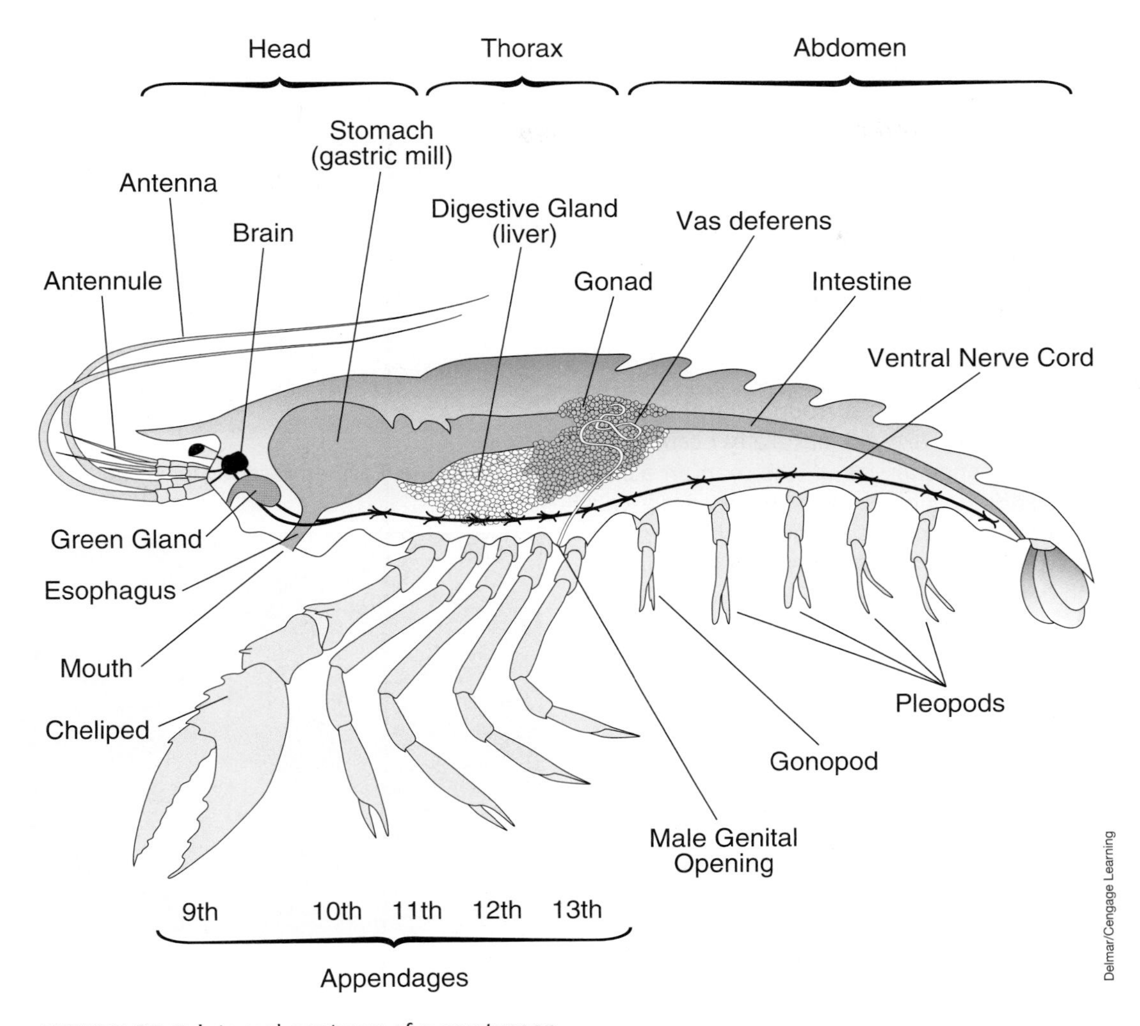

FIGURE 25-5 Internal anatomy of a crustacean.

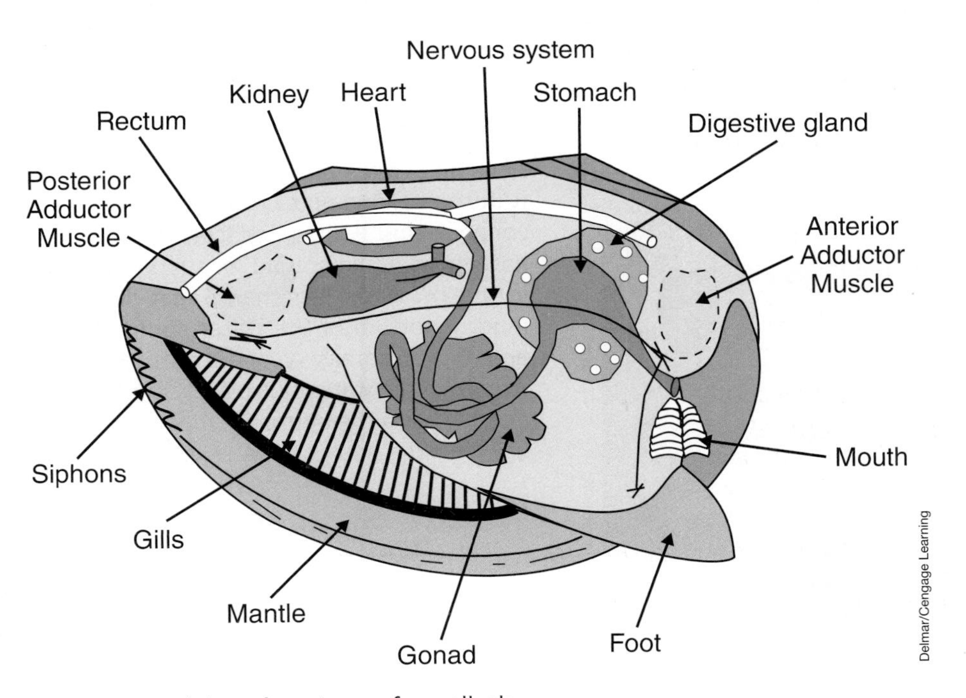

FIGURE 25-6 Internal anatomy of a mollusk.

Aquaculture Production Systems

Aquaculture is based on the need for a **water facility**. Examples of water facilities include ponds, rivers, lakes, streams, tanks, vats, and aquariums. These facilities are developed to resemble the natural environment and accommodate each specific type of organism.

Ponds

Ponds are a contained water area surrounded by human-built dams or levees; they are not naturally occurring bodies of water. They range in size of less than an acre up to 50 acres. Ponds do not drain or have an active moving water source, so management of water is important. Ponds are typically rectangular in shape and range from 3–5 feet in depth (see Figure 25-7).

Raceways

Raceways are long and narrow water structures with flowing water. Raceways are common in fish hatcheries for breeding purposes and are used to increase the number of fish that may be stocked in other bodies of water during fishing seasons. They may be built of different types of materials and are usually built in naturally occurring waterways. They have the advantage of holding a larger number of fish as flowing water creates more oxygen. Pumps may be needed to correct the water flow, and electricity costs can be expensive (see Figure 25-8).

Tanks, Vats, and Aquariums

Tanks, vats, and aquariums are smaller bodies of water that may be used to house aquatic species. **Tanks** are large, round areas (see Figure 25-9). Vats are similar in shape but smaller in size. **Aquariums** are much smaller in size and typically used to raise ornamental fish. All three may be made of glass, wood, fiberglass, metal, or concrete. They require proper water flow and oxygenation.

FIGURE 25-7 Crawfish ponds.

FIGURE 25-8 Earthen raceway for trout.

FIGURE 25-9 Tanks for raising catfish.

Pens and Cages

Pens and **cages** are small structures placed in bodies of water to confine aquacrops. A pen is a structure used to connect land with water (see Figure 25-10). They may have posts in the ground for support and are usually kept in shallow water less than 3 feet deep. They rest in the muddy bottom of the water source to provide nesting and breeding areas. A cage is a structure that floats on the surface of the water. It is usually made of net material and doesn't touch the bottom of the water bed. It is commonly used in large bodies of water to confine aquatic species.

Water Sources

There are three types of water sources used in aquaculture production. The first is **freshwater**. Freshwater has little to no salt content and is the most

FIGURE 25-10 Ocean net pens.

common water source in North America. Some examples of freshwater sources include wells, streams, and surface runoff areas. Another water source is **saltwater**. Water is measured by its **salinity**, or salt content. Salt content is measured in **ppt** or parts per thousands; saltwater is 16.5 ppt or greater. Examples of saltwater sources are oceans and seas. The last type of water source is called **brackish water**. This is a mixture of freshwater and saltwater. These sources are found where streams enter larger bodies of water such as oceans and seas. The salt content is higher than freshwater but less than saltwater. Only a few species of aquatics are capable of living in this type of environment. Crabs are the most common.

Water Quality

Water temperature is another factor when raising aquatics. The temperature of water is important to the organism's metabolism and regulation of the body temperature. **Cool water** species grow best in 50–60 degree water. If the water temperature is too low or high it will affect the overall health and growth of the animal. Examples of cool water species are salmon and trout. **Cold water** is best for species that require temperatures under 50 degrees. **Warm water** is considered between 65 and 85 degrees. Species in water can survive at both the high and low ends of water temperatures. Warm water species are the most abundant. Table 25-2 lists common aquatic species and the water temperature requirements for their health.

TABLE 25-2

Common Species and Water Temperatures

Catfish	75–85 degrees
Tilapia	75–85 degrees
Bass	77–88 degrees
Crawfish	65–85 degrees
Atlantic char	40–50 degrees
Salmon	50–60 degrees
Trout	50–60 degrees

Other factors that must be considered for water needs include the quality and possible pollutants, pH levels, mineral contents, gases, and living plants and animals in the water. Water management is necessary in some aquatic production sources where surface water and wells may cause contamination. **Surface water runoff** is excess water drainage from precipitation. The water must be pollution free and there are legal regulations on the usage of this type of water. **Well water** is pumped from aquifers deep within the earth where water is found naturally. This is a source of high-quality water free of pollutants, but due to the need for permits to drill and the amount of labor involved, this source of water is expensive. It is also usually cold to cool water. A more common natural source of water is a **spring**, which is a natural opening in the earth that provides a natural clean water source. Most springs develop at high elevations but are not reliable as these water sources may easily dry up.

Water quality involves how suitable water is for a particular source or use. Living organisms require fresh, clean water that is pollutant free and has a sufficient amount of **dissolved oxygen** (DO). *DO* is oxygen found in water for living organisms to use for breathing. It is found in a free, gaseous form rather than in a molecular form such as humans breathe. Plants are involved in providing the transfer of oxygen into the water. Information on water quality and management pertaining to pH levels, alkaline and acidic water, and water filters is discussed in Chapter 14 with regard to ornamental fish. This type of production is very similar and should be reviewed for more information on water management and control.

Aquatic Reproduction and Breeding

Fish and aquatic animals reproduce and grow through production cycles. A **production cycle** is when the young are full grown and have reached market size. The cycles follow the seasons of the year. Aquatic reproduction season is in the spring; the young are born and grow over the summer. Most species take two years to reach adulthood. Aquatic animals have four production cycles: broodfish and spawning, hatching and raising fry, producing fingerlings, and growing to adult size or market weight.

Broodfish and Spawning

The first production cycle includes **broodfish**, which are mature breeding fish used for reproduction. Broodfish may be male or female. The broodfish produce young so that fish may be stocked and used as food fish for other aquatics. Breeding, which is called **spawning**, usually begins around 3 years of age. Adult fish are separated from the young, as many species of fish and aquatic animals will eat their young. Spawning occurs when pairs or groups of fish are isolated in a water source. Broodfish actively lay a group of eggs called **spawn**. The spawn are laid in a nest with the eggs varying in size, from as small as a plant seed. Female broodfish produce thousands of eggs per spawning. Many fish can produce 2,000 eggs per pound of body weight (see Figure 25-11). Spawning

FIGURE 25-11 Trout eggs.

FIGURE 25-12 Fertilization of salmon eggs.

seasons vary with species. Catfish spawn when the water becomes warmer in the late spring to early summer. Coho salmon will spawn once and then die. After spawning, some fish—males or females—will care for the nest. Other species are **mouth brooders**, which means they incubate the eggs in their mouth.

Artificial fertilization

Artificial fertilization and incubation is used with some species, especially those that die after spawning. This is done by **expressing** the female, or placing pressure on the abdomen and squeezing to remove eggs. The male is expressed in the same manner so that sperm, known as **milt**, is removed. Both are then mixed in a container for fertilization and hatching to occur (see Figure 25-12). This is done at a **hatchery** where eggs are fertilized and hatched after incubation, which is the period between spawning and hatching. During incubation, the most important factors are water temperature and circulation. Specialized equipment is needed, as eggs are incubated in trays, jars, or troughs (see Figure 25-13).

Hatching and Raising Fry

Fry, or newborn fish larvae, are then hatched. Fry are very small and delicate and must be kept in ideal conditions for growth and to prevent disease. The fry will hatch and a **sac fry**, or yolk sac, is attached for a day after the hatch (see Figure 25-14).

TERMINOLOGY TIP

Fry is the term for young fish larvae, but salmon young are called **alevin**.

FIGURE 25-13 incubation of trout eggs.

Fry are fed finely ground fish meal that is organ based, such as heart or kidney mixed with plankton, yeast, and chicken yolks made into a **slurry**, or soft soup-like substance. The fry grow to an inch long and are then moved to a new location. At this stage they become **fingerlings**, or immature fish greater than an inch in size but not at stocking or market size. As they grow they are moved to large tanks or small ponds.

TERMINOLOGY TIP

Fingerlings are young immature fish over an inch in size, but salmon fingerlings are called **smolt**.

Producing Fingerlings

Fingerlings should be fed a high-protein diet around 36% that is compounded as a commercial granule feed. It takes one growing season to meet stocking size. Fish species need to be at least 5 inches to be stocked in a natural water source.

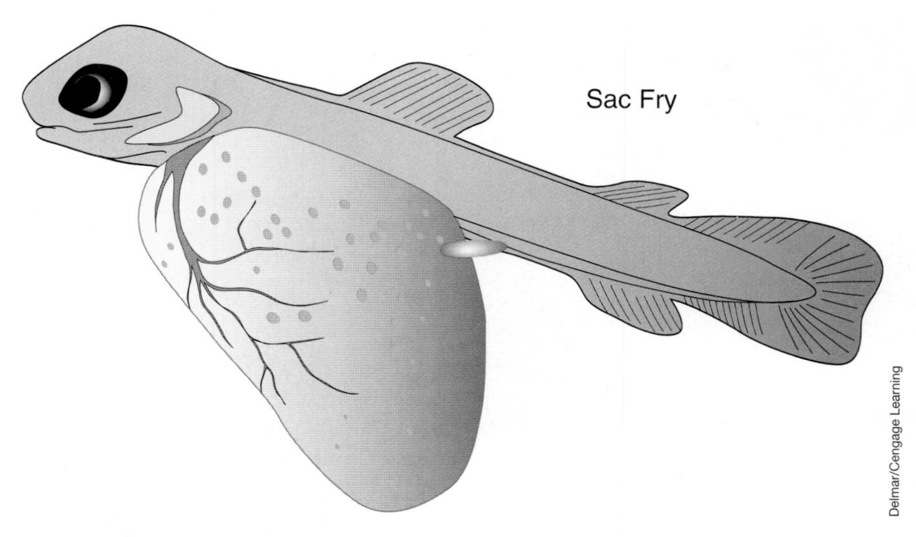

FIGURE 25-14 Fry with yolk sac.

Readying Fish for Market

Food fish production includes raising fingerlings to market size, and they should be fed 3% of their body weight to gain weight as quickly as possible. This type of production cycle requires 28–32% protein and typically takes 6 months to harvest and reach market weight.

Breeds or Species

There are many different species and breeds of aquatic animals that are found natively in the United States and many species that are raised for production purposes. The species are divided into the following groups: finfish, ornamental fish, mollusks, crustaceans, and aquatic mammals. Some popular production methods within the United States pertaining to finfish include salmon and catfish; crustaceans and mollusks include shrimp, clams, and oysters.

Breed Selection

Species and breed selection are based on many of the same factors as livestock. Successful production with aquatics species means consideration of the following:

- Environmental needs
- Nutritional needs
- Reproductive habits
- Adaptability to housing
- Disease and health resistance
- Market demand and profitability

The ability to reproduce and survive is an important factor in aquatic production. Aquatic species require specific handling, care, and knowledge to be a successful operation.

Nutrition

The nutritional needs of aquatics must meet their energy needs for swimming and regulating their body temperature. Most aquatics require a balanced daily ration of protein, carbohydrates, fats, vitamins, and minerals. Protein is necessary for growth and should be provided as an animal protein source, such as shrimp, fish, or chicken and with at least 28% protein. Carbohydrates are used for energy and include soybeans, corn, and grains for at least 10% of the diet. Fats are needed for waste production and are provided from by products at 4–15% of the diet. Vitamins and mineral supplements are added to the diet in such items as calcium, iron, and sodium. Aquatic animals may eat a type of water plant nutrient called **plankton**, which grows naturally in the water. Some vegetation sources that may be added into commercial diets include rice, plant stems, leaves, and grains. Many commercial diets are formulated for specific species. Diets come in several formulas that are made to float in the water and easily break apart. **Meal** is finely ground food for small fish and aquatic animals. **Crumbles** are flakes or blocks of food. **Pellets** are meal bound into larger particles. Pellets vary in size and are made to float. Feeding rates vary with species and life stage and many require feeding in a range from every hour to 1–2 times a day. All aquatics and fish should be fed 3% of their body weight. Fry are fed every hour to

FIGURE 25-15 A type of automated self-feeder used in trout production.

several times a day as they continue to grow. Fingerlings are fed 2–3 times a day. Food fish and broodfish are fed 1–2 times daily. Many aquaculture programs feed on an automatic feeder system. This type of feeder provides a specific amount of food on demand as the fish touches a "trigger" that releases the food (see Figure 25-15). Some feeders are set on timers to release a specific amount of food at a certain time of the day.

Common Diseases

Aquatic species are prone to several types of infectious and noninfectious diseases and conditions. Several factors may be involved with these diseases, such as the environmental conditions, nutritional program, water quality, and health status of the animal.

Fin rot is a bacterial condition in which the fin tissues begin to erode or slough away. The bacterial infection begins by causing the fragile tissue to become red and inflamed. Factors that may introduce the bacteria include poor water sources, poor water management, stress, and wounds caused by aggressive fish. Treatment may include a bath immersion in salts if the species is salt tolerant, antibiotic therapy, and use of topical medications.

> **TERMINOLOGY TIP**
>
> A *topical* is a product, such as a medication, applied onto the skin or hair coat on the outside of the body.

Ich is a common fish and aquatic parasitic disease caused by a protozoan. This disease is commonly called **whitespot** as it causes white spots to occur on the skin and fins of the body. This parasite is introduced into the water by poor water sources or other species that have contaminated the water. Stress and poor nutrition also make a fish more susceptible to the disease. The best prevention is to isolate all new species of aquatic animals and plants. Treatment is by parasitic medications added to the water for a 10–14 day period.

Lymphocytosis is a viral disease that affects certain species of fish and is commonly known as **cauliflower disease**, due to the white to gray growths that develop over the skin and fins. This virus has no treatment, is rarely fatal, and is caused by poor water sources and management and stress. Some species will not show signs but are carriers of the virus, spreading it to other fish.

Fish fungus is known as **cotton-wool disease** due to the growths that develop over the face, gills, or eyes and resemble cotton. This condition is caused by fungal spores that contaminate a water source. Fish should be isolated and treatment includes antifungal medications placed in the water and salt bath immersion for salt-tolerant species. Salt should be used at 2 grams per liter of water.

Cloudy eye is caused by a bacterial infection that may affect one or both eyes. This condition primarily occurs in tropical fish. It may spread to the nervous system, causing damage. This bacterial infection requires proper water quality, changing water frequently, and filtering water to prevent the spread of bacteria. Treatment is through antibacterial medications and adding salt to the water.

Ascites, also called **dropsy**, is caused by a viral or bacterial infection that produces body swelling due to fluid buildup and a pinecone-like appearance to the scales. This condition is caused by poor water quality and a buildup of ammonia in the water. Treatment is with antibacterial antibiotics and salt water immersion.

Swim bladder disorder is common in goldfish and koi and causes fish to have difficulty swimming to the surface and in different depths of water. This condition may be caused by poor diet, bacterial infection, and genetic deformities. Some fish will not recover, but treatments that may help are diet changes, feeding flakes instead of pellets, and changing the water quality.

It is important to prevent diseases from spreading and control disease through good management practices, such as isolating new and sick fish, monitoring water quality, feeding proper diets, and treating the water with chemicals to prevent disease.

SUMMARY

Aquatic animals are typically raised for human consumption. The most widely produced animals in aquaculture are fish, crustaceans, and mollusks. These animals may be raised in freshwater, saltwater, or brackish water. The water source and quality are important to production of aquatic animals. Aquaculture is becoming a popular area within the veterinary industry.

Key Terms

abductor muscle string muscle that allows a shell to open and close

alevin young salmon

arthropods crustacean with an external skeleton

aquacrops commercially produced plant or animal living in or around a water source

aquaculture production of living plants, animals, and other organisms that grow and live in or around water

aquarium small-sized, rectangular area that holds water to contain small fish

aquatic pertaining to water

ascites fluid buildup in the abdomen of a fish that caused by a virus or bacterial infection; commonly called dropsy

barbs also called barbels; devices on the fish that detect food, danger, and movement

bivalves mollusks with two shells

brackish water a mixture of freshwater and saltwater

broodfish mature male and female breeding fish used for reproductive purposes

cage structure made to float on the water to hold aquatics

carapace thorax, abdomen, and limbs

caudal fin the tail fin that acts as a propeller

cauliflower disease nickname for lymphocytosis

cloudy eye bacterial infection in one or both eyes of fish

cold water water temperature below 50 degrees

cool water water temperature between 50 and 60 degrees

cotton-wool disease nickname for fish fungus

crumbles flakes or blocks of fish food

crustacean aquatic animal with an outer body shell and legs

decapod crustacean with five pairs of legs

dissolved oxygen (DO) oxygen found in water for living organisms to breathe

dropsy nickname for ascites

exoskeleton external cartilage on the outside of the body

expressing placing pressure on the abdomen to remove eggs or sperm

fingerling immature fish greater than an inch in size

fin rot bacterial condition that causes the fin tissue to wear away or deteriorate

fins structures that act as arms and legs and allow movement in the water

fish aquatic species that have scales, gills, and fins

fish fungus fungal spores that develop in growths on the face, eyes, and gills and resemble cotton; commonly called cotton-wool disease

food fish production raising fingerlings to market size to be sold as food products

freshwater water with little to no salt content

fry newborn fish larvae

gills structures that act as lungs as water passes over them

hatchery place where fish are fertilized, bred, hatched, and raised

head part of the body containing the eyes, mouth, and gills

ich common aquatic parasite caused by a protozoan that produces white spots on the skin; commonly called whitespot

intensity biomass the number of species in a volume of water

lymphocytosis viral disease that causes white to gray growths to develop over the skin; commonly called cauliflower disease

meal fine ground food for young fish and aquatics

milt fish sperm

mollusk aquatic species that have only a thick, hard shell

mouth brooders fish that incubate their eggs inside the mouth

operculum gill covering that divides the head from the trunk

pellets fish meal bound into larger particles

pen structure that connects land to water

pincers pair of structures on the head used for smelling, feeling, and protection

plankton natural plant food source that grows in water

pond contained water area surrounded by human-built dams and levees

ppt parts per thousands measurement that determines the amount of salt in water

production cycle the time for young aquatic species to reach full market size

production intensity assurance that the size of the body of water will hold the appropriate amount of food sources and be large enough for a high production level of products

raceways long and narrow water structures with flowing water

sac fry yolk sac attached to a fry for a day after the hatch

salinity the amount of salt in water

saltwater water with a salt content of at least 16.5 ppt

scales individual segments that cover the body and act as skin

slurry soft soup-like substance fed to fry

smolt immature salmon greater than an inch in size

spawn group of fish eggs in a nest

spawning breeding or reproduction in fish

spring natural opening in the earth that provides a clean natural water source

surface water runoff excess water drainage from rainfall or precipitation

swim bladder structure that acts like a small balloon that inflates and gives the fish buoyancy to float and balance in all levels of water

swim bladder disorder condition common in goldfish and koi that causes a fish difficulty swimming to the surface and in depths of water

tail part of the body containing the caudal fin that allows movement and guidance in the water

tank large round areas of water

topical product applied to the outside of the body

trunk part of the body that contains the body, fins, and anus

univalve mollusks with one shell or part of a shell

valve structure in mollusks that acts like gills

vat smaller, circular areas that hold water

vertebrates animal with a backbone

warm water water between 65 and 85 degrees

water facility areas built to house aquatic animals in natural environments

well water water pumped from aquifers from deep within the earth where water occurs naturally

whitespot nickname for ich

REVIEW QUESTIONS

1. What common species of aquatic animals are produced in the United States?
2. What are the general requirements for aquaculture production?
3. Explain production intensity.
4. What are the three types of water?
5. What types of production facilities are available for aquaculture or aquacrops?
6. What is dissolved oxygen?
7. What are the types of production systems in aquaculture?
8. What is ich?

Clinical Situation

Jimmy, a veterinary assistant working in an aquaculture facility, is doing a daily monitoring of the tanks and vats. He is checking the water temperature and water quality of each tank. Jimmy notices that some fish in several tanks are dead and floating at the top of the water source.

On better examination, Jimmy notices that several tanks have fish that look sick and diseased. He begins noting the tank numbers, signs of disease, and the number of fish that have died. He also decides to take some water samples of the tanks containing diseased and dead fish.

- Did Jimmy handle this situation correctly? Why or why not?
- What would you do as a veterinary assistant in this situation?
- What is the importance of properly handling a food fish situation?

Section III

General Anatomy and Disease Processes

26

The Structure of Living Things

Courtesy of Image Source

Objectives

Upon completion of this chapter, the reader should be able to:

- ☐ Define anatomy, physiology, and pathology
- ☐ Identify structural units of living things
- ☐ Describe the functions of cells and cell division
- ☐ Identify the function of various types of tissues
- ☐ Explain the purpose of organ systems

Introduction

Knowledge and understanding of normal anatomy is critical to the veterinary professional who assists in restraint, handling, and care of a variety of animals. **Anatomy** is the study of the internal and external body structure. **Physiology** is the study of the functions of those structures and how they work. A sound understanding of what is normal will aid in the identification of disease processes when something abnormal occurs. **Pathology** is the study of disease processes.

While animal species vary in appearance, there are similarities in external body structures and the functions of those structures. Sometimes the structures share a common name and other times the name for the structure may be specific to a particular species of animal. Each species of animal was discussed in the production chapters of this text. It is recommended that you refer to each chapter for an external anatomy diagram and proper terminology of the external body parts related to a specific species. All animals have a head, body, and tail area. Some animals have feet, paws, or hooves. It is important that veterinary assistants become familiar with the terminology of external anatomical structures of the species they commonly work with.

Cells

The most basic structural unit of an animal is the **cell**. Cells are microscopic in size, and each living thing is made up of trillions of cells. Cells come in a variety of shapes and sizes and each cell has a specific function.

The Structure of a Cell

A cell consists of many **organelles** (small units within the cell responsible for a specific function; see Figure 26-1). A cell's structure includes a **cell membrane**, also called a cell wall or plasma membrane, that holds the cell together. The **cytoplasm** is the fluid part of the cell that allows the internal structure of the cell to move. The **nucleus** is the brain of the cell and is located in the center of the cell. Within the nucleus is the **nucleolus**, which forms genetic material. The **mitochondria** are the part of the cell that makes energy and is often referred to as the "powerhouse of the cell." The **ribosomes** are structures within the cell that make protein. The **lysosomes** are structures that digest food and proteins. Cells allow for **metabolism**, a reaction that occurs within the body and allows chemicals to break down and be used by the body. Cells serve many functions within the body (see Table 26-1).

Cell Division

Cell division occurs when a cell splits into two cells. Two types of cell division occur within the body: mitosis and meiosis. **Mitosis** is cell division of **somatic cells** (nonreproductive), which aid in the body's growth and repair. This process involves several phases (see Figure 26-2).

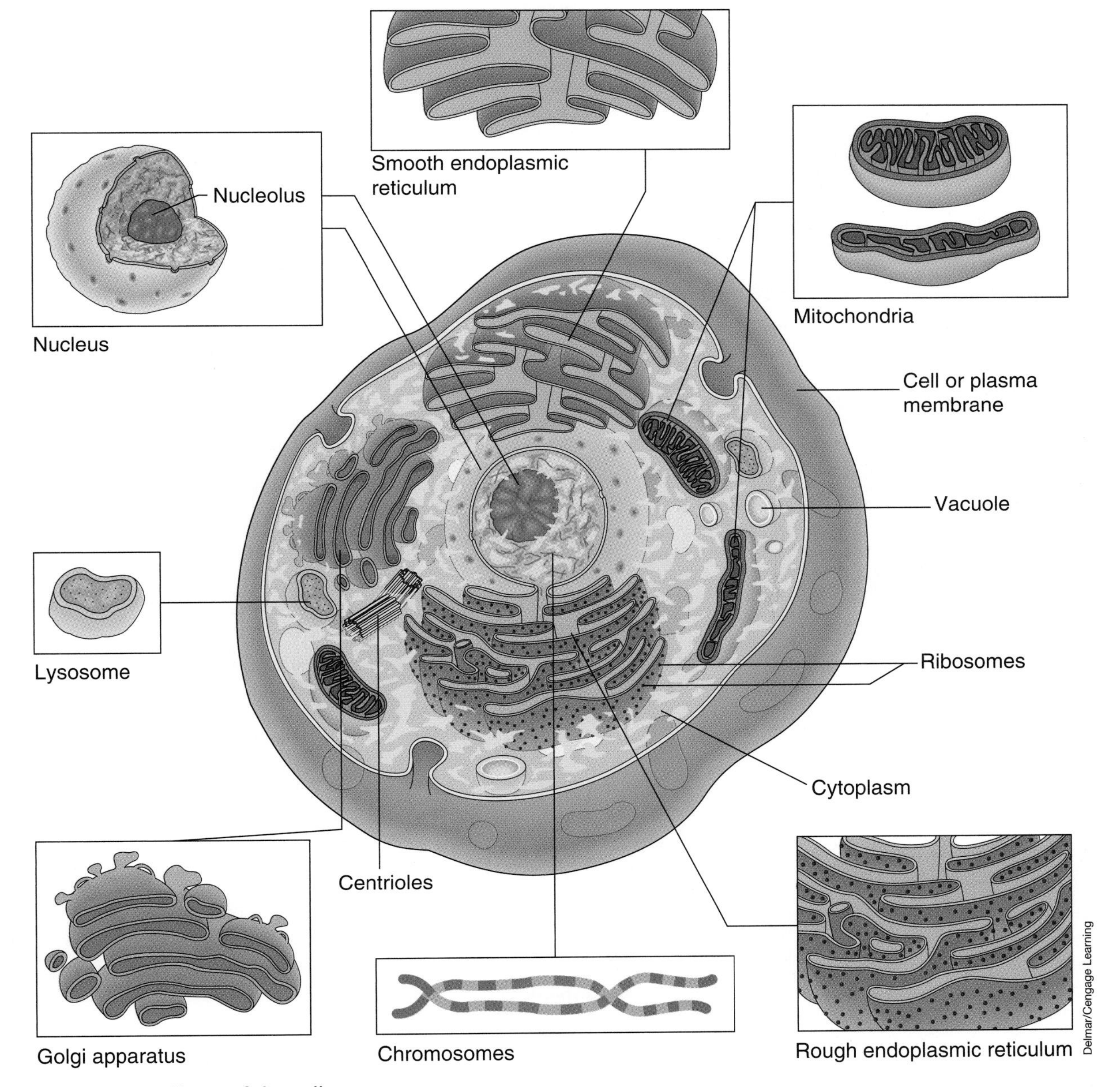

FIGURE 26-1 Parts of the cell.

TABLE 26-1	
Cell Functions	
Active Transport	A process in which a substance found in a lower concentration will move to an area where it is higher in concentration
Anabolism	A process in which smaller particles combine to form larger particles
Catabolism	A process in which larger particles break down into smaller particles
Diffusion	A process in which a substance found in a higher concentration will move to an area where it is lower in concentration
Endocytosis	Process in which a cell can take in a particle
Extra-Cellular Fluid (ECF)	Fluid found outside the cell, such as blood
Homeostasis	Maintenance and balance of body processes
Osmosis	The movement of a substance through and across cell membranes
Phagocytosis	Process by which dead cells and waste materials are eaten or removed from the body

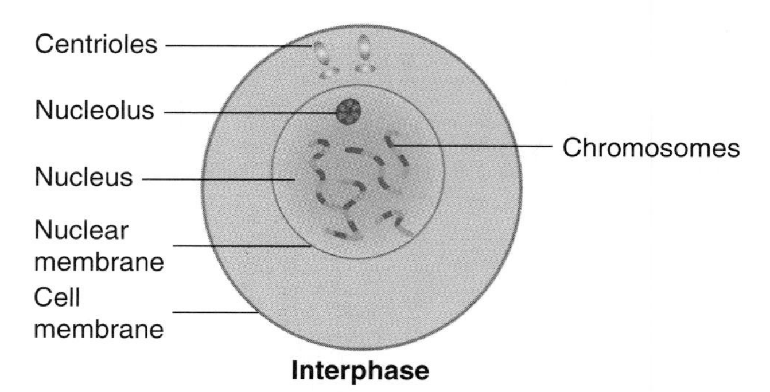

The nucleus and nuclear membrane are distinct. The chromosomes are contained in the nucleus in a thread-like mass.

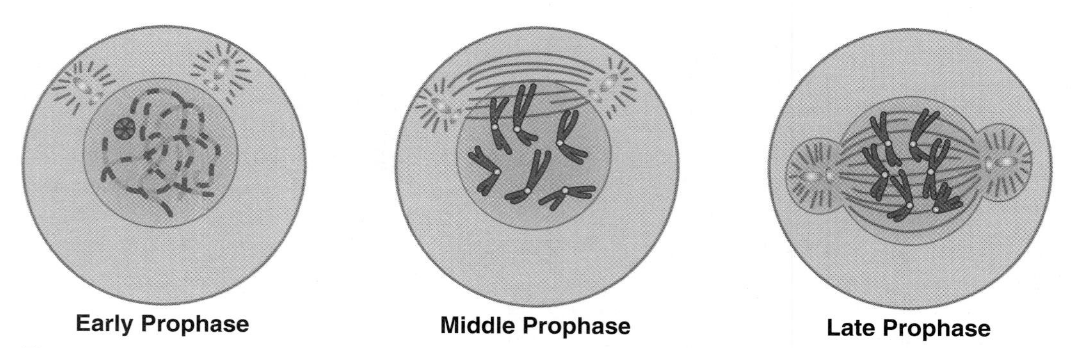

The centrioles begin to move toward opposite poles creating fibers between them. The chromosomes begin to condense. The nuclear membrane becomes less distinct.

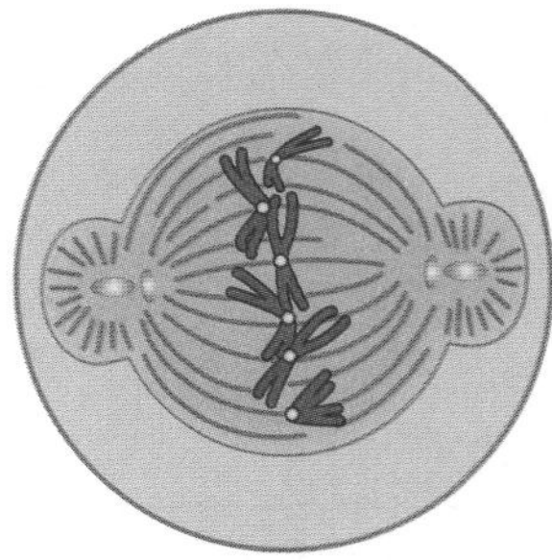

The chromosomes begin to attach to the fibers between the centrioles.

FIGURE 26-2 The phases of mitosis. (*Continues*)

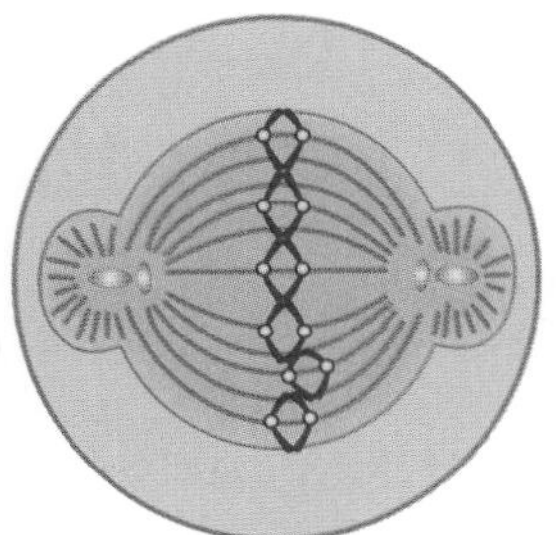

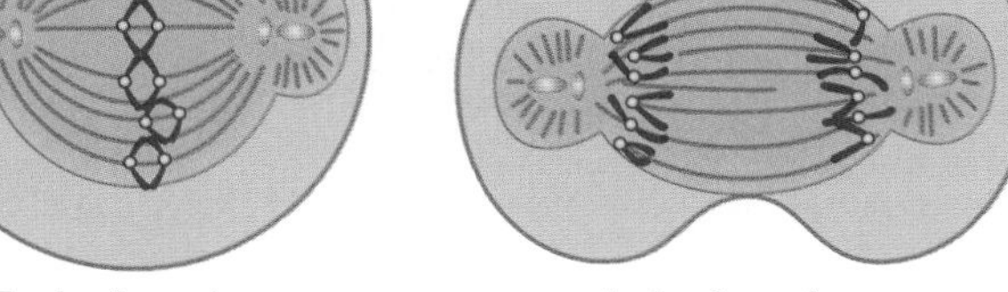

Early Anaphase **Late Anaphase**

The chromosomes separate and move to opposite poles.

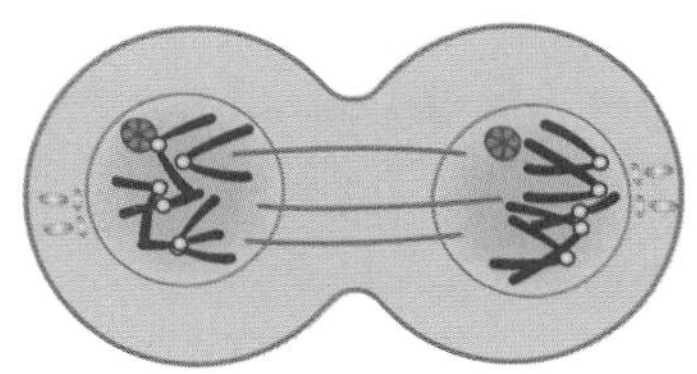

Telophase

The chromosomes are at opposite poles and a nuclear membrane begins to form.

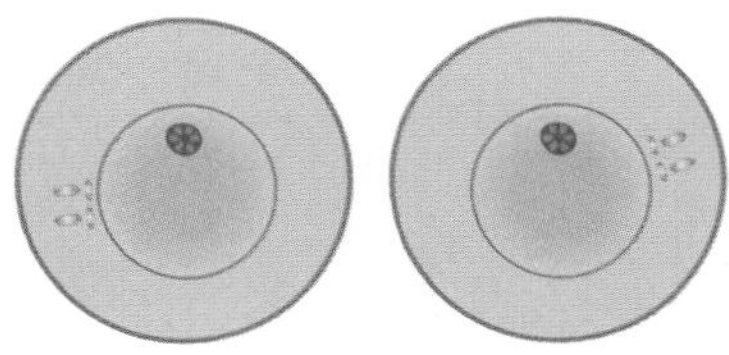

Interphase

Cell division is complete. Two new cells are identical to the original.

FIGURE 26-2 (*Continued*)

One cell divides into two cells identical to the original cell. **Meiosis** is cell division for breeding and reproduction. This process involves a cell that divides into four different cells, called **daughter cells** (see Figure 26-3). Each daughter cell has half the number of chromosomes of the parent cell. The process of fertilization provides the other half of the chromosomes. All new cells contain identical genetic materials from the parents. The stages of cell division are outlined in Table 26-2.

> **TERMINOLOGY TIP**
>
> *Fertilization* is the union of the sperm to the egg to create an embryo.
> A *chromosome* is the genetic material inherited from each parent and passed to the offspring.

Meiosis I in Males

Prophase I
Chromosomes begin to condense and chromosomes similar in structure pair and cross over.

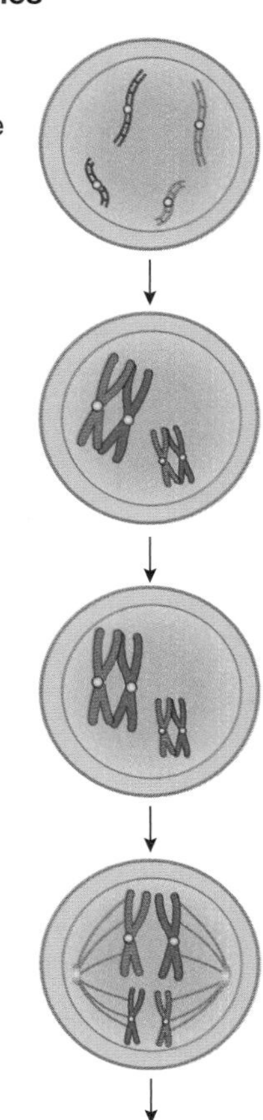

Metaphase I
Spindle fibers attach to chromosomes and chromosomes line up in center of cell.

Anaphase I
Chromosomes start to move to opposite ends of cell as spindle fibers shorten.

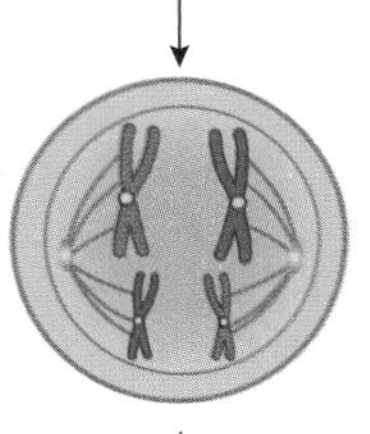

Telophase I
Chromosomes reach opposite ends of cell and nuclear membrane forms.

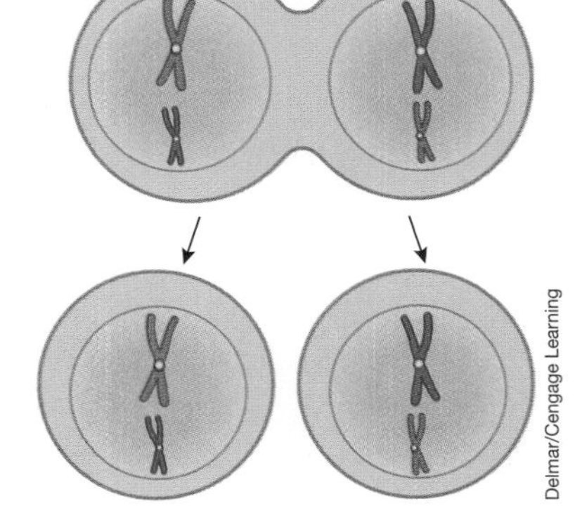

Cytokinesis
Cell division occurs producing two identical cells.

FIGURE 26-3 The phases of meiosis. (*Continues*)

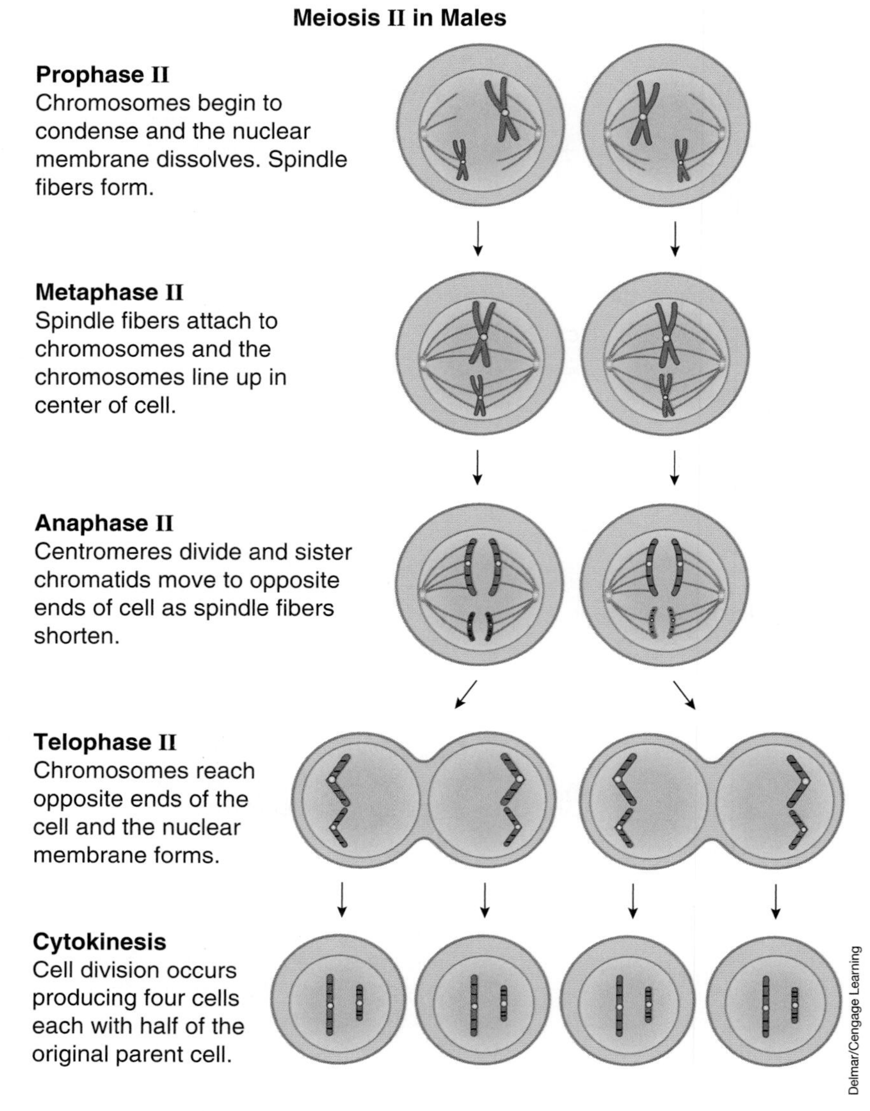

Meiosis II in Males
Prophase II
Chromosomes begin to condense and the nuclear membrane dissolves. Spindle fibers form.
Metaphase II
Spindle fibers attach to chromosomes and the chromosomes line up in center of cell.
Anaphase II
Centromeres divide and sister chromatids move to opposite ends of cell as spindle fibers shorten.
Telophase II
Chromosomes reach opposite ends of the cell and the nuclear membrane forms.
Cytokinesis
Cell division occurs producing four cells each with half of the original parent cell.
Delmar/Cengage Learning

FIGURE 26-3 *(Continued)*

TABLE 26-2

Stages of Mitosis and Meiosis

Interphase	Cells in their normal state
Prophase	Chromatin forms and begins to form an X shape
Metaphase	A spindle forms at the center
Anaphase	Chromosomes split
Telophase	Divide to create two or four new cells

Enzymes are chemicals that change other chemicals and are part of the metabolism process. Enzymes break down cell components to allow the cells to be absorbed and used in the body. This is common with how medications work within the body.

Tissues

Tissues are a group of cells that are alike in structure. Tissues perform specific functions, such as forming skin and bone. The study of tissues is called **histology** and is an area of study in veterinary medicine. There are four types of animal tissues: epithelial, connective, muscular, and nervous. **Epithelial tissue** covers the body's surface, lines internal organ structures, and serves as a protective layer. **Connective tissue** holds and supports body structures by connecting cells, such as ligaments and tendons (see Figure 26-4). **Muscular tissue** allows movements of body parts (see Figure 26-5). **Nervous tissue** contains particles that respond to a stimulus and cause a reaction in the body.

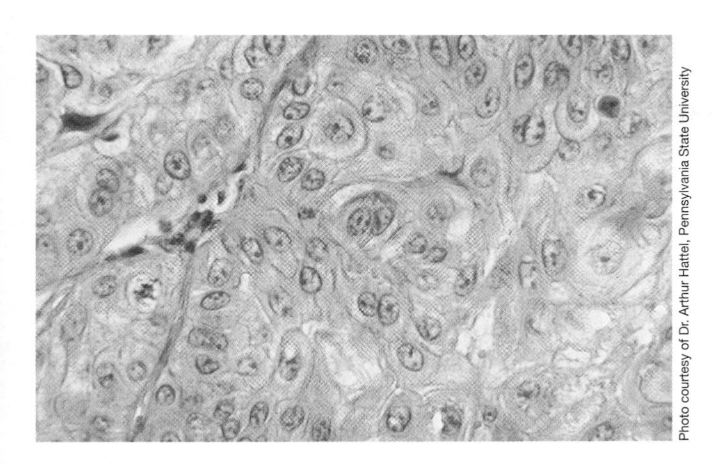

Photo courtesy of Dr. Arthur Hattel, Pennsylvania State University

FIGURE 26-4 Photomicrograph (high power) of cancer in the skin of a horse. There are variations in the shape and size of the nucleus and cells.

Organs

A group of similar tissues form **organs**, such as the liver, heart, skin, kidneys, pancreas, and so on. Each organ performs a specific function. Organs do not function separately; rather, they act together to perform specific related functions. This group of organs working together is called an organ system. Each organ system serves a particular function in maintaining life. Every living thing is made up of many organ systems. Each system coordinates to make up a complete living, functioning organism.

Disease and Injury

Body systems and their functions can be disrupted by disease processes or injury. **Infection** is the invasion of a foreign substance causing disease. Infection usually results in inflammation. **Inflammation** is a protective response to an injury and will result in pain, swelling, and redness.

External forces can cause trauma, such as a fall from a height. Trauma can result in tissue damage or injury. Emergency management of trauma is necessary to prevent complications such as excessive blood loss or infection.

Cell division may occur rapidly and in some cases develop in localized areas called **tumors**. A tumor may be **benign** or not cancerous or it may be **malignant**,

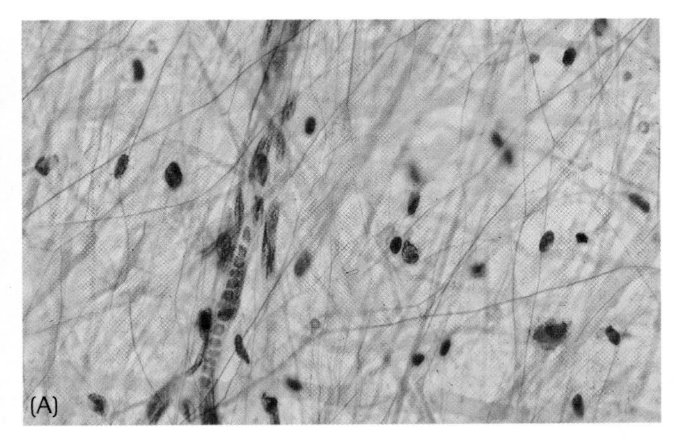

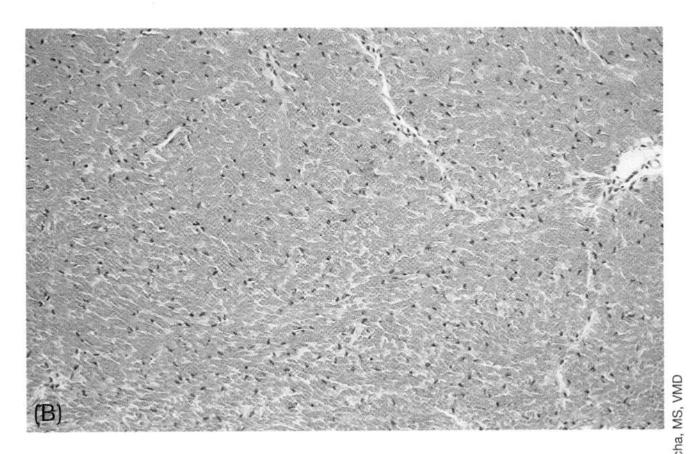

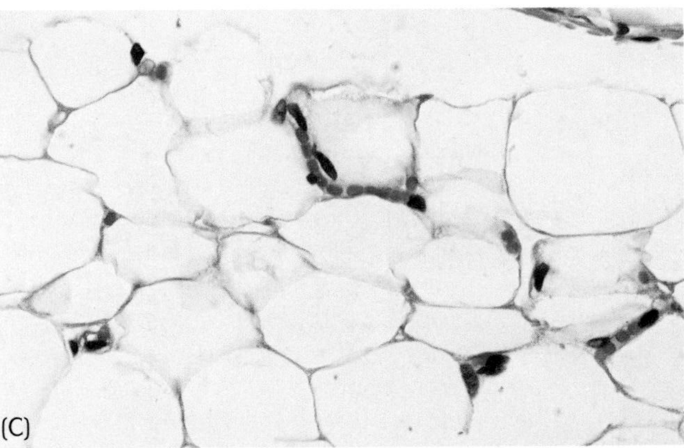

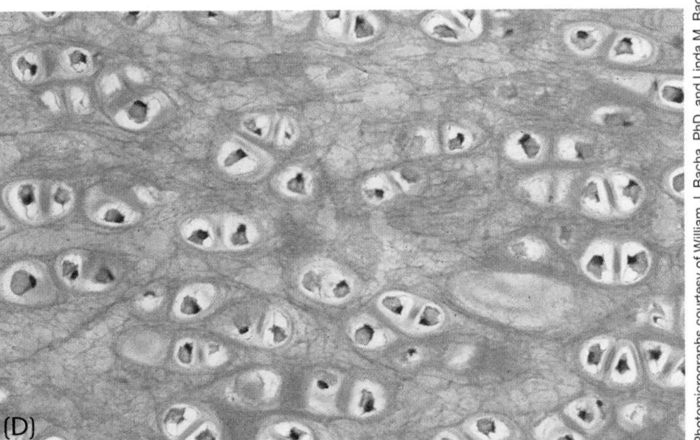

FIGURE 26-5 Types of connective tissue: (A) mesentery (loosely arranged connective tissues attached to abdominal organs) of cat, which shows collagen fibers (pink) and elastic fibers (thin dark fibers); (B) densely packed elastic fibers found in ligament of a sheep; (C) adipose (fat) tissue; (D) cartilage found in a pig.

meaning cancer causing. Veterinary medicine deals with determining if a tumor is cancerous and how it should be treated. Cancerous cells have a large nucleus, have an abnormal spindle, and when stained to view under a microscope the cells look clumped (see Figure 26-6).

SUMMARY

Cells are the most basic structure of living organisms. Cells divide and multiply forming tissues. Tissues form organs, which form specific functions. Groups of organs work together to provide all vital life functions for living things.

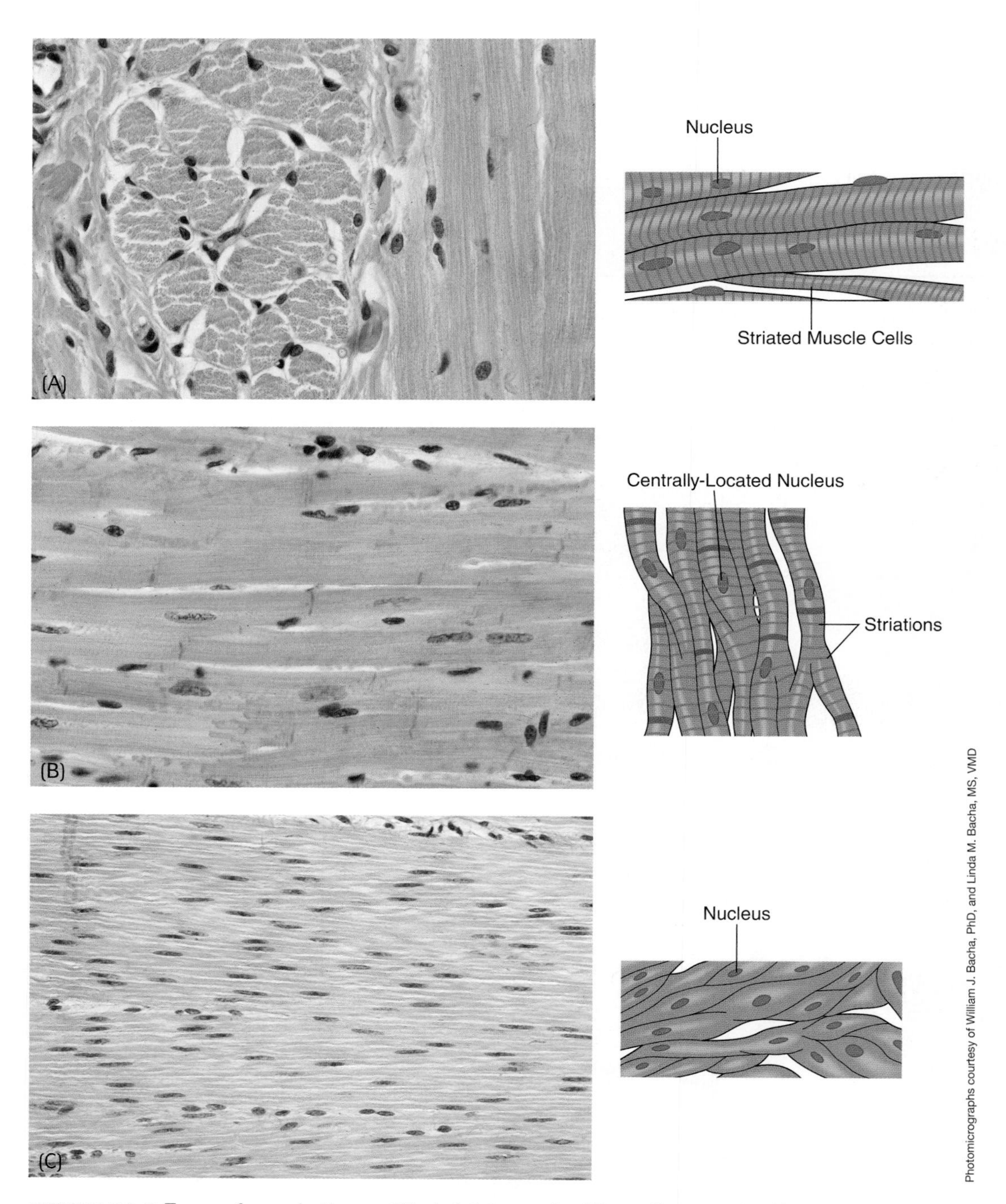

FIGURE 26-6 Types of muscle tissue: (A) skeletal muscle; (B) cardiac muscle; (C) smooth muscle.

Key Terms

anatomy the study of the internal and external body structures and parts of an animal

benign noncancerous tumor

cell basic unit of life and the structural basis of an animal

cell division process that occurs when a cell splits into two cells

cell membrane cell wall that forms the outside of a cell and holds the structure together

connective tissue holds and supports body structures by connecting cells

cytoplasm the liquid or fluid portion of a cell that allows the internal cell structures to move

daughter cells four cells that are created from one cell during meiosis and each cell has half the number of chromosomes of the parent cell

enzyme chemical that changes other chemicals during metabolism

epithelial tissue protective layer that cover the body's organs and acts as a lining of internal and external structures

histology the study of tissues

infection invasion of the body by foreign matter causing illness and disease

inflammation protective mechanism causing redness, pain, and swelling

lysosomes the structures within the cell that digest food and proteins

malignant cancerous tumor

meiosis cell division for breeding and reproductive processes

metabolism reaction within the body that allows chemicals to break down and be used by the body

mitochondria "powerhouse of the cell" and the part of the cell that makes energy

mitosis cell division that divides one cell into two cells and aids in animal growth and tissue repair

muscular tissue allows movement of body parts

nervous tissue contains particles that respond to a stimulus and cause a reaction in the body

nucleolus located within the nucleus and forms genetic material

nucleus brain of the cell that is located in the center of a cell

organ a group of tissues

organelle structures that are found within a cell

pathology the study of cells and diseases

physiology the study of the functions of body structures and how they work

ribosomes structures within the cell that make protein

somatic cells cells that divide in mitosis and help in growth and repair of the body

tissue a group of cells similar in structure and function

tumor cell division and development in localized areas

REVIEW QUESTIONS

1. What are cells that are alike in size, shape, and function called?
2. What is the type of tissue that lines internal structures called?
3. What is the study of tissues called?
4. What structure is known as the "powerhouse of the cell"?
5. Explain the term "homeostasis".
6. What are the parts of a cell?
7. What are the stages of mitosis?
8. What is the difference between anatomy and physiology?
9. What is the difference between a benign and a malignant tumor?
10. Give examples of the various types of tissues and describe their functions.

Clinical Situation

Kate, a veterinary assistant at Green Lane Animal Clinic, is preparing for a surgery on "Callie" Ames, a 12 yr old SF Cocker Spaniel. Kate has read the medical history and notes that "Callie" has a 4 cm mass on her L lateral abdomen that Dr. Smith is scheduled to surgically remove. The tumor is to be biopsied at an outside lab to determine the cause and if it is malignant.

Kate notes that "Callie" is to have presurgical blood work and that the owner has left several questions with the receptionist. The note from the owner states Mrs. Ames would like to know what will happen if the mass is cancerous and what care "Callie" will need when she is sent home.

- What items and supplies will Kate need to prepare for "Callie's" surgical procedure?
- What are some treatments to consider if the tumor is cancerous?
- What are some factors that should be addressed on "Callie's" home care instructions?
- Who should discuss the lab results with Mrs. Ames?

27

Delmar/Cengage Learning

The Musculoskeletal System

Objectives

Upon completion of this chapter, the reader should be able to:

- Name the structures that make up the musculoskeletal system
- State the functions of the musculoskeletal system
- Identify types of joints and where they are found in the body
- Describe common diseases of the musculoskeletal system

Introduction

The **musculoskeletal system** consists of bones, muscles, tendons, and ligaments. The skeleton makes up the bony structure of the animal, providing support and protection for vital organs. The muscular system allows the animal to move. The skeleton is adaptable to bend and grow.

Bones

Bone is a hard, active tissue that consists mostly of calcium and forms the skeleton of the animal. Bones give the animal support and structure while protecting the internal organs. Bones also serve as a mineral reserve to the body, storing calcium and phosphorous for use when necessary and providing nutrients to aid in red blood cell formation. Young and growing animals need calcium and other essential minerals in their diets to allow the bones to form and grow.

The number of bones varies with the age and species of the animal. Bone shapes are described as long, short, flat, or irregular. Bone structure is classified as either **compact** or **cancellous** bone. Compact bone is a thick tissue that forms the outer layer of bone. This part of the bone can be repaired, and it gives elasticity and rigidity to a bone. Cancellous bone is sponge-like and softer and located inside the end of bones. Bones are formed through the process of **ossification.** A hollow center within the bone is called the **medullary cavity** and produces blood cells within a fluid known as **bone marrow**. Additional structures of bone are defined in Table 27-1. Figure 27-1 illustrates the anatomical features of a long bone.

TABLE 27-1

Other Structures of Bone	
Osteoblast	Particles that begin ossification in young, developing bones
Osteocyte	Bone cells that begin to develop mature bone
Osteocast	Mature bone particles that form minerals and compact bone
Periosteum	Thin connective tissue covering outer bone
Endosteum	Thin connective tissue covering inner bone

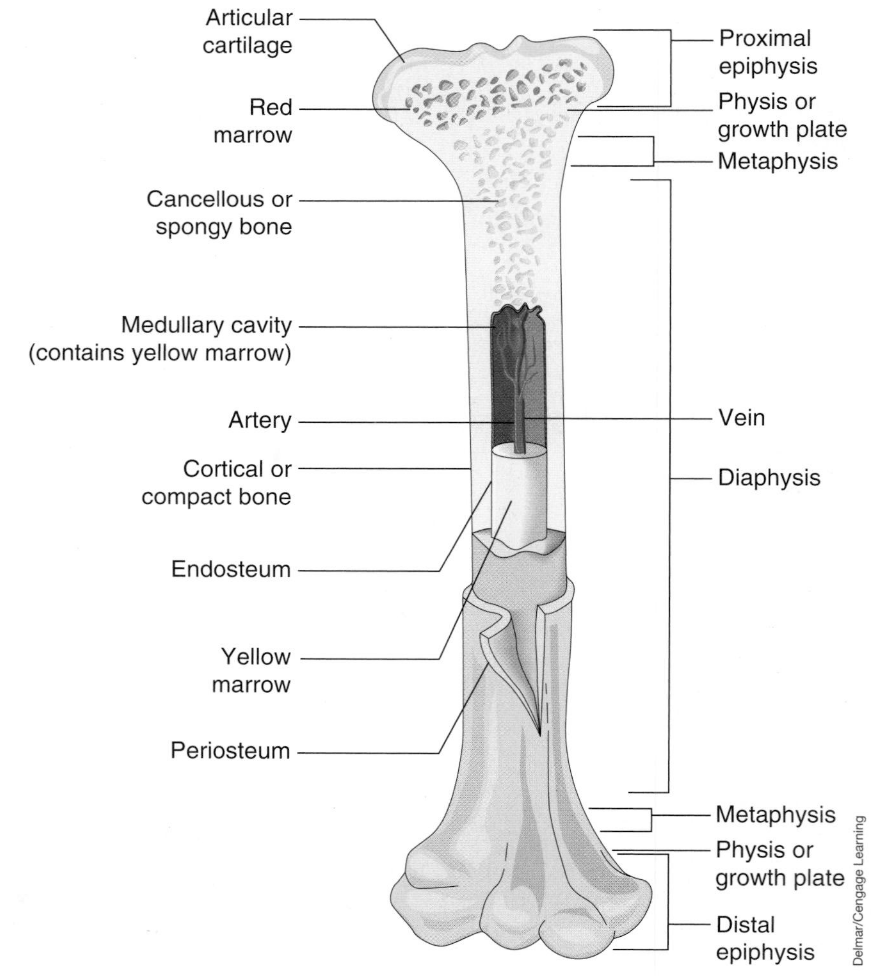

FIGURE 27-1 Anatomy of a long bone.

Joints

The area where two bones meet forms a **joint.** Joints allow the animal to bend in areas with a level-like movement. There are three main types of joints located in the body. A **fibrous joint** is a fixed joint with little to no movement. The area where two bones of a fibrous joint meet is called a **suture**. This is a fine line that connects the bones, but little to no movement occurs. An example of this type of joint occurs throughout the skull. A **cartilage joint** connects at the end of a joint where **cartilage** forms to protect and cushion the area where two bones meet. When an animal is young and growing this area is called the **growth plate**; over time and as the animal ages, the cartilage joint becomes bone. The third type of joint is a **synovial joint**. This type of joint is movable and may be a **hinge** joint that opens and closes in a one-way movement, a **pivot** joint that rotates around a fixed point, or a **ball and socket** joint that rotates in numerous directions. Various joints are illustrated in Figure 27-2 and listed in Table 27-2.

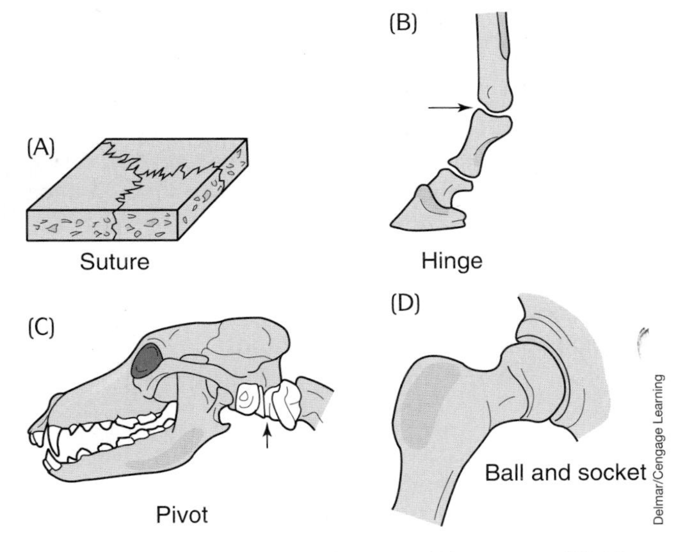

FIGURE 27-2 Examples of joints: (A) suture; (B) hinge; (C) pivot; (D) ball and socket.

TABLE 27-2

Types of Joints and Locations

LOCATION	TYPE OF JOINT
Skull	Fibrous joint
Growth plates at end of bones	Cartilage joint
Elbow	Synovial joint (hinge)
Vertebrae	Synovial joint (pivot)
Hip	Synovial joint (ball and socket)

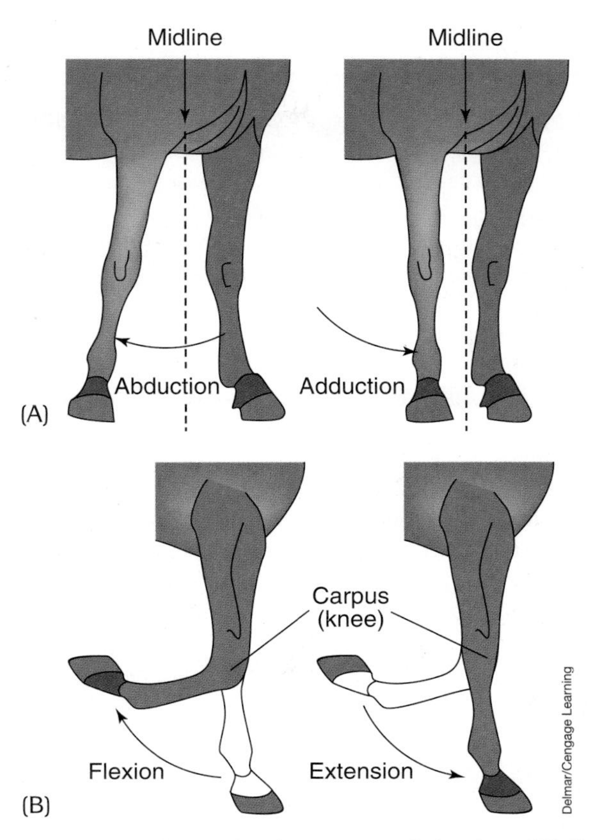

FIGURE 27-3 (A) adduction versus abduction; (B) flexion versus extension.

Joints move in different directions depending on the type of joint. Joints that bend move with **flexion** or closing causing the joint to shorten and **extension** or opening causing the joint to lengthen. Animals also are capable of joint movements known as **abduction** and **adduction**. *Abduction* is the movement away from the body and *adduction* is movement toward the body (see Figure 27-3).

Muscles and Connective Tissues

Muscles are located throughout an animal's body and allow movement. They attach to different locations of the body and serve as protection and provide the animal strength. **Ligaments** are fibrous strands of tissue that connect one bone to another bone and serve as support to all areas of the body. They also allow bending and movement of the limbs. **Tendons** are fibrous strands of tissue that connect a bone to a muscle. They also aid in movement of the limbs.

Axial Skeleton

The skeleton is divided into two parts, the **axial skeleton** and the **appendicular skeleton**. The axial skeleton contains the bones of the body that lie perpendicular or lengthwise. This includes the skull, spinal column, ribs, and sternum. The **skull** holds and protects the brain and is made up of many bones. The **maxilla**, or upper jaw, and **mandible**, or lower jaw, make up part of the skull (see Figure 27-4). Small, open spaces located within the skull and nasal canal are called **sinuses**. The **spinal column** contains the **vertebrae**, which are individual bones that surround and protect the **spinal cord** (see Figure 27-5). The spinal column extends from the base of the skull to the end of the tail and provides the animal with movement. It may also be referred to as the vertebral column or spine. The spinal column is divided into sections according to the body location.

> **TERMINOLOGY TIP**
>
> The term *vertebrae* is plural and refers to the entire spinal column. The singular form is *vertebra* and refers to one bone.

The first section of the spinal column is called the **cervical** vertebrae and extends over the head and neck area. The first vertebra located at the base of the skull is the **atlas** and is the first cervical vertebra that allows the up and down nodding motion of the head to occur. The second cervical vertebra is the **axis** and allows for the rotation and shaking motion of the head. The second section of vertebrae is called the **thoracic** vertebrae and lie over the shoulders and chest area. Each **rib** attaches to an individual thoracic vertebra and protects the heart and lungs. The **sternum**, or breastbone, helps to protect the organs of the chest. The next section of the spinal column is the **lumbar** vertebrae, which lie over the lower back or **loin** area. The **sacral** vertebrae form over the pelvis or rump area. Several vertebrae are fused together directly over the pelvis at the highest point of the spine in an area called the **sacrum**. A firm joint on each side of the sacrum connects to the **pelvis**, or hip bone. The last section of the spinal cord is the **coccygeal** vertebrae and forms the tail area. Much of this section doesn't contain the spinal cord. Figure 27-6 illustrates the various parts of the axial skeleton.

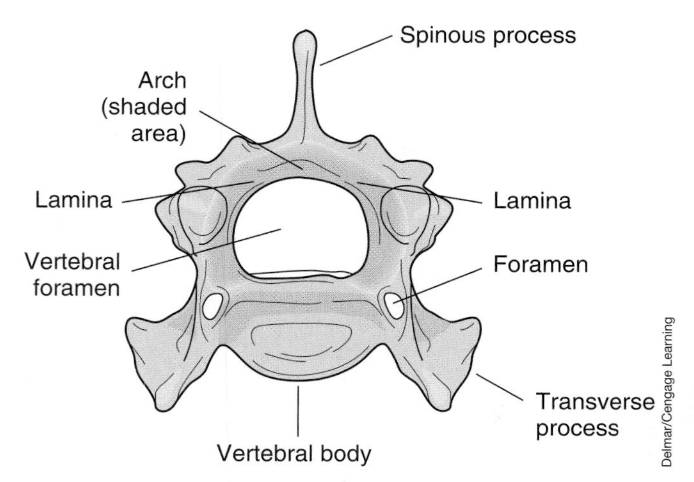

FIGURE 27-5 Vertebra, a bone of the spinal column.

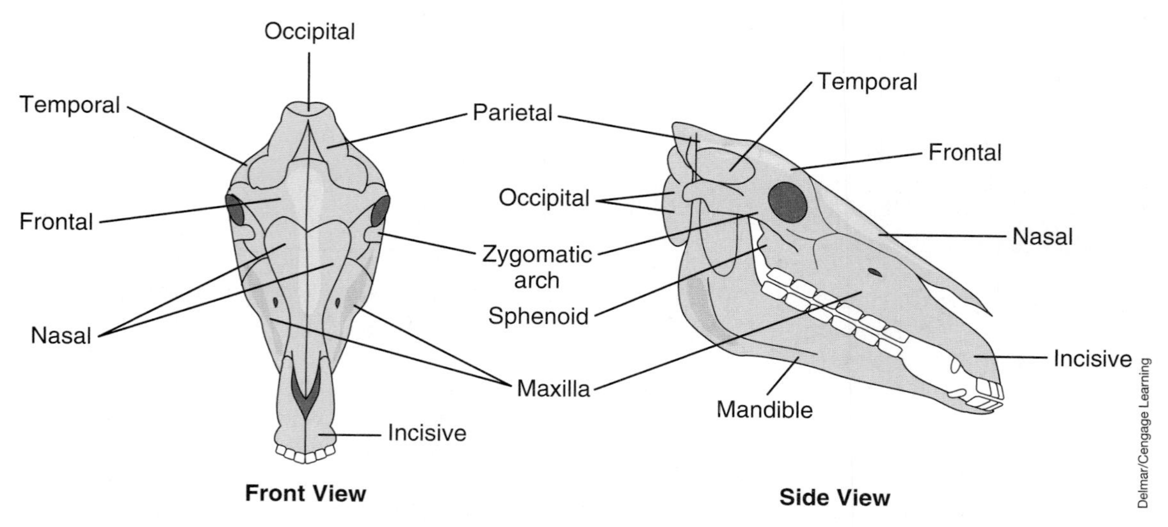

FIGURE 27-4 Selected bones of the skull and face.

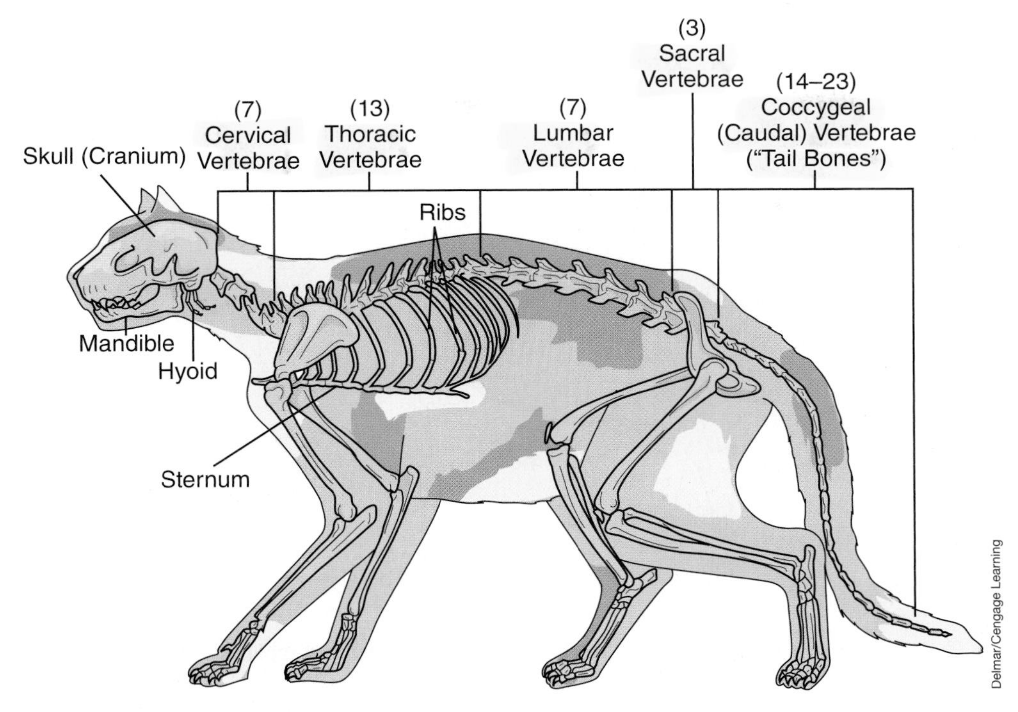

FIGURE 27-6 Parts of the axial skeleton.

Appendicular Skeleton

The appendicular skeleton is the part of the body that contains the appendages or limbs. The foreleg or **forelimb** is the front leg of an animal. The front limbs include the **scapula**, commonly called the shoulder blade; the **olecranon**, known as the elbow; the **humerus** or the large upper bone of the forelimb; and the **radius** and **ulna**. The *radius* is the larger bone located in the lower front limb. The *ulna* is a smaller bone and lies directly behind the radius. Both bones are fused together at the top and bottom and connect to the humerus to form the elbow joint. The lower leg forms the **carpal** bones, similar to the wrist bones in humans. There are several small carpal bones that are arranged in two rows collectively called the **carpus** joint. The **metacarpal** bones form the long bones of the front feet. In hoofed animals, these bones form the **cannon** bone. Horses have two small bones that form at the back of the cannon bone called **splint bones** that can become inflamed with large amounts of work due to the body absorbing movement over the front legs. The **phalanges** form the toes or digits. The short inside bone that forms the first digit and acts as an opposable thumb is called the **dewclaw**. The hind limb or the **pelvic limb** forms the rear leg of an animal. The pelvis is formed by three sections, the **ilium**, **ischium**, and **pubis**. The pelvis holds the rear limb in place by the **acetabulum**, or ball and socket joint of the hip. The **femur**, or thigh bone, forms the large bone of the upper rear leg. The **patella** forms the joint between the upper and lower bones of the rear leg; this is commonly called the kneecap. The bones in the lower leg are called the **tibia** and **fibula**. The *tibia* is the larger bone in the lower rear leg and lies in the front; the *fibula* is smaller and lies to the rear of the leg. The point of the rear leg where the tibia and fibula meet is commonly called the **hock**. The hock forms several bones similar to the ankle bone called **tarsal** bones and the collective term for the entire joint is the **tarsus**. The **metatarsal** bones form the long bones of the rear feet and the digits of both the front and rear limbs are all called the *phalanges*. Figure 27-7 illustrates bones that make up the appendicular skeleton.

Common Musculoskeletal Diseases and Conditions

Hip dysplasia is a common genetic condition of large-breed dogs. The ball and socket joint of the pelvis and femur becomes diseased and does not sit properly (see Figure 27-8). This condition varies in degrees of severity. The socket of the pelvis is shallow and the ball of the femur **subluxates**, or comes partially out of the joint. Common signs of the condition include reluctance to lie down or rise, inability to move on stairs, and lameness and pain in mild to severe stages. Some breeds

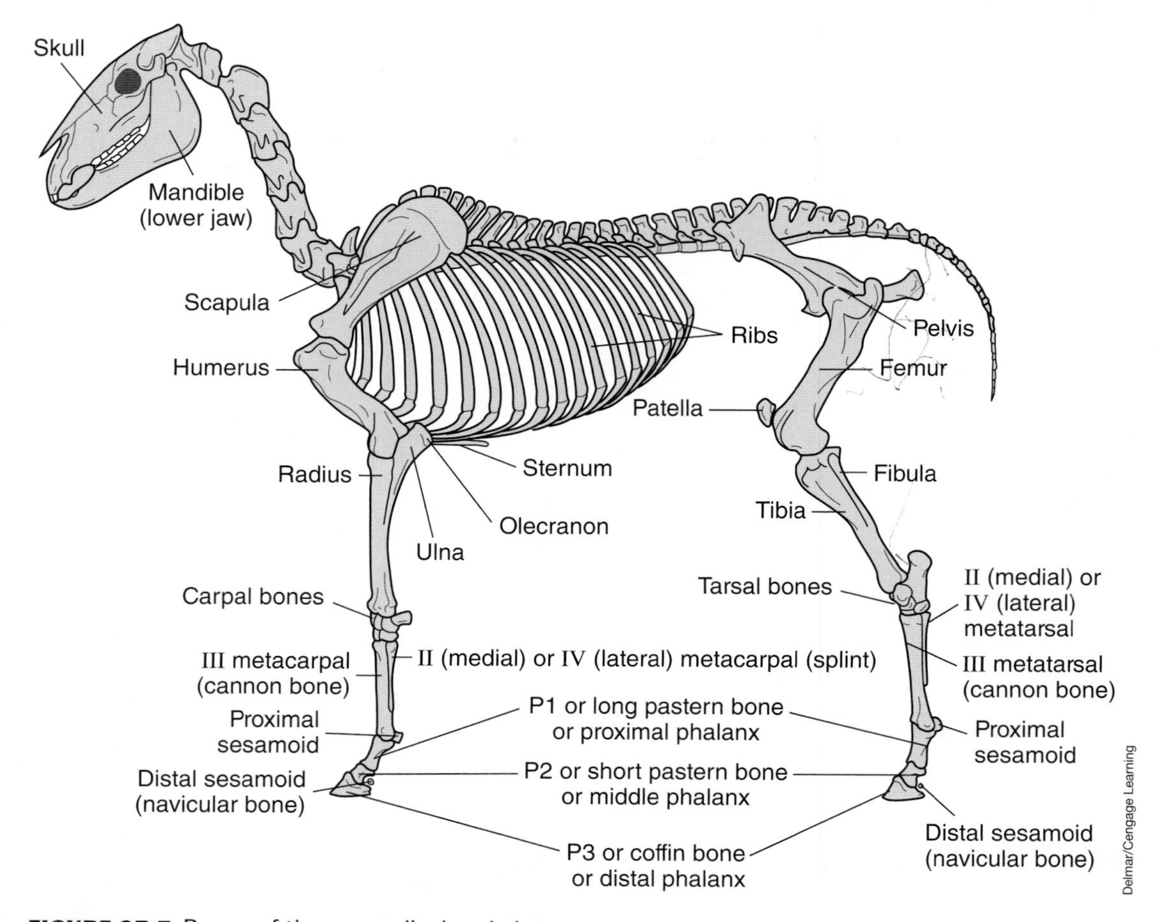

FIGURE 27-7 Bones of the appendicular skeleton.

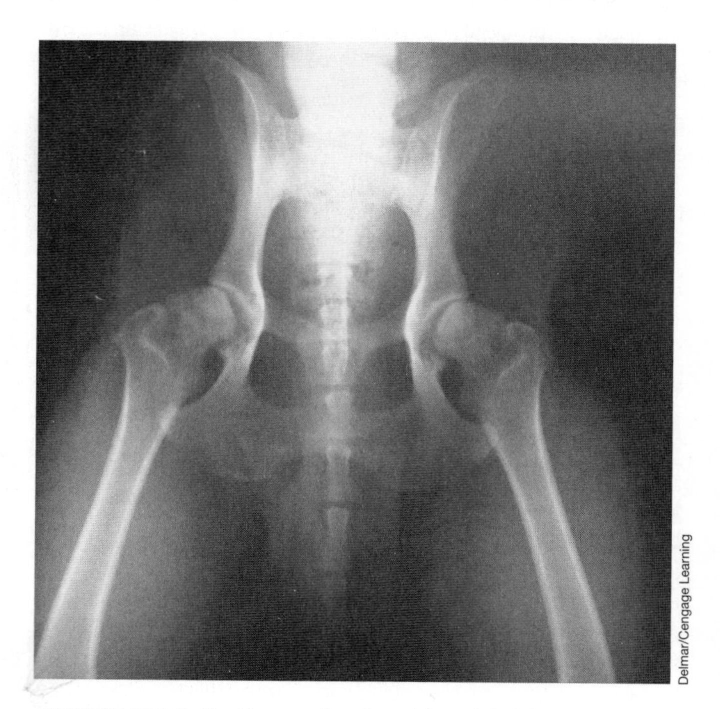

FIGURE 27-8 Radiograph of a dog with hip dysplasia. Note the shallow acetabulum and the subluxation (partial dislocation) of the femur.

are more predisposed for the condition and large-breed puppies that grow too quickly are more prone to this condition. Anti-inflammatory medications may be given to relieve pain and inflammation of the condition. Severe stages of hip dysplasia may require surgical hip replacement. Careful breeding of selective dogs will decrease the occurrence of hip dysplasia. Breeders should be evaluating their breeding dogs for hip dysplasia by having a hip X-ray certification through the Orthopedic Foundation of Animals (OFA) or through the University of Pennsylvania's PennHIP program.

Dog Breeds with Genetic Disposition to Hip Dysplasia

- German Shepherd
- Rottweiler
- Labrador Retriever
- Great Dane
- Golden Retriever
- Saint Bernard

Arthritis is a common condition of older animals in which the joints become inflamed and other changes in the joints may be noted. This condition causes lameness and pain and worsens over time. Signs of arthritis in animals include the following:

- Favoring a limb or limping
- Difficulty sitting or standing
- Sleeping more
- Seeming to have stiff or sore joints
- Hesitancy to jump, run, climb stairs
- Weight gain
- Decreased activity or less interest in play/activity
- Attitude or behavior changes
- Being less alert

Pain relievers and anti-inflammatory medications as well as dietary supplements may be helpful in reducing pain. Many arthritis medications are now available to aid the joints of affected animals.

Intervertebral disc disease or **IVD** is another common occurrence in dogs, but there are more injuries in certain breeds of long-backed dogs, such as the Basset Hound, Dachshund, and Corgi. IVD is caused when pressure occurs between the discs due to inflammation from an injury. The inflammation and swelling place pressure between the affected vertebrae, causing pain on the spinal nerves and possible trauma to the spinal cord. Most common signs of the condition include severe pain and possible paralysis. The paralysis can affect the back, head, neck, and limbs. The paralysis may affect the ability of the animal to control urination and defecation. Steroids may be used to help decrease inflammation and muscle relaxers may be necessary to reduce pain.

Cruciate ligament tears are a common injury to any active dog but tend to be more complicated in large-breed dogs. It is possible for any type of animal to have a ligament injury. Ligaments are located throughout the limbs, but the most common ligament injury is to the cruciate ligament, which lies over the patella. The cruciate ligament provides stability to the kneecap and is commonly torn during vigorous exercise or trauma to the knee. Lameness and sometimes complete loss of use of the limb will result. Diagnosis is by the determination of a **cranial drawer sign**, which indicates a torn ligament by moving the knee forward and back like a drawer opening and closing (see Figure 27-9). The tear may also involve the **meniscus**, another ligament that lies over the knee, with both ligaments forming an X shape and crossing over the patella. Restricted movement, pain medications, and surgical repair are necessary to reconstruct and repair the ligament tear.

Bone **fractures** cause a bone to break (see Figure 27-10). A bone may fracture in many different ways. A bone may come partially out of place causing a subluxation, or it may completely move out of place causing a **dislocation** or **luxation**. A **simple fracture** occurs when a bone has a single break but stays in place. A **comminuted fracture** is a bone that breaks in several different locations and causes fragments of the bone to occur. A **compound fracture** is a break in the bone that causes the bone to break through and penetrate the skin. Fractures can be repaired in several ways, but the bone must be realigned and not allowed to move for it to heal. Some fractures can be healed with the use of a **splint**, which is a support applied with padding and bandages to immobilize the injury. A **cast** is a material placed on the area of the

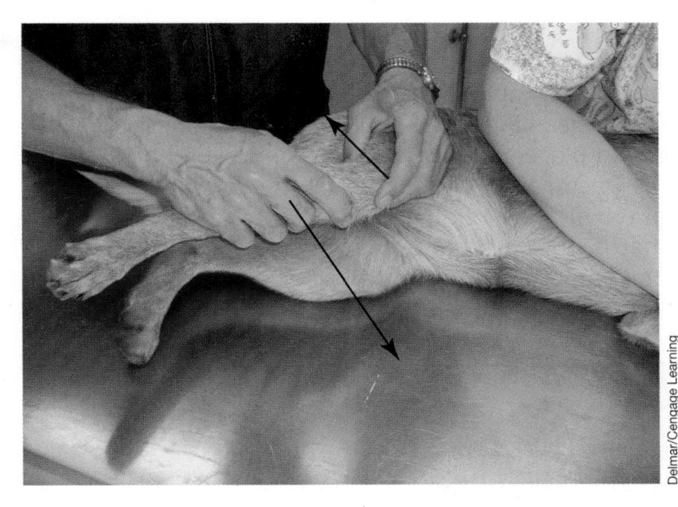

FIGURE 27-9 A veterinarian tests for cranial drawer sign in a dog's knee joint. The arrows indicate the direction of force being applied.

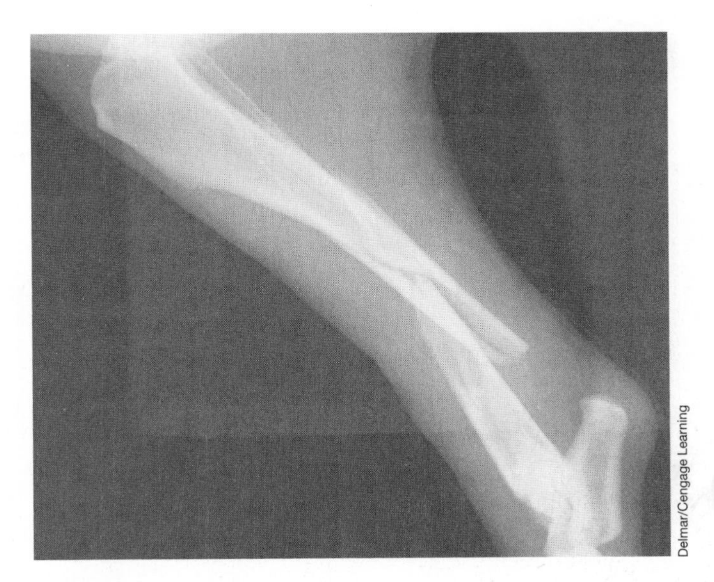

FIGURE 27-10 Radiograph of a fractured leg.

body where the fracture has occurred and is made of a hard substance to keep the bone in place. Surgical repair may include the need for an **intramedullary** or **IM pin** (see Figure 27-11). This is a stainless steel pin or rod placed into the center of a broken bone to hold the fracture site together while the bone heals. A **bone plate** may be required and is a surgical steel plate that is placed around the fracture site that holds the bone together with surgical screws (see Figure 27-12).

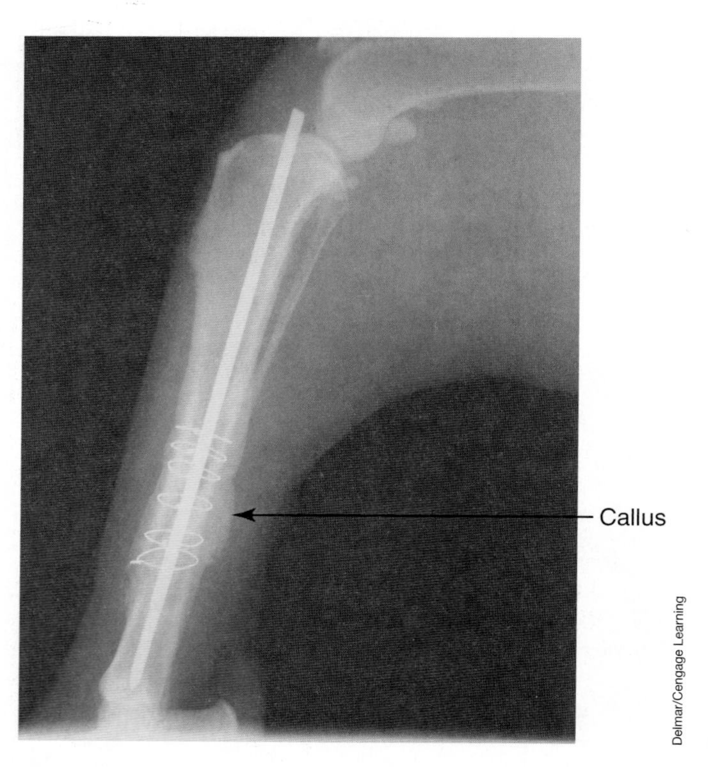

FIGURE 27-11 Radiograph showing placement of intramedullary pin.

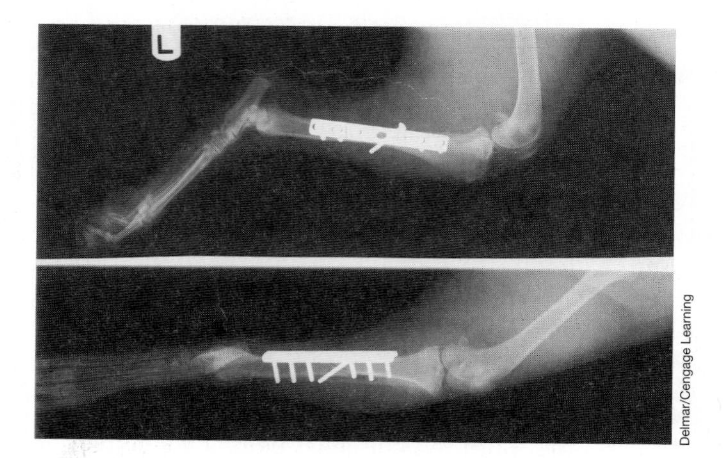

FIGURE 27-12 Radiograph showing use of bone plate to treat a fracture.

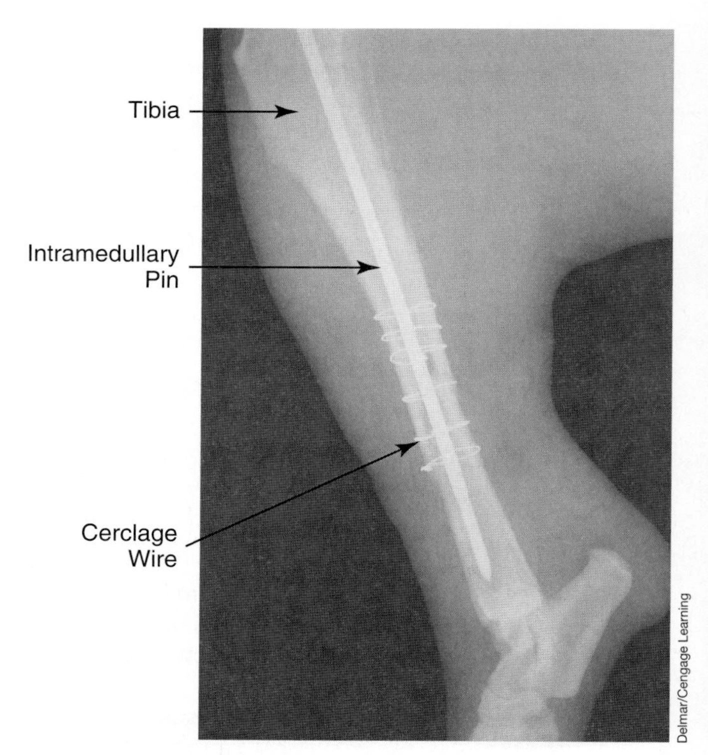

FIGURE 27-13 Radiograph showing use of cerclage wire in treatment of a fracture.

If bone fragments occur they may be held together with the use of **cerclage wire** to support the fragments as they heal (see Figure 27-13). Bone healing is similar to bone development. Cartilage rebuilds and forms a **callus** over the fracture site. This area becomes thickened and within two weeks osteoblasts begin to develop, and gradually the callus is replaced by bone. The duration of the complete healing depends on the fracture site, type of trauma, and healing care. Any bandage changes and postoperative care are essential for proper healing.

MAKING THE CONNECTION

For additional reading relating to disorders of the musculoskeletal system, see the following:

Chapter 19—Laminitis in horses
Chapter 21—Porcine Stress Syndrome
Chapter 22—Foot rot and Polyarthritis
Chapter 23—Joint ill

SUMMARY

The musculoskeletal system consists of bones, muscles, joints, and tendons. The axial skeleton consists of the skull, spine, and rib cage. The appendicular skeleton consists of the upper and lower extremities. This system provides support and protection for vital organs. It also allows the animal to have movement. Common disorders of the musculoskeletal system include disorders affecting joints. Trauma can also result in bone fractures. Musculoskeletal injuries may present with muscle pain or weakness, paralysis, and deformity. The animal may be less active and gaining weight.

Key Terms

abduction movement away from midline or the axis of the body

acetabulum area of the pelvis where the femur attaches to form the hip joint

adduction movement toward midline or the axis of the body

appendicular skeleton contains bones of the body that hang; the limbs or appendages

arthritis inflammation of the joint

atlas first cervical vertebrae located at the base of the brain that allows for up and down motions of the head

axial skeleton contains bones of the body that lie perpendicular or lengthwise

axis second cervical vertebrae that allows for rotation and shaking motions of the head

ball and socket joint that rotates in numerous directions

bone hard, active tissue that consists mostly of calcium and forms the skeleton of the animal; gives the body support, structure, and protection

bone marrow fluid within the medullary cavity where blood cells are produced

bone plate surgical steel plate placed around a fractured bone that holds the bone together with surgical screws

callus cartilage that rebuilds and thickens over a fractured bone that has healed

cancellous bone sponge-like, softer layer of bone located inside the end of bones

cannon long bones in hoofed animals located above the ankle joint

carpal lower leg that is similar to the wrist

carpus joint formed by several carpal bones and arranged in two rows in the area of the wrist

cartilage forms at the end of bones to protect and cushion the bone

cartilage joint connects at the end of a joint where two bones meet and acts as a cushion

cast material made of a hard substance and placed over a broken bone to keep the bone in place as it heals

cerclage wire surgical wire used to hold fragments of broken bone together while healing

cervical first section of the spinal column located over the neck area

coccygeal last section of the vertebrae that lies over the tail area

comminuted fracture bone breaks in several different locations and causes fragments

compact bone thick tissue that forms the outer layer of bone

compound fracture a break in a bone that causes the bone to break through and penetrate the skin

cranial drawer sign in and out movement or motion of the knee when the cruciate ligament has been torn

cruciate ligament tear common knee injury to large-breed dogs that tears the ligament over the patella

dewclaw the short inside bone that forms the first digit that acts as an opposable thumb

dislocation a bone that completely moves out of place

endosteum thin connective tissue covering inner bone

extension bending causing a joint to open and to lengthen

(Continues)

femur large upper bone of the rear leg; thigh bone

fibrous joint fixed joint with little to no movement

fibula smaller bone that lies in the lower rear leg and in the back of the limb

flexion bending causing the joint to close and to shorten

forelimb front leg of an animal

fracture broken bone

growth plate area of the bones in young animals that allow the bones to grow and mature as the animal ages and the cartilage joint turns to bone

hinge joint that opens and closes

hip dysplasia common genetic condition of large-breed dogs where the ball and socket joint of the femur and pelvis becomes diseased

hock common term for the point of the rear leg where the tibia and fibula meet

humerus large upper bone of the forelimb

ilium first section of the pelvis

IM pin intramedullary pin

intervertebral disc disease (IVDD) common injury of the back in long-backed breeds where inflammation and swelling occur when one vertebra puts pressure on another

intramedullary pin stainless steel pin or rod placed into the center of a broken bone to keep the bone in place as it heals

ischium second section of the pelvis

joints where two bones meet and allow a bending motion

ligament fibrous strands of tissue that attach bone to bone

loin the lower back or lumbar area

lumbar third section of the vertebrae that lie over the lower back

luxation a bone that comes completely out of place when broken

mandible lower jaw

maxilla upper jaw

medullary cavity hollow center within bone where blood cells are produced

meniscus ligament that lies directly over the patella and forms an X shape over the cruciate ligament

metacarpal bones that form the long bones of the feet in the forelimbs

metatarsal bones that form long bones of the feet in the rear limbs

muscle structures located throughout the animal's body that attach to different locations and serve to protect, allow bending and movement, and aid in the strength of the animal

musculoskeletal system body system consisting of bone, muscles, tendons, and ligaments

olecranon the elbow

ossification the formation of bone

patella knee cap that forms joint of the upper and lower rear leg

pelvic limb rear leg of an animal

pelvis hip bone

periosteum thin connective tissue covering outer bone

phalanges toes or digits

pivot joint that rotates around a fixed point

pubis third section of the pelvis

radius larger bone on the front of the lower forelimb

rib bone that attaches to individual thoracic vertebrae and protects the heart and lungs

sacral the fourth section of the vertebrae that lies over the pelvis area

sacrum area over the pelvis that is fused together and forms the highest point of the hip joint

scapula shoulder blade

simple fracture a break in the bone that may be complete or incomplete, but does not break through the skin; also known as a closed fracture

sinuses small, open spaces of air located within the skull and nasal bone

skull the bone that holds and protects the brain and head area

spinal column extends from the base of the skull to the end of the tail and allows movement; also called the spine

spinal cord canal of nerves that run through the center of the vertebrae located in the spinal column

splint support applied to a broken bone with bandage and padding to allow a bone to heal

splint bone two small bones on the back of a horse's cannon bone that may become inflamed from large amounts of work and stress on the legs

sternum breastbone that protects the organs of the chest

subluxate bone partially out of the joint

suture where two fibrous joints meet; fine line where little or no movement occurs

synovial joint moveable joint at the area where two bones meet

tarsal the point of the rear leg where the tibia and fibular meet; ankle area

tarsus joint of the ankle of the rear limb

tendon fibrous strands of tissue that attach bone to muscle

thoracic second section of vertebrae located over the chest area

tibia larger bone in the lower rear leg and lies in front

ulna smaller bone on the back of the lower forelimb

vertebrae individual bones of the spine that surround and protect the spinal cord

REVIEW QUESTIONS

1. What minerals are commonly stored in bones?
2. What does the term “flexion” mean?
3. What are the fibrous bands of tissue that connect one bone to another bone called?
4. In which section of the spinal column are the atlas and axis bones located?
5. What is the common name for the bone at the point of the rear leg where the tibia and fibula meet?
6. What are the parts of the musculoskeletal system?
7. Name the parts of the axial skeleton.
8. Name the parts of the appendicular skeleton.
9. Name and describe the musculoskeletal condition that large-breed dogs are prone to.
10. Name and describe a common disorder often affecting older animals.

Clinical Situation

© iStock/Waldemar Dabrowski

"Jake" Anderson, a 6 yr old NM English Springer Spaniel is rushed into Bell Animal Hospital. The dog has been HBC and is not bearing any weight on his R front limb. He is very painful. The veterinary assistant, Mark, helps transport "Jake" into an exam room and notes severe swelling over the leg and thinks the leg appears to be broken. Dr. Johns does a PE and some blood work, which appears to be normal, except for the leg injury. Mark and the vet tech, Julie, are asked to take radiographs of the leg. They perform two x-ray views and take the x-rays to Dr. Johns for evaluation.

Dr. Johns determines the R front radius and ulna are fractured. It appears to be a simple fracture over both bones. Dr. Johns takes the x-rays into the exam room to discuss his findings with the owner.

- What treatment options might Dr. Johns discuss with the owners?
- What supplies or equipment may be necessary to treat "Jake" when the vet is done talking to the owners?
- What home care may "Jake" require when he is discharged?

28

The Digestive System

Objectives

Upon completion of this chapter, the reader should be able to:

- State the structures that make up the digestive system
- Describe the function of the digestive system
- Differentiate between the monogastric, ruminant, nonruminant, and avian digestive systems
- Name and describe common disorders associated with the digestive system

Introduction

The **digestive system** or **gastrointestinal system** (GI) is responsible for **digestion** of food. *Digestion* is the breaking down of food particles into nutrients that are used by the body to allow the animal to live. Nutrients are converted to a form that cells can use and are transported through the digestive system.

TERMINOLOGY TIP

The digestive system is the common name for the **alimentary canal**, which is the veterinary medical term for the GI system.

Teeth

Food enters the mouth and is chewed through the use of the teeth. Animal teeth are adapted and structured according to the type of food they eat. Carnivores are animals that eat a meat-based diet. Their teeth are made for grasping food, tearing it up, and chewing it into smaller particles. Dogs and cats are carnivores and have long **canine** teeth, also known as fangs, that are used to tear apart food. **Herbivores** are plant-based food eaters and their teeth are structured for grinding down plant particles. Cows, sheep, goats, and horses are examples of herbivores. **Omnivores** are animals that eat both a plant- and meat-based diet. Their teeth are adapted for a variety of foods. Pigs and humans are omnivores.

Teeth are made of the hardest substance found in the body called **enamel**. Enamel is the covering of the tooth known as the **crown** and protects it from damage. The crown is the top part of the tooth that lies above the gum line and is visible in the mouth. The second layer of tooth is called the **dentin** and is similar to bone. The **root** is the portion of the tooth that is located below the gum and holds the tooth in place with one or more roots. Within the center of a tooth is the **pulp cavity**, which holds the nerves, arteries, and veins. Figure 28-1 illustrates the anatomical structure of the tooth.

Teeth are either **deciduous** or **permanent**. Deciduous teeth are known as baby teeth and are developed in the newborn animal and appear curved in shape. They are shed around the time the animal reaches maturity and replaced by the permanent teeth or the adult set, which are more straight in shape. The teeth are arranged in the mouth in a specific way known as **dentition** (see Figure 28-2 and Table 28-1). The teeth are arranged the same in every domestic animal, but the number of teeth differs according to the species. The front teeth located on the upper and lower jaw bones are the **incisors**. They are used to bite and grasp food. The canines or fangs are the next set of teeth and are the longest and sharpest teeth in the mouth. They are used to tear food apart. The next sets of teeth are the **premolars** and are wider teeth used to grind down and crush food. Dogs and cats have an upper 4th premolar and a lower 1st molar that

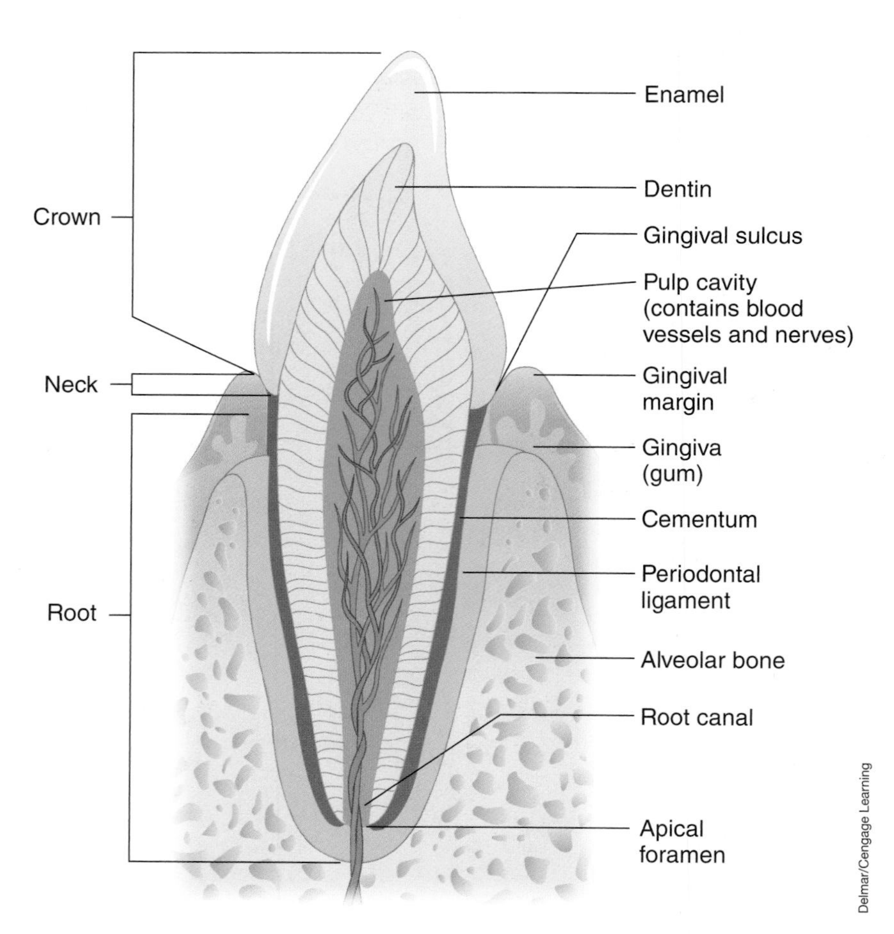

FIGURE 28-1 Anatomical structure of teeth.

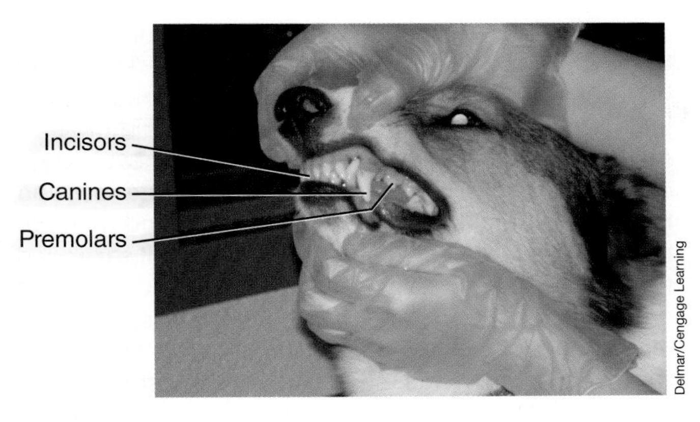

FIGURE 28-2 Dentition in dog.

TABLE 28-1

Dentition of Adult Domestic Animals

Dog	2 (I 3/3 C 1/1 P 4/4 M 2/3) = 42
Cat	2 (I 3/3 C 1/1 P 3/2 M 1/1) = 30
Horse	2 (I 3/3 C 1/1 P 3-4/3 M 3/3) = 24-42
Goat	2 (I 0/3 C 0/1 P 3/3 M 3/3) = 32
Sheep	2 (I 0/3 C 0/1 P 3/3 M 3/3) = 32
Pig	2 (I 3/3 C 1/1 P 4/4 M 3/3) = 44
Cow	2 (I 0/3 C 0/1 P 3/3 M 3/3) = 32
Rabbit	2 (I 2/1 C 0/0 P 3/2 M 3/3) = 28

Note: I—Incisor; C—Canine; P—Premolar; M—Molar

have a tendency to become infected and abscessed. This tooth is called a **carnasial tooth** and when abscessed causes a swelling in the check under the eye. The last sets of teeth are the **molars**. The molars are large teeth located in the back of the mouth. Horses use their teeth in a grinding motion that breaks down food. Over time, the teeth wear with age, helping to determine a horse's age due to the shape of the tooth and the amount of surface wear.

The Mouth

Animals have four **salivary glands** within the mouth that produce **saliva**, a fluid used to break down food and line the digestive tract for ease of swallowing and moving food along the intestinal tract (see Figure 28-3). The **tongue** is a muscle that lies within the mouth and is used to hold food within the mouth. The tongue is where saliva begins to mix with food. Some animals have **papillae** or hair on the tongue that act as taste buds. The next part of the digestive tract is the throat or the area at the back of the mouth called the **pharynx**. Food passes into the **esophagus**, the tube from the throat that passes food into the stomach. The entire digestive system is lined with a thin connective tissue called **mucosa**, which helps ease the passing of food. As food enters the stomach and intestinal tract, it moves through the system in wavelike

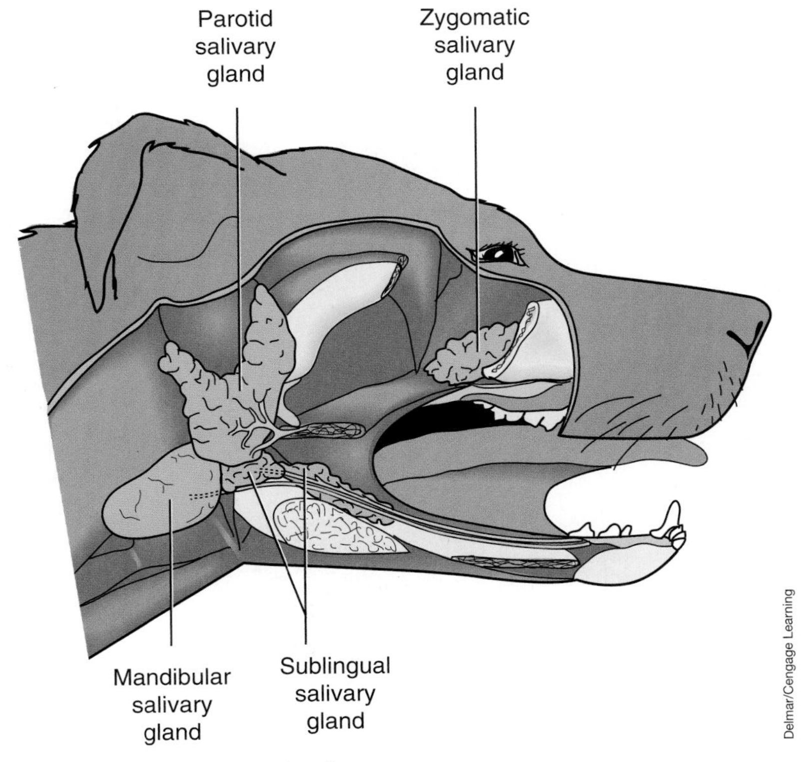

FIGURE 28-3 Salivary glands.

motions called **peristalsis**. These contractions move food through the entire GI system.

Stomach Systems

There are four types of animal digestive systems: monogastric, ruminant, nonruminant, and avian. Each system serves the animal by digesting particular types of food.

Monogastric Systems

The **monogastric** stomach is a single, simple sac that is divided into three parts: the **cardia**, the **body**, and the **pylorus** (see Figure 28-4). The cardia is the entrance to the stomach and filters food into the entrance through the opening called the **fundus**. The body of the stomach enlarges as food enters. The pylorus is the exit passage of the stomach. When the stomach is empty, it lies in folds known as **rugae**. Sphincter muscles allow food to enter and exit the stomach and peristalsis motions continue to move the food through the stomach as digestion occurs. Stomach acids are produced by the **liver**, the organ that lies behind the stomach and creates **bile,** secretions that help break down food to aide in digestion and absorption of food.

Food then moves from the stomach into the small intestine where most of the digestion and absorption occurs. The small intestine is actually longer than the large intestine, but the width is smaller. The small intestine is divided into three sections: duodenum, jejunum, and ileum. The **duodenum** is the short first section of the small intestine. The **jejunum** is the second or middle section of the small intestine. The last and final section is the **ileum**. The small intestine then enters into the large intestine by way of a small sac called the **cecum**. This pouch aides in further digestion of food and begins the widening of the large intestine, commonly called the **colon**. The cecum does not play a vital role in monogastric animals. The large intestine is also divided into three sections: the **ascending**, **transverse**, and **descending** colon. The ascending is the first section, the transverse is the second or middle section, and the descending is the last section.

The stomach and intestinal system are lined with different types of connective tissues. The **peritoneum** is a thin, clear lining of the entire abdomen. The **mesentery** is a connective tissue that extends from the peritoneum and carries blood vessels and nerves to the small intestine. The **omentum** is a thin lining that surrounds all abdominal organs. Several organs are responsible for digestion and metabolism occurring in the digestive system. The **pancreas** lies next to the intestinal tract and secretes enzymes that help aid in digestion of food. The three types of digestive enzymes produced by the pancreas are **trypsin**, which digests proteins; **amylase**, which breaks down starches; and **lipase**, which breaks

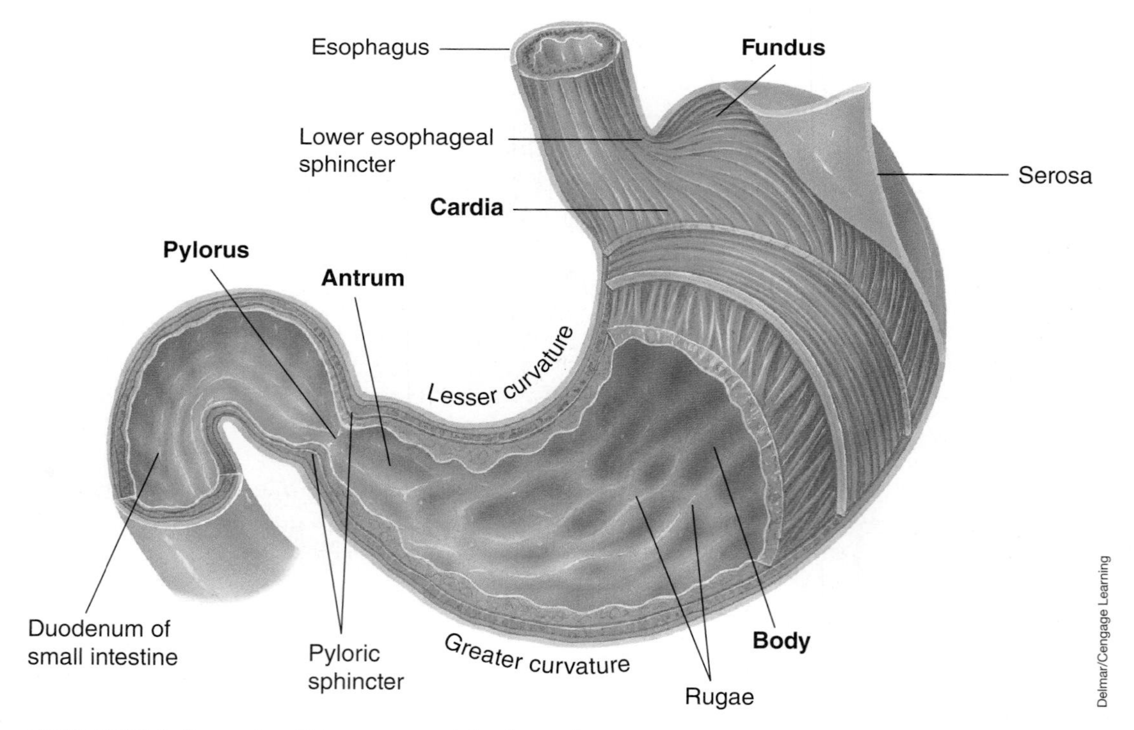

FIGURE 28-4 Example of the monogastric stomach.

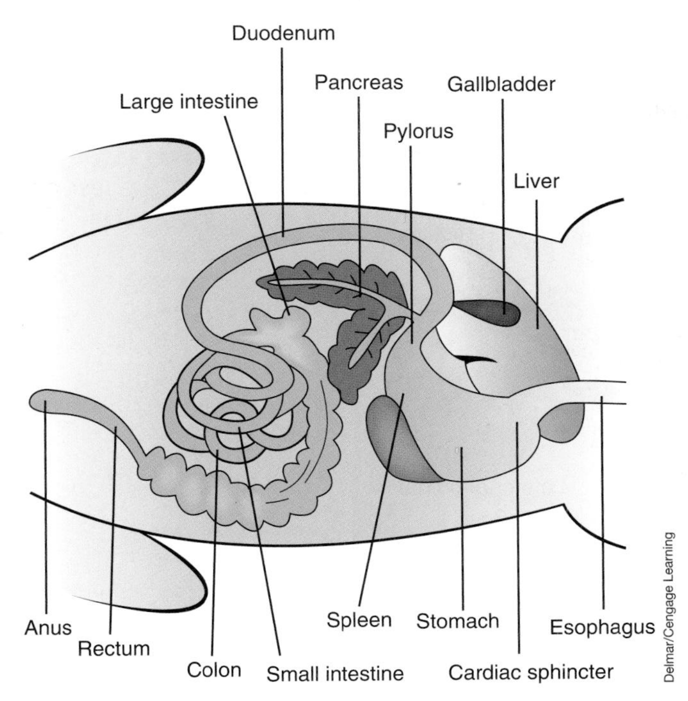

FIGURE 28-5 Structures of the monogastric digestive system.

down fats. The pancreas is also the organ that helps maintain blood sugar levels and produces **insulin** that is released into the bloodstream and helps to regulate the body's use of blood sugar, also called **glucose**. The liver lies under the pancreas and is constantly producing bile, which is a digestive enzyme that is transported to the gall bladder where it is stored for later use. Bile helps break down and digest fat. Many drugs and other toxins are metabolized and broken down by the liver. When too much or too little blood glucose is produced, the body finds it difficult to regulate and results in a condition known as **diabetes**.

Figure 28-5 illustrates the organs associated with the monogastric digestive system.

The Ruminant System

Another type of stomach in the digestive system is the **ruminant** stomach (see Figure 28-6). A ruminant is an animal, such as a cow, sheep, or goat, that has one large stomach with multiple compartments or sections. The ruminant stomach occupies about ¾ of the abdomen. Ruminant animals have no front teeth and are known as "cud chewers." **Cud** is the mixture of grass or hay sources and saliva. Food is broken down by the ruminant system and regurgitated for further breakdown to allow digestion to occur. **Regurgitation** is the process of bringing food into the mouth after it has been swallowed. The first section of the ruminant stomach is called the **rumen**. The rumen acts as a storage vat and soaks food to soften it during a process called **fermentation**, which allows food particles to break down by the use of bacteria for easier digestion. The rumen consists of about 80% of the stomach and is the largest compartment. The second section is called the **reticulum**. This compartment acts as a filter and composes about 5% of the stomach. The reticulum looks similar to a honeycomb pattern. This helps to trap dangerous materials that are not digestible and prevents them from passing through the digestive tract. Items that are not capable of passing this compartment are regurgitated for further breakdown. The third compartment is the **omasum**. The omasum composes about 8% of the stomach and absorbs nutrients and water. It also helps to grind down roughage and grains as they pass through the system. The last section is the "true" stomach called the **abomasum**. The true stomach is about 10% of the GI system and acts similar to the monogastric stomach. As food passes into each compartment and is regurgitated to be re-chewed for further breaking down, re-salivation, and re-swallowing. This occurrence continues until food is able to pass into the abomasum. Ruminants are known to chew with open mouths and air easily enters the rumen, forming gas. As the gas builds up, belching occurs to rid the rumen of the gas in a process called **eructation**.

The Nonruminant System

Another type of digestive system is the **nonruminant system**. This system is noted in horses, rodents, and rabbits. This system is similar to the monogastric digestive tract in appearance except it has a well-developed intestinal tract with a large cecum located between the small and large intestines (see Figure 28-7). Plant fibers are broken down and digested in the cecum. Small amounts of healthy bacteria also break down plant and hay fibers through fermentation. Nonruminant animals tend to eat slowly and require the need of digestion to occur through most of the day. A horse's intestinal system is about 70 feet in length, with the cecum measuring about 4 feet. Nonruminant animals are not capable of vomiting, as they do not have a **gall bladder**, which is an organ that stores bile. Nonruminant animals make bile in their intestinal systems and have no need for a gall bladder. This is also a factor as to why they can't vomit.

The Avian System

The last type of digestive system is the **avian system** (see Figure 28-8). This is a specialized GI system in birds and poultry species characterized by including

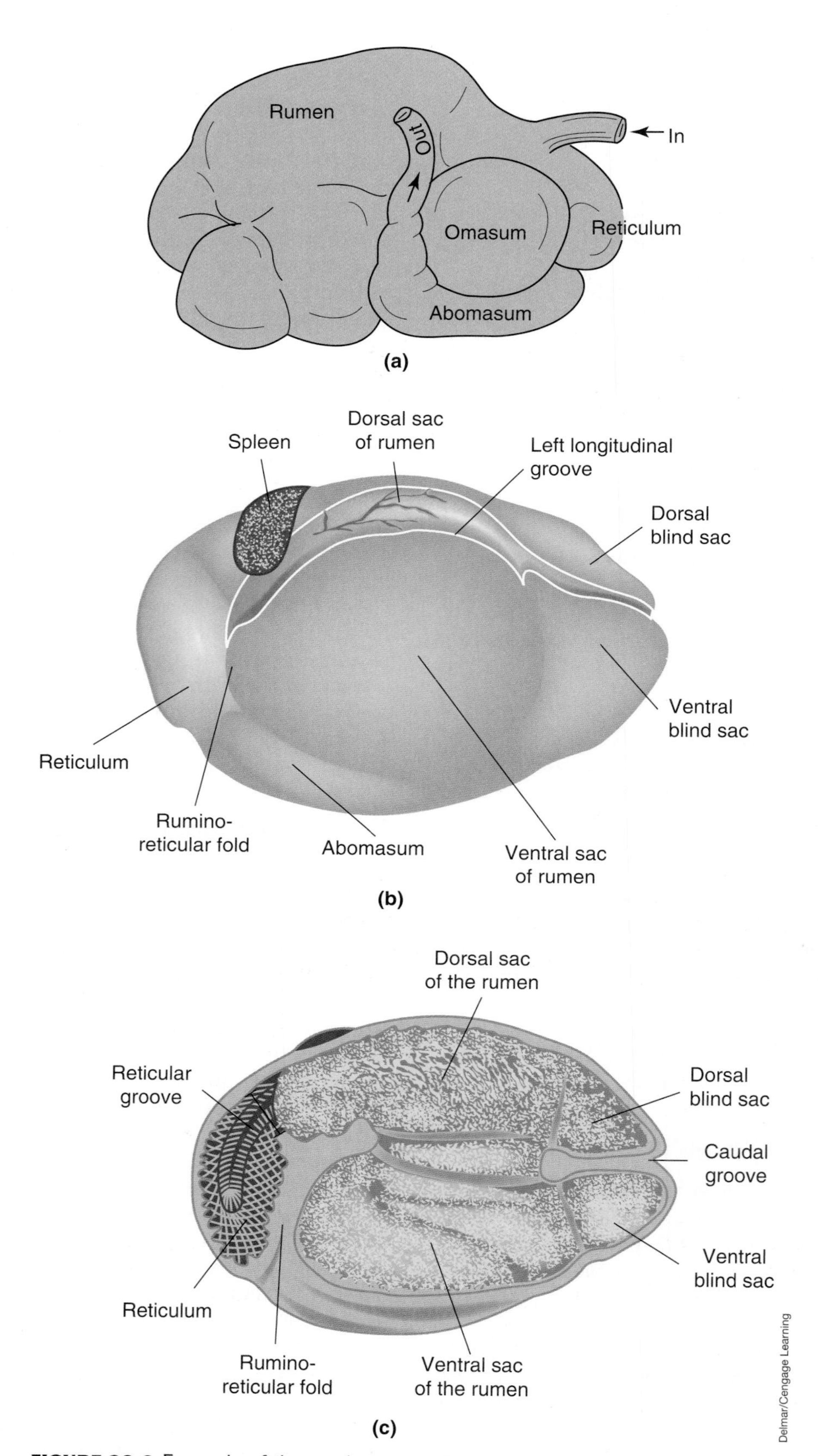

FIGURE 28-6 Example of the ruminant stomach.

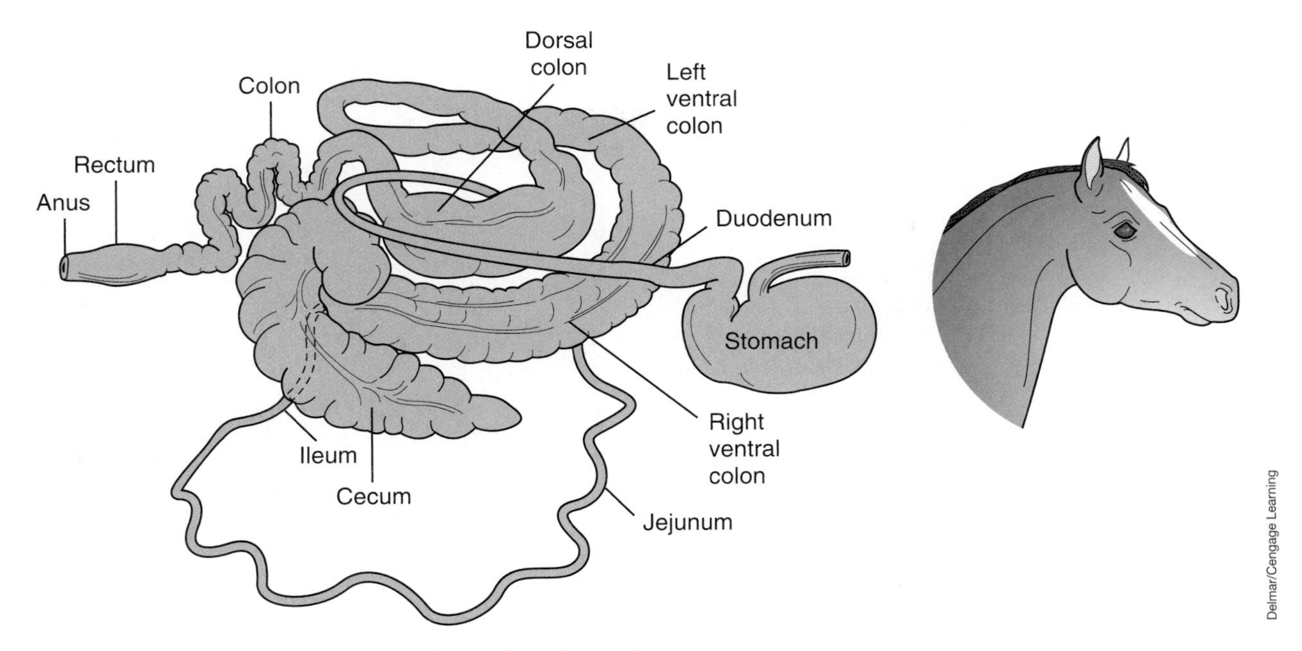

FIGURE 28-7 Example of a non ruminant digestive system.

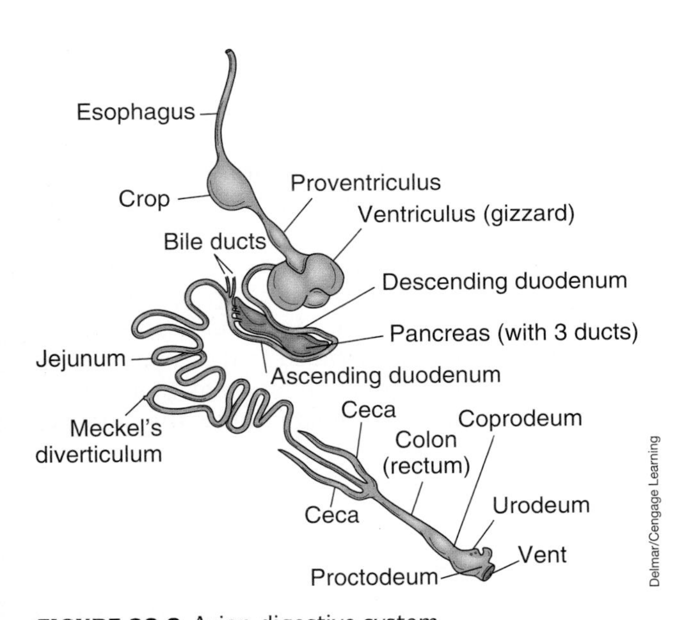

FIGURE 28-8 Avian digestive system.

several organs not found in other species that are used to break down and grind hard food particles. Avian mouth parts have no teeth and form what is called a **beak**. Salivation occurs to help soften food for swallowing. The esophagus passes food into a small holding tank called the **crop**. The crop softens and stores food for later use and slows digestion. The crop can be palpated to determine if digestion is occurring. Similarly to ruminants, birds can regurgitate food from the crop to be filtered and broken down for later digestion. When the food particles are small enough to pass into the stomach, the food passes into two specialized organs located in the stomach: the **proventriculus** and the **gizzard**. The proventriculus acts like the monogastric stomach and begins the digestion process by releasing excretions to soften food. The gizzard is a muscular organ located after the proventriculus and grinds down hard food particles, such as bones or seed shells. The food filters through small and large intestines similar to other animal species. The end of the digestive tract is the **cloaca**, which is the area where waste materials pass. Avian species pass both urates and feces at the same time from the cloaca. The cloaca is commonly called the **vent** and is the external area similar to the rectum.

Common Digestive Conditions

Digestive upset can be caused by many factors. The digestive systems reaction to this upset is typically **vomiting** or **diarrhea**. *Vomiting* is the act of bringing up partially or undigested food that has been in the stomach. Vomiting occurs in monogastric animals only. Ruminants and avians have a similar normal action in regurgitation, but this is due to how the animal needs to have food sources filtered and broken down for digestion. *Diarrhea* is the occurrence of waste materials becoming soft and watery, which can lead to an uncontrollable bowel movement. Animals that have occurrences of vomiting or diarrhea should be examined to determine the cause.

Dehydration

Both vomiting and diarrhea can lead to **dehydration**, the loss of fluids in the body. Dehydration can be established by the evaluation of **skin turgor**. This is the area of skin at the base of the neck and over the shoulders and back that is elastic-like. The hydration status can be determined by lifting the skin up and monitoring how quickly it "snaps" back in place. The normal skin turgor should snap back in place immediately like elastic. When dehydration occurs the skin turgor will stay "tented" or appear not to go back in its normal position over the neck and shoulders. The hydration status can also be determined by evaluating the gums or **mucous membranes (mm)** (see Figure 28-9). The color and amount of moisture can help identify if dehydration is occurring. Dehydration can also be seen in the eye sockets of animals and the eyes will appear sunken into the head.

FIGURE 28-10 Administering SQ fluids under the skin to treat dehydration.

> **TERMINOLOGY TIP**
>
> The gums or mm can be labeled normal or moist, **tacky** or slightly dry, and dry or no moisture.

Dehydration may cause the need to replace the body with **intravenous** (IV) or **subcutaneous** (SQ) fluids. Intravenous fluids are given into the vein by placement of an IV catheter, which is a small, sterile plastic tube that is inserted into the vein. This procedure is typically done by the veterinarian or veterinary technician and is necessary for severely dehydrated animals that may need long-term fluid therapy. Subcutaneous fluids are given under the skin, usually at the base of the neck or an area where the skin is plentiful and elastic-like (see Figure 28-10). This is a procedure that may be done by the veterinary assistant and requires a bag of fluids, a fluid line, and an appropriate-sized needle. This type of fluid replacement is for animals that are not too dehydrated and is a short-term therapy. Fluid replacement must be decided by the veterinarian. The type of fluids will depend on the amount of dehydration and the cause of the dehydration. It will also depend on the animal's overall physical health. Many fluids are sodium based as thirst control is provided to the animal. When vomiting occurs, the veterinarian may require an animal to have food and water withheld to allow the digestive system a period of rest.

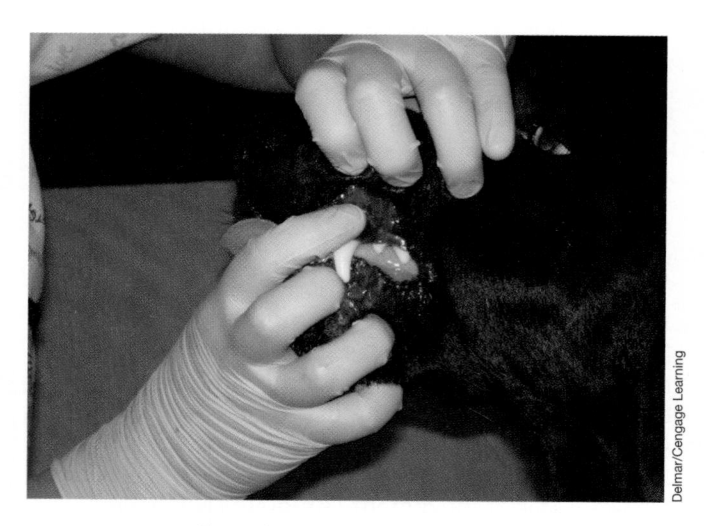

FIGURE 28-9 Examination of mucous membranes to determine hydrations status.

> **TERMINOLOGY TIP**
>
> The abbreviation **NPO** means "nothing per os" or nothing by mouth, meaning no food or water. The opposite of this abbreviation is **PO** or "per os" or by mouth.

Fluids can be given to replace water in the body until the animal is ready to eat and drink. When sodium-based fluids are given, it encourages the animal to drink. **Sodium chloride** or NaCl is a saltwater fluid. **Normal saline** is an isotonic solution, meaning it has the same concentration level of salt, and is often used to trigger an animal's brain to drink water. **Lactated Ringer's**

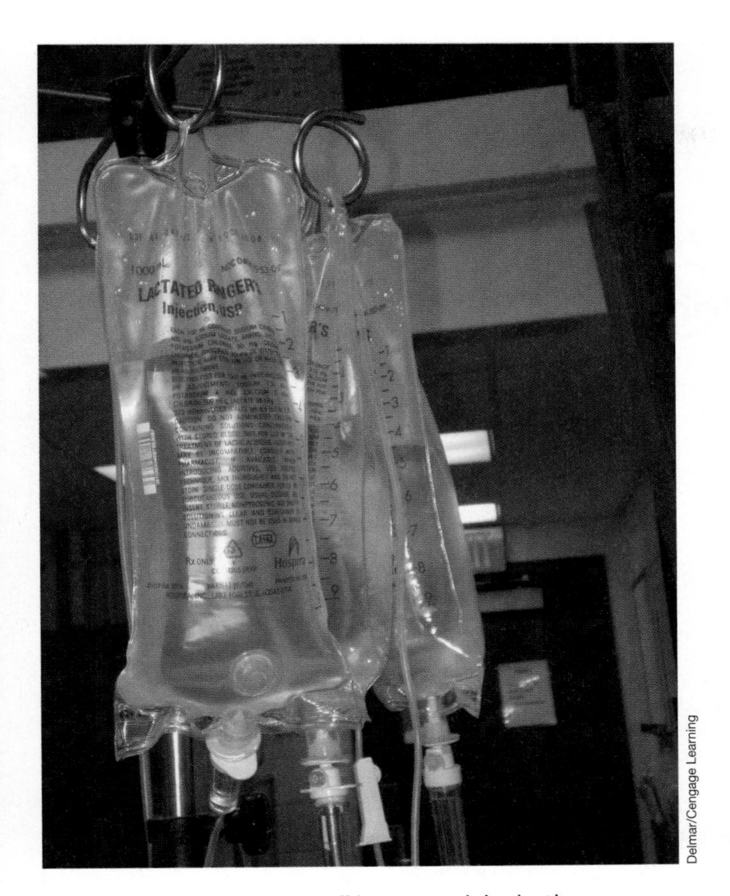

Delmar/Cengage Learning

FIGURE 28-11 Starting an IV to treat dehydration.

Solution (LRS) is a fluid that is commonly used to replace fluid loss (see Figure 28-11). The dehydration status helps to identify how much replacement fluid is necessary for an animal. Table 28-2 shows the amount of dehydration and the clinical signs. The calculation used to determine the amount of replacement fluids is the percentage of dehydration multiplied by the animal's body weight in kilograms (kg) multiplied by 1,000. This determines the replacement volume of fluids in milliliters (ml).

% dehydration × body of weight in kg × 1,000
= ml replacement volume

TABLE 28-2

Dehydration Percentages

6–7% dehydration	Moist mm; slight skin turgor; slightly sunken eyes
8–9% dehydration	Tacky mm; obvious sunken eyes; slow skin turgor with noted tenting
10–12% dehydration	Dry mm; very sunken eyes; very slow and very tented skin tugor

Constipation

Constipation is another occurrence in the digestive system that can cause little to no bowel movement production. A constipated animal may show signs of stomach pain, straining to produce a bowel movement, and decreased appetite. Many factors can cause constipation, such as poor diet, eating items that are not digestible, not drinking enough, or a low-fiber diet. Constipated animals should be evaluated by a veterinarian to determine the cause. Patients should be treated by the veterinarian according to how constipated they are. Some animals may require an **enema**, or a procedure that passes fluids into the rectum and colon to soften feces to allow the animal to have a bowel movement. Many times this is done using warm, soapy water. Animals may need **stool softeners** or **laxatives** given by mouth to help keep the waste materials soft for ease of passing. A high-fiber diet may also be recommended to produce normal bowel movements.

Colic

Colic is a common condition in horses. Colic is a severe stomach pain that may be caused by many factors; sometimes the factor remains unknown (see Table 28-3). Horses are not able to vomit, and when they begin showing signs of colic it is important to prevent them from rolling. The stomach pain caused by colic can cause a horse to lie down and roll to try to relieve the pain and pressure. When a horse rolls, it is likely that the stomach and intestinal tract can rotate upon itself. This is a condition known as **intussusception**. This causes the intestine to telescope upon itself, thus cutting off the circulation through the digestive tract (see Figure 28-12). When the circulation is cut off to an area of the body, the tissue begins to die, and this can cause a fatal condition in the animal. Horses, being nonruminant animals, must slowly digest food particles throughout the entire day. When colic or intussusception occurs this prevents digestion from occurring.

TABLE 28-3

Causes of Colic

Ingesting large amounts of grain or pasture	Ingesting fresh grass or hay sources
Stress	Excessive gas
Internal parasites	Excessive amounts of water after exercise or high temperatures
Dehydration	Ingesting sand
Sudden change in diet	Constipation or impaction
Abdominal tumors	Hernia
Medications	Unknown origins

Skills Competency

Converting pounds (# or lbs) to kilograms (kg)

Equipment and Supplies:

- Scale
- Pen or pencil
- Paper
- Calculator

1. Determine the weight of the animal in pounds using a scale. Record the weight of the animal in pounds.
2. Set up the formula of determining kg by using the following:

   weight in pounds divided by 2.2 = _______ kg

   # or lbs/2.2 = _______ kg
3. Record the body weight in pounds and kg in the medical record.

Converting kilograms (kg) to pounds is done in reverse.

1. Record the weight of the animal in kilograms (kg).
2. Set up the formula of determining # or lbs by using the following formula:

   weight in kg multiplies by 2.2 = _______ # or lbs

   kg × 2.2 = _______ # or lbs

EXAMPLE:

20# animal = 9.0 kg
20# / 2.2 = 9.0 kg

EXAMPLE:

10 kg animal = 22 pounds
10 kg × 2.2 = 22 #

Colic is an emergency and a veterinarian should be notified immediately. A physical exam and rectal palpation are necessary to determine how severe the condition may be. Feces should be monitored for amount and consistency to determine if the animal is capable of passing stool. The pain is typically treated and the cause may or may not be determined. A common pain reliever used for colic is Banamine. Intussusception is corrected through surgical repair, but the longer the condition has occurred the less likely the surgery will be successful. Signs of colic include frequent lying down and rising, rolling, **flanking** (looking at or biting at the sides of the abdomen), kicking at the sides of the abdomen, little to no bowel movement production, anorexia, depression, and little to no drinking.

Bloat

Bloat is a condition that affects dogs and some ruminants, such as cattle, sheep, and goats. Bloat causes the abdomen to become swollen and painful

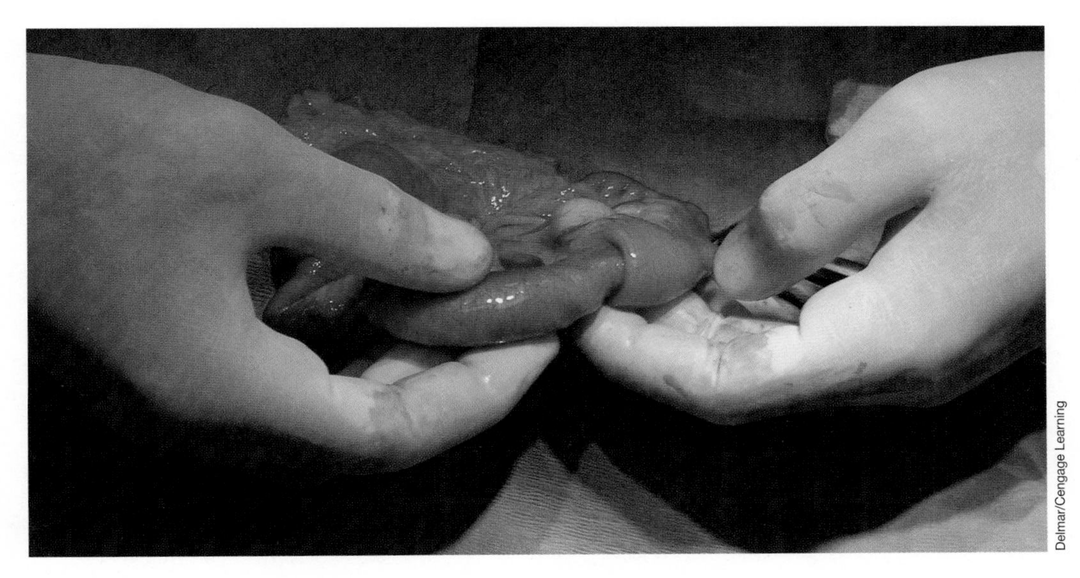

FIGURE 28-12 Intraoperative photograph of intussusception, where a region of the intestine has telescoped into another section.

due to air and gas within the stomach and intestinal tract. Bloat in dogs is more common in large and giant breeds, but it can occur in any type of dog. Dogs that become bloated should be handled as an emergency and seen immediately by a veterinarian. Dogs may become bloated for several reasons, usually by ingesting food very quickly, exercising shortly after eating, getting into garbage or food they are not accustomed to eating, or drinking excessive amounts of fluids. This condition is called **gastric dilation**. Dogs that become bloated could progress to a condition known as **gastric dilation volvulus** or GDV, in which the stomach or intestinal tract rotates and causes a similar condition to that of horses that colic. Bloat or gastric dilation may be treated by placing a stomach tube into the GI system to remove excessive air or gas. Dogs that have GDV require immediate surgery to correct the rotation. Cattle and ruminants that become bloated usually cause the rumen to become gas filled, and eructation can't occur to remove the gas. Bloat develops causing a **distended** or swollen abdomen. This typically occurs on the left side of the rumen (see Figure 28-13). Two types of bloat may occur in ruminant animals: free gas or frothy bloat. **Free gas** accumulates in the dorsal rumen and the animal usually shows signs of choking when the esophagus becomes obstructed with ingesta, causing the gas to not be able to escape. The rumen stops contracting and regurgitation ceases. **Frothy bloat** is caused by gas being trapped in small bubbles within the rumen. This is typical of animals that eat lush green pastures and build up gas that becomes trapped. Large

FIGURE 28-13 Bloat in a cow. Note the distended appearance on the left side.

ruminants are typically treated by the veterinarian by placing a **trocar** within the rumen to allow the gas to escape and relieve the pressure on the animal's stomach. A trocar is a plastic or metal instrument with a pointed, sharp end that is placed within the rumen through the abdominal wall and that allows the air to drain (see Figure 28-14).

Ruminants can also have dilation volvulus occur in which the stomach rotates and becomes displaced. This requires a surgical incision into the side of the abdominal wall in which the veterinarian can place an arm and rotate the stomach back to its place.

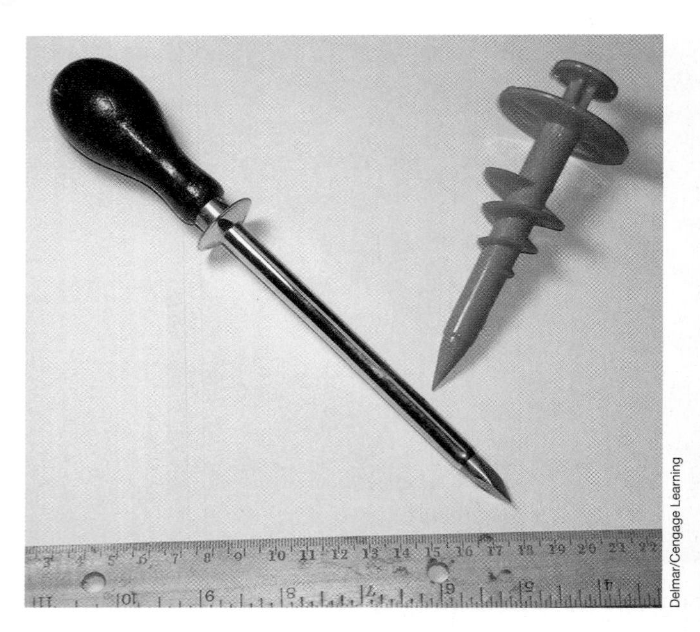

FIGURE 28-14 Examples of trocars.

Foreign Body Obstructions

Foreign body obstructions may occur in any type of animal that ingests an item that is not digestible or may become impacted within the digestive system. Items that are ingested may be toys, household items, or items found outside (see Figure 28-15). Diagnosis of a foreign body within the digestive tract may be done through radiographs or through the use of **oral barium study**. Oral barium is given by mouth and radiographs are taken over time as the barium passes through the digestive system. The barium is **radiopaque** and fluoresces as it passes through the GI system (see Figure 28-16). The radiographs will help determine if an object is moving through the system or if it has become obstructed and requires surgical removal. A foreign body obstruction may cause the intestinal tract to telescope upon itself causing an intussusception. This causes the blood supply to be cut off. When a foreign body is surgically removed from the stomach or intestinal tract and the blood supply has been cut off causing an area of dead or **necrotic** tissue to occur, there may be a need for an **anastomosis** surgery. This is the surgical removal of an area of dead tissue along the intestinal tract, which requires a resection of the intestines.

Signs of an animal with a foreign body obstruction include the following:

- Vomiting
- Retching or gagging
- Diarrhea
- Anorexia

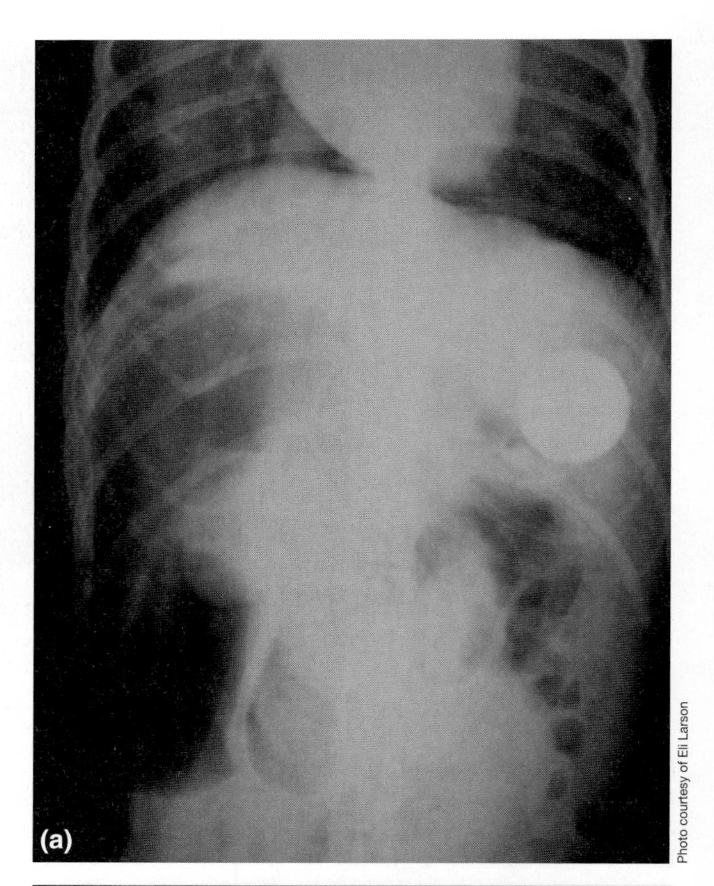

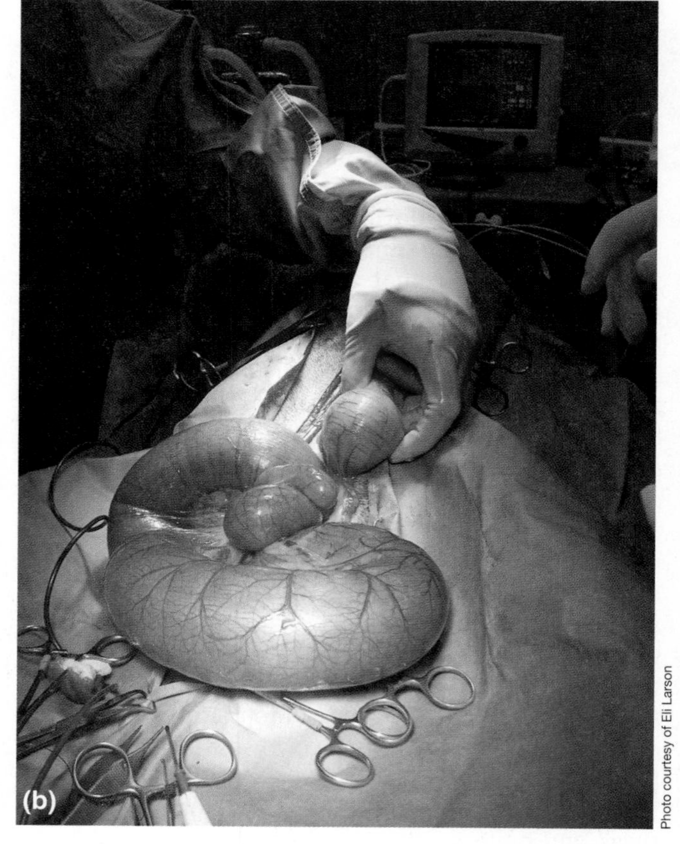

FIGURE 28-15 Dog with an intestinal obstruction (golf ball).

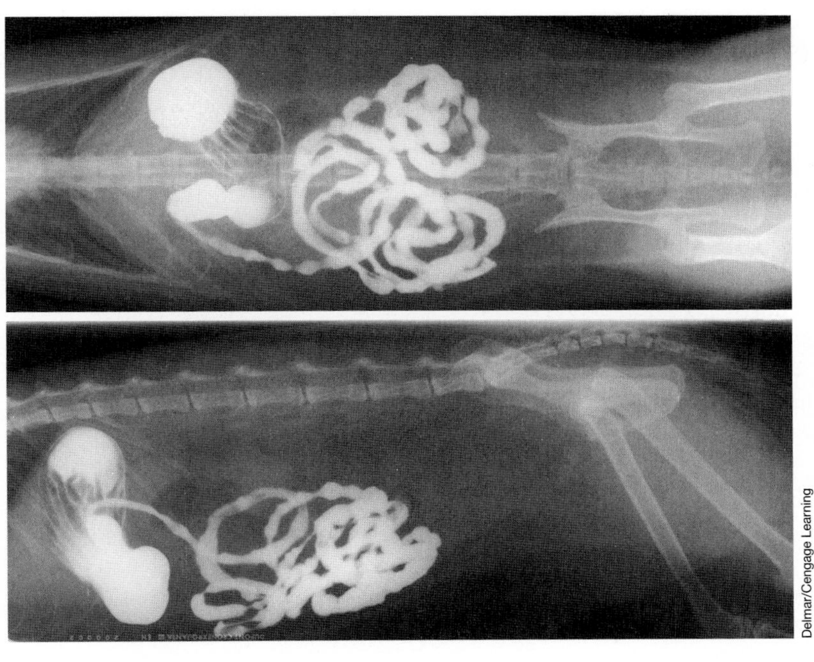

Delmar/Cengage Learning

FIGURE 28-16 Radiograph showing barium series in a cat. The bright white area is the barium within the gastrointestinal tract.

- Abdominal pain
- Dehydration
- Depression

SUMMARY

The digestive system provides nutrients to the body. Four different types of digestive systems can be found in animal species: monogastric, ruminant, nonruminant, and avian. The variation in the types of digestive systems found in animals is related to the types of foods the animals eat. Some of the common conditions veterinarians will treat related to the digestive system are vomiting, diarrhea, constipation, dehydration, bloat, colic, and foreign body obstructions. Some of these conditions present as medical emergencies and immediate care is required to save the animal's life.

MAKING THE CONNECTION

Additional information on disorders affecting the digestive system can be found in the following:

Chapter 10—fatty liver disease
Chapter 12—Tyzzer's disease
Chapter 18—bovine viral diarrhea, campylobacter, bloat, grass tetany, and ketosis
Chapter 19—colic
Chapter 21—swine dysentery
Chapter 22—enterotoxemia, Johne's disease, and lamb dysentery
Chapter 23—bloat

Key Terms

abomasum last section of the ruminant stomach that acts as the true stomach and allows food to be digested

alimentary canal veterinary medical terminology for the GI system

amylase enzyme produced by the pancreas that breaks down starches

anastomosis surgical removal of a dead area of tissue along the digestive tract and re-sectioning the area back together

(*Continues*)

ascending colon first section of the large intestine

avian system specialized digestive system of birds

beak avian mouth with no teeth that forms an upper and lower bill

bile yellow fluid secretion that helps break down food for digestion and absorption of food

bloat condition that causes the abdomen to become swollen and painful due to air and gas within the intestinal tract

body central part of the stomach that expands as food enters

canine teeth also known as fangs that are used to tear apart food

cardia entrance of the stomach that filters food

carnasial tooth upper 4th premolar and lower 1st molar in dogs and cats that tends to become infected and abscessed

cecum the small sac that lies between the small and large intestines

cloaca end of the digestive tract where waste materials pass

colic condition in horses that causes severe stomach pain

colon common term for the large intestine

constipation occurrence in the digestive tract that can cause little to no bowel movement

crop small sac that acts as a holding tank for food as it is passed from the esophagus in birds

crown the upper part of the tooth that lies above the gum line

cud mixture of grass sources and saliva that is chewed and regurgitated to break down food for digestion

deciduous baby teeth that are developed in newborn animals and eventually shed when adulthood is reached

dehydration loss of fluids in the body

dentin second layer of teeth similar to bone

dentition the way teeth are arranged in the mouth

descending colon third or last section of the large intestine

diabetes condition that is produced when too much or too little blood sugar is produced and the body finds it difficult to regulate

diarrhea process of waste materials and feces become soft and watery

digestion breaking down food particles into nutrients to be used by the body to allow an animal to live

digestive system the body system that contains the stomach and intestines

distended swollen, as in the abdomen

duodenum short, first section of the small intestine

enamel hardest substance in the body that covers and protects the teeth

enema procedure of passing fluids into the rectum to soften feces to produce a bowel movement

eructation gas buildup where belching occurs to rid the rumen of air

esophagus tube that passes food from the mouth to the stomach

fermentation process of soaking food that allows bacteria to break down food for easier digestion

flanking looking at or biting at the sides of the abdomen due to stomach pain, often in a colic situation in horses

foreign body obstruction an animal ingests a foreign object that is not digestible and it becomes impacted within the intestinal tract

free gas air accumulates in the dorsal rumen of a ruminant's stomach causing the animal to choke when the esophagus becomes obstructed with food and saliva, causing the gas to not be able to escape

frothy bloat caused by gas being trapped within small bubbles within the rumen causing the abdomen to become swollen and painful

fundus opening of the stomach

gall bladder organ that stores bile

gastric dilation veterinary term for the condition known as bloat in which air or gas fills the stomach, causing the abdomen to become swollen and painful

gastric dilation volvulus (GDV) condition where the stomach and intestinal tract rotate after becoming swollen due to air or gas in the GI tract, causing the intestinal tract's circulation to be cut off

gastrointestinal system (GI) the digestive system that contains the stomach and intestines

gizzard muscular organ located after the proventriculus in birds that grinds down hard food substances

glucose veterinary terminology for blood sugar

herbivores animals that eat plant-based food sources

ileum third or last section of the small intestine

incisors the front teeth located in the upper and lower jaws

insulin chemical produced by the liver that is released into the bloodstream and regulates the body's blood sugar

intravenous (IV) given within the vein

intussusception condition when the stomach or intestine telescopes upon itself, cutting off circulation of the digestive tract

jejunum second or middle section of the small intestine

Lactated Ringer's Solution (LRS) fluid of lactic acid that is commonly used to replace fluids, as in dehydration

laxative veterinary term for stool softeners or medicine given to soften feces to produce a bowel movement

lipase enzyme produced by the pancreas that breaks down fats

liver organ behind the stomach that makes bile and produces glucose

mesentery connective tissue from the peritoneum and carries blood vessels and nerves to the small intestine

molars last set of teeth that are large and located in the back of the mouth

monogastric digestive system of an animal with one simple stomach

mucosa thin connective tissue that lines the intestinal tract

mucous membranes (mm) the gums

necrotic dead area of tissue

nonruminant system digestive system similar to monogastric animals with a larger, well-developed cecum for breaking down plant fibers

normal saline solution with the same concentration level as salt

NPO nothing by mouth, as in food or water

omasum third section of the ruminant stomach that absorbs nutrients and water

omentum thin lining that surrounds the organs within the abdomen

omnivores animals that eat both plant- and meat-based food sources

oral barium study barium solution given by mouth to pass through the digestive system to allow X-rays to be taken over time to view internal structures of the GI tract

pancreas organ that lies next to the stomach and secretes enzymes that aid in digestion

papillae hair on the tongue that acts as taste buds

peristalsis wave like motion of the stomach that moves food through the intestine in contractions

peritoneum clear thin lining of the abdomen

permanent adult teeth that are formed after the deciduous teeth are shed

pharynx throat or area at the back of the mouth

PO by mouth

premolars wider teeth at the back of the mouth used to grind and tear food

proventriculus acts as a monogastric stomach and begins the digestion process in birds by releasing excretions to soften food

pulp cavity center of the tooth that holds the nerves, veins, and arteries

pylorus exit passageway of the stomach

radiopaque solution that fluoresces and allows radiation to pass through to view internal body structures during X-rays

regurgitation process of bringing food into the mouth from the stomach to break it down

reticulum second section of the ruminant stomach that acts as a filter for food

root part of the tooth located below the gum line that holds the tooth in place

rugae folds within the stomach when it is empty

rumen first section of the ruminant stomach that acts as a storage vat and softens food for fermentation

ruminant animal with a digestive system that has a stomach with four sections or compartments

(*Continues*)

saliva fluid that helps soften and break down food for ease of swallowing and digestion

salivary glands area within the mouth that produces saliva

skin turgor process of evaluating an animal for dehydration by lifting the skin over the base of the neck or shoulder blades

sodium chloride NaCl; saltwater fluid

stool softeners medication given to produce a bowel movement by softening feces

subcutaneous (SQ) given under the skin

tacky slightly dry, as in the gums

tongue muscle within the mouth used to hold food within the mouth

transverse colon second or middle section of the large intestine

trocar plastic or metal pointed instrument placed into the rumen of a ruminant animal that has bloated to relive the air pressure on the animal's stomach

trypsin enzyme produced by the pancreas that digests proteins

vent external area of an avian that passes waste materials; also called cloaca and similar to the rectum

vomiting process of bringing up partially or undigested food that has been in the stomach of monogastric animals

REVIEW QUESTIONS

1. The involuntary muscle action of the digestive system is called:
2. What is an animal that eats a meat-based diet called?
3. What are the teeth used to grind down food called?
4. What is the enzyme that breaks down starches called?
5. What is a common gastrointestinal condition in horses that causes severe pain and may be fatal called?
6. What are the four types of digestive systems?
7. List two examples of species with each type of digestive system.
8. Compare and contrast bloat and colic.
9. List the organs associated with the digestive system.
10. Describe the function of the ruminant digestive system.

Clinical Situation

© iStockphoto/Eric Lam

"Harley" is a 4 yr NM lab known for ingesting clothing, toys, and anything he can get in his mouth. He has been vomiting for 2 days and is lethargic. He appears depressed and in some discomfort when Allie, the vet assistant, takes him into an exam room. "Harley" has a normal RR and HR but has an elevated temperature of 103.0 degrees.

Dr. Osborne completes a PE, blood work and orders radiographs. The x-rays show a buildup of gas in the stomach and small intestine, but no foreign body is noted. Dr. Osborne suggests that "Harley" be admitted for a barium study.

Allie and Kathleen, another vet assistant, admit "Harley" into the ICU. The owner has several questions for the assistants as they place "Harley" into a kennel. Her main concern is asking about the barium study procedure.

- What should the assistants explain to the owner about the barium study?
- What should be discussed with the owner if the vet finds surgery is necessary?
- What supplies and equipment will Allie and Kathleen need to complete the barium study radiographs?
- What steps should the assistants take in admitting "Harley" into the ICU?

29

Delmar/Cengage Learning

Circulatory System

Objectives

Upon completion of this chapter, the reader should be able to:

- State the structures that make up the circulatory system
- Describe the functions of the circulatory system
- Identify and describe the function of the various types of blood cells
- Describe blood flow through the heart
- Describe common disorders of the circulatory system

Introduction

The **circulatory system** is essential for life and involves the heart, blood, arteries, veins, and capillaries. The functions of the circulatory system include oxygen flow, blood circulation, and transport of nutrients, waste removal, and the movement of hormones.

Blood

The circulatory system begins with blood flow. **Blood** is composed of 40% cells and 60% **plasma**. *Plasma* is formed of various proteins: albumin, globulin, and fibrinogen. **Albumin** draws water into the bloodstream and helps in providing hydration to the body. **Globulins** provide antibodies to help prevent disease. **Fibrinogen** aids in clotting blood.

> **TERMINOLOGY TIP**
>
> The term to **clot** or clotting is the process of blood flow stopping, as in a cut or puncture that causes bleeding.

Hematology is the study of blood and an essential practice in the veterinary facility. The veterinary assistant serves an important role in setting up for blood collection, running blood analysis samples, restraining patients for blood collection, and completing blood testing kit samples. Blood samples are collected by the veterinarian or technician. Blood tube samples should be labeled with the patient's name, owner's name, date, and test procedure requested. Some testing procedures require the blood tube to be spun at high speeds using a piece of equipment called a **centrifuge** (see Figure 29-1). The centrifuge uses speed to separate blood elements, causing the cells to separate from the liquid portion of blood, known as **serum**. Serum is the liquid portion of a blood sample that is used to analyze chemistry values to determine the functions of the body's organs. The blood is composed of red cells and white cells. **Red blood cells** (RBCs) are known as **erythrocytes** and are the most abundant blood cell in the body (see Figure 29-2A). Their main function is to transport oxygen throughout the body. The red blood cells are produced in the bone marrow through the process of **erythropoiesis**. Red blood cells are constantly being produced and replaced. The main component of red blood cells that allows for oxygen transport and produces iron that allows red blood cells to be replicated or continue being produced is a protein called **hemoglobin**. **White blood cells** (WBCs) are called **leukocytes** and are the body's main defense against infection. The blood cells are commonly evaluated under a microscope to determine the type of cell and how many are noted in the blood. (Figure 29-2 B, C, D) There are five types of white blood cells found in an animal's bloodstream: neutrophils, lymphocytes, eosinophils, monocytes, and basophils. Each type of white blood cell serves a specific function. **Neutrophils** are the most commonly seen white blood cells and destroy any microorganisms in the tissues. A mature neutrophil has a nucleus with segments or divisions. An immature neutrophil is called a **band cell** and indicates that there is an infection in the body. Band cells have a nucleus that is shaped like the letter U (see Figure 29-2B). **Lymphocytes** make up the largest part of the bloodstream and aid in immune functions and help to fight diseases by producing antibodies in the blood. Lymphocytes have one large single nucleus that makes up most of the blood cell (see Figure 29-2C). **Eosinophils** fight against allergic reactions, help control inflammation, and help to prevent parasite infections within the body. They look similar to neutrophils in that they have a segmented nucleus, but they have a large number of granules in the cytoplasm that give them a dotted appearance (see Figure 29-2D). **Monocytes** are the largest white blood cells and help the neutrophils by removing organisms, dead cells, and foreign particles. As monocytes age they become **macrophages**, which are cells that eat and destroy organisms at different locations within the body. **Basophils** are cells that are involved with allergic reactions, and they have a segmented nucleus with granules that stain very dark (see Figure 29-2E). The granules contain **histamine**, which is released during an allergic reaction. Another type of cell found within the blood is a **platelet**. Platelets aid in clotting time of blood after an injury occurs to a blood vessel, causing the vessel to constrict. Platelets attach to the vessel site and plug the hole and help decrease bleeding. Platelets are also known as **thrombocytes** and appear in a blood smear as small dots. When not enough platelets are formed in the blood, a condition called **anemia** occurs. This causes a low red blood cell count that doesn't allow the red blood cells to replenish. **Autoimmune disease** may occur when the animal's red blood cells are destroyed by its immune system.

Delmar/Cengage Learning

FIGURE 29-1 Centrifuge.

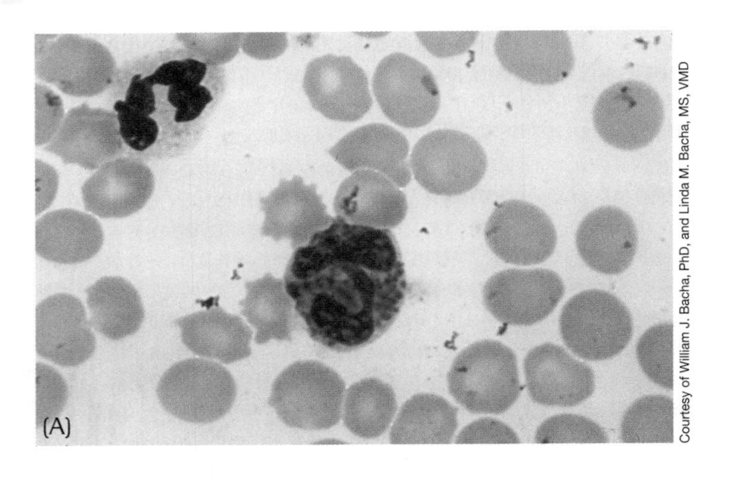

Courtesy of William J. Bacha, PhD, and Linda M. Bacha, MS, VMD

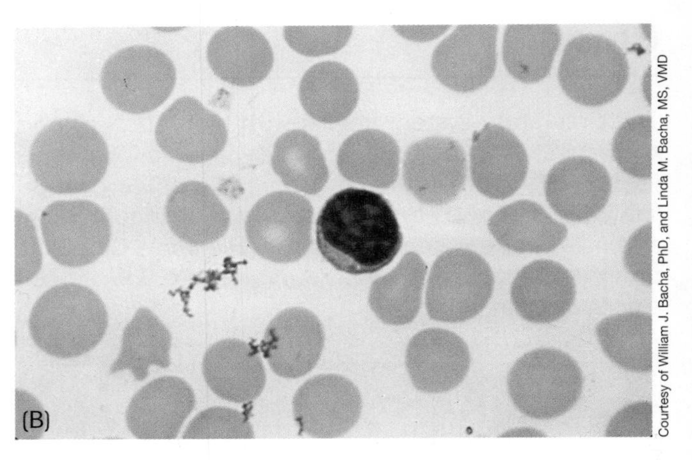

Courtesy of William J. Bacha, PhD, and Linda M. Bacha, MS, VMD

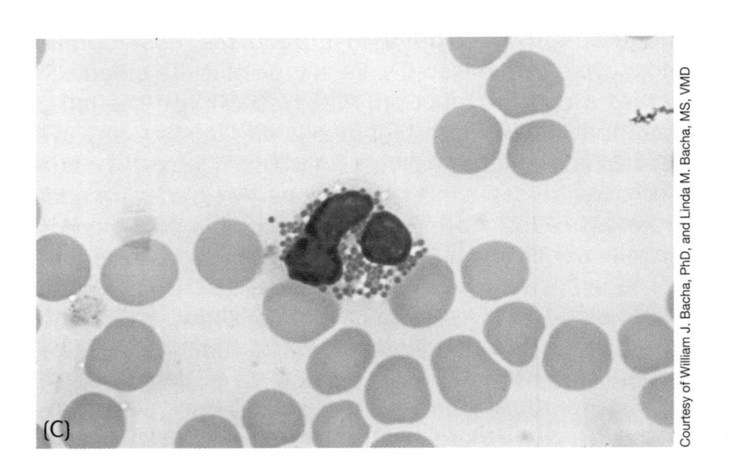

Courtesy of William J. Bacha, PhD, and Linda M. Bacha, MS, VMD

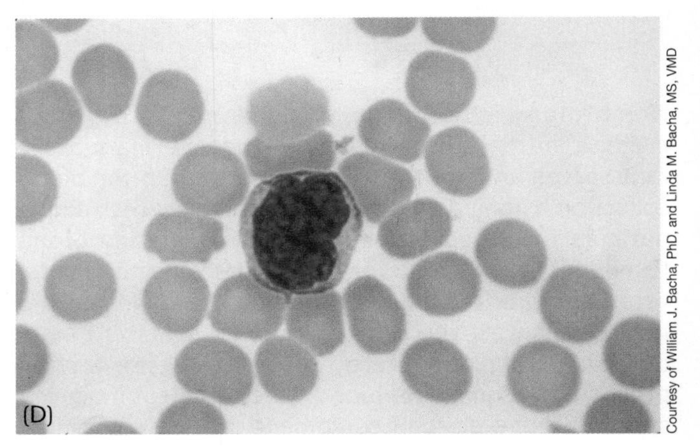

Courtesy of William J. Bacha, PhD, and Linda M. Bacha, MS, VMD

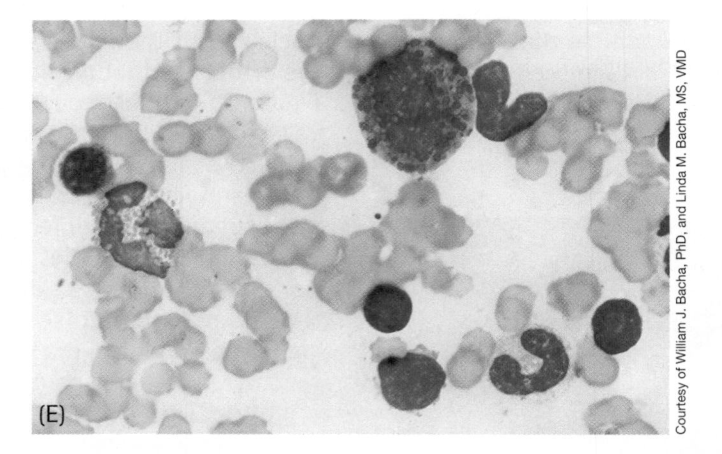

Courtesy of William J. Bacha, PhD, and Linda M. Bacha, MS, VMD

FIGURE 29-2 (A) A blood smear from a dog. Numerous red blood cells are visible: The large, darkly stained cell with granules is a basophil. The other white blood cell with lightly stained cytoplasm and a segmented nucleous is a neutrophil. (B) A bone marrow smear from a dog. Multiple white blood cells are visible, including an immature neutrophil or band cell. (C) A blood smear from a dog showing a small lymphocyte. Two platelets are also visible. The dark spots are deposits of stain. (D) A blood smear from a dog showing an eosinophil. The cell has granules and a nucleus similar in shape to that of a basophil. (E) A blood smear from a dog showing a monocyte.

TERMINOLOGY TIP

Terms used in blood work evaluation to determine the diagnosis of an animal's condition include the following:

- hyper—above normal or excessive
- hypo—below normal or reduced
- emia—refers to the blood
- penia—refers to a blood count that is below normal
- cytosis—refers to a blood count that is above normal

The Heart

The **heart** is an organ with four chambers and is located in the chest between two lungs. The wall of the heart is made of thick cardiac muscle called **myocardium**. A thin inner layer of muscle is called the **endocardium**. The outside thin covering of the myocardium is called the **epicardium**. The outside of the heart is lined by a sac called the **pericardium**, which is a thin membrane that covers, protects, and maintains the beating action of the heart (see Figure 29-3).

Blood Flow

Blood flows throughout the body allowing for **systemic circulation** and oxygenation. The systemic flow of blood delivers nutrients to the entire body. Oxygen is delivered to the lungs and is exchanged with carbon dioxide. **Veins** are located throughout the body and are vessels that carry blood to the heart. **Arteries** are vessels located throughout the body that carry blood away from the heart. Veins are much thinner and less muscular than arteries, and they contain valves that keep blood flowing in one direction. Arteries have thick, muscular walls and can be used to determine the **pulse** of an animal. The pulse is the **heart rate**, which is the pumping action of blood through the heart. Blood flows into the heart through a large vessel called the **vena cava**.

As blood flows into the heart, it enters the thin walled chamber of the **right atrium** (see Figure 29-4). It then passes into the **right ventricle**, a thicker walled chamber that pumps blood to the lungs. The atrium and ventricle are separated by the **AV valve**, which opens and closes to allow blood to flow through in one direction. The **pulmonary arteries** carry oxygen-poor blood and prevent blood flow back to the heart. The pulmonary veins allow blood to flow back to the heart and into the **left atrium**, a thin walled chamber, and into the **left ventricle**, the thickest walled chamber of the heart, also separated by an AV valve. The **aorta** is the large vessel that allows blood flow out of the heart and back into systemic circulation. The aorta is an artery that prevents blood flow back to the heart and is located in the midchest and abdomen. Table 29-1 lists the normal heart rates for various species.

As blood flows through the heart, the body produces **blood pressure** (BP) that can be measured for contraction and relaxation. The contraction is caused by **vasoconstriction**, in which the diameter of the vessel decreases and causes the blood pressure to increase. Relaxation is caused by **vasodilatation**, in which the vessel diameter increases and causes the blood pressure to decrease. Blood pressure measures one complete contraction of the heart in the cardiac cycle. The contraction

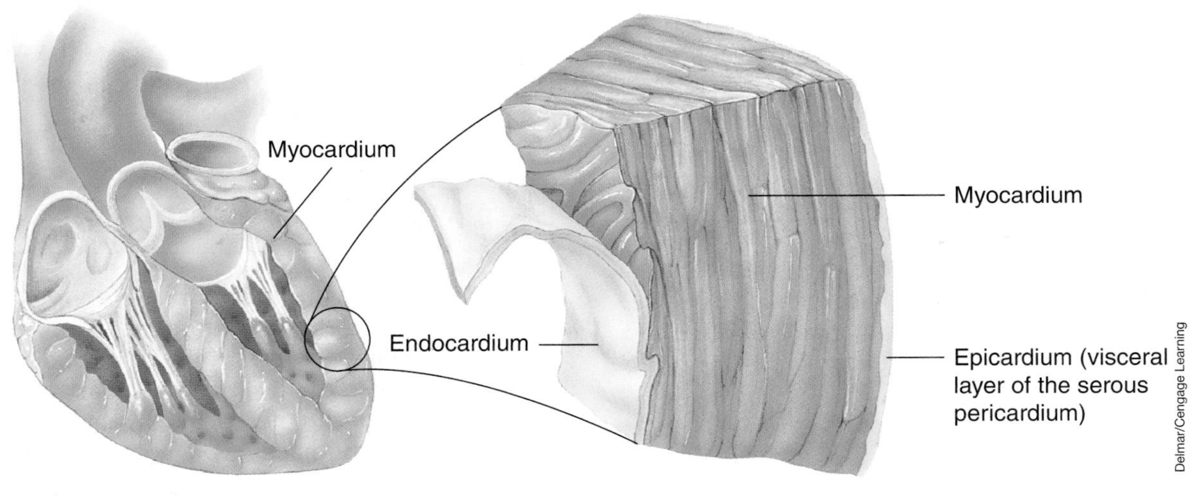

FIGURE 29-3 Layers of the heart.

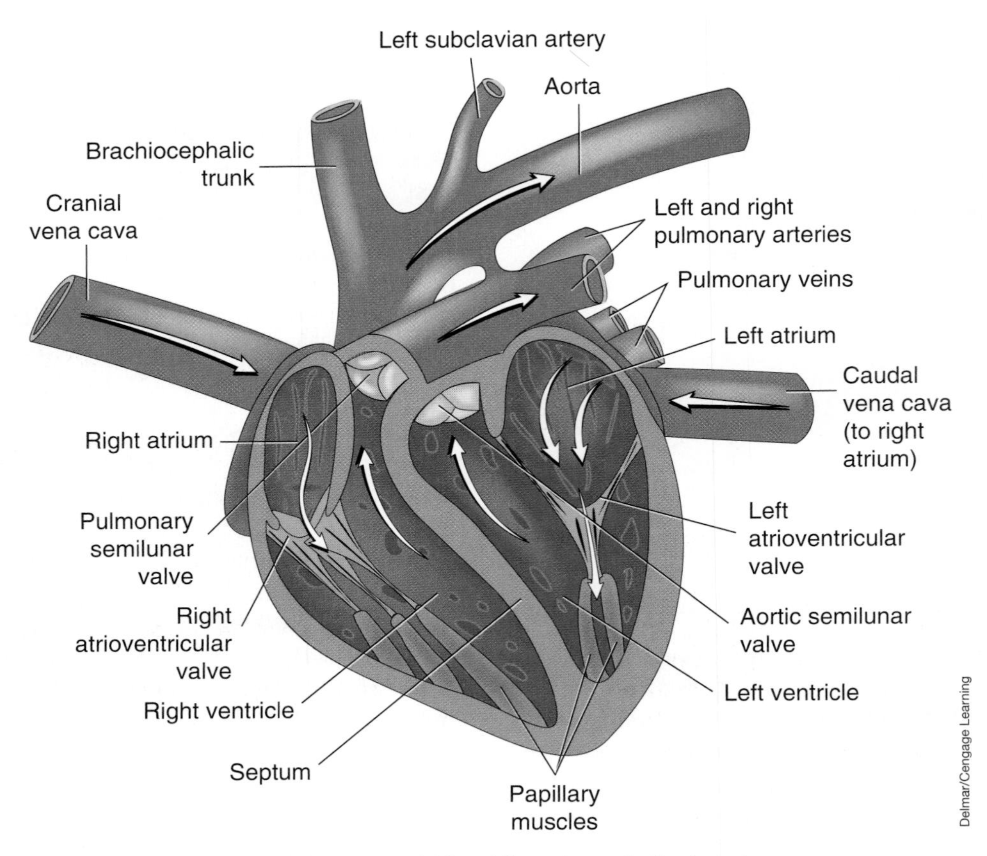

FIGURE 29-4 Internal structure and blood flow through the heart.

TABLE 29-1

Normal Heart or Pulse Rates

SPECIES	NORMAL HEART RATE (BEATS PER MINUTE)
Dog	60–120 bpm
Cat	160–240 bpm
Rabbit	130–325 bpm
Guinea Pig	240–250 bpm
Horse	30–40 bpm
Cow	60–80 bpm
Pig	60–80 bpm
Chicken	200–300 bpm
Sheep or Goat	70–80 bpm

phase is called **systole**, which is the top number that is normally higher. The relaxation phase is called **diastole**, which is the bottom number that is normally lower. Blood pressure is measured in mm of mercury or Hg.

Example
Blood pressure = 120/80
120 is systole
80 is diastole

Heart Sounds

The heart sounds are created by the closing of valves. Heart sounds can be heard through the use of a **stethoscope**, an instrument that is used to listen to the heart, lungs, and chest (see Figure 29-5). Heart sounds are controlled by the **pacemaker** of the heart, known as the **SA node**. The pacemaker system controls the heart's rhythm. The first heart sound, known as the "lub," is created by the closure of the AV valves. The second heart sound, or the "dub," is the closure of the aorta and pulmonary valve.

Electrocardiography

The veterinary facility needs to be able to monitor heart sounds and rhythms of animals that are under anesthesia or may have existing heart conditions. This may be done with the use of **electrocardiography** or the evaluation of the electrical currents of the heart through the use of machines. One type of equipment used to evaluate the electricity of the heart is an **electrocardiogram**, an EKG or ECG (see Figure 29-6).

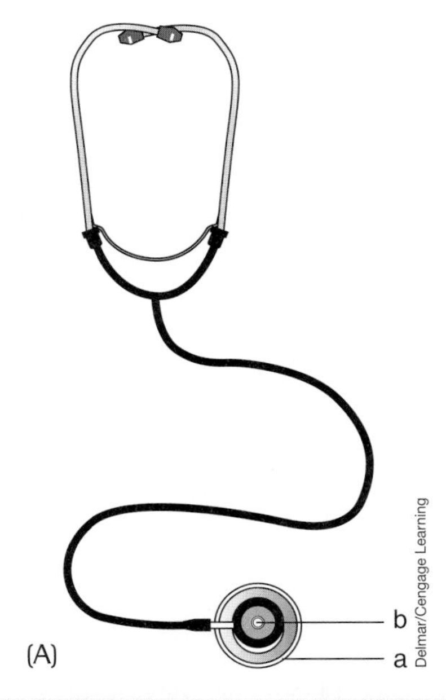

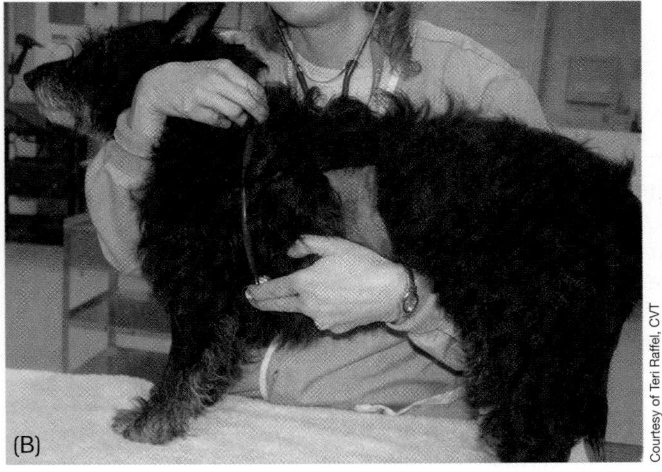

FIGURE 29-5 (A) Stethoscope; (B) listening to heart sounds.

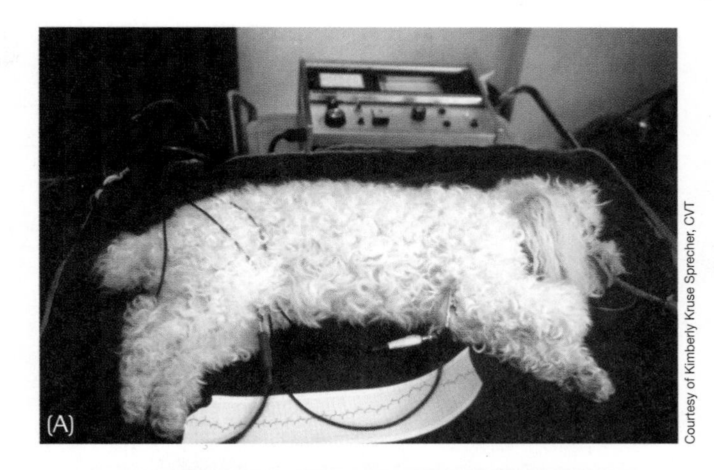

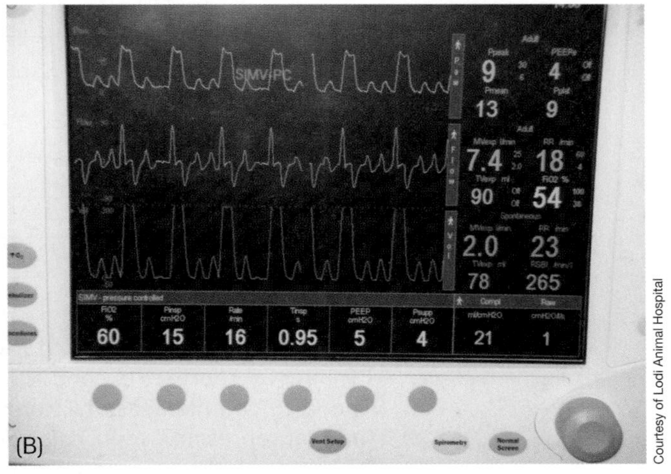

FIGURE 29-6 (A) EKG monitor; (B) electrocardiograph of a dog.

An electrocardiogram is a machine that records the electrical currents of the heart. The rhythm of the heart is shown on a screen that is depicted in waves or lines that peak according to the heartbeats. Each peak is identified by a letter that relates to the heart's activity and functions. The EKG helps veterinarians identify problems with the heart. A normal heart rate and rhythm is called a **sinus rhythm**.

The first peak is called the **P wave**, which forms at the SA node (pacemaker) and reflects the flow of electricity and blood through the atrium and that a contraction has occurred. The QRS wave or **QRS complex** is a series of peaks that show that the electrical current and blood have flowed through the AV node and have caused a contraction in the ventricles. The **T wave** is the last peak and the most important part of the EKG. It shows that an electrical impulse has traveled through the entire heart and has completed the contraction and re-polarization of the heart (see Figure 29-7).

TERMINOLOGY TIP

The term **re-polarize** means to re-set or re-start, and when describing the heart it means to get ready for the next electrical impulse to start.

Common Diseases and Conditions of the Circulatory System

Many abnormalities can occur with the heart. These conditions may be diagnosed through the use of physical exam, radiology, EKG, or ultrasound evaluation.

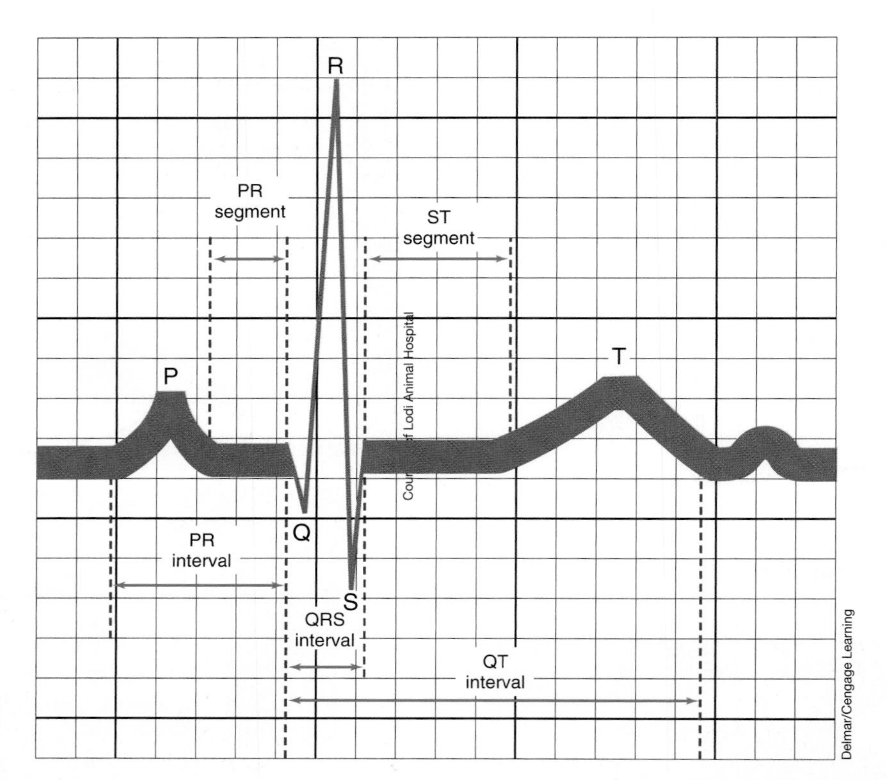

FIGURE 29-7 Anatomy of an electrocardiogram. The first deflection, the P wave, represents excitation (depolarization) of the atria. The PR interval represents conduction through the atrioventricular valve. The QRS complex results from excitation of the ventricles. The QT Interval represents ventricular depolarization and re-polarization. The ST segment represents the end of ventricular depolarization to the onset of ventricular repolarization. The T wave results from recovery (re-polarization) of the ventricles.

Some common types of abnormal heart conditions include a **heart murmur**, which is caused by an abnormal valve that produces an abnormal flow of blood that creates a "swishing" noise upon auscultation.

> **TERMINOLOGY TIP**
>
> **Auscultation** or to auscultate means to listen to, as in using a stethoscope to auscultate the heart.

Arrhythmias

An **arrhythmia** is a change in the heart's rhythm or rate. **Tachycardia** refers to the heart beating faster than normal. **Bradycardia** is the term that refers to the heart beating slower than normal. When **cardiac arrest** occurs, this means the heart is not contracting appropriately and is similar to a heart attack in humans. Cardiac arrest may be from **atrial fibrillation**, commonly called A-fib. This is a condition that occurs when the pacemaker of the heart or SA node is not working. **Ventricular fibrillation**, commonly called V-fib, is a condition causing the ventricles to fire electrical currents rapidly and is the most serious cause of cardiac arrest. When this occurs the heart goes into a condition called **asystole**, meaning the heart stops contracting and heart failure occurs.

Shock

Shock is a condition that occurs when an animal does not have enough blood and oxygen reaching the tissues. This may occur with any type of trauma or heart condition. Shock is a medical emergency and care should be provided immediately.

SUMMARY

The cardiovascular system consists of the heart, blood, and blood vessels. The function of the system is to transport oxygen and nutrients to all body structures through the blood. Arteries carry blood away from the heart and veins carry blood back to the heart. Diseases and disorders of the cardiac system include alterations in the heart's rate and rhythm. Severe blood loss resulting from trauma is a medical emergency.

Key Terms

albumin part of the blood that draws water into the blood stream and provides hydration

anemia low red blood cell count where the red blood cells are not replenished

aorta large vessel that allows blood to flow out of the heart and back into systemic circulation

arrhythmia change in the heart's rate or rhythm

artery vessel that carries blood away from the heart

asystole heart stops contracting and heart failure occurs

atrial fibrillation A-fib; condition that occurs when the SA node or the pacemaker of the heart is not working

auscultation the procedure of listening to a body part

autoimmune disease animal's red blood cells are destroyed by the immune system

AV valve opens and closes between the atrium and ventricle to allow blood to flow through each chamber of the heart

band cell immature neutrophil that indicates an infection within the body; shaped like a U

basophil white blood cell with a segmented nucleus and granules that stain very dark and aid in allergic reaction control

blood red liquid within the circulatory system that contains 40% of cells in the body and transports oxygen through the body

blood pressure (BP) the heart's contractions and relations as blood flows through the chambers

bradycardia heart beating slower than normal

cardiac arrest the heart is not contracting appropriately

centrifuge piece of equipment that uses high speeds to separate liquid portions from solids

circulatory system body system essential for life that includes the heart, blood, veins, arteries, and capillaries with the functions that include oxygen flow, blood circulation, and transport of nutrients, waste removal, and the movement of hormones

clot formation of a plug in the blood to stop bleeding

diastole relaxation phase of blood pressure or the lower number, which is normally lower

electrocardiogram (ECG or EKG) a machine that controls the electrical currents of the heart

electrocardiography the evaluation of the electrical currents of the heart through the use of machines

endocardium thin inner muscle layer of the heart

eosinophil white blood cells that fights against allergic reactions, controls inflammation, and protects the body from parasite infection and have a large nucleus with segmented granules within the cytoplasm

epicardium outer thin covering of the myocardium

erythrocytes veterinary term for red blood cells

erythropoiesis the production process of red blood cells within the bone marrow

fibrinogen protein that aids in clotting blood

globulin provide antibodies to help prevent disease

heart organ with four chambers located in the chest between the lungs

heart murmur an abnormal valve that produces an abnormal flow of blood that creates a "swishing" noise upon auscultation

heart rate pumping action of the blood through the heart creating a beating action

hematology the study of blood

(Continues)

hemoglobin main component of red blood cells that allows oxygen transport and produces iron to allow cells to multiply

histamine chemical released during an allergic reaction

left atrium thin walled chamber of the heart that passes from the pulmonary artery into the left side of the heart

left ventricle thickest walled chamber of the heart

leukocytes veterinary term for white blood cells

lymphocytes white blood cells that make up the largest part of the bloodstream and aid in immune functions that protect the body from disease; have one large single nucleus that makes up most of the cell

macrophage cells that eat and destroy organisms throughout the body

monocyte largest white blood cell that helps neutrophils rid the body of wastes and cell debris

myocardium thick muscle that forms the wall of the heart

neutrophils most common white blood cell that destroys microorganisms within tissues that has a nucleus with segments

pacemaker system of the heart that controls heart sounds and rhythm

pericardium thin membrane that covers, protects, and maintains the beating action of the heart

plasma formed of various proteins in the body and consists of 60% of the blood system; forms a solid portion of blood

platelet blood cell used to clot blood by constricting the vessels to stop bleeding

pulmonary artery carry oxygen poor blood and prevent blood flow back to the heart

pulse the heart rate of an animal or each beat created from the heart

P wave first peak on an EKG that forms at the SA node (pacemaker) and reflects the flow of electricity and blood through the atrium and a contraction has occurred

QRS complex series of peaks that show that the electrical current and blood have flowed through the AV node and have caused a contraction in the ventricles

red blood cells (RBC) most abundant cells within the body that are produced in the bone marrow and transport oxygen throughout the body

re-polarize re-set or re-start the heart and means to get ready for the next electrical impulse to start

right atrium thin walled chamber of the heart where blood enters

right ventricle thicker walled chamber of the heart that pumps blood to the lungs

SA node the term for the pacemaker of the heart

serum liquid portion of blood

shock condition that occurs when an animal does not have enough blood and oxygen reaching the tissues

sinus rhythm a normal heart rate and rhythm

stethoscope instrument used to listen to the heart, lungs, and chest

systemic circulation blood system that delivers nutrients and oxygen to all areas of the body

systole contraction phase of blood pressure or the top number that is normally higher

tachycardia heart beating faster than normal

thrombocyte veterinary term for platelets

T wave last peak and the most important part of the EKG; shows that an electrical impulse has traveled through the entire heart and has completed the contraction and re-polarized the heart

vasoconstriction diameter of a vessel decreases as blood pressure increases

vasodilatation diameter of a vessel increases as blood pressure decreases

vein vessels that carry blood to the heart

vena cava large vessel that transports blood into the heart from systemic circulation

ventricular fibrillation V-fib; condition causing the ventricles to fire electrical currents rapidly and is the most serious cause of cardiac arrest

white blood cells (WBC) five types of blood cells that protect the body from infection and aid in immune system protection

REVIEW QUESTIONS

1. What is the outer layer of the heart called?
2. What is the blood vessel that transports blood to the right atrium called?
3. What is the device used to measure the electrical activity of the heart called?
4. What is the blood vessel that carries blood away from the heart toward the lungs called?
5. What is the name of the protein that aids in blood clotting called?
6. Name the organs of the circulatory system.
7. Name the three major types of blood cells.
8. Describe the path of blood through the heart.
9. Explain the difference between tachycardia and bradycardia.
10. Differentiate between an artery and a vein.

Clinical Situation

© iStock/Derek Dammann

Dr. Jansen and Dale, a veterinary assistant from Ashton Oak Equine Center are on a farm call to examine a Thoroughbred named "Roscoe". The horse has been lethargic, weak after exercise, and had a decreased appetite for several days.

The vet completes a PE and determines the horse has pale mucous membranes, a decreased HR and RR, is 5–10% dehydrated, appears weak and to have lost some weight since the last exam. The vet is suspicious of parasites or an auto immune disease, since some of the signs appear to be anemia.

Dr. Jansen explains to the owner that anemia causes a low red blood cell count causing less oxygen to be transported to the tissues. Thus, the body is not producing enough red blood cells. Upon completing the exam, Mark assists the vet in obtaining some blood samples and a fecal sample for further diagnosis.

- What are some possible causes for the anemia?
- What are some factors that could prevent the horse from having anemia?
- What treatment options might Dr. Jansen discuss with the owner?

30

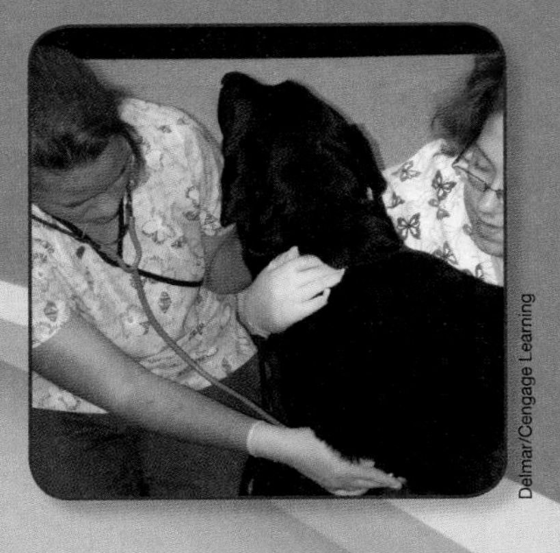

The Respiratory System

Objectives

Upon completion of this chapter, the reader should be able to:

- State the structures that make up the respiratory system
- Describe the function of the respiratory system
- Name and describe common disorders associated with the respiratory system

Introduction

The **respiratory system** works closely with the circulatory system in serving to supply the body with oxygen and rid the body of carbon dioxide (see Figure 30-1). This provides an exchange of gases between the animal and the environment.

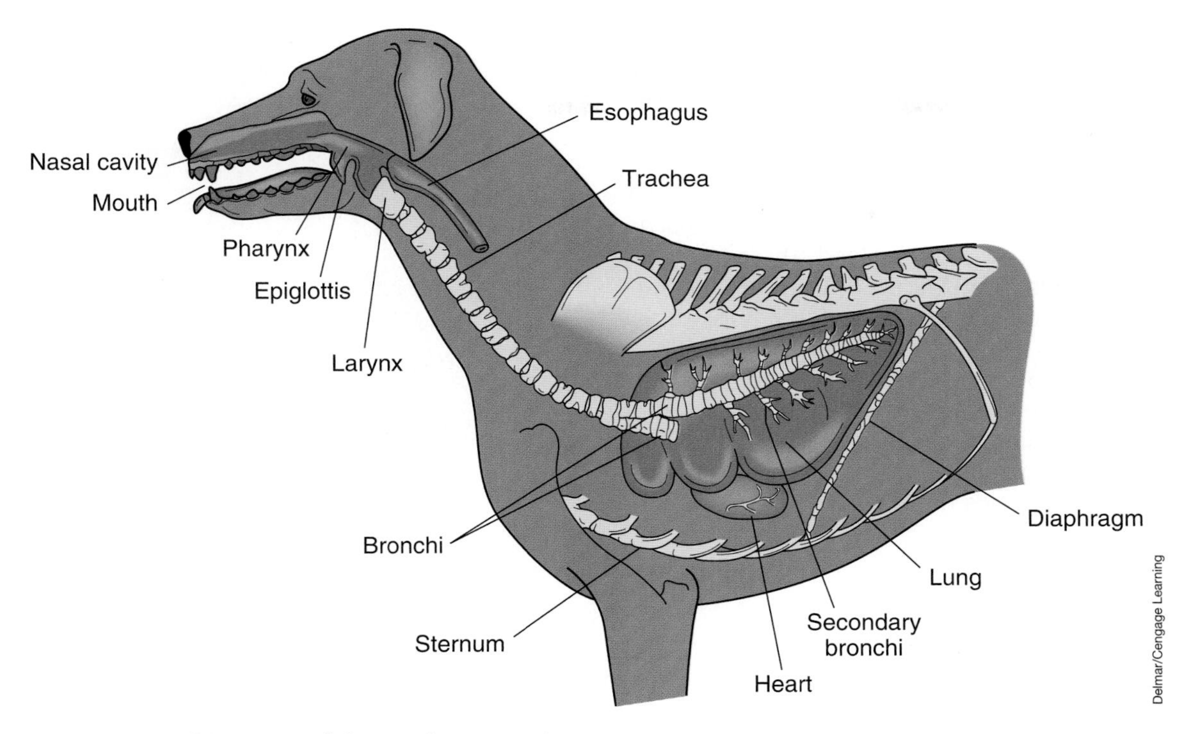

FIGURE 30-1 Structures of the respiratory system.

Structures of the Respiratory System

The respiratory system begins at the nostrils or **nares**. Each nostril is covered by a lining of mucous membranes and has numerous blood vessels. They also contain thin bonelike scrolls or plates of cartilage. The nostrils create an airway and help trap dust and foreign particles within the upper airway to prevent entry into the lungs. The nostrils have the ability of raising the temperature of cold air before entering the lungs. The mouth also serves as an airway that leads to the respiratory system. The **pharynx** or throat area is the common passageway for air and food. The **larynx** lies within the throat and is cartilage that opens into the airway. The **glottis** opens into the larynx and has a covering called the **epiglottis** that prevents food from entering the airway and lungs during swallowing. The **trachea** is the windpipe and is the airway that needs to be established during anesthesia. A tube called an **endotracheal tube** is placed in the trachea to control anesthesia and keep the animal sedated during surgery (see Figure 30-2). Endotracheal tubes or E-tubes are also used to establish an airway when an animal is not breathing or in cardiac arrest. It is important to inflate the cuff on the end of the tube to prevent anesthesia gas from escaping or any objects, such as vomit or saliva, from entering the lungs.

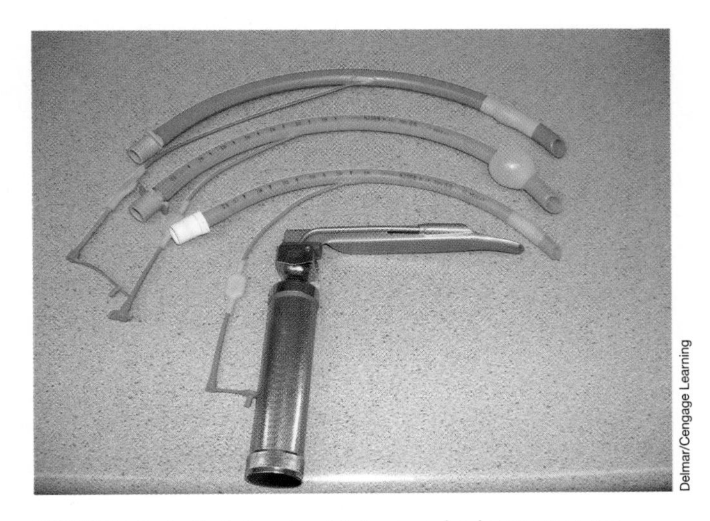

FIGURE 30-2 Endotracheal tubes and a laryngoscope.

TERMINOLOGY TIP

Anesthesia is the process of placing an animal under a loss of consciousness to perform surgery. Anesthesia prevents pain sensation.

The trachea is made of cartilage and forms a tube that connects the pharynx to the lungs. The base of the trachea forms two branches that lead to each lung. The branches are called **bronchi.** The bronchi are large passages connected to the lungs through tiny tubes called **bronchioles**. They resemble tree branches that connect through tiny sacs of air called **alveoli**. The alveoli exchange oxygen and carbon dioxide in the blood. These branches connect to the lungs (see Figure 30-3). Each **lung** surrounds the heart and inflates with air during breathing. There are two lungs, which are the main organs of respiration (see Figure 30-4). They are elastic-like sacs that act like balloons, filled with air, and the lungs act like a sponge and absorb water vapor. The lining of the lungs is called the **pleura**, a double membrane covering of the **pleural cavity**, which holds the lungs. The pleura contain

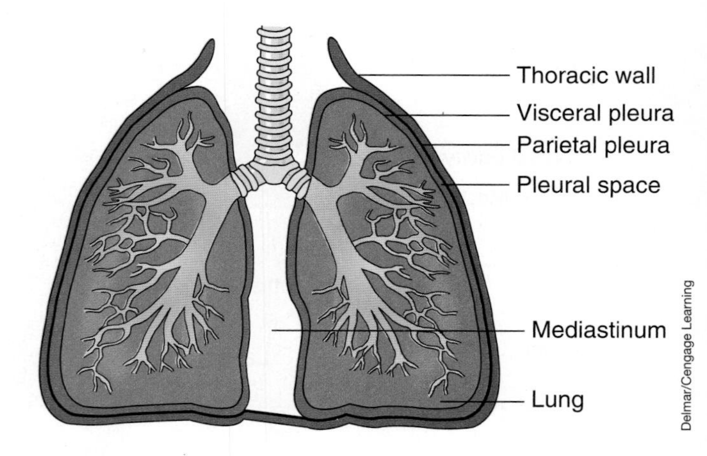

FIGURE 30-4 Respiratory structures of the thoracic cavity.

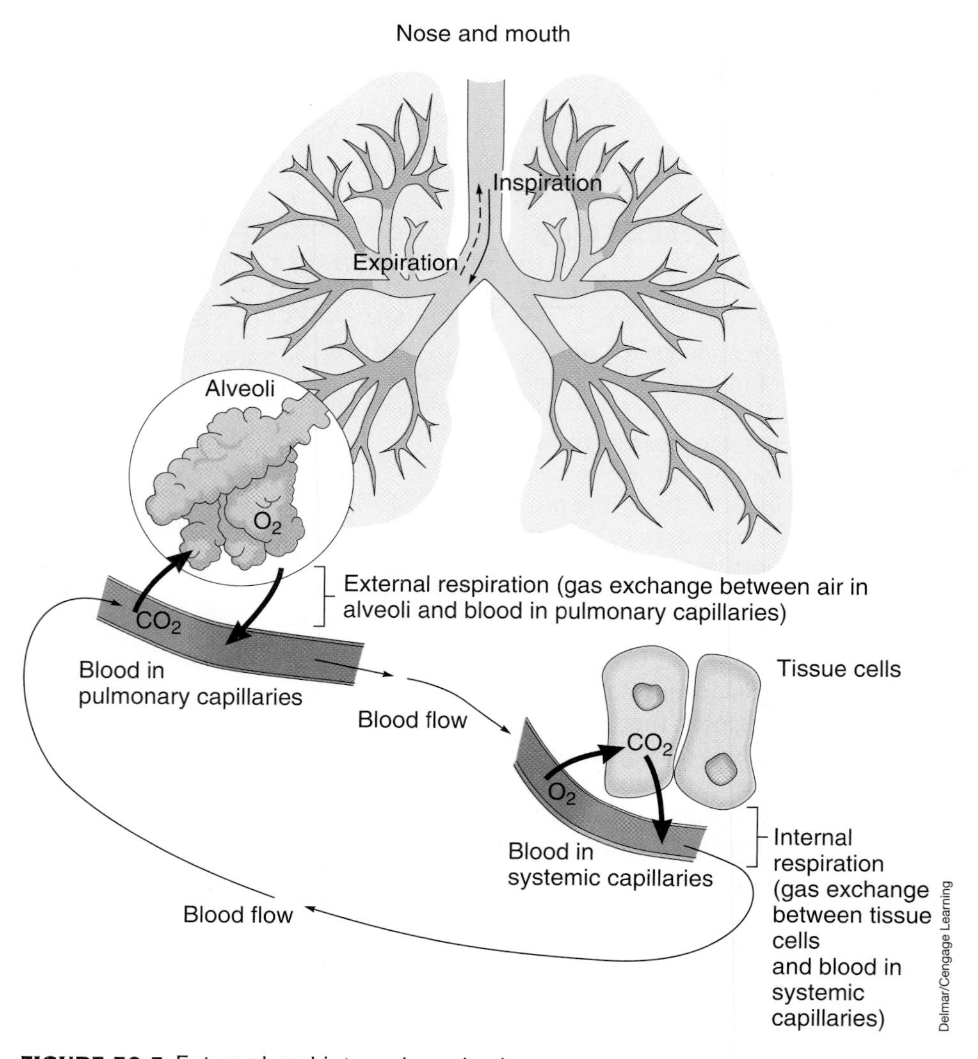

FIGURE 30-3 External and internal respiration.

a fluid that allows the lungs to function and exchange oxygen. This allows the process of respiration to occur.

Respiration

Respiration is the act of inhalation and exhalation creating a breath. **Inhalation** is the act of inhaling or taking in air to the lungs with the nostrils and trachea opening. The rib muscles expand or relax and the diaphragm contracts. The **thoracic cavity** or chest cavity enlarges and air rushes in (see Figure 30-5).

> **TERMINOLOGY TIP**
>
> The ***diaphragm*** is the area of the chest that surrounds the ribs and protects the lungs. It serves the ability to expand and relax during the breathing process.

Exhalation or to exhale is the act of air flowing out of the lungs. Relaxation of the trachea and nostrils occurs, causing the rib muscles to contract and the size of the thoracic cavity to decrease as air flows out. Respiration or breathing is controlled by the respiratory center of the brain by an involuntary action. The respiratory rate (RR) is controlled and affected by several factors, including excitement, body temperature, exercise and activity, fever or pain, and oxygen levels in the blood. Table 30-1 lists normal respiratory rates of some species.

Diseases and Conditions

Respiratory disorders are common and are often the result of bacterial or viral infections. Some respiratory conditions may be the result of trauma. Some of the most common disorders seen in veterinary practice are discussed here.

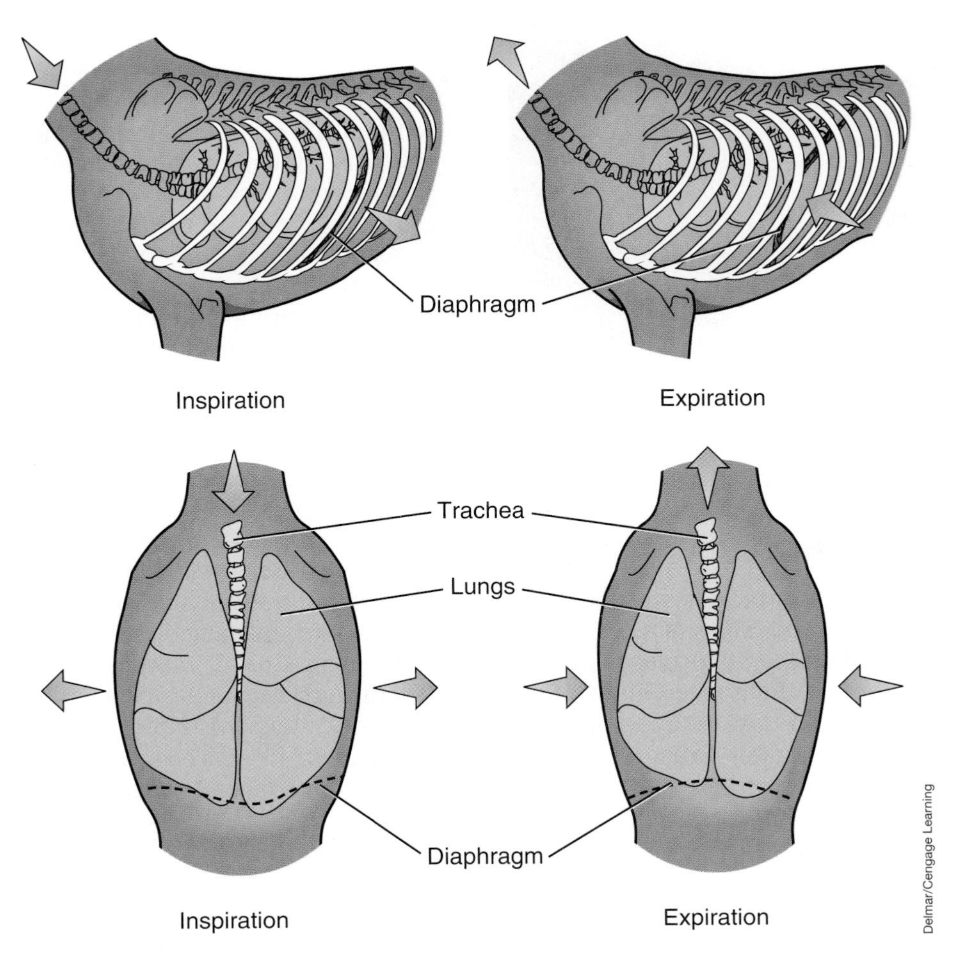

FIGURE 30-5 The mechanics of breathing.

TABLE 30-1

Normal Respiratory Rates of Animals (breaths per minute)

Dog	16–20 br/min
Cat	22–26 br/min
Rabbit	30–60 br/min
Guinea Pig	42–105 br/min
Horse	12–15 br/min
Cow	30–35 br/min
Pig	12–15 br/min
Chicken	36–40 br/min
Sheep/Goat	20–22 br/min

Delmar/Cengage Learning

FIGURE 30-6 A young kitten suffering from an upper respiratory infection. Note the sore eyes and thick nasal discharge.

Upper Respiratory Infection

A common disorder of the respiratory system is an **upper respiratory infection** or URI. Signs of a URI include sneezing, coughing, nasal discharge, ocular discharge, wheezing, and difficulty breathing (see Figure 30-6). **Sneezing** is a common action that removes dust and foreign particles from the respiratory tract through the use of physical force.

Pneumonia

Animals can also develop a condition of the lungs called **pneumonia** (see Figure 30-7). This is a disease that causes the lungs to become inflamed. There are two types of causes of pneumonia: bacterial and viral. Advanced stages of pneumonia can develop into a **pleural friction rub**. This is an abnormal lung sound that develops from pneumonia. It may cause a crackling noise called **crackles** or **rales**, which sound like cellophane paper being folded.

Delmar/Cengage Learning

FIGURE 30-7 Heifer suffering from pneumonia as exhibited by thick nasal discharge and labored breathing.

Asthma

Asthma is a condition in which the lungs have difficulty taking in enough oxygen. Asthma can lead to a condition where a lack of oxygen causes the animal's mucous membranes to turn blue called **cyanosis**. During an asthma attack the sound of **wheezing** can be heard. Medications may be used to open up the bronchioles and lungs during these situations. These medications are known as **bronchodilators**.

Kennel Cough (Bordetella)

Bordetella or tracheobronchitis, commonly called kennel cough, occurs in the respiratory tract of animals. This bacterium causes a severe and chronic cough mostly in dogs. Some dogs will also develop a nasal discharge. Contact and transmission often occurs with the transfer of nasal and coughing discharge. Vaccination is typically done using a nasal vaccine sprayed into the nostrils. Several vaccines are available and may be administered in various ways. It is important to vaccinate dogs that are frequently kenneled, are show dogs, or are taken to locations where many dogs gather, such as dog parks.

Shipping Fever

Shipping fever is a condition that occurs in livestock, such as horses and cattle, that are transported. Shipping fever causes a respiratory infection that affects the lungs and chest cavity, causing a severe problem that can be fatal. Shipping and transporting livestock can be stressful, can cause overheating conditions where bacteria commonly grow, and poor sanitation within a trailer can cause the disease to spread rapidly through livestock. Antibiotics and proper sanitation are key to treating the disease. Signs of the disease in livestock include fever, nasal discharge, and coughing. A vaccine is available in cattle and can be given three to four weeks prior to transport. Proper trailer ventilation and sanitation are the best methods to prevent the disease.

Diaphragmatic Hernia

A condition of the respiratory system caused by severe trauma, such as a hit by car (HBC), is called a **diaphragmatic hernia**. This affects the diaphragm by causing a tear in the chest muscle that internal organs may protrude through. It causes the inability of the lungs to expand properly and labored and difficult breathing to occur. Surgical correction is necessary to repair the tear and replace any organs to their original position. It can be an emergency if the tear in the muscle is extensive and affects numerous chest and abdominal organs.

Heaves

Horses are affected by several respiratory conditions. One condition is called **heaves** and is caused by an allergy usually related to dust. Heaves is a common term for COPD, or **chronic obstructive pulmonary disease**. This is a lifelong condition that may be controlled by antihistamines and steroids. Horses that are affected may need to be housed outside or in an area with good ventilation, as dust and moisture can cause serious breathing difficulty. They may also need to have their hay source watered down to reduce the amount of dust spores that occur in the air. Horses with COPD tire easily, have labored and difficulty breathing, and exhibit nasal discharge, coughing, and wheezing.

Roaring

Another common condition in horses is called **roaring**. Roaring is a condition that causes a sound (like a roar) due to the larynx only opening a small amount due to a trauma on the nerves of the throat area. It may be controlled with activity limitations and medications. Severe trauma and breathing problems may be corrected surgically.

MAKING THE CONNECTION

For additional information on the respiratory system, please see the following chapters:

Chapter 9—rhinotracheitis, Feline calcivirus, feline infectious peritonitis
Chapter 12—respiratory disease in rabbits, pasteurella
Chapter 18—infectious bovine rhinotracheitis
Chapter 21—pneumonia
Chapter 24—Newcastle disease, infectious bronchitis

SUMMARY

The respiratory system is the body system that regulates breathing and the exchange of oxygen and carbon dioxide in the body. This oxygen exchange is critical to maintaining life processes. Most of the diseases of the respiratory system are related to viral or bacterial invasion. Trauma can also be a source of respiratory injury.

Key Terms

alveoli tiny sacs of air at the end of the bronchioles

asthma condition of lungs that have difficulty taking in enough oxygen

bordetella tracheobronchitis; commonly called kennel cough

bronchi branches at the base of each lung connected by large passages

bronchioles tiny tubes that connect to the lungs and bronchi and resemble tree branches

bronchodilator medications may be used to open up the bronchioles and lungs

(*Continues*)

chronic obstructive pulmonary disease (COPD) respiratory condition similar to heaves and asthma that causes difficulty breathing

crackle noise that sounds like cellophane

cyanosis blue color of mouth and gums due to a lack of oxygen

diaphragm area of the chest that surrounds the ribs, heart, and lungs and protects the chest organs

diaphragmatic hernia condition that affects the diaphragm by causing a tear in the chest muscle, and internal organs may protrude through

endotracheal tube E-tube; tube placed into the windpipe to create and establish an airway during anesthesia or CPR

epiglottis covering or flap that prevents food from entering the airway and lungs during swallowing

exhalation act of exhaling or leaving air out of the lungs

glottis opening into the larynx

heaves respiratory condition in horses due to an allergic reaction usually to dust

inhalation act of inhaling or taking in air to the lungs

larynx cartilage that opens into the airway and lies within the throat area

lungs organs that surround the heart and inflate with air for breathing

nares the nostrils or the nose

pharynx throat area for the passage of food or air

pleura double membrane covering that lines the lungs

pleural cavity double membrane covering that holds the lungs

pleural friction rub abnormal lung sound that develops with pneumonia and causes a crackling noise

pneumonia condition of the lungs causing inflammation and may be of viral or bacterial sources

rale noise or crackle caused by the lungs during pneumonia

respiration the process of breathing through inhalation and exhalation

respiratory system part of the body that controls oxygen circulation through the lungs and the body

roaring respiratory condition in horses that sounds like a roar due to the larynx only opening a small amount due to a trauma on the nerves of the throat area

shipping fever a condition that occurs in livestock, such as horses and cattle, that are transported

sneezing act of removing dust or other particles from the respiratory tract through physical force

thoracic cavity the chest

trachea the windpipe or tube-like airway established during anesthesia

upper respiratory infection (URI) condition causing sneezing, coughing, nasal discharge, ocular discharge, wheezing, and difficulty breathing

wheezing sound produced during asthma that causes difficulty breathing

REVIEW QUESTIONS

1. What system works closely with the respiratory system in the transport of oxygen to organs?
2. Which of the following contains organs of the respiratory system?
   a. periosteum, medullary cavity, cartilage
   b. fundus, mesentery, peritoneum
   c. pericardium, plasma, erythrocytes
   d. pharynx, pleura, trachea
3. What abnormal lung sound develops from pneumonia?
4. What type of medication is used to open the bronchioles and lungs?
5. What respiratory condition may be related to a dust allergy?
6. Describe the process of respiration.

Clinical Situation

Courtesy of Photo Disc

Dr. Joseph and Amanda, vet assistant, at Redmont Animal Hospital, are on a farm call at a local dairy farm. Mr. Calvins is a local farmer who has a large herd of Holstein dairy cattle and has a cow that he has been treating for about a week. The cow has appeared sick to him and he has been giving injectable penicillin that he purchased at a local farm supply store.

Amanda takes a history of the cow while Dr. Joseph gets his equipment. Mr. Calvins says the cow has been sick about a week with signs of labored breathing, poor appetite, and decreased milk production. Amanda also notes in the history the penicillin information on the bottle.

Dr. Joseph completes a PE and upon auscultation of the chest, hears a crackling sound in the lungs. The cow also has an increased HR and RR, as well as a slight fever.

- What does Dr. Joseph likely determine is the cow's medical problem?
- Was self treatment by the farmer a good idea? Why or why not? Explain.
- How might have this situation been prevented?

31

The Endocrine System

Objectives

Upon completion of this chapter, the reader should be able to:

- State the structures that make up the endocrine system
- Describe the functions of the endocrine system
- Describe common disorders of the endocrine system

Introduction

The body needs to coordinate and integrate all of its functions into one harmonious whole. This is called **homeostasis**. The maintenance of homeostasis involves growth, maturation, reproduction, and metabolism. The body system responsible for this coordination is the **endocrine system**.

Structures of the Endocrine System

The endocrine system is composed of numerous glands throughout the body that secrete hormones into the bloodstream and **excrete** chemicals to rid the body of wastes (see Figure 31-1). These chemicals and hormones are passed in the body through blood and tissues. The **chemicals** and **hormones** provide changes in the body that regulate growth, regulate sexual production and development, metabolize nutrients in cells, and maintain a homeostasis in the body, allowing for a balance in the body systems.

Endocrine Glands

The endocrine glands secrete hormones directly into the bloodstream. They are then transported to all areas of the body. The endocrine glands are the following:

- Pituitary
- Thyroid
- **Parathyroid**
- Adrenal
- Thymus
- Pancreas
- Pineal
- Gonads (ovaries in females and testes in males)

Pituitary gland

The main endocrine gland is the **pituitary gland**. The pituitary gland works with the **hypothalamus** to control the endocrine system and create a link with the nervous system. The pituitary gland lies at the base of the brain and has an anterior and posterior lobe. The **posterior lobe** connects to the hypothalamus. The hypothalamus develops from brain tissue while an animal is in the embryo stage. The hypothalamus serves as a reservoir for hormones and allows for the release and regulation of hormones. The posterior or back lobe releases two peptide hormones: oxytocin and antidiuretic hormone (ADH). **Oxytocin** plays a vital role in **parturition**, or the labor process. The hormone releases and causes the muscles of the uterine wall to contract and milk production to begin in the mammary glands. An injectible form can be given to increase these functions. **Antidiuretic hormone** or ADH is a hormone that promotes urine formation, water absorption, controls blood pressure, and changes control in water content. A condition known as **diabetes insipidus** affects the water content in the bloodstream, causing the urine to become very dilute.

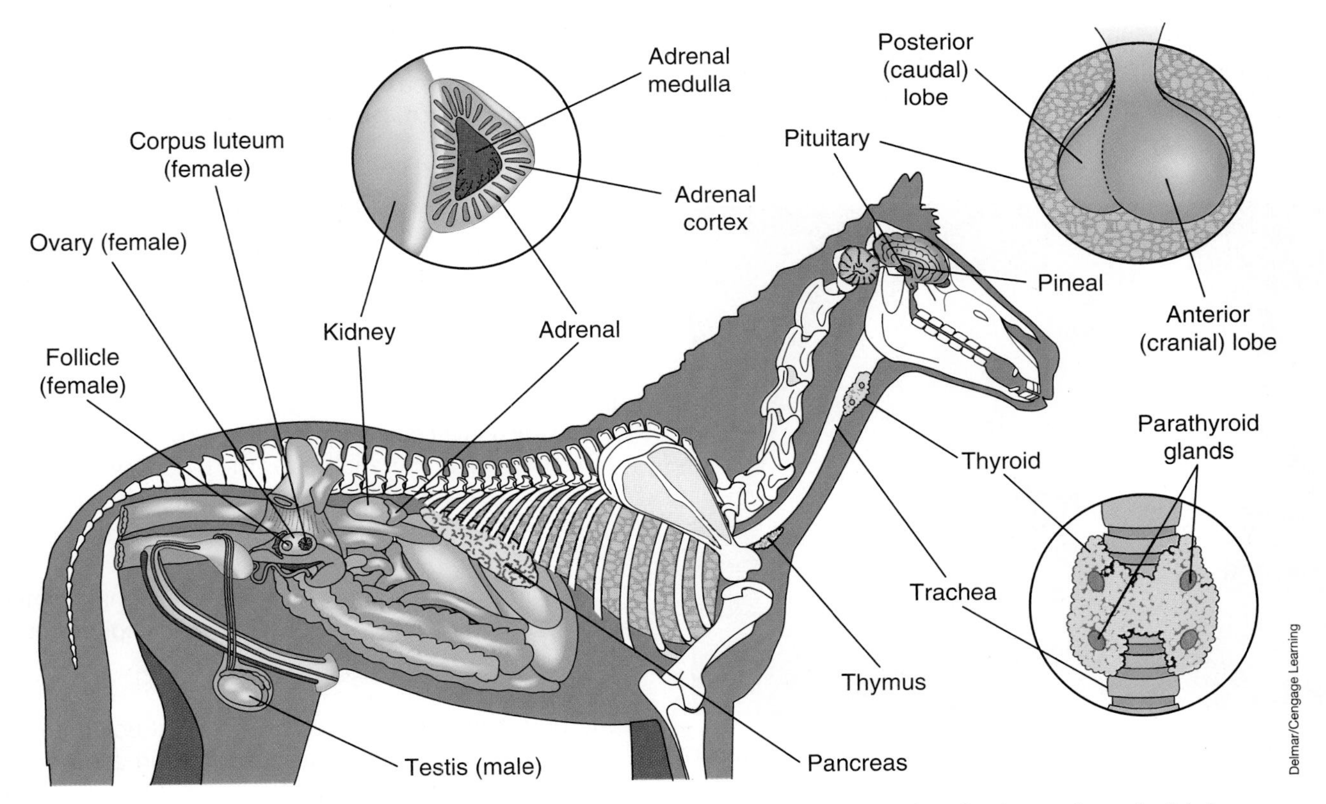

FIGURE 31-1 Locations of the endocrine glands. The relative locations of the endocrine glands are shown in this horse.

This decreases the ADH causing the animal to have polydipsia and polyuria.

TERMINOLOGY TIP

Polyuria means excessive and uncontrolled urination; **polydipsia** means uncontrolled thirst. The terms are abbreviated PU and PD and often occur simultaneously.

The anterior lobe of the pituitary gland is controlled by the hypothalamus. Its main function is to release hormones into the bloodstream and produce and regulate growth hormones within the body. The major growth hormone **somatotropin** increases protein synthesis within the body, thus increasing the size of the animal. The anterior lobe also produces the hormone **prolactin**, which controls milk production. The pituitary gland is often called the "master gland" as it controls the **thyroid gland**, which regulates and secretes hormones within the body (see Figure 31-2).

Thyroid gland

The thyroid gland can become overactive or underactive in animals, causing many conditions that can affect animals. The thyroid gland produces the **thyroid stimulating hormone** (TSH) and tells the thyroid gland when to produce **thyroxine**, which is the hormone that controls the actions of the thyroid gland and cell metabolism. The two hormones that occur in the bloodstream that help to identify normal or abnormal thyroid gland production are called T3 and T4. Both are important in diagnosis of thyroid gland disease. The **T3** hormone is more potent and a more active form of the hormone. The **T4** hormone is converted into tissues and breaks down fats and helps control cholesterol. The thyroid gland is located in the neck and contains two lobes, one on each side of the trachea. It is the only endocrine gland that can be palpated when it is enlarged.

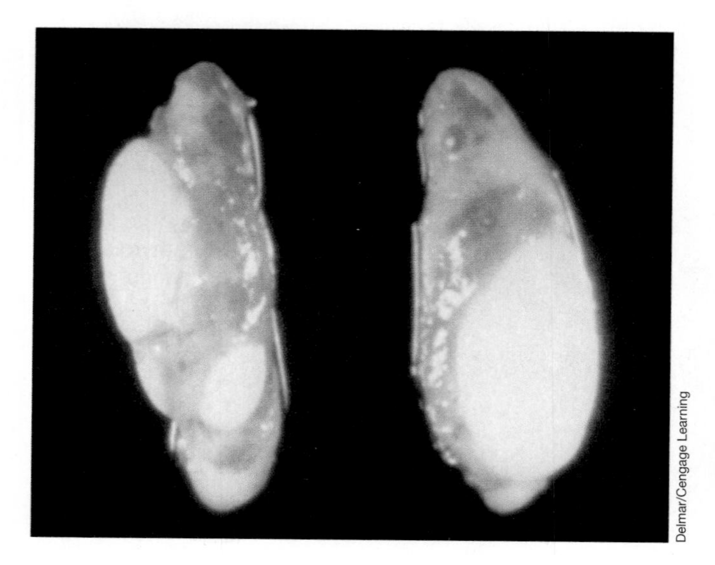

FIGURE 31-2 Thyroid and parathyroid glands from a cat. The white areas are very enlarged parathyroid glands. Normal parathyroid glands would not be visible.

Parathyroid glands

The parathyroid glands are located on the surface of the thyroid gland. They secrete parathormone, which helps regulate blood calcium and phosphorus levels.

Adrenal glands

The **adrenal glands** are also a part of the endocrine system and lie cranial to the kidneys. The adrenal glands produce and release adrenaline and other hormones. **Adrenaline** is a chemical that is released by the nervous system in times of stress to create a response in an animal's fight or flight instinct reaction. There is a release of a chemical called **epinephrine**, which is short-acting. A longer acting hormone chemical is **norepinephrine**, which increases the heart rate, blood pressure, blood flow, blood glucose, and metabolism. The adrenal glands also regulate the function of **ACTH** or adrenaocorticotropic hormone. ACTH is a hormone that controls blood pressure, releases cholesterol, and also controls the body's production of steroids. Stress is a factor of ACTH production and increases the amount of ACTH in the body, which in turn decreases the production rate.

Pancreas

The **pancreas** is another organ within the endocrine system. It serves a dual role in endocrine and exocrine production and functions. The pancreas produces and regulates insulin production. Insulin is regulated by the **blood glucose** level, also known as blood sugar. The release of insulin following a meal is created by a response from the pancreas. This response causes sugar metabolism, which is a conversion of insulin to glucose to glycogen to fat. Sugar metabolizes as the blood level decreases as food digests, and this is repeated after each meal.

Thymus

The **thymus** is a gland in young animals. It has immunologic functions through maturation of t-lymphocytes.

Pineal gland

The pineal gland is located in a central portion of the brain. The functions of the pineal gland are not fully understood. It plays a role in the regulation of body rhythms.

Gonads

The gonads are the glands associated with reproduction.

Hormones

Hormones are made up of four major chemical groups that regulate parts of the body. The **fatty acids** control hormones involved in estrus in the female animal's heat cycle. **Steroids** occur naturally in the body and regulate chemicals, such as cholesterol, in the body that control essential life functions. **Amino acids** are the simplest hormones and control the thyroid gland functions. **Peptides** are the largest hormones and control the proteins in the body. Hormone functions are delivered to target cells located throughout the body and fit into each cell like a lock and key. The body will send a signal for the necessary hormone function, which responds by creating an **enzyme**. Enzymes are chemicals that create a reaction that changes within the body. The enzymes create and release the hormones that are needed until the function is complete. An example of hormone and enzyme function is insulin production as the blood sugar elevates. The blood sugar or glucose level begins to drop, causing insulin secretion to slowly stop. Glucagon hormone is released to increase the blood glucose level. Diabetes results when this function is not properly controlled by the hormones in the body.

Hormones and Estrus

The endocrine system is active in producing and controlling hormones involved in male and female reproduction. **Estrus** hormones are produced to control the heat cycle of female animals. One estrus hormone is called **luteinizing hormone (LH)**, which allows the production of testosterone, allows for ovulation to occur, and forms the corpus luteum (CL) during the reproduction cycle. **Ovulation** is the release of the egg in the female to allow for reproduction to occur when sperm is present. The **corpus luteum (CL)** forms in the female during reproduction and is present through the entire gestation period. This is what maintains the pregnancy.

> **TERMINOLOGY TIP**
>
> The term **gestation** means the length of pregnancy of an animal. Each species of animal has a different gestation period.

Another estrus hormone, the **follicle stimulating hormone (FSH)**, allows for sperm production, regulates the female estrus cycle, forms the follicle during the breeding process, and produces estrogen. **Estrogen** is the hormone that begins the estrus cycle. Both the LH and FSH hormones are controlled by the hypothalamus. Another hormone, **gonadatropin (GnRH)**, is produced to regulate and maintain a normal estrus cycle.

> **TERMINOLOGY TIP**
>
> *Gonadatropin* is often called chorionic gonadatropin or Gonadatropin Releasing Hormone, abbreviated GnRH.

Diseases and Conditions

Many diseases and conditions are caused by the glands that secrete chemicals in the body and are part of the endocrine system. These conditions often cause notable signs and symptoms in animals, such as weight problems, skin problems, and changes in thirst.

Hypothyroidism

Hypothyroidism is a condition that decreases the thyroxine hormone production from the thyroid gland. This condition causes the body to slow down, causing weight gain, lethargy, hair and skin problems, weakness, and intolerance to cold. Skin conditions that are common in this disease include **alopecia** (hair loss) and **dermatitis** (inflammation of the skin). Treatment is through a lifetime therapy of thyroxine supplement. This condition is common in dogs and is inherited in many breeds.

> ***Dog Breeds Prone to Hypothyroidism***
>
> - Boxer
> - Great Dane
> - Golden Retriever
> - Dachshund
> - Doberman
> - Schnauzer

Hyperthyroidism

Hyperthyroidism is the disease that causes an increased thyroxine production in the thyroid gland. This condition occurs most commonly in elderly cats.

Signs of this disease include weight loss, a ravenous appetite, polyuria (PU), polydipsia (PD), increased activity, increased heart rate, and an enlarged, palpable thyroid gland. As the disease progresses, signs of vomiting and diarrhea may occur. Treatment options include a **thyroidectomy**, or the surgical removal of the thyroid gland, or thyroxine therapy and supplementation. Cats that have had a thyroidectomy usually require medication supplementation to prevent the condition from becoming hypothyroidism. A caution with the surgery procedure should be noted to the owner. Any damage or accidental removal of the parathyroid gland will cause severe **hypocalcemia**, or a decrease in blood calcium. Another treatment option is a **radioactive iodine** treatment, known as Radiocat. This involves iodine being injected into the bloodstream to treat any areas that are overactive in thyroxine production, causing damage to the thyroid tissue. Less active areas of tissue are left undamaged. Another medication therapy option is **Methimazole**, an oral medication supplement that blocks the synthesis of thyroxine and requires the need for adjusting the dosage for the lifetime of the cat.

Diabetes

Hyperglycemia or **diabetes mellitus** is a condition that causes a high blood glucose level (see Figure 31-3). Animals with this condition tend to have an increased appetite but lose weight, have an increased thirst, increased water consumption, and increased urination (PU/PD). Animals that develop diabetes have cells that deteriorate and break down the pancreas, which in turn cause the body's immune system to attack the cells. It is estimated that about 75% of the cells are destroyed when signs of the disease begin to occur. The treatment is supplementing insulin into the body given by an injection with an insulin needle measured in IUs or insulin units. Diabetes can be challenging and difficult to regulate and control. It requires a regular and consistent diet that may be a therapy diet necessary to control weight. The insulin dose is measured based on the blood glucose levels. The goal of controlling diabetes is to keep insulin levels consistent and as close to normal on a daily basis, and it requires regular attention and blood level evaluations. The long-term use of insulin can cause nerve damage to the kidneys and can lead to kidney failure. Side effects of the disease include vision problems leading to **cataracts**, or the opacity of the lens, or retinal problems. Insulin overdose can occur resulting in **hypoglycemia** or a low blood glucose level. Signs of low blood glucose include weakness, lethargy, ataxia, a trance-like appearance, seizures, and possible coma. Young newborn animals can be treated with sugar water or Karo syrup placed on the gums. Severe cases will need treatment with IV fluids and glucose supplements.

FIGURE 31-3 Cat with diabetes mellitus.

Hyperadrenocorticism

Hyperadrenocorticism, often called **Cushing's disease**, involves a production problem within the adrenal gland (see Figure 31-4). This results in a pituitary gland tumor when the ACTH hormone is increased or an adrenal gland tumor when cortisol hormone production is increased. This disease can be **iatrogenic**, meaning it may be caused by humans, such as in the situation of using steroids. Diagnosis is done by measuring cortisol levels. This is completed through a dexamethasone suppression test. A baseline blood sample is acquired and then the corticosteroid **dexamethasone** is administered to the animal by injection. A blood sample is then collected 4–8 hours post-injection. Normal results will show a decreased ACTH and cortisol levels. If a pituitary tumor is present, there will be an increased ACTH level and no change in cortisol levels. If an adrenal gland

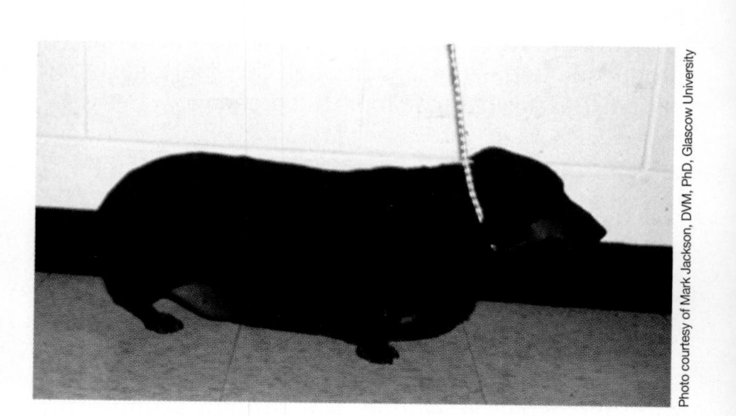

FIGURE 31-4 Dog with hyperadrenocorticism.

tumor is present, there will be no change in ACTH levels and an increase in cortisol levels. Signs of Cushing's disease include the following:

- PU/PD
- Increased appetite
- Thin skin and hair coat
- Panting
- Swollen abdomen
- Weakness
- Lethargic
- Hair loss

Treatment of Cushing's disease includes surgical tumor removal and medication supplementation to decrease the cortisol level in the bloodstream. This is a lifelong therapy that may lead to a decreased production of steroids within the adrenal glands if the glands fail to function. This condition is known as **hypoadrenocorticism** or Addison's disease. Signs of this condition include lethargy, weakness, weight loss, vomiting, diarrhea, poor appetite, and GI problems. **Hyponatremia** may occur, which is a decrease in sodium (Na) and chlorine (Cl) production within the blood. **Hyperkalemia** may also occur, which is an increase of potassium (K) in the blood. Diagnosis is with the use of an ACTH stimulation test. A baseline cortisol blood sample is obtained and then an injection of ACTH gel is administered. The cortisol blood level is sampled 1–2 hours post-injection. Normal results will show an increased cortisol level, and signs of Addison's disease will show no change in the cortisol level. Treatment of Addison's includes long-lasting steroid therapy and IV fluids.

MAKING THE CONNECTION

Chapter 18—hypocalcemia and ketosis
Chapter 23—hypocalcemia

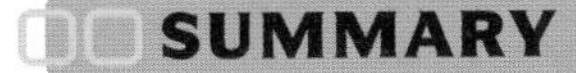

SUMMARY

The endocrine gland works with the nervous system to regulate homeostasis in the body. This is accomplished through the secretion of hormones, which act on specific tissues and organs to produce a desired effect.

Key Terms

ACTH adrenaocorticotropic hormone released by the adrenal gland that controls blood pressure, releases cholesterol, and regulates the body's production of steroids

adrenal gland located cranial to the kidneys and produces and releases adrenaline and other hormones

adrenaline chemical that is released by the nervous system in times of stress to create a response in an animal's fight or flight instinct reaction

alopecia hair loss

amino acid simplest hormone that controls thyroid gland functions

antidiuretic hormone (ADH) hormone that promotes urine formation, water absorption, controls blood pressure, and changes control in water content

blood glucose blood sugar or insulin produced by the pancreas

cataracts opacity of the lens in the eye

chemical a change in the body that affects growth, sexual reproduction, and development

corpus luteum (CL) known as the yellow body and forms during the gestation period of a female to maintain pregnancy

Cushing's disease common term for hyoeradrenocorticism

dermatitis inflammation of the skin

Dexamathasone corticosteroid administered by injection to determine pituitary problems

diabetes insipidus condition that affects the water content in the bloodstream, causing the urine to become dilute

diabetes mellitus condition that causes high blood glucose levels

endocrine system excretory system that rids the body of waste materials

enzyme chemical reactions that change within the body and create and release hormones

(Continues)

epinephrine short-acting chemical released during the fight or flight response

estrogen female reproductive hormone that begins the estrus cycle

estrus the heat cycle that releases hormones for reproduction

excrete to remove and rid the body of waste

fatty acid controls hormones involved in estrus in the female animal's heat cycle

follicle stimulating hormone (FSH) estrus hormone that allows for sperm production, regulates the female estrus cycle, forms the follicle during the breeding process, and produces estrogen

gestation length of pregnancy

gonadotropin (GnRH) hormone produced and released to maintain a normal estrus cycle

homeostasis a balance within the body

hormone a chemical change in the body that regulates growth, regulates sexual production and development, metabolizes nutrients in cells

hyperadrenocorticism increased production problem within the adrenal gland that results in a pituitary gland tumor when the ACTH hormone is increased or an adrenal gland tumor when cortisol hormone production is increased; also called Cushing's disease

hyperglycemia high blood glucose level

hyperkalemia increased potassium in the blood

hyperthyroidism condition that increases the thyroxine hormone production from the thyroid gland and common in cats

hypoadrenocorticism decreased production of steroids within the adrenal glands if the glands fail to function; also called Addison's disease

hypocalcemia decrease in blood calcium

hypoglycemia low blood glucose level

hyponatremia decrease of sodium and chlorine in the blood

hypothalamus develops from brain tissue while an animal is in the embryo stage, and the gland serves as a reservoir for hormones and allows for the release and regulation of hormones

hypothyroidism condition that decreases the thyroxine hormone production from the thyroid gland and common in dogs

iatrogenic condition caused by humans

luteinizing hormone (LH) estrus hormone that allows the production of testosterone, allows for ovulation to occur, and forms the corpus luteum (CL) during the reproduction cycle

Methimazole medicine for cat to treat hyperthyroidism; blocks the synthesis of thyroxine

norepinephrine long-acting chemical hormone that increases the heart rate, blood pressure, blood flow, blood glucose, and metabolism

ovulation release of the egg during estrus in the female to allow reproduction to occur when united with sperm

oxytocin hormone that releases and causes the muscles of the uterine wall to contract and milk production to begin in the mammary glands

pancreas organ with a dual role in endocrine and exocrine production and functions; produces and regulates insulin production

parathyroid gland gland located below the thyroid gland

parturition the labor process

peptide largest hormone that controls proteins in the body

pituitary gland lies at the base of the brain and controls hormone release in the endocrine glands

polydipsia increased thirst; PD

polyuria increased urine production; PU

posterior lobe back lobe of the pituitary gland that controls peptide hormones

prolactin hormone that controls milk production

radioactive iodine Radiocat; treatment involving iodine being injected into the bloodstream to treat any areas that are overactive in thyroxine production, causing damage to the thyroid tissue

somatotropin growth hormone that increases protein synthesis in the body, causing an increase in the animal's size

steroid occurs naturally in the body and regulates chemicals, such as cholesterol, in the body that control essential life functions

thymus a gland in young animals that has an immunologic function

thyroidectomy surgical removal of the thyroid gland

thyroid gland "the master gland"; controls, secretes, and regulates hormone production within the body and is located in the neck area

thyroid stimulating hormone (TSH) hormone produced by the thyroid gland that controls the chemical thyroxine

thyroxine hormone that controls the actions of the thyroid gland and cell metabolism

T3 most potent and active thyroid gland hormone measured in the bloodstream to diagnosis thyroid problems

T4 thyroid hormone converted into tissues and breaks down fats and helps control cholesterol

REVIEW QUESTIONS

1. What organ is known as the "master gland"?What is the pituitary hormone that is necessary to govern metabolism called?
2. What is the hormone responsible for stimulating ovulation called?
3. Hypoadrenocorticism, a disorder caused by deficient adrenal cortex production of glucocorticoid, is known as what other condition?
4. What is an abnormally low blood glucose level called?
5. What is the difference between hypothyroidism and hyperthyroidism?
6. What are the stages of estrus?

Clinical Situation

Amelia, a veterinary assistant at Small Creatures Vet Clinic, is answering the phones. She is talking to Mrs. Carl about her cat, "Ruby", a 16 yr SF DSH. "Ruby" has been having some problems such as weight loss, excessive drinking, and increased vocalization. She also notes that she sees "Ruby" in the litterbox more frequently than normal.

Amelia schedules an appointment with Dr. Hall and asks Mrs. Carl to bring "Ruby" right in. Mrs. Carl asks her if she should be concerned and Amelia replies "With "Ruby's" age and clinical signs, this may be an emergency.

Photo by Isabelle Francais

- What may be wrong with "Ruby"?
- What other signs may be present with "Ruby" on a PE?
- What will Dr Hall most likely suggest to the owner?

32

The Renal System

Delmar/Cengage Learning

Objectives

Upon completion of this chapter, the reader should be able to:

- State the structures that make up the renal system
- Describe the function of the renal system
- Name and describe common disorders associated with the renal system

Introduction

Food is transformed through the process of digestion, absorption, and metabolism. The blood and lymph fluids transport products of digestion to the tissues. The cells of the tissues use the oxygen and food they need for growth and repair. The waste products are then taken away to be removed from the body. The primary body system involved in this process is the **renal system**. The functions of the renal system include excreting waste products, such as nitrogen; breaking down proteins and amino acids; producing ammonia and urea; regulating the body's water balance; regulating chemicals, such as potassium, sodium, and chloride; producing hormones that control blood pressure; and producing a chemical called **creatinine**. *Creatinine* is a chemical hormone filtered by the kidneys and indicates the kidney function levels within the blood.

Structures of the Renal System

The structures of the urinary system include the kidneys, ureters, bladder, and urethra. The renal system controls urinary production to rid the body of waste products within the bloodstream (see Figure 32-1). Urine is formed in the kidneys, flows through the ureters to the bladder, is stored in the bladder, and flows through the urethra and outside the body.

Kidneys

The **kidneys** are two organs located in the dorsal abdomen on either side of the spine. The kidneys produce urine. They are reddish brown in color and are bean shaped and smooth (see Figure 32-2). Cows are the only animal species that have lobed kidneys, rather than a smooth surface (see Figure 32-3). Each kidney has a **renal artery** and **renal vein** that are located in the center of the kidney and allow blood flow through the renal system. About 20–25% of the body's blood flows through the kidneys.

The kidney's anatomy includes the **cortex**, which is the outer section of the structure. The **medulla** is the center section of the kidney. The **renal pelvis** is the inner-most section and filters blood and urine through the veins and arteries. The **nephron** is the unit of measurement of the kidney that produces urine and is

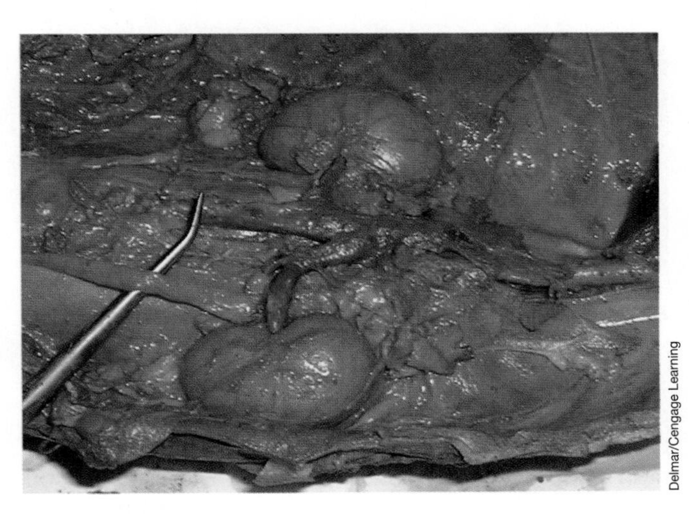

FIGURE 32-2 External appearance of the feline kidney. This photograph is of a prepared specimen. The blue vessels are the renal veins and caudal vena cava injected with latex. The ureter passes over the probe.

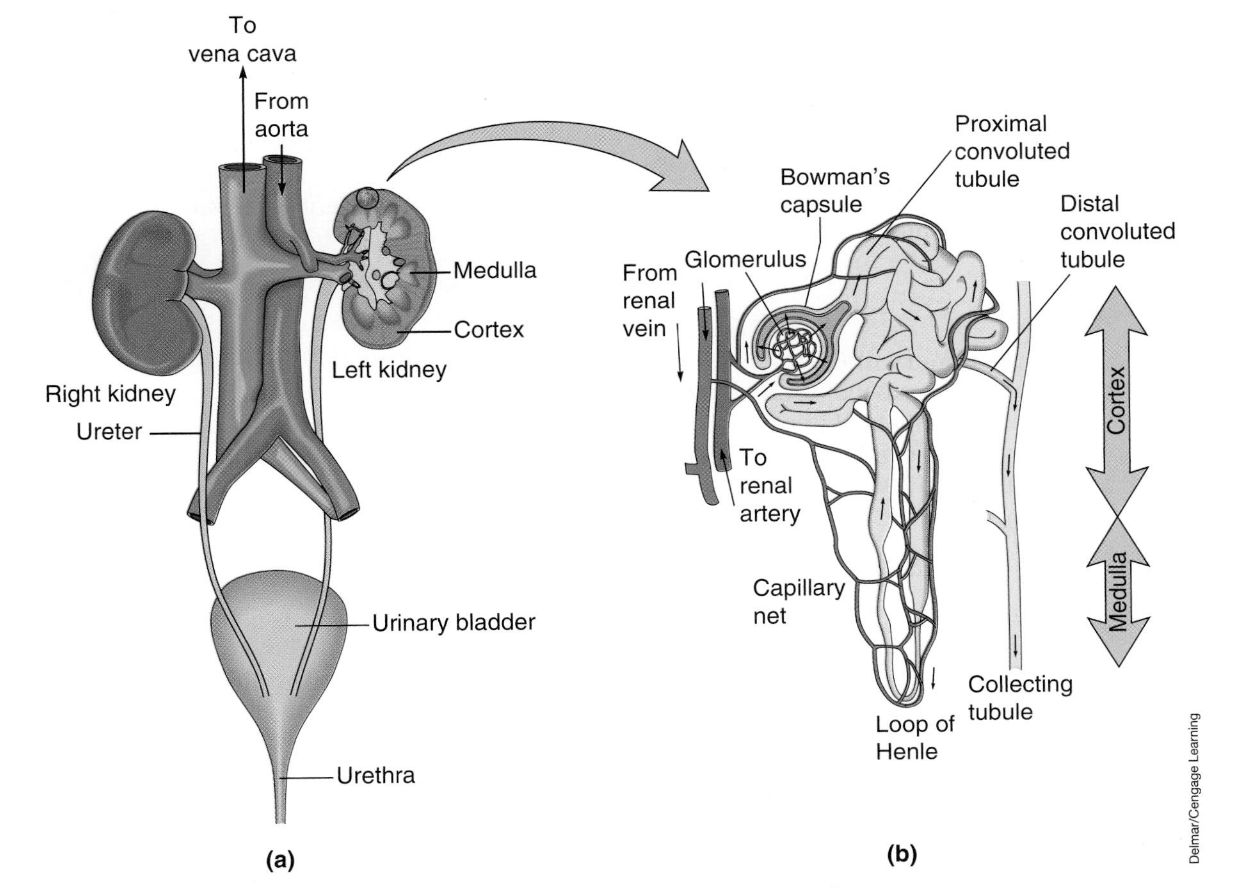

FIGURE 32-1 (A) Structures of the renal system; (B) Structures of a nephron.

FIGURE 32-3 External appearance of a bovine kidney.

produced at the renal pelvis through the tubes that filter blood into the kidneys. This tube is called the **Bowman's capsule**, which is enclosed and wrapped around several capillaries known as the **glomerulus**. Urine production begins through **osmosis** or water being absorbed by the body. Water mixes with the blood and absorbs sodium, causing the water to become diluted and concentrated. This begins the process of urine formation, which involves 80% of the body's water.

Ureters

In the center of each kidney is a tube that filters urine out to the bladder. This tube is called the **ureter**. Each ureter is made of smooth muscle and pushes urine out of the kidneys through contractions.

Urinary Bladder

The ureter connects to the **urinary bladder**, which is a hollow organ that holds urine by expanding and stretching. The bladder holds large amounts of urine based on the size of an animal. For example, a 25-pound dog has a bladder that can hold around 100 ml of urine.

Urethra

The bladder connects to the urethra, which is a muscular tube that allows urine to flow to the outside of the body for elimination. Males typically have a longer and narrower urethra than females. Male animals are more prone to urinary obstructions due to this anatomy. Male dogs have a bone within the urethra known as the **os penis**. The end of the urethra is typically S-shaped and is the site of the narrowest portion of the tube. Urine is passed to the outside of the body through the **penis**. The female urethra is shorter and wider. Female animals are more susceptible to urinary **incontinence** or uncontrolled bladder leakage.

Common Disease and Conditions

The process of eliminating waste from the body is critical to maintaining health and homeostasis.

Urinary Incontinence

Urinary incontinence is the uncontrolled leaking of urine from the bladder. This condition is more frequent in females that are seniors and have been spayed. It commonly occurs during sleeping and causes accidents to occur in the household due to the inability to control the sphincter muscles within the bladder. Many animals respond to treatment with medication of a natural hormone supplement of estrogen to help control the bladder. Females are also more prone to urinary tract infections (UTI) called **cystitis** or inflammation of the urinary bladder. Bacteria migrates into the urethra through contaminates from the environment or increased licking of the **vulva** or **prepuce** areas. Signs of cystitis include **hematuria** (blood in the urine), increased or frequent urination, accidents within the house, and **dysuria** (difficult or painful urination).

TERMINOLOGY TIP

The *vulva* is the external female anatomy where urine is excreted; the *prepuce* is the skin covering the male anatomy of the penis where urine is excreted.

Urinary Blockage

Urinary blockage is caused by several conditions. One condition is **uroliths** or bladder stones that form within the urinary bladder, ureter, or urethra and cause an obstruction (see Figure 32-4). Diet is the common source for bladder stone formation. Radiographs can be helpful in diagnosis of the condition, and many animals may need surgical removal of the bladder stones, known as a **cystotomy**. The uroliths should be analyzed for the components to allow proper treatment with medications or dietary control to prevent further occurrences (see Figure 32-5). Dalmatians are a breed of dog that is known for a genetic urolith condition from uric acid bladder stones. Dalmatians, as a breed, can't metabolize uric acid as their bodies do not create enough enzymes to allow the metabolism. This creates a buildup of uric acid that forms into bladder stones. This condition can be controlled by a specialized diet.

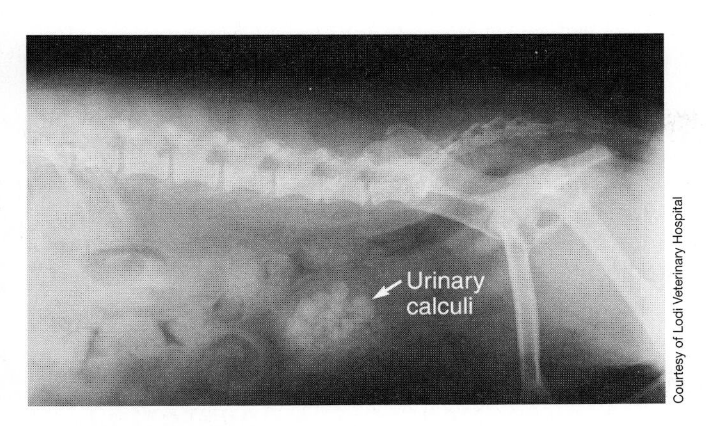

FIGURE 32-4 Radiograph of the urinary bladder showing urinary calculi in the bladder.

FIGURE 32-5 Bladder stones.

Feline Urologic Syndrome

Cats are as a species of animal known for urinary tract disease. Male cats that are castrated have an increased incidence for a condition called **feline urologic syndrome** or FUS. Male cats that are overweight or obese are more at risk for this condition. The condition causes cats to become blocked and they are not able to urinate. This is an emergency situation that must be treated quickly. Cats have urine that tends to build up **sediment**, which is sand-like **crystals** that form in the urine. The sediment easily causes a blockage within the urethra. Cats with FUS must be treated within 24–48 hours as the inability to urinate causes kidney failure to develop. Cats with FUS are sedated, and a urinary **catheter** is used to flush out the bladder and urethra. A urinalysis should be evaluated for treatment options. Typically, veterinarians will keep FUS cats catheterized for 48 hours to allow urination and flushing of sediment along with fluid therapy and medication treatment for an underlying infection. The catheter is removed and the cat is monitored for an additional 24 hours for proper urination. Many FUS cats can be placed on a specialized diet to reduce blockages. Those cats that have increased occurrences of FUS blockages may be candidates for a **perineal urethrostomy** or PU surgery. This surgery essentially removes the urethra tube and makes the male cat more anatomically like a female, eliminating the narrow and long tube that causes sediment buildup and blockage. Signs of FUS include the following:

- Vocalization due to pain
- Painful abdomen
- Hard palpable bladder
- Dehydration
- Weakness and lethargy
- Panting and increased HR and RR

Toxicity

Antifreeze toxicity or **ethylene glycol** poisoning is a condition that affects the renal system and causes a severe condition in an animal. Dogs and cats may ingest antifreeze that is found outside due to its fruity taste and smell, making it inviting to animals. Ethylene glycol is the active ingredient in antifreeze and is not capable of being metabolized by the kidneys. This buildup becomes toxic to the renal system, causing it to fail and shut down. About 80% of animals that ingest antifreeze will die. Time is of the essence with this condition. Signs of antifreeze toxicity include central nervous system (CNS) conditions, such as ataxia or loss of body control, disorientation, inability to stand, weakness, difficulty breathing, oliguria or anuria, PU/PD, painful abdomen, vomiting, seizures, coma, and death. Bloodwork and urinalysis evaluation help in determining the diagnosis and treatment of antifreeze toxicity, with the main therapy being that of fluids to flush out the renal system.

TERMINOLOGY TIP

The **central nervous system** or CNS is the part of the body under involuntary brain control, such as the brain stem and spinal cord. **Ataxia** is a term used to note an animal that has the loss of control over its body causing wobbliness and inability to control movement. **Oliguria** is decreased amounts of urine production, whereas **anuria** is no urine production.

Renal Failure

Renal failure may occur as an **acute** condition, which is one that occurs suddenly and is short term. It may also be a **chronic** condition in which the signs appear over a long time. Chronic renal failure is typical of senior pets due to age and other disease conditions that cause problems with the renal system. Signs of chronic renal failure include increased thirst (PD), poor appetite, weight loss, loss of urinary control (incontinence), vomiting, and diarrhea. Blood work typically reveals an increased blood urea nitrogen (BUN) level, increased creatinine, and increased phosphorus levels. Anemia is also seen on a complete blood count or CBC. Fluid therapy along with dietary control with reduced protein and phosphorus are the usual treatment. Routine monitoring of the blood levels helps with managing the condition.

SUMMARY

The renal system's main responsibility is the removal of wastes from the body. The removal of wastes is achieved by the filtering of blood. This mainly occurs in the kidney. The renal system also helps to maintain proper balance of body fluids, which aids in the maintenance of homeostasis.

MAKING THE CONNECTION

Chapter 18—leptospirosis
Chapter 21—leptospirosis

Key Terms

acute short term occurring condition

antifreeze toxicity poisoning that affects the renal system and shuts down the system causing renal failure; also known as ethylene glycol toxicity

anuria no urine production

ataxia term used to note an animal that has the loss of control over its body, causing wobbliness and inability to control movement

Bowman's capsule tubes located at the renal pelvis that filter blood into the kidneys from excessive amounts of dietary components

catheter thin plastic or rubber tube that is passed into the urinary bladder or bloodstream

central nervous system (CNS) part of the body under involuntary brain control, such as the brain stem and spinal cord

chronic long-term condition

cortex outer section of the kidney structure

creatinine chemical hormone filtered by the kidneys and indicates the kidney function levels within the blood

crystals solid formations within a urine sample that form from increased components within the urinary tract or renal system

cystitis term for a bladder infection or urinary tract infection

cystotomy surgical incision into the urinary bladder to remove urinary bladder stones

dysuria difficult or painful urination

ethylene glycol toxicity poisoning by antifreeze that causes renal failure

feline urologic syndrome (FUS) condition that commonly affects neutered male obese cats in which they become blocked and are unable to urinate

glomerulus capillaries located in a bundle at the renal pelvis

hematuria blood in the urine

incontinence uncontrolled bladder leakage

kidney two organs located in the dorsal abdomen located on either side of the spine that produce urine and are reddish brown in color and are bean shaped and smooth

medulla center section of a kidney

nephron unit of measurement of the kidney that produces urine

oliguria decreased amounts of urine production

osmosis water absorption by the body through membranes

os penis bone within the urethra of male dogs

penis male sex organ and organ that allows urine to be excreted externally

perineal urethrostomy (PU) surgical creation of a new opening at the urethra and penis to eliminate the occurrence of urinary blockages

prepuce the skin covering the male anatomy of the penis where urine is excreted

renal artery located in the center of the kidney and allows blood flow out of the renal system

renal pelvis inner-most section of the kidney that filters blood and urine through the veins and arteries

renal system body system that involves the kidneys and urinary production to rid the body of waste products within the bloodstream

renal vein located in the center of the kidney and allows blood flow into the renal system

sediment sand-like buildup in the urine composed of crystals and casts

ureter tube made of smooth muscle that pushes and filters urine out of the kidneys through contractions

urolith term for a bladder stone that forms within the urinary bladder

vulva external female anatomy where urine is excreted

REVIEW QUESTIONS

1. What is a UTI? What are the signs of a UTI? What is urinary incontinence? Explain. What is the medical term for excessive urination?
2. The pathway of urine formation is:
   a. kidney, ureter, urethra, bladder
   b. ureter, pelvis, urethra, bladder
   c. kidney, urethra, bladder, ureter
   d. kidney, ureter, bladder, urethra
3. What does the term "oliguria' mean?
4. What are the parts of the urinary system in both male and female animals?
5. What are the components of a urinalysis?
6. List and describe the functions of the renal system.
7. Explain the difference between urinary incontinence and urinary blockage.
8. Describe how urine is formed.

Clinical Situation

Photo by Isabelle Francais

Kelly, a vet assistant at High Pines Animal Hospital, is performing a UA sample on "Max", a 4 yr NM Dalmatian. Kelly has just recently been trained to run lab work samples. "Max's" owner, Mr. Kline, brought the urine sample in this morning and Kelly placed it in the fridge until she has a chance to analyze the sample. "Max" has a history of bladder stones and frequent UTI's.Kelly notes the color of the urine is brown and appears cloudy and somewhat thick. She performs a clinical stick analysis with the following results that she provides to the vet.

Color: brown
Appearance: cloudy, thick, viscous
SG: 1.245
Glucose: NEG
Ketones: NEG
Bilirubin: TRACE
Protein: TRACE to +
Blood: POS + + +
pH: 8.5

- What is most likely wrong with "Max"?
- What other parts of a UA might Kelly complete?
- What factors may be involved with the sample and how the owner collected the specimen?

33

Reproductive System

Objectives

Upon completion of this chapter, the reader should be able to:

- State the structures that make up the reproductive system
- Describe the functions of the reproductive system
- Identify and describe the differences between male and female reproduction
- Describe common disorders of the reproductive system

Introduction

The **reproductive system** is responsible for producing offspring. Both male- and female-specific organs are required to complete offspring production. The animal reproductive system, in general, is basically similar in many species. The reproductive organs, whether male or female, are called **genitals** or **genitalia**.

The Female Reproductive System

In the female, the **vulva** is the external beginning of the reproductive system. The vulva is the external opening of the **vagina**, or the internal female reproductive organ that allows for breeding, labor, and urination processes to occur. At the base of the vagina is the **urethra**, which is a narrow tube that connects to the urinary bladder. The end of the vagina holds the **cervix**, which is the opening to the uterus. The **uterus** is the main area for allowing an embryo to begin the gestation phase. The uterus is Y-shaped and has two tubes that are called **uterine horns**. It is a large, hollow organ that expands through the gestation period to allow the young to grow. Some animal species carry single or multiple young in the uterine horns, and other species carry the embryo or embryos within the uterus. Each end of the horn tapers into a small tube called an **oviduct**, which holds the ovary and is the transport system for sperm and eggs. The **ovary** is the primary female reproductive organ and produces eggs in the female through the process of meiosis. There are two ovaries, one located at the end of each uterine horn. Some variations exist between species; for example, chickens have only one fully developed left ovary. The ovary contains **follicles** that are tiny structures where the eggs are produced. Each animal species has a set number of follicles and eggs that are present when the animal is born. The eggs are used slowly throughout the life of the animal. As an egg matures, it will move to the edge of the ovary for release. The time frame when an egg is released is called **ovulation**. Figure 33-1 shows female reproductive organs in a dog and cow.

Some species of animals, such as avian species, some fish, and some reptiles, have a reproductive system for laying eggs rather than bearing live young. The egg develops in the **infundibulum**, which is a funnel within the uterus. Species that lay eggs have an oviduct that has four sections or segments: the **ovum** (egg yolk), **albumen** (egg white), **isthmus** (shell membrane), and the uterus. The external reproductive area where the egg is laid is called the **cloaca** or **vent**. The time frame of ovulation until an egg is laid is about 24 hours. Figure 33-2 shows the reproductive organs in a bird.

Estrogen is the main female reproductive hormone that causes behavior changes in females during the estrus cycle and also allows follicle production and development to occur. **Progesterone** is a hormone that is produced to maintain pregnancy.

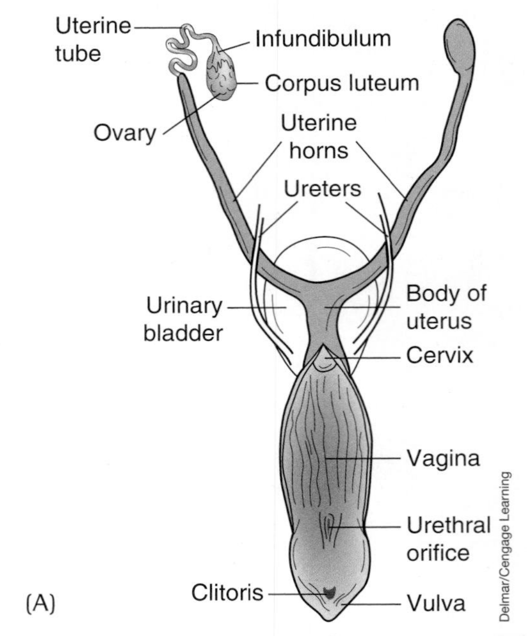

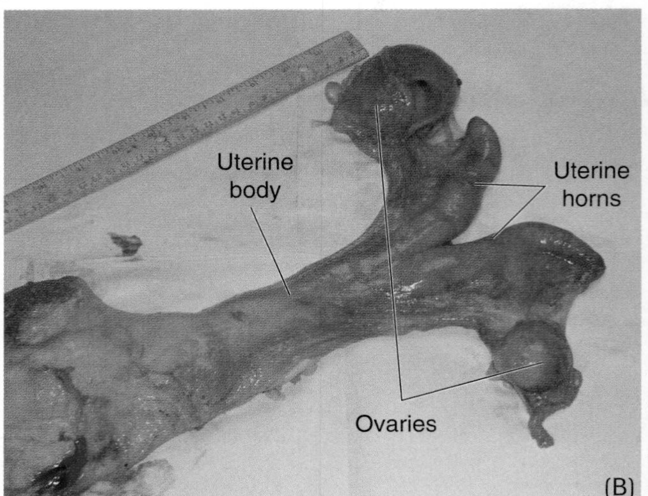

FIGURE 33-1 (A) Reproductive tract of a female dog; (B) reproductive tract of a cow.

The Male Reproductive System

The male reproductive system begins with the external structure known as the **penis**. The penis is the male reproductive organ used in breeding and is also part of the urinary and renal system. The penis is covered and protected by skin called the **prepuce**. The primary reproductive organs in males are called **testicles**. The testicles produce sperm and are held in place by a sac known as the **scrotum**. The scrotum consists of skin with two lobes

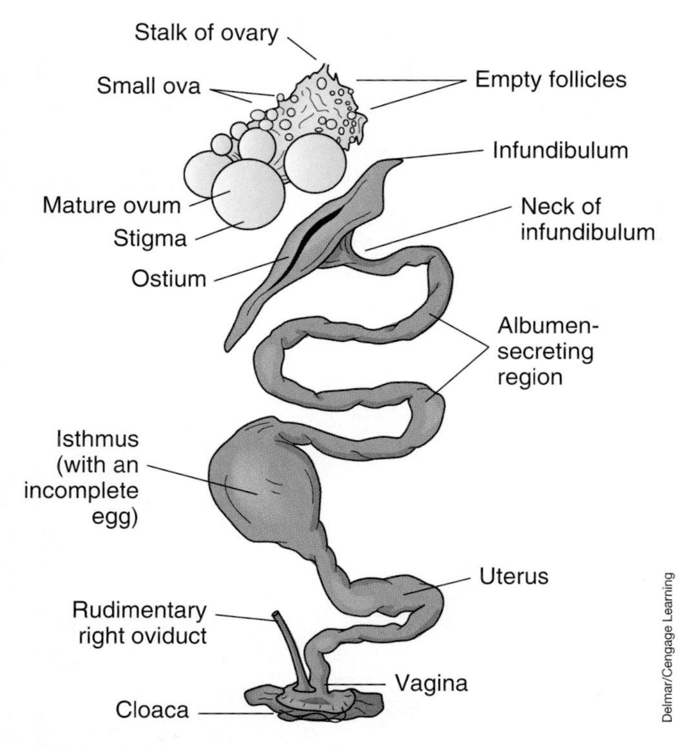

FIGURE 33-2 Reproductive tract of a bird.

that hold and protect each testicle. The scrotum is temperature controlled for sperm production and must be cooler than the body for best sperm production. **Sperm** production begins at puberty and continues throughout the life process. Sperm is the male reproductive cell. Sperm is formed in the **seminiferous vesicles**, which are tubes located within the testicles. Sperm passes through the **epididymus** and into the **vas deferens**, the transport systems for sperm from the testicles to the outside of the body. Both organs also carry **semen**, which is fluid that transports the sperm during breeding. In mammals, the sperm travels through the urethra. In avian species and reptiles, the sperm travels through the cloaca. Avian species do not have a true penis due to an undeveloped organ. In fish species, the sperm travels through the urogenital pores, which also passes feces and urates. Figure 33-3 shows the reproductive system in a stallion. **Testosterone** is the main male reproductive hormone that produces male characteristics and is essential for sperm production. Male characteristics developed by testosterone include the following:

- Large body size
- Muscles and powerful structure
- Aggressive tendencies
- Deep vocal sounds

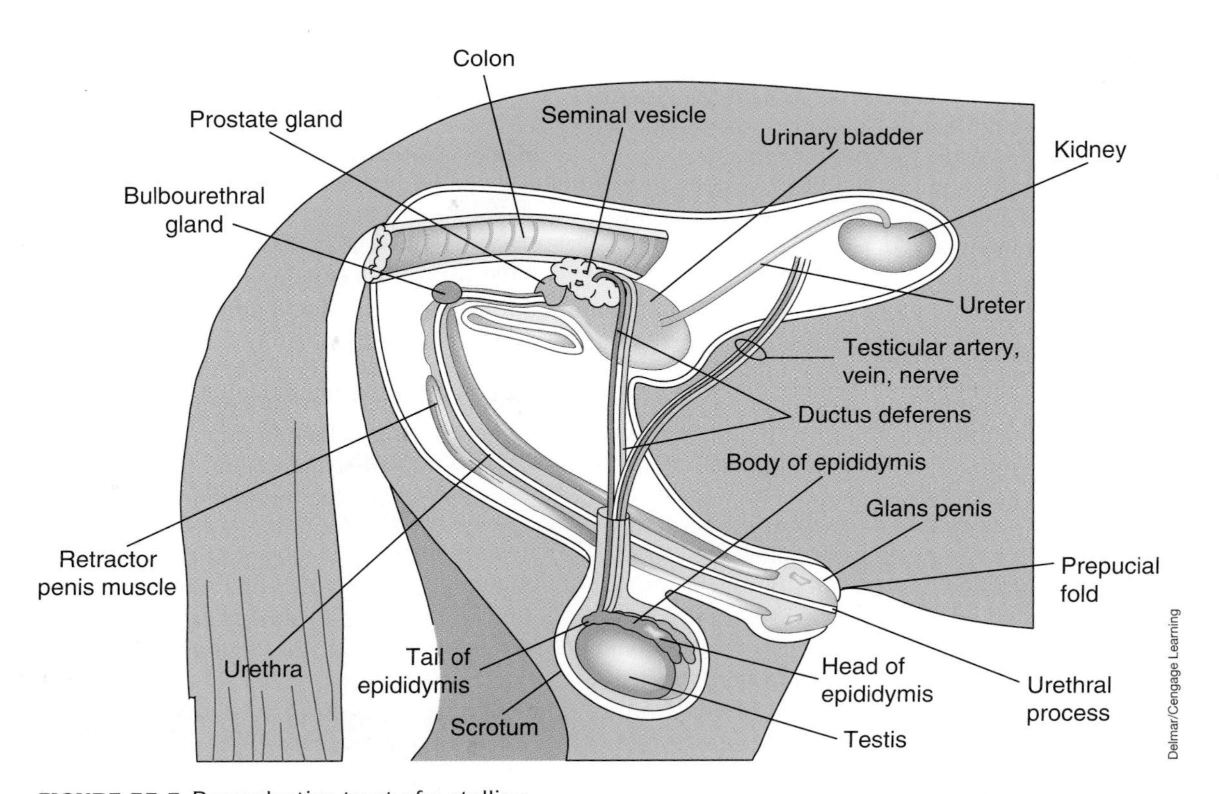

FIGURE 33-3 Reproductive tract of a stallion.

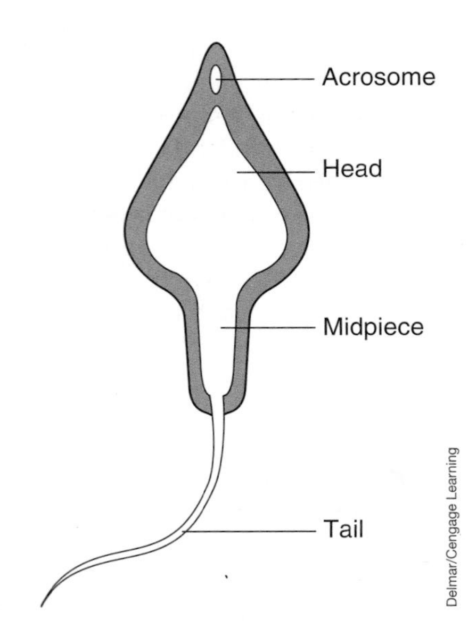

FIGURE 33-4 Parts of sperm.

Semen

Semen morphology is the process of determining if sperm is normal and adequate for breeding. Sperm cells multiply by meiosis. The sperm have a head and a tail, known as the **flagella** (see Figure 33-4). The flagella allow for movement and motility. Chromosomes are found in the sperm's head and are what determine the DNA of the offspring. A normal sperm cell has one head and one straight tail. Sperm quality is noted through the use of a microscope, and all sperm should move rapidly. Sexual reproduction occurs when the egg and sperm unite. This requires a male and female of the same species. When egg and sperm unite, the process is called **fertilization** (see Figure 33-5).

Fertilization

In mammals, fertilization occurs directly in the reproductive tract. In avian species, the female will develop an egg and pass it externally, and incubation must then occur to allow the egg to hatch. **Incubation** is the process of increasing the temperature to allow the embryo to grow. Fish fertilization occurs outside of the reproductive tract through either eggs, as live young, or as mouth brooders. **Mouth brooders** carry their eggs in their mouth until hatching. Fertilization leads to **conception**, which is the process of creating new life in the form of an embryo. An **embryo** is a cell that develops into a newborn animal. Pregnancy occurs and is known as the **gestation** or the length of time until the animal is born. **Hormones** are chemicals that are produced by the body that allow for sexual reproduction to occur in both males and females.

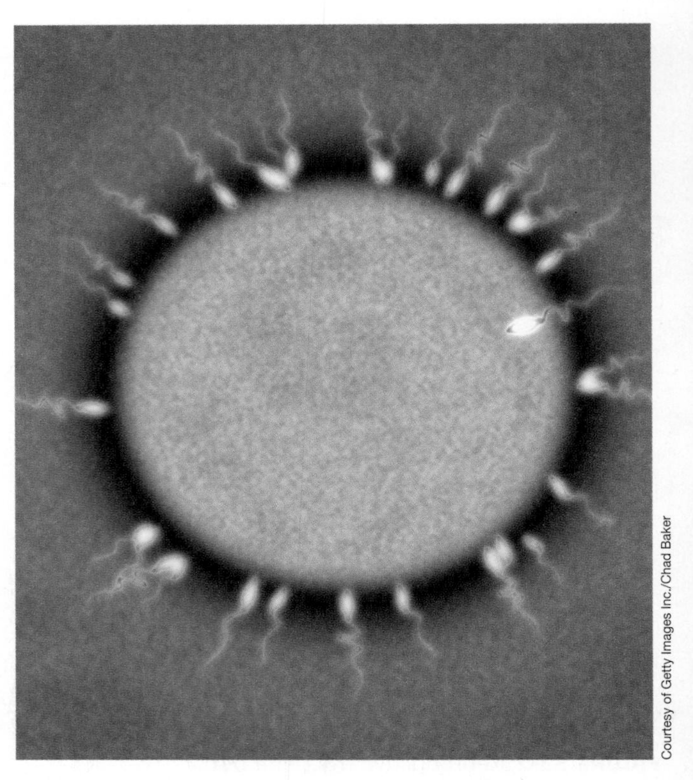

FIGURE 33-5 The outer membrane of an egg must be dissolved before the sperm can enter the egg.

Estrus

Estrus cycle is also called the heat cycle in animals. This is the period of time that animals are receptive to breeding. During the heat cycle a female animal is bred. There are four stages within the heat cycle, and each animal species has a different length of cycle and duration. Many animals have a seasonal estrus, such as many large animal species, which have an estrus cycle that depends on the length of the day. **Proestrus** is the stage that occurs just prior to the heat cycle in an animal that is not pregnant. Progesterone is released in this stage. **Estrus** is the time when the female animal is receptive to breeding or mating with a male. Some estrus cycles are as short as 12 hours and others as long as 14 days. The estrus cycle is triggered by the release of estrogen. Signs of estrus vary from animal to animal, but many female animals in estrus will have a swollen vulva, a change in behavior, vaginal discharge, increased urination, and increased vocal behaviors. **Metestrus** is the stage just following the heat cycle and when ovulation occurs. Some species are **induced ovulators** and must be mating for the release of the egg to occur. Examples of induced ovulators are cats and rabbits. Some species release one egg during ovulation, whereas other species release multiple eggs. Table 33-1 summarizes the types of ovulation that occur in various species.

TABLE 33-1

Animal Species Ovulation Types

Dog	Multiple eggs released during ovulation
Cat	Induced ovulation with multiple eggs released
Horse	One egg released during ovulation
Cow	One egg released during ovulation
Sheep	One egg released during ovulation
Pig	Multiple eggs released during ovulation
Goat	One egg released during ovulation
Rabbit	Induced ovulation with multiple eggs released

TERMINOLOGY TIP

Induced ovulation is the occurrence of the release of the egg at the time sperm is introduced. This causes immediate fertilization to occur at the time of mating.

Metestrus produces the **luteinizing hormone (LH)**, which allows an egg to develop and release. This hormone triggers the **corpus luteum (CL)** to develop an egg. The CL allows for the follicle to enlarge in size. During this stage, the CL is called the "yellow body" due to its color and size. The follicle grows in size as a response to the **follicle stimulating hormone (FSH)**. After ovulation occurs, breeding must occur to ensure pregnancy. Metestrus typically lasts 3–4 days, and at the end of the stage the follicle will collapse and estrogen production will decrease. **Diestrus** is the stage when pregnancy occurs and is characterized by a functional CL that releases progesterone to maintain pregnancy. This occurs over a 9–12 day time frame and prepares the uterus for gestation. **Proestrus** is the end stage of the heat cycle of an animal that has become pregnant. When gestation is occurring, the CL becomes smaller in size and hormone production decreases. If breeding does not occur or pregnancy is not maintained, the animal will go out of the estrus cycle and remain in anestrus until the next scheduled estrus occurs. **Anestrus** is the time when an animal is not in the estrus cycle. Figure 33-6 illustrates the estrus cycle in a cow. Table 33-2 summarizes estrus cycles in various species.

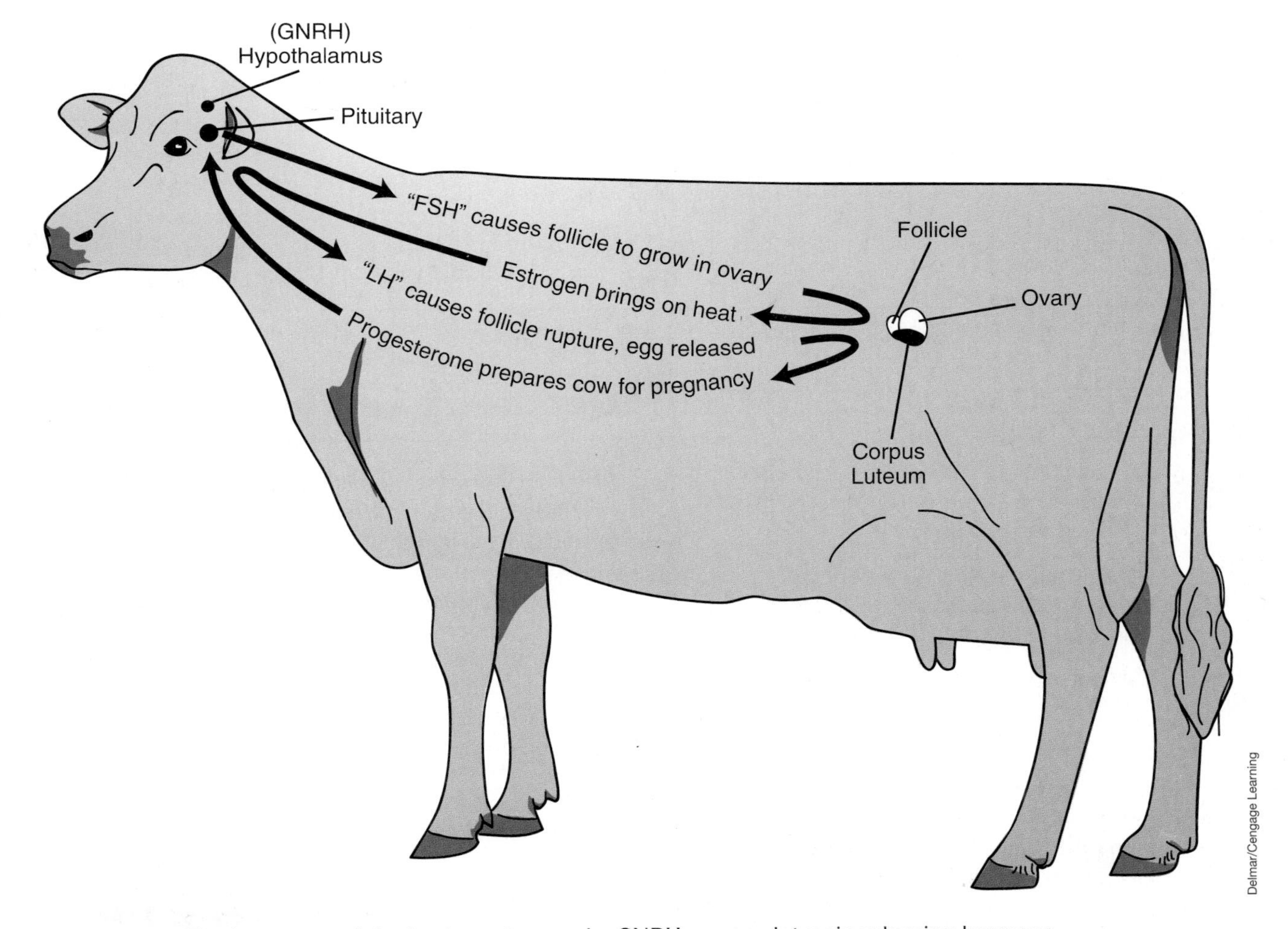

FIGURE 33-6 The hormones of the bovine estrus cycle. GNRH = gonadotropin-releasing hormone.

TABLE 33-2

Estrus Cycles in Various Species

SPECIES	CYCLE OCCURRENCES	LENGTH OF ESTRUS	OVULATION TIME	SIGNS OF ESTRUS
Dog	Cycle twice a year*	9–10 days	1–2 days after estrus	Swollen vulva Vulva discharge **Tail flagging** **Standing heat**
Cat	Polyestrus—cycle every 14–21 days	4–5 days	24 hours after mating (induced ovulation)	**Lordosis** Friendly Vocal
Horse	Cycle every 21 days	6–10 days	1–2 days prior to the end of estrus	Tail flagging **Winking** Increased urination Vaginal discharge Swollen vulva Vocal
Cow	Cycle every 21 days	12–18 hours	10–14 hours after estrus	Standing heat Restless Mounting other cows
Goat	Cycle every 21 days	30–40 hours	End of cycle	Rapid tail wagging Tail held in air Mounting other goats
Sheep	Cycle every 17 days	24–36 hours	Late estrus	Minimal signs Difficult to detect Mounting other sheep
Pig	Cycle every 21 days	40–72 hours	Mid-estrus	Swollen vulva Standing Heat Lordosis
Rabbit	Constant cycle—polyestrus	polyestrus	8–10 hours after mating (induced ovulation)	Aggression Lordosis

*The exception in the dog is the Basenji, which only has one estrus cycle per year.

TERMINOLOGY TIP

Many terms relate to breeding animals. The following are several common terms:

Lordosis—exhibiting a prayer position where the front end of the body is lowered to the ground with the back end held high in the air

Polyestrus—many heat cycles occur at regular intervals

Standing Heat—the female animal will stand for breeding when a male is present

Tail Flagging—the raising of the tail held high in the air

Winking—movement of the vulva with frequent vaginal discharge and urination

Gestation and Incubation

Pregnancy or gestation is the period of development of the fetus in the uterus from conception to birth. Gestation periods vary in length in different species (see Table 33-3). Signs of pregnancy include increased appetite, weight gain, and increased size of mammary glands.

In egg-bearing animals, once an egg is laid the embryo continues to grow inside the egg. This period of growth is known as the incubation period. This period varies by species (see Table 33-4). Proper temperature and humidity are important to incubation.

The Labor Process

The **parturition** process, also known as birth or labor, has three stages. During the labor process the levels of progesterone decrease and estrogen increases.

TABLE 33-3

Gestation Length of Various Species

SPECIES	GESTATION LENGTH
Dog	Average 63 days
Cat	Average 63 days
Horse	Average 330 days
Cow	Average 285 days
Goat	Average 50 days
Sheep	Average 150 days
Pig	Average 114 days
Rabbit	Average of 30 days

TABLE 33-4

Incubation Table

Chicken	21 days
Duck	28 days
Turkey	28 days
Goose	28–30 days
Pheasant	24 days

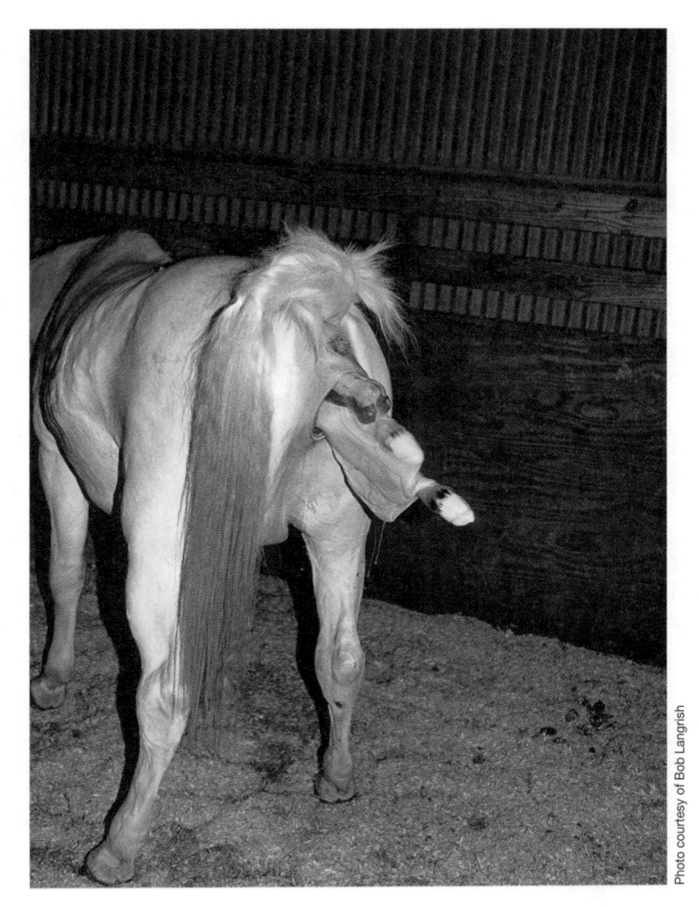

Photo courtesy of Bob Langrish

FIGURE 33-7 Birth of a foal.

The estrogen hormone prepares the newborn for the delivery process. Labor begins when cortisone is released by the fetus. The first stage of labor begins with **contractions** of the uterus as the fetus begins to move around. Signs of first-stage labor include restlessness, anxiety, nervousness, and abdominal discomfort. The cervix begins to dilate, or open, and causes the tissues to soften. The second stage of labor occurs when the fetus moves into the pelvic opening of the birth canal. Strong abdominal contractions increase as each fetus enters the vaginal area for birth. The fetus exits the birth canal during this stage (see Figure 33-7). Most female animals will lick at the newborn immediately after birth to help clean and dry the baby and also to stimulate breathing. This process removes the fetal membranes over the mouth and nose. After the active labor stage is complete, the third stage occurs, and the placental tissues are expelled from the uterus. The **placenta** is the tissue and membrane collection that connect the fetus to the mother to allow nutrients to be absorbed during the gestation time (see Figure 33-8). The placenta is also the collection area for fetal wastes. After the placenta is removed, the uterus begins to shrink to its normal size through a process called **involution**. This usually occurs 1–3 weeks after birth depending on animal species.

Semen Collection

Semen collection is the procedure using specialized veterinary equipment to collect sperm from male animals to be used in breeding practices. Semen collection is done in numerous species and breeds of animals, especially in horses, dogs, and cattle. Collection methods will vary between each species of animal, but proper and sterile collection methods are important to the success of the practice. The most common collection method is with the use of an **artificial vagina** (AV), which is a sleeve or tube used to collect the semen of an animal during the ejaculation process. It is important that veterinary assistants and other professionals have advanced handling and restraint skills when working with breeding animals. This includes knowledge of and experience in body language and behavior of the species. Once semen is collected it should be evaluated for any abnormal appearance. It must then be cooled slowly, preserved, and properly stored. Storage and shipping of semen may be cooled or frozen. **Cooled semen** must be shipped and used within a 24–48 hour time frame for sperm survival. **Frozen semen** may be stored in liquid nitrogen and may be preserved for up to 40 years.

Delmar/Cengage Learning

FIGURE 33-8 Placenta.

Artificial Insemination

Artificial insemination (AI) is a procedure that is popular in many animal species and breeds and is used in many different situations. This procedure is the practice of breeding female animals through the use of veterinary equipment rather than the traditional mating method (see Table 33-5). Timing is of the most vital importance when using AI methods. The female must be in heat and near ovulation, and determining this requires proper estrus cycle evaluation and observation. The most common failure of the AI procedure is failure to properly detect and determine the ovulation of the animal. There are many reasons to use AI, such as the male animal located in another part of the country, the male and female are not compatible, the animal is too aggressive to safely breed, or the female's estrus cycles are difficult to regulate. The first step in the artificial insemination process is having quality sperm. The sperm should be evaluated on arrival for motility and quality. Cooled and frozen semen must be warmed to 95–98 degrees. Fresh semen should be used within 24 hours. The insemination process should be done at intervals during at least two times prior to ovulation. This requires knowledge of the species and proper timing of the cycle.

Some AI procedures place the semen in the uterus while others place the semen within the cervix. This depends on the animal species and where the embryo grows within the reproductive process. Many female animals will stand for the AI procedure while other species may need to be sedated for surgical AI placement. It is important to prevent the female from urinating or leaking any semen from the vagina for 30 minutes after the AI process is completed. An X-ray or ultrasound procedure is done after gestation has started to ensure the AI breeding process has occurred and to determine the number of offspring.

Neutering

Neutering is the surgical removal of the reproductive organs. The **ovariohysterectomy (OHE)** or spay is the surgical procedure in females that removes the uterus

TABLE 33-5

Common AI Procedures

Dogs	Estrus smear to evaluate cells and stage in cycle	AI during last stages when cornified epithelial cells are present	AI tube placed in cervix
Horses	Palpate or ultrasound to determine stage in cycle	AI on days 3, 5, and 7 in every-other-day method	AI speculum used with caution due to delicate tissue; AI tube placed into uterus
Cows	Most estrus activity occurs at night	AI during the last 2/3 or 1–2 hours after estrus ends	Palpate to feel ovaries; place AI tube in cervix
Sheep	Rare; difficult to detect estrus	AI twice during estrus cycle	AI tube placed in uterus
Pigs		AI 24 hours after estrus cycle begins; AI again 40–48 hours after estrus begins	AI tube placed in cervix
Turkeys	Used due to difficulty breeding	Once a week for 2 inseminations	Storage glands for semen in hens

and ovaries. A surgical incision is placed on the midline of the abdomen, or the center of the belly near the umbilicus or navel. This procedure is done to prevent overpopulation of animals, eliminate aggressive or difficult behaviors, and prevent cancer development in animals. It is not recommended to perform surgery in females that are in estrus as increased bleeding may occur, causing complications.

The **castration** surgery is the removal of the testicles in male animals (see Figure 33-9). It is done by making a small incision near the testicles over the scrotum. This is a surgical procedure under anesthesia in small animals. Sutures or staples are usually used to close the incision. In large animals, such as cows and horses, castration is usually done standing with sedation. The area is left open to heal. Both surgeries require postoperative care in reduced activity, monitoring for swelling and infection, and suture removal if necessary.

Some male animals may not have both or one testicle descend into the scrotum. This is known as cryptorchid. A **unilateral cryptorchid** is one retained testicle, and a **bilateral cryptorchid** occurs when both testicles are retained. Retained testicles may be in the abdomen, inguinal canal, or under the skin near the scrotum. Most testicles are smaller in size when they don't descend properly into the scrotum. They are also at an increased risk for tumor development and can cause pain. Cryptorchid animals are predisposed for a genetic tendency to pass this condition on to their offspring.

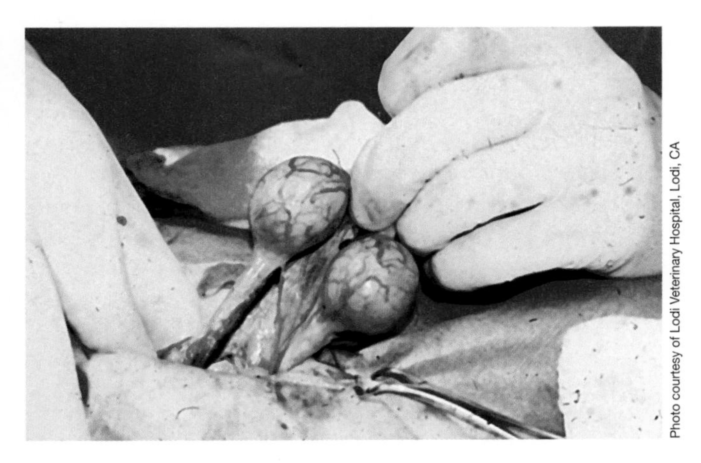

Photo courtesy of Lodi Veterinary Hospital, Lodi, CA

FIGURE 33-9 Canine castration.

Reproductive Problems

Female animals may develop a uterine infection known as a **pyometra**, where the uterus fills with pus. This condition is an emergency and must be treated immediately. The buildup of bacteria in the system causes a toxic affect when it gets into the bloodstream, causing a **septic** condition. An immediate ovariohysterectomy should be done to remove the infected uterus. Another reproductive condition in females is a **prolapsed uterus**. This is more common in cattle but can occur in any species of female that is post-labor (see Figure 33-10). It has an increased occurrence in females that have large newborns or numerous young in which labor requires excessive straining. Difficult deliveries involving **dystocia** also may cause a uterus to

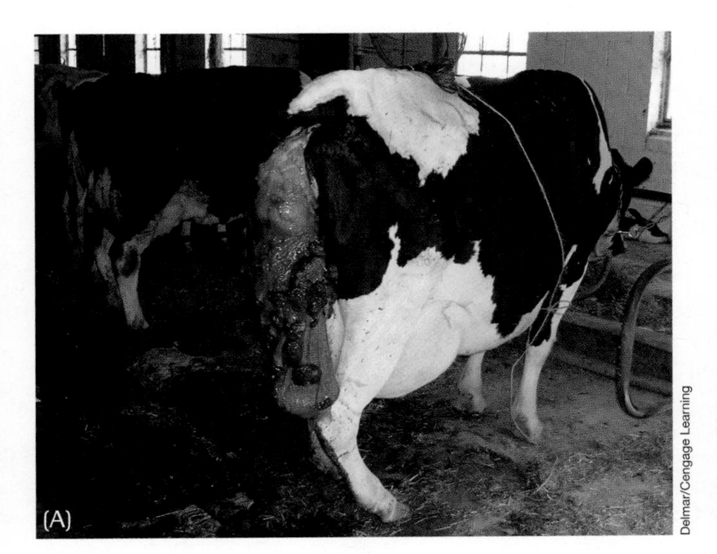

FIGURE 33-10 (A) Cow with a prolapsed uterus; (B) close-up of prolapsed uterus showing placentomes.

prolapse. The uterus and vaginal tissues separate from the body and allow the uterus body to be passed outside the body due to the stretching of the vagina and vulva. This is a life-threatening emergency situation as the exposure to the outside of the body can cause severe septic infection, or possible tearing can cause a uterine artery to hemorrhage.

Ovarian cysts also occur in intact females and may occur on the follicles or ovaries of the reproductive system. They usually are larger than 1 cm in size and begin when the follicle or ovary doesn't release an egg. This condition typically causes the female animal to be **sterile**. Signs of ovarian cysts include irregular and decreased estrus cycles. Many females with cysts also produce an excessive amount of estrogen.

Mammary Gland Problems

Mammary gland tumors are a common occurrence in females with intact reproductive systems. These tumors may be cancerous and should be evaluated by a veterinarian. Swelling may occur on one or more mammary glands. The surgical removal of the mammary glands is known as a **mastectomy**. There is a decreased occurrence in females that have had an OHE prior to having the estrus cycle begin.

SUMMARY

The reproductive system allows animals to breed and create new life. Male and female reproductive organs are required for fertilization and incubation to take place. Reproduction is an important part of the animal industry. Disorders of the reproductive system can cause sterility, rendering an animal incapable of giving birth. Some reproductive disorders can be life-threatening.

MAKING THE CONNECTION

All chapters in Section III contain information on reproduction and breeding of various animals species.

Key Terms

albumin egg white in an animal that lays eggs

anestrus time when a female animal is not actively in the heat cycle

artificial insemination (AI) the practice of breeding female animals through the use of veterinary equipment rather than the traditional mating method

artificial vagina (AV) sterile sleeve or tube used to collect the semen of an animal during the ejaculation process

bilateral cryptorchid both testicles are retained and do not descend into the scrotum

castration surgical removal of the testicles in male animals

cervix end of the vagina that opens to the uterus

cloaca external reproductive area where egg-laying animals pass urine, feces, and eggs; also called the vent

conception process of creating a new life that forms an embryo

contraction first stage of labor when the fetus begins to move in the uterus to prepare for birth by moving into the birth canal

cooled semen process of transporting and storing sperm for use within 24–48 hours due to cool temperatures

corpus luteum (CL) known as the "yellow body" of the ovary that forms to allow an ovary to increase in size as an egg develops

diestrus stage of the heat cycle when pregnancy occurs and is characterized by a functional CL that releases progesterone to maintain pregnancy

dystocia difficult labor

embryo cell that develops into a newborn life

epididymus tube that transports sperm from the seminiferous vesicles to the vas deferens

estrogen main female reproductive hormone that causes behavior changes in females during the estrus cycle and also allows follicle production and development to occur

estrus the time when the female animal is receptive to breeding or mating with a male

estrus cycle heat cycle in female animals when the female is receptive to mating with a male

fertilization uniting of egg and sperm to allow an embryo to form

flagella tail of sperm that allows movement

follicle tiny structures within the ovary that enlarge with egg development and allow ovulation to occur

follicle stimulating hormone (FSH) hormone released to allow a follicle to grow and allow ovulation to occur

frozen semen method of shipping and storing sperm in liquid nitrogen; may be frozen and preserved for up to 40 years

genitalia reproductive organs

genitals reproductive organs

gestation length of pregnancy

hormone chemicals that are produced by the body that allow for sexual reproduction to occur in both males and females

incubation the process of increased temperature to allow an embryo to grow

induced ovulation occurrence of the release of the egg at the time sperm is introduced when fertilization occurs at mating

infundibulum funnel within the uterus where an egg develops in an egg-laying animal

involution the process of the uterus shrinking after labor

isthmus shell membrane of an animal that lays eggs

lordosis exhibiting a prayer position where the front end of the body is lowered to the ground with the back end held high in the air

luteinizing hormone (LH) allows an egg to develop and be released in the ovary during metestrus

mastectomy surgical removal of the mammary glands

metestrus stage of the heat cycle following estrus when ovulation occurs

mouth brooder fish that carry their eggs in their mouth for incubation to protect and allow hatching

ovarian cyst small masses that develop on the ovary of an intact female animal, causing eggs not to develop and sterility

ovariohysterectomy (OHE) term for a spay; surgical removal of the ovaries and uterus

ovary the part of the uterine horn that allows egg development and is the primary female reproductive organ in egg growth

oviduct small tube on the end of the uterine horn that is tapered and holds the ovary

ovulation the release of the egg from the ovary

ovum egg yolk in an animal that lays eggs

parturition the labor process

penis the male external reproductive organ used in breeding and mating

placenta tissue and membrane collection that connect from the fetus to the mother to allow nutrients to be absorbed during gestation

polyestrus many heat cycles occur at regular intervals throughout the breeding season

prepuce the skin that covers and protects the penis in male animals

proestrus stage of the heat cycle before estrus that releases the progesterone hormone prior to mating

progesterone hormone produced to maintain pregnancy

prolapsed uterus condition when the uterus and vaginal tissues separate from the body and allow the uterus body to be passed outside the body

pyometra severe infection of the uterus in female intact animal that causes the uterus to fill with pus

reproductive system the organs in the body responsible for both male and female breeding and reproduction

scrotum skin over the two testicles that protects and controls the temperature for sperm production

semen fluid that transports sperm during reproduction

semen collection procedure using specialized veterinary equipment to collect sperm from male animals to be used in breeding practices

semen morphology process of determining if sperm is adequate for transport and breeding by analyzing the quality and quantity

(*Continues*)

seminiferous vesicle tubes located within the testicle that transport sperm to the testicles

septic buildup of bacteria in the system causing a toxic affect when it gets into the bloodstream

sperm male reproductive cell developed in the testicle

standing heat the female animal will stand for breeding when a male is present

sterile not able to reproduce

tail flagging holding the tail high in the air during the estrus cycle

testicle the primary reproductive organ in the male that holds sperm

testosterone main male reproductive hormone that produces male characteristics and is essential for sperm production

unilateral cryptorchid one retained testicle that has not descended into the scrotum

urethra narrow tube that connects the male and female reproductive organs to the urinary bladder

uterine horn two parts of the uterus that hold single or multiple embryos that expand as the embryos grow

uterus large, hollow reproductive organ that holds an embryo during the gestation period and is Y-shaped

vagina internal female reproductive organ that allows mating and labor to occur

vas deferens small tube that transports sperm from the testicles to the outside of the body

vent the external area where eggs are laid in egg-bearing animals; also called the cloaca

vulva external beginning of the reproductive system in the female and where urine is excreted

winking movement of the vulva with frequent vaginal discharge and urination

REVIEW QUESTIONS

1. What is the heat cycle in females known as?
2. What is the term used for both male and female external reproductive organs?
3. What is the veterinary medical term for a spay?
4. What are the male hormones? Describe each.
5. What is the time frame when an egg is released in a female called?
6. Explain the estrus cycle.
7. Describe the stages of labor.
8. How does fertilization and incubation of an egg occur?
9. Explain artificial insemination and why it may be practiced.
10. Explain neutering and why it may be performed.

Clinical Situation

Photo by Isabelle Francais

Leslie, a veterinary assistant at Whiskers and Tails Veterinary Clinic, is setting up for a C-section surgery. A 3 yr F English Bulldog is scheduled for surgery. Leslie reviews the patient's medical record history and notes that the dog has had a C-section a year ago and had a normal routine surgery with 6 puppies. The medical record notes show that the dog was seen about 3 weeks ago and radiographs were performed, showing 8 puppies in the uterus.

Dr. Dawson and the vet tech, Sara, arrive in the prep area and prepare to anesthetize the patient. Leslie continues to set up for surgery.

- What equipment and supplies will Leslie need to provide for the surgical procedure?
- What concerns and factors should the veterinary medical team consider with this patient?
- What client education may be provided to the owner of the animal?
- What factors in this situation may or may not show the owners of this dog are responsible breeder's?

34

Delmar/Cengage Learning

The Immune System

Objectives

Upon completion of this chapter, the reader should be able to:

- State the structures that make up the immune system
- Describe the functions of the immune system
- Describe common disorders of the immune system

Introduction

The **immune system** is responsible for keeping the body healthy and protecting animals from disease. **Immunity** is the term used to mean *protection*. Immunity begins at birth when an animal begins nursing from its mother and acquiring **antibodies** from the first 24 hours of milk known as **colostrum**. Antibodies are specific proteins produced to protect a newborn against disease and attack cells relating to any diseases, known as **pathogens**. Antibodies protect young animals from possible disease until vaccines are begun to help build up the immune system. They form as chains within the bloodstream.

Vaccines

Vaccinations are placed within the body as **antigens**, which are foreign materials used to create an immune response. Vaccinations are started in young animals as a series of **booster shots** that slowly build up the immune system and help prevent infection of the body by becoming resistant to diseases. Vaccine booster shots provide a primary response and a secondary response. The **primary response** of a vaccine provides antigens that need 3 to 14 days to build an immune response that begins producing antibodies. The **secondary response** of a vaccine provides a quick, repeated exposure to an antigen that creates immunity to prevent disease development. Some booster shots are required 2 to 3 times to build up an immune response. This depends on the type of vaccine, age, species of animal, and the disease type.

> **TERMINOLOGY TIP**
>
> A *vaccine* is used to build up immunity and resistance to disease within the body. Vaccines are also called *shots*, *vaccinations*, *boosters*, and *immunizations*.

Vaccine types include modified live and killed vaccines. **Modified live vaccines** are made from altered antigens created from disease pathogens that place small amounts of a disease into the animal's body (see Figure 34-1). **Killed vaccines** are manufactured from dead pathogens of a disease and placed into the animal's body in an inactive form. Immunity, thus, provides different types of protection to the body. **Active immunity** is developed from exposure of a pathogen through the process of vaccination. **Passive immunity** is developed through antibodies acquired from one animal or source to another. Passive immunity can be provided through colostrum, commercial diets or formulas, and plasma donated from related animals.

> **TERMINOLOGY TIP**
>
> **Plasma** is a part of the blood that is spun down and separated from the serum. The plasma contains natural immunity antibodies that may be given to the newborn in cases when the young have not received adequate amounts of colostrum.

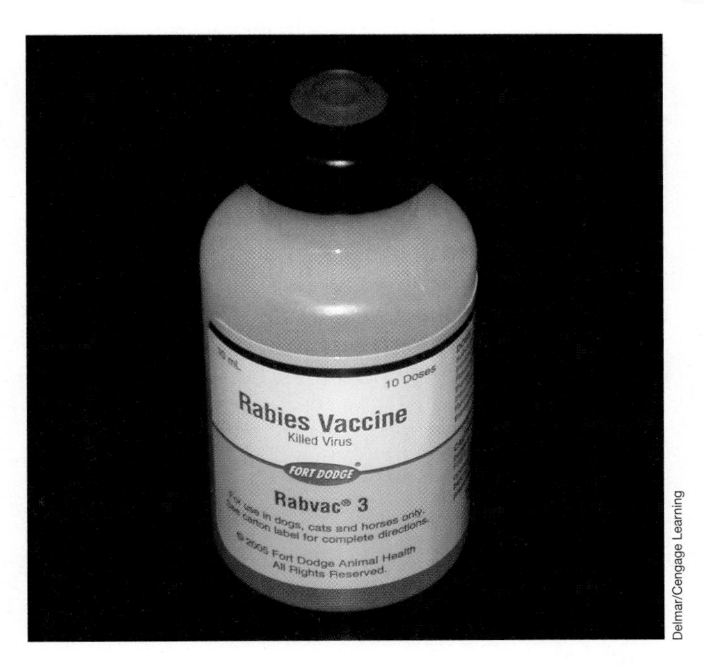

FIGURE 34-1 Rabies is a disease that can be controlled with a vaccine.

Vaccination Routes

Vaccines are given by various routes and methods. The most common routes are **subcutaneous (SQ)** or under the skin and **intramuscular (IM)** or into the muscle. Vaccination programs are the most important part of the clinical health of an animal and its immune system. Some vaccines may be labeled for use as **intranasal (IN)** or by applying in the nostril. This is a common route of vaccine administration of the bordetella or "kennel cough" vaccine. Another route is **intraocular** or into the eye as in an eye drop. *Lingual* or *oral* routes mean by mouth. **Intradermal (ID)** means into the dermis layer of skin. Figure 34-2 illustrates the variations in routes of injections.

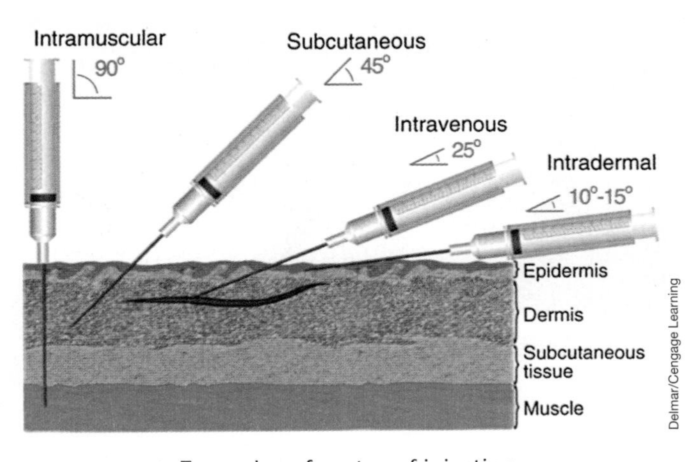

FIGURE 34-2 Examples of routes of injection.

Infection

The body's immune system controls body temperature (see Table 34-1) and helps prevent **infection**. Infection is the immune system's response to an abnormal condition. The immune system also helps to control and destroy diseases. **Inflammation** is a sign of infection within the body. Inflammation causes white blood cells to build up in the site that is affected and may cause pus to form, redness around the area, warm to hot skin temperatures, increased body core temperatures, edema, and pain.

TERMINOLOGY TIP

Edema is a buildup of fluid under the skin (see Figure 34-3).

TABLE 34-1

Normal Body Temperatures	
Dog	101–102 degrees
Cat	101–102.5 degrees
Cow	101–101.5 degrees
Horse	99–100 degrees
Pig	102–102.5 degrees
Goat	102 degrees
Sheep	103 degrees
Rabbit	102–104 degrees

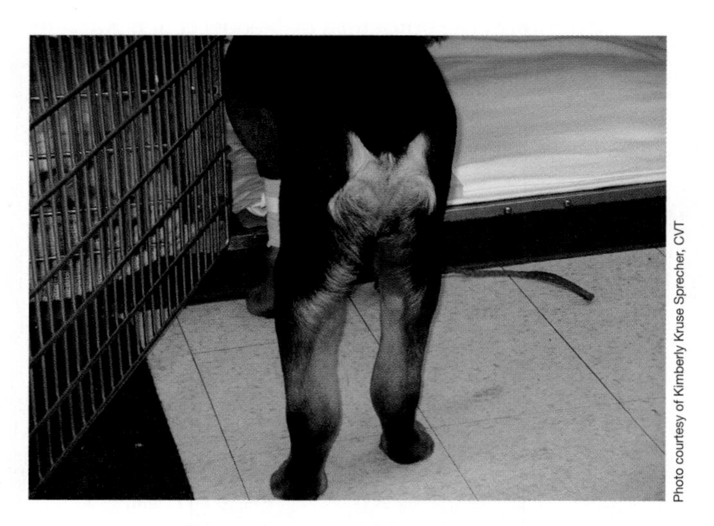
Photo courtesy of Kimberly Kruse Sprecher, CVT

FIGURE 34-3 Edema (swelling) in the hind legs of a dog.

Allergies

Allergies are a common clinical condition with animals. Allergies develop from **allergens**, which are pollens in the air that cause an allergic reaction. Some examples of common allergens are ragweed, trees, grass, or flowers. Animals that have allergies have a **hypersensitivity** within their immune system to certain allergens. This sensitivity causes a release of a chemical known as **histamine** when an allergy reaction occurs. This response may cause such allergy signs as itchiness, eye or nasal discharge, sneezing, rubbing at eyes, or redness to the skin. Many allergy reactions are treated with **antihistamines** or drugs used to prevent and control the allergy reaction. Skin allergies are known as **atopy**. Skin allergies may be reactions to items in the air, the household, the outside environment, or the animal's diet. Atopy is a secondary skin infection from scratching. Skin testing may be done to determine the cause of the atopy. Skin injections consist of **intradermal** injections placed into the layers of the skin. A positive reaction to an allergen will respond with the skin becoming swollen and red at the site of the injection. Another method of allergy testing is through **blood titer** levels. The titer is a measured amount of antibody within the bloodstream. Blood titer levels can be measured with a simple blood draw to detect diseases and allergies within an animal's body. High amounts of blood titer levels mean that an allergy or a pathogen is the cause.

ELISA Testing

Some clinical tests that are used in-house to detect pathogens and diseases are **ELISA** tests, which are enzyme-linked immunosorbent assays. These simple tests are used to measure antigen or antibody levels in a blood sample. Many ELISA tests come in a test kit that contains all the necessary items to run a blood sample in the facilities lab. Most tests can be completed within 5 to 10 minutes. They are relatively inexpensive and easy to run. Some examples of ELISA test kits are heartworm tests, feline leukemia tests, and FIV or FIP tests. Most of these test kits are called **SNAP tests** as they are packaged as a piece of equipment that when mixed with the blood is snapped to activate the test (see Figure 34-4).

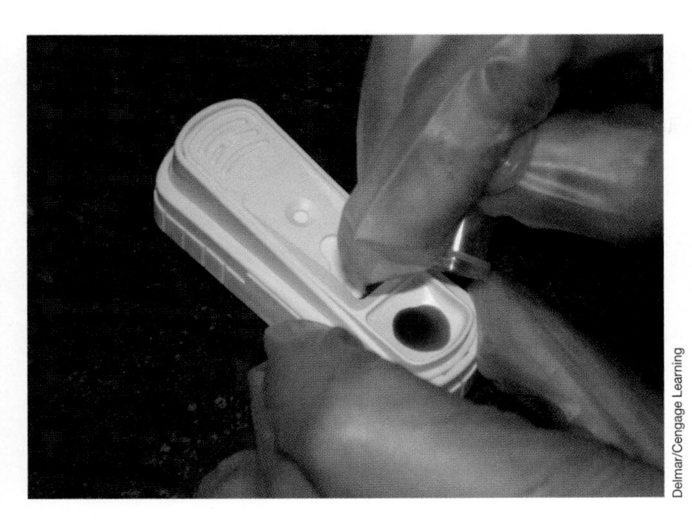
Delmar/Cengage Learning

FIGURE 34-4 SNAP test.

SUMMARY

The immune system functions to protect the body from harmful substances. The immune system produces antibodies that respond to invasion of the body by a foreign substance. Immunity can be provided through the use of vaccinations.

Key Terms

active immunity developed from exposure of a pathogen through the process of vaccination

allergen pollens in the air that cause an allergic reaction

antibody specific proteins produced to protect a newborn against disease and attack cells relating to any diseases

antigen foreign materials used to create an immune response

antihistamine drugs used to prevent and control an allergic reaction

atopy skin or food allergy causing a skin infection

blood titer measured amount of antigen within the bloodstream

booster series vaccines placed into the immune system to build up protection and immunity over an amount of time

colostrum the mother's first 24 hours of milk after labor, which pass antibodies to the young

edema buildup of fluid under the skin

ELISA enzyme-linked immunosorbent assays; simple test used to measure an antigen or antibody level within the blood

histamine chemical released during an allergic reaction

hypersensitivity increased reaction to allergens causing an allergic reaction

immune system responsible for keeping the body healthy and protecting an animal from disease

immunity term used to mean protection and begins at birth during the nursing process

infection the immune system's response to an abnormal condition

inflammation causes white blood cells to build up in the site that is affected and may cause pus to form, redness around the area, warm to hot skin temperatures, increased body core temperatures, edema, and pain

intradermal (ID) injection given into the layers of skin

intramuscular (IM) injection given into the muscle

intranasal (IN) given into the nasal cavity or nostrils

intraocular into the eyes

killed vaccine manufactured from dead pathogens of a disease and placed into the animal's body in an inactive form

modified live vaccine made from altered antigens created from disease pathogens that place small amounts of a disease into the animal's body

passive immunity developed through antibodies acquired from one animal or source to another, as in colostrum, formula, or plasma

(*Continues*)

pathogen a disease

plasma contains natural immunity antibodies that may be given to the newborn in cases when the young have not received adequate amounts of colostrum

primary response provides antigens that need 3 to 14 days to build an immune response that begins producing antibodies

secondary response provides a quick repeated exposure to an antigen that creates immunity to prevent disease development

SNAP test test kit equipment that when mixed with the blood is snapped to activate the test

subcutaneous (SQ) injection given under the skin

vaccination placed into the body to build up resistance in the immune system to disease

REVIEW QUESTIONS

1. Explain the difference between antibodies and pathogens.
2. Explain the difference between a live vaccine and a killed vaccine.
3. Describe the primary and secondary response to a vaccine.
4. Explain the difference between inflammation and infection.
5. Describe allergies and how they can be treated.

Clinical Situation

Courtesy of Photo Disc

Dr. Davis, a large animal veterinarian, is at a farm call to examine a young sheep that has recently been banded for castration. The banding was done by the owner about a month ago. The sheep has not been doing well for several days. When Dr. Davis examines the sheep, he notes the sheep's stance is very wide, stiff and reluctant to move. Dr. Davis performs a PE and notes when he attempts to perform an oral exam that the sheep is unable to open its mouth. The area of the testes also had developed an open wound due to the poor circulation from the band. Bacteria had most likely entered the wound and caused the clinical signs that Dr. Davis was seeing. Treatment of the sheep at this stage would not be successful.

- What is most likely wrong with the sheep?
- How could this condition have been prevented?
- What other signs may Dr. Davis see in this sheep as the condition progresses?
- What options might Dr. Davis give the owner?

35

Delmar/Cengage Learning

The Nervous System

Objectives

Upon completion of this chapter, the reader should be able to:

- State the structures that make up the nervous system
- Describe the functions of the nervous system
- Describe common disorders of the nervous system

Introduction

The **nervous system** is the body's control network. It coordinates and controls body activity. It does this by detecting and processing internal and external information and responding in an appropriate manner.

The nervous system is made up of **neurons**, or specialized cells within the nervous system that control impulses. The body has three types of neurons: sensory, motor, and interneurons. The **sensory neurons** deliver signals from the central nervous system (CNS). **Motor neurons** deliver signals from the central nervous system to the muscles to give a response. The **interneurons** deliver signals from one neuron to another from the brain to the spinal cord.

Structures of the Nervous System

The major structures of the nervous system are the brain, spinal cord, peripheral nerves, and sensory organs. The nervous system has two major divisions: the **central nervous system** and the **peripheral nervous system**.

Central Nervous System

The central nervous system or CNS consists of the brain and spinal cord. The major organ in the nervous system is the **brain**. The brain controls the actions of the nervous system and most of the body. The brain is divided into three regions or sections: the cerebrum, cerebellum, and brain stem. The **cerebrum** controls the voluntary movements of the body and thought processes and is the largest region of the brain. The **cerebellum** provides control over coordination and movement. The **brain stem** controls functions that maintain life. Within the brain stem is a small area called the **medulla oblongata**, which controls the body's functions, such as the heart rate, respiratory rate, and blood pressure. Damage to this region causes instant death. The **midbrain** area controls the senses, which include sight, smell, hearing, taste, and sensation. Damage to the midbrain causes a **coma** condition, where there is a loss of consciousness and lack of awareness. Other organs that are housed within the brain include the **thalamus**, which is located at the top of the brain stem; the **hypothalamus**, which is located in front of the thalamus and controls hormone production; and the **pituitary gland**, which is located between the thalamus and hypothalamus and controls hormone release and body functions. The brain stem attaches to the **spinal cord**. Figure 35-1 shows the structures of the brain.

The spinal cord begins at the base of the brain and continues to the sixth or seventh lumbar vertebrae. The spinal cord is held in place by the vertebrae and is a hollow tube made of strands and fibers that run through the middle of the vertebrae. Nerves, called **tracts**, attach and run through the spinal column, allowing for body sensations. These are sensory nerves that cause pain and allow for pain detection. The spinal cord nerves also allow reflexes and coordination functions. The spinal cord serves as a pathway for impulses to go to and from the brain.

> **TERMINOLOGY TIP**
>
> The lumbar vertebrae lie over the lower back and are designated L when noting spinal cord and vertebral conditions. The other vertebrae include cervical (C), thoracic (T), sacral (S), and coccygeal (Co) (see Figure 35-2).

The Peripheral Nervous System

The peripheral nervous system or PNS consists of the cranial and spinal nerves, as well as the autonomic nervous system.

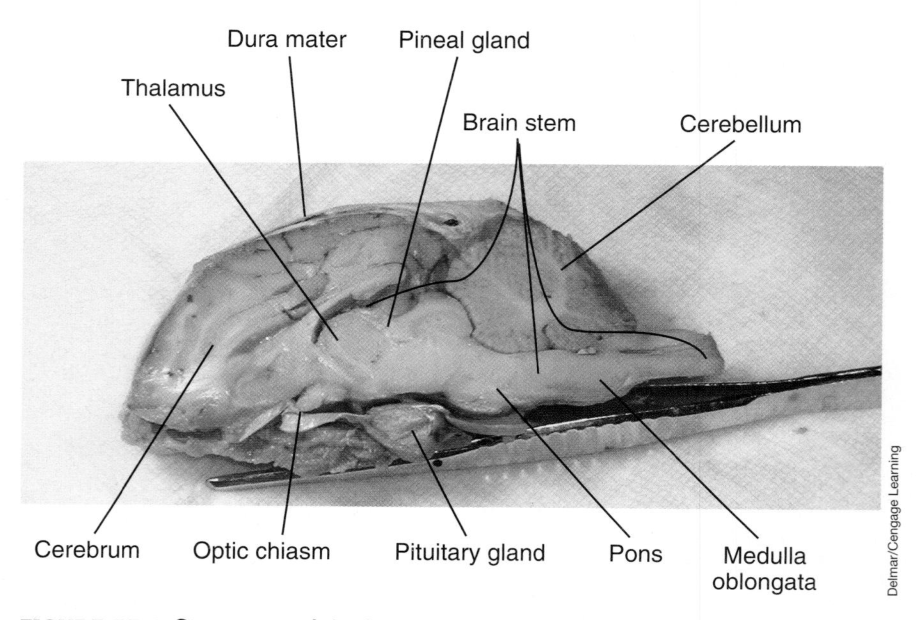

FIGURE 35-1 Structures of the brain.

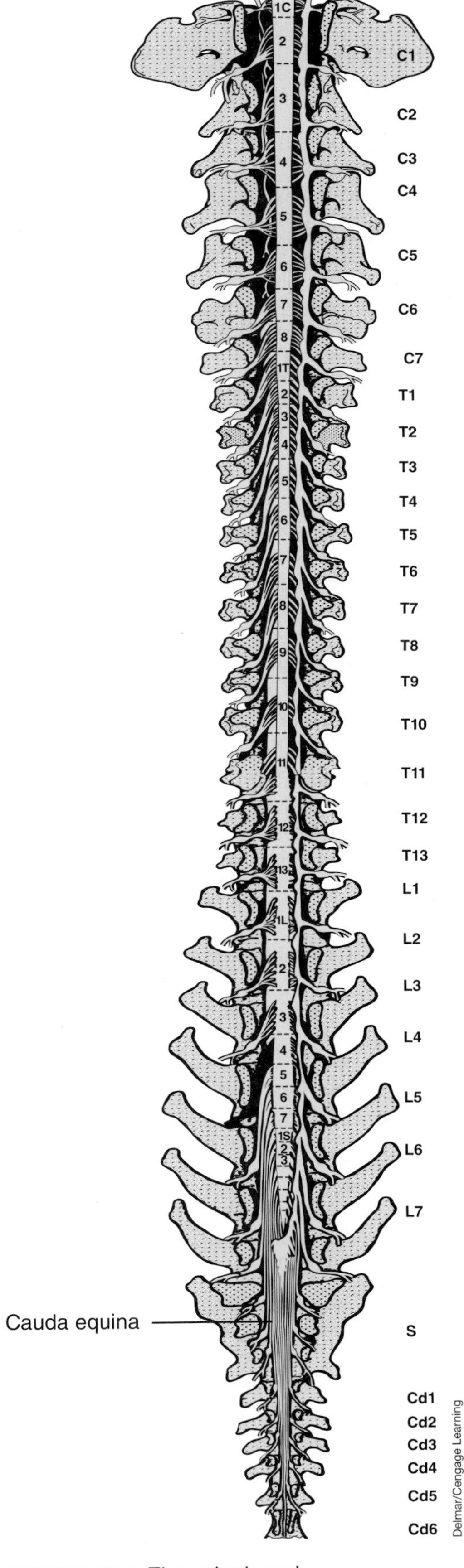

FIGURE 35-2 The spinal cord.

There are 12 pairs of cranial nerves. They originate from under the surface of the brain. The cranial nerves are named for the function they serve and are represented by Roman numerals (see Figure 35-3 and Table 35-1).

TABLE 35-1

Cranial Nerves

CRANIAL NERVE	NAME	FUNCTION
I	olfactory	conducts sensory impulses from the nose to the brain (smell)
II	optic	conducts sensory impulses from the eyes to the brain (vision)
III	oculomotor	sends motor impulses to the external eye muscles (dorsal, medial, and ventral rectus; ventral oblique; and levator superioris) and to some internal eye muscles
IV	trochlear	sends motor impulses to one external eye muscle (dorsal oblique)
V	trigeminal	three branches: ophthalmic = sensory to cornea; maxillary = motor to upper jaw; mandibular = motor to lower jaw
VI	abducent	motor innervation to two muscles of the eye (retractor bulbi and lateral rectus)
VII	facial	motor to facial muscles, salivary glands, and lacrimal glands and taste sensation to anterior two-thirds of tongue
VIII	acoustic or vestibulo-cochlear	two branches: cochlear = sense of hearing; vestibular = sense of balance
IX	glossopharyngeal	motor to the parotid glands and pharyngeal muscles, taste sensation to caudal third of tongue, and sensory to the pharyngeal mucosa
X	vagus	sensory to part of the pharynx and larynx and parts of thoracic and abdominal viscera; motor for swallowing and voice production
XI	accessory	accessory motor to shoulder muscles
XII	hypoglossal	motor to the muscles that control tongue movement

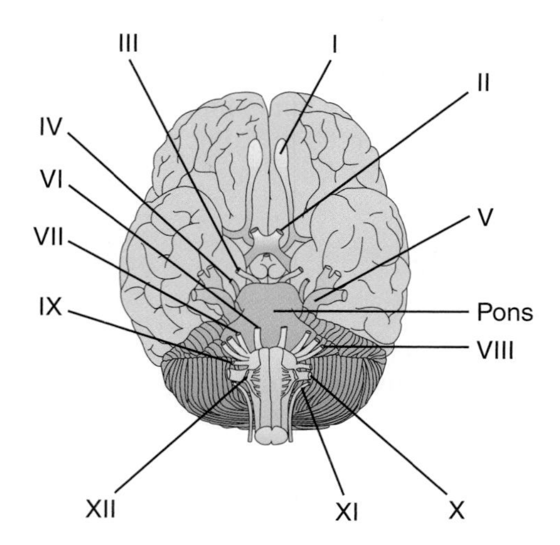

FIGURE 35-3 Cranial nerves.

The spinal nerves arise from the spinal cord and are paired. Each segment of vertebrae has two nerve branches: the dorsal root and the ventral root. The **dorsal root** contains sensory nerves and the **ventral root** contains motor nerves.

The autonomic nervous system is part of the peripheral nervous system and functions to innervate smooth muscle, cardiac muscle, and glands. It is further divided into the parasympathetic system and the sympathetic system. The **sympathetic system** is responsible for the "fight or flight" reaction, increased vital signs, and certain drug reactions. The **parasympathetic system** is responsible for opposite reactions, such as decreased vital signs, regulating the body system back to normal, and controlling peristalsis. The main nerve in these systems is the **vagus nerve**.

TERMINOLOGY TIP

Fight or flight is a natural instinct of animals in which they take care of themselves by either running away or protecting themselves by fighting. **Vital signs** are the body signs such as heart rate (HR), respiratory rate (RR), and temperature (T). **Peristalsis** is the wave-like action of food digesting in the stomach.

Coordinated Function of the CNS and PNS

The CNS controls the brain and spinal cord by receiving signals from the PNS. The PNS controls the nerves, detects a stimulus, send signals to the CNS, and causes a response or action to occur. The body is made up of electrical units of measurements called **volts**. These volts create a force within a cell in specific amounts. Volts help in creating **reflexes**, which are units of function that produce a movement. A signal is sent to the brain to create a reflex, and the movement occurs. Reflexes may be voluntary or involuntary responses. **Voluntary reflexes** occur when an animal asks its body to perform a function, such as a dog jumping up to greet its owner or a horse kicking at another horse. **Involuntary reflexes** occur without thinking and serve as many of the body system functions, such as the heart beating and breathing. Figure 35-4 summarizes the divisions of the nervous system and their coordinated function.

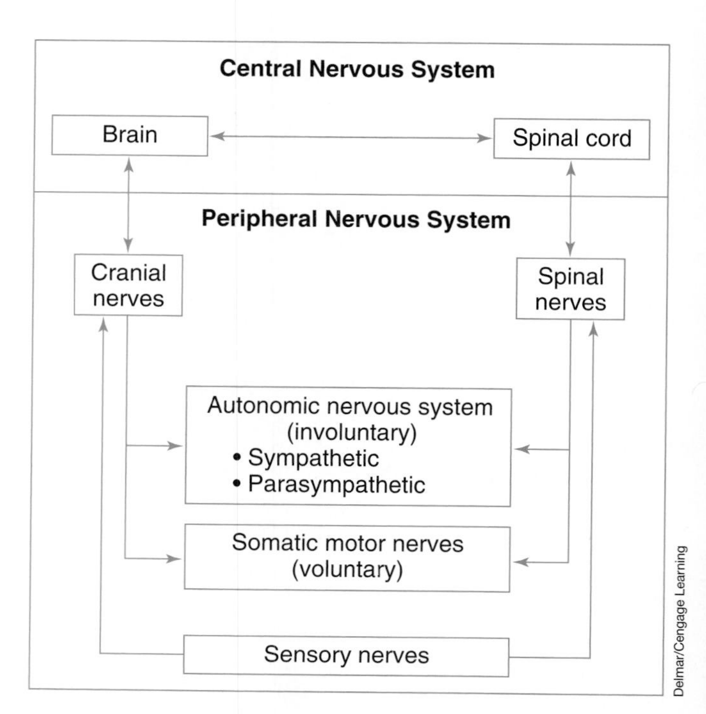

FIGURE 35-4 Divisions of the nervous system.

Receptors

Receptors are found within the nerves located throughout the body and allow for detecting changes within the body and the environment. The receptors cause vessels to dilate or constrict depending on the body's temperature. Skin receptors help control the body temperature of an animal; pain receptors detect pain; pressure receptors determine feeling and sensations; and mechanical receptors determine **proprioception** or movement. **Chemoreceptors** are chemical receptors that allow animals to smell, taste, and detect sounds. The taste buds in the mouth have four types of receptors: bitter, sour, salty, and sweet.

> **TERMINOLOGY TIP**
> The term *proprioception* relates to the body's ability to recognize body position for posture or movement.

Testing

The **cranial nerve test** is done to determine how well the animal's reflexes and nervous system are working. This exam helps to determine if a brain-related injury has occurred as a result of some trauma.

The **spinal nerve reflexes** should also be evaluated for proprioceptive reflexes of the muscles and tendons. The animal should stand to be evaluated. One test to determine proper body balance is turning one foot to the opposite surface to see if it can place it in the correct position. Normal animals will right the foot immediately. Nerve damage causes a prolonged response to appropriately position the foot. The **knee jerk reflex** is also used to assess spinal nerves. This involves the use of the **reflex hammer**, a tool used to tap the area of the knees and joints to elicit a response. The patella ligament is tapped lightly to determine any reflex movements (see Figure 35-5). Normal reflexes should have the motor nerves extend the knee joint. A lack of response is a sign of nerve damage to the lower lumbar region of the vertebrae.

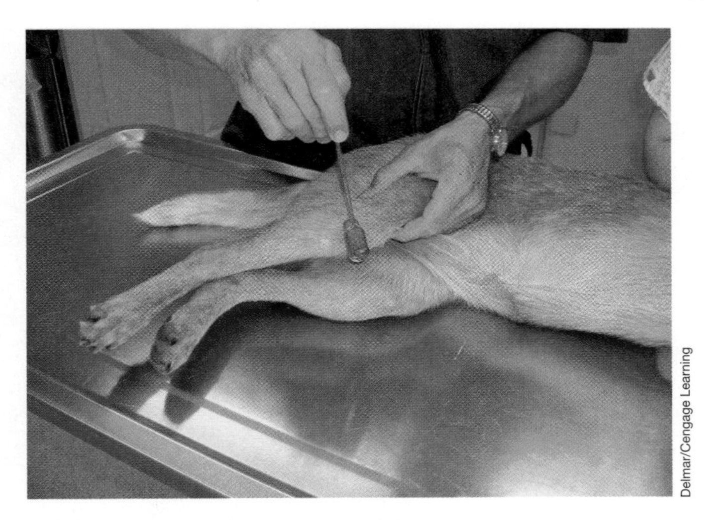

FIGURE 35-5 A veterinarian tests the patellar reflex on a dog. By striking the patellar tendon, receptors sense the sudden lengthening and quickly contract the muscle in response.

Common Diseases and Conditions

Diseases of the nervous system can originate from invasion of bacterial or viral organisms, other disease processes, or traumatic injury.

Intervertebral Disc Disease

Many animals and breeds are predisposed to a condition that causes pressure to be placed on an area of the vertebral column, causing compression on the discs. This is called **intervertebral disc disease** or IVDD. Enough pressure can cause compression of the spinal cord (see Figure 35-6). This condition causes severe back pain and in severe cases partial or complete **paralysis**. In dogs, the long-backed breeds are more susceptible to injury. Some of these breeds include the Basset Hound, Dachshund, and Corgi.

> **TERMINOLOGY TIP**
> The term *paralysis* means a loss of motor skills. This may be partial loss in an area of the body or complete loss of motor skills throughout the body.

Epilepsy

Epilepsy is the medical term for seizures. A **seizure** is a loss of voluntary control of the body with an amount of unconsciousness causing uncontrolled violent body activity (see Figure 35-7). Seizures may occur due to disease, heredity, or toxin-related response. Many animals have abnormal behavior noted prior to seizure activity and may also be noted after the seizure. Signs of seizures include the following:

- Paddling motion of legs
- Opening and closing of the jaw (chewing gum seizures)

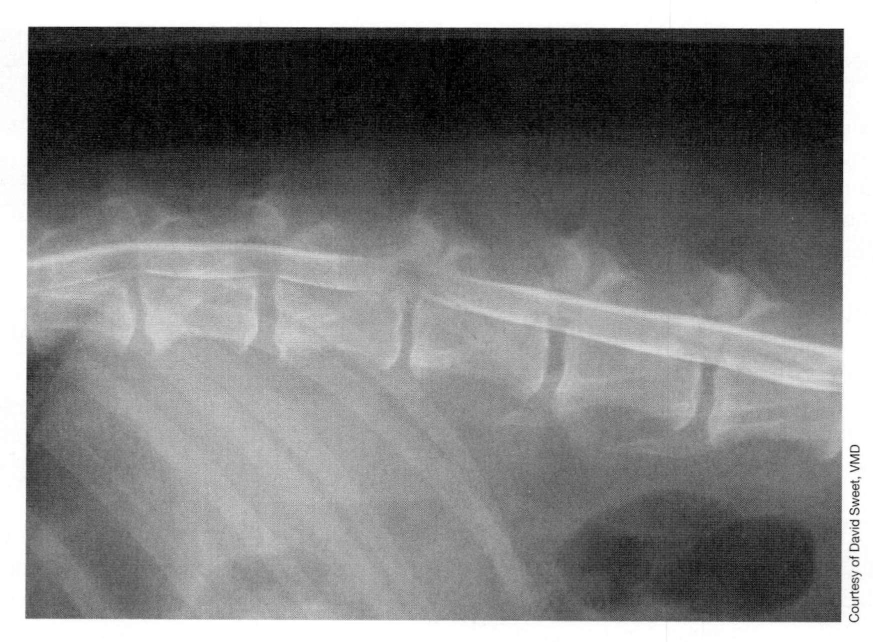

Courtesy of David Sweet, VMD

FIGURE 35-6 Myelogram showing intervertebral disc disease. Dye injected into the spinal column shows the compression of the spinal cord.

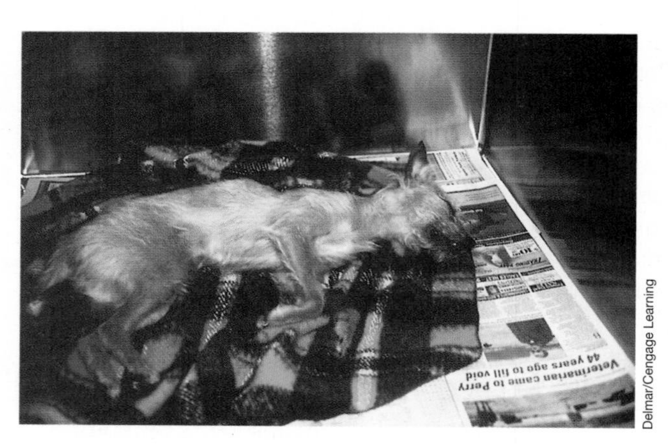

Delmar/Cengage Learning

FIGURE 35-7 A seizing dog. During a seizure, dogs typically are unresponsive, lying on their sides, paddling their legs, and chomping their jaws.

- Muscle twitches
- Excessive drooling/salivation
- Uncontrolled urination
- Uncontrolled defecation
- Vocalization—crying, barking, howling
- Loss of consciousness
- Stiff appearance to body
- Ocular dilation

The severity of signs will depend on the length and severity of the seizure activity. Care should be taken for any animal suffering a seizure. Animals may tend to bite as they have no control over their nervous system. They should be monitored around stairs or on furniture as seizure activity can cause trauma and further injury. Each seizure activity should be timed; 30 minutes of seizure activity becomes an emergency situation. It is important for a veterinarian to examine any animal that has a seizure to determine the cause of the episode. Some common causes of seizures may include the following:

- Epilepsy
- Hereditary factors
- Trauma or nervous system injury
- Tumor—brain or nervous-system related
- Toxins or chemicals
- Infectious diseases affecting the nervous system

Hereditary epilepsy is the most common occurrence of seizure activity. There is no underlying cause as the seizures are caused by genetic factors passed from one or both parents. Genetic seizure conditions usually begin to occur between 6 months and 5 years of age. Several breeds of dogs are known for genetic epilepsy: Cocker Spaniel, Beagle, Dalmatian, Labrador Retriever, and Springer Spaniel. Epilepsy begins with infrequent occurrences and increases with age and time. Medication treatments may be used to control frequent and severe episodes. Some examples of drugs used to control and prevent seizures include Phenobarbital and potassium bromide (KBr). These medicines must be monitored over time for the need to adjust doses and to assess liver functions as this is the main organ in which these drugs are metabolized. Toxicity and seizures can occur when animals ingest chemicals such as organophosphates, which are ingredients in insecticides and weed killers. These chemicals are toxic to the nervous system and often produce a seizure activity response as they block normal activity between the CNS and PNS.

Rabies

Rabies is a deadly virus that can affect any mammal or human. Rabies is usually transmitted by the saliva of an infected animal transferred into the bloodstream. Most cases of rabies occur in wild animals, with bats, skunks, fox, and raccoons being the most common. Rabies vaccinations should be given to all dogs, cats, ferrets, and livestock species to help control and prevent the rabies virus. Vaccines should be kept up to date, and state and local rabies laws should be evaluated as many rabies laws exist for certain animal species and may be different from location to location. Limiting animals to the indoors, especially companion animals, will help decrease the potential for exposure. Also, attraction of wild animals to the property is discouraged, so not feeding wild animals and not leaving out garbage will decrease the risk of wild animals coming in contact with pets and livestock. Signs of rabies in animals may vary. Animals may become aggressive, but many wildlife species become calm and friendly, venturing out during the daytime and approaching people. Other signs include excessive salivation, foaming at the mouth, lack of coordination, stumbling, circling, and other changes in behavior. Any pet or livestock suspected of being bitten by another animal should be seen immediately by a veterinarian. The rabies virus is fatal and testing is done through use of the brain tissues, so any animal expected of having rabies must be euthanized and the brain submitted to a State Laboratory testing facility for testing.

West Nile Virus

West Nile virus is a concern for birds and horses. West Nile is transmitted through mosquito bites and causes inflammation or swelling of the brain and spinal cord. The life cycle starts when an infected bird is bitten by a mosquito and is then carried to a human, bird, or horse and transmitted through another blood meal. Signs of infection may only take 5–15 days. A West Nile vaccine is available for prevention. Signs of West Nile virus include the following:

- Stumbling or tripping
- Muscle weakness or twitching
- Partial paralysis
- Loss of appetite
- Depression or lethargy
- Head pressing or tilt
- Impaired vision
- Wandering or circling
- Inability to swallow
- Inability to stand up
- Fever
- Convulsions
- Coma
- Death

SUMMARY

The nervous system is the body's control center. It regulates and coordinates body functions as well as movement. There are two divisions of the nervous system: the central nervous system and the peripheral nervous system. Disorders of the nervous system can arise from other disease processes, invasion of viruses or bacteria, or traumatic injury.

MAKING THE CONNECTION

Chapter 8: rabies
Chapter 9: rabies
Chapter 21: pseudo rabies
Chapter 24: Marek's disease

Key Terms

brain the major organ in the nervous system that controls the body's actions

brain stem part of the brain that controls functions that maintain life

central nervous system (CNS) controls the brain and spinal cord by receiving signals from the PNS

cerebellum part of the brain that controls coordination and movement

cerebrum part of the brain that controls the voluntary movements of the body and thought processes and is the largest region of the brain

chemoreceptor chemical receptor that allows an animal to taste, smell, and detect sounds

coma damage to the midbrain that causes a loss of consciousness and awareness

cranial nerve test reflex test of the eye where a hand is moved quickly toward the eye without touching the eye or any hair around the eye

dorsal root nerve branch of the vertebrae that contain sensory nerves

epilepsy veterinary term for seizures

fight or flight natural instinct of animals in which they take care of themselves by either running away or protecting themselves by fighting

hypothalamus organ within the brain located in front of the thalamus that controls hormone production

interneurons cells that deliver signals from one neuron to another from the brain to the spinal cord

intravertebral disc disease (IVDD) injury to the spine that causes pressure on the discs and/or spinal cord and may cause partial or complete paralysis

involuntary reflex occurs without the need for thinking and serves as a necessary body function, such as the heart beating

knee jerk reflex reflex used to assess the spinal nerves

medulla oblongata part of the brain that controls the body's involuntary functions

midbrain part of the brain that controls the senses

motor neuron cells that deliver signals from the central nervous system (CNS) to the muscles to give a response

nervous system the body's control network primarily controlled by the brain

neurons specialized cells within the nervous system that control impulses

paralysis partial or complete loss of motor skills

parasympathetic system spinal cord system responsible for opposite reactions, such as decreased vital signs, regulating the body system back to normal, and controlling peristalsis

peripheral nervous system (PNS) controls the nerves, detects a stimulus, send signals and informs the CNS, and causes a response or action to occur

peristalsis wave-like motions of the stomach that allow digestion of food

pituitary gland organ located between the thalamus and hypothalamus that controls hormone releases in the body

proprioception relates to the body's ability to recognize body position for posture or movement

rabies a deadly virus that can affect any mammal, including humans; usually transmitted by the saliva of an infected animal transferred into the bloodstream

receptor found within the nerves located throughout the body and allows for detecting changes within the body and the environment

reflex units of function that produce a movement created by volts

reflex hammer tool used to tap the area of the knees and joints to elicit a response

seizure loss of voluntary control of the body with an amount of unconsciousness causing uncontrolled violent body activity

sensory neuron cells that deliver signals from the central nervous system (CNS)

spinal cord strands of fibers that attach the brain stem to the PNS to send and receive signals through the body; begins at the base of the brain and continues to the lumbar vertebrae

spinal nerve reflexes evaluated for proprioceptive reflexes of the muscles and tendons

sympathetic system spinal cord system that is responsible for the "fight or flight" reaction, increased vital signs, and certain drug reactions

thalamus organ within the brain located at the top of the brain stem

tract nerves that attach to the spinal cord and run through the body, providing sensations and feelings

vagus nerve main nerve within the spinal cord system

ventral root nerve branch of the vertebrae that contains motor nerves

vital sign body signs such as heart rate (HR), respiratory rate (RR), and temperature (T)

volt electrical units of measurement within the body that create a force within cells

voluntary reflex occurs when an animal asks its body to perform a desired function, such as walking or running

West Nile virus a mosquito-borne disease that causes inflammation or swelling of the brain and spinal cord

REVIEW QUESTIONS

1. What is the division of the autonomic nervous system that is concerned with body functions under emergency or stress called?
2. What type of nerve carries impulses away from the CNS and toward muscles?
3. What is the largest portion of the brain involved with thought and memory?
4. What part of the brain is associated with muscle movement?
5. What type of reflexes are involved in the heart and breathing?
6. Describe the structures of the central nervous system and explain their functions.
7. Describe the structures of the peripheral nervous system and explain their functions.
8. Differentiate between the sympathetic and parasympathetic divisions of the autonomic nervous system.
9. Describe how an impulse travels to cause a response.
10. Differentiate between the cranial nerves and the spinal nerves.

Clinical Situation

Photo by Isabelle Francais

"Sally", a 3 yr SF Cocker Spaniel, was brought into the emergency clinic due to multiple seizure episodes. Alice, the vet assistant, obtained information from the owner and noted in the medical record that "Sally's" seizures began the night before. As Alice was taking the vital signs of the dog, "Sally" began to have another seizure. She immediately called for the veterinarian, Dr. Howe. "Sally" was lying on her side with her legs thrashing wildly. Her jaws were opening and closing quickly. The seizure episode lasted around 60 seconds.

The owner was very upset and concerned and continued to say that "Sally" had 4 episodes over night lasting from 1–2 minutes. Dr. Howe performed a PE and drew some blood work, which showed normal results. Dr. Howe concluded that "Sally" had epilepsy.

- What causes epilepsy?
- What signs may occur to alert the owner that seizure activity may be occurring?
- What signs or symptoms may occur to an animal after a seizure?
- What treatments may be offered for "Sally"?
- What precautions might Dr. Howe discuss with "Sally's" owners about epilepsy in dogs?

36

Delmar/Cengage Learning

The Sensory System

Objectives

Upon completion of this chapter, the reader should be able to:

- ☐ State the structures that make up the sensory system
- ☐ Describe the functions of the sensory system
- ☐ Describe common disorders of the sensory system

Introduction

The sensory system consists of those structures and organs associated with touch, vision, hearing, smell, and taste. The function of the senses is to receive stimuli from sensory receptors and transmit those impulses to the brain for interpretation. This chapter will cover the senses of sight and hearing.

The Eye

Vision receptors within the eye allow the lens to focus on light-sensitive cells and allow for day and night vision in animals. The eye is made of several parts that work together to allow vision to occur. There are three layers of the eye: retina, sclera, and iris (see Figure 36-1). The **retina** is the inner layer of the eye that contains two types of cells that aid in seeing color and determining depth. The **rods** are the cells that allow the eye to detect light and depth and make up 95% of the cells within the eye. The **cones** are cells that allow an animal to see certain colors. Both rods and cones work together to allow each animal species to have its developed eyesight. Behind the retina is the **posterior chamber** of the eye, which holds the nerves, blood vessels, and optic vessels. The middle layer of the eye is called the **iris**. The iris gives the eye its color and holds the pupil in the center, which dilates and constricts in darkness and light to allow vision to occur. Light allows the pupil to **constrict** (close) and darkness allows the pupil to **dilate** (open). In front of the iris is the **anterior chamber**, which holds the **vitreous humor**, a fluid-filled material behind the iris and lens that is somewhat jelly-like in consistency. This fluid regulates the pressure within the eye. The outer white layer of the eye is the **sclera**. The sclera is covered by a clear layer called the **cornea**. This is often a common site for scratches and trauma in the eye.

Reflex Testing

The **menance response test** is a reflex test of the eye where a hand is moved quickly toward the eye without touching the eye or any hair around the eye. A normal response to this reflex is to blink or close the eye. An abnormal response would show no change in the eye and not elicit a blink. The **pupillary light response test** uses a light source shown into the animal's eyes to note constriction or shrinking of the **pupil** and dilation or expanding when the light source is removed from the eye. Both pupils should react in the same manner. Normal reflexes show that the optic nerves are working correctly. The eyes should also be examined for any signs of **nystagmus**, which is a condition where the eyes jump back and forth in rhythmic jerks due to damage in the inner ear, brain stem, or cranial nerves. This movement causes the animal to have symptoms similar to motion sickness, such as vomiting and lack of coordination. Normal cranial nerves will show no signs of uncontrolled ocular movement.

Ear

The **ear** is an organ that enables hearing and helps to maintain balance. The ear is divided into the outer, inner, and middle portions. The outer ear consists of the pinna and external auditory canal. The **pinna** is the external portion of the ear that catches sound waves and transmits them to the external auditory canal. The **external auditory canal** is the tube that transmits sound to the tympanic membrane. The middle ear consists of the tympanic membrane, the auditory ossicles, Eustachian tube, oval window, round window, and the tympanic bulla. The **tympanic membrane** is tissue that separates the outer and inner ear structures. The **auditory ossicles** are three bones of the inner ear (malleus, stape, and incus) that transmit sound vibrations. The **Eustachian tube** is a narrow duct that leads from the ossicles to the nasopharynx and serves

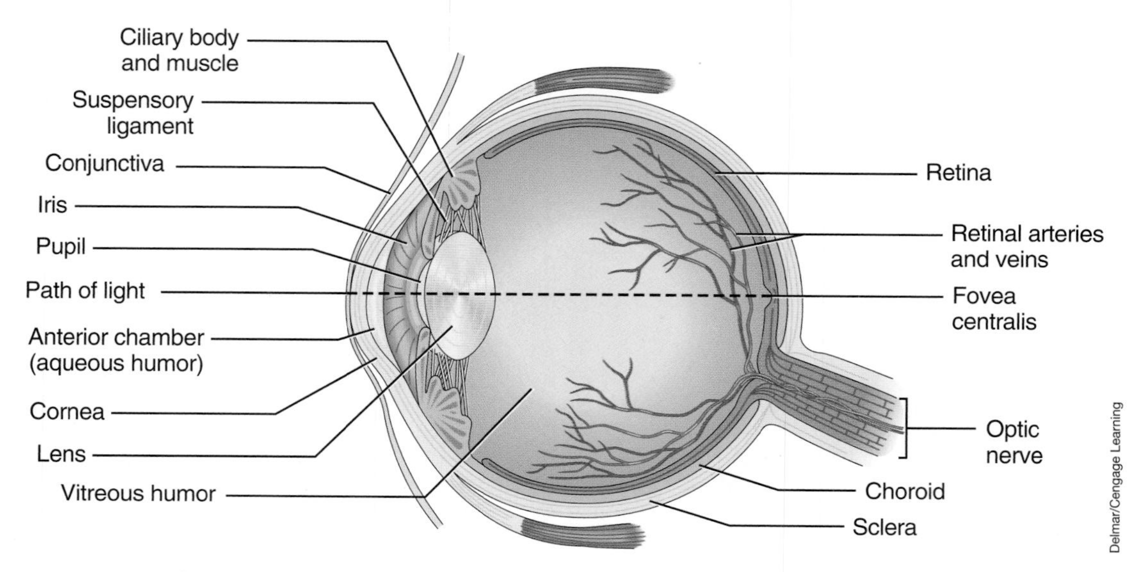

FIGURE 36-1 Structures of the eyeball.

to maintain air pressure in the middle ear. The **oval window** is a membrane that separates the middle and inner ear. The **round window** is a membrane that receives sound waves, and the **tympanic bulla** is an osseous chamber at the base of the skull. The inner ear consists of the bony labyrinth, which is divided into three parts: the vestibule, semicircular canals, and the cochlea. The **vestibul**e contains receptors for balance and position. The **semicircular canals** contain sensory cells that detect changes in position. The **cochlea** is a spiral-shaped passage that vibrates and relays vibrations that allow sound to be heard. The structures of the ear are illustrated in Figure 36-2.

Hearing

Sound waves enter the ear through the pinna, travel through the auditory canal, and strike the tympanic membrane. The tympanic membrane vibrates and moves the ossicles. The ossicles conduct the sound waves through the middle ear. Sound vibrations reach the inner ear through the round window and displace fluid within the structures of the inner ear. Cells of the inner ear initiate a nerve impulse that is relayed to the brain.

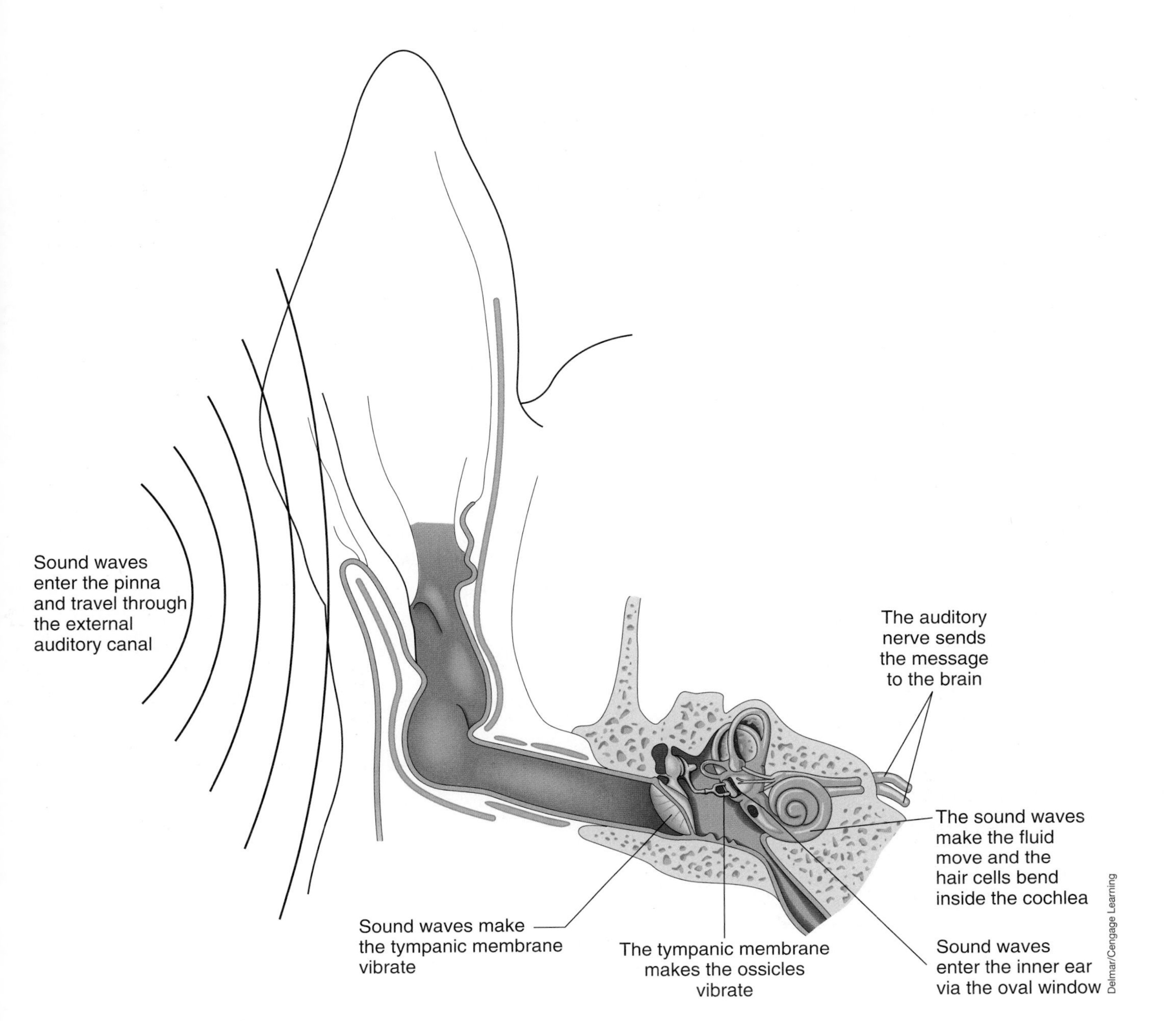
Sound waves enter the pinna and travel through the external auditory canal
The auditory nerve sends the message to the brain
The sound waves make the fluid move and the hair cells bend inside the cochlea
Sound waves make the tympanic membrane vibrate
The tympanic membrane makes the ossicles vibrate
Sound waves enter the inner ear via the oval window
Delmar/Cengage Learning

FIGURE 36-2 Structures of the ear.

Equilibrium

The ear also aids in **equilibrium**, which is a state of balance. **Static equilibrium** is controlled by the organs of the inner ear. The structures of the inner ear bend with the movement of the head in response to gravity. **Dynamic equilibrium** is controlled by the semicircular canals. These structures maintain balance in response to rotational or angular movement of the head.

SUMMARY

The eyes and the ears are organs of the sensory system. The eyes allow sight and control the ability to see in dark and light situations. The ears allow hearing and help to maintain equilibrium.

Key Terms

anterior chamber area of the eye in front of the iris that holds the vitreous humor and regulates the pressure in the eye

auditory ossicle bones of the middle ear that transmit sound vibrations

cochlea spiral-shaped passages in inner ear that receives sound vibrations and initiates an impulse to the brain for translation

cones cells in the eye that allow animals to see colors

constrict to close (as in the pupil in the eye)

cornea clear outer layer of the eye that is commonly easily damaged or scratched

dilate to open (as in the pupil in the eye)

dynamic equilibrium maintaining balance in response to rotational or angular movement.

ear sensory organ that enables hearing

equilibrium state of balance

Eustachian tube narrow duct that leads from middle ear to nasopharynx and maintains air pressure

external auditory canal tube that transmits sound from the pinna to the tympanic membrane.

iris middle layer of the eye that gives an eye its color and holds the pupil

menance response test a reflex test of the eye where a hand is moved quickly toward the eye without touching the eye or any hair around the eye

nystagmus condition where the eyes jump back and forth in rhythmic jerks due to damage in the inner ear, brain stem, or cranial nerves

oval window membrane that separates the middle and inner ear

pinna external portion of the ear

posterior chamber area behind the eye that holds the nerves

pupil area in the center of the eye that opens and closes with light responses

pupillary light response test light shown into the animal's eyes to note constriction or shrinking of the pupil and dilation or expanding when the light source is removed from the eye

retina inner layer of the eye that contains two types of cells that aid in seeing color and determining depth

rods cells that allow the eye to detect light and depth and make up 95% of the cells within the eye

round window membrane that receives sound waves through fluid passed through the cochlea

sclera outer white layer of the eye

semicircular canals sensory cells that generate nerves impulses to regulate position

static equilibrium maintenance of the position of the head relative to gravity

tympanic bulla osseous chamber at the base of the skull

tympanic membrane eardrum; tissue that separates the outer and middle ear

vestibule portion of inner ear that contains specialized receptors for balance and position

vitreous humor jelly-like fluid material located behind the iris and lens that controls the pressure of the eye

REVIEW QUESTIONS

1. What is the term for "a state of balance"?
2. What part of the ear separates the outer or external ear from the middle ear?
3. What is the colored muscular layer of the eye that surrounds the pupil?
4. What is the region of the eye where nerve endings of the retina gather to form the optic nerve called?
5. What is the name of the spiral-shaped passage that leads from the oval window to the inner ear?
6. Explain the pupillary light response.
7. Explain the process of hearing.
8. Explain how the inner ear aids in maintenance of equilibrium.

Clinical Situation

Photo by Isabelle Francais

"Sophie", a 10 yr SF Persian, is being examined by Dr. Lister. Mrs. George, the cat's owner, has noticed changes in "Sophie's" eyes recently. Dr. Lister notes that both eyes have developed a thin, whitish appearance to both lenses. Mrs. George tells the vet that the cat has been having difficulty jumping up on furniture and often walks into items.

Dr. Lister performs a PE on "Sophie" and notes in the progress notes the bilateral opaque appearance to the lenses. The pupil responses are decreased.

Dr. Lister discusses some treatment options with Mrs. George and states that he recommends a CBC and blood chemistry panel due to the patient's age, and that often cataracts in cats are caused from another underlying condition. Mrs. George agrees to the blood work and allows the vet tech, Amanda, to take "Sophie" to the prep area for blood collection.

- What causes or factors may begin cataracts in cats?
- What treatment options may be discussed with Mrs. George?
- What is most likely the cause of "Sophie's" cataracts?

37

Animal Nutrition

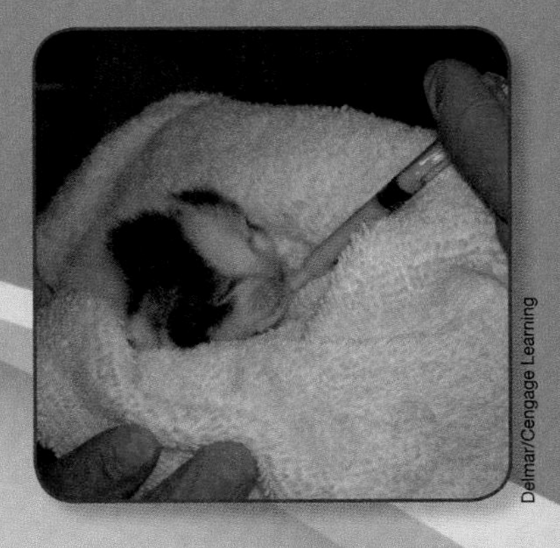
Delmar/Cengage Learning

Objectives

Upon completion of this chapter, the reader should be able to:

- Discuss the major nutrient needs of animals
- Describe the purpose of quality animal nutrition
- Contrast feed requirements with digestive systems types and structures
- Describe the ways that animals use nutrition and nutrients
- Discuss and describe the types of animal feed
- Explain how animals are fed
- Discuss the importance of rations

Introduction

Animal nutrition is a part of veterinary medicine that provides knowledge and scientific evidence about food that is both a specific quality and quantity for different animals. Many researchers, scientists, and veterinary professionals devote large amounts of time to improving animal nutrition and health care. Every species of wild or domesticated animal has its own specific nutritional requirements for a balanced diet. Producers want animals to grow big and strong as quickly as possible, pet owners want animals to be healthy and happy, and veterinary professionals want to provide proper information for every aspect of animal health, especially one as important as nutrition. Veterinary assistants can provide detailed information to clients regarding appropriate nutrients and diets. They may also help owners select favorable diets.

Nutritional Needs of Animals

All animals require food to grow, live, reproduce, and work. **Animal nutrition** is the science of determining how animals use food in the body and all body processes that transform food into body tissues and energy for activity. Each animal species has needs and requirements based on the animal's environment, work and activity level, age and stage of development, genetic makeup, and health. A **nutrient** is any single class of food or group of like foods that aids in the support of life and makes it possible for animals to grow or provides energy for physiological processes in life.

Animals require a specific amount of food in their daily needs. This amount is known as a **ration** and is the total amount of food an animal needs within a 24-hour time frame. A ration may be met as one amount or in divided amounts throughout the day. A **balanced ration** is a diet that contains all the nutrients required by an animal in correct and specific amounts. Nutrients may be broken down, digested, absorbed, and used by the body. Some nutrients may be given in decreased or increased amounts that may be harmful to the animal's health or growth. Therefore, it is important to have an understanding of the correct rations for various species of animals.

Nutrients

Nutrients are the components of food that are meant to sustain life. There are six essential nutrients that animals require: water, carbohydrates, fats, proteins, vitamins, and minerals.

Water

Water makes up more than 75% or more of an animal's body. Newborn animals may have up to 90% of their body weight as water. Water provides several functions within the body, including body temperature control, body shape maintenance, transporting nutrients within the body's cells, aiding in digestion of food, breaking down food particles, and serving as a carrier for waste products. Water makes up the major part of all body fluids, such as urine, blood, feces, sweat, and lung vapors. Water is the most critical nutrient in an animal's diet. An animal can live longer without food than without water. Animals require 3 pounds of water for every 1 pound of food they consume. The amount an animal needs beyond its food is determined by the amount of activity, gestation, or lactation occurrences and the environment. Water loss through the process of **dehydration** causes a serious problem especially when 10% or more fluid is lost by the body. At 20% of water loss an animal will die. The following factors contribute to water loss in an animal:

- Diarrhea
- Vomiting
- Urination
- Panting
- Sweating
- Defecation
- Lactation
- Gestation

Water allows nutrients to be transported through the body and also helps to break down cells into smaller parts through a chemical process called **hydrolysis**. This means that water is added to a molecule in a process that breaks down nutrients into smaller particles, allowing for them to be transported in the body.

Carbohydrates

Carbohydrates are nutrients that provide energy for body functions and allow for body structure formation. Carbohydrates are the largest part of an animal's food supply, being about 75% of the diet. Carbohydrates include starches, sugars, and fiber materials. They are made of combinations of carbon, hydrogen, and oxygen. These nutrients are not stored in the body and are required within a food source on a daily basis. Carbohydrates are capable of being converted to fats. Some of the functions of carbohydrates include maintaining the body's blood sugar levels, creating lactose in milk, storing fat, and completing metabolism. **Starches** are plant or grain materials that provide fiber and bulk in an animal's diet. Starches convert to glucose (sugar) during digestion. Examples of starches are cereal grains, silage, oats, and corn (see Figure 37-1).

Sugars are another type of carbohydrate that forms the simplest example of a nutrient in an animal's diet. Sugars include such food items as fruits and milk. Sugars are further classified as **simple** or **monosaccharides** and **double** or **disaccharides**. A simple sugar is glucose and a double sugar is sucrose or table sugar. A long chain of simple sugars forms a **polysaccharide**. These carbohydrates form the building blocks of other major nutrients and are easily digested within the stomach and intestine. **Fiber** is material from plant cells that is left after other nutrients are digested. Fiber helps produce positive bacteria and aids in digestion of food particles. Examples of fiber include hay and grass (see Figure 37-2). Fiber helps slow down the process of digestion and helps protect the lining of the stomach and intestinal tract.

FIGURE 37-1 Silage is an example of carbohydrates that can be provided in the diet of livestock.

Fats

Fats or **lipids** furnish animals with a concentrated source of energy. Fats are found in every cell within the animal body and contain the highest amount of energy of all nutrients. Fats function within the body by providing insulation from the environment, protection for vital organs, energy reserves, and flavor in food. Fats form cholesterol, steroids, and other hormones found naturally within the body. When fat is absent from the diet, it can lead to hair and skin problems. In increased amounts, it can lead to obesity and other health problems. Many diets contain **fatty acids**, which are oils that are products of fat sources that may be used as nutrients or supplements within a diet. Fatty acids are helpful in skin and hair coat hydration. Fats are measured by **calories** or the unit of measurement that defines the energy in food. Animal feed is measured in **kilocalories**, which are the amount of energy to raise 1 gram of water by 1 degree, and is written as a kcal. Calories and kilocalories are measured with a calorimeter, which is an instrument that measures the amount of heat released by a food product. This measurement determines the length of metabolism, which in turn predicts the amount of energy in the food particle. This is the information that is placed on food labels for nutrition information. Examples of fatty acids include fish oil, linseed oil, fish meal, and vegetable oil. Fats form solid particles and oils or acids form liquid particles. One benefit for animal feeds is that fat increases the **palatability** of food, or how good it tastes and if it is eaten by any animal. Fatty acid sources may be **essential** or **nonessential**. Essential fatty acids are necessary in a diet and produce natural hormones needed within the body. Nonessential fatty acids are not necessary within the body and are used as additions to a diet.

FIGURE 37-2 Pasture provides fiber for some species of animals such as horses.

Proteins

Proteins are nutrients that are essential in growth and tissue repair. Protein is useful in the forming and development of muscles, internal organs, skin, hair, wool, feathers, hoofs, and horns. Protein forms the basis of the structure and function of cells. It is also vital to growth and development of young animals and in reproduction and breeding. Protein-based diets aid animals that require weight gain for health and market situations. Proteins contain **amino acids**, the building blocks of proteins that form in chain-like structures. There are 23 amino acids available in nutritional foods, and 10 are essential to the health needs of animals. Essential amino acids are used to make other amino acids required by the body for life. Proteins and amino acids work together to help food digest in the body through each animal's digestive system. Protein sources may come from soybean meal, skim milk, fish meal, or alfalfa hay (see Figure 37-3). Proteins are listed as a percentage on food labels based on the **biological value**, which describes the quality of the source. The higher the biological value of the protein, the more easily digested the food. Many large and small animal food sources provide protein through such ingredients as eggs, milk, fish, meat, and corn.

Protein deficiency tends to be one of the most common nutritional problems in animals. Animals that do not receive adequate amounts of protein will have both development and health problems that may affect their lives. Young animals require larger amounts of protein

FIGURE 37-3 Proteins are added to foods of animals.

than adult animals. Signs of protein deficiency include the following:

- Poor or slow growth
- Anorexia or decreased appetite
- Anemia
- Poor appearance and coat
- Low birth weight
- Low milk production
- Dental problems
- Edema
- Brittle bones

Minerals

Minerals are needed in every area of the body but are found mostly in the bones and teeth. Minerals are used by the body based on the animal's needs and mineral availability. Calcium is a mineral that makes up about 50% of the body's mineral source and is found in many organic body fluids, such as blood and tissues. The body's functions rely on mineral balances to maintain regular rates and duties, such as the heart rate and respiratory rate. There are two types of minerals: macrominerals and microminerals. **Macrominerals** are minerals needed in large amounts, such as calcium, phosphorus, and iron. **Microminerals** are minerals needed in small amounts, such as sodium, potassium, and magnesium. These minerals are often called **trace minerals**.

Vitamins

Vitamins are nutrients needed in small amounts for life and health maintenance. Vitamins help provide a defense against disease (vitamin E), promote growth (vitamin D) and reproduction (vitamin B12), and contribute to the general health of the animal by regulation of body functions. Vitamins react with enzymes to allow body functions to occur. Some vitamins act as **antioxidants** within the body's system to boost the immune system. There are two types of vitamins: **fat soluble** and **water soluble**. Fat soluble vitamins are stored in fat and released when needed within the body. Vitamins A, D, E, and K are fat soluble. Vitamin A is the most common vitamin *not* available in commercial animal foods. Fat soluble vitamins can be given in excessive amounts, resulting in a toxic reaction that poisons the system and that can cause damage to the kidneys and heart. Water soluble vitamins are not stored within the body, are dissolved by water, and are therefore needed in daily doses by the body. Vitamins C and D are water soluble. Vitamin D is commonly acquired by natural sunlight. Vitamins are usually supplied to the body through the use of **supplements**, which are additives placed into the diet in solid or liquid form when needed by the animal.

Animal Nutrition and Concentrations

The **concentration** of food is dependent on how it is delivered to the animal. Food concentration is based on **dry matter**, or the amount of nutrients without the water. The concentration is determined by the amount of food fed divided by the percentage of dry matter. Dry matter is critical in how much the animal needs to be fed. The concentration helps determine the amount of food to feed and helps ease digestion. **Digestion** is the breakdown of food from large molecules to smaller molecules for use by the body.

Types of Diets

Animal foods come in a variety of types (see Figure 37-4). Different species as well as ages of animals, the activity of the animal, and the health status of an animal, all must be considered when selecting a type of diet to feed an animal. Factors to consider when planning an animal's diet include the following:

- age
- environment
- species
- size
- health condition
- breed
- medical history

Courtesy of Brain Yacur and Guilderland Animal Hospital

FIGURE 37-4 There are a wide variety of pet foods on the market to choose from.

Photo courtesy of Catherine Wein

FIGURE 37-5 Puppies should be fed a diet that will encourage growth.

Growth Diets

Young animals may be placed on a **growth diet** (see Figure 37-5). Growth diets are specialized and formulated to increase the size of muscles, bones, organs, and body weight of young offspring. Each animal species will have specific needs and requirements. Livestock and meat production animals are raised in programs that provide quick body mass increases in a short amount of time as a goal for market weight and profitability. Growth diets include large components of such nutrients as protein, vitamins, and minerals.

Maintenance Diets

Maintenance diets are given to adult animals that are in their prime age and health condition. These animals may be working or competing, and the maintenance diet goal is to keep that animal at a specific ideal weight; the same weight and health condition means maintenance. These types of diets require energy to work and be active, to maintain the body's temperature, and for good health. Maintenance diets consist of high fats and carbohydrates and small amounts of proteins, vitamins, and minerals.

Reproductive Diets

Reproduction diets are given to breeding animals for additional nutrient needs. Specialized reproduction diets are used to increase the energy needs of female animals that begin the lactation phase and those developing embryos. The first **trimester** of a pregnancy in animals is the most critical time for nutrition. This is the first three months of pregnancy when the embryos require larger amounts of protein, vitamins, and minerals for proper development. Male breeding animals require a reproductive diet to allow quality sperm production. Breeding animals that don't get enough adequate nutrition will show signs of poor overall health, underweight offspring, and possible abortion.

TERMINOLOGY TIP

The term **abortion** refers to lost fetus in a pregnant female animal. In humans this is called a *miscarriage*.

Lactation Diets

Lactation diets are provided to females that have completed the gestation phase and are currently producing milk for young to nurse. These diets require large amounts of water and high amounts of protein, vitamins, and minerals. The most important minerals are calcium and phosphorus. These nutrients improve the milk's quality and quantity.

Work Diets

The **work diet** is typically based on livestock that use a large amount of energy for some type of work or activity. All activity requires energy to provide the necessary amounts of nutrients to fuel work. Examples of work include plow animals, draft or wagon animals, race horses, hunting animals, and show or competition animals. Work diets include increased carbohydrates, fats, vitamins, and minerals.

Reduced Calorie Diets

Reduced calorie diets are often for animals that are overweight or less active due to a health complication. These diets are specific for an animal's low energy needs and are formulated with decreased amounts of carbohydrates, fats, and proteins and moderate vitamins and minerals for normal body functions.

Senior Diets

Another diet that is specific to age and health is the **senior diet**. This type of food is formulated for geriatric or senior animals over a specific age for their species that may require certain increased nutrients and some decreased nutrients. Senior diets are usually low in carbohydrates and fats, moderate in protein for healthy bone and skeletal mass, and increased in vitamins and minerals to protect the body and immune system as it ages. Most of the reduced calorie and senior diets are species-specific and formulated for animal types based on digestive systems.

Feeding Animals

Veterinary medicine requires all veterinary professionals to have some basic knowledge of animal nutrition. This is helpful in determining quality diets and proper amounts to feed. **Feed** is what animals eat to obtain nutrients and nutrition. **Feedstuff** is the ingredients in animal food that help determine the nutrient contents, such as corn meal, barley, or oats. Animal feed is divided into classes based on the most numerous type of nutrient: forages, pastures, concentrates, and supplements.

Forages

Forages are also called roughages, which are high in fiber and are plant-based sources, such as grass, stems, or leaves. Most forage is low in protein with the exception of alfalfa hay. Hay and grass forage sources tend to be relatively inexpensive and usually plentiful dependent on the weather.

Pastures

Pastures are grown as either **temporary** or **permanent** pasture sources. Temporary pastures are planted each season by reseeding the grass source, such as corn, hay, or millet seed. Permanent pastures are grass sources that re-grow each year without the need to be replanted or seeded, including clover or bluegrass (see Figure 37-6).

FIGURE 37-6 Permanent pastures don't require replanting year after year.

Concentrates

Concentrates are provided to an animal as an additional nutrient source when the primary food source is not adequate or abundant. Concentrates are high in protein and energy and tend to be fed when other food sources, such as hay, are not of the best quality. These foods include many of the bagged commercial animal feeds from Purina, Hills, or Agway.

Supplements

Supplements are the last class. They are provided when necessary as a diet additive during specific health or conditional requirements. Supplements may be a vitamin or mineral or a mineral block, such as salt.

Ideal Weight

When feeding an animal its required food source for adequate nutrients, it is important that the foodstuff be researched and of high quality. When feeding for ideal nutritional value, it is best to feed the animal according to the manufacturer's recommendations and the animal's ideal weight. The **ideal weight** is the breed standard based on the animal's age, species, breed, purpose or use, and health condition.

Example

An animal that is 80 lbs, but considered overweight for its age and breed, should be assessed by the veterinarian. The veterinarian recommends the animal lose 10 lbs, so the animal's ideal weight is 70 lbs. The owner would feed the animal based on the 70 lb ideal weight range according to the food label and manufacturer's recommendations.

Body Condition Scoring

Weight is measured by an ideal body appearance called the **body condition scoring**. This is a rating of how an animal appears in looks based on an ideal weight. An animal that is larger than an ideal weight may appear to be overweight or obese. These animals will have excess amounts of fat and will appear wide over the body and hips (see Figure 37-7). On a scale from 1 to 10, 5 is average and considered an ideal weight. The numbers 7 to 10 are overweight and the numbers below 4 are underweight and thin (see Figure 37-8). Animals that are thin lack body fat and substance (see Figure 37-9).

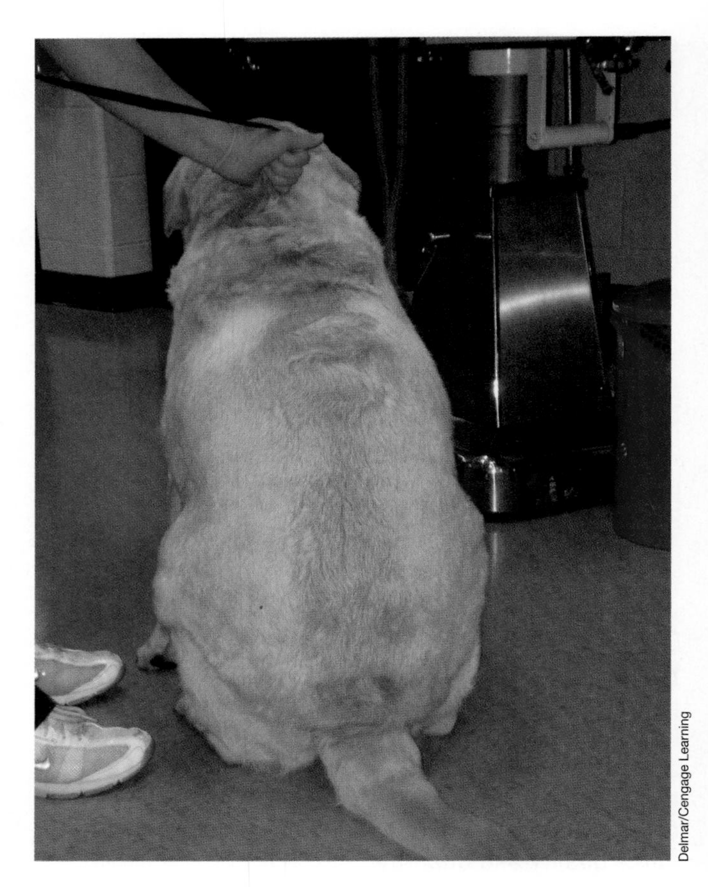

FIGURE 37-7 Example of a dog that is overweight.

Food Analysis

Food analysis is completed by animal feed companies and requires specific information that is tested and documented safe for the species of animal the food is intended for. Food analysis is the process of determining the nutrients in food and prepared mixes to assure it serves as a balanced ration. The food analysis is legally required to be placed on the animal food label. Items that are displayed include the dry matter, crude protein, ash, fat, crude fiber, nitrogen-free extracts, and various other nutrients.

The nutritional information of animal food is prepared by the National Research Council (NRC) and provides all feedstuff ingredients including an international feed number or IFN. This information is provided to formulate proper rations of how much to feed according to the size of the animal. The following are required on all feed labels (see Figure 37-10):

- Product name
- Nutrient list
- Bar code
- Manufacturer name and address

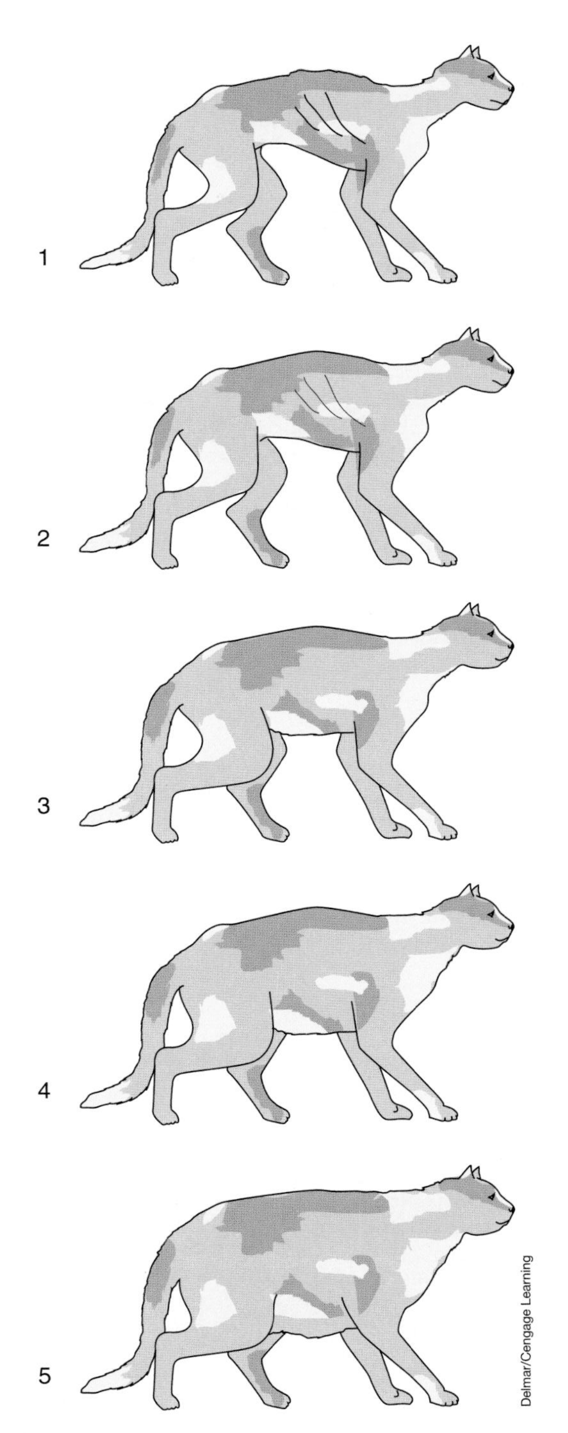

FIGURE 37-8 Body condition scoring: (1) emaciated (2) thin (3) ideal (4) overweight (5) obese.

Feeding Schedules

Feeding animals may be done through **free choice** or **scheduled feedings**. Free choice may also be called free access. This is a popular feeding method for large herds of livestock. Free choice allows animals to eat whenever they want food. Many times the food is provided in accessible feeders or bowls and is always available for the animal. This is common with hay or grass sources for livestock. Water is also commonly allowed free choice. A scheduled feeding is more common for companion animals or livestock that are housed separately or in reduced numbers. Scheduled feedings have a set amount of food given at specific times during the day. This is the best method for ideal nutrition to make sure each animal receives and obtains its adequate food amount. Most animals on a feeding schedule are fed at least 2 to 3 times a day. The feedstuff and schedule are based on the management and practice of the animal program. This includes individual rations given to each animal. Some larger producers and animal owners may purchase specialized equipment that is computerized and set to release certain amounts of food at specified times throughout the day. This equipment is called an **automatic feeder** and may also be available in water form, as an automatic waterer. Factors to balance rations include the following:

FIGURE 37-9 Example of a dog who is emaciated due to neglect.

- Knowledge of the needs of the animal species
- Knowledge of the food and nutrient contents
- Research and documentation
- Knowledge of proper nutrition

FIGURE 37-10 Example of a pet food label.

SUMMARY

Nutrition is the process by which animals eat and use their food for survival. The food is used for various reasons according to several factors involving the individual animal. Nutrients are substances found within the food of an animal and are necessary for animals to function properly. Each animal species has different nutritional requirements. It is essential for veterinary assistants to have a basic understanding of animal nutrition and the needs of animals regarding a quality diet and proper ration.

Key Terms

abortion a pregnant female animal loses the fetus

amino acids building blocks of proteins that form in chain-like structures

animal nutrition science of determining how animals use food in the body and all body processes that transform food into body tissues and energy for activity, and the process which by animals grow, live, reproduce, and work

antioxidants vitamins that boost the body's immune system

automatic feeder specialized equipment that is computerized and set to release certain amounts of food at specified times throughout the day

balanced ration diet that contains all the nutrients required by an animal in correct and specific amounts

biological value percentage on a food label that describes the quality of the food source

body condition scoring rating of how an animal appears in looks and weight based on its ideal body weight

calories unit of measurement that defines the energy in food

carbohydrates nutrients that provide energy for body functions and allow for body structure formation and make up the largest part of an animal's diet needs

concentrates food sources that are provided to an animal as an additional nutrient source when the primary food source is not adequate or abundant

concentration determined by the amount of food fed divided by the percentage of dry matter and how it is delivered to the animal

dehydration water loss by the body

digestion the process of breaking down food from larger particles into smaller particles for use by the body in order to function

disaccharide a double sugar

double sugar two particles of carbohydrates that form the building blocks of nutrients and are called disaccharides

dry matter the amount of nutrients in a food source without the water content

essential fatty acids nutrients necessary in a diet and produce natural hormones needed within the body

fats concentrated source of energy, also called lipids

fat soluble vitamins that are stored in fat and released when needed within the body and include vitamins A, D, E, and K

fatty acids oils that are products of fat sources that may be used as nutrients or supplements within a diet

feed materials that animals eat to obtain nutrients and nutrition

feedstuff ingredients in animal food that determine nutrient contents

fiber material from plant cells that is left after other nutrients are digested

food analysis the process of determining the nutrients in food and prepared mixes to assure it serves as a balanced ration

forages plant-based sources of nutrition that are high in fiber; also called roughages

free choice feeding method of animals that allows animals to eat when they want food; also called free access

growth diet specialized food formulated to increase the size of muscles, bones organs, and body weight of young offspring

hydrolysis chemical process of breaking cells down into smaller particles

ideal weight breed standard based on an animal's breed, age, species, purpose, or use and health status

kilocalories amount of energy to raise 1 gram of water by 1 degree; written as a kcal

lactation diet specialized food provided to breeding females that have completed the gestation phase and are currently producing milk for their offspring

lipids concentrated source of energy that contains the highest amount of energy of all nutrients

macrominerals minerals needed in large amounts, such as calcium

maintenance diet specialized diet fed to animals that may be working or competing; the goal is to keep the animal at a specific and constant ideal weight

microminerals minerals needed in small amounts, such as iron

minerals nutrients needed in every area of the body but are found mostly in the bones and teeth; used by the body based on the animal's needs and mineral availability

monosaccharide a simple sugar

nonessential fatty acids nutrients that are not necessary within the body and are used as additions to a diet when necessary.

nutrient any single class of food or group of like foods that aids in the support of life and makes it possible for animals to grow or provides energy for physiological processes in life

palatability how well food tastes and is eaten by an animal

permanent pasture grass source that re-grows each year without the need of planting or seeding

polysaccharide long chain of simple sugars

proteins nutrients that are essential in growth and tissue repair

ration total amount of food an animal needs within a 24-hour time frame

reduced calorie diet specialized food given to animals that are overweight or less active due to their health status

reproduction diet specialized food given to breeding animals for additional nutrient needs

scheduled feeding a set amount of food given at specific times during the day

senior diet specialized food given to geriatric or older animals over a specific age for their species that require certain nutrient levels due to age

(*Continues*)

simple sugar single particle of carbohydrates that form the building blocks of nutrients and are called monosaccharides

starches plant or grain materials that provide fiber and bulk in an animal's diet and converts to glucose or sugar during digestion

sugar type of carbohydrate that forms the simplest type of a nutrient

supplements additives placed into the diet in solid or liquid form when needed by the animal

temporary pasture grass source that does not re-grow each year and needs to be planted or seeded

trace minerals minerals needed in small amounts; also called microminerals

trimester the first three months of pregnancy when additional nutrition must be provided

vitamins nutrients needed in small amounts for life and health maintenance

water nutrient that makes up 75% of the body and provides several functions within the body, including body temperature control, body shape maintenance, transporting nutrients within the body's cells, aiding in digestion of food, breaking down food particles, and serves as a carrier for waste products

water soluble vitamins that are not stored within the body, are dissolved by water and therefore needed in daily doses by the body to work; these include vitamins C and D

work diet specialized food given to animals that use a large amount of energy for work or an activity

MAKING THE CONNECTION

For nutrition related to specific species, review the content in each chapter in Section III of this book. Also review Chapter 28: The Digestive System

REVIEW QUESTIONS

1. What is animal nutrition?
2. What is the importance of proper animal nutrition?
3. What is a ration?
4. What is a balanced ration?
5. What are nutrients?
6. What are the six nutrients necessary for animals?
7. What is the function of lipids?
8. What is the function of protein?
9. What is the difference between vitamins and minerals?
10. What ways are animals fed?
11. What are the types of diets available for animals?

Clinical Situation

Photo by Isabelle Francais

Betsy, a veterinary assistant at Happy Dog Vet Hospital, is discussing some nutritional information with Mr. Good, who owns "Snoopy," a 10 yr old NM Beagle that is used for hunting and field trials. Mr. Good has noticed that "Snoopy" has been losing weight over the last few months. His medical records show that he lost 4 pounds since his visit last year. Mr. Good is concerned about the food he is feeding "Snoopy." "Do you think he should be on another type of food?" he asks.

"Well, that depends, there are several options that you may want to consider. You have to think about 'Snoopy's' age, his activity level, and his health. It appears from his medical record that he has been in good health. Dr. Wilson will be examining him today, so you may want to talk to him about a diet change."

"What do you think he should eat?" asks Mr. Good.

"Well, if he were my dog, I'd feed him a high-quality diet that has a lot of fat to put some weight back on him quickly." replies Betsy.

- Did Betsy handle educating the client correctly?
- What options may you have considered to Mr. Good?
- What are some factors in selecting an ideal food for "Snoopy"?

Competency Skill

Feeding a Hospitalized Animal

Objective:

Ability to properly prepare a diet for an animal under the care of the veterinarian.

Preparation:

- Appropriate feedstuffs
- Container for food

(*Continues*)

- Measuring devices
- Cleaning products

Procedure:

1. Prepare the appropriate diet. Consult with the veterinarian for specialized diet needs.
2. Determine the amount of food to be consumed in 24 hours.
3. Place the food in the cage or animal holding area. Place in a clean bowl or bucket.
4. Remove the bowl or bucket when the animal is done eating.
5. Clean and disinfect the bowl or bucket.

Competency Skill

Watering a Hospitalized Animal

Objective:

Ability to properly prepare water for an animal under the care of the veterinarian.

Preparation:

- Water
- Container
- Cleaning materials

Procedure:

1. Provide water as appropriate.
2. Refill the water bowl or bucket throughout the day.
3. Monitor the amount of water that is consumed by the patient.
4. If the water container is spilled, clean and dry the area as soon as the spill is noticed.
5. Replace the water bowl or bucket.
6. Clean the bowl or bucket as needed.

Competency Skill

Feeding and Watering the Presurgical or Pre-Anesthesia Patient

Objective:

Ability to properly prepare a diet and provide water for an animal undergoing surgical procedures.

Preparation:

- Appropriate charts
- Appropriate foodstuffs
- Water
- Containers
- Cleaning materials

Procedure:

1. Withhold food from certain species of animals for 12 hours before anesthesia.
2. Withhold water 8 hours prior to anesthesia.
3. Water can be provided postoperatively when the patient is fully awake and standing.
4. Food should be offered within hours in ruminant and nonruminant animals.
5. Food should be offered within 12–24 hours after anesthesia in monogastric animals.

38

Microbiology and Parasitology as Disease Processes

Objectives

Upon completion of this chapter, the reader should be able to:

- ☐ Describe the various types of microorganisms that contribute to disease processes
- ☐ Discuss and describe common animal external parasites
- ☐ Discuss and describe common animal internal parasites
- ☐ Identify the importance of zoonosis in humans

Introduction

The health of an animal can be altered by invasion of microorganisms from its environment. There are also smaller insects and animals that may play a role in passing disease onto companion or production animals. These parasites can only survive by feeding off of the host animal, and as a result they pass disease to the animal. These types of diseases occurring in production animals can greatly impact the success of the business and health of humans who ingest products from these animals. A sound understanding of how microorganisms and parasites contribute to disease will enable the veterinary assistant to identify these situations and help to provide care and control of the disease processes.

Microbiology

The study of microorganisms is called **microbiology**. Many microorganisms are beneficial and can be used to make antibiotics and food products. However, some microorganisms are harmful; these are called **pathogens**. Pathogens can cause disease that may be contagious or non-contagious. Contagious diseases are easily spread between animals and humans. Non-contagious diseases are not passed to others.

Contagious diseases may be spread through direct or indirect contact. **Direct contact** means through contact with an animal or a body fluid. **Indirect contact** means spread through ways other than touching an infected animal, such as airborne or through bedding sources.

Types of Microorganisms

Disease may be defined as a change in structure or function within the body. It is usually identified when abnormal symptoms occur in the health or behavior of an animal. There are five primary disease-causing microorganisms: viral, bacterial, fungal, protozoan, and rickettsial.

Viruses

Viral diseases or viruses are caused by particles that are contagious and spread through the environment. Viruses can only live inside a cell. They cannot get nourishment or reproduce outside of a cell. Viral diseases are often not treatable with antibiotics and must run the course of the illness and may be maintained through supportive therapy or care. **Supportive care** or therapy includes such treatment as fluid replacement, medications to decrease vomiting and diarrhea, and pain medication. Common viruses include influenza strains, HIV, and West Nile virus.

MAKING THE CONNECTION

See the following chapters:

Chapter 8—discussion of viral disease
Chapter 9—discussion of panleukopenia, rhinotracheitis, feline calicivirus, feline leukemia, feline immunodeficiency virus, and rabies
Chapter 10—discussion of pisittacosis
Chapter 14—discussion of lymphocytosis
Chapter 18—discussion of bovine viral diarrhea
Chapter 19—discussion of equine infectious anemia, equine encephalomyletis, and equine influenza
Chapter 21—discussion of brucellosis and pseudo rabies
Chapter 22—discussion of blue tongue and soremouth
Chapter 23—discussion of goat pox
Chapter 24—discussion of Marek's disease, Newcastle disease, infectious bronchitis, avian pox, and avian influenza
Chapter 25—discussion of lymphocytosis and ascites

Bacterial Disease

Bacterial diseases are spread through single-celled organisms called *germs* and are located in all areas of an animal's environment. Bacteria need an environment that will provide food for them to survive. Bacterial disease can cause diarrhea, pneumonia, sinusitis, and various infections.

MAKING THE CONNECTION

Chapter 8—discussion of bacterial disease
Chapter 12—discussion of Pasturella and Tyzzer's disease
Chapter 13—discussion of salmonella
Chapter 14 and Chapter 25—discussion of fin rot, cloudy eye, ascites, and swim bladder disease
Chapter 18—discussion of leptospirosis and Black Leg
Chapter 21—discussion of leptosirosis and pneumonia
Chapter 22—discussion of foot rot, mastitis, actinomycosis, enterotozemia, and lamb dysentery
Chapter 23—discussion of joint ill
Chapter 24—discussion of fowl cholera

Fungal Disease

Fungal diseases are spread by simple-celled organisms or spores that grow on the external body and other areas of the environment, mostly in moist and humid conditions. A fungal disease generally occurs in an animal that is immunologically impaired.

MAKING THE CONNECTION

Chapter 8—discussion of fungal disease
Chapter 14 and Chapter 25—discussion of fish fungus

Protozoal Disease

Protozoan diseases are the simplest forms of life caused by single-celled organisms that are known as parasites and that may live inside or outside the body. Most protozoa feed off of dead or decaying matter and cause infection in animals that ingest them.

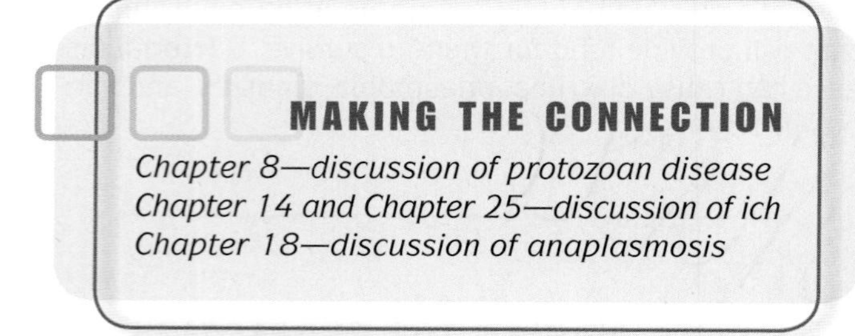

Rickettsial Disease

Rickettsial diseases are spread by biting insects, such as fleas and ticks. Rickettsia need to be in a living cell to reproduce.

Parasitology

Parasitology is the study of organisms that live on or in other organisms to survive. Parasites may invade the internal or external parts of an animal. Most parasites have one location that they live in and feed off the **host**, or the animal that is infected. Parasites go through a life cycle in which they are born as **larvae** and continue through growth stages until they become adults and are capable of reproduction. Each parasite has a different life cycle and growth stage. Many parasites invade the digestive tract, skin, or muscles of an animal. They may enter the animal's body system through many different methods. One may be through ingestion, as in food or water sources. Other routes of contamination include contact with soil, through penetration of the skin, through nursing young, and contact with infected feces. Parasites can be controlled and prevented through proper sanitation and disinfection methods.

Internal Parasites

Common internal parasites that occur in small and large animals include roundworms, whipworms, hookworms, heartworms, coccidian, tapeworms, and strongyles. Many other types of internal parasites occur in various species of animals and should be evaluated in a fecal analysis (see Figure 38-1). Many internal parasites invade the intestinal system, causing vomiting and diarrhea.

Roundworms

Roundworms are also called **ascarids** and are the most common intestinal parasite in small and large animals. They occur most often in young animals. All roundworms live in the small intestine of animals. There are several different species of roundworms. Female roundworms can lay up to 200,000 eggs a day.

Toxascaris leonina has the simplest life cycle. After an animal ingests infective eggs, the eggs hatch and the **larvae** mature within the small intestine. The adult female worm lays eggs that are passed in the feces. The eggs become infective after remaining in the environment for at least 3–6 days. Animals become infected if they eat something contaminated with infected feces.

Toxocara canis have a more complicated life cycle and an effective way of making sure its species will be passed from generation to generation. They are very hardy and resistant. An animal can acquire a *T. canis* infection several ways: ingestion of eggs, ingestion of a transport host, or by larvae entering the animal while in the uterus or through the milk. The larvae migrate through the circulatory system and go either to the **respiratory** system or other organs or tissues in the body. If they enter body tissues, they can **encyst** (become walled off and inactive). They can remain encysted in tissues for months or years.

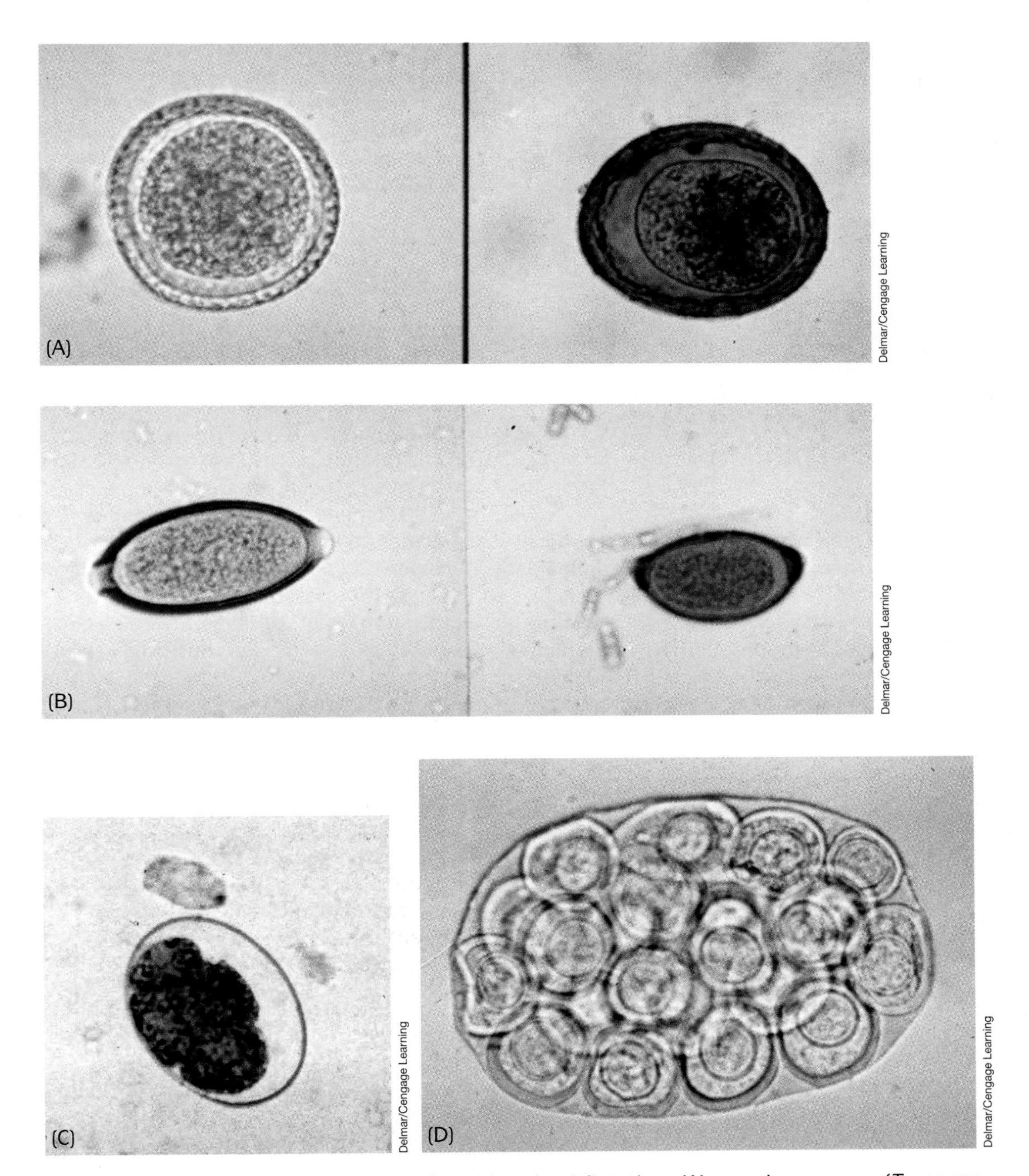

FIGURE 38-1 Parasite eggs commonly found in a fecal flotation: (A) roundworm eggs (*Toxocara canis*, left; *toxacaris leonina*, right); (B) whipworm (*Trichuris vulpis*, left) and lungworm (*Capillaria aerophilia*); (C) hookworm (*Ancylostoma*); (D) tapeworm (*Dipylidium caninum*).

Toxocara cati are similar to *T. canis*. The infective eggs are swallowed, and the larvae hatch and penetrate the stomach wall. From there the larvae migrate through the **liver**, other tissues, and lungs. Some larvae may encyst in the tissues. Larvae that enter the lungs are coughed up and swallowed. The larvae mature in the stomach and small intestine, and the adult female worms start laying eggs.

Parascaris equorum in large animals are swallowed in contaminated hay or water, and they hatch in the intestinal tract. The young worms burrow through the intestinal wall, taking about a week to make their way to the lungs. From there the young worms travel up the trachea to the mouth, to be swallowed a second time. They mature in the intestine in two to three months, then lay eggs that are passed in the feces to start the cycle anew.

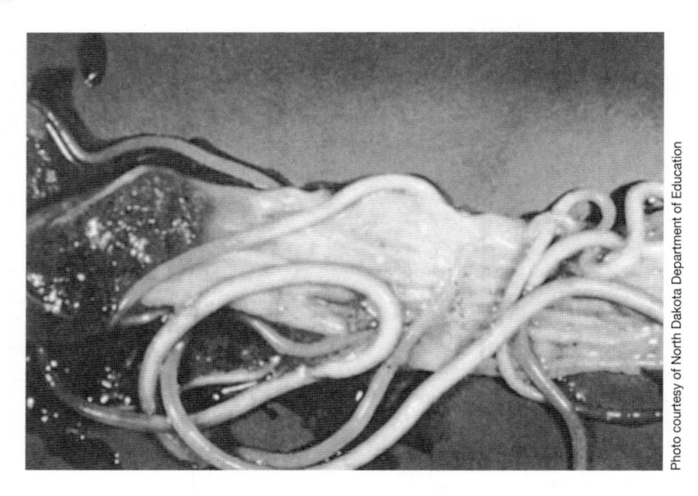

Photo courtesy of North Dakota Department of Education

FIGURE 38-2 Common roundworm found in dogs (*Toxicana canis*).

Small animals infected with roundworms will show signs of vomiting, diarrhea, a bloated stomach appearance, and visible roundworms in the feces, which appear like long thin pieces of pasta (see Figure 38-2). Large animals will show signs of abdominal pain, coughing and diarrhea, as well as visible roundworms in the feces that may be 6–12 inches in length. The eggs of roundworms are circular shaped with dark circular centers when viewed under a microscope. Table 38-1 summarizes types of roundworms and their common hosts.

Whipworms

Whipworms are another common intestinal parasite, occurring mostly in dogs. Whipworms get their name from the whip-like shape of the adult parasite. The front end of the worm is thin and the back end of the worm is thick, similar to a whip handle. Whipworms live in the large intestine of infected animals. ***Trichuris vulpis*** species infect animals through ingesting food or water contaminated with whipworm eggs. The eggs are swallowed, hatch, and in a little less than three months, the **larvae** mature into adults in the cecum and large intestine where they burrow their mouths into the intestinal wall and feed on blood. Adult worms lay eggs that are passed in the feces. The eggs must remain in the soil for about a month to mature and be capable of causing infection. The eggs are resistant and hardy and difficult to control. Animals infected with whipworms may exhibit diarrhea, weight loss, possible blood in the feces, and anemia. Whipworms are not visible in the feces and are diagnosed through a fecal analysis under a microscope. The eggs of whipworms are football shaped (see Figure 38-1B).

Hookworms

Hookworms are another common intestinal parasite of dogs and occasionally cats. Hookworms have either teeth-like structures or cutting plates with which they attach themselves to the wall of the intestine and feed on the animal's blood. Hookworms can cause a skin disease in humans called **cutaneous larval migrans**. Infections of the intestines in people can also cause severe abdominal pain. ***Ancylostoma caninum*** live in the small intestine of their host animal, where they attach themselves and feed on the host's blood. The adults lay eggs that pass out in the feces. In 2–10 days, the eggs hatch and the larvae are released. These larvae are excellent swimmers that travel through raindrops or dew on leaves and vegetation and wait for a suitable host animal to come along. The larvae enter a host either by being ingested or by burrowing through the host's skin. Animals infected with hookworms show signs of vomiting, diarrhea, anemia, weakness, black darkened feces, a dull poor coat appearance, and occasional coughing. Hookworms are not visible in the feces and are diagnosed with a fecal analysis using a microscope. Hookworm eggs are oval and clear (see Figure 38-1C).

Heartworms

Heartworm disease was first identified in the United States in 1847 and occurred most frequently in the southern United States. In recent years, heartworm disease has been found throughout the United States. The movement of infected animals that could serve as sources of infection for others is probably a significant contributing factor to heartworms spreading throughout North America. The actual number of infected dogs

TABLE 38-1

Roundworm Hosts

ROUNDWORM	PRIMARY HOST	TRANSPORT HOST
Toxascaris leonina	Dog, cat, fox, other wild carnivores	Small rodents
Toxocara canis	Dog, fox	Small rodents
Toxocara cati	Cat	Small rodents, beetles, earthworms
Parascaris equorum	Horse	Small rodents, flies

and cats in the United States is unknown. Heartworm disease is spread by the mosquito and is of most concern in dogs and cats and occasionally ferrets. ***Dirofilaria immitis*** belong to the same class as roundworms and look similar in structure, like long, thin pieces of pasta. *D. immitis* affects the heart and circulatory system of infected animals. Adult heartworms in the heart lay tiny **larvae** called **microfilaria**, which then live in the bloodstream. These microfilaria enter a mosquito when it sucks blood from an infected animal. In 2–3 weeks, the microfilaria develop into larger larvae in the mosquito and migrate to the mosquito's mouth. When the mosquito bites another animal, the larvae enter the animal's skin. The larvae grow and after about 3 months finish their migration to the heart, where they grow into adults, sometimes reaching a length of 12–14 inches (see Figure 38-3 and Figure 38-4). The time from when an animal was bitten by an infected mosquito until adult heartworms develop, mate, and lay microfilaria is about 6 months in dogs and 8 months in cats. Dogs that have heartworm disease will show signs of exercise intolerance, difficulty breathing, coughing, decreased appetite, weight loss, and lethargy. Cats typically show no signs and may die suddenly. Some cats will show similar signs to that of dogs. Simple heartworm test kits are commonly used in diagnosing the disease. Treatment of heartworm disease is a concern; when the heartworms die within the heart and/or bloodstream, this can cause blood flow to be blocked and may cause the pet to die. Prevention with monthly medication is the key to this disease.

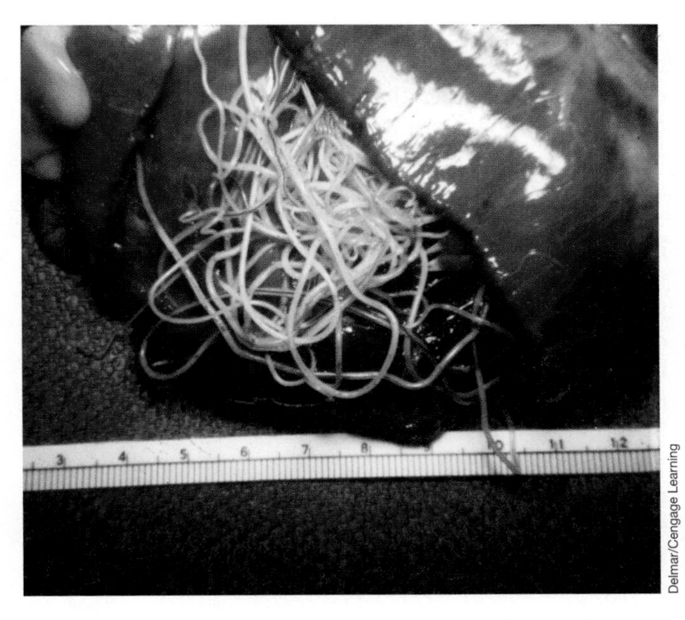

FIGURE 38-4 Heartworms can plug up the heart of an animal.

Strongyles

Large and small **strongyles** are a common parasite of large animals, especially horses. Large strongyles are a group of internal parasites, also known as bloodworms or redworms. Eggs in manure hatch into larvae that are consumed by the grazing animal. The larvae mature in the intestinal tract and burrow out into blood vessels

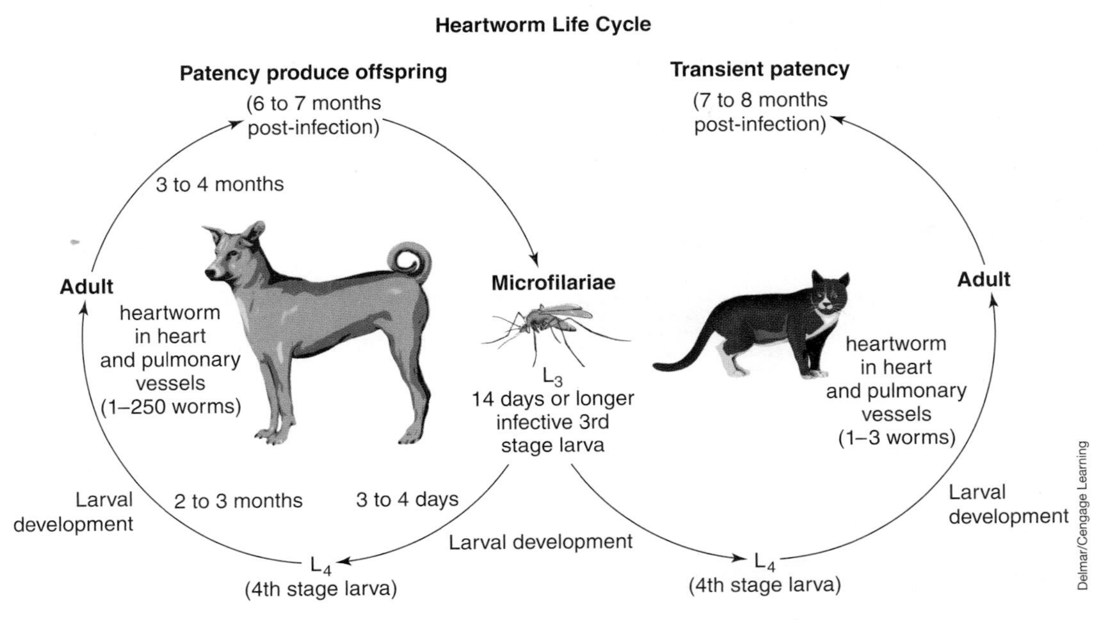

FIGURE 38-3 The life cycle of heartworm.

where they migrate throughout various organs and eventually back to the intestine. The larvae can cause extensive damage to the lining of blood vessels. Signs of large strongyles are weight loss, anemia, and abdominal pain and, in extreme cases, sudden death. Small strongyles differ from large strongyles in several ways. First, small strongyles do not migrate through tissues as do large strongyles. Second, small strongyle larvae may become encysted. This means that they burrow into the intestinal wall and lay dormant waiting for the proper conditions to emerge. During this encysted period, unlike adult parasites, small strongyle larvae are not susceptible to most dewormers. Signs of small strongyles include diarrhea, weight loss, poor growth, poor coat quality, and abdominal pain. Frequent deworming is recommended with all livestock. Yearly fecal analysis should be done to identify any internal parasites (see Figure 38-5). Strongyle eggs appear as large, long, oval shapes under a microscope.

Coccidia

Coccidia or **coccidiosis** is a common protozoan internal parasite that is a simple one-celled organism that occurs in all mammals. Most animals infected with coccidia are infected by ***Iospora canis***, although several species of coccidian do occur. Animals with coccidia will shed **oocysts**, or the single-celled eggs in the feces. The eggs live within the intestinal tract and the infected animal may show no signs of illness. Other animals may develop chronic diarrhea, and blood and mucous may be present. Sometimes coccidiosis is difficult to diagnose even on fecal analysis. The microscopic cells appear small and circular and may be missed on a fecal smear. Most animals become infected with coccidian by contaminated food, water, or soil sources. Bird droppings are the most common source of the parasite.

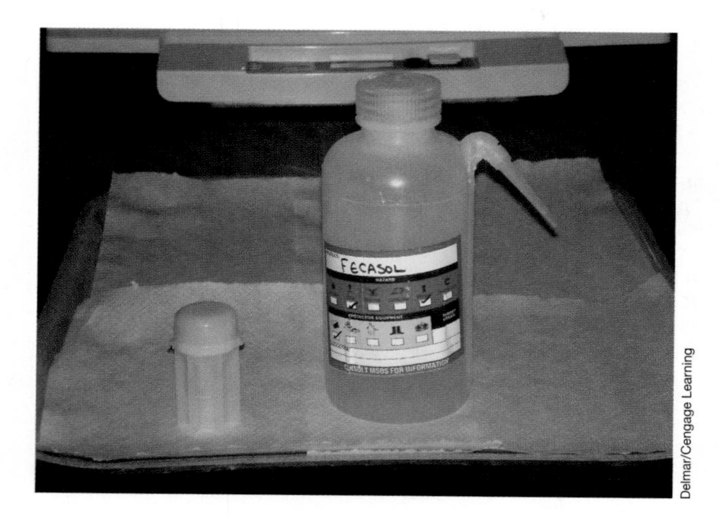

Delmar/Cengage Learning

FIGURE 38-5 Fecal floatation materials.

Giardia

Giardia or giardiasis is a protozoan or single-celled organism that lives in the intestinal tract of animals. Animals become infected from ingestion of contaminated water or soil sources. This is a common occurrence in water sources that are outside or are contaminated when pipes break. Many people will be provided boil-water advisory information when giardia is present in a water source. Animals frequently get diarrhea when they have giardia present in the feces. Other signs of giardia include vomiting, weight loss, and poor overall appearance.

Tapeworms

Tapeworms are flat worms that are segmented or have individual parts that grow and shed as the parasite ages. Tapeworms live within the small and large intestine. The segments, known as **proglotids**, may often be seen near the anus of the dog or cat. These segments may move if recently passed, or if dried, they look like grains of uncooked rice or cucumber seeds (see Figure 38-6). Tapeworm infections are usually diagnosed by finding these segments on the animal. Most animals develop tapeworms from ingesting a flea, ingesting a rodent, or ingesting a mite. ***Dipylidium caninum***, which can be up to 20 inches long, lives in the small intestine. The segments, full of eggs, are passed in the feces. While warm, the segments are active, but as they dry, they break open and liberate the eggs inside. Either an adult louse or a flea **larva** ingests the eggs. The egg develops into an immature form in the insect. When a dog or cat eats the insect, the immature form develops into an adult worm and the life cycle is completed (see Figure 38-7). Large and small animals may become infected with ***Taenia*** species, which have a similar life cycle (see Figure 38-8). Fleas may begin the life cycle in small animals. Mites living in a pasture may consume tapeworm eggs from the feces of infested animals. Grazing animals may then swallow the mites and become infested with tapeworms. Signs of tapeworms include poor hair coat, abdominal discomfort, or visible segments in the feces.

External Parasites

Several external parasites invade all types of small and large animals. Mites, fleas, ticks, lice, and biting flies are common, especially in warmer weather. External parasites may live in the hair coat, on the skin, or within the ear canals.

Fleas

Adult female fleas can lay up to 50 eggs a day. The eggs are laid on the host where they fall off to hatch in the environment. Eggs incubate best in high humidity and

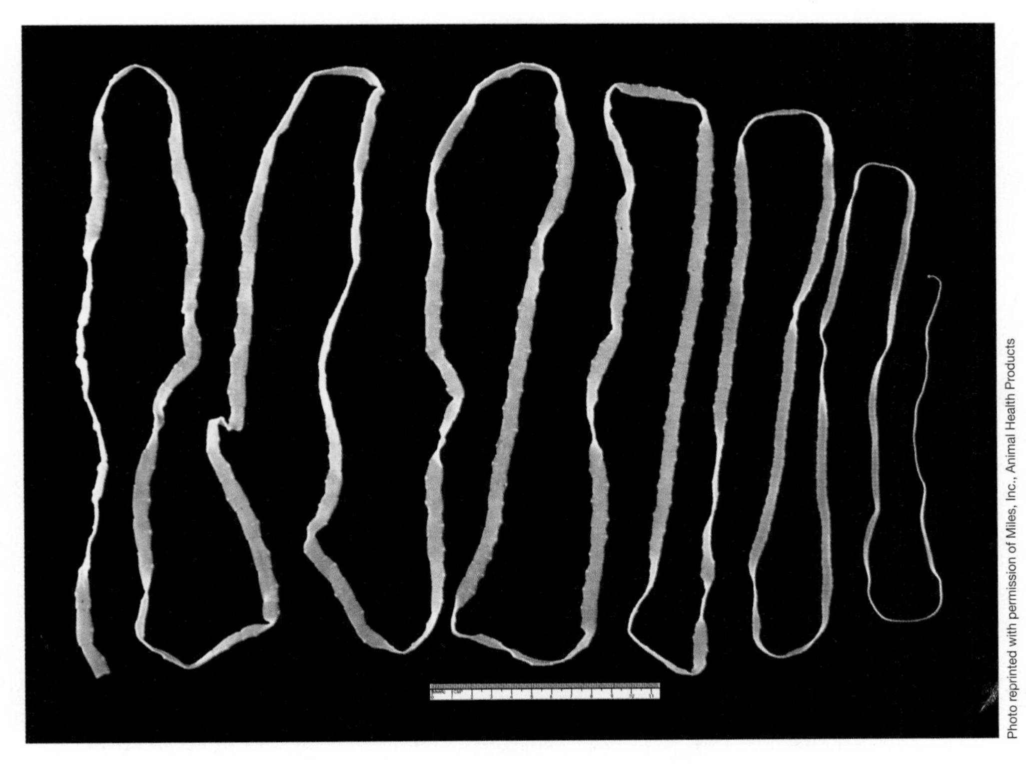

Photo reprinted with permission of Miles, Inc., Animal Health Products

FIGURE 38-6 Tapeworm; note the segmented body.

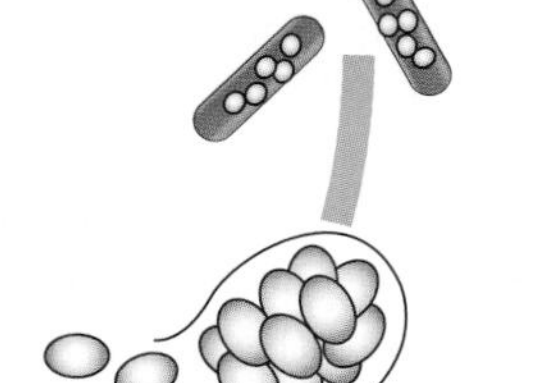

Life cycle of:
Dipylidium caninum
(dog/cat-flea tapeworm)
INTERMEDIATE HOST:
The Common Flea
1. Pet passes egg-filled tapeworm segments in feces
2. The segments rupture, releasing individual eggs that are eaten by flea larvae
3. During grooming or biting, pet ingests tapeworm-carrying fleas
4. Tapeworm matures in pet
5. Adult tapeworm releases mature, egg-filled segments and, without treatment, the cycle begins again
An important point to remember about this tapeworm species is that any pet that has fleas also has a very real possibility of having tapeworms.

Delmar/Cengage Learning

FIGURE 38-7 Life cycle of flea tapeworm (*Dipylidium caninum*).

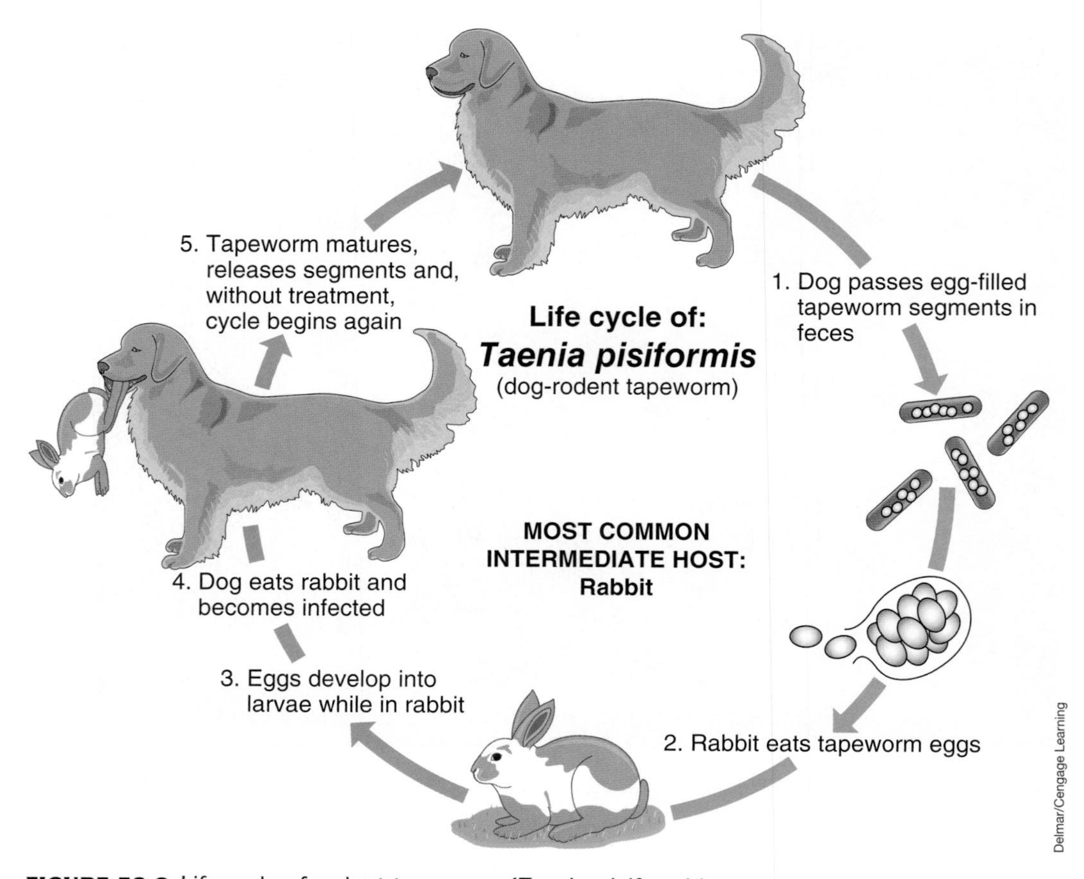

FIGURE 38-8 Life cycle of rodent tapeworm (*Taenia pisiformis*).

temperatures of 65–80 degrees. The eggs will hatch around 10 days and become the larvae stage. Larvae are like little caterpillars crawling around and grazing on the flea dirt that is generally in their vicinity. This is the stage when large animals pick up tapeworms while they graze. Larvae go through several molt stages and then become **pupae**, where they spin a cocoon-like structure where inside they become an adult flea. This stage can remain dormant several months to a year before emerging as an adult flea. Under ideal temperatures and conditions, this stage may occur in as little as 21 days from the time eggs were hatched. The pupae emerge when they feel vibrations and heat of a host to feed off of. Adult fleas may survive up to several months without a food source, which is a blood meal (see Figure 38-9). Once a female flea has a blood meal, it will produce eggs within 24–48 hours and then continually lay eggs until it dies. The average life cycle of an adult flea is 4–6 weeks. Fleas are easily noted on an animal by applying a flea comb to the coat. The veterinary assistant may also look for flea "dirt," which is dark black to brown small particles that look like the animal is dirty. This flea "dirt" is actually the feces of the flea. If the veterinary assistant isn't sure if it is flea "dirt" or real dirt from the outside environment, place some of the particles on a paper towel or other light colored surface and apply a drop of water to the particles. If the particles turn red rust to brown in color, they are flea feces, which are the blood meal of the host animal. Owners can also use a warm towel placed on a high animal access area, such

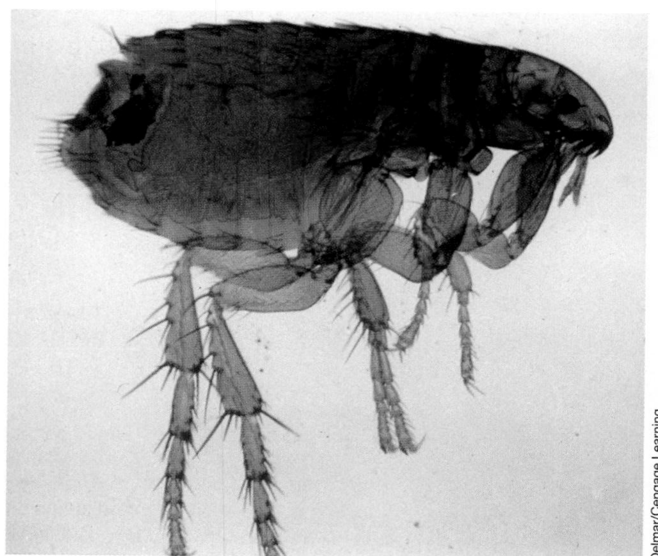

FIGURE 38-9 Flea.

as a living room, with the lights turned out for several minutes. When the lights are turned on, the owners can inspect the towel for fleas. Fleas will quickly seek heat if there are in the household. Signs of an animal having fleas include itching and scratching of the skin, biting at the skin, hair loss, scabs, and visible fleas or "dirt." Fleas can cause severe skin irritation and itching, and in some animals, the flea saliva causes an allergic reaction when an animal is bitten. This is called **flea allergy dermatitis** or **FAD**. Signs of FAD include itching and scratching of the skin, biting at the skin, hair loss, scabs or bumps, and red areas on the skin. Fleas can also cause an animal to have **anemia** or a loss of blood. An animal that has anemia from flea bites will have pale gums, weakness, and be lethargic. Animals that ingest a flea or a stage of the life cycle of the flea are susceptible to developing tapeworms. The most common sign of tapeworms is itching around the anal area or visible segments shedding from the tapeworm noted around the rectum or on bedding. Another less common condition caused by flea bites is **plague**. This disease affects animals and humans and causes high fever, dehydration, and enlarged lymph nodes that can eventually lead to death. ***Bartonella*** is a bacteria spread by fleas, most common in cats. Cats transmit the organism when they are infested by fleas, scratch themselves, and get infected flea dirt in their claws, and scratch a person or another cat with their dirty claws. Cats can also harbor *Bartonella* in their mouths and transmit the infection via bites or groom their feet and self-infect their claws. Cats also are prone to a disease called **feline infectious anemia** or **hemobartonella**. This disease is caused by the bite of a flea with bacteria that infects the animal's red blood cells and immune system. Infected cats are often pale, jaundiced, anemic, lethargic, and have a fever.

Ticks

Ticks are arthropods that seek heat and movement (see Figure 38-10). Ticks undergo four stages of life until they become adults. There are several species of ticks that transmit different types of serious diseases. Adult female ticks will lay eggs depending on how much of a blood source they have from a host. The larger the blood meal, the more eggs the female will produce. After an adult has completed a blood meal, it will drop off of the host and lay its eggs in the environment. Eggs are laid from several days to several weeks. The female dies shortly after laying the eggs. The eggs then develop into the larvae stage and then molt into a nymph stage, where they then become an adult tick. During each stage the tick will need a blood meal from a host to develop into the next life cycle stage. The entire life cycle process from egg to adult typically takes 2 years. Ticks will feed off any animal species or humans. When they have a blood meal they can transmit viral, bacterial, and rickettsial diseases. The most common tick species in animals are the American Dog Tick, the Deer Tick, the Brown Dog Tick, the Lone Star Tick, and the *Ixodes* species. The American Dog Tick transmits a disease known as **Rocky Mountain Spotted Fever**. Signs include a fever, joint pain (arthrodynia), depression, and anorexia. This disease can be fatal if not treated. The Lone Star Tick transmits a disease known as **Ehrlichia**. This bacterial disease causes a fever, joint pain, depression, anemia, anorexia, weight loss, and some skin swelling. If not treated, this disease may be fatal. The Deer Tick, also known as the Black Legged Tick, transmits a common disease called **Lyme disease**. This disease causes joint pain, lameness, fever, depression, anorexia, lethargy, and swelling of the joints and in more severe cases may lead to severe liver, cardiac, and neurological problems. Some animals will show no signs of the disease in the early stages of infection. Another less common disease caused by the Deer Tick is **Babesia**. This disease causes signs of anemia, jaundice (yellow coloring of the skin and **mucous membranes**), fever, and vomiting, and severe stages can cause kidney failure. This disease may be difficult to control and relapses often occur. Table 38-2 summarizes the various types of disease caused by ticks.

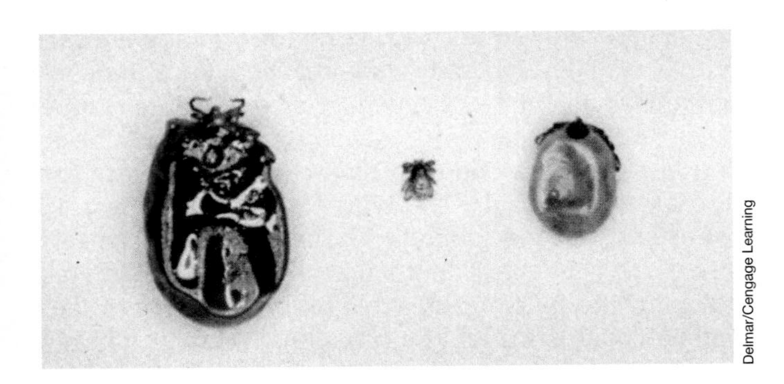

FIGURE 38-10 Ticks.

Lice

All animals and humans get lice. Each animal species, including humans, gets lice that are transmitted only to that species; thus, lice are species specific. Humans can't get lice from dogs, cats, horses, or other animals, and people can't transfer lice to animals. Animal lice are almost motionless, and some species of lice will attach to the hair by sticking and feeding off the hair and skin cells. Other species of lice bite and remove blood from the animal. Animal lice are gray to light beige in color and are wingless parasites that are about a twelfth of an inch in length. They are small in size but can be seen with the naked eye (see Figure 38-11). A piece of tape can be used to remove them for better identification; the lice and eggs easily stick to a piece of tape. Lice can

TABLE 38-2

Tick Species and Disease

TICK SPECIES	APPEARANCE	DISEASE POTENTIAL
American Dog Tick	■ Chestnut brown color ■ Light to white colored markings on back	Rocky Mountain Spotted Fever
Brown Dog Tick	■ Shades of light to dark brown ■ Solid color with no markings	None
Deer Tick (Black Legged Tick)	■ Dark brown body color ■ Black legs	Lyme disease Babesia
Lone Star Tick	■ Reddish brown color ■ White spot or "star" on back	Ehrlichia (Ehrlichiosis)
Ixodes cookie (Groundhog tick)	■ Light brown to tan body ■ Dark brown head and neck area	Encephalitis (rare)

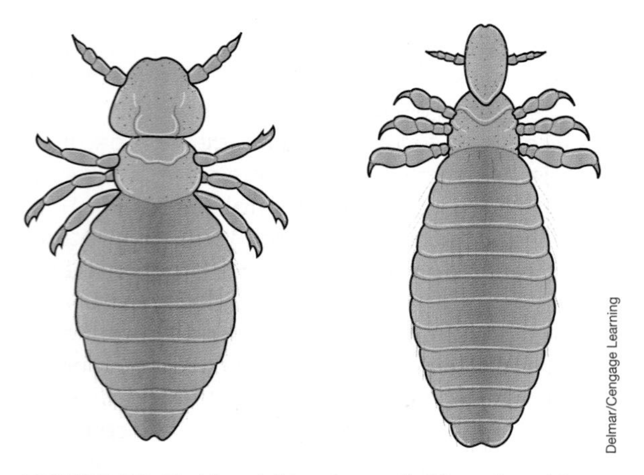

FIGURE 38-11 Lice: biting louse (left) and sucking louse (right).

spread by direct contact or contact with bedding and areas that animals lie in frequently. Lice eggs, called **nits**, may fall off and be passed to another animal of the same species. Grooming supplies and clippers are a common source of infection if sanitation is not done properly. Lice cause severe itching, areas of hair loss, and anemia, and sometimes adult lice and nits can be seen visibly. Lice can lay up to 100 eggs and the life cycle to adults takes about 21 days.

Mosquitoes

Mosquitoes are small flying insects that bite humans and animals and have a blood meal to survive. When they bite, they are capable of spreading diseases from animal to animal. One disease transmitted by mosquitoes is heartworm disease. Mosquitoes transmit microfilaria, or immature heartworms, from animal to animal when they have a blood meal from an infected animal.

West Nile virus is also a mosquito-borne disease that causes inflammation or swelling of the brain and spinal cord. Birds and horses are at the greatest risk.

Biting Flies

Biting flies can cause animals, especially large animals, lots of irritation. One disease that is of concern from biting flies, especially the horsefly, is **equine infectious anemia** or **EIA**. Equine infectious anemia, also known as swamp fever, is a viral disease for which there is no vaccine and no cure. This is of huge concern from state to state and many horses that travel are required to have a **Coggins test**, which screens for the EIA virus. The Coggins test is performed by a state-approved laboratory through a blood sample. Many states have regulations on horses requiring a negative Coggins test, and veterinary assistants should be familiar with state laws on equine travel. The best route of prevention is through the use of fly sprays to help reduce the amount of biting flies that have contact with a horse. Manure and waste control and removal are helpful.

Mites

Mites come in many forms and invade several areas of an animal's body. Mites may live on the skin or hair coat, or they may develop and invade the ears. **Ear mites** are common in all animals (see Figure 38-12). Ear mites are tiny infectious organisms resembling microscopic ticks. Infection usually produces a characteristic dry black ear discharge, which is dry and crusty and is commonly said to resemble coffee grounds. The mite lives on the surface of the ear canal skin, though it sometimes migrates out onto the face and head of its host. Eggs are laid and hatch after 4 days of incubation. The larva hatches from the egg, feeds on ear wax and skin oils for about a week, and then goes through several molts to become

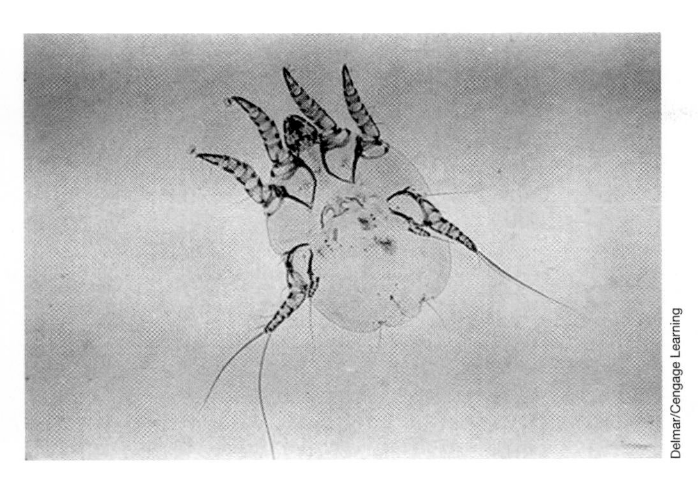

Delmar/Cengage Learning

FIGURE 38-12 Ear mite (Otedectes mite).

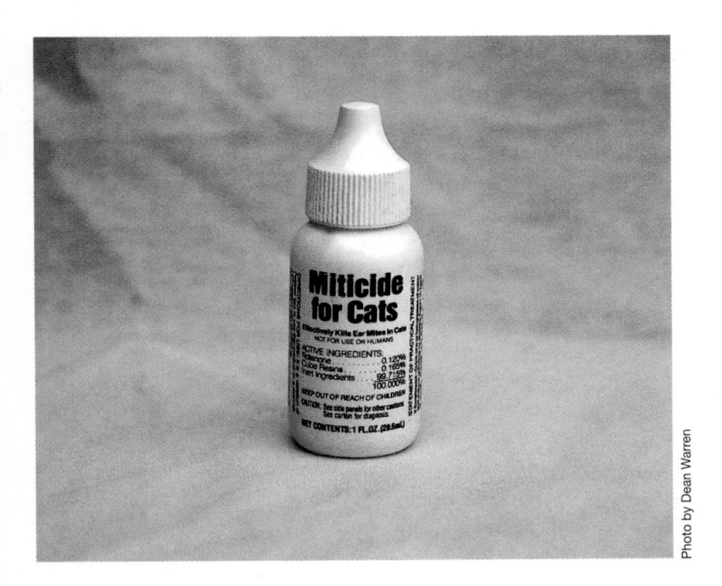

Photo by Dean Warren

FIGURE 38-13 Ear drop products for control of mites.

an adult mite. The entire life cycle takes about 3 weeks to complete. Adult ear mites live about 2 months. Ear mite infections can be easily treated by cleaning the ear canals and applying medications that kill the ear mites (see Figure 38-13). Other signs of ear mites in animals include scratching the ears, shaking the head, and open sores within the ear.

Mange mites are another common occurrence, especially in dogs. Mange mites live on the skin and hair coat. There are two types of mange mites: sarcoptic mange and demodectic mange. **Sarcoptic mange** is the name for the skin disease caused by infection with the *Sarcoptes scabei* mite (see Figure 38-14). Mites are not insects; instead, they are more closely related to spiders. They are microscopic and cannot be seen with the naked eye. Adult mites live 3–4 weeks in the host's skin. After mating, the female burrows into the skin and deposits 3–4 eggs in a tunnel behind her. The eggs hatch in 3–10 days,

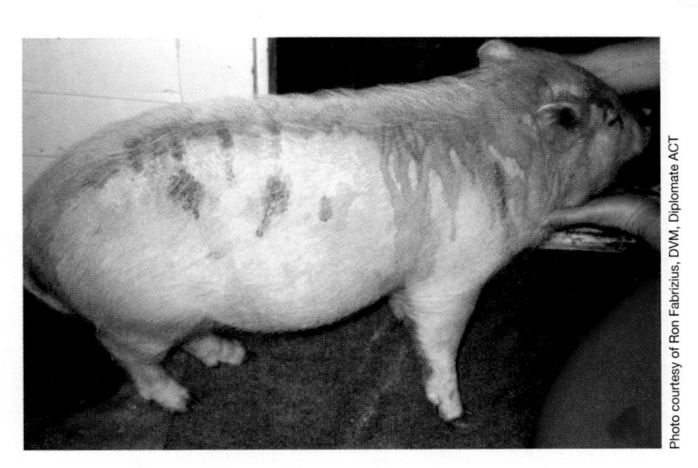

Photo courtesy of Ron Fabrizius, DVM, Diplomate ACT

FIGURE 38-14 Alopecia in a pig with sarcoptic mange.

producing a larvae that then molts into the adult mite within 10–15 days. This mite is zoonotic and causes the condition known as **scabies** in humans. Sarcoptic mange is highly contagious by direct contact or contact with infected bedding or other items such as grooming tools that have not been properly disinfected. Signs of sarcoptic mange in animals include excessive itching, hair loss, damage to the hair coat, open lesions on the skin, and skin infection. Diagnosis is typically through a skin scraping and identifying the mites under a microscope. **Demodectic mange** is another mite that invades the skin, and all dogs raised normally by their mothers possess this mite as mites are transferred from mother to pup via cuddling during the first few days of life. Most dogs live without any problems with their mites, never suffering any consequences. However, in some dogs the mites can become overpopulated and cause severe skin irritation. This mange is not contagious. Signs of the disease include localized hair loss, increased itching, and dry areas of skin. A skin scraping is used to identify the microscopic mites that are shaped like cigars. Both mange mites are treatable with medications.

Ringworm (Dermatophytosis)

Dermatophytosis, commonly known as **ringworm**, is a fungal disease of animals and can be transmitted to humans (see Figure 38-15). This condition can be passed from any infected animal to another species or person. It is the most common contagious skin disease of animals. Dogs, cats, and all livestock species commonly get ringworm from the environment. Spores from infected animals can be shed into the environment and live for up to 24 months. Humid, warm environments encourage growth of the fungus. Spores can be on brushes, bedding, furniture, stalls, or anything that has been in contact with an infected animal or the animal's hair. Ringworm is transmitted by direct contact of spores, which may be on the animal or in the environment.

Delmar/Cengage Learning

FIGURE 38-15 Holstein heifer with ringworm.

Skin lesions may vary in appearance and size. The classic sign is a small round area with no hair and scaly, dry skin located in the center. The lesion may start as a small spot and continue to grow. The area may or may not become itchy. In some cases, the fungus can spread over the body to other locations. Most common sites in animals are on the head, face, ears, and tail. The most common method for diagnosis is by skin scraping and looking at the hair and skin under a microscope. Another method is by use of a Wood's lamp, which fluoresces the area with a black light. Several species of the ringworm fungus will glow a fluorescent color when exposed to a Wood's lamp. However, many species do not show up with this method. Oral and topical treatment is used in curing the animal, but the environment must also be disinfected to prevent reoccurrence.

Deworming Programs

Small and large animals alike should be on a regular deworming program. Large animals should be wormed every 6–8 weeks due to the increased exposure to parasites. Small animals should be wormed yearly or as recommended by the veterinarian. Dewormers should be given according to the common parasites that affect the species of animal. Many dewormers are available for treatment of these parasites. Unfortunately, none of the over-the-counter or prescription dewormers will kill all of these parasites. Thus, dewormers must be chosen according to the type of parasite that is present. Table 38-3 summarizes the recommendations by the Companion Animal Parasite Council (CAPC) and details control methods for common internal (GI) parasites for dogs and cats in the United States. Table 38-4 lists common deworming products used for companion animals.

TABLE 38-3

Companion Animal Deworming Recommendations

1. Administer year-round broad-spectrum heartworm medications.
2. Deworm puppies and kittens at age 2, 4, 6, and 8 weeks of age and then monthly until 6 months of age.
3. Deworm nursing dogs and cats at the same time as their offspring.
4. Conduct fecal examinations 2 to 4 times per year for adult animals, depending on health status and lifestyle factors.

TABLE 38-4

Common Companion Animal Deworming Products

PRODUCT NAME	PARASITES TREATED
Drontal	Roundworms, hookworms, whipworms, and tapeworms
Albon	Coccidia
Droncit	Tapeworms
Heartgard	Heartworms, roundworms, and hookworms
Interceptor	Heartworms, roundworms, hookworms, and whipworms.
Nemex	Roundworms and hookworms
Panacure	Roundworms, hookworms, whipworms, and some species of tapeworms
Proheart	Heartworms and hookworms
Revolution	Heartworms, fleas, ear mites, ticks, roundworms, mange, and hookworms
Sentinel	Heartworms, roundworms, hookworms, whipworms, and flea eggs
Strongid T	Roundworms, hookworms, whipworms, tapeworms

Large animal deworming recommendations include worming livestock every 2 months for common internal gastrointestinal (GI) parasites. Additional recommendations for parasite control include the following:

- Remove manure daily from stalls and weekly from pastures.
- Be sure pastures and paddocks are well-drained and not overpopulated.
- Compost manure rather than spreading it on fields where animals graze.
- Use a feeder for hay and grain and avoid ground feeding.
- Initiate effective fly control programs.
- Routinely examine all animals for signs of infestation.
- Establish a parasite prevention and monitoring program with your veterinarian.

MAKING THE CONNECTION

All chapters in Section II discuss parasites and parasitic disease specific to the species discussed in the chapter. Please review these content areas in those chapters for more information on parasites.

Neoplasia (Cancer)

Neoplasia, commonly called cancer, is common in all animals, but there are many types of cancers that affect every area of an animal's body. The rate of cancer increases with age. Certain animal species and breeds are more prone to certain types of cancer, such as osteosarcoma or bone cancer (see Figure 38-16). In general, animals get cancer at roughly the same rate as humans. Cancer accounts for almost half of all animal deaths over the age of 10. Some cancer is easily preventable, such as with early neutering. When an animal is suspected of having cancer, such diagnostics as blood work, X-rays, ultrasound, and biopsy (tissue sample) can be helpful in diagnosis and possible early treatment. Common signs of cancer in animals include the following:

- Abnormal swellings that persist or continue to grow
- Sores that do not heal
- Weight loss
- Loss of appetite
- Bleeding or discharge from any body opening
- Offensive odor
- Difficulty eating or swallowing
- Hesitating to exercise or loss of stamina
- Persistent lameness or stiffness
- Difficulty breathing, urinating, or defecating
- Overall poor general appearance
- Depression

Each type of cancer requires specific care and some early stages of cancer may be treatable, while others may be difficult to treat. Treatment of cancer may include **chemotherapy** (treatment with chemicals), surgical removal, radiation, **cryosurgery** (freezing tissue), medications, or a combination of treatments (see Figure 38-17). Table 38-5 lists commonly occurring cancers in animals.

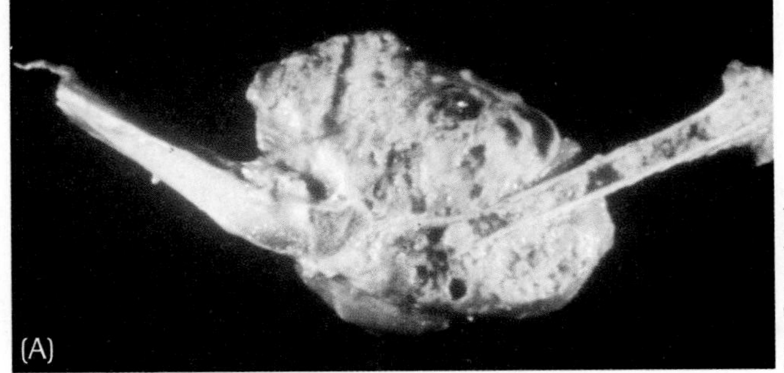

Delmar/Cengage Learning

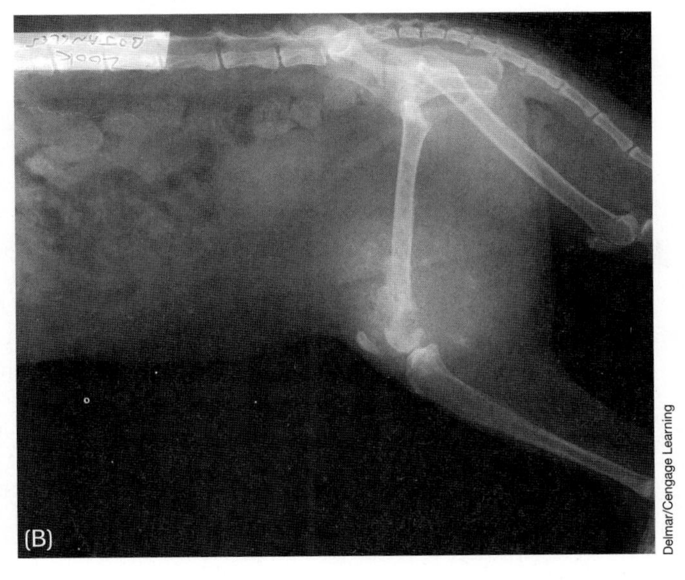

Delmar/Cengage Learning

FIGURE 38-16 (A) An osteosarcoma, a malignant bone tumor (B) Radiograph of an osteosarcoma.

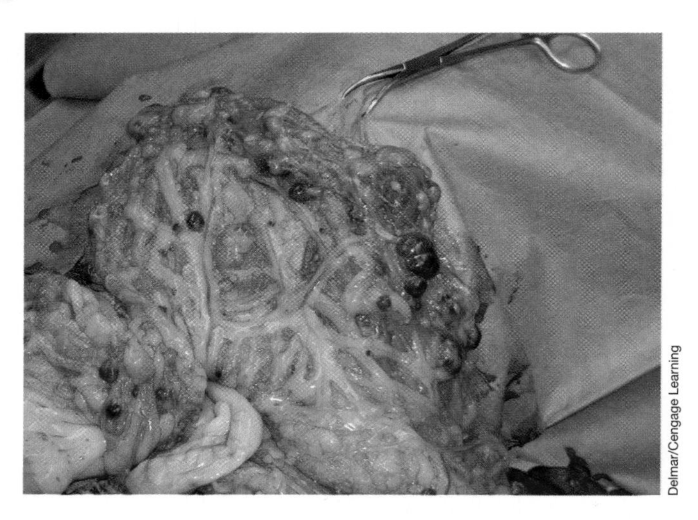

FIGURE 38-17 Intraoperative photograph showing numerous dark red tumors in the omentum of a dog with a splenic tumor. The tumors had also spread to the liver.

Zoonosis

Diseases that are transmitted from animals to humans are called **zoonoses**. Many animals infected with a zoonotic disease may not show any signs of illness. The large majority of zoonoses are not clinically significant. However, it is important for veterinary professionals to be aware of zoonotic disease and take precautions to prevent the spread or contraction of such diseases. The most common zoonotic diseases of significance are discussed here.

Toxoplasmosis

Toxoplasma gondii is a single-celled organism that can occur in any mammal but is shed through the feces of cats, which are the only animals that pass the Toxoplasmosis through the feces. This disease is of a concern in humans, especially pregnant women, who may contract the disease, which may cause abortions and congenital defects in the fetus. Cats that have this condition shed oocysts (egg-like immature particles) in their feces, and the oocysts must be in the environment 1–5 days before they are infective (see Figure 38-18). Cats pass the eggs in their feces only 2–3 weeks after becoming infective. The oocysts can survive several years and are difficult to remove from the environment. Cats generally become infected through eating rodents or birds. Signs of animals having toxoplasmosis include depression, anorexia, fever, and lethargy. Prevention is the key, so for humans it is recommended that pregnant women not clean or handle litter boxes during pregnancy. When cleaning litter, working outside in gardens or soil, and children playing in sandboxes should wash their hands immediately on going inside. These are common sites for cats to eliminate and pass oocysts.

Brucellosis

Brucellosis is a reproductive disease that can affect all mammals. It is of most concern in dogs and cattle. This is a contagious disease spread through vaginal discharge, usually through breeding programs. In cattle this disease is also known as "Bang's disease." This disease causes increased abortions and infertility problems, usually with the animal becoming sterile. When humans come in contact with milk from a cow that has brucellosis, they will become infected with a disease called undulant fever, which causes a high fever, making brucellosis a public health issue. There is no treatment for brucellosis, so testing of breeding animals prior to reproduction is essential. A vaccine is available in cattle. Testing in breeding dogs is vital for

TABLE 38-5

Common Cancers in Animals

CANCER TYPE	COMMON SPECIES	OCCURRENCE	SIGNS
Bone cancer	Large/giant breed dogs	Leg bones near the joints	Lameness, swelling, pain
Skin cancer	Dogs; horses	Body and limbs	Change in color, swelling, drainage, pain
Lymphoma	Dogs, cats, horses, cattle	Enlarged lymph nodes throughout the body, spleen, bone marrow	Palpable lymph nodes, anorexia, pain, weight loss, lethargy
Abdominal cancer	Dogs, cats, horses, rodents	Stomach, intestines	Weight loss, abdominal swelling, pain
Mammary cancer	Dogs, cats	Female uterus, mammary glands	Swelling, palpable hard mammary glands, pain
Head and neck cancer	Dogs, horses	Mouth, nose, throat	Bleeding, odor, oral masses, difficulty eating and drinking, difficulty breathing, facial swelling, pain

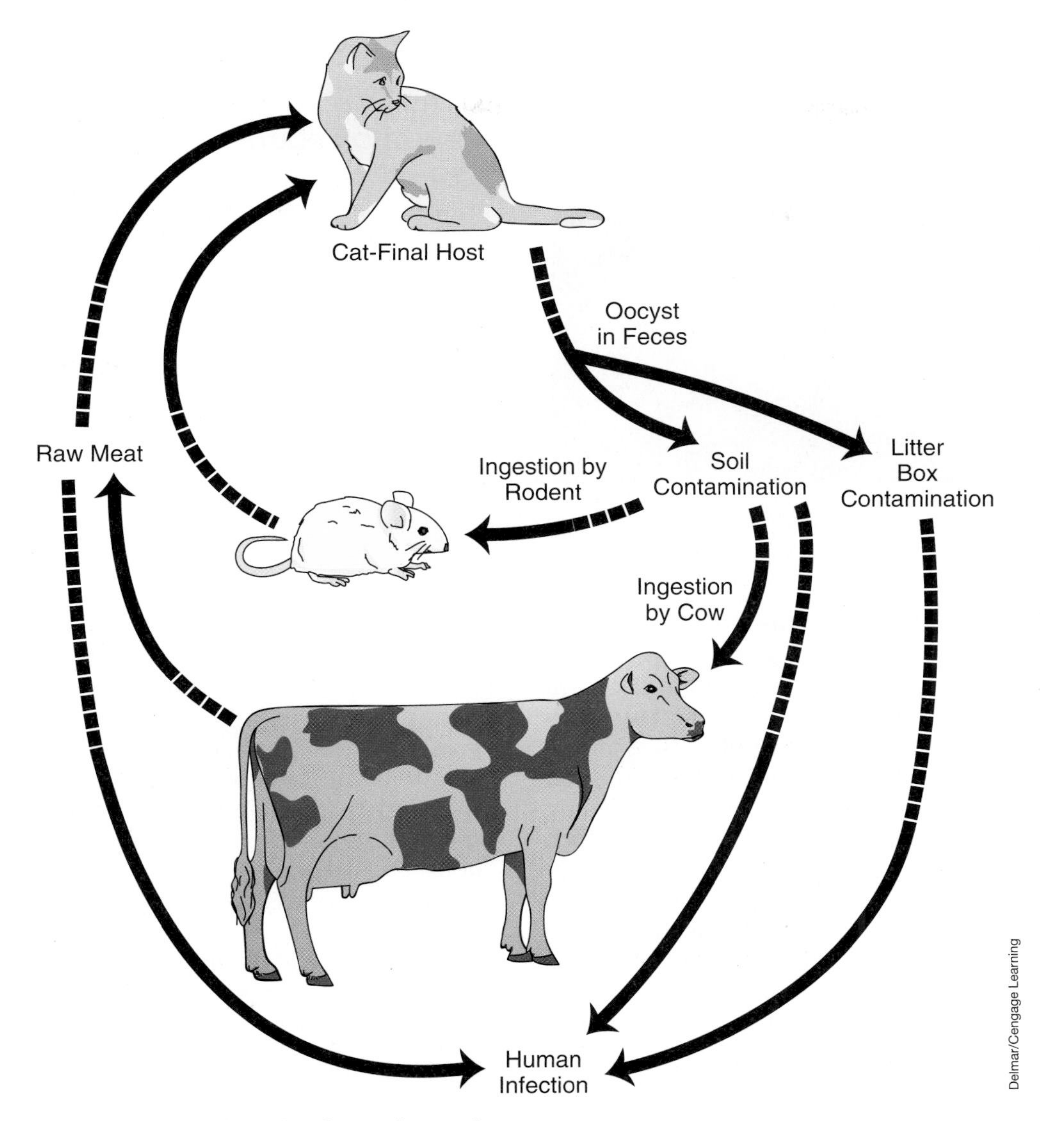

FIGURE 38-18 Life cycle of toxoplasmosis.

control and prevention. The organism becomes infective in about 3 weeks.

Encephalitis

Encephalitis, commonly called "sleeping sickness," is a viral disease that affects horses and people. There are several strains of the disease that are of concern in the horse, such as Eastern Equine Encephalitis (EEE), Western Equine Encephalitis (WEE), and Venezuelan Equine Encephalitis (VEE). This disease is spread by mosquitoes and the virus causes central nervous system infections. Birds are the reservoir host, which means a mosquito bites an infected bird that then spreads the disease on to horses. No treatment is available, but a vaccine is available to prevent these strains of disease. Signs of encephalitis in horses include possible blindness, walking aimlessly, walking into things, fever, anorexia, depression, a sleepy appearance, difficulty swallowing, grinding teeth, paddling motion with front limbs, and eventually death.

Maintaining Good Animal Health

Many factors must be considered in maintaining ideal animal health. The environment is one of the essential factors. Vaccination programs are the best method

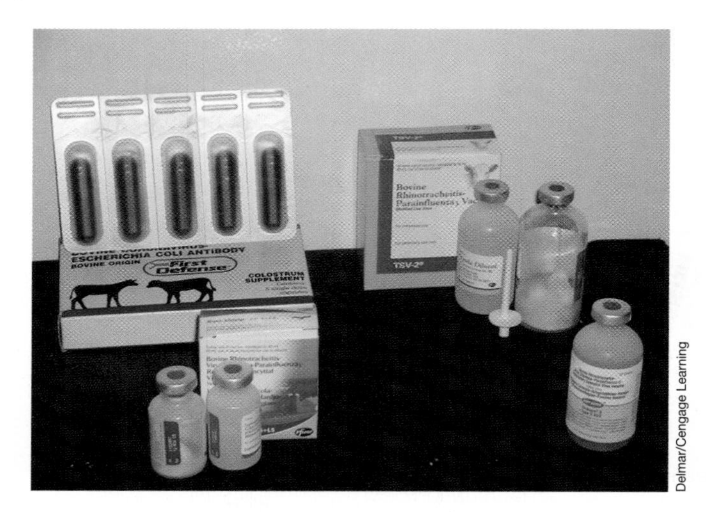

Delmar/Cengage Learning

FIGURE 38-19 Examples of vaccines.

TABLE 38-6

Common Injection Routes

INJECTION ROUTE	LOCATION
SQ—subcutaneous	Under the skin
IM—intramuscular	Into a muscle
IV—intravenous	Into a vein
ID—intradermal	Into the skin (within the layers)
IN—intranasal	Into the nasal cavity
IP—intraperitoneal	Into the peritoneum or body cavity
IO—intraosseous	Into the bone

of lowering disease risks (see Figure 38-19). Knowing the common diseases that affect a species of animal or are at high risk in a given area will help to decrease the chances that animals will become infected with disease. If animals do become infected, it is essential to isolate any sick animals, especially those housed in large groups, to decrease the spread of disease. Young animals can be preconditioned to reduce stress, which will lower the chances of disease. **Preconditioning** prepares young animals for possible stress factors, such as learning to load and travel, learning to be around large numbers of people or animals, and being in a show environment. Other factors in keeping animals healthy include a clean and sanitary environment, proper nutritional program, proper space, a yearly physical exam and health program, and a deworming program.

Treating Disease

When animals do become sick and need veterinary treatment, veterinary assistants should be familiar with medication treatments and routes of medication administration. Medications are called **biologicals**, which are drugs used to treat disease. They are purchased from a pharmacy or a pharmaceutical company that manufactures and sells medications. Some biologicals are **antibiotics**, a group of medications used to kill organisms, such as bacteria. **Pesticides** are a group of medications used to treat internal and external parasites. Some of these pesticides are known as **anthelmintics**, or dewormers. Medications may be given by many different routes, such as injectible, oral, or topical. Injections are given into the body in various ways (see Table 38-6). One route of injection is subcutaneous (SQ), or under the skin. Another method is intramuscular (IM), or into the muscle. Another common injection is the intravenous (IV) injection.

Veterinary assistants are commonly supervised to perform SQ and IM injections. Oral medication is given by mouth. Topical medications are placed on the external body, such as the skin and hair coat. Some medications are additives that are mixed in with the food or water.

SUMMARY

Animal diseases are an important part of veterinary medicine, and veterinary assistants must have an understanding of normal and abnormal animal health. Some diseases are contagious, while others do not spread from animal to animal. Some diseases are zoonotic and create a public health hazard. Certain animal species are of concern as they may be used for human consumption. Many factors must be considered when dealing with animal health and diseases. It is essential to know how to properly handle an animal that is ill and understand the concerns with animal diseases.

Key Terms

Ancylostoma caninum hookworm species that live in the small intestine of their host animal, where they attach themselves and feed on the host's blood

anemia loss of blood causing blood cells not to replenish

anthelmintic a dewormer used to treat parasites in animals

antibiotic medication used to treat infection or disease

ascarids veterinary term for roundworms

babesia less common disease caused by the Deer Tick that causes signs of anemia, jaundice (yellow coloring of the skin and mucous membranes), fever, and vomiting, and more severe stages can cause kidney failure

bacterial class of disease spread through single-celled organisms called germs and are located in all areas of an animal's environment

bartonella bacteria caused by a flea bite that affects cats within their bloodstream and infects the red blood cells and immune system

biological drugs used to treat disease

brucellosis reproductive disease of mammals spread through breeding

chemotherapy therapy treatment with chemicals or drugs

coccidiosis common protozoan internal parasite that is a simple one-celled organism that occurs in all mammals

Coggins test blood test in equines that screens for equine infectious anemia

cryosurgery procedure involving freezing parts of tissue

cutaneous larval migrans hookworm infection in humans caused by penetrating the skin

demodectic mange mite that invades the skin; all dogs raised normally by their mothers possess this mite as mites are transferred from mother to pup via cuddling during the first few days of life

dermatophytosis fungal disease caused by an external parasite

Dipylidium caninum tapeworm species that can be up to 20 inches long and lives in the small intestine

direct contact contact with an animal or a bodily fluid from that animal

Dirofilaria immitis heartworm species that belong to the same class as roundworms and look similar in structure, like long, thin pieces of pasta

disease poor health that is a disturbance or change in an animal's function and structure

ear mites microscopic parasites that live within the ear canal causing severe itching and a dark, crusty discharge

ehrlichia bacterial disease caused by the Lone Star Tick and causes a fever, joint pain, depression, anemia, anorexia, weight loss, and some skin swelling

encephalitis viral disease in horses and people that is called sleeping sickness; several strains are spread by mosquitoes and affect the nervous system

equine infectious anemia (EIA) highly infectious disease of horses also known as swamp fever; a viral disease for which there is no vaccine and no cure

feline infectious anemia another name for hemobartonella, which causes an infection within the red blood cells of cats

flea allergy dermatitis (FAD) allergic reaction to a flea bite caused by the flea's saliva that affects the animal's skin

fungal class of disease spread by simple-celled organisms or spores that grow on the external body and other areas of the environment, mostly in moist and humid conditions

heartworm disease disease spread by the mosquito; of most concern in dogs and cats and occasionally ferrets, causing long white worms to build up in the heart

hookworms common intestinal parasite of dogs and occasionally cats that have either teeth-like structures or cutting plates with which they attach themselves to the wall of the intestine and feed on the animal's blood

host the animal that is infected by a parasite and that the parasite feeds off of

indirect contact spread through ways other than touching an infected animal, such as airborne or through bedding sources

(*Continues*)

Iospora canis most common coccidia species affecting animals

larvae immature stage of a parasite

Lyme disease bacterial disease transmitted by the bite of a tick that causes joint inflammation in affected animals

microbiology the study of microscopic organisms

microfilaria tiny immature heartworm larvae that live in the bloodstream

mucous membrane (mm) the gums of an animal and their color

neoplasia veterinary term for cancer

nits lice eggs

oocysts single-celled eggs

Parascaris equorum roundworm species in large animals that are swallowed in contaminated hay or water; hatch in the intestinal tract

parasitology the study of organisms that live off of other organism to survive

pathogens diseases and what causes them

pesticide chemical used to treat external parasites

plague disease that affects animals and humans and causes high fever, dehydration, and enlarged lymph nodes; can eventually lead to death

preconditioning building up an animal's body to prevent stress from occurring

proglotids segments that shed off a tapeworm and appear like small white pieces of flat rice

protozoan class of disease; the simplest forms of life caused by single-celled organisms that are known as parasites that may live inside or outside the body

pupae immature larvae stage that go through several molting stages during growth by spinning a cocoon

rickettsial class of disease spread by biting insects, such as fleas and ticks

ringworm fungal condition that affects the skin of humans and mammals and is highly contagious

Rocky Mountain Spotted Fever rickettsial disease transmitted by the American Dog tick and causes signs that include a fever, joint pain (arthrodynia), depression, and anorexia

roundworms visible long white worms that are the most common intestinal parasite in small and large animals

sarcoptic mange name for the skin disease caused by infection with the *Sarcoptes scabei* mite that causes a zoonotic mite skin condition; burrow into the skin causing infection

scabies sarcoptic mange condition in humans

strongyles common intestinal parasite in large animals that are long, thin, and white in appearance

supportive care therapy or treatment that includes such treatment as fluid replacement, medications to decrease vomiting and diarrhea, and pain medicine

taenia genus of tapeworm that affects companion animals and livestock

tapeworms intestinal flat worms that are segmented or have individual parts that grow and shed as the parasite ages

Toxocara canis roundworm species with a more complicated life cycle and an effective way of making sure its species will be passed from generation to generation

Toxocara cati roundworm species similar to T. *canis* that may infect the lungs or abdominal tissue

Toxascaris leonina roundworm species with the simplest life cycle

Toxoplasma gondii single-celled organisms that can occur in any mammal but is shed through the feces of cats, which are the only animals that pass it through the feces

Trichuris vulpis whipworm species infect animals through ingesting contaminated food or water

viral class of disease caused by particles that are contagious and spread through the environment

whipworms common intestinal parasite, occurring mostly in dogs, that gets its name from the whip-like shape of the adult parasite

zoonosis disease passed from animals to humans

REVIEW QUESTIONS

1. What are signs of normal animal health?
2. What are signs of abnormal health in an animal?
3. What are common internal parasites in animals?
4. What are common external parasites in animals?
5. What are some ways to treat or control parasites?
6. Name several diseases that are zoonotic.
7. What is preconditioning?
8. What are the routes of injections?

Clinical Situation

Courtesy of Photo Disc

The Brickell family has just purchased a farm. They called the Alpine Veterinary Clinic to ask some questions about the sanitation of the farm and when they can start transporting their animals to the location. Amanda, a veterinary assistant at the facility, is talking with Mr. Brickell about his concerns. "I was wondering if you can tell me about the types of animals that were raised on this farm previously."

"Well, the previous owners had some cattle, some horses, and several small animals. We saw most of their livestock and pets. I am just concerned about moving my animals here in case there are any health concerns I need to worry about. Is it possible to talk to the vet about this?"

"Sure, I will take your phone number and have Dr. Daniels call you."

- What are some concerns the new owner should be worried about?
- What precautions might the veterinarian suggest he take with the farm?
- When might the veterinarian recommend he move his animals to the farm?

Section IV

Clinical Procedures

39

Animal Behavior

Objectives

Upon completion of this chapter, the reader should be able to:

- ☐ Describe types of behavior
- ☐ Identify and recognize normal animal behavior
- ☐ Identify and recognize abnormal animal behavior
- ☐ Describe common animal behavior problems in companion animals
- ☐ Describe common animal behavior problems in large animals
- ☐ Discuss animal behavior and veterinary medicine

Introduction

Animal behavior relates to what an animal does and why it does it. The types of behaviors exhibited are varied. Some are genetically determined, or instinctive, while others are learned behaviors. Animals learn behaviors from their mothers, other animals, and people.

Instinctive Behaviors

Instinctive behaviors occur naturally to an animal in a reaction to a stimulus. Many behaviors in the early life of companion animals are instinctual. Kneading action of the paws of a puppy or kitten on the mother's mammary glands is an instinctual behavior that literally determines the young animal's survival. This particular behavior stimulates an equally instinctual behavioral response for the mother—the lactating response. Hormones are released from the brain thereby, releasing the milk from the mammary ducts to the teats. This lactating response is seen in human beings as well.

Many instinctive behaviors can be observed between a mother and her newborns. Calling out to the mother as the young animal's eyes are still closed and nestling in with the rest of the litter maintaining the body heat of such a small individual are instinctive behaviors (see Figure 39-1). The mother instinctively licking the young keeps the young's coat clean and aids in the process of elimination of feces and urine. This licking action is comforting to the young and in the case of the cat elicits the purring response that is characteristic only of the cat.

FIGURE 39-1 Kittens will snuggle together to maintain body warmth.

Marking is another instinctive behavior in animals that has both social and sexual purposes. **Pheromones** present in the urine give all who encounter the marks left by others a variety of messages, including sexual receptiveness, eating habits, age, and overall health. Mating behaviors are also instinctual in nature. Female animals in estrus give off pheromones that indicate the female is ready to breed. Male animals in return detect these odors and respond by locating the female and breeding with her.

Predatory behaviors are another example of instinct. Animals understand that in order to survive they must eat and food must be hunted.

Submission and dominance postures are also examples of behaviors that are instinctive (see Figure 39-2). These postures allow animals to know who is the leader or that they don't pose a threat.

Instinctive behaviors include behaviors that are acquired through an animal's genetic makeup, which is the set of instructions that build the parts of the body, including the brain. Each species' brain is built on a somewhat different pattern. When happy, a dog may wag its tail, a cat purrs, and a human smiles, while a turtle shows no outward signs at all. These differences are due to the different brain structures inherent to each species. The brain and other parts of the nervous system generate behaviors with a response to environmental input and hormones. Some of the behaviors may be simple, immediate responses to simple stimuli, including the knee jerk reflex and the cry of pain when an animal is hurt.

Maternal instincts are also a behavior to recognize. Many mothers with newborns will become protective and may be more defensive and difficult to handle. When possible, allow young animals to stay close to the mother when handling.

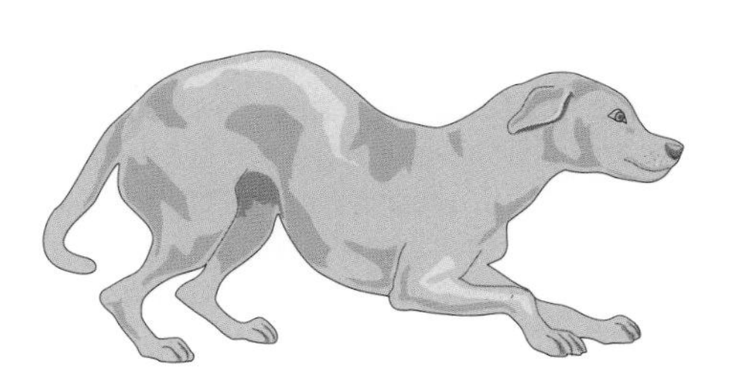

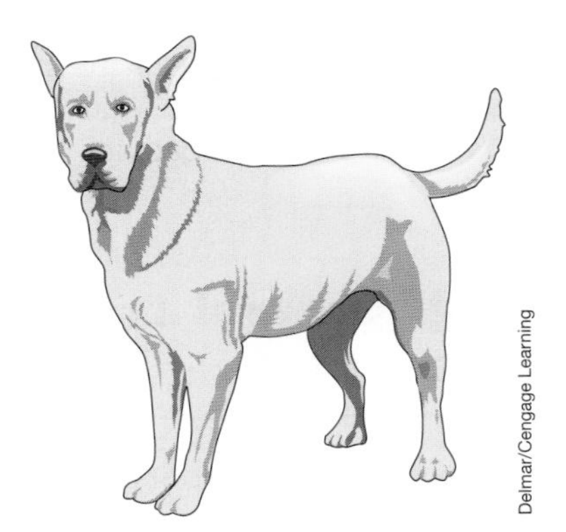

FIGURE 39-2 Dominance and submissive postures in dogs. The dominant dog has a raised head and tail. The submissive dog has a lowered head and neck as well as a lowered tail.

Learned Behaviors

Learning is the modification of behavior in response to specific experiences. **Learned behaviors** of animals can be classified in various ways. Understanding how an animal learns will help those in the veterinary field to better serve customers with animals presenting with behavior problems. Some common behavior modification techniques include conditioning, modeling, and imprinting.

Conditioning

Animals can learn to associate one stimulus with another. This is a behavior that is learned through **conditioning**. Conditioning is the process of teaching an animal an action in relationship to another action. An example of conditioning is rewarding a dog for coming when a clicker sound is given. The dog hears the clicking sound and has learned that when it approaches its owner, it receives a reward such as food.

Modeling

Another learned behavior is **modeling**. Modeling is when the animal learns a behavior through watching other animals conduct the behavior. For example, in a pack animal such as the wolf, hunting behaviors fit this category.

Imprinting

Finally, **imprinting** is a learned behavior due to a process that must occur within a distinct, usually short, time period. It involves an attachment to an object that will emit adult behaviors and can be generalized to all examples of the object. For example, imprinting in a newborn horse can be achieved by rubbing its entire body and lifting and touching its feet to allow it to become accustomed to being touched and handled. These training techniques, if done properly, make an animal much easier to handle and adjust to changes in the environment. Much of this learned behavior allows animals to trust people.

Distinguishing Normal from Abnormal Behaviors

Over 40% of animal owners have reported behavioral problems in their pets. These may range from elimination problems to aggression. Animals with behavior problems tend to be passed from home to home, only developing more problems in the process. Some animal behavior problems are due to improper handling or training techniques while others may be genetically related. It is important to recognize that not all "normal" animal behaviors are acceptable to owners. It is helpful to understand an animal's natural instincts, which will allow the veterinary assistant to provide suggestions for outlets for normal behaviors the pet owner and pet may enjoy. Appropriate management and regular training will help to raise an emotionally well-balanced companion whose company will be enjoyed for years. Animal behavior problems can be solved with appropriate training and guidance, which is likely to start at the veterinary clinic. Understanding normal versus abnormal animal behaviors is a great starting point for being able to train an animal for undesired habits. For example, some animals are naturally **coprophagic**, or eat their own waste materials and feces. In some animals, this is a sign of nutritional problems, where in others they need the nutrients in their feces for adequate nutrition. Table 39-1 lists normal and abnormal behaviors of a variety of animal species.

Abnormal animal behaviors may be common for one species and uncommon for another. It is important to learn to observe an animal's body language and determine if the situation provoked the behavior or if the animal became stressed and the attitude changed in relationship to the situation. When animals become too aggressive to handle, this is a safety issue and can be a liability for an animal owner. Seeking professional help with difficult-to-handle animals may be the best solution for certain animal behaviors.

Most livestock rely on their sense of smell, hearing, and eyesight. Animals can react to odors and sounds that people may not be able to detect. Odors can cause behavior reactions that may trigger defense instincts, especially in mothers with newborns. Animals have sensitive ears and sounds could potentially frighten or excite them. Many animals, such as cattle and sheep, see most objects in black and white. Other animals, such as dogs, cats, and horses, see some colors. Some animals are near-sighted and others far-sighted. Some animals, such as horses, have limited vision and can only see objects in front of them, whereas cattle have a large visionary field and can see all areas around them. These well-developed senses can cause an animal to react in a manner that is unpredictable. Many animals have a specific hierarchy in a herd or group where they are equals or have a specific role or placement within the group (see Figure 39-3).

TABLE 39-1

Normal and Abnormal Animal Behaviors

ANIMAL SPECIES	NORMAL BEHAVIORS	COMMON ABNORMAL BEHAVIORS
Dog	Barking, marking, chewing, digging, jumping	Aggression, biting, growling, food aggression, destructive behaviors, house soiling, coprophagia
Cat	Chasing, pouncing, jumping, stalking, purring, hissing, hunting, grooming, scratching	House soiling, chewing, excessive self-grooming, destructive scratching
Rabbit	Grooming, coprophagia (eating feces), marking, growling, thumping	Biting, striking, hiding, screaming
Bird	Screaming, mimicking, preening feathers, biting, beak grinding	Excessive screaming, aggression, territorial, feather picking, self-destruction, destructive chewing
Horse	Kicking, biting, flattening ears, whinnying	Bucking, rearing, charging, destructive biting, aggression, destructive chewing
Cow	Pushing, tail swishing, charging, pawing ground	Aggression, kicking, biting
Goat	Butting, jumping, climbing, chewing	Aggression, destructive chewing, biting
Sheep	Butting, bumping, herd instinct, vocal, playful	Aggression, biting
Pig	Biting, squealing, bumping	Aggression, charging

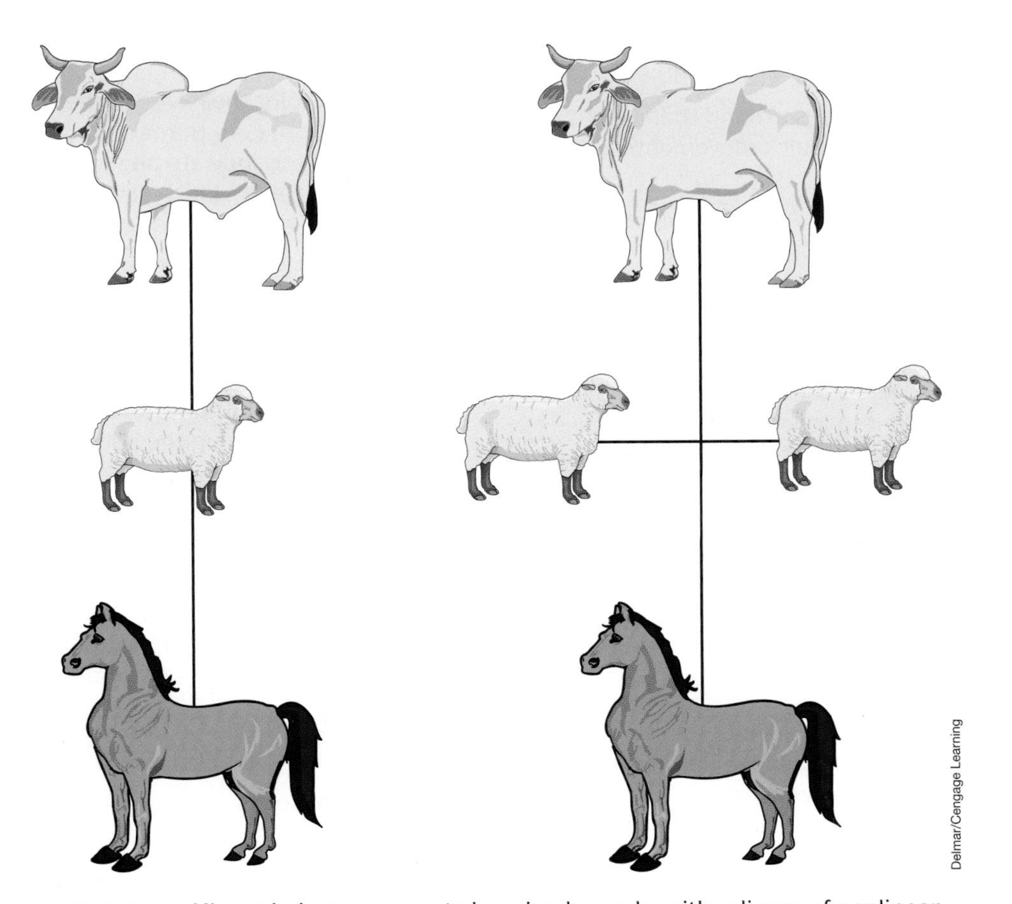

FIGURE 39-3 Hierarchal arrangements in animals can be either linear of nonlinear.

Livestock Handling and Behavior

Livestock behavior, in general, is predictable. Many livestock species have their own behavior patterns, but overall some guidelines can be used in working around livestock:

- Most livestock respond well to calm, quiet, and gentle handling.
- Livestock are easily scared or frightened with changes in environment or daily routine.
- Livestock groups have a social order; dominant animals should be fed first and in enough space for the herd.
- Many livestock, such as cattle and sheep, are herd animals and should not be isolated.
- Fast or aggressive movements may cause livestock to react dangerously.

Cattle Behavior

Some tips on cattle behavior that should be considered during handling, restraint, or general work within their pastures or stall areas include the following:

- Cattle can see 360 degrees around them and can easily be startled by quick movements (see Figure 39-4).
- Cows may seem content in a pasture but are one of the most nervous domesticated herd animals.
- Cattle are easily startled by strange noises, people, or changes in the environment.
- Always announce your presence near a cow in a quiet and calm manner; gently touch a cow when next to it rather than allowing it to touch or move into you.
- When working with a cow, allow it to settle down and become used to the environment.
- Cows tend to kick forward and to the side rather than back.
- Always leave yourself an exit when working in an enclosed area with a cow.

Horse Behavior

Some tips on equine behavior that should be considered during handling, restraint, or general work within their pastures or stall areas include the following:

- Horses have blind spots in which they can't see well, so work quietly and calmly around a horse.
- Allow a horse to know where you are at all times by moving slowly and talking gently.

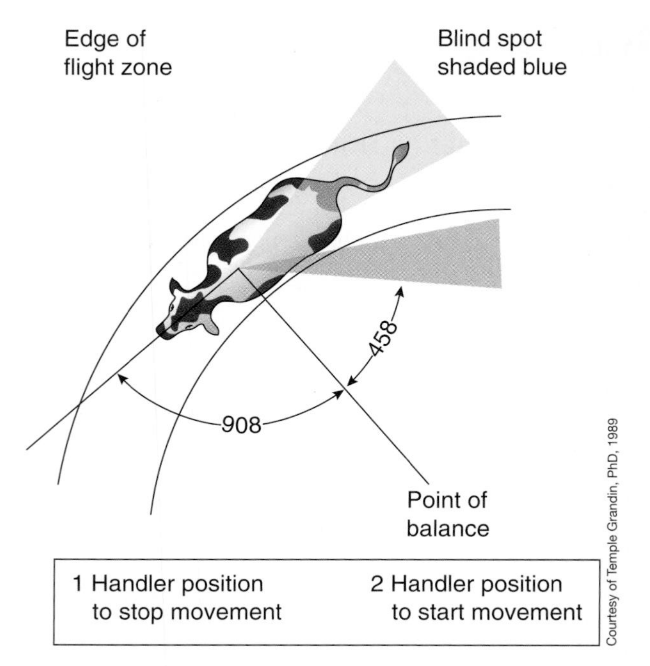

FIGURE 39-4 The flight zone of a cow. Note the presence of a blind spot and the positions where the handler can influence the animal's movement.

- Horses point their ears in the direction they are focused on. Ears that are flattened back or pinned on the head indicate anger and may mean a horse is going to kick or bite. Horses' ears present different types of messages (see Figure 39-5).
- Being nervous around a horse will make the horse nervous and difficult.
- Approach horses from their left shoulder, moving confidently and calmly, and watching the body language.
- Horses that run toward you may be chased away by raising and waving the hands and arms above your body.

Swine Behavior

Some tips on swine behavior that should be considered during handling, restraint, or general work within their pastures or stall areas include the following:

- Sows are very protective of their young.
- Moving pigs is best done with panels or chutes.
- Young piglets should be isolated from the sow when veterinary attention is necessary.
- A panel should be used to protect your lower body when moving swine.
- Pigs will bite when threatened or scared and typically go for the legs and feet.

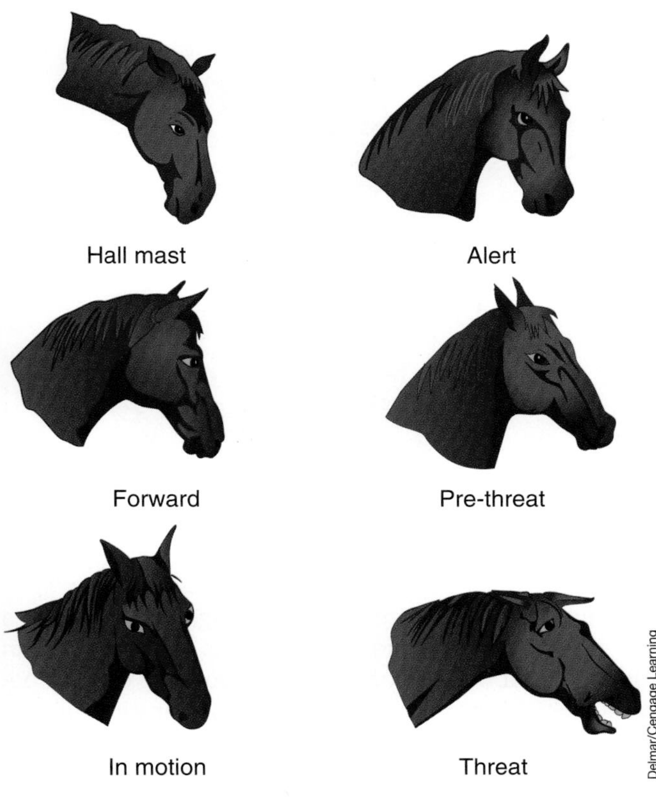

FIGURE 39-5 The ear positions of horses communicate various messages.

Sheep and Goat Behavior

Goats and sheep are animals that are flighty and panic easily. Some tips on sheep and goat behavior that should be considered during handling, restraint, or general work within their pastures or stall areas include the following:

- Sheep and goats are herd animals.
- Females are protective of their young.
- Sheep and goats will commonly butt when threatened.
- Sheep can be immobilized by sitting them on their rumps.

Common Behavior Problems in Animals

In veterinary medicine, it is important that any animal that is showing signs of a behavior problem be examined to rule out any physical abnormalities or conditions that could cause the behavior. When a veterinarian has ruled out a physical problem, then it must be determined what the behavior problem is and if it is normal or abnormal for the species. The veterinarian may have experience in animal behavior and offer some **modification techniques** used to retrain the animal and get rid of inappropriate behaviors. Some behavior problems are difficult to modify, while others may be easy handled.

Inappropriate Elimination (House Soiling)

Many indoor companion animals have **inappropriate elimination** problems, commonly called house soiling. These behaviors may be due to a physical problem, such as a urinary tract infection, and these should first be ruled out by a veterinarian. When a behavior problem is established, it should be handled by an animal behavior specialist or a veterinary professional trained in animal behavior problems. Elimination problems in dogs may be from poor housebreaking techniques. A good place to start with dogs that have house-soiling issues is to retrain them in housebreaking guidelines, similar to training a puppy. It is highly recommended that all animals in the household be neutered as this will decrease the chances of a house-soiling problem. Cats tend to have more elimination problems than dogs. Cats may eliminate or spray areas to mark territory. Many cats begin eliminating in the house due to issues with the litter box. It is important to determine if a cat is having difficulties accepting the litter box and possibly its location in the house. This is a behavior problem in which the cat chooses not to use the litter box for any one of a variety of reasons, electing to use an alternative area for elimination of urine, feces, or both. Affected cats simply avoid the litter box and select a quiet spot behind a chair or in the corner of a room instead. Some common reasons a cat may dislike the litter box are the following:

- Too few litter boxes
- Inappropriately positioned boxes (damp cellar, high-traffic area)
- Inconvenient location (basement, second level)
- Hooded box (most cats dislike hoods)
- Box too dirty (not scooped/cleaned often enough)
- Box too clean (cleaned with harsh-smelling chemicals, such as bleach)
- Liners (some cats are intimidated by plastic liners)
- Plastic underlay (convenient for the owner but not always accepted by the cat)
- Wrong type of litter (crystals or pellets)
- Litter not deep enough
- Animosity between cats in the house (competition/guarding)
- Difficulty getting into/out of the box, especially in elderly, arthritic cats

Cats with soiling problems can sometimes be easily accommodated, and other times other behavior modifications, such as behavior or anxiety medications, must be prescribed by a veterinarian. The causes of anxiety-related problems can be difficult to determine and may need additional therapy and treatment. Stress can be a large part of anxiety, such as changes in the household. Stressors such as moving, changes in routine, holidays, or changes in the makeup of the family can result in inappropriate elimination. Reducing these stressors or decreasing their impact on the household will benefit the cat.

Barking

Barking is a mode of communication that seems to be more common in dogs than other canine species. Most dogs bark, howl, and whine to some degree. Sometimes this may be the result of human encouragement. Certain breeds have been bred to bark as part of their watchdog or herding duties, such as the Rottweiler or German Shepherd. Barking is used to alert or warn others and defend a territory, to seek attention or play, to identify oneself to another dog, and as a response to boredom, excitement, being startled, lonely, anxious, or teased.

It is important that a dog understand when barking is appropriate and when it is not. This is commonly achieved by training the dog to know when to stop barking. It is important to not reward the dog's barking behavior. Excessive barking in dogs is a common behavior problem. Before barking can be corrected, it is important to determine why the dog is vocalizing in the first place. Training techniques for excessive barking may be learned through dog trainers or animal behavior specialists. It is important that the training be followed consistently.

Howling is used as a means of long-range communication in many different circumstances. Howls are more often associated with wolves, but dogs howl too. Wolves often howl to signify territorial boundaries, locate other pack members, coordinate activities such as hunting, or attract other wolves for mating. Dogs may howl as a reaction to certain stimuli such as sirens or other loud noises. Growling can occur in different activities. It is used to threaten, warn, defend, in aggression, and to show dominance. But growling is also used in play as well. The body posture should be able to tell the difference. Growls during aggression are accompanied by a stare or snarl, and the growling dog often remains stationary. Play-growls occur in combination with a happy tail and a play bow to signal willingness to play. These dogs are often moving and jumping about to entice play.

Aggression

Animal aggression can result from many factors, including sexual or mating behaviors, genetic disposition, breed disposition, and improper training or handling. An animal that becomes aggressive poses a threat to society. Some warning signs may be noted in animals that become aggressive depending on the species, such as growling, hissing, biting, or kicking (see Figure 39-6). Aggressive behaviors may be caused by the following:

FIGURE 39-6 Signs of aggression in a cat may include hissing.

- Fear—A fearful or frightened animal may bite.
- Territorial instincts—An animal may feel the need to protect property, yard, people, and other animals by vocalization or attacking.
- Possessiveness—An animal may protect its food, toys, and bed from other animals by vocalization of attacking.
- Dominance—A more dominant animal may show aggression toward an animal of lower status to maintain a hierarchy within a species.
- Pain—An animal in pain may bite.
- Mating or sexual behavior—Males intact around females in estrus may bite.

Aggression must be handled carefully depending on the size of the animal, the gender, and the type of aggression. It is important that veterinary assistants know proper handling and safety techniques when working with animal species and have some training and experience in what to do with aggressive species.

Digging

Dogs are known to cause destructive digging problems in yards (see Figure 39-7). Digging can be related to many factors that include wanting to escape the confines of the yard, trying to stay cool in the summer or warm in the winter, predator instinct of wildlife, separation

FIGURE 39-7 Digging can be a problem behavior common in some dogs.

FIGURE 39-8 Cribbing.

anxiety, boredom, curiosity by digging at unfamiliar objects in the soil, aggression at an object located outside of the yard, and displaced aggression or digging out of frustration at not being able to get to an object. Some cases of digging can be resolved by using a screen or other distracting items to block the view of objects. Placing the animal in an area with limited access to soil or ground can limit digging. Providing toys and other items for play can redirect animals from digging. Cases of separation anxiety may need medication for behavior modification.

Biting

Any animal has the potential to bite especially when scared or threatened. This is a reflex inherited by animals in the wild and becomes a defensive reaction. Some animals may show signs of biting, such as growling, hissing, or bearing their teeth, while others may show no signs at all. Many animals are known for biting severely, such as dogs, cats, birds, and horses. These animals can cause severe injuries and in cases of horses or large birds can amputate fingers or ears. It is important when animals bite to determine why. Some animals are naturally aggressive and bite out of anger. Other animals bite as a mechanism to be left alone. Yet others may bite as a reaction to pain or injury. To solve a biting problem, it is essential to determine why the animal is biting. Many biting issues can be resolved through training modification techniques.

Chewing

Large animals that chew excessively are common, such as horses and goats. The consumption and chewing of wooden fences, walls, and boards may be a nervous habit, boredom, or a nutritional deficiency. Chewing can cause damage to buildings and fences and is not good for the health of the animal, possibly resulting in broken teeth, cuts in the mouth, or colic. Some horses will chew on an area and then suck in air in a condition called **cribbing** (see Figure 39-8). This is a vice or bad habit that can easily be picked up by horses due to stress or modeling.

Euthanasia

Animals sometimes become too much to handle or too aggressive to safely be around people. One last resort to behavior problems with animals, especially those that become overly aggressive around people, is euthanasia. This process requires discussion and approval by the veterinarian and may be the only resolution in a behavioral situation. It is important to consider the animal when making this decision. This decision must be taken only as a last chance after all other options have been considered. When animals have behavior problems and are taken to shelters, auctions, or other dispersal areas, these behaviors may become worse in the stress of a new environment and change. Aggressive livestock can also be potentially life threatening in cases of anger and difficulty handling.

Animal Behaviorists

Veterinary animal behaviorists are specialized veterinarians that have a special interest and experience in animal behavior problems. Veterinary behaviorists are trained and licensed to diagnose and treat problems in animals, whether they are medical or behavioral. Being veterinarians, these behaviorists can diagnose medical problems that may be contributing to the animals' behavioral problems. A veterinary behaviorist is also licensed to prescribe drugs and is familiar with behavior modification medications, such as tranquilizers and antidepressants, and their uses and side effects. It is important to understand that many animal behavior problems are actually normal behaviors that are being performed out of context or inappropriately from an owner's perspective. Understanding what an animal is doing and the reason why is often a big relief and a crucial step in resolving the problem. Such knowledge also helps owners develop the patience and understanding necessary to implement the recommended behavior modification strategies. Veterinarians may recommend or suggest a consultation with a veterinary behavior specialist if the behavior problem is too complicated to treat or diagnose.

SUMMARY

Veterinary medicine often sees clinical cases relating to animal behavior problems. Determining what is normal versus abnormal animal behaviors in specific species is necessary to provide answers and solutions to clients. Animals may have problems that go beyond the care of veterinarians and require behavioral specialists or animal trainers that have more experience and knowledge in species and problems. It is important that animal owners understand that the beginning point of evaluating an animal for problems is a physical exam to rule out medical issues before pursuing behavioral modification treatment.

Key Terms

animal behavior manner that relates to what an animal does and why it does it

conditioning the process of teaching an animal an action in relationship to another action

imprinting process of learning through an attachment to an object that will emit adult behaviors and can be generalized to all examples of the object

inappropriate elimination commonly called house soiling; when the animal urinates or defecates in the house

instinctive behaviors behaviors that occur naturally to an animal in a reaction to a stimulus

learned behavior the response to a stimulus that creates a manner by means of watching another animal or person

learning the modification of behavior in response to specific experiences

modeling when an animal learns a behavior through watching other animals conduct the behavior

modification techniques training used to retrain the animal and get rid of inappropriate behaviors

pheromones hormone odors present in the urine giving all who encounter the mark left by others a variety of messages, including sexual receptiveness, eating habits, age, and overall health

veterinary animal behaviorist specialized veterinarians that have a special interest and experience in animal behavior problems

REVIEW QUESTIONS

1. How do animals learn behaviors?
2. What are the differences between conditioning, modeling, and imprinting?
3. What are instinctive behaviors? Give five examples of instinctive behaviors.
4. What are signs of normal animal behavior? Give three examples of species and their normal behaviors.
5. What are signs of abnormal animal behaviors?
6. What is the definition of modification technique?
7. What are veterinary animal behavior specialists?
8. What are the types of aggression in animals?

Clinical Situation

© iStock/Attila Kornlos

Diane, a veterinary assistant at Kitty Haven Veterinary Clinic, is discussing some questions that Mrs. Harris has concerning her three indoor-only cats. Mrs. Harris is a client who calls frequently asking questions, but she often refuses to make appointments to have the cats examined.

"Oh, Diane, I am having an awful time with the cats. They are going potty all over the house. I just can't stand it and the smell is terrible. What can I do to stop this?"

"Well, Mrs. Harris, we really need to see the cats for an exam and determine if there is a medical problem that may be causing this to occur. Do you know if it is one cat doing this or all of the cats?"

"I am not sure; I just know I can't stand it anymore. I will do anything to stop it."

"Then let's start by setting up an appointment to see the cats and Dr. Miller will examine each of them and go from there."

- What would you do if Mrs. Harris doesn't want to set up an exam appointment?
- What are some reasons you can give Mrs. Harris as to why this may be occurring?
- If the problem is determined to be behavioral elimination, what are some suggestions to try at home?
- What are some other treatments Dr. Miller may suggest?

40

Basic Veterinary Restraint and Handling Procedures

Objectives

Upon completion of this chapter, the reader should be able to:

- Discuss the safety concerns with proper restraint and handling
- Understand the equipment needs for animal restraint
- Understand the circumstances necessary for animal restraint
- Properly tie knots used in animal restraint
- Properly restrain small companion animals, such as dogs and cats
- Properly restrain large animals, such as cattle and horses
- Properly restrain exotic animals, such as birds and reptiles
- Properly restrain rodents, such as mice and rats

Introduction

Veterinary assistants, as well as the entire veterinary staff, should be trained in the proper safety and handling of animals for restraint procedures. This is important as many veterinary patients may not be willing to have certain procedures done and their behaviors may change, making for an unsafe environment for both the animal and the staff. Therefore, it is extremely important that the veterinary staff learn how to properly restrain all types of patients. Practice is the most important part of training the staff in animal restraint and handling techniques.

Interpreting an Animal's Body Language and Behavior

It is important for veterinary staff members to have knowledge and experience in working around the animals that are seen in the facility. Each animal species has instincts and behavior characteristics that allow it to respond to the people around it, its environment, its mental and health status, and its natural instincts. When the environment of an animal changes, an animal's behavior pattern also may change as a response. A trip to the veterinary facility is an example of a change of environment that an animal may respond positively or negatively to. Animals visiting the veterinary facility may exhibit characteristics indicating that they are happy, scared, or angry. This requires the veterinary assistant to gain knowledge and experience with the normal *body language* in the species they work with. Body language is communication by the animal about how it feels toward other animals, people, and its environment. Body language will help determine how easy or difficult an animal may be during handling and restraint.

FIGURE 40-1 A happy animal is relaxed, alert, and comfortable.

The Happy Animal

The happy animal displays normal behavior for animals. A happy animal will be relaxed, alert, stand, sit or lie with a comfortable appearance, and have ears up and forward (see Figure 40-1). Various species will have other body language that displays that they are happy. A happy animal is easier to handle than one that is scared or angry. However, the veterinary assistant should be alert to a change in behavior depending on the procedures being performed.

FIGURE 40-2 An animal that is scared will lower its body and put its ears back and close to its head.

The Scared Animal

The scared animal has the potential to be difficult to handle and become aggressive if not dealt with in the correct manner. Animals that are scared tend to have a stiff stance, shake or tremor from nervousness, avoid direct eye contact, lay their ears flat or back on the head, and lower their body or tail to the ground (see Figure 40-2). They may become **submissive**, meaning they give in to humans due to an instinct that makes them feel threatened. This may lead to fear biting or some other protective instinct that makes them injure a human out of a natural instinct to protect themselves.

The Angry Animal

The angry animal is known as aggressive. **Aggression** is a behavior that makes an animal difficult and unsafe to handle. An animal may be aggressive for several different reasons. Aggressive animals may display body language such as a stiff stance, bearing teeth, head lowered to the ground, staring, and tail raised (see Figure 40-3). Aggressive animals may fall into the category of dominance aggression, fear aggression, territorial aggression, or redirected aggression. **Dominance aggression** refers to the "pack" animal instinct and social status within a group. **Fear aggression** refers to the defense reaction to being harmed and the instinct for an animal to protect itself. Many animals are ruled by the "fight or flight" instinct, fight being protection and flight being to run away. **Territorial aggression** refers to an animal's protective nature of its environment, such as an owner, offspring, or food. **Redirected aggression** refers to an animal's predator instinct where the animal turns its aggressive behaviors on the owner. This is the most serious situation involving animal aggression. It is important, once again, to note that the veterinary staff must gain experience and knowledge in normal animal behavior and normal body language. It is very difficult to work with a species of animal without the knowledge of what is normal or abnormal when handling the animal safely.

MAKING THE CONNECTION

Refer to Chapter 39 for more information regarding animal behavior and aggression. Behavior relating to various animal species is covered in each chapter in Section III.

Photo courtesy of iStock

FIGURE 40-3 An aggressive animal will bear its teeth and may vocalize or attack.

Restraint Considerations

Restraint means to hold back, check, or suppress an action and to keep something under control using safety and some means of physical, chemical, or psychological action (see Figure 40-4). Various degrees of restraint are used by the veterinary staff to allow an animal to be controlled for procedures for safety of both the animal and people. Restraint may be of physical control, using the hands or body or a piece of equipment; of chemical control, using a sedative or tranquilizer; or of psychological control, in which the voice, eyes, or other body language of a person may help to aide in the control of an animal.

TERMINOLOGY TIP

A **sedative** or **tranquilizer** is a medication given to an animal to keep it calm during certain stressful procedures or circumstances.

Veterinary assistants should learn animal restraint for the species they work with through experience and practice. The goal of any restraint situation is to minimize the effects of the handling by properly controlling the animal safely and with the least amount of stress possible. This is sometimes referred to by the phrase, "less is more." It is important for all veterinary professionals to use good judgment, contain their temper, and be alert of the animal at all times. Working as a team is a responsibility for safety and focus (see Figure 40-5). All people involved with animal restraint should use the guidance of those that have experience.

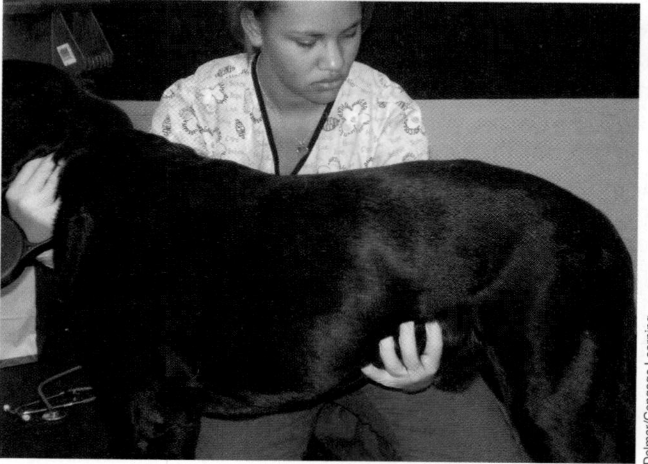

Delmar/Cengage Learning

FIGURE 40-4 The veterinary assistant is maintaining physical restraint of this dog.

Delmar/Cengage Learning

FIGURE 40-5 Teamwork is important when working with animals that require restraint.

Animal Safety

Animal safety must be considered of every patient. Many animals are accustomed to humans, but those that are not or are removed from their territory will become easily stressed. This is important in very young or very old animals. Young animals have small and brittle bones that are growing and forming and should be handled with care. Older animals may be arthritic and in pain and should also be handled with care. Rough handling may delay recovery or could lead to death. Pregnant animals must also be treated with caution during times of stress. Every time restraint of animals is necessary, the safety of the animal and the staff must be considered. Never allow non-veterinary staff or animal owners to restrain an animal; any mishap could potentially be a legal issue for the staff or veterinary facility.

Restraint Equipment

When animals become difficult and attempt to bite, scratch, or kick, it is necessary to sometimes use other means of control for safety measures. Examples of equipment that may be used in these situations include muzzles, anti-kick bars, hobbles, or stanchions. Overly aggressive behaviors must be handled depending on the species of animal, situation involved, and the behavior pattern of the animal. The hands can be used to soothe and calm an animal when they are placed in locations that are safe. The veterinary assistant should consider his or her hands as fragile instruments that can easily be injured when placed in the wrong area. Most restraint equipment is designed for use on a specific species with most being used to distract the animal from the pressure or pain of a procedure while they are being worked on. Equipment can cause injury if not properly used.

Muzzles

A muzzle is a device placed over the mouth and nose of an animal to prevent it from biting (see Figure 40-6). Muzzles are commonly made for dogs, cats, and horses. Muzzles may be made of nylon, leather, wire, or basket materials. Muzzles can also be made with gauze, tape, or leashes.

Delmar/Cengage Learning

FIGURE 40-6 Muzzles can be used to prevent an animal from being able to bite.

Competency Skill

Applying a Commercial Muzzle on Dogs and Cats

Objective:

To properly and safely apply a muzzle to prevent the animal from biting.

Preparation:

muzzle

Procedure for Dogs:

1. Hold the muzzle with the narrow side up and the wide side down. Grasp the muzzle on each side.
2. Stand to the side of the dog and place the muzzle carefully over the nose of the dog. The muzzle should fit snugly (see Figure 40-7A).
3. Place each end behind the ears and snap in place like a seat belt.
4. Pull the end snugly (see Figure 40-7B).
5. Release the clasp by pinching and pulling open.
6. Hold one end of the muzzle and allow it to slip off the nose of the dog.

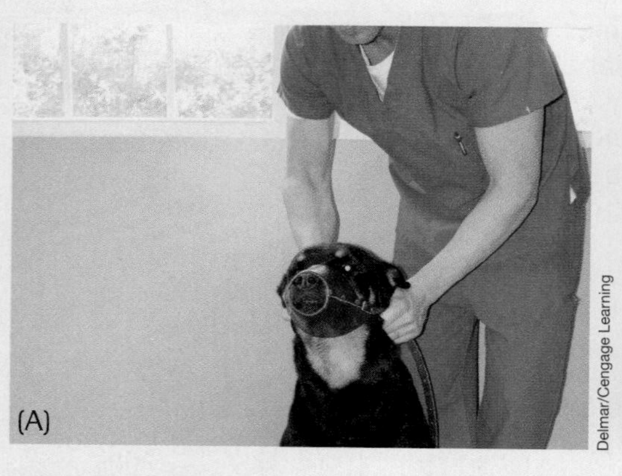

(A)

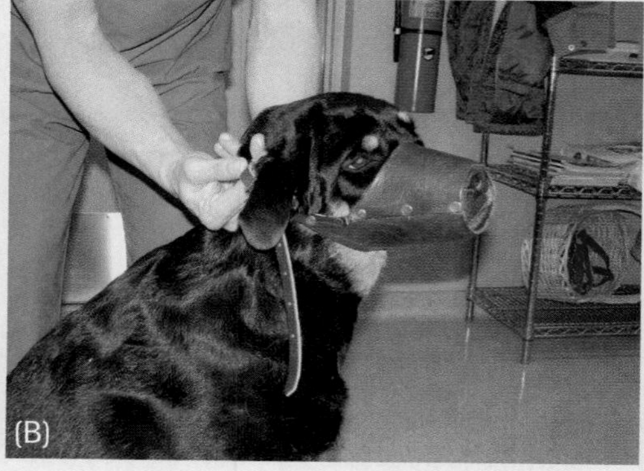

(B)

FIGURE 40-7 Placing a muzzle on a dog: (A) the muzzle is placed on the dog with the straps pulled to the back of the head; (B) the straps are clasped and pulled snugly.

Procedure for Cats:

1. Hold the muzzle with the wide side upward and the narrow side down.
2. Stand to the side or the rear of the cat and gently slip the muzzle over the face, including the eyes (see Figure 40-8A).
3. Secure the back of the muzzle behind the ears by closing the Velcro flap. Make sure the flap is pulled snugly and secure (see Figure 40-8B).
4. Restrain the limbs to prevent the cat from scratching off the muzzle.
5. Release the Velcro flap and pull the muzzle off the face.

Delmar/Cengage Learning

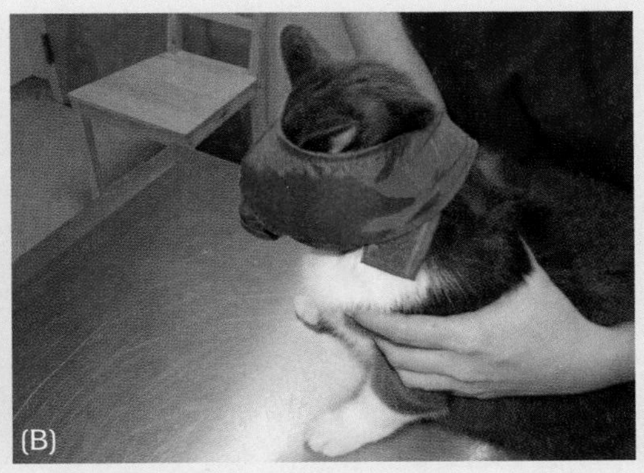

Delmar/Cengage Learning

FIGURE 40-8 Applying a muzzle to a cat: (A) cover the cat's face with the muzzle pulling the straps to the back of the head; (B) secure the straps.

Competency Skill

Applying a Gauze Muzzle

Objective:

To properly and safely apply a muzzle to prevent the animal from biting.

Preparation:

gauze

(Continues)

Procedure:

1. Make a loop using a piece of heavy gauze long enough to fit the dog (see Figure 40-9A).
2. Slip the loop over the dog's nose by standing to the side of the dog. Pull the loop tightly (see Figure 40-9B).
3. Take each side under the jaw and tie in a square knot under the chin (see Figure 40-9C).
4. Take each side behind each ear and tie in a knot with a bow behind the ears (see Figure 40-9D).
5. Untie the bow and slip the muzzle off the nose of the dog.

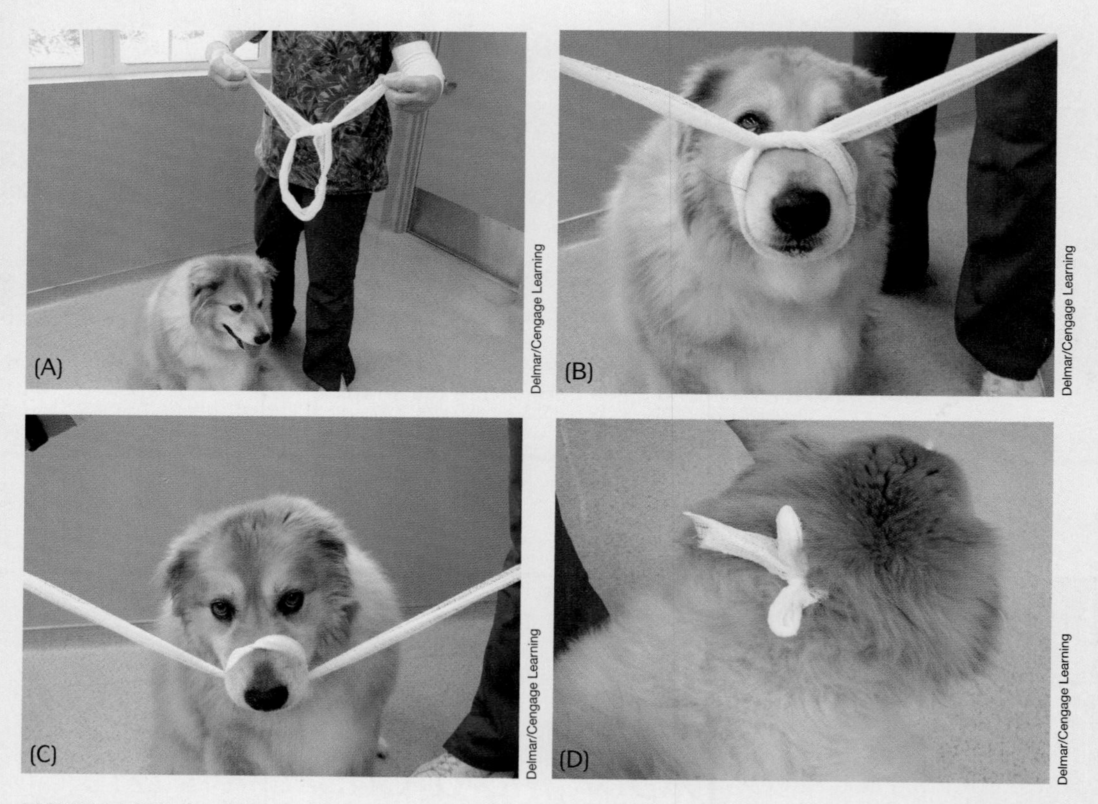

FIGURE 40-9 Applying a gauze muzzle: (A) make a loop large enough to fit over the dog's muzzle; (B) Tighten the gauze on the muzzle; (C) tie with a square knot under the muzzle; (D) tie with a bow behind the dog's head.

Competency Skill

Applying a Tape Muzzle

Objective:

To properly and safely apply a muzzle to prevent the animal from biting.

Preparation:

tape

Procedure:

1. Make a muzzle using a piece of 1-inch tape that is long enough to fit the dog.
2. Fold the entire piece of tape onto itself to make a long piece of nonstick tape.
3. Make a large loop with the tape.
4. Slip the loop over the dog's nose by standing to the side of the dog. Pull the loop tightly.
5. Take each side under the jaw and tie in a square knot under the chin.
6. Take each side behind each ear and tie in a knot with a bow behind the ears.
7. Untie the bow and slip the tape muzzle off the nose of the dog.

Competency Skill

Applying a Leash Muzzle

Objective:

To properly and safely apply a muzzle to prevent the animal from biting.

Preparation:

leash

Procedure:

1. Make a loop using a commercial nylon leash long enough to fit the dog (see Figure 40-10A).
2. Slip the loop over the dog's nose by standing to the side of the dog. Pull the loop tightly (see Figure 40-10B).

(Continues)

3. Take each side under the jaw and tie in a square knot under the chin.
4. Take each side behind each ear and tie in a knot with a bow behind the ears (see Figure 40-10C and D).
5. Untie the bow and slip the leash muzzle off the nose of the dog.

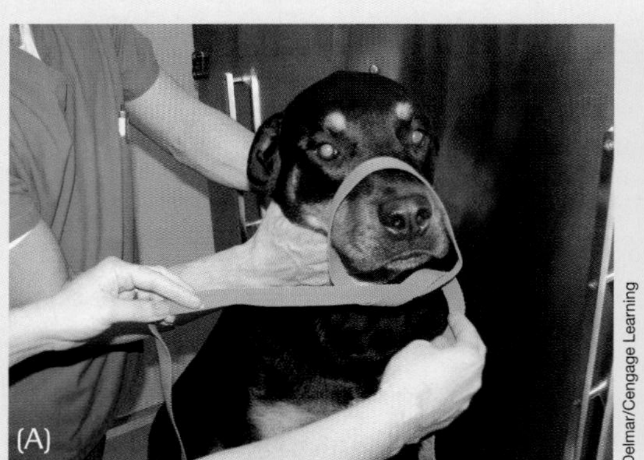

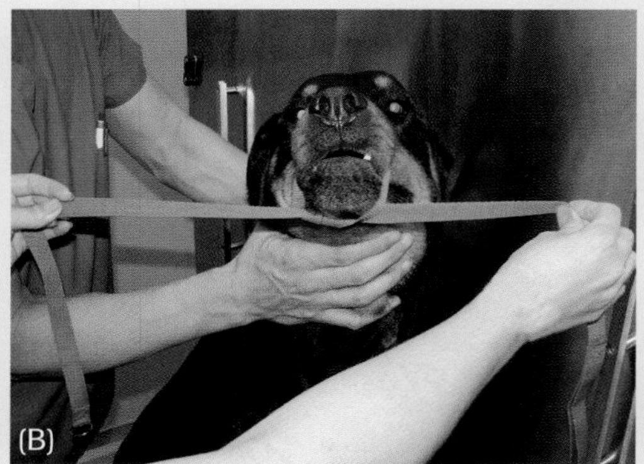

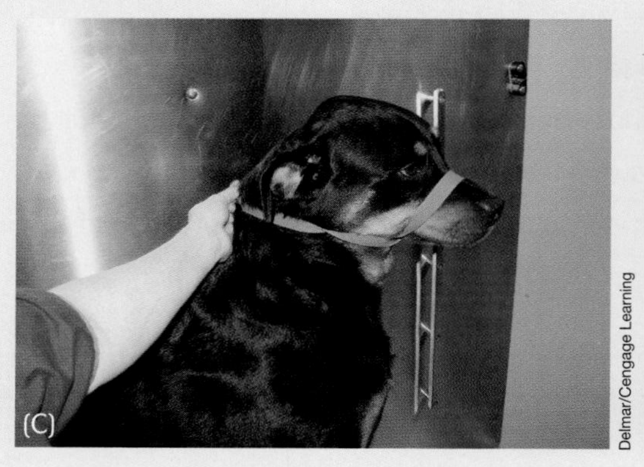

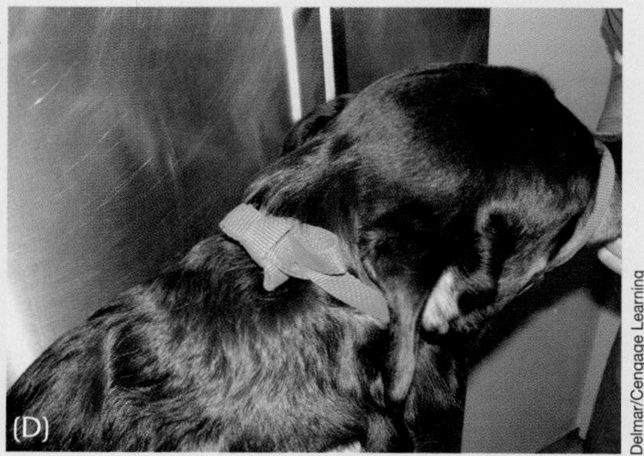

FIGURE 40-10 Applying a leash muzzle to a dog: (A) make a loop large enough to go around the dog's face; (B) tighten the loop around the dog's face; (C, D) tie the leash behind the dog's head.

Towels

Towels can be used to restrain small animals by keeping them wrapped and contained within the material (see Figure 40-11). Many animals are more comfortable wrapped and secure, especially when they feel hidden.

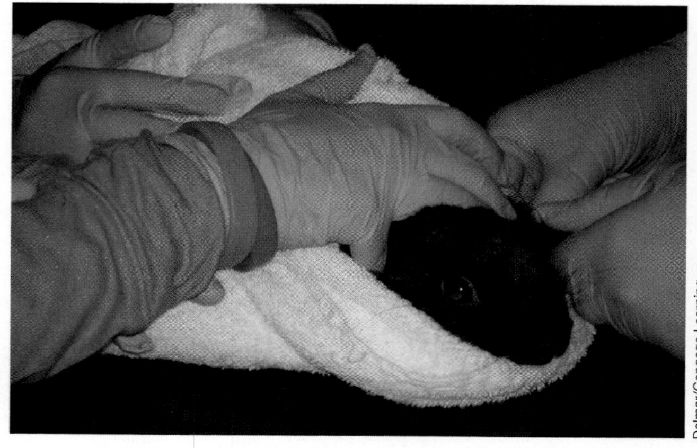

FIGURE 40-11 A towel can be used to restrain some animals and make them feel more secure.

Squeeze Cages

Squeeze cages may also be used with small animals and help to contain an animal without placing a person's hands directly on the animal but in a cage with open areas used to sedate an animal or give vaccines.

Competency Skill

Use of a Towel to Restrain a Cat

Objective:

To successfully use a towel to calm and restrain a cat.

Preparation:

heavy towel

Procedure:

1. Gently apply a towel over the cat's body and head.
2. Reach under the towel and scruff the cat.
3. Pull the cat's body toward your body to keep secure.
4. Keep the towel over the cat's head to keep the cat comfortable.

Competency Skill

Use of a Squeeze Cage

Objective:

To safely and properly restrain an animal using a squeeze cage.

Preparation:

squeeze cage

Procedure:

1. Open the top of the squeeze cage.
2. Place the cat inside the cage and close the lid.
3. Use the side flaps to pull the cat forward against the side of the cage.
4. Release the sides of the cage.
5. Open the lid and remove the cat from the cage.

Leather Welding Gloves

Leather welding gloves may also serve as protection for the hands when handling certain species (see Figure 40-12).

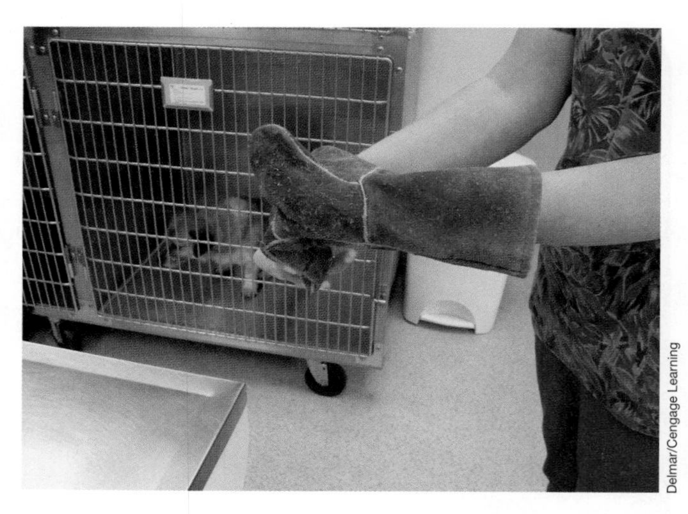

FIGURE 40-12 Leather welding gloves can be worn to protect the hands when working with some animals.

Anti-Kick Bars

Anti-kick bars are used to immobilize cattle when giving injections or medications.

Hobbles

A hobble can limit the movement of an animal by tethering its legs. They are made of various materials such as leather, nylon, or rope. They look similar to a set of handcuffs that are placed on the animal's legs.

Competency Skill

Use of Welding Gloves

Objective:

To properly use welding gloves to restrain animals for procedures.

Preparation:

welding gloves

Procedure:

1. Apply the welding gloves to each hand.
2. Grasp the animal firmly over the neck with one hand and around the body with the free hand. The scruff technique may be used in cats.
3. Use caution to avoid injuring the animal with too harsh of a grip. Use caution to not let the animal get away.
4. A towel can also be used to keep the animal covered and quiet.

Competency Skill

Use of the Anti-Kick Bar on a Cow

Objective:

To properly and safely restrain a cow.

Preparation:

kick bar

Procedure:

1. Locate the metal kick bar.
2. Stand behind the cow or at the front and side of the cow.
3. Avoid standing to the side of the rear legs.
4. Position the bar around the cow's flanks.
5. Tighten down over the flanks.

Competency Skill

Placement of Hobbles on Horse

Objective:

To safely restrain a horse and keep it from wandering too far too quickly.

Preparation:

hobbles

Procedure:

1. Locate the hobbles.
2. Stand to the side of the horse using caution to not be kicked.
3. Place the left hobble around the lower rear leg and buckle in place.
4. Move to the right side of the horse and place the right hobble around the lower rear leg and buckle in place.

Stanchions or Head Gates

Head gates and stanchions are structures that are used to contain large animals so that procedures can be performed safely. A head gate is a large, crate-like structure that an animal, such as a cow, can be loaded into. At the front of the structure the device has bars that lock around the animal's head, securing it. The sides can then be adjusted on some models to prevent the animal from moving around in the structure. On some, side panels can be removed to allow the handler or veterinarian access to the animal to administer procedures such as injections.

Competency Skill

Use of a Stanchion or Head Gate

Objective:

To safely restrain a cow for various medical treatments or procedures.

Preparation:

head gate or stanchion

Procedure:

1. Open the head gate or stanchion chute.
2. Push the cow forward into the chute.
3. As the cow reaches the front of the chute, allow the head to enter the gate.
4. Close the gate at the base of the neck at the shoulders (see Figure 40-13).
5. If the chute has side walls, they should be squeezed in toward the cow to prevent movement.

FIGURE 40-13 Use of a head gate.

Planning the Restraint Procedure

The best place for restraint is an area with enough room for the animal and restrainers and an area that is clean, dry, and well lit. This may be an exam room or a barn stall. Several considerations should be made when planning restraint areas. It is best to discuss a plan of restraint with everyone involved in handling the animal. This communication allows everyone to work safely together. Move any costly equipment so as not to damage items. Allow for a nonslip area for workers and animals. Temperature should be considered when working outside with animals. Some species may become hyperthermic easily when handled roughly or stressed in warm weather. This must be taken into consideration in warm areas and with certain procedures. Early morning may be the best time when heat is a factor in handling animals. Always allow for a shaded area, use fans, and provide water when heat is a factor. In cold weather, hypothermia may become a factor. Anesthetized animals have difficulty maintaining their body temperature. Doors and windows should be monitored when working with unruly animals. Always plan what should be done if an animal happens to get away from a restrainer. Areas to hold and contain livestock should be inspected for safety before use. Regardless of how well a restraint is planned, unpredictable circumstances may cause problems. It is best to always anticipate what may happen and have a backup plan if something should occur. This allows for safe planning and awareness. Do not become stressed in events when animals are difficult as animals can sense these emotions and may become more difficult to control.

Restraint Knots

Knots are made from one to two pieces of rope material where one section of the rope prevents slipping of another. Knots allow animals to be tied and restrained for a temporary amount of time. Ropes should be inspected for tears and the ability to hold the animal. Ropes may be used to tie animals in a standing position or in a lying position. It is important that the knots that are used to tie an animal are properly placed and tied.

Square Knot

A **square knot** is commonly used to secure an animal and is a nonslip knot that doesn't come untied or easily loosen. The saying "right over left, left over right" is the common method for making a square knot. Two ropes may be used to make a square knot and when complete it looks like two intertwined loops and is easily untied when the opposite ends are pushed together. A single rope can be used to make the square knot by passing one shorter end behind the opposite longer end and folding a loop and pushing it through the loop created by the ends of the rope (see Figure 40-14).

Delmar/Cengage Learning

Delmar/Cengage Learning

FIGURE 40-14 Square knot: (A) loop in the short end held over the long end or the rope attached to the animal's halter; (B) Place the fold of rope into the loop.

Competency Skill

How to Tie a Square Knot

Objective:

To tie a knot that will not slip or easily loosen.

Preparation:

two pieces of rope

Procedure:

1. Take one piece of rope and apply the right over the left.
2. Then apply the left over the right.
3. The end product will appear intertwined.
4. Untie by pushing the opposite ends together.

Reefer's Knot

A **reefer's knot** is a single bow knot that allows a non-slip, quick-release tie. It is the same as the square knot with the exception that the second throw is made upon itself, creating a hold that can easily be untied by pulling the end of the rope (see Figure 40-15). This is a common tie for large animals to prevent them from injuring their heads and necks during restraint.

FIGURE 40-15 Reefer's knot: (A) Form a loop with the short end crossing over the long end. (*Continues*)

FIGURE 40-15 Reefer's knot: (B) wrap the short end of the rope around the long end; (C) pass the short end through the loop and pull in the direction indicated by the arrow. *(Continued)*

Competency Skill

How to Tie a Slip Knot or Quick-Release Knot

Objective:

To tie a knot that can be quickly and easily removed.

Preparation:

rope, tie area

Procedure:

1. Place the rope over a pole or tie area.
2. Allow about 2 feet of rope for the animal to move slightly.
3. Make a loop in the rope close to the pole or tie area.
4. Pass the loop under the lead rope near the pole and pass it through the loop. This should look like a pretzel.
5. Pull the loop to tighten the knot.
6. Untie by pulling the end of the lead rope for a quick release.

Competency Skill

How to Tie a Reefer's Knot

Objective:

To tie a knot that can be quickly and easily removed.

Preparation:

rope, tie area

Procedure:

1. Place the rope around a pole or tie area.
2. Make a loop in the end of the long end so the short end overlaps the long end.
3. Pass the short end of the rope up through the loop.
4. Reach under the long end of rope and grasp the short end so that it wraps around the long end.
5. Pass the short end of rope back through the loop in the opposite direction of the first pass.
6. Pull securely.

Half Hitch

A **half hitch** is a tie that makes a loop around a stationary location, such as a post or fence. One pass is made around a location with the next pass going through the resulting loop and pulling up on the pinch ends of the hitch between the standing part of the rope and the item it is tied to (see Figure 40-16). The half hitch is commonly used to secure an animal to a surgery table.

Restraint Positions

Common restraint techniques in small animals include standing restraint, sitting restraint, sternal recumbency, lateral recumbency, dorsal recumbency, and restraint for blood collection. The term **recumbency** means a lying position. **Standing restraint** is used to keep an animal standing for a procedure and preventing it from sitting or lying down. **Sitting restraint** is used to keep an animal in a sitting position for ease of completing a procedure. **Sternal recumbency** is placing the animal on its chest for restraint. **Lateral recumbency** is placing the animal on its side for restraint and may be done in left or right lateral recumbency. Left lateral recumbency is the left side down and right is the right side down in contact with a table or the floor. **Dorsal recumbency** is used to place the animal on its back for restraint and is a common restraint during surgical procedures and radiology techniques.

FIGURE 40-16 Half hitch: (A) pull in direction shown by arrow; (B) second half knot.

Competency Skill

How to Tie a Half Hitch

Objective:

To tie a knot that will securely restrain an animal to a particular site.

Preparation:

rope, pole or tie area

Procedure:

1. Place the rope over a pole or tie area.
2. Pass the short end of rope under the long end and then back over the top, thus creating a closed loop around a tie area.
3. Push the rope down between the toe area and the loop that was formed.
4. Pull the loop tight.
5. Pass the short end of rope over and under the end forming a loop.
6. Pass the short end up through the loop and pull it tight.

Standing Restraint of Dogs and Cats

Objective:

To safely restrain a dog while the dog is in a standing position.

Preparation:

No special equipment is required. Small dogs may require only one restrainer; medium to large dogs will require two or more restrainers.

Procedure for a Dog:

1. Dog should be grasped in a bear hug around the head and neck with one arm.
2. The free arm should be placed under the dog's stomach and the dog raised to a standing position (see Figure 40-17).
3. Do not allow the dog to sit down.

Procedure for a Cat:

1. Scruff the cat with the dominant hand and support the abdomen with the other hand (see Figure 40-18).
2. Keep the cat's body close to your body for better control.
3. Monitor the cat for changes in body language and divert the cat's attention as necessary.

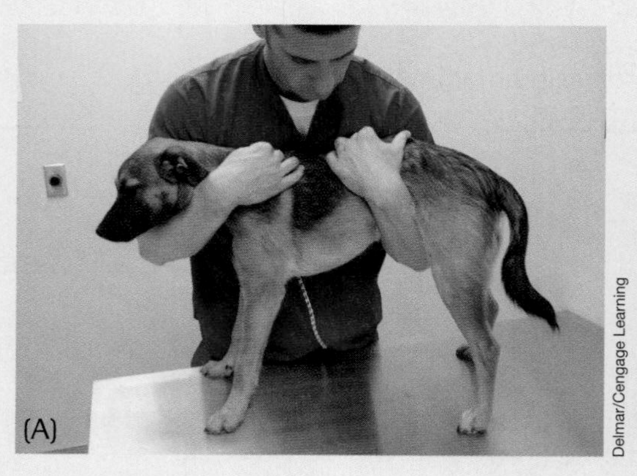

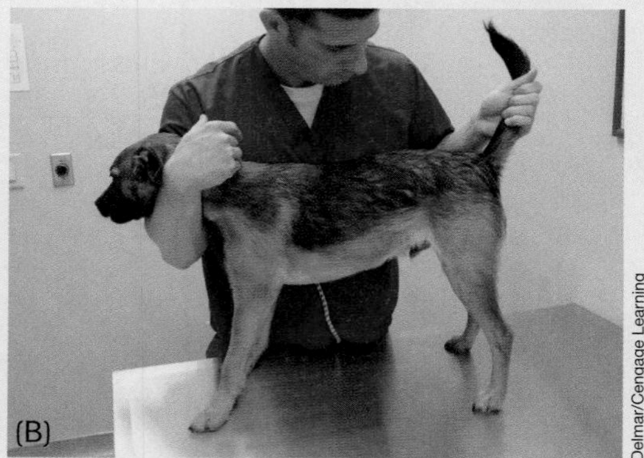

FIGURE 40-17 (A) Standing restraint; (B) Standing restraint holding the tail up.

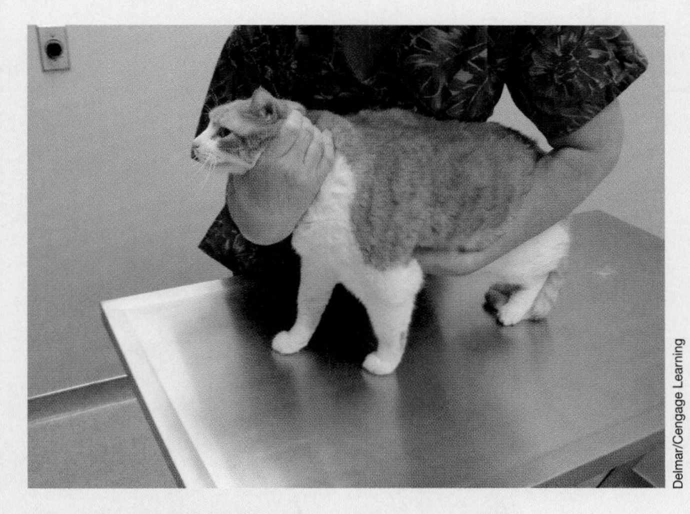

FIGURE 40-18 Standing restraint of the cat.

Competency Skill

Sitting Restraint

Objective:

To safely restrain a dog in the sitting position.

Preparation:

No special equipment is required. More than one person may be necessary to properly restrain a larger animal.

Procedure:

1. Dog should be grasped in a bear hug around the head and neck with one arm.
2. The free arm should pull the dog's body close to the restrainer and gently place pressure over the back to place the dog in a sitting position (see Figure 40-19).
3. Small dogs may require only one restrainer; medium to large dogs will require two or more restrainers.

(*Continues*)

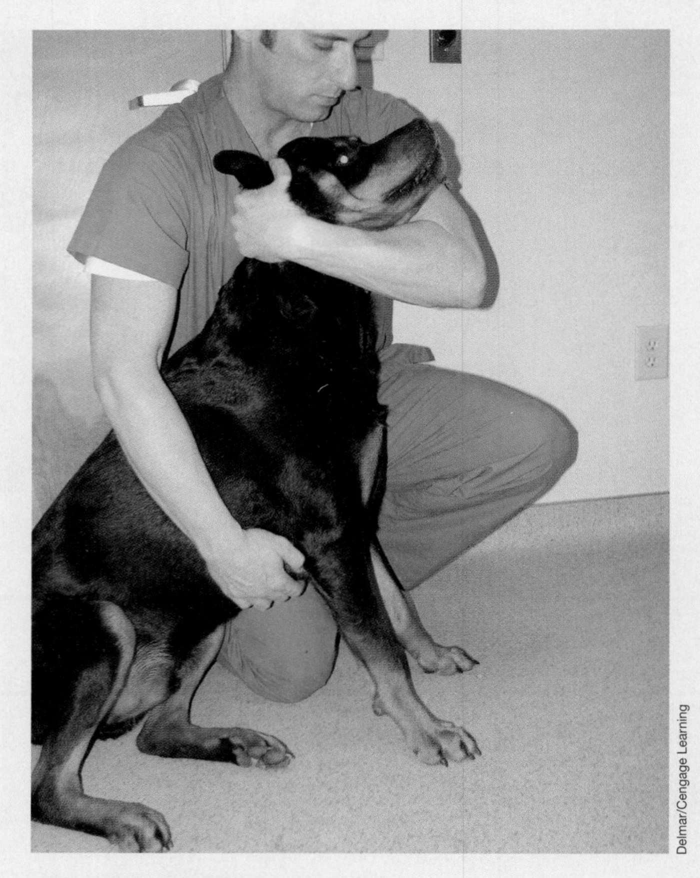

FIGURE 40-19 Sitting restraint.

Competency Skill

Sternal Recumbency of Dogs and Cats

Objective:

To properly and safely restrain a dog or cat for procedures.

Preparation:

No special equipment is necessary. Small dogs may require only one restrainer; medium to large dogs will require two or more restrainers.

Procedure for Dogs:

1. The dog should be in the standing position.
2. Apply pressure over the hips, forcing the dog to sit.
3. Slide one hand under the front legs while applying pressure over the shoulders with the free hand.
4. Maintain pressure over the back and shoulders until the dog is lying on its chest.
5. Restrain the head in a bear hug with one arm.
6. Use the free arm to apply continued pressure over the back to prevent standing (see Figure 40-20).

FIGURE 40-20 Sternal recumbency.

Procedure for Cats:

1. The cat should be in standing restraint with the scruff method applied with the dominant hand.
2. Gently place the free hand over the back near the hips.
3. Reach around the cat's body with the free hand and gently use the elbow to place the cat on its chest and restrain both front and rear limbs to keep the cat from scratching (see Figure 40-21).
4. A towel can be helpful in aggressive or scared cats.

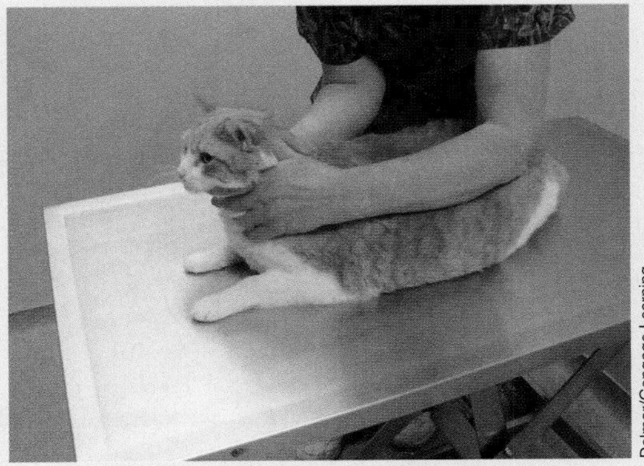

FIGURE 40-21 Sternal recumbency of the cat.

Competency Skill

Dorsal Recumbency

Objective:

To properly and safely restrain a dog or cat for procedures.

Preparation:

No special equipment is necessary. Small dogs may require only one restrainer; medium to large dogs will require two or more restrainers.

1. The dog or cat should be placed in lateral recumbency as outlined.
2. The dog or cat is gently rolled onto its back with the front legs extended forward and the back legs extended backward (see Figure 40-22).
3. The head should be kept between the front legs and held securely.
4. The rear legs should be kept still to avoid injury to the back.

Cats may be restrained with one or two people depending on the attitude of the cat. The wrist and forearm can be used to hold down the head, but caution must be used to monitor the cat's mouth and avoid being bitten!

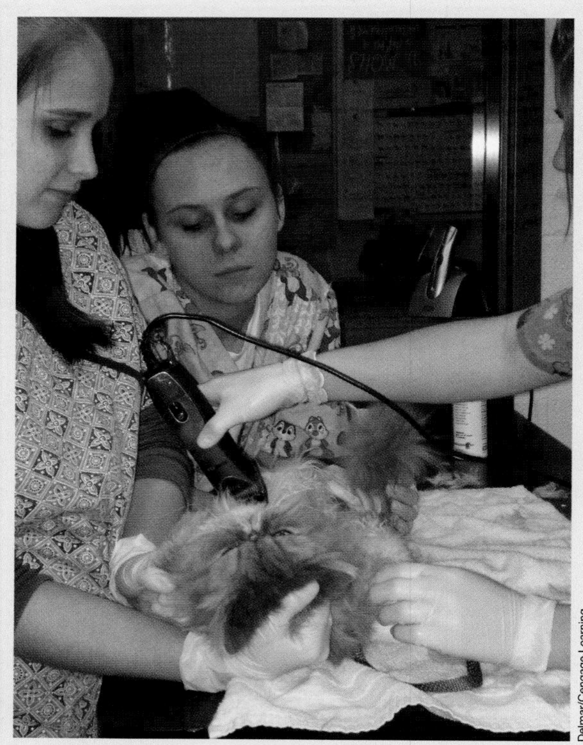

FIGURE 40-22 Dorsal recumbency.

Lateral Recumbency

Objective:

To properly and safely restrain a dog or cat for procedures.

Preparation:

No special equipment is necessary. Small dogs may require only one restrainer; medium to large dogs will require two or more restrainers.

1. Determine which side of the dog must be placed down on the floor or table.
2. Stand on the side of the animal that is to be the down side.
3. Reach both arms over the back and chest of the dog and firmly grasp the front and rear paws (see Figure 40-23A).
4. With a sudden jerk, flip the dog onto its side. Caution should be used to avoid having the dog's head hit any objects or the floor.
5. The front and rear limbs that are on the down side should be firmly grasped to prevent the dog from standing.
6. The upper body of the restrainer should be used to apply pressure over the back and chest of the dog to prevent standing.
7. The elbow should apply pressure over the neck of the dog to avoid biting (see Figure 40-23B).

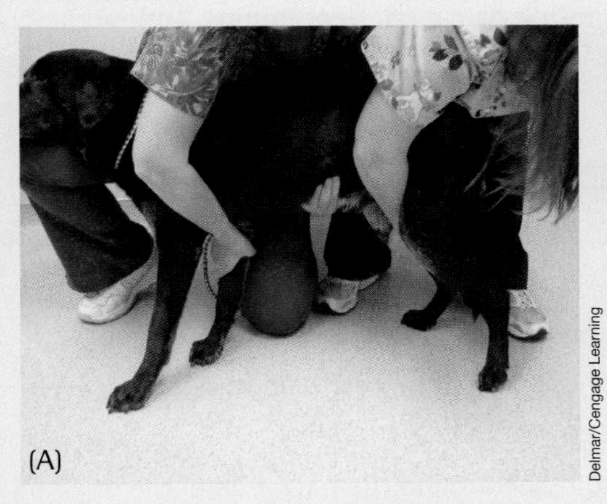

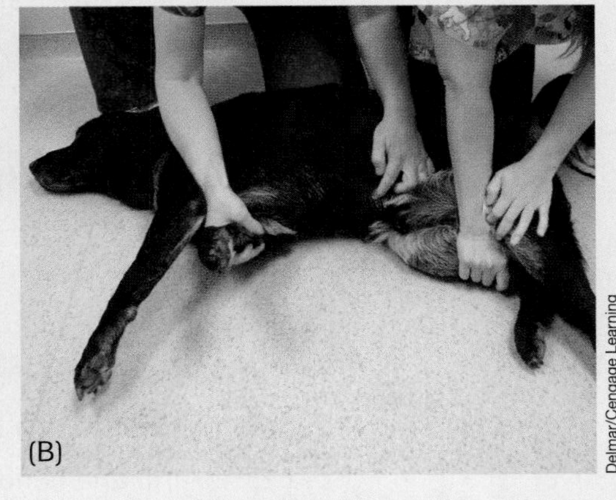

FIGURE 40-23 Placing a dog in lateral recumbency: (A) place the arms over the dog's back and grasp the legs that will be on the floor side; (B) gently lift and lower the dog to the floor and use pressure of your body to maintain control to prevent the dog from standing.

MAKING THE CONNECTION

Many terms refer to directional locations on an animal's body and are useful in restraint procedures. Review Chapter 1 for discussion of anatomical and directional terminology.

Restraint for Blood Collection

Blood collection is commonly performed in small animals in one of three locations: the cephalic vein, the jugular vein, or the saphenous vein. The **cephalic vein** is located in the medial aspect of the front limbs and is called cephalic **venipuncture**. The restraint for blood collection in the cephalic area is typically done in sitting or sternal recumbency. The **jugular vein** is located on either side of the neck in the lower throat area. The restraint for jugular venipuncture is commonly sitting or sternal recumbency. The **saphenous vein** is located on the lateral surface of the rear limbs just proximal to the hock. The restraint commonly used for saphenous venipuncture is lateral recumbency.

Competency Skill

Cephalic Venipuncture Restraint for Dogs and Cats

Objective:

To properly and safely restrain a dog or cat while obtaining a blood sample from the cephalic vein.

Preparation:

No special equipment is necessary.

Procedure for Dogs:

1. Place the dog in sitting recumbency.
2. Stand on the opposite side of the limb to be restrained.
3. Restrain the head using a bear hug.
4. Use the free arm and hand to reach around the back side of the dog and hold the leg at the elbow.
5. Extend the elbow forward.
6. Apply pressure around the elbow by making a fist and allowing the vein to pop up over the front limb.
7. Keep the leg extended throughout the procedure (see Figure 40-24).

8. When the venipuncture is complete, gently apply pressure to the puncture site until bleeding stops or a bandage is applied.
9. If blood is on the fur, clean with hydrogen peroxide and rinse fur with water and dry.

Procedure for Cats:

1. Place one hand around the head and under the jaw to prevent biting. Use the free hand to extend the elbow of the leg that is being used for blood collection.
2. Grasp the elbow and apply pressure to hold off the cephalic vein at the top of the leg (see Figure 40-25).
3. Place a thumb over the puncture site after blood collection to stop bleeding or until a bandage is applied.
4. Clean the fur with hydrogen peroxide and wash the area with water.

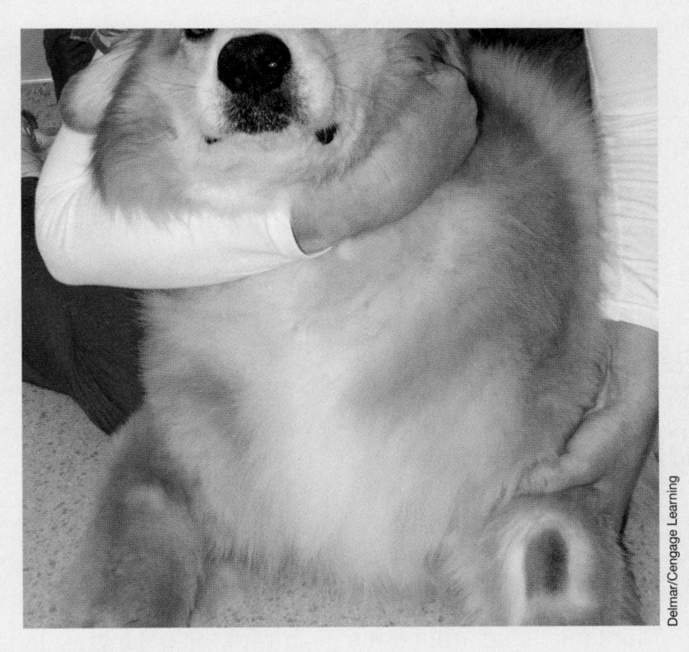

FIGURE 40-24 Cephalic venipuncture in a dog.

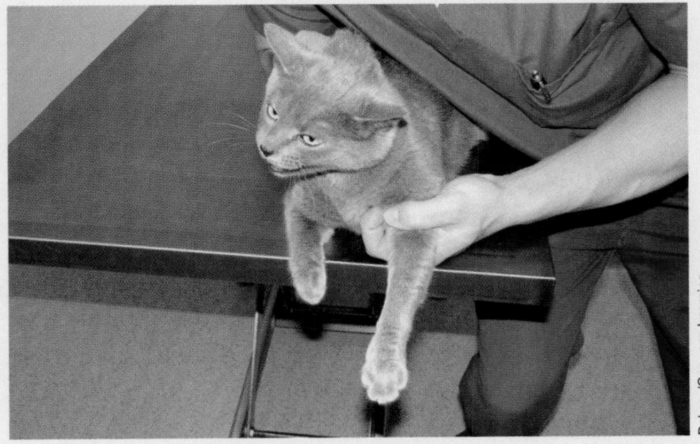

FIGURE 40-25 Cephalic venipuncture in a cat.

Competency Skill

Jugular Venipuncture Restraint for Dogs and Cats

Objective:

To properly and safely restrain a dog or cat while obtaining a blood sample from the jugular vein.

Preparation:

No special equipment is necessary.

Procedure for Dogs:

1. A large dog should be placed on the floor in sitting or sternal recumbency. A small dog should be placed on a table with the front legs extended over the edge.
2. Collars should be removed to prevent injury.
3. The head is extended upward and away from the person collecting blood. The muzzle is held closed to prevent biting (see Figure 40-26).
4. The free arm is held over the body to prevent moving.
5. Once the needle is removed from the jugular vein, the restrainer should place a finger over the collection site to prevent bleeding.
6. Clean the fur with hydrogen peroxide and rinse fur with water and dry.

Procedure for Cats:

1. Place one hand around the head with the palm grasping the lower jaw to prevent biting.
2. With the free hand hold the front limbs placing the fingers between each paw and holding firmly to prevent scratching (see Figure 40-27).
3. Extend the front limbs over the edge of a table.
4. Extend the head upward securely.
5. Place a finger over the puncture site after blood collection to stop bleeding.
6. Clean the fur with hydrogen peroxide and wash the area with water.

FIGURE 40-26 Jugular venipuncture in a dog.

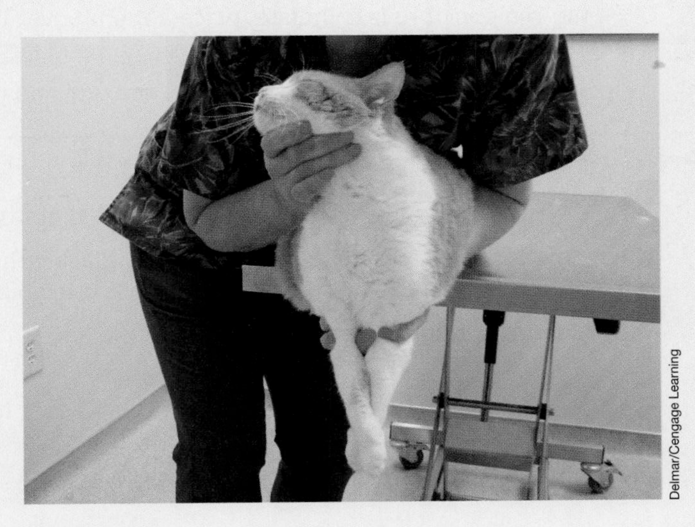

FIGURE 40-27 Jugular venipuncture in a cat.

Competency Skill

Saphenous Venipuncture Restraint in Dogs and Cats

Objective:

To properly and safely restrain a dog or cat while obtaining a blood sample from the saphenous vein.

Preparation:

No special equipment is necessary.

Procedure for Dogs:

1. Place the dog in lateral recumbency according to the limb being used. The hind limb being used is the one that should be on top.
2. Restrain the head in a bear hug.
3. Modify the restraint for lateral recumbency by placing the hand restraining the hind limbs around the top of the leg below the knee or thigh area.
4. Maintain pressure over the lateral surface of the limb so that the saphenous vein will pop up (see Figure 40-28).
5. Once the venipuncture is complete, place a thumb over the blood collection site and maintain pressure until the bleeding stops or a bandage is placed.
6. If blood is on the fur, clean with hydrogen peroxide and rinse fur with water and dry.

Procedure for a Cat:

1. Place the cat in lateral recumbency using the scruff and stretch technique.
2. Modify the restraint for lateral recumbency by placing the hand restraining the hind limbs around the top of the leg below the knee or thigh area (see Figure 40-29).
3. Maintain pressure over the lateral surface of the limb so that the saphenous vein will pop up.
4. Once the venipuncture is complete, place a thumb over the blood collection site and maintain pressure until the bleeding stops or a bandage is placed.
5. If blood is on the fur, clean with hydrogen peroxide and rinse fur with water and dry.

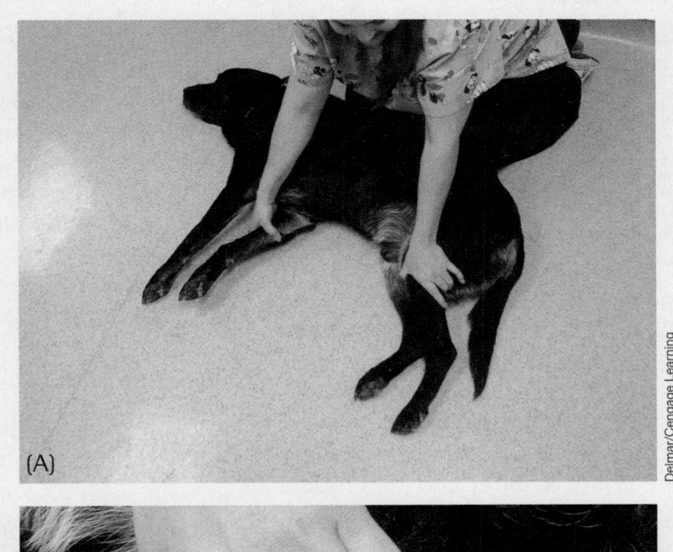

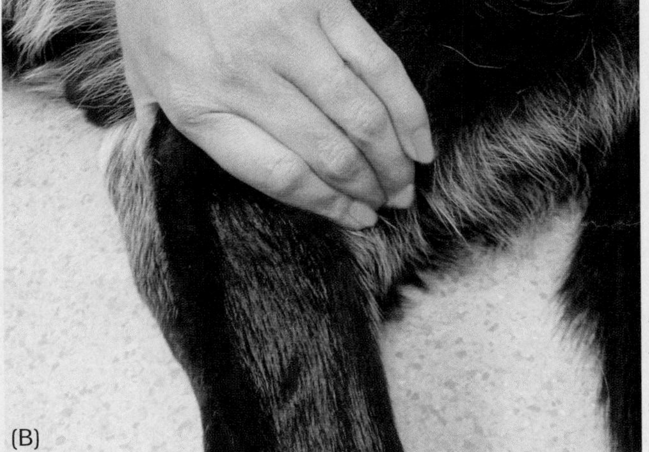

FIGURE 40-28 A and B Saphenous venipuncture in a dog.

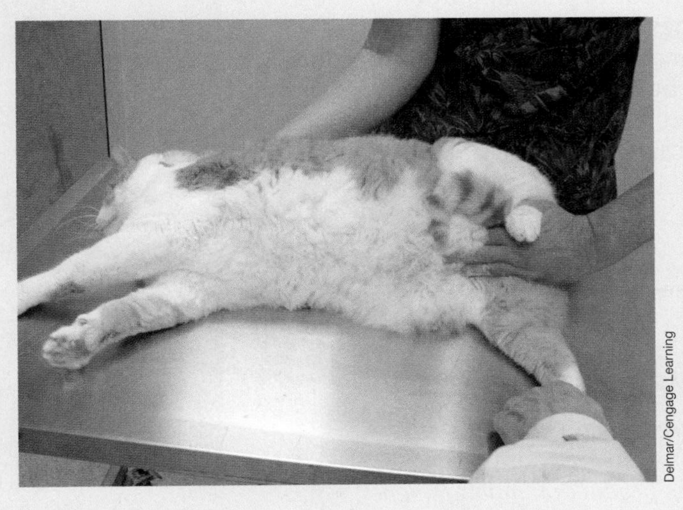

FIGURE 40-29 Saphenous venipuncture in a cat.

Restraint Procedures and Techniques

There are various methods of restraint for large and small animals. Many of these procedures include using equipment that is meant to protect both the animal and the restrainer and veterinary staff. Basic restraint begins with removing an animal from its cage, stall, or pen. This is done with a type of head control device specific to the animal species. The control device should be properly fitted and be placed over the animal's head or neck area according to the animal husbandry and production system methods discussed in previous chapters. For example, dogs are commonly controlled with the use of a leash and collar, whereas horses are commonly controlled with the use of a halter and lead rope. Many basic components are used for handling animals for physical exams.

Small Animal Restraint

Small animals typically are restrained for exams and veterinary procedures in which the animal is held in a safe manner through body control. Injuries that commonly occur to veterinary staff during small animal restraint include bites and scratches. Small animals include cats, dogs, rodents, rabbits, ferrets, reptiles, and birds. When working with small animals, it is helpful to talk and reassure the animal and use several techniques that may divert its attention from procedures that are being done. Such diversions that may be helpful include talking to the animal, making soothing and calm noises, lightly blowing in the face, lightly rubbing the temple area, or a combination of these techniques. Sometimes it is necessary to be creative in distracting an animal's attention from a procedure that is being done to be able to complete the restraint safely. Monitoring and learning animal body language is helpful in safe restraint.

> **MAKING THE CONNECTION**
>
> *Species specific restraint techniques have been covered in previous chapters on individual species production management practices. Please see chapters in Section III.*

Cats

Cats tend to be one of the most difficult domesticated animals to control during restraint when they become upset and aggressive from stress. Cats are able to bite and scratch and due to increased bacteria on the teeth and nails can cause severe skin infections quickly and easily. Caution should always be taken if a scratch or bite ensues and immediate care of the wound should be completed. It is recommended that any cat wounds that appear infected be seen by a physician.

One of the most important parts of restraining cats is to properly and safely restrain and have control over the head. Towels placed over a stressed cat may allow it to feel less threatened as cats feel more secure when they are hidden under an object (see Figure 40-30). Cats that become aggressive may be muzzled with special cat muzzles that cover the mouth and eyes. **Cat bags** are also available to control the limbs and allow better control over the head, which is the only body part exposed when a cat is placed in the cat bag (see Figure 40-31). Cats may also be placed in squeeze cages or anesthesia chambers for better control and less need for hands-on restraint. **Squeeze cages** are wire boxes with small

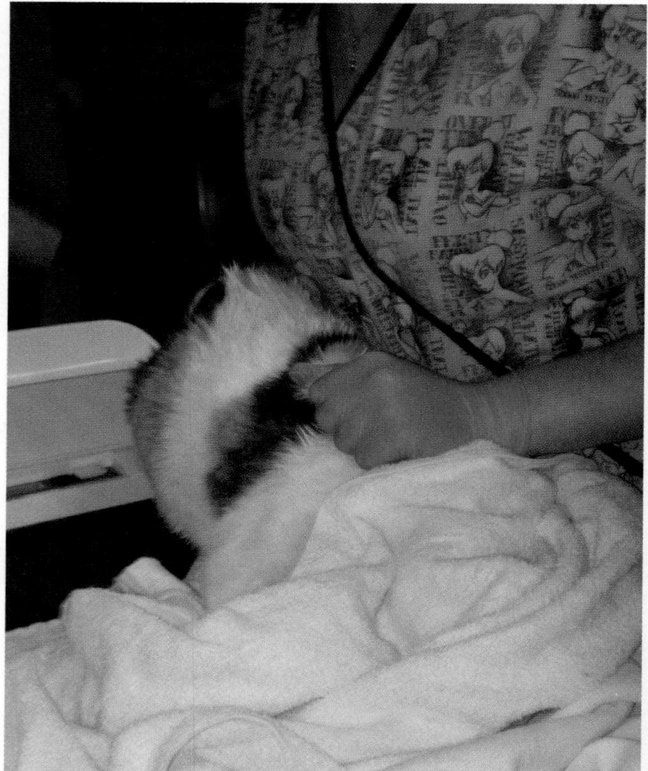

FIGURE 40-30 A towel can be used to aid in removing a cat from a cage.

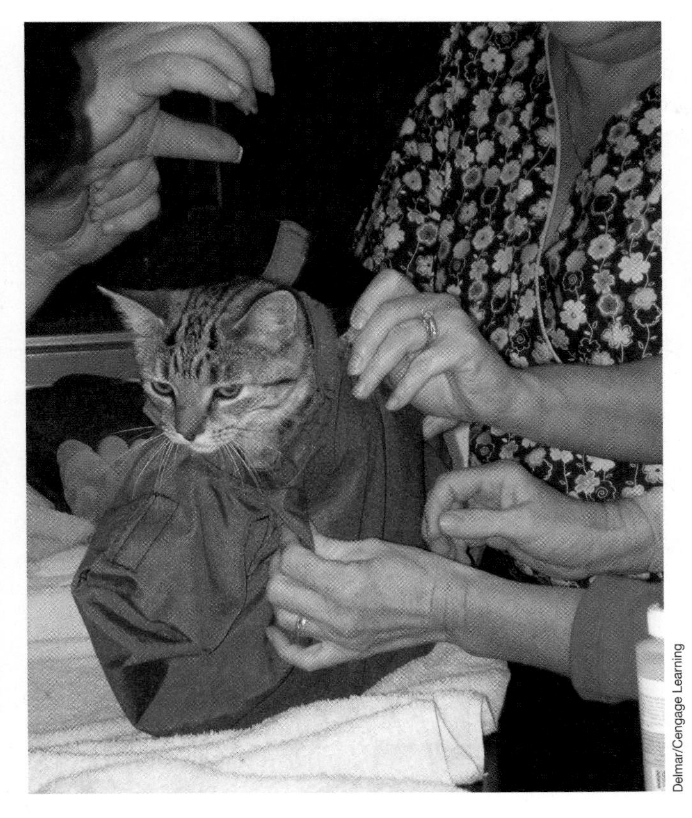

FIGURE 40-31 A cat bag can be used to help control and restrain the cat for procedures.

slats that allow injections to be given to a cat, such as vaccines or sedatives. They have handles that pull a cat to the side of the box for better access. An **anesthesia chamber** is used to sedate a cat or small mammals under gas anesthesia that is pumped into the glass or plastic chamber. Holes in the chamber allow the anesthesia machine to attach to the chamber and filter gas and oxygen in to induce sedation. These chambers can also be used to examine and observe a cat without the need for hands-on restraint. They can further be used to sedate a difficult or aggressive cat if need be.

Some hands-on restraint techniques that work specifically in cats include the scruff and the stretch. The **scruff technique** is used on the base of the neck over the area where the skin is elastic and able to be grasped with a fist. This may divert the attention of the cat from the procedure that is being done. It gives some control over the head, but safety must be taken; a cat can easily turn and bite even with the best grip. The **stretch technique** is done by scruffing the cat with one hand while in lateral recumbency and using the free hand to hold the rear limbs and pull them dorsally, thus putting the cat in a stretch position.

Head Restraint of a Cat

Objective:

To safely restrain a cat by manually holding its head.

Preparation:

No specific equipment is required.

Procedure:

1. Place one hand around the side of the head while standing over and to the rear of the cat.
2. Place the thumb on the top of the head and the fingers under the chin to hold the mandible (lower jaw) shut.

(Continues)

See Figure 40-32. This technique can be done with the left or right hand.

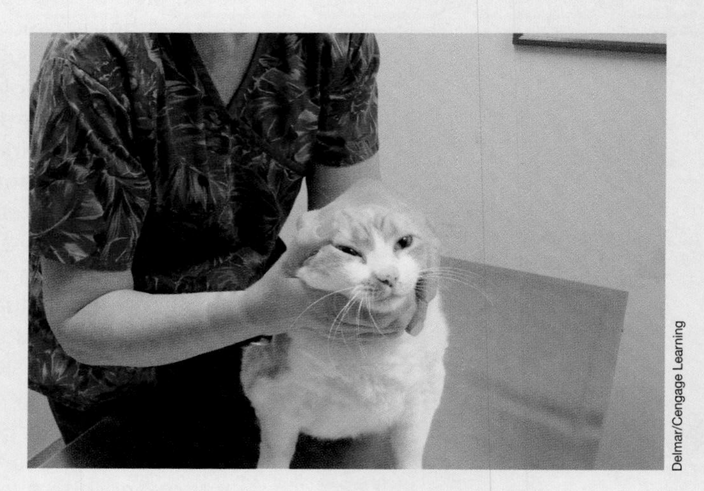

FIGURE 40-32 Restraint of the cat's head.

Competency Skill

Use of a Cat Bag

Objective:

To properly use a cat bag to restrain and control a cat during veterinary procedures.

Preparation:

cat bag

Procedure:

1. Open a cat bag and place it on a table.
2. Place the cat on top of the open cat bag.
3. While scruffing the cat, pull the bag over the body of the cat.
4. Zip the bag shut carefully to avoid not catching any hair in the zipper.
5. Close the bag over the entire body with the head exposed.
6. Control the head to prevent biting.
7. Open the areas in the cat bag to perform procedures.

See Figure 40-31.

Competency Skill

Use of an Anesthesia Chamber

Objective:

To safely and properly use an anesthesia chamber for restraint or observation of an animal.

Preparation:

anesthesia chamber

Procedure:

1. Remove the lid of the anesthesia chamber.
2. Place the cat inside the chamber and place lid securely on top and latch sides.
3. Make sure the openings over the top of the chamber are open and clear.
4. Open the lid of the chamber and remove the cat.

Competency Skill

Scruff Technique

Objective:

To safely and successfully use the scruff technique to restrain a cat.

Preparation:

No special equipment is required.

Procedure:

1. Use the dominant hand to grasp the loose skin over the base of the neck (see Figure 40-33).
2. Use the free hand to hold the rear limbs to prevent scratching.
3. Use a towel over the cat if the cat is difficult or trying to scratch with the front limbs.

(Continues)

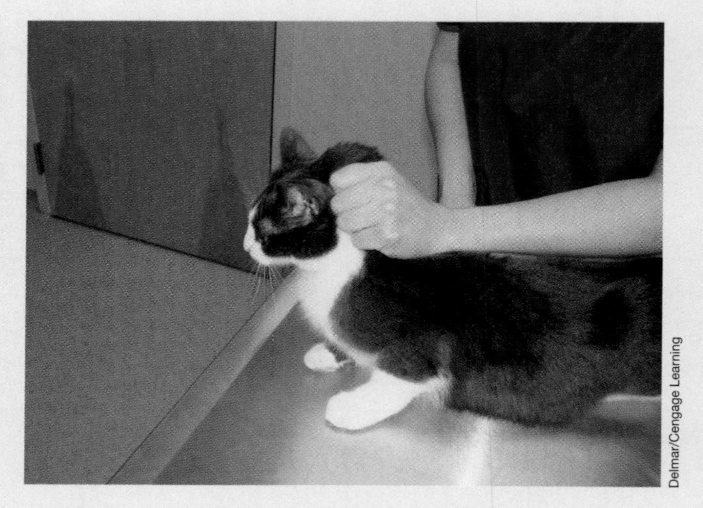

FIGURE 40-33 Cats can be restrained using a scruff technique.

Competency Skill

The "Stretch" Technique

Objective:

To safely and successfully use the stretch technique to restrain a cat.

Preparation:

No special equipment is required.

Procedure:

1. Grasp the cat firmly at the scruff of the neck using one hand.
2. With the free hand grasp the two rear legs by slipping a finger in between each rear paw.
3. Roll the cat to the side as in lateral recumbency.
4. Use the forearm of the hand holding the scruff to support the body and keep the cat from moving (see Figure 40-34).

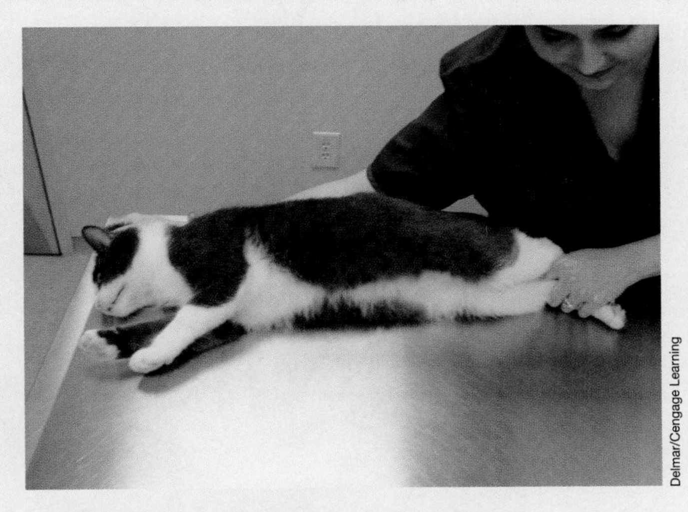

FIGURE 40-34 Another common restraint technique used for cats is the stretch technique.

Competency Skill

Use of the "Kitty Taco"

Objective:

To successfully use a towel to calm and restrain a cat.

Preparation:

heavy towel

Procedure:

1. Open a heavy towel on a table.
2. Place the cat on top of the open towel.
3. Scruff the cat and pull one side of the towel over the body of the cat and tuck under the cat.
4. Take the opposite side of the towel and wrap around the cat to make a taco-like appearance. The cat's four limbs will be enclosed in the towel and the head will be exposed (see Figure 40-35).
5. Properly control the head to prevent biting.

(*Continues*)

This restraint may be used to examine the head and mouth and administer medications. One limb may be removed from the towel to examine or complete a nail trim.

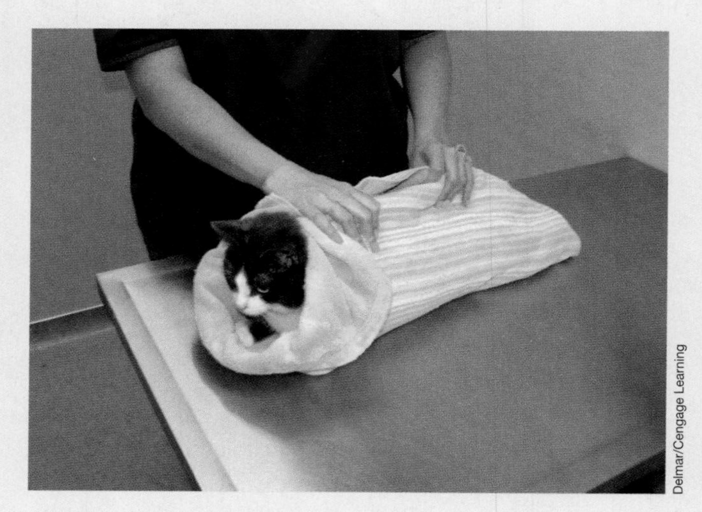

FIGURE 40-35 Use of a towel to create a Kitty Taco.

Competency Skill

Removing a Cat from a Cage or Carrier

Objective:

To safely and successfully remove a cat from a cage or carrier for examination.

Preparation:

towel

Procedure:

1. Close all windows and doors prior to restraint.
2. Open the cage or door and allow the cat to move to the front or step out.
3. If the cat is not willing to come to the front of the cage, reach into the area and scruff the cat with the dominant hand.

4. Use the free hand and place the hand under the abdomen and lift the cat to your body.
5. A towel may be used for scared or aggressive cats to feel more secure. Use the towel by placing it over the cat and then follow the above steps (see Figure 40-36).

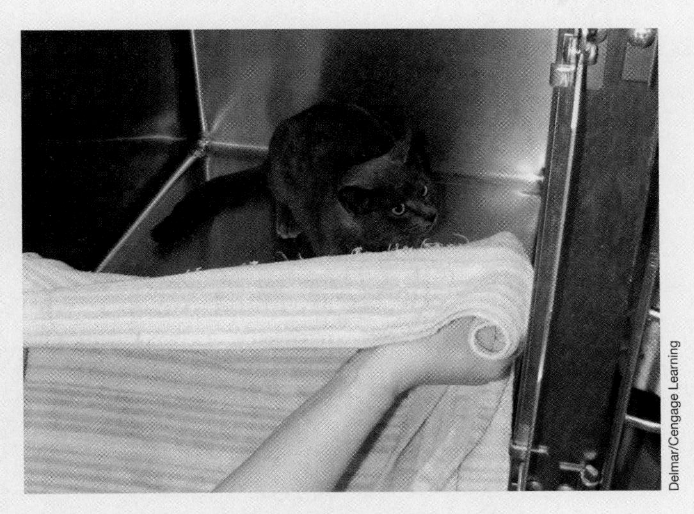

FIGURE 40-36 A towel can be placed over the cat to safely remove the cat from a cage.

Competency Skill

Carrying a Cat

Objective:

To safely carry a cat to an examination room or cage.

Preparation:

No special equipment is required.

Procedure:

1. Carry a calm and happy cat by placing one hand on the front of the body to control the head and front limbs.
2. Place the other hand under the abdomen and rump to control the rear limbs.
3. Pull the cat to your body for support.

(*Continues*)

If the cat is aggressive:

1. Place the dominant hand over the scruff of the cat's neck and apply the scruff technique.
2. Use the free hand to lift the cat under the rump and hold onto the rear limbs to prevent scratching.
3. Lift the cat in the air with the cat's limbs away from the body.

See Figure 40-37. A towel can be placed over the cat and the same steps followed.

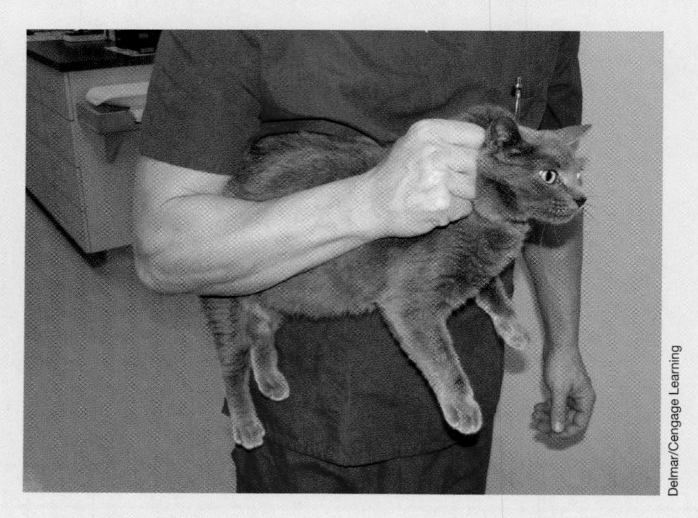

FIGURE 40-37 Technique for carrying an aggressive cat.

Dogs

Dogs that are difficult may be captured and restrained using a **rabies pole** or **snare pole** (see Figure 40-38). This pole is long and placed inside a cage or kennel to capture a dog around the neck. A noose on the end of the pole acts as a collar and the pole acts as a leash. The end of the pole controls the noose by pulling in or out on the pole to loosen or tighten the noose to fit the head of the dog. Once the noose is on the dog, it can be safely removed from the cage or kennel and walked to a safe area for further restraint or sedation. When walking a dog with a snare pole, the dog should be kept in front of the restrainer and pushed along, rather than pulled. Pulling a dog can cause severe head and neck injuries.

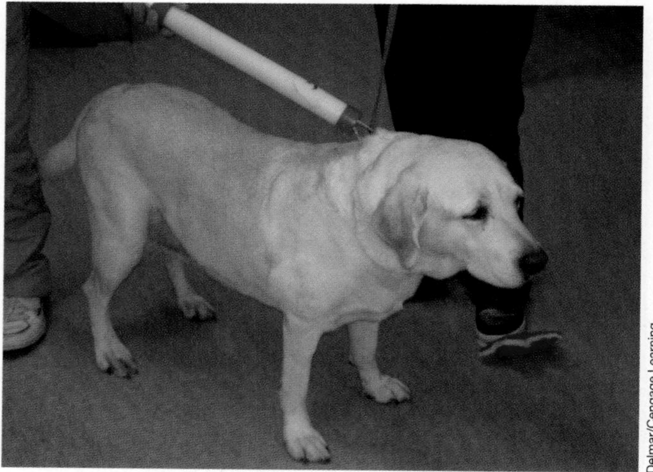

FIGURE 40-38 Use of a rabies control to restrain a dog.

Competency Skill

Removing a Dog from a Cage or Kennel

Objective:

To safely and successfully remove a dog from a cage or kennel.

Preparation:

leash

Procedure:

1. Place a leash in one hand with a large loop open and ready to place over the dog's head (see Figure 40-39A).
2. Open the cage door enough to slip the hand holding the leash into the cage (see Figure 40-39B).
3. Slip the leash over the neck of the dog and gently tighten the leash around the patient's neck (see Figure 40-39C).
4. Open the door and allow the dog to exit the cage (see Figure 40-39D).
5. Keep the dog to your side while maintaining a slight tension on the leash.

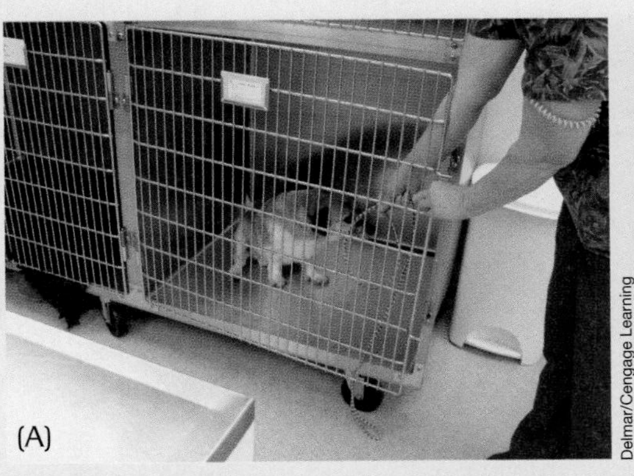

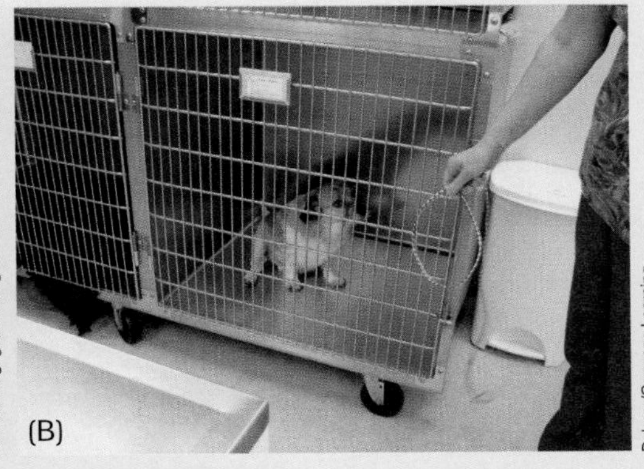

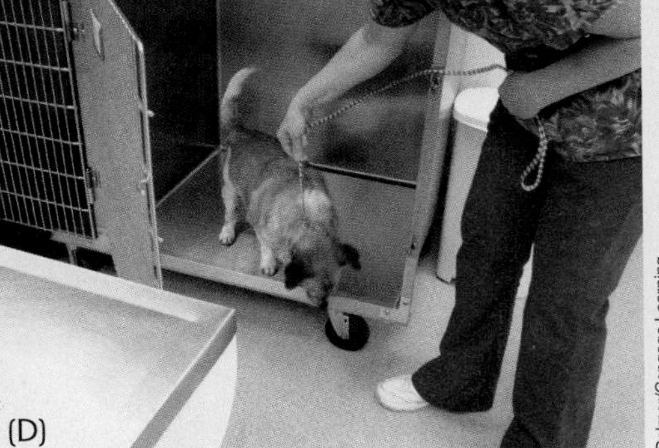

FIGURE 40-39 (A) make a large loop at the end of the leash; (B) open the door of the cage wide enough to place the hand with the leash in the cage; (C) place the loop over the dog's head and gently tighten; (D) allow the dog to exit the cage.

Competency Skill

Walking a Dog in the Heel Position

Objective:

To safely walk a dog in the veterinary facility.

Preparation:

leash

Procedure:

1. Place a leash in one hand with a large loop open and ready to place over the dog's head.
2. Slip the leash over the head while standing to the slide of the dog.
3. Gently tighten the leash around the patient's head.
4. Stand to the left side of the dog.
5. Walk forward with the dog at the left side of the body (see Figure 40-40).

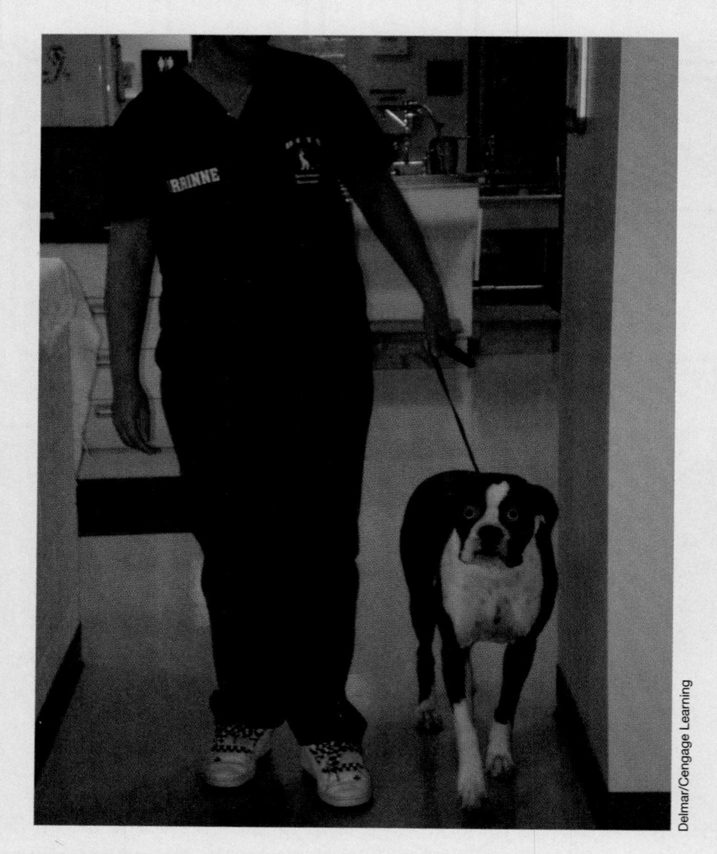

FIGURE 40-40 Walking the dog in the heel position.

Competency Skill

Applying a Commercial Leash to a Dog

Objective:

To properly and safely use a leash to control a dog.

Preparation:

leash

Procedure:

1. Place a leash in one hand with a large loop open and ready to place over the dog's head.
2. Slip the leash over the head while standing to the slide of the dog.
3. Gently tighten the leash around the patient's head (see Figure 40-41).
4. Stand to the left side of the dog.

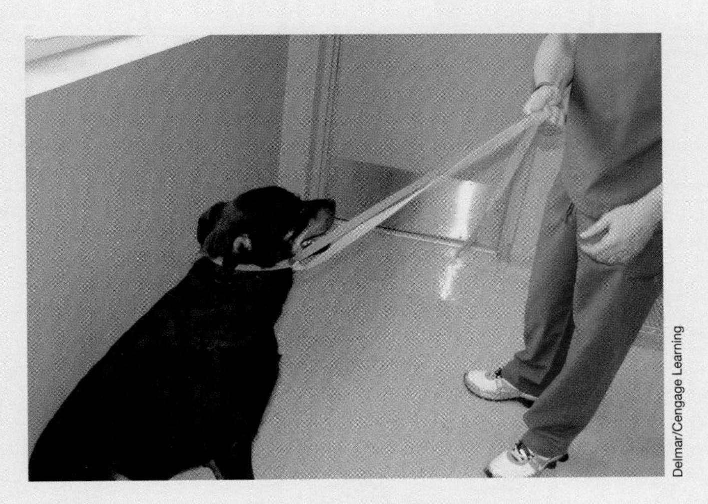

FIGURE 40-41 Application of a leash on a dog.

Competency Skill

Dorsal-Ventral (DV) Recumbency

Objective:

To properly and safely place an animal in a **dorsal-ventral** recumbency position.

Preparation:

No special equipment is necessary.

Procedure:

1. The dog or cat should be in the standing position.
2. Cats should be scruffed.
3. Apply pressure over the hips, forcing the dog or cat to sit.
4. Slide one hand under the front legs while applying pressure over the shoulders with the free hand.
5. Maintain pressure over the back and shoulders until the animal is lying on its chest (see Figure 40-42).
6. Restrain the head in a bear hug with one arm with a dog. Cats should be scruffed.
7. Use the free arm to apply continued pressure over the back to prevent standing. Small dogs may require only one restrainer; medium to large dogs will require two or more restrainers. Cats may be restrained with one or two people.

This is a common position for X-rays when the light beam enters the dorsal area (back) first and the ventral area (stomach) second.

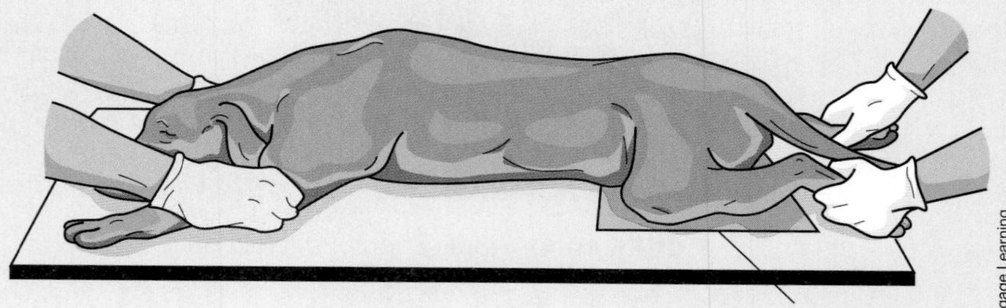

FIGURE 40-42 Dorsal-ventral recumbency.

Competency Skill

Ventral-Dorsal (VD) Recumbency

Objective:

To properly and safely place an animal in a **ventral-dorsal** recumbency position.

Preparation:

No special equipment is necessary.

Procedure:

1. The dog or cat should be placed in lateral recumbency as outlined.
2. The dog or cat is gently rolled onto its back with the front legs extended forward and the back legs extended backward.
3. The animal's head should be kept between the front legs and held securely.
4. The rear legs should be kept still to avoid injury to the back (see Figure 40-43).

Small dogs may require only one restrainer; medium to large dogs will require two or more restrainers. Cats may need two restrainers.

This is a common position for X-rays when the light beam enters the ventral area (stomach) first and the dorsal area (back) second.

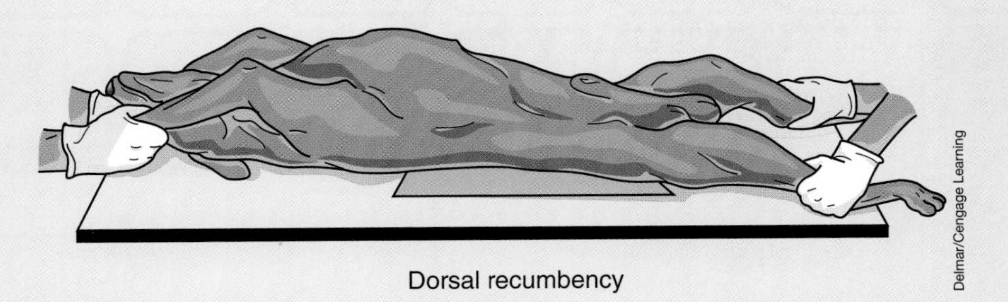

FIGURE 40-43 Ventral-dorsal recumbency.

Competency Skill

Returning a Dog to a Cage or Kennel

Objective:

To safely move a dog from a procedure area to the cage or kennel.

Preparation:

leash

Procedure:

1. Open the cage door completely.
2. Place the animal in the cage with the dog facing you.
3. Close the door with only one arm remaining in the cage.
4. Gently slip the loop of the leash over the dog's neck and pull the leash from the cage.
5. Completely close the cage door and latch securely.
6. Check the door to make sure it is properly closed.

Competency Skill

Handling a Fractious or Scared Dog

Objective:

To properly and safely handle a dog that is behaving in an unruly manner.

Preparation:

leash, muzzle

Procedure:

1. Work calmly and slowly with the scared dog.
2. Talk with a soft and gentle voice and reassure the dog.

3. Move slowly toward the dog from the side. Do not approach from the front or rear of the dog.
4. Gently apply a leash using the loop method over the neck.
5. Continue to talk softly to the dog.
6. Slowly walk the dog from the area.
7. Use a muzzle as necessary.

Handling of Other Small Animals

Many of the techniques discussed for the dog and cat can be used or slightly adapted for use in other small animals. The scruff technique works well on ferrets. Birds and small mammals can be handled and secured by wrapping them in towels. Some small animals have some specific needs that the veterinary assistant should become familiar with. Rabbits, for instance, can become paralyzed if their hind limbs are not properly supported.

Competency Skill

Lifting a Rabbit

Objective:

To properly and safely lift a rabbit.

Preparation:

No special equipment is necessary.

Procedure:

1. When lifting a rabbit, keep its feet forward and its back toward your body.
2. Grasp the scruff of the neck with one hand and the hind legs with the other hand.
3. Extend the rear legs and lift the rabbit.
4. Use the palm of the hand that is holding the rear legs to support the hindquarters and body (see Figure 40-44).
5. Keep the back against your body.

(*Continues*)

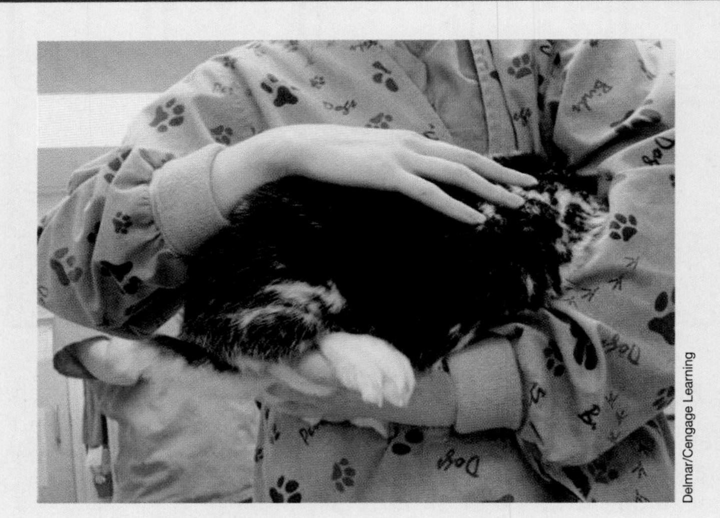

FIGURE 40-44 Properly holding a rabbit.

Large Animal Restraint

Large animals include livestock such as horses, cattle, goats, swine, and sheep. Large animal behavior is typically based on these species being in groups or herds of animals. Livestock have the instincts of prey animals and many times when people approach them for restraint, if they are not used to humans or being handled, typical behavior is that of prey. They will use the "fight or flight" instinct as part of their reaction to restraint. Horses tend to be more willing to be handled by people as they have usually been trained and worked with by humans. Some may become frightened by certain movements or smells. Other livestock, such as cattle, sheep, and swine, may not be as accustomed to human handling and should be restrained with caution and an experienced hand. The large size of livestock can be a safety issue, and many species of livestock are capable of kicking, rearing up into the air, biting, and using their large bodies to injure people. This is a safety situation that requires veterinary staff to be experienced in large animal behaviors and species-specific handling techniques. Many times large animals that are aggressive or deemed dangerous to handle should be sedated for the safety of the staff and the animal.

Competency Skill

Application of a Halter

Objective:

To properly and safely apply a halter to lead a large animal such as a cow or horse.

Preparation:

halter

Procedure for the Cow:

1. Stand on the left side of the cow.
2. Place the nose band of the halter over the face and around the nose.
3. The headstall should then be placed over the back of the head and behind the ears (see Figure 40-45).
4. Make sure the halter is a good fit and doesn't slip off.
5. Apply a lead to the bottom of the halter.

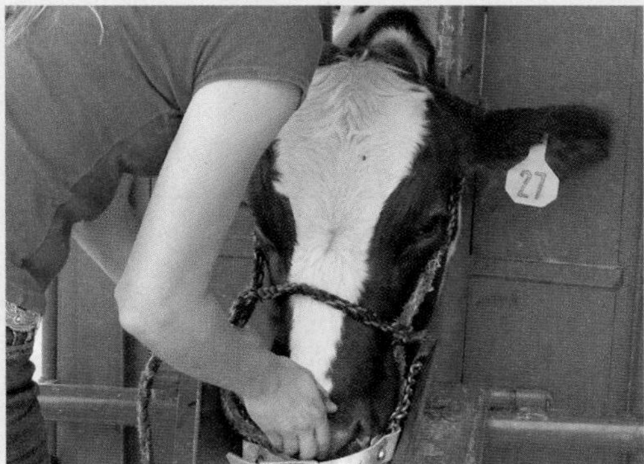

Delmar/Cengage Learning

FIGURE 40-45 Haltering a cow.

Procedure for a Horse

1. Stand on the left side of the horse.
2. Place the nose band of the halter over the face and around the nose.
3. The headstall should then be placed over the back of the head and behind the ears.
4. Secure the halter in place (see Figure 40-46).
5. Make sure the halter is a good fit and doesn't slip off.
6. Apply a lead to the bottom of the halter ring.
7. Hold the lead loops in the left hand and hold the lead close to where it connects to the halter in the right hand.

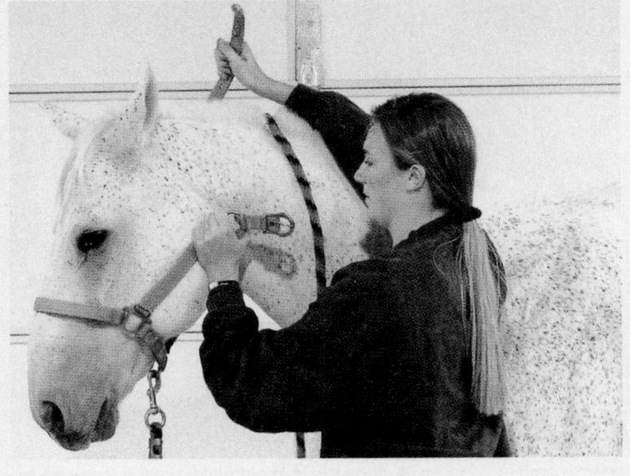

Delmar/Cengage Learning

FIGURE 40-46 Haltering a horse.

Competency Skill

Leading a Cow or a Horse

Objective:

To safely lead a horse or a cow from location to location.

Preparation:

halter, rope, pole or prod

Procedure:

1. Hold the lead loops in the left hand and hold the lead close to where it connects to the halter in the right hand.
2. Stand on the left side of the horse or cow near the shoulder (see Figure 40-47).
3. Stand about 1 foot from the horse or cow.
4. Face the same direction as the horse and click with the tongue to ask the horse to move forward. Cattle may need to be pulled or prodded with a pole.
5. Walk the horse or cow forward and watch the front feet to avoid being stepped on.

FIGURE 40-47 Leading a horse.

Competency Skill

Jogging a Horse on a Lead

Objective:

To safely exercise a horse while on a lead.

Preparation:

rope, halter

Procedure:

1. Hold the lead loops in the left hand and hold the lead close to where it connects to the halter in the right hand.
2. Stand on the left side of the horse near the shoulder.
3. Stand about 1 foot from the horse.
4. Face the same direction as the horse and click with the tongue to ask the horse to move forward.
5. Jog with the horse in a forward motion and watch the front feet to avoid being stepped on.

Competency Skill

Tying a Cow

Objective:

To safely and securely tie a cow for restraint.

Preparation:

rope

(*Continues*)

Procedure:

1. Pull the cow's head to the side.
2. Place the lead over a bar or chute.
3. Apply a quick release knot by making one half of a bow tie.
4. Pull the end of the rope to quickly untie.

Competency Skill

Tying a Horse

Objective:

To safely and securely tie a horse for restraint.

Preparation:

rope

Procedure:

1. Place the lead over a pole or tie area.
2. Allow about 2 feet of rope for the horse to move slightly.
3. Make a loop in the lead rope close to the pole or tie area.
4. Pass the loop under the lead rope near the pole and pass it through the loop. This should look like a pretzel.
5. Pull the loop to tighten the knot.
6. Untie by pulling the end of the lead rope for a quick release.

Competency Skill

Cross Tying a Horse

Objective:

To safely and securely tie a horse for restraint.

Preparation:

rope

Procedure:

1. Walk the horse to the cross tie area.
2. Place the left cross tie on the lateral cheek ring of the halter.
3. Move in front of the horse to the right side.
4. Place the right cross tie on the lateral cheek ring of the halter.

Competency Skill

Longing a Horse

Objective:

To safely exercise a horse.

Preparation:

rope

Procedure:

1. Place a long longe line onto the horse's halter by snapping securely.
2. Allow the horse to move away from you in a large circle.
3. Work the horse at a walk, jog, and lope in both directions to allow the vet to assess any lameness problems.

(*Continues*)

4. The horse should continue to move in large circles around the handler (see Figure 40-48).
5. Hold the end of the longe line firmly in one hand.

FIGURE 40-48 Lunging a horse.

Competency Skill

Tail Tie Technique

Objective:

To safely and properly place a tail tie to aid in maneuvering of a horse.

Preparation:

rope

Procedure:

1. Stand to the rear of the horse and slightly to the side.
2. Grasp the tail about 1/3 of the way from the tail base just below the tail bone.
3. Use both hands to gently lift the tail upward.

4. Place a rope around the tail making a loop.
5. Make a slip knot and pull the end of the rope toward the head (see Figure 40-49).
6. Tie the end of the rope to the front leg or neck using a slip knot tie.

FIGURE 40-49 Tail tie.

Competency Skill

The Twitch Restraint

Objective:

To use a device to elicit minor pain to distract an animal from procedures.

Preparation:

twitch device

(*Continues*)

Procedure:

1. Stand to the left side of the horse's neck and head.
2. Place the end of the twitch loop over the left wrist.
3. Hold the twitch handle with the right hand.
4. Grasp the horse's upper lip with the left hand and press the edges together.
5. Quickly slip the twitch loop over the lip so that it lies around the upper lip (see Figure 40-50).
6. Tighten the twitch loop by twisting the handle clockwise before letting go of the lip.
7. Hold the twitch in the left hand and the halter in the right.
8. Release the twitch by untwisting and removing from the lip.

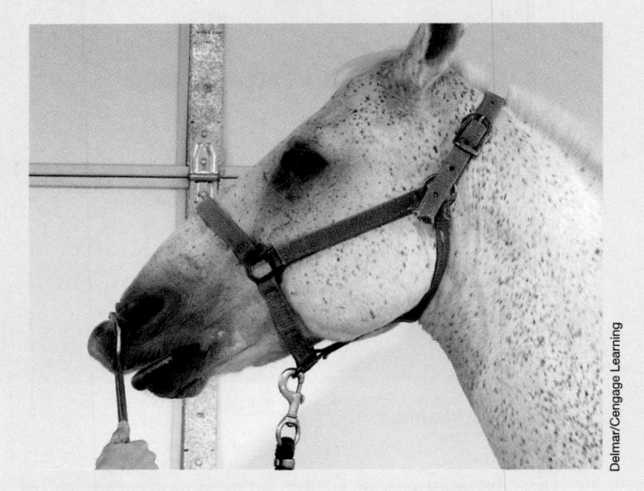

Delmar/Cengage Learning

FIGURE 40-50 Application of a twitch.

Competency Skill

Use of a Chain Shank

Objective:

To apply an increased amount of restraint.

Preparation:

chain shank

Procedure:

1. Stand to the left side of the horse's neck and head.
2. Place the chain through the left lateral cheek ring.
3. Run the chain over the nose, under the chin, or under the lip (see Figure 40-51).
4. Move in front of the horse to the right side.
5. Place the chain through the right lateral check ring and snap in place.
6. Move in front of the horse to the left side to control the horse. Hold the chain and lead close to where it connects to the halter.

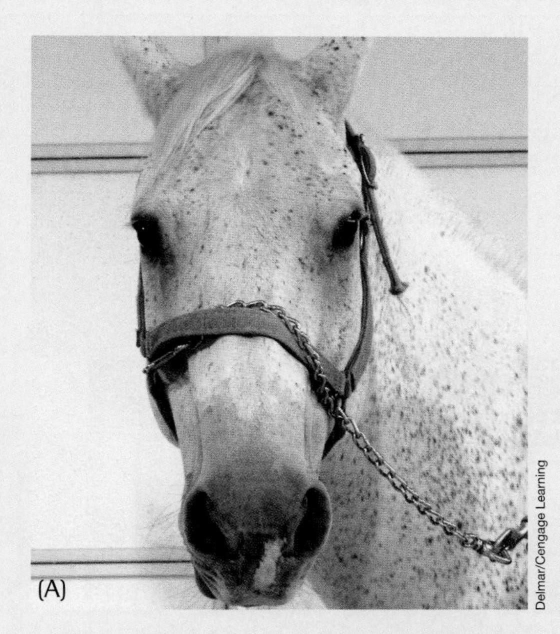

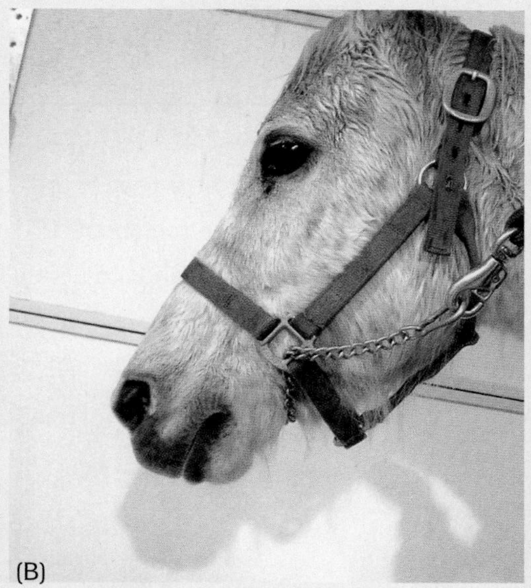

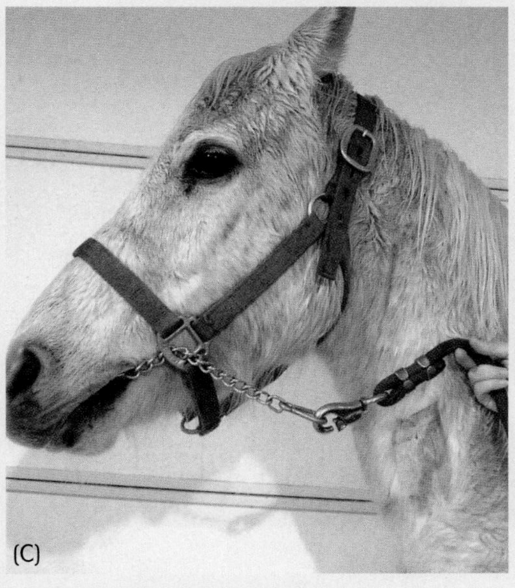

FIGURE 40-51 A to C Chain shank.

Competency Skill

The Tail Jack Restraint

Objective:

To properly and safely restrain a cow.

Preparation:

No special equipment is required.

Procedure:

1. Stand to the rear of the cow and just behind the legs.
2. Grasp the tail about 1/3 of the way from the tail base.
3. Use both hands to gently lift the tail upward (see Figure 40-52).

FIGURE 40-52 Tail jack restraint.

Competency Skill

The Tail Switch Restraint

Objective:

To properly and safely restrain a cow.

Preparation:

No special equipment is required.

Procedure:

1. Stand to the side and as close to the legs as possible.
2. Use both hands to grasp the tail switch.
3. Rotate the tail gently by the switch and turning it under the cow.
4. Make a small circle with the end of the tail.

Competency Skill

Restraint of a Large Animal for Blood Collection or an Injection

Objective:

To properly and safely restrain a large animal for the purpose of collecting a blood sample or to give a medication by injection.

Preparation:

halter

(*Continues*)

Procedure:

1. Apply a halter and lead to the animal.
2. Stand to the left side of the animal near the shoulder.
3. Hold the animal close to the halter with the left hand.
4. Place the right hand just behind the eye to distract the animal.
5. Reach to the neck area and pinch the skin with the right hand to further distract the animal.

MAKING THE CONNECTION

Some species of livestock may require specialized restraint. Refer to the chapters on specific large animal production for techniques.

Competency Skill

Restraint of a Foal

Objective:

To properly and safely restrain a foal.

Preparation:

No special equipment is required.

Procedure:

1. Hold one arm and hand around the neck to control the head and front of the body.
2. The arm may be moved lower to hold the chest as necessary.
3. Place the other arm around the rump holding the foal in the arms (see Figure 40-53).
4. The tail may be lifted up over the rump toward the head to control a foal from moving.

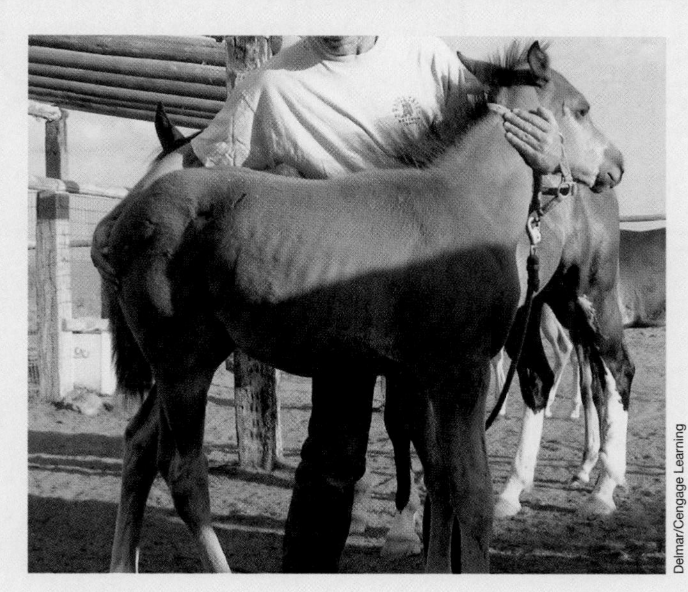

FIGURE 40-53 Cradling a foal.

SUMMARY

Restraining small animals must be done with several safety issues considered at all times. The safety of the animal and the restrainers is most important. The restrainer must focus on the animal's body language during all procedures. The environment must be monitored for safety issues. Anticipation of possible occurrences must be considered to provide a safe restraint setting. Animals can be unpredictable and behaviors may change with the slightest amount of stress, especially in a veterinary setting. Allowing for a pleasant experience can sometimes be helpful when working with animals. Large animal restraint requires a great deal of practice and skill. It also requires knowledge and experience in species behavior. Large animals can cause severe injury and possibly death in situations where handling and restraint are difficult. Care and handling of animals may require additional restraint equipment that must be used properly to ensure safety of the staff and the animal. Sedation may also be a possibility.

Key Terms

aggression behavior that makes an animal angry and difficult and potentially unsafe to handle

anesthesia chamber restraint equipment used to sedate a cat or small mammals under gas anesthesia that is pumped into the glass or plastic chamber

body language communication by the animal about how it feels toward other animals, people, and its environment

cat bag material that covers the body and limbs of a cat to control it for various procedures; only exposed part is the head

cephalic vein blood vessel located on the medial front limb

dominance aggression behavior that refers to the pack instinct of an animal and its social status within a group

(*Continues*)

dorsal recumbency restraint position with an animal lying on its back

dorsal-ventral (D-V) position where the animal is restrained on its chest with the X-ray beam traveling through the dorsal aspect first and the ventral aspect second

fear aggression refers to the defense reaction to being harmed and the instinct for an animal to protect itself

half hitch tie that makes a loop around a stationary location, such as a post or fence

jugular vein blood vessel located in the neck or throat area

knot tying two pieces of rope to make them not slip and to contain an animal

lateral recumbency restraint position of the animal lying on its side

rabies pole long pole used to capture a dog around the neck with a noose on the end of the pole that acts as a collar and a pole that acts as a leash; also called a snare pole

recumbency to lie as in a restraint position

redirected aggression refers to an animal's predator instinct where the animal turns its aggressive behaviors on the owner

reefer's knot single bow knot that ties in a nonslip but quick release tie

restraint to hold back, check, or suppress an action and to keep something under control using safety and some means of physical, chemical, or psychological action

saphenous vein blood vessel located on the lateral thigh of the rear limb just proximal to the hock

scruff technique used on the base of the neck over the area where the skin is elastic and able to be grasped with a fist technique

sedative medication used to calm an animal

square knot commonly used to secure an animal and is a nonslip knot that doesn't come untied or easily loosen

sitting restraint restraint used to keep an animal sitting on its rump and prevent it from standing

snare pole long pole with a noose on the end that acts as a leash and collar to capture a difficult dog; also called rabies pole

squeeze cage wire boxes with small slats that allow injections to be given to a cat, such as vaccines or sedatives

standing restraint restraint used to keep an animal standing on its feet and prevent it from sitting or lying down

sternal recumbency restraint position with the animal lying on its chest

stretch technique scruffing the cat with one hand while in lateral recumbency and using the free hand to hold the rear limbs and pull them dorsally in a stretch

territorial aggression refers to an animal's protective nature of its environment, such as an owner, offspring, or food

tranquilizer medication used to calm an animal during stress

venipuncture practice of placing a needle into a blood vessel

ventral-dorsal (V-D) position where the animal is restrained on its back with the X-ray beam traveling through the ventral aspect first and the dorsal aspect second

REVIEW QUESTIONS

1. What is animal behavior?
2. What is the difference between innate and learned behavior?
3. What is body language?
4. What is a square knot?
5. What are some similarities in restraining small and large animals?
6. What are some differences in restraining small and large animals?

Clinical Situation

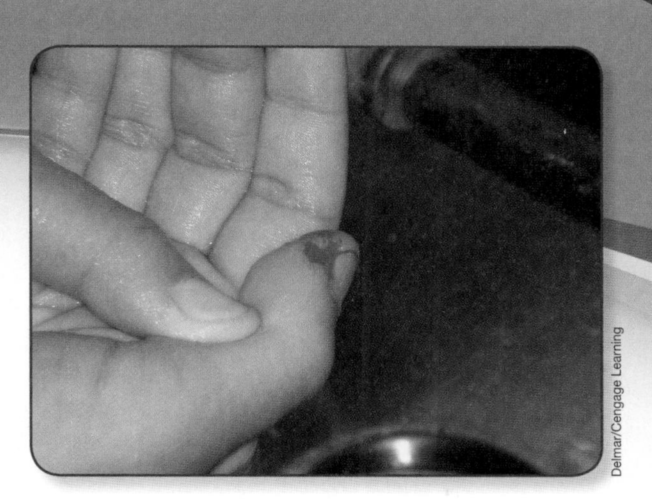
Delmar/Cengage Learning

Lila, the veterinary assistant at Companion Animal Vet Clinic, was working in the veterinary kennel ward. She began going down the cage rows, giving water to those animals that were allowed to have water. She noted that one kennel had some waste material that needed to be cleaned. When she opened the cage and began to remove the waste, "Precious," a nervous Bichon Frise, began growling and raising his upper lip. Lila gently reached toward the dog and patted him on the head.

"It's okay, 'Precious,' no one is going to hurt you," she said quietly.

The dog snapped at her hand, causing a scratch with his teeth. Lila quickly closed the cage door and went to clean the wound. The scratch was not deep, but to be safe, she checked with Dr. Todd to see if there was anything to be concerned about.

- How did Lila handle the situation correctly?
- What did Lila do incorrectly?
- What preventative measures should have Lila taken in this situation?

41

Delmar/Cengage Learning

Veterinary Safety

Objectives

Upon completion of this chapter, the reader should be able to:

- State the concerns of human and animal safety in the veterinary industry
- Discuss the safety hazards within the veterinary industry
- Describe and discuss OSHA guidelines in veterinary medicine
- Explain the use of MSDS
- Complete the OSHA log and accident report paperwork
- Identify safety signs and equipment
- Discuss safety plan guidelines in a veterinary facility

Introduction

Veterinary safety practices are important to the well-being of the veterinary staff, clients, and patients within the veterinary facility. Safety issues within the veterinary facility include handling and restraint of animals, OSHA rules and regulations, handling human medical emergencies, exposure to radiation and anesthetic gases, physical safety hazards, and sanitation concerns. It is important that the veterinary assistant knows the hazard concerns within the veterinary industry. This chapter discusses safety related to physical hazards and how to develop a safety plan within the facility.

Safety in the Veterinary Facility

The veterinary facility is required to have emergency plans prepared for multiple types of safety issues. All staff members are required to be properly trained in safety practices and to maintain a copy of any safety plan created in the veterinary facility. Training and yearly safety updates are required for all new and current employees.

Fire and Safety Plans

All veterinary facilities should develop a fire safety plan and other evacuation details that allow the staff members to know what to do during an emergency, such as a fire, severe weather, or other disaster situations. Emergency numbers for police, fire, and other emergency response teams should be within easy reach of each telephone in the facility. A fire plan should be enacted that outlines evacuation routes for people and animals and responsibilities of each staff member during the emergency. All fire extinguishers should be checked regularly for proper dating and assurance that they are in working order (see Figure 41-1). A fire exit plan should be posted near every entrance and exit of the facility and be clearly marked and easy to read. Fire alarms should be checked regularly for proper working order (see Figure 41-2). A meeting location that is a safe distance from the facility should be established for the entire staff. Routine fire drills should be practiced to ensure the staff is knowledgeable in the plan and procedures.

Professional Dress

Every veterinary health care team member should be appropriately dressed to reduce contamination of clients, employees, and patients. This includes wearing proper dress attire outlined by the personnel manual of each facility and the duties and responsibilities for each staff member's role in the facility. Veterinarians, veterinary technicians, veterinary assistants, and kennel staff typically wear scrub uniforms that are easily cleansed and changed when necessary to reduce

FIGURE 41-1 All staff should know the location of the fire extinguishers in the facility.

FIGURE 41-2 The facility should have fire alarms that are in the proper working order.

the spread of contamination and disease throughout the facility. Scrub uniforms entail a well-fitting top that covers the upper body and well-fitting pants that cover the lower body (see Figure 41-3). Lab coats are another common item worn by employees handling animals and other contaminants. When scrubs or lab coats become soiled, they are easily changed and laundered to prevent the spread of disease. Several uniforms and lab coats should be available for each staff member working with animals.

Footwear should include a closed-toe shoe that fits well over the entire foot to prevent injury from sharp items and animals. This may include nonslip sneakers, which are easily laundered when necessary. It may also include work boots used around large animals for more foot protection and ease of cleansing.

Loose jewelry, such as earrings, bracelets, and necklaces, should be avoided as they can easily catch on items or animals and cause severe injury to a person or patient. Rings should be removed from the fingers to avoid injury and damage. Waterproof watches should be kept snug on the wrist and be easy to clean if contaminated. Many facilities have regulations and rules on specific body-piercing locations and the visibility of tattoos, which will be outlined in the personnel manual. Each employee should have a clear understanding of the employee and employer expectations.

FIGURE 41-3 Veterinary staff members typically wear neat and properly fitting scrubs that can be easily changed and cleaned.

Competency Skill

Professional Dress

Objective:

To present oneself in a manner that is professional and will serve to reduce the spread of infection.

The following guidelines should be followed when working in a veterinary facility.

- All clothing should be well fitting, including the scrub top and scrub pants.
- Shoes should be nonslip and fully enclose the foot.
- Exam gloves should be worn for patient and human safety.
- Do not wear long or fake nails as they hold bacteria and increase the possibility of infection and injury to other people and animals. The nails should be natural and not extend beyond the end of the finger.

- Do not wear loose, large, or floppy jewelry, such as bracelets, necklaces, rings, or hoop earrings. These items can be caught on or bitten by animals, causing injury to the person or the animal.
- Limit watches to waterproof items.
- Hair should be pulled back out of the face to allow easy visibility and a clean working area.
- Tattoos and facial/body piercings are not professional and should not be visible to clients.

Personal Protective Equipment

Exam gloves should be available to protect the hands from contaminants and should be used when handling any animal that may be potentially infectious or cause a person to become injured or contaminated in any way. Exam gloves should be changed before handling of a different animal. Hands should be washed thoroughly using warm water and antibacterial soap after the handling of each animal.

Personal Protective Equipment

Objective:

To know what PPE is available and the proper uses and needs for it.

Follow the guidelines below for proper use of PPE:

- Wear exam gloves at all times when working with chemicals, animals, and hazardous materials.
- Wear goggles or safety glasses when mixing chemicals, using the dremel, or other tasks that may risk eye injury.
- Wear face masks and shields when brushing teeth, working on dentals, using high-powered equipment, or when a risk of splashing items in the face is possible.
- Wear ear plugs when working in noisy areas, such as kennels.
- Wear aprons or gowns when bathing animals or working in surgical assisting and isolation wards.

Competency Skill

Glove Removal

Objective:

To properly perform the removal of gloves.

Procedure:

1. Firmly pinch the outer surface of one glove about 2 inches below the cuff (see Figure 41-4A).
2. Pull the grasped glove downward and pull the hand out of the glove. The removed glove will be inside out (see Figure 41-4B).
3. Holding on to the removed glove with the fingertips, gather it into the palm of the other gloved hand.
4. Firmly pinch the outer surface of the second glove about 2 inches below the cuff.
5. Pull the gloved hand upward removing the hand from the glove. The remaining glove will be inside out with the first glove inside it (see Figure 41-4C).
6. Discard both gloves in the medical waste container.
7. Wash hands with disinfectant soap immediately. Dry hands thoroughly. Apply hand sanitizer.

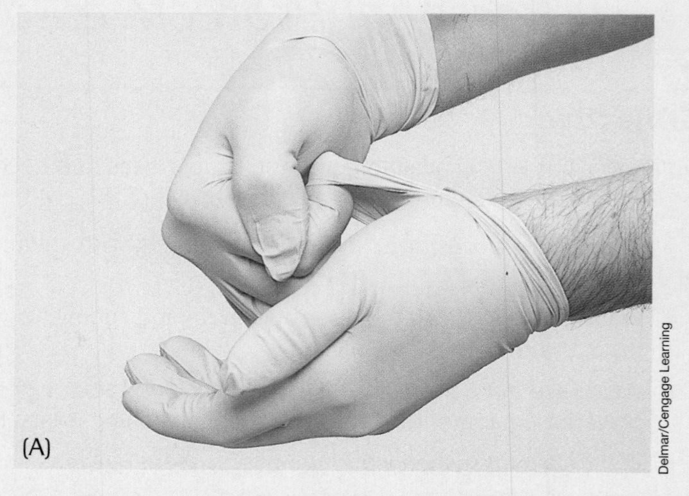

FIGURE 41-4 Proper removal of examination gloves. (*Continues*)

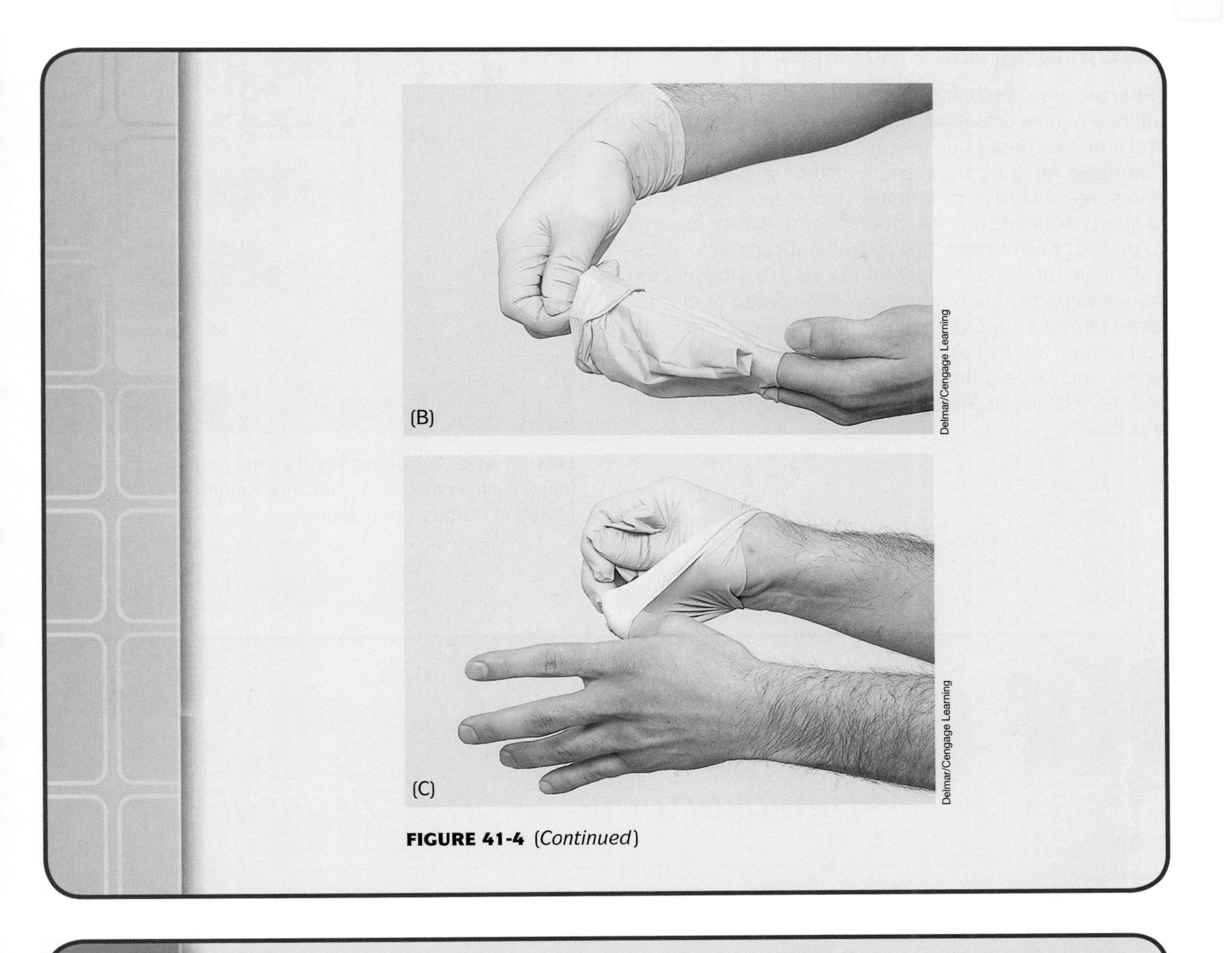

FIGURE 41-4 *(Continued)*

Competency Skill

Hand Hygiene

Objective:

To properly clean the hands to prevent spread of infection.

Follow these guidelines for proper hand hygiene:

- Wash hands after removing gloves.
- Wash hands after handling each animal.
- Wash hands using warm water and antibacterial soap, cleansing each surface of the hand for 5 minutes.

Instruments and Equipment

"**Sharps**" are sharp instruments and equipment that can injure a human or animal and cause a wound or cut that transmits a contagious disease due to contamination. Examples of sharps materials include glass, needles, and surgical blades. Veterinary staff members must be properly trained to work around these safety hazards to prevent health risks. Any instrument capable of causing a puncture wound must be placed in a **sharps container** to prevent contamination and spread of disease (see Figure 41-5). Other safety hazards that commonly cause injuries to people and animals include heavy gates and large animal restraint equipment. Fingers may be pinched or the equipment may potentially crush a person.

FIGURE 41-5 Any object that has the potential to cause an injury such as a scratch, puncture wound, or cut should be placed in a sharps container.

Competency Skill

Biohazard and Sharps Disposal

Objective:

To identify the sharps container and properly dispose of hazardous materials into it.

Follow these guidelines when disposing of hazardous wastes contaminated with biohazardous materials:

- Locate all biohazard containers (see Figure 41-6).
- Place needles, blades, and any sharp materials that may have infectious waste in a sharps container.
- Place all body fluids, including blood, urine, and feces, in medical waste bags.
- Place all surgical drape material in medical waste bags.
- Dispose of sharps containers and medical waste in proper location for incineration or medical waste pickup.

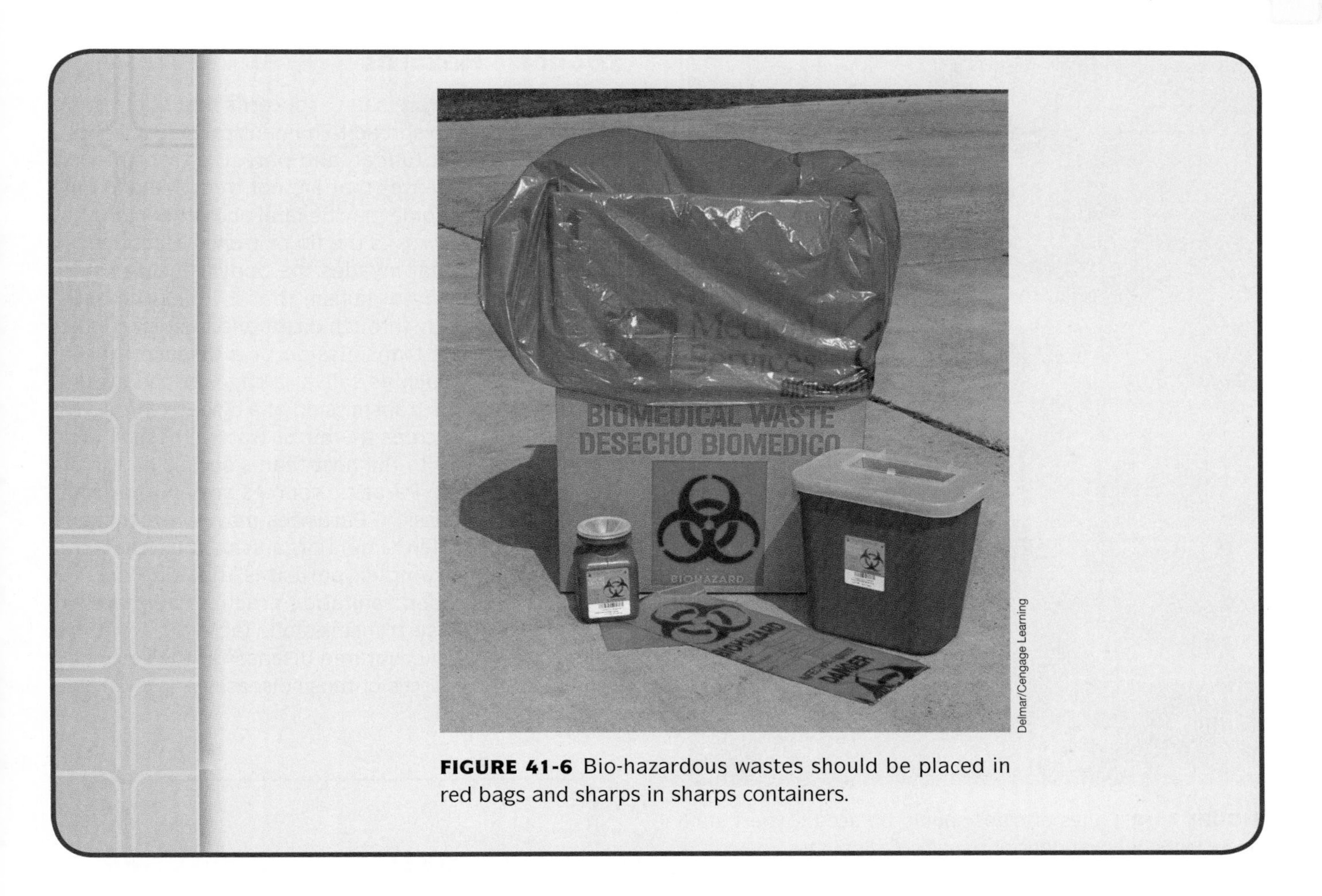

FIGURE 41-6 Bio-hazardous wastes should be placed in red bags and sharps in sharps containers.

Physical Hazards

Physical hazards are safety concerns that can cause physical harm to a human or animal's body. Examples of physical hazards include animal bites, back injuries from lifting heavy objects, falls on wet surfaces, and radiation exposure.

Animal bites and scratches should be cleaned with warm water and soap and a bandage applied until medical attention is given to the injury (see Figure 41-7). Floors that are wet should have a wet floor sign placed in any areas where people are walking and may slip and fall (see Figure 41-8).

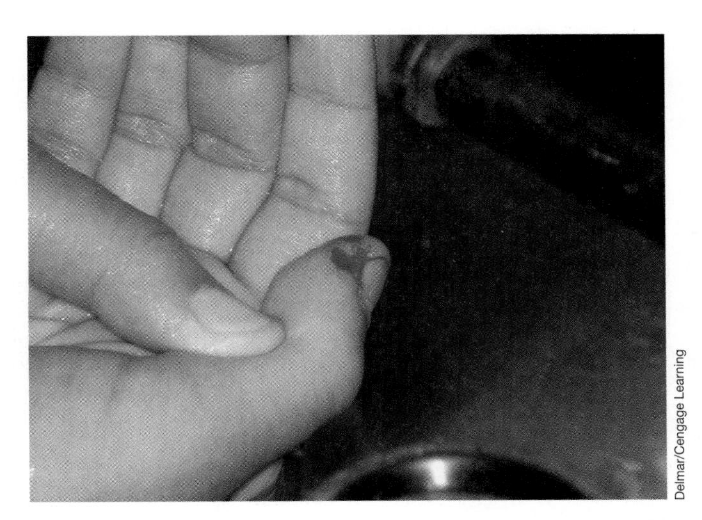

FIGURE 41-7 Thoroughly clean and administer first aid to any wound obtained while on the job. Make sure facility protocols and procedures are followed in regard to care and reporting of the incident.

Chemical Hazards

Chemical hazards are safety concerns that may cause injury to the skin, lungs, eyes, or other areas due to exposure. Chemical hazards include chemotherapy drugs, cleaning agents, insecticides, and anesthetic gases.

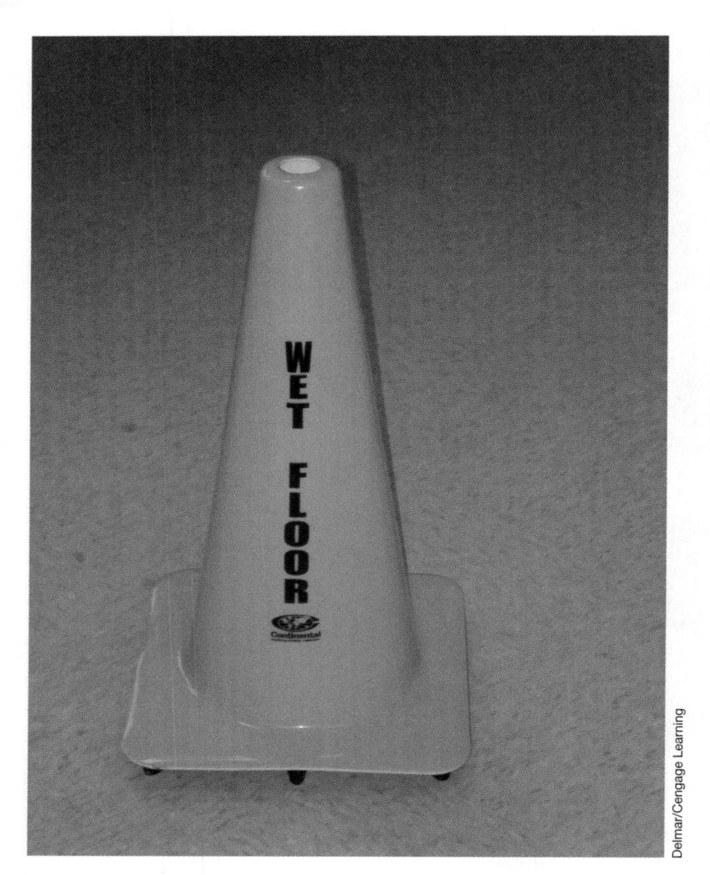

Delmar/Cengage Learning

FIGURE 41-8 Cones or signs should be used to alert staff and clients of a wet floor or danger of slipping.

Many hazardous chemicals are used in the veterinary industry. These chemicals may cause severe injury through direct contact or through inhalation of vapors. Some vaccines are potential risks as they may cause disease to a person who is accidentally exposed. Some antibiotics that are handled may also cause exposure to chemicals that are hazardous.

TERMINOLOGY TIP

Chemotherapy is the use of chemical agents to treat forms of cancer.

Biological Hazards

Biological hazards are safety concerns that pose a risk to humans and animals through contamination of living organisms through body tissues and fluids. Biological hazards include blood, urine, feces, and live vaccines. **Medical waste** is a concern as living tissue and fluids may have contagious and infectious organisms that may cause zoonotic disease. Medical waste may be surgical drapes, bandage materials, urine-soaked bedding, or saliva.

Zoonotic Hazards

Zoonotic hazards are safety concerns that allow contagious organisms to spread to humans causing infections, viruses, bacterial, fungal, and parasitic transmission. A **virus** causes diseases that are not treatable and must run their course. Some may be fatal and others may not. An example of a virus is the flu or rabies. **Bacteria** is a living organism that invades the body, causing illness. A **fungus** is a living organism that invades the external area of the body through direct and indirect contact. **Direct contact** is transmission of a disease through a direct source, such as saliva, a cage, or the ground. **Indirect contact** is transmission of a disease through an indirect source such as the air or water. These sources may be secondary to the host that is spreading the disease or contagion. Parasitic sources are another route of disease transmission. **Parasites** may be **protozoans**, which are one-celled simple organisms, or **rickettsial,** which are more complex parasites that bite such as fleas and ticks. Good sanitation practices help prevent and control disease transmission. Table 41-1 outlines some commonly encountered diseases in the veterinary facility and the sources of those diseases.

MAKING THE CONNECTION

Review Chapter 38 on microbiology for additional content related to zoonosis.

TABLE 41-1

Examples of Diseases and Sources

DISEASE NAME	DISEASE TYPE
Brucellosis	Bacterial
Canine Distemper	Viral
Ringworm	Fungal
Coccidia	Protozoan
Rabies	Viral
Lyme Disease	Bacterial
Rocky Mountain Spotted Fever	Rickettsial
Leptosporosis	Bacterial
Tetanus	Bacterial
Parvovirus	Viral
Hepatitis	Viral
Toxoplasmosis	Protozoan
Tick Paralysis	Rickettsial

OSHA Guidelines and Regulations

The **Occupational Safety and Health Act (OSHA)** of 1970 provides for safe and healthy work environments and conditions for all employees. The OSHA guidelines were created by the U.S. Department of Labor and state that all employers are required to inform their employees of potential workplace hazards and risks so that everyone can work safely in their profession. OSHA regulates many safety aspects of business in all industries within the United States. Veterinary practices are regulated for the following areas:

- Sanitation, cleanliness, and safety
- Provision of clearly marked, adequate sized, and unobstructed exits
- Ensuring all compressed gas cylinders for anesthesia and oxygen purposes are properly marked and in working order
- Provision of strict guidelines when handling hazardous chemicals
- Requiring protective devices for eyes, face, hands, feet, respiratory, and skin in situations posing any form of physical hazard
- Providing adequate toilet facilities and clean water
- Maintenance of operating fire extinguishers and sprinkler systems, as well as fire exit plans
- Requiring protective guards over all hazardous machinery
- Providing hazard-free electrical equipment
- Providing information to employees regarding any and all potential hazards, including MSDS binders and OSHA accident reports

Material Safety Data Sheets (MSDS)

The **right-to-know station** is the area where all OSHA binder information and the **MSDS safety sheets** should be kept to allow all employees access to information regarding any hazard within the facility. MSDS are material safety data sheets that provide information published by manufacturers of various products that have the potential to harm humans within the facility (see Figure 41-9). These may be chemicals or physical products. Parts of the MSDS include the following:

- Product Name
- Manufacturer Information

Material Safety Data Sheet

I—Product Identification

Company Name: We Wash Inc.

Address: 5035 Manchester Avenue
Freedom, TX 79430

Tel. No.: (314) 621-1818
Nights: (314) 621-1399
CHEMTREC: (800) 424-9343

Product Name: Spotfree
Synonyms: Warewashing Detergent

Product No.: 2190

II—Hazardous Ingredients of Mixtures

Material:	(CAS#)	% by Wt.	TLV	PEL
According to the OSHA Hazard Communication Standard, 29 CFR 1910.1200, this product contains no hazardous ingredients.		N/A	N/A	N/A

III—Physical Data

Vapor pressure, mm Hg: N/A
Evaporation rate (ether = 1): N/A
Solubility in H_2O: Complete
Freezing point F: N/A
Boiling point F: N/A
Specific gravity $H_2O = 1$ @25C: N/A

Odor: Mild chemical odor

Vapor density (air = 1) 60–90F: N/A
% Volatile by wt.: N/A
pH @ 1% Solution 9.3–9.8
pH as distributed: N/A
Appearance: Off-white granular powder

Courtesy of POL Consultants

FIGURE 41-9 Sample MSDS. (*Continues*)

IV—Fire and Explosion

Flash point F: N/AV Flammable limits: N/A

Extinguishing media: The product is not flammable or combustible. Use media appropriate for the primary source of fire.

Special firefighting procedures: Use caution when fighting any fire involving chemicals. A self-contained breathing apparatus is essential.
Unusual fire and explosion hazards: None known.

V—Reactivity Data

Stability: Conditions to avoid: None known.

Incompatibility: Contact of carbonates or bicarbonates with acids can release large quantities of carbon dioxide and heat.

Hazardous decomposition products: In fire situations, heat decomposition may result in the release of sulfur oxides.

Conditions contributing to hazardous polymerization: N/A

Spotfree
VI—Health Hazard Data

Effects of overexposure (medical conditions aggravated/target organ effects)

A. Acute (primary route of exposure) Eyes: Product granules may cause mechanical irritation to eyes.
Skin (primary route of exposure): Prolonged, repeated contact with skin may result in drying of skin.
Ingestion: Not expected to be toxic if swallowed; however, gastrointestinal discomfort may occur.

B. Subchronic, chronic, other: None known.

VII—Emergency and First Aid Procedures

Eyes: In case of contact, flush thoroughly with water for fifteen minutes. Get medical attention if irritation persists.
Skin: Flush and dry Spotfree from skin with flowing water. Always wash hands after use.
Ingestion: If swallowed, drink large quantities of water and call a physician.

VIII—Spill or Leak Procedures

Spill management: Sweep up material and repackage, if possible.
Spill residue may be flushed to the sewer with water.
Waste disposal methods: Dispose of in accordance with federal, state, and local regulations.

IX—Protection Information/Control Measures

Respiratory: None needed. Eye: Safety glasses. Glove: Not required.

Other clothing and equipment: None required.

Ventilation: Normal.

X—Special Precautions

Precautions to be taken in handling and storing: Avoid contact with eyes. Avoid prolonged or repeated contact with skin. Wash thoroughly after handling. Keep container closed when not in use.
Additional information: Store away from acids.

Prepared by: D. Martinez Revision date: 04/11/__

Seller makes no warranty, expressed or implied, concerning the use of this product other than indicated on the label. Buyer assumes all risk of use and/or handling of this material when such use and/or handling is contrary to label instructions.

While Seller believes that the information contained herein is accurate, such information is offered solely for its customers' consideration and verification under their specific use conditions. This information is not to be deemed a warranty or representation of any kind for which Seller assumes legal responsibility.

FIGURE 41-9 (*Continued*)

- Hazard Ingredients/Identity Information
- Physical/Chemical Characteristics
- Fire and Explosion Hazard Data
- Reactivity Data (PEL—Permissible Exposure Limit) in Radiology
- Health Hazard Data
- Precautions for Safe Handling and Use
- Control Measures

The OSHA binder should include safety plans that contain an evacuation plan in the event of an emergency; locations of water, gas, and electric shutoffs; locations of fire extinguishers; and emergency telephone numbers that include the police, fire department, and emergency service companies. All chemicals, disinfectants, and other products within the facility should be labeled with hazardous warning labels that identify which body systems they may be toxic to. Hazardous labels should identify the product name, what body systems or organs it affects, what PPE equipment is necessary when working with the item, and include a picture that identifies it as hazardous (see Figure 41-10). In addition, the facility should have emergency and personal protection equipment such as safety glasses or goggles, latex and vinyl gloves, ear plugs, and radiation gear. A person should be designated to oversee the safety training within the practice of all new employees and design a yearly OSHA safety meeting that reviews all updated information regarding employee safety. It is the employee's responsibility to read and understand the facility's practice safety manual and regulations and know where all safety equipment is kept and how it is used. It is also the employee's responsibility to learn how to properly read the MSDS safety sheets. Many practices have documents and accident reports that must be filed as soon as an injury occurs. Failure to report an incident may result in a violation of legal requirements that may delay the processing of necessary insurance claims and benefits.

Table 41-2 lists safety equipment that should be in each veterinary facility.

Controlled Substances

Controlled substances are drugs that have the potential for abuse and addiction and must be kept in a locked cabinet by law. These medications are regulated by the **Drug Enforcement Administration (DEA)**.

SAMPLE

CHEMICAL INVENTORY FORM

Date updated ______________________

Dental office ______________________

Chemical Name	Hazard Class				Physical State	Manufacturer	Comments
	(H)	(F)	(R)	(P)			

(H) Health	(F) Fire Hazard	(R) Reactivity	(P) Protection
0—Minimal	0—Will not burn	0—Stable	A—Goggles
1—Slight	1—Slight	1—Slight	B—Goggles/gloves
2—Moderate	2—Moderate	2—Moderate	C—Goggles/gloves/clothing
3—Serious	3—Serious	3—Serious	D—Goggles/gloves/clothing/mask
4—Extreme	4—Extreme	4—Extreme	E—Goggles/gloves/mask
			F—Gloves
			G—Face shield/gloves

Delmar/Cengage Learning

FIGURE 41-10 Each MSDS binder should contain a list of the chemicals and the health dangers they may impose.

TABLE 41-2

Safety Equipment in the Veterinary Facility

Fire extinguishers	Safety glasses/goggles
Exam gloves (latex and vinyl)	Wet floor signs
Warning labels	Biohazard signs
Radiation hazard signs	Lead apron, gloves, and thyroid shield
Ear plugs	Anesthetic gas scavenger system
Eye wash station	Sharps container
Medical waste bags	

Medications that are controlled are listed in schedules according to their potential for abuse, ranging from schedule I to schedule V, with V being the lowest potential and I being the highest potential for abuse or addiction (see Table 41-3).

Under DEA laws and regulations, veterinarians are required to obtain a DEA license to purchase and dispense any controlled substance. The license must be maintained yearly with the DEA, the registration number must be placed on all prescriptions, a **controlled substance log** must be maintained accurately at all times and be kept on file for 2 years, and all **scheduled drugs** must be locked at all times. The controlled substance log requires information that must be recorded and accurate and be available for inspection at any time. Table 41-4 outlines the required legal information that must be kept on every controlled substances prescribed by the veterinarian.

TABLE 41-3

Drug Schedules

Schedule I: no medical use—high abuse (illegal substances, e.g., marijuana)

Schedule II: accepted medical use—high abuse (e.g., morphine)

Schedule III: accepted medical use—medium abuse (e.g., Phenobarbitol [seizures])

Schedule IV: accepted medical use—low abuse (e.g., Tylenol with codeine)

Schedule V: accepted medical use—very low abuse (e.g., liquid cough medications)

TABLE 41-4

Controlled Substance Log Required Information

Drug name	Drug strength
Date of use	Client name
Patient name	Amount used
Amount remaining	Initial of handler(s)

Competency Skill

OSHA Veterinary Hospital Plan

Objective:

To become aware of the facilities safety plans and know what role to play when a plan needs to be activated.

Preparation:

OSHA plan, safety binder, or policies and procedures manual; MSDS binder

Procedure:

1. Review the veterinary OSHA plan.
2. Identify all hazards in the work area (physical, chemical, zoonotic, and biological).

3. Locate the MSDS binder.
4. Review the safety labeling on hazardous items.
5. Learn the emergency procedures for the hospital (accidents, hazardous materials spills, fire evacuation, gas releases).
6. Review emergency procedures for patient evacuation.
7. Locate the fire extinguishers.
8. Locate all PPE (personal protective equipment).
9. Locate where to dispose of sharps containers and medical waste.

Competency Skill

Lifting Heavy Objects

Objective:

To properly lift objects without injury.

Follow these guidelines when lifting heavy objects:

- Use more than one person if the object is over 50 pounds.
- To lift, each person should be in position to lift simultaneously.
- Bend at the knees and keep the back straight.
- Lift the object and move to the desired location.

SUMMARY

It is essential that all veterinary facility employees are properly trained in safety protection and have working knowledge of all safety plans. The MSDS safety sheets should be readily and easily accessible to all employees. Additional information and OSHA guidelines are available online at http://www.osha.gov or at 1-800-321-OSHA (6742) or at the U.S. Department of Labor, OSHA, 200 Constitution Avenue, Washington, DC 20210.

Key Terms

bacteria living organism that invades the body causing illness

biological hazards safety concerns that pose a risk to humans and animals through contamination of living organisms through body tissues and fluids, such as needles or surgery blades

chemical hazards safety concerns of products that may cause injury such as burns or vapors that cause injuries to the eye, lungs, or skin

chemotherapy treatment using chemical agents that help with cases of cancer

(*Continues*)

controlled substances drugs that have the potential for abuse and addiction and must be kept in a locked cabinet by law

controlled substance log area where medicines that are scheduled drugs are kept when dispensed and must be maintained accurately and kept of file for 2 years

direct contact transmission of a disease through a direct source, such as saliva or the ground

Drug Enforcement Administration (DEA) agency that regulates controlled substances and requires veterinarians to obtain a license to dispense scheduled drugs

fungus living organism that invades the external area of the body through direct or indirect contact

indirect contact transmission of a disease through an indirect source such as the air or water

medical waste storage and trash for items such as surgical drapes or bandage material contaminated by body fluids and living tissues that may be infected with contagious diseases

MSDS safety sheets material safety data sheets; information published by manufacturers of various products that have the potential to harm humans within the facility

OSHA Occupational Safety and Health Administration; agency that provides for safe and healthy work environments and conditions for all employees

parasites living organisms that may occur on the internal or external area of the body and feed off the animal, causing disease

personal protective equipment (PPE) equipment used for safety purposes and protection

physical hazards safety concerns that may cause physical injuries such as bites, kicks, scratches, or back injuries caused by lifting heavy objects

protozoans one-celled simple organisms or parasites

rickettsial more complex multi-cellular organism that are tick- or flea-borne

right-to-know station area in the facility where the OSHA and MSDS binders are kept with safety and product information

scheduled drugs medicines with the potential for abuse and categorized from I to V according to strength; I is highest potential and V lowest potential for abuse

sharps sharp instruments and equipment that can injure a human or animal and cause a wound or cut that transmits a contagious disease due to contamination

sharps container area that holds sharp items, such as needles, surgical blades, and glass, that may spread disease

virus causes a disease that is not treatable and must run it course

zoonotic hazards safety concerns that allow contagious organisms to be spread to humans causing infections, viruses, bacterial, fungal, and parasitic transmission

REVIEW QUESTIONS

1. What are the four types of hazards found in the veterinary industry?
2. What does OSHA stand for?
3. What is OSHA's importance in the veterinary industry?
4. What are the four classes of zoonotic infection sources?
5. What is the difference between direct and indirect contact? Give examples of each.
6. What are MSDS?
7. What are the parts of the MSDS?
8. What is the DEA?
9. What is the importance of the DEA in veterinary medicine?
10. What is a controlled substance?
11. What information must be legally kept in the controlled substance log?

Clinical Situation

© iStock/Ralph Huyskens

Jessica, a new veterinary assistant at Glenmoor Vet Clinic, was checking the post-op surgery patients in the afternoon. Dr. Miles, the veterinarian who performed the daily surgeries, asked her to check each patient for any signs of pain or bleeding. Jessica carefully looked at each animal, opening and closing each cage, observing each animal for signs of pain. She also looked at each animal's surgical incision. She did not wear gloves and did not wash her hands between each patient. She also decided to offer each animal small amounts of food and water. One small Beagle, named "Daisy," appeared extremely sleepy. She gave her a new towel and patted her head. Another patient, a cat named "Rascal," who was declawed that day, was attempting to bite at his bandages. Jessica giggled, saying "Oh, 'Rascal,' are you trying to chew your booties?" She wrapped the cat in a large towel, thinking this would keep him from thinking about the bandages. She quietly turned out the lights, tuned on a radio, and left the post-op ward.

- What did Jessica do correctly?
- What did Jessica do incorrectly?
- What should Jessica have done in the surgery ward?

42

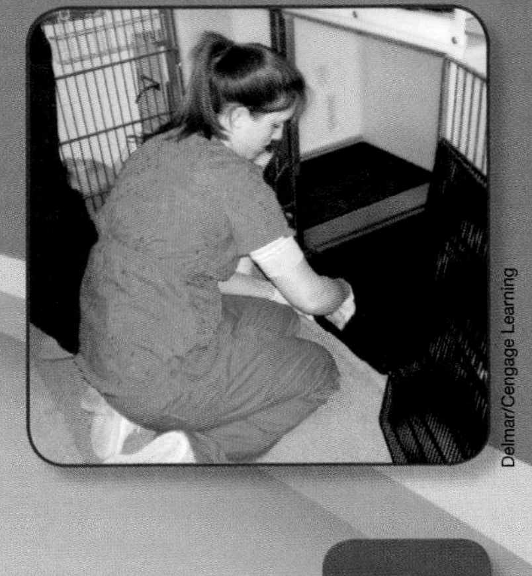

Veterinary Sanitation and Aseptic Technique

Objectives

Upon completion of this chapter, the reader should be able to:

- Explain the types of sanitation used in the veterinary industry
- Discuss methods of sanitation in a veterinary facility
- Describe common disinfectants used in a veterinary facility
- State the considerations in selecting a disinfectant
- Discuss areas of importance in keeping clean in the veterinary facility
- Identify factors of cleanliness in the veterinary facility
- Describe aseptic techniques
- Discuss the importance of the isolation ward

Introduction

The care and cleanliness of the veterinary facility serves as a safety measure to staff, clients, and patients. Preventing the spread of disease is of utmost importance. Sanitation, disinfection, and aseptic techniques govern the standards of proper cleanliness in a veterinary care facility. All staff members should have proper knowledge of how to clean, disinfect, sanitize, and sterilize the areas of the veterinary facility. When clients enter the facility, the smell and appearance will be their first impression, and first impressions are everything to the client.

Methods of Sanitation

There are many methods of sanitation used in a veterinary facility. **Sanitation** is the process of keeping an area clean and neat. This includes appearance and odor control. The type of sanitation will depend on the location in the facility, the purpose of the area, and the type of chemical or product used for sanitation purposes. First, it is important to understand the levels of cleaning. **Cleaning** is the process of physically removing all visible signs of dirt and organic matter such as feces, blood, and hair. **Disinfecting** is the process of destroying most microorganisms on nonliving things by physical or chemical means (see Figure 42-1). **Sterilizing** is the process of destroying *all* microorganisms and viruses on an object using chemicals and/or extreme heat or cold under pressure.

Physical Cleaning

Physical cleaning is the most common method of sanitary control within the veterinary facility. It involves the use of a chemical with a cleaning object, such as a mop, sponge, or washcloth. This action may include dusting, mopping, or cleaning up urine or feces within a cage. Physical cleaning is the use of the hands to remove dirt, debris, and organisms from all surfaces of the veterinary facility. This includes disinfecting areas to prevent the spread of disease (see Figure 42-2).

FIGURE 42-1 Disinfection of animal cages.

FIGURE 42-2 Mopping is one method used to physically clean the facility.

Sterilization

Sterilization is the process of killing all living organisms on a surface. This is typically used in the exam rooms, treatment area, and surgical suite to ensure the tables and instruments are free of disease. There are several classifications of sterilization.

Cold Sterilization

Cold sterilization is the process of soaking items in a disinfectant chemical until they are cleaned for reuse. A **cold tray** holds a chemical that acts as a sterilizing agent (see Figure 42-3). Items such as nail clippers, brushes, endotracheal tubes, and some surgical scissors may be prepared through cold sterilization techniques. These are items used repeatedly in the facility and require a simple and quick sanitation method. Table 42-1 lists common cold tray solutions.

FIGURE 42-3 Cold tray used for sterilization.

Dry Heat or Incineration

Dry heat may be used with the use of a flame by exposing an item to extreme heat or through incineration. An **incinerator** is used to burn the remains of items that have the potential to spread disease. **Incineration** may be necessary with biological hazards, medical wastes, or animal carcasses that may be contagious or zoonotic. This method requires items to be burned to ashes to prevent the spread of disease.

Autoclave

Another common method for sterilizing items, especially surgical equipment, is the use of an autoclave. An **autoclave** is a piece of equipment in the form of a sealed chamber in which objects are exposed to heat and steam under pressure at extremely hot temperatures to kill all living organisms (see Figure 42-4).

Radiation and Ultrasound

Another method of sterilization may be through the use of radiation or ultrasound. **Radiation** is using ultraviolet or gamma rays that radiate and kill living organisms. **Ultrasound** is passing high-frequency sound waves through a solution to create a vibration that scrubs an object to remove debris. This is commonly done using an ultrasonic cleaner machine where items soak in a solution that vibrates to remove dirt and debris.

FIGURE 42-4 Autoclave.

TABLE 42-1

Common Cold Tray Solutions

SOLUTION NAME	USE	CONTACT TIME	DILUTION
Chlorhexidine Solution	Dilute with water for wiping, soaking, mopping, and spraying on external animate or inanimate objects.	5 minutes	1 ounce or 2 tablespoons per gallon of water
Nolvasan Solution	Dilute with water for wiping, mopping, and spraying on or soaking inanimate objects.	10 minutes	1 ounce or 2 tablespoons per gallon of water
Cetylcide-G	Dilute with water for soaking inanimate objects.	40 minutes	60 ml per gallon water; 15 ml per quart water
Pink Germicide	Dilute with water for soaking inanimate objects.	4 minutes	10 ml per quart water

Filtration

Another method of sanitation is the use of filtration within a facility. **Filtration** is removing particles from the air using a physical barrier and is common in lab areas or research facilities. This usually entails a separate room that is entered prior to entering the area where animals are housed or contained. This room is pressurized to prevent organisms from entering the facility. It is common for people entering a filtration area to change into sterile clothing and PPE equipment prior to entering the housing area.

Veterinary Sanitation Chemicals and Cleaners

There are many common chemicals in veterinary facilities that are used for cleaning, disinfecting, and sterilization. Chemicals should always be handled with care and consideration. Some chemicals are harmful to people, animals, or both. Certain chemicals may have vapors that are harmful if inhaled. Other chemicals may cause burns if contact is made with the skin or eyes. These are reasons to be careful when mixing or handling any chemical. It is important to always read the bottle label or MSDS to determine the safety of the chemical. Chemicals should only be handled or used when wearing exam gloves. When mixing or diluting chemicals, it is important to read all labels and directions before using the materials. Some cleaning agents may be used in one location but not in another. A chemical solution that is used on the floor may be harmful to an animal and should not be used to clean a cage or exam table.

Antiseptics are solutions that destroy microorganisms or inhibit their growth on living tissue and are effective disinfecting agents. Disinfecting agents have a variety of properties that alter their uses and effectiveness. Some of these properties include the following:

- Spectrum of activity—what the agent will kill such as viruses, bacteria, or fungus
- Concentration of the solution—the strength of the solution. Some agents must be diluted before use and some can be used at full strength
- Contact time of the solution—how long the disinfectant should sit before being cleaned from the surface
- Appropriate surface uses—the types of items the agent may be used to clean
- Inhibiting factors—uses that should be avoided when using the disinfectant
- Toxic effects—the hazardous effects the agent may have to humans or animals

TERMINOLOGY TIP

Spectrum refers to a wide variety of factors.

TERMINOLOGY TIP

Dilution means to lessen in strength by adding another component, such as water. This requires adequate measuring skills.

Some of the common agents used in the veterinary facility are summarized in Table 42-2.

TABLE 42-2

Common Veterinary Cleaners and Disinfectants

CLEANER	TYPE OF AGENT	USE
Nolvasan Solution (Chlorhexadine)	Disinfectant	Used on inanimate objects. Dilute with water at 1:40 dilution. Used in cold sterilization. Two-day activity. Inhibited by soaps.
Nolvasan Scrub (Chlorhexidine)	Disinfectant and antiseptic	Used on humans and animals. Used for surgical preps. Sterile prep. Two-day activity. Inhibited by soaps.

(Continues)

TABLE 42-2 (*Continued*)

Clorox or Chlorine Bleach	Cleaner disinfectant	Caution with odors and vapors. Toxic to skin. Treats all living organisms.
Pink Germacide	Disinfectant	Used on inanimate objects. Dilute with water. Used in cold sterilization. Toxic to skin.
Instrument Milk	Cleaner and disinfectant	Used on surgical equipment and instruments. Used in cold sterilization.
Roccal-D	Disinfectant	Dilute with water. Toxic to skin.
Citrus II	Cleaner and disinfectant	Dilute with water. Pleasant smelling. Used on inanimate objects.
Ethyl Alcohol or Isopropyl Alcohol	Disinfectant and antiseptic	Requires long-term contact to cause sterilization. Requires 70% concentration. Used on living and nonliving sites.
Iodine or Betadine	Disinfectant and antiseptic	Use with gloves due to staining effect on skin. Used on living and nonliving sites. 2% solution—1:10 dilution for skin contact—1:100 dilution for tissue lavage. Surgical scrub and solution. Inactive with alcohol or organic matter. Four- to 6-hour activity.
Quaternary Ammonia	Disinfectant	Toxic to skin. Toxic vapors. Caution mixing with other chemicals. Inanimate objects.
Formaldehyde	Disinfectant	Preservative of tissues. Used less due to highly toxic affects on skin and toxic vapors. Potential for causing cancer.
Hydrogen Peroxide	Antiseptic	Used to clean areas on patients, especially blood and skin wounds. Used at 3–20% concentration.

Competency Skill

Dilution of a Substance or Disinfectant

Objective:

To properly work with disinfecting agents to obtain the best possible effects of cleanliness in the facility.

Equipment needed:

disinfectant, PPE, containers, dilutant

Procedure:

1. Assemble all disinfectants in one location.
2. Wear eye protection and gloves when handling chemicals.
3. Assemble secondary containers, such as buckets or bottles.
4. Dilute product as label indicates. Use a syringe to measure proper amount.
    - 1 teaspoon = 5 ml
    - 1 tablespoon = 15 ml
    - 1 ounce = 30 ml
5. Add water to the container.
6. Place dilatants into the container without overfilling.
7. Replace lids on containers, wipe from bottles any spillage, and place in storage location.

Hand Hygiene

Proper hygiene includes hand-washing (see Figure 42-5). For animals and other staff members to be safe and sanitary, it is important that everyone in the veterinary health care field practice proper hand-washing techniques. The most common method of spreading disease is through direct hand contact.

Hand-washing should take place:

- After visiting the restroom
- After coughing
- After sneezing
- After removing exam gloves
- After touching or handling each animal
- After handling money
- After contact with people (shaking hands)
- After contact with visibly sick people
- After touching phones
- After touching door handles
- After using a computer keyboard

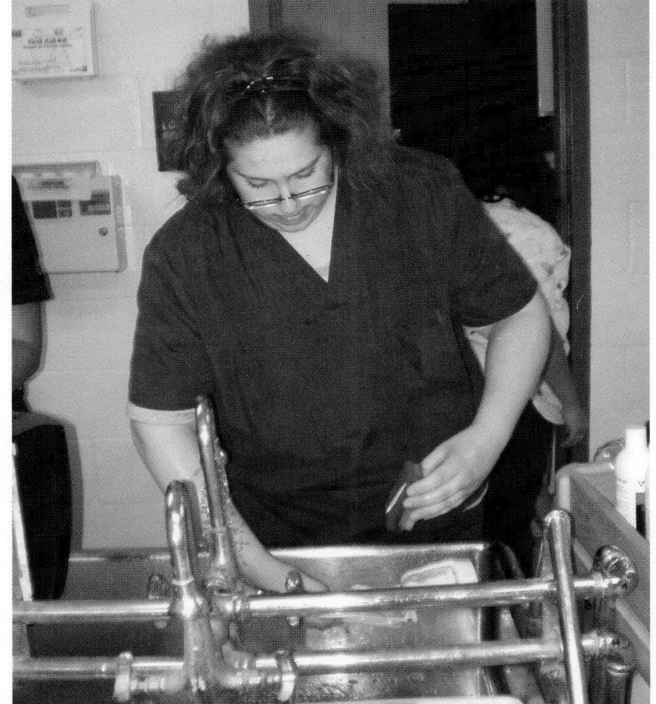

FIGURE 42-5 Proper hand-washing is the key to prevention of the spread of infection.

The proper hand-washing procedure begins with an antibacterial soap and the use of hand sanitizer (see Figure 42-6). Staff members should avoid wearing jewelry, having artificial nails, and allowing nails to grow beyond the end of the finger, as this will decrease the possibility of contamination. Due to frequent hand-washing, the use of moisturizers is recommended to keep hands soft and skin healthy. Signs should be placed throughout the facility to remind all staff members to wash their hands.

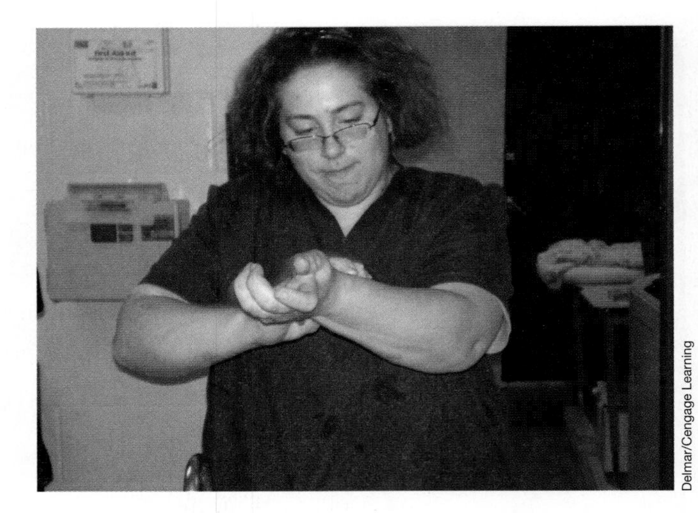

FIGURE 42-6 An antibacterial solution should be used initially when washing the hands.

Competency Skill

Hand-Washing

Objective:

To properly and effectively clean the hands to prevent cross contamination of oneself, patients, and clients.

Equipment needed:

antibacterial soap, sink with warm water, paper towels, hand sanitizer, hand lotion

Procedure:

1. Gather paper towels.
2. Wet hands with warm water.
3. Place antibacterial soap into the palm of one hand. Use an amount the size of a nickel.
4. Use vigorous friction by rubbing both hands together and lather both hands to the wrist consistently for 15–30 seconds.
5. Cover all surfaces of the hand, including the back of the hands and between the fingers.
6. Rinse hands thoroughly allowing the water to run downward off the skin.
7. Dry hands completely with paper towels.
8. Use a paper towel to turn off faucets and open doors.
9. Clean up area.

Housekeeping and General Cleaning

The most important aspect of general cleaning in a veterinary facility is to practice good hygiene and odor control. Cleanliness must be a priority to every staff member. If an area looks clean but smells dirty, then the cleanliness has not been achieved. Many areas of the veterinary clinic need daily cleaning and routine disinfecting. Some areas will need to remain sterile for the safety of staff and patients. Areas of high traffic and animal contact require repeated methods of cleaning and sanitation. Cleaning protocols will vary from facility and area of the hospital. It is important that all staff members know proper disinfecting and cleaning protocols for the entire hospital.

Many practices develop a daily and weekly cleaning routine. Daily cleaning is conducted on areas and objects that require strict cleansing (see Figure 42-7). Every staff member should practice good hygiene and clean up after himself or herself, including the equipment and supplies that he or she uses. It is everyone's responsibility to help keep the facility clean on a daily basis. This career requires keeping an open mind of the duties that may be required as a veterinary professional. Many veterinary assistants are required to keep the facility clean, the animals clean, and equipment maintained. This may include animal bathing and grooming, cage and kennel disinfecting, and lawn care. It is important that the veterinary assistant be familiar with the facility's cleaning and disinfecting methods and chemicals available for use.

It is important to note some rules for cleaning:

- Work from the top to the bottom when cleaning or disinfecting items.
- Work from the back of a space toward the door or entrance.
- Use towels, beds, or newspaper as needed to keep animals and cages clean.

All veterinary team members should practice the "clean as you go" attitude to keep the facility clean and safe. This means putting items back where they were originally located, replacing items after use, tidying up an area after a procedure, dusting and sweeping, and cleaning up messes that may occur. This practice helps minimize the spread of disease. Many factors can be considered in helping decrease the spread of disease to animals and humans. These factors include the following:

- Limit area where animals eliminate (urinate/defecate).

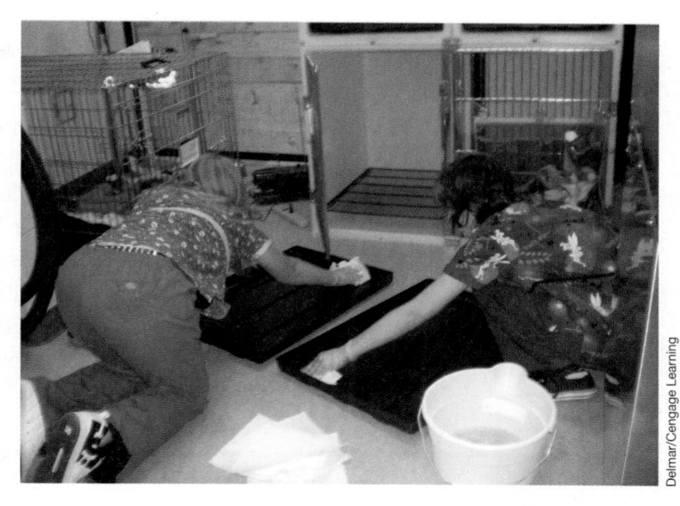

FIGURE 42-7 Animal enclosures must be cleaned daily.

- Routinely clean and change bedding in stalls and cages.
- Remove uneaten food.
- Change and clean water sources.
- Change clothing when contaminated.
- Treat animals in isolation last.
- Clean as you go.
- Use proper odor control.
- Clean rugs and chairs.
- Wipe down walls and doors.
- Wear gloves when handling animals.
- Keep a second set of clothes and scrubs to replace contaminated work clothes.
- Clean parking lots and walk areas.

Cleaning must be done properly to allow proper disinfection to follow. Cleaning a surface depends on whether it is a smooth or rough surface. Also, the wash ability of the surface must be considered. Some items are best vacuumed or mopped for cleaning. Fabrics may need laundering for proper cleaning. Hair is the most common issue within the veterinary clinic. Hair must be removed from all surfaces for proper disinfecting to take place. Hair may float around the clinic causing contamination or the spread of disease. Sweeping, vacuuming, dusting, and mopping should be a routine part of the daily cleaning schedule. Windows and glass doors should be cleaned as well. Table 42-3 lists the various areas that must be cleaned in the veterinary facility.

TABLE 42-3

Veterinary Areas to Clean

Exam tables	Countertops	Sinks and faucets
Wash tubs	Desk tops	Floors
Walls	Light switches	Cages/kennels
Door handles	Scales	Drawer handles
Windows and glass	Cabinets and shelves	Restrooms
Instruments	Equipment	Computer keyboards
Isolation ward	Surgical suite	X-ray table
Telephones	Lab area	Air filters

Competency Skill

Cage or Kennel Cleaning

Objective:

To properly clean animal enclosures to eliminate the risk or spread of infection.

Equipment needed:

disinfectant, waste containers, PPE, buckets, sponges

Procedure:

1. Remove all bedding, toys, or bowls.
2. Place dirty bedding into laundry area using appropriate laundry detergent.
3. Disinfect all toys and bowls appropriately with soapy disinfectant and water and allow soaking.
4. Remove all feces, urine, or other dirt materials with paper towels and discard in the garbage or medical waste.
5. Wash all sides of the cage or kennel walls, including door, grates, and latches, with appropriate disinfectant and warm water. Use a washcloth soaked in disinfectant.
6. Use a stiff brush for any items on walls or other areas that are not easily removed with a washcloth.
7. Allow the disinfectant to stand on the cage items for the allotted contact time.
8. Rinse with warm water.
9. Dry all items.
10. Disinfect or launder all cleaning items.
11. Wipe all light switches, doors, and walls of the kennel area daily.
12. Replenish cleaning supplies as necessary.

Competency Skill

Cleaning Surfaces and Supplies

Objective:

To understand how to properly maintain and clean all surfaces in the veterinary facility.

Equipment needed:

disinfectants, buckets, sponges, PPE

Procedure:

1. Evaluate the areas and surfaces that need to be cleaned.
2. Evaluate the cleaners and disinfectants that may be safely used on the areas of the facility.
3. Dust all areas to remove debris and hair.
4. Sweep or vacuum the floor to remove debris and hair.
5. Mop all floors and place wet floor signs for safety.
6. Dry all wet areas to prevent streaks from forming.

Competency Skill

Laundry

Objective:

To properly launder and maintain cleanliness of the linens and clothing in the veterinary facility.

Equipment needed:

detergent, washing machine, dryer

Procedure:

1. Sort laundry materials and size of the laundry load.
2. Pre-treat all laundry in washing machine by soaking.
3. Set washing cycle for load size and proper temperature, usually warm to hot water.
4. Add laundry detergent that disinfects bedding and other items.
5. Place items in dryer when washer cycle is complete.
6. Fold items properly and store in closed cabinet.

Exam Room Sanitation

Exam room sanitation is an important part of safety and disease control. It is important for not only the exam area to be sanitary but also the equipment, tools, and supplies to be clean and properly disinfected, and staff members should be using sanitary practices and proper hand-washing techniques.

The exam room should be cleaned after every patient that is seen in the room and at the end of the day. This includes sweeping up hair and debris. It is best to clean from the top of the room to the bottom of the room and from dry areas to wet areas. This will prevent dust, debris, and bacteria from spreading to cleaned areas. Areas of the exam room that contain the most dirt and debris should be cleaned first. Begin cleaning at a starting point and work clockwise around the room. This will allow every area to be cleaned. When using a spray disinfectant on tables and countertops, one should spray and wipe using an ample amount of disinfectant and allowing enough contact time as noted on the label to provide proper sanitation. When using mixed chemicals with cloths or sponges, make sure the area is cleaned well and dried to prevent streak marks. Always clean using an up-and-down motion or side-to-side motion to ensure all surfaces have been covered. Chemicals should be safe to use around animals and produce little to no odor that may cause respiratory irritation.

The trash should be cleaned up on a regular basis to prevent the spread of disease and odors from forming. Spray and wipe out the empty trash cans and replace each can with a new liner. Place any sharp items that have been used in the hazardous waste or sharps container. When the sharps containers are full, make sure they are closed securely and place them in a location for proper removal.

A cold tray should be located in each exam room to allow tools and equipment that may be safely submerged in disinfectant to be sanitized. When tools are placed in cold trays they should first be wiped free of any blood, debris, or bodily waste materials using a disinfectant. The tool should then be placed in the cold tray and fully submerged. Soak time should be followed according to the disinfectant label. Most soak times are short with an average of 5 minutes. After soaking, the tool should be removed from the cold tray and washed with warm water and then dried well to prevent any rusting. The cold tray chemicals should be changed on a regular basis depending on how much they are used.

All areas of the exam room floor should be swept or vacuumed and cleared of debris prior to mopping. A bucket or large push bucket with ringer should be used for mopping. This allows proper dilution of chemicals in water and decreases the chance of cross-contamination in another area. A mop bucket should be used only in exam rooms and in no other areas of the facility to prevent spread of disease. A wet floor sign should be placed near the area prior to mopping. The solution should be placed in the mop bucket and warm water added to the bucket in appropriate amounts. Place the mop into the cleaning solution and wring it out so that it is only damp. Begin at the farthest corner of the exam room and mop the floor area lengthwise along the base of the room. Then use figure-eight motions along the remaining area of the floor. Apply more disinfectant as the mop becomes dry and repeat as necessary. Work toward the door and allow the room to air dry. When the bucket water becomes visibly soiled, it should be changed immediately. When finished mopping, empty the mop bucket in a safe area. Wring out the mop and allow the mop to air dry. All mops should be rinsed thoroughly and laundered on a regular basis. Never use a dirty or odorous mop to clean.

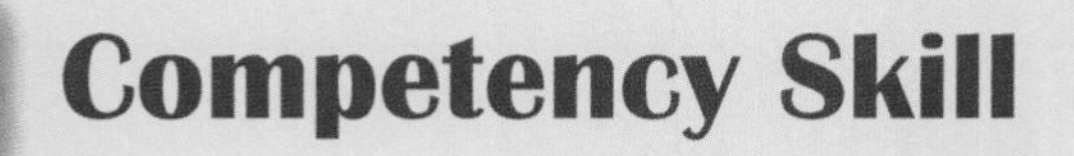

Competency Skill

Cleaning Exam Rooms and the Surgical Suite

Objective:

To properly maintain cleanliness and sterility in examination rooms and surgical suites.

Equipment needed:

disinfectants, buckets, sponges, PPE, waste containers

Procedure:

1. Gather all disposable materials and place in garbage or medical waste container.
2. Clean areas of hair and other debris.
3. Apply disinfectant solution to all areas of animal and human contact. Dry with paper towels.
4. Sweep or vacuum floor.
5. Spot clean or mop floor. Place wet floor sign for safety.

Competency Skill

Cleaning Surgical Instruments

Objective:

To properly clean and sterilize surgical instruments and prevent the spread of infection.

Equipment needed:

disinfectants, sterilization tools, PPE

Procedure:

1. Place all instruments in a cold tray solution.
2. Use a wire brush to scrub all debris from items and rinse with instrument milk or cleaner.
3. Rinse with distilled water.
4. Use ultrasonic cleaner if large amounts of debris. Place items in cleaner and soak basket in solution.
5. Follow recommendations of manufacturer.
6. Remove from solution and rinse with distilled water.
7. Dry instruments.
8. Use instrument milk to lubricate items by immersing for 30 seconds.
9. Lay on towel with all hinged instruments open to dry.

Aseptic Techniques

The most important principle in the veterinary facility is maintaining aseptic technique. **Asepsis** and the practice of **aseptic technique** is the practice of keeping a sterile environment and keeping the environment disease- and contaminant-free. This is especially vital in the surgical suite. Aseptic technique governs how the facility is cleaned, how equipment and instruments are cleaned, and how surgical and medical procedures are preformed. A break in aseptic technique leads to possible infection, disease, and potential patient death.

Sterile techniques include washing the hands frequently, wearing gloves when handling animals or other possible contaminants, and cleaning all surfaces with disinfectants to prevent spread of disease. When a human causes the spread of disease and contamination of an animal, this is known as a **nosocomial infection**. A nosocomial infection may occur with unsterile surgical practices, contamination of a healthy animal due to unsafe sterile practices with hands or equipment not being cleansed, or allowing contagious animals to be in contact with healthy animals.

Isolation Ward

Hospital housing is arranged according to the type of cases and patient care needs. This may include healthy patients, outpatients, surgical patients, medical patients, and contagious patients. Contagious patients are kept away from all healthy patients in the **isolation ward**. This separate

Competency Skill

Aseptic and Sterile Techniques

Objective:

To properly maintain asepsis in the veterinary facility.

Equipment needed:

disinfectants, PPE, appropriate waste containers

Procedure:

1. Treat every patient as if it were contagious.
2. Wear scrubs or a lab coat to protect clothing and patients.
3. Wear mask, goggles, foot covers, and hair covers as necessary.
4. Disinfect everything a patient or human uses or is in contact with.
5. Place contagious animals in isolation.
6. Place all disposable items in garbage or medical waste.
7. Disinfect all non-disposable items by soaking in a cold tray or cleaning and sterilizing.
8. Place used bedding in the laundry area for immediate attention.
9. Wash hands after removing exam gloves. Use an antibacterial soap and hand sanitizer.
10. Change scrubs when in contact with contagious animals or any body fluids.

housing groups similar patients, making it safe for all animals and staff. Each ward should have its own medical supplies, cleaners, disinfectants, and equipment. The isolation ward should have items that are never removed from this location. The laundry and bedding should be disinfected separately from other laundered items. Signs should be kept on all cages and doors detailing the types of contagious diseases of the patients that are housed in the ward. All staff members that enter should wear exam gloves, sterile gowns, face masks, shoe covers, and hair covers (see Figure 42-8). All items should be discarded and either thrown away or appropriately disinfected after each use in the isolation ward.

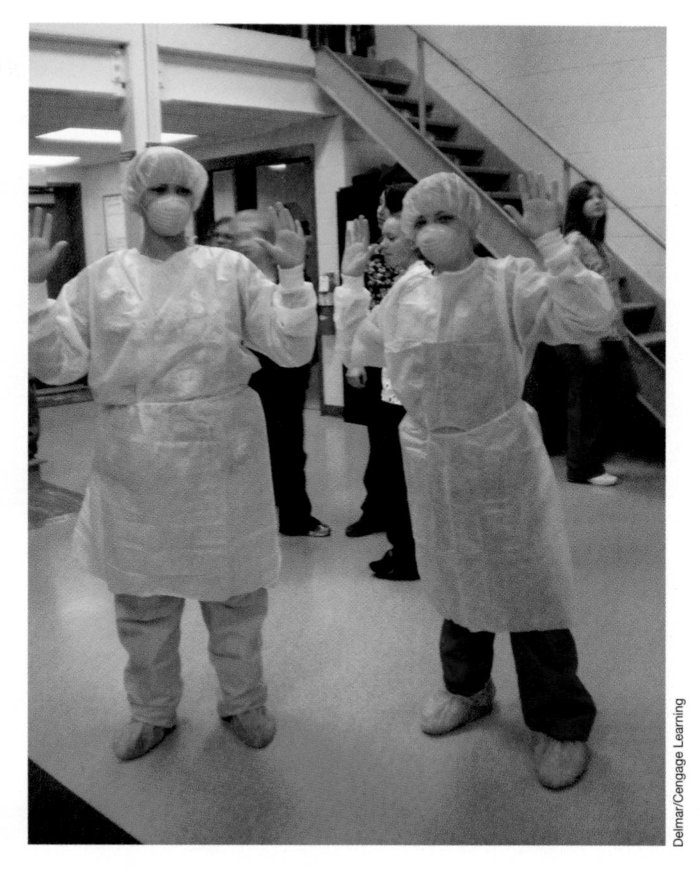

Delmar/Cengage Learning

FIGURE 42-8 Proper PPE must be worn when working within an isolation ward.

SUMMARY

Sanitation involves cleaning and disinfecting the veterinary facility. This controls the direct spread of organisms on surfaces, in the air, and on other objects within the facility. All members of the veterinary staff must follow standard sanitation procedures. Failure to maintain a high standard of disinfecting and sanitation will result in harm to the patients, loss of clients, a decrease in business, and the loss of employee jobs. Every staff member must have a high level of awareness for the potential transmission of diseases from patient to patient and patient to people. Everyone must be diligent in proper cleaning and sanitation techniques and follow protocols for aseptic procedures.

Key Terms

antiseptics solutions that destroy microorganisms or inhibit their growth on living tissue and are effective disinfecting agents

asepsis the practice of keeping a sterile environment and keeping the environment disease- and contaminant-free

aseptic technique governs how the facility is cleaned, how equipment and instruments are cleaned, and how surgical and medical procedures are preformed

autoclave a piece of equipment in the form of a sealed chamber in which objects are exposed to heat and steam under pressure at extremely hot temperatures to kill all living organisms

cleaning the process of physically removing all visible signs of dirt and organic matter such as feces, blood, and hair

cold sterilization the process of soaking items in a disinfectant chemical until they are necessary for reuse

cold tray holds a chemical that acts as a sterilizing agent

dilution to lessen in strength by adding another component, such as water

disinfecting the process of destroying most microorganisms on nonliving things by physical or chemical means

dry heat method of sterilization through the use of a flame by exposing an item to extreme heat or through incineration

filtration removing particles from the air using a physical barrier; common in lab areas or research facilities

incineration the burning of infectious materials or animal carcasses

incinerator a device used to burn the remains of items that have the potential to spread disease

(*Continues*)

isolation ward separate housing that groups similar patients, making it safe for all animals and staff

nosocomial infection when a human causes the spread of disease and contamination of an animal

physical cleaning the most common method of sanitary control within the veterinary facility

radiation using ultraviolet or gamma rays that radiate and kill living organisms

sanitation the process of keeping an area clean and neat

spectrum wide variety of factors

sterile techniques washing the hands frequently, wearing gloves when handling animals or other possible contaminants, and cleaning all surfaces with disinfectants to prevent spread of disease

sterilization killing all living organisms on a surface

sterilizing destroying all microorganisms and viruses on an object using chemicals and/or extreme heat or cold under pressure

ultrasound passing high-frequency sound waves through a solution to create a vibration that scrubs an object to remove debris

REVIEW QUESTIONS

1. What are the three levels of cleaning?
2. What are methods of sterilization?
3. What are the factors to consider when using disinfectants?
4. What are some chemicals and disinfectants used in the veterinary hospital?
5. What is aseptic technique?
6. What areas should be cleaned in a veterinary facility?

Clinical Situation

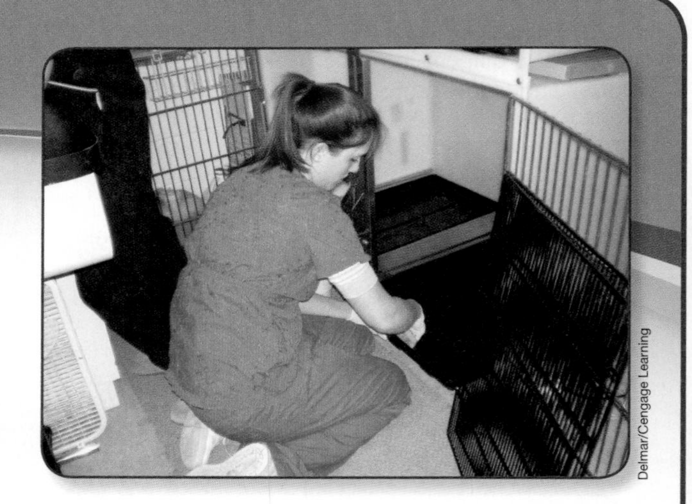
Delmar/Cengage Learning

Judy, a veterinary assistant, and Leslie, a veterinary technician, have been working the entire morning cleaning the exam rooms and the surgery area. Judy has been working on disinfecting the exam rooms and equipment. Leslie has been sterilizing the surgical suite and autoclaving instruments for the next day's surgery schedule.

"I am out of disinfectant in the exam rooms," says Judy. "What else can I use to disinfect the rooms?"

"Just get some Lysol out of the closet." says Leslie. "That should work."

Judy locates the Lysol and begins cleaning the exam areas and equipment with Lysol. She notices several other disinfectants in the closet, but the lemon-scented Lysol smells so much better. When she is done, she takes the mop and bucket of water and Lysol mixture she has used in the exam rooms to Leslie and begins mopping the surgery floor.

- Did Leslie and Judy follow proper cleaning and disinfecting guidelines?
- How would have you handled cleaning the facility?
- Is Lysol a good cleaning agent? Explain.

43

Delmar/Cengage Learning

The Physical Examination and Patient History

Objectives

Upon completion of this chapter, the reader should be able to:

- Demonstrate how to obtain a complete patient history
- Explain the importance of the patient history
- Discuss the tools and supplies necessary for a physical examination
- Explain the importance of the physical examination
- Describe normal and abnormal body system signs
- Discuss each body system evaluated in the physical examination

Introduction

The physical examination is an essential evaluation of an animal's overall health. The complete and accurate results of a physical exam begin with a full patient history of an animal's past medical problems and other health factors that will determine if an animal is healthy or if the animal is showing signs of disease. Veterinary assistants are a vital part of obtaining a patient history. Some veterinarians also expect the support staff to be able to record vital signs and complete a pre-physical exam based on the owner's chief complaints and reason for visiting the veterinary facility.

The Patient History

The **patient history** is background information of past medical, surgical, nutritional, and behavioral factors that have occurred over the life of the animal. A patient history should be obtained with every new animal and updated with each patient visit to the veterinary facility. A thorough knowledge of the history of an animal is necessary. Frequently a careful description of the history or onset of a problem will simplify the diagnosis. Most owners will not provide additional information if they are not asked the appropriate questions, so the questions are important to the patient history. While it is usually best to deal with the person who spends a major portion of the day with the animal, information from other family members should be included if it contributes to a better understanding of the problem. Specific areas of history that should be included are as follows:

- Signalment (age, gender, breed)
- Past illnesses or diseases
- Past surgeries or medical procedures
- Previous injuries
- Genetic history of parents and other offspring
- Chief complaint
- Signs observed by the owner
- Changes that have occurred in the animal (behavior, attitude, exercise)
- Duration of signs
- Vaccine history
- Nutritional history

The medical history should be noted in brief comments within the chart, either on a patient history sheet or within the progress notes. It is best to not ask a leading question. Instead, it is best to ask "yes" or "no" questions to grasp an idea of any changes that have occurred with the animal. Asking open-ended questions will then encourage owners to give a more descriptive answer about what they are seeing with their pet. When the opportunity arises, ask the client to explain in detail what they have seen in the animal. Table 43-1 lists the common questions used to obtain accurate patient histories in veterinary practice.

The veterinary assistant should record the patient history according to what he or she interprets from the owner. It is important to note the **chief complaint**, or the reason the animal is being seen, as this is the most important issue to address. It is also important to establish how long the chief complaint has been occurring. Other areas of importance on the patient history include obtaining the patient's name, age or date of birth, gender, color, and breed or species. This information is known as the animal's **signalment**. This information provides statistics on the animal that may help indicate problems and also identify the animal. Vaccine history should include the last vaccination dates and the types of vaccines that were given. Nutrition is important to recognize, so the veterinary assistant should obtain diet information, how much is fed and how often, and if any changes in eating have occurred. Some problems are frequently associated with certain breeds, so it is important to establish an awareness of particular genetic problems associated with parents, grandparents, siblings, or progeny. *Progeny* refers to the offspring of the animal and genetic problems are easily passed on to the offspring. While the degree of a genetic disease will vary from animal to animal, the problem may be present to some degree in the presenting animal. Such information from an owner can be helpful in diagnosing a disease or condition. In some cases, especially behavior-related situations, it is ideal to obtain a home environment history, such as if the animal is indoor, outdoor, or both and what type of cage or confinement the animal has. Some other areas to consider with some patients are travel history, any medications that an animal is currently taking, and other animals in the household or that the pet comes in contact with.

The Physical Examination

A **physical examination (PE)** is an observation of each body system to determine abnormal problems that can cause a health problem. This includes examining each body system of the animal from head to tail. The PE is

TABLE 43-1

Examples of Patient History Questions

PATIENT HISTORY	CORRECT QUESTION	INCORRECT QUESTION
Appetite	"Has the appetite changed recently?"	"You haven't observed any change in appetite, have you?"
Chief complaint	"What is the reason for today's visit?"	"What do you need done today?"
Behavior	"Has there been a change in behavior?"	"There hasn't been a change in behavior, has there?"

one of the most important skills completed on an animal. It is also one of the most challenging, as animals can't talk and tell us what is wrong or where it hurts. A proper PE can detect early signs of illness before they become serious health issues. A yearly physical exam should be completed on every patient to recognize any changes in health (see Figure 43-1). As animals age, the frequency of a PE may be increased to every 6 months to detect changes. Animals that are having anesthesia or surgical procedures should have a PE to determine they are healthy enough for the procedure.

A consistent method should be developed and used with every patient exam. This helps to maintain an effective exam and not miss an area of importance. It also allows the staff member to assess areas quickly and easily. Although it is important to focus on the chief complaint, all other areas must be evaluated. A full PE should be given regardless of the immediate health problem. In some cases, multiple animals that need to be seen immediately for emergency situations may present to the facility. In these cases, the support staff may need to perform a **triage**, which is a quick general assessment of an animal to determine how quickly the animal needs a veterinarian's attention.

Tools for the Physical Examination

It is important to have all the PE tools in the areas of the facility where physical exams are typically completed (see Figure 43-2). This would include the exam rooms and the treatment area. All tools should be maintained and kept in working order; this would include checking any batteries, electrical cords, and switches. In some cases, certain species may require additional restraint and safety devices that should be available in case of handling problems. This would include muzzles, towels, welding gloves, anti-kick devices, and hobbles, depending on the animal species. Certain procedures that may also need to be completed during a physical exam may require the additional need for tools and supplies. Anticipation of these items is important in saving time and testing an animal's patience. It is also important to make certain there is enough space available for the PE in accordance to the size of the animal. Some animals may require additional restrainers and enough people should be available to help as necessary. The tools and the exam location should be prepared and disinfected prior to an animal entering the exam area. It is important to perform the PE in a quiet area that is comforting to the animal.

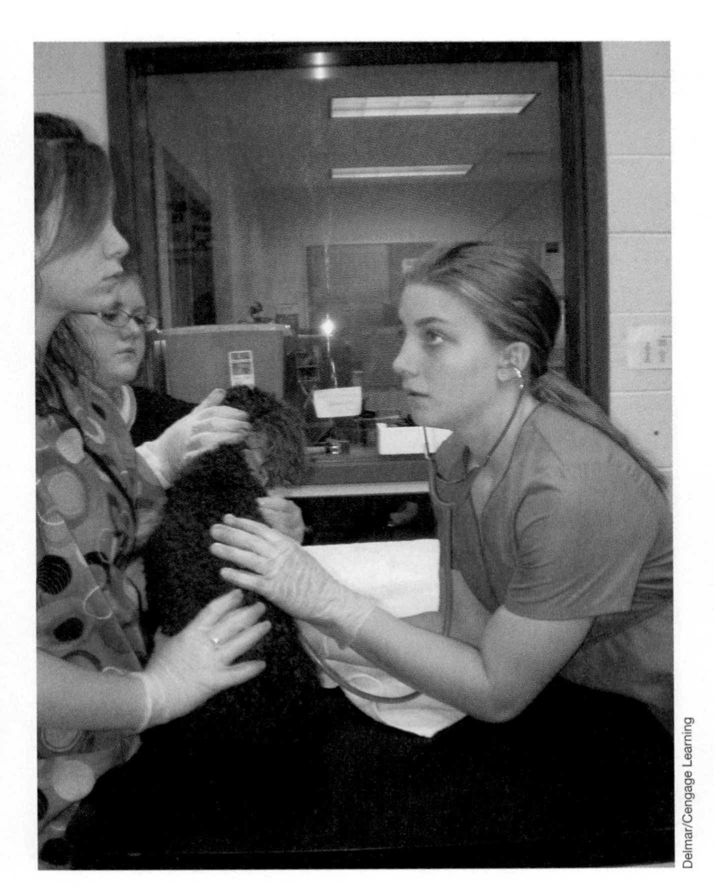

FIGURE 43-1 Physical examinations are an important part of maintaining a pet's health.

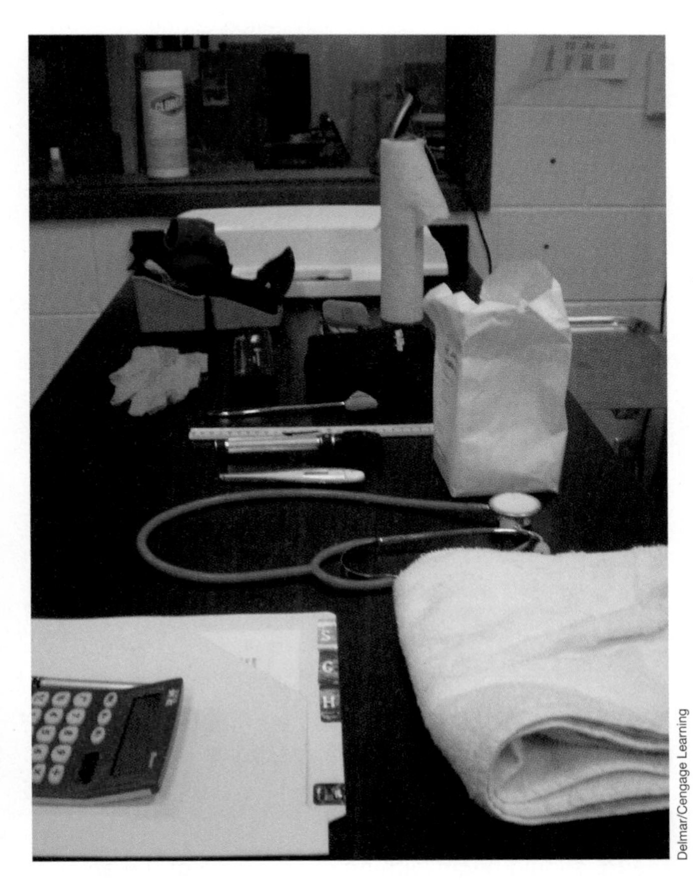

FIGURE 43-2 All physical examination tools should be set up before beginning any examination.

TABLE 43-2

Tools and Supplies for the Physical Examination

TOOL	FUNCTION
Stethoscope	Instrument used to examine and listen to the heart, chest, and lungs
Exam gloves	Protection to cover the hands to prevent contamination of the skin or transfer of organisms to other animals
Paper towels	For quick cleanup of any messes or spills
Restraint equipment	To safely handle an animal
Vaseline or KY gel	Lubricant used to ease discomfort when performing rectal exams or using a rectal thermometer
Otoscope	Instrument used to examine the ears
Medical chart	Documentation of the examination findings and record of patient history
Cotton-tipped applicators	Long, thin, wooden sticks with a cotton end used to clean areas of the body or apply medications; commonly called Q-tips
Gauze sponges	soft square mesh pads used to clean areas of an animal
Clock, timer, or watch	Used to count respirations and pulse rates
Ophthalmoscope	Instrument used to examine the eyes
Blue/black pen	Used to record information in a medical chart
Reflex hammer	Instrument used to tap areas of the body to test the animal's reflex responses
Cotton balls	Small, soft, white circular balls used to clean areas of an animal
Penlight	Instrument that provides a light source to observe the mouth or other body areas where there is limited light
Calculator	Used to determine drug dosages
Scale	Equipment used to weigh an animal
Thermometer	Instrument used to determine an animal's internal body temperature

Table 43-2 lists the tools commonly used in the physical examination and their functions.

Observation

The PE should begin with an observation of the animal. This is done before a hands-on exam is performed. An **observation** is an inspection of an animal at a distance to see what it looks like, acts like, or how it moves naturally. The veterinary assistant should observe the animal as it walks to the exam area, observing the gait for any limping or signs of lameness. This is also an appropriate time to observe the animal's overall behavior for signs of aggression, shyness, fear, dominance, or calmness. This is important when the hands-on exam must begin, to predict how the animal will react. The observation helps to note how the animal acts prior to stressing it out and potentially becoming excited or agitated by the hands-on assessment. Some other areas to note on observation include the following:

- Mental status (e.g., alert, focused)
- Conformation of the animal (body shape)
- Body condition score (e.g., thin, obese)
- Neurological problems (e.g., wobbling, unable to control movements, head tilt)
- Overall general appearance (e.g., weak, labored breathing, depressed)

On completing an observation, the veterinary assistant should socialize and greet the animal by allowing the animal to approach and sniff a hand. This will allow the animal to familiarize itself with the person and allow the person to determine the animal's behavior and acceptance.

Weight

The weight of an animal should be established when the animal enters the exam area. A scale should be provided according to the size of the animal. Small animals may be weighed on a feline scale or a pediatric baby scale (see Figure 43-3). Large dogs may be weighed on a floor

FIGURE 43-3 Small animals can be weighed on a feline or pediatric scale.

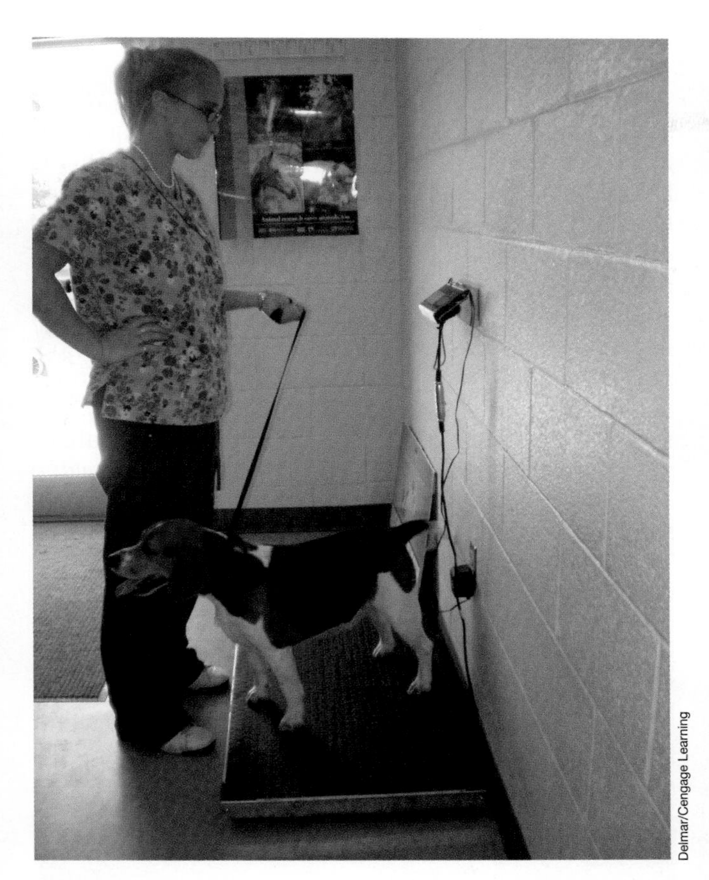

Delmar/Cengage Learning

FIGURE 43-4 Large animals can be weighed on a floor scale.

scale, and livestock should be weighed on a livestock scale or a weight tape measurement (see Figure 43-4). The weight should be recorded and assessed according to species and breed, as well as age, gender, purpose or use, and health status to determine any abnormal conditions.

The Oral Cavity

The **oral cavity** includes the gums, mouth, teeth, tongue, and throat. Examining the mouth area should include examining the teeth by lifting the lip, noting any signs of tartar on teeth, stains, odor, or broken teeth. The mucous membrane (mm) or gum color can be noted at this time (see Figure 43-5). Pink is the normal color of healthy gums. The capillary refill time (CRT) can be established by pressing on the gums and noting the time it takes for the color to return to the tissues. A normal CRT is 1–2 seconds. Then, safely open the mouth to examine the tongue and throat area for any abnormal appearance or objects (see Figure 43-6). If an animal appears to be aggressive, do not attempt to pry open the mouth as this may cause the animal to bite.

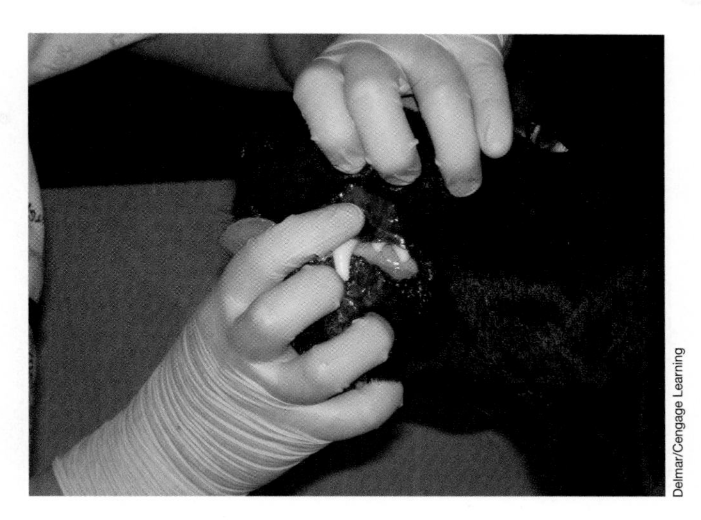

Delmar/Cengage Learning

FIGURE 43-5 When examining the oral cavity, note the color of the mucus membranes.

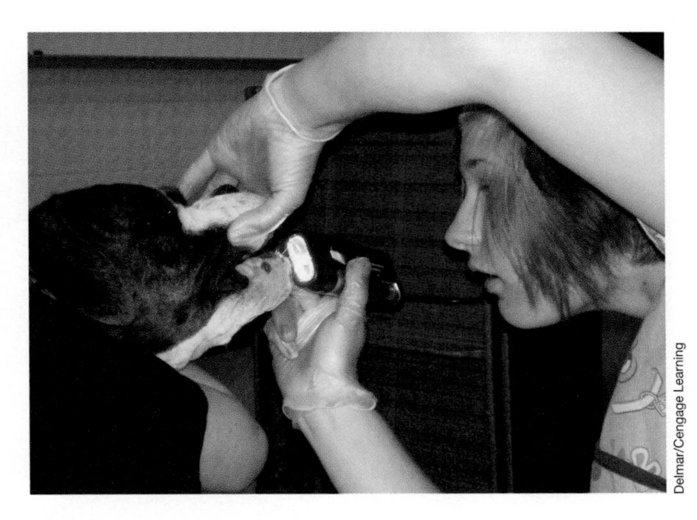

Delmar/Cengage Learning

FIGURE 43-6 Look inside the mouth to note any abnormalities or objects.

The Head Exam

The **head exam** includes examining the eyes, ears, and nose. The eyes should be examined using an ophthalmoscope to note any signs of ocular discharge, redness, squinting, scratches, sores, and pupil reflexes in response to a light source (see Figure 43-7). When a light is placed into the pupil of the eyes by shining it directly into the eyes, the pupil should **constrict** or close. When the light source is removed, the pupil should **dilate**, or open wide. It is important to note even pupil size and reflexes on each eye. The ears should be examined using an otoscope and any signs of ear wax discharge, redness, sores, and odor should be noted (see Figure 43-8). The nasal area or nostrils should be examined with a

FIGURE 43-7 The eyes should be examined using an ophthalmoscope to note any signs of ocular discharge, redness, squinting, scratches, sores, and pupil reflexes in response to a light source.

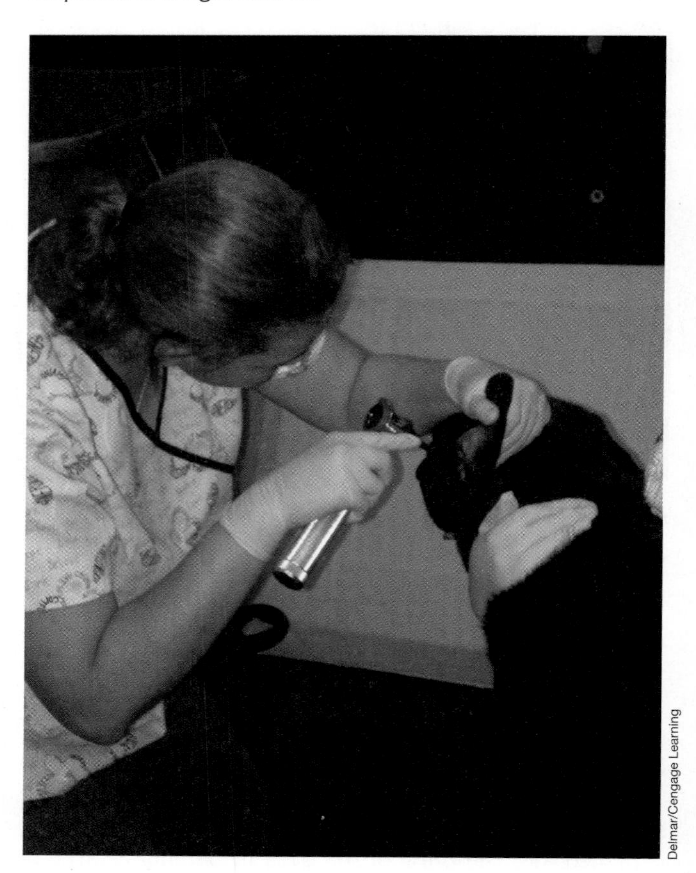

FIGURE 43-8 The ears should be examined using an otoscope and any signs of ear wax discharge, redness, sores, and odor should be noted.

penlight or the otoscope to evaluate the nasal passage for discharge, foreign objects, or swelling. The head, face, and neck should be examined for signs of sores, hair loss, or areas of swelling.

The Thoracic Cavity

The thorax or chest area should be evaluated with a stethoscope by listening to the heart, chest, and lungs for any abnormal sounds (see Figure 43-9). This is known as an **auscultation**. A breathing pattern should be established and any abnormal signs noted, such as **dyspnea** (difficulty breathing), chest movement (normal, shallow, deep), and abnormal sounds. Abnormal sounds may include wheezing (sounds like musical notes) or **rales** or **crackles** (sounds like cellophane paper). Heart and lung sounds should be evaluated for abnormal sounds, such as a **murmur** (often heard as a swishing sound), an **arrthymia** (abnormal heartbeat), or any muffled sounds that may be caused by fluid buildup in the chest. A heart rate (HR) and respiratory rate (RR) should be evaluated and recorded (see Figure 43-10). The femoral pulse should be evaluated for strength and quality with any signs of weakness or bounding noted.

Skin and Integumentary System

The **integumentary system** includes the skin, hair, pads, horns, beak, nails, hooves, and feathers. The skin and hair coat should be evaluated for signs of **alopecia**

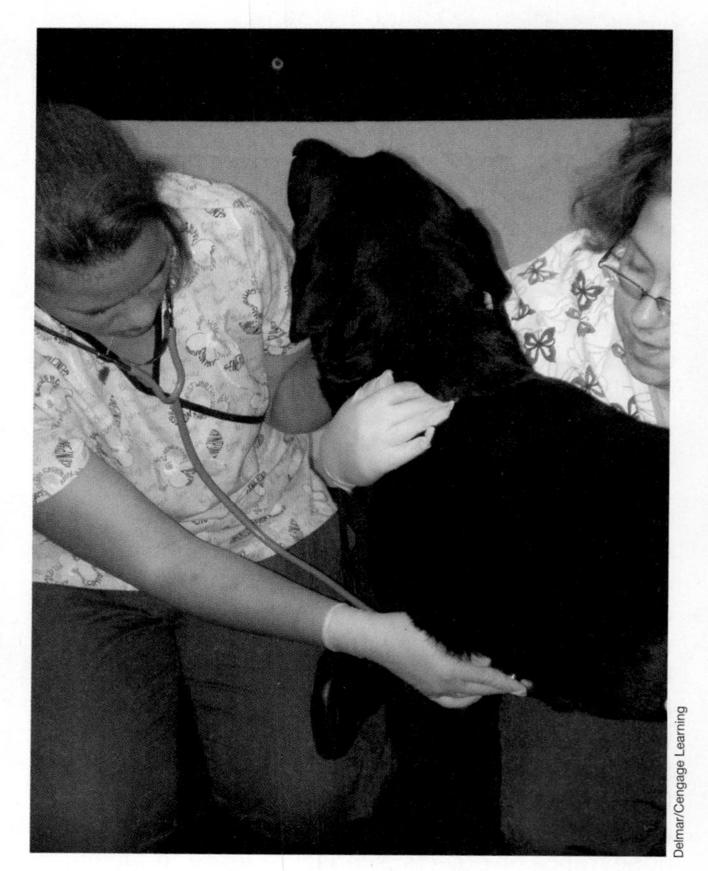

FIGURE 43-9 The thorax or chest area should be evaluated with a stethoscope by listening to the heart, chest, and lungs for any abnormal sounds.

FIGURE 43-10 Heart rate (HR) and respiratory rate (RR) should be evaluated and recorded.

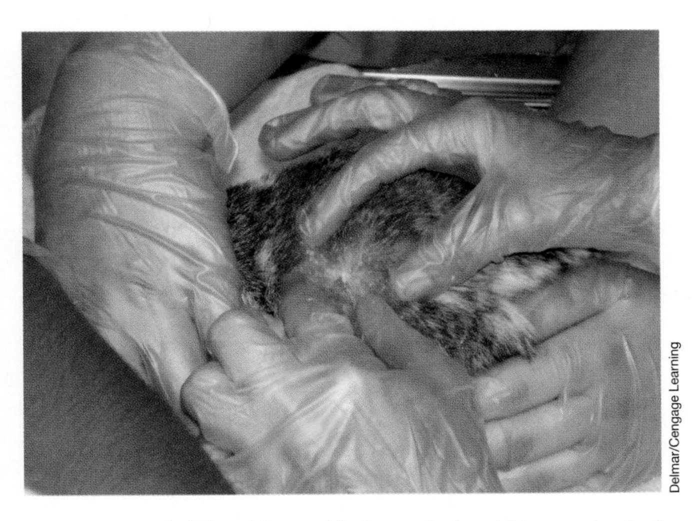

FIGURE 43-11 The skin and hair coat should be evaluated for signs of **alopecia** (hair loss), dandruff or flaky skin, dryness, excessive oil, matting, redness, sores, or signs of parasites.

(hair loss), dandruff or flaky skin, dryness, excessive oil, matting, redness, sores, or signs of parasites (see Figure 43-11). External parasites include fleas and dirt, ticks, mites (ear and skin), lice, and fly eggs called maggots. Signs of hair loss may potentially be a fungus or any other condition that is contagious. The hair-coat quality should be evaluated for signs of cleanliness or unkempt appearances. The **hydration status** should be evaluated over the neck and shoulder area of the skin, observation of the eye sockets, and gum moistness. Loss of elasticity of the skin (skin turgor) is the first sign of dehydration. This is called tenting and should be reported in subjective forms of body weight percentages from 0–20%. The nail bed, hooves, and pads, as well as the beak, horns, and antlers, should be evaluated for signs of injury, dryness, or wounds (see Figure 43-12).

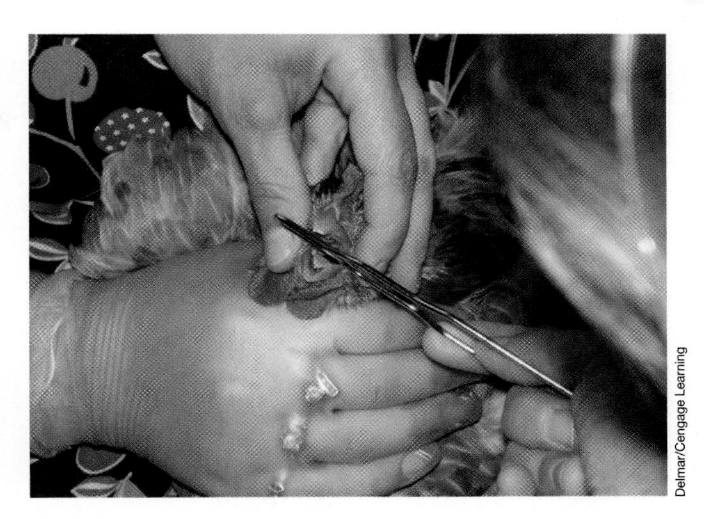

FIGURE 43-12 Checking the beak for injury or abnormalities.

Feathers should be evaluated for similar signs relating to the hair and skin quality.

The Nervous System

The **nervous system** is evaluated throughout the entire physical exam. Areas to note are the response of pupil reflexes, the gait, as well as other body reflexes, such as limbs, paws, and digits. Reflex test functions that can easily be evaluated include the joint reflexes, which can be assessed using the reflex hammer to see the **knee jerk reaction**. This is the reaction to a slight tap over the knee joint that causes the knee to jerk in response. The limbs should be evaluated by bending at each joint to note any pain or swelling (see Figure 43-13). The paws can be assessed by placing the top surface of the paw in contact with the ground and noting that the animal immediately rights the paw in the proper position. The digits can be pinched to elicit a response that should be to pull away from the pressure (see Figure 43-14). This is called the **toe pinch response**. These reflexes allow the assistant to determine proper nervous system function. Small animals that have questionable nervous system signs may be evaluated using the **wheel barrow technique**, which is slightly lifting the animal's hind end off the ground and pushing the body forward in a "wheelbarrow" motion. The animal should walk forward in a normal nervous system function.

The Musculoskeletal System

The **musculoskeletal system** includes the bones, joints, and muscles. The gait can be used to determine any signs of lameness or pain. Areas of the limbs and body should be evaluated for signs of swelling, heat, injury or wounds, atrophy, and pain. The muscles are areas in the body where movement occurs and should be assessed

Delmar/Cengage Learning

FIGURE 43-13 The limbs should be evaluated by bending at each joint to note any pain or swelling.

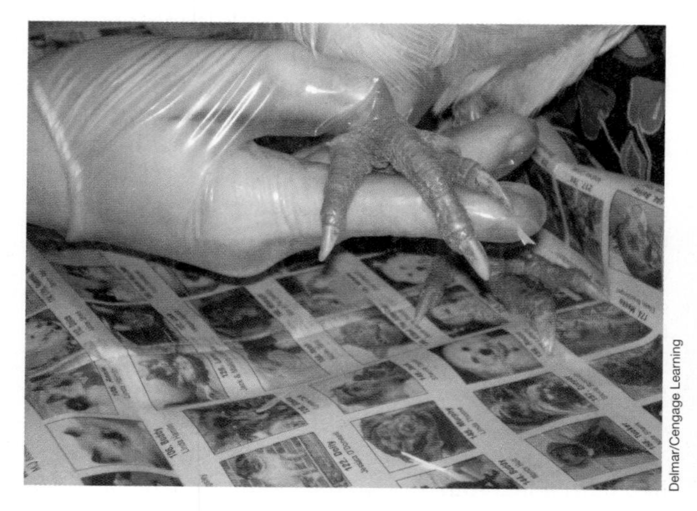

Delmar/Cengage Learning

FIGURE 43-14 Toe pinch response.

for quality. These areas are prone to pain, swelling, and **atrophy**, which is the shrinking of the muscle when it is not being used. The joints of the body should be evaluated for signs of swelling, pain, inflammation, or arthritis.

The Abdominal System

The **abdominal system** includes the stomach, intestines, and internal organs, such as the liver, pancreas, and spleen. The abdomen should be palpated (to feel) for any signs of masses, distention, pain, and bruising. The stomach area should be palpated gently and slowly without grabbing any locations that may feel abnormal (see Figure 43-15). Using both hands and carefully allowing the fingers to move gently over the abdomen will allow the assistant to feel any signs of masses, feces in the intestine, fluid, gas, or possible fetuses in a pregnant female. Large animals should be handled carefully to prevent injury from kicking or striking.

The Urogenital System

The **urogenital system** includes the kidneys, urinary bladder, external genitals, mammary glands, and rectal area. Palpation of the dorsal abdomen will allow evaluation of the kidneys and bladder. It is difficult to palpate the kidneys of some animal species. The mammary glands should be palpated for signs of masses, tumors, nodules, or cysts and any sign of milk discharge or pain and swelling. Female animals that have recently been bred or were in labor may show signs of mastitis, including heat, redness, and discharge. Lactating animals should have their milk expressed to determine normal appearance. Female animals should have the vulva examined for signs of discharge, swelling, inflammation, and masses. Male animals should have the prepuce, penis, and testicles evaluated for discharge, inflammation, swelling, and tumors. Urine samples should be collected and evaluated as deemed necessary (see Figure 43-16). The rectal area should be evaluated for signs of diarrhea, size of anal glands, odor from the anal glands or rectum, and any masses (see Figure 43-17). The rectal temperature of the animal can be

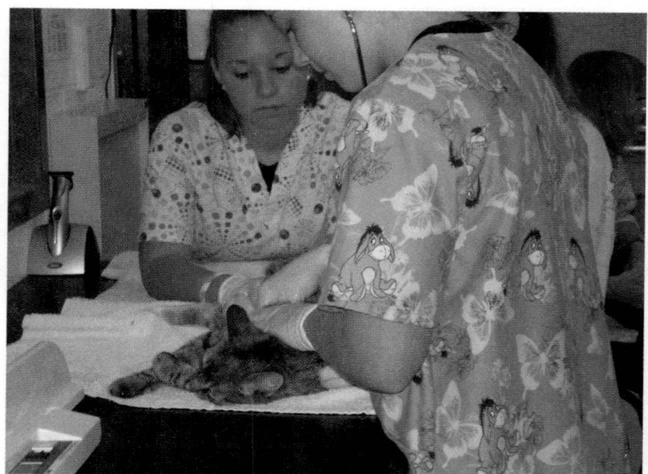

Delmar/Cengage Learning

FIGURE 43-15 Palpate the abdomen for signs of swelling or discomfort.

FIGURE 43-16 Obtaining a urine sample from a dog.

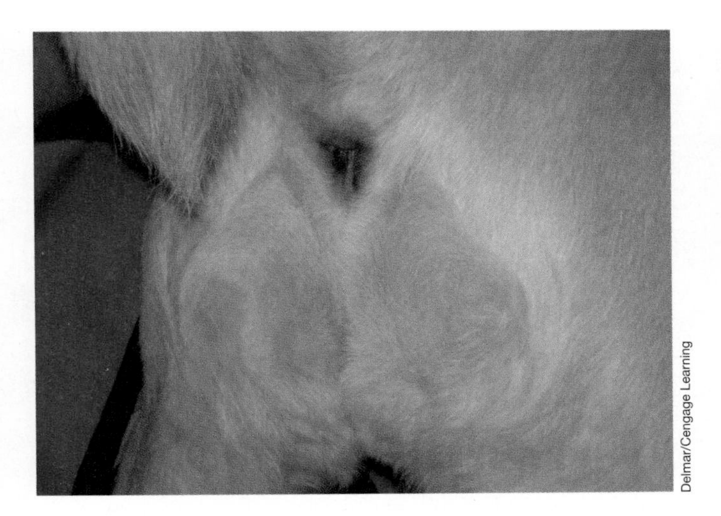

FIGURE 43-17 The rectal area should be evaluated for signs of diarrhea, size of anal glands, odor from the anal glands or rectum, and any masses.

completed at the end of the exam by lubricating a digital or mercury thermometer with Vaseline or KY gel to allow the ease of passing the tool rectally and minimizing any discomfort to the animal (see Figure 43-18). It normally takes about 60 seconds for the thermometer to complete the reading. It is ideal to end with this so as not to stress or agitate the animal and then have a difficult animal during the rest of the PE.

Recording the Physical Examination

Many veterinary facilities have a physical exam (PE) report form that is used to note each body system of an animal. This form should be completed by the veterinarian, veterinary technician, and veterinary assistant as their duties allow (see Figure 43-19). Any abnormal findings may be noted in the progress notes so follow-up reports can be monitored. It is important to note that when learning to complete PEs, one should concentrate on the hands-on skills and have another person record the findings until the veterinary assistant becomes experienced and familiar with each body system. This will allow fewer mistakes and missing and abnormal notations in the medical record. When the assistant becomes familiar and experienced in evaluating patients, he or she should be able to complete a full PE and record the findings at the end of the exam, when the exam gloves can be removed and hands sanitized prior to recording in the medical record. This avoids potential contamination of the exam area and medical records. When the medical chart is completed, take a few moments to review the notes and make certain all items were covered, results were recorded accurately, and spelling and grammar are correct. It is important to use descriptive words and proper veterinary terminology when noting any abnormal occurrences. Any problems noted that are a possible emergency should be immediately reported to the veterinarian for further evaluation. The evaluating veterinarian will then review the medical record and continue prompt assessment and examination of the patient.

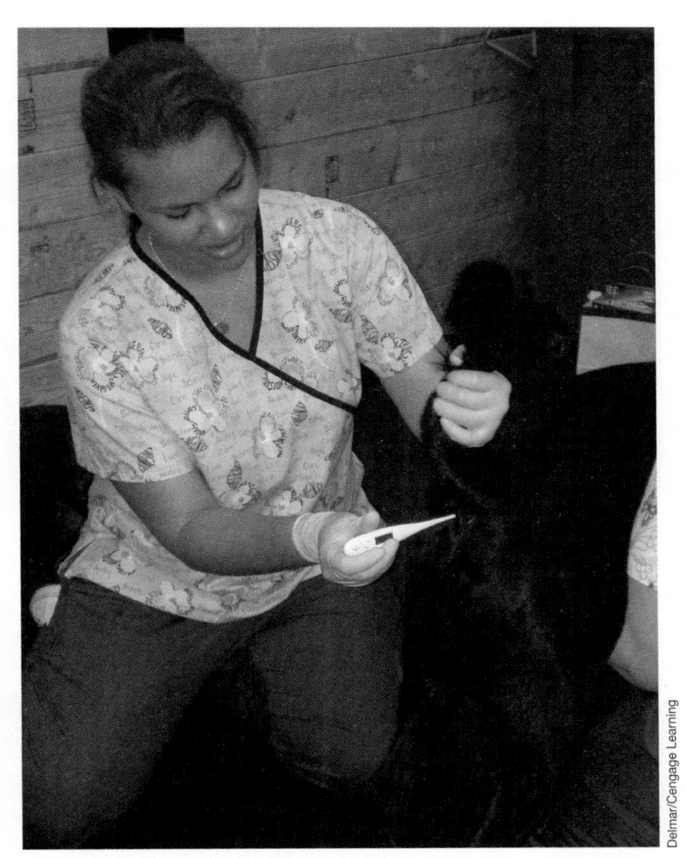

FIGURE 43-18 Taking a rectal temperature.

FIGURE 43-19 Completion of a physical examination report.

SUMMARY

Completing a thorough physical examination even when there are no obvious problems allows veterinary staff to determine whether a problem is occurring or locate signs of diseases or changes in a patient. In addition, in animals with known abnormalities, such as fatty tumors or heart murmurs, regular exams allow the staff to chart the course of disease and determine whether it is progressive or remains the same. The veterinary assistant and support staff play a vital role in assessing the patient's overall appearance and condition.

Key Terms

abdominal system includes the stomach, intestines, and internal organs, such as the liver, pancreas, and spleen

alopecia areas of hair loss

atrophy the shrinking of the muscle when it is not being used

auscultation process of listening to the heart, chest, and lungs for any abnormal sounds using a stethoscope

chief complaint the reason the animal is being seen by the veterinarian

constrict to close

crackles abnormal sounds within the chest that sounds like cellophane paper

dilate to open

dyspnea difficulty breathing

head exam includes examining the eyes, ears, and nose

hydration status area evaluated over the neck and shoulders, eye sockets, and gums to determine how much water loss has occurred

integumentary system includes the skin, hair, pads, horns, beak, nails, hooves, and feathers

knee jerk reaction the reaction to a slight tap over the knee joint that causes the knee to jerk in response to assess the nervous system

murmur abnormal heart sound that presents as a swishing sound

musculoskeletal system includes the bones, joints, and muscles

nervous system assessment of the brain and spinal cord through the response of pupil reflexes, the gait, as well as other body reflexes, such as limbs, paws, and digits

observation an inspection of an animal at a distance to see what it looks like, acts like, or how it moves naturally

oral cavity includes the gums, mouth, teeth, tongue, and throat

patient history background information of past medical, surgical, nutritional, and behavioral factors that have occurred over the life of the animal

physical examination (PE) an observation of each body system to determine abnormal problems that can cause a health problem

rales abnormal sounds within the chest that sound like musical notes

signalment the patient's name, age or date of birth, gender, color, and breed or species

toe pinch response digits can be pinched to elicit a response that should be to pull away from the pressure to assess the nervous system

triage a quick general assessment of an animal to determine how quickly the animal needs a veterinarian's attention

urogenital system includes the kidneys, urinary bladder, external genitals, mammary glands, and rectal area

wheel barrow technique slightly lifting the animal's hind end off the ground and pushing the body forward in a "wheelbarrow" motion to assess the nervous system

REVIEW QUESTIONS

1. What is the importance of the patient history?
2. What are the parts of a patient history?
3. What is the importance of the physical exam?
4. What are the body system parts of a physical exam?
5. Why is an observation performed first in a physical exam?
6. What is meant by the term *signalment*?
7. What is the chief complaint?
8. What is a triage? Why is it performed?

Clinical Situation

Melissa, a veterinary assistant at the Ponderosa Veterinary Clinic, is setting up for a dog exam in one of the hospital exam rooms. Melissa has worked in the facility only for 2 weeks and has been taking notes of the duties she must complete and certain expectations of the veterinarians and staff. Melissa has checked the appointment scheduled, retrieved the medical chart for the patient, and located all of the necessary paperwork for the office visit and physical exam. The appointment notes that Charles Delmont is bringing in a dog named "Taffy" at 4:00 p.m. for a PE and anal gland expression. "Taffy" is a 4-year-old SF Pomeranian.

Melissa reviews the medical chart and notes that "Taffy" has a WILL BITE sticker on the front of the chart. "Taffy" also has a history of matted hair around the rectum and skin problems. Several times she has had fleas and ticks and alopecia.

- What tools and supplies should Melissa prepare for the PE?
- What information should Melissa ask the owner for a patient history?
- Does Melissa need to do anything else to prepare the exam room for "Taffy" and Mr. Delmont?

44

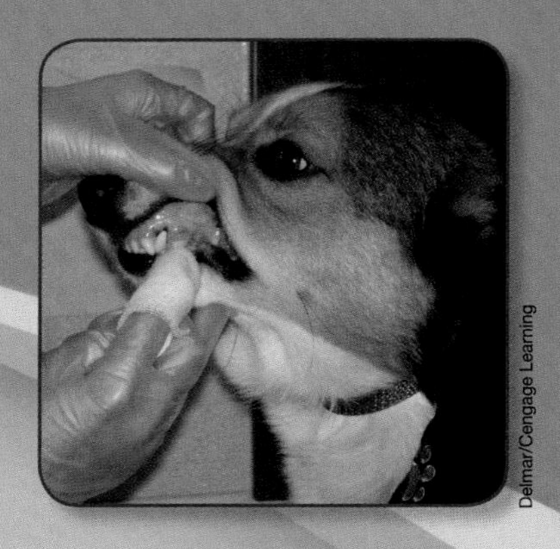
Delmar/Cengage Learning

Examination Procedures

Objectives

Upon completion of this chapter, the reader should be able to:

- Demonstrate common examination procedures
- Discuss and describe common procedures with clients
- Use and explain common veterinary terminology and abbreviations relating to examination procedures

Introduction

Examination procedures must be understood by the entire veterinary health care team. Even though the patient's liability and responsibility lie with the attending veterinarian, the veterinary team and support staff must be trained and reliable professionals that can perform auscultation, obtain vital signs, provide client education, and assist the veterinary technician and veterinarian in performing common examination procedures. It is important to note that cleanliness, good hand hygiene, and PPE or personal protective equipment should be used when necessary. The veterinary assistant plays a vital role in the examination room.

Vital Signs

Animals are evaluated for normal health status by taking their **vital signs**, which are the signs of life that are measurements to assess the basic functions of the body. Vital signs include the heart rate, respiratory rate, temperature, blood pressure, mucous membrane color, capillary refill time, and weight of the animal. The first three vital signs are often referred to as the **TPR.** The TPR refers to the temperature, pulse, and respiration.

Temperature Assessment

The patient's temperature is an essential part of an animal's exam and health status. The body temperature can allude to signs of infection, disease, environmental conditions, and other factors. The **endothermic** temperature is evaluated in an animal that regulates its body temperature internally or within the body. An **ectothermic** temperature describes an animal that regulates its body temperature externally or through its environment (see Table 44-1).

Hypothermia is a condition where the body temperature is below normal causing the animal to become cold, and it may be a sign of environmental exposure to cold climates, metabolic disease, or shock. **Hyperthermia** is when the body temperature is above normal and causes the animal to have a **fever**, which is an excessive amount of heat in the body. Hyperthermia can be a sign of infection, toxicity, or heat stroke. **Heat stroke** is a common occurrence in areas that have high environmental temperatures and can be a serious condition for animals. Heat stroke may be considered when body temperatures reach or exceed 105 degrees and can produce severe damage to internal body organs, the nervous system, and the blood supply.

The core body temperature is best evaluated by taking a rectal temperature using a rectal thermometer (see Figure 44-1). The animal should remain standing or in a sternal position so as not to break the thermometer or cause injury to the animal's rectal area. The rectal thermometer is applied carefully within the rectum by lubricating the end of the thermometer that has a silver metal tip. The thermometer only need be inserted

TABLE 44-1

Examples of Endothermic and Ectothermic Species

ENDOTHERMIC SPECIES			ECTOTHERMIC SPECIES	
Dog	Cow	Mouse	Snake	Frog
Cat	Goat	Rabbit	Lizard	Turtle
Horse	Sheep	Rat	Fish	Toad

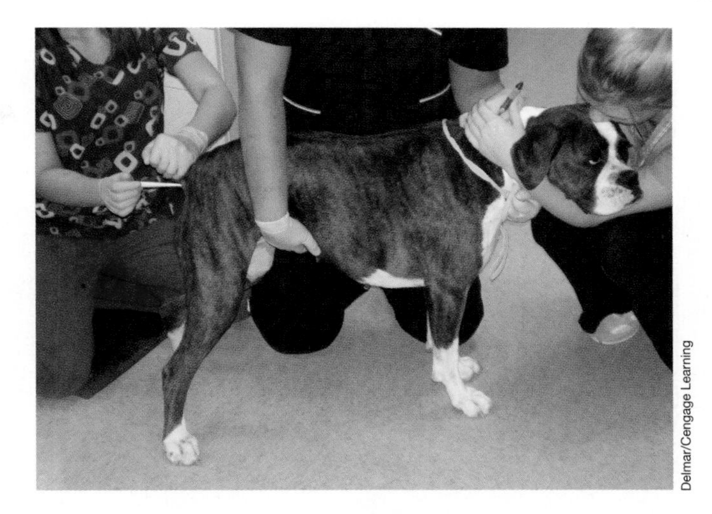

Delmar/Cengage Learning

FIGURE 44-1 Rectal temperature is the most accurate measurement of body temperature.

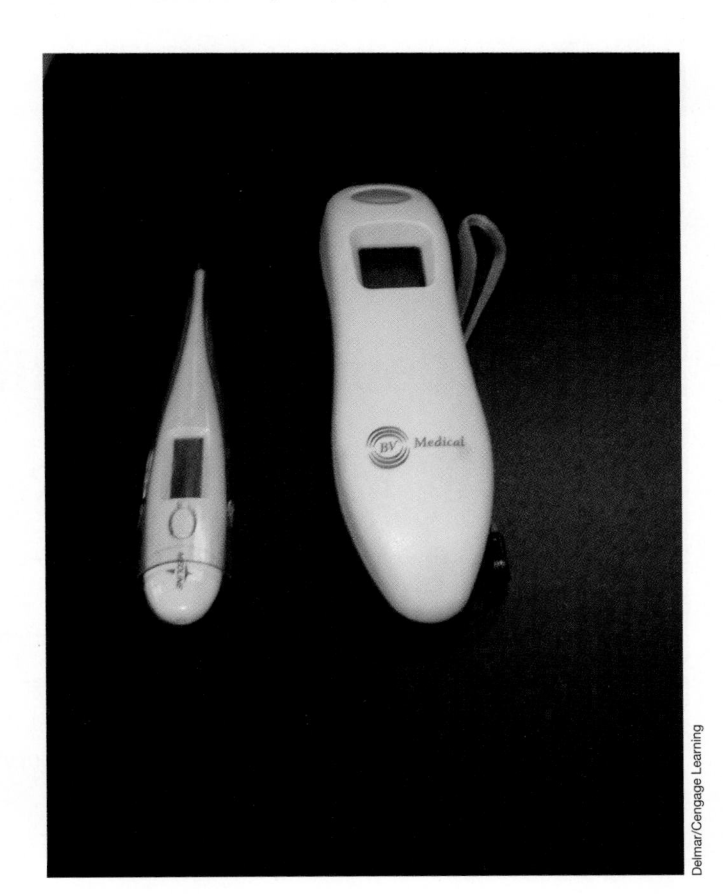

Delmar/Cengage Learning

FIGURE 44-2 Digital rectal thermometer.

¼ inch to ½ inch within the rectum. Some rectal thermometers are digital and will alert with an alarm sound when the temperature is recorded (see Figure 44-2). Some thermometers are mercury based and have to be read after a set amount of time to determine the temperature (see Figure 44-3). Mercury thermometers are a safety hazard in that if broken, mercury is a toxic substance that will erode and burn. Care must be taken when handling mercury thermometers. Never touch mercury or glass with bare hands. Prior to using a mercury thermometer it must be carefully shaken down to decrease the mercury from the previous use. The mercury line in the glass thermometer is visible as a silver thin line located in the center of the glass when the number line is seen. This is also the area where the temperature reading is noted. The mercury will stop at the highest point of the internal body temperature and should be read on the number scale at that point. Ear thermometers fit into the external ear canal and record the body temperature of the ear. Table 44-2 lists common normal body temperatures in various species.

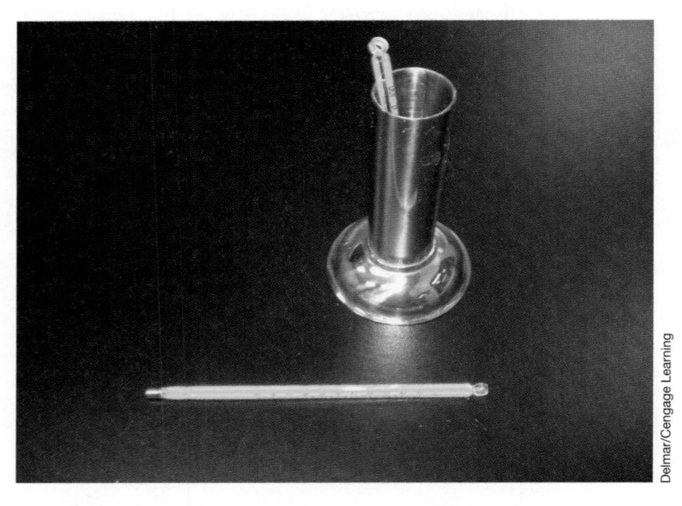

FIGURE 44-3 Mercury rectal thermometer.

TABLE 44-2

Normal Core Body Temperatures

SPECIES	NORMAL CORE BODY TEMPERATURE
Dog	101–102 degrees
Cat	101–102.5 degrees
Rabbit	102–104 degrees
Guinea Pig	100–103 degrees
Horse	99–100.5 degrees
Cow	101–101.5 degrees
Sheep	102–102.5 degrees
Goat	103–103.5 degrees
Pig	102–102.5 degrees
Chicken	105–106 degrees
Hamster	101–102 degrees

Competency Skill

Taking a Rectal Temperature

Objective:

To properly take a rectal temperature to accurately measure core body temperature.

Equipment Needed:

rectal thermometer (digital or mercury), Vaseline or KY gel lubricant, exam gloves, gauze sponges, paper towels

Procedure:

1. Apply exam gloves.
2. Shake down mercury thermometer with a flick of the wrist or turn on and calibrate digital thermometer.
3. Apply a small amount of Vaseline or KY gel to a gauze sponge.
4. Place the metal tipped end of the thermometer into the lubricant.
5. Lift the animal's tail and secure one hand at the base of the tail. Keep animal in standing position or sternal position.

6. With the other hand gently insert the lubricated end of the thermometer into the animal's rectum.
7. Allow digital thermometers to beep when done or mercury thermometers to be in place for at least 1 minute.
8. Remove thermometer gently from rectum and clean end of thermometer with a clean gauze sponge.
9. Read the thermometer for the core body temperature.
10. Record the temperature in degrees in the medical record.
11. Disinfect the thermometer and place in a safe area.
12. Clean up supplies and disinfect the work area.

Pulse

The **pulse** is the number of times the heart beats per minute. The pulse is taken by palpating an artery. The **heart rate** (HR) is the number of times the heart relaxes and contracts per minute. The heart rate is taken by auscultating the heart with a stethoscope (see Figure 44-4). Table 44-3 lists normal heart rates for various species.

Respiratory Rate

The respiration or **respiratory rate** (RR) measures the breathing of an animal and evaluates how many breaths an animal takes in a minute. Respiration is one total inhale and one total exhale (see Figure 44-5). Table 44-4 lists the normal average respiratory rates in various species.

TABLE 44-3

Normal Heart Rates

SPECIES	HEART RATES (bpm)
Dog	70–180
Cat	170–240
Rabbit	130–325
Guinea Pig	240–350
Horse	35–45
Cow	60–70
Sheep	60–80
Goat	70–80
Pig	60–80
Chicken	200–400
Hamster	250–500

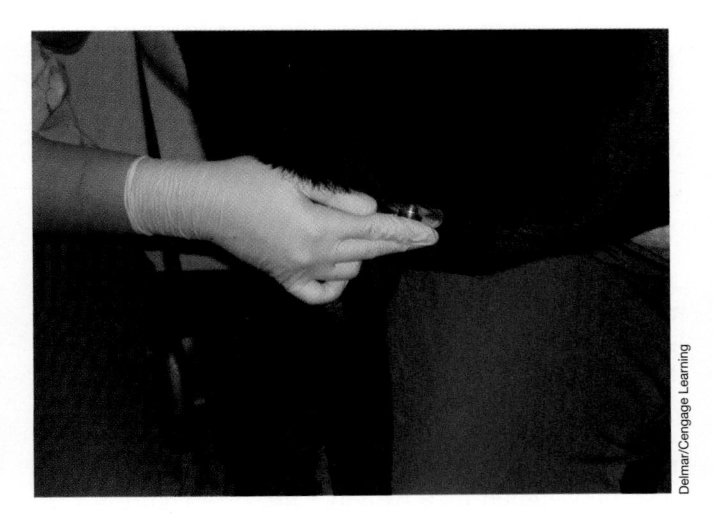

Delmar/Cengage Learning

FIGURE 44-4 Listening to the heart to determine heart rate.

Delmar/Cengage Learning

FIGURE 44-5 Counting the dog's respiratory rate.

TABLE 44-4
Normal Respiratory Rates

SPECIES	RESPIRATORY RATES (BREATHS/MINUTE)
Dog	16–20
Cat	20–30
Rabbit	32–60
Guinea Pig	40–150
Horse	8–16
Cow	10–30
Sheep	12–20
Goat	12–20
Pig	8–15
Chicken	15–30
Hamster	35–135

TERMINOLOGY TIP

The heart rate or pulse may be recorded **P** or **HR** and is labeled **bpm**, which stands for beats per minute. The respiration or respiratory rate is recorded **R** or **RR** and is labeled **br/min**, which means breaths per minute. The weight may be recorded as **wt**. or the pound sign (#) to refer to pounds. Pounds are also abbreviated lb.

Blood Pressure

Blood pressure is the tension exerted by blood on the arterial walls. Blood pressure is measured by a sphygmomanometer (blood pressure cuff). This device measures the amount of pressure exerted against the walls of the blood vessels. Systolic pressure occurs when the ventricles contract. Diastolic pressure occurs when the ventricles relax. *Hypertension* is the term used to describe high blood pressure while *hypotension* is the term used to describe blood pressure that is too low.

Mucous Membranes

The **mucous membrane** color is the color of an animal's gums. The gums should be a nice light pink color. Gum color such as grey, blue, brick red, or white is abnormal and should be reported to the veterinarian (see Figure 44-6).

Capillary Refill Time

The **capillary refill time** (CRT) evaluates how well an animal's blood is circulating in the body. This is done by placing a finger on the gums with a small amount of pressure, allowing the area to turn white. Releasing the finger and pressure allows the color to return to the gum area (see Figure 44-7). The time it takes for this to occur is the CRT. A normal CRT time is 1–2 seconds. Times that are longer than this should be reported to the veterinarian.

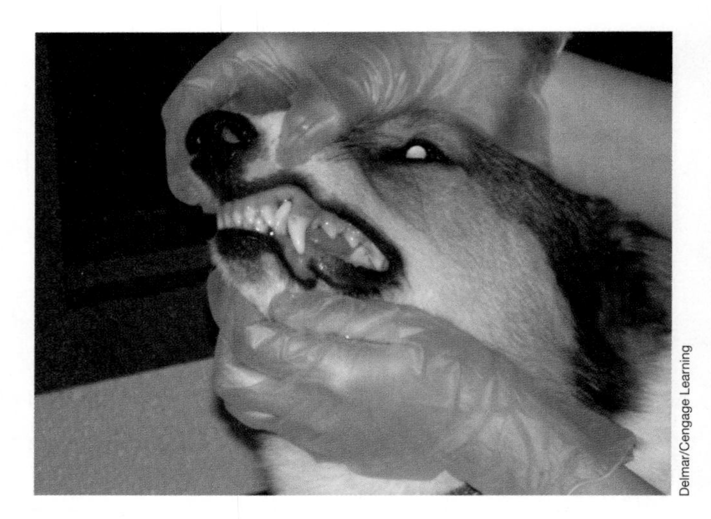
Delmar/Cengage Learning

FIGURE 44-6 Checking mucous membranes.

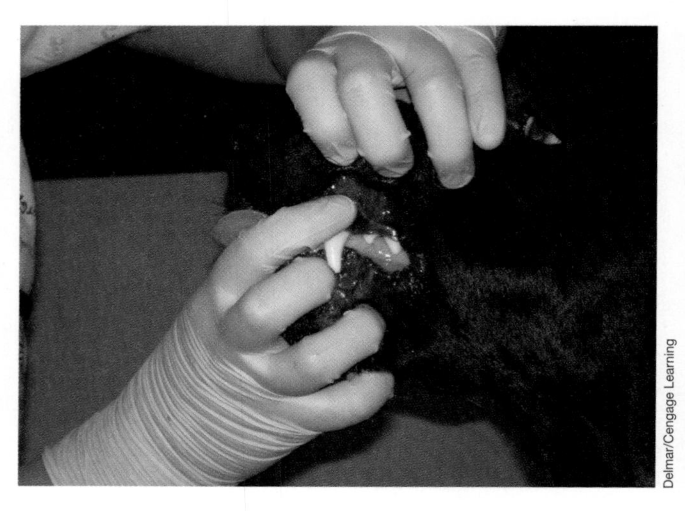
Delmar/Cengage Learning

FIGURE 44-7 Performing a capillary refill test.

Weight Assessment

An animal should be weighed and evaluated for obesity or thin body conditions. An animal is weighed using a size-appropriate scale or weight tape. There are scales for small animals, medium-sized animals, and livestock. These scales are typically meant for an animal to stand or sit on and weigh the body mass of the animal through digital measurement (see Figure 44-8). Livestock can also be weighed by a **weight tape** (see Figure 44-9). This is the use of a measurement tape around the girth of an animal that gives an estimate of its body mass.

FIGURE 44-8 Weight is an important measurement in assessing the health status of animals.

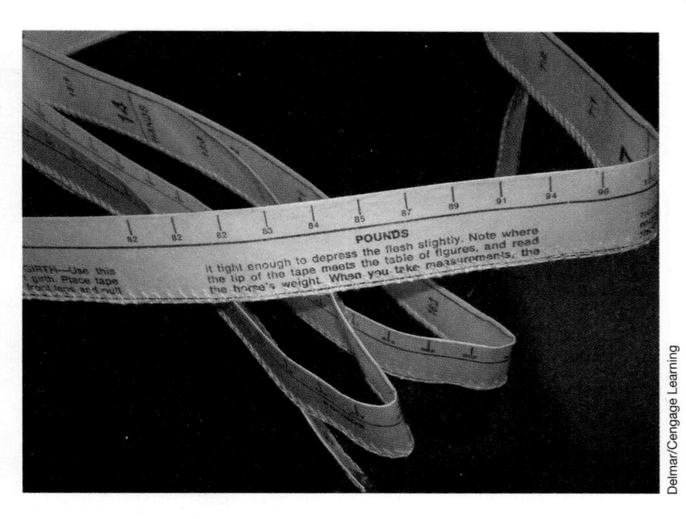

FIGURE 44-9 The weight of livestock can be measured using a weight tape.

Competency Skill

Weighing an Animal Using a Scale

Objective:

To properly use a scale to obtain an accurate measurement of weight.

Equipment Needed:

scale

Procedure:

1. Turn on the scale and balance it prior to placing an animal on the scale.
2. Allow the animal to walk on the scale or place the animal on the scale.
3. Maintain the animal by allowing it to stand, sit, or lie quietly and calmly on the scale with as little movement as possible (see Figure 44-10).
4. Read the digital weight measurement on the scale screen.
5. Remove the animal from the scale.
6. Turn off the scale.
7. Record the animal's weight in the medical record.
8. Disinfect the scale.

(*Continues*)

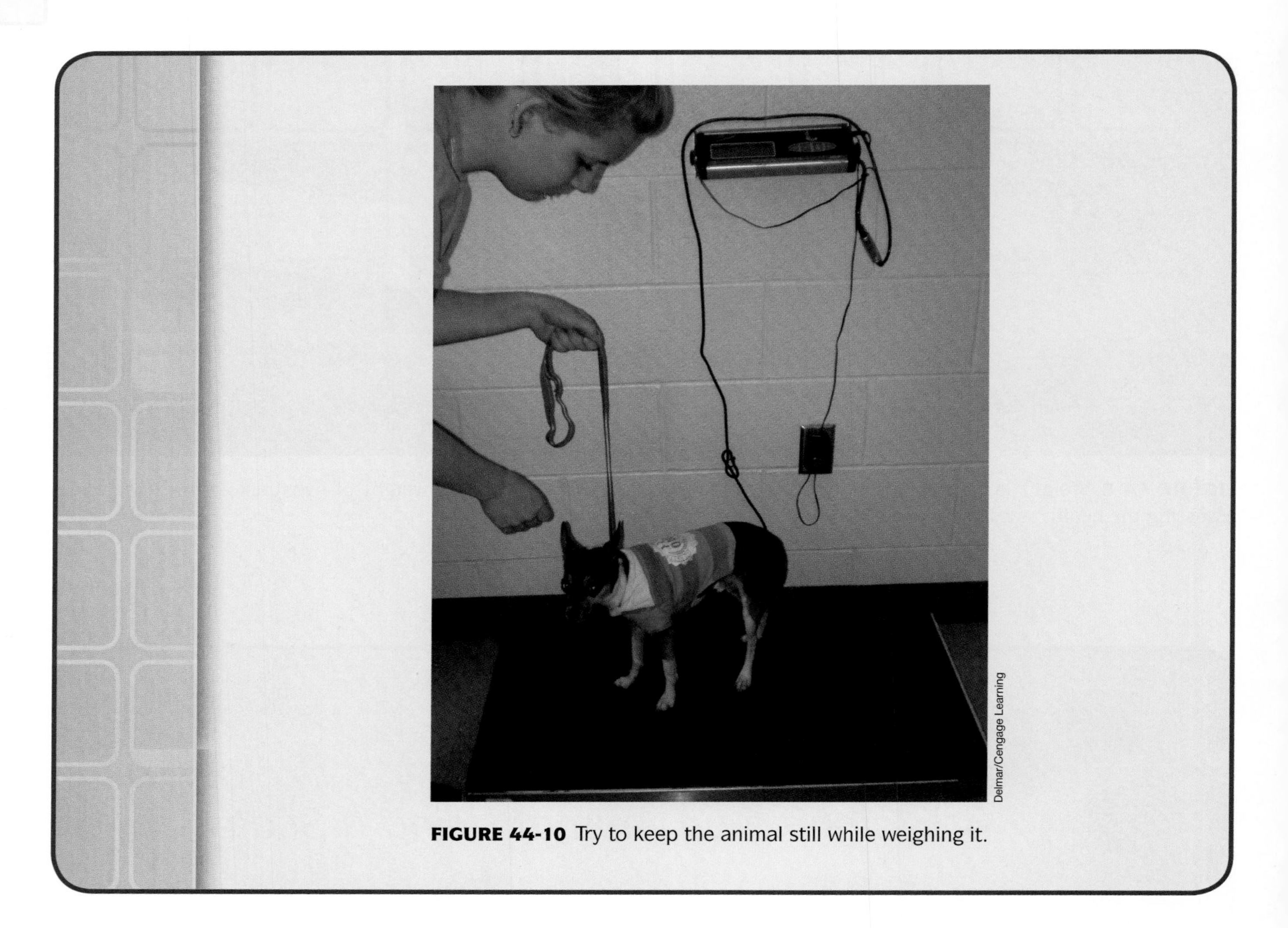

FIGURE 44-10 Try to keep the animal still while weighing it.

Competency Skill

Weighing an Animal Using a Weight Tape (Livestock)

Objective:

To properly use a weight tape to accurately measure the weight of large animals and livestock.

Equipment Needed:

weight tape

Procedure:

1. Tie or restrain the large animal.
2. Unfold or unroll the entire weight tape. Note the side of the tape that measures the weight of an animal.
3. Place the weight tape around the girth of the animal just behind the front limbs and encircling the back and chest area.
4. Read the measurement at the end of the tape where the ends meet with the tape snugly around the girth.
5. Remove the tape and record the measurement in the medical record.

Auscultation

Auscultation of the chest cavity includes listening to the heart and lungs. A stethoscope is used to listen to the chest area by placing the end of the stethoscope that appears bell-like in appearance on the animal's heart area. The heart area can be located by taking the left front leg and pulling it back to the chest or thorax area, and where the elbow meets the chest is the heart location. This area is also near the ribs. The stethoscope should be moved to at least three areas of the chest to listen to the heart and lungs to rule out abnormal sounds (see Figure 44-11). The same procedure should be completed on the right side of the chest as well. The heart can also be listened to through the sternum or breast bone on the lower ventral chest area. A strong heart beat should be noted.

Irregular heart sounds such as decreased rhythm may be a sign of a mass or fluid located in the chest area. **Bradycardia** is the term for a decreased or slow heart rate and **tachycardia** is the term for an increased or fast heart rate. It may also be a sign of heart disease. Irregular lung sounds may be present in sounds such as crackles or wheezing. Crackles, sometimes referred to as rales, sound like crumpling cellophane and many times mean fluid buildup in the chest. Wheezing is a raspy sound that produces a whistle-like noise that is common in narrow or constricted airways. Some sounds may be normal for breeds that have brachycephalic (flattened or shortened) appearing nasal passages such as pugs, bulldogs, and Himalayans.

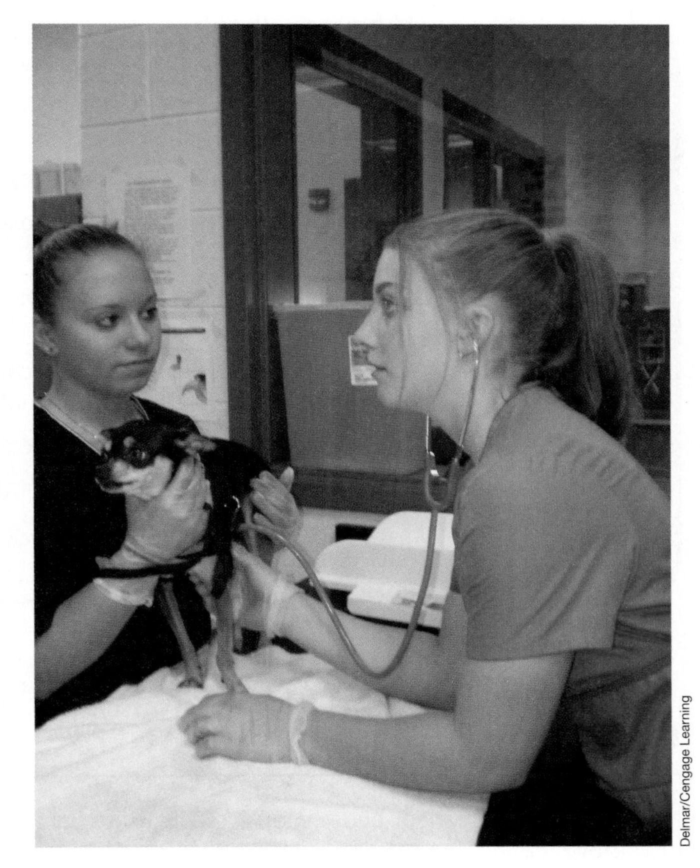

FIGURE 44-11 A stethoscope is used to assess heart and lung sounds.

Competency Skill

Auscultation Procedure

Objective:

To properly use the stethoscope to listen to heart sounds when performing a physical examination.

Equipment Needed:

stethoscope, exam gloves, quiet location

Procedure:

1. Apply exam gloves.
2. Locate a quiet place to listen to the chest cavity.
3. Locate the heart by taking the front limb and placing the elbow at the lateral chest area. The area where the elbow meets the chest is the heart location.
4. Place each end of the stethoscope that moves outward into each ear canal.
5. Apply the bell end of the stethoscope to the lateral chest area and listen to the heart sounds.
6. Move the stethoscope to at least three areas of the chest and listen for abnormal sounds.
7. Move the stethoscope slightly caudal to the heart and listen to three areas of the lung field and listen for abnormal sounds.
8. Repeat these steps on the opposite side of the chest.
9. Record any abnormal sounds of the heart and lungs.
10. Disinfect the end of the stethoscope and the ends of the ear probes.

Eye Exam

Normal eyes appear clear and responsive. The white part of the eye, called the sclera, should have small blood vessels present. The cornea, or the colored part of the center of the eye, should be clear. Any abnormal signs, such as a scratch, redness, swelling, discharge, or jaundice color, should be noted. **Jaundice** is a yellow coloration to the skin or mucous membranes and may be a sign of liver disease or other illness. A light source should be placed on the eye with a penlight or ophthalmoscope. Any abnormal signs such as poor pupil response, sensitivity to light, squinting, ocular discharge, or irregular eye movements should be noted (see Figure 44-12). Pupillary light response (PLR) may be normal or absent. When the PLR is normal it is called **direct PLR**. When the PLR is abnormal it is called **indirect PLR**. Uneven-sized pupils are a condition known as **anisicoria**. Rapid and irregular eye movements that may cause the eyes to bounce from side to side or in circular motions is a condition known as **nystagmus**. This may suggest a balance problem, such as an ear infection or a CNS disease. Blood noted inside the eye is a condition called **hyphema**.

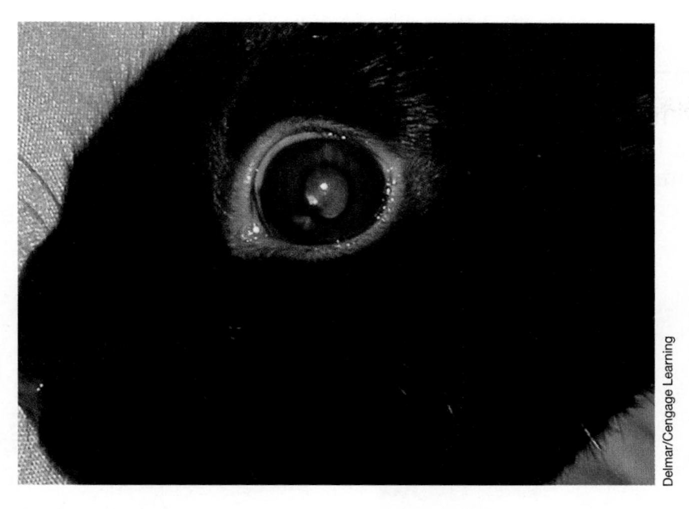

FIGURE 44-12 When examining the eye, note any abnormalities.

Competency Skill

Use of an Ophthalmoscope

Objective:

To properly use the ophthalmoscope to assess the eyes and pupillary response to light.

Equipment Needed:

ophthalmoscope, exam gloves

Procedure:

1. Apply exam gloves.
2. Elevate the animal's head with one hand under the chin area to stabilize any movements. Each eyelid may need to be held open to observe the eye structures.
3. Turn on the light source of the ophthalmoscope.
4. Place the ophthalmoscope about 2–3 inches from the eye and look into the eye using the eyepiece located on the center of the ophthalmoscope.
5. Observe the opposite eye.
6. Note any abnormal signs of the eyes.
7. Turn off the light source of the ophthalmoscope and place it in a safe location.
8. Record any notes in the medical record.

Ears

The ears should be examined using the otoscope (see Figure 44-13). Generally, the ear canal and the ear flap, also known as the pinna, should be a light pink color and free of wax and discharge (see Figure 44-14). Changes in the appearance of the ear should be noted, including any odor, redness, swelling, discharge, or debris. A common condition of the ear flap in animals that may scratch or shake the ears frequently is called an **aural hematoma**. A hematoma is a blood-filled swelling caused by the rupture of a blood vessel. The area fills up with blood, causing pressure on the area.

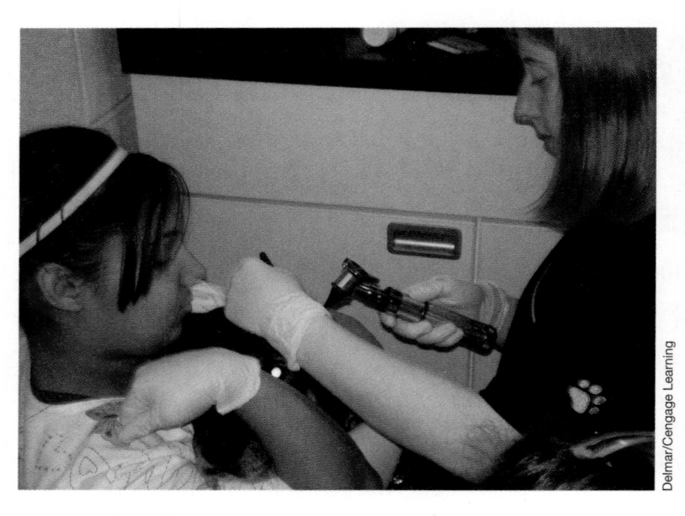

FIGURE 44-13 Examination of the ear is done using an otoscope.

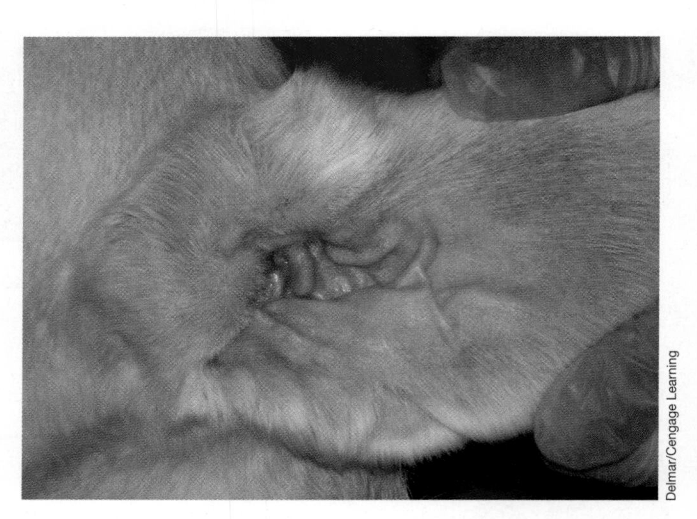

FIGURE 44-14 Note the normal color that should be observed on examination of the ear.

Competency Skill

Use of the Otoscope

Objective:

To properly use the otoscope to assess the ears during a physical examination.

Equipment Needed:

Otoscope, exam gloves, otoscope cones

Procedure:

1. Apply exam gloves.
2. Place a clean cone on the end of the otoscope and make sure it is secure.
3. Turn on the light source for the otoscope.
4. Pull back an ear flap to expose the internal ear canal. Hold the ear flap with one hand.

5. Using the free hand, place the otoscope cone near the internal ear canal.
6. Observe the ear canal by looking into the otoscope eyepiece in the center of the instrument.
7. Note any abnormal appearance.
8. Repeat the above steps with the opposite ear.
9. Turn off the otoscope light source.
10. Remove the cone and disinfect the cone.
11. Place the cone and otoscope in a safe location.
12. Record any abnormal findings in the medical record.

Dental Examination

The dental exam should be performed to note any signs of dental disease or tooth problems (see Figure 44-15). Signs of dental disease include broken teeth, decayed teeth, inflamed gums, oral odor, and tartar or stains of the teeth. **Plaque** is a soft buildup of material found over the surfaces of the teeth that is composed of bacteria, food materials, and saliva. It may be easy to remove by brushing the teeth but often returns quickly (see Figure 44-16). Tartar or **calculus** is a mineralized plaque material that appears brown to dark yellow in color and is difficult to remove. The **gingiva** or gums are the soft tissue that surround the teeth and are generally pink in color or may be dark pigments in certain breeds of animals. **Gingivitis** is the inflammation of the gums and is the first stage of dental disease and causes red, inflamed gums that are swollen and often bleed easily when touched. Animals that have dental disease may require anesthesia to clean the teeth through the process of a dental **prophylaxis**, which is the term for the professional cleaning of a patient's teeth.

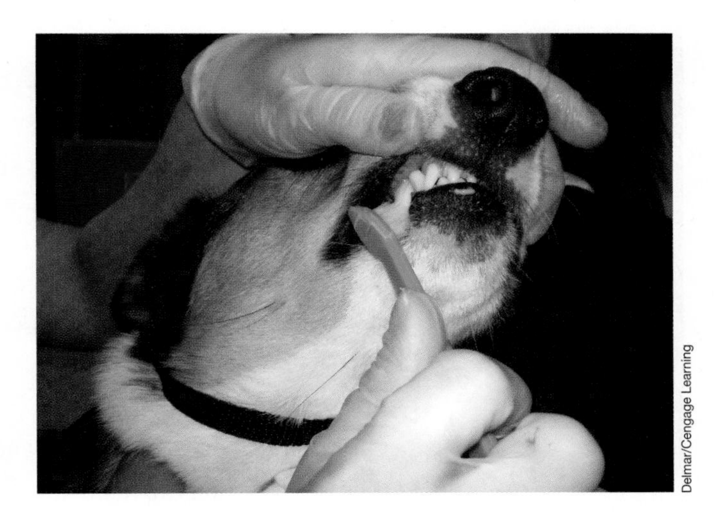

FIGURE 44-16 Brushing may help to prevent plaque buildup and promote good oral health.

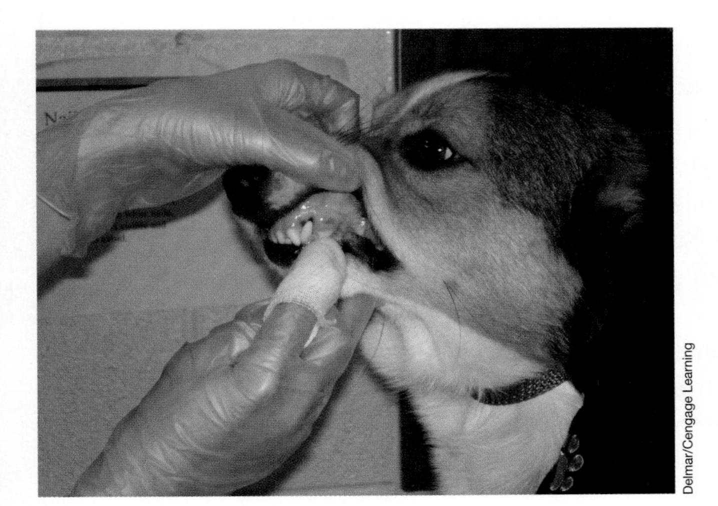

FIGURE 44-15 Examination of the oral cavity and dentition is an important part of the physical examination.

A common condition in animals is a **tooth abscess**. The roots of teeth can abscess when a tooth is broken or damaged. This causes an infection within the root and gingival tissues. Injuries can occur when the animal chews on hard items. A root infection can also be caused by severe dental disease. The **carnassial tooth** is a common site for root abscess. The carnassial teeth are the largest teeth in an animal's mouth and are the upper fourth premolars and the lower first molars. These teeth may have 3–4 roots that may be affected. The abscess of the root causes severe facial swelling and pain, typically located below an eye as the teeth are located close to the sinuses and eye sockets.

Another common dental condition is excessive amounts of gingival tissue growth called **gingival hyperplasia**. These masses are common in some dog breeds, such as the Boxer, Great Dane, Doberman, and Dalmatian. Some animals, such as horses and rodent species, will require teeth to be floated or trimmed. **Floating** is the process of filing down long, sharp edges of teeth that can prevent horses from eating properly and can be painful during chewing or riding. Rodents' and rabbits' teeth continuously grow and may need to be trimmed for them to properly eat and not have the teeth grow into areas of the mouth causing irritation and pain.

Competency Skill

Dental Exam

Objective:

To properly perform an examination and assessment of the teeth and oral cavity.

Equipment Needed:

exam gloves, gauze sponges, dental probe, dental mirror

Procedure:

1. Apply exam gloves.
2. Carefully approach the animal and observe the facial area for any swelling.
3. Carefully restrain the head and gently lift the upper lip to observe the teeth. Note any abnormal appearance or odor.
4. Repeat the same process on the opposite side of the mouth.
5. The mouth can be opened by applying pressure at the area where the maxilla and mandible meet. Once open carefully and gently pull the tongue out to one side to allow viewing of the mouth.
6. Animals may bite down on the tongue causing wounds, so caution should be used when examining the mouth.
7. Note any abnormal appearances or odors in the medical chart.
8. The dental probe may be used to measure depths of tooth injuries or openings. It may also be used to check debris and tartar buildup.
9. The gauze sponges may be used to wipe away any bleeding, saliva, or debris from the teeth.
10. Clean up the area and disinfect any tools.

SUMMARY

Exam room procedures entail a large amount of a veterinary facility's daily case load. Various animals and appointments will occupy exam areas on a daily basis with the potential for contagious organisms to come in contact with animals, staff members, and clients. Understanding of common exam room procedures and proper exam room sanitation will ensure a clean and happy veterinary environment.

Key Terms

anisicoria uneven pupil size

aural hematoma ear filled with blood due to a blood vessel bursting

bradycardia decreased or slow heart rate

calculus mineralized plaque material that appears brown to dark yellow in color and is difficult to remove from teeth

capillary refill time (CRT) evaluates how well an animal's blood is circulating in the body by pressing on the gums and monitoring how long it takes for the color to return to the area

carnassial tooth the largest teeth in an animal's mouth and are the upper fourth premolars and the lower first molars; common abscess sites

direct PLR normal pupil light response assessment

ectothermic temperature evaluated in an animal that regulates its body temperature externally or through its environment

endothermic temperature evaluated in an animal that regulates its body temperature internally or within the body

fever elevated body temperature due to excessive heat

floating filing down teeth that are overgrown and sharp

gingiva soft gum tissue

gingival hyperplasia excessive amounts of gingival tissue growth

gingivitis inflammation of the gums

heart rate (HR) evaluates how many beats the heart produces in a minute

heat stroke serious condition where the body overheats at temperatures of 105 degrees or more

hyperthermia condition where body temperature is above normal and causes the animal to have a fever

hyphema blood noted inside the eye

hypothermia condition where body temperature is below normal causing the animal to become cold

indirect PLR abnormal pupil light response assessment

jaundice yellow coloration to the skin or mucous membranes

mucous membrane (mm) the gums of an animal and their color

nystagmus condition that causes the eyes to bounce from side to side or in circular motion

plaque soft buildup of material found over the surfaces of the teeth that is composed of bacteria, food materials, and saliva

prophylaxis professional cleaning of a patient's teeth

pulse the heart rate

respiratory rate (RR) measures the breathing of an animal and evaluates how many breaths an animal takes in a minute

tachycardia increased or elevated heart rate

tooth abscess infection of the tooth root

TPR refers to the temperature, pulse, and respiration

vital signs the signs of life that are measurements to assess the basic functions of an animal's body and include the heart rate, respiratory rate, temperature, blood pressure, mucous membrane color, capillary refill time, and weight of the animal

weight tape the use of a measurement tape around the girth of an animal that gives an estimate of its body mass

REVIEW QUESTIONS

1. What is the difference between a scale and a weight tape?
2. What is the difference between hypothermia and hyperthermia?
3. What is the importance of safe use of a mercury thermometer?
4. When obtaining the rectal temperature of a dog, what should the normal temperature be? What should the normal rectal temperature of a horse be?
5. Explain how to locate the heart in the chest area of an animal.
6. What is the term for an increased or elevated heart rate? A decreased or slow heart rate?
7. What are some signs of dental disease?
8. What is the importance of proper hand washing?
9. What are some abnormalities that may be noted when using an ophthalmoscope?
10. What is the proper way to mop an exam room floor?

Clinical Situation

Delmar/Cengage Learning

Brandon, the veterinary assistant at Woodbine Animal Clinic, is cleaning up the exam room to prepare it for the next patient. Brandon notes when he enters the room that there are several used needles and syringes, empty vaccine vials, dirty paper towels, and several exam tools including an otoscope, thermometer, and a pair of bandage scissors on the exam counter. The table is dirty and soiled with what appears to be urine. The floor has some areas of wetness and visible blood. Brandon realizes he has a lot to clean up to prepare the room for the next patient.

- What protocol should Brandon use?
- What should Brandon do first in preparing the room?
- How should the client be handled if the room needs thorough cleaning?

45

Delmar/Cengage Learning

Hospital Procedures

Objectives

Upon completion of this chapter, the reader should be able to:

- ☐ Discuss the proper admission and discharge practices of hospitalized patients
- ☐ Describe how to record observations of hospitalized patients
- ☐ Assess and determine the handling of emergency situations
- ☐ Explain how to tell the gender of animal species
- ☐ Explain how to tell the approximate age of an animal
- ☐ Demonstrate how to properly mix and prepare a vaccine
- ☐ Demonstrate wound cleaning techniques
- ☐ Demonstrate basic bandaging techniques
- ☐ Discuss proper procedures for feeding and watering hospitalized patients
- ☐ Perform proper sanitation of cages of hospitalized patients
- ☐ Discuss proper isolation ward practices

Introduction

Working within a veterinary facility requires being knowledgeable and experienced in hospital skills and being able to recognize an emergency situation when an animal is presented to the facility or is a hospitalized patient. The veterinary assistant must be aware of the needs of the patients and veterinary staff to provide successful work within the hospital treatment area. Several procedures may be necessary of hospitalized patients and the veterinary assistant will be a part of the preparation, procedure, and follow-up care.

The Hospitalized Patient

Animals that are admitted to the hospital may require constant care, thorough observation, critical care and support, or routine and regular treatments. It is important that the veterinary assistant know the proper intake procedures for hospitalized patients, how to note the treatments or procedures that are required, and where the patient should be placed for care. It is also necessary for the assistant to know the case history and when the patient is ready to be discharged so that the owner is given accurate and appropriate instructions for home care.

Admitting the Hospitalized Patient

The client should be aware of what treatments, procedures, and necessary care are required of the patient prior to admitting the animal to the hospital. Appropriate waiver forms and estimates should be provided to the client and then the client may leave (see Figure 45-1). In some cases, this may be difficult for some people and they may need to be reassured that their pet is in excellent care and comfort to ease the separation from the client. The patient, medical record, cage card, and treatment instructions should be placed in the treatment area where all information is kept on hospitalized patients. Some facilities also place a **neck band** immediately on the patient for further identification. The neck band and cage card should contain patient information and reason for hospitalization. Table 45-1 summarizes the various forms of patient identification that may be used in the hospital.

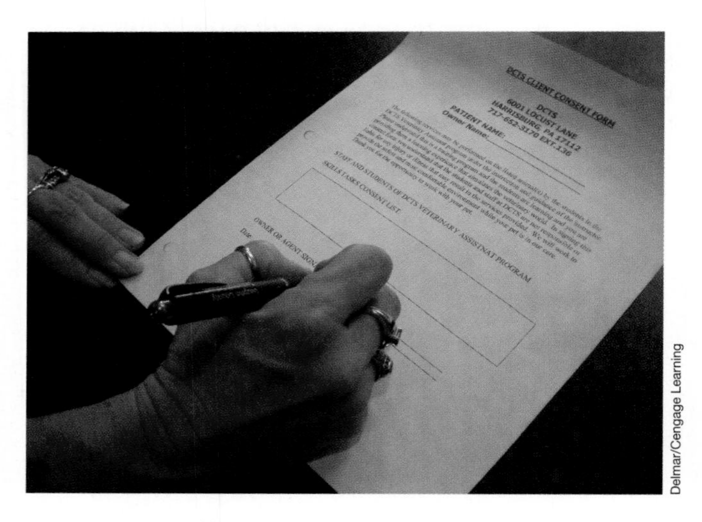

Delmar/Cengage Learning

FIGURE 45-1 Clients must sign a consent form before leaving the patient at the facility.

The cage, kennel, or holding area should be prepared based on the size of the animal and the type of care that is needed. Towels, beds, blankets, or newspapers may be used to line cages for small animals and companion animals. Clean bedding should be provided in a stall for large animals. When the patient is placed in the enclosure and properly secured, the assistant should locate the **hospital treatment board** and medical record bin. The hospital treatment board is an essential and valuable tool in the veterinary clinic (see Figure 45-2). This board contains patient information and displays treatments that need to be completed or have been completed and general health information that is easily accessible to the entire veterinary and support staff. The information captured on the hospital treatment board includes the following:

- Client name
- Phone number
- Patient name
- Patient age
- Patient breed
- Patient gender
- Patient weight
- Reason for hospitalization
- Treatment needs or procedures
- Medications and dosages
- Warnings or other vital information

Many facilities record all hospitalized patients by the medical record filing method so that the medical chart is easily located and follows the same labeling procedure for accuracy. Facilities that file records by client's last name or client number would place the information on the board in the same manner to identify the client, patient, and medical chart (see Figure 45-3).

TABLE 45-1

Identification Information for Hospitalized Patients

Cage card	Neck band	Patient name	Patient age
Species	Breed	Color/Markings	Gender
Phone number	Reason for hospitalization	Client name	Personal items (toys, blanket, etc.)

Delmar/Cengage Learning

FIGURE 45-2 Recording information on the hospital treatment board.

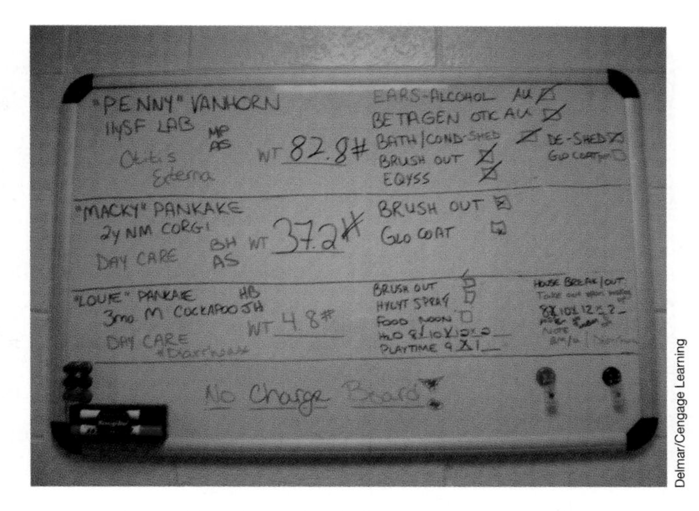

Delmar/Cengage Learning

FIGURE 45-3 Note how the information is recorded on this hospital treatment board.

Every hospitalized patient should have a treatment sheet provided with the veterinarian's treatment instructions written on it. It is important that every staff member document every treatment, procedure, and observation on a daily basis as the work is completed.

Discharging the Hospitalized Patient

When the veterinarian feels the patient no longer needs constant care and is able to safely go home, it is important that written home care instructions are provided in the medical chart. This will be a guideline for the veterinary assistant and support staff in providing

Competency Skill

Admitting the Hospitalized Patient

Objective:

To efficiently and effectively process the admission information for patients that are to be hospitalized.

Preparation:

medical record, estimate, consent form or waiver, cage or holding area according to patient size and care needs (bedding, water, feed, litter box, etc.)

(*Continues*)

Procedure:

1. Review patient treatment plan and estimate with the client.
2. Obtain a signature from the client on a consent form or waiver.
3. Assure the client the patient will be well cared for.
4. Take the patient to the treatment area, along with the medical record.
5. Place a neck band and cage card on the cage for identification (see Figure 45-4).
6. Place patient in cage or holding area.
7. Close cage and check that cage is secure.
8. Place patient information on the hospital treatment board.
9. Place medical record in treatment bin.

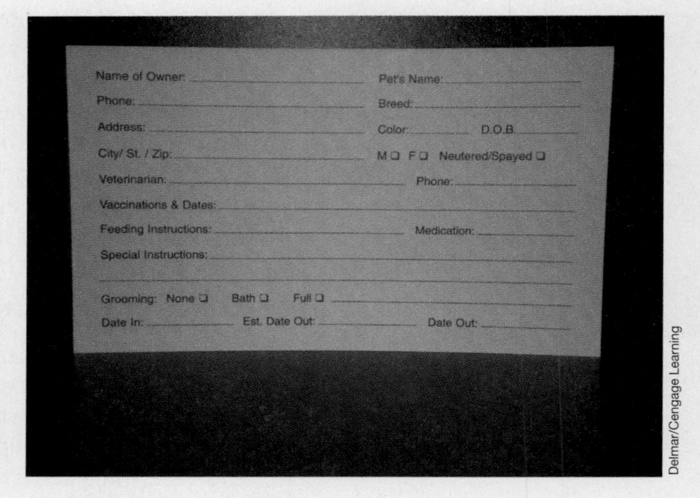

Name of Owner: ____________ Pet's Name: ____________
Phone: ____________ Breed: ____________
Address: ____________ Color: ________ D.O.B. ________
City/ St. / Zip: ____________ M ☐ F ☐ Neutered/Spayed ☐
Veterinarian: ____________ Phone: ____________
Vaccinations & Dates: ____________
Feeding Instructions: ____________ Medication: ____________
Special Instructions: ____________
Grooming: None ☐ Bath ☐ Full ☐ ____________
Date In: ________ Est. Date Out: ________ Date Out: ________

Delmar/Cengage Learning

FIGURE 45-4 Cage card.

information on patient care to the client. Compliance of home care is improved when the owner has the appropriate information and instructions for home care. The medical record should include home care instructions, any medications that need to be dispensed and sent with the pet, any pending laboratory tests, and an invoice. Once the prescription is filled, the assistant should review the dosage and directions for giving the medicine. The owner should be told when any pending lab test results will be available, if a recheck appointment is necessary, and any health conditions to monitor at home. It is important to ask each client if he or she has questions regarding the patient or the home care instructions. The veterinary assistant should stress to the owner to call if he or she has any problems or questions (see Figure 45-5). The client and patient should be escorted to the front desk or reception area for invoicing and checkout procedures. The assistant should note the conversation in the medical chart.

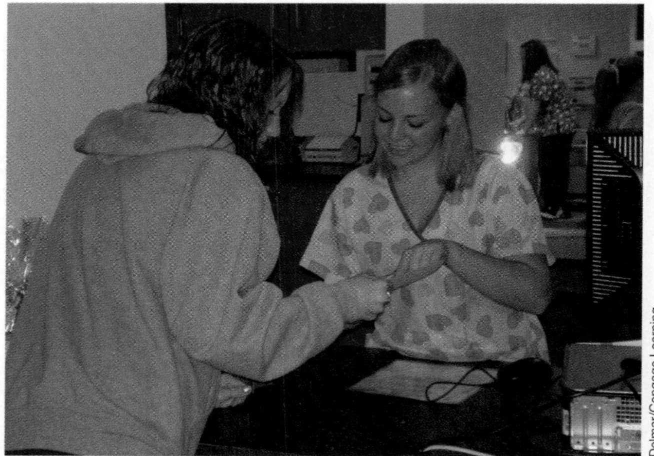

Delmar/Cengage Learning

FIGURE 45-5 The veterinary assistant should go over all discharge instructions with the client and allow the client time to ask questions.

Competency Skill

Discharging a Hospitalized Patient

Objective:

To efficiently and effectively process the discharge information for patients that are able to return home after hospitalization.

Preparation:

medical record, home care instructions, medication, invoice, patient, and any personal items

Procedure:

1. Complete any prescriptions.
2. Review home care instructions, medications, and other information with the client.
3. Demonstrate how to administer medications.
4. Provide information on lab tests or follow-up visits.
5. Review the patient discharge form with written instructions.
6. Obtain the patient for the client.
7. Escort the client to the front desk.
8. Give the medical record and invoice to the receptionist.
9. Inform the receptionist if a follow-up visit is needed.
10. Thank the client for coming and ask him or her to call if any problems or questions arise.

Recording Observations

It is essential for all staff members to learn to observe patients. This is a necessary skill from the moment a patient enters the facility until it is discharged. **Observation** is a key factor in successful patient care and treatment. Observation includes watching and noting an animal's behavior, appearance, mental status, and overall health (see Figure 45-6). These signs should be compared with what is normal for the species and breed of the animal. Any changes that occur should be monitored and noted in the medical chart. Some changes may be subtle and can be easily overlooked. Observations include visual monitoring, smelling odors, palpating changes, and auditory monitoring. Examples of observations that may be helpful in treating a patient include noting odorous gas or diarrhea, a tense abdomen, or ocular discharge. Daily observations should be noted in the medical record.

FIGURE 45-6 Observation skills are critical to the health and safety of the patient while under the care of the veterinarian.

Evaluating Emergency Situations

Emergencies are a fact of life in all veterinary facilities. Patients may arrive in serious or critical condition from injuries, hemorrhaging, shock, respiratory distress, or poisoning cases. An **emergency** is a situation that requires immediate life-saving measures. When emergencies arrive or occur within the hospital, it is important that all staff members work together as a team under the leadership of the attending veterinarian to provide adequate emergency care. The success of handling emergencies relies on the staff working as a team and staying calm under pressure. The veterinarian and veterinary technician should be providing hands-on care while the veterinary assistants are responsible for locating supplies, medications, and emergency equipment for use in the treatment area. A **crash cart**, which is a movable table that holds emergency equipment and supplies, should be easily accessible during a serious patient condition. All items are prepared ahead of time for emergency situations. An effective and efficient veterinary assistant should be able to complete the following:

- Locate the crash cart or emergency kit
- Locate emergency equipment that may not be located on the crash cart or emergency kit due to size or limited supply
- Maintain emergency equipment, drugs, and supplies on the crash cart or kit
- Update the emergency supplies as newer equipment and fresh drugs are ordered
- Identify common emergency equipment
- Locate common general supplies
- Restrain and position patients properly
- Provide skin preparation as necessary
- Perform simple in-house laboratory procedures

Table 45-2 lists the common emergency equipment the veterinary assistant should be familiar with.

Once the patient has been stabilized, the assistant should provide veterinary care and patient monitoring as needed under the veterinarian and veterinary technician. All staff members should be trained on what to do during an emergency so that each member of the veterinary team has a job to perform and is useful in these situations. Monthly staff meetings can be devoted to practicing such emergency procedures and protocols according to the veterinarian. Knowledge of common emergency procedures is invaluable during an emergency as this is not a time to learn and be trained.

Signs of an animal that is in a life-threatening emergency include the following:

- No signs of a heartbeat
- No signs of breathing
- An animal that is not alert
- An animal that is having difficulty breathing
- An animal that has a low body temperature and is pale in gum color
- An animal that has an excessively high temperature
- An animal that is hemorrhaging excessively

It is important that assistants have knowledge of emergency equipment and how to clean and maintain its care. A regular maintenance schedule should be followed to make sure all items are in working order. All drugs should be monitored for proper dating and a checklist made for all items.

Common Hospital Procedures

The veterinary assistant should be able to perform basic hospital procedures. These procedures include determining the gender of an animal, applying a basic bandage, mixing up a vaccine or medication, basic wound care, determining the approximate age of an animal, and feeding and watering procedures.

TABLE 45-2

Common Emergency Equipment

Laryngoscope and blade (see Figure 45-7)	Various endotracheal tube sizes (see Figure 45-8)	Heat pad or blanket	Clippers and blades
Instrument packs	IV stand	IV bags	IV lines
Pulse oximeter	Oxygen cage	Oxygen tank	Anesthesia machine
Stomach tube	IV catheter	Bandage materials	Urinary catheter
Stethoscope	Penlight	Syringes	Needles
EKG	Blood pressure monitor	Ambu bag	Emergency drugs

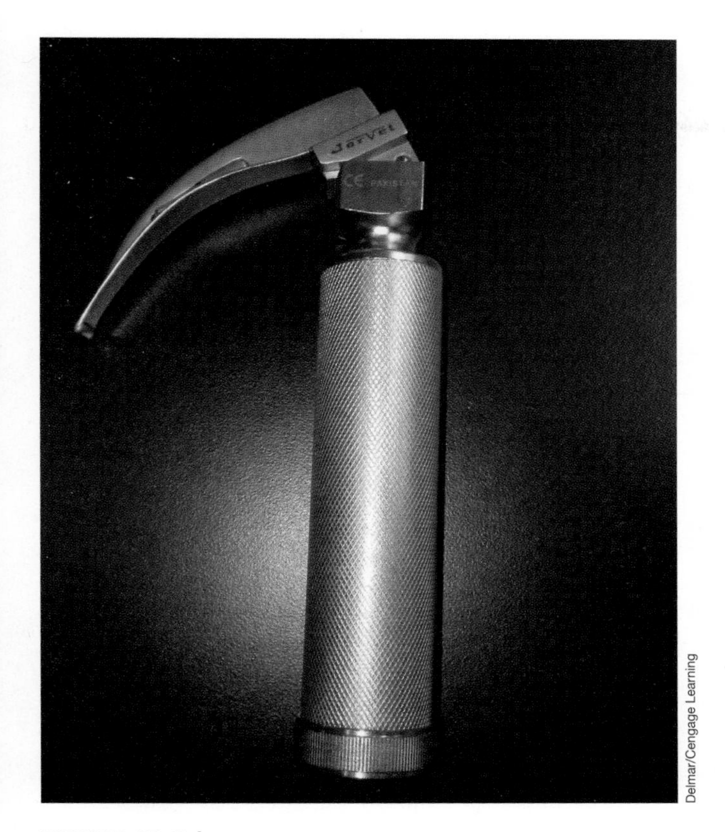

Delmar/Cengage Learning

FIGURE 45-7 Laryngoscope.

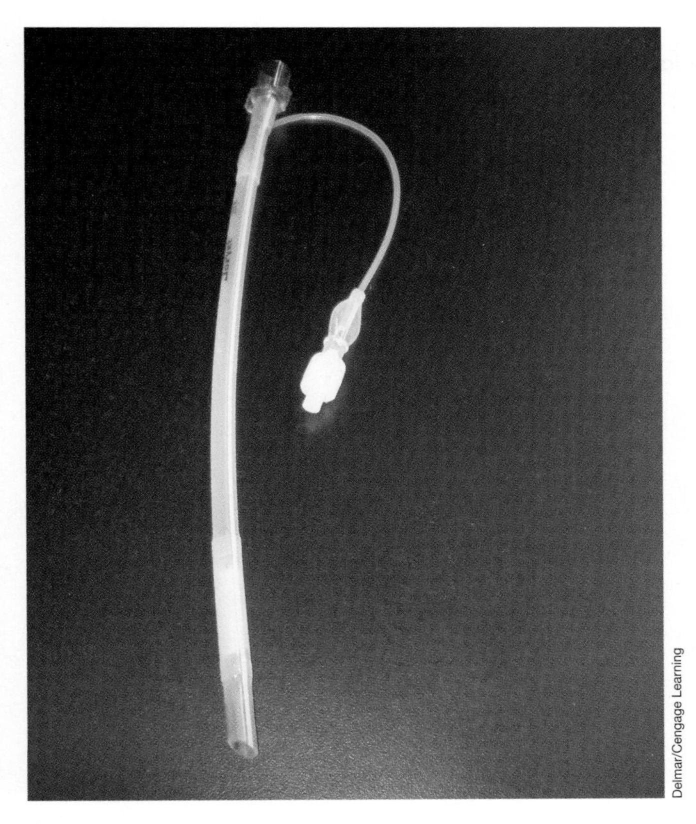

Delmar/Cengage Learning

FIGURE 45-8 Endotracheal tube.

Determining Animal Gender

Although it may seem humorous and odd, some owners may not know the gender or may incorrectly identify the gender of an animal. In some animals, such as kittens, rabbits, and pocket pets, it may be more difficult to identify the sex. Some young animals may not have developed reproductive organs to easily identify their gender. Some animals may have been neutered, as well. The gender of most pocket pets, rabbits, and cats can be identified by the shape of the genital opening and the distance between the anus and genital area, commonly called the **anogenital distance**. Some animals can be sexed by this distance from the rectum or anus and the genital area where urination occurs. Many male rodent species and cats have a wider space than in the female (see Figure 45-9). Gently lift the tail at its base and move it upward to get a visual appearance. Table 45-3 summarizes the shape of the genitals in various species.

Determining Approximate Animal Age

The age of an animal can be approximated in part by examining its **dentition**. The dentition is the arrangement of teeth in the mouth. The eruption time of deciduous and permanent teeth occurs at a specific time in each animal species. The deciduous teeth, or the baby teeth, are sharper and smaller than the adult teeth, which are larger. Also, as animals grow the teeth become more widely spaced as the jaw grows, when the baby teeth are pushed out and replaced by the adult teeth. The appearance of the teeth will also help determine age in an animal (see Figure 45-10). Signs of staining, tartar, and general wearing will be noted in older animals. In the case of horses, the shape of the angle of

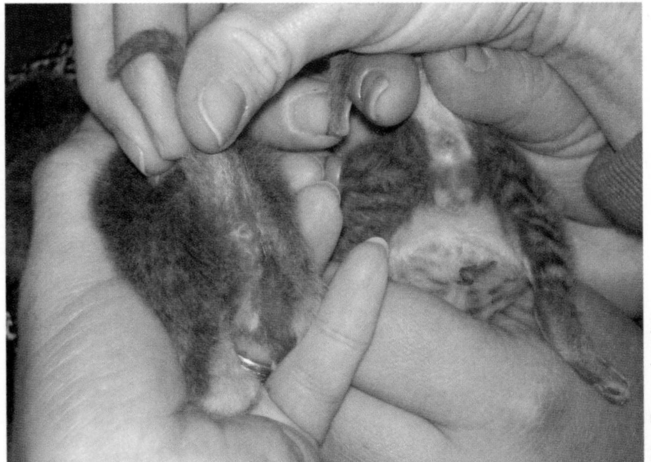

Delmar/Cengage Learning

FIGURE 45-9 Determining sex in kittens. The female kitten is on the left and the male is on the right.

TABLE 45-3

Genital Shape of Species

SPECIES	MALE GENITAL SHAPE	FEMALE GENITAL SHAPE
Mouse	Penile shaft	Vaginal orifice or opening with flap of skin
Rat	Outward projection	Closed
Rabbit	Round opening	Slit opening
Hamster	Round opening	Long, slit-like opening
Guinea Pig	I-shaped; able to expose penis	Y-shaped opening
Cat	Round; period shaped	Slit; comma shaped

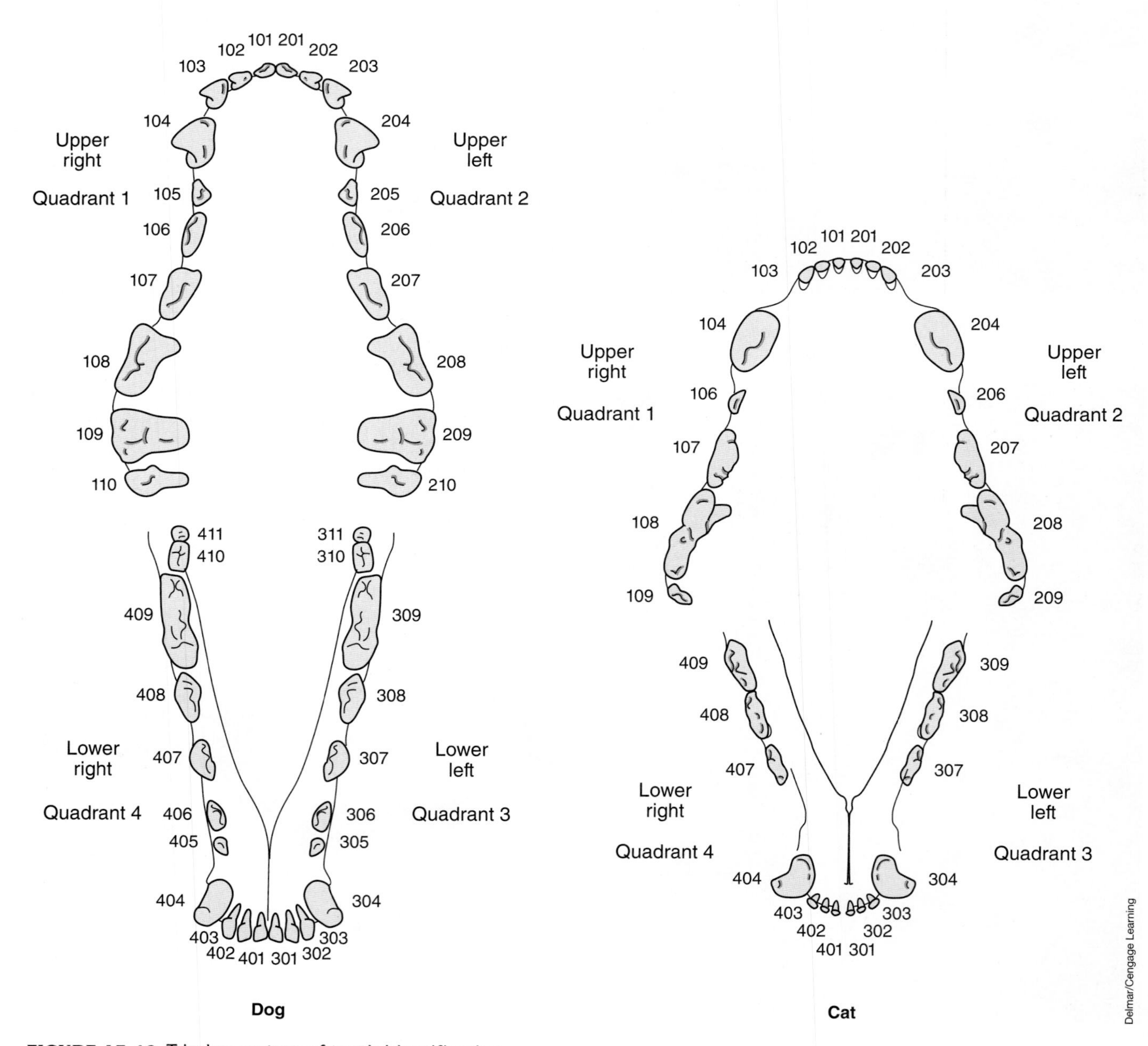

FIGURE 45-10 Triadan system of tooth identification.

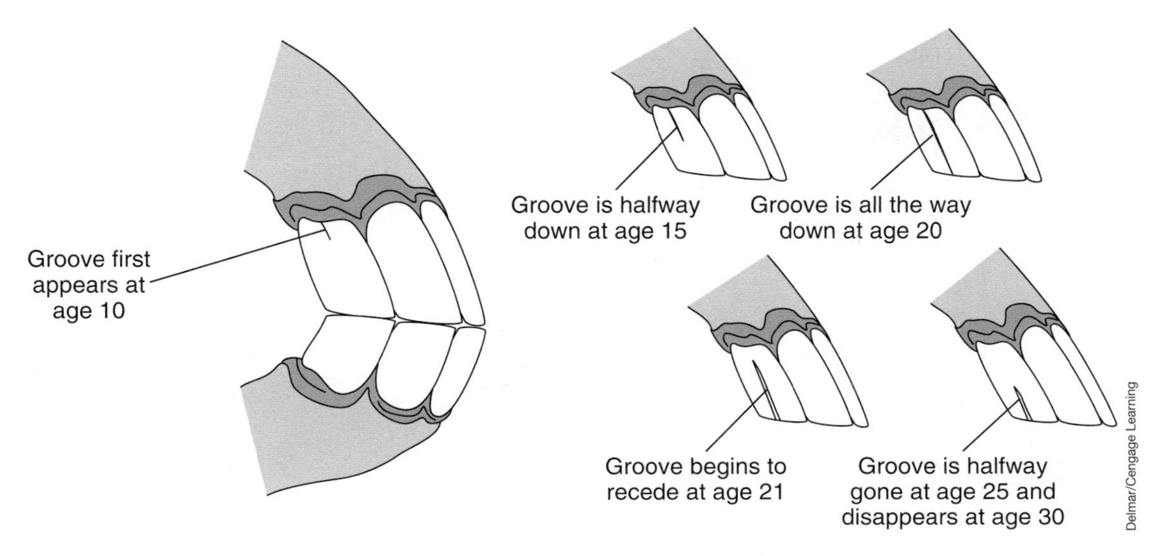
Groove first appears at age 10
Groove is halfway down at age 15
Groove is all the way down at age 20
Groove begins to recede at age 21
Groove is halfway gone at age 25 and disappears at age 30
Delmar/Cengage Learning

FIGURE 45-11 Grooves to determine age in horse teeth.

the teeth changes and the cups (center spot in the teeth) disappear with age (see Figure 45-11). Younger horses have more round appearing teeth, middle-aged horses have triangular shaped teeth, and senior horses have rectangular shaped teeth. Dogs and cats typically begin losing their deciduous teeth between 3 and 6 months of age and have all of the adult teeth between 6 and 8 months of age. Table 45-4 presents a comparison of the dentition in various species.

Preparing Vaccines

Vaccinations are a regular and routine occurrence in the veterinary hospital. All animals require booster vaccines while young or beginning a new type of vaccine and are placed on a regular vaccine program throughout the adult life. Animals that appear ill or diseased should not be vaccinated until they are healthy. Vaccines come either in single or multiple doses. Single doses may be premixed or in two vials, one with a liquid **diluent** and one with a powder (see Figure 45-12). The two vials must be combined to form a viable vaccine. The liquid diluent must be removed with a syringe and needle and may be an active vaccine component or sterile water. The diluent is then inserted into the powder vial and inverted to mix the components. Always mix the diluent provided for the vaccine, never one from another vaccine type. The vaccine is then ready to be administered in the animal species that it is labeled for. Once a vaccine is reconstituted, it must be used within an hour unless the label indicates otherwise. Vaccines are considered biologicals and must be kept refrigerated (see Figure 45-13). Multiple-dose vials will indicate the amount of vaccine required for each animal. This is usually 1 ml but the label should be read for the directions. Multiple doses are premixed and ready for use. Care must be taken not to contaminate the vial as this will make the entire contents unusable.

TABLE 45-4

Comparison of Dentition between Species*

		DECIDUOUS				PERMANENT			
		INCISORS	CANINES	PREMOLARS	MOLARS	INCISORS	CANINES	PREMOLARS	MOLARS
Dogs	Upper	3	1	3	3	1	4	2	
	Lower	3	1	3	3	1	4	3	
Horse	Upper	3	0	3	3	1	3 or 4†	2	
	Lower	3	0	3	3	1	4	3	
Cattle	Upper	0	0	3	0	0	3	3	
	Lower	3	1	3	3	1	3	3	

*The chart lists the number of teeth present on one side of the mouth. The total number of teeth is double.

†The mare (female horse) often does not have the canine tooth, and the first premolar is often absent.

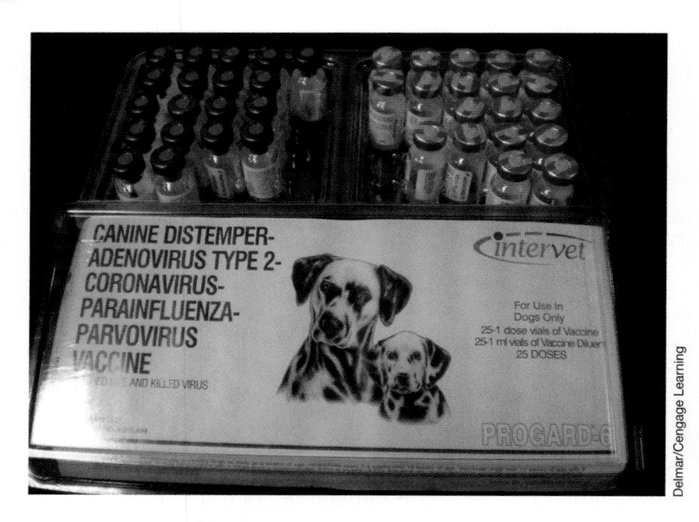

Delmar/Cengage Learning

FIGURE 45-12 Vaccines.

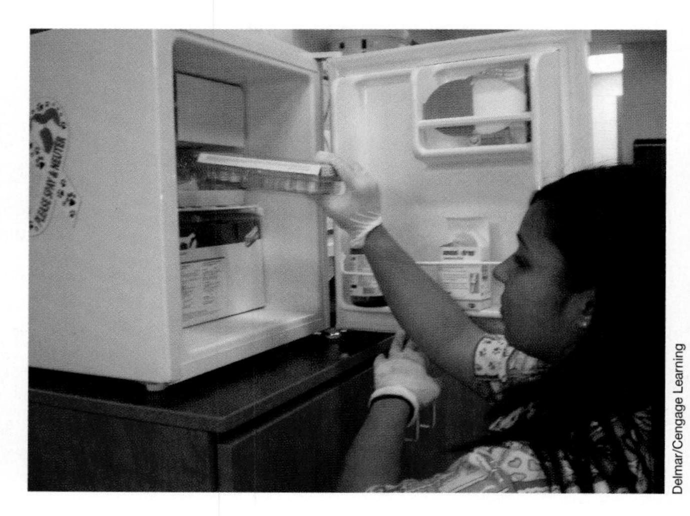

Delmar/Cengage Learning

FIGURE 45-13 Vaccines must be kept refrigerated.

Competency Skill

Mixing Vaccines

Objective:

To properly prepare vaccinations to be administered to patients.

Preparation:

vaccine for the appropriate species, an appropriate-size syringe and needle (usually 1–3 ml), alcohol, cotton balls, exam gloves (see Figure 45-14)

Procedure:

1. Apply exam gloves.
2. Soak a cotton ball with alcohol.
3. Wipe the top of each vial with the alcohol.
4. Obtain the liquid diluent vial and invert the bottle. (Multiple dose vials volume should be withdrawn according to label; see Figure 45-15.)
5. Insert the needle into the center of the rubber vial (see Figure 45-16).
6. Remove entire contents of vial into the syringe. For a multiple dose vial remove only the desired amount as specified on the label (see Figure 45-17).
7. Inject entire contents of liquid diluent into the powder vial.
8. Gently invert bottle 4–5 times until thoroughly mixed.
9. Withdraw the contents into the syringe and remove the needle from the vial.

10. Recap the needle.
11. Label each syringe with the peel-off label or a piece of tape indicating the contents.
12. Place the syringe in the treatment area.
13. Clean up work area.

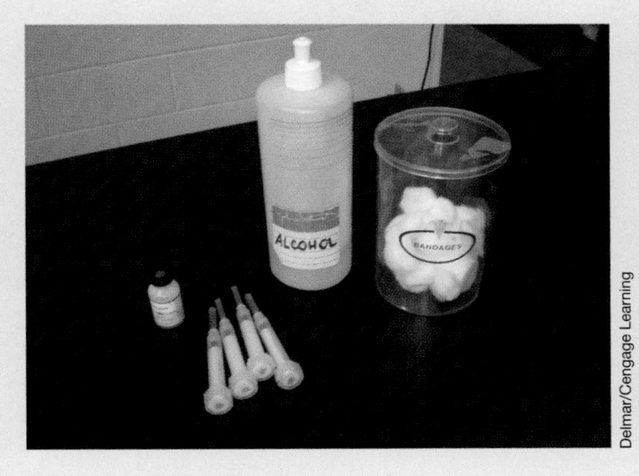

Delmar/Cengage Learning

FIGURE 45-14 Preparation for mixing vaccines.

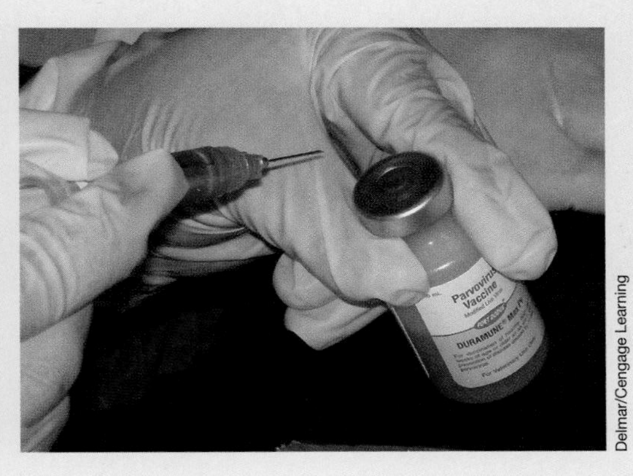

Delmar/Cengage Learning

FIGURE 45-15 Look at the label for the correct amount to withdraw when working with a multiple dose vial.

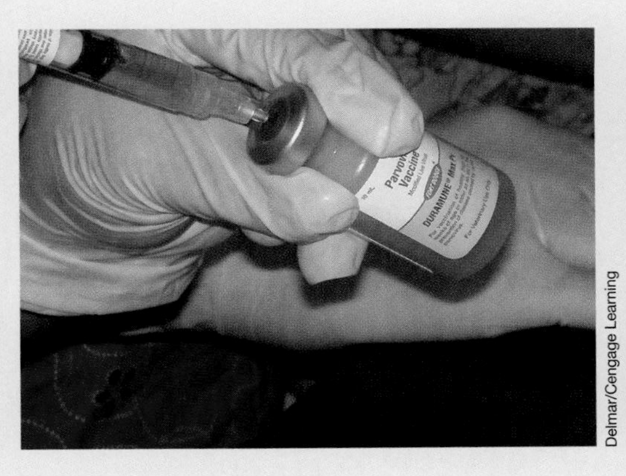

Delmar/Cengage Learning

FIGURE 45-16 Insert the needle into the rubber top of the vial.

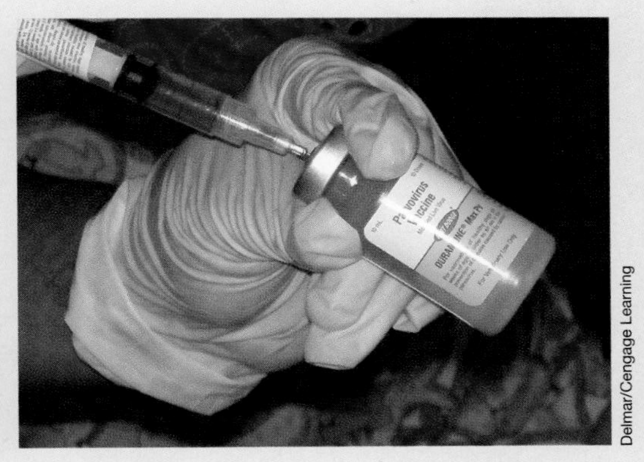

Delmar/Cengage Learning

FIGURE 45-17 Only withdraw the amount indicated for a single dose when using a multiple dose vial.

Wound Cleaning and Care

Wounds may be as superficial as a lesion or skin irritation or may be deep within the flesh as a cut or puncture (see Figure 45-18). It is important that the wound be evaluated by the veterinarian to determine treatment options and the possibility of suturing. Many wounds may be on areas where excessive amounts of hair are located, and these areas will need to be clipped away for better visibility and treatment care. Dependent on the treatment involved with a wound, the area will need to be cleaned and disinfected. The area may be contaminated with debris and bacteria, so it is best to flush the site after the hair has been clipped away. This will also flush out any hair that may have gotten into the site. A surgical scrub should be used to cleanse and disinfect the area, such as Nolvasan or chlorhexidine scrub (see Figure 45-19). Wound care will depend on the location of the wound, the size and depth of the wound, and how well the wound is managed. Antibiotics are common treatments to protect the animal against infection.

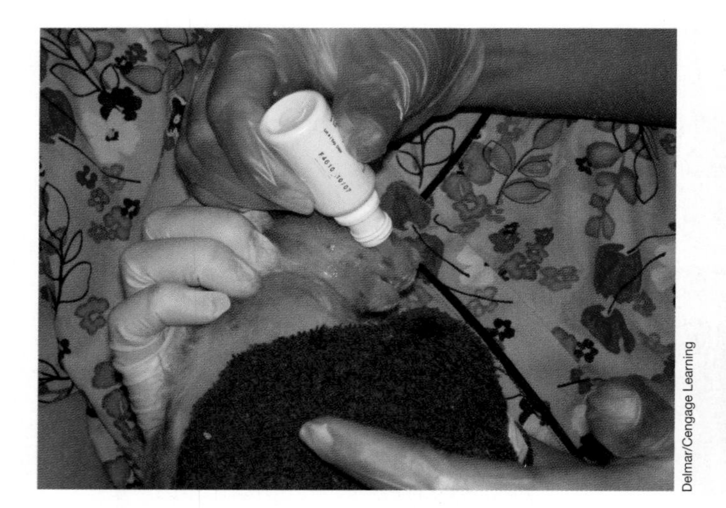

Delmar/Cengage Learning

FIGURE 45-18 Skin lesions are a type of wound.

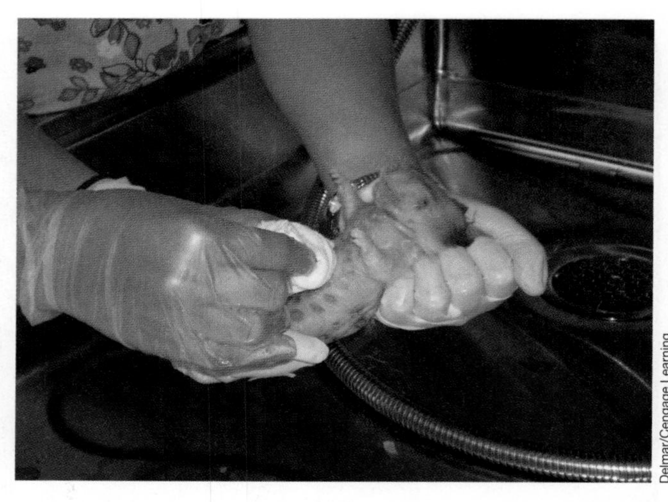

Delmar/Cengage Learning

FIGURE 45-19 Clean the wound with a surgical scrub to disinfect the area.

Competency Skill

Wound Cleaning

Objective:

To properly clean and prep the site of a wound for the veterinarian to assess and provide treatment for it.

Preparation:

exam gloves, clippers, KY gel, vacuum, gauze sponges, nonabsorbent pad, water bowl, sterile saline, betadine, Nolvasan or chlorhexidine scrub, 50 ml syringe and large gauze needle, sponge, forceps, towels

Procedure:

1. Apply KY gel over the wound area.
2. Clip the hair from the wound area going against the growth of the hair coat.
3. Clip wide margins with straight edges.
4. Vacuum loose hair from the area.
5. Place towels around the area of the animal where the wound is located.
6. Gently flush the wound with sterile saline using a 50 ml syringe.
7. Repeat the flushing until the wound is free of dirt and debris.

8. Begin the scrubbing process by applying sterile water and Nolvasan or chlorhexidine scrub on gauze sponges. Use a water bowl to soak the sponges. Begin at the margin of the wound, scrubbing at the center of the wound and working to the clipped edges. Repeat as necessary for 3–4 cycles.
9. Begin scrubbing at the center of the wound in circular movements working outward to the hair coat. Repeat 3–4 times.
10. Use a new gauze sponge with each cleaning.
11. Apply a surgical disinfectant, such as betadine (iodine), over the wound.
12. Place a sterile nonabsorbent pad or gauze sponge over the site to keep it clean.

Bandaging and Bandage Care

Bandages may be applied to an animal for several reasons, such as wound dressings, protection, immobilization (to prevent movement), and for covering splints. Bandages consist of several layers, each with a specific purpose. The veterinary assistant should have a general knowledge of the materials and equipment necessary for bandaging techniques and a basic knowledge of how to apply a simple protective bandage (see Figure 45-20). The general materials needed for bandaging include the following:

- Rolled cotton
- Adhesive tape (1″–2″)
- Rolled gauze or kling
- Co-flex or vet wrap bandage
- Non-adhesive pad
- Bandage scissors

The **primary bandage layer** is the layer closest to the skin. The primary layer should be soft and provide padding to prevent rubbing and skin irritation and act as a protective covering of the area. This may include cotton rolls, cotton bandages, or other soft fiber materials wrapped around the area. If a wound or suture line is present, the area should be covered with a nonabsorbent protective pad. The bandage layer application should be tight enough to remain in place but not so tight as to cut off circulation. Bandage wrapping should begin at the center of the area to be bandaged and wrapped downward or forward of the area and then reversed to cover the upward or backward area. This allows the bandage to cover all areas and provides several primary layers of padding and protection. Many times the bandage layers are applied at a 45-degree angle, which may help prevent bandage slipping. Each layer should lie smoothly against the previous layer with no wrinkling present and with overlap of the layer before it.

The **secondary bandage layer** is a thin fiber material that clings to the first layer and holds the primary layer in place. It typically consists of a non-adherent wrap, such as gauze or kling, extending proximally and then distally beyond the edges of the primary layer (see Figure 45-21).

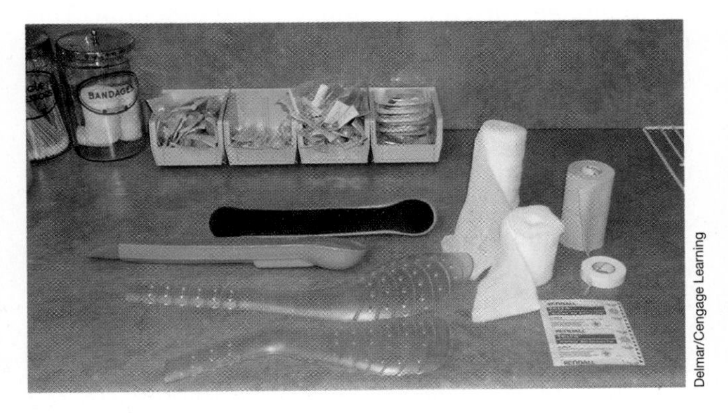

Delmar/Cengage Learning

FIGURE 45-20 A selection of bandage material and splints.

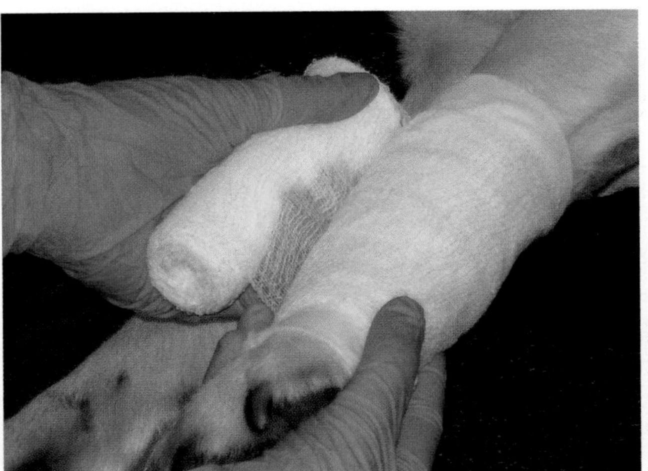

Delmar/Cengage Learning

FIGURE 45-21 The secondary layer of gauge is placed over the primary layer of the bandage.

The **tertiary bandage layer** is the final layer that acts as the outer covering that holds the bandage on the animal and is usually waterproof to protect the underlying layers. It should extend proximally and distally beyond the edges of the secondary layer. The material for this layer is usually self-adherent, meaning it sticks to itself, such as co-flex or vet wrap (see Figure 45-22). The edges should be wide enough to prevent slipping but not so extensive that it causes discomfort to the patient. Adhesive tape can be applied to the outer edges of the last layer of bandage on the proximal and distal ends to also help prevent the bandage from slipping (see Figure 45-23). Half of the tape piece should attach to the bandage and half to the hair of the animal. It is important to note that bandage layers that are placed too tight may cut off areas of circulation, causing serious damage to tissue.

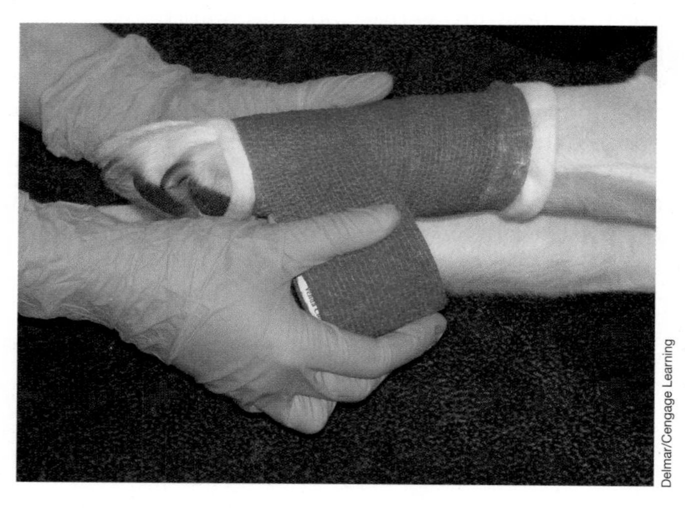

Delmar/Cengage Learning

FIGURE 45-22 The tertiary bandage layer is usually waterproof and helps to protect and keep the underlying layers clean.

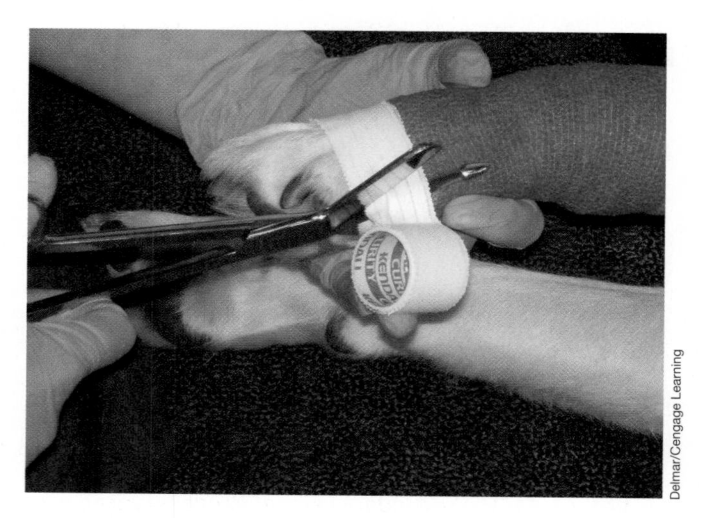

Delmar/Cengage Learning

FIGURE 45-23 Tape can be applied to the ends of the bandage to help keep it in place.

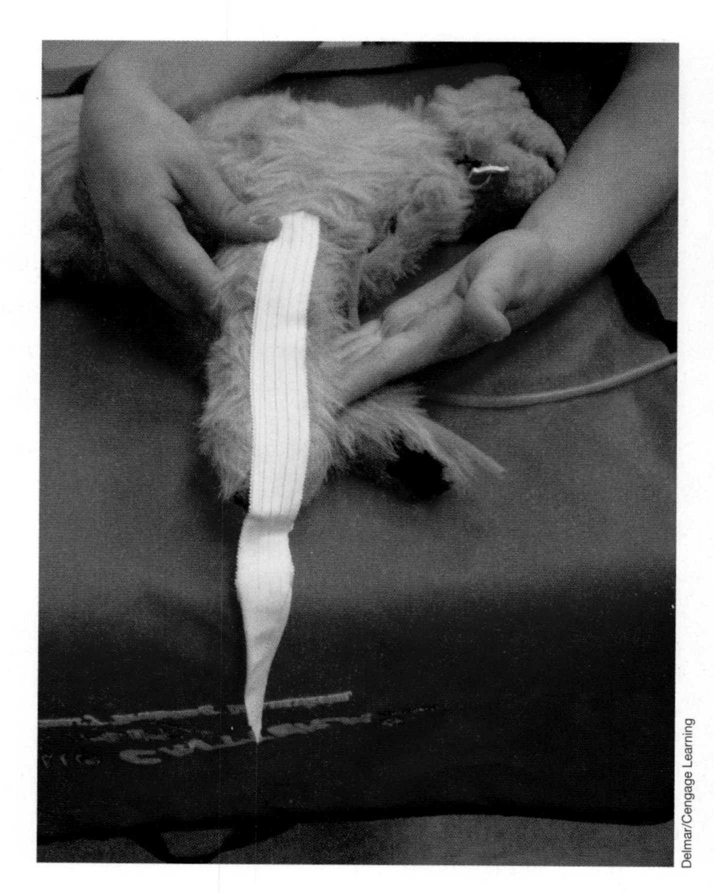

Delmar/Cengage Learning

FIGURE 45-24 Application of adhesive tape to prevent slippage of the bandage.

Bandage application and type depend on the purpose of the bandage and the location to which they are applied. If slipping is a concern, such as on the limbs, adhesive tape strips may be applied directly onto the skin on opposite sides of the limb (see Figure 45-24). The strips should extend beyond the end of the limb. The bandage is then applied and the strips are incorporated into the secondary bandage layer. These tape strips are referred to as **stirrups**.

Bandage care and cleanliness are essential. This means that the bandage should remain dry and free of debris. This may mean frequent bandage changes for areas of wounds or if a bandage becomes soiled. Animals are frequent lickers and chewers of bandages so care must be met to prevent the animal from damaging or removing the bandage. Sprays that have nontoxic but bad-tasting chemicals may be applied to the outer layer to limit these behaviors. Examples of this type of spray include Bandguard and Bitter Apple. More drastic measures may be required for difficult patients, such as applying an E-collar around the head and neck to prevent chewing. Bandage care may also include restricted exercise, monitoring for signs of swelling or slipping, applying a plastic bag over the bandage to protect from wetness, and monitoring for odor.

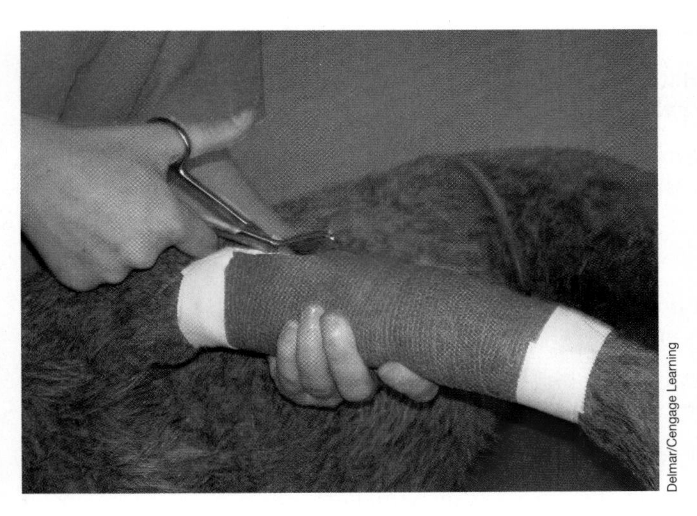

FIGURE 45-25 Use bandage scissors to cut away the bandage materials for removal.

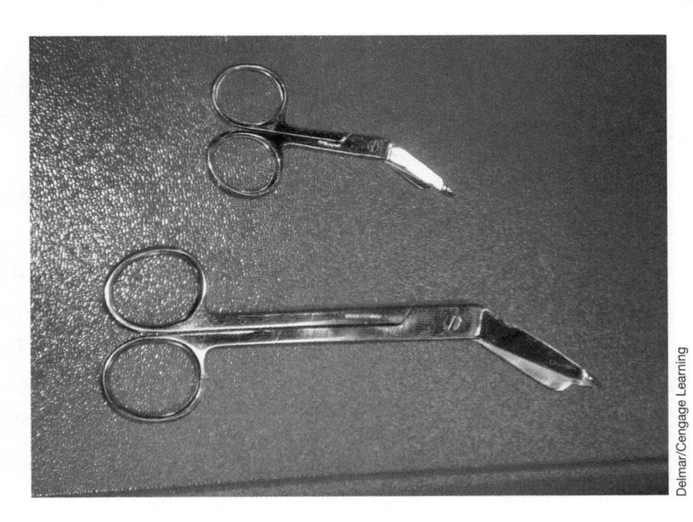

FIGURE 45-26 Bandage scissors.

Bandage removal is completed when the veterinarian determines the bandage may be removed. **Bandage scissors** are used to cut away the layers of the bandage (see Figure 45-25). Bandage scissors have an angled end called a dog leg (see Figure 45-26). This angled end of the scissor blades prevents cutting into the animal during bandage removal. The bandage should be removed from the proximal end working distally and making certain that the end of the bandage scissors is always pointed upward and in contact with the bandage layers.

Competency Skill

Bandage Application

Objective:

To properly place a bandage on a wound.

Preparation:

Cotton roll, gauze roll or kling, 1″–2″ adhesive tape, Vet wrap or Co-flex bandage, bandage scissors, exam gloves

Procedure:

1. Apply exam gloves.
2. Apply a strip of 1″ or 2″ tape along the cranial aspect of a limb and one strip on the caudal aspect of a limb to prevent slipping. The tape strips should extend beyond the end of the limb.

(Continues)

3. Hold the roll of cotton in one hand while using the other hand to hold the wrap. *Do not* place too tight, cutting off circulation.
4. Apply the primary layer using a cotton roll by applying distally to the end of the area and then proximally to cover the entire surface. Apply the cotton roll at a slight 45-degree angle.
5. Apply the cotton roll smoothly and evenly, preventing wrinkling.
6. Apply each layer of cotton so that it overlaps the previous layer. Continue wrapping to desired padding and protection amount.
7. Apply the gauze roll as the secondary layer distally to the end of the area and then proximally to cover the entire surface. *Do not* place too tight, cutting off circulation.
8. Apply the gauze roll at a slight 45-degree angle.
9. Apply each gauze layer smoothly and evenly, preventing wrinkling.
10. Apply each layer of gauze so that it overlaps the previous layer. Pull each strip of tape upward into the gauze roll layers.
11. Apply to desired amount with a slight extension beyond the end of the primary layers.
12. Apply the vet wrap tertiary layer by applying distally to the end of the area and then proximally to cover the entire area.
13. Apply the vet wrap at a slight 45-degree angle.
14. Apply the vet wrap smoothly and evenly, preventing wrinkling.
15. Apply each layer of vet wrap so that it overlaps the previous layer.
16. Apply to desired amount with the edges of the bandage extending slightly beyond the secondary layer. *Do not* place too tight, cutting off circulation.
17. Apply 1″ or 2″ adhesive tape to the proximal end and the distal end of the bandage edges. Attach half of the tape layer to the bandage and half of the tape layer to the hair.
18. A small amount of tape can be placed along the last edge of the vet wrap at the end of the bandage opening. You may make a triangle end of the vet wrap bandage to easily locate the opening.

Competency Skill

Bandage Removal

Objective:

To properly remove the bandage without causing any additional trauma.

Preparation:

bandage scissors, exam gloves

Procedure:

1. Apply exam gloves.
2. Work from the proximal end to the distal end of the bandage.
3. Place the long, blunt blade of the bandage scissors against the skin and slightly under the bandage edge.
4. Keep the blade flat against the skin and the end raised slightly upward in contact with the bandage.
5. Place the bandage layers between the scissor blades.
6. Begin cutting proximally. Peel the bandage layers away from the patient using a firm motion.
7. Unpeel or cut layers moving toward the distal portion of the bandage.
8. Remove stirrups if present when the entire bandage has been removed.
9. Gently remove each layer of bandage.
10. Notify the veterinarian when the bandage has been removed.
11. Clean up work area.

Feeding and Watering Hospitalized Patients

There are several factors influencing patient feeding and watering in the veterinary hospital. The health status of the patient, procedures or surgeries to be completed, medications, and the patient's appetite must be factored into the patient's needs. Some animals may be placed on NPO or no food or water by mouth. This may be due to the patient's condition, disease, or surgical needs. If an animal is allowed to eat and drink, it must be determined which diet is appropriate for the patient. Age and illness will influence the type of diet. Patients with certain conditions will need specialized veterinary diets or foods that may be prescribed by the attending veterinarian. Typically, heart failure patients need a low-salt diet and renal failure patients need a low-protein diet. The diet should be noted in the chart so that consistency during the hospital stay is followed.

Some animals may feel comfortable eating in a veterinary setting. Other animals may feel insecure to eat in a caged setting or outside of their normal environment. This needs to be taken into consideration when trying to encourage patients to eat. Some animals may not be used to the type or consistency of food. For example, some dogs eat only canned or dry food and may not eat another type. Some horses or cattle may eat a specific type of hay and not want another variety. In small animals, canned foods may be warmed up to increase the palatability. Gentle heating of food also increases the aroma or smell, making it more appealing. Mixing small amounts of water can sometimes increase the acceptance. Dry food can be watered down to make it softer in texture. In large animals, water or applesauce can be added to grains to make it more appealing. Foods may also be force fed to animals if necessary. **Force feeding** is placing the food in the animal's mouth and making it swallow. This is done by several methods. One should gently opening the animal's mouth and place a small amount of food in the mouth using a finger against the roof of the mouth. Another method is to liquefy the diet and feed it through a syringe or feeding tube. The syringe is filled with food and the end of the syringe or the feeding tube is placed into the side of the mouth near the back teeth. The nose of the patient is tipped slightly upward. The end of the syringe is pushed to move the food contents into the mouth and the patient is allowed to swallow between amounts. The feedings are broken down into several meals throughout the day. Food should be fed slowly to prevent the patient from aspirating food into the lungs. **Aspiration** means swallowing food that enters the lungs rather than the stomach. Some patients may begin to eat on their own after a few amounts of food are passed. Some animals will fight the process of force feeding. It may be necessary to use a **nasogastric tube** placed into the stomach to feed the animal (see Figure 45-27). This is done by the veterinarian or veterinary technician. The nasogastric tube enters the nostril and passes through the pharynx and into the esophagus until the tip is placed into the stomach. Liquefied food can then be administered through the tube as needed. The amount of food required should be evaluated by the veterinarian to make certain the patient's total calorie needs are met. Animals that are capable of eating should be monitored for the amount of food they eat. Animals should never go longer than 24 hours without eating. Feeding hospitalized patients requires creativity, persistence, and patience.

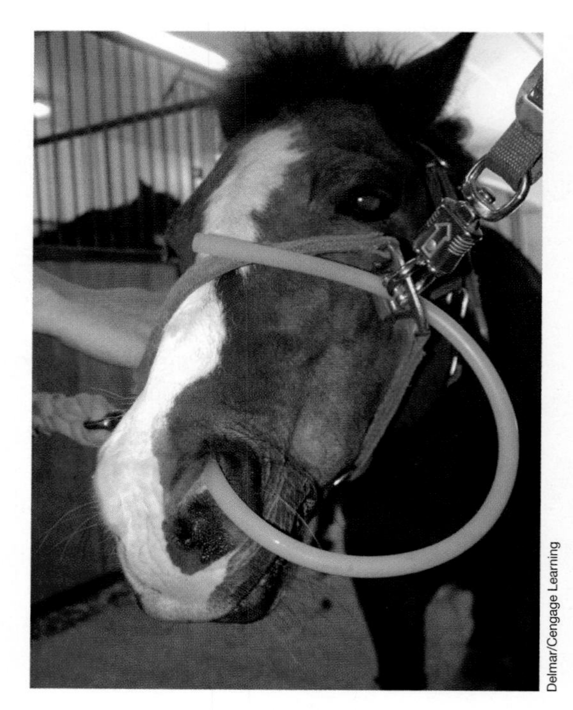

Delmar/Cengage Learning

FIGURE 45-27 Nasogastric feeding tube.

Competency Skill

Feeding Hospitalized Patients

Objective:

To properly meet the dietary needs of the hospitalized patient through feeding and watering.

Preparation:

select the appropriate variety of food, note how much food is to be given to the patient, food bowl, bucket, or other utensils, exam gloves, microwave or heat source

Procedure:

1. Prepare the approved diet.
2. Determine the amount of food to be fed.
3. Offer the food to the animal. If the animal doesn't seem interested in eating, apply a small amount of water to the food. For soft foods, gently warm the contents to room temperature.
4. If the patient is not interested, place a small amount on the fingertip.

5. Gently open the patient's mouth and place the food against the roof of the mouth.
6. Allow the patient to swallow the food.
7. Repeat slowly, allowing the patient time to swallow.
8. Repeat until the food is refused or the animal eats the amount.

Competency Skill

Force Feeding Hospitalized Patients

Objective:

To properly administer nutritional products to the animal by force.

Preparation:

select the appropriate variety of food, note how much food is to be given to the patient, syringe, feeding tube, exam gloves, microwave or heat source

Procedure:

1. Apply exam gloves.
2. Prepare the food by liquefying the appropriate amount with warm water.
3. Apply the food into a syringe. A feeding tube can be applied to the end of the syringe if needed.
4. Place the syringe in the side of the patient's mouth toward the back teeth.
5. Tilt the head slightly upward.
6. Pull the corner of the lip outward to form a small pouch.
7. Gently administer the liquid into the mouth between the cheek and back molars.
8. Allow the food to be swallowed.
9. Repeat the process throughout the day until the total daily calorie needs are met.

Water is another important nutrient required by animals. Hospitalized patients may or may not be allowed to have water available to them. Animals may become dehydrated if they are not consuming enough water. Water should be available in adequate amounts if allowed. Hospitalized patients readily spill their water, so cages should be monitored or bowls placed on metal rings to prevent spilling. Some patients will require more water than others. Water should be provided as much as needed and kept fresh and clean.

SUMMARY

Hospitalized patients require specialized care, constant observation, and proper sanitation. Many patients are cared for by the veterinary assistant. Hospitalized patients may be contagious, in a serious health condition, or require many treatments throughout the day. The veterinary assistant must know what is being done with the patients, what is normal or abnormal for each patient, and what treatments are to be done by the assistant. This requires knowledge, experience, and a planned working schedule to meet the diverse needs of every patient.

Key Terms

anogenital distance distance between the area of the vulva or penis and the rectum by which gender can be determined

aspiration swallowing food that enters the lung field

bandage covering applied to a wound, fracture, or area of injury to prevent movement

bandage scissors scissors with an angled blunt end used to carefully remove a bandage from an animal

crash cart movable table that holds emergency equipment and supplies that is easily accessible during a serious patient condition

dentition the arrangement of teeth within the mouth

diluent liquid mixture of a vaccine

emergency situation that requires immediate life-saving measures

force feeding placing food within a patient's mouth and forcing it to swallow

hospital treatment board area in the facility where hospitalized patients are listed along with the skills that are to be completed on them

nasogastric tube tube passed through nostrils into the stomach to force feed a patient

neck band paper or plastic slip placed around the neck that identifies the animal

observation watching and noting an animal's behavior while hospitalized

primary bandage layer first bandage layer that acts as padding

secondary bandage layer holds the first layer in place

stirrups pieces of tape at the end of a bandage on a limb that hold the bandage in place and prevent slipping

tertiary bandage layer third bandage layer that acts as a protective covering

REVIEW QUESTIONS

1. What is the importance of the hospital treatment board?
2. What information should be placed on a hospital treatment board?
3. What are some veterinary assistant duties and responsibilities in emergency situations?
4. What are some common emergency tools and supplies?
5. What are the three layers of a basic bandage?
6. What is the purpose of each layer?
7. What items are needed for applying a basic bandage?
8. What information should be addressed for basic bandage care?

Clinical Situation

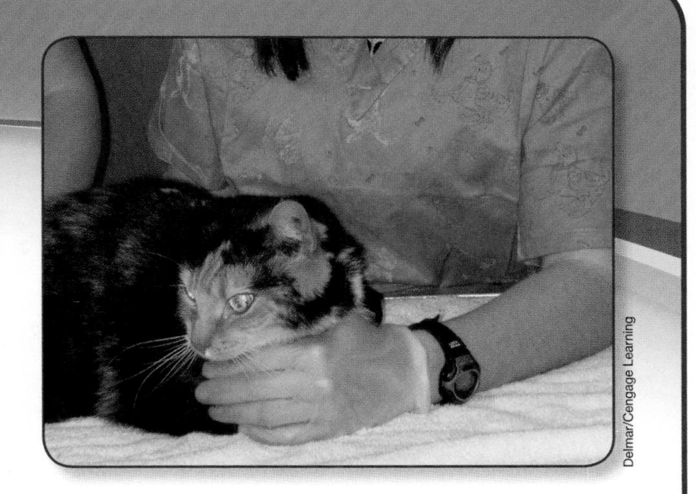
Delmar/Cengage Learning

Alicia, a veterinary assistant at Kind Heart Pet Hospital, is working as the exam room vet assistant. Dr. Moore has asked her to set up the treatment area for a bandage application for a cat that has an injured leg. The cat had X-rays that show no fracture, but the leg is swollen and sore and the vet feels the cat should not use the limb for a week until it has some time to heal and the swelling is reduced. Alicia goes into the treatment area to set up the supplies and equipment and to assist Dr. Moore in applying the bandage.

- What supplies will be needed for this treatment situation?
- What may be needed if the cat is in pain and difficult to handle?
- When complete, what home care instructions might Dr. Moore have Alicia prepare for the owners?

46

Grooming Procedures

Objectives

Upon completion of this chapter, the reader should be able to:

- ☐ Discuss and describe common grooming procedures performed in the veterinary facility
- ☐ Maintain grooming equipment and tools through proper sanitation and care
- ☐ Perform a brush-out of patients, including patients that may be matted
- ☐ Demonstrate proper maintenance and care of clippers and blades
- ☐ Demonstrate how to properly trim, clip, and shave patients using clippers and appropriate blades
- ☐ Demonstrate how to properly clean normal ears of a patient
- ☐ Demonstrate how to properly trim nails of a patient
- ☐ Demonstrate how to properly express anal sacs using external expression techniques
- ☐ Demonstrate how to properly bathe patients
- ☐ Demonstrate how to properly dry patients

Introduction

Grooming is a large part of the veterinary assistant's duties and responsibilities in a veterinary facility. **Grooming** is the care of an animal's external body, including hair coat, ears, nails, and anal glands. There is a difference in grooming care provided by a professional groomer, whose training is of an entirely different skill set and perspective, than that of the veterinary assistant. The focus of grooming in a veterinary facility includes improving the well-being of the patient, but not necessarily appearance. The main goal in grooming tasks in the veterinary facility is improving the overall health and comfort of the animal.

Grooming Skills in the Veterinary Facility

Many patients require hair care due to poor overall hair coat quality, appearance, and lack of care. Long-haired animals may become matted, have waste materials collected on their hair coat, have hair become attached to the rectum or urogenital areas, have overgrown nails, enlarged anal sacs, or dirty ears. Many of these factors can cause an unhappy, unhealthy, and uncomfortable animal. Routine care and maintenance may be provided by the veterinary facility, usually by the assistant.

A variety of grooming tools should be available at the veterinary facility. All tools that have been in contact with an animal should be properly disinfected to prevent transmission of external parasites or other contagious diseases. Grooming items tend to be the most common spread of parasites and disease in a veterinary facility.

Brushing and Combing

Brushing and combing of an animal's hair coat is required for all animals, especially those that may have long and thick hair. **Brushing** is the act of cleaning the hair coat using a soft brush (see Figure 46-1). This removes dead hair and dirt materials. **Combing** is the act of cleaning the hair with a thin comb that helps remove tangles and mats, as well as other debris and foreign substances located in the hair. All animals should have regular brushing and combing of the hair. Combing is important before a bath to untangle any hair or mats that may have developed (see Figure 46-2). Many shampoos and conditioners can't penetrate a matted hair coat and this may only cause further tangling and matting to occur. Mats in the hair coat may cause sores on the skin and are uncomfortable to animals. Brushes and combs come in various sizes, shapes, and material textures (see Figure 46-3). Any combs or brushes with wire teeth, such as a slicker brush, should be used with caution as they can scratch or irritate the skin or pull on mats and tangles, causing discomfort. Some brushes are so soft that they don't penetrate the hair and get through the hair coat. The teeth of combs and brushes may be close together or farther apart. Combs with teeth close together, such as a flea comb, are meant to pull at debris from the coat (see Figure 46-4). Teeth

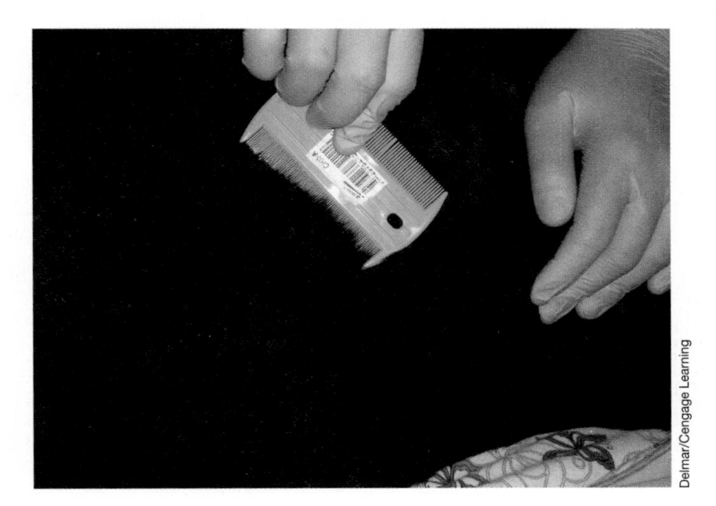
Delmar/Cengage Learning

FIGURE 46-2 Combing helps remove matted hair.

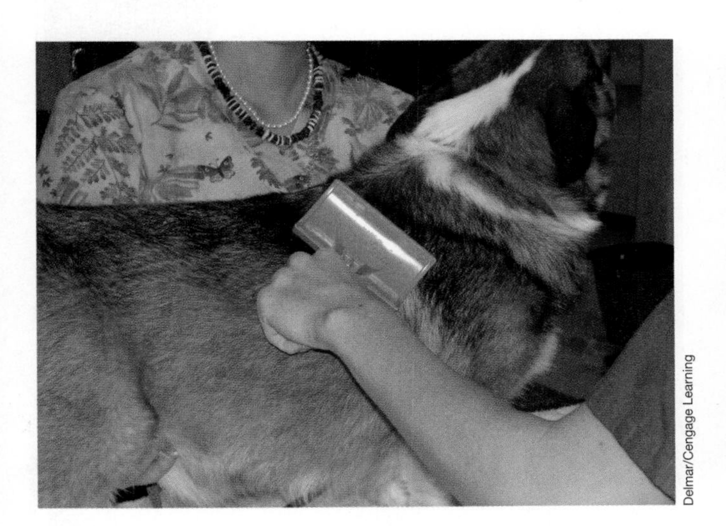
Delmar/Cengage Learning

FIGURE 46-1 Brushing helps maintain a healthy hair coat.

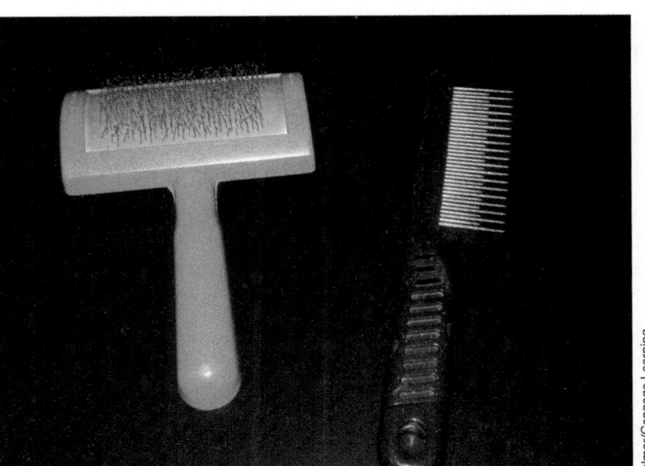
Delmar/Cengage Learning

FIGURE 46-3 Types of brushes and combs.

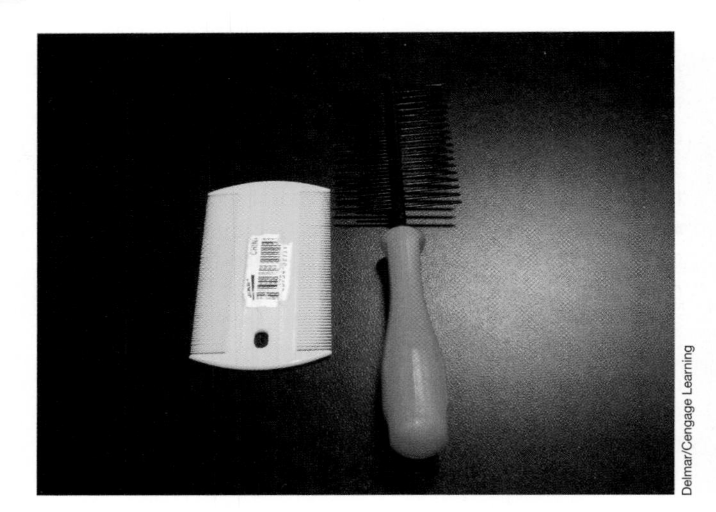

FIGURE 46-4 Flea comb (left) and detangler comb (right).

FIGURE 46-5 Mat splitter.

that are farther apart are meant to break apart mats and tangles. Combs and brushes must meet the need of each individual patient.

Basic brushing and combing should begin at the back of the body and work forward, brushing with the direction of the hair coat. Also, it is best to begin at the feet and work upward. Any mats should be broken down by working across the mat with a wide-toothed comb until the mat is free from the underlying skin and hair. **Mats** are areas of hair that are interwoven together and form a large clump that is irritating to an animal. A **mat splitter** may be used to break down and separate mats into smaller ones, but care must be taken as these tools are sharp and can injure the person, animal, or hair coat (see Figure 46-5). It is best to work with a comb and gently work the areas of the mat using patience, as some animals will become irritated with excessive pulling on the hair coat. Severely matted areas may need to be clipped out with electric clippers (see Figure 46-6). When clipping out individual mats over the hair coat it may result in a choppy appearance to the hair coat. When clipping mats it is best to use the lowest blade number possible. For example, a number 5 blade leaves a longer coat length than a number 10 blade. When using clippers, it is important that the clipper blade be kept flat or parallel to the skin and hair coat (see Figure 46-7). The blades should not be tipped or angled inward toward the hair coat. When mats have been removed, it is important to note the condition of the skin under the matted area. These areas may become inflamed, irritated, have open wounds, have signs of parasites, or dry skin. Any observation of abnormal skin should be reported to the veterinarian.

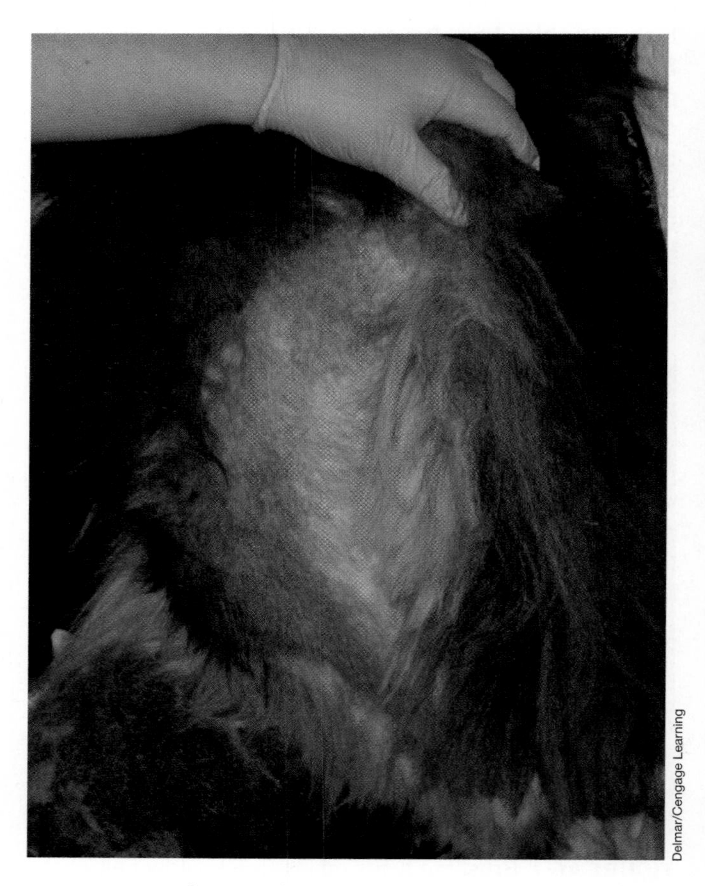

FIGURE 46-6 A severely matted hair coat may need to be shaved.

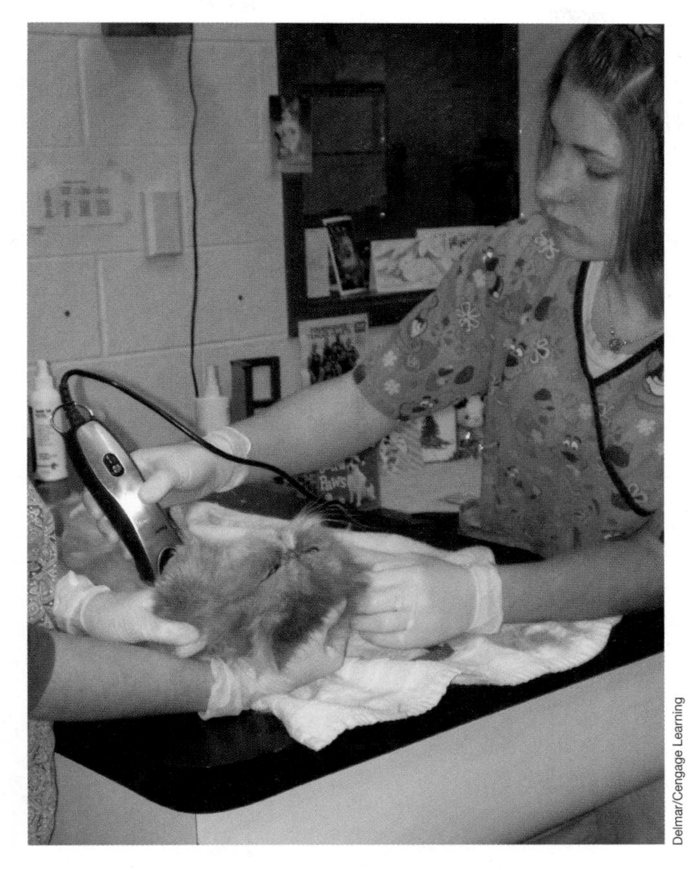

Delmar/Cengage Learning

FIGURE 46-7 The clippers must be kept parallel to the skin or hair coat.

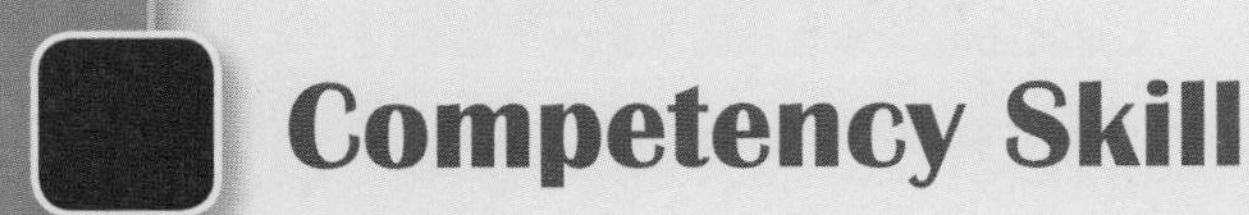

Brushing or Combing

Objective:

To properly brush or comb various hair coats as a health maintenance technique.

Preparation:

appropriate brush (thin spaced teeth), appropriate comb (wide spaced teeth), scissors, clippers, clipper blades in assorted sizes, exam gloves, grooming table (if needed)

Procedure:

1. Begin combing medium to long hair areas over the back and sides of the animals, working from the tail forward toward the head (see Figure 46-8).
2. Part the hair and expose the skin to allow the comb to work through all hair.

(*Continues*)

3. Comb legs working from the feet proximally working up the leg.
4. Comb the medial aspect of each limb while standing on the opposite side of the body.
5. Comb lateral areas of the limb closest to you.
6. Comb the axillary (armpit) and groin areas with the patient standing. Do not pull excessively in these areas, as they are tender.
7. Comb the tail by working cranially from the tip.
8. Comb the ventral body last working from the distal area to the proximal area.
9. Comb out or clip any mats as necessary. Be patient working on mats, being careful not to pull excessively on the hair.
10. Brush over the body to remove loose hair, brushing in the direction of the hair coat.
11. Clean and disinfect all tools and the work area.

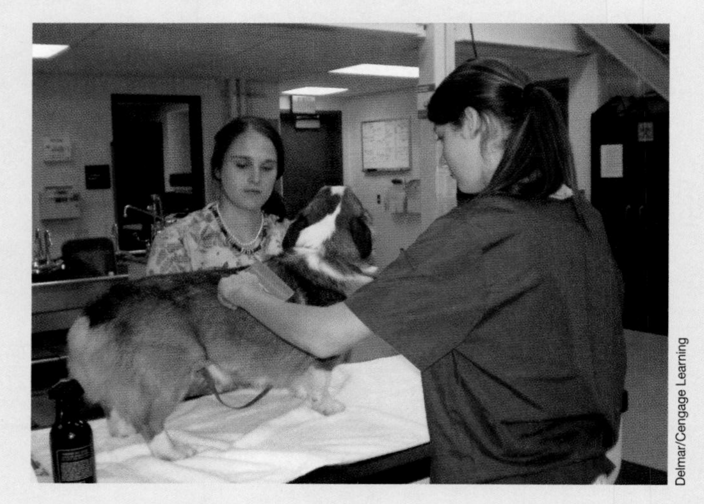

FIGURE 46-8 When brushing, work from the tail forward.

Bathing

Bathing is done to clean the skin and hair coat of the animal or to apply medicated shampoos or dips to the skin and hair coat. Bathing removes dirt and debris from the skin and hair with the use of shampoo, conditioner, and water (see Figure 46-9). **Dipping** is the process of applying a chemical pesticide or medication to the skin and hair coat to treat a specific condition, such as mites or fleas. Dips usually remain on the skin and hair coat for a time period to allow them to work as specified.

Bathing should begin after the animal has been completely combed and brushed and is free of mats. Other grooming needs, such as anal gland expression, ear cleaning, or nail trims, should be performed prior to bathing. Large cotton balls should be placed into each ear canal to prevent water from entering the ear and

FIGURE 46-9 Bathing removes dirt and debris.

potentially causing an inner ear infection. A protective ophthalmic ointment should be applied to the eyes to protect them from shampoo, conditioner, or any medications that may irritate the eyes (see Figure 46-10). A common ointment used is Puralube.

All supplies and tools should be located and prepared prior to placing the patient in the bathing area, either the wash tub or wash stall. Tools used to bathe a patient are listed in Table 46-1. A shampoo and conditioner or medicated shampoo should be selected based on the needs of the patient. Medications should be prescribed by the veterinarian. Towels should be available to dry the patient. A cage should be set up for drying or a clean area should be provided for tying large animals for drying. Washcloths, mitts, sponges, or other absorbable washing material should be used to bathe the animal, especially the head and face area. Aprons or water-resistant smocks should be worn to prevent excessive wetness of scrubs. Goggles or glasses may be worn to prevent water from getting into the eyes. All drains should have a protection to prevent hair and other debris from getting into the drain. Make certain the water is warm and comfortable for the patient.

TABLE 46-1

Bathing Tools

Protective eye ointment	Cotton balls	Towels	Washrags, sponges
Warm water	Dryer	Aprons/smocks	Goggles
Exam gloves	Shampoo	Conditioner, detangler	medications

The patient should be placed in the bathing area with a leash or halter and lead for added control. One end may be held or tied to the bath area to prevent the animal from escaping the bath area. The patient should never be left unattended when tied in a bath area. Begin the bath by thoroughly soaking the entire coat with warm water (see Figure 46-11). The shampoo should be mild, lathered into the hair coat, and rubbed deep into the hair and skin with a massaging motion. All areas of the body should be lathered well including between the digits, around the rectum and the genital

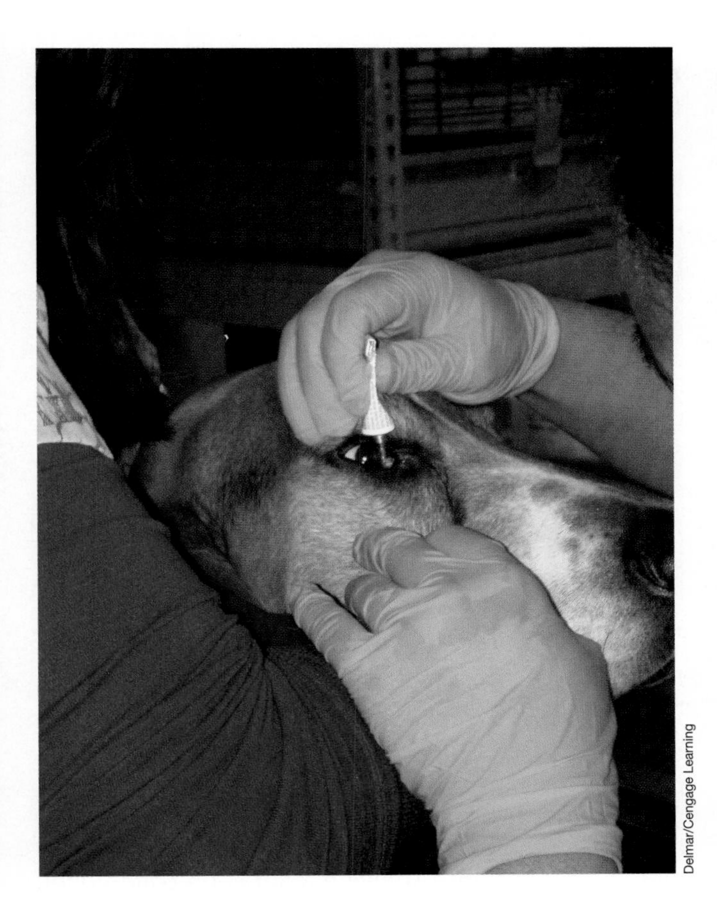

Delmar/Cengage Learning

FIGURE 46-10 An eye ointment is used to protect the eye from cleaning products used when bathing.

Delmar/Cengage Learning

FIGURE 46-11 Wet the entire coat with warm water to begin the bathing process.

area, the axillary areas, and behind the ears. The face area should be washed using a washcloth with a small amount of shampoo applied to a wet cloth. Care must be used around the eyes to prevent any shampoo from entering and burning the eyes. Once the patient is fully lathered, allow the shampoo to stand on the body for about 5 minutes to thoroughly clean the skin and hair coat. Any medicated shampoos that are used should be handled according to the label; many require a standing time in contact with the hair and skin. While soaking, it is important to continue massaging the shampoo into the skin for the time indicated. After shampooing is completed, begin rinsing the patient with warm water. All areas should be rinsed well, especially in creases and folds of the skin. The conditioner is then applied much in the same manner as the shampoo. Conditioners do not generally lather up like shampoos, so it is important to use enough conditioner in all areas of the hair coat. A comb or massage brush can be used to work through the hair coat when the conditioner is on to help make certain the conditioner makes appropriate contact. The conditioner must be rinsed well from the entire hair coat. Any dips should be applied according to the label directions. Rinse all materials from the hair coat using the hands to remove excessive amounts of water by squeezing as much water off the hair as possible. Towels should be applied around the animal and rubbed over the body to begin the drying process (see Figure 46-12). Blow drying using handheld dryers, cage dryers, or high-powered vacuum-type dryers is a convenient way to dry the coat (see Figure 46-13). It is important to comb out the hair coat as the animal is drying as this helps prevent the animal from overheating, reduces tangling of hair, softens the hair coat, and decreases wet odor smell in the animal. As in combing and brushing the animal's coat, work from the distal end to the

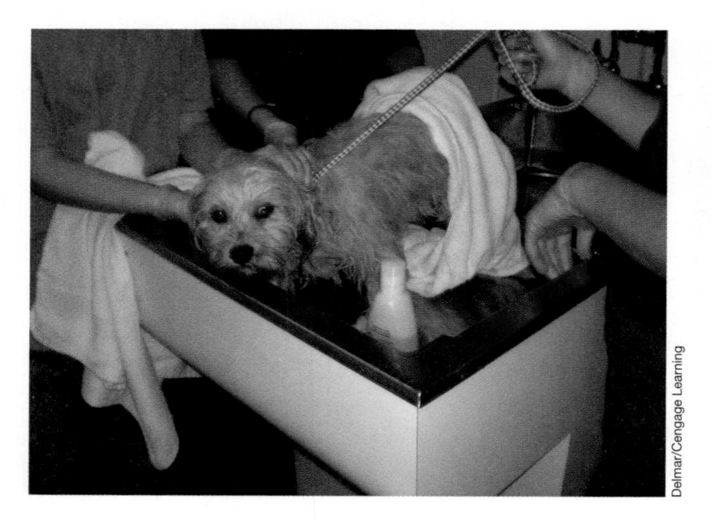

FIGURE 46-12 Use towels to begin the drying process and to remove excess water from the hair coat.

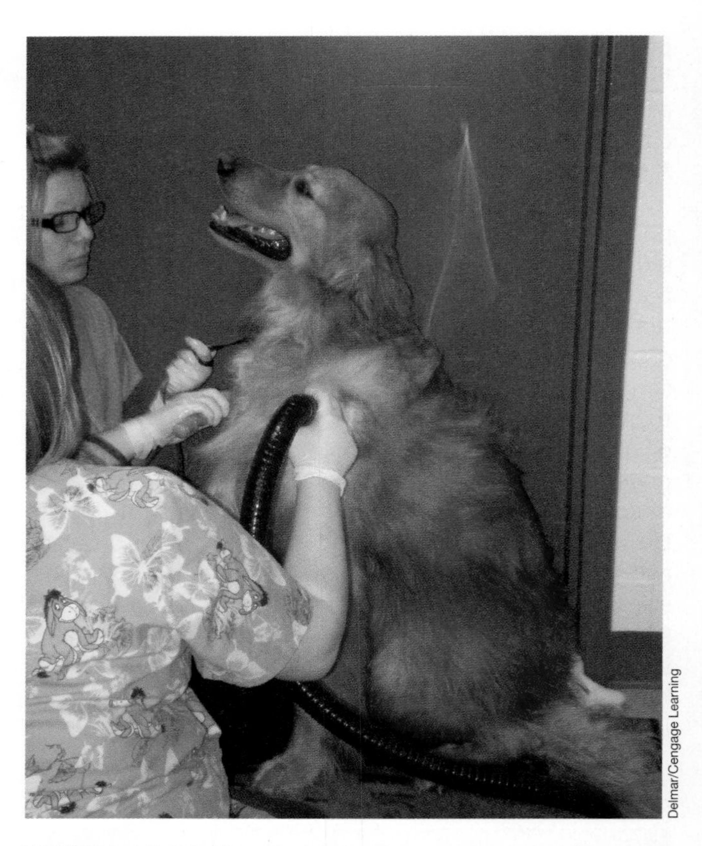

FIGURE 46-13 Handheld dryers can be used to more thoroughly dry the coat.

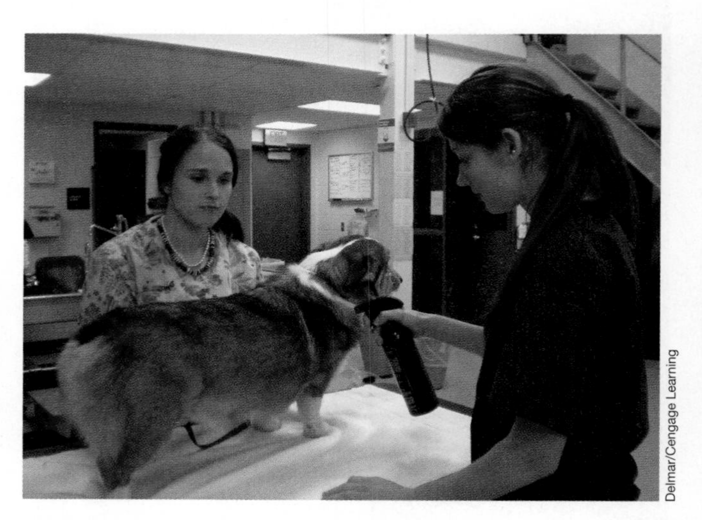

FIGURE 46-14 Applying a spray conditioner helps keep the hair coat neat and odor free.

proximal area and caudal to cranial. It is important to make certain all areas have been completely dried, such as between the digits, under the tail, at the abdomen, and in the ear flaps. Applying a spray-on conditioner also helps with keeping the hair coat neat and odorless (see Figure 46-14). When hand drying, do not leave the heat of the dryer on one spot very long. When cage drying, check on animals every 5 minutes to make certain they are not overheating. Use low temperatures to dry and make certain the dryer is not too close to the animal to prevent skin burns. Cotton balls can be removed from the ears.

After the patient is dry, a final comb-out and brush-out should be done and a spray-on conditioner applied to the hair coat. The patient should be placed in a clean area to avoid becoming soiled. The bath area should be cleaned and disinfected. It is important to remove all hair from the bath area and drain. A mop should be used to dry any floor areas that may be wet. Towels should be laundered and all items including shampoo and conditioner bottles cleaned and put away. Hair should be vacuumed and the dryer disinfected and put away.

Competency Skill

Bathing Procedure

Objective:

To properly bathe an animal in a safe manner for effective health maintenance.

Preparation:

shampoo/conditioner/medicated shampoo, protective ophthalmic ointment, cotton balls, towels, washcloths, sponges, comb/brush, leash, apron/smock, exam gloves, goggles, dryer

Procedure:

1. Apply exam gloves.
2. Place leash on patient.
3. Place cotton balls in both ear canals.
4. Apply eye ointment to both eyes.
5. Place patient in wash area.
6. Secure patient in wash area.
7. Wet entire hair coat thoroughly with warm water.
8. Apply shampoo by lathering well through coat. For medicated shampoos, read and follow all labels. Allow to stand 5 minutes.
9. Rinse entire coat well.
10. Apply conditioner by massaging into hair coat. Allow to stand 5 minutes.
11. Rinse entire coat well.
12. Apply dip, if necessary, by following all directions.
13. Rinse well.
14. Squeeze excessive hair from coat.
15. Wrap patient in towels and begin towel drying.

(*Continues*)

16. Remove patient from wash area.
17. Begin drying process.
18. Place in cage with cage dryer set on low to moderate heat checking every 5 minutes for overheating.
19. Hand dry at low temperature moving over entire coat.
20. Comb or brush through hair coat as it dries.
21. Apply spray-on conditioner with combing or brushing.
22. When completely dry, apply final spray-on coat conditioner.
23. Place animal in clean cage.
24. Clean up work area and disinfect all tools.

Clipping, Trimming, and Shaving

Clipping of the hair is typically the process of removal of small amounts of hair from one or several areas of the patient. **Trimming** refers to the process of removing a specific amount of hair from one or more locations. **Shaving** is the process of taking down an amount of hair close to the skin. Clipping and trimming may be done using scissors or electric clippers (see Figure 46-15). In veterinary medicine this usually refers to areas of the body that need care and maintenance, such as the rectal area, genital area, or facial area. Clipping and trimming around the rectum is common in animals that may have long hair that may collect around the rectum causing a buildup of feces (see Figure 46-16). This causes a sanitary problem that may cause bad odor or soiling of furniture. The genital area may need trimming for similar reasons. Many male and female animals may urinate with urine collecting around the hair of the genital area. Trimming and clipping hair from these sites is common in care

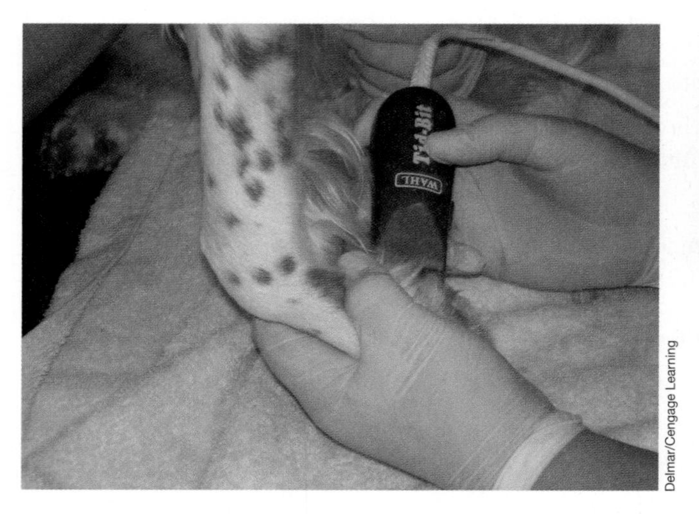

Delmar/Cengage Learning

FIGURE 46-15 Electric clippers are used to clip and trim the hair coat.

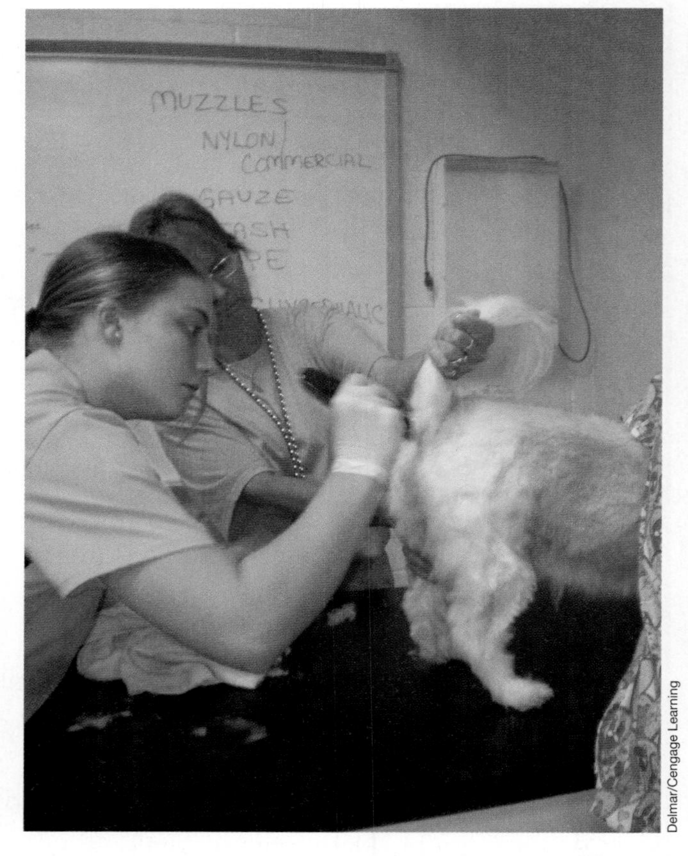

Delmar/Cengage Learning

FIGURE 46-16 Many animal breeds may require trimming of the hair coat around the rectum and genital areas to prevent urine and feces from building up on the hair coat.

for patients. Shaving is another skill in veterinary medicine that is necessary for several reasons. One is the removal of mats and another is shaving a surgical area to properly prepare it for a surgical procedure.

When clipping or trimming hair, blunt scissors should be used. Blunt scissors have a non-sharp end so that the possibility of injury is decreased around an animal. Smaller electronic clippers may be used but should be used carefully so that areas are not trimmed too short (see Figure 46-17). To prevent trimming areas too short while using electronic clippers, it is recommended that veterinary assistants use **clipper guards**, which fit over the end of the clippers. The clipper guards come in various sizes from ¼ inch to 2 inches to allow uniform trimming (see Figure 46-18). Clipper guards should be placed over a #40 clipper blade. When trimming a hair coat to a shorter length, work against the hair coat. This is best done by using a slicker brush and brushing against the hair coat allowing the area to stand. Use long, even strokes when trimming long-haired coats. Shaving animals is typically done with surgery patients or animals that may have severely matted coats or skin conditions. When shaving an animal it is best to know how short the area must be shaved. Surgical patients will need the entire surgical preparation surface shaved to the skin, free of all hair and debris. This is typically done with a #40 or #50 surgical clipper blade. Shorter shaves may be done with smaller-number clipper blades, which have wider spaces between the teeth, causing less hair to be removed. Clipper guards can also be used to keep the area uniform. When using clippers, one must keep the clipper blade surface parallel to the animal's body and flat going with the hair coat. In surgical preparation, shave in the direction of the hair and then against the hair growth to remove all hair from the desired site.

FIGURE 46-17 Electric clippers.

FIGURE 46-18 Clipper guards.

Clipper and Clipper Blade Care and Maintenance

Clippers and blades receive a large amount of hard use. To prolong their performance, several things may be done to maintain clippers and blades. It is best to read the manufacturer's manual to know the proper use and care of the clippers. A backup clipper should always be available in case of damage or overheating. Electronic clippers come in a variety of sizes, shapes, and speeds. Some may be cordless and others have electrical cords. Corded models should be checked on a regular basis for any damage to the cord or plug. Cordless models are kept in a charger and should be disinfected and returned to the charger immediately after each use. Some clippers have multiple speeds.

One area on the clipper that should be checked after each use is the air filter, which may easily become blocked by hair and debris. This will cause the motor to overheat and possibly become damaged. The instruction manual should note regular maintenance checks. Cleaning the clippers after each use and keeping clipper blades free of hair and debris is essential to prolonging the life of the clippers. When using the clippers, always use a cooling spray every

Delmar/Cengage Learning

FIGURE 46-19 Cooling spray and blade cleanser.

5 minutes and a lubricating oil every 15 to 20 minutes to prevent the clippers from overheating and excessive wear and tear (see Figure 46-19). Never use an oil lubricant other than the clipper oil. This means turning off the clippers to apply a spray that cools down the clipper motor and lubricating oil to keep the clippers working properly. Items such as WD-40 will ruin the clippers. After using a blade, remove it from the clipper body and clean the teeth using a heavy wire brush (see Figure 46-20). Blade wash should be used to soak the blade. Blades should be stored face down to prevent tooth damage. Blades should be monitored for damaged or missing teeth. Broken teeth can cause skin tears and irritation. The blades should be dried well after each cleaning. Any wet areas on clippers or blades will cause corrosion. Blades will become dull with excessive use. Avoid using clippers on wet and dirty areas of the coat. Blades can be sharpened by professionals or using commercial blade sharpeners. Extra sharpened blades in various sizes should be available. The clipper blade attachment area of the clippers should be cleaned after each use by using a wire brush and clipper blade wash. Again, dry all areas well to prevent corrosion. Several times a year, the clipper head should be removed and clipper

Delmar/Cengage Learning

FIGURE 46-20 Wire brush used to clean clipper blades.

Delmar/Cengage Learning

FIGURE 46-21 It is necessary to remove the head of the clipper to thoroughly clean the device to maintain it properly.

grease applied to the drive gear (see Figure 46-21). This should be noted in the instruction manual. Some clippers may need to be serviced by the manufacturer if damaged or not working properly.

Ear Cleaning

The ear canals of animals may need to be cleaned on a regular basis. Some animals have hair growing within the ear canal as part of the breed or species characteristic. Some people will pluck out this hair while others will shave it out to remove it (see Figure 46-22). Check with the veterinarian for a preference. If the hair is to be plucked, try to remove as much hair from the inside ear flap with your fingers. The hair growing within the ear canal can be removed using a pair of forceps (see Figure 46-23). If clipping the ear flap and canal,

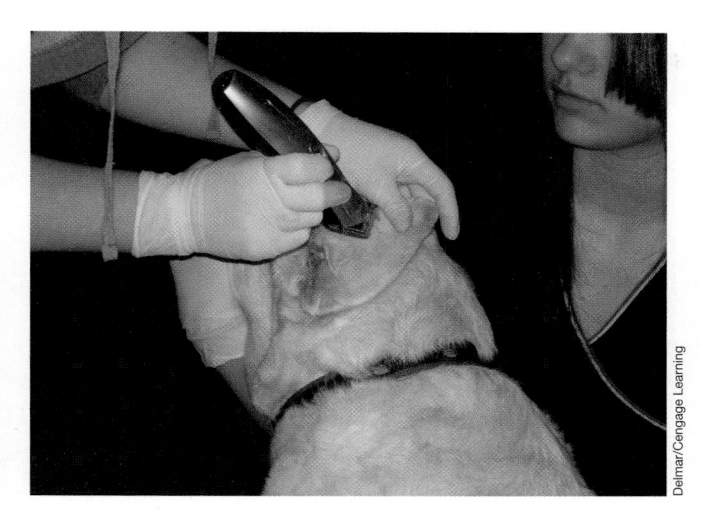

FIGURE 46-22 Some breeds may require the shaving of the inside of the ear to keep the ear canal clean.

FIGURE 46-23 Hair inside the ear can also be removed using forceps to pluck it.

use caution not to damage the folds of the ear. Blades should be kept flat on the ear surface. The ear opening is shaped like a "U" whereas the ear canal is shaped like an "L." The eardrum, or **tympanic membrane**, is located at the end of the "L". This area is easy to puncture during cleaning. It is also easy to pack ear wax and discharge further into the canal around this area. It is not recommended to use cotton-tipped applicators in the internal ear canal. They may be used to clean the folds of the external ear flap. The **pinna**, or ear flap, can be cleaned using a cotton ball that has alcohol or a commercial ear-cleaning solution presoaked on the cotton ball. Wipe all areas of the ear flap and external ear canal. If signs of infection are present, such as redness, inflammation, strong odor, or discharge, notify the veterinarian.

Ears that are filled with wax or discharge should be cleaned by placing a small amount of ear cleaning solution or ear antiseptic into the internal ear canal (see Figure 46-24). Pull the ear fold caudally to open the ear canal (see Figure 46-25). After applying the

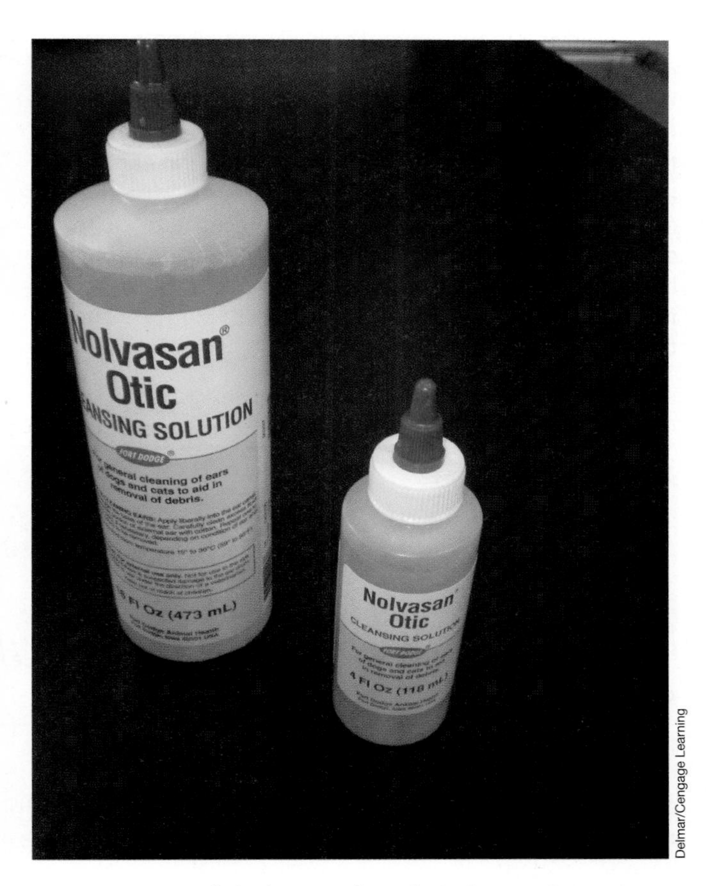

FIGURE 46-24 Solution used to clean internal structures of the ear.

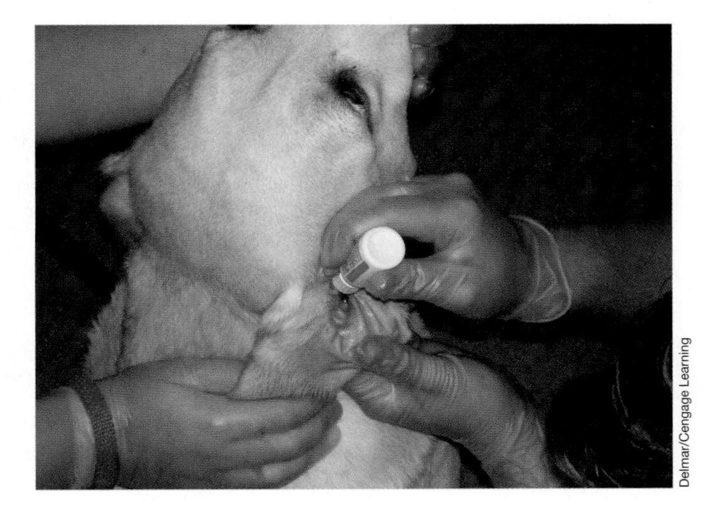

FIGURE 46-25 When instilling ear solutions, pull the ear flap caudally to open the ear canal.

solution, fold the ear flap back into place and gently massage the base of the ear and cartilage to allow the solution to loosen debris. When done correctly, you should hear a sloshing sound as the fluids soften and loosen ear debris. Place a cotton ball or gauze sponge over the ear canal opening, and tip the patient's head sideways ventrally. This will help any excess fluid drain out onto the cotton ball or sponge. Wipe the ear canal to remove excess discharge. Repeat this several times until the fluid is clean and the ear canal is free of debris. Dry the ear canal with a cotton ball or gauze sponge reaching into the canal with the fingertip. Make certain all ear cleaning solution has been removed from the ear as moisture buildup can promote bacterial growth and an ear infection. An otoscope can be used to examine the internal ear canal (see Figure 46-26).

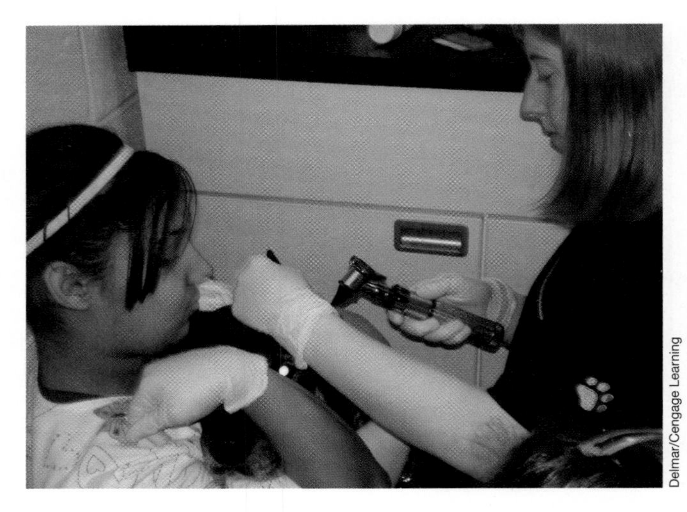

FIGURE 46-26 Use an otoscope to check the ear for buildup of wax or debris.

Competency Skill

Ear Cleaning Procedure

Objective:

To properly trim or shave hair around the ears and remove excessive wax or debris from the ear canal.

Preparation:

cotton-tipped applicator, cotton balls or gauze sponges, alcohol/ear cleaning solution, forceps, exam gloves

Procedure:

1. Apply exam gloves.
2. Examine ears using an otoscope.
3. Contact veterinarian if signs of redness, odor, inflammation, pain, or discharge are present.
4. Pluck or shave any hair from ear flap or ear canal if necessary.
5. Clean outer pinna and ear flap with alcohol or ear solution on cotton ball or gauze sponge. Wipe away from the internal ear canal.
6. Clean the outer pinna and ear flap creases and folds with cotton-tipped applicators. Work away from the internal ear canal.
7. Repeat until clean.

8. Fill ear canal with ear cleaning solution by pulling flap caudally.
9. Massage the base of the external ear canal and cartilage.
10. Place cotton ball or gauze at external ear canal opening and tilt head. Drain fluid from ear and wipe away any discharge.
11. Repeat until fluid is clean.
12. Wipe out external ear canal using a cotton ball or gauze sponge.
13. Check internal ear canal for moisture using an otoscope.

Nail Trimming

The nails on animals can easily become overgrown. They may require regular trimming so that they don't begin to catch on rugs, blankets, or other items or split from overgrowth. Sometimes long nails will begin to curl and grow into the foot pad, causing severe pain and discomfort. The purpose of nails is for protection or defense and traction on smooth surfaces. In cats, they are also used for climbing. Some animals have hoofs, which need to be trimmed and keep from overgrowth as well. This section focuses on small animal nail trimming, as most hoofed animals are maintained by a farrier, who is a professional trained in trimming hoofed animals. The anatomy of the nail includes the **nail bed**, the area where the nail growth occurs and is located at the end of the digit near the hair growth of the foot. Within the nail bed lies the **quick**, or the blood and nerve supply to the nail. The outer nail covering is composed of **keratin** similar to human nails. Keratin is a protein that allows the nail to grow and strengthen. The goal of the nail trim is to cut the nail just distal to the quick (see Figure 46-27). Some nails are white or unpigmented, whereas others are black to brown in color and contain pigment. In white nails it is much easier to see and determine where the quick lies. The quick is a faint pink triangular area within the end of the nail (see Figure 46-28). In dark-colored nails it is not easy to see the quick. Small cuts of nail must be done in dark nails to look for the dark spot in the nail bed that surrounds the quick. At this point you should stop trimming as it is the near the quick. Most animals use the hind digits more for traction and movement than the front digits, so typically the front nails need more taken off. Dewclaws may also be present and have no contact with the ground and typically become long and tend to curl in toward the skin and hair coat. Every animal should be checked for signs of dewclaws on all four limbs. Some cats are known for being polydactyl and may have extra toes and digits that need to be trimmed.

If the quick is cut into, the nail will bleed. This is painful to the animal as the nerve supply also lies within the nail bed. When trimming the nails, always have **styptic powder** available to apply to any bleeding nail beds. The styptic powder is a yellow loose powder that packs into the nail bed and clots the blood. There are special

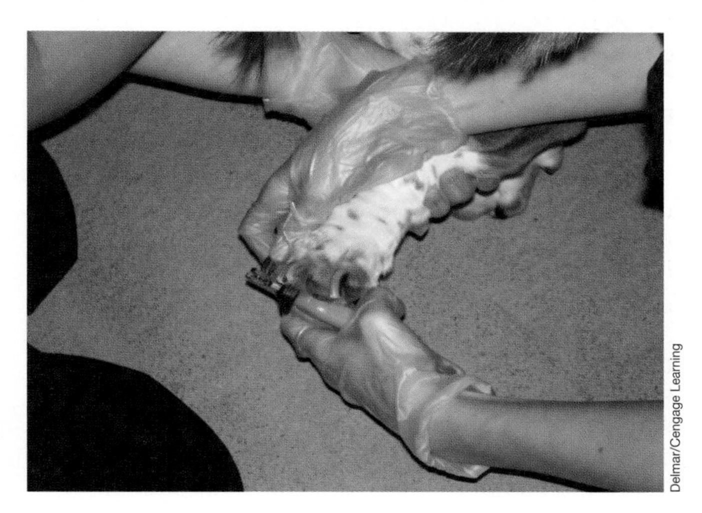

FIGURE 46-27 Nails should be trimmed just distal to the quick.

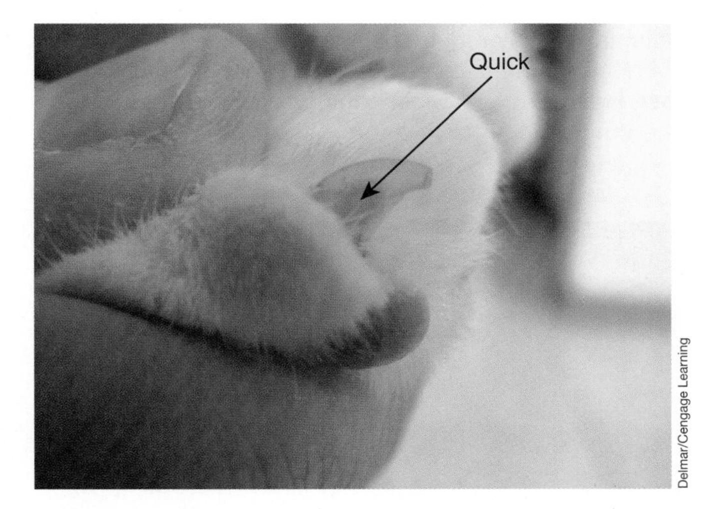

FIGURE 46-28 Quick of a cat nail.

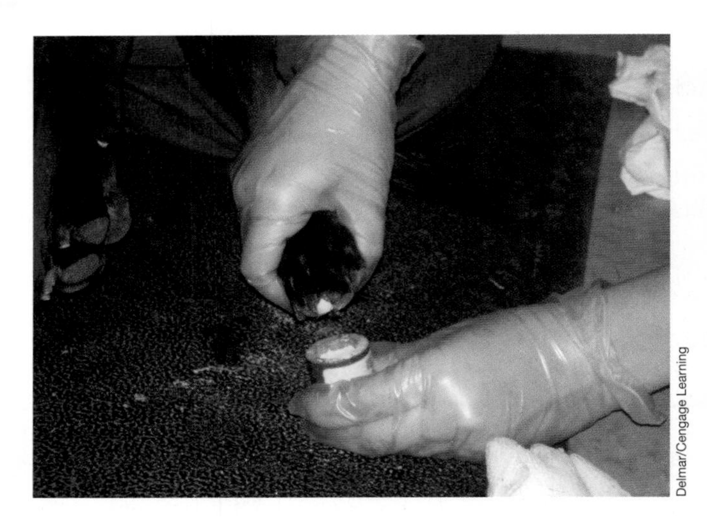

FIGURE 46-29 Styptic powder can be used to stop bleeding that may occur when nails are trimmed.

well containers that can hold the powder or the powder can be placed on a gauze sponge (see Figure 46-29). Caution must be used to not allow the powder to have contact with any skin or pads of people or animals, as the powder is caustic and burns materials. Sometimes the powder will not provide proper clotting and a wooden **silver nitrate stick** is necessary. These are slim, long, wooden applicators with silver nitrate located on the end (see Figure 46-30). The silver nitrate chemical burns the nail bed to clot the bleeding. Caution must also be used when using these sticks due to the caustic risks. Animals that have severely overgrown nails may need to be sedated to trim the nails, as they may need to be cut back into the quick, which is painful for the animal and may require pain medication.

Nail trimmers come in a variety of shapes and sizes (see Figure 46-31). Most commonly used types include the White clippers, which are scissor-like and useful for

FIGURE 46-30 Silver nitrate sticks.

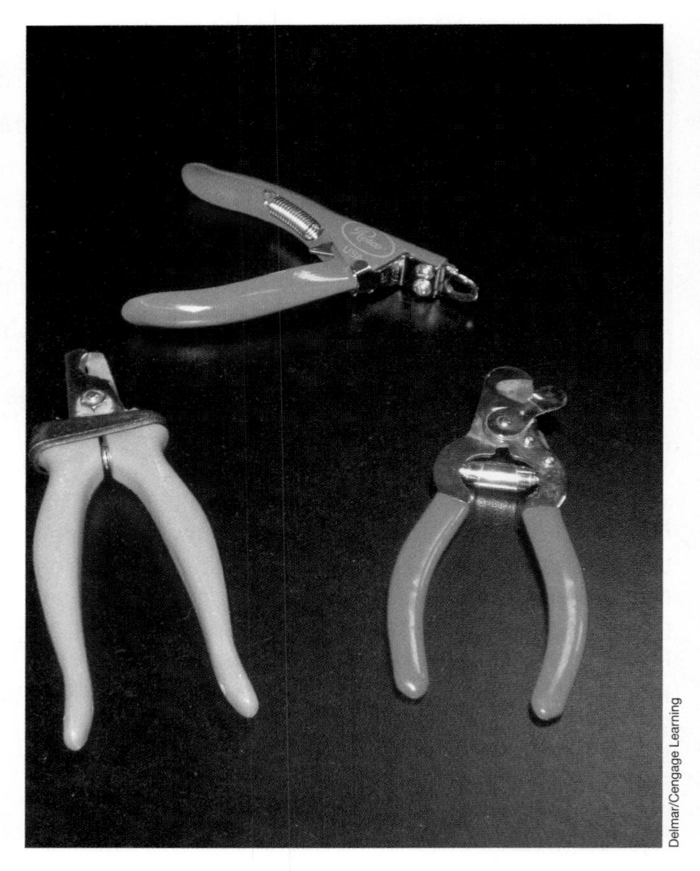

FIGURE 46-31 Nail trimmers.

cats, small dogs, birds, and pocket pets. They are useful in trimming overgrown nails that curl into pads. Another common type of clipper is the handheld professional trimmers that have a blade and scissor-like handle that come in a variety of sizes depending on the animal. A less-used nail trimmer is the Rescoe, or guillotine nail clipper. A sharp blade in the center moves to cut the nail. Extreme care must be used with this type of clipper as in small animals it could amputate a toe. Human nail trimmers can be used on small thin nails. The dremel tool is also useful with animals that may not sit well for nail trims. The dremel is a grinding tool with a sandpaper surface that rounds and smoothes the end of the nail without as much possibility as hitting the quick (see Figure 46-32).

When trimming or dremeling nails, the foot should be held and pressure applied to the digit with the nail being trimmed (see Figure 46-33). This will extend the nail and hold it in place to prevent too much from being cut. The nail is then trimmed. The animal can be in a standing or sitting position and smaller pets may need to be held for the nails to be trimmed. It is best to begin at the rear paws and trim away from the face to allow the animal time to settle down. The front legs are then trimmed pulling the limbs back away from the head. This allows animals that do not like their nails trimmed to feel less threatened.

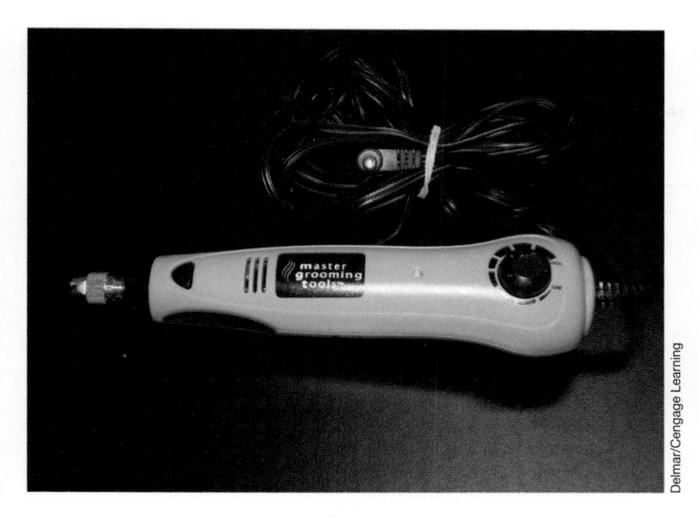

Delmar/Cengage Learning

FIGURE 46-32 Dremel tool used to trim nails.

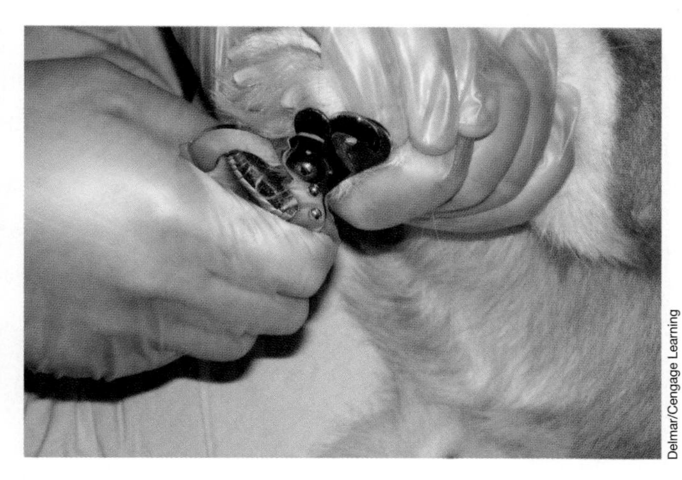

Delmar/Cengage Learning

FIGURE 46-33 Proper method of holding the foot when trimming nails.

Competency Skill

Nail Trimming Procedure

Objective:

To properly trim nails.

Preparation:

nail trimmers, dremel, styptic powder, silver nitrate sticks, powder well, gauze sponge, exam gloves

Procedure:

1. Apply exam gloves.
2. Locate the proper trimmers or dremel tool.
3. Grasp a rear foot with one hand (see Figure 46-34).
4. Apply pressure to each digit to extend the nail (see Figure 46-35).
5. Use the other hand to clip or file the nail just distal to the quick. Dark nails that are difficult to see the quick should be trimmed in small amounts until the dark center circle of the quick is noted.
6. Repeat with each nail, remembering to check for dewclaws, if present.
7. If bleeding occurs, use the styptic powder by applying the nail into the yellow powder and applying pressure using a gauze sponge. Severe bleeding nails should have a silver nitrate stick applied to the end of the nail to control bleeding.

(*Continues*)

8. Trim each foot pad beginning with the rear limbs and then moving to the front limbs.
9. Check all nails for signs of bleeding after completing the trim.

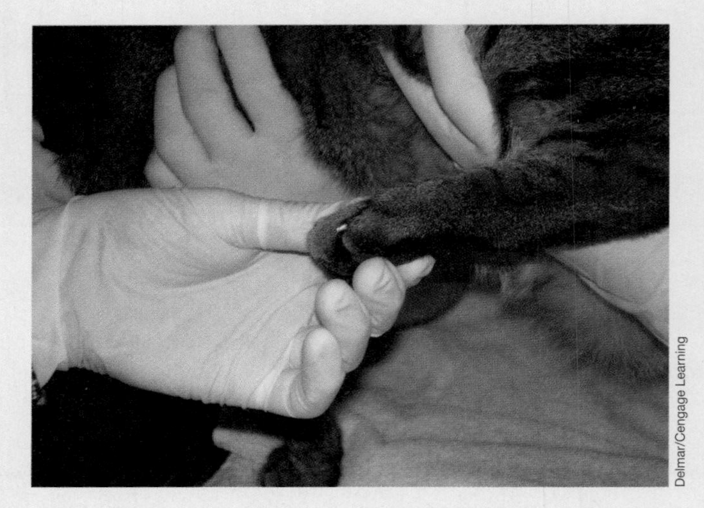

FIGURE 46-34 To begin trimming nails, start with a rear limb and grasp the foot firmly.

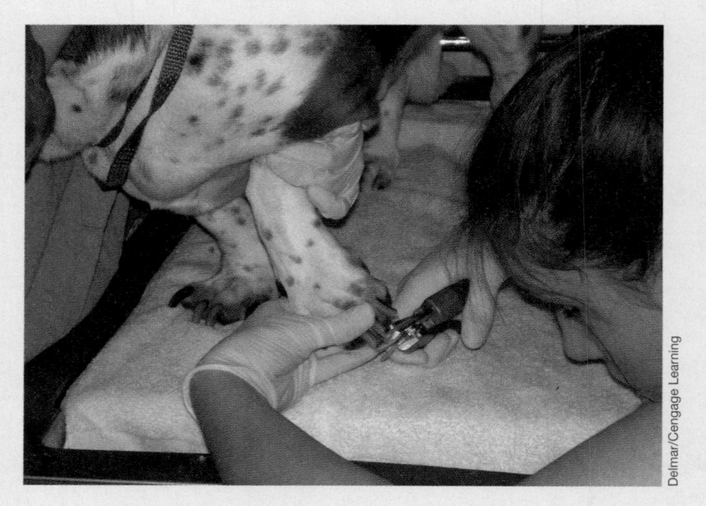

FIGURE 46-35 Apply pressure to the digit to extend the nail.

Expressing Anal Glands

Anal glands or sacs are located on either side of the rectum and are the scent glands of dogs and cats (see Figure 46-36). The sacs themselves lie ventrally and slightly anterior to the skin around the rectum at the 4 o'clock and 8 o'clock positions. A duct travels to a small opening that allows the glands to be expressed. The material that builds up in the sacs is secreted into the sac every time the animal has a bowel movement. If the material becomes thickened and is not removed properly, the sacs may become abscessed and painful. Signs that the anal glands should be expressed include the dog or cat scooting across the floor on its hind end, excessive licking at the rectum, and odor noted from the rectum.

There are two techniques for expressing anal glands. The first technique that should be used by the veterinary assistant is the external expression of

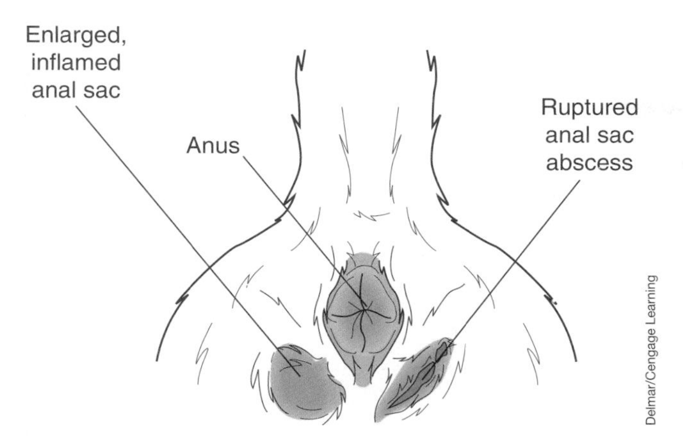

FIGURE 46-36 Location of the anal glands.

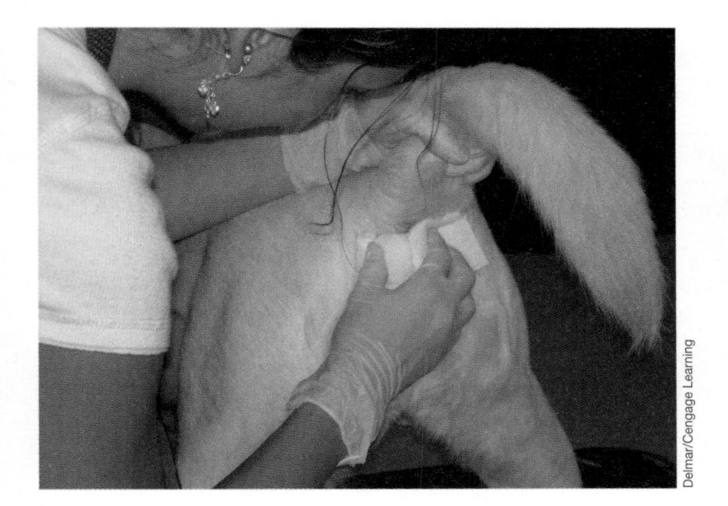

FIGURE 46-37 External expression of the anal glands.

the anal glands (see Figure 46-37). Gloves are worn while the thumb and index finger are used to gently palpate the ventral and lateral sides of the anus. The tail of the animal is held upward and out of the way. Stand to the side of the animal after locating the anal glands, and place a paper towel over the area and gently press the fingers medially while squeezing each sac against each other with slight pressure. The fingers can be moved in an upward and then outward semicircular motion while squeezing the glands against each other. This will allow the release of the anal secretions through each gland duct. The secretions will be caught in the paper towel. Note how much secretion is expressed and the color and consistency. After completing the procedure, place the paper towel and gloves in the trash. Apply a clean pair of gloves and then clean the anus and hair around the anus with a damp paper towel or gauze sponges soaked in warm, soapy water. A pet safe cologne or coat conditioner can be applied to the hair coat (see Figure 46-38).

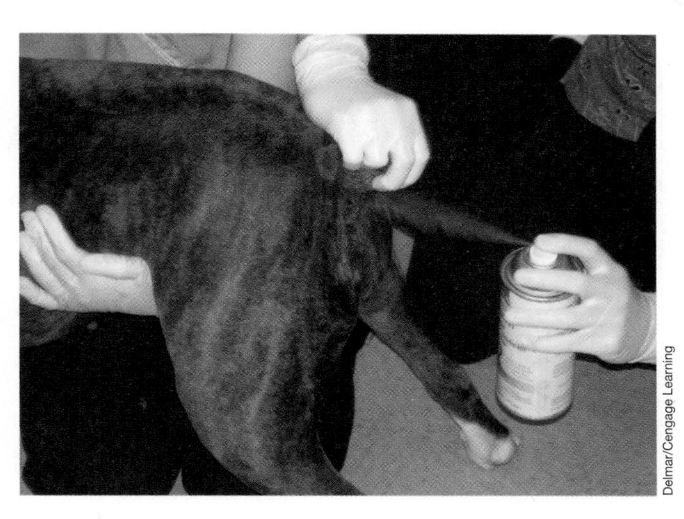

FIGURE 46-38 After expression of the anal glands, the animal should be cleaned and a scented spray can be used.

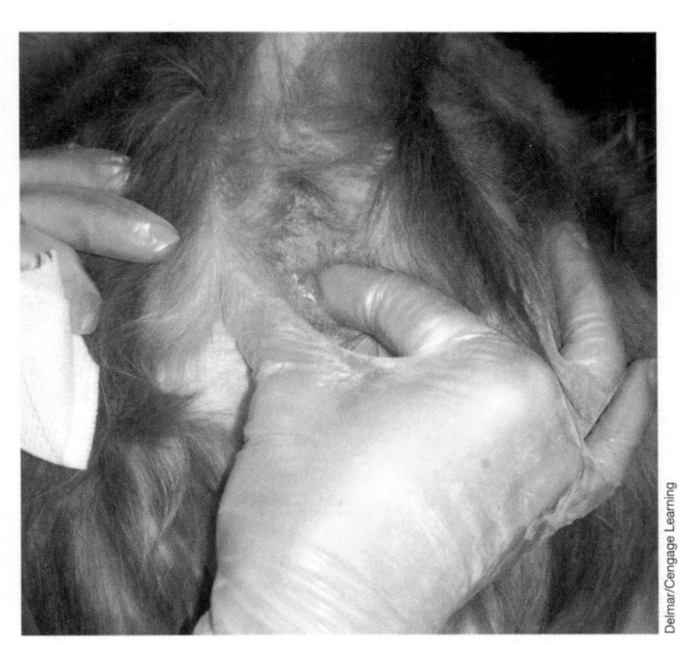

FIGURE 46-39 Internal expression of the anal glands.

If the anal glands are difficult to express using the external anal gland expression technique, the veterinary technician or veterinarian should be alerted and an internal expression method should be used (see Figure 46-39). This technique allows greater pressure to the anal sacs when the secretions may be thickened or a duct is blocked. If the secretions are expressed, it is important to note any abnormal findings, such as signs of blood, yellow or pus-like secretions, red or swollen anal glands, or severe pain in the anal gland area. These may be signs of an infection or abscess.

Competency Skill

External Anal Gland Expression Procedure

Objective:

To safely and effectively express the anal glands using the manual method.

Preparation:

exam gloves, Vaseline or KY gel lubricant, paper towels, gauze sponges, small bowl, soap or scrub, cologne or conditioning spray

Procedure:

1. Apply exam gloves and stand to the side of the animal, lifting the tail upward. A small amount of lubricant can be applied to the paper towel for comfort.
2. Place a paper towel over the anus and using the thumb and index finger locate and palpate the anal sacs, located laterally and slightly ventrally, at 4 o'clock and 8 o'clock.
3. Place the thumb over the lateral aspect of one gland and the index finger over the lateral aspect of the other gland.
4. Gently squeeze the fingers together in an upward and outward, semicircular motion, as if massaging the area.
5. Note the amount and appearance of the secretions.
6. Place paper towels in the trash.
7. Replace exam gloves if soiled.
8. Clean the rectum area and the hair around the anus with clean paper towels or gauze sponges soaked in warm, soapy water.
9. Dry the area.
10. Apply cologne or conditioning spray.

SUMMARY

Veterinary assistants play an important role in grooming and maintaining the care and cleanliness of hospitalized patients. Animals may require veterinary care that the owners have not been able to provide at home or have neglected. Common basic grooming skills, such as nail trims, ear cleaning, and bathing and anal gland expression, are helpful to many clients that can't properly provide this care at home. The veterinary assistant should be able to complete these duties easily and efficiently.

Key Terms

anal glands scent glands located on either side of the rectum

bathing process of cleaning the skin and hair coat of the animal or to apply medicated shampoos or dips to the skin and hair coat

brushing the act of cleaning the hair coat using a soft brush

clipper guard tool that fits over the end of the clippers to prevent trimming too short

clipping the process of removal small amounts of hair from one or several areas

combing the act of cleaning the hair with a thin comb that helps remove tangles and mats

dipping process of applying a chemical pesticide or medication to the skin and hair coat

grooming the care of an animal's external body including, hair coat, ears, nails, and anal glands

keratin a protein that allows the nail to grow and strengthen

mat areas of hair that are interwoven together and form a large clump

mat splitter tool used to break down and separate mats into smaller ones

nail bed area where the nail growth occurs and is located at the end of the digit near the hair growth of the foot

pinna external ear flap

quick the blood supply to the nail

shaving the process of taking down an amount of hair close to the skin

silver nitrate stick slim, long, wooden applicators with silver nitrate located on the end that is used to clot a bleeding nail by burning the quick

styptic powder yellow loose powder that packs into the nail bed and clots the blood

trimming the process of removing a specific amount of hair from one or more locations

tympanic membrane veterinary term for the ear drum located in the inner ear

REVIEW QUESTIONS

1. What tasks are involved in the grooming care of an animal?
2. What is the difference between combing and brushing?
3. What is meant by trimming, clipping, and shaving patients?
4. What items and supplies are necessary for bathing?
5. What are signs of an ear infection?
6. What is the importance of styptic powder or silver nitrate sticks?
7. Why do dewclaws often become overgrown?
8. What are signs of a dog or cat needing anal glands expressed?

Clinical Situation

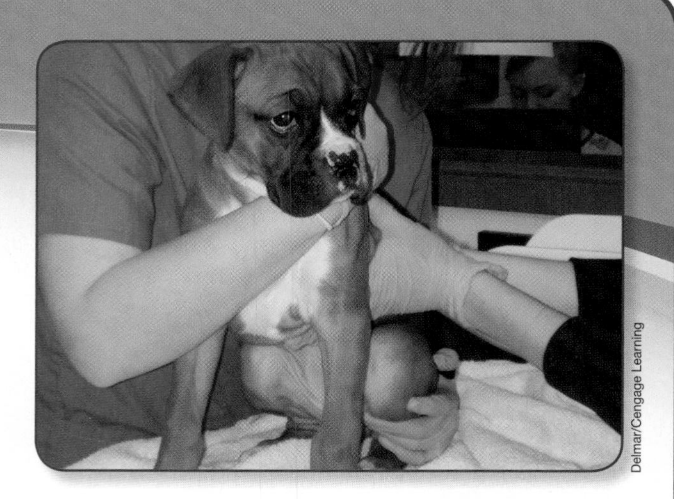
Delmar/Cengage Learning

Rhonda, the veterinary assistant, and Kelly, the veterinary technician at Harland Veterinary Hospital, are working with Mrs. Cramer's dog, "Tilly." "Tilly" is a 5-year-old SF Boxer that is hyper and tends to get excited at the veterinary clinic. "Tilly" needs her ears cleaned frequently, as the ears are not cropped and floppy, and she tends to have a lot of wax and discharge in her ears. When there is a lapse in the cleanings, she is prone to ear infections. Mrs. Cramer frequently visits the clinic for routine ear cleanings.

"I would really like to clean 'Tilly's' ears at home. It would save me driving to the clinic so much. Is it possible for me to clean her ears at home?"

"Sure, you would just need to have the necessary supplies and know the correct procedure."

"Wonderful, maybe that will make 'Tilly's' ears less of a problem."

- What should you discuss with Mrs. Cramer about home ear cleaning?
- What items will you suggest she have at home for cleaning "Tilly's" ears?
- What signs of problems should Mrs. Cramer know about that should alert her to get a veterinary appointment?

47

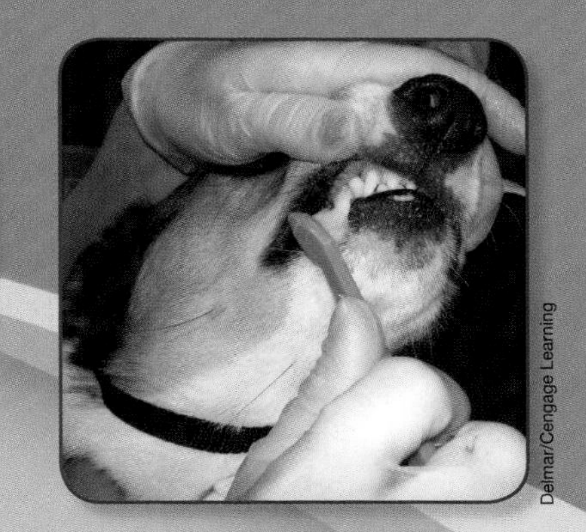

Veterinary Assisting Procedures

Objectives

Upon completion of this chapter, the reader should be able to:

- State the need for dental care for patients and perform basic teeth-brushing techniques
- Discuss and perform the proper method of filling a syringe and labeling the syringe
- Discuss and perform subcutaneous, intranasal, and intramuscular injections
- Demonstrate how to administer SQ fluids
- Explain the importance of monitoring IV catheters and fluids
- Discuss the importance of socialization and exercise of a hospitalized patient
- Explain the euthanasia procedure
- Demonstrate veterinary laundering techniques

Introduction

Veterinary assisting to animal patients is similar to nursing care in human patients. This is a service and trained area essential to caring for and preventing illness and disease in hospitalized patients when the owners are not capable of caring for them at home. Veterinary assistant skills are essential to the health of the patient. Nursing care provided to patients is a team effort by the veterinarian and support staff. Veterinary nursing skills include dental care, providing injections for the health of the patient, monitoring IV fluids and catheters, and understanding the needs and process of euthanasia. Injuries and wounds are as much a disease condition warranting nursing care as is parvovirus and heartworm disease.

Dental Care

Dental care in animals is important and has an increased demand in all types of species, such as companion animals, large animals, and pocket pets. Several aspects of dental care fall into the hands of the veterinary assistant. The veterinary technician performs dental care under anesthesia, and the veterinarian performs oral surgery and extractions. Preventative dental care and teeth brushing are an important part of the veterinary assistant's responsibility (see Figure 47-1). All of these factors make animal dental care an essential part of veterinary medicine and patient care. Promoting ideal dental health begins with educating clients on home dental care needs including daily brushing of teeth, feeding a dry diet, and providing routine veterinary dental care such as dental prophylaxis or floating. It is also important to educate owners on how to properly keep teeth strong and healthy and also prevent dental injury. Proper toys and dental chews should be provided, such as KONGs or veterinary dental bones (see Figure 47-2). Items such as raw bones, chains, or rocks should not be chewed on as they may damage and injure teeth and gums. Young animals should be monitored for proper tooth eruption as deciduous (baby) teeth are shed and permanent (adult) teeth begin alignment. A yearly physical exam should include a dental exam by the veterinarian.

Delmar/Cengage Learning

FIGURE 47-2 Chew toys promote good dental health.

Daily Dental Care and Brushing

Daily dental care should include feeding a hard, dry diet that promotes dental health. Hard treats and dental toys or bones are highly recommended to further allow healthy teeth and provide animals that chew with appropriate items to prevent tooth injury. Daily dental care also means brushing the teeth in the proper manner. Owners think of dental care similar to their own teeth; however, there are several modifications in pet dental care that can cause difficulty or harm to the animal. The veterinary assistant should teach owners how to properly brush their pet's teeth. Teaching the proper method and necessary tools and supplies will help owners accept the needs of the pet's dental care and make the process easier. When owners have difficulty brushing their pet's teeth, they soon give up and stop dental care altogether. Animals should never have human toothpaste placed in their mouth, as human toothpaste has certain enzymes that are not digestible in animals. This may cause severe irritation or potential toxicity. Veterinary **dentifrices**, or toothpaste, come in a variety of flavors that increase the animal's acceptance to brushing of the teeth (see Figure 47-3). These dentifrices have safe, digestible enzymes that easily break down in the animal's body. Flavors that are enjoyed by pets include fish, beef, malt, and poultry.

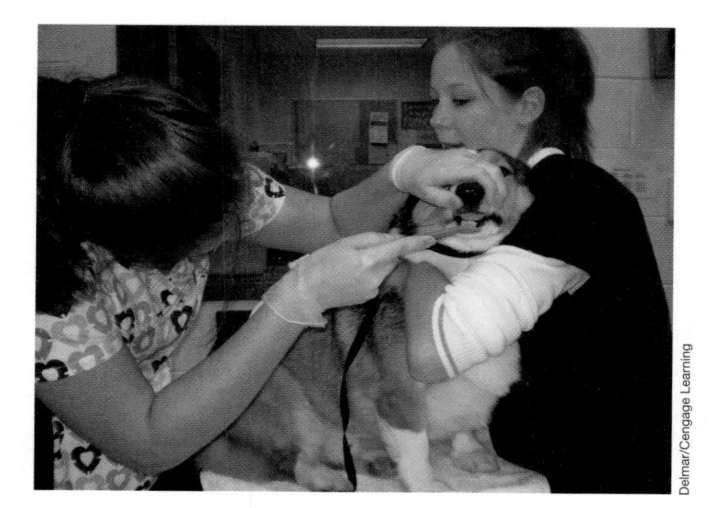

Delmar/Cengage Learning

FIGURE 47-1 The veterinary assistant can perform teeth brushing.

The toothbrush should also be a specialized instrument that fits the shape and size of the animal. Human toothbrushes are not ideal for all animals, such as cats

FIGURE 47-3 Proper tools for dental care in pets.

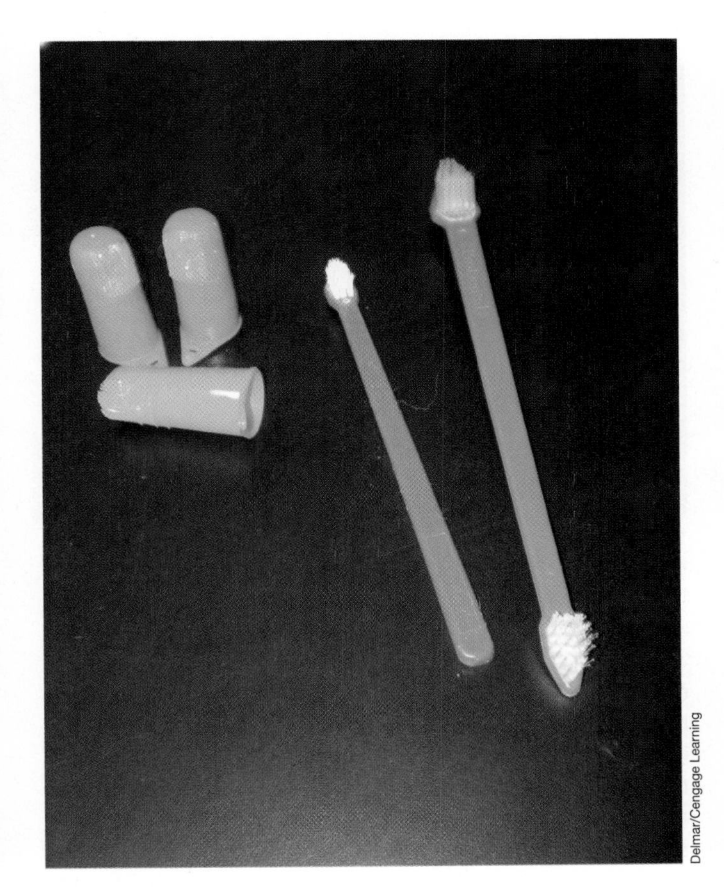

FIGURE 47-4 Toothbrushes and fingerbrushes.

and small dogs. Pet toothbrushes may have a small round head with soft bristles and a short narrow handle that fits comfortably into the mouth of an animal. Various sizes are available depending on the size of the animal. A **fingerbrush** is a small tool that fits on the end of a finger and is a plastic, thimble-like device that has small, soft bristles that are rubbed over the teeth (see Figure 47-4).

The brushing procedure is similar to brushing one's own teeth with the exception that the lingual surface, or inner aspect of the teeth near the tongue, is not brushed. Only the buccal, labial, and occlusal surfaces of the teeth are brushed. The **buccal** surface of the teeth is located on the outer area near the cheek (see Figure 47-5). The **labial** surface of the teeth is the front area that is covered by the lips (see Figure 47-6). The **occlusal** surface is the top area of the teeth. It is difficult and not safe to open an animal's mouth to brush the inside surfaces of the teeth. Injury may be caused to the person or the animal. The areas of importance are easily accessed by lifting the lips to expose the outer tooth surface (see Figure 47-7). Animals will accept teeth brushing with time and consistency, as well as patience. Begin with a small amount of toothpaste placed on the tip of a finger or piece of gauze to allow the animal to smell and possibly taste it.

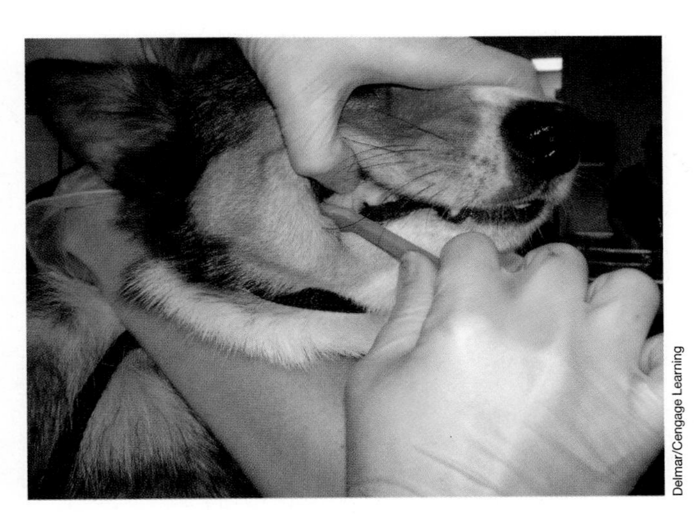

FIGURE 47-5 Buccal surface of the teeth.

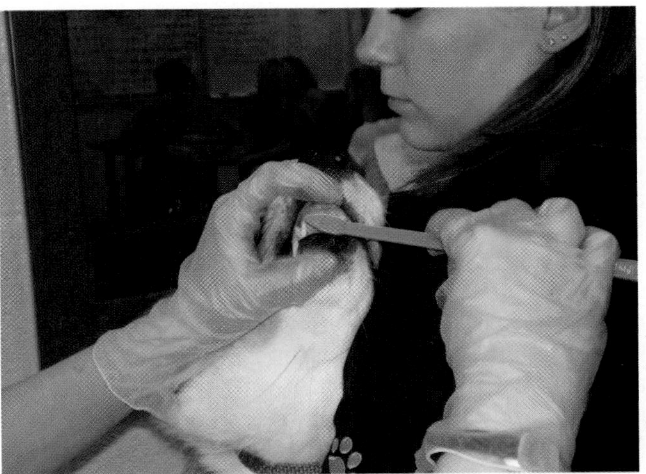

FIGURE 47-6 Labial surface of the teeth.

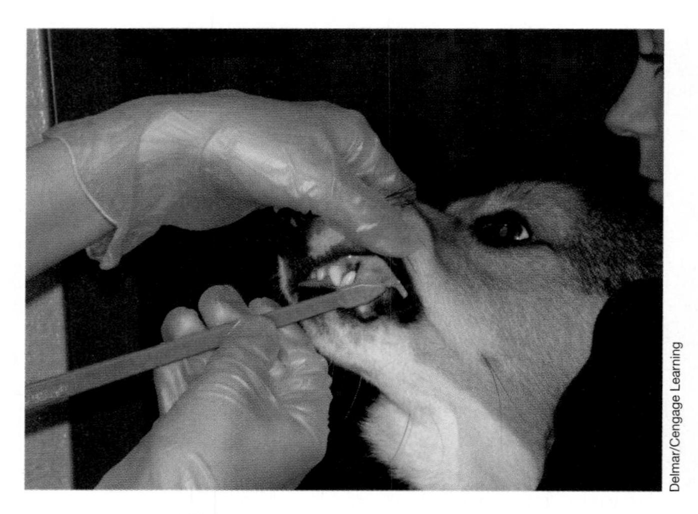

FIGURE 47-7 Lift the lips to expose the outer tooth surface.

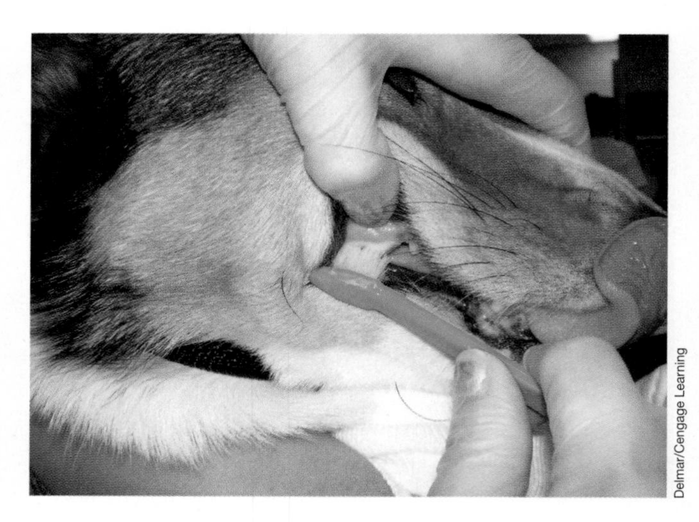

FIGURE 47-9 Proper angle for brushing the teeth.

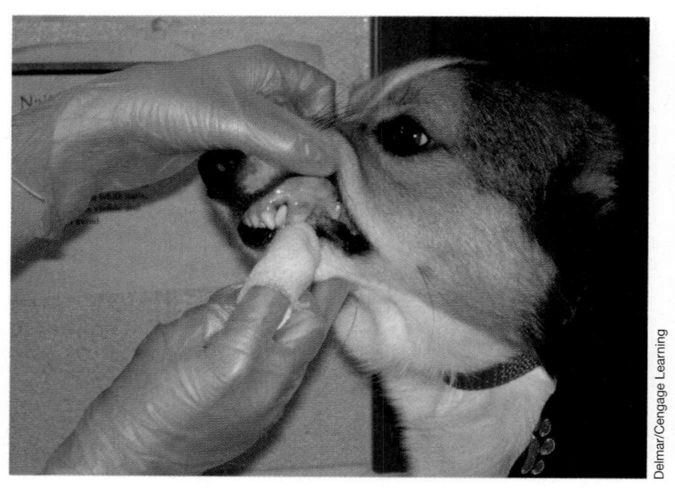

FIGURE 47-8 Use of gauze to rub toothpaste onto the surface of the teeth.

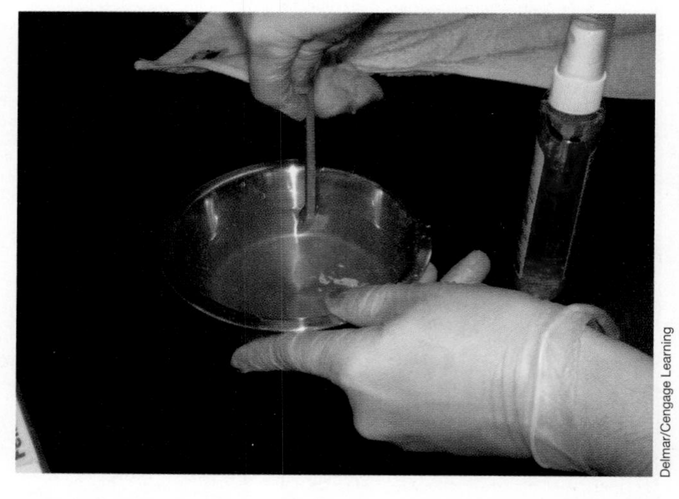

FIGURE 47-10 Use a small bowl of water to rinse the toothbrush.

A small amount of toothpaste can be rubbed over the outer buccal surface by using the fingertip or gauze and using a rubbing motion on the tooth (see Figure 47-8). As the pet accepts this, the owner can then proceed to using a toothbrush. By carefully lifting one area of the lip and working on both the upper and lower aspects of the teeth, the teeth can be brushed. The toothbrush should be held at a 45-degree angle over the tooth and brushed with a circular motion (see Figure 47-9). This process can be repeated until all outer aspects of the upper and lower teeth are brushed. Toothpaste should be applied as needed. A small amount of water in a bowl can be used to rinse the brush before applying more toothpaste (see Figure 47-10).

Pets should be introduced to teeth brushing at an early age and the veterinary assistant should show owners how to properly brush the pet's teeth. This helps the process become more pleasurable to the animal and the owner. Some patients may require daily teeth brushing and dental care. Sample kits including a toothbrush, fingerbrush, and pet-safe toothpaste should be sent home to offer compliance. A dental model may be used to show owners the proper brushing technique. Owners may need to be encouraged to provide home dental care. It is also important to discuss with the owner the possible need for **dental prophylaxis** or professional cleaning by scaling and polishing the teeth. This may need to be done once a year or more depending on the health and age of the animal.

Large animals may require regular floating of the teeth to prevent sharp edges from forming, causing difficulty eating. Floating is an advanced technique that requires additional training and experience. Floating equipment often includes a tooth float and file blade and mouth gag (see Figure 47-11). Horses commonly require floating when sharp edges form on the teeth that possibly cut into the gums, cheek, or tongue. When this condition progresses, horses may have difficulty chewing hay and grass and often drop grain out of the mouth during chewing.

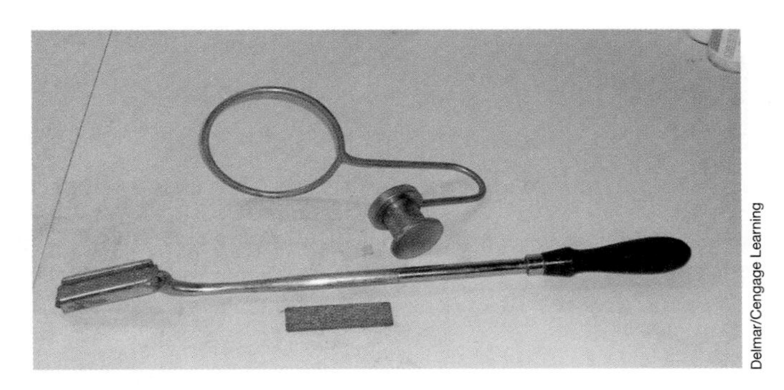
Delmar/Cengage Learning

FIGURE 47-11 A tooth float is used to file the teeth of horses. A gag is used to hold the mouth open for the procedure.

Competency Skill

Teeth Brushing

Objective:

To properly provide dental care and cleansing of pet's teeth.

Preparation:

toothbrush or fingerbrush, pet-safe toothpaste, gauze sponges, small bowl with fresh water, dental spray, exam gloves

Procedure:

1. Apply exam gloves.
2. Place a small amount of toothpaste on the end of the index finger or piece of gauze.
3. Allow patient to smell or taste the toothpaste.
4. Rub the toothpaste over a tooth area by lifting the upper lip. Repeat as necessary until patient is accepting of the process.
5. Apply a small amount of toothpaste to the toothbrush or fingerbrush.
6. Hold the mouth closed and lift the upper lip to expose the outer tooth surfaces.
7. Begin brushing the central incisors, holding the brush at a 45-degree angle.
8. Work posteriorly to the back of the mouth ending at the molars.
9. Rinse the toothbrush with water and apply more toothpaste as needed.
10. Brush the left and right upper and lower dental arcade.
11. If the patient is willing, slip the toothbrush between the upper and lower arcades. Rotate the bristles so they contact the inner surface of the upper arcade.

(*Continues*)

12. Repeat on the lower arcade.
13. Brush both surfaces of the upper and lower lingual arcade.
14. Clean the patient's face when done.
15. Clean up the work area and disinfect all items.

Dental Prophylaxis

The professional dental prophylaxis should be provided by the veterinary technician. It is important for the veterinary assistant to have an understanding of the importance of the procedure to educate clients. Some veterinary assistants may be asked to perform certain aspects of the dental prophylaxis under direct veterinary supervision. Several courses are available for advanced training in dental techniques and should be evaluated for requirements and state practice act guidelines. The dental prophylaxis includes placing the animal under general anesthesia. A dental exam should be completed first, noting any damaged or broken teeth, signs of gingivitis, excessive plaque, or any signs of missing or loose teeth (see Figure 47-12). A veterinarian may be asked to check any areas that are diseased or to extract teeth. The prophylaxis should then begin with scaling all surfaces of the upper and lower dental arcade. Then all tooth surfaces should be polished. This process allows for the best dental care for the patient and should be discussed with all clients. Each animal should be evaluated for dental health care needs.

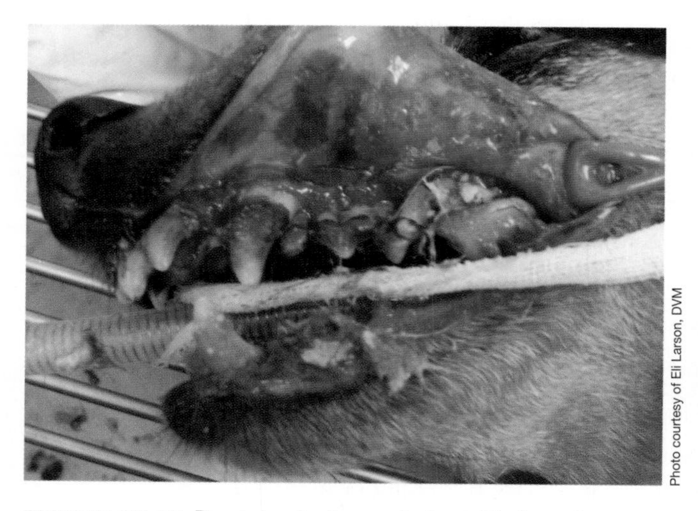

FIGURE 47-12 Dental calculus and gingivitis in a dog.

Administering Injections

Some veterinary practices will allow veterinary assistants to prepare and administer certain injections. The subcutaneous, intranasal, and intramuscular injections may be administered in some states, depending on the veterinary practice act and the tasks delegated within the facility, by the veterinary assistant (see Figure 47-13). The **subcutaneous (SQ)** injection is administered under the skin. The **intranasal (IN)** injection is administered into the nose, usually in drops placed into the nostril rather than using a needle. The **intramuscular (IM)** injection is administered into a muscle. Intravenous (IV) and intradermal (ID) injections are given by the veterinary technician and veterinarian, respectively.

Filling a Syringe

Several things must be considered before filling a syringe. The syringe and needle size are the first items to consider when preparing for an injection. Select a syringe that has a volume slightly larger than the dose being administered. Vaccines typically hold 1 ml, so

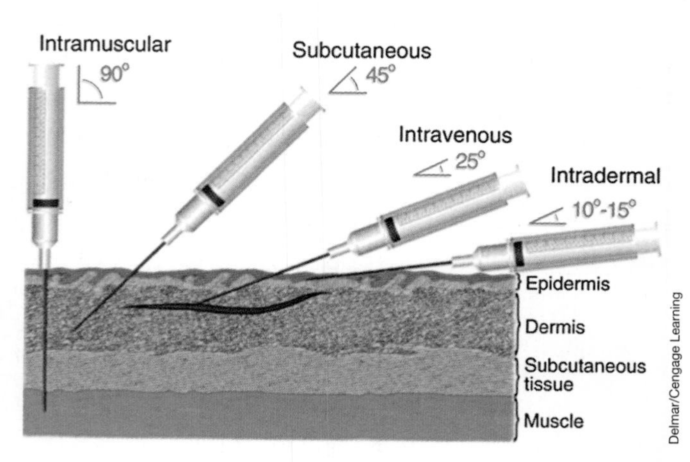

FIGURE 47-13 Examples of injection routes for drug administration.

an appropriate-sized syringe would be a 3 ml syringe (see Figure 47-14). When the syringe volume size is slightly larger than the needed amount to be administered, it allows for space to remove any air bubbles that may be drawn into the syringe. It also allows space for aspiration. The **aspiration** process of a syringe is when the plunger is drawn back slightly to make certain no blood vessel has been accidently penetrated prior to administering an injection. This is done immediately prior to injecting any medication into a patient, regardless of the route, to avoid accidental injection of a medication into the bloodstream. The needle size or **gauge** should be determined by patient size, thickness of the liquid being administered, and the rate at which the injection is being administered. A gauge is measured by the diameter of the needle. The greater the diameter of the needle, the lower the gauge size. As an example, an 18 gauge needle has a greater diameter than a 25 gauge needle. The more **viscous**, or thick, a liquid drug is, the lower the size gauge needle. The more rapidly a drug may be administered, the lower the size gauge needle. Cats have smaller veins and muscles than horses, so the gauge size selected should be appropriate for the injection type and the animal species. The needle length is also based on the type of injection and depth the medication will be administered. A short length needle may be used in thin-skinned animals, such as cats and rodents, and a longer needle may be used when an injection is given into a muscle. The syringe needs to be handled with care to avoid any possible contamination. It is important to clean any injection vial with alcohol prior to applying a needle into the vial. Medications and vaccines should never be mixed in the same syringe unless otherwise directed by the label, as many different substances may create an unwanted chemical reaction. It is important to always label any syringes with the contents and initial it by the person preparing the syringe. The syringe should be labeled as in Table 47-1.

When ready to withdraw the contents into a syringe, the vial may create a vacuum making it somewhat difficult to remove the substance from the container. When this occurs, detach the syringe from the needle so that it is still penetrating the top of the vial. This will allow some air to enter the container and eliminate the vacuum. Securely attach the syringe into the needle and continue to withdraw the necessary amount of medication. The vial should be held upside down with one hand and the other hand controlling the syringe, which is pointing upward within the vial. The needle should penetrate the rubber stopper of the vial at the level of the medication. The plunger of the syringe should be pulled back to withdraw the proper amount of contents. Then the needle is withdrawn from the vial and the side of the syringe should be tapped gently with a snap of the fingertip to remove any air bubbles from the syringe. A gentle push on the end of the plunger may also help expel air bubbles. Recap the needle by gently sliding the needle into the needle cover and attach an appropriate label for the medication or vaccine. Always check the patient record or hospital treatment board to make certain the correct dose and medication is being administered. Read every drug label at least three times to make certain the correct drug is being used.

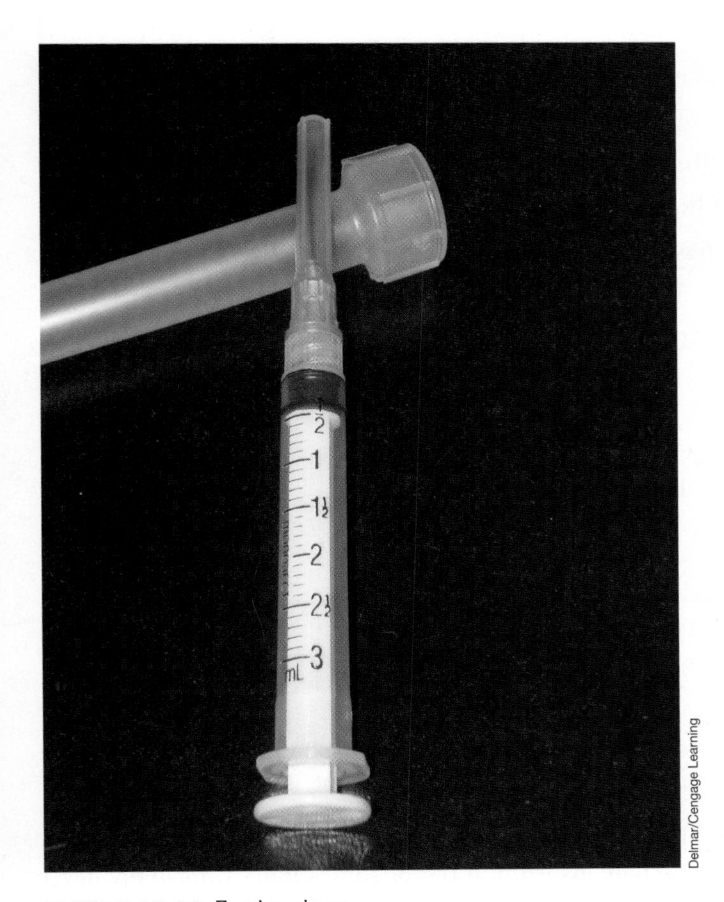

Delmar/Cengage Learning

FIGURE 47-14 3 ml syringe.

TABLE 47-1

Labeling a Syringe

- Drug or Vaccine Name/Type
- Amount or dose prepared
- Date
- Patient Name
- Initial of person preparing syringe

Competency Skill

Filling a Syringe

Objective:

To properly prepare a syringe with an accurate dose of medication.

Items Needed:

proper-sized syringe, proper needle gauge, proper needle length, vial of medication or vaccine, alcohol, cotton balls, adhesive tape, exam gloves

Procedure:

1. Determine the drug or vaccine and amount to be placed in the syringe.
2. Apply exam gloves.
3. Select the proper-sized syringe, needle length, and gauge.
4. Prepare a label with the drug or vaccine name, amount to be withdrawn, date, patient name, and veterinary assistant's initial.
5. Place the label on the distal barrel of the syringe.
6. Prepare a cotton ball saturated with alcohol.
7. Place the cotton ball on the top of the vial and wipe the rubber stopper area.
8. Place the vial upside down in one hand with the fingers curling around the vial securely.
9. Uncap the needle and insert the needle into the rubber top of the vial.
10. Withdraw the proper volume.
11. Remove the needle from the vial.
12. Gently tap or snap the edge of the syringe to remove any air bubbles, or slightly expel the air by pushing the end of the plunger.
13. Recap the needle.

Subcutaneous Injections

The SQ injection is the easiest to administer and is the most frequently used injection for vaccines and antibiotics. Small animals have loose skin over the base of the neck and between the shoulder blade area, making an easy site for the SQ injection (see Figure 47-15). The site to be injected should be cleaned with alcohol. One hand should hold the syringe and the free hand should pinch the skin over the shoulder blades and lift gently to form a triangle, or tent. The needle is inserted at the base of the triangle or tent formation parallel to the body. If the needle is short it may need to be fully inserted; if the needle is long it may only need to be partly inserted. Caution should be taken to avoid passing the needle outside of the skin. To determine proper needle placement, release the skin and feel for the end of the needle under the skin using the index finger. If the needle is entirely under the skin, aspirate for proper placement. If no blood is noted, administer the injection. Withdraw the needle and briefly rub the area and praise the patient.

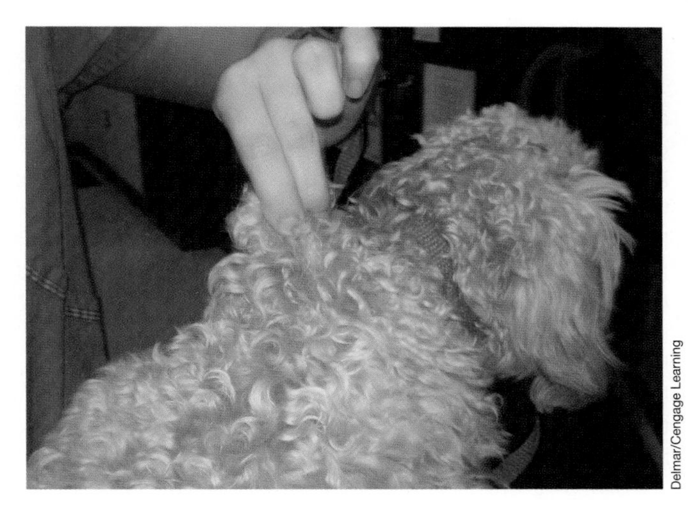

Delmar/Cengage Learning

FIGURE 47-15 Site for subcutaneous injections.

Competency Skill

Administering a Subcutaneous Injection

Objective:

To properly administer a subcutaneous injection.

Preparation:

proper-sized syringe, proper gauge and length needle, alcohol, cotton balls, exam gloves

Procedure:

1. Apply exam gloves.
2. Lift the skin between the shoulder blades using the thumb and forefinger of one hand. Form a triangle or tent with the skin.
3. Wipe the area with an alcohol-soaked cotton ball.
4. Using the other hand, insert the syringe into the skin at the base of the tent or triangle parallel to the body.
5. Once the needle is placed, release the skin.
6. Use the free hand to palpate the needle below the skin to check for accurate placement, noting the needle is not through the skin.
7. Aspirate the end of the plunger, looking for any signs of blood entering the syringe. If no blood enters the syringe, administer the injection.
8. Withdraw the needle and place in sharps container.
9. Rub the injection site with one hand and praise the patient.

Intramuscular Injections

The IM injection is placed deeper into the body through a muscle. There are many muscle sites on an animal's body to place IM injections. The common sites for IM injections in small animals are in the **quadriceps** or **hamstring** group of muscles and the **epaxial** muscles located over the bands of muscles along either side of the spinal column near the back end of the animal. The quadriceps muscle is located on the hind limb at the cranial part of the thigh, whereas the hamstring muscle is located on the hind limb at the mid- to distal part of the thigh. In large animals, IM injections are administered in the brisket muscle over the front chest between the front limbs, the neck muscles, or the hamstring muscles.

IM injections that are given into the rear limb should be given by one hand stabilizing the limb so the animal doesn't readily move the leg. Caution must be taken when giving an IM injection in the hamstring or quadriceps area of the thigh as the sciatic nerve may be easily injected, which can cause irreversible damage and potential paralysis. The fingers should be placed on the medial aspect of the leg and the thumb on the lateral aspect at the mid-thigh. The free hand should hold the syringe, and the femur area should be palpated for correct muscle placement. The needle should be placed anterior to the femur bone approximately halfway between the hip and the knee and parallel to the femur. The site should be cleaned using an alcohol-soaked cotton ball. When injecting the needle, place the syringe slightly downward as it enters the muscle to control the depth of the needle placement (see Figure 47-16). Avoid any bone placement. Aspirate the needle to ensure a blood vessel has not been entered. If there is no sign of blood, slowly inject the substance. Withdraw the needle and gently massage the injection area. If multiple or repeated IM injections are required, it is recommended to use alternating rear limbs to decrease pain and trauma to areas of the limbs.

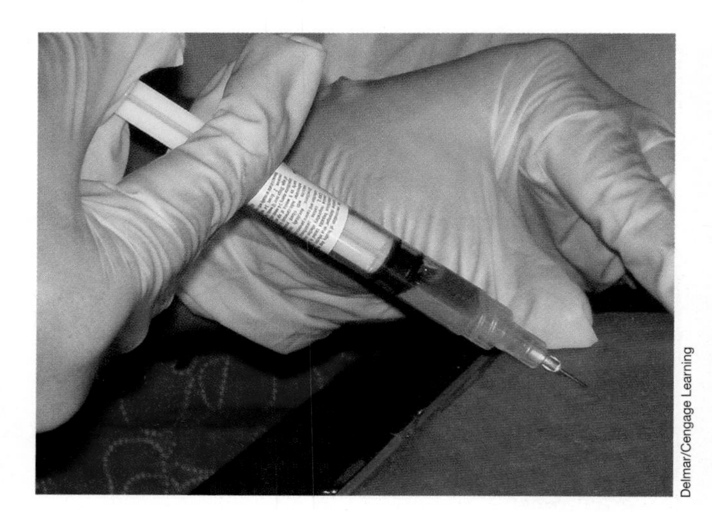

Delmar/Cengage Learning

FIGURE 47-16 Correct angle for an IM injection.

Competency Skill

Administering an Intramuscular Injection

Objective:

To properly administer an intramuscular injection.

Preparation:

proper-sized syringe, proper gauge and length needle, alcohol, cotton balls, exam gloves

Procedure:

1. Apply exam gloves.
2. Place one hand with the fingers located medially along the middle of the femur. The thumb is placed on the lateral aspect of the mid-thigh. The IM injection may also be

given into the epaxial muscles located over the back end of the animal along either side of the spinal column.

3. Rub an alcohol saturated cotton ball over the injection site.
4. Insert the needle cranially and parallel to the femur.
5. Direct the needle below the skin and into the muscle mass in front of the femur between the knee and hip joints. Do not go through the muscle or touch the femur bone.
6. Aspirate the plunger on the syringe. If no blood is noted, inject the substance slowly.
7. Withdraw the needle and place in a sharps container.
8. Massage the area where the injection was given and praise the patient.

Large animal injections may commonly be placed in the neck muscles located at the base of the neck. This area forms a triangle shape from the muscles and ligaments in this area where the shoulder meets the crest of the neck. The center of the triangle is the best IM injection site location. The pectoral muscles or brisket area over the lower front chest located at the top of the front limbs is another possible IM injection site in large animals. When placing the needle into the muscle of a large animal, use a quick and decisive thrust that is placed deep into the muscle location. The site should be aspirated to make sure no blood vessel has been entered.

Intranasal Injections

The IN injection is administered by placing drops of liquid into the nasal cavity or nares of an animal. Many respiratory medications and vaccines are given by the IN route. Small animals may be in standing, sitting, or sternal recumbency, and large animals should be standing. The most important aspect of administering the IN injection is elevating the head and neck to prevent the nasal drops from being forced back out of the nostrils. A firm grip must be kept on the head for best control. The nose should be pointed upward long enough for the drops to coat the mucous membranes of the nares. Many animals do not like or tolerate nasal drops, so care must be taken when administering IN injections.

Competency Skill

Administering an Intranasal Injection

Objective:

To properly administer an intranasal injection.

Preparation:

proper-sized syringe, exam gloves

(*Continues*)

Procedure:

1. Apply gloves.
2. Extend the head and neck of the patient with the nose pointed upward.
3. Apply several drops to each nostril.
4. Keep the head and neck extended to allow drainage of the substance on the mucous membranes.
5. Blow on the nasal area to eliminate sneezing.
6. Allow head to relax, and discard syringe or dropper in sharps container.

Fluid Administration

Fluids are administered to patients for fluid replacement therapy due to illness or dehydration. Some patients may need small amounts of fluid replacement given as an SQ injection. Other animals that are very sick and require large amounts of fluids may need an IV catheter and IV fluid therapy. SQ fluid therapy may be administered by the veterinary assistant. IV catheter placement is completed by the veterinary technician. However, the veterinary assistant should be able to monitor the catheter and IV fluid line while the patient is undergoing treatment.

Subcutaneous Fluid Therapy

SQ fluids are placed into the loose skin at the base of the neck of small animals that require small amounts of fluid replacement. This is usually less than a 200 ml volume of fluid. These fluids are placed under the skin and slowly absorbed by the body. Common reasons for SQ fluids include treatment for kidney disease, urine collection, and minimal vomiting and diarrhea cases. The most common type of SQ fluid used for therapy is **Lactated Ringers Solution (LRS)**.

SQ fluids are given using an IV bag with an attached IV line that is attached to the proper-sized needle according to the animal's size and species, just as in injections. The IV line or extension set is a flexible plastic tubing that connects to the fluid bag. The fluid bag is made of vinyl material that is clear. Most animals tolerate being given SQ fluids, which are similar to an SQ injection. One location can hold about 5–10 ml per pound body weight as a volume of fluid. Average absorption time of SQ fluids is between 6 and 8 hours. The fluid bag should be opened and an IV drip set applied to the end of the bag, which is kept sealed by a plug that must be pulled and removed (see Figure 47-17). The bag should be inverted when the plug is removed to prevent fluids from leaking. The sharp end of the IV set should be inserted into the open fluid bag at the unplugged site. The IV drip set control located on the plastic tubing should be maneuvered closed to prevent fluids from leaking from the tubing. The bag is then placed upright and hung on an IV pole. The tubing may be opened to allow fluids and air to flow out of the bag. A fluid bag has two ports located on the end: the injection port and the spike port. The **injection port** is used to remove and insert additional substances using a needle and syringe. The **spike port** is the site where the drip set tubing is placed. Most fluid bags hold a 1,000 ml fluid volume.

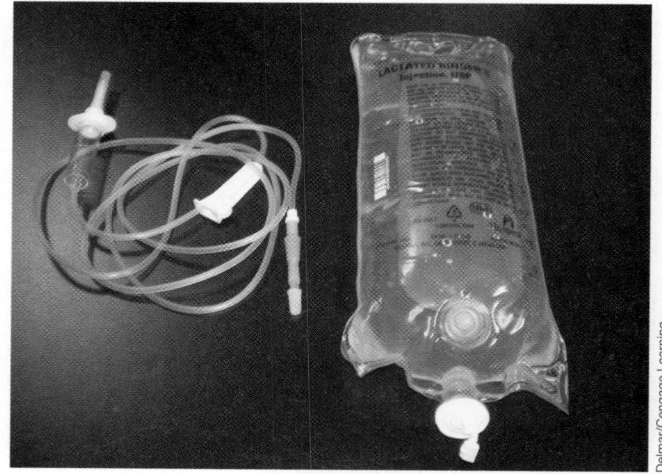

Delmar/Cengage Learning

FIGURE 47-17 IV fluid and administration set.

Competency Skill

Preparing an IV Fluid Bag and IV Drip Set

Objective:

To properly prepare an IV bag and tubing for medication administration.

Preparation:

IV fluid bag, IV drip set, IV stand or pole, appropriate-sized needle

Procedure:

1. Wash hands well, as gloves are not necessary with this task.
2. Remove an IV fluid bag from the plastic cover.
3. Remove an IV drip set from the plastic cover.
4. Close the IV drip set wheel located on the plastic tubing.
5. Remove the cover over the white spike of the IV drip set located at the end of the tubing, which is sharp and should be handled with care.
6. Remove the spike port cover on the IV fluid bag. Pull firmly to remove.
7. Hold the spike port in the left hand to guide the spike straight into the port. Push firmly. If the spike is pushed in at an angle, it may puncture the bag of fluids.
8. Hang the bag on an IV pump or IV poie.
9. Open the IV drip set clamp to allow fluids and air to be removed from the bag and tubing.
10. Close the IV drip set clamp when all air bubbles have been removed.
11. Apply the appropriate-sized needle to the end of the IV drip set tubing.

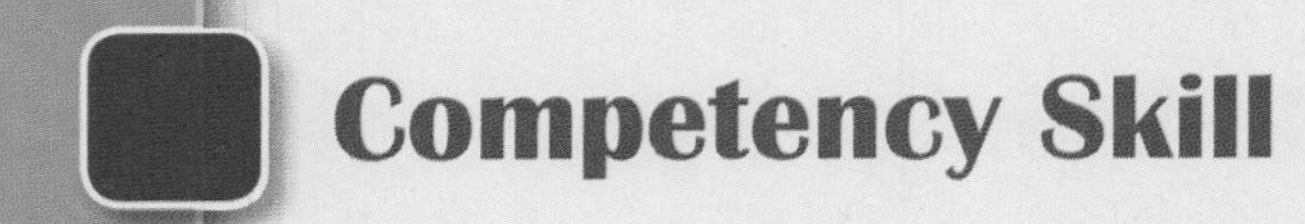

Competency Skill

Administering SQ Fluids

Objective:

To properly administer SQ fluids.

Preparation:

IV fluid bag, IV drip set, IV stand or pole, appropriate-sized needle, exam gloves

Procedure:

1. Apply gloves.
2. Tent the skin over the base of the neck between the shoulder blades as in an SQ injection (see Figure 47-18).

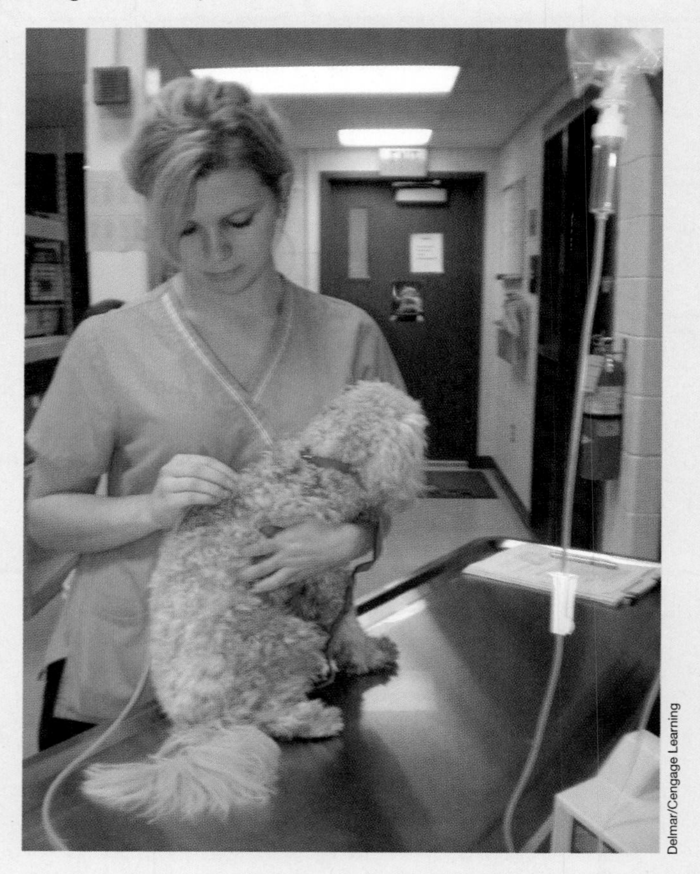

FIGURE 47-18 Preparation for the administration of SQ fluids.

3. Insert the needle into the base of the tented area slightly angled downward and just under the skin.
4. Insert along the long axis of the skin fold.
5. Let go of the tented area of skin.
6. Hold the needle in place with one hand.
7. Use the free hand to turn on the wheel of the IV drip set clamp. The higher the bag is placed on the pole, the faster the fluid rate. The bag can be squeezed gently to increase fluid flow.
8. Give the appropriate volume of fluid as directed by the veterinarian.
9. When complete, close the clamp on the IV drip set.
10. Remove the needle and apply pressure over the injection site.
11. Some fluid will leak from the injection site.
12. Clean work area and disinfect supplies.
13. Place needle in sharps container.

IV Catheter Monitoring

An **IV catheter** is a small plastic piece of equipment that is placed within a vein to administer fluids and medications directly into a patient's bloodstream (see Figure 47-19). This task is done by the veterinary technician or veterinarian as it requires advanced skills in vein location and properly inserting the catheter into the area, which is relatively small. The veterinary assistant is responsible for maintaining and monitoring the IV catheter patency after it has been placed. **Patency** refers to the proper flow and purpose of the item and that it remains intact and usable. Most IV catheters are placed in the cephalic vein; however, large animals and larger breed dogs may have a jugular catheter. The veterinary assistant should be able to recognize where the catheter is placed, that it is patent and flowing properly, and that the fluid rate is proper and flowing. A bandage is placed to cover and protect the catheter and the bandage should be checked on a daily basis for signs of leaking, disconnection, bleeding, or swelling. This may require a daily bandage change to assess the site. Some catheter sites may become inflamed and require antibiotic or antiseptic around the placement site. The catheter site should then be re-bandaged. **Phlebitis** may develop around the site and should be monitored for signs of redness, swelling, pain, and inflammation. The IV catheter should be replaced every 3 to 4 days if the treatment is lengthy.

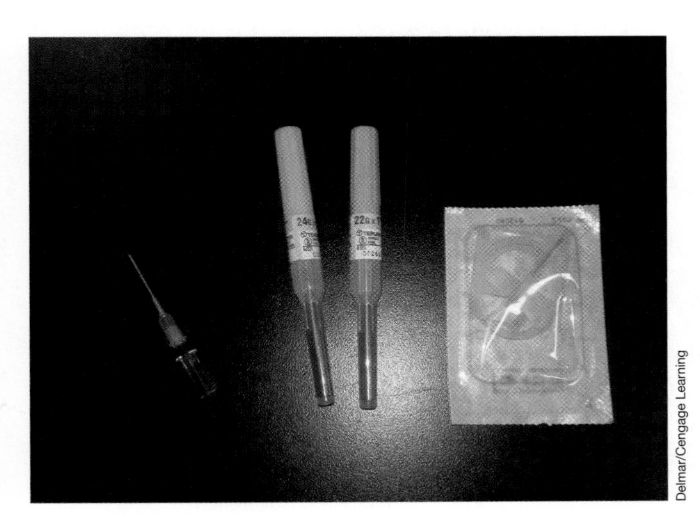

FIGURE 47-19 Intravenous catheter.

One of the most common problems with the IV catheter is patency, which may be caused from the IV tubing kinking or a clot occluding the vein and preventing a free flow of fluids. Catheters may require flushing with several milliliters of fluid or **heparinized saline.** Heparin is a drug that prevents blood from clotting and is mixed with saline to flush into the catheter site to prevent a blood clot from occluding the catheter. This is done every time the catheter is detached from the IV set or prior to medication being administered directly into the catheter. Some animals with cephalic vein IV catheters may kink the tubing by pulling the leg in a direction that prevents fluid flow. A splint or bandage can be placed to help maintain the flow of fluids. Animals that chew or bite at the catheter site or bandage may need an E-collar to prevent the animals from damaging or removing the catheter.

If **pitting edema** is noted, or swelling around the catheter site causing the fluid to be outside of the vein, the IV fluids should be stopped and a new catheter placed at another vein site.

IV Fluid Monitoring

IV fluids require constant and consistent monitoring by the veterinary assistant. Frequent observations will help detect improper fluid flow or occlusion. The veterinarian should determine the volume to be delivered to each patient on a daily basis. The amount of fluids to be delivered over the course of a day is called the **flow rate**. When a total volume is determined, the veterinary assistant may note how much fluid is in an IV bag at the start of treatment by placing a piece of adhesive tape along the edge of the bag or container lengthwise where no information is written. The initial volume and time are marked in pen along the tape at the level of the fluid line. The final volume desired and when the volume should be completed are marked on the opposite end of the tape at the proper fluid level. Ideally, an **infusion pump** should be available to adjust the rate and regulate the fluid flow (see Figure 47-20). This equipment is used to provide a constant flow of fluid at a specific rate throughout the day. The infusion pump has an alarm that sounds when fluids are no longer passing through the tubing, alerting when an obstruction occurs or when a bag is empty.

There are two types of IV drip lines that are used depending on patient fluid volume. A **macrodrip line** is a regular IV drip that delivers 15 drops of fluid per milliliter or 15 drops/ml. A **microdrip line** is a small line that delivers 60 drops/ml (see Figure 47-21). After the veterinarian has determined the total volume a patient requires in a day or 24-hour period, the veterinary assistant may determine how much fluid should be given each hour. This is done by taking the total volume and dividing by 24, the number of hours in a day. This number gives you the amount of fluid to be delivered to a patient each hour. Table 47-2 shows some examples of fluid rate calculations.

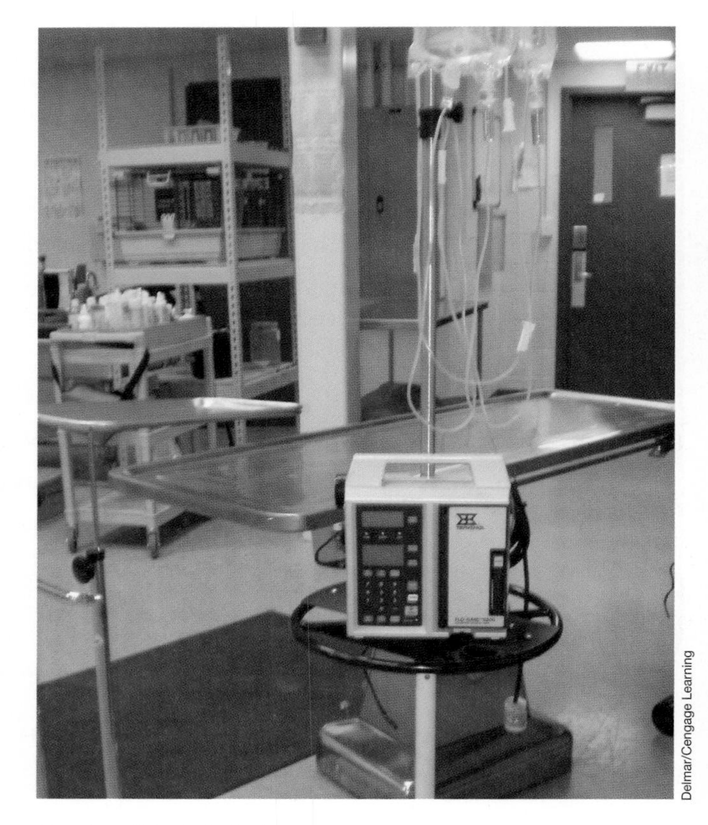

FIGURE 47-20 Infusion pump.

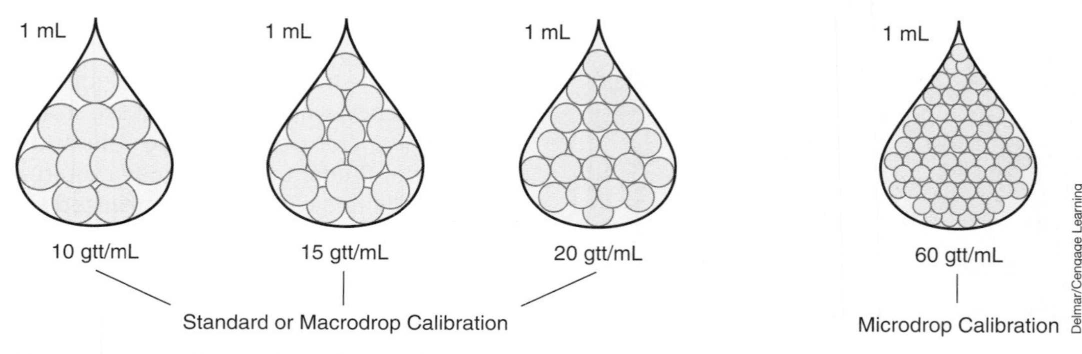

FIGURE 47-21 Comparison of drop size in macrodrip and microdrip infusions.

TABLE 47-2

Fluid Rate Calculation Examples
Example 1
Total volume: 1000 ml
1000 ml / 24 hours per day = 41.6 ml/hr
Example 2
Total volume: 3500 ml
3500 ml / 24 hours per day = 145.8 ml/hr
Example 3
Total volume: 9000 ml
9000 ml / 24 hours per day = 375 ml/hr

Competency Skill

Catheter Patency

Objective:

To properly assess the IV catheter and site of administration to ensure patency of the line.

Preparation:

3 ml syringe, sterile saline, LRS solution, heparin

Procedure:

1. Observe fluid flow of catheter through IV tubing and IV bag.
2. If fluid flow has stopped, determine the occlusion cause.
3. Extend the patient's leg by extending the foot and elbow forward.
4. If flow returns, place a splint or bandage to extend the limb.
5. If no flow returns, remove the bandage over the catheter. Note any swelling, redness, or irritation.
6. If no swelling is noted and the catheter appears to be in the vein, detach the IV tubing from the catheter.
7. Flush a combination of saline, LRS, and a small amount of heparin into the catheter using a 3 ml syringe.
8. If the fluid flows through the catheter, reattach the tubing and monitor fluid flow.
9. When the fluid flow is appropriate, re-bandage the catheter site.
10. If no fluid flow appears, contact the veterinary technician or veterinarian.

Competency Skill

Infusion Pump Monitoring

Objective:

To properly assess the IV infusion pump to maintain patency of the line.

Preparation:

Infusion pump, 3 ml syringe, sterile saline, LRS solution, heparin

Procedure:

1. Check tubing to see that it is properly placed in the infusion pump mechanism. See that there are no kinks or obstructions.
2. If the clamp is controlling the fluid flow, make certain the clamp is open.
3. Press the tubing under the clamp to make certain fluids can flow through the tubing.
4. If no fluid is flowing, flush a combination of saline, LRS, and a small amount of heparin into the catheter using a 3 ml syringe.
5. Monitor the catheter site for signs of swelling.
6. Note if the rate of fluid flow is correct and the pump is working.

Competency Skill

Monitoring Fluid Flow Rate

Objective:

To properly assess the infusion pump and administration set for proper flow rate.

Preparation:

infusion pump, IV bag and tubing

Procedure:

1. Note drops per milliliter multiplied by the total number of milliliters equals the total number of drops to be administered: 15 drops/ml × 300 ml = 4,500 total drops to be delivered.

2. Total volume of fluids delivered per day divided by 24 hours equals how much fluid should be administered per hour: 300 ml / 24 = 12.5 ml/hr
3. Total number of drops to be administered divided by total number of minutes over which to be administered equals the drops per minute: 4,500 drops/240 minutes (4 hours × 60 minutes/ 240 minutes) = 18.75 drops or approximately 19 drops/ minute

Competency Skill

Catheter Bandage Replacement

Objective:

To properly assess the IV catheter and administration site and replace the bandages as appropriate.

Preparation:

cotton roll, gauze roll, vet wrap, adhesive tape, bandage scissors, exam gloves

Procedure:

1. Apply exam gloves.
2. Remove all layers of bandage.
3. Do not touch tape holding catheter in place.
4. Examine catheter for swelling or bleeding. Note tubing patency.
5. Flush catheter as necessary.
6. If catheter is patent, re-bandage by applying a cotton roll as the primary layer.
7. Apply a gauze roll layer as the secondary bandage layer.
8. Place a loop of IV tubing without any kinks into the tertiary bandage layer.
9. Apply the vet wrap as the tertiary bandage layer.
10. Apply tape to the outer area as necessary.
11. Contact the veterinarian or veterinary technician if the catheter site is not normal or patent.

Socialization and Exercise of Patients

Animals are accustomed to human interaction and when hospitalized may not receive enough human contact. It is important that the veterinary assistant provide positive social interaction to all hospitalized patients during treatment, cage cleaning, or during exercise. This includes petting animals, talking to the patients, using a gentle and soothing voice, or holding them if possible (see Figure 47-22). These positive actions help put patients at ease.

Exercise may be provided to some hospitalized animals depending on their condition and treatment needs. Exercise is important as some animals will not urinate in the confines of the cage, such as dogs. This requires a brief walk several times a day outside of the facility (see Figure 47-23). A designated walking area should be determined to control the spread of parasites and disease and allow proper sanitation measures. Always place the appropriate leash or control needs on the patient. Areas inside the facility may be used for exercise and play areas. The time an animal is outside of its cage will vary from patient to patient. Be careful to indicate when an animal is roaming free within an area. Interaction of patients is an important part of their recovery and treatment.

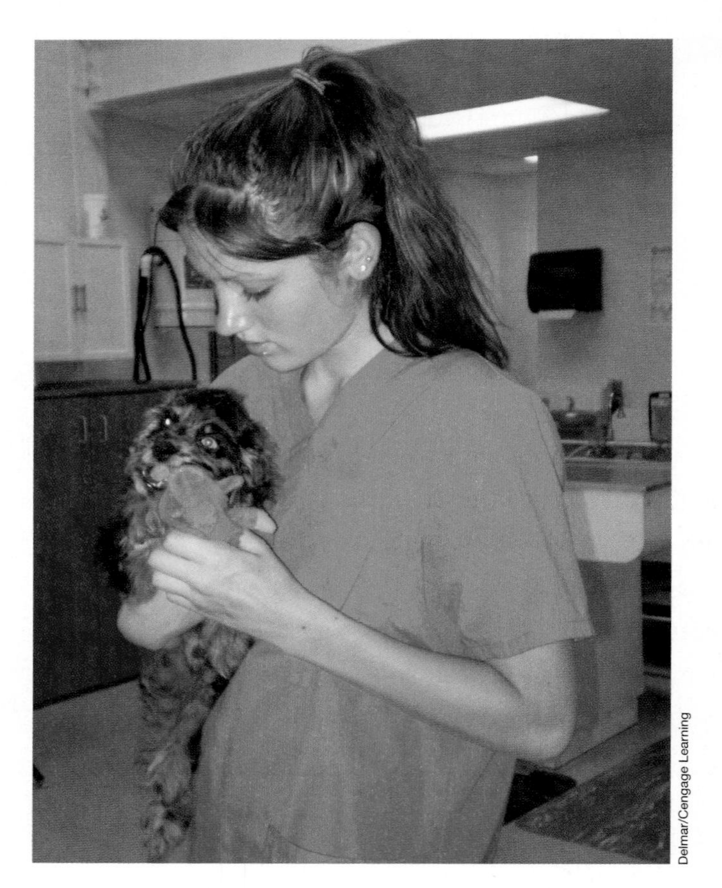

FIGURE 47-22 It is important for hospitalized patients to receive human contact and attention.

FIGURE 47-23 Some animals will need to be walked while in the care of the veterinary facility.

Euthanasia

Euthanasia is the process of putting an animal to sleep using humane methods by means of a painless death. Some hospitalized patients may not recover and continue to become ill, and the owners may decide to end the animal's suffering by means of euthanasia. This decision is based on personal values, religious beliefs, and previous experiences along with the veterinarian's guidance and recommendations. This process is often difficult for the owners as well as the veterinary staff.

The process of the euthanasia should be as pain-free and stress-free as possible. During this time, the veterinary staff should be supportive and sympathetic to the owner's needs and values. It is important to remember the need for patient and client confidentiality and limit discussions about the case with only those team members involved with the patient care. The owner should be provided a waiver that is signed to allow permission for the procedure to be done. The preparation for euthanasia should include discussing the procedure and sequence of events with the client. It is also important to determine what the owner would like done with the animal's remains. Many facilities offer burial or cremation services and some owners may opt not to take the animal's body or complete burial at home. Several items need to be considered in the preparation for euthanasia:

- Will the client be present with the animal during the procedure?
- Would the client like to pay before the procedure or be billed?
- Inform the client of legal restrictions in certain areas for private burial.
- Determine how the client would like to care for the animal's remains, such as pet cemetery burial or cremation.
- Inform the client of the costs of the services.
- Complete all necessary paperwork and consent forms prior to the procedure.

An exam room should be prepared with a box of tissues, a blanket or towel, chairs for the owner and family members, and any other needs for the family. A cadaver bag should be placed in an area of the room that is out of sight of the owner, such as a drawer or closet. The drug of choice should be available to the veterinarian with an accurate weight of the animal. Some facilities administer sedatives to calm an animal prior to the procedure. If the procedure requires IV catheter placement, the catheter and materials should be placed on the exam counter.

Competency Skill

Preparation for Euthanasia

Objective:

To properly assist with and prepare for administration of euthanasia procedures.

Preparation:

blankets or towels, tissue box, chairs or seats for the family, cadaver bag, tape, ID tag, sedatives or tranquilizers, euthanasia solution, appropriate-sized syringe and needle

Procedure:

1. Prepare an exam room or area for the euthanasia procedure.
2. Place a blanket or towels on the exam table or floor depending on patient size.
3. Place a box of tissues on the exam counter.
4. Place chairs near the area where the euthanasia will be performed.
5. Place a cadaver bag in a drawer or cabinet with a roll of adhesive tape and an ID tag to label the bag with the patient name and client name.
6. Place any sedatives that may be used to sedate and calm the patient on the counter.
7. Place the euthanasia solution on the exam counter.
8. Place an appropriate-sized syringe and needle on the exam counter near the euthanasia solution.

The euthanasia procedure itself should be performed in a quiet area with the patient and family members placed in the area immediately on arrival to the facility. The patient and family should be made as comfortable as possible. Any paperwork should be completed prior to the procedure and placed in the patient's medical record. Many family members may ask about the procedure so that they understand what to expect during the process. It is important to make certain the owners are prepared for the procedure and have some understanding of what the veterinarian will be doing and how the patient will react. The veterinarian may discuss the procedure with the owners and family members or may ask the assistant to answer any questions. It is important that the family know the veterinarian will perform the procedure and the assistant will be available to restrain the patient and assist as needed. If a sedative or tranquilizer is administered, the patient and family members will remain in the exam area until the medication takes effect. The veterinarian will administer any sedatives and the assistant should periodically check on the patient every few minutes during this period to determine when the drug has taken effect. The assistant should make certain the owners are comfortable and if anything can be done for them. If an IV catheter is required, the veterinary technician will do so with the assistant restraining the patient. If a simple venipuncture is used to administer the euthanasia solution, the assistant will restrain the patient. Once the procedure is ready to begin, the veterinarian will prepare the necessary amount of solution and administer the injection. Once the entire amount has been administered, the patient's body will relax completely. The needle is then withdrawn and placed in a sharps container. The patient should be placed in lateral recumbency and the veterinarian will check for residual heartbeats or signs of a pulse and reflexes. During this time there may be some **agonal** respirations, which are gasps of breaths while the respiratory system shuts down. The patient's pupils will dilate and the mucous membranes will become **cyanotic** or bluish gray in color. It is important to note that the patient may commonly lose control of the bladder and bowel functions. When no heartbeat or pulse is detected, the patient is pronounced deceased. Owners' reactions will be variable and the staff should be prepared at this point to handle a variety of grief reactions from relief to hysteria. Make the owners feel as comfortable as possible and ask if they would like to remain with the patient alone to say good-bye. The owners may wish to remove any collar to keep as a memory. Some clients may benefit from information on a pet loss group or grief support number. When the owner has finished visiting with the pet, he or she is then escorted outside by a side door if possible. If the owner is not taking the patient's body along, the assistant should then enter the exam area to place the body in a cadaver bag and label the bag with an ID tag containing the patient name and owner name and what is to be done with the remains. The bag should be closed with adhesive tape and the remains placed in the appropriate area of the facility. The assistant should make certain that the sedatives and euthanasia solution are returned to the pharmacy area and any controlled substances are recorded in the controlled substance log. The exam area should be disinfected and cleaned.

Laundering Materials

Laundry disinfection and cleaning is an important part of the practice of a sanitation and disease control plan. Many items must be laundered in a facility after being used with an animal. These items include the following:

- Towels, blankets, and other bedding materials
- Muzzles and cat bags
- Collars and leashes
- Scrubs, surgical gowns, lab coats, and other clothing materials
- Surgical towels and pack covers

Sorting the laundry materials is an important part of the laundering technique. In most cases, laundry should be pretreated by soaking in water with detergent or bleach materials as necessary. Laundry should be sorted according to surgical materials, regular hospital materials, and contagious items. The three types should not be mixed as this will help prevent and control spread of disease. Any items with bloodstains, urine, feces, or other bodily fluids should be soaked and treated in water and detergent or bleach mixtures for at least 30 minutes before washing. Items that are bloodstained should also be pretreated with hydrogen peroxide to remove stains. All contagious items are presoaked in bleach and warm water.

Washing is more effective at high temperatures and with high quality industrial strength detergents. A fabric softener may be used during the rinse cycle or in the dryer using sheets. Each load of laundry should be set on the proper load size and monitored for overloading. The washer must be cleaned and disinfected on a regular basis to remove hair, debris, and detergent residues. Lint should be removed from filters after each load. Surgery laundry should be folded immediately and placed in an area to complete packs and re-sterilize materials. Other hospital laundry should be folded and stored in clean and dry locations.

Competency Skill

Laundry Procedure

Objective:

To properly clean and sanitize all soiled items that may be laundered.

Preparation:

laundry baskets (3), buckets (3), industrial strength laundry detergent, bleach, fabric softener, hydrogen peroxide, exam gloves, washer, dryer

Procedure:

1. Sort laundry into baskets labeled surgery, hospital, and contagious.
2. Place items needing soaking in buckets with detergent or bleach.
3. Pretreat any blood stains with hydrogen peroxide.
4. Pretreat each laundry load in the washing machine.
5. Set washer cycle on hot and appropriate load size.
6. Add fabric softener in rinse cycle.
7. Dry according to load size.
8. Disinfect and clean washer.
9. Clean lint filters.
10. Disinfect and clean dryer.
11. Fold laundry.
12. Put laundry away in proper clean area.

SUMMARY

The proper handling and care of a hospitalized patient is essential for the veterinary assistant. In order for the assistant to properly provide this care, one must be able to recognize the needs of the patient and the needs of the veterinary care. Such needs include dental care, veterinary assistant nursing care, proper set up and maintenance of fluid therapy, understanding of euthanasia procedures, and the proper need for laundering of bedding and hospital materials.

Key Terms

agonal gasps of breaths while the respiratory system shuts down

aspiration process of a syringe when the plunger is drawn back slightly to make certain no blood vessel has been accidently penetrated prior to administering an injection

buccal surface of the teeth located on the outer area near the cheek

cyanotic bluish gray color of the mucous membranes due to a lack of oxygen

dental prophylaxis professional cleaning by scaling and polishing the teeth under general anesthesia

dentifrices veterinary term for toothpaste used on animals

epaxial located above the axis of a body part, such as above the spine.

euthanasia the process of putting an animal to sleep using humane methods

fingerbrush a small toothbrush that fits on the end of a finger and is a plastic, thimble-like device that has small, soft bristles that are rubbed over the teeth

flow rate amount of fluids to be delivered over the course of a day

gauge the size of a needle measured by the diameter of the needle

hamstring muscle located on the hind limb at the mid- to distal part of the thigh

heparinized saline drug that prevents blood from clotting and is mixed with saline to flush into the catheter site to prevent a blood clot from occluding the catheter

infusion pump equipment is used to provide a constant flow of fluid at a specific rate throughout the day

injection port area on an IV bag used to remove and insert additional substances using a needle and syringe

intramuscular (IM) injection given into the muscle

intranasal (IN) medication administration into the nasal passage

IV catheter a small plastic piece of equipment that is placed within a vein to administer fluids and medications directly into a patient's bloodstream

labial surface of the teeth located on the front area that is covered by the lips

lactated ringers solution (LRS) most common fluid used to hydrate an animal

macrodrip line a regular IV drip that delivers 15 drops of fluid per milliliter or 15 drops/ml

microdrip line a small line that delivers 60 drops/ml

occlusal the top surface area of the teeth

patency refers to the proper flow and purpose of the item and that it remains intact and usable

phlebitis area around the IV catheter site causing redness, swelling, pain, and inflammation

pitting edema swelling around the catheter site causing the fluid to be outside of the vein

quadriceps muscle located on the hind limb at the cranial part of the thigh

spike port the site on an IV bag where the drip set tubing is placed

subcutaneous (SQ) injection given under the skin

subcutaneous fluids fluids placed under the skin and slowly absorbed by the body

viscous thick liquid substance

REVIEW QUESTIONS

1. What are dental care responsibilities of the veterinary assistant?
2. What tools and supplies are necessary for brushing a patient's teeth?
3. What type of injections may be completed by the veterinary assistant?
4. What does the term *aspiration* mean?
5. What is the difference between an 18 gauge needle and a 24 gauge needle?
6. What items should be placed on a syringe label?
7. What is the difference in using SQ fluid therapy versus IV fluid therapy?
8. What items should be monitored with patients that have IV catheters?
9. What are ways to socialize hospitalized patients?
10. What items are necessary when preparing for euthanasia?

Clinical Situation

The veterinary assistant, Kathryn, is monitoring a patient's IV fluids during the cat's hospital stay. Kathryn notes the fluids are not flowing correctly and, at times, no fluids are dripping from the IV bag. Kathryn checks the hospital treatment board and the medical record and notes the cat should be receiving 1,400 ml of fluid a day. She is concerned since the fluids are obviously not being received correctly.

- What should Kathryn do?
- What can be done in this situation to help treat the cat correctly?

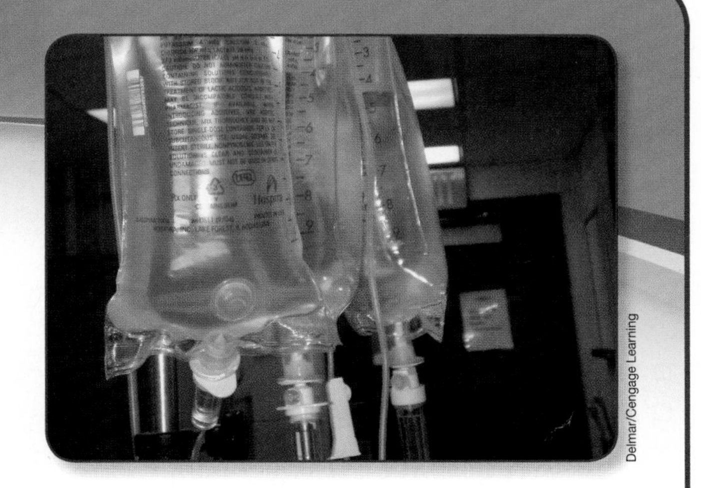

Delmar/Cengage Learning

- How much fluid volume should the cat receive on an hourly basis?
- What maybe causing the fluid flow problem?

48

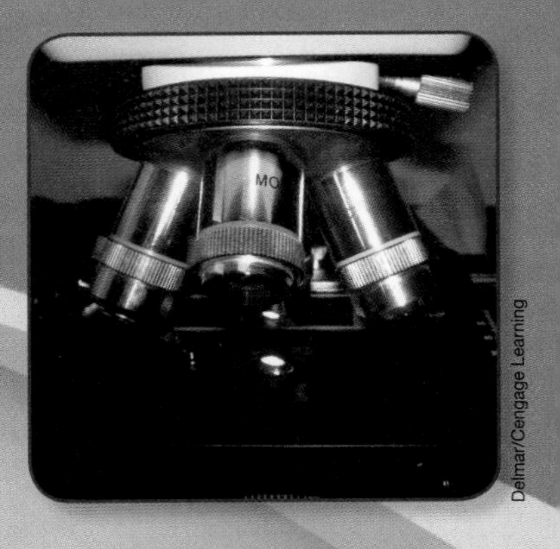

Laboratory Procedures

Objectives

Upon completion of this chapter, the reader should be able to:

- Identify common laboratory equipment used in a veterinary facility
- Demonstrate proper collection of a fecal sample
- Demonstrate how to properly conduct a gross fecal examination
- Demonstrate how to properly prepare a fecal smear
- Demonstrate how to properly set up a fecal floatation
- Demonstrate how to properly prepare a blood sample
- Demonstrate how to properly prepare a usable blood smear
- Demonstrate how to properly stain a blood smear using Wright's stain
- Demonstrate how to properly complete a CBC
- Demonstrate how to properly complete a blood chemistry sample using in-house analyzers
- Demonstrate how to properly determine a plasma protein or total protein sample using a refractometer
- Demonstrate how to properly use assorted serologic testing kits
- Demonstrate how to properly collect a voided urine sample
- Demonstrate how to properly conduct a gross urine examination
- Demonstrate how to properly determine urine specific gravity
- Demonstrate how to properly understand and conduct urine chemistries using test strips
- Demonstrate how to properly prepare urine sediment for microscopic examination
- Demonstrate how to properly collect a sample for culture and sensitivity
- Demonstrate how to properly prepare a gram stain smear
- Demonstrate how to properly set up equipment and materials for a veterinary necropsy

Introduction

Laboratory skills are meant to aid the veterinarian in determining a patient diagnosis and thus determining the necessary treatment. Many testing procedures are done both inside and outside the veterinary facility. Most laboratory testing begins after the initial patient history and physical exam. The veterinarian will determine the testing needs of each patient, and the veterinary technician and assistant will properly work together to prepare samples, conduct in-house laboratory tests, and record laboratory data in medical records. The patient's well-being depends on the entire veterinary health care team to provide accurate and timely results. Laboratory testing technology changes and updates regularly in veterinary medicine. Many protocols, equipment, and tests may have unique factors in each veterinary facility. These factors require flexibility in all staff members and understanding of the importance of continuing education and learning to keep up with the changes in veterinary medicine. It is important to remember safety when handling lab samples to ensure both protection of the sample as well as the protection of the person handling the samples. Many samples may potentially contaminate an area or a person or patient, so wearing appropriate personal protective equipment, such as exam gloves and goggles, is essential.

Veterinary Laboratory Equipment

Many laboratory tests and services are performed within the veterinary facility and are referred to as **in-house testing**. Some tests that may not be completed in-house are often sent to a commercial laboratory referred to as a **reference lab**. The equipment that is available for the test determines which services are offered in-house or through reference labs.

Microscope

Microscopes are a basic and key component to all in-house veterinary labs (see Figure 48-1). They are extremely useful for basic testing procedures and can help identify and diagnose problems quickly and accurately. Many veterinary facilities will have a binocular microscope and a research microscope that allows a variety of magnification powers to view different test samples. They also allow slides to be observed under **oil immersion**, which is the use of a specialized oil substance that is placed over a sample to view the contents. It is important that the veterinary assistant has a general and clear understanding of how to appropriately use a microscope. All microscopes should be covered when not in use. Each microscope will have several **lens objectives** that offer a variety of magnifications to view slide samples (see Figure 48-2). These objectives offer a variety of different powers (see Table 48-1). A slide should be placed on the **stage**, which is the flat section under the lens. A slide should always be viewed using the lowest power magnification first and then moving to a higher power. A slide is viewed by looking through the **eye piece**. To focus each lens to view a sample, the **focus knobs** should be turned gently and slowly while looking through the eye piece. When the microscope is on, a light source will be present. The slide sample should be moved up and down and side to side to view the entire sample. This motion is done by moving the **diaphragm mechanism**. All lenses should be cleaned after each use, and this is accomplished by using lens paper and a small amount of lens cleaner or alcohol to wipe each lens. It is important to follow the manufacturer's instructions when using and cleaning any microscope. All slides and **coverslips**, which fit over the sample on the slide, should be clean and have no cracks or chipped edges.

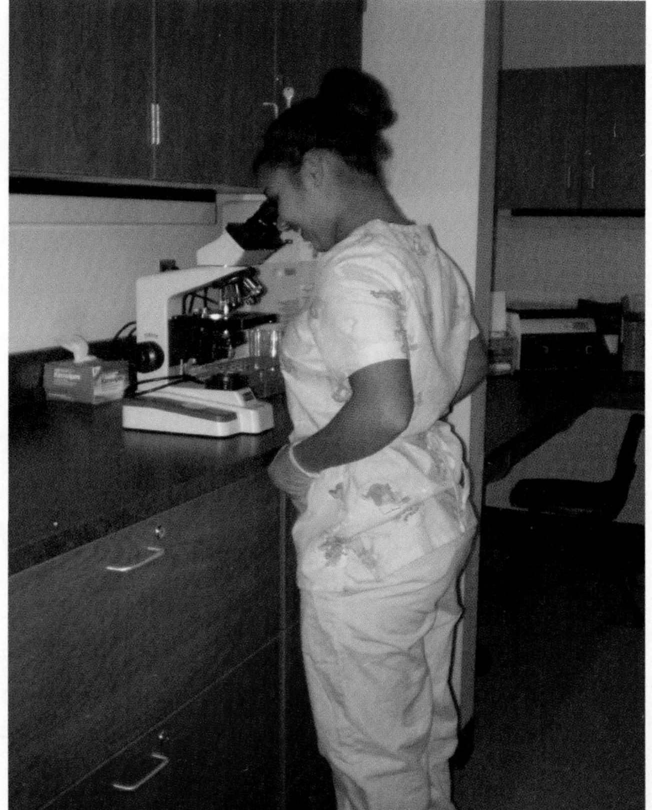

FIGURE 48-1 It is important to learn the proper use of the microscope.

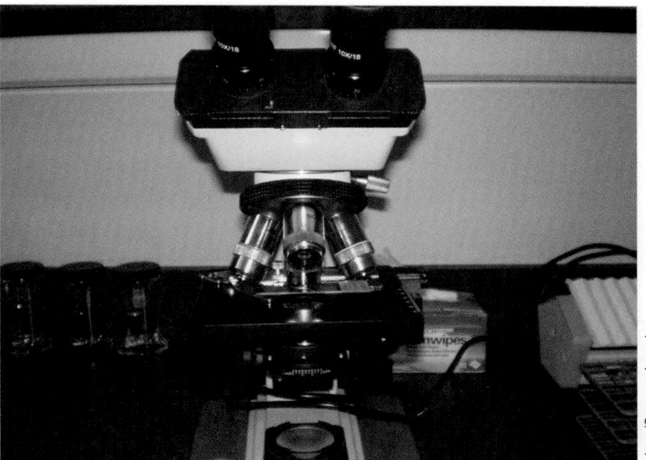

FIGURE 48-2 Note the different lens objectives that can be used on this microscope.

TABLE 48-1

Magnification Powers

	EYE PIECE	OBJECTIVE	MAGNIFICATION
Low Power	Green	10 ×	100 ×
Medium Power	Blue	30 ×	300 ×
High Power	Yellow	40 ×	400 ×
Oil Immersion	Red	100 ×	1000 ×

Competency Skill

Use of a Microscope

Objective:

To properly understand the mechanics of the microscope and use it in practice.

Preparation:

microscope, microscope slide, coverslip, lens paper, immersion oil

Procedure:

1. Remove the plastic cover from the microscope. The arm of the microscope should face the user.
2. Plug in the microscope and turn on the power source including the light.
3. Place a slide on the stage of the microscope, securing it in place (see Figure 48-3).
4. Place the projection lens on a low power.

Delmar/Cengage Learning

FIGURE 48-3 Be sure to secure the slide in place when placing it on the stage of the microscope.

5. Use the adjustment knob to lower the objective power while looking through the eyepiece (see Figure 48-4).
6. Look at the stage through the eyepiece while focusing the slide. Adjust the focus accordingly. Raise the objective arm away from the slide when complete.
7. Apply a drop of immersion oil into the center of the slide specimen if using the oil immersion objective lens.
8. Lower the immersion objective slowly into the immersion oil.
9. Slowly adjust the focus knob while looking in the eyepiece.
10. When complete, raise the objective arm away from the slide.
11. Remove the slide from the microscope stage.
12. Use a kim wipe or lens paper piece to gently cleanse the objective.

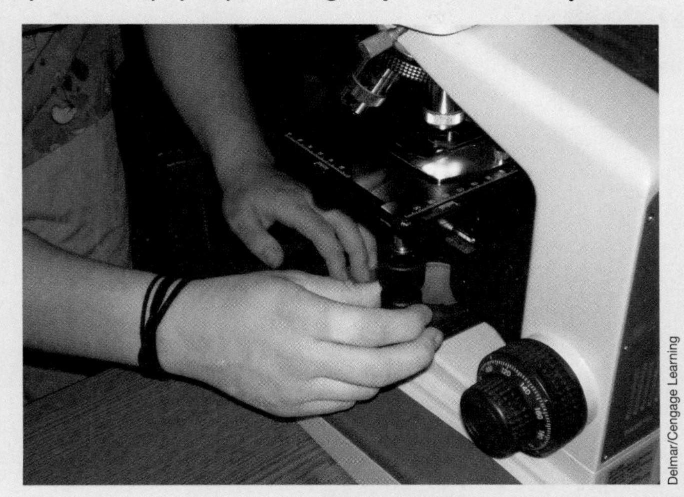

Delmar/Cengage Learning

FIGURE 48-4 Lower the objective power while looking in the eyepiece.

Centrifuge

A **centrifuge** is another common veterinary tool used to spin lab samples at a high rate of speed and force (see Figure 48-5). This device is used to separate or concentrate materials suspended in a liquid form. Gravity helps to separate the different forms. Centrifuges come in a variety of shapes and sizes and many facilities have two, one used to spin microhematocrit tubes and one standard machine used to spin larger volumes of fluid samples. All centrifuges have a variety of **rotors,** or wheels, that spin at a variety of speeds depending on the sample type (see Figure 48-6). Each time the centrifuge is used, it must be balanced before turning it on. Samples that may be used in a centrifuge include blood tubes of various sizes, **centrifuge tubes** that have **conical** or pointed ends and may hold urine or fecal samples, or **microhemocrit tubes**, which are thin, small glass tubes that hold blood (see Figure 48-7).

Delmar/Cengage Learning

FIGURE 48-5 Centrifuge.

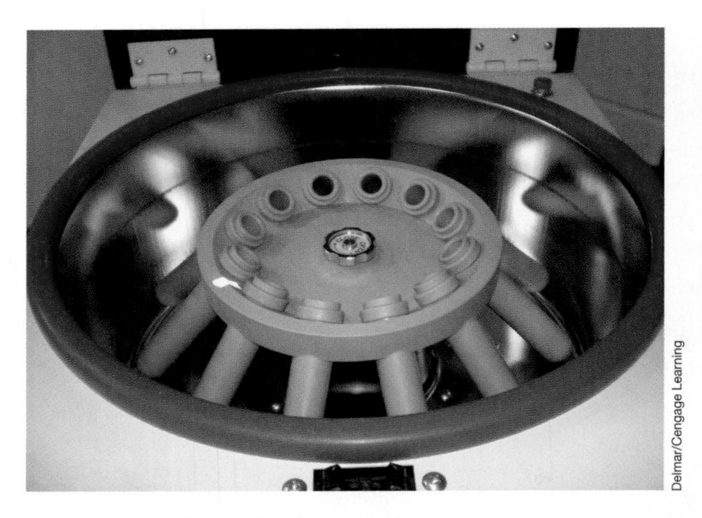

FIGURE 48-6 Rotors in the centrifuge.

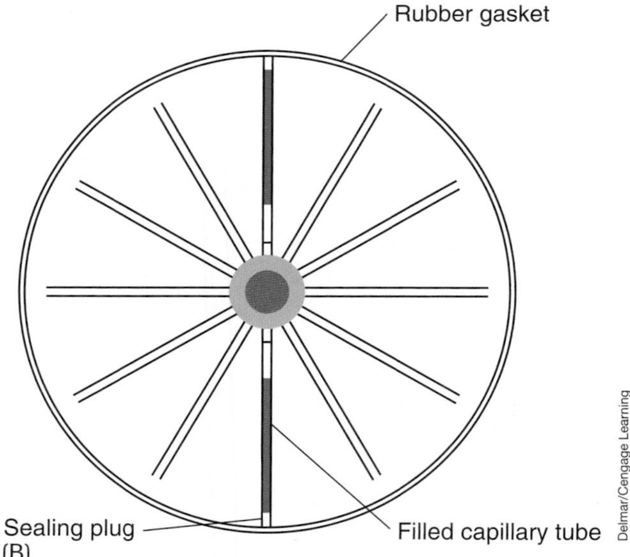

FIGURE 48-7 (A) Microhematocrit centrifuge; (B) proper placement of sealed capillary tubes in a microhematocrit centrifuge.

FIGURE 48-8 After placing a sample in the hematocrit, the rotor should be closed and secured.

When a sample is placed within a rotor, the lid should be closed and secured (see Figure 48-8). The machine should be set on the proper test sample setting and speed and the timer set for the required spinning time. The centrifuge spins at a high rate of speed; no one should stand directly over the centrifuge while it is in use and one should never try to stop or slow down the rotors.

Refractometer

A **refractometer** is a tool used to measure the weight of a liquid and determine a liquid's pH level (see Figure 48-9). The veterinary refractometer has two scales that are viewed in a similar fashion to a prism. This tool is handheld and lightweight. A liquid sample is placed on the prism plate and the cover is closed (see Figure 48-10). The refractometer is then held to a light source and viewed through the eye piece to determine the specific gravity and pH level of the liquid (see Figure 48-11). Each scale is labeled, one on the right

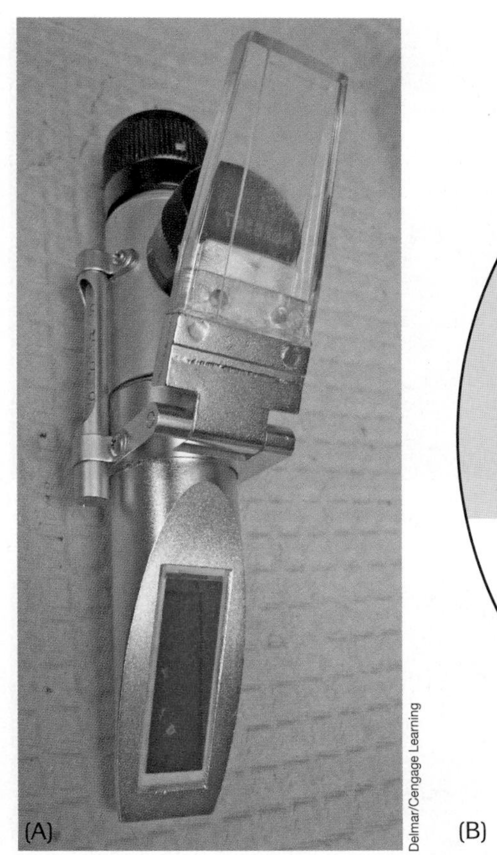

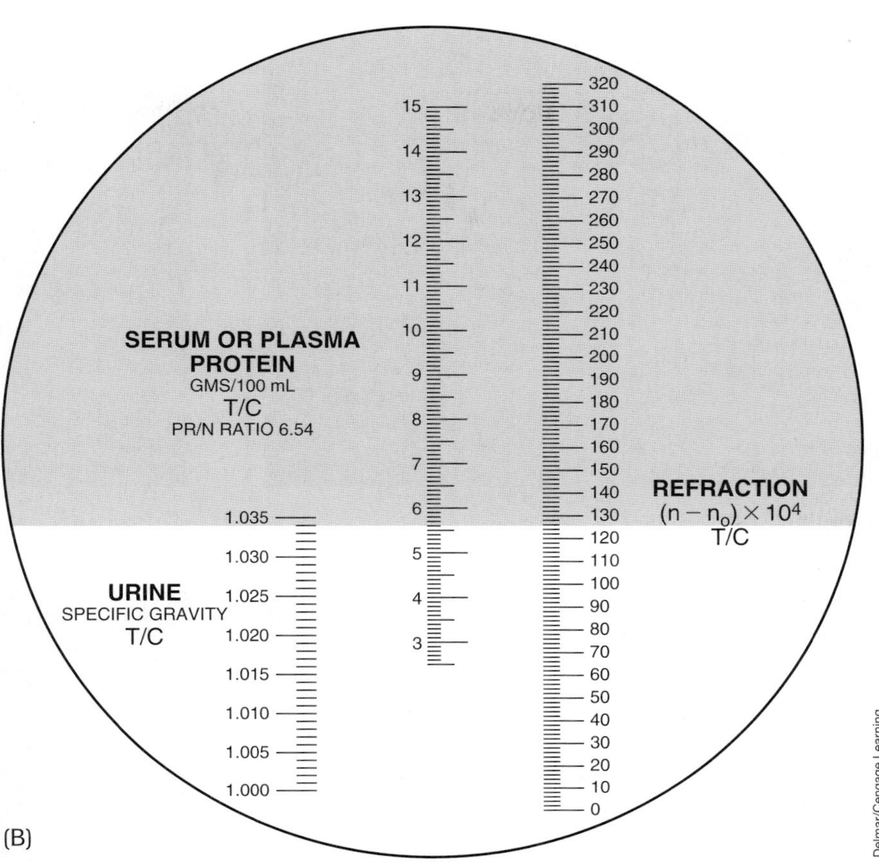

FIGURE 48-9 (A) Refractometer; (B) this refractometer scale shows a urine specific gravity of 1.034 (lower left scale) and serum or plasma protein of 5.6 (middle scale).

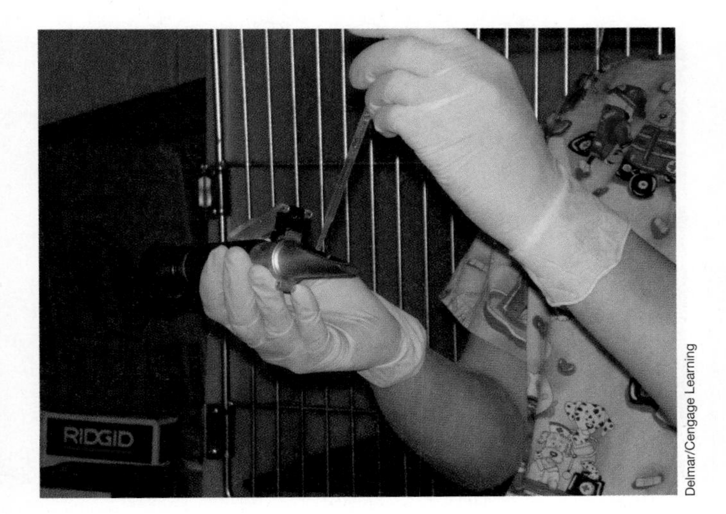

FIGURE 48-10 Adding liquid sample to the refractometer.

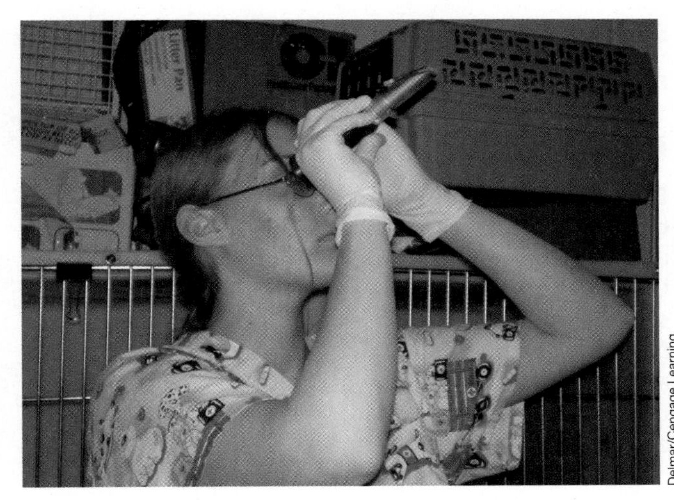

FIGURE 48-11 Viewing the sample in the refractometer.

side of the prism and one located on the left side. The tool should be cleaned after each use by using lens paper and cleaner or alcohol (see Figure 48-12). The manufacturer's instructions should be followed for cleaning and use. It is important to calibrate the refractometer on a regular basis using distilled water, which weighs zero. If the tool is not properly calibrated it will not provide accurate results.

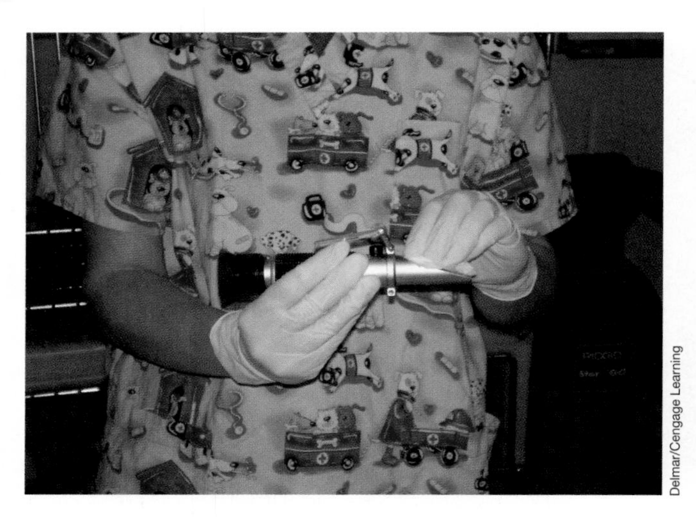

FIGURE 48-12 Cleaning the refractometer.

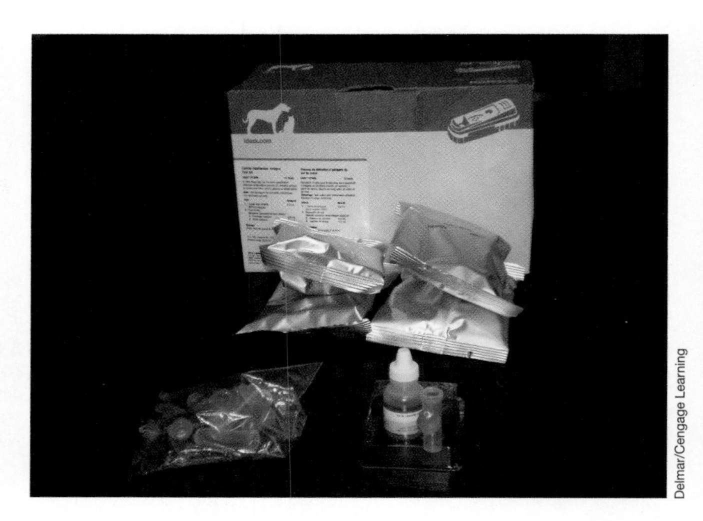

FIGURE 48-14 Serological test kit.

FIGURE 48-13 Blood rocker.

Blood Chemistry Analyzers

Blood chemistry analyzers are machines that run blood samples and measure routine blood chemistries and electrolytes, including **complete blood counts** or CBCs. Blood chemistries and CBCs may be run using samples of whole blood, serum, or plasma. The manufacturer's instructions should be followed in how to use and maintain the equipment, as there are a variety of types of analyzers. Many blood chemistry machines are easy to run, efficient, and accurate and provide results quickly for patient diagnosis. Blood samples are commonly placed on a blood rocker, which allows the blood to mix until it is analyzed (see Figure 48-13).

Serological Test Kits

Commercial serologic test kits are commonly used in a veterinary facility to provide quick and accurate results of common viruses and diseases (see Figure 48-14).

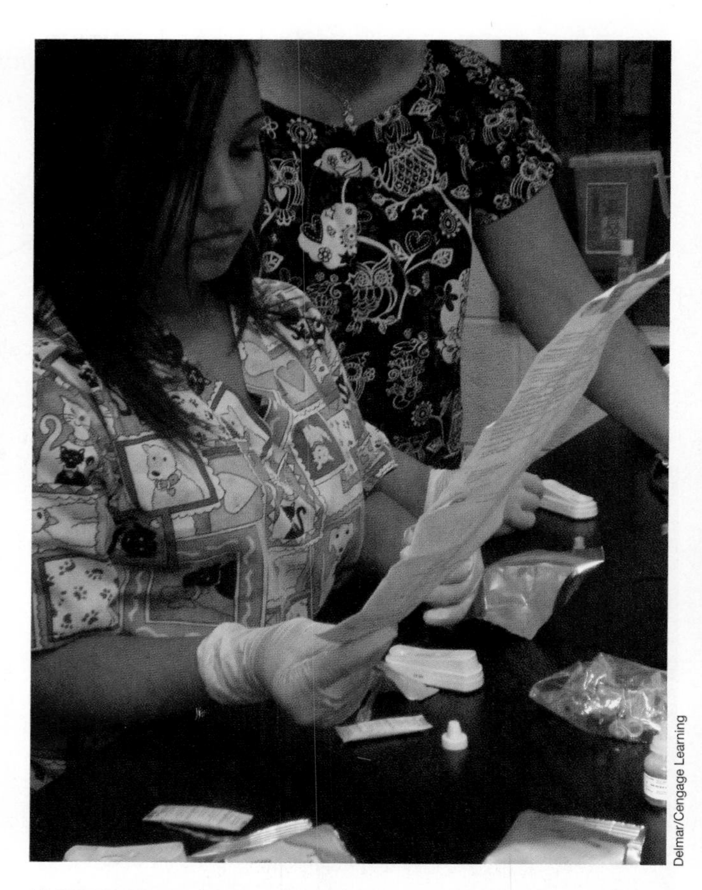

FIGURE 48-15 Instructions that come with each kit should be followed.

There are an increasing number of test kits available by several manufacturers to detect a variety of diseases. These test kits usually have a set number of individual tests along with the items necessary to perform them. Step-by-step instructions accompany each test kit and should be followed to conduct the test and interpret the results (see Figure 48-15). The instructions should be kept with the test kit, which is usually kept refrigerated and warmed to room temperature prior to use. A variety of **reagents** or chemicals are used to run each individual test in the entire kit and should be kept with the test kit and never thrown out until the entire kit has been used (see Figure 48-16).

Stains, test strips, and other chemical reagents are also commonly used in a veterinary lab. Veterinary assistants should be familiar with the types of items used in the facility and how to properly interpret the results. Over time, some stains and reagents tend to crystallize or evaporate and should be monitored for age and expiration date. Some items are kept in **Coplin jars**, which hold the chemical for repeated use (see Figure 48-17). These items may need to be changed on a regular basis as the materials become aged and collect debris. Never add a new chemical to an old chemical. The items should be entirely discarded and renewed. Keep all items tightly capped and stored appropriately.

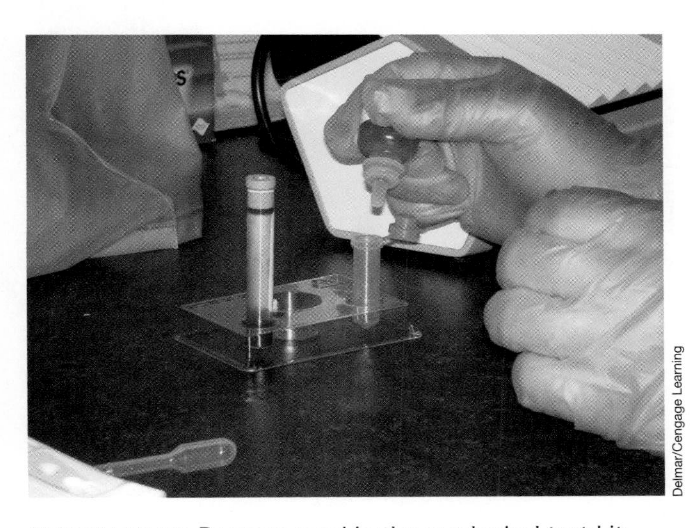

FIGURE 48-16 Reagent used in the serological test kit.

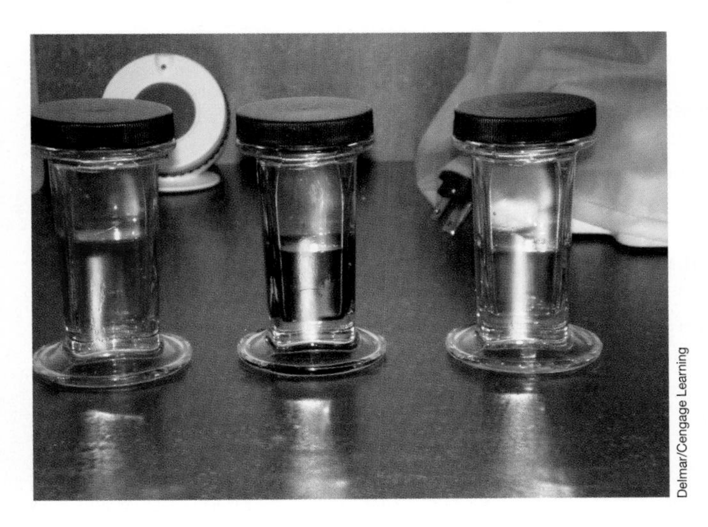

FIGURE 48-17 Coplin jars.

Recording Laboratory Results

All laboratory results should be recorded or placed within each patient's medical record. Some laboratory tests are kept in logs in the lab area for easy and quick reference. As soon as test results are completed, they should be recorded in the file. The veterinarian should be notified to interpret the results for patient diagnosis. All results are given in units of numeric value and should include normal reference ranges in parenthesis to the right of the results. Many reference values vary from laboratory and textbook. Some test results are provided in positive or negative form. Laboratory log books should be maintained and kept neat and clean and ideally in a three-ring binder with clear plastic sleeves. Medical charts should have an area to place all laboratory reports according to date, with newest being on top.

The Fecal Sample

Fecal samples are used to diagnose internal parasites and the presence of blood within the stool sample. A fecal sample is assessed using a small amount of an animal's stool or bowel movement materials. Changes in the fecal sample aid in determining animal health issues and signs of disease.

The fecal sample begins with the collection, which may be done at the facility or brought in with the animal's owner. The fecal sample should be placed within an airtight container or bag, which is labeled with the client name, patient name, sample test, and date. Samples that are collected fresh and not immediately tested should be kept refrigerated for freshness. At least 1–2 grams of feces is needed to complete a fecal analysis.

The first part of the fecal analysis sample begins with the **gross examination** or visible observation of the feces. The observations are recorded in the patient's medical record and possibly in a laboratory fecal log. The gross fecal observation should include the following:

- Color
- Consistency
- Odor
- Presence of blood ranging from dark black to bright red
- Presence of observable parasites
- Presence of mucous
- Presence of foreign materials or debris

The next step is performing a **fecal smear** preparation by placing a small amount of the fecal sample onto a microscope slide. This is usually done using an applicator stick to transfer a small amount of the sample directly onto a slide by rolling the material lengthwise along the slide (see Figure 48-18). Samples that are dry may require 1–2 drops of saline solution to mix with the feces before applying to the slide. A thin layer of feces should be placed across the slide surface. The veterinarian or veterinary technician should then read the fecal smear using a microscope.

The **fecal floatation** technique is done to sample the fecal material to determine if any parasite eggs are within the sample. This is done based on the concept that fecal ova (eggs) are lighter in weight than the solution they are submerged within. The result causes any ova to float to the top of the solution and become attached to a coverslip or slide applied over the sample. The slide or coverslip is then viewed under the microscope for evaluation by a veterinarian or veterinary technician.

Fecal samples require fecal floatation solution, a bottle of Lugol's solution, and new methylene blue stain. Solutions commonly used to prepare a fecal sample are listed in Table 48-2. Slides and coverslips should be available, along with applicator sticks, floatation system kits, centrifuge tubes, and a laboratory space designated for fecal samples (see Figure 48-19). A small tray may be used to decrease contamination and spread of disease within the lab. Many parasites and diseases within a fecal sample are zoonotic to humans, and exam gloves should be worn at all times. The area should be cleaned and disinfected after every fecal sample analysis is conducted.

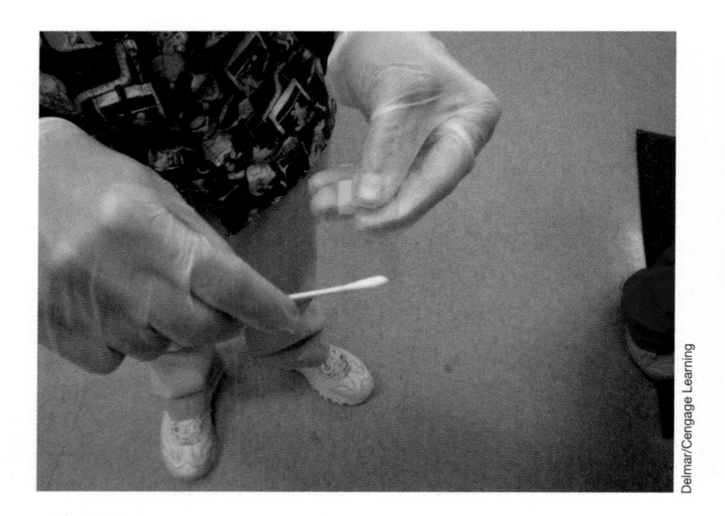

Delmar/Cengage Learning

FIGURE 48-18 Preparation of a fecal sample slide.

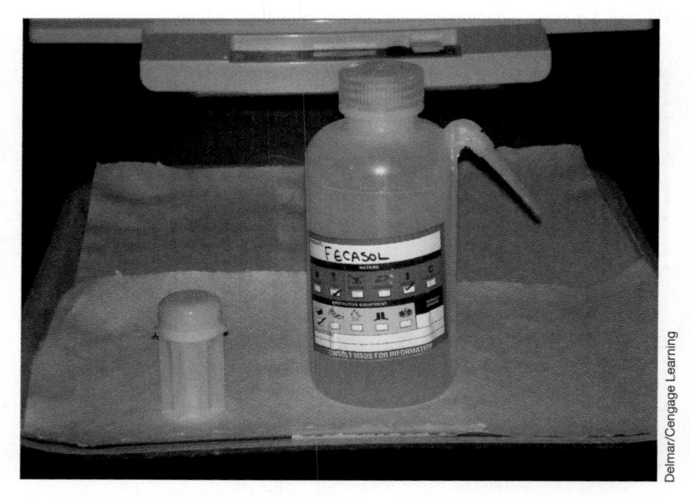

Delmar/Cengage Learning

FIGURE 48-19 Preparation needed for fecal floatation.

TABLE 48-2

Fecal Sample Solutions

Solution	Description
Lugol's Solution	Iodine tincture solution applied to fecal smear to determine parasites, commonly oocysts and single-celled organisms
New Methylene Blue	Blue stain solution applied to fecal smear to note bacteria and single-celled organisms
Saline	Saltwater solution used in fecal smears to determine ova
Sodium nitrate	Most common solution; used in commercial fecal floatation kits; relatively expensive
Zinc sulfate	Used to diagnose Giardia cysts; used with centrifuging fecal samples
Sucrose	Granulated sugar solution used when centrifuging fecal samples; more expensive and sticky
Magnesium sulfate	Epsom salts; inexpensive, but crystallizes quickly and easily causing egg ova distortion
Sodium chloride	Table salt solution; causes cloudy appearance to sample; inexpensive

Fecal Analysis Procedure

Objective:

To properly view and prepare a fecal sample.

Preparation:

exam gloves, fecal sample, tray, laboratory work area, fecal kit/container/centrifuge tube, fecal floatation solution, microscope, microscope slide, coverslip, applicator sticks, wax pencil, Lugol's solution, new methylene blue, saline, centrifuge, timer

Procedure:

1. Apply exam gloves.
2. Conduct gross fecal examination. Record results.
3. Prepare fresh fecal smear.
4. Label slide with wax pencil or pencil (patient name, client name, date).
5. Place small amount of feces in center of slide using applicator stick.
6. Add one drop of saline solution.
7. Mix well and spread thin layer lengthwise over slide.
8. Place coverslip over smear.
9. Place at microscope for evaluation.
10. Prepare another fecal smear as above.
11. Place 1 drop Lugol's solution or New Methylene blue to slide.
12. Place coverslip over slide.
13. Place at microscope for evaluation.
14. Set up fecal floatation (commercial fecal solution kits such as Ovassay or Fecalyzer).
15. Label the container with patient name, client name, and date.
16. Place small amount of fecal sample in container.
17. Fill the container halfway with the fecal floatation solution.
18. Mix the sample and solution well.
19. Insert any filter and lock container in place.
20. Fill the vial to the top and allow solution to dome upward without overfilling (see Figure 48-20).
21. Apply a coverslip to the top of the solution to form a vacuum (see Figure 48-21).

(*Continues*)

Delmar/Cengage Learning

FIGURE 48-20 Preparation of the fecal floatation.

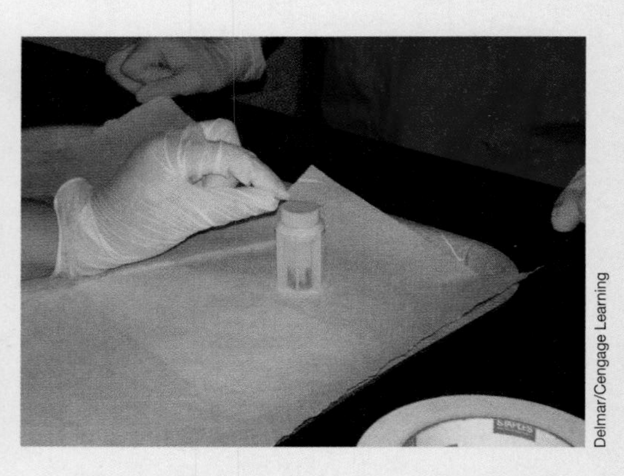

Delmar/Cengage Learning

FIGURE 48-21 Placing a slide cover slip on the top of the fecal floatation preparation.

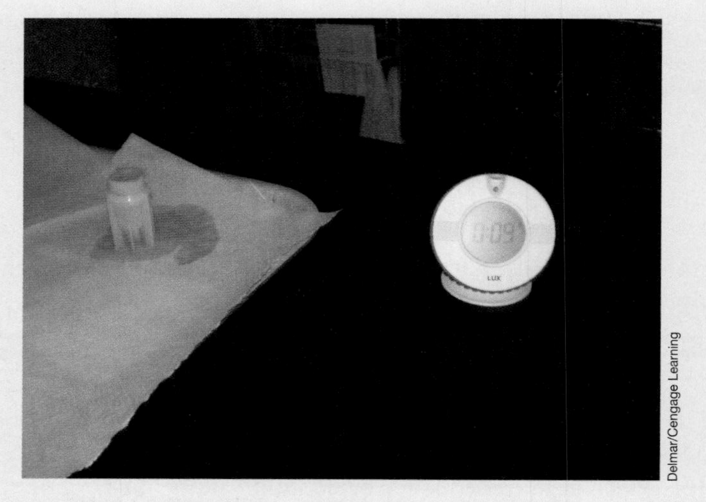

Delmar/Cengage Learning

FIGURE 48-22 Time the sample.

22. Place timer on 15 minutes to allow ova to float (see Figure 48-22).
23. When timer sounds, remove coverslip and apply to microscope slide.
24. Place slide at microscope for evaluation.
25. Fecal floatation with centrifuge (if not using commercial kit).
26. Mix fecal sample in centrifuge tube using 1 tsp of feces and small amount of water to soften feces. Mix using applicator stick.
27. Add fecal floatation solution to the top of tube allowing upward dome to form.
28. Place a coverslip on the tube to form a vacuum.
29. Place in centrifuge and spin 3–5 minutes on slow speed.
30. Remove coverslip and apply to a microscope slide.
31. Place slide at microscope for evaluation.
32. Clean work area well.
33. Dispose of feces in biohazard waste container.

Blood Chemistry Procedures

Blood collection procedures are often performed to collect blood samples for various tests both in-house and for reference labs (see Figure 48-23). **Vacutainer tubes** are necessary to place blood samples in for future sampling. These tubes are commercial tubes that have a vacuum created to place a needle and syringe into a rubber plunger to apply a blood sample (see Figure 48-24). The blood sample is then housed within the tube and prepared for analysis. A variety of tubes are available for specific testing requirements. Most frequently used tubes are red top tubes, tiger strip top tubes, and lavender top tubes. Blood tubes should be handled gently and carefully to prevent damage to the sample. After a sample has been collected from a patient, it is best to apply the blood to the vacutainer tube by removing the top and applying the blood sample directly into the tube. This prevents **hemolysis**, which is the rupture of red blood cells causing a pink coloration to develop in the plasma or serum (see Figure 48-25). This discoloration often interferes with test results.

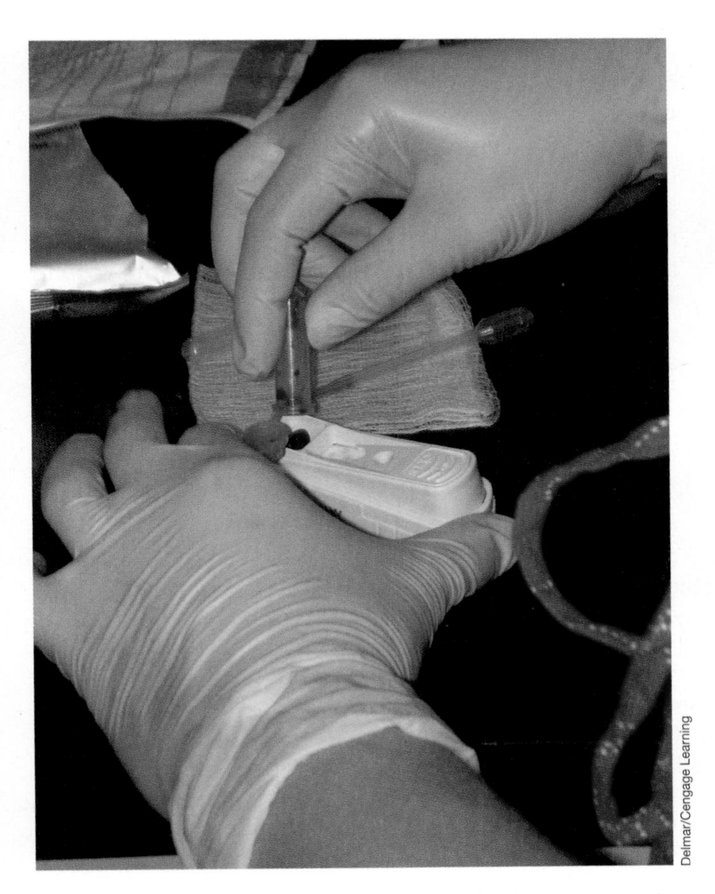
Delmar/Cengage Learning

FIGURE 48-23 Blood samples are a common diagnostic procedure.

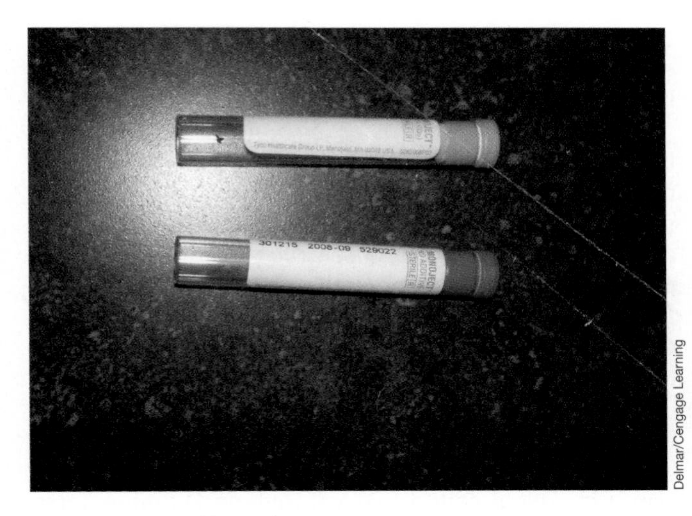
Delmar/Cengage Learning

FIGURE 48-24 Vacutainers.

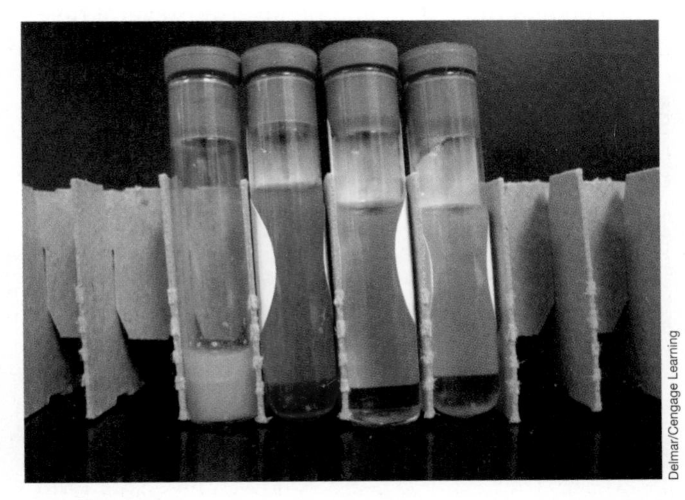
Delmar/Cengage Learning

FIGURE 48-25 The second tube from the left shows the pinkish, blood-tinged coloration that occurs when red blood cells are damaged on improper handling of the specimen.

Each blood test type requires a specific amount of blood, serum, or plasma to run the sample. The testing requirements should be noted before blood collection. A general rule is to collect at least 2 and ½ times the amount of whole blood needed for serum or plasma test samples. Blood samples that are not able to be evaluated immediately should be refrigerated to preserve freshness and sample accuracy. Blood samples that require centrifuging should be allowed to clot at least 30 minutes prior to spinning but no more than 60 minutes. Each vacutainer tube should be labeled with the patient name, client name, testing requirements, and date. A laboratory test log should be kept to record appropriate tests. Any outside reference tests should be prepared and paperwork completed and the lab notified for pick-up.

Most testing samples require **blood serum**, which is the liquid portion of the whole blood sample. Serum blood samples are usually placed within red top or tiger stripe top blood tubes. The samples are slowed to clot and are then spun within a centrifuge for 10 minutes to produce serum. **Blood plasma** samples may be needed and require the blood sample to be frozen and then centrifuged to obtain plasma. Serum is most commonly used in blood chemistry profiles.

Whole blood samples are common for CBC testing and are placed within lavender top tubes (see Figure 48-26). The lavender top tubes contain a chemical called EDTA (Ethylene Diamine Tetra Acetate) that prevents the blood from forming a clot, called an **anticoagulant**. When a whole blood sample is placed in a tube, it should be inverted several times to allow the EDTA to react with the blood. Table 48-3 lists the various types of blood collection tubes and their uses.

FIGURE 48-26 A whole blood sample placed in a blood rocker before testing.

The Complete Blood Count

The complete blood count or CBC is a group of tests that evaluate the different types of white blood cells (WBCs). The CBC is composed of the following:

- Examination of a blood smear
- Differential white blood cell count and morphology
- Platelet estimation and count
- Red blood cell morphology
- Examination of blood for blood-borne parasites
- Packed cell volume (PCV)
- Plasma or total protein (TP)
- Total WBC count
- Mean corpuscular volume (MCV)
- Hemoglobin concentration

Many of these tests are conducted in-house and are simple to run and performed by the veterinary assistant. Some facilities may need to send them to outside reference labs. In-house blood analyzer machines efficiently provide results to these tests and several can be run by hand.

The CBC requires unclotted whole blood collected in a lavender top tube. The lavender top tube should be filled as much as possible. An in-house sample is run on a CBC analysis machine, for which the assistant should be proficient in understanding the results and sampling methods. The results should be printed out and placed in the patient's medical chart. The results should be provided to the veterinarian for further diagnosis and treatment.

TABLE 48-3

Blood Collection Tubes

Tube	Description
Red Top Tube	Sterile, no anticoagulant or additives. Contains gel separator. Collection of serum for chemical or serological and bacteriologic studies. May be used for any procedure requiring serum.
Tiger Stripe Top Tube	Contains no silicone, gel separators, anticoagulants, or additives of any kind. Can be used for collection of serum.
Lavender Top Tube	Sterile, contains EDTA as the anticoagulant. Primarily for collection of hematology studies, blood bank procedures, and certain chemistries with whole blood.
Green Top Tube	Sterile, contains lithium heparin as the anticoagulant. For collection of other miscellaneous studies. Electrolytes, glucose, blood urea nitrogen (BUN) can be performed more quickly than from a red top.
Light Blue Top Tube	Sterile, contains sodium citrate solution as the anticoagulant. Tube calibrated to hold only 4.5 ml of blood. Primarily for collection of coagulation studies.
Gray Top Tube	Sterile, contains potassium oxalate and sodium fluoride as the anticoagulant. Used for the collection of glucose and lactate samples. Not suitable for enzymes or electrolytes.

Blood Smear

The **blood smear**, also known as a blood film, is used to look at the morphology of blood cells. The **morphology** includes the cell structure, shape, color, and appearance in numbers. The blood smear is made by the assistant and is made by using one slide, acting as a spreader, to pull a small drop of blood across a clean slide making it into a thin film (see Figure 48-27). This is done by holding the spreader slide at a 45-degree angle. Two methods may be used to make a blood film: pushing or pulling the spreader slide. Each method requires consistent pressure applied during the entire spreading process. A suitable film is thin and tapers to a **feathered edge** before the far end of the slide (see Figure 48-28). The spreader slide ensures 2 straight edges parallel to the long edges of the slide. Immediately after the film is spread, it is dried quickly by fanning the slide in the air.

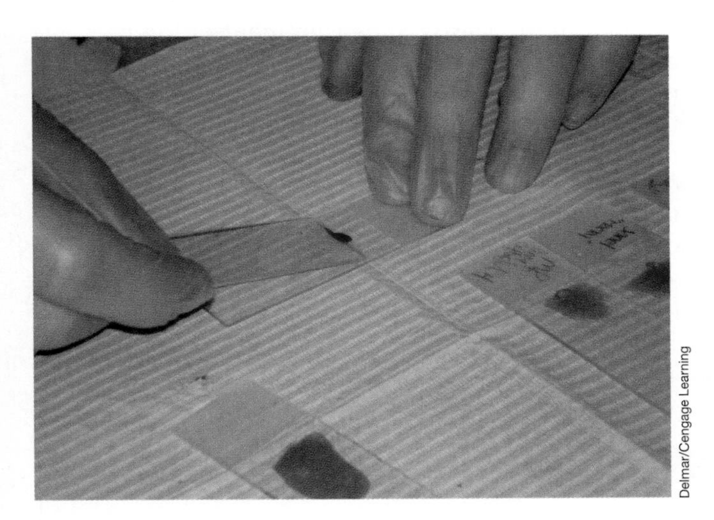

FIGURE 48-27 Creating a blood smear.

FIGURE 48-28 The edge of the blood smear should taper to a feathered edge.

The feathered edge is the area where the veterinarian or veterinary technician reads and interprets a blood slide. It must be layered and forms a slight circular motion at the end of the slide. Some recommendations to prevent errors in blood slides include the following:

- Use slides that are clean and free of breaks and chipped areas; no residue should be on the edges of any slide.
- Do not use large amounts of blood.
- Use a very small drop of blood at the center of one end of the slide.
- Place the spreader slide centrally into the drop of blood.
- Always keep the spreader slide at a 45-degree angle.
- Keep a constant and even pressure on the spreader slide.
- Do not push or pull the spreader slide slowly; this makes a poor feathered edge.
- Make a quick motion when using the spreader slide.

When the slide is completely dry, it should be stained with Wright's stain, which is a three-step staining method using commercial mixed solutions that allow blood cells to be easily viewed under a microscope. The solutions may also be in a commercial kit known as a Diff Quick set. These include three solutions labeled A, B, and C, which fix the stain with a red and blue coloration (see Figure 48-29).

When preparing blood smear slides, it is best to make two slides: one is stained and viewed and the other remains unstained in case of further testing. Both slides should be labeled on the frosted end using a wax pencil or regular pencil with the patient name, client name, and date (see Figure 48-30). Immersion oil should be

FIGURE 48-29 Stains used for blood smears.

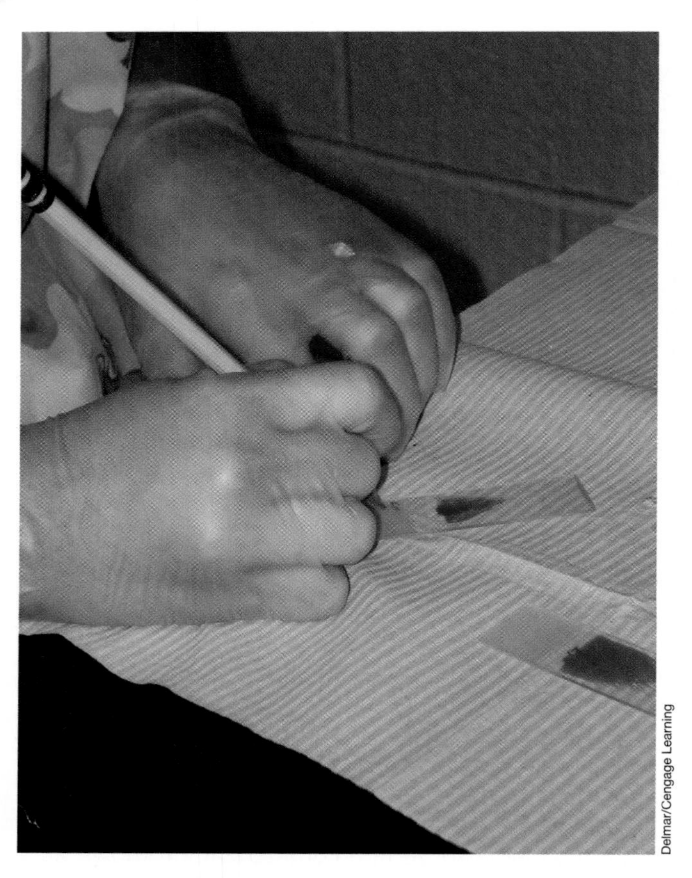

FIGURE 48-30 Properly label all blood smear slides.

placed next to the microscope and the slide placed on a paper towel near the microscope for evaluation.

The blood smear, when stained, will allow the red blood cells to appear a pink, red, or salmon color. The white blood cells will stain the nuclei with a dark blue to purple coloration. The platelets stain dark blue to violet in color. The cytoplasm and granules will vary in color from pink to purple. When evaluating the stained slide with the naked eye, it should have an overall even coloring of purple with pink hues. The stains should be maintained in separate Coplin jars marked with the appropriate order of use, such as A, B, and C. The jars should be filled at least 3/4 full to allow slides to be covered by the stain. Distilled water should be used to rinse slides during staining.

Packed Cell Volume

The **packed cell volume** or PCV is also referred to as a **hematocrit**. This is a measurement of the percentage of red blood cells in whole or unclotted blood. The red blood cells, called **erythrocytes**, carry oxygen to the bloodstream. The PCV test is simple, rapid, and requires a very small amount of blood. The equipment includes a centrifuge with a microhematocrit rotor, hematocrit tubes, and clay. The microhematocrit centrifuge has a lid that screws onto the rotor to prevent tubes from emerging. The rotor must be properly balanced. Plain tubes may be used with whole blood

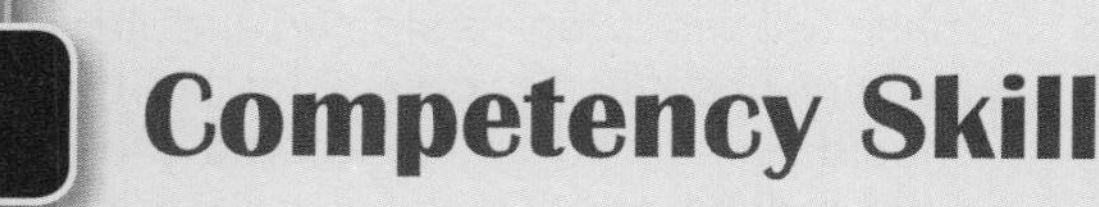

Blood Smear Procedure

Objective:

To properly prepare a blood smear for microscopic evaluation.

Preparation:

microscope, microscope slides, wax pencil or pencil, blood sample, syringe, exam gloves

Procedure:

1. Apply exam gloves.
2. Place a small drop of blood on the end of a slide using a clean syringe or from the lip of the blood top tube.
3. Apply a second slide as the spreader slide directly in front of the blood drop.

4. Place the spreader slide at a 45-degree angle.
5. Gently pull or push the spreader slide lengthwise along the slide.
6. Keep constant and even pressure on the spreader slide.
7. Use a quick motion, making an even feathered edge.
8. Repeat this method until an even feathered edge is formed (see Figure 48-31).
9. After the slide is completed, air dry by fanning or set aside to dry.
10. Make a second blood slide using the same steps.
11. Set slides aside to stain.

FIGURE 48-31 A properly and improperly prepared blood smear.

Competency Skill

Blood Smear Stain Procedure

Objective:

To properly prepare a blood smear with the appropriate stain.

Preparation:

blood smear slides (dried), Wright's stain or Diff Quick set, forceps, distilled water, immersion oil, tray, paper towels, exam gloves

(*Continues*)

Procedure:

1. Apply gloves.
2. Open each Coplin jar, marked A, B, and C.
3. Using forceps, the slide should be immersed completely inside jar A (the eosinate or red color stain) that applies the stain fixative. Time immersion according to kit directions.
4. Lift slide from jar and tilt and drain onto a paper towel, dabbing the end of the slide several times to remove excess stain.
5. Immerse slide in distilled water, lifting up and down to rinse.
6. Using forceps, the slide should be immersed inside jar B (the polychrome or blue color stain) (see Figure 48-32).Time immersion according to kit directions.
7. Lift slide from jar and tilt and drain onto a paper towel, dabbing the end of the slide several times to remove excess stain.
8. Immerse slide in distilled water, lifting up and down to rinse.
9. Using forceps, the slide should be immersed inside jar C (clear stain). Time immersion according to kit directions.
10. Lift slide from jar and tilt and drain onto a paper towel, dabbing the end of the slide several times to remove excess stain (see Figure 48-33).
11. Wipe off back of the slide using a paper towel.
12. Allow to air dry completely.
13. Place on a paper towel near microscope.
14. Place a bottle of immersion oil near the slide.

FIGURE 48-32 Staining technique for a blood smear.

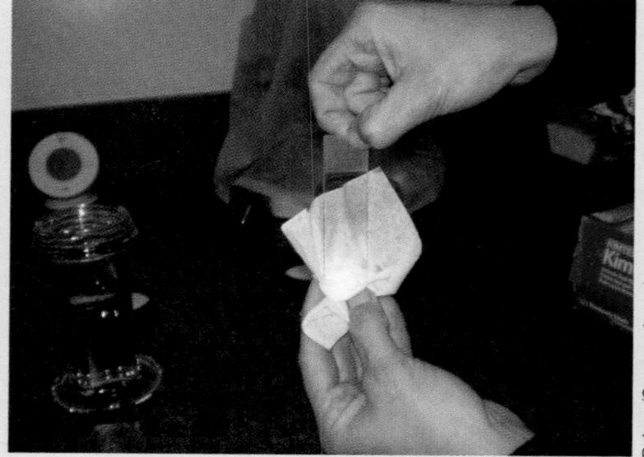

FIGURE 48-33 Allow slide to drain onto a paper towel.

that contains an anticoagulant, such as EDTA, and untreated whole blood directly from a syringe should use a heparinized hematocrit tube. Hematocrit tubes are sometimes called capillary tubes.

Blood is drawn into the hematocrit tube by **capillary action**, meaning the blood drains into the tube through gravity. One end of the tube is placed into the blood sample and is allowed to rise through the tube until it is 3/4 to completely full. Some tubes have a black fill line indicating how much blood should be placed, although the mark is not necessary for the test to be accurate. Place a finger over the end of the tube and wipe clean using lens paper. Quickly and firmly apply the end of the tube into a clay sealant. This seals the end of the tube preventing blood from leaking. The tube is then placed into the centrifuge rotor and balanced with another capillary tube filled with water. Multiple PCV tubes can be run in one rotor. The tubes should be placed opposite each other to balance the rotor and the clay end of the tubes should be facing outward, which is important because if not properly placed, the sample will be lost as it spins. Then apply and tighten the metal lid and set the centrifuge to the proper setting and time.

When the centrifuge has completed spinning, the rotor is opened and the tube is ready to read. The tube will have a dark red layer composed of red blood cells near the clay-sealed end. Just above this line is a yellow to clear area of plasma. Between both layers is a tiny area of visible white, known as the **buffy coat**, composed of white blood cells and platelets (see Figure 48-34).

To determine the PCV level or the percent of red blood cells in the patient's sample, a microhematocrit scale is used (see Figure 48-35). The scale can be a chart-like paper or it may be a plastic form that fits over the centrifuge. Using either type of scale, align the sealant at zero and the top of the plasma at 100. The line running through the interface of red blood cells and the buffy coat layer is the percentage of red blood cells of the patient. This test helps the veterinarian determine if anemia is present in an animal. The percentage should be recorded in the medical chart.

Plasma Protein

Plasma protein, also called **total protein** (TP), measures the ratio of protein within the blood and helps veterinarians determine the hydration status and inflammation occurrence of patients. The TP is determined using a refractometer or total solids meter. The TP sample is evaluated after the PCV test has been

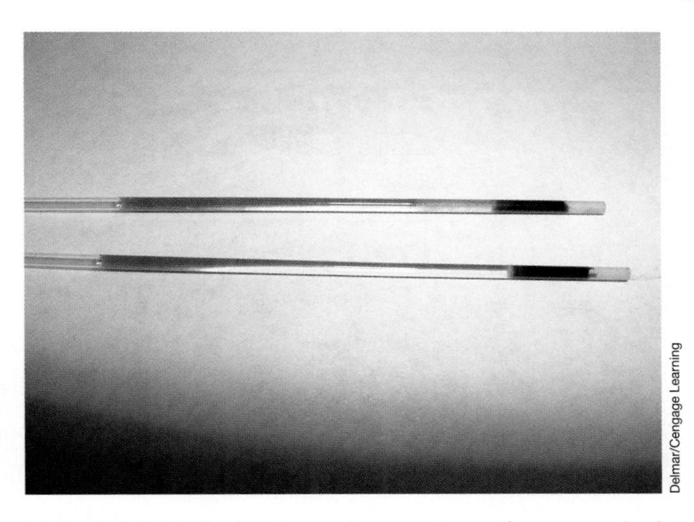
Delmar/Cengage Learning

FIGURE 48-34 A blood sample spun to perform a packed cell volume. Note the yellow color of the serum showing icterus or jaundice.

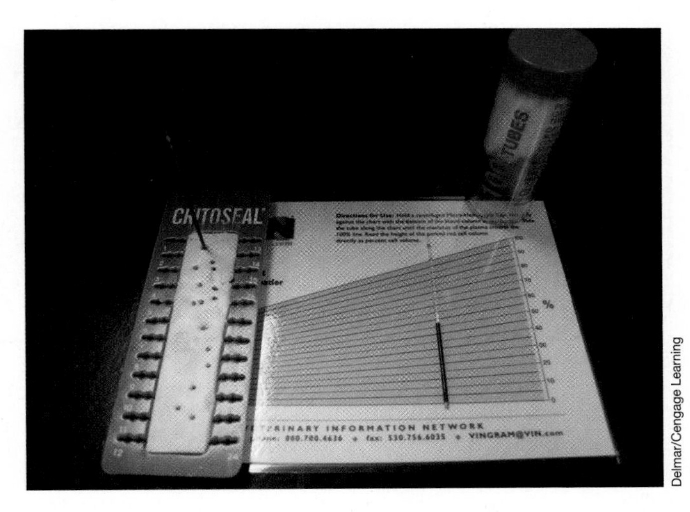
TUBES
%
Delmar/Cengage Learning

FIGURE 48-35 Hematocrit scale.

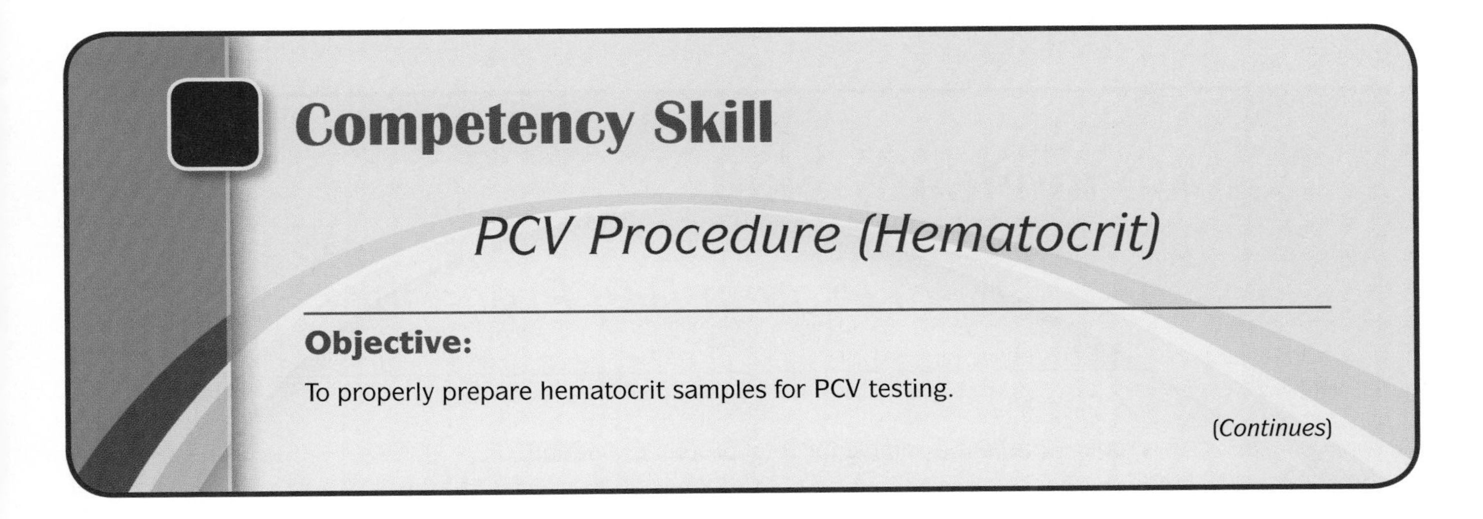
Competency Skill
PCV Procedure (Hematocrit)
Objective:
To properly prepare hematocrit samples for PCV testing.
(Continues)

Preparation:

capillary tube, hematocrit tube or microhematocrit tube, clay sealant, lens paper, exam gloves, centrifuge, microhematocrit scale, blood sample

Procedure:

1. Apply gloves.
2. Fill one capillary tube with a blood sample.
3. Apply a finger to the end of the tube that was in the blood tube.
4. Wipe the tube clean with lens paper.
5. Quickly apply the end of the tube firmly into the clay sealant.
6. Fill one capillary tube with water and apply end to clay sealant.
7. Apply the microhematocrit rotor to the centrifuge.
8. Place both tubes on opposite sides within the centrifuge rotor. Both tubes should have the clay sealant facing outward.
9. Apply lid to the rotor and secure tightly.
10. Place setting on hematocrit and time according to manufacturer's instructions.
11. After the centrifuge has stopped spinning, remove the tube with the blood sample.
12. Place the tube on the microhematocrit scale. Align the top of the sealant at zero and the top of the plasma at 100.
13. Read the line crossing the red blood cells and the buffy coat.
14. Record the percentage of the PCV in the medical chart.

determined. The microhematocrit tube holding the spun-down blood is used to evaluate the TP by using the plasma located above the buffy coat layer in the tube. The clear to slightly yellow coloration at the top area of the tube holds the plasma. The tube is broken just above the buffy coat to allow a drop of plasma to be applied to the refractometer prism. The cover plate is lowered and the refractometer is held toward a light source to read the scale for serum or plasma protein, often labeled TP. The scale is read where the dark and light colors intersect on the scale. Record the results in the medical chart in g/dL (deciliter). This is a simple and quick test that the veterinary assistant should be able to proficiently perform.

Competency Skill

Total Protein (TP) Procedure

Objective:

To properly prepare a sample for total protein evaluation.

Preparation:

capillary or microhematocrit tube (spun), refractometer, lens paper, exam gloves

Procedure:

1. Apply exam gloves.
2. Open the refractometer lid.
3. Break the capillary tube containing the patient's blood just above the buffy coat in the blood plasma (clear to light yellow fluid).
4. Apply one drop of fluid to the refractometer.
5. Close the lid to the refractometer.
6. Direct the refractometer toward a light source.
7. Focus the eye piece and read where the light and dark areas interface on the TP scale.
8. Record the results in g/dL in the medical record.
9. Place tube in medical waste container.
10. Clean refractometer using alcohol and lens paper.

Blood Chemistry Samples and Electrolytes

Blood chemistry samples and electrolytes are often used to diagnose diseases and conditions of ill patients. Some of these tests may be run in-house and others may be sent to reference labs. In-house blood analyzers allow access to rapid results. Many pre-anesthetic blood panels are performed on patients prior to surgery, as well as yearly health profiles on senior or geriatric pets. Many of these blood chemistries are useful in determining developing conditions before a patient becomes clinically ill. Approximately 10% of veterinary patients have a diagnosed condition determined during routine wellness or pre-anesthetic blood work.

Most in-house veterinary labs consist of analyzers used for serum blood chemistries, electrolytes, and hematology evaluation. Blood serum is obtained from a blood sample spun down in the centrifuge to allow serum fluid to be analyzed. **Electrolytes** determine the balance of elements in the body, specifically potassium, sodium, and chloride. **Hematology** refers to the study of the blood, specifically with a CBC. Many analyzer machines may run a series of tests called **panels** or individual tests. These tests provide chemical information on specific organ functions that help evaluate a patient's health. It is important to note that blood machines are constantly upgrading in systems and software, so it is expected that current equipment may require frequent upgrades to further allow testing capabilities in-house. Testing becomes faster and more dependable, is reduced in testing errors, and often becomes easier to use. Veterinary assistants should be proficiently trained in operating, cleaning, and maintaining blood machine equipment. There are many different veterinary blood machine brands and much information to be learned about them. The assistant should be able to understand and determine the following general information on blood machines:

- How to obtain and print out patient results
- How to determine reference value ranges for the machine's results
- How to properly calibrate machines and keep records of the calibration process
- How to change empty and expired reagent chemicals used in the machine
- How to properly run the testing panels
- How to troubleshoot problems that may occur with the machine and who to contact with manufacturer and technical problems

It is important that assistants know how to operate and maintain the blood machines. The machine must be kept clean, and follow recommended directions for maintenance. The validity of results is dependent on the skill of the person performing the testing samples. Equipment must be turned on properly, allowed to warm up, calibrated and cleaned on a regular basis, and serviced by the manufacturer as necessary. The directions and testing instructions should be kept in a binder near the blood machines to refer to as necessary.

Serologic Testing Kits

Several manufacturers produce testing kits that help diagnose conditions and diseases quickly and easily within the veterinary facility (see Table 48-4). These tests range in species and type of sample needed. Some samples may require whole blood, serum, or a fecal sample. Many of these tests involve antibody or antigen presence that reacts in the test kit. Some serology test kits test for one specific disease and others test for multiple diseases. Most of the test kits offer results on a positive or negative basis. The majority of testing kits provide results within 10 to 15 minutes. The veterinary assistant is essential in performing these tests quickly and accurately.

Test kits come packaged individually in boxes with all of the necessary materials required to run each test. Instructions are provided in a step-by-step process of how to set up and read the test. Some kits require being kept at room temperature where others require refrigeration. The kits and the chemicals provided have expiration dates that should be monitored. Reagents within a test kit should always be kept with the kit and never removed, replaced, or thrown away. Most of the serologic test kits are based on the same type of principle and procedure, often called a **SNAP test**. A SNAP test is named for the snap action that begins the reaction to determine the test result (see Figure 48-36).

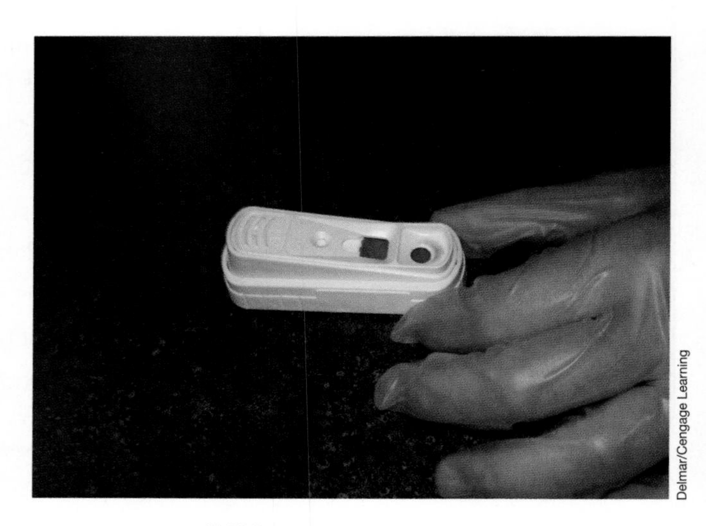

FIGURE 48-36 SNAP test.

TABLE 48-4

Types of Serologic Test Kit

TEST TYPE	MANUFACTURER	DISEASES TESTED
SNAP FeLV	Idexx	Feline Leukemia Virus
SNAP FIV/FeLV Combo	Idexx	Feline Immunodeficiency Virus; Feline Leukemia Virus
SNAP Giardia	Idexx	Feline Giardia; Canine Giardia
SNAP Heartworm	Idexx	Canine Heartworm Disease
SNAP Parvo	Idexx	Canine Parvovirus
PetCheck Heartworm	Idexx	Canine Heartworm Antigen
SNAP Foal IgG	Idexx	Equine IgG Antibodies
SNAP Antibiotic Residue	Idexx	Ruminant/Dairy Penicillin; Tetracycline, Aflatoxin, Gentmicin, Sulfamethazine
SNAP cPL	Idexx	Canine Pancreatitis
SNAP Feline Triple	Idexx	Feline Immunodeficiency Virus; Feline Leukemia Virus; Feline Heartworm Disease
SNAP 3Dx	Idexx	Heartworm disease, *Ehrlichia canis* and Lyme disease
SNAP 4Dx	Idexx	Heartworm disease, *Ehrlichia canis* and Lyme disease, *Anaplasma phagocytophilum*
D-Tec CB	Synbiotics	Canine Brucellosis
Witness Relaxin	Synbiotics	Canine and Feline Pregnancy test
FeLV Assure	Synbiotics	Feline Leukemia Virus
OvuCheck	Synbiotics	Canine Ovulation
Solo Step	Heska	Feline and Canine Heartworm Disease

The Urine Sample

The urine sample is an essential part of an animal's health, especially when urinary problems exist. The veterinary assistant's role in the urine sample involves collecting voided samples, restraint during other urine collection methods, gross examination of urine samples, using clinical strip tests, and preparing the microscopic urine examination. The complete urine exam is called the **urinalysis.**

Urine Collection

There are several methods for collecting urine, depending on the urine test and condition of the animal. Most samples that are collected are **voided** samples, meaning they are collecting as the animal is urinating. This is the only type of urine collection method the veterinary assistant may perform. The voided sample should be collected **midstream** shortly after the urination begins and just prior to the process ending (see Figure 48-37). It is important that when collecting a urine sample, the container must be sterile. Dog urine can be collected using a plastic sterile collection cup and in difficult animals or those low to the ground, a wire attachment with a loop can be made to hold the container and give the assistant an extension without interfering with the patient's normal urination process. A similar method can be used in collecting urine samples from large animals. Cats should be provided a pelleted non-absorbable litter that doesn't soak up the urine when a cat eliminates in the litter pan. Some cats may need an empty litter pan. The urine sample is then transferred into a sterile container. It is best to not use urine samples collected off of the floor or a cage area if at all possible, as these samples will be contaminated and give inaccurate results.

Other urine collection methods include palpating the bladder and expressing the urine by applying external pressure, which is common in cats. The veterinarian or veterinary technician expresses the bladder, and the assistant should hold the container and catch the midstream sample. Some animals may need to be **catheterized** for urine collection. This is applying a long, thin rubber or plastic sterile tube inserted into the bladder through the external urinary opening. The assistant restrains the patient while the catheter is placed by the veterinarian or technician. A syringe is then attached to the end of the catheter and urine is aspirated. In a female animal, the catheter is placed through the vulva to enter the urethra and in a male animal the catheter is passed into the penis and then through the urethra (see Figure 48-38). A male animal is much easier to catheterize than a female animal. Aseptic technique is required for catheter placement and urine collection. The animal may be restrained in standing recumbency, lateral recumbency, or dorsal recumbency depending on the veterinarian's preference.

A urine sample may also be collected via **cystocentesis**, which is the surgical puncture into the bladder using a needle to collect a urine sample (see Figure 48-39). This is typically used to collect a sterile sample checked for bacteria or other microorganisms, such as a culture

FIGURE 48-37 Midstream urine collection.

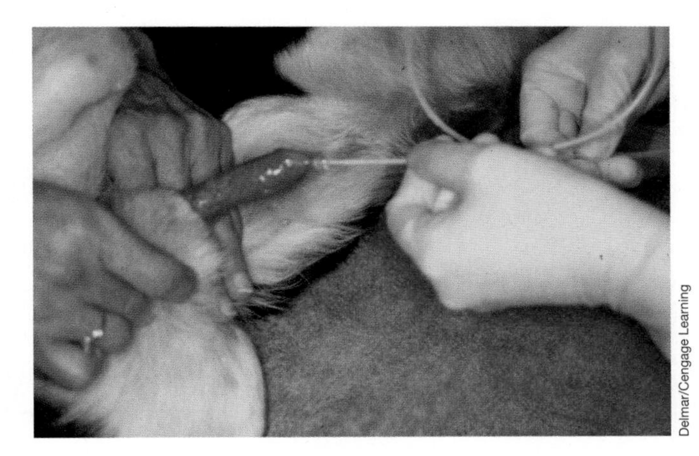

FIGURE 48-38 Catheterization of the male.

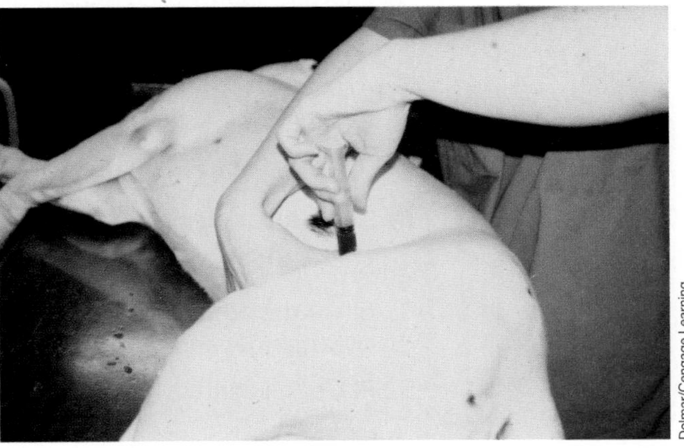

FIGURE 48-39 Cystocentesis of a dog.

and sensitivity (C/S). A **culture and sensitivity** determines if a bacterial issue is occurring and what type of bacteria is causing the health condition and what antibiotic should be used in treating the problem. A cystocentesis avoids bacterial contamination and avoids the urine from coming in contact with the urethra and causing contamination. This technique requires advanced skills on the part of the veterinarian or technician as a needle attached to a syringe is passed through the abdominal wall directly into the bladder. Urine is aspirated into the syringe. The assistant restrains the patient, usually in lateral or dorsal recumbency.

Gross Examination of Urine

The urine exam is divided into several parts: physical appearance, chemical properties, and microscopic examination. The veterinary assistant is responsible for completing the physical examination, clinical test strip evaluation, and preparing the sample for microscopic examination. The sample should be completely evaluated within 30 minutes of collection.

The physical appearance of the urine sample includes noting the color, clarity and consistency, odor, and the presence of foam. Most veterinary facilities have a urinalysis laboratory sheet that can easily be completed by filling in the parts of the urinalysis. Normal urine color in most animal species varies in shades of yellow. The color of urine can range from colorless to red or brown. Certain medications and supplements can alter the color of urine, as can certain types of diseases. Most animal species urine is clear, although some may be cloudy and be normal for the type of animal. The clarity can range from clear to cloudy to **flocculent**, which contains large amounts of particles called sediment suspended in the urine. The odor may not be diagnostic and may range in varying degrees by animal species. Certain large animals that may

Competency Skill

Collecting a Voided Urine Sample

Objective:

To properly obtain a midstream urine sample from a patient.

Preparation:

exam gloves, sterile container, label, pen, wire cup holder and extension

Procedure:

1. Apply gloves.
2. Leash or control the patient and take the animal outside.
3. Walk around the area, monitoring the patient for signs of urine elimination.
4. When the patient assumes the urination posture, prepare to slip the sterile container under the patient into midstream urine.
5. Withdraw the container prior to the patient ending urination.
6. Allow the patient to finish urinating.
7. Return the patient to the cage or enclosure.
8. Clean the outside of the sterile container and label the container with patient name, client name, and date.
9. Place the container in the refrigerator if it is not able to be tested immediately.

have more bacteria present in the urine tend to have ammonia odor noted in their urine. Another odor that may be detected in animal urine is due to the presence of ketones, which produce a sweet, fruit-like smell. When urine is shaken, it should produce a small amount of white foam. High levels of protein may create a large volume of foam, and bile pigments noted in the urine will produce a foam that is greenish-yellow in color.

Urine Specific Gravity

The urine **specific gravity** (SG), or the weight of the liquid, is determined using a refractometer. A drop of the urine sample is placed on the prism of the refractometer. The specific gravity should be noted on the urine SG scale on the refractometer. The SG number is recorded as a decimal, such as 1.025. The SG should be recorded on the urinalysis report.

Chemical Test Strips

The urine chemical test strips are used to evaluate the chemical properties of a urine sample using a **reagent strip**, also called a **chem strip** or **dip stick** (see Figure 48-40). Each test is on a long thin plastic strip separated by individual square pads containing a chemically treated paper. There may be from one to 10 tests on the strip depending on the manufacturer and test type. The bottle containing the testing strips provides a label interpreting the color changes and results. Each test should be noted on the urinalysis sheet. The test strip is aligned with the label to determine what the color change means, and note and record the result. The manufacturer's instructions should be read and followed. The urine strip should be saturated in urine, usually by applying a drop of urine to each square pad with a plastic applicator or needle and syringe. Each test will take a specific amount of time until it should be read. All values should then be recorded on the urinalysis sheet. Some clinical urine test strips are recorded in numbers and others in values such as +1, +2 pr +3 or negative (−).

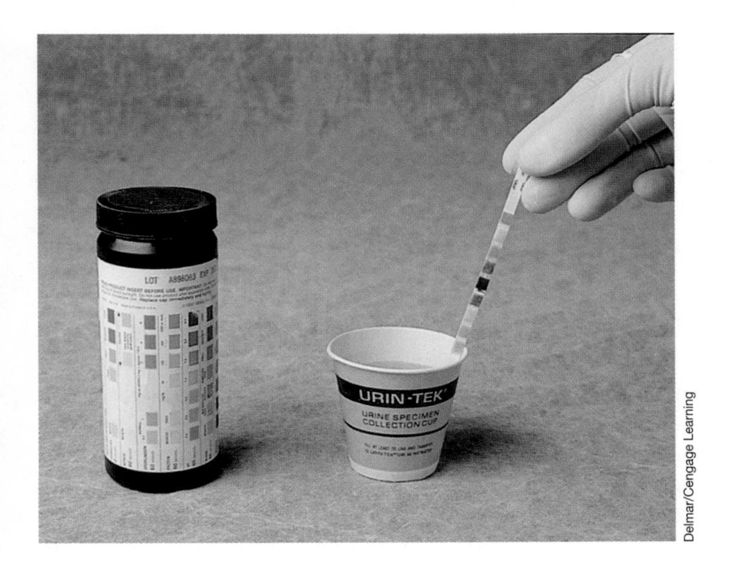

FIGURE 48-40 Urinalysis using chemical reagent strip.

The preparation of urine for microscopic examination should then begin by placing the urine in a conical or tapered end centrifuge tube, similar to a test tube. The same amount of water should be placed in a second centrifuge tube and both tubes placed on opposite sides in the centrifuge. The correct rotor size should be selected for the tubes. It is important to balance each tube placed in the rotor. The centrifuge should be placed on the urine spin cycle, which is typically between 1,500 and 2,000 rpm for 5 minutes. When the centrifuge has completed spinning, the test tube holding the urine should be removed and decanted over a sink area. The **decant** process is pouring the urine out of the tube and allowing only the sediment on the bottom to remain. A drop of urine should cling to the bottom of the tube within the sediment. A drop of urine sediment stain, such as new methylene blue or a commercial urine stain, should be added to the sample, and a gentle tap on the test tube should help to mix the sample. The most common commercial stain is Sedi-stain. Some veterinarians may ask for two slides, one that is unstained and one stained. A clean slide should be labeled with the patient name, client name, and date. A drop of urine should be placed in the center of the slide. A coverslip is placed over the drop. Place the slide next to the microscope for evaluation by a veterinarian or veterinary technician. Figure 48-41 shows crystals, cells, and cast as seen in microscopic evaluation of a urine sample.

Gram Stain

Gram stain tests are evaluated to determine presence of bacteria and type of bacteria in a sample. **Gram positive** and **Gram negative** bacteria may be identified and the number of bacteria used to determine the diagnosis of the animal. Gram positive bacteria stain purple and Gram negative bacteria stain red. The bacteria may be in the shape of **rods** (oblong) or **cocci** (round). **Budding yeast** cells may also be noted, appearing as a budding flower. The sample may be feces, urine, or other body fluid such as pus or discharge. The sample must be collected aseptically so as to not introduce any organisms that are not presently occurring in the sample. If contaminants are introduced into the sample, the results will be invalid and make diagnosis difficult. The sample should be collected using a sterile cotton swab. The cotton swab tip is placed into

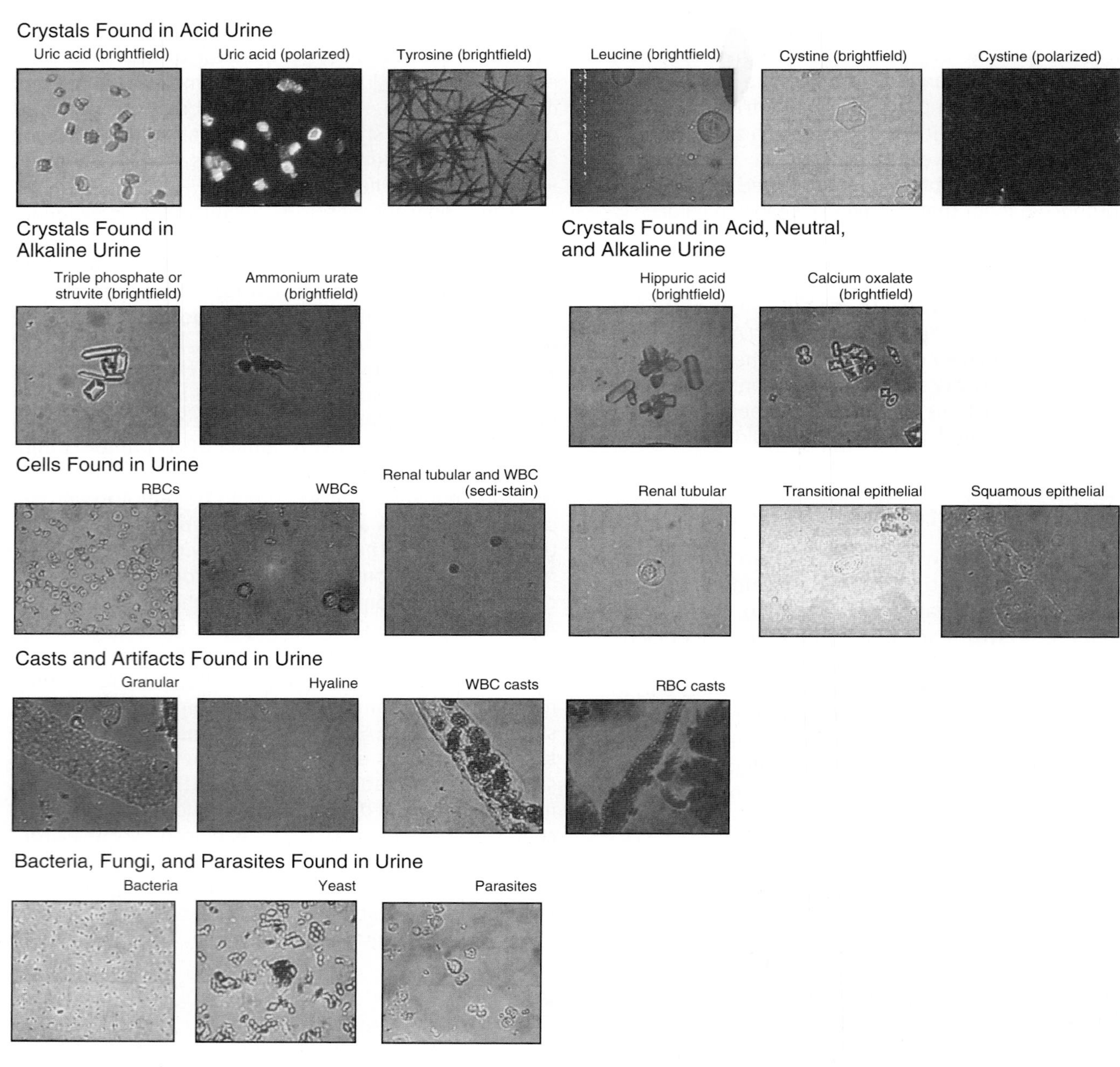

FIGURE 48-41 Crystals, cells, and casts found in urine.

the sample area and then transferred to a microscope slide, where a thin sample is spread over the center of the slide. The slide must be air dried or heat fixed prior to staining. When dry, a gram stain kit is used to stain the slide. The kit consists of three stains, the crystal violet or purple stain, Lugol's iodine or orange stain, the saffranin stain or red stain and the decolorizer as a rinse. The stains should be placed in Coplin jars for easy staining. The veterinary assistant should be able to collect a sterile sample and be able to perform a Gram stain preparation and alert the veterinarian or technician when the slide is ready to examine.

Culture and Sensitivity

A culture and sensitivity (C/S) is a test done through a sterile process to determine if any bacterial or fungal growth occurs on a sample. This requires a liquid or solid sample using a sterile cotton swab that is then placed within an applicator that may or may not contain a **medium** or gel where organism growth may occur. The cotton swab should not touch any site other than the sample area and the medium. A cap should be placed over the swab, which may entirely be submerged in the gel or the stick broken off. The container may be a culturette system, agar plate, or

Competency Skill

Urine Sample Procedure

Objective:

To properly prepare a urine sample for analysis.

Preparation:

sterile urine container, exam gloves, pen, conical test tubes, test tube rack, centrifuge, clinical urine test strips, pipette, microscope slides, cover slips, urine stain, refractometer, lens paper, alcohol, paper towels, tray and work area

Procedure:

1. Apply gloves.
2. Conduct physical appearance examination.
3. Note color, transparency, consistency, odor, foam content.
4. Perform the specific gravity test by using the refractometer. Record value.
5. Clean refractometer with alcohol and lens paper.
6. Locate a clinical urine test strip kit.
7. Apply a drop of urine onto each pad of the test strip using a pipette.
8. Time each test according to the direction panel.
9. Read each test pad as indicated. Align each test strip with correct label.
10. Record results based on the urine color chart on the bottle label.
11. Prepare urine sediment.
12. Fill one conical test tube with urine.
13. Fill another conical test tube with the same amount of water.
14. Place tubes opposite each other in a standard centrifuge.
15. Set centrifuge for urine spin and recommended speed and time.
16. After centrifuge stops, decant urine.
17. Place one drop of non-stained urine on microscope slide.
18. Place cover slip over drop.
19. Mix one drop of urine stain with sediment.
20. Tap bottom of tube to mix.
21. Place one drop of urine sediment onto a microscope slide (see Figure 48-42).
22. Place cover slip over drop.

(*Continues*)

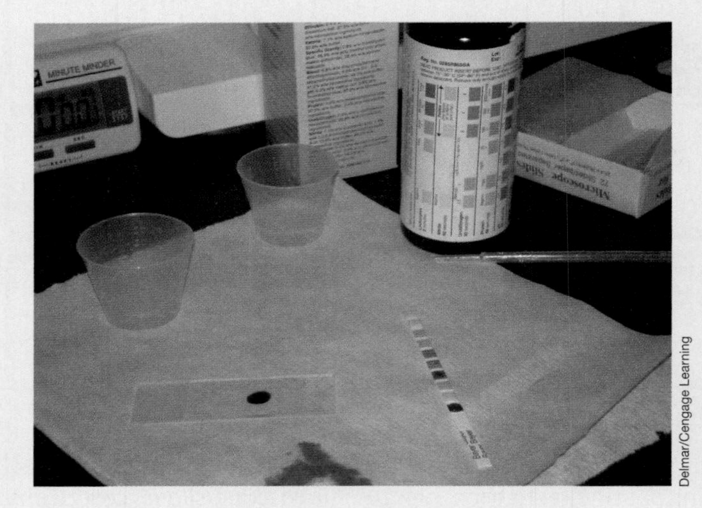

Delmar/Cengage Learning

FIGURE 48-42 Preparation of a urine sample slide.

23. Place slides near microscope for evaluation.
24. Clean work area.

Competency Skill

Gram Stain Procedure

Objective:

To properly prepare a sample as a Gram stain.

Preparation:

exam gloves, sterile cotton swab, microscope slide, open flame or match for heat source, Gram stain kit, staining rack, paper towels, immersion oil

Procedure:

1. Apply gloves.
2. Collect sample using a sterile cotton swab.
3. Apply thin layer of sample to clean microscope slide by rolling sterile swab over the center of the slide.
4. Place over open flame to dry.
5. Apply crystal violet stain for 30 seconds.
6. Gently rinse slide with tap water.

7. Apply Lugol's iodine solution for 30 seconds.
8. Gently rinse slide with tap water.
9. Wash slide with decolorizer for 10 seconds or until purple color is no longer present.
10. Apply saffranin stain for 30 seconds.
11. Gently rinse with tap water.
12. Air dry or blot dry using a paper towel.
13. Place slide near microscope.
14. Place immersion oil near slide.

Competency Skill

Bacterial or Fungal Culture Procedure

Objective:

To properly prepare a sample for a culture and sensitivity test.

Preparation:

Agar plate, DTM bottle or culturette tube, exam gloves, sterile cotton swab

Procedure:

1. Apply gloves.
2. Apply sterile cotton swab to sample site.
3. Place swab in transport media (culturette tube, agar plate or DTM bottle)
4. Place patient name, client name, and date on sample.
5. Send culturette tube to lab for analysis.
6. Close DTM bottle lid so that it fits loosely to allow ventilation.
7. Place DTM bottle in well-lit area and monitor for fungal growth or color change.
8. Place agar plate in incubator upside down.
9. Check daily for growth.

DTM bottle. The DTM means dermatophyte test medium. Agar plates are used to hold the sample by sweeping or rolling the cotton swab over the plate. This must be done gently so the agar gel is not damaged. Agar plates and DTM tests should be placed in a safe place for 4–6 weeks to monitor growth and color change, looking for the presence of bacteria or fungus. The top of the bottle should not be secured tightly so that ventilation may occur.

Competency Skill

Necropsy Setup

Objective:

To properly prepare for a necropsy procedure.

Preparation:

waterproof apron, goggles or face shield, exam gloves, heavy rubber gloves, mask, specimen containers, pathology forms, slides, large sharp knife, ronguers, scalpel, scalpel blade, forceps, sharp scissors, needle holder, suture material, suture needle, Cadaver bag, tape, ID tag

Procedure:

1. Apply gloves.
2. Place cadaver body over a sink on a drainage rack.
3. Place the animal in left lateral recumbency.
4. Place all tools and supplies next to the work area.
5. When necropsy is complete, place cadaver in cadaver bag.
6. Tape bag securely closed.
7. Label bag with client name, patient name, date, and way to dispose of remains.
8. Place in refrigerator or holding area for pickup.
9. Clean work area by disinfecting table, sink, and supplies.
10. Place all lab samples in lab area with forms prepared.
11. Place specimens in area for pickup.

Necropsy Procedure

The **necropsy** procedure is done by examining the body of a deceased animal to determine the cause of death. It is similar to a human autopsy. A necropsy procedure is performed shortly after death since prolonging the time will result in a breakdown of tissues and inaccurate results.

The body is examined and all organs are viewed to note any abnormalities. Tissue samples are collected and sent to a veterinary pathologist for microscopic exam. The packaging and shipping of these samples is often a duty of the veterinary assistant. This requires the assistant to know what reference lab is used and how the sample is to be submitted. Many labs provide containers and paperwork for submitting samples. Certain tests require special instructions and should be read to determine handling of the sample. The submission form should provide client name, patient name, and other identification information as well as a patient history and the type of tissue samples included. Most tissue samples are placed in **formalin** to preserve the tissue. The necropsy procedure is often performed in the treatment area away from general facility activities. It is best to place the animal on a rack over a sink for ease of area cleanup. The tools and supplies the veterinarian will need include the following:

- Scalpel blade and handle
- Forceps
- Sharp scissors
- Ronguers (bone cutters)

- Large, sharp knife
- Needle holders
- Suture needle
- Suture material
- Slides
- Sample containers

The veterinarian should wear an apron, mask, goggles, and heavy rubber gloves to prevent contamination. The necropsy procedure is thorough and detailed. Key observations are made much like in a physical exam. In the case of rabies suspects, the head must be sent for rabies testing immediately on death or euthanasia of the animal. The packaging and shipping should follow the state veterinary lab requirements.

On completing a necropsy, all tissues and organs should be replaced into their respective body cavities, and the animal is sutured closed. The cadaver body is placed in a cadaver bag for disposal. The entire work area should be disinfected and all tools rinsed and re-sterilized.

SUMMARY

The veterinary laboratory has a variety of testing procedures and clinical evaluations that are performed by the veterinary assistant. These practices, procedures, and protocols require knowledge of the procedure and the tools and equipment that are necessary to perform. The equipment must be used properly, cleaned, and kept well maintained. Protective equipment must be used when obtaining and analyzing samples. The veterinary assistant plays an essential role in all aspects of lab medicine and obtaining results.

Key Terms

anticoagulant substance that prevents the blood from forming a clot

blood chemistry analyzer machine that runs blood samples that measure routine blood chemistries and electrolytes

blood plasma blood sample that is frozen and then centrifuged to obtain plasma

blood serum liquid portion of the whole blood sample

blood smear blood film placed on a slide that is used to look at the morphology of blood cells

budding yeast bacteria that are in the shape of a budding flower

buffy coat the layers within a hematocrit or PCV that form a tiny area of visible white composed of white blood cells and platelets

capillary action the action of blood draining into the tube through gravity

catheterized a long, thin rubber or plastic sterile tube inserted into the bladder through the external urinary opening to collect a urine sample

centrifuge veterinary tool used to spin lab samples at a high rate of speed and force used to separate or concentrate materials suspended in a liquid form

centrifuge tube glass or plastic tube that holds samples within a centrifuge

chem. strip long, thin plastic strip separated by individual square pads containing a chemically treated paper

cocci round bacteria shape

commercial serologic test kit commonly used test kit in a veterinary facility to provide quick and accurate results of common viruses and diseases

complete blood count (CBC) blood test

conical pointed shaped end

Coplin jar glass container that holds chemicals for use in staining slide samples

coverslip thin piece of glass that fits over the sample on the slide

culture and sensitivity (C/S) test that determines if a bacterial issue is occurring and what type of bacteria is causing the health condition and what antibiotic should be used in treating the problem

cystocentesis surgical puncture into the bladder using a needle to collect a urine sample

decant process of pouring the urine out of the tube and allowing only the sediment on the bottom to remain in a tube

diaphragm mechanism part of the microscope that allows the slide sample to be moved both up and down and side to side to view the entire sample

(*Continues*)

dip stick long, thin plastic strip separated by individual square pads containing a chemically treated paper

electrolytes determine the balance of elements in the body, specifically potassium, sodium, and chloride

erythrocytes red blood cells

eye piece the portion of a microscope that you look through

feathered edge the staggered area at the end of the slide where the veterinarian or veterinary technician reads and interprets a blood slide

fecal floatation technique done to sample the fecal material to determine if any parasite eggs are within the sample by placing feces within a liquid

fecal sample used to diagnose internal parasites and the presence of blood within the stool sample

fecal smear preparation by placing a small amount of the fecal sample onto a microscope slide

flocculent contains large amounts of particles called sediment suspended in the urine and appearing cloudy

focus knob the portion of the microscope that allows better visualization of the sample

formalin chemical used to preserve tissue samples

Gram negative bacteria that stains red

Gram positive bacteria that stains purple

Gram stain tests are evaluated to determine presence of bacteria and type of bacteria in a sample

gross examination visible observation of the feces

hematocrit measurement of the percentage of red blood cells in whole or unclotted blood; also called a PCV

hematology the study of blood

hemolysis the rupture of red blood cells causing a pink coloration to develop in the plasma or serum

in-house testing laboratory samples analyzed within the veterinary facility

lens objective microscope viewer that offers a variety of different powers

medium gel where organism growth may occur

microhematocrit measurement of the percentage of red blood cells in whole or unclotted blood; also called a PCV

microhematocrit tube thin, small glass tubes that hold blood within a centrifuge

midstream urine collected shortly after the urination begins and just prior to the process ending

morphology the cell structure, shape, color, and appearance in numbers

necropsy procedure done by examining the body of a deceased animal to determine the cause of death

oil immersion the use of a specialized oil substance that is placed over a sample in order to view the contents

packed cell volume (PCV) measurement of the percentage of red blood cells in whole or unclotted blood; also called a hematocrit

panel individual blood test

plasma protein measures the ratio of protein within the blood and helps veterinarians determine the hydration status and inflammation occurrence in patients; also called total protein (TP)

reagent chemicals are used to run each individual test in the entire kit

reagent strip chemical test strip used to analyze urine or blood on a long, thin plastic strip separated by individual square pads containing a chemically treated paper

reference lab laboratory samples analyzed outside of the facility in a commercial lab

refractometer tool used to measure the weight of a liquid and determine a liquid's pH level

rods oblong bacteria shape

rotors wheels in a centrifuge that spin at a variety of speeds depending on the sample type

SNAP test a serologic test

specific gravity (SG) the weight of a liquid

stage part of the microscope that is the flat section under the lens

total protein (TP) measures the ratio of protein within the blood and helps veterinarians determine the hydration status and inflammation occurrence in patients; also called plasma protein

urinalysis break down of urine components to determine a diagnosis

vacutainer tube tube used to place blood samples in for future sampling

voided urine sample collected as the animal is urinating

whole blood blood sample placed in a lavender top tube to prevent clotting of the sample

REVIEW QUESTIONS

1. What is the difference between an in-house test and a reference lab test?
2. What are some important pieces of lab equipment used in the veterinary facility and what do they do?
3. What are the components of performing a fecal sample?
4. What is the difference between whole blood, serum, and plasma?
5. What is the importance of the feathered edge in a blood slide?
6. What are the components performed in a urinalysis?
7. What are the ways to collect a urine sample from an animal?
8. What is the importance of the Gram stain?
9. What lab tests are performed using a refractometer?
10. What lab tests are performed using a hematocrit tube?

Clinical Situation

Photo by Isabelle Francais

The veterinary assistant, Andrea, is attempting to collect a urine sample on a patient named "Kimba." "Kimba" is a 3-year-old SM Doberman that has a history of eliminating in the house. Each time that Andrea walks the dog outside to collect a urine sample, the dog either will not eliminate or attempts to eliminate, and when Andrea tries to place a container for collection, the dog stops urinating.

- What may be clinical conditions to consider with "Kimba"?
- How can Andrea try to collect a voided urine sample?
- What other ways may a urine sample be obtained?

49

Delmar/Cengage Learning

Radiology Procedures

Objectives

Upon completion of this chapter, the reader should be able to:

- ☐ Explain the practice of radiation safety within the veterinary facility
- ☐ Demonstrate how to maintain a radiology log of radiographs taken within the facility
- ☐ Describe directional terms and abbreviations pertaining to radiology
- ☐ Demonstrate proper animal restraint and placement during radiographic procedures
- ☐ Demonstrate how to use the caliper to measure the thickness of a body part in centimeters
- ☐ Demonstrate how to use a technique chart properly
- ☐ Demonstrate how to set and determine proper settings on an X-ray machine
- ☐ Demonstrate how to use identification markers for X-ray films
- ☐ Properly develop X-ray films using a hand developing tank
- ☐ Properly develop X-ray films using an automatic developer
- ☐ Properly load and unload film from a cassette
- ☐ Properly clean a film cassette
- ☐ Properly file patient films
- ☐ Properly clean and maintain the veterinary darkroom
- ☐ Explain the importance of ultrasound and endoscopy equipment

Introduction

Radiology is the study of radiation and the proper way to take radiographs, commonly called **X-rays**. **Radiation** is a safety hazard in the veterinary industry and requires proper training and safety methods to allow a safe working environment for staff. The veterinary assistant plays an important role in properly performing radiographic procedures and other diagnostic techniques. **Diagnostic imaging** is the term used for performing radiographs, ultrasounds, and endoscopy. These tools are helpful in allowing the veterinarian to properly diagnose the patient. It is important that the assistant be familiar with these procedures and the proper handling and setup of the valuable equipment used in these procedures.

Radiation Safety

The veterinary assistant should have a basic understanding of radiation, the hazards of working around radiation, and how to properly reduce exposure and enact safety measures when working in radiology. Regulations regarding equipment and personal protective safety are governed by OSHA and the State Department of Health. Each state has various state laws governing radiation exposure and radiation safety principles.

Radiation causes generalized damage to cells throughout the body with reproductive cells, thyroid cells, and eye cells being easily damaged and of concern. Certain body cells are more sensitive to radiation than others. Damage from radiation occurs with small exposure limits over a lengthy time frame or from large exposure limits over a short period of time. It is important to note that radiation is invisible and is not able to be felt, smelled, or tasted, so exposure is unknown.

Following safety measures greatly reduces the risk of radiation exposure. This is achieved through the use of the ALARA method or "as low as reasonably achievable" method of taking radiographs. The lowest amount of radiation exposure will increase the safety of the patient and staff when obtaining X-rays.

The **dosimeter**, or X-ray badge, measures the amount of radiation a person is exposed to (see Figure 49-1). It should be worn every time a radiograph is performed or a person is working around or near radiation. It measures the levels of radiation exposure to each staff member and calculates the level of exposure of each person. The standard exposure level for a person is 0.005 Sievert per year for professionals within an occupation using radiation. The general public can only receive 1/10 of this amount and should never be in access of the radiology area.

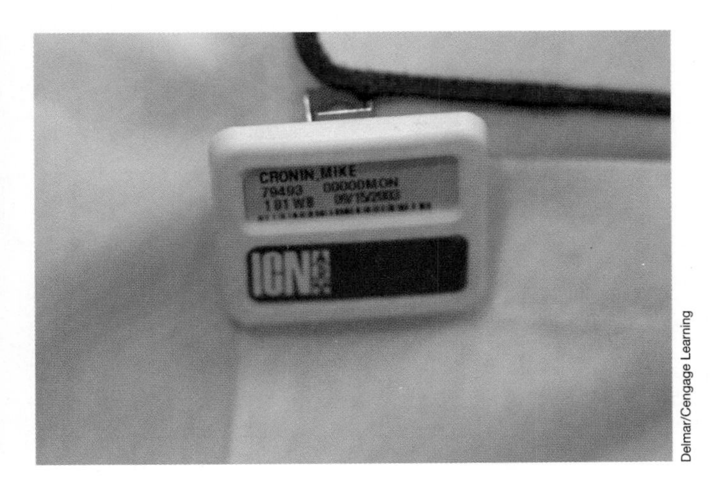

Delmar/Cengage Learning

FIGURE 49-1 Dosimeter.

All professionals should be educated in not placing any body part directly into the X-ray beam where radiation is strongest. All individuals must be at least 18 years of age to work with radiation and pregnant women should not be near any areas of radiation.

Radiation protection begins with the use of protective equipment whenever there is exposure to radiation. This includes the use of a lead apron, lead gloves, thyroid shield, and lead glasses. This equipment should be stored in the radiology area by hanging on a wall as the lead items may not be folded. Folding the lead items may cause cracks and tears in the lead material, which would reduce their level of protection. Ideally, the items should hang on specialized hangers on the wall with the gloves on a glove rack for proper ventilation after each use. This equipment should be radiographed on a regular basis to inspect for leaks and damage that may occur in the lead shielding. The dosimeter should be stored outside of the radiation area. When dressing for radiographs, the thyroid shield should be placed over the neck first. When applying the lead apron it should fit properly over the shoulders and chest and tie in or Velcro in easily. It should cover the entire front of the person. The thyroid shield should completely cover the neck and thyroid gland area and be tucked under the gown. The goggles should then be applied over the eyes and fit well without easily sliding off during movement. The dosimeter (X-ray badge) should fit onto the front of the gown, usually clipped onto the neck or pocket of the gown. It should face outward and lie flat.

All X-ray equipment should be maintained and cleaned after each use. The **cassettes** that hold the film have a **screen** that lines the cassette and should be cleaned every month using a screen cleaner to prevent a buildup of dust and debris (see Figure 49-2). Any cracks or damage in the screen must be noted as the cassette will need to be replaced. The exterior of the cassette is cleaned after each use and may be cleaned with a mild disinfectant if exposed to an animal or with soap and water. An **X-ray log** should be maintained that notes each film taken and the exposure settings of the machine (see Figure 49-3). The patient's name, client's name, date, X-ray number, X-ray position, thickness of body measurement, and area exposed should be noted in the log. The X-ray machine should be serviced by a field representative on a regular basis to make certain the equipment is in proper working order and properly calibrated. Dosimeters should be evaluated on a regular basis to determine employee exposure rates.

When working with patients, it is helpful to reduce as much radiation exposure as possible. If the patient is easily restrained and capable of safely being X-rayed, the smallest size cassette should be used to allow less radiation exposure. This will allow the **collimator** to reduce the field size and prevent scatter radiation.

Delmar/Cengage Learning

FIGURE 49-2 Screen on X-ray cassette.

The collimator sets the size of the X-ray beam that produces radiation to take the picture, based on cassette size. The setting allowing the least amount of radiation should be used. Avoid any retakes of films by using the correct exposure settings, proper patient positioning, and correct area to be radiographed. Items such as sandbags, foam wedges, and a commercial positioner may be used to help keep the patient still. Animals that are in pain or not cooperative may require anesthesia or sedation to make the radiograph process easier. Some states have regulations regarding whether the staff may be in the same room as the patient when the radiograph is being taken.

Strict safety measures must be implemented and followed by every veterinary team member to help reduce radiation exposure. Once radiation damage occurs it is not reversible and can cause serious illness or injury. Simple safety precautions during every radiographic procedure will help prevent unnecessary exposure to staff, clients, and patients.

Radiology Terminology

There are many terms that relate to the equipment, procedures, and positioning of patients when discussing radiology. It is important for veterinary assistants to understand basic radiology terms and understand positioning when working in radiology. This is important to proper film exposure and development and patient diagnosis.

X-ray Terms

When a radiograph or X-ray is taken and developed, there are several terms that should be understood and known to appropriately provide notations in the X-ray log and medical chart. X-rays may be developed and must be properly exposed, avoiding such circumstances as causing an X-ray to become radiopaque or radiolucent. **Radiopaque** means appearing white to light gray in color and may indicate hard tissue such as bone or

X-RAY LOG													
Date	X-Ray #	Client Name	Patient Name	Species	Sex	Age	Weight	View	CM	KVP	mAs	Body part	Initials
								-----	----	-----			-------
								-----	----	-----			-------
								-----	----	-----			-------
								-----	----	-----			-------
								-----	----	-----			-------
								-----	----	-----			-------
								-----	----	-----			-------
								-----	----	-----			-------
								-----	----	-----			-------
								-----	----	-----			-------

Delmar/Cengage Learning

FIGURE 49-3 X-ray log.

may be an exposure factor in that the machine settings were too low or the patient was not properly measured. **Radiolucent** means appearing black or dark gray in color and may indicate soft tissue or air and may be an exposure factor relating to increased exposure settings or improper patient measurement.

X-ray Machine Terminology

When taking X-rays, the machine must properly be set and used. This requires knowledge of the procedure and understanding of terms. The patient should be measured using the caliper to determine the body part thickness in centimeters (cm) (see Figure 49-4). This helps determine machine settings that will properly allow the setting of the **kilovoltage peak (kVp)** and **milliamperage (mA)**. The kVp is the strength of the X-ray beam. Thus, this is the strength of radiation. The mA represents the number of X-ray beams based on time. Thus, this is the amount of radiation released during a set time. The settings of the machine will be based on the thickness of the animal's body part to be X-rayed and thus should be measured accurately and settings properly calculated.

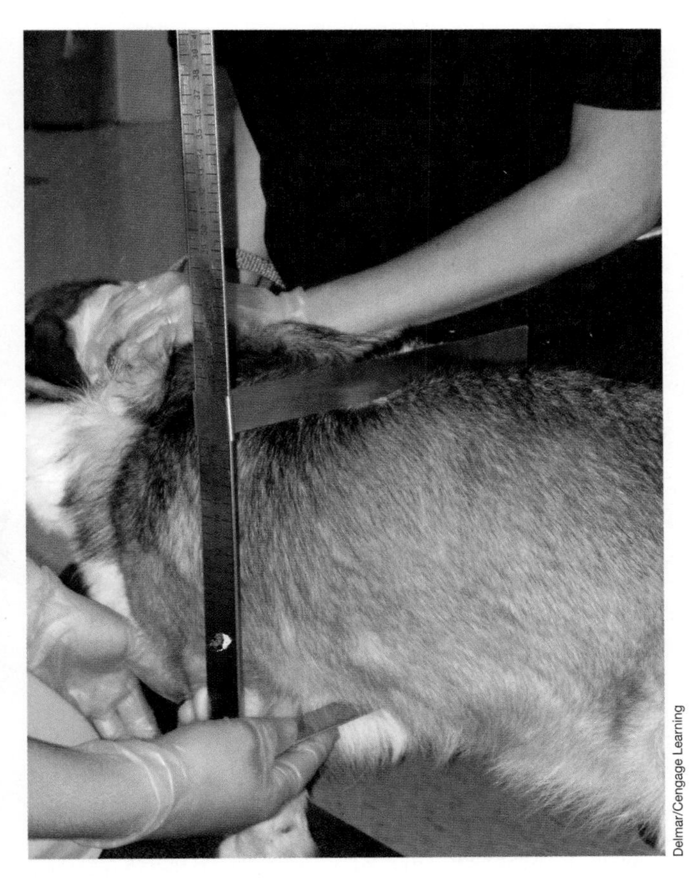

Delmar/Cengage Learning

FIGURE 49-4 Measuring the patient to determine proper settings for taking an X-ray.

X-ray Positioning Terms

X-rays usually have several positioning types depending on the animal and body part being evaluated. The most common X-ray positions are the lateral, ventrodorsal, dorsoventral, and oblique views. The **lateral (LAT)** view is an animal positioned on its side. The X-ray beam passes the animal from side to side. The **ventrodorsal (V-D)** view is the animal positioned on its back with the X-ray beam going through the ventral area first (stomach) and the dorsal area second (back). The **dorsoventral (D-V)** view is the animal positioned on its stomach with the X-ray beam going through the back first and the stomach second. The **oblique** view is used on areas that need to be placed at an angle to prevent double exposure from other body organs. This is common with skull X-rays to prevent an area from causing shadows on the film. It is important to note the left and right sides of an animal on an X-ray view so that the proper side of the animal is identified by the film. An **AP** view or **anterior posterior** view is often used on limbs and common in large dogs and large animals. This means the X-ray beam penetrates the anterior portion of the body area as the entrance and the posterior portion of the body as the exit point.

Diagnostic Terms

There are several diagnostics performed in addition to general radiograph films. Radiographic **contrast mediums** are used to study parts of the body and films are taken in sequences to monitor the contrast medium as it moves through the area. A contrast medium is a material or substance administered into the body to show structures on an X-ray film that would otherwise be difficult to view. **Barium sulfate** is an example of a contrast medium used in a **barium study**. Barium sulfate is a solution administered by mouth or rectally and used to fluoresce body parts of the digestive system. A barium study is a series of films taken as the material passes through the digestive tract. A **lower GI** is a contrast film of the lower digestive tract organs and is often given via an enema, which is a large amount of a substance passed into the colon via the rectum. This is called a **barium enema.** An **upper GI** is a contrast study of the structure of the upper digestive tract in which the barium is given orally.

Veterinary Animal Restraint

When an animal is positioned and restrained during a radiographic procedure, the correct positioning is important. A lateral view requires the proper side down depending on the reason for the X-ray and the location

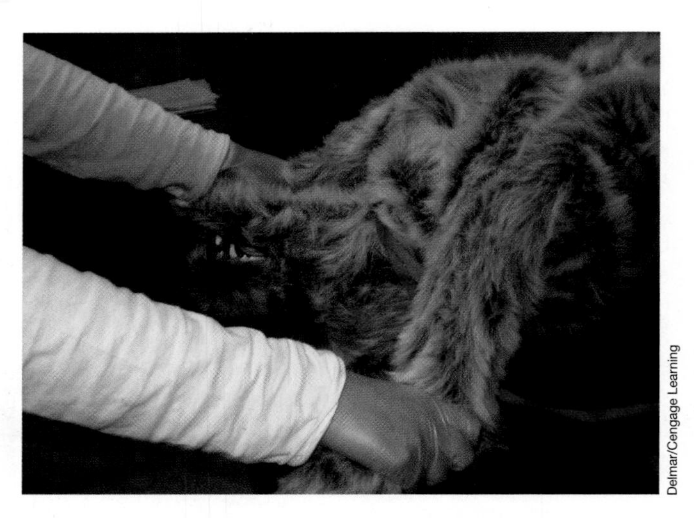

FIGURE 49-5 Proper positioning of the front limbs for a ventrodorsal X-ray.

of the film and body part. Most patients are placed in right lateral recumbency unless otherwise noted by the veterinarian. The limbs are pulled forward or backward depending on the area to be viewed. A V-D film places the patient on its back with the front limbs extended parallel to each other and beyond the patient's head. The head should be stabilized between each leg (see Figure 49-5). The hind limbs are parallel and pulled firmly away from the patient's body until they are straight. A D-V film places the animal on its sternum or chest, with the front limbs pulled forward with the head stabilized between each front leg. The hind limbs are placed comfortably under the patient and out of view of the body part being X-rayed. The X-ray beam is centered directly over the center of the body part being radiographed. After the proper position is determined the veterinary assistant and technician must maneuver the patient into position. The use of sandbags, a trough positioner, or foam wedge may keep the patient in place. The goal is to obtain and maintain as normal a position as possible while preventing movement or superimposed body areas that may create a shadow. A minimum of two films should be taken to ensure proper film positioning, calculation, and development. All items should be set up and prepared prior to placing the animal on the X-ray table. This reduces patient and staff stress and allows the team to focus on the task at hand.

Radiology Log

The radiology or X-ray log is a legally required document that must be kept updated and accurate and should be kept within the radiology area. The X-ray log serves several purposes including meeting the standards of the veterinary profession, as a comparison in techniques used on patient follow-up films, and by recording the quality of each film taken as a reference to locate the film or improve film quality. The log is usually maintained in a binder with a functional blue or black pen secured to the binder. Individual entries are made during each radiographic procedure. The format is similar in all logs and is composed of columns and rows where information is entered for each patient. X-ray log papers may be created by the hospital or purchased commercially through an office supply store. Each state practice act outlines the required information kept within the veterinary X-ray log. These items include the following:

- Date
- X-ray or radiograph number
- Client name
- Patient name or identification
- Breed or species
- Gender
- Age
- Weight
- Body location of radiograph or study
- Thickness of body part in centimeters (cm)
- X-ray view or position
- Kilovoltage peak (kVp)
- Milliamperage (mA)
- Seconds of exposure time
- Quality of the film
- Veterinary diagnosis and comments

Patient Measurement

Patients are measured to ensure that the X-ray beam can sufficiently penetrate the thickness of the tissue requiring an X-ray. **Calipers** are used to measure the thickness of a body part (see Figure 49-6). This device or instrument is a simple ruler-like tool that slides on a fixed bar or arm that moves parallel to it. The upright portion along the moveable bar is marked in **centimeters** (cm) and inches. The cm scale is to be used when measuring for X-rays. This is due to the techniques chart that is based on cm readings and settings for the X-ray machine. The caliper is placed over the thickest part of the area to be X-rayed. The moveable bar is allowed to move and fit lightly around the area being measured (see Figure 49-7). The cm scale should be read at the point just *below* the moveable arm. Reading the scale above the arm will provide an inaccurate measurement. Table 49-1 shows some examples of caliper readings for various species and body parts.

The caliper should be kept close to the X-ray log book so it is easily accessible to note the proper cm measurements and necessary views. Correct use of the caliper leads to quality films and less error.

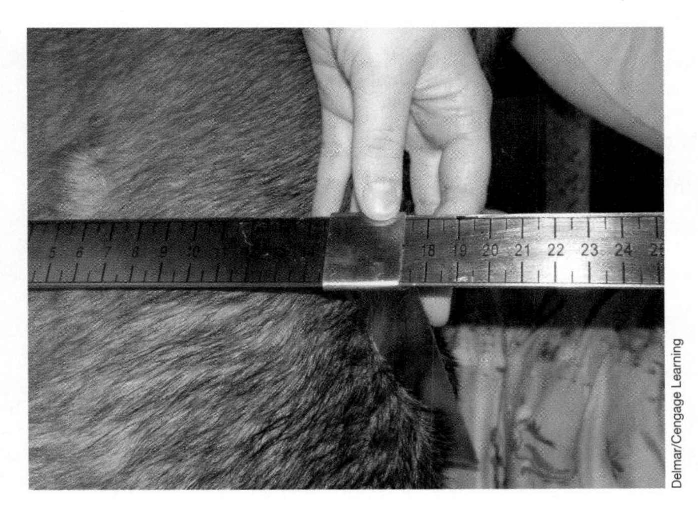

FIGURE 49-6 Calipers are used to measure the thickness of the body part to be X-rayed.

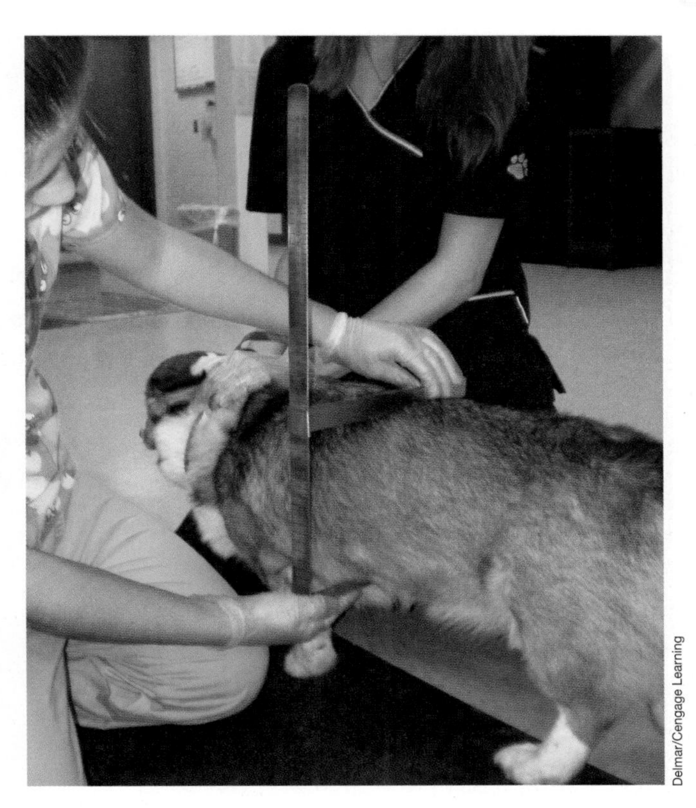

FIGURE 49-7 Use of a caliper to measure the thickness of a body part to be X-rayed.

TABLE 49-1

Example of Caliper Readings

BODY PART	SPECIES/BREED	MEASUREMENT (CM)	VIEW
Carpal joint	Feline/DSH	2 cm/2 cm	LAT/V-D
Chest	Canine/Dachshund	8 cm/10 cm	LAT/V-D
Skull	Canine/Mastiff	15 cm/11 cm	Oblique/D-V
Abdomen	Ferret	3 cm/5 cm	LAT/V-D
Tarsal Joint	Equine	6 cm/ 9 cm	LAT/AP

Competency Skill

Caliper Measurement Procedure

Objective:

To properly measure the thickness of a body part to ensure an accurate and useful X-ray is obtained.

(*Continues*)

Preparation:

caliper, X-ray views requested, X-ray log book, pen

Procedure:

1. Place the fixed portion of the caliper at the site where the X-ray beam is to exit.
2. Slide the moveable arm over the site the beam is to enter.
3. Move the caliper along the body area to the thickest portion of tissue.
4. Read the centimeter (cm) scale at the portion of the moveable arm closest to the body. Read at the area just below the arm.
5. Record the measurement in cm along with the view in the X-ray log.
6. Repeat for any additional views.
7. Clean the caliper after each use by disinfecting.
8. Put caliper in storage area.

Technique Chart

An X-ray **technique chart** is a listing of settings on the X-ray machine based on the thickness of the area to be radiographed. It is set up in rows and columns and lists the thickness in centimeters (cm). Once the thickness of a body part is determined, the technique chart is used to determine the machine settings. This is done by noting the measurement in the column listed cm and moving across the row to determine the kVp, mA, and exposure time in seconds or mAs. Technique charts will be different for a grid technique and a tabletop technique. The **grid technique** is the plate that holds the film below the X-ray table and causes the **X-ray tube** holding the radiation source to be lower. The film is not in contact with the patient. The **tabletop technique** is the distance between the X-ray tube and the top of the X-ray table surface and the film cassette is placed on top of the table and in contact with the patient.

Technique charts are formulated for a specific machine. The constants of the machine are predetermined before the chart is created. These factors include the following:

- The distance from the source of the X-ray beam to the film. This is usually between 36 and 40 inches. This is a fixed distance based on the machine and changes when the grid is used versus the tabletop technique.
- A filter is located between the window of the X-ray tube and the collimator, which is usually between 2 to 2.5 mm in thickness. The filter absorbs the scatter radiation to prevent excessive exposure amounts.
- The grid height and width within the table.
- Film speed that depends on exposure time. Fast film requires less exposure time but lacks definition and detail, whereas slow film requires greater exposure time but produces greater detail.
- **Intensifying screens** that are located within the film cassette to help produce a better exposure to the film. They are rated according to speed and relate to exposure time with high speeds requiring less exposure to radiation. Slow film speeds have greater detail but cause blurry appearances with any movement.

A technique chart should be developed for both a grid technique and a tabletop technique. Most common procedures include a tabletop technique used for views of the skull, extremities, and avian and rodent or exotic animals. The grid is used for larger animals and the thorax, abdomen, and chest.

Machine Setting Procedures

The control panel of the machine has several dials, switches, and knobs. It is important to first locate the on/off power switch. This turns the machine on and off. It is important to make certain the X-ray machine is on during use. Some machines have a buzzing sound and others are silent. The three important selector knobs used for setting the machine are the kVp, mA, and the timer. Some machines have the mA and timer linked and also provide the mA setting. The goal is to use the highest mA setting and the shortest exposure time. The knobs should be set to meet the readings on the technique chart based on the patient's body measurement. It is important for the assistant and technician to locate the exposure button on the console panel, which may also be in the form of a foot pedal that enables the staff to take a picture while simultaneously restraining the patient in the proper X-ray position. When the exposure button is pressed, a red light flashes or a buzzer sounds to indicate the emission of radiation. The age of the machine will also determine how the machine is set up and used. All settings should be double checked and confirmed for accuracy. All staff members working in radiology should be properly trained and experienced in taking radiographs.

Film Identification

Radiographs are a part of a patient's medical record and are also considered legal documents. Most X-rays are filed separately from the medical record but must be properly labeled and identified to locate and review. Each film that is taken should be permanently marked with patient and client information, as well as the veterinary facility information. Each state requires specific information to be placed permanently on the film, meaning it must be placed on the film prior to X-ray exposure or X-ray development. Information that must be on the label includes the following:

- Hospital name, address, and phone number
- Veterinarian's name
- X-ray number
- Client name
- Patient name
- Date

The film should also be marked with a directional label or marker to note which side of the animal is being viewed. This is usually in the form of a lead L or R placed on the proper side of the cassette film prior to X-ray exposure. The markers identify the left and right side of the patient when using a V-D, D-V, or AP view and the down side when using a LAT or oblique view. Make certain the marker is placed properly on the film cassette and within the area of the X-ray beam. Some facilities use lead tape that contains the patient information labeled on it and placed on the X-ray film cassette to be exposed onto the film during the X-ray. Other facilities use an electronic film identification printer that stamps the information onto the film prior to film development. If an electronic flasher or printer system is used, an area must be blocked on the film cassette prior to exposure to prevent that area from being occluded.

Developing Film

There are two methods for developing film in the veterinary facility. The methods are **manual developing** or **automatic film processing**. The older method is manual developing through using **hand tanks**. Manual development is cheaper to set up and maintain than an automatic processing system, but takes longer to develop a film than the automatic method. The manual method also requires temperature control of the hand tanks, which hold the chemicals used to develop the film. The automatic film processor is a costly piece of equipment that has advantages to hand developing. The most noted is the speed of development and quality of the film. The solutions are temperature controlled by the machine and reduce errors and increase the film quality and life of the film. The time for a person to develop the film is significantly reduced. Professional maintenance is also required with the automatic processor unit. Table 49-2 highlights the pros and cons of manual versus automatic developing.

TABLE 49-2

Pros and Cons of Development Methods

	PROS	CONS
Manual Developing	■ Low cost ■ Easy to use ■ No backup system required ■ Low maintenance	■ Lengthy developing time ■ Must maintain chemical temperatures ■ Must stir and prepare chemicals ■ Less life span of developed film
Automatic Processing	■ High speed of development ■ Temperature controlled chemicals ■ Reduced developing errors ■ Increased life span of film ■ Higher quality film	■ High cost ■ Machine warm-up and setup time ■ Requires maintenance and servicing ■ Requires backup method if equipment breaks or fails

Film processing involves several steps independent of the method of developing. These steps include film developing, film **rinsing**, film **fixing**, and film drying. **Processing** involves being in a darkroom that has no light source or light leaks from the outside area. Light will ruin the film and make it undevelopable. The darkroom is equipped with a **safe light**, which is a red light that is low in intensity and a filter that doesn't damage the film. The entrance door should be secured firmly to prevent anyone from entering during film development. A light or sign should be placed on the door when the room is in use. The undeveloped film is removed from the cassette before it is developed. A new film is replaced within the cassette.

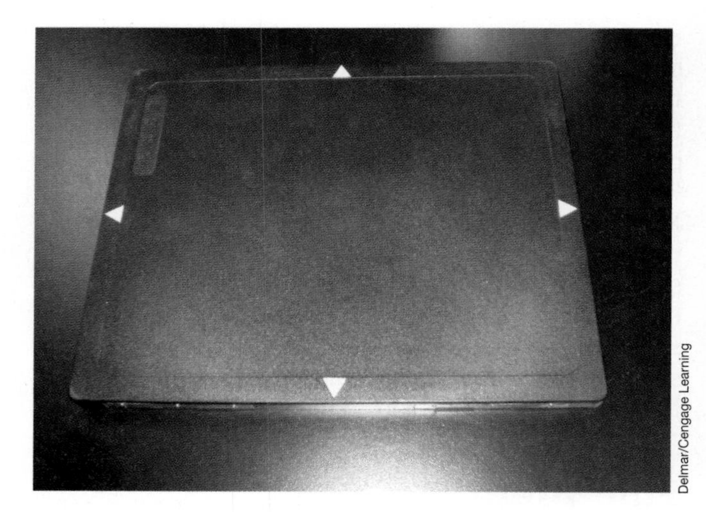

Delmar/Cengage Learning

FIGURE 49-8 Film cassette.

Film Cassettes

Cassettes hold the film that is used to take an X-ray and prevent the film from being exposed to light (see Figure 49-8). They have latches that secure the cassette tightly to keep light from entering. It is important to work in the darkroom when loading and unloading film from a cassette. The cassette should be placed face down so the back latches can be unlocked. The cassette is then turned over face up and the top open. Film is removed and should be handled with care. Only handle the film at the corners and as carefully as possible, avoiding dropping the film to prevent streaks, smudges, dirt marks, static electricity, and light exposure. When film is refilled into the cassette, it is removed from the film box, which is kept in a film bin that is light-tight. The appropriate film size should be selected and the lid removed. Each sheet of film is covered by a paper or plastic cover that protects the film. A single sheet of film should be carefully removed and placed in the cassette. The cassette is closed and properly locked. The cassette is ready for use or storage in its customary place. A cassette should never be left without film.

Competency Skill

Unloading Film from a Cassette

Objective:

To properly remove film from a cassette while maintaining the integrity of the film.

Preparation:

film cassette, film, darkroom, safe light

Procedure:

1. Place the cassette upside down and unlock.
2. Turn cassette face up and open cover.
3. Remove film from cassette by one corner, grasping firmly with the fingertips.
4. Leave the cassette open while developing film.

Competency Skill

Loading Film to a Cassette

Objective:

To properly prepare a cassette to maintain the integrity of the film.

Preparation:

film cassette, film, darkroom, safe light

Procedure:

1. Place cassette face down and unlock.
2. Open cassette.
3. Place the film box containing the same size film next to cassette.
4. Remove the box lid.
5. Open paper or plastic cover and remove one sheet of film.
6. Place film in cassette and close. Lock cassette.
7. Fold paper or plastic cover over film box and place in storage bin.
8. Place cassette in storage area.

Competency Skill

Cleaning a Cassette

Objective:

To properly maintain equipment for use in obtaining X-rays.

Preparation:

film cassette, mild soap, warm water, paper towels

(*Continues*)

Procedure:

1. Open cassette.
2. Apply small amount of warm water and soap to paper towel and wipe cassette screens on both surfaces.
3. Dry each screen surface.
4. Allow cassette to remain open until dry.

Film Hangers

Film hangers are used to secure film for manual processing and development. These tools hold the film onto a metal frame with clips and hold the film as it is submerged in chemicals during developing (see Figure 49-9). The film is removed from the cassette and the proper hanger size is selected to hold the film. The film hanger is held upside down as the film is attached to clips. Once the bottom clips are secured into the corners of the film, rotate the film and secure the film to the top clips by pulling them downward and snap onto the corners. The film holder should secure all four corners of the film. The film is then handled by holding the bar at the top of the hanger. This bar allows the film to hang in the tank during the developing process.

Delmar/Cengage Learning

FIGURE 49-9 Film hanger.

Manual Developing

Developing tanks are required for hand-processing films. These tanks may be metal or heavy plastic and hold large volumes of chemicals. The solutions require periodic changing at which time the chemicals are drained. The tanks are cleaned well to remove bacteria and algae, rinsed, and then refilled. The frequency of this practice depends on how much use occurs and any debris and contaminants that may get into the chemicals. The tanks contain the developer solution, the fixer solution, and a wash tank that contains water. The **developer solution** is usually located on the left-hand side of the tank. The developer solution develops the X-ray film. The **fixer solution** is usually located on the right-hand side and is a smaller tank compared to that of the water. The fixer solution fixes the X-ray film and helps maintain the life span of the film, allowing it to stay readable. The water rinse tank is usually located to the far right side and is larger than the other two tank sections. When the processing method is complete, the film should hang to dry.

Automatic Film Processing

The automatic film processor unit should be started early to allow it to warm up and the chemicals to reach the desired temperature. All chemical tanks should be

Competency Skill

Manual Film Processing

Objective:

To properly develop X-ray film using the manual method.

Preparation:

film cassette, film, darkroom, safe light, automatic film marker, patient identification card, pen, hand developing tanks, developer solution, fixer solution, water, timer, thermometer

Procedure:

1. Make patient identification card and place in automatic film marker.
2. Stir developer and fixer solution and make certain proper temperatures are noted.
3. Place film cassettes face down in the darkroom and unlock back.
4. Place film cassettes face up and open top.
5. Remove film by one corner.
6. Place unexposed corner of film into automatic film marker over card and lower top to expose film corner.
7. Begin attaching film to bottom corner of film hanger; attach to both clips.
8. Rotate hanger and attach top clips to both corners.
9. Place film hanger in the developer on the left side of the processing tank.
10. Lift film up and down several times.
11. Set times for proper time required for developer.
12. While film is in developer, refill film into cassette.
13. Put cassette away.
14. Place film box in storage bin.
15. When timer sounds, remove film from developer and rinse in water tank by moving film hanger up and down several times.
16. Place film hanger in fixer solution and lift up and down several times.
17. Set times for proper time required for fixer.
18. Cover tanks with the correct lids.
19. Exit darkroom.
20. When timer sounds, return to darkroom and remove film from fixer and rinse in water tank.
21. Place in water for 30 minutes. After 30 minutes, hang film to dry.
22. When film is dry, unclip the film from the hanger.
23. Notify veterinarian when film is ready to read.

checked for proper levels. The roller racks should be removed and washed on a regular basis. The process of using the automatic process for film developing is simple. The film cassette is taken to the darkroom and the exposed film removed from the cassette. The film should be grasped by the edges and placed on the feeder tray of the processor. The film is aligned with the roller bar and placed against it. The film will be pulled into the processor and developed automatically. The completed film will be removed on the opposite end of the feeder tray dry and ready to view. X-rays are viewed on a viewing box that illuminates the film and allows the veterinarian to make a diagnosis (see Figure 49-10).

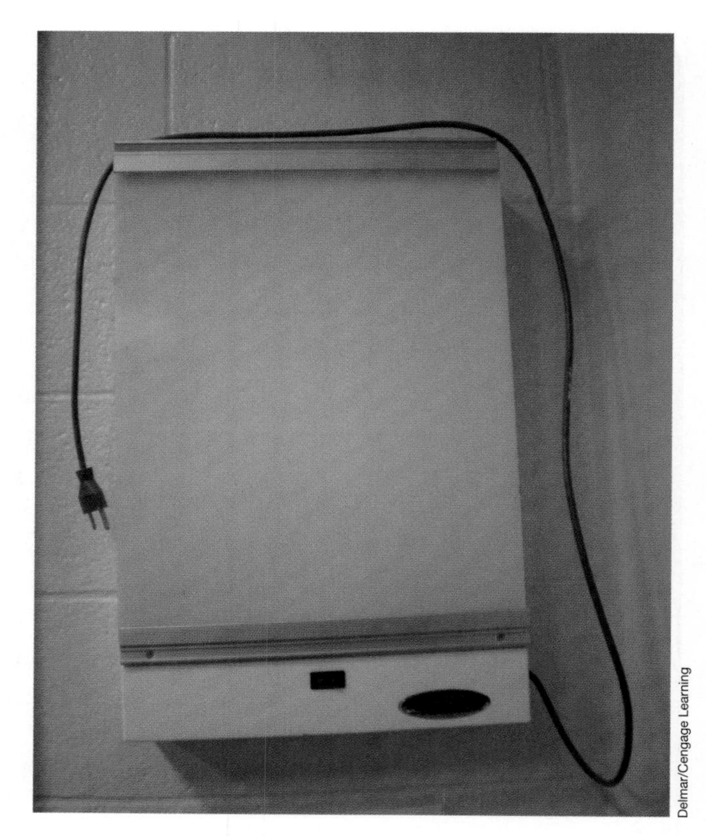

FIGURE 49-10 X-ray view box.

Digital Radiology

Many facilities are beginning to switch from using chemical processing to produce radiographs to using digital radiology. Digital X-rays are transferred onto a computer disc rather than on a film. This type of diagnostic imaging has multiple advantages including ease of use, production of high-quality images, affordability, and improved neatness and cleanliness of images. This radiology practice is quite common in veterinary medicine and in the coming years will be considered the norm in radiology.

Competency Skill

Automatic Film Processing

Objective:

To properly develop X-ray film using the automatic method.

Preparation:

film cassette, film, darkroom, safe light, automatic film marker, patient identification card, pen, automatic processor, developer solution, fixer solution

Procedure:

1. Make certain the processor is ready to use.
2. Make certain all chemicals are replenished.
3. Place cassettes face down in darkroom and unlock.

4. Turn cassettes over and open cover to remove film.
5. Grasp film by one corner and align film with feeder tray.
6. Gently place film against roller bar.
7. The film will enter the processor automatically.
8. While the film is being processed, refill the film cassette. Replace film box in storage bin.
9. Remove film from opposite end of feeder tray.
10. Give film to veterinarian to be read.

Filing Film

Radiographs must be filed just as medical records are filed for ease of location. X-rays are too large to be filed with the medical record and are often stored in a nearby file area in large protective X-ray folders or paper envelopes. Several systems have been used to file radiographs, such as alphabetical or numeric filing. Most facilities file the radiographs by the X-ray number and then store them in numeric order. All X-rays for the patient are stored in one file envelope. Once a film has been developed, an envelope and number should be recorded in the patient's medical record and a file folder labeled with the X-ray and patient information. The information label should include the following:

- X-ray number
- Patient name
- Client name
- Veterinarian name
- Dates of radiographs
- Type of study or view
- Diagnosis

The file folder is labeled properly, the film placed within the file folder, and the folder then properly filed in the storage location. Filing should be done accurately as a misfiled folder creates disorder in a clinic.

Darkroom Care and Maintenance

The darkroom is an area that should be kept clutter free and clean at all times to help improve film quality and proper developing methods. The darkroom should be checked on a regular basis for any light leaks from the outside. The door should be observed for cracks around the door. The safety light should be checked by closing the door and keeping the darkroom dark with only the safe light source illuminated. An unexposed film should be placed on the counter with a metal object, such as a paper clip or nail over the film. The film should be exposed for 2 minutes and then the film should be developed. If the object appears on the film, the safety light is not working properly or some other light source is entering the room. This test evaluates the proper lighting environment of the darkroom.

The counter workspace should be cleaned on a daily basis and all items placed in their storage locations for ease of working in the dark. There should be two areas to the darkroom: a wet side and a dry side. The wet side holds the processor and hand developing tanks. The dry side is the counter workspace where cassettes are loaded and unloaded with film. The film storage boxes should be stored in a bin under the work area to decrease light exposure.

Floors and work areas may become wet and towels should be kept in the area to clean up leaks and spills. A mop should be used to clean the floor as spills occur. A cleaning regimen should be developed for darkroom maintenance and materials and chemicals monitored for reordering. Figure 49-11 shows an example of a darkroom maintenance schedule.

Ultrasound Diagnostics

Ultrasound is a diagnostic tool using ultrasonic sound waves to view images of internal organs and structures. The sound waves bounce off the patient and create an echo within the tissues and respond back to the ultrasound machine and are projected on a screen (see Figure 49-12). The stronger the ultrasound signal during the return phase, the brighter and whiter the image.

Tasks	Cleaning Schedule	Date	Initials
Clean processor	Daily		
Clean tanks	Weekly		
Replenish solutions	Daily, as needed		
Clean counters	Daily		
Clean/mop floor	Daily		
Inventory supplies	Weekly		
Check light leaks	Monthly		
Check safe light	Monthly		

Delmar/Cengage Learning

FIGURE 49-11 Darkroom maintenance schedule.

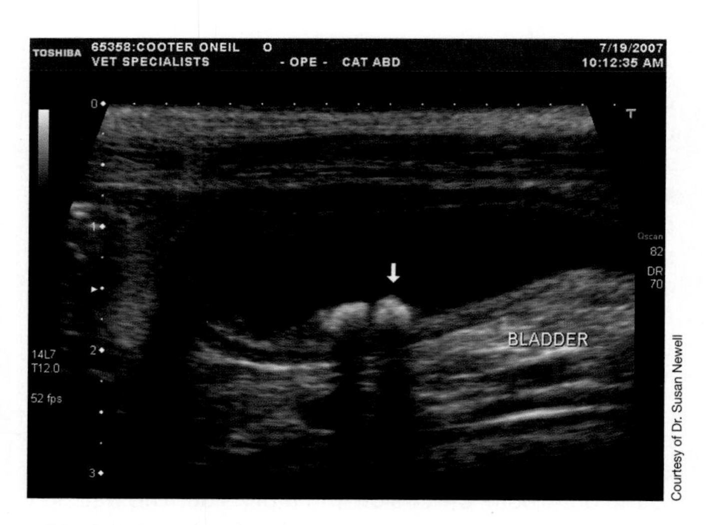

Courtesy of Dr. Susan Newell

FIGURE 49-12 An ultrasound image taken during examination of a cat's abdomen.

The veterinary assistant plays the role of preparing and restraining the patient for the ultrasound procedure. The quality of the image is affected by how the patient is prepared and how calm the animal is kept during restraint. The image quality may be decreased if air, hair, dirt, or other debris is not properly removed making an unsterile environment. The **transducer** is the wand tool used to scan the area being examined. The area where the transducer is placed should be clipped as that for a surgery using a #40 or #50 clipper blade. The area is gently washed with sterile scrub materials and dried well to enhance the contact area of the transducer. A small amount of transducer gel is applied to the area to help visualize internal structures. The patient should be positioned depending on the area being examined. Abdominal examination in small animals requires dorsal or lateral recumbency. Cardiac examination requires lateral recumbency. Many large animal ultrasounds are done through standing recumbency.

The equipment should be safely placed in the examination or treatment area, plugged in, turned on, and prepared for use. The equipment should be set up prior to the patient being brought into the area. When the ultrasound is complete, the equipment should be disinfected properly according to machine instructions. A printout of the ultrasound, or the **sonogram**, is made, the image is dated, and the report placed in the patient's file. The sonogram is the image of the internal structures. Some images may be recorded on videotape or CD. The ultrasound diagnostic report should note this and the tape or disc dated and stored within the medical record or other storage area.

Endoscopy

Endoscopy is the procedure of visually examining the interior of the body by using an endoscope or a tool used to view the inside of the body. The **endoscope** is made of a bundle of fine glass rods moved through the body to project an image using a light source. The use of the endoscope is limited by its diameter and length relative to the structure being passed. A small opening within the endoscope allows for other diagnostic tools and instruments to pass for further use of sampling. Such items as biopsies are completed with an endoscope. The procedure requires sedation or anesthesia.

The glass rods within the endoscope are fragile and may break easily. The endoscope should be stored within a padded case provided by the manufacturer. The endoscope should be stored in a straight position, disinfected and dried after each use, and stored in a dry area.

The veterinary assistant should be able to set up the endoscope and monitor screen, understand the proper cleaning and disinfection method, and understand storage factors. If tissue samples or biopsies are taken during the procedure, the assistant should prepare lab work forms and containers for sampling. The reference lab should be contacted for any special instructions of handling guidelines.

SUMMARY

Radiology procedures are an important part of the veterinary assistant's duties. The proper need for training and experience are necessary in areas of radiation safety, maintaining radiology logs and films, proper directional term understanding, proper set up and maintenance of the x-ray machine, proper film development and handling and proper patient restraint during radiology procedures. Radiology is an area that is changing every day and the need to maintain knowledge and education in this field is recommended.

Key Terms

anterior posterior (AP) the X-ray beam penetrates the anterior portion of the body area as the entrance and the posterior portion of the body as the exit point

automatic film processing equipment that develops and dries the film automatically

barium enema a large amount of substance passed into the colon via the rectum that fluoresces

barium study a series of films taken as the material passes through the digestive tract

barium sulfate an example of a material or substance administered into the body to show structures on an X-ray film that would otherwise be difficult to view

caliper device or instrument with a simple ruler-like tool that slides on a fixed bar or arm that moves parallel to it and is used to measure the thickness of a body part

cassette tool that holds X-ray film

centimeters (cm) measurement used in measuring body parts for X-rays

collimator part of the machine that sets the size of the X-ray beam that produces radiation to take the picture

contrast medium material or substance administered into the body to show structures on an X-ray film that would otherwise be difficult to view

developer solution chemical that develops the film

diagnostic imaging the term used for performing radiographs, ultrasounds, and endoscopy

dorsoventral (D-V) view is the animal positioned on its stomach with the X-ray beam going through the back first and the stomach second

dosimeter X-ray badge that measures the amount of radiation a person is exposed to

endoscope fiber-optic instrument used to visualize structures inside the body

endoscopy the procedure of visually examining the interior of the body by using an endoscope or a tool used to view the inside of the body

film hanger used to secure film for manual processing and development

fixer solution chemical that fixes the X-ray film and helps maintain the life span of the film allowing it to stay readable

fixing process of using a solution to maintain the life of an X-ray film

grid technique the plate that holds the film below the X-ray table

hand tanks large containers that hold chemicals and water from developing film

intensifying screen located within the film cassette to help produce a better exposure to the film

kilovoltage peak (kVp) the strength of the X-ray beam

lateral (LAT) an animal positioned on its side

lower GI a contrast film of the lower digestive tract organs and is often given via an enema

manual developing developing X-rays by hand using chemical tanks

milliamperage (mA) represents the number of X-ray beams based on time

oblique view used on areas that need to be placed at an angle to prevent double exposure from other body organs

processing process of developing film that involves being in a darkroom that has no light source or light leaks from the outside area

radiation safety hazard that allows X-rays to be produced to take a radiograph

(*Continues*)

radiology the study of radiation

radiolucent appearing black or dark gray in color and may indicate soft tissue or air and may be an exposure factor relating to increased exposure settings or improper patient measurement

radiopaque appearing white to light gray in color and may indicate hard tissue such as bone or may be an exposure factor in that the machine settings were too low or the patient was not properly measured

rinsing using water to remove any chemicals from an X-ray film

safe light a red light that is low in intensity and a filter that doesn't damage the film

screen area that lines the film cassette

sonogram the printout of an ultrasound recording

tabletop technique the distance between the X-ray tube and the top of the X-ray table surface and the film cassette is placed on top of the table and in contact with the patient

technique chart a listing of settings on the X-ray machine based on the thickness of the area to be radiographed

transducer the instrument used with an ultrasound that scans the body and transmits the waves back to the screen

ultrasound diagnostic tool using ultrasonic sound waves to view images of internal organs and structures

upper GI a contrast study of the structure of the upper digestive tract in which the barium is given orally

ventrodorsal (V-D) animal positioned on its back with the X-ray beam going through the ventral area first (stomach) and the dorsal area second (back)

X-ray common term for radiograph

X-ray log records the patient's name, client's name, date, X-ray number, X-ray position, thickness of body measurement, and area exposed

X-ray tube part of the machine that holds the radiation source

REVIEW QUESTIONS

1. What are safety equipment items that are used in the radiology area?
2. What is the kVp?
3. What is the mA?
4. What is the difference between a V-D view and a D-V view?
5. What information is kept in an X-ray log?
6. What tool is used to measure a patient? What is the correct way to use this tool to measure the animal?
7. What is the significance of the technique chart?
8. What is the difference between manual and automatic developing?
9. What information must legally be placed on a film?
10. What is an ultrasound?
11. What is an endoscope?

Clinical Situation

Delmar/Cengage Learning

Molly, a veterinary technician, and Lee, a veterinary assistant, are working in radiology at the Best Friends Vet Hospital. They have several patients to X-ray, including two large dogs, a difficult cat, and a rabbit. The facility has both an automatic processor and manual hand tanks. The vet would like the animals restrained for the X-ray views and would like to complete each view without the use of sedatives or anesthesia if necessary.

- What would you do to prepare for the X-rays?
- In what order would you complete the X-rays?
- What restraint items may be useful with the patients?
- How would you utilize the developing equipment to process the films?

50

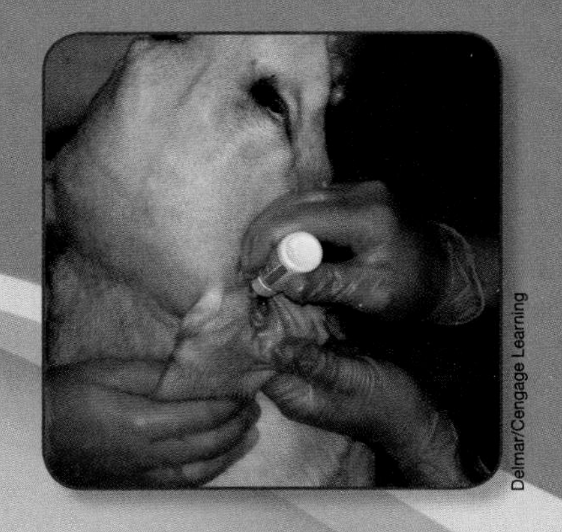

Pharmacy Procedures

Objectives

Upon completion of this chapter, the reader should be able to:

- Explain how to interpret a prescription
- Explain how to read labels correctly
- Demonstrate how to properly label a dispensing container
- Calculate the proper quantity of medication to dispense
- Demonstrate how to properly use a pill counting tray to count medications
- Explain how to process and log controlled substances according to DEA regulations
- Demonstrate how to administer oral medications to a patient
- Demonstrate how to administer aural medications to a patient
- Demonstrate how to administer topical medications to a patient
- Demonstrate how to administer ophthalmic medications to a patient
- Demonstrate how to properly store medications in the pharmacy according to manufacturer labels
- Discuss the importance of accuracy with pharmacy skills

Introduction

The **pharmacy** is the area where medications are stored and prepared for veterinary patients. Pharmacy skills are an essential part of the veterinary assistant's duties and must be accurate in several major areas, including reading the veterinarian's prescription, understanding common abbreviations used in the veterinary pharmacy, properly identifying the drug of choice, and properly calculating the amount of medication to dispense to a patient. Accuracy is essential and mandatory when working with medications and prescriptions.

Reading a Prescription

Veterinary prescriptions, often abbreviated Rx, are written in a combination of words and abbreviations. The **prescription** is the type of medication, amount of medication, and directions for use of the medication. The prescription is determined and prepared by the veterinarian. There are several parts to the prescription including the following (see Figure 50-1):

- Medication name
- Medication strength
- Method of administration
- Amount to be administered
- Frequency of administration
- Length of use
- Amount to be dispensed
- Special instructions
- Number of refills
- Veterinarian name

It is essential for the assistant to understand all parts of the prescription and have knowledge of common veterinary pharmacy abbreviations, as discussed in Chapter 1. The patient must legally have a VCPR established within 6 months to a year for the veterinarian to legally prescribe medications.

Competency Skill

Reading a Prescription

Objective:

To properly read and interpret a prescription written by the veterinarian.

Preparation:

patient file, prescription

Procedure:

1. Identify the drug name.
2. Identify the drug strength.
3. Determine how the medication is to be administered (oral, aural, ophthalmic, topical, etc.).
4. Determine how much is administered.
5. Determine the frequency of administration (SID, BID, TID, etc.).
6. Determine how long the medication should be used.
7. Determine how much medicine is to be dispensed.
8. Note any special instructions.
9. Determine the number of refills.
10. Recheck all items for accuracy.
11. Recheck any math used to determine items.

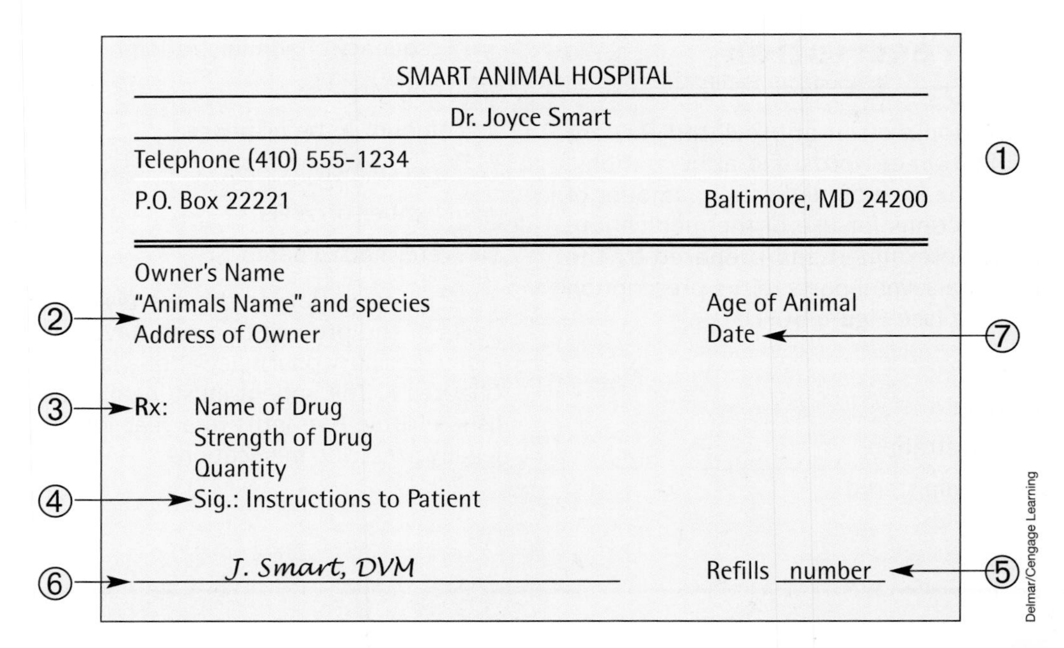

FIGURE 50-1 Parts of the prescription: (1) name, address, and phone number of veterinarian; (2) client's name and address; species and patient's name; (3) name, strength, and quantity of drug; (4) instructions for giving drug to patient; (5) number of refills; (6) veterinarian signature; (7) date script was written.

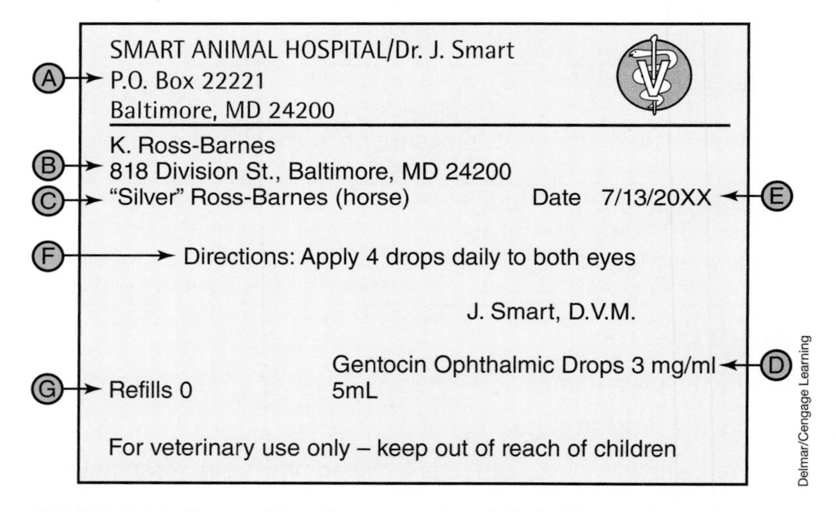

FIGURE 50-2 Parts of the dispensed drug label: (A) veterinarian name and address; (B) client name; (C) animal name and species; (D) drug name, strength, and quantity; (E) date ordered; (F) directions for use; (G) refill information.

Labeling a Prescription

The purpose of a **prescription label** is to indicate to the owner how to properly give the medication and identify the medication within the container (see Figure 50-2). Many facilities use printed or handwritten labels and place individual labels on each container. When a label is prepared it should be legible if handwritten, clean and understandable, accurate, and contain all required label information. Legally, the following information is required on a pharmacy label:

- Veterinary facility name, address, phone number
- Name of prescribing veterinarian
- Client name
- Client address if controlled substance

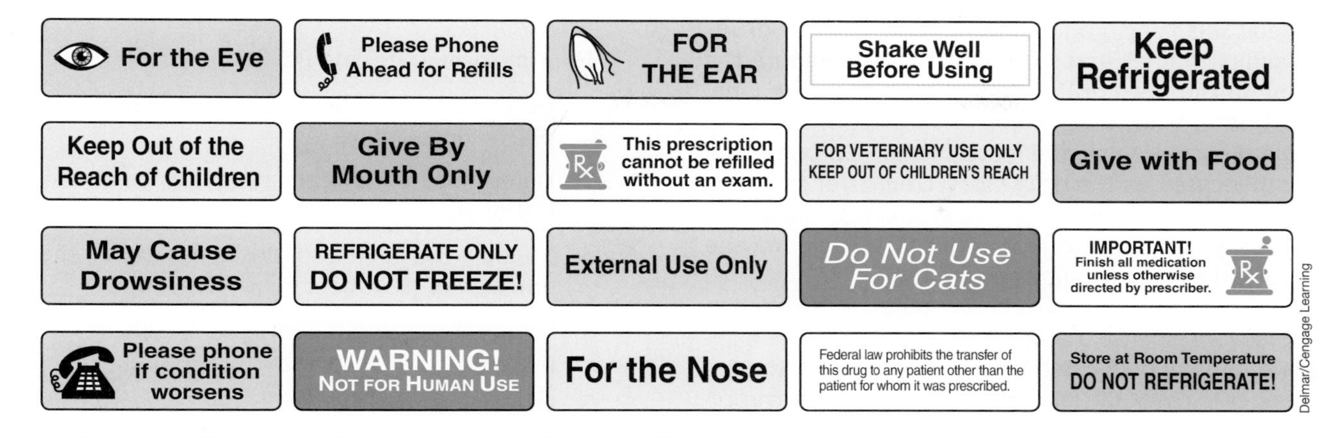

FIGURE 50-3 Warning labels placed on prescription medication.

- Patient name or ID number
- Medication name
- Medication strength
- Quantity dispensed
- Expiration date of medication
- Number of refills
- Amount per treatment or use
- Route of use
- Frequency of treatments
- Length or duration of use
- Special instructions

Additional labels may be required for the container and are often added as warning labels in sticker form, such as "Keep refrigerated," "Give with food," or "Shake well" (see Figure 50-3). Every container should be labeled with "For Veterinary Use Only" as a legal documentation of how the drug is meant to be used. When labeling the directions for use on the medication, it is imperative that plain and understandable basic words are used. Never use pharmacy terms or abbreviations on a label. Recheck each label three times for accuracy and discuss the directions with the client and make certain he or she has no questions on the use of the drug. When applying the label to the container, make certain it is centered and straight and no folds are located over any information.

Dispensing Medications

Every medication should be dispensed in a childproof container (see Figure 50-4). The only exception for this is if a client requests a non-childproof container due to physical reasons making it difficult to open such a lid. It is important to note that children or pets can easily get into containers and safety is an issue. It is important to

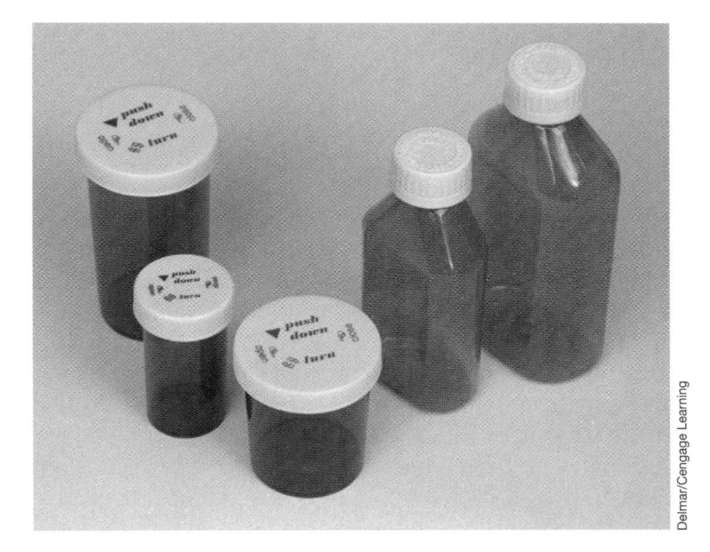

FIGURE 50-4 Various child-resistant caps.

document in the medical record when a non-childproof container is requested and dispensed.

Medication Containers

Plastic vials and bottles are commonly used with a twist-off or snap-off lid. The size of the container is measured in **drams**. The proper size vial or bottle must be selected to hold the required amount of medicine to be dispensed. Vials and bottles are usually amber in color to protect the medication from degrading due to light exposure. Vials are used for tablets, capsules, and powders. Bottles are used for liquid medications.

Drug Identification

Accuracy in selecting the proper medication to be dispensed is essential. The proper dosage and strength must also be accurate. A patient's condition may worsen or not improve if the wrong medicine or strength is selected. This is also a legal ramification. Many drugs have similar names and spellings. Always compare the spelling of the prescription with that of the drug label on the bottle. Many drugs come in a variety of strengths and the prescription should match with the proper drug strength located on the drug label. Drug strengths may be written in milligrams (mg), kilograms (kg), milliliters (ml), cubic centimeters (cc), or grains (gr). Some may be a combination such as mg/kg or mg/tablet.

All drugs selected should be evaluated for expiration dates. Never dispense a drug that has expired. The expiration date may be located on the drug label or the lid. Remember to check the following information for accuracy:

- Right patient
- Right drug
- Right strength
- Right quantity
- Right frequency

Amount to Dispense

The amount to be dispensed may be listed on the prescription or may need to be calculated by the veterinary assistant. This requires knowledge of pharmacy abbreviations and basic math. The amount to be dispensed is based on the prescription provided by the veterinarian. This requires knowledge of how much to give per treatment, how may treatments per day, and how long the duration of the treatment will be. Some medications may require determining a dose if the actual drug strength used is not equal to the required amount to be given. Recheck all dosages and dispensing amounts three times for accuracy.

Example 1:
Rx: Amoxicillin 250 mg PO BID × 14

1. Determine how often the drug is given (BID = twice a day).
2. Determine how long the drug is to be given (14 days).
3. Multiply the number of doses per day by the duration of treatment (2 × 14 = 28).
4. Locate the strength of drug as closely as possible to the prescription (250 mg).
5. Determine how many units per dose are needed. (250 mg tablet per dose).
6. Multiply the number of tablets per dose by the number of doses (1 × 28 = 28).This is the number to be dispensed (28).
7. Determine the amount per dose needed. This is done by dividing the units per dose by the units per tablet available. This example receives one tablet that is 250 mg.

Example 2:
Rx: Amoxicillin 250 mg PO SID × 10 d1.

1. Determine how often the drug is given (SID = once a day).
2. Determine how long the drug is to be given (10 days).
3. Multiply the number of doses per day by the duration of treatment (1 × 10 = 10).
4. Locate the strength of drug as closely as possible to the prescription (500 mg).
5. Determine how many units per dose are needed. (250 mg tablet divided by 500 mg tablet = 0.5 tablet).
6. Multiply the number of tablets per dose by the number of doses (0.5 × 10 = 5). This is the number to be dispensed (5).
7. Determine the amount per dose needed. This is done by dividing the units per dose by the units per tablet available. This example receives one half (0.5) of a 500 mg tablet once a day and requires dispensing five tablets.

Pill Counting Tray

When the amount to be dispensed has been determined, the assistant should locate the medication, and if tablets or capsules are used, the **pill counting tray** should be used for counting the required quantity. The pill counting tray is a device that has a flat area for medicine to be placed and a channel or funnel-like area to place medicine that has been counted and is to be dispensed to a patient (see Figure 50-5). This is helpful in limiting the amount of contact with the medicine and helps determine the proper amount to be dispensed is accurate.

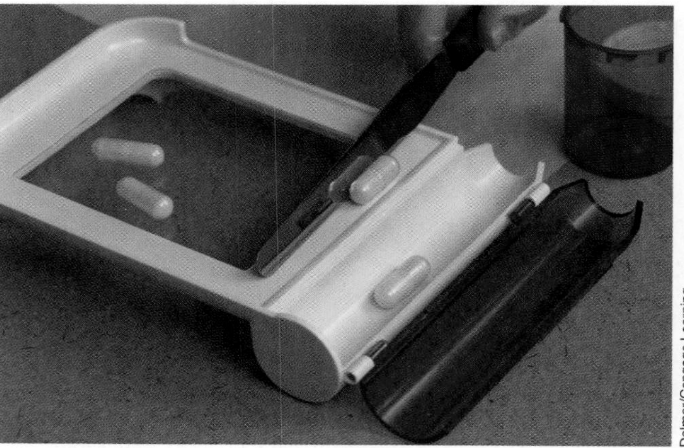
Delmar/Cengage Learning

FIGURE 50-5 Pill counting tray.

Competency Skill

Pill Counting Tray Procedure

Objective:

To properly prepare a prescription using a counting tray.

Preparation:

proper medication bottle, pill counting tray, spatula or tongue depressor, medicine vial or container

Procedure:

1. Place the pill counting tray on the pharmacy counter with the channel to the left and the open plate in front of you.
2. Pour the medication tablets or capsules onto the tray plate.
3. Open the channel cover.
4. Using a spatula or tongue depressor push groups of five tablets or capsules into the channel.
5. When you have counted the desired amount of medicine, close the channel cover. Lift the tray and place the channel spout into the medicine vial or container.
6. Tilt the tray to pour the medicine into the vial or container.
7. Place the vial on the counter.
8. Place the medicine bottle on the pharmacy shelf.
9. Clean the pill tray using water.

Types of Medications

Medications are divided into three categories: **over-the-counter drugs (OTC)**, **prescription drugs**, and **controlled substances**. OTC drugs are nonprescription drugs and can be purchased by anyone at anytime, commonly in human pharmacies. Prescription drugs can only be ordered and prescribed by a licensed veterinarian. Controlled substances are prescription drugs that have the potential for abuse or addiction and are regulated by the **Drug Enforcement Agency (DEA)**. The DEA has specific regulations and guidelines for ordering, storing, and dispensing controlled drugs. A veterinarian must have a **controlled substance license** to prescribe these drugs. This license is a separate number from the veterinary license and must be kept posted within the pharmacy. A controlled substance is also referred to as a **scheduled drug**, as the class of a controlled drug is categorized by schedule type from Schedule I to Schedule V (see Table 50-1).

Controlled Substances

Handling controlled substances is common among all veterinary health care team members. Several guidelines and regulations must be known regarding controlled substances and how they are handled, stored,

TABLE 50-1

Scheduled Drug Classes

SCHEDULE TYPE	EXPLANATION OR EXAMPLE
Schedule I	No medical use; illegal drugs, e.g., cocaine; limited use in research only; not found in veterinary facilities.
Schedule II	Potential for severe addiction, e.g., morphine; no refills allowed.
Schedule III	Moderate potential for addiction, e.g., Hycodan; limited to 5 refills every 6 months.
Schedule IV	Low potential for addiction, e.g., Valium; limited to 5 refills every 6 months.
Schedule V	Low potential of addiction or abuse, e.g, Robitussin; no limits on refills.

and dispensed. The only person who may order and prescribe controlled substances is a licensed veterinarian who has a DEA controlled substance license. If a veterinarian in the facility doesn't have a DEA license, he or she may not prescribe controlled drugs. These licenses are issued by the DEA and renewed every 3 years. The license should be posted within the pharmacy area within a reasonably visible area.

Controlled substances must legally be stored behind two locks. This means they must be stored within a locked box within a locked safe unit. The outer locked area must be permanently fixed in place securely. The access to the controlled substance storage location should be limited to one or two people within the facility. This controls the access to medications and potential theft.

Any controlled drug must legally be logged into the **controlled substance log** kept within the pharmacy. A written entry is required for every substance used or dispensed by the facility. The log must be recorded using blue or black ink and must be a permanently bound book maintained for at least 2 years with requirements on record keeping varying from state to state. Information necessary to be entered into the log includes the following:

- Name of controlled substance
- Drug strength
- Drug form (tablet, capsule, liquid, injection, etc.)
- Quantity dispensed
- Quantity on hand
- Date dispensed
- Time dispensed
- Client name and address
- Patient name
- Animal breed/species
- Initials or signature of person dispensing medication

A running inventory account of all controlled substances is legally required by each facility. The amount dispensed and the amount left on hand must be accurate and accounted for. A regular inventory should be maintained and be accurate according to the controlled substance log.

Educating Client on Medication Use

The veterinarian may discuss medications with the client but it is up to the staff members to make certain the client completely understands how to give the medicine, when, and how much. Certain terminology may confuse a client, or so much information is provided at one time that people may easily become confused and may be afraid to question or ask about a topic. There is always the potential for a client to misunderstand or have incomplete information, which may serve as a failure for proper pet health. This may delay or impede a patient's recovery. The veterinary assistant should be able to review all medication instructions before the client and patient leaves the facility. The client should also be urged to call with any questions or problems.

Some clients will receive a handout about how to give the medication with examples or diagrams showing hand placement. It is important that the assistant demonstrate the procedure for the client and ask if he or she has any questions. A written copy or review should include the following information:

- Why medication is being administered
- How medication is to be administered
- How much is to be administered
- When the medication is to be administered

Administering Oral Medications

Oral medications are administered by mouth. Oral medications may be tablets, capsules, or liquids. The form of the medication administered may be based on several factors. The species of animal depends on how the mouth is opened and what instrument may

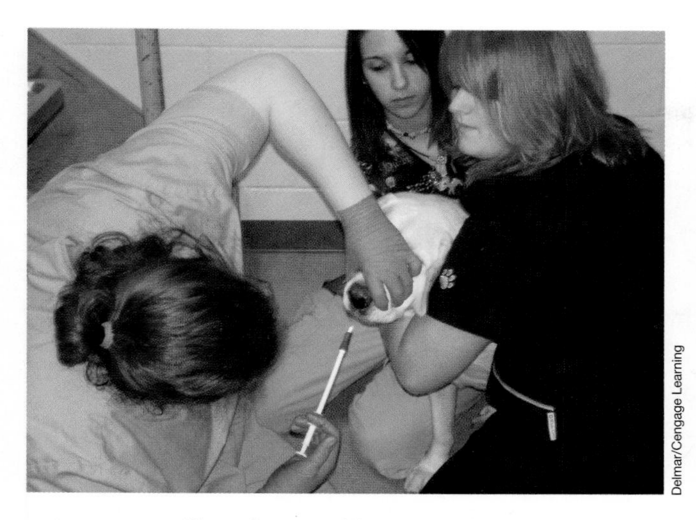

FIGURE 50-6 Use of a pet piller to administer medication to a dog.

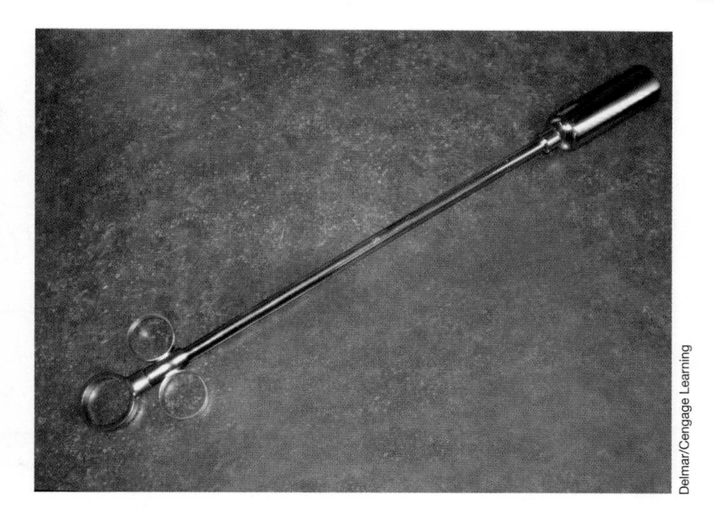

FIGURE 50-7 Balling gun.

be used to administer the medication. Small animals, such as dogs or cats, will require the mouth to be opened and the medicine placed at the back of the mouth and swallowed. Large animals, such as horses or cattle, require more power to hold and administer medications. Some animals may require pet piller or dosing syringes to aid in giving medicine by mouth. A **pet piller** is a small device, usually made of plastic, that has a long thin handle with a plunger on the end. The plunger holds the medicine and the handle is used to throw the medicine in the back of the throat (see Figure 50-6). Large animals may require a **balling gun**, which is a metal device with a long handle that has a plunger at the base, used to hold medicines (see Figure 50-7). The disposition of an animal also plays a role in how medication is given. When administering medicine, it is important to consider the type of drug used. Tablets and capsules may be coated with butter, shortening, or vegetable oil to make them slippery and easier to pass down the throat. They may also be coated with foods that tempt them to eat, such as peanut butter or a piece of meat. Some animals may eat the food item and spit out the medicine, so caution must be used in this method. Tablets may be crushed into a fine powder and mixed with food or water and administered as a liquid medication. This is often done using a mortar and pestle, which grinds down medications to a powder-like substance (see Figure 50-8). Liquids may be given using a syringe without a needle (see Figure 50-9). This allows proper measurement of the volume to be given and allows for slow administration to prevent the animal from spitting out the medicine. Pastes may be given from the original container, which is usually squeezed from the tube and is placed between the upper and lower teeth of the animal.

FIGURE 50-8 A mortar and pestle can be used to grind medications.

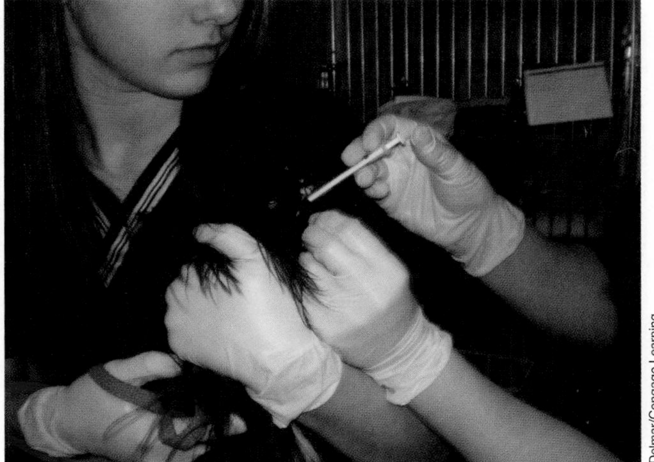

FIGURE 50-9 Use of a syringe to administer medications.

Competency Skill

Administering Oral Medications

Objective:

To properly administer a medication vial the oral route.

Preparation:

proper medication (right patient, right drug, right strength, right dose, right time), pet piller or balling gun, syringe, exam gloves

Procedure:

Tablets or capsule (small animal)

1. Apply gloves.
2. Elevate head upward.
3. Open the mouth at the side by pressing between the upper and lower jaw joint by cheek (see Figure 50-10).
4. Place medicine between index finger and thumb of free hand.
5. Place middle finger of same hand at front of mouth and apply pressure to open mouth wide.
6. Drop the tablet or capsule at the back of the throat.
7. If safely able to use index finger of free hand, push the medicine deeply into the throat (see Figure 50-11).

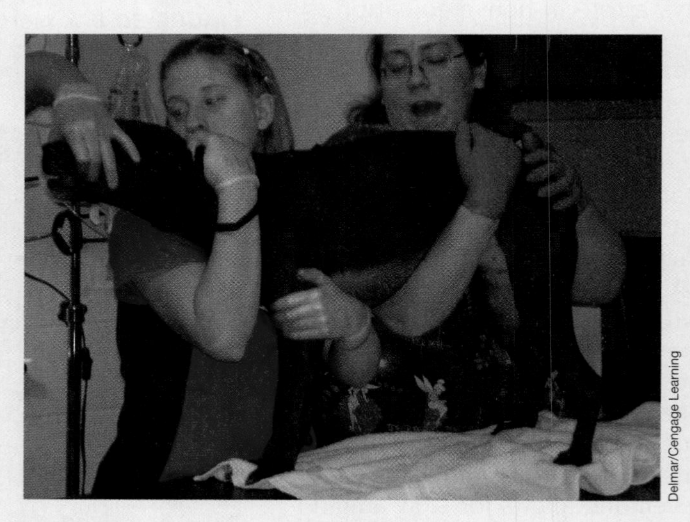

FIGURE 50-10 Press on the joint between the upper and lower jaw to open the mouth.

FIGURE 50-11 Gently press medication deep into throat if possible to do so safely.

8. Close the mouth and hold until the patient swallows.
9. Gently blow on face or rub throat to stimulate swallowing.
10. Monitor for any signs of the medication not being ingested.

Pet Piller or Balling Gun

1. Apply gloves.
2. Place pill in end of plunger of pet piller or balling gun.
3. Apply the end of the piller or balling gun into the side of the mouth between the upper and lower teeth.
4. Once in mouth, push plunger to throw medicine to the back of the throat.
5. Remove balling gun or piller and immediately elevate head upward.
6. Blow on nose or rub neck to stimulate swallowing.
7. Check to make sure medicine was swallowed.
8. Disinfect items and put away.

Liquids and Pastes

1. Apply gloves.
2. Elevate head upward.
3. Insert syringe or paste tube into the side of the mouth between upper and lower teeth.
4. Press the plunger to the desired amount into the back of the throat.
5. Remove syringe or tube and continue elevating head upward.
6. Blow on nose or rub throat to stimulate swallowing.
7. Keep mouth held shut until swallowing occurs.
8. Disinfect any tools and put away.

Administering Aural Medications

Aural medications are placed within the ear canal and may be used for such treatments as ear infections, ear cleaners, or ear mites. The opening of the ear canal structure includes the haired area of the **tragus**, located cranially on the ear flap (see Figure 50-12). The area located caudally and with no hair is called the **pinna**. The internal ear canal is shaped in an "L" fashion where at the beginning or exterior area the structure is wider and as the canal gets closer to the eardrum it gets shorter. The ear canal can be palpated through the skin, below the opening of the external pinna, and extending down toward the lower jaw. The ear canal is made of cartilage and when palpated feels similar to a rubber hose or tube.

Ears may be treated for many reasons, most commonly after developing an ear infection, which is more common in long-eared dogs and those with floppy ears. These ears do not allow for good air ventilation and moisture and bacteria may easily build up within the canal. Some dog breeds, such as the poodle or Bichon Fries, have hair that grows within the ear canal and may prevent proper ventilation. All animals are susceptible to **ear mites**, a parasite that occurs deep within the ear canal. Flies and mosquitoes may bite the ear flaps, especially in livestock, causing crusting and bleeding around the ear edges. Wax and other debris may become lodged within the internal ear canal, causing trauma and irritation. Dogs and cats commonly inflict bite wounds over the ear flaps causing damage to the pinna and surrounding tissue that may develop into a **hematoma**. This may also occur due to excessive head shaking and may be a secondary trauma to an ear infection. When an animal shakes its head severely it may cause a blood vessel to rupture and the pinna to fill with blood.

Medications placed into the ear are usually in liquid form, either as a drop-like solution or ointment (see Figure 50-13). Some medicine is packaged in tubes and others in bottles with droppers. Many tubes are meant for multiple uses. It is often necessary for the ear to be cleaned prior to aural medications being administered. When ears are a concern during a diagnosis, the veterinary assistant is asked to restrain the animal for an exam. Medications may be administered after the cleaning has occurred. The veterinary assistant may be asked to show clients how to properly administer aural medications at home. It is important that gloves be worn to prevent contamination. When a multiple-use ear dropper or container is used, the top should be cleaned after each use using alcohol.

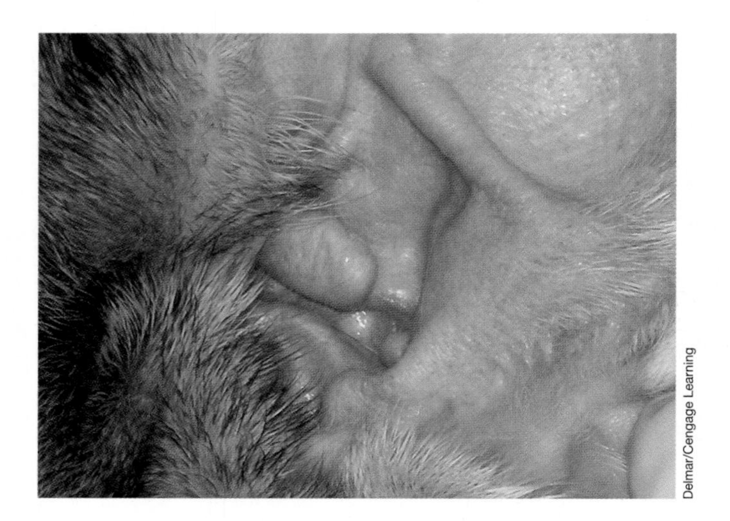

FIGURE 50-12 External anatomy of the ear.

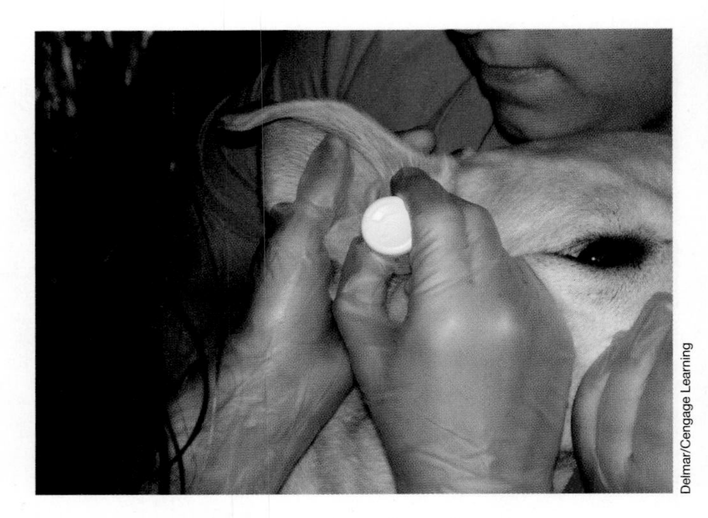

FIGURE 50-13 Administration of ear drops.

Administering Topical Medications

A **topical** medication is a substance applied to the outside of the hair or skin coat on an external body surface. Topical medications may include antiseptics to clean the skin surface, flea and tick preventions, or wound treatments (see Figure 50-14). Some areas of the body must be cleaned prior to a topical being administered. Areas of wounds or abrasions may have scabs or crusts that need to be soaked prior to topical medication application. This is best done by soaking with warm water and a surgical scrub, sterile saline, or betadine solution. This may take several minutes and soakings may need to be repeated. The area then must be dried. Application of topical medicines is then completed at the appropriate site following the veterinarian's recommendations.

Some topical medications come in a single-use container and others are in a large volume meant for multiple use. Topical medications in large volumes used in the treatment area should be kept sterile by removing the required amount from the container with an item that is clean and has not touched the

Competency Skill

Administering Aural Medication

Objective:

To properly administer medications via the aural route.

Preparation:

ear medication, exam gloves

Procedure:

1. Apply gloves.
2. Clean ears as recommended by the veterinarian. This procedure is discussed in Chapter 36.
3. Place medication dropper or tip deep inside the ear canal only to the depth of the start of the vertical portion of the "L" shape.
4. Apply the proper amount of medication into the ear as noted by the veterinarian. This is usually in drops that are counted according to the veterinarian's label.
5. Remove the medication dispenser from the ear.
6. Massage the base of the outside of the ear canal. This will create a swishing sound caused by the medication moving around the ear canal.
7. Wipe any solution that may have leaked onto the outside of the ear flap or hair.
8. Disinfect the medication dispenser with alcohol and place in appropriate area.

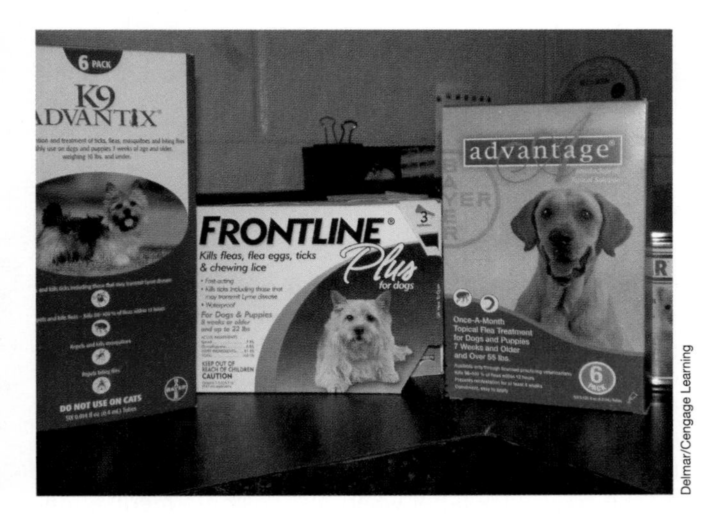

FIGURE 50-14 Flea and tick products are topical medications.

patient. This may be done using a spatula, tongue depressor, or other similar item. Areas of wounds on the skin should be treated carefully to prevent further damage or trauma. Healing tissue and wounds are fragile.

Flea and tick topical medications should be applied according to the label instructions. There are a variety of products available that offer monthly treatments. Each product treats specific animals for specific uses. Some topicals treat fleas only, whereas others treat fleas and ticks and possibly mosquitoes or other biting insects. Gloves must be worn when applying these topicals as they include a variety of chemicals that may be harmful to human skin. Most of these products require separating the hair and placing the entire contents in one area of the skin or in multiple areas of skin.

Competency Skill

Administering Topical Medications

Objective:

To properly apply topical medications for safe and effective use.

Preparation:

exam gloves, topical medication, tongue depressor

Procedure:

1. Apply gloves.
2. Clean area as necessary. This is discussed in Chapter 45.
3. Use a tongue depressor to transfer an amount of topical medication if the container is for multiple uses.
4. Apply the ointment onto the area in a circular motion, starting at the center of the wound and gently working outward.
5. Do not contaminate the medication by touching items that touched the animal. Flea and tick topicals should be applied according to the instructions.
6. Separate hair from skin and apply as directed.
7. Clean area and put items away.

Administering Ophthalmic Medications

Ophthalmic medications are ointments or solutions applied to the eyes (see Figure 50-15). This may be for treatment of an eye condition or used prior to bathing and grooming to protect the cornea from damage. Ophthalmic medicines are usually packaged for a single patient. In the veterinary treatment area, some items may be used on multiple animals and must be kept sterile. After removing the cap, wipe the end of the applicator with a cotton ball or gauze sponge with a small amount of alcohol. After using the medication, repeat this process. Avoid contact of the applicator with the animal's eye or other body parts when applying the medicine into the eyes. Touching the surface with the tip will not only contaminate it but may also cause trauma to the eye. Wear gloves when applying eye medicines. The eye will need to be held open to better view the eye and place the substance (see Figure 50-16). Eye drops can be dropped into the eye by holding the bottle above the open eye. Ointments should be applied by a thin layer over the lower eyelid and allow the animal to blink several times to move the substance throughout the eye.

Medication Storage

Medications have an expiration date and should be monitored for proper dating prior to dispensing. All medicines in the pharmacy should be stored with oldest dates first to use up the products that will expire first (see Figure 50-17). Drugs expiring last are placed at the back of the shelf. The expiration date is located on the label of the bottle. This date must be included on the pharmacy label when dispensing a medication. Any outdated items are removed from the shelf.

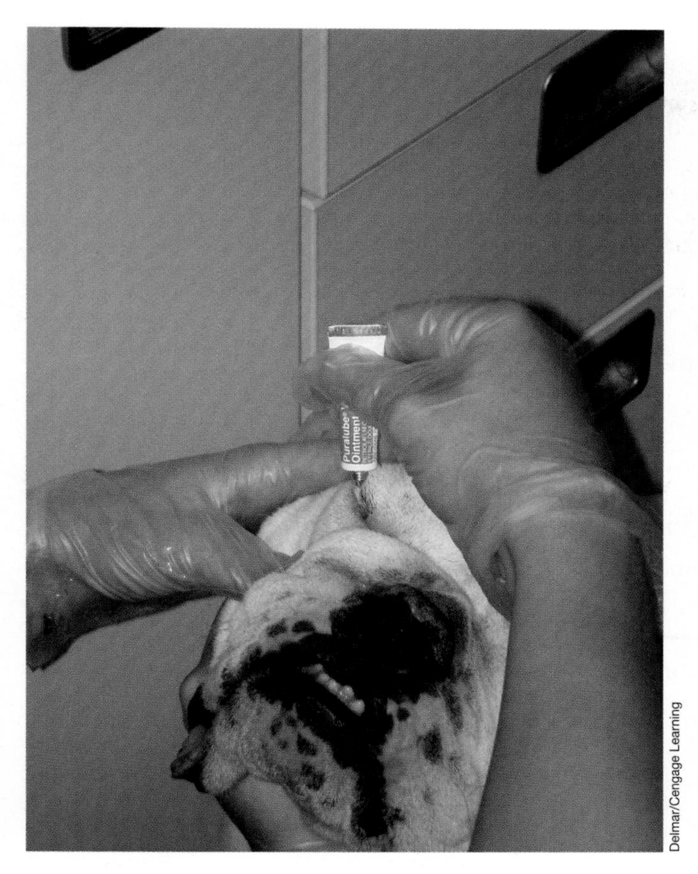

FIGURE 50-15 Application of ophthalmic medications.

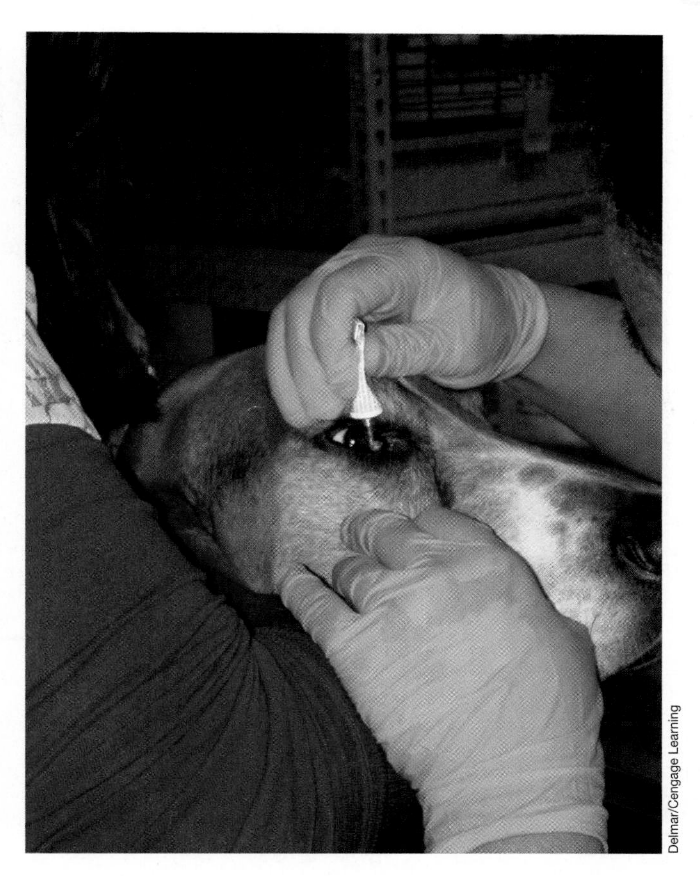

FIGURE 50-16 Hold the eye open to properly apply an ophthalmic medication.

Competency Skill

Administering Ophthalmic Medication

Objective:

To properly administer medication to treat eye ailments.

Preparation:

exam gloves, ophthalmic medicine, gauze sponges

Procedure:

1. Apply gloves.
2. Wipe any discharge from the patient's eye using a gauze sponge.

(Continues)

3. Open the end of the ophthalmic medicine and hold in one hand.
4. Using the free hand, use the index finger and thumb to pull the upper and lower lids apart to open the eye.
5. The thumb pulls the lower lid down and the index finger pulls the upper lid upward.
6. The other finger may rest on the head of the animal.
7. Gently tilt the head upward.
8. Apply the drops or ointment gently into the eye, counting each drop and applying the proper amount. Do not touch the surface of the eye with the dispenser.
9. Apply the ointment over a thin layer on the lower lid. Do not touch the surface of the eye with the dispenser.
10. Release the eyelids.
11. Allow the animal to blink to move the medication throughout the eye.
12. Clean the dispenser with a small amount of alcohol on a gauze sponge.

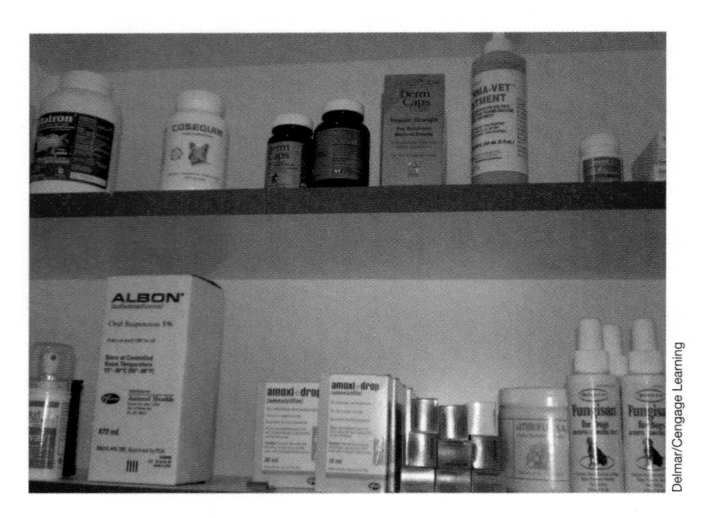

FIGURE 50-17 Proper storage of medications at the veterinary facility.

Drugs typically arrive in an individual container. Each medicine has a drug insert within the container that details the actions and use of the drug. Some medications may have the insert attached to the outside of the bottle. Storage information is also located on the drug insert and details how the medication is stored. Some items will need to be stored at room temperature and others may require refrigeration. Biologicals, especially vaccines, are stored under refrigeration. A separate refrigerator should be used for pharmacy items. Some drugs must be stored in a dark place and kept from direct light. These drugs will lose their **efficacy**, or the strength and life of the drug, when exposed to light sources. All drugs without information on storage will most likely be kept at room temperature and in a dry location.

Most veterinary pharmacies keep medications stored on shelves in alphabetical order for ease of locating. All controlled substances must be kept locked at all times. These items must be inventoried similar to other items in the hospital. When the items are opened, they should be sealed tight and replaced to their proper pharmacy location. Multiple bottles of the same medicine should be stored in a separate area to make certain they are not opened before the previous bottle is used. The pharmacy shelves must be cleaned on a regular basis to keep dust and debris from the bottles.

SUMMARY

Pharmacy techniques and skills require accuracy. When handling and dispensing medicines, it is important to make sure the right patient, right medicine, right strength, and right dose should be determined. If unsure of any item on a prescription, check with the veterinarian for clarification. It is best to ask the veterinarian to clarify what is written rather than risk incorrectly filling a prescription. It is important when calculating or dispensing medications to always check your work at least three times for accuracy.

Key Terms

aural pertaining to the ears

balling gun a metal device with a long handle that has a plunger at the base, used to hold medicines to administer to large animals

controlled substance prescription drugs that have the potential for abuse or addiction; also called scheduled drug

controlled substance license paper given by the DEA to allow a veterinarian to prescribe scheduled drugs

controlled substance log written entry is required for every substance used or dispensed by the facility

dram the measurement size of vials or bottles

Drug Enforcement Agency (DEA) government agency that sets regulations and guidelines for ordering, storing, and dispensing controlled drugs

ear mites microscopic parasites that live within the ear canal

efficacy the strength of the drug

hematoma the rupture of a blood vessel causing a fluid filled pocket of blood

over-the-counter drugs (OTC) medications that do not require a prescription

pet piller small device, usually made of plastic, that has a long thin handle with a plunger on the end to administer medications to small animals

pharmacy the area where medications are stored and prepared for veterinary patients

pill counting tray tool used to count out tablets or capsules to send home with a client

pinna the ear flap

prescription (Rx) type of medication, amount of medication, and directions for use of the medication prepared by the veterinarian

prescription drug medication prescribed by the veterinarian

prescription label label to indicate to the owner how to properly give the medication and identify the medication within the container

scheduled drug prescription drugs that have the potential for abuse or addiction; also called controlled substance

topical a drug that is administered to the skin

tragus opening of the ear canal structure; includes the haired area

REVIEW QUESTIONS

1. What items should a prescription include to fill it?
2. What items must be included on a label to dispense a medicine?
3. What items should be checked at least three times for accuracy when dispensing medicines?
4. Complete the following pharmacy questions based on the information given:

   2 caps Cephalexin 500 mg BID × 14

   a. What does the prescription call for?
   b. How many items should be dispensed for this dosage?
   c. What total mg amount is given per day?

   ½ tab Amoxicillin 100 mg TID × 10

   a. What does the prescription call for?
   b. How many items should be dispensed for this dosage?
   c. What total mg amount is given per day?
5. Why is a pill counting tray used?
6. What items are recorded in a controlled substance log?
7. What is the difference between a pet piller and a balling gun?
8. What is an aural medicine?
9. What is a topical medicine?
10. What is the importance of storing items out of a direct light source?

Clinical Situation

Delmar/Cengage Learning

Amy, a veterinary assistant at Seaside Vet Clinic, is working in the pharmacy to dispense hospitalized patient medications. She has several prescriptions to fill for Dr. Andrews. Amy has to fill the following prescriptions:

Advantage Canine for a 50# dog, apply 1 dose topically once a month for 6 months

2 capsules Doxycycline SID for 3 weeks

1 tablet Cefa-Tabs BID for 1 week

1 ml Amoxi-Drops BID for 10 days

- What information does Amy need to fill each prescription?
- How should each prescription be written for the client?
- How many items of each prescription will Amy send with the patient?

51

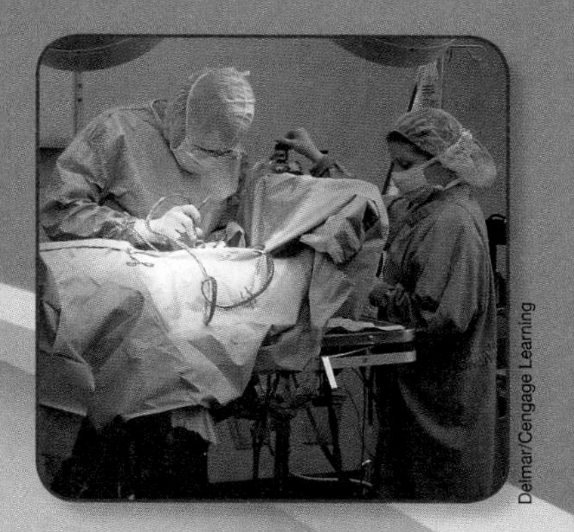
Delmar/Cengage Learning

Surgical Assisting Procedures

Objectives

Upon completion of this chapter, the reader should be able to:

- ☐ Describe how to maintain asepsis during surgical procedures
- ☐ Explain how to maintain and record in a surgical log book
- ☐ State the importance of a sterile surgical suite
- ☐ Demonstrate how to assist with anesthesia preparations and inductions
- ☐ Demonstrate how to properly restrain the patient for intubation
- ☐ Demonstrate how to properly clip and prepare the patient for surgery
- ☐ Demonstrate how to properly handle and open sterile surgical packs
- ☐ Demonstrate how to properly prepare surgical drapes, gowns, instrument packs, and towel packs for surgical procedures
- ☐ Demonstrate how to properly use an autoclave for sterilization procedures
- ☐ Demonstrate how to properly assist the surgeon in gowning
- ☐ Demonstrate how to properly position the patient for surgery
- ☐ Describe how to place and maintain patient monitors
- ☐ Explain the planes of anesthesia
- ☐ Demonstrate how to properly monitor a patient during anesthesia and post-operative recovery
- ☐ Demonstrate how to set up, maintain, and disconnect the anesthesia machine
- ☐ Describe the difference between rebreathing and non-rebreathing anesthesia systems
- ☐ Demonstrate how to properly refill the vaporizer and soda lime components of the anesthesia machine
- ☐ Identify and locate components on the anesthesia machine
- ☐ Select appropriate-sized rebreathing bags for the patient
- ☐ Demonstrate how to properly test the functions of the anesthesia machine
- ☐ Explain how to make adjustments to the anesthetic flow of gases during induction, maintenance, and recovery phases
- ☐ Describe the anesthesia and surgical report
- ☐ Demonstrate how to properly glove and gown for surgical assisting
- ☐ Demonstrate how to properly extubate a patient during recovery
- ☐ Explain how to review the postoperative care of a patient
- ☐ Complete a suture removal appointment
- ☐ Clean and maintain the surgical suite

Introduction

Surgical assisting requires the veterinary assistant and all other staff members to understand and maintain **aseptic techniques**. This is essential in the surgical suite. This requires knowledge and attention to the patient and the staff members working in the surgical area. **Asepsis** governs the patient, the health care team, the surgical suite, the instruments, and all items housed within the surgical area. Cleaning and sanitation are a vital part of keeping the surgery suite sterile and aseptic. Any break in asepsis may lead to a potentially life-threatening patient infection, delayed healing, or patient death.

Surgical Asepsis

The goal of surgical aseptic technique is to prevent any organisms from entering into a patient via the surgical incision, inhalation, or IV anesthesia. The first barrier of entry is through the skin. Lengthy surgeries pose more of a potential risk for infection, as well as surgeries involving deep body cavities, immune-compromised patients, and orthopedic surgeries. Many preparation techniques have been developed to ensure patient safety.

Proper disinfecting and sterilization techniques help reduce the number of living organisms in the surgical environment. Any item that is not sterile is not usable. The goal is to maintain the utmost sterility of every object that comes in contact with the patient during surgery. This would include anesthetic equipment, the surgery table, surgical prep equipment, surgical instruments, and staff members.

Another area that must be considered in maintaining asepsis is the surgical suite's ventilation and air flow. The surgical suite should have a separate ventilation area from the rest of the hospital. Surgical doors should be kept closed during all surgery procedures. Staff member traffic should be kept at a minimum to reduce the risk of spreading organisms.

Only sterile surgical procedures should be completed in the surgery room. This means such procedures as flushing abscesses, unblocking urinary obstructed cats, suturing wounds, and performing dentals should be done in the treatment area to reduce the spread of living organisms.

Surgical Log Book

All aspects of the surgery should be noted in a surgical log and report. This includes presurgical documentation, surgical notes, and postoperative recovery stages. The surgical consent form, surgical fee estimate, anesthesia report, surgical report, and recovery report are forms that should be included in each patient's medical record. The **surgical log** is a record of details recording all surgeries performed in the facility (see Figure 51-1). Similar to the radiology log, this book is used to meet legal and state regulations in noting all procedures completed within the facility. The information within the surgery log should include the following:

- Surgery date
- Patient name or number
- Client name
- Breed/species
- Gender
- Weight
- Procedure(s) performed
- Pre-anesthetic medications administered, dosage, and route of use
- Anesthetic administered, dosage, and route of use
- Surgical assessment score
- Technician and assistant initials
- Veterinarian's initials
- Length of surgery time
- Laboratory specimens taken
- Surgical comments

The surgery log book is kept in the surgical suite and entries are recorded prior to surgery and at the completion of the surgery.

Anesthesia Log Book

An **anesthesia log** book may also be kept in addition to or included in the surgery log book. The anesthesia log details the patient's status through induction, maintenance, and recovery stages. This should also include TPR and BP every 5 minutes, fluid type, and amount received during the procedure, pain medicines used, and oxygen flow rates. These items should be noted every 5 to 15 minutes throughout the surgical procedure. Items that should be recorded in the anesthesia log should include the following:

- Pre-anesthetic drug and dosage
- Anesthetic drug and dosage
- Route of drug use (IV, IV, SQ)
- Time medications given
- Pain relief drugs and dosage

ANESTHESIA/SURGERY LOGBOOK

Date	Patient	Client	Procedure(s) Performed	Drugs Administered	Dosage	Route(s) of Administration	Length of Procedure	Surgeon(s)	Anesthetist

Comments:

____

____

____

____

____

____

____

FIGURE 51-1 A sample anesthesia/surgery log book.

- Vital signs
- Complications
- Length of anesthesia
- Technician initials

The anesthesia log is helpful in determining what medications were used during each patient's procedure and any adjustments that may be necessary on follow-up procedures. The log is also legal documentation of the anesthesia used during surgery.

Surgical Suite Maintenance

The surgical suite should be the cleanest area of the veterinary facility. The surgery area should only be used for the surgical procedure. Patient preparation should be completed in the treatment area and the patient carefully moved to the surgical suite to reduce risk of contamination. The surgery room should be kept closed and sealed to prevent spread of organisms. Only surgical staff should be in the room during surgeries. The surgery room itself should have separate cleaning equipment from all other areas of the hospital. Sanitation should occur after each surgery and a final sanitation and disinfection procedure completed at the end of every surgery day.

Ceiling Sanitation

The ceiling of the surgical suite should be spot-cleaned daily. The entire ceiling should be mopped once a week using a sponge mop and bucket used only within the surgical suite. The ceiling should be cleaned prior to other areas, as dust particles and hair or other debris that have collected on the ceiling may fall onto other objects during cleaning. Dry vacuuming may be necessary depending on the ceiling surface material. When vacuuming areas within the surgery area, use a clean filter and bag each time to prevent spread of organisms. Ventilation fan filters should be changed on a weekly basis.

Wall Sanitation

The walls of the surgical suite should be spot-cleaned after each surgery. Each wall surface should be mopped down using a sponge mop and bucket on a daily basis. A vacuum may be used as necessary as outlined in the ceiling sanitation section. Spot-cleaning of walls should be done with a disinfecting cleaner and paper towels.

Counter and Shelf Sanitation

Counters, tabletops, shelves, sinks, and waste containers should be disinfected on a daily basis. Items that require spot-cleaning should be cleaned between each surgery using a disinfectant and paper towels. All flat surfaces and edges should be wiped down and kept clean. All items that have been used and are disposable should be placed within a medical waste container. The medical waste container should be emptied each day or when full between surgeries.

Floor Sanitation

Floors should be mopped on a daily basis and may require mopping between surgeries. The surgical mop and bucket should only be used in the surgery area. It may be necessary to label these items "for surgery room use only." Never use a regular hospital mop and bucket that have been used in other areas of the facility. This is an easy way to contaminate the surgery room from highly contagious sites.

Surgical floor mopping should include the dual mop method. This process involves one bucket containing fresh warm water for mop rinsing. A second bucket should contain a disinfectant solution for actual floor mopping. The mop is placed in the disinfectant solution and wrung out well. A section of floor is mopped in the furthest area of the surgical room. The mop is then placed in the rinse water and wrung out. It again goes into the disinfectant solution and is wrung out and mopping continues. This process continues until the entire floor area is mopped. At the end of the mopping, the mop is rinsed and disinfected and wrung out well and hung to dry. Empty both buckets immediately and refill them for each use. The mop head should be washed weekly. This should be done in a washing machine with hot water and bleach. Multiple mop heads should be available for regular changing and cleaning purposes.

Equipment

All surgical equipment should be cleaned and disinfected following the manufacturer's recommendation. Any surgical equipment that is located within the surgery suite may not be able to be cleaned depending on the part of the equipment and its power source. It is important to read the cleaning and care guidelines for all items. Permanent fixtures, such as surgical lights, should be wiped down and cleaned daily. The surgery table must be disinfected after each use and should include cleaning the top surface, edges, bottom surface, and base (see Figure 51-2). The same procedure applies to any

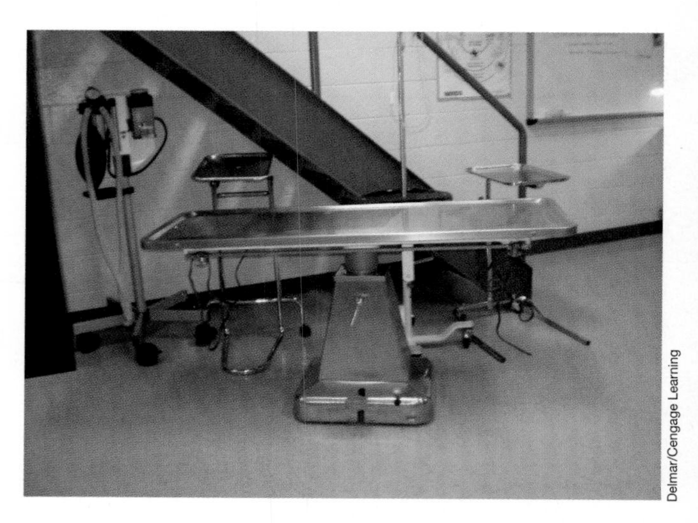

FIGURE 51-2 The surgical suite, table, and equipment must be cleaned and maintained on a daily basis.

Mayo stand, or the instrument tray that is elevated to hold the surgeon's surgical instruments and supplies. Surgical ties should be washed weekly unless they are soiled or contaminated by bodily fluids, in which case they should be laundered immediately. Any surgical positioner and restraint devices should be disinfected after each use. It is important to note that any disinfectant used in the surgical suite be safe for patient contact as most items are used in close contact to the patient.

Pre-anesthetic Patient Care

Any patient undergoing a surgical procedure should be admitted into the hospital early in the day to complete necessary presurgical blood work and physical exams prior to the surgery. All consent forms and fees, as well as an estimate and review of procedures, should be provided to the client to complete as soon as the patient is admitted. Emergency surgical procedures are worked into the schedule dependent on how soon the patient needs to be attended to. The early admission also allows patients to **fast**, or not have any food or water, prior to surgery depending on the animal species and required fasting time. For some species this means a 12-hour fast and for others fasting is not required. This is dependent on the animal's digestive system. Dogs and cats typically are fasted 12 hours prior to surgery. Rodents, cattle and other ruminants, and horses do not require fasting. Monogastric animals are fasted so that no contents are in the stomach, reducing the chances of vomiting during anesthesia.

The physical exam for surgery patients should include an evaluation of each body system, evaluation of the vital signs, and the overall health status of the patient. The veterinary assistant should evaluate the vital signs and the veterinarian should complete the PE evaluation.

Presurgical blood work is frequently performed and may be required by the facility or the type of surgical procedure involved (see Figure 51-3). The information provided by the blood work helps to determine the anesthetics that will be administered to the patient and the anesthesia classification of the patient. The classification of each patient is a presurgical assessment score that rates the patient in a class between I and V. The age and physical status of the patient serve as a guideline for the classification. The older the patient, the higher the score and the more surgical protocols that will be in place during the procedure. Table 51-1 summarizes the classification.

The choice of presurgical blood testing, pre-anesthetic drugs, anesthesia, and surgical monitoring will vary according to the classification assessment. Each facility will have its own protocol and doctor preference for presurgical and surgical procedures. The choice of pre-anesthetic medications will depend on several factors including the following:

- Species/breed
- Age
- Anesthesia

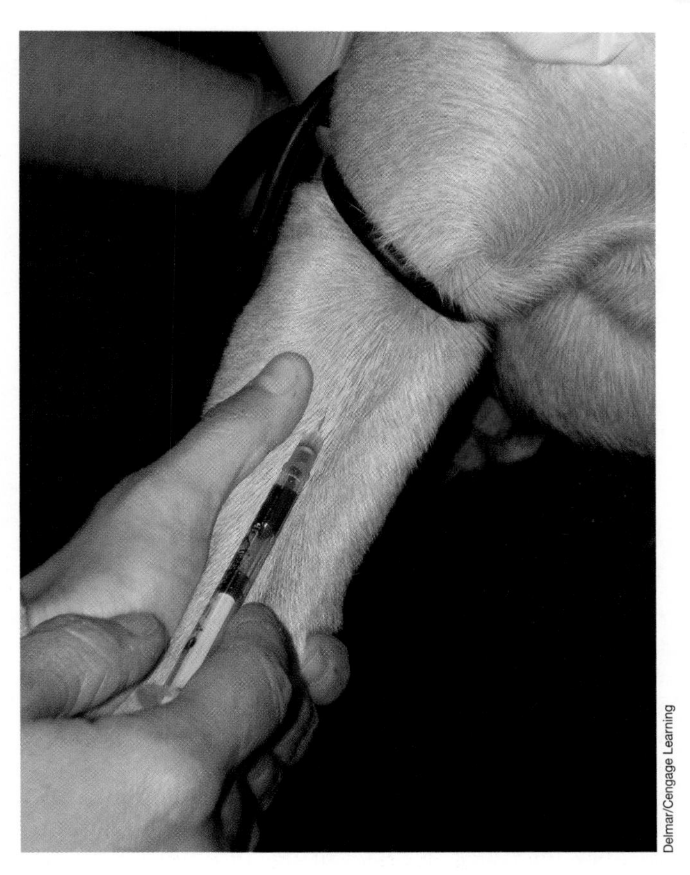

FIGURE 51-3 A blood draw is an important step in the presurgical assessment.

TABLE 51-1

Anesthesia Classification Assessment

Class	Description
Class I	Minimal Risk: Normal, healthy patient
Class II	Slight Risk: Slight systemic disease with no clinical signs; may include very young, geriatric, underweight or obese patients
Class III	Moderate Risk: Moderate systemic disease; slight clinical signs noted
Class IV	High Risk: Severe systemic disease; severe clinical signs with a threat to life
Class V	Grave Risk: Moribund patient; may die with or without surgery within a 24-hour period

- Classification/score
- Surgical procedure
- Length of surgery
- Veterinarian preference
- Patient weight

The purpose of pre-anesthetic drugs is to calm patients, reduce pain, reduce the amount of anesthesia necessary for the surgery, and reduce side effects of other drugs necessary during the procedure. All medications used for presurgical anesthetics should be recorded in the patient chart and any controlled substance noted in the log.

Fluid Therapy

A surgical protocol that may be used for every surgical patient or on surgical case basis is fluid therapy. It is important that all surgical patients have an IV catheter placed during the pre-anesthesia preparation. This provides an access to the patient's bloodstream for easy administration for anesthesia medications and pain medicines. It is also an assurance to the bloodstream should any complications occur during anesthesia or surgery for quick access during an emergency. The IV catheter is commonly used as a fluid therapy site to provide fluids during surgery. Most IV catheters are placed in the cephalic vein by the veterinary technician and attached to an IV fluid bag and line. The fluids of choice are usually **isotonic crystalloid** fluids, such as Lactated Ringer's Solution (LRS). In small patients or young patients, a 5% dextrose solution may be used or added to the LRS. The veterinarian should select any fluids being used with the patient. The flow rate of the fluids is typically 10–20 ml/kg/hr. Dehydrated patients will require a higher flow rate. Patients with renal or cardiac damage should receive a reduced rate of flow, which should be established by the veterinarian. The veterinary assistant should be able to set up an IV bag and line along with the proper-sized IV catheter necessary for the technician to place the catheter. The assistant should also be able to properly monitor the fluid rate and flow.

Anesthesia Induction

The veterinary assistant will provide restraint for all patients being induced with anesthesia. The **induction phase** is the time when the patient is being given anesthesia to make it sleep and remain unconscious and free of sensation during the surgery. The veterinarian or technician will administer the anesthesia. The assistant should properly restrain the patient depending on the type of anesthesia being administered. Anesthesia may be given as an injectible, such as IV or IM, or may be given as an inhalant, in which the patient inhales the anesthetic gases. The inhalant route may cause the animal to struggle and enough staff members should be available for additional restraint. The inhalant route uses an anesthesia machine to continue to keep the patient anesthetized (see Figure 51-4). The vital signs should be monitored throughout the procedure. This is often done using a pulse oximeter, which evaluates the heart rate and oxygen levels (see Figure 51-5). Once an animal has been induced and has achieved the brief period of unconsciousness, the patient should be intubated with an endotracheal tube. **Intubation** is the process of placing a tube into the trachea to establish an airway and allow the patient to continue to inhale gases that keep the patient under general anesthesia as long as necessary. The **endotracheal tube** is a flexible tube that should fit snuggly into the trachea.

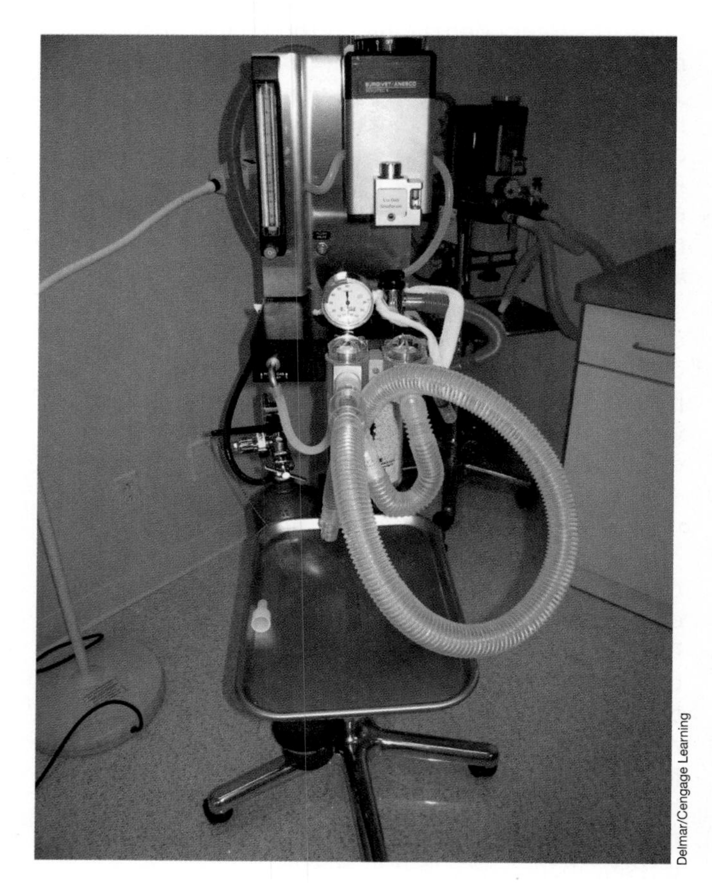
Delmar/Cengage Learning

FIGURE 51-4 A machine used to provide inhalation anesthesia.

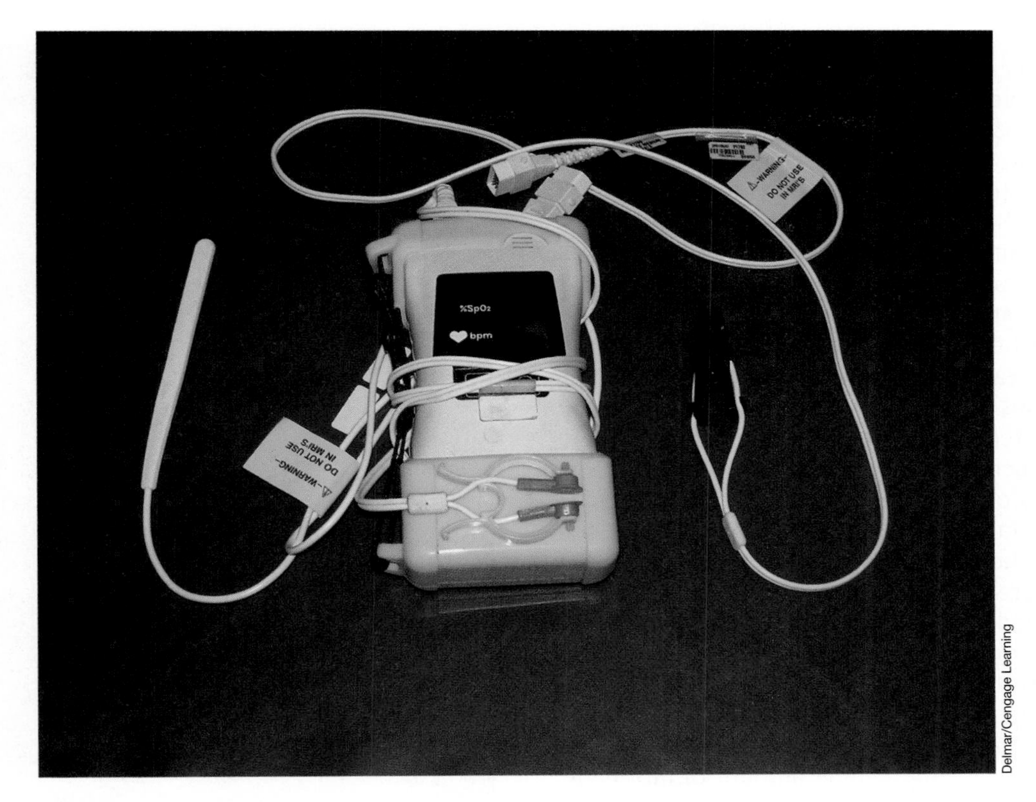

FIGURE 51-5 Pulse oximeter.

The **plane** of anesthesia determines how awake or how asleep a patient is under anesthesia. A **blink reflex** can be evaluated to determine which plane of anesthesia a patient is in. The blink reflex should not be visible in an adequate plane of anesthesia for surgery. An eye ointment lubricant is placed into each eye to protect the cornea and prevent the eyes from drying out during the anesthesia and surgical procedure.

Intubation Procedure

The veterinarian or veterinary technician performs the endotracheal tube placement as the assistant restrains for the procedure. Intubation consists of passing a tube through a patient's mouth and pharynx and into the trachea. This establishes an airway and allows the patient to be attached to the anesthesia machine during a procedure. It also prevents the patient from inhaling and aspirating saliva, vomit, or other fluids into the lungs. The endotracheal tube comes in a variety of lengths and diameters to fit securely into the trachea of various animal species (see Figure 51-6). The selected size should be the approximate diameter of

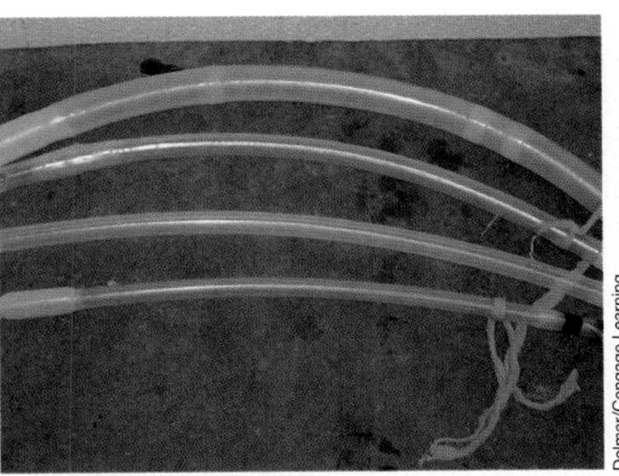

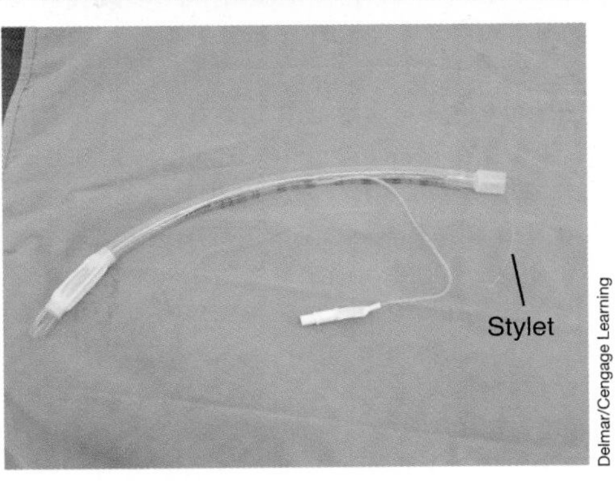

FIGURE 51-6 Endotracheal tubes.

the patient's trachea. The tube is made of heavy-duty plastic or rubber material that is slightly curved in the shape of the trachea. One end is slightly tapered to allow ease of placement into the trachea. The opposite end has an adapter that attaches to the anesthesia machine. Some tubes have a **cuff** located about 3/4 of the way down the tube that is inflatable and allows the tube to be inflated with air to create a tight seal around the trachea to prevent any moisture or material from entering the lung field. The cuff is attached to the endotracheal tube by a thin tube running parallel to the endotracheal tube. When inflated, the internal and external parts of the cuff look like a balloon. The cuff is inflated using a syringe that pushes air into the cuff. Too much air can be placed into the cuff causing the tube to overinflate, so care must be taken to not irritate or damage the trachea. The tube and the cuff should be checked prior to use to make certain no tears or leaks are present. It is important for the assistant to understand how the cuff and tube work during the intubation process, as the cuff must be deflated prior to **extubation** or endotracheal tube removal from the patient.

Actual preparation of the endotracheal tube should consist of placing a small amount of lubricant, such as KY gel, on a gauze sponge. The end of the endotracheal tube is lubricated for ease of passage into the trachea. Some veterinarians prefer a topical anesthetic lubricant that is placed directly on the trachea prior to tube placement. An example of this type of solution is lidocaine gel. This is most commonly used in cats due to a very active reflex that opens and closes the trachea during intubation. Lidocaine gel helps relax the reflex.

The patient is placed in sternal or lateral recumbency with the head and neck extended upward to allow the throat to be viewed during placement. The mouth is held open with the tongue pulled forward and down over the patient's front teeth. This requires gloves and the tongue can be held using a gauze sponge to prevent slipping. The veterinarian or technician will use a **laryngoscope** to place the tube into the trachea (see Figure 51-7). This is a tool made of heavy metal that has a light source on the end to help light the airway view. The laryngoscope is placed into the throat area and a blade that extends off the tool is used to hold down the cartilage of the larynx. As the patient breathes, the larynx reflex opens and closes and the laryngoscope helps in visualization of the laryngeal cartilage and trachea. Once the tube is placed within the trachea, it should be checked for proper placement. This may be done by feeling the end of the endotracheal tube during a breath to feel the air expired from the patient or when attached to the anesthesia machine, note that the breathing is in

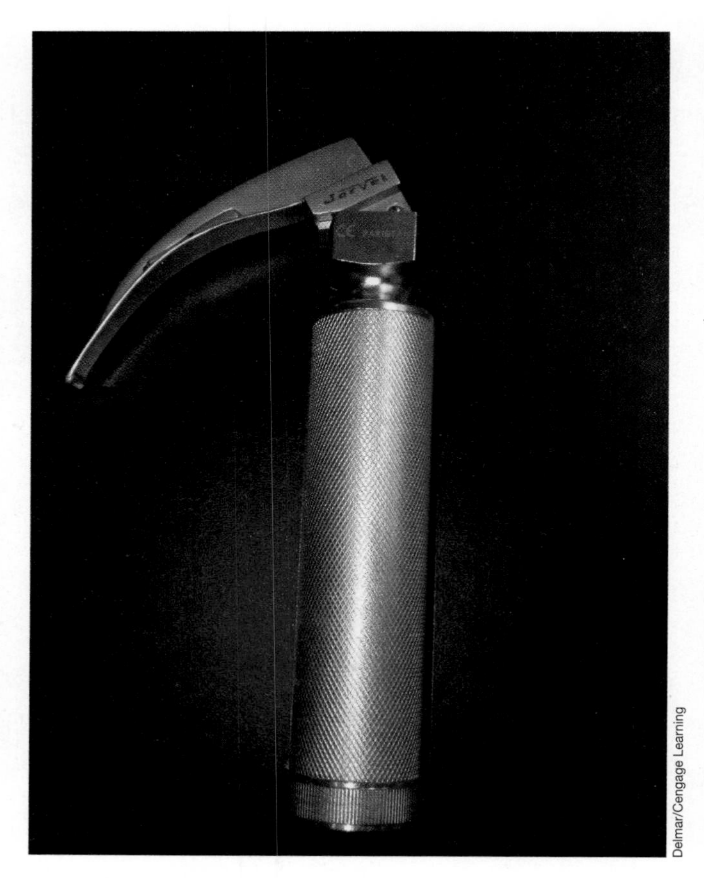

FIGURE 51-7 Laryngoscope.

coordination with the bag on the machine. As soon as the tube is in place, it should be secured by placing a piece of rolled gauze above the adapter lip of the endotracheal tube in a square knot and both sides secured behind the patient's canine teeth and tied over the patient's nose or secured in a bow behind the patient's ears. This prevents the tube from accidental displacement. At this point, the assistant releases the patient's mouth and inflates the cuff and then rotates the patient into lateral recumbency.

The patient is then attached to the anesthesia machine by attaching the hose to the end of the tube adapter. The patient is then placed on the machine for inhalation anesthesia, which is maintained throughout the procedure. Various vital sign monitors are then placed on the patient according to veterinarian preference and availability.

Patient Monitors

Patient monitors may include such equipment as a pulse oximeter, a temperature probe, a respiratory monitor, an esophageal stethoscope, blood pressure monitor, and an

Competency Skill

Restraint for Intubation

Objective:

To properly restrain a patient for safe insertion of an endotracheal tube.

Preparation:

gauze sponge, lubricant (KY gel or lidocaine gel), proper-sized endotracheal tube, gauze roll, laryngoscope, 3 or 5 ml syringe, exam gloves

Procedure:

1. Apply gloves.
2. Place patient in sternal or lateral recumbency.
3. Extend the head and neck and open the mouth widely.
4. Apply a gauze sponge to the tongue and pull forward and down between the front teeth.
5. Follow the veterinarian's or technician's positioning needs.
6. Once the endotracheal tube is placed, close the mouth and hold the end of the tube.
7. Tie the end of the tube with a piece of rolled gauze.
8. Place the ends of the gauze roll on either side of the face and tie on top of nose or behind the ears. Tie the ends in a bow for a quick release.
9. Inflate the tube cuff using a 3–5 ml syringe and close the cuff.
10. Place the patient in lateral recumbency.

electrocardiogram (ECG or EKG). These monitors are important for detailing a patient's vital signs and ensuring the animal is stable throughout the anesthesia procedure. These monitors can also detect the plane of anesthesia an animal is in and if more or less inhalant is necessary.

The **pulse oximeter** is a device used to measure a patient's vital signs by indirectly measuring the oxygen saturation within the blood and any changes in the blood volume. The tool is commonly called a *pulse ox* and is attached to the patient with a probe that may be placed on the tongue, pulse point over a paw, or a rectal probe. The screen readout shows the oxygenation rate at all times and some units provide a heart rate, temperature, and respiratory rate. The monitored signal bounces in time with the heartbeat because the arterial blood vessels expand and contract with each heartbeat. Many pulse ox devices have a printer that allows the vital signs to be printed out for each patient and the data placed in the medical record. Normal oxygenation rates are around 98%. This value determines how much oxygen the patient is receiving. Values below 90% indicate cyanosis and the veterinarian should be notified immediately.

An **esophageal stethoscope** is a device placed into the esophagus next to the endotracheal tube and emits the sound of the heartbeat throughout a procedure. This type of monitor is an inexpensive tool that consists of a tube placed through the mouth and down the esophagus to the level of the heart. This special tube can be attached to a device that amplifies the sound of the heart, which is then transmitted through the speaker on the device.

A **blood pressure monitor** or **sphygmomanometer** is used to measure an animal's blood pressure during surgery. This will also help determine tissue perfusion as well as depth of anesthesia. If the pet is in a light surgical plane (under light anesthesia), it may have an increase in blood pressure when the surgeon starts the surgery or manipulates the tissues. The surgeon also monitors the blood loss that is occurring during the procedure, since that also affects tissue perfusion. The blood pressure monitor measures the pressure through the use of a cuff similar to that used in humans. The BP cuffs come in various sizes to accommodate the different species of animals.

An **electrocardiogram** (ECG or EKG) monitor is placed onto a patient through metal leads called **electrodes** that measure the electrical activity of the heart. An EKG measures the electric currents generated by the heart. It is used to monitor the heart rate, rhythm, and changes in the nerve impulses in the heart. Continuous monitoring with an EKG allows early recognition of electrical changes associated with disorders of conduction in the heart and arrhythmias that may need to be treated. Lead II is the most common lead used for monitoring surgical patients.

Surgical Preparation

All surgical patients should be prepared for surgery in a specific way that makes the patient's surgical area sterile and aseptic. This begins with making the patient urinate if possible to prevent contamination of urine during surgery. The animal's bladder should be palpated and expressed before the preparation. The preparation of the surgical site should be completed in the treatment area. The incision area is clipped and a **surgical margin** should extend 2 to 4 inches beyond the anticipated incision borders (see Figure 51-8). Long-haired breeds should have any hair near the surgical margin trimmed to a length that prevents contamination of the incision area. The clipped area should be kept neat and even and smooth over all areas where hair is removed (see Figure 51-9). A surgical clipper blade of #40 or #50 should be used. The blade is held flat against the skin to avoid clipper burns or knicks in the skin. The clipper is used first with the growth of the hair coat and then against the hair growth to clip as short as possible. The surgical clip should provide a smooth area of skin with no remaining hair over the site. Some areas of certain animal species may require plucking rather than surgical clipping due to delicate tissue that may be damaged or torn by clipper blades. This is common in male tomcats and rabbits that are being castrated. When orthopedic surgeries are prepared on a limb, the entire circumference of the limb should be clipped free of hair. When all areas of the hair are clipped, a vacuum

FIGURE 51-8 Shaving the flank of the cow to prepare a surgical margin.

should be used to clean up all hair. Shop vacuums and built-in vacuum systems are common in collecting hair. It is also useful to gently run the vacuum over the patient's prep site to remove any loose hair.

The skin preparation then begins with an initial scrub to rinse off any remaining hair over the surgical site. The **surgical scrub** process then begins with sterile solutions placed onto the surgical margin (see Figure 51-10). The most common surgical scrub solution is chlorhexidine, such as Nolvasan scrub. Scrub solution contains a soapy component that when mixed with water helps to break down oils, dirt, and debris on the skin to better cleanse the area. The scrub process is done multiple times by using scrub-soaked gauze sponges to cleanse the area beginning in the center of the site and working clockwise in a circular motion and working to the outer edge of the hair coat. This allows the area to be cleaned without touching a previously wiped area. Always work from the incision site outward toward the remaining hair. This process is repeated as often as necessary until the gauze sponges are no longer showing signs of dirt or debris. An alcohol-soaked

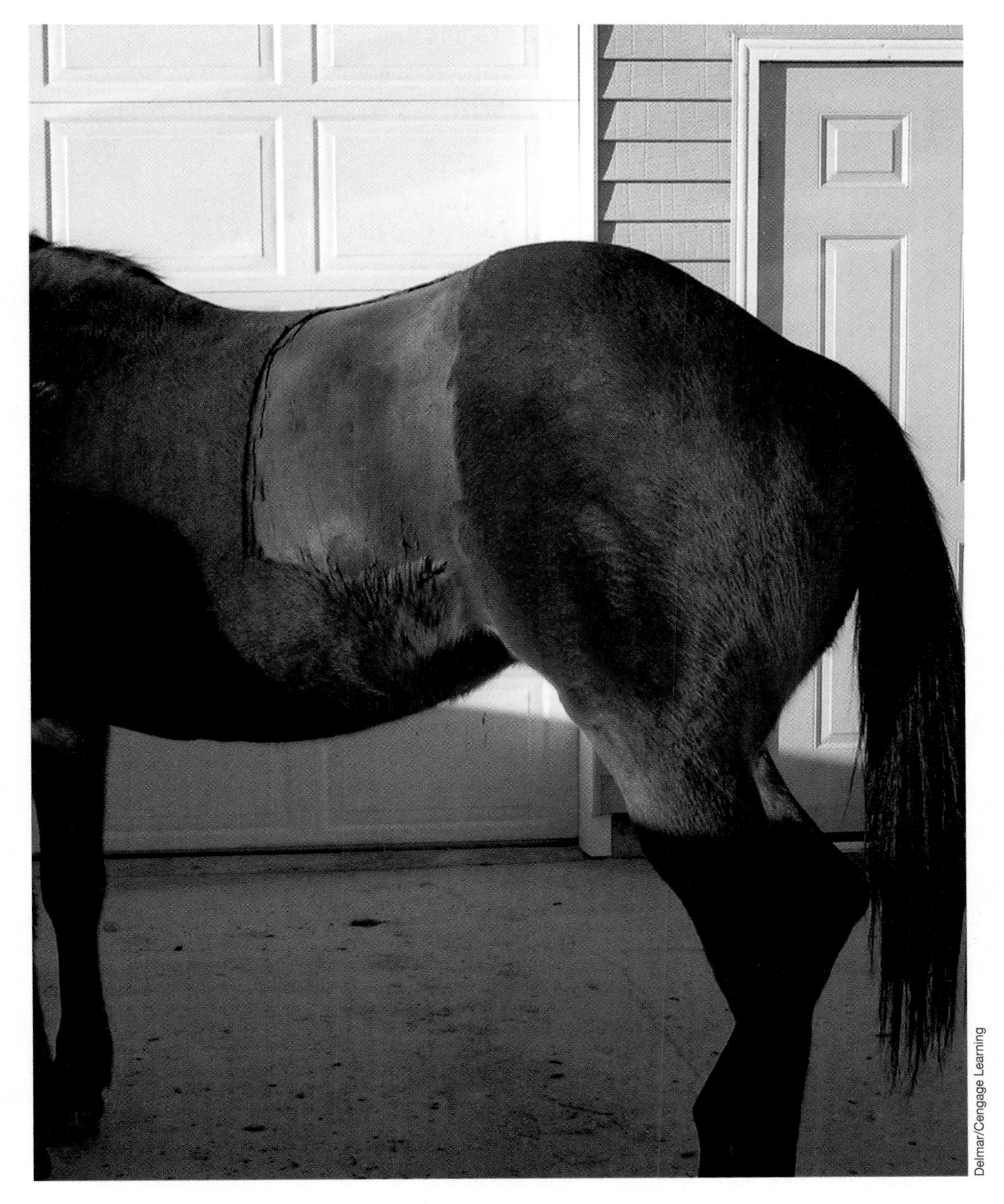

FIGURE 51-9 Flank of horse shaved in preparation for surgery.

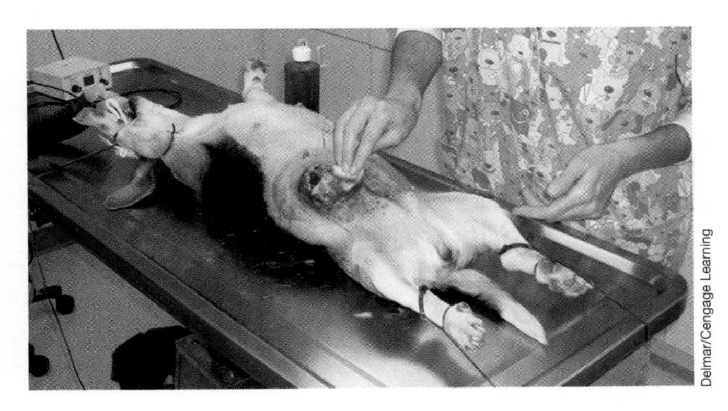

FIGURE 51-10 Surgical scrub used to prepare the surgical site.

sponge may then be used in the same manner by scrubbing three times. A surgical antiseptic is then applied to the skin, such as iodine or betadine solution, and the patient may then be moved to the surgical area. A gauze sponge or several sponges can be placed over the area to prevent contamination. A second antiseptic spray may be completed when the patient is in place on the surgery table.

The patient should be properly placed onto the surgical table without contaminating the preparation area and tied onto the surgery table using rope ties. A final scrub may be completed as necessary. Any surgical monitors should then be applied to the patient and turned on for monitoring purposes.

Competency Skill

Surgical Preparation

Objective:

To properly prepare the surgical site.

Preparation:

clippers, #40 or #50 surgical blades, vacuum, exam gloves, gauze sponges, surgical scrub, water, alcohol, antiseptic (iodine or betadine)

Procedure:

1. Apply gloves.
2. Clip the patient using #40 or #50 clipper blades and clippers.
3. Clip the incision site and 2 to 4 inches around the incision site according to the veterinarian's direction. Trim any long hair that may fall into the surgical margin.
4. Vacuum any loose hair that was removed from the patient. Vacuum patient, table, and floor areas.
5. Rinse the surgical site with a gauze sponge and water.
6. Using a gauze sponge soaked in water and surgical scrub, begin the sterile scrub at the center of the incision site and move in a circular motion outward toward the unclipped area.
7. Repeat this process until the gauze sponges show no dirt or debris from the clipped area.
8. Repeat the process using alcohol-soaked gauze sponges and repeat three times.
9. Repeat surgical scrub and water cleaning three more times.
10. Apply antiseptic spray over the incision area.
11. Apply gauze sponges over the incision area to move the patient to the surgical suite.
12. Carefully move the patient to the surgery table and properly position and tie onto the table using the rope ties.
13. Repeat the surgical scrub and water cleaning three more times.
14. Apply another light spray of antiseptic and allow to dry on the patient's skin.

Surgical Pack Preparation

Surgical pack preparation is essential for all surgical procedures. The process must be completed with a sterile and aseptic technique. The process begins with the surgeon's items being properly opened and available for the surgery procedure. The surgeon needs the following nonsterile items:

- Hair cover or cap
- Surgical mask
- Booties or shoe covers

The veterinary surgeon requires the following items and supplies that must be sterile for each surgery:

- Scrub pack containing bristle brush
- Sterile linen hand towels
- Surgical scrub
- Surgery glove pack with correct-sized gloves
- Sterile gown pack

Surgery Packs

All packs should be sterilized with an indicator strip and tape that ensures the items have been properly sterilized. The indicator strip is located inside the pack and changes color when the heat process has reached the interior objects. The autoclave tape that seals the pack has strips that turn dark in color when heat has provided proper sterilization (see Figure 51-11). Each pack that is prepared should be closed with autoclave tape that is turned under to form a tab for easy opening. The tape should be labeled with the pack contents, date autoclaved, and the initials of the person preparing the pack. The packs should be carefully opened in the surgery suite and placed on a mayo stand. When opening the pack, the tab should be pulled and the pack placed on the instrument stand. Then the outer pack cover should opened with two fingers, using care not to contaminate any items inside the pack. The left and right sides are opened and the bottom part of the pack cover pulled down to expose the surgical instruments.

Surgery packs will include the following items that are sterilized separately:

- Instrument packs (spay pack, castration pack, wound pack, orthopedic pack, etc.)
- ½ surgical drape
- ¼ surgical drape
- Towel packs
- Individual instrument envelopes
- Bowel packs

The surgical pack is put together in a specific way that may vary from facility to facility and be based on the veterinarian's recommendation or preference. Surgical packs include the instruments for each individual pack type but should also include basic supplies necessary for all surgeries (see Figure 51-12). These items are required for every type of surgery. Surgery pack supplies include the following:

- Gauze sponges
- Laparotomy towel
- Suture material
- Surgical blade

The surgical pack is prepared in a way that when opened it should be easily accessible to the veterinarian and be kept neat and organized for the surgical purpose. Typically, instruments are placed in the center of the pack on top of the laparotomy towel and gauze sponges (see Figure 51-13). The **laparotomy**

FIGURE 51-11 Surgical pack.

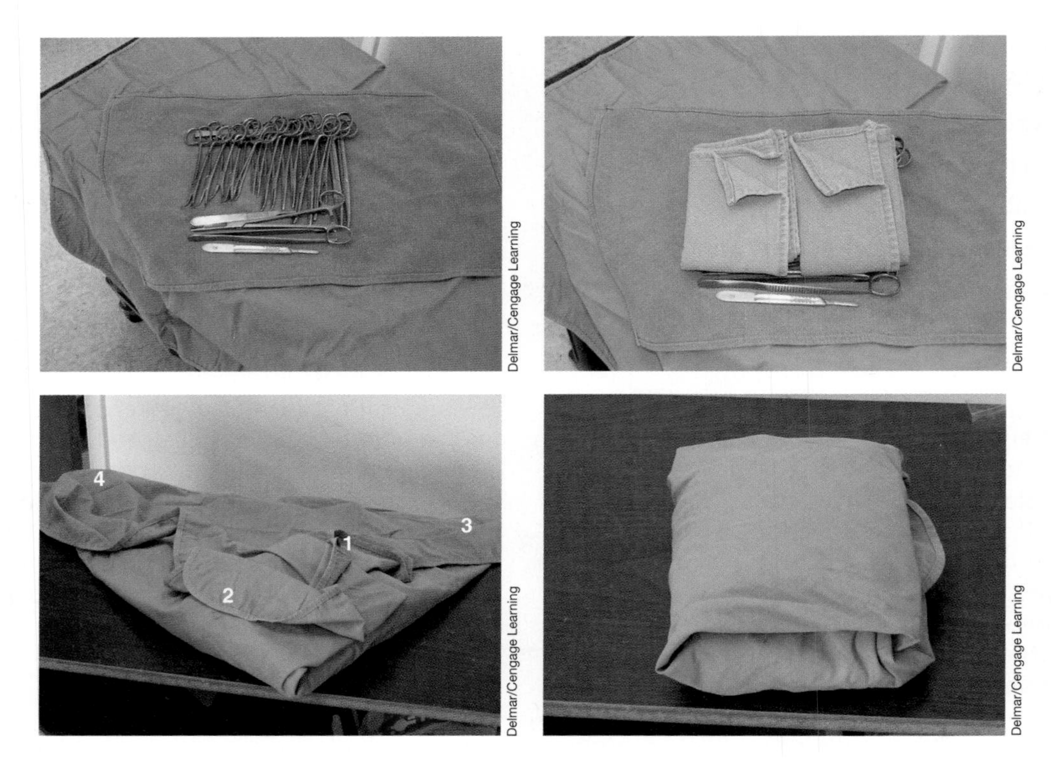

FIGURE 51-12 Contents and proper folding of the surgical pack.

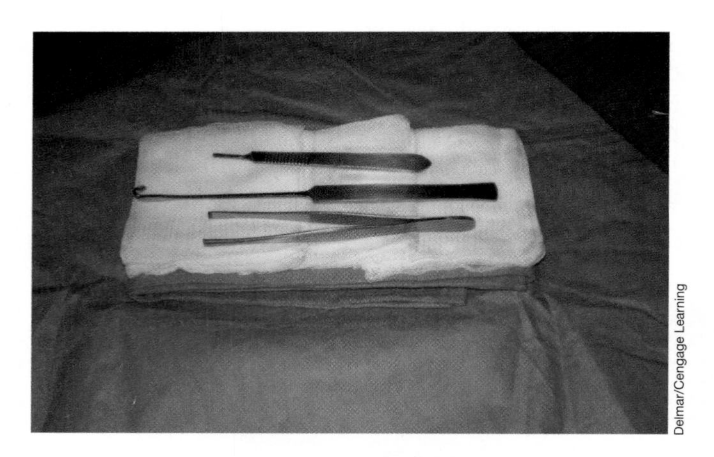

FIGURE 51-13 Proper placement of items in surgical pack.

towel is used to control bleeding during surgery and as a hand towel for the surgeon. The gauze sponges are placed on top of the towel and should be counted and consistent in every pack. This assures when the surgery is complete and packs are cleaned that every gauze sponge is accounted for and has not been accidentally left inside of a patient. The instruments are then placed on top of the gauze and the pack is wrapped in the proper method.

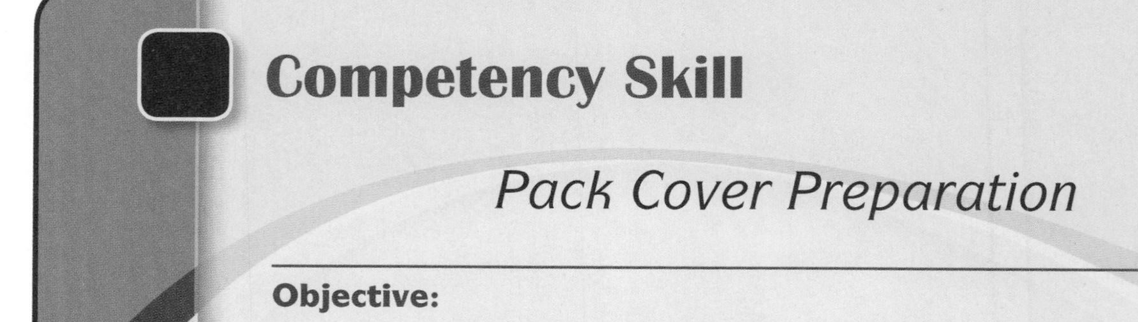

Competency Skill

Pack Cover Preparation

Objective:

To properly complete and fold a surgical pack for the sterilization process.

Preparation:

surgical drape, scissors, permanent marker, autoclave tape

Procedure:

1. Cut a piece of surgical drape about 12 inches in length.
2. Open the drape fully.
3. Place the drape in the shape of a diamond on the work surface.
4. Place the pack item or surgical pack in the center of the pack cover.
5. Fold the bottom flap upward over the contents. Hold flap in place.
6. Fold the left flap inward and make a small dog ear at the end of the flap. Hold flaps in place.
7. Fold the right flap inward and make a small dog ear at the end of the flap. Hold flaps in place.
8. Make a small crease at the side of each side of the upper flap. This should make an envelope shape.
9. Fold the contents upward and snuggly tighten the pack in place within the cover.
10. Fold again and make a dog ear folded inward on the end of the drape.
11. Apply a piece of autoclave tape with a flap tucked under the tape.
12. Write the date, pack contents, and initials of the preparer on the tape with a permanent marker.

Competency Skill
Surgical Pack Preparation

Objective:

To properly complete and fold a surgical pack.

Preparation:

surgical drape, scissors, surgical instruments (will vary between packs), indicator strip, permanent marker, autoclave tape, gauze sponges, laparotomy towel

Procedure:

1. Remove one arm's length of a piece of surgical drape or approximately 3 feet of drape.

(*Continues*)

2. Open the drape up fully.
3. Place the drape flat on the work surface.
4. One-fourth of the bottom of the drape should be folded upward twice in a fan fold and the fold be located on the back of the drape.
5. One-fourth of the top of the drape should be folded downward twice in a fan fold and the fold be located on the back of the drape.
6. The laparotomy towel is folded lengthwise in half and then in four square fanfolds and placed in the center of the drape.
7. Apply three stacks of 4 x 4 gauze sponges of equal amounts on top of the laparotomy towel. The total number of sponges should be consistent in each pack.
8. An indicator strip should be applied to the top of the sponges face up.
9. The spay hook, tissue forceps, and surgical blade handle should be placed on top of the sponges.
10. The four towel clamps should be linked together and held in one hand.
11. The curved hemostats should be placed in the hand in the opposite direction on top of the towel clamps.
12. The straight hemostats should be placed in the hand in the opposite direction on top of the curved hemostats.
13. The scissors are placed in the hand in the opposite direction on top of the straight hemostats.
14. The needle holders are placed in the hand in the opposite direction on top of the scissors.
15. The entire stack of instruments is then placed on top of the three flat instruments and held in place with one hand (see Figure 51-14).
16. While holding the instruments in place, fold the bottom half of the drape upward over the instruments. Continue to hold instruments in place.

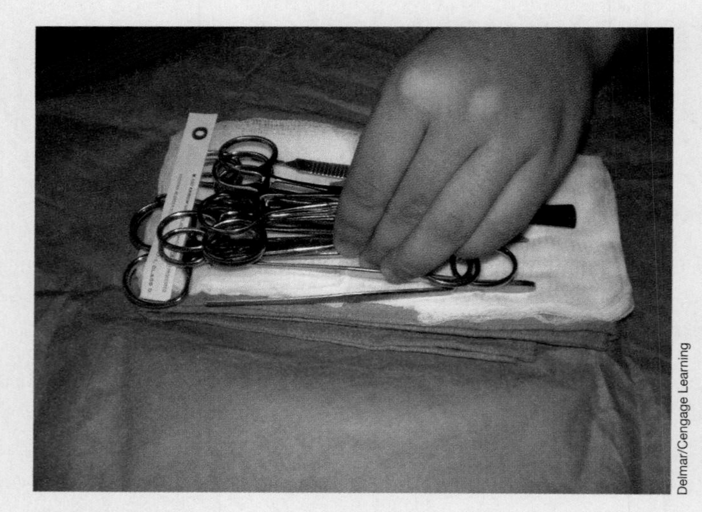

FIGURE 51-14 Hold the instruments that make up the surgical pack in one hand.

17. Fold the top half of the drape downward over the instruments.
18. Continue to hold the middle of the pack in place.
19. Fold the left side of the drape inward over the pack. Make a crease in the side of the fold.
20. Fold half of the left flap backward in a fan fold and place a dog ear fold on one corner of the drape.
21. Fold the right side of the drape inward over the pack. Make a crease in the side of the fold.
22. Fold half of the right flap backward in a fan fold and place a dog ear fold on one corner of the drape.
23. Place the surgical pack on top of a pack cover. Complete the fold of the pack cover.

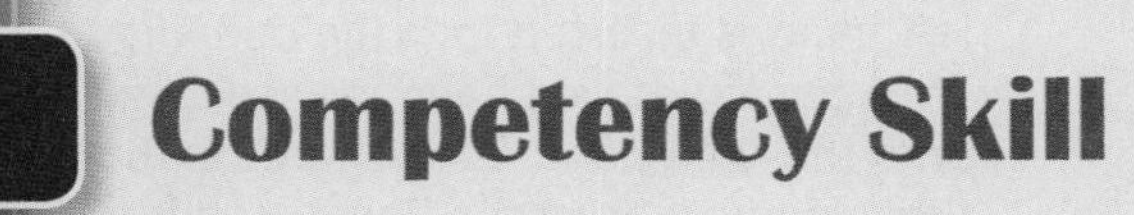

One-Half or One-Fourth Drape Preparation

Objective:

To properly prepare surgical drapes.

Preparation:

surgical drape, scissors, indicator strip, permanent marker, autoclave tape

Procedure:

1. Cut a piece of surgical drape that is ½ a yard or ¼ of a yard in length.
2. Open the drape fully.
3. Place the drape on the work surface.
4. Fold the drape in half lengthwise.
5. Fold the drape in half again lengthwise.
6. Fold equal amounts of the drape in fan folds creating four equal parts.
7. Place the drape in the center of a pack cover.
8. Place an indicator strip on top of the drape.
9. Fold the pack cover.

Scrub Packs

The **scrub pack** contains the necessary items to properly prepare the surgeon and technician or assistant for surgery. This is a surgical scrub for staff members to ensure they are sterile and aseptic for the procedure. The contents in a scrub pack include the following:

- A hard bristle brush
- Sterile hand towels
- Surgical scrub
- Surgery gloves

The bristle brush should be placed on top of a stack of folded hand towels. The towels should be folded in half lengthwise and in a fan fold. The end of the towel fold should result in a neat small square form. The number of towels should be counted and used consistently in each scrub pack. A sterile indicator strip should be placed between the hand towels and the brush. The scrub pack should be carefully opened in the same manner as the surgery pack and placed near the sterile scrub area, usually within the treatment area. The surgical scrub should be kept near the sink location for hand washing and must be kept in a sterile container (see Figure 51-15). Individual bristle brushes and hand towels should be sterilized separately in case of staff members needing to re-scrub or items that may be dropped or contaminated.

The **glove pack** is usually a prepackaged disposable sterile glove pack that is used one time and then discarded. The proper-sized surgery gloves should be opened by pulling down the outer plastic cover and left in the scrub area.

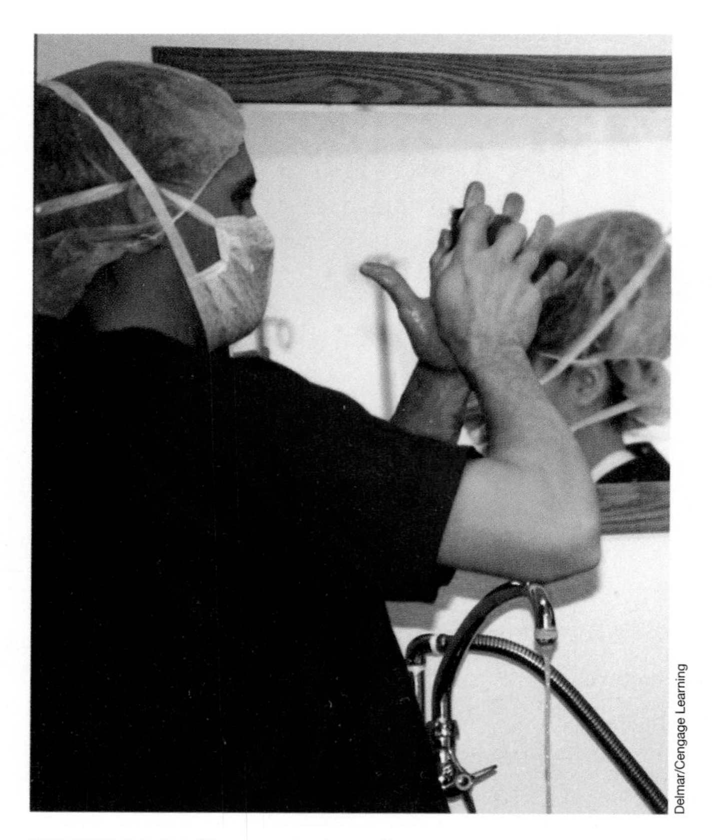

FIGURE 51-15 The surgical scrub.

Gloving Techniques

The veterinary assistant should be aware of gloving techniques as they may be asked to assist in surgery. Gloving begins after the sterile hand scrub is completed and the hands have been dried. Each person should know his or her glove size. The gloves are opened and ready for use. All rings and jewelry items should be removed from the hands and fingernails should not extend beyond the end of the finger. The inner glove pack should be opened once the person has completed the sterile hand scrub.

Open gloving is done by removing the left glove using the right hand by grasping it and the edge of the cuff. The left hand is inserted into the opening as far as possible. Then the left gloved hand is used to guide the right glove onto the right hand. This is done by holding the gloves at the folded cuff. Use the fingers to lift the cuff upward and invert the right cuff over the gown sleeve. Pull the glove fully into place on the right hand. Then do the same with the left glove so that at no time is the outer surface of either glove in contact with the skin.

Closed gloving is done when the gown is placed with the hands still within the sleeves so that no skin is visible and potentially contaminating the gloves. One glove is grasped at the cuff edge through the sleeve. The other side of the cuff is grasped with the other hand still in the sleeve. The hand is then slid further down the gown sleeve and into the glove. The same procedure is repeated with the other glove. The cuffs of the gown are completely covered by the cuffs of the sterile gloves. When both hands are completely gloved, grasp them together by locking hands and adjust the gloves as necessary.

When the gloving procedure is completed, the staff member should immediately keep the elbows elevated and the hands together and not touch any nonsterile surface. Immediately move to the surgical area to assist the surgeon.

Gown Pack

The **gown pack** holds the gown that is worn by the surgeon and veterinary staff assisting in the surgical area. Gowns may be disposable and meant for one-time use or they may be reusable and require laundering after each use and then may be sterilized. The gown must be folded in the proper method to assure when opened and removed from the pack cover it is the proper way for applying without causing contamination.

Competency Skill

Gown Pack Preparation

Objective:

To properly prepare gowns in a sterile pack.

Preparation:

surgical drape, surgical gown, scissors, permanent marker, autoclave tape

Procedure:

1. Place the surgical gown on a flat surface.
2. Place the gown front side up.
3. Place the sleeves across the front of the gown. The sleeves should slightly overlap in the center of the gown.
4. Fold the gown in half lengthwise with the center of the gown forming a fold.
5. Fold the gown in half lengthwise again.
6. Tuck the loose ends of the neck and back ties into the folds of the gown.
7. Begin at the hem of the gown and fan fold the gown in small square sections.
8. Continue folding the gown throughout its length until the neck line is reached. This should form a neat square shape.
9. Fold the top corner of the gown into a dog ear.
10. Place the gown on top of a pack cover.
11. Place a sterile indicator strip on top of the gown.
12. Fold the pack cover.

Competency Skill

Opening a Pack

Objective:

To properly open a sterile pack without contamination of the contents.

(*Continues*)

Preparation:

surgical pack

Procedure:

1. Place sterilized pack on mayo or instrument stand in the surgical area.
2. Open the pack by pulling the tab in the autoclave tape.
3. Open the pack by pulling the top flap upward. Do not touch any part of the pack inside.
4. Open the right flap outward by pulling the dog ear flap. Do not touch any part of the pack inside.
5. Open the left flap outward by pulling the dog ear flap. Do not touch any part of the pack inside.
6. Leave the pack for the surgeon.

Sutures and Surgical Blades

The veterinary assistant should know the types and sizes of suture material and surgical blades and where the supplies are kept. These items come prepackaged individually and are already sterile (see Figure 51-16). These items are needed by the surgeon and should be placed onto the Mayo or instrument stand after opening the surgical pack and providing the surgery drape materials. Scalpel blades are numbered to indicate their size and shape. Each surgeon will have a preference to size depending on the patient and procedure. The blades are wrapped in a foil, peel-apart package that is opened by peeling the edges away from the blade and allowing the blade to carefully drop onto the pack surface. The blade should be dropped onto an open area for easy visibility.

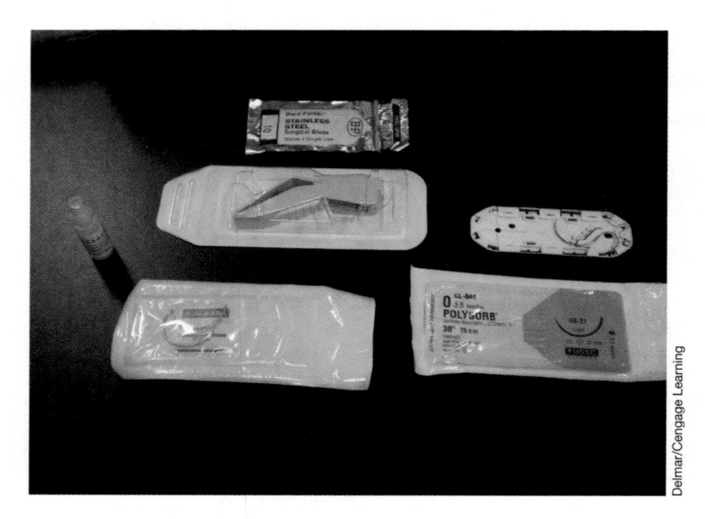

FIGURE 51-16 Suture materials.

When the surgery is complete, make certain the scalpel blade is removed from the handle. Grasp the narrow end of the handle and using a pair of forceps grasp the blade over the hooks of the handle and lift the end upward. When the blade is raised from the hooks, it can then be slid from the handle. Caution should be used when handling the blade as it is very sharp, and they should be discarded after each use by placing them in the sharps container.

Sutures come in a variety of materials, sizes, and lengths. They are dated and housed within a chemical substance to keep them sterile and soft for use. Different types of suture material are used on different tissues. There are two classes of suture in the form of absorbable and non-absorbable materials. **Absorbable suture** is suture that is broken down within the body over a period of time and is absorbed by the body. This type of suture is used both internally and externally. **Non-absorbable suture** is not broken down within the body and remains intact for long periods. This type of suture is commonly used on the external body. Non-absorbable sutures are generally used to close skin and must be removed after wound healing is complete. There are some procedures where non-absorbable sutures may be placed internally but they cannot be removed. Some sutures have a needle already attached and are called **swedged on needles** and other suture material is on a **suture reel** and must be thread into a needle.

TABLE 51-2

Suture Size

EXAMPLES OF SUTURE SIZES FOR USE IN ANIMALS				
10-0–8-0	7-0–5-0	4-0–3-0	2-0–0	1–2
Microsurgery; corneal	Ophthalmic; neurosurgery; ureters	Skin and subcutaneous; bowel; bladder	Abdominal Fascia; Stomach; Hernia	Rib Retention; Cutaneous Stents

Suture size is similar to needle size in that the smaller the number, the larger the needle size and diameter of the suture material (see Table 51-2). The larger the number, the smaller the needle and suture size. Larger whole numbers indicate larger sutures, so a number 2 suture is larger than 1. However, larger numbers followed by an "**ought**" (0) are smaller. A 5-0 is smaller than 2-0. The smallest sutures that are used for ophthalmic and microsurgery range down to 12-0 size.

Surgical staples are a fast, easy way to close the skin. They are useful but require application using a staple gun, and there is a specific device required for removing them. **Surgical glue** is useful for closure of small wounds or surgical sites and is commonly used in feline declaws. This is a sterile type of Krazy glue that eventually breaks down in the body.

Gowning the Surgeon and Assistants

The surgeon and surgical assistants must put on a mask, cap, or hair cover and booties or shoe covers. Next the surgical staff should perform a surgical scrub preparation. On completing the surgical hand scrub, the hands are dried using a sterile hand towel. The gown pack should already be open and ready for the surgical staff. Then the gown is picked up by the folded dog ear at the neck line of the gown. The gown is held up and away from the body to prevent contamination. Care should be made not to allow the gown to have contact with the floor. The person places his or her hands through each armhole. The person then applies the sterile gloves with either an open or closed technique. This is when an assistant should stand behind the person and, without touching the external area of the gown and touching only the ties, should tie in the staff member. The ties should be placed in a bow. Many gowns have ties at the base of the neck and at the waist. Never touch any part of the gown other than the ties.

Patient Surgical Positioning

The patient positioning is dependent on the surgical procedure and the veterinarian's preference. Many surgeries require the animal to be placed in dorsal recumbency on the surgery table (see Figure 51-17). This is completed by keeping the patient straight and centered on the table. Each limb of the patient is then tied into the table hooks. The front limbs are pulled forward toward the head, which is centered between the front limbs. The rear limbs are pulled backward toward the tail, which is centered between the rear limbs.

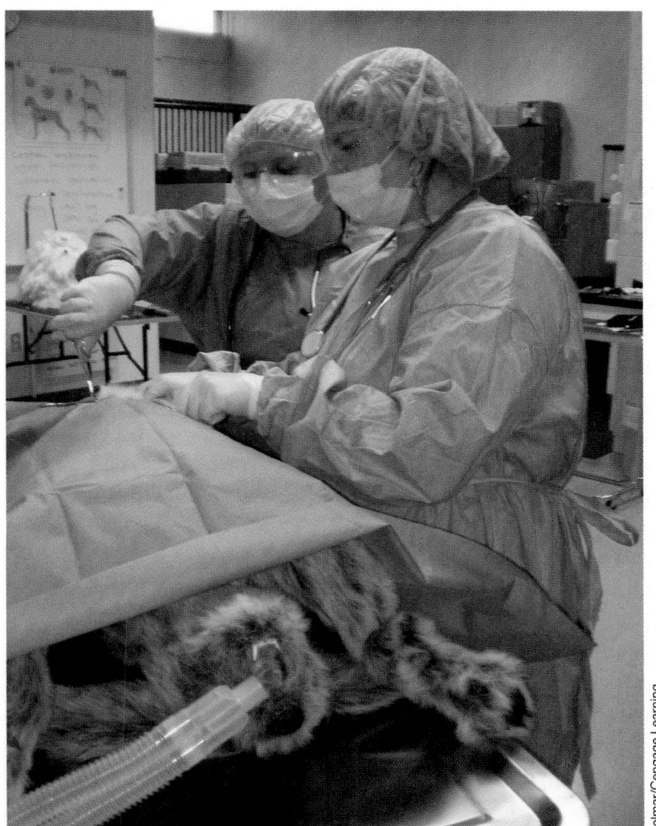

FIGURE 51-17 Patient in dorsal recumbency.

TABLE 51-3

Common Procedure Positions

Spay or OHE	Dorsal Recumbency
Dog Castration	Dorsal Recumbency
Cat Castration	Sternal or Lateral Recumbency
Abdominal Surgery	Dorsal Recumbency
Orthopedic Surgery	Lateral or Dorsal Recumbency
Ear Crop or Head Surgery	Sternal Recumbency
Declaw or Dewclaw Surgery	Lateral Recumbency

The ties may be placed in vice-like devices or on hooks in which figure eights are made to hold the limbs. Other common surgical positions include sternal and lateral recumbency. The limbs should be secured as naturally as possible for the surgical procedure. Table 51-3 lists positioning for some common procedures.

The surgery table should be placed at the proper height for the procedure and the veterinarian. Some surgeons may prefer to sit or stand and the table should be properly adjusted for the necessary height. Some patients may require a heating pad that is placed between the patient and the table. Other patients may require a V-trough positioner to keep the animal centered on the table.

Surgical lights should be positioned directly over the sight when the incision will be made. Final light positioning should be done after the animal is draped and the surgeon is in place to direct any light needs.

Anesthesia Planes

Anesthesia planes or depths are measured by evaluating the patient's involuntary reflexes. The veterinary assistant may be asked to monitor reflexes to determine the depth of anesthesia. The **palpebral reflex** is evaluated by tapping the fingertip lightly over the surface of the eyelid. If a strong to slight blink reflex is noted, the patient is in a light plane of anesthesia and the inhalant gas should be increased for surgical purposes. The blink reflex will decrease and eventually disappear as the plane of anesthesia deepens. The **pedal reflex** is used on the digit or the skin between digits, which is pinched to determine reflex response. A patient that moves or withdraws the foot is in a light plane of anesthesia. The reflex weakens and disappears as the plane deepens. Muscle tone is noted by the plane level and is moderate when in a light plane and lessens as the plane level deepens.

Plane I is during the induction phase when a patient becomes disoriented for a brief period and then passes into Plane II. Plane II is the excitement phase when a patient displays involuntary movements and vocalization. This stage passes quickly and may also be visible in the recovery phase. Light sometimes affects this plane of anesthesia and placing the patient in a lightly lit or dark area is helpful. Plane III is a light depth of anesthesia. This is the plane ideal for surgery where vital signs remain normal, pupils are regular in size and rotate medially, and mild to no reflexes are present. Plane IV is an overdose of anesthesia and a deeper level in which the patient could potentially die. No reflexes are noted, vital signs are decreased and pupils are centrally fixed and dilated. At this plane, cardiac arrest is possible.

The Anesthesia Machine

The veterinary assistant should have a basic knowledge and understanding of the parts of the anesthesia machine (see Figure 51-18). The most critical part of the

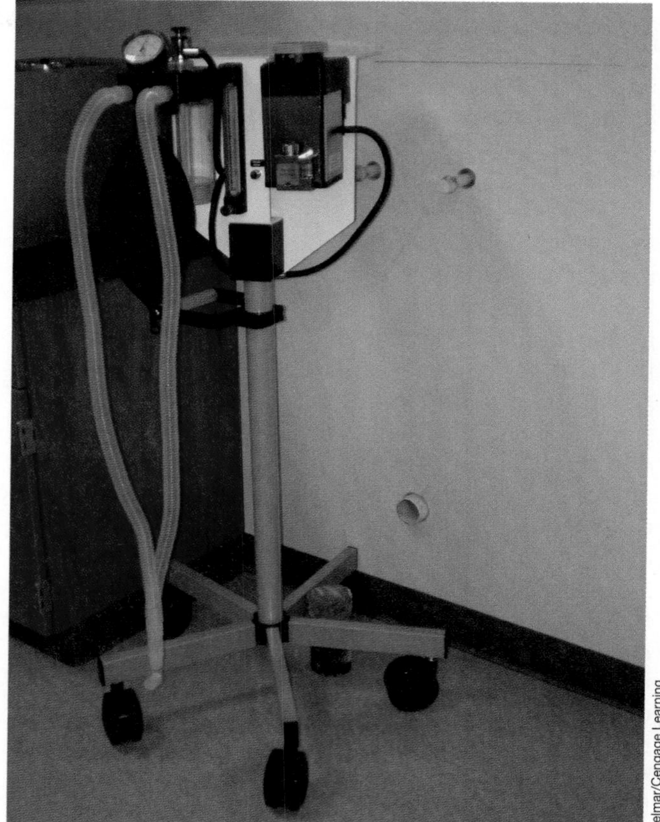

FIGURE 51-18 The anesthesia machine.

machine is the **vaporizer**. The vaporizer converts the liquid anesthesia into a gaseous form, which the patient inhales. Each vaporizer is meant for a specific gas inhalant. A vaporizer that is meant for isoflurane cannot be used with halothane or nitrous oxide. The vaporizer has a dial that adjusts and controls the delivery rate of the gas. It is essential for all staff members to know how to turn the dial on and off, as well as adjust the dial between number settings. The delivery of gas is measured in percentage volumes. When the liquid gas form is used it must be replenished into the vaporizer. This level must be monitored during anesthesia. The liquid level should be noted prior to sedating an animal. During the refill process, limited people should be around as the gaseous odor may escape. All anesthetic gases are toxic and care should be used when working around the machine. Any woman who is pregnant should not work with anesthesia. Escaped gases that are inhaled cause a potential health issue such as renal damage or liver damage when inhaled over time. Short-term exposure can cause lightheaded signs or dizziness.

The **oxygen tanks** are located under the machine or in separate cylinders next to the machine (see Figure 51-19). These **cylinders** are metal containers that hold compressed forms or liquid to keep patients oxygenated and breathing during anesthesia. Oxygen cylinders are green in color and come in a variety of sizes. Each is designated as to its size and use by a series of numbers and letters. When you look at an oxygen cylinder, you will see a number posted on the side; this number denotes the number of liters of oxygen the cylinder holds. The letters A, B, C, D, and E are used to designate the size of the cylinder. An A cylinder is one of the smaller cylinders and holds only about 34 liters of compressed oxygen; an E cylinder is one of the larger, holding approximately 680 liters of compressed air. Large cylinders are often mounted outside of the building or in low-traffic areas and deliver oxygen to the rooms where the anesthesia machines are located by hoses that descend from the wall or ceiling and attach to the machine. Often, you will see the letter M displayed on an oxygen cylinder. This indicates that this oxygen cylinder is for medical use only. The oxygen tank has a **pressure valve** that indicates how much oxygen is left in the tank. The valve should be checked daily and empty tanks removed and a full tank opened and connected to the machine. It is essential that oxygen never runs out during a procedure. Tags may be placed on the tank as to their status, which should indicate "full," "empty," or "in use." In some facilities, nitrous oxide gas is used as an anesthetic gas and is supplied in a compressed cylinder that is blue in color. The oxygen **flow meter** is located on the machine near the vaporizer and is adjusted for the proper rate of oxygen flow to the patient. The flow meter is calculated in liters/minute. The rate of flow is determined by the patient size, method of administration, and the type of breathing system used. A common rate of flow is 30 ml/kg/minute.

Example

Weight in Pounds	Weight in kg	O_2 Flow Rate
22 lb	22# / 2.2 = 10 kg	30 ml × 10 kg = 300 ml
50 lb	50# / 2.2 = 22.7 kg	30 ml × 22.7 kg = 681 ml

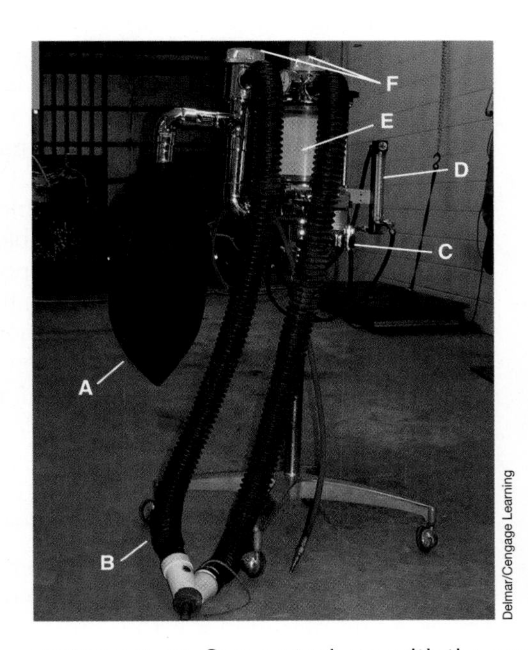

FIGURE 51-19 Oxygen tank use with the anesthesia machine.

When a patient exhales during breathing, carbon dioxide (CO_2) is produced and must be absorbed without entering the facility. The CO_2 is absorbed in a **canister** that contains granules called **soda lime** that collect and absorb the gas. Soda lime begins as a white color and as CO_2 is absorbed the granules change color, such as pink, blue, or purple depending on type. When 2/3 of the canister contents have changes in color, the canister should be changed.

The bag that connects to the machine and inflates and deflates as the patient breaths is called a **rebreathing bag**. The size of the bag depends on the patient size and the **tidal volume** required by the animal. The average tidal volume is 10 ml/kg. The bag size is determined by multiplying the tidal volume by 6 to determine the size of bag required on the machine.

Example

A 22-pound dog weighs 10 kg

10 kg × 10 ml/kg = 100 ml

100 ml × 6 = 600 ml

Bag sizes range from 0.5 liter to 5 liters. In the example, 0.5 liters equals 500 ml so a 0.5 liter bag would be too small, so the next size bag would be a 1 liter bag, which is the appropriate size for the patient. Bags are made of rubber material and slip easily onto the steel or metal tubing of the anesthesia machine. The movement of the bag is observed to evaluate the patient's rate and depth of breathing. In some cases the patient may need assisted breathing by compressing the bag with the hands in a regular and rhythmic method. The bag should be removed and cleaned after each use. It should be dried completely before reuse.

The **pop-off valve** is used to release excess amounts of gas when the pressure becomes too high within the machine system. This valve is usually located at the top of the machine. The **scavenger hose** is attached to the pop-off valve to prevent gas from leaking into the room. The scavenger hose should be connected to an outside area of the building where the gases escape into the outside air.

The hose that connects to the patient is made of **corrugated tubing** that is plastic and thin and allows the oxygen and anesthetic gas to flow to the patient. The tubing is connected to the patient's endotracheal tube with a **"Y" connector**. One end of the tubing is attached to the inhalation valve and the other end of the tube is attached to the exhalation valve. After each use, the hose should be rinsed and hung to dry. The hose must be dry prior to the next use.

Some animals may require the use of an **anesthesia chamber** for induction methods. An anesthetic chamber is a glass or plastic square or rectangular box that is clear and has two openings in which the inhalation and exhalation tubes on the anesthesia machine connect to the chamber. This allows gas and oxygen flow into the chamber to induce the patient for anesthesia. Once the patient is in a proper plane, the patient may be intubated and placed on the machine. The chambers are large enough for cats, rodents, reptiles, and very small dogs. **Anesthesia masks** may be used in larger patients and are placed over the muzzle and face area and must fit snuggly for the animal to breathe in the gases for induction. The masks are cone shaped and made of plastic and rubber components. Once a patient is in a level plane, the patient may be intubated and placed on the machine. The mask must be cleaned after each use and properly dried.

Breathing Systems

There are two types of breathing systems used with the anesthesia machine and determined by patient size. In the **non-rebreathing system** the canister and breathing bag are not used. This system is used for small patients such as rodents, birds and reptiles, or small puppies and kittens that are too small to move the breathing bag with their respirations. The most common system type is called a **Bain system** and uses a single corrugated or plastic tube that connects to the patient. Another hose passes through the center of the tube that contains the gas mixture that the patient inhales. The outer tube passes the exhaled gases. The pop-off valve is located at the end of the tube and is opened to allow release of pressure. The flow rate is between 100 ml and 300 ml based on patient size.

The **rebreathing system** allows the patient to rebreathe anesthetic gas after it has passed through the CO_2 absorbing canister. There is a closed and semi-closed system. The closed system provides enough oxygen to meet the patient requirements and the pop-off valve is kept closed. This is more economical as oxygen is not wasted. The semi-closed system has the pop-off valve open with a high oxygen flow rate.

The veterinary assistant should be able to locate the parts of the anesthesia machine and be able to adjust the knobs and valves as appropriate. It is important to know the oxygen flow and the type of systems used based on patient size. The appropriate-sized bag should be able to be selected. All manufacturer information should be reviewed to understand how each machine functions.

Post-anesthetic Care

Once the animal has been removed from the anesthesia machine, the patient should be moved to the recovery area and placed in a warm, quiet, darkened, and comfortable area to wake up. The patient should be placed in lateral recumbency with the head and neck extended. The endotracheal tube is still in place and should be untied but left intact until the animal is safely awake enough for extubation. It is important to monitor the patient's reflexes, especially the **swallow reflex**. The moment the patient begins to swallow is when the endotracheal tube should be extubated. This begins with the patient licking with the tongue or having increased jaw tone. This is a critical time to note because if the tube is extubated too soon, the patient may have difficulty breathing or aspirate materials into the lungs. If the tube is removed after the animal is waking up, the tube may be bitten or severed causing an obstruction in the throat. When the patient is ready for extubation, the cuff should be opened and deflated and the tube gently reversed out of the throat.

As the animal recovers, the heart rate and respiratory rate will increase. The patient will begin to move around and try to sit up or stand. As the patient wakes up, some movements may be exaggerated and noise may cause an excited response. During this time, the patient may thrash around and should be monitored as there is always a chance of injury. Vital signs should be monitored until the patient is mentally alert and aware of its surroundings. Some patients may require additional heat sources. Animals that are recumbent longer than 30 minutes should be rotated every 15 to 20 minutes to prevent the development of bed sores or pneumonia. Well-padded cages should be used for animals during recovery.

Postoperative Patient Care

Postoperative care is completed after the patient has recovered and continues until the patient is discharged. This is the time frame when the patient is monitored for pain, signs of infection, or hemorrhage. Once the incision is closed, the patient is disconnected from the anesthesia machine and the incision area is cleaned with hydrogen peroxide. Do not wipe the incision itself, only areas around the site. The area should be dried with a clean gauze sponge. The veterinarian may prescribe antibiotics or pain medications that should be given to the patient. The patient should be moved to the recovery area. Some patients may require IV fluids, which should be placed on an IV pole or on an infusion pump. The veterinary assistant should monitor the recovery phase to make certain the patient's vital signs are normal and the patient is comfortable.

Pain is sometimes difficult to assess and identify. Signs of pain may include the following:

- Restlessness
- Whining or vocalization
- Reluctance to move
- Chewing or biting at incision
- Anorexia
- Change in behavior

Pain control provided before and after surgery is imperative in veterinary medicine. The veterinarian may have a pain protocol for surgical procedures or may prescribe specific medications. The veterinary technician and assistant must see that the medications are administered properly and as often as necessary. Most surgical patients are sent home with pain relief medication.

Food is usually withheld from monogastric animals for a 12-hour period and water is often given in small amounts the evening after the surgery. When food and water are begun they are given in small amounts as they may cause digestive upset. Large animals and ruminants are often given small amounts of food and water as soon as they are standing and alert. The surgical patient should be eating and drinking within 24 hours after surgery unless the procedure involved the GI tract or mouth.

When the patient is ready for discharge, the home care instruction sheet should be prepared according to the procedure and veterinarian instructions. The sheet should be reviewed with the owner at the time of discharge. Most surgical patients will require a **suture removal** appointment that may be scheduled with the technician or assistant. The suture removal may be between 7 and 14 days depending on the surgical procedure. The incision is monitored for any signs of infection or damage, which is noted within the medical record. Any abnormal signs with the incision should be reported to the veterinarian for observation. If the incision is normal, the sutures are cut individually with **suture removal scissors** (see Figure 51-20). These scissors have one blade that

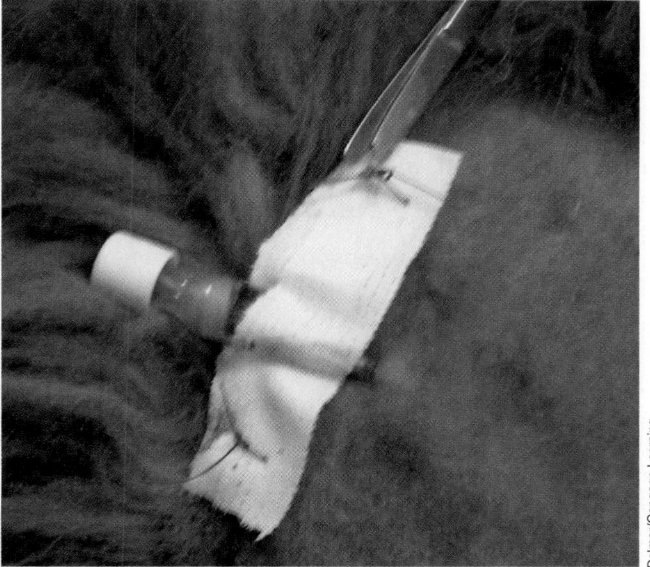

Delmar/Cengage Learning

FIGURE 51-20 Suture removal with the use of suture removal scissors.

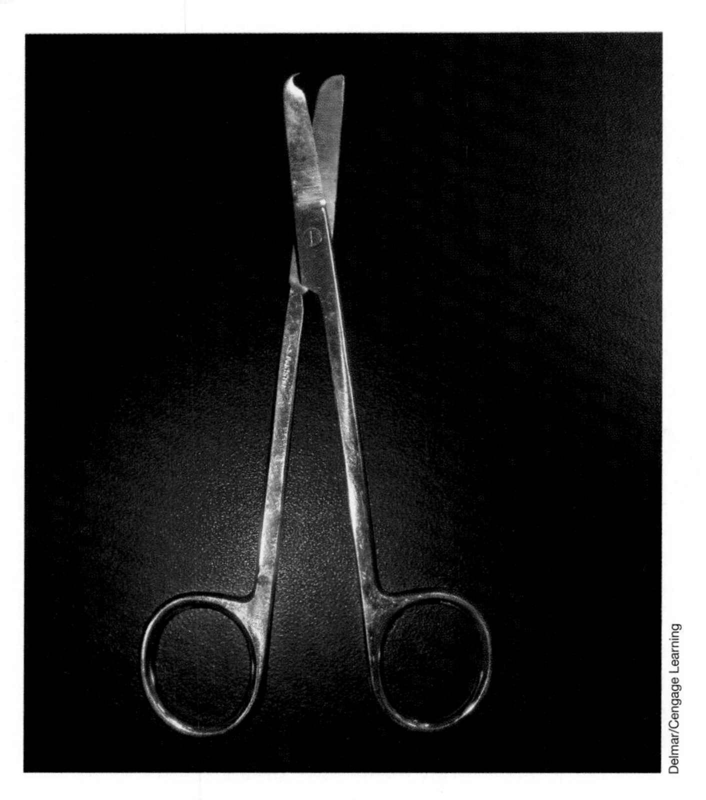
Delmar/Cengage Learning

FIGURE 51-21 Suture removal scissors.

has a small hook shape on the end to better get under the suture (see Figure 51-21). The hook area is placed under the suture and then cut. The assistant will then gently pull each suture from the skin by holding and pulling at the tie area. Staples are removed with a **staple remover**, similar to what humans use in offices. The staple remover is placed under the staple and squeezed to open up the staple, which is then gently removed from the site.

Sterilization Techniques

The most common method for surgical sterilization of packs and instruments is the **autoclave** (see Figure 51-22). It is not used on heat sensitive materials that are made of plastic, rubber, or other materials that may melt. Autoclaves come in a variety of shapes and sizes. **Distilled water** is heated in the chamber to allow high temperatures and pressures to produce the necessary heat to sterilize items. The distilled water is sterile water that is placed in the water reservoir to the fill line. Tap water should never be used as it will cause a mineral buildup within the equipment and clog the heat source. The higher the temperature, the quicker organisms are destroyed. Pressure forces steam into the interior

Competency Skill

Suture or Staple Removal

Objective:

To properly remove sutures or staples at a surgical site.

Preparation:

exam gloves, suture removal scissors, mosquito hemostats, staple remover

Procedure:

1. Apply gloves.
2. Inspect incision for signs of infection or damage.
3. Direct veterinarian to inspect incision as necessary.
4. Slip suture removal scissor blade under each knot and cut between the knot and skin.
5. Use a finger or mosquito hemostat to pull the knot through the skin and remove.

6. Slip the staple remover under the staple and squeeze the handle.
7. Remove the staple and place in sharps container.
8. Clean up and disinfect work area.
9. Disinfect all equipment.

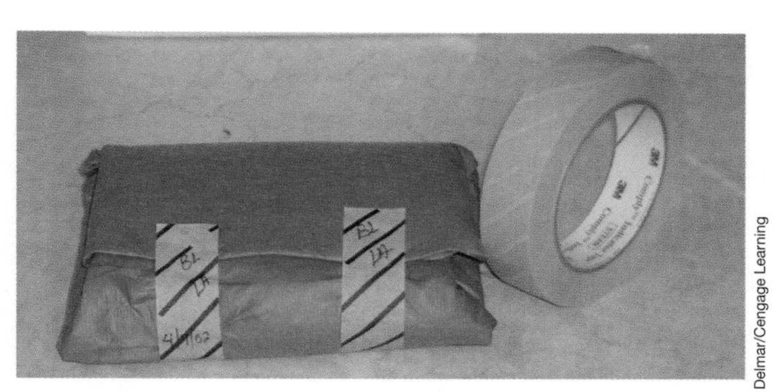

FIGURE 51-22 Autoclave (top) and autoclave tape (bottom).

packs and then destroys any living organism. Materials should not be packed into the autoclave too tightly as this will prevent steam from entering all sites. Materials are packed onto trays that sit within the autoclave chamber. Once the packs are placed in the autoclave, a lever or button is released to fill the interior chamber, located below the trays, with the distilled water. When the area is filled, the door is closed and should be properly sealed. The heat source is turned on and the temperature set. Temperature settings and times will vary with equipment. A chart of directions should be located with the autoclave to note temperature and settings for various materials. The preheat cycle begins and is similar to an oven. Indicator lights will turn on and an audible sound will be noted as the autoclave begins to heat. The timer should be set. The entire cycle should be allowed to function and observation may be required especially if the device requires decreasing the temperature according to the type of machine. When the cycle is complete, an alarm will sound. At this time, the machine should be turned off and the pressure allowed to decrease. When the pressure has properly decreased, the door can be partially opened to allow the steam to air out. Caution must be taken as to not be burned by the steam or the heat of the device. Never make direct skin contact with the autoclave. The door should be left partially ajar until the contents have cooled and dried. When the items are ready to be removed, they should be properly stored and the autoclave door closed.

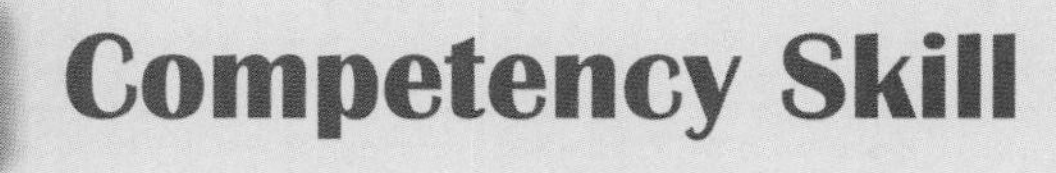

Competency Skill

Use of an Autoclave

Objective:

To properly fill and remove contents from an autoclave for sterilization.

(Continues)

Preparation:

autoclave, distilled water, surgical packs

Procedure:

1. Fill the water holding tank with distilled water. Add distilled water to the fill line.
2. Close the holding tank.
3. Open the door to the autoclave.
4. Remove trays.
5. Place material on trays without stuffing items into the area.
6. Place trays into chamber.
7. Allow water chamber to fill with water.
8. Close door.
9. Set temperature and timer.
10. Allow to preheat.
11. Set temperature and timer for autoclave cycle.
12. When timer sounds, turn off autoclave and allow cooling.
13. When pressure has reduced, open the door partially to allow steam to escape.
14. Allow contents to cool and dry.
15. Unpack autoclave and store items in proper area.
16. Close door.

Surgical Instruments

Veterinary assistants should be able to identify common surgical tools, know how to clean and maintain the life of the instruments, and have an understanding of what the instruments are used for and how they work.

Cleaning Surgical Instruments

Instruments must be cleaned after each surgical use (see Figure 51-23). Surgery produces blood materials, waste materials, and other debris that may build up on instruments. Rust can occur on different types of metal instruments if not properly cleaned, dried, or maintained. It is best to use special instrument cleaners and distilled water to prevent deterioration of instruments and thus prolong their use.

FIGURE 51-23 All surgical instruments must be properly cleaned.

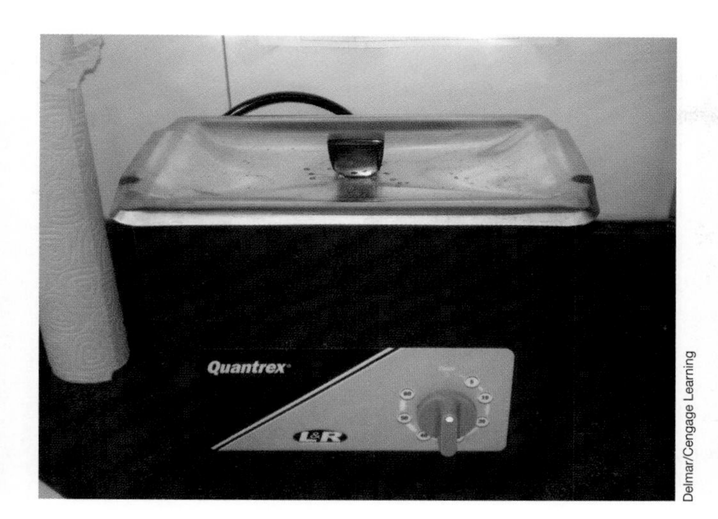

Delmar/Cengage Learning

FIGURE 51-24 Ultrasonic cleaner.

Immediately after a surgery, all instruments in a pack should be removed and soaked in a disinfectant solution safe for the instruments. Organic debris should not be allowed to collect and dry on the instruments. The instruments should soak for at least 10 minutes and then be cleaned using a wire brush and manual scrubbing. Chlorhexidine solution and pink germicide are common instrument disinfectants. Instruments can be hand cleaned and particular attention should be paid to the hinged areas and crevices. Mechanical cleaning is another method for cleaning instruments. This is done by using an **ultrasonic cleaner** (see Figure 51-24). Instruments are placed within a wire basket and the basket is seated within the cleaner, which holds a disinfectant solution. The instruments should always be placed in the basket in the open position and ideally spaced out. The cleaner is plugged in and turned on and allowed to vibrate and cleanse for at least 5–10 minutes. The instruments should be fully submerged in the solution. After the cleaning cycle is complete, the instruments are rinsed with distilled water. The instruments should be thoroughly dried. They should then be dipped into a solution of **instrument milk** for 30 seconds. The instrument milk acts as a lubricant to keep the instruments protected from corrosion. Allow the instruments to drain and air dry before being placed in the surgical pack.

Scalpel Blade Handle

Scalpel blade handles are small and flat and a blade is placed on the end to make a surgical incision (see Figure 51-25).

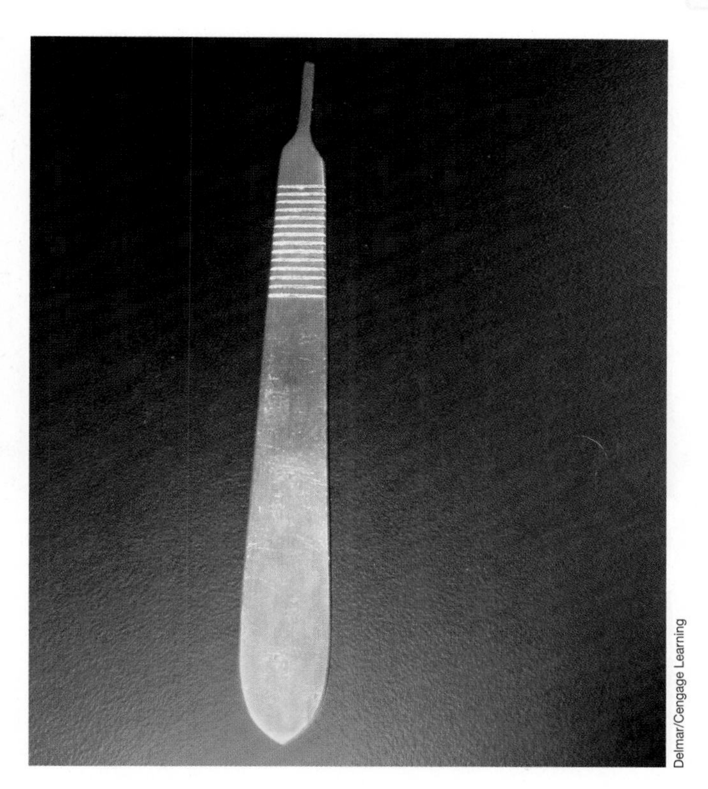

Delmar/Cengage Learning

FIGURE 51-25 Scalpel blade handle.

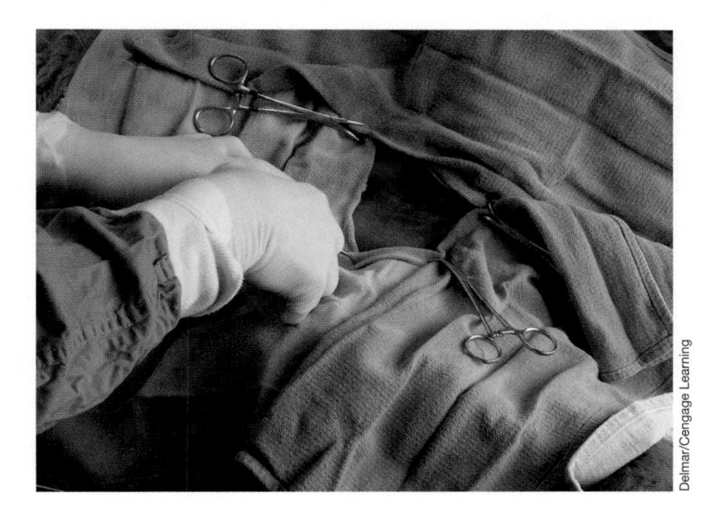

Delmar/Cengage Learning

FIGURE 51-26 Use of towel clamps.

Towel Clamps

The **towel clamps** are used to hold the surgical drape in place over the patient during the surgery (see Figure 51-26). The towel clamps attach lightly to the skin of the animal and the four corners of the drape. They are locking forceps in various sizes with curved,

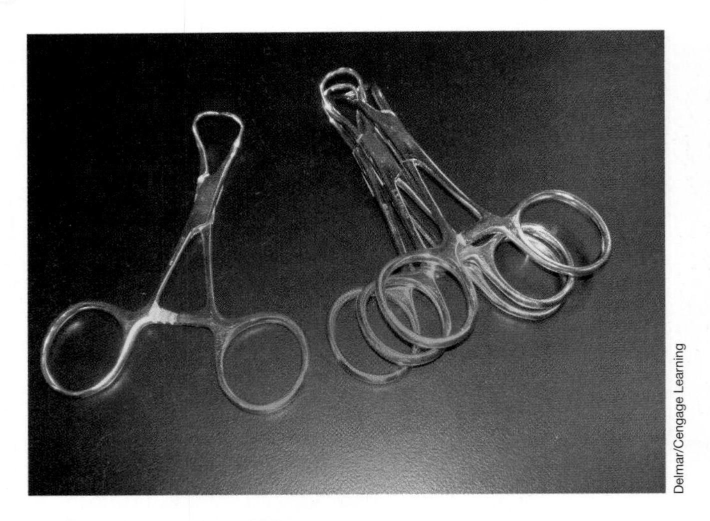

FIGURE 51-27 Towel clamps.

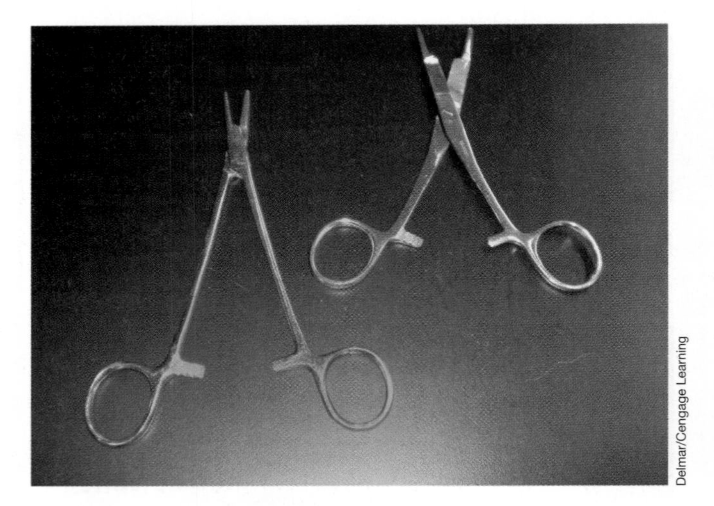

FIGURE 51-28 Needle holders.

sharp, pointed tips (see Figure 51-27). Occasionally towel clamps are used to hold tissue. The most common type of towel clamp is the Backhaus Towel Clamp. Four towel clamps should be placed in all surgical packs.

Needle Holders

The **needle holders** are used to hold a needle while placing sutures into a patient during surgery. They are hinged and lock in place to hold the needle (see Figure 51-28). Some types of needle holders are clamps only and other are a combination of clamp and scissors. Mayo-Hegar needle holders are heavy instruments with mildly tapered jaws that have clamps only and no cutting blades. Olsen-Hegar needle holders include both a locking jaw and scissor combination for holding the needle and cutting the suture (see Figure 51-29). Care must be taken when suturing to not accidentally cut off the needle from the suture material.

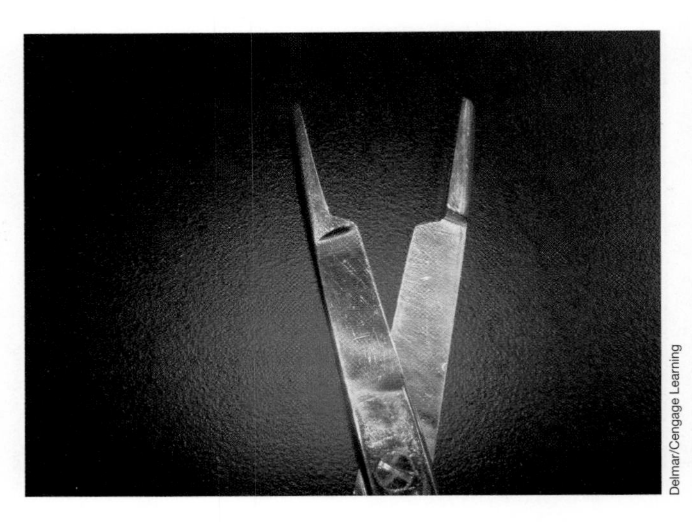

FIGURE 51-29 Head of the Olsen-Hegar needle holder.

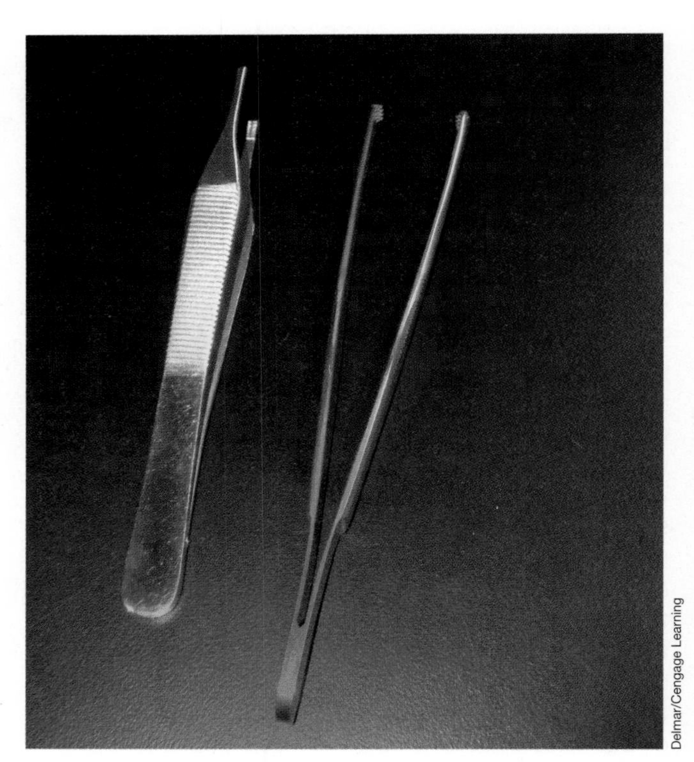

FIGURE 51-30 Tissue forceps.

Tissue Forceps

Tissue forceps are used to grasp tissue and have teeth that grip the tissue (see Figure 51-30). **Dressing forceps** are used to grasp and hold tissue but have no teeth, but rather a smooth surface. Forceps have two handles held together at one end with a spring-like action that opens and closes. **Rat tooth forceps** have large teeth, similar to a rat's, that hold large thick tissues and

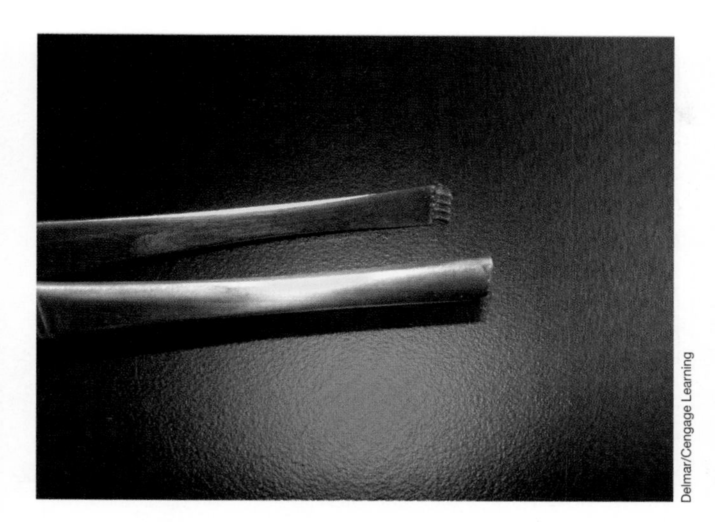

FIGURE 51-31 Rat-toothed forceps.

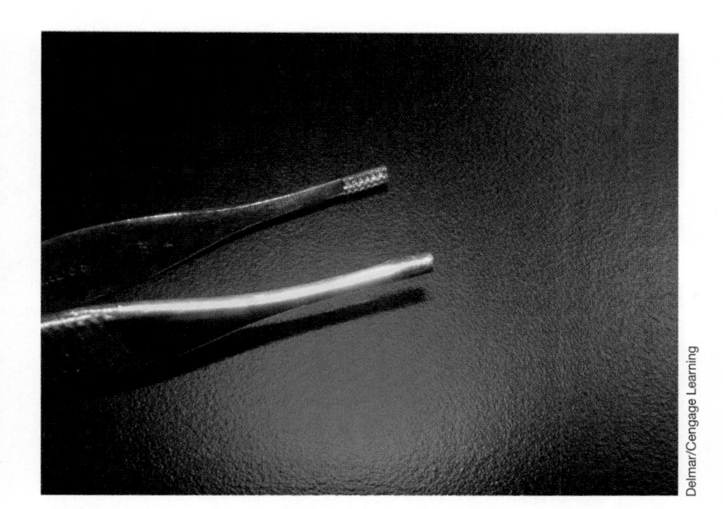

FIGURE 51-32 Adson Brown tissue forceps.

skin (see Figure 51-31). **Adson Brown tissue forceps** have several small and delicate teeth that are used for light handling of delicate tissue (see Figure 51-32). **Allis tissue forceps** are intestinal forceps that lock in place to hold and grasp tissue (see Figure 51-33). They have small interlocking teeth used to grasp bowel tissue, intestinal tissue, and skin. When locked, Allis forceps have an opening that is wide to allow thicker tissues to be clamped. **Babcock tissue forceps** are also intestinal forceps but less traumatic than Allis forceps. They have broad, flared ends with smooth tips. They are used to hold intestines and bladder tissue that is delicate and easily damaged. **Sponge forceps** may be straight or curved and are used to hold gauze sponges or other absorbent material to clear away bleeding and other debris (see Figure 51-34). They may have smooth or serrated edges.

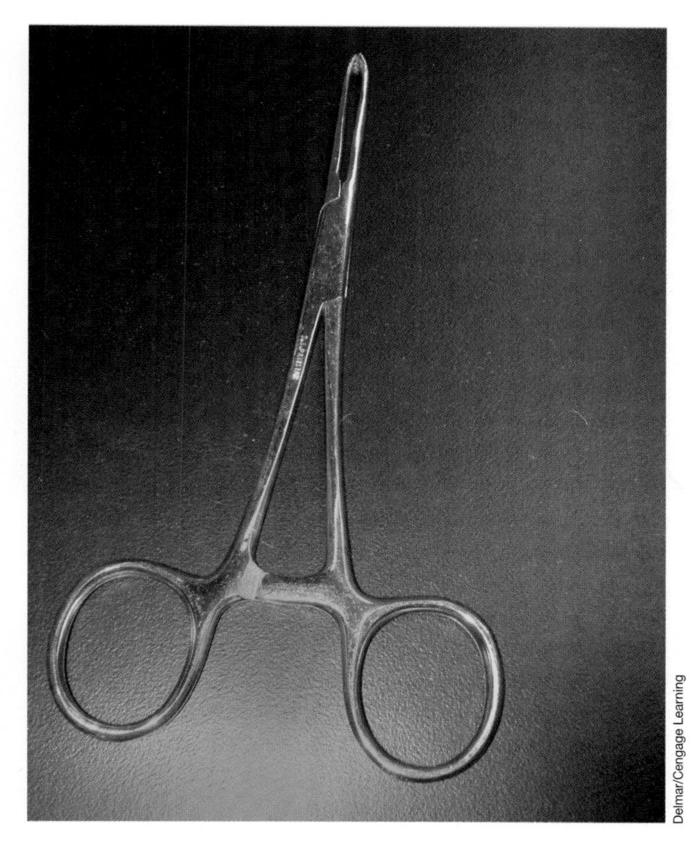

FIGURE 51-33 Allis tissue forceps.

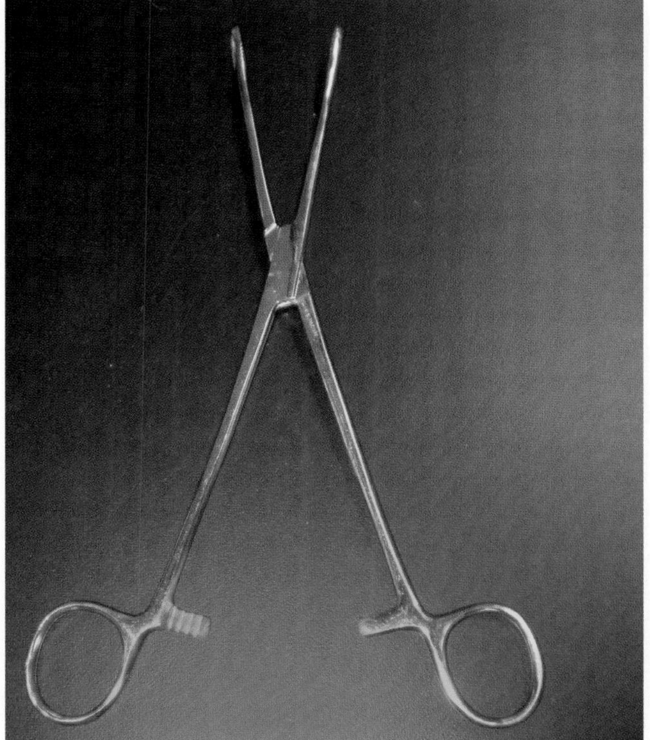

FIGURE 51-34 Sponge forceps.

Hemostatic Forceps

Hemostatic forceps or hemostats are hinged locking instruments that are designed to clamp and hold off blood vessels (see Figure 51-35). Hemostats may be straight or curved. **Mosquito hemostats** are small in size with a fine tip that is used to hold delicate blood vessels (see Figure 51-36). **Kelly hemostats** are larger in size and are partially serrated to better hold and grasp blood vessels (see Figure 51-37). **Crile hemostats** are similar in size and shape to Kelly forceps except they are fully serrated. The serrations are located within the interlocking jaws and are located vertically along the blades. **Carmalt hemostats** are much heavier and used to hold off the ovarian stump during a spay surgery. The interlocking jaws are fully serrated lengthwise.

Intestinal Forceps

Intestinal forceps are used during abdominal surgeries. **Doyen intestinal forceps** or clamps are non-crushing intestinal forceps that have serrations lengthwise. They hold intestinal vessels and tissues without causing any damage. They are used for short periods of time to tie off larger vessels.

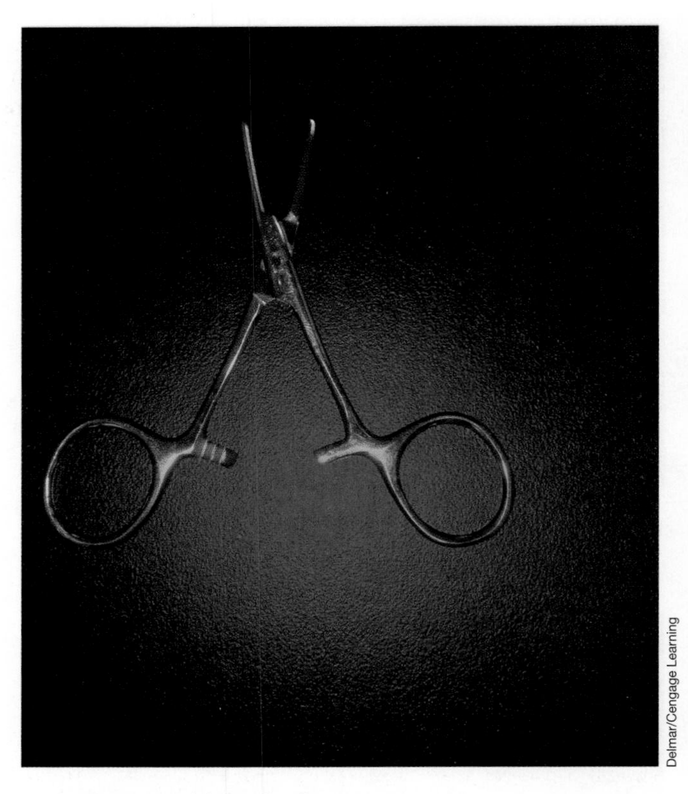
Delmar/Cengage Learning

FIGURE 51-36 Mosquito hemostats.

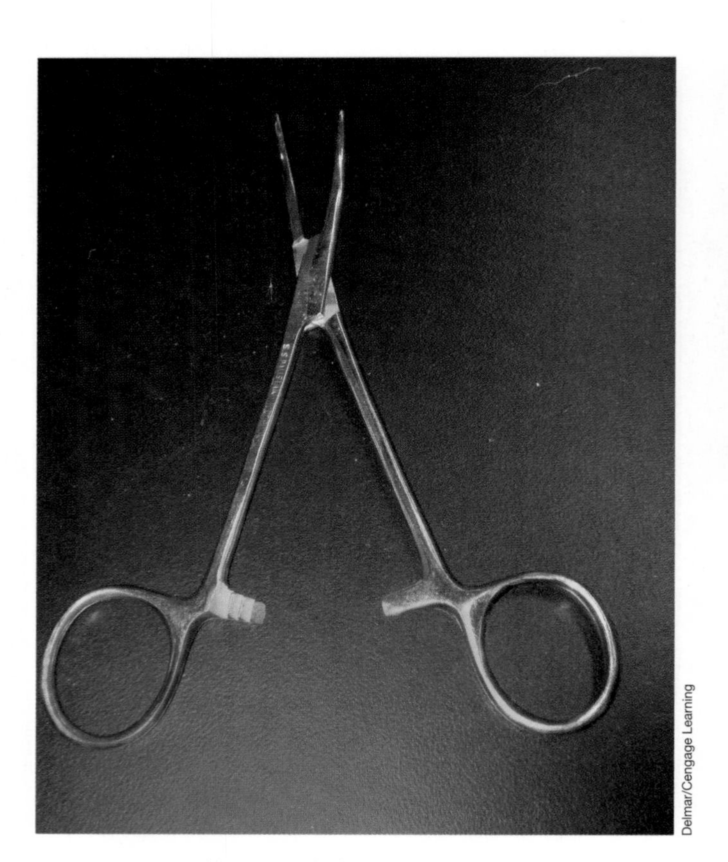
Delmar/Cengage Learning

FIGURE 51-35 Hemostatic forceps.

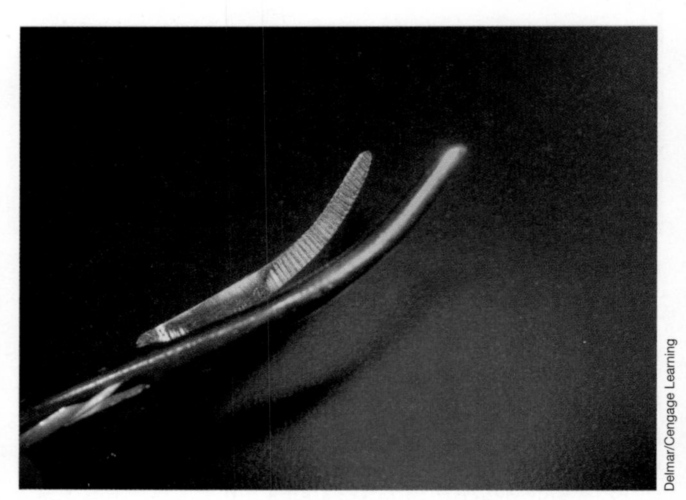
Delmar/Cengage Learning

FIGURE 51-37 Kelly hemostats.

Scissors

Scissors are used to remove tissues during surgery. They may have sharp ends that are pointed or blunt ends that are rounded. They may also have straight or curved blades. **Mayo scissors** are used to cut heavy tissues and sutures (see Figure 51-38). They are proportionate in size and when lying on a flat surface curve upward without contact at the end of the blades. **Metzenbaum scissors** are more delicate, thinner in size, and have long

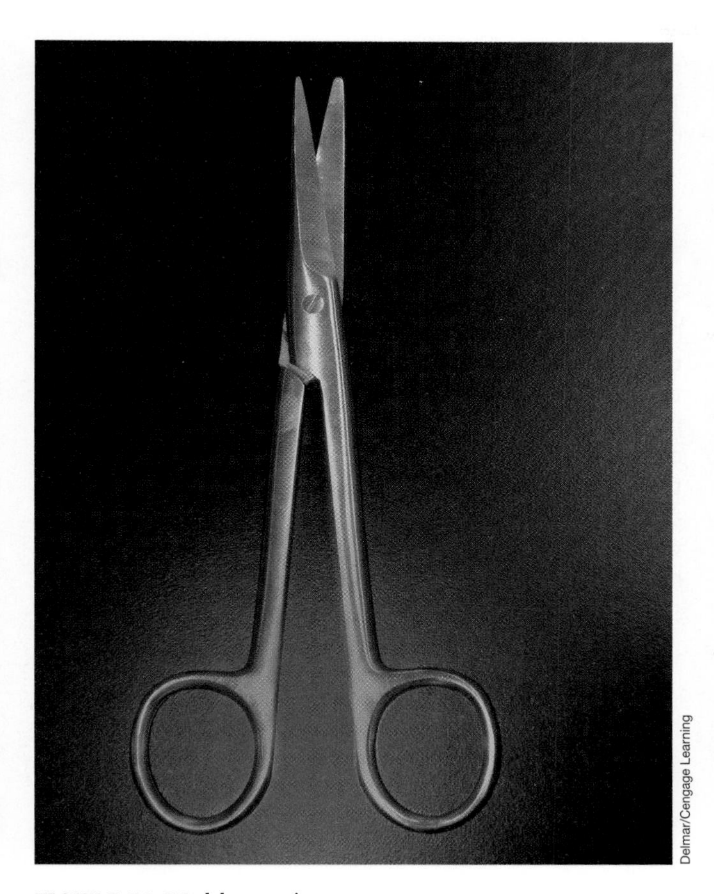

FIGURE 51-38 Mayo scissors.

handles in relationship to the blade length. They are used to cut and separate delicate tissues. Suture removal scissors have a sharp tip with a section in the blade used to get under the suture material and cut it for removal.

Retractors

Retractors may be handheld or self-retracting and are used during surgeries to hold open an animal's body cavity for exploration. Senn and Holman retractors are both handheld. **Senn retractors** have blades at each end that look like fingers on the end of a hand. They may be blunt or sharp in shape. **Holman retractors** are levers that are flat with a pointed appearance that look similar to an arrow (see Figure 51-39). They have holes in the handle and are commonly used during orthopedic procedures. Self-retracting retractors include the Weitlaner and **Gelpi retractors**. The ends hold tissues apart for better visualization within a body cavity. The **Weitlaner retractor** has rake-like teeth on the ratchet ends and can be blunt or sharp (see Figure 51-40). The Gelpi retractor has single points on the end that are sharp.

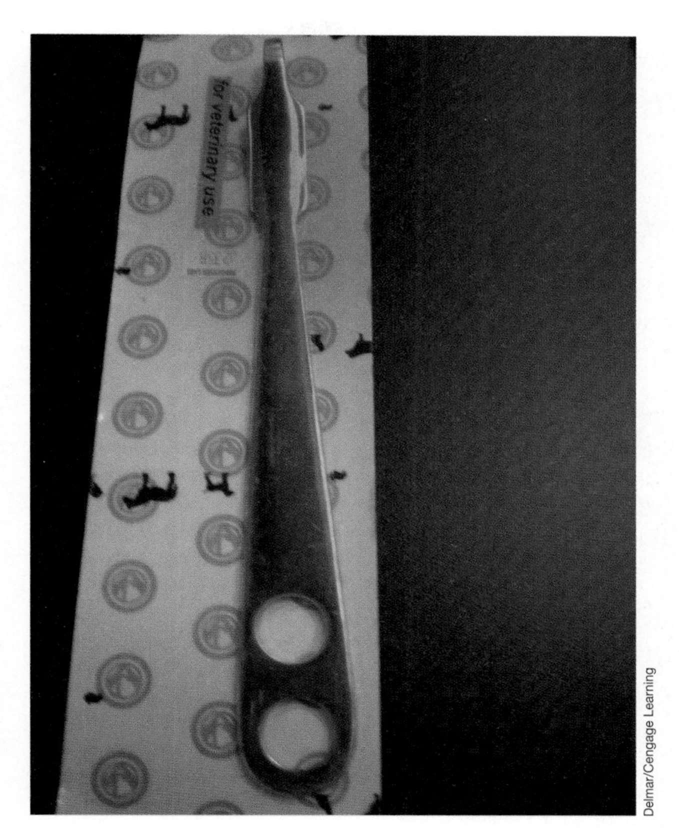

FIGURE 51-39 Holmann retractor.

FIGURE 51-40 Weitlaner retractor.

FIGURE 51-41 Spay hook.

Spay Hook

The **spay hook** is used to locate the uterus and uterine horns in small female animals. It is a long, thin, metal device with a hook on the end (see Figure 51-41).

SUMMARY

Many aspects are involved with surgical assisting. Surgical preparation involves understanding care and maintenance of surgical equipment and instruments, patient preparation and care, surgeon care and assistance, a clean and aseptic surgical environment, and postoperative care of the patient. The goal of all aspects within the surgical suite is to maintain a sterile and aseptic environment that protects the patient and the veterinary health care team.

Key Terms

absorbable suture material that is broken down within the body over a period of time and is absorbed by the body

Adson Brown tissue forceps instrument with several small and delicate teeth that are used for light handling of delicate tissue

Allis tissue forceps intestinal forceps that lock in place to hold and grasp tissue and when locked, have an opening that is wide to allow thicker tissues to be clamped

anesthesia chamber a glass or plastic square or rectangular box that is clear and has two openings in which the inhalation and exhalation tubes on the anesthesia machine connect to the chamber

anesthesia log book that details the patient's status through induction, maintenance and recovery stages during an anesthesia procedure

anesthesia mask cone-like device used in larger patients; placed over the muzzle and face area and must fit snuggly for the animal to breathe in the gases for induction

asepsis the process of a clean, sanitary, and sterile environment

aseptic technique the process of obtaining a sterile and organism-free environment

autoclave most common method for surgical sterilization of packs and equipment by using a large box-like area to heat the items

Babcock tissue forceps intestinal forceps that are less traumatic and have broad, flared ends with smooth tips

Bain system uses a single corrugated or plastic tube that connects to the patient and another hose through the center of the tube that contains the gas mixture that the patient inhales

blink reflex evaluation of the eye to determine which plane of anesthesia a patient is in

blood pressure monitor device used to assess the pumping action of the heart

canister area that absorbs carbon dioxide and contains granules to absorb it

Carmalt hemostats much heavier and used to hold off the ovarian stump during a spay surgery with interlocking serrations located lengthwise

closed gloving process done when the gown is placed with the hands still within the sleeves so that no skin is visible and potentially contaminating the gloves

corrugated tubing plastic and thin hose and allows the oxygen and anesthetic gas to flow to the patient by connecting from the machine to the endotracheal tube

Crile hemostats similar in size and shape to Kelly forceps except they are fully serrated with serrations located within the interlocking jaws and vertically along the blades

cuff located about 3/4 of the way down the endotracheal tube that is inflatable and allows the tube to be inflated with air to create a tight seal around the trachea to prevent any moisture or material from entering the lung field

cylinder metal containers that hold compressed forms or liquid to keep patients oxygenated and breathing during anesthesia

distilled water sterile water that is placed in the water reservoir to the fill line

Doyen intestinal forceps clamps that are non-crushing intestinal forceps that have serrations lengthwise and hold intestinal vessels and tissues for short time periods without causing any damage

dressing forceps instrument used to grasp and hold tissue but have no teeth, but rather a smooth surface

electrocardiogram monitor placed onto a patient through metal leads called electrodes that measure the electrical activity of the heart; also called an ECG or EKG

electrode metal leads that connect a patient to the EKG unit

endotracheal tube a flexible tube that should fit snuggly into the trachea

esophageal stethoscope a device placed into the esophagus next to the endotracheal tube and emits the sound of the heartbeat throughout a procedure

extubation the process of the endotracheal tube removal from the patient

fast to not give food or water or anything by mouth prior to surgery

flow meter located on the machine near the vaporizer and is adjusted for the proper rate of oxygen flow to the patient

Gelpi retractor self-retracting instrument with single points on the end that are sharp

glove pack a prepackaged disposable sterile glove pack that is used one time and then discarded

gown pack holds the sterile gown that is worn by the surgeon and veterinary staff assisting in the surgical area

hemostatic forceps hinged locking instruments that are designed to clamp and hold off blood vessels; also called hemostats

Holman retractor levers that are flat with a pointed appearance that look similar to an arrow with holes in the handle and are commonly used during orthopedic procedures

induction phase the time when the patient is being given anesthesia to make them sleep and remain unconscious and free of sensation during the surgery

instrument milk solution that acts as a lubricant to keep the instruments protected from corrosion

intubation the process of placing a tube into the trachea to establish an airway and allow the patient to continue to inhale gases that keep the patient under general anesthesia as long as necessary

isotonic crystalloid fluids given to an animal that is dehydrated and create a balance of elements within the body

Kelly hemostats larger in size and are partially serrated to better hold and grasp blood vessels

laparotomy towel sterile towel used to control bleeding during surgery and as a hand towel for the surgeon

laryngoscope a tool made of heavy metal that has a light source on the end to help light the airway view

Mayo scissors instrument used to cut heavy tissues and sutures that are proportionate in size and when lying on a flat surface curve upward without contact at the end of the blades

Mayo stand the instrument tray that is elevated to hold the surgeon's surgical instruments and supplies

Metzenbaum scissors instrument delicate and thinner in size and have long handles in relationship to the blade length and are used to cut and separate delicate tissues

(*Continues*)

mosquito hemostats small in size with a fine tip that is used to hold delicate blood vessels

needle holder instrument used to hold a needle while placing sutures into a patient during surgery

non-absorbable suture material not broken down within the body and remains intact for long periods

non-rebreathing system system used for very small patients such as rodents, birds and reptiles, or small puppies and kittens that are too small to move the breathing bag with their respirations

open gloving process done by removing the left glove using the right hand by grasping it and the edge of the cuff

oxygen tank located under the machine or in separate cylinders next to the machine

palpebral reflex evaluated by tapping the fingertip lightly over the surface of the eyelid

pedal reflex used on the digit or the skin between digits, which is pinched to determine a reflex response

plane determines how awake or how asleep a patient is under anesthesia

pop-off valve used to release excess amounts of gas when the pressure becomes too high within the machine system

pressure valve indicates how much oxygen is left in the tank

pulse oximeter a device used to measure a patient's vital signs by indirectly measuring the oxygen saturation within the blood and any changes in the blood volume

rat tooth forceps instrument with large teeth, similar to a rat's, that hold large thick tissues and skin

rebreathing system bag that connects to the machine and inflates and deflates as the patient breathes and allows the patient to rebreathe anesthetic gas after it has passed through the CO_2 absorbing canister

retractor handheld or self-retracting and are used during surgeries to hold open an animal's body cavity for exploration

scavenger hose tube attached to the pop-off valve to prevent gas from leaking into the room and pushes gases outside of the facility to escape within the outside air

scrub pack sterile supply that contains the necessary items to properly prepare the surgeon, technician, or assistant for surgery for washing hands

Senn retractor retractor blades located at each end that look like fingers on the end of a hand may be blunt or sharp in shape

soda lime granules within the canister that absorb the carbon dioxide

spay hook instrument used to located the uterus and uterine horns in small female animals that has a long, thin, metal device with a hook on the end

sphygmomanometer device used to measure an animal's blood pressure during surgery; will also help determine tissue perfusion as well as depth of anesthesia; also called a blood pressure monitor

sponge forceps straight or curved and used to hold gauze sponges or other absorbent material to clear away bleeding and other debris

staple remover device placed under the staple and squeezed to open up the staple, which is then gently removed from the site

surgery pack sterile instrument prepared for surgical procedure with tools and supplies required for the procedure

surgical glue sterile substance used for closure of small wounds or surgical sites and is commonly used in feline declaws

surgical log a record of details recording all surgeries performed in the facility

surgical margin area of the patient that should extend 2 to 4 inches beyond the anticipated surgical incision borders

surgical scrub process then begins with sterile solutions placed onto the surgical margin to clean the patient

surgical staples fast and easy way of using a device that closes the skin with metal pieces

suture reel container of suture material that must be cut and thread into a needle

suture removal removal of sutures within 7–10 days after surgery

suture removal scissors scissors that have one blade that has a small hook shape on the end to better get under the suture to remove

swallow reflex the patient licking with the tongue or having increased jaw tone during extubation of the endotracheal tube

swedged on needle suture material that has a needle already attached

tidal volume amount of oxygen required by the patient

tissue forceps instrument used to grasp tissue and have teeth that grip the tissue

towel clamp instrument used to hold the surgical drape in place over the patient during the surgery

ultrasonic cleaner instruments are placed within a wire basket and the basket is seated within the cleaner, which holds a disinfectant solution and creates vibrations to clean the items

Weitlaner retractor self-retracting instrument with rake-like teeth on the ratchet ends and can be blunt or sharp

vaporizer converts the liquid anesthesia into a gaseous form, which the patient inhales

Y connector plastic device that connects the intake and outtake hose from the anesthesia machine to the endotracheal tube of the patient

REVIEW QUESTIONS

1. What is the importance of asepsis?
2. What is the goal of an aseptic surgical area?
3. What items are recorded in an anesthesia log?
4. Why is pre-anesthetic medication given to patients?
5. What is intubation?
6. How does an assistant know when to extubate a patient?
7. What vital sign monitors are found in the surgical suite?
8. What items must be sterile when working in the surgical area?
9. What items are found in a surgical pack?
10. What are signs of pain in animals?
11. What factors should veterinary assistants know about the anesthesia machine?

Clinical Situation

Sally, a veterinary assistant at Kitty Kat Klinic, is working in the surgical suite with Dr. Sprouse. Sally has been trained to open the surgical packs and prepare the surgical monitors and equipment, as well as clean and sterilize the items used in surgery. Today, Sally has noticed that the surgical pack she is beginning to open has a small tear in it. She also notices that the autoclave tape indicator is not very dark.

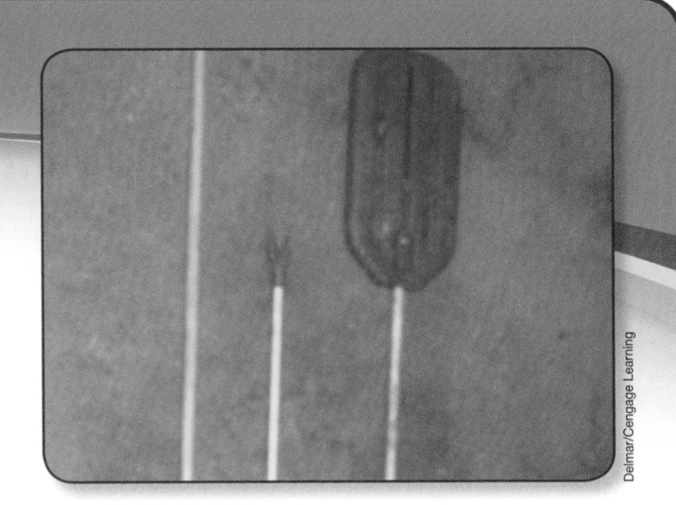

Delmar/Cengage Learning

- What should Sally do?
- What may have caused these occurrences?
- What may prevent these situations in the future?

Appendix A
AVMA Accredited Veterinary Schools

ALABAMA

Auburn University
College of Veterinary Medicine
104 J. E. Greene Hall
Auburn University, AL 36849-5517
(334) 844-4546
Status: Full Accreditation
Last Evaluation: 2005
Next Evaluation: 2012
SAVMA Chapter: Yes
http://www.vetmed.auburn.edu

Tuskegee University
School of Veterinary Medicine
Tuskegee, AL 36088
(334) 727-8174
Status: Limited Accreditation
Last Evaluation: 2008
Next Evaluation: 2013
SAVMA Chapter: Yes
http://tuskegee.edu

CALIFORNIA

University of California
School of Veterinary Medicine
Davis, CA 95616-8734
(530) 752-1360
Status: Full Accreditation
Last Evaluation: 2004
Next Evaluation: 2011
SAVMA Chapter: Yes
http://www.vetmed.ucdavis.edu

Western University of Health Sciences
College of Veterinary Medicine
309 E Second Street—College Plaza
Pomona, CA 91766-1854
(909) 469-5628
Status: Limited Accreditation
Last Evaluation: 2008
Next Evaluation: 2010
http://www.westernu.edu/cvm.html

COLORADO

Colorado State University
College of Veterinary Medicine and Biomedical Sciences
Fort Collins, CO 80523-1601
(970) 491-7051
Status: Full Accreditation
Last Evaluation: 2008
Next Evaluation: 2015
SAVMA Chapter: Yes
http://www.cvmbs.colostate.edu

FLORIDA

University of Florida
College of Veterinary Medicine
PO Box 100125
Gainesville, FL 32610-0125
(352) 392-2213
Status: Full Accreditation
Last Evaluation: 2008
Next Evaluation: 2015
SAVMA Chapter: Yes
http://www.vetmed.ufl.edu

GEORGIA

University of Georgia
College of Veterinary Medicine
Athens, GA 30602
(706) 542-3461
Status: Full Accreditation
Last Evaluation: 2006
Next Evaluation: 2013
SAVMA Chapter: Yes
http://www.vet.uga.edu

ILLINOIS

University of Illinois

College of Veterinary Medicine
2001 South Lincoln Avenue
Urbana, IL 61802
(217) 333-2760
Status: Full Accreditation
Last Evaluation: 2006
Next Evaluation: 2013
SAVMA Chapter: Yes
http://www.cvm.uiuc.edu

INDIANA

Purdue University

School of Veterinary Medicine
1240 Lynn Hall
West Lafayette, IN 47907-1240
(765) 494-7607
Status: Full Accreditation
Last Evaluation: 2004
Next Evaluation: 2011
SAVMA Chapter: Yes
http://www.vet.purdue.edu

IOWA

Iowa State University

College of Veterinary Medicine
Ames, IA 50011
(515) 294-1242
Status: Full Accreditation
Last Evaluation: 2006
Next Evaluation: 2010
SAVMA Chapter: Yes
http://www.vetmed.iastate.edu

KANSAS

Kansas State University

College of Veterinary Medicine
Manhattan, KS 66506
(785) 532-5660
Status: Full Accreditation
Last Evaluation: 2003
Next Evaluation: 2010
SAVMA Chapter: Yes
http://www.vet.ksu.edu

LOUISIANA

Louisiana State University

School of Veterinary Medicine
Baton Rouge, LA 70803-8402
(225) 578-9900
Status: Limited Accreditation
Last Evaluation: 2005
Next Evaluation: 2010
SAVMA Chapter: Yes
http://www.vetmed.lsu.edu

MASSACHUSETTS

Tufts University

School of Veterinary Medicine
200 Westboro Road
North Grafton, MA 01536
(508) 839-5302
Status: Full Accreditation
Last Evaluation: 2004
Next Evaluation: 2011
SAVMA Chapter: Yes
http://www.tufts.edu/vet

MICHIGAN

Michigan State University

College of Veterinary Medicine
G-100 Veterinary Medical Center
East Lansing, MI 48824-1314
(517) 355-6509
Status: Full Accreditation
Last Evaluation: 2005
Next Evaluation: 2012
SAVMA Chapter: Yes
http://cvm.msu.edu

MINNESOTA

The University of Minnesota

College of Veterinary Medicine
1365 Gortner Avenue
St. Paul, MN 55108
(612) 624-9227
Status: Full Accreditation
Last Evaluation: 2007
Next Evaluation: 2014
SAVMA Chapter: Yes
http://www.cvm.umn.edu

MISSISSIPPI

Mississippi State University
College of Veterinary Medicine
Mississippi State, MS 39762
(662) 325-3432
Status: Full Accreditation
Last Evaluation: 2007
Next Evaluation:2014
SAVMA Chapter: Yes
http://www.cvm.msstate.edu

MISSOURI

University of Missouri–Columbia
College of Veterinary Medicine
Columbia, MO 65211
(573) 882-3877
Status: Full Accreditation
Last Evaluation: 2006
Next Evaluation: 2013
SAVMA Chapter: Yes
http://www.cvm.missouri.edu

NEW YORK

Cornell University
College of Veterinary Medicine
Ithaca, NY 14853-6401
(607) 253-3700
Status: Full Accreditation
Last Evaluation: 2003
Next Evaluation: 2010
SAVMA Chapter: Yes
http://www.vet.cornell.edu

NORTH CAROLINA

North Carolina State University
College of Veterinary Medicine
4700 Hillsborough Street
Raleigh, NC 27606
(919) 513-6210
Status: Full Accreditation
Last Evaluation: 2007
Next Evaluation: 2014
SAVMA Chapter: Yes
http://www.cvm.ncsu.edu/

OHIO

The Ohio State University
College of Veterinary Medicine
1900 Coffey Road
Columbus, OH 43210-1092
(614) 292-1171
Status: Full Accreditation
Last Evaluation: 2006
Next Evaluation: 2013
SAVMA Chapter: Yes
http://www.vet.ohio-state.edu

OKLAHOMA

Oklahoma State University
Center for Veterinary Health Sciences
Stillwater, OK 74078
(405) 744-6595
Status: Full Accreditation
Last Evaluation: 2003
Next Evaluation: 2010
SAVMA Chapter: Yes
http://www.cvm.okstate.edu

OREGON

Oregon State University
College of Veterinary Medicine
Corvallis, OR 97331
(541) 737-2098
Status: Full Accreditation
Last Evaluation: 2007
Next Evaluation: 2014
SAVMA Chapter: Yes
http://www.vet.orst.edu

PENNSYLVANIA

University of Pennsylvania
School of Veterinary Medicine
3800 Spruce Street
Philadelphia, PA 19104-6044
(215) 898-5438
Status: Full Accreditation
Last Evaluation: 2009
Next Evaluation: 2016
SAVMA Chapter: Yes
http://www.vet.upenn.edu

TENNESSEE

University of Tennessee

College of Veterinary Medicine
2407 River Drive
Knoxville, TN 37996
(865) 974-7262
Status: Full Accreditation
Last Evaluation: 2008
Next Evaluation: 2010
SAVMA Chapter: Yes
http://www.vet.utk.edu

TEXAS

Texas A&M University

College of Veterinary Medicine & Biomedical Sciences
College Station, TX 77843-4461
(979) 845-5051
Status: Full Accreditation
Last Evaluation: 2009
Next Evaluation: 2016
SAVMA Chapter: Yes
http://www.cvm.tamu.edu

VIRGINIA

Virginia Tech

Virginia-Maryland Regional
College of Veterinary Medicine
Blacksburg, VA 24061-0442
(540) 231-7666
Status: Full Accreditation
Last Evaluation: 2007
Next Evaluation: 2014
SAVMA Chapter: Yes
http://www.vetmed.vt.edu

WASHINGTON

Washington State University

College of Veterinary Medicine
Pullman, WA 99164-7010
(509) 335-9515
Status: Full Accreditation
Last Evaluation: 2003
Next Evaluation: 2010
SAVMA Chapter: Yes
http://www.vetmed.wsu.edu

WISCONSIN

University of Wisconsin–Madison

School of Veterinary Medicine
2015 Linden Drive West
Madison, WI 53706-1102
(608) 263-6716
Status: Full Accreditation
Last Evaluation: 2008
Next Evaluation: 2015
SAVMA Chapter: Yes
http://www.vetmed.wisc.edu

Calgary, Alberta

University of Calgary

Faculty of Veterinary Medicine
Calgary, Alberta T2N 4N1
CANADA
Phone: 403-210-3961
Status: Letter of Reasonable Assurance
Last Evaluation: 2009
Next Evaluation: 2013
http://www.vet.ucalgary.ca

Ontario

University of Guelph

Ontario Veterinary College
Guelph, ON N1G 2W1
CANADA
Phone: 519-824-4120
Status: Full Accreditation
Last Evaluation: 2009
Next Evaluation: 2016
SAVMA Chapter: No
http://www.ovc.uoguelph.ca/

Prince Edward Island

University of Prince Edward Island

Atlantic Veterinary College
550 University Avenue
Charlottetown, PE C1A 4P3
CANADA
Phone: 902-566-0882
Status: Full Accreditation
Last Evaluation: 2004
Next Evaluation: 2011
SAVMA Chapter: Yes
http://www.upei.ca/~avc

Quebec

Université de Montréal

Faculté de médecine vétérinaire
C.P. 5000
Saint Hyacinthe, PQ J2S 7C6
CANADA
Phone: 450-773-8521
Status: Full Accreditation
Last Evaluation: 2006
Next Evaluation: 2012
SAVMA Chapter: No
www.medvet.umontreal.ca

Saskatchewan

University of Saskatchewan

Western College of Veterinary Medicine
52 Campus Drive
Saskatoon, SK S7N 5B4
CANADA
Phone: 306-966-7447
Status: Full Accreditation
Last Evaluation: 2003
Next Evaluation: 2010
SAVMA Chapter: No
http://www.usask.ca/wcvm

Australia

Murdoch University

Division of Veterinary and Biomedical Sciences Murdoch
WESTERN AUSTRALIA
Phone: 61 8 9360 2566
Status: Full Accreditation
Last Evaluation: 2009
Next Evaluation: 2016
http://wwwvet.murdoch.edu.au

The University of Melbourne

Faculty of Veterinary Science
Werribee, VIC Australia
Phone: 61 3 9731 2261
Status: Full Accreditation
Last Evaluation: 2006
Next Evaluation: 2013
http://www.vet.unimelb.edu.au

The University of Sydney

Faculty of Veterinary Science
NSW AUSTRALIA
Phone: 61 2 9351 6936
Status: Full Accreditation
Last Evaluation: 2005
Next Evaluation: 2012
http://www.vetsci.usyd.edu.au/

England

University of London

The Royal Veterinary College
Royal College Street
London NW1 OTU
ENGLAND
Phone: 44 (0)20 7468-5000
Status: Full Accreditation
Last Evaluation: 2005
Next Evaluation: 2012
http://www.rvc.ac.uk

Ireland

University College Dublin

School of Agriculture, Food Science & Veterinary Medicine
Belfield, Dublin 4 Ireland
Phone: 353 1 716 6100
Status: Full Accreditation
Last Evaluation: 2007
Next Evaluation: 2014
http://www.ucd.ie/agfoodvet

The Netherlands

State University of Utrecht

Faculty of Veterinary Medicine
PO Box 80.163
3508 TD Utrecht
THE NETHERLANDS
Phone: 31 30 253-4851
Status: Full Accreditation
Last Evaluation: 2007
Next Evaluation: 2014
http://www.vet.uu.nl

New Zealand

Massey University College of Sciences

Institute of Veterinary, Animal, and Biomedical Sciences
Palmerston North
NEW ZEALAND
Phone: 64 6 350-5714
Status: Full Accreditation
Last Evaluation: 2007
Next Evaluation: 2014
http://ivabs.massey.ac.nz

Scotland

University of Glasgow

Faculty of Veterinary Medicine
Glasgow, Scotland G61 1QH
Phone: 44 (0)141 330-5700
Status: Full Accreditation
Last Evaluation: 2006
Next Evaluation: 2013
http://www.gla.ac.uk/faculties/vet

The University of Edinburgh

Royal (Dick) School of Veterinary Studies
Summerhall
Edinburgh EH9 1QH
SCOTLAND
Phone: 44 131 650-1000
Status: Full Accreditation
Last Evaluation: 2008
Next Evaluation: 2015
http://www.vet.ed.ac.uk

Appendix B AVMA Accredited Veterinary Technology Programs

ALABAMA

Jefferson State Community College

Veterinary Technology Distance Learning Program
2601 Carson Road
Birmingham, AL 35215-3098
205-856-8519
www.jeffstateonline.com/vet_tech/
(Christie Wallace, DVM Director)
Associate of Applied Science
Initial Accreditation: March 16, 2007
PROVISIONAL ACCREDITATION

ARIZONA

Anthem College

1515 E. Indian School Road
Phoenix, AZ 85014
602-279-9700
www.anthem.edu
(Jeremy Prupas, DVM, Director)
Associate in Science
Initial Accreditation: March 3, 2008
PROVISIONAL ACCREDITATION

Kaplan College

Veterinary Technology Program
13610 N. Black Canyon Hwy., Suite 104
Phoenix, AZ 85029
602-548-1955
http://getinfo.kaplancollege.com/KaplanCollegePortal/KaplanCollegeCampuses/Arizona/PhoenixEast/
(James Flom, DVM, Program Director)
Associate of Occupational Studies
Initial Accreditation: August 19, 2005
PROVISIONAL ACCREDITATION

Mesa Community College

Veterinary Technology/Animal Health Program
1833 W. Southern Avenue
Mesa, AZ 85202
480-461-7488
www.mc.maricopa.edu
(Jill Sheport, DVM, Director)
Associate of Applied Science
Initial Accreditation: February 3, 2006
PROVISIONAL ACCREDITATION

Penn Foster College

Veterinary Technician Distance Education Program
14624 N. Scottsdale Road, Suite 310
Scottsdale, AZ 85254
800-275-4410
www.pennfostercollege.edu
(Margi Sirois, EdD, Director)
Associate in Science
Initial Accreditation: January 5, 2006
PROVISIONAL ACCREDITATION

Pima County Community College

Veterinary Technology Program
8181 E. Irvington Road
Tucson, AZ 85709-4000
520-206-7414
FAX 520-206-7803
www.pima.edu
(Daniel DeNoon, DVM, Director)
Associate of Applied Science
Initial Accreditation: January 29, 2003
FULL ACCREDITATION

ARKANSAS

Arkansas State University-Beebe
1000 Iowa Street
PO Box 1000
Beebe, AR 72012
501-882-6263
www.asub.edu
Kristie Coley, DVM Director
Associate in Applied Science
Initial Accreditation–June 10, 2009
PROVISIONAL ACCREDITATION

CALIFORNIA

California State Polytechnic University-Pomona
College of Agriculture
Animal Health Technology Program
3801 W. Temple Ave.
Pomona, CA 91768
909-869-2136
www.csupomona.edu
(James Alderson, DVM, Coordinator)
Bachelor of Science
Initial Accreditation: April 1996
FULL ACCREDITATION

Cosumnes River College
Veterinary Technology Program
8401 Center Pkwy.
Sacramento, CA 95823
916-691-7355
http://ww2.losrios.edu/crc/x77.xml
(Christopher Impinna, DVM, Director)
Associate in Science
Initial Accreditation: April 1975
FULL ACCREDITATION

Foothill College
Veterinary Technology Program
12345 El Monte Rd.
Los Altos Hills, CA 94022
650-949-7203
FAX 650-949-7375
www.foothill.fhda.edu
(Karl Peter, DVM, Director)
Associate in Science
Initial Accreditation: April 1977
FULL ACCREDITATION

Hartnell College
Animal Health Technology Program
156 Homestead Ave.
Salinas, CA 93901
831-755-6855
www.hartnell.cc.ca.us
(Sharon McKinney-Radel, RVT, Director)
Associate in Science
Initial Accreditation: April 1981
TERMINAL ACCREDITATION

Los Angeles Pierce College
Veterinary Technology Program
6201 Winnetka Ave.
Woodland Hills, CA 91371
818-347-0551
www.macrohead.com/rvt
(Elizabeth White, RVT, Director)
Associate in Science
Initial Accreditation: December 1975; April 1993
FULL ACCREDITATION

Mt. San Antonio College
Animal Health Technology Program
1100 N. Grand Ave.
Walnut, CA 91789
909-594-5611
www.mtsac.edu
(Jean Hoffman, RVT, Director)
Associate in Science
Initial Accreditation: April 1977
FULL ACCREDITATION

Western Career College-Citrus Heights Campus
7301 Greenback Lane, Suite A
Citrus Heights, CA 95621
916-722-8200
www.westerncollege.edu
Ms. Patti Hebert, RVT, Program Director
pattih@westerncollege.com
Associate in Science
Initial Accreditation: June 16, 2006
FULL ACCREDITATION

Western Career College–Pleasant Hill
Veterinary Technician Program
380 Civic Drive, #300
Pleasant Hill, CA 94523
925-609-6650
www.westerncollege.com/ph_campus.asp
(Andrea Monroe, DVM, Director)
Associate in Science
Initial Accreditation: March 31, 2004
FULL ACCREDITATION

Western Career College-Sacramento
Veterinary Technology Program
8909 Folsom Blvd.
Sacramento, CA 95826
916-361-1660
www.westerncollege.com/ph_campus.asp
(Janelle Emmett, DVM, Program Director)
Associate in Science
Initial Accreditation: August 20, 2004
FULL ACCREDITATION

Western Career College–San Jose
Veterinary Technician Education Program
6201 San Ignacio Ave.
San Jose, CA 95119
408-360-0840
www.westerncollege.edu
(Barbara McManus, RVT, Interim Director)
Associate in Science
Initial Accreditation: December 1, 2006
PROVISIONAL ACCREDITATION

Western Career College–San Leandro
Veterinary Technology Program
170 Bayfair Mall
San Leandro, CA 94578
510-276-3888
www.westerncollege.com/ph_campus.asp
(Laura Jamison, RVT, Program Director)
Associate in Science
Initial Accreditation: August 18, 2004
FULL ACCREDITATION

Western Career College–Stockton Campus
Veterinary Technician Education Program
1313 West Robinhood Drive, Suite B
Stockton, CA 95207
209-956-1240 x 44116
www.westerncollege.edu
(Jamie Larson, RVT, Interim Program Director)
tyeatts@westerncollege.com
Associate in Science
Initial Accrediation: June 14, 2006
PROVISIONAL ACCREDITATION

Yuba College
Veterinary Technology Program
2088 N. Beale Rd.
Marysville, CA 95901
530-741-6962
www.yccd.edu/yuba/vettech/index.html
(Scott Haskell, DVM, Director)
Associate in Science
Initial Accreditation: April 1978
FULL ACCREDITATION

CANADA

Ridgetown College, University of Guelph
Veterinary Technology Program
Main Street East
Ridgetown, Ontario N0P 2C0
519-674-1666
www.ridgetownc.on.ca
(Dr. Irene Moore, Program Director)
Diploma
Initial Accreditation: February 4, 2005
FULL ACCREDITATION

AHT/VT Programs in Canada
Canadian recognition

At its June 2006 meeting, the AVMA Executive Board approved a recommendation that the AVMA recommends that veterinary technician credentialing (i.e., licensing, registration, or certification) entities in the United States recognize graduates of Canadian Veterinary Medical Association (CVMA)-accredited veterinary technology programs as eligible for credentialing. In turn, the CVMA recommends that Canadian provincial licensing bodies recognize graduates of AVMA-accredited veterinary technology programs as being eligible for licensure. As always, eligibility for licensure/registration/certification of veterinary technicians is the purview of each state and provincial credentialing agency.

COLORADO

Bel-Rea Institute of Animal Technology
1681 S. Dayton St.
Denver, CO 80231
800-950-8001
www.bel-rea.com
(Nolan Rucker, DVM, President)
Associate of Applied Science
Initial Accreditation: April 1975
FULL ACCREDITATION

Colorado Mountain College
Veterinary Technology Program
Spring Valley Campus
3000 County Rd. 114
Glenwood Springs, CO 81601
970-945-8691
www.coloradomtn.edu
(Jeff Myers, DVM, Coordinator)
Associate of Applied Science
Initial Accreditation: December 1975
FULL ACCREDITATION

Community College of Denver
Veterinary Technology Program
1070 Alton Way, Bldg. 849
Denver, CO 80230
303-365-8300
www.ccd.rightchoice.org
(Shannon Burkhalter, CVT, Director)
Associate of Applied Science
Initial Accreditation: March 2002
FULL ACCREDITATION

Front Range Community College
Veterinary Research Technology Program
4616 S. Shields
Ft. Collins, CO 80526
970-226-2500
www.frcc.cc.co.us
(Elizabeth Bauer, DVM, Director)
Associate of Applied Science
Initial Accreditation: April 1996
FULL ACCREDITATION

Pima Medical Institute-Colorado Springs
3770 North Citadel Drive
Colorado Springs, CO 80910
Phone: 719-482-7462
www.pmi.edu
Christina LeMay, CVT Director
Associate of Occupational Science
Initial Accreditation–August 7, 2009
PROVISIONAL ACCREDITATION

CONNECTICUT

NW Connecticut Community College
Veterinary Technology Program
Park Place East
Winsted, CT 06098
860-738-6490
www.nwctc.commnet.edu/vettech/
(Sheryl Keeley, CVT, or Nancy Marchetti, DVM, Coordinator)
Associate in Science
Initial Accreditation: March 2000
FULL ACCREDITATION

Quinnipiac University
Veterinary Technology Program
Mt. Carmel Ave.
Hamden, CT 06518
203-582-8958
www.quinnipiac.edu
(Steve Carleton, DVM, Director)
Bachelor of Science
Initial Accreditation: April 1980
TERMINAL ACCREDITATION

DELAWARE

Delaware Technical and Community College
Veterinary Technology Program
PO Box 610, Route 18
Georgetown, DE 19947
302-855-5918
FAX 302-858-5460
www.dtcc.edu
(Valerie Quillen, DVM, Department Chair)
Associate in Science
Initial Accreditation: May 1, 2002
FULL ACCREDITATION

FLORIDA

Brevard Community College
Veterinary Technology Program
1519 Clearlake Rd.
Cocoa, FL 32922
321-433-7594
FAX 321-634-4565
www.brevard.cc.fl.us
(Laura Earle-Imre, DVM, Director)
Associate in Science
Initial Accreditation: November 1999
FULL ACCREDITATION

Hillsborough Community College
Veterinary Technology Program
PO Box 31127
Tampa, FL 33631-3127
813-757-2157
www.hccfl.edu
(Vincent Centonze, DVM-Director)
Associate in Science
Initial Accreditation-August 1, 2008
PROVISIONAL ACCREDITATION

Miami-Dade College
Veterinary Technology Program
Medical Center Campus
950 NW 20th Street
Miami, FL 33127
305-237-4473
FAX 305-237-4278
www.mdc.edu/medical/AHT/Vet/default.asp
(Lois Sargent, DVM, Coordinator)
Associate in Science
Initial Accreditation: February 21, 2003
FULL ACCREDITATION

St. Petersburg College
Veterinary Technology Program
Box 13489
St. Petersburg, FL 33733
727-341-3652
www.spcollege.edu/hec/vt/
(Richard Flora, DVM, Director)

On-Campus Program
Associate in Science
Initial Accreditation: April 1978
FULL ACCREDITATION

Distance Learning Program
Associate in Science
Bachelor of Science
www.spcollege.edu/bachelors/vtech.
php?program = vtech
Initial Accreditation: November 1995
FULL ACCREDITATION

GEORGIA

Athens Technical College
Veterinary Technology Program
800 US Highway 29N
Athens, GA 30601
706-355-5107
www.athenstech.edu
(Carole Miller, DVM, Director)
Associate of Applied Science
Initial Accreditation: February 28, 2003
FULL ACCREDITATION

Fort Valley State University
Veterinary Technology Program
1005 State University Drive
Fort Valley, GA 31030
478-825-6353
FAX 478-827-3023
www.fvsu.edu/
(Seyedmehdi Mobini, DVM, Interim Director)
Associate of Applied Science
Bachelor of Science
Initial Accreditation: November 1978; (4 yr)
September 2001
FULL ACCREDITATION

Gwinnett Technical College
Veterinary Technology Program
5150 Sugarloaf Pkwy.
Lawrenceville, GA 30043
770-962-7580
FAX 770-962-7985
www.gwinnetttechnicalcollege.com
(Bonnie Ballard, DVM, Director)
Associate of Applied Science
Initial Accreditation: March 2000
FULL ACCREDITATION

Ogeechee Technical College
Veterinary Technology Program
1 Joe Kennedy Blvd.
Statesboro, GA 30458
912-688-6037
www.ogeecheetech.edu
(Dr. Janice Grumbles, Program Director)
Initial Accreditation: August 12, 2005
Associate of Applied Science
PROVISIONAL ACCREDITATION

IDAHO

College of Southern Idaho
Veterinary Technology Program
315 Falls Ave.
Twin Falls, ID 83303-1238
208-733-9554 ext. 2408
FAX 208-736-2136
www.csi.cc.id.us
(Jody Rockett, DVM, Director)
Associate of Applied Science
Initial Accreditation: March 1999
FULL ACCREDITATION

ILLINOIS

Joliet Junior College
Agriculture Sciences Department
1215 Houbolt Road
Joliet, IL 60431
815-280-2746
FAX 815-280-2741
www.jjc.cc.il.us
(R. Scott Keller, DVM, Coordinator)
Associate of Applied Science
Initial Accreditation: March 2001
FULL ACCREDITATION

Parkland College
Veterinary Technology Program
2400 W. Bradley Ave.
Champaign, IL 61821
217-351-2224
www.parkland.edu
(Paul Cook, DVM, Director)
Associate of Applied Science
Initial Accreditation: November 1976
FULL ACCREDITATION

Rockford Business College
4230 Newburg Rd.
Rockford, IL 61108
815-965-8616
www.rockfordbusinesscollege.edu
(Melissa Hagemann, CVT, Director)
Associate of Applied Science
Initial Accreditation: February 1, 2008
PROVISIONAL ACCREDITATION

Southern Illinois Collegiate Common Market (SICCM)
Includes schools at: John A. Logan College at Carterville; Kaskaskia College at Centralia; Rend Lake College at Ina; Shawnee Community College at Ullin; Southeastern Illinois College at Harrisburg; Southern Illinois University at Carbondale; Southern Illinois University at Edwardsville
3213 South Park Avenue
Herrin, IL 62948
Jamie Morgan, CVT-Director
618-942-6902
www.siccm.com
Associate of Applied Science
Initial Accreditation-February 6, 2009
PROVISIONAL ACCREDITATION

Vet Tech Institute at Fox College
18020 Oak Park Ave.
Tinley Park, IL 60477
(Dr. Jordan Siegel, Program Director)
708-636-7700
www.foxcollege.edu
Associate of Applied Science
Initial Accreditation: October 22, 2007
PROVISIONAL ACCREDITATION

INDIANA

Brown Mackie College-Michigan City
325 East US Highway 20
Michigan City, IN 46360
Barbara Kaufman, RVT-Program Director
800-519-2416
www.brownmackie.edu
Associate of Science
Initial Accreditation-March 20, 2009
PROVISIONAL ACCREDITATION

Brown Mackie College-South Bend
3454 Douglas Rd.
South Bend, IN 46635
Phone: 574-237-0774
www.brownmackie.edu
Jolynn Rudman, DVM Director
Associate of Science
Initial Accreditation–September 18, 2009
PROVISIONAL ACCREDITATION

Harrison College (formerly Indiana Business College)
School of Veterinary Technology
6300 Technology Center Drive
Indianapolis, IN 46220
317-873-6500
www.harrison.edu
(William Finney, DVM-Director)
Associate of Applied Science
Initial Accreditation-October 31, 2008
PROVISIONAL ACCREDITATION

Purdue University
School of Veterinary Medicine
Veterinary Technology Program
West Lafayette, IN 47907
765-494-7619
www.vet.purdue.edu/vettech/index.html
(Pete Bill, DVM, Director)

On-Campus Program
Associate in Science
Bachelor of Science
Initial Accreditation: November 1976
FULL ACCREDITATION

Distance Learning Program
Associate in Science
Initial Accreditation: July 31, 2002
FULL ACCREDITATION

The Vet Tech Institute at International Business College at Fort Wayne
Veterinary Technology Program
5699 Coventry Lane
Fort Wayne, IN 46804
260-459-4500
www.vettechinstitute.edu/camp_wayne.php
(Stephanie Coates, DVM, Director)
Associate of Applied Science
Initial Accreditation: February 23, 2007
PROVISIONAL ACCREDITATION

The Vet Tech Institute at International Business College–Indianapolis
7205 Shadeland Station
Indianapolis, IN 46804
317-813-2300
www.ibcindianapolis.edu
(Peg Villanueva, DVM, Director)
Associate of Applied Science
Initial Accreditation: February 29, 2008
PROVISIONAL ACCREDITATION

IOWA

Des Moines Area Community College
Veterinary Technology Program
2805 SW Snyder Dr., Suite 505
Ankeny, IA 50023
800-362-2127
www.dmacc.edu
(Frank Cerfogli, DVM, Program Director)
Associate of Applied Science
Initial Accreditation: March 14, 2005
FULL ACCREDITATION

Iowa Western Community College
Veterinary Technology Program
2700 College Road, Box 4-C
Council Bluffs, IA 51502
712-325-3431
http://iwcc.cc.ia.us/programs/departments/veterinary-tech.asp
(Dan DeWitt, DVM, Program Director)
Associate of Applied Science
Initial Accreditation: March 2, 2007
PROVISIONAL ACCREDITATION

Kirkwood Community College
Animal Health Technology Program
6301 Kirkwood Blvd., SW
Cedar Rapids, IA 52406
319-398-4978
www.kirkwood.edu
(Anne Duffy, RVT, Coordinator)
Associate of Applied Science
Initial Accreditation: April 1989
FULL ACCREDITATION

KANSAS

Colby Community College
Veterinary Technology Program
1255 S. Range
Colby, KS 67701
785-460-5466
http://www.colbycc.edu/?m=4&s=80&l=163
(Jennifer Martin, DVM, Director)
Associate of Applied Science
Initial Accreditation: November 1974
FULL ACCREDITATION

KENTUCKY

Brown Mackie College–Louisville
3605 Fern Valley Road
Louisville, KY 40219
Ms. Robin Butler, LVT-Director
502-968-7191
www.brownmackie.edu
Associate of Science
Initial Accreditation-January 23, 2009
PROVISIONAL ACCREDITATION

Morehead State University
Veterinary Technology Program
25 MSU Farm Dr.
Morehead, KY 40351
606-783-2326
www.morehead-st.edu
(Scott Rundell, DVM, Coordinator)
Associate of Applied Science
Initial Accreditation: November 1977
FULL ACCREDITATION

Murray State University
Animal Health Technology Program
Department of Agriculture
100 AHT Center
Murray, KY 42071
270-762-7001
breathitt.murraystate.edu/bvc/
(Terry Canerdy, DVM, Director)
Bachelor in Science
Initial Accreditation: November 1986
FULL ACCREDITATION

LOUISIANNA

Delgado Community College
615 City Park Avenue
Building 4, Room 301
New Orleans, LA 70119-4399
504-671-6234
http://www.dcc.edu/campus/cp/ahealth/vet_tech/
(Dr. James Hurrell, Program Director)
Associate of Applied Science
PROVISIONAL ACCREDITATION effective
September 22, 2006

Northwestern State University of Louisiana
Veterinary Technology Program
Department of Life Sciences
225 Bienvenu Hall
Natchitoches, LA 71497
318-357-5323
www.nsula.edu
(Brenda Woodard, DVM, Director)
Associate in Science
Initial Accreditation: April 1981
FULL ACCREDITATION

MAINE

University College of Bangor
Veterinary Technology Program
85 Texas Avenue, 217 Belfast Hall
Bangor, ME 04401-4367
207-262-7852
www.uma.maine.edu/bangor
(Robert Gholson, DVM, Director)
Associate in Science
Initial Accreditation: March 2002
FULL ACCREDITATION

MARYLAND

Essex Campus of the Community College
Of Baltimore County
Veterinary Technology Program
7201 Rossville Blvd.
Baltimore, MD 21237
410-780-6306
FAX 410-780-6145
www.ccbcmd.edu
(Jack Stewart, RVT, Coordinator)
Associate of Applied Science
Initial Accreditation: April 1978
PROBATIONARY ACCREDITATION

MASSACHUSETTS

Becker College
Veterinary Technology Program
964 Main Street
Leicester, MA 01524
508-791-9241
FAX 508-892-8155
www.beckercollege.com
(James Q. Knight, DVM, Director)
Associate in Science
Bachelor of Science
Initial Accreditation: November 1978
FULL ACCREDITATION

Holyoke Community College
Veterinary Technician Program
303 Homestead Ave.
Holyoke, MA 01040-1099
413-538-7000
www.hcc.mass.edu
(Walter Jaworski, DVM, Coordinator)
Associate in Science
Initial Accreditation: April 1989
FULL ACCREDITATION

Mount Ida College
Veterinary Technology Program
777 Dedham St.
Newton, MA 02459
617-928-4545
FAX 617-928-4072
www.mountida.edu
(Tracy Blais, CVT, Director)
Associate in Arts
Bachelor of Animal Science
Initial Accreditation: November 1985
FULL ACCREDITATION

North Shore Community College
Veterinary Technology Program
1 Ferncroft Road
Danvers, MA 01923
978-762-4000
www.northshore.edu
(Karen Komisar, DVM, Director)
Associate of Applied Science
Initial Accreditation: March 28, 2007
FULL ACCREDITATION

MICHIGAN

Baker College of Cadillac
Veterinary Technology Program
9600 East 13th Street
Cadillac, MI 49601
231-775-8458
www.baker.edu
(Tina Burke, DVM, Coordinator)
Associate of Applied Science
Initial Accreditation: May 9, 2003
PROVISIONAL ACCREDITATION

Baker College of Clinton Township
34950 Little Mack Ave.
Clinton Township, MI 48035-4701
586-790-9430
www.baker.edu
(Marianne Tear, LVT, Director)
Associate of Applied Science
Initial Accreditation: November 7, 2007
PROVISIONAL ACCREDITATION

Baker College of Flint
Veterinary Technology Program
1050 W. Bristol Road
Flint, MI 48507
800-964-4299 or 810-766-4153
FAX 810-766-2055
www.baker.edu
(Aaron Walton, DVM, Director)
Associate of Applied Science
Initial Accreditation: November 20, 2002
FULL ACCREDITATION

Baker College of Jackson
Veterinary Technology Program
2800 Springport Road
Jackson, MI 49202
517-789-6123
www.baker.edu
(Katrina Bowers, LVT, Director)
Associate of Applied Science
Initial Accreditation: March 21, 2006
PROVISIONAL ACCREDITATION

Baker College of Muskegon
Veterinary Technology Program
1903 Marquette Avenue
Muskegon, MI 49442
800-937-0337 or 231-777-5275
FAX 231-777-5201
www.baker.edu
(Marcy O'Rourke, DVM, Department Chair)
Associate of Applied Science
Initial Accreditation: October 18, 2002
FULL ACCREDITATION

Baker College of Port Huron
Veterinary Technology Program
3403 Lapeer Road
Port Huron, MI 48060-2597
810-985-7000
www.baker.edu
(Norman Bayne, DVM, Director)
Associate of Applied Science
Initial Accreditation: February 2, 2007
PROVISIONAL ACCREDITATION

Macomb Community College
Veterinary Technician Program
Center Campus
44575 Garfield Rd.
Clinton Township, MI 48044
586-286-2096
FAX 586-286-2098
www.macomb.cc.mi.us
(Lori Renda-Francis, LVT, Director)
Associate of Applied Science
Initial Accreditation: April 1979
FULL ACCREDITATION

Michigan State University
College of Veterinary Medicine
Veterinary Technology Program
A-10 Veterinary Medical Center
East Lansing, MI 48824
517-353-7267
www.cvm.msu.edu/vettech
(Helene Pazak, DVM, Director)
Certificate
Bachelor of Science
Initial Accreditation: August 1973
FULL ACCREDITATION

Wayne County Community College District
Veterinary Technology Program
c/o Wayne State University
Div. of Laboratory Animal Resources
540 E. Canfield
Detroit, MI 48201
313-577-1156
FAX 313-577-5890
www.dlar.wayne.edu/vtp
(Karen Hrapkiewicz, DVM, Director)
Associate of Applied Science
Initial Accreditation: November 1976
FULL ACCREDITATION

MINNESOTA

Argosy University
Twin Cities
Veterinary Technician Program
1515 Central Parkway
Eagan, MN 55121
888-844-2004
www.argosy.edu
(Paula Lind, CVT, Interim Director)
Associate of Applied Science
Initial Accreditation: November 1981
FULL ACCREDITATION

Duluth Business University
Veterinary Technology Program
4724 Mike Colalillo Dr.
Duluth, MN 55807
800-777-8406
www.dbumn.edu
(Deserie Bremer, DVM, Program Director)
Associate of Applied Science
Initial Accreditation: August 6, 2004
PROBATIONARY ACCREDITATION

Globe University
Veterinary Technology Program
8089 Globe Drive
Woodbury, MN 55125
651-714-7360 or 800-231-0660
FAX 651-730-5151
http://www.msbcollege.edu/campus-locations/woodbury-mn/
(Brian Hoefs, DVM, Interim Program Director)
Associate of Applied Science
Bachelor of Science
Initial Accreditation: February 14, 2003
FULL ACCREDITATION

Minnesota School of Business–Blaine
Veterinary Technology Program
3680 Pheasant Ridge Dr. NE
Blaine, MN 55449
763-225-8000
www.msbcollege.edu
(Mary Mortenson, CVT, Program Director)
llewandowski@msbcollege.edu
Associate of Applied Science
Bachelor of Science
PROVISIONAL ACCREDITATION effective June 2, 2006

Minnesota School of Business–Plymouth
Veterinary Technology Program
1455 County Road 101 North
Plymouth, MN 55447
763-476-2000
www.msbcollege.edu
(Carrie Padget, CVT, Interim Director)
Associate of Applied Science
Initial Accreditation: February 4, 2004
FULL ACCREDITATION

Minnesota School of Business–Rochester
2521 Pennington Dr. NW
Rochester, MN 55901
507-536-9500
www.msbcollege.edu
(Julie Legred, CVT, Director)
Associate of Applied Science
Initial Accreditation:February 15, 2008
PROVISIONAL ACCREDITATION

Minnesota School of Business–St. Cloud
Veterinary Technology Program
1201 2nd Street South
Waite Park, MN 56387
320-257-2000
www.msbcollege.edu
(Sarah Bjorstrom, CVT, Director)
Associate of Applied Science
Initial Accreditation: January 12, 2007
PROBATIONARY ACCREDITATION

Minnesota School of Business–Shakopee
Veterinary Technology Program
1200 Shakopee Town Square
Shakopee, MN 55379
952-345-1200
(Ms. Teresa Reuteler, CVT, Program Director)
Associate of Applied Science
Initial Accreditation: June 8, 2005
PROVISIONAL ACCREDITATION

Ridgewater College
Veterinary Technology Dept.
2101 15th Ave., NW
Willmar, MN 56201
320-235-5114
FAX 320-222-8260
www.ridgewater.mnscu.edu
(Allen Balay, DVM, Director)
Associate of Applied Science
Initial Accreditation: November 1993
FULL ACCREDITATION

Rochester Community and Technical College
Animal Health Technology Program
851 30th Avenue SE
Rochester, MN 55904-4999
800-247-1296
www.rctc.edu
(Dr. Kimberly Rowley, Program Director)
Associate of Applied Science
Initial Accreditation: January 28, 2005
PROVISIONAL ACCREDITATION

MISSISSIPPI

Hinds Community College
Veterinary Technology Program
1100 PMB 11160
Raymond, MS 39154
601-857-3456
FAX 601-857-3577
www.hindscc.edu/Departments/Agriculture/VeterinaryTech.aspx
(Bobby Glenn, DVM, Director)
Associate of Applied Science
Initial Accreditation: April 1990
FULL ACCREDITATION

MISSOURI

Crowder College
601 LaClede Avenue
Neosho, MO 64850
417-455-5772
www.crowder.edu
Stephanie Watson, DVM, Program Director
swatson@crowder.edu
Associate of Applied Science
FULL ACCREDITATION
effective October 20, 2006

Jefferson College
Veterinary Technology Program
1000 Viking Dr.
Hillsboro, MO 63050
636-942-3000
FAX 636-789-4012
www.jeffco.edu
(Dana Nevois, RVT, Director)
Associate of Applied Science
Initial Accreditation: April 1978
FULL ACCREDITATION

Maple Woods Community College
Veterinary Technology Program
2601 NE Barry Rd.
Kansas City, MO 64156
816-437-3235
FAX 816-437-3239
www.mcckc.edu/vettech
(Chris Morrow, DVM, Director)
Associate of Applied Science
Initial Accreditation: April 1975
FULL ACCREDITATION

Sanford Brown College–St. Peters
Veterinary Technology Program
100 Richmond Center Blvd.
St. Peters, MO 63376
(Dr. Amy Wolff, Program Director)
888-769-2433
www.sanford-brown.edu
Associate of Applied Science
Initial Accreditation: August 31, 2007
PROVISIONAL ACCREDITATION

Sanford Brown College–Fenton
Veterinary Technology Program
1345 Smizer Mill Road
Fenton, MO 63026
(Ms. Angela Hutchinson, RVT)
888-769-2433
www.sanford-brown.edu
Associate of Applied Science
Initial Accreditation: October 11, 2007
PROBATIONARY ACCREDITATION

The Vet Tech Institute at Hickey College
Veterinary Technology Program
2780 N. Lindbergh
St. Louis, MO 63114
314-434-2212
www.hickeycollege.edu
(Angela Scherer, DVM-Director)
Specialized Associate Degree in Veterinary Technology
Initial Accreditation-September 26, 2008
PROVISIONAL ACCREDITATION

NEBRASKA

Nebraska College of Technical Agriculture
Veterinary Technology Program
RR3, Box 23A
Curtis, NE 69025
308-367-4124
www.ncta.unl.edu
(Barbara Berg, LVT, Department Chair)
Associate of Applied Science
Initial Accreditation: August 1973
FULL ACCREDITATION

Northeast Community College
Veterinary Technician Program
801 E. Benjamin Ave.
Norfolk, NE 68702-0469
402-371-2020
FAX-402-844-7400
www.northeastcollege.com
(Mike Cooper, DVM, Director)
Associate of Applied Science
Initial Accreditation: November 1997
FULL ACCREDITATION

Vatterott College
Veterinary Technician Program
11818 I Street
Omaha, NE 68137-1237
402-392-1300
www.vatterott-college.com
(Joni Brunssen, DVM, Director)
Associate of Applied Science
Initial Accreditation: November 1978
FULL ACCREDITATION

NEVADA

The College of Southern Nevada
Veterinary Technology Program
6375 W. Charleston Blvd.
Las Vegas, NV 89146-1164
702-651-5852
www.ccsn.nevada.edu
(Dennis Olsen, DVM, Director)
Associate of Applied Science
Initial Accreditation: March 8, 2006
FULL ACCREDITATION

Pima Medical Institute
Veterinary Technician Program
3333 E. Flamingo Road
Las Vegas, NV 89121
702-458-9650
www.pmi.edu
(Dennis Lopez, LVT, Director)
Occupational Associate
Initial Accreditation: March 10, 2006
PROVISIONAL ACCREDITATION

Truckee Meadows Community College
Veterinary Technology Program
7000 Dandini Blvd.
Reno, NV 89512
775-850-4005
(Ms. Wendi Ford, LVT, Program Director)
Associate of Applied Science
Initial Accreditation: September 16, 2005
PROBATIONARY ACCREDITATION

NEW HAMPSHIRE

Great Bay Community College (formerly New Hampshire Community Technical College)
Veterinary Technology Program
277 Portsmouth Ave.
Stratham, NH 03885-2297
603-772-1194
FAX 603-772-1198
www.greatbay.edu
(Margaret MacGregor, VMD, Director)
Associate in Science
Initial Accreditation: April 1998
FULL ACCREDITATION

NEW JERSEY

Bergen Community College
School of Veterinary Technology
400 Paramus Road
Paramus, NJ 07652
201-612-5389
www.bergen.cc.nj.us
(Harriet Terodemos, RVT-Coordinator)
Associate of Applied Science
Initial Accreditation-October 1, 2008
PROVISIONAL ACCREDITATION

Camden County College
Animal Science Technology Program
PO Box 200
Blackwood, NJ 08012
856-227-7200
www.camdencc.edu
(Margaret Dorsey, VT, Coordinator)
Associate in Animal Technology
Initial Accreditation: November 1978
FULL ACCREDITATION

NEW MEXICO

Central New Mexico Community College
Veterinary Technology Program
525 Buena Vista SE
Albuquerque, NM 87106
505-224-5043
(Dr. Bonnie Snyder, Program Director)
Associate of Applied Science
Initial Accreditation: June 17, 2005
FULL ACCREDITATION

San Juan College
Veterinary Technology Distance Learning Program
4601 College Blvd.
Farmington, NM 87402
505-566-3182
www.sjc.cc.nm.us/pages/1.asp
(David Wright, DVM, Director)
Associate of Applied Science
Initial Accreditation: November 8, 2006
FULL ACCREDITATION

NEW YORK

Alfred State College
Veterinary Technology Program
Agriculture Science Building
Alfred, NY 14801
607-578-3009
FAX 607-587-4721
www.alfredstate.edu
(Melvin Chambliss, DVM, Director)
Associate of Applied Science
Initial Accreditation: September 2001
FULL ACCREDITATION

La Guardia Community College
The City University of New York
Veterinary Technology Program
31-10 Thomson Ave.
Long Island City, NY 11101
718-482-5470
www.lagcc.cuny.edu
(Robin Sturtz, DVM, Director)
Associate of Applied Science
Initial Accreditation: April 1983
FULL ACCREDITATION

Medaille College
Veterinary Technology Program
18 Agassiz Cr.
Buffalo, NY 14214
716-884-3281
FAX 716-884-0291
www.medaille.edu
(Joseph Savarese, DVM, Director)
Associate in Science
Bachelor of Science
Initial Accreditation: November 1996
FULL ACCREDITATION

Mercy College
Veterinary Technology Program
555 Broadway
Dobbs Ferry, NY 10522
914-674-7530
www.mercy.edu/acadivisions/natscivettech/vettech.cfm
(Jack Burke, DVM, Director)
Bachelor of Science
Initial Accreditation: April 1987
FULL ACCREDITATION

State University of New York–Canton
Agricultural & Technical College
Health Sciences & Medical Technologies
Veterinary Science Technology Program
34 Cornell Drive
Canton, NY 13617
315-386-7410
www.canton.edu
(Mary O'Horo-Loomis, DVM, Director)
Associate of Applied Science
Initial Accreditation: November 1979
FULL ACCREDITATION

State University of New York–Delhi
College of Technology
Veterinary Science Technology Program
156 Farnsworth Hall
Delhi, NY 13753
607-746-4306
www.delhi.edu
(Dawn Dutka, DVM, Interim Director)
Associate of Applied Science
Bachelor in Business Administration
Initial Accreditation: (2 yr) December 1975; (4 yr) March 2002
FULL ACCREDITATION

State University of New York–Ulster (Ulster County Community College)
Veterinary Technology Program
Cottekill Road
Stone Ridge, NY 12484
800-724-0833, ext. 5233
www.sunyulster.edu
(Dr. Beth Alden, Program Director)
Associate of Applied Science
Initial Accreditation: December 3, 2004
FULL ACCREDITATION

Suffolk Community College
Veterinary Science Technology Program
Western Campus
Crooked Hill Rd.
Brentwood, NY 11717
631-851-6289
www.sunysuffolk.edu
(Elia Colon-Mallah, DVM, Director)
Associate of Applied Science
Initial Accreditation: April 1996
FULL ACCREDITATION

NORTH CAROLINA

Asheville-Buncombe Technical Community College
Veterinary Medical Technology Program
340 Victoria Road
Asheville, NC 28801
828-254-1921 ext. 273
www.abtech.edu/ah/vet/
(Dianne Cotter, LVT, Director)
Associate of Applied Science
Initial Accreditation: March 3, 2006
FULL ACCREDITATION

Central Carolina Community College
Veterinary Medical Technology Program
1105 Kelly Dr.
Sanford, NC 27330
919-775-5401
FAX 919-775-1221
www.ccarolina.cc.nc.us
(Paul Porterfield, DVM, Director)
Associate of Applied Science
Initial Accreditation: November 1974
FULL ACCREDITATION

Gaston College
Veterinary Medical Technology Program
201 Hwy. 321 South
Dallas, NC 28034-1499
704-922-6200
FAX 704-922-6440
www.gaston.cc.nc.us
(Kristine Blankenship, DVM, Director)
Associate of Applied Science
Initial Accreditation: November 1996
FULL ACCREDITATION

NORTH DAKOTA

North Dakota State University
Veterinary Technology Program
Van Es Laboratories
Fargo, ND 58105
701-231-7511
FAX 701-231-7514
vettech.ndsu.nodak.edu
(Tom Colville, DVM, Director)
Bachelor of Science
Initial Accreditation: May 1978
FULL ACCREDITATION

OHIO

Brown Mackie College-Cincinnati
1011 Glendale-Milford Rd.
Cincinnati, OH 45215
513-672-1969
www.brownmackie.edu
Nancy Killeen, RVT Director
Associate of Science
Initial Accreditation–June 19, 2009
PROVISIONAL ACCREDITATION

Columbus State Community College
Veterinary Technology Program
550 E. Spring St.
Columbus, OH 43216
614-287-3685
www.cscc.edu
(Brenda Johnson, DVM, Director)
Associate of Applied Science
Initial Accreditation: May 1974
FULL ACCREDITATION

Cuyahoga Community College
Veterinary Technology Program
11000 Pleasant Valley Rd.
Parma, OH 44130
216-987-5450
www.tri-c.cc.oh.us
(Kathy Corcoran, DVM, Director)
Associate of Applied Science
Initial Accreditation: April 1998
FULL ACCREDITATION

Kent State University-Tuscarawas
School of Veterinary Technology
330 University Drive NE
New Philadelphia, OH 44663
330-339-3391
www.tusc.kent.edu
(Ron Southerland, DVM-Director)
Associate of Applied Science
Initial Accreditation-November 7, 2008
PROVISIONAL ACCREDITATION

UC Raymond Walters College
Veterinary Technology Program
9555 Plainfield Road
Blue Ash, OH 45236
513-936-7173
FAX 513-558-0734
www.rwc.uc.edu
(Jennifer Wells, DVM, Director)
Associate of Applied Science
Initial Accreditation: April 1975
FULL ACCREDITATION

Stautzenberger College–Brecksville
Veterinary Technology Program
8001 Katherine Blvd.
Brecksville, OH 44141
440-846-1999
http://www.sctoday.edu
(Ms. Heather Machles RVT, Program Director)
Associate of Applied Science
Initial Accreditation-December 8, 2006
FULL ACCREDITATION

Stautzenberger College-Maumee
Veterinary Technology Program
1796 Indian Wood Circle
Maumee, OH 43537
419-866-0261
FAX 419-866-0261 x 237
www.stautzen.com
(Laurie Pfaff-Cherry, RVT, Director)
Associate of Applied Science
Initial Accreditation: April 1996
FULL ACCREDITATION

Vet Tech Institute at Bradford School
2469 Stelzer Road
Columbus, OH 43219
614-416-6200
www.bradfordschoolcolumbus.edu
Angela Beal, DVM, Program Director
abeal@bradfordschoolcolumbus.edu
Associate of Applied Science
PROVISIONAL ACCREDITATION effective August 11, 2006

OKLAHOMA

Murray State College
Veterinary Technology Program
One Murray Campus
Tishomingo, OK 73460
580-371-2371
FAX 580-371-9844
www.msc.cc.ok.us
(Carey Floyd, DVM, Director)
Associate of Applied Science
Initial Accreditation: November 1980
FULL ACCREDITATION

Oklahoma State University–Oklahoma City
Veterinary Technology Program
900 N. Portland Ave.
Oklahoma City, OK 73107
405-945-9112
www.osuokc.edu
(Sally Henderson, DVM, Coordinator)
Associate of Applied Science
Initial Accreditation: March 2000
FULL ACCREDITATION

Tulsa Community College
Veterinary Technology Program
7505 W. 41st St.
Tulsa, OK 74107
918-595-8212
FAX 918-595-8216
www.tulsacc.edu
(Jan Weaver, DVM, Coordinator)
Associate of Applied Science
Initial Accreditation: March 2000
FULL ACCREDITATION

OREGON

Portland Community College
Veterinary Technology Program
PO Box 19000
Portland, OR 97219
503-244-6111
www.pcc.edu
(Brad Krohn, DVM, Division Chair)
Associate of Applied Science
Initial Accreditation: April 1989
FULL ACCREDITATION

PENNSYLVANIA

Harcum College
Veterinary Technology Program
750 Montgomery Ave.
Bryn Mawr, PA 19010-3476
610-526-6055
FAX 610-526-6031
www.harcum.edu
(Nadine Hackman, VMD, Director)
Associate in Science
Initial Accreditation: April 1976
FULL ACCREDITATION

Johnson College
Veterinary Science Technology Program
3427 N. Main Ave.
Scranton, PA 18508
1-800-2WE-WORK or 570-342-6404
FAX: 570-348-2181
www.johnsoncollege.com
(Rosemary Cook, CVT, Chair)
Associate in Science
Initial Accreditation: November 1995
FULL ACCREDITATION

Lehigh Carbon & Northampton Community Colleges
Veterinary Technology Program
3835 Green Pond Rd.
Bethlehem, PA 18020
610-861-5548
FAX 610-861-4132
www.lccc.edu
(Susan Stadler, DVM, Director)
Associate of Applied Science
Initial Accreditation: March 1999
FULL ACCREDITATION

Manor College
Veterinary Technology Program
700 Fox Chase Road
Jenkintown, PA 19046
215-885-2360
www.manorvettech.com
(Joanna Bassert, VMD, Director)
Associate in Science
Initial Accreditation: April 1992
FULL ACCREDITATION

The Vet Tech Institute
Veterinary Technician Program
125 Seventh Street
Pittsburgh, PA 15222
412-391-7021 or 800-570-0693
FAX 412-232-4345 x 218
www.vettechinstitute.com
(Barbra Karner, VMD, Director)
Associate in Specialized Technology
Initial Accreditation: June 14, 2002
FULL ACCREDITATION

Sanford-Brown Institute (formerly Western School of Health and Business Careers)
Veterinary Technology Program
421 7th Avenue
Pittsburgh, PA 15219
412-281-2600
www.westernschool.com
(Cynthia Watkins, CVT, Program Director)
Associate of Specialized Technology
Initial Accreditation: September 22, 2004
FULL ACCREDITATION

Wilson College
Veterinary Medical Technology Program
1015 Philadelphia Ave.
Chambersburg, PA 17201
717-264-4141
FAX 717-264-1578
www.wilson.edu
(Freya S. Burnett, CVT, Director)
Bachelor of Science–College for Women
Initial Accreditation: May 1984
FULL ACCREDITATION

PUERTO RICO

University of Puerto Rico
Veterinary Technology Program
Medical Sciences Campus
PO Box 365067
San Juan, PR 00936-5067
787-758-2525, ext. 1051 or 1052
www.cprsweb.rcm.upr.edu
(Elizabeth Rivera, DVM, Director)
Bachelor of Science
Initial Accreditation: April 1996
FULL ACCREDITATION

SOUTH CAROLINA

Piedmont Technical College
Newberry Campus
540 Wilson Road
Newberry, SC 29108
Sylvia MacFarlane, RVT-Director
803-276-9000
www.ptc.edu
Associate in Health Science
Initial Accreditation-December 5, 2008
PROVISIONAL ACCREDITATION

Tri-County Technical College
Veterinary Technology Program
PO Box 587
Pendleton, SC 29670
864-646-8361
www.tctc.edu
(Peggy Champion, DVM, Director)
Associate in Allied Health Science
Initial Accreditation: April 1979
FULL ACCREDITATION

Trident Technical College
Veterinary Technology Program
1001 South Live Oak Drive
Moncks Corner, SC 29461
843-899-8011
www.tridenttech.edu
(Paul Kerwin, DVM, Director)
Associate in Allied Health Sciences
Initial Accreditation: March 2002
FULL ACCREDITATION

SOUTH DAKOTA

National American University
Allied Health Division
Veterinary Technology Program
321 Kansas City St.
Rapid City, SD 57701
800-843-8892
www.national.edu/veterinary_tech.html
(Margaret Behrens, DVM, Director)
Associate in Applied Science
Initial Accreditation: April 1981
FULL ACCREDITATION

TENNESSEE

Chattanooga State Technical Community College
4501 Amnicola Highway
Chattanooga, TN 37406-1097
423-697-4400
www.chattanoogastate.edu
(Susie Matthews, DVM, Director)
Associate of Applied Science
Initial Accreditation: March 28, 2008
PROVISIONAL ACCREDITATION

Columbia State Community College
Veterinary Technology Program
P.O. Box 1315, Health Sciences 105
Columbia, TN 38401
931-540-2722
www.coscc.cc.tn.us
(Boyce Wanamaker, DVM, Director)
Associate of Applied Science
Initial Accreditation: April 1979
FULL ACCREDITATION

Lincoln Memorial University
Veterinary Technology Program
Cumberland Gap Pkwy.
LMU Box 1659
Harrogate, TN 37752
423-869-6278
FAX 423-869-7151
www.lmunet.edu
(Mary Hatfield, LVMT, Director)
Associate of Applied Science
Bachelor of Science
Initial Accreditation: November 1987
FULL ACCREDITATION

TEXAS

Cedar Valley College
Veterinary Technology Program
3030 N. Dallas Ave.
Lancaster, TX 75134
972-860-8127
FAX 972-860-8057
ollie.dcccd.edu/vettech
(Steven Grubbs, DVM, Director)

On-Campus Program
Associate of Applied Science
Initial Accreditation: November 1978
FULL ACCREDITATION

Distance Learning Program
Associate of Applied Science
Initial Accreditation: September 2001
FULL ACCREDITATION
(Kelly Alan Black, DVM, Director)

McLennan Community College
Veterinary Technology Program
1400 College Drive
Waco, TX 76708
254-299-8750
www.mclennan.edu/departments/workforce/vtech/
(Jennifer Garretson, DVM, Director)
Associate of Applied Science
Initial Accreditation: January 23, 2004
FULL ACCREDITATION

Midland College
Veterinary Technology Program
3600 N. Garfield
Midland, TX 79705
432-685-4619
FAX 432-685-6431
www.midland.edu
(Kerry Coombs, DVM, Director)
Associate of Applied Science
Initial Accreditation: November 1992; April 1995
FULL ACCREDITATION

Palo Alto College
Veterinary Technology Program
1400 W. Villaret Blvd.
San Antonio, TX 78224-2499
210-486-3355
FAX 210-486-3356
www.accd.edu
(Fonzie Quance-Fitch, DVM, Director)
Associate of Applied Science
Initial Accreditation: April, 1998
FULL ACCREDITATION

Sul Ross State University
School of Agriculture & Natural Resource Sciences
Veterinary Technology Program
P.O. Box C-114
Alpine, TX 79830
432-837-8205
FAX-432-837-8409
www.sulross.edu
(Darwin R. Yoder, DVM, Director)
Associate Degree
Initial Accreditation: November 1977
PROBATIONAL ACCREDITATION

Lone Star College–Tomball

Veterinary Technology Program
30555 Tomball Pkwy.
Tomball, TX 77375-4036
281-351-3357
FAX-281-351-3384
wwwtc.nhmccd.cc.tx.us/
(George W. Younger, DVM, Director)
Associate of Applied Science
Initial Accreditation: April 1990
FULL ACCREDITATION

The Vet Tech Institute of Houston

4669 Southwest Freeway, Suite 100
Houston, TX 77027
713-629-1500
www.vettechinstitute.edu
(Catherine Huff, RVT-Director)
Associate of Applied Science
Initial Accreditation-October 21, 2008
PROVISIONAL ACCREDITATION

UTAH

Utah Career College

Veterinary Technician Program
1902 West 7800 South
West Jordan, UT 84088
801-304-4224
FAX 801-676-2220
www.utahcollege.com/ushs/veterinary_technician.html
(Alex Arangua, DVM, Program Director)
Associate of Applied Science
Initial Accreditation: August 22, 2003
PROVISIONAL ACCREDITATION

Utah Career College–Layton

869 West Hill Field Road
Layton, UT 84041
Phone: 801-542-8314
http://www.utahcollege.edu/campus-locations/layton-ut/
Michael Poulsen, CVT Director
Associate of Applied Science
Initial Accreditation - September 4, 2009
PROVISIONAL ACCREDITATION

VERMONT

Vermont Technical College

Veterinary Technology Program
Randolph Center, VT 05061
802-728-3391
www.vtc.vsc.edu
(Amy Woodbury-St. Denis, DVM, Director)
Associate of Applied Science
Initial Accreditation: April 1991
FULL ACCREDITATION

VIRGINIA

Blue Ridge Community College

Veterinary Technology Program
Box 80
Weyers Cave, VA 24486
540-234-9261
FAX 540-234-9066
www.br.cc.va.us
(Stuart Porter, VMD, Director)
Associate of Applied Science
On-Campus Initial Accreditation: November 1976
Distance Learning Initial Accreditation (VA residents only): November 1999
FULL ACCREDITATION

Northern Virginia Community College

Veterinary Technology Program
Loudoun Campus
1000 Harry Flood Byrd Hwy.
Sterling, VA 20164-8699
703-450-2525
FAX 703-450-2536
www.nv.cc.va.us

On Campus Program
(Leslie Sinn, DVM, Head)
Associate of Applied Science
Initial Accreditation: April 1980
FULL ACCREDITATION

Distance Learning Program
Associate of Applied Science
(Terri Champney, DVM, Director)
Initial Accreditation: February 20, 2004
FULL ACCREDITATION

WASHINGTON

Pierce College Ft. Steilacoom

Veterinary Technology Program
9401 Farwest Dr., SW
Lakewood, WA 98498
253-964-6668
www.pierce.ctc.edu
(Salvador Hurtado, DVM, Director)
Associate in Veterinary Technology
Initial Accreditation: May 1974
FULL ACCREDITATION

Pima Medical Institute-Seattle

Veterinary Technology Program
9709 Third Ave. NE, Suite 400
Seattle, WA 98115
(Dr. Carol Mayer, Program Director)
800-477-PIMA
www.pmi.edu
Occupational Associate Degree
Initial Accreditation: October 2, 2007
PROVISIONAL ACCREDITATION

Yakima Valley Community College

Veterinary Technology Program
P.O. Box 22520
Yakima, WA 98907-1647
509-574-4759
FAX 509-547-4751
www.yvcc.edu
(Susan Wedam, DVM, Coordinator)
Associate of Applied Science
Initial Accreditation: March 1999
FULL ACCREDITATION

WEST VIRGINIA

Carver Career Center and the Bridgemont Community and Technical College

4799 Midland Drive
Charleston, WV 25306
Dawn DeMoss, RVT-Director
304-348-1965 ext. 117
http://kcs.kana.k12.wv.us/carver/
Associate of Applied Science
Initial Accreditation-February 27, 2009
PROVISIONAL ACCREDITATION

Pierpont Community & Technical College

Veterinary Technology Program
1201 Locust Ave.
Fairmont, WV 26554
304-367-4589
www.fscwv.edu
(Anna Romano, RVT, Coordinator)
Associate of Applied Science
Initial Accreditation: April 1981
PROBATIONAL ACCREDITATION

WISCONSIN

Madison Area Technical College

Veterinary Technician Program
3550 Anderson
Madison, WI 53704
608-246-6100
www.madison.tec.wi.us/matc
(Jane Clark, DVM, Director)
Associate of Applied Science
Initial Accreditation: November 1974
FULL ACCREDITATION

Moraine Park Technical College

Veterinary Technology Distance Learning Program
235 N. National Ave.
Fond du Lac, WI 54936
920-922-8611
www.morainepark.edu
(Laura Lien, BS, CVT, Program Director)
Associate of Applied Science
Initial Accreditation: September 7, 2007
PROVISIONAL ACCREDITATION

WYOMING

Eastern Wyoming College

Veterinary Technology Program
3200 W. "C" St.
Torrington, WY 82240
800-658-3195, ext. 8268
www.ewc.wy.edu
(Patti Sue Peterson, AHT, Director)
Associate in Applied Science
Initial Accreditation: April 1976
FULL ACCREDITATION

Appendix C
Wildlife Services and Resources

National Wildlife Rehabilitators Association (NWRA),
2625 Clearwater Rd,
Suite 110, St.
Cloud, MN 56301
(320) 230-9920
www.nwrawildlife.org

U.S. Fish and Wildlife Services
www.usfw.org

Wildlife Rehabilitation Centers and Rehabilitators Directory
http://www.southeasternoutdoors.com/wildlife/rehabilitators/directory-us.html

Glossary

abdominal system includes the stomach, intestines, and internal organs, such as the liver, pancreas, and spleen

abduction movement away from midline or the axis of the body

abductor muscle string muscle that allows a shell to open and close

abomasum last section of the ruminant stomach that acts as the true stomach and allows food to be digested

abortion a pregnant female animal that has lost the fetus

abscess bacterial infection that causes pus to build up in a localized area

absorbable suture material that is broken down within the body over a period of time and is absorbed by the body

acceptance emotion when a person understands and accepts that a pet has passed away

account history client's previous invoices that can be viewed at any time to see past payments and charges

acetabulum area of the pelvis where the femur attaches to form the hip joint

acid rain polluted rainwater

acidity water pH levels below 70

ACTH adrenaocorticotropic hormone released by the adrenal gland that controls blood pressure, releases cholesterol, and also the body's production of steroids

actinobacillosis disease of the head and jaw that causes tumor-like lumps of yellow pus to build up within the jaw; also called "lumpy jaw"

actinomycosis bacterial infection of the head and neck lymph nodes and causes hard lesions to form over the tongue; also called "wooden tongue"

active immunity developed from exposure of a pathogen through the process of vaccination

acute condition that occurs short term

ad lib giving something, such as food or water, as much as necessary or so that it is available at all times

adduction movement toward midline or the axis of the body

adrenal gland located cranial to the kidneys; produce and release adrenaline and other hormones

adrenaline chemical that is released by the nervous system in times of stress to create a response in an animal's fight or flight instinct reaction

Adson Brown tissue forceps instrument with several small and delicate teeth that are used for light handling of delicate tissue

aeration oxygen or air created in the water by making bubbles

aggression behavior that makes an animal angry and difficult and potentially unsafe to handle

aggressive behavior of an animal exhibiting an attempt to do harm such as biting

aging allowing a water source to stand in a container at least 5 days

agonal gasps of breaths while the respiratory system shuts down

agouti a mixture of two or more colors aiding in digestion of food, breaking down food particles, and serves as a carrier for waste products

air cell empty area located at the large end of an egg that provides oxygen

albumin (1) part of the blood that draws water into the blood stream and provides hydration; (2) egg white that surrounds the yolk

alevin young salmon

alfalfa hay high in protein

algae plant-based source of food that grows spontaneously in water when left in the sunlight

alimentary canal veterinary medical terminology for the GI system

alkalinity water pH levels above 70

allergen pollens in the air that cause an allergic reaction

alligators large reptiles raised for meat, skin, by-products, and for release into the wild

Allis tissue forceps intestinal forceps that lock in place to hold and grasp tissue and when locked, have an opening that is wide to allow thicker tissues to be clamped

alopecia hair loss

alpacas ruminant mammals similar to llamas but smaller and that originated in South America; they have silky fine coats and are raised for their hair

alphabetical filing system organization of medical records by patient or client identification number and color coding

alveoli tiny sacs of air at the end of the bronchioles

Amazons large breed of cage bird from South or Central America that are talkative, entertaining, and trainable

American Animal Hospital Association (AAHA) sets the standards of small animal and companion animals hospitals

American Association for Laboratory Animal Science (AALAS) association developed for the exchange of information and humane care of laboratory animals used in research

American class chicken class developed in North America for egg and meat production

American Medical Association (AMA) agency that represents the human side of medicine and conducts many studies that benefit the advancement of human medicine and surgical procedures

American Veterinary Medical Association (AVMA) accredits college programs for veterinarian and veterinary technician state licensure and sets the standards of the veterinary medical industry

Americans with Disabilities Act (ADA) agency that governs the accommodations of people with physical and mental problems to allow them to work more efficiently

amino acids building blocks of proteins that form in chain-like structures

amphibian animal with smooth skin and spends part of its life on land and in the water

amylase enzyme produced by the pancreas that breaks down starches

anal glands scent glands or sacs located on each side of the rectum

anastomosis surgical removal of a dead area of tissue along the digestive tract and resectioning the area back together

anatomy the study of the internal and external body structures and parts of an animal

Ancylostoma caninum hookworm species that live in the small intestine of their host animal, where they attach themselves and feed on the host's blood

anemia low red blood cell count due to inability of body to make or replicate blood cells

anesthesia chamber a glass or plastic square or rectangular box that is clear and has two openings in which the inhalation and exhalation tubes on the anesthesia machine connect to the chamber

anesthesia log book that details the patient's status through induction, maintenance and recovery from anesthesia procedures including the patient's name, species, type of anesthesia, and amount used

anesthesia mask cone-like device used in larger patients and are placed over the muzzle and face area and must fit snuggly for the animal to breathe in the gases for induction

anestrus time when a female animal is not actively in the heat cycle

Angora goat group of goats that produce hair

Animal Abuse Laws rules that govern the neglect and abuse of animals and typically regulated by a humane society or dog warden

animal behavior manner that relates to what an animal does and why it does it

animal behaviorists people who study animals to collect data on their behavior in captivity

animal models live animals used to assist in gaining knowledge and information on human and veterinary medicine by studying behavior, diseases, procedures, treatments, and the effects of medications on the body

animal nutrition science of determining how animals use food in the body and all body processes that transform food into body tissues and energy for activity and the process which by animals grow, live, reproduce, and work

animal refuge large protected area of land where exotic animals are housed in a manmade setting that resembles their habitat

animal rights rules that govern how animals are handled and cared for, especially in research facilities

animal tissue living part of an animal that is cultured in a lab and grown for specific needs

Animal Welfare Act rules governing how animals are handled and cared for and takes a large interest in monitoring research facilities

anisicoria uneven pupil size

anogenital the area located around the far stomach between the rear legs and the base of the tail

anogenital distance distance between the area of the vulva or penis and the rectum in which gender can be determined

anorexia not eating

anterior chamber area of the eye in front of the iris that holds the vitreous humor and regulates the pressure in the eye

anterior posterior (AP) the X-ray beam penetrates the anterior portion of the body

anthelmintic medication also known as a dewormer, used to treat and prevent parasitic infection

antibiotic medication used to treat infection or disease

antibody specific proteins produced to protect a newborn against disease and attack cells relating to any diseases

anticoagulant substance that prevents the blood from forming a clot

antidiuretic hormone (ADH) hormone that promotes urine formation, water absorption, controls blood pressure, and changes control in water content

antifreeze toxicity poisoning that affects the renal system and shuts down the system causing renal failure; also known as ethylene glycol toxicity

antigen foreign materials used to create an immune response

antihistamine drugs used to prevent and control an allergic reaction

antioxidants vitamins that boost the body's immune system

antiseptics solutions that destroy microorganisms or inhibit their growth on living tissue; effective disinfecting agents

anuria no urine production

aorta large vessel that allows blood to flow out of the heart and back into systemic circulation

appendicular skeleton contains bone of the body that hang; the limbs or appendages

applied research research done for a specific purpose, such as developing a vaccine or medication

appointment cards a written reminder for clients that includes a date and time to remember the appointment scheduled for their pet

aquacrops commercially produced plant or animal living in or around a water source

aquaculture production of living plants, animals, and other organisms that grow and live in or around water

aquarium basic container that holds water and fish

aquarium maintenance schedule (ams) list of important duties that should be completed and the frequency necessary for an adequate fish tank

aquatic animal that lives in the water or pertaining to water

arboreal animal that lives in trees

arrhythmia abnormal heartbeat

artery vessel that carries blood away from the heart

arthritis inflammation of the joint

arthropods crustacean with an external skeleton

artificial incubation providing a controlled temperature in which eggs hatch, such as an incubator

artificial insemination (AI) the practice of breeding female animals through the use of veterinary equipment rather than the traditional mating method

artificial vagina (AV) sterile sleeve or tube used to collect the semen of an animal during the ejaculation process

ascarid intestinal parasite, long white worm, roundworm

ascending colon first section of the large intestine

ascites fluid buildup in the abdomen of a fish that caused by a virus or bacterial infection; commonly called dropsy

asepsis technique of keeping a sterile environment

aseptic sterile environment condition

aseptic technique the process of obtaining a sterile and organism-free environment

Asiatic class chicken class developed in Asia for show and ornamental breeds

aspiration (1) process of pulling the plunger back on a syringe slightly to make certain no blood vessel has been accidently penetrated prior to administering an injection; (2) swallowing food that enters the lung field

assessment veterinary diagnosis of problem

associate degree (AS) a two-year degree program

asthma condition of the lungs that have difficulty taking in enough oxygen

asystole heart stops contracting and heart failure occurs

ataxia an animal that has the loss of control over its body causing wobbliness and inability to control movement

atlas first cervical vertebrae located at the base of the brain that allows for up and down motions of the head

atopy skin or food allergy causing a skin infection

atrial fibrillation A-fib; condition that occurs when the S-A node or the pacemaker of the heart is not working

atrophy the shrinking of the muscle when it is not being used

auditory ossicle bones of the middle ear that transmit sound vibrations

aural pertaining to the ears

aural hematoma ear filled with blood due to a blood vessel bursting

auscultation process of listening to the heart, chest, and lungs for any abnormal sounds using a stethoscope

autoclave piece of equipment in the form of a sealed chamber in which objects are exposed to heat and steam under pressure that is used at extremely hot temperatures to kill all living organisms

autoimmune disease animal's red blood cells are destroyed by the immune system

automated inventory system system of counting all supplies and products within the facility by hand and keeping a list of how many items are in stock

automatic feeder specialized equipment that is computerized and set to release certain amounts of food at specified times throughout the day

automatic film processing equipment that develops and dries the film automatically

AV Valve opens and closes between the atrium and ventricle to allow blood to flow through each chamber of the heart

avian influenza commonly called bird flu and occurs as a naturally occurring virus within the intestinal tract and affects poultry and humans

avian pox virus of poultry spread by mosquitoes and lower the immune system slowly

avian system specialized digestive system of birds

axial skeleton contains bones of the body that lie perpendicular or lengthwise

axis second cervical vertebrae that allows for rotation and shaking motions of the head

Babcock tissue forceps intestinal forceps that are less traumatic and have broad, flared ends with smooth tips

babesia less common disease caused by the Deer tick that causes signs of anemia, jaundice (yellow coloring of the skin and mucous membranes), fever, vomiting; more severe stages can cause kidney failure

bachelor of science (BS) a four-year college degree program

back-grounding system production system that raises calves to market size for profit

back rubbers equipment used to allow cows to rub their back on a rolling bar that applies fly control chemicals

bacon type hog hogs raised for the large amount of meat, known as bacon, on their body

bacteria living organism that invades the body causing illness

bacterial class of disease spread through single-celled organisms called germs and are located in all areas of an animal's environment

Bain system uses a single corrugated or plastic tube that connects to the patient and another hose through the center of the tube that contains the gas mixture that the patient inhales

balanced ration diet that contains all the nutrients required by an animal in correct and specific amounts

ball and socket joint that rotates in numerous directions

balling gun a metal device with a long handle that has a plunger at the base, used to hold medicines to administer to large animals

band cell immature neutrophil that indicates an infection within the body; shaped like a U

band large group

bandage covering applied to a wound, fracture, or area of injury to prevent movement

bandage scissors scissors with an angled blunt end used to carefully remove a bandage from an animal

banding method of castrating male livestock by using a tight rubber band placed over the testicle area to cut off circulation and cause the tissue to fall off

Bang's disease common name for brucellosis

barbs also called barbels; devices on the fish that detect food, danger, and movement

bargaining stage of grief that allows the person an attempt to resolve the pet's problem by any means possible

barium enema a large amount of substance passed into the colon via the rectum that fluoresces

barium study a series of films taken as the material passes through the digestive tract

barium sulfate an example of a material or substance administered into the body to show structures on an X-ray film that would otherwise be difficult to view

barrow young castrated male pig

bartonella bacteria caused by a flea bite that affects cats within their bloodstream and infects the red blood cells and immune system

basic research research studied in a lab setting to determine knowledge and understanding in life processes and diseases

basophil white blood cell with a segmented nucleus and granules that stain dark and aid in allergic reaction control

bathing process of cleaning the skin and hair coat of the animal or to apply medicated shampoos or dips to the skin and hair coat

beak avian mouth with no teeth that forms an upper and lower bill

beak trim procedure to shorten or round the sharp edges of the beak

beard hairs located under the wattle in turkeys

beefalo cross breed of bison and domesticated beef cattle to produce beef animals with healthier and tastier meat products

beef cattle cows specifically raised for meat

Belgian Flemish draft horse breed that is light brown in color with a lighter mane and tail

benign noncancerous tumor

bilateral cryptorchid both testicles are retained and do not descend into the scrotum

bile yellow fluid secretion that helps break down food for digestion and absorption of food

billy common name for the adult male goat

biological drugs used to treat disease

biological filtration process of providing bacteria and other living organisms in the water to change harmful materials into forms that are safe

biological hazard safety concerns that pose a risk to humans and animals through contamination of living organisms

biological oxygenation the use of plants to create oxygen in the water

biological value percentage on a food label that describes the quality of the food source

biology the study of life

bird flu highly contagious virus that affects avian species and humans; also known as avian influenza

birdkeepers people who raise and care for caged birds

birds of prey large birds that are hunters and eat small mammals and rodents; also protected by law and are not legal to hunt

bison ruminant mammals closely related to cattle

bit equipment placed in a horse's mouth when riding a horse

bivalve mollusks with two shells

blink reflex evaluation of the eye to determine which plane of anesthesia a patient is in

bloat condition that causes the abdomen to become swollen and painful due to air and gas within the intestinal tract

blood red liquid within the circulatory system that contains 40% of cells in the body and transports oxygen through the body

blood chemistry analyzer machine that runs blood samples that measure routine blood chemistries and electrolytes

blood feathers growing feathers that have a blood vessel in the center

blood glucose blood sugar or insulin produced by the pancreas

blood plasma blood sample that is frozen and then centrifuged to obtain plasma

blood pressure (BP) the heart's contractions and relations as blood flows through the chambers

blood pressure monitor device used to assess the pumping action of the heart

blood serum liquid portion of the whole blood sample

blood smear blood film placed on a slide that is used to look at the morphology of blood cells

blood titer measured amount of antigen within the bloodstream

blue tongue virus spread by gnats that weakens the immune system and causes secondary infections to occur

boar adult male pig of reproductive age

body central part of the stomach that expands as food enters

body condition scoring rating of how an animal appears in looks and weight based on its ideal body weight

body language communication by the animal about how it feels toward other animals, people, and its environment; the use of mannerisms and gestures that are observed and tell how a person feels

bone hard, active tissue that consists mostly of calcium and forms the skeleton of the animal; give the body support, structure, and protection

bone marrow fluid within the medullary cavity where blood cells are produced

bone plate surgical steel plate placed around a fractured bone and holds the bone together with surgical screws

booster series of vaccines given multiple times to build up the immune system

booster series vaccines placed into the immune system to build up protection and immunity over an amount of time

bordetella tracheobronchitis; commonly called kennel cough

bottom dwelling fish fish that live and feed on the bottom or floor of a body of water

bovine veterinary term for a cow

Bovine viral diarrhea (BVD) virus in cattle causing diarrhea and respiratory signs

Bowman's capsule tubes located at the renal pelvis that filter blood into the kidneys

brachycephalic short, nonexistent nasal area, giving a flat face appearance

brackish water a mixture of freshwater and saltwater

bradycardia decreased or slow heart rate

brain the major organ in the nervous system that controls the body's actions

brain stem part of the brain that controls functions that maintain life

branding the use of extreme heat or cold temperatures to mark the skin with a number or symbol

bray loud vocal sound of a mule or donkey

breeding stock the male and female animals of a species raised specifically for reproductive purposes

broiler young chick between 6–8 weeks of age and under 4 pounds used for meat

broiler egg production system where chicks are raised for large production numbers and selected for broiler programs

broiler production system used to produce the most amount of meat as possible

bronchi branches at the base of each lung connected by large passages

bronchioles tiny tubes that connect to the lungs and bronchi and resemble tree branches

bronchodilator medications may be used to open up the bronchioles and lungs

brood a group of reptiles or amphibians

broodfish mature male and female breeding fish used for reproductive purposes

broodmare female horse used for reproduction and breeding purposes

browse woody type of plant

brucellosis reproductive disease that is highly contagious and has no treatment; also known as Bang's disease

brumation process in reptiles similar to hibernation when their body temperature decreases

brushing the act of cleaning the hair coat using a soft brush

buccal surface of the teeth located on the outer area near the cheek

buck adult male goat

bucking lifting and kicking action of the hind limbs

budding yeast bacteria that are in the shape of a budding flower

budgerigar commonly called a parakeet; small bird breed that comes in a variety of bright colors with bars located over the wings and back

budgie abbreviated name for a budgerigar or parakeet

buffy coat the layers within a hematocrit or PCV that form a tiny area of visible white composed of white blood cells and platelets

bull adult male cow of breeding age

burrow to dig a tunnel into an area

butting the action of using the head to hit an object or a person

by-products the parts of animals that are not the main profit or production source, such as the hooves, antlers, hair coat, or internal organs

by-products tissue tissue that is removed from an animal when it is used for other purposes in research or for human consumption

cage structure made to float on the water to hold aquatics

cage card identification and location of patient while in care of facility

calculus mineralized plaque material that appears brown to dark yellow in color and is difficult to remove from teeth

calf young newborn cow of either gender

California Mastitis Test (CMT) test used to diagnose and measure the level of cell infection within the mammary glands

calicivirus highly contagious virus that affects the upper respiratory system of cats and may cause a head tilt

caliper device or instrument with a simple ruler-like tool that slides on a fixed bar to allow a patient's body part to be measured for a radiograph

callus cartilage that rebuilds and thickens over a fractured bone that has healed

calories unit of measurement that defines the energy in food

calving the labor process of cows

campylobacter bacterial infection that affects the intestinal tract or reproductive tract in cattle

canaries small breed of bird in solid bright colors that originated in the Canary Islands and is known for its singing ability

cancellous bone sponge-like, softer layer of bone located inside the end of bones

candling technique of using a light to view the internal structures of an egg

canine teeth also known as fangs that are used to tear apart food

canine distemper viral infection may be fatal to puppies; dogs who recover may have long-term neurological affects

Canis familiaris genus and species of domesticated dog

canister area that absorbs carbon dioxide and contains granules to absorb it

cannibalism the act of rodents eating their young

cannon long bones in hoofed animals located above the ankle joint

canter three-beat running gait of English horses

capillary action the action of blood draining into the tube through gravity

capillary refill time (CRT) evaluates how well an animal's blood is circulating in the body by pressing on the gums and monitoring how long it takes for the color to return to the area

capon male castrated chicken over 6 pounds and 6 months of age

caprine veterinary term for goats

captivity an animal that is housed in an enclosure or with people

carapace thorax, abdomen, and limbs

carbohydrates nutrients that provide energy for body functions and allow for body structure formation and make up the largest part of an animal's diet needs

carcass dead body of an animal

cardia entrance of the stomach that filters food

cardiac arrest the heart is not contracting appropriately

Carmalt hemostats much heavier and used to hold off the ovarian stump during a spay surgery with interlocking serrations located lengthwise

carnasial tooth upper fourth premolar and lower first molar in dogs and cats that tends to become infected and abscessed

carnivore meat eating

carotene protein that promotes the health of organs

carpal lower leg that is similar to the wrist

carpus joint formed by several carpal bones and arranged in two rows in the area of the wrist

carrier (1) animals that are infected by a disease but do not show signs of disease; (2)box-like object that safely holds a cat for transport

carrying capacity the number of animals living in an area that can be supported and sustained by that area

cartilage forms at the end of bones to protect and cushion the bone

cartilage joint connects at the end of a joint where two bones meet and acts as a cushion

cashmere fine down undercoat of goats that is warm and used in clothing

cassette tool that holds X-ray film

cast material made of a hard substance and placed over a broken bone to keep the bone in place as it heals

castration surgical removal of the testicles in male animals

cat bag restraint equipment that holds the body and limbs of the cat and controls for certain procedures, such as a nail trim

Cat Fancier's Association (CFA) cat breed registry association that promotes the health and responsible breeding of purebred cats

cat muzzles restraint equipment placed over the face and mouth of a cat to prevent biting

cataracts opacity of the lens in the eye

catheter thin plastic or rubber tube that is passed into the urinary bladder or bloodstream

catheterized a long, thin rubber or plastic sterile tube inserted into the bladder through the external urinary opening to collect a urine sample

caudal fin the tail fin that acts as a propeller

cauliflower disease nickname for lymphocytosis

caustic chemicals used to burn horn buds to prevent horn growth

cautery unit handheld equipment that heats to high temperatures to burn through nails and clot them at the same time

cecum area within the intestinal tract located between the small and large intestine that helps break down roughage and protein

cell basic unit of life and the structural basis of an animal

cell division process that occurs when a cell splits into two cells

cell membrane cell wall that forms the outside of a cell and holds the structure together

centimeters (cm) measurement used in measuring body parts for X-rays

central nervous system (CNS) part of the body under involuntary brain control, such as the brain stem and spinal cord

central processing unit (CPU) the brain of the computer that carries out instructions asked by the software

centrifuge piece of equipment that uses high speeds to separate liquid portions into solids

centrifuge tube glass or plastic tube that holds samples within a centrifuge

cephalic vein blood vessel located on the medial front limb

cephalic venipuncture blood collection from vein inside the front of the legs

cerclage wire surgical wire used to hold fragments of broken bone together while healing

cerebellum part of the brain that controls coordination and movement

cerebrum part of the brain that controls the voluntary movements of the body and thought processes and is the largest region of the brain

certified veterinary technician (CVT) a veterinary technician that has passed a state licensing exam

cervical first section of the spinal column located over the neck area

cervix end of the vagina that opens to the uterus; birth canal

chain shank chain placed around the head and halter of a horse for added control

chalaza part of the egg that anchors the yolk to the egg white

chammy by-product of sheep skin used as a leather product for polishing

channel a chosen route of communication

cheek pouches area within the mouth where food is stored

chelonians veterinary term for turtles

chem strip long, thin, plastic strip separated by individual square pads containing a chemically treated paper

chemical a change in the body that affects growth, sexual reproduction, and development

chemical filtration using special chemicals such as ozone and activated charcoal to keep the water clear and from becoming a yellowish color

chemical hazard safety concerns of products that may cause injury such as burns or vapors that cause injuries to the eye, lungs, or skin

chemical restraint use of sedatives or tranquilizers to calm and control an animal

chemoreceptor chemical receptor that allows an animal to taste, smell, and detect sounds

chemotherapy treatment using chemical agents that help with cases of cancer

chick young chicken

chicken most important type of poultry and largest industry with four different classes

chief complaint the reason the animal is being seen by the veterinarian

Child Labor Laws laws and guidelines that regulate children working and the conditions in which they work in

chronic long-term condition

chronic obstructive pulmonary disease (COPD) respiratory condition similar to heaves and asthma that causes difficulty breathing

circulatory system body system essential for life that includes the heart, blood, veins, arteries, and capillaries with the functions that include oxygen flow, blood circulation, and transport of nutrients, waste removal, and the movement of hormones

cleaning physically removing all visible signs of dirt and organic matter such as feces, blood, and hair

client the owner of the animal

clinical research research that is conducted in a lab setting and focuses on human and veterinary related issues

clipper guard tool that fits over the end of the clippers to prevent trimming too short

clipping the process of removal of small amounts of hair from one or several areas

cloaca external reproductive area where egg-laying animals pass urine, feces, and eggs; also called the vent

closed gloving process done when the gown is placed with the hands still within the sleeves so that no skin is visible and potentially contaminating the gloves

Clostridium bacteria that causes tetanus and is found naturally in manure and soil

clot stop or control bleeding

cloudy eye bacterial infection in one or both eyes of fish

cloven shaped split in the toe of the hoof that separates the hoof into two parts

clutch group of eggs

Clydesdale Scottish draft horse breed that is dark in color with white face and legs and is famous as the Budweiser mascot

coat conditioner spray placed on hair coat to eliminate drying of skin and hair

cocci round bacteria shape

coccidia fungal disease

coccidiosis common simple protozoan internal parasite that is a simple one-celled organism that occurs in all mammals; is spread through water, food, and soil contamination

coccygeal last section of the vertebrae that lies over the tail area

cochlea spiral-shaped passages in inner ear that receives sound vibrations and initiates a impulse to the brain for translation

cock adult male chicken

cockatiels small bird breed that is a combination of yellow, gray, white, and orange colors and is easy to train

cockatoos large bird breed that is commonly white and has a large speaking ability

cockerel male chicken under a year of age

coffin bone bone located within the hoof joint that rotates when laminitis occurs

Coggins test blood test in equines that screens for EIA or equine infectious anemia

cold housing building with no heat where air circulates out moisture and usually houses the entire herd as a group

cold sterilization soaking items in a disinfectant chemical until they are necessary for reuse

cold tray holds a chemical that acts as a sterilizing agent

cold water water temperature below 50 degrees

colic condition in horses that causes severe stomach pain

collimator part of the machine that sets the size of the X-ray beam that produces radiation to take the picture

colon common term for the large intestine

color code method used to file medical records for visual ease of use

colostrum milk produced by female within 24 hours of giving birth; provides antibodies that protect the immune system

colt young male horse

coma damage to the midbrain that causes a loss of consciousness and awareness

comb flesh-like projection on the top of the head

combing the act of cleaning the hair with a thin comb that helps remove tangles and mats

combining vowel a letter, usually a vowel, that is placed between the prefix and suffix that makes the word easier to pronounce

commercial medium rabbits are between 8 and 12 pounds in size

commercial business grade type of cow that is a mixed breed

commercial serologic test kit commonly used test kit in a veterinary facility to provide quick and accurate results of common viruses and diseases

Common Laws regulations and rules based on legal violations

communication process the four essential steps to relaying information to others including the sender, the receiver, the message, the channel and feedback

compact bone thick tissue that forms the outer layer of bone

companion fish fish kept in people's homes as stress relievers and as entertainment

complete blood count (CBC) blood test

compound fracture a break in a bone that causes the bone to break through and penetrate the skin

computer inventory system method of monitoring supplies and products within the facility and knowing when items need to be reordered

computer systems computerized models that mimic animal behaviors and internal or external structures; also known as a simulated system

computerized appointment book software item used to schedule appointments and maintain medical information on the veterinary computer system and allows future use of information in an organized manner

concave inward appearance

concentrates food sources that are provided to an animal as an additional nutrient source when the primary food source is not adequate or abundant

concentration determined by the amount of food fed divided by the percentage of dry matter and how it is delivered to the animal

conception process of creating a new life that forms an embryo

conditioning the process of teaching an animal an action in relationship to another action

cones cells in the eye that allow animals to see colors

confidential information that is private and not to be shared by anyone outside of the facility without permission

confidentiality maintenance of privacy of medical information

confinement method system of raising sheep in an aseptic environment for human consumption purposes

conformation the body's shape and form

conical pointed shaped end

connective tissue holds and supports body structures by connecting cells

consent form permission given by owner to treat animal

constipated unable to pass waste materials and feces

constipation inability to have a bowel movement

constrict to close (as in the pupil in the eye)

consumable products supplies used on a daily basis that are not considered in the client invoice, such as paper towels, gauze sponges, and cotton balls

contagious capable of spreading disease

continuing education (CE) seminars and classes for employees that train and educate veterinary health care members on topics in the profession

contraction first stage of labor when the fetus begins to move in the uterus to prepare for birth by moving into the birth canal

contraction phase second phase of the labor process that causes wave-like motions in the uterus to move the fetus into the birth canal

contrast medium material or substance administered into the body to show structures on an X-ray film that would otherwise be difficult to view

controlled substance drugs that have the potential for abuse and addiction and must be kept in a locked cabinet by law

controlled substance license paper given by the DEA to allow a veterinarian to prescribe scheduled drugs

controlled substance log area where medicines that are a scheduled drug are kept when dispensed and must be maintained accurately and kept of file for 2 years

conures medium bird breed that is brightly colored from Central America; active and playful

cool water water temperature of 50–60 degrees

cooled semen process of transporting and storing sperm for use within 24–48 hours due to cool temperatures

copier machine equipment used to make exact copies of paper or other documents

Coplin jar glass container that holds chemicals for use in staining slide samples

cornea clear outer layer of the eye that is commonly damaged or scratched

corpus luteum (CL) known as the "yellow body" of the ovary that forms to allow an ovary to increase in size as an egg develops

corrugated tubing plastic and thin hose and allows the oxygen and anesthetic gas to flow to the patient by connecting from the machine to the endotracheal tube

cortex outer section of the kidney structure

cotton wool disease; nickname for fish fungus

courtesy the emotion of placing one's needs and concerns before your own

coverslip thin piece of glass that fits over the sample on the slide

cow adult female cow of breeding age

cow-calf system production system that raises cattle to breed them as adults

cow kicking the direction of cows kicking to the side with their rear legs

crackles abnormal sounds within the chest that sounds like cellophane paper

cranial drawer sign in and out movement or motion of the knee when the cruciate ligament has been torn

cranial nerve test reflex test of the eye where a hand is moved quickly toward the eye without touching the eye or any hair around the eye

crash cart movable table that holds emergency equipment and supplies that is easily accessible during a serious patient condition

creatinine chemical hormone filtered by the kidneys and indicates the kidney functions levels within the blood

credit application paperwork completed by clients to allow them to obtain a line of credit to pay for veterinary services

creep feeders equipment used to feed young calves that prevents adult cattle from eating the food

crest top of an animal's head

Crile hemostats similar in size and shape to Kelly forceps except they are fully serrated with serrations located within the interlocking jaws and vertically along the blades

crop small sac that acts as a holding tank for food as it's passed from the esophagus in birds

cross-breeding mating of two different breeds of related animal species and producing an animal with a combination of characteristics from each animal

cross ties short ropes placed on either side of the horse to allow safe movement while working around the animal

crown the upper part of the tooth that lies above the gum line

cruciate ligament tear common knee injury to large breed dogs that tears the ligament over the patella

crumbles flakes or blocks of fish food

crustacean aquatic animal with an outer body shell and legs

cryosurgery procedure involving freezing parts of tissue

crystals solid formations within a urine sample that form from increased components within the urinary tract or renal system

cud mixture of grass sources and saliva that is chewed and regurgitated to break down food for digestion

cuff located about ¾ of the way down the endotracheal tube that is inflatable and allows the tube to be inflated with air to create a tight seal around the trachea to prevent any moisture or material from entering the lung field

culled to remove from the herd

culture and sensitivity (C/S) test that determines if a bacterial issue is occurring

cursor blinking line on the computer screen that determines where typing is occurring

Cushing's disease common term for hyoeradrenocorticism

cusp center of the tooth that wears with age

cutability quality and quantity of meat from a beef animal

cutaneous larval migrans hookworm infection in humans caused by penetrating the skin

cyanosis blue color of mouth and gums due to a lack of oxygen

cyanotic bluish gray color of the mucous membranes due to a lack of oxygen

cylinder metal containers that hold compressed forms or liquid to keep patients oxygenated and breathing during anesthesia

cystitis term for a bladder infection or urinary tract infection

cystocentesis surgical puncture into the bladder using a needle to collect a urine sample

cystotomy surgical incision into the urinary bladder to remove urinary bladder stones

cytoplasm the liquid or fluid portion of a cell that allows the internal cell structures to move

dairies buildings where cows are milked

dairy cattle cows raised specifically for milk

Dairy Herd Improvement (DHI) program maintains records of dairy herd information

dairy goat group of goats that produce milk

dam female parent of the horse

daughter cells four cells that are created from one cell during meiosis and each cell has half the number of chromosomes of the parent cell

debeak procedure to remove the tip of a chick's beak to prevent pecking damage to other chicks

decant process of pouring the urine out of the tube and allowing only the sediment on the bottom to remain in a tube

decapod crustacean with five pairs of legs

deciduous baby teeth that are developed in newborn animals and eventually shed when adulthood is reached

declaw the surgical removal of the digits or claws on the feet of a cat

deer ruminant animals that graze on pasture and have antlers

defensive aggression natural behavior that occurs when a cat is attempting to protect itself from people or other animals

defensive reaction behavior when an animal protects itself from danger

dehorning process of removing the horns to prevent injury to people and other animals

dehydration water loss by the body

demodectic mange mite that invades the skin; all dogs raised normally by their mothers possess this mite as mites are transferred from mother to pup via cuddling during the first few days of life

denial the emotion that occurs when a person will not accept a pet's death

dental prophylaxis professional cleaning by scaling and polishing the teeth under general anesthesia

dentifrices veterinary term for toothpaste used on animals

dentin second layer of teeth similar to bone

dentition the arrangement of teeth within the mouth

depression the emotion of sadness where a person becomes so sad that he or she can't handle the normal functions of daily life

dermatitis inflammation of the skin

dermatophytosis fungal disease caused by an external parasite

descending colon third or last section of the large intestine

designer breed the cross breeding of two breeds to form a new breed not yet recognized by the AKC

developer solution chemical that develops the film

dewclaw first digit on the paw; doesn't make contact with the ground

dewlap flap of skin under the chin

Dexamathasone corticosteroid administered by injection to determine pituitary problems

diabetes condition that is produced when too much or too little blood sugar is produced and the body finds it difficult to regulate

diabetes insipidus condition that affects the water content in the bloodstream, causing the urine to become dilute

diabetes mellitus condition that causes high blood glucose levels

diagnostic imaging the term used for performing radiographs, ultrasounds, and endoscopy

diaphragm area of the chest that surrounds the ribs, heart, and lungs and protects the chest organs

diaphragm mechanism part of the microscope that allows the slide sample to be moved both up and down and side to side to view the entire sample

diaphragmatic hernia condition that affects the diaphragm by causing a tear in the chest muscle, and internal organs may protrude through

diarrhea waste materials and feces become soft and watery

diastole relaxation phase of blood pressure or the lower number, which is normally lower

diestrus stage of the heat cycle when pregnancy occurs and is characterized by a functional CL that releases progesterone to maintain pregnancy

digestion the process of breaking down food from larger particles into smaller particles for use by the body

digestive system the body system that contains the stomach and intestines

dilate to open (as in the pupil in the eye)

dilated the occurrence of widening or opening, such as the pupils

dilation phase first phase of the labor process causing the birth canal to open and expand

diluent liquid mixture of a vaccine

dilution to lessen in strength by adding another component, such as water

dimorphism the occurrence of having both male and female body parts and appearances

dip stick long, thin, plastic strip separated by individual square pads containing a chemically treated paper

dipping process of applying a chemical pesticide or medication to the skin and hair coat

Dipylidium caninum tapeworm species that can be up to 20 inches long and lives in the small intestine

direct contact (1) contact with an animal or a bodily fluid from that animal; (2) transmission of a disease through a direct source, such as saliva or the ground

direct PLR normal pupil light response assessment

Dirofilaria immitis heartworm species that belong to the same class as roundworms and look similar in structure, like long, thin pieces of pasta

disaccharide a double sugar

disease poor health that is a disturbance or change in an animal's function and structure

disinfecting destroying most microorganisms on nonliving things by physical or chemical means

dislocation a bone that completely moves out of place

displaced abomasums condition in cattle that causes the stomach to rotate out of place

dissect separate into pieces, or break down into parts to identify the meaning of a word

dissolved oxygen (DO) oxygen found in water for living organisms to breathe

distended swollen, as in the abdomen

distilled water sterile water that is placed in the water reservoir to the fill line

diurnal animals that are active during the day and sleep at night

diversity difference

DNA test test using blood or a feather that helps determine a bird's gender

dock tail area

doe adult female goat

domestication taming of an animal to coexist peaceably with humans

domestics cats with unknown parentage that are of two or more breeds and not registered

dominance aggression behavior that refers to the pack instinct of an animal and its social status within a group

donkey member of the equine family known as a burro or ass

dorsal recumbency restraint position with an animal lying on its back

dorsal root nerve branch of the vertebrae that contain sensory nerves

dorsal-ventral (D-V) position where the animal is restrained on its chest with the X-ray beam traveling through the dorsal aspect first and the ventral aspect second

dorsoventral (D-V) view is the animal positioned on its stomach with the X-ray beam going through the back first and the stomach second

dosimeter X-ray badge that measures the amount of radiation a person is exposed to

double sugar two particles of carbohydrates that form the building blocks of nutrients and are called disaccharides

down soft feather covering that grows under primary feathers

Doyen intestinal forceps clamps that are non-crushing intestinal forceps that have serrations lengthwise and hold intestinal vessels and tissues for short time periods without causing any damage

draft animals type of animal that is muscular, powerful, and strong and used for pulling heavy loads

draft capacity force an animal must exert to move or pull at a constant rate of 1/10 of its body weight

draft hitch class contest based on the number of animals used to pull a wagon and how the team works together

draft horse large and muscular horse standing over 17 hands in height

drake adult male duck

dram the measurement size of vials or bottles

dremel tool used to grind nails

drenching method of giving medicine by mouth with a dose syringe to large numbers of animals

dressing the amount of meat produced on one pig

dressing forceps instrument used to grasp and hold tissue but have no teeth, but rather a smooth surface

drop-off appointment a set time for the client to bring in an animal and leave the patient for a set amount of time for exams, surgeries, and other procedures according to the veterinarian's schedule

drop-off time set time when an animal is expected at the veterinary facility for a later exam or surgery

dropsy nickname for ascites

Drug Enforcement Administration (DEA) agency that regulates controlled substances and requires veterinarians to obtain a license to dispense scheduled drugs

dry cow period time that milk production is stopped to allow the reproductive system to rest

dry heat method of sterilization through the use of a flame by exposing an item to extreme heat or through incineration

dry matter the amount of nutrients in a food source without the water content

dual purpose breed breed that serves more than one purpose; in the case of cattle, both milk- and meat-producing breed

duckling young ducks

ducks small water birds used for meat, feathers, and down

duodenum short, first section of the small intestine

dust bath method of grooming of the chinchilla in which the animal rolls in a dusty medium to condition its skin and hair coat

dwarf rabbits are less than 2 pounds as adults

dynamic equilibrium maintaining balance in response to rotational or angular movement

dyspnea difficulty breathing

dystocia difficult labor

dysuria difficult or painful urination

ear sensory organ that enables hearing

ear marking notches made on the edges of the ear flaps to identify cattle

ear mites microscopic parasites that live within the ear canal

ear notching practice of making a small cut on the ears to identify the pig number and litter number

ear plucking removal of hair from the ear canal

ear tag identification tags with numbers applied to the ears by piecing the ear flap

"earthquake" technique procedure used in training a caged bird to stand on the hand and be held that teaches a bird not to bite; when a bird attempts to bite the hand is shaken slightly to cause a balance check

ecdysis process of snakes shedding their skin

ecology the study of the environment that animals live in

ectothermic temperature evaluated in an animal that regulates its body temperature externally or through its environment

edema buildup of fluid under the skin

efficacy the strength of the drug

egg bound a female bird that is attempting to lay an egg that has become stuck in the cloaca

egg depositers fish lay eggs on areas such as plants, rocks, or the bottom of a water source for protection

egg production system used to produce a quality egg for human consumption

egg scatterers the fish spontaneously scatter their eggs around the tank as they swim

egg-laying fish reproduce when the female fish releases eggs that are fertilized by the male's sperm

eggshell outer calcium layer that protects an egg

ehrlichia bacterial disease caused by the Lone Star tick and causes a fever, joint pain, depression, anemia, anorexia, weight loss, and some skin swelling

elastrator instrument used to stretch bands over the testicles for castration

electrocardiogram monitor placed onto a patient through metal leads called electrodes that measure the electrical activity of the heart; also called an ECG or EKG

electrocardiography the evaluation of the electrical currents of the heart through the use of machines

electrode metal leads that connect a patient to the EKG unit

electrolytes determine the balance of elements in the body, specifically potassium, sodium, and chloride

ELISA enzyme linked immunosorbent assays; simple test used to measure an antigen or antibody level within the blood

elk ruminant animal, much like a deer; grazes on pasture and has antlers

emaciated severe weight loss and dehydration

emasculator surgical tool used to cut the spermatic cord during a castration procedure

embryo cell that develops into a newborn life

emergency situation that requires immediate life-saving measures

emergency services an appointment that must be seen immediately and is a life or death situation

empathy the emotion of being able to understand another's feelings

emu large flightless bird native to Australia, smaller than the ostrich

enamel hardest substance in the body that covers and protects the teeth

encephalitis viral disease in horses and people that is called sleeping sickness and have several strains that are spread by mosquitoes; affects the nervous system

endangered animal species that are decreasing in numbers in the wild and are close to becoming extinct

endocardium thin inner muscle layer of the heart

endocrine system excretory system that rids the body of waste materials

endoscope fiberoptic instrument used to visualize structures inside the body

endoscopy the procedure of visually examining the interior of the body by using an endoscope or a tool used to view the inside of the body

endosteum thin connective tissue covering inner bone

endothermic temperature evaluated in an animal that regulates its body temperature internally or within the body

endotracheal tube (E-tube) tube placed into the windpipe to create and establish an airway during anesthesia or CPR

enema procedure of passing fluids into the rectum to soften feces to produce a bowel movement

English class chicken class developed in England for better meat production

enterotoxemia bacterial disease from overeating that causes toxins to build up within the rumen and are not capable of being absorbed by the body

enzyme chemical reactions that change within the body and create and release hormones

eosinophil white blood cell that fights against allergic reactions, controls inflammation, and protects the body from parasite infection; have a large nucleus with segmented granules within the cytoplasm

epaxial referring to the posterior area of a patient

epicardium outer thin covering of the myocardium

epididymus tube that transports sperm from the seminiferous vesicles to the vas deferens

epiglottis covering or flap that prevents food from entering the airway and lungs during swallowing

epilepsy veterinary term for seizures

epinephrine short-acting chemical released during the fight or flight response

epithelial tissue protective layer that cover the body's organs and acts as a lining of internal and external structures

equilibrium state of balance

equine veterinary term for the horse

equine encephalomyelitis viral disease of horses found in different strains transmitted by mosquitoes; also called sleeping sickness

equine infectious anemia (EIA) highly infectious disease of horses also known as swamp fever; a viral disease for which there is no vaccine and no cure; transmitted by mosquitos

Equus caballus genus and species of the horse

eructation gas buildup where belching occurs to rid the rumen of air

erythrocytes red blood cells

erythropoiesis the production process of red blood cells within the bone marrow

esophageal stethoscope a device placed into the esophagus next to the endotracheal tube and emits the sound of the heartbeat throughout a procedure

esophagus tube that passes food from the mouth to the stomach

essential fatty acids nutrients necessary in a diet and produce natural hormones needed within the body

estimate the approximate cost of services

estrogen main female reproductive hormone that causes behavior changes in females during the estrus cycle and also allows follicle production and development to occur

estrus stage of the heat cycle when the animal is receptive to being bred

estrus cycle heat cycle in female animals when the female is receptive to mating with a male

ethical the act of doing what is right

ethics rules and regulations that govern proper conduct

ethylene glycol toxicity poisoning by antifreeze that causes renal failure

eustachian tube narrow duct that leads from middle ear to nasopharynx and maintains air pressure

euthanasia the process of humanely putting an animal to sleep

evert turn inside out

ewe the adult female sheep capable of reproduction

exam room area where patients receive routine exams

excrete to remove and rid the body of waste

exemptions fees determined when an employee claims a set of yearly wages from their income taxes

exhalation act of exhaling or leaving air out of the lungs

exoskeleton external cartilage on the outside of the body

exotic animal animal not native to the area where it is raised and may be rarely found in its natural habitat

exotic animal dealer people that buy and sell exotic animals to private collectors and zoos

expel to remove

expiration to breathe out and expand the chest

expressing placing pressure on the abdomen to remove eggs or sperm

extension bending causing a joint to open and to lengthen

external auditory canal tube that transmits sound from the pinna to the tympanic membrane

extinct animal species that are no longer found in their natural habitats

extubation the process of the endotracheal tube removal from the patient

eye piece the portion of a microscope that you look through

Fair Labor Standards Act agency that governs the age of children working and the duties they can perform

family size herd a group of fewer than 100 head of cattle

farm flock method most popular sheep system that involves market lamb production for meat sources as the primary system and wool production as the secondary system

farming demonstration draft animals used in plowing and farming activities

farrier professional who trims and shoes horses hooves; also known as a blacksmith

farrow to finish production system where sows farrow out and the piglets are raised to market weight

farrowing the labor process of pigs

fast to not give food or water or anything by mouth prior to surgery

fat soluble vitamins that are stored in fat and released when needed within the body and include vitamins A, D, E, and K

fats concentrated source of energy; also called lipids

fatty acids oils that are products of fat sources that may be used as nutrients or supplements within a diet

fatty liver disease condition of birds in which large amounts of fat are deposited in the liver

fax machine equipment use to copy and send documents via a telephone line

fear aggression refers to the defense reaction to being harmed and the instinct for an animal to protect itself

feathered edge the staggered area at the end of the slide where the veterinarian or veterinary technician reads and interprets a blood slide

feathers long hair over the lower legs of horses located just above the hoof

fecal egg analysis test to determine the type of parasite found in fecal matter of animal

fecal floatation technique done to sample the fecal material to determine if any parasite eggs are within the sample by placing feces within a liquid

fecal sample used to diagnose internal parasites and the presence of blood within the stool sample

fecal smear preparation by placing a small amount of the fecal sample onto a microscope slide

feed materials that animals eat to obtain nutrients and nutrition

feedback the return message that is sent by the receiver

feedlot size the ideal weight of calves that increases size and profit

feedstuff ingredients in animal food that determine nutrient contents

feline distemper virus called panleukopenia that causes a decreased white blood cell count in cats

Feline Immunodeficiency Virus (FIV) commonly called feline AIDS virus; fatal virus that attacks the immune system of cats

feline infectious anemia another name for hemobartonella, which causes an infection within the red blood cells of cats

feline infectious peritonitis (FIP) fatal virus of cats that has a dry and wet form that affects the lungs and chest cavity

feline leukemia fatal virus in cats that affects the white blood cells and immune system

feline urologic syndrome (FUS) condition that commonly affects neutered male obese cats in which they become blocked and are unable to urinate

Felis catus genus and species of cats

FeLV abbreviation for feline leukemia virus

femur large upper bone of the rear leg; thigh bone

feral animal that is wild and not domesticated to live in an area maintained by humans

fermentation process of soaking food that allows bacteria to break down food for easier digestion

fertile egg with an embryo present due to a male fertilizing the egg

fertility ability to reproduce

fertilization uniting of egg and sperm to allow an embryo to form

fertilize the ability of a male to mate with a female to create an egg with an embryo

fever elevated body temperature due to excessive heat

fiber material from plant cells that is left after other nutrients are removed

fibrinogen protein that aides in clotting blood

fibrous joint fixed joint with little to no movement

fibula smaller bone that lies in the lower rear leg and in the back of the limb

fight or flight natural instinct of animals in which they take care of themselves by either running away or protecting themselves by fighting

filly young female horse

film hanger used to secure film for manual processing and development

filtration removing particles from the air or water using a physical barrier; common in lab areas or research facilities

filtration system water system that cleans the water source

finches a small variety of songbird that comes in a variety of colors and patterns and is easy to raise and breed

fingerbrush a small tooth that fits on the end of a finger and is a plastic, thimble-like device that has small, soft bristles that are rubbed over the teeth

fingerling immature fish greater than an inch in size; fish that are two years of age

finishing system production system where pigs are fed to market size

fin rot bacterial condition that causes the fin tissue to wear away or deteriorate

fins structures that act as arms and legs and allow movement in the water

fish aquatic species that have scales, gills, and fins

fish fungus fungal disease known as "cotton-wool disease" due to the appearance of tufts of growth that occur on the skin of the face, gills, or eyes that resemble cotton or wool

fixed-office hours appointments seen throughout the day when clients arrive and the patients are seen in the order they enter the facility during set times

fixer solution chemical that fixes the X-ray film and helps maintain the life span of the film, allowing it to stay readable

fixing process of using a solution to maintain the life of an X-ray film

flagella tail of sperm that allows movement

flanking looking at or biting at the sides of the abdomen due to stomach pain, often in a colic situation in horses

flaxen color similar to blonde

flea external parasite; wingless, feeds off blood of animal

flea allergy dermatitis (FAD) allergic reaction to a flea bite caused by the flea's saliva that affects the animal's skin

flexion bending causing the joint to close and to shorten

flight feathers adult feathers located over the edges of the wings

float instrument used to file down teeth; also called a rasp

floating filing down teeth that are overgrown and sharp

flocculent contains large amounts of particles called sediment suspended in the urine and appearing cloudy

flock a group of poultry

floss material within a filter made of gravel, charcoal, and fiber materials

flow meter located on the machine near the vaporizer adjusted for the proper rate of oxygen flow to the patient

flow rate amount of fluids to be delivered over the course of a day

flow schedules appointments scheduled based on each patient being seen at a specific time by the veterinarian during the day, such as an appointment scheduled for 1:00 p.m.

foal young newborn horse of either gender

foaling the labor process of horses

focus knob the portion of the microscope that allows better visualization of the sample

follicle tiny structures within the ovary that enlarge with egg development and allow ovulation to occur

follicle stimulating hormone (FSH) hormone released to allow a follicle to grow and allow ovulation to occur

food analysis the process of determining the nutrients in food and prepared mixes to assure it serves as a balanced ration

Food and Drug Administration (FDA) agency that sets manufacturing standards of food additives and medications used in animals, especially for human consumption

food fish production raising fingerlings to market size to be sold as food products

foot rot bacterial condition of the hoof that decays or damages the hoof tissue

forage hay or pasture source

forager animal that eats grass and pasture

forages plant-based sources of nutrition that are high in fiber; also called roughages

force feeding placing food within a patient's mouth and forcing it to swallow

forced molting the process of appliying a light source to avian species to allow them to shed their flight feathers each season and grow new feathers

forcep tool used to remove hair or parasite from body

foreign body obstruction condition when an animal ingests a foreign object that is not digestible and becomes impacted within the intestinal tract

forelimb front leg of an animal

Form W-4 Employee Withholding Allowance Certificate; form filled out according to the employee wage bracket and marital status for tax purposes

formalin chemical used to preserve tissue samples

founder common term for laminitis

fountain structures made of plastic, fiberglass, or concrete that hold fish species and must be watertight

fowl another term for poultry

fowl cholera bacterial infection of poultry that causes a purple coloration to the head

fracture broken bone

free choice feeding method of animals that allows animals to eat when they want food; also called free access

free gas air accumulates in the dorsal rumen of a ruminant's stomach causing the animal to choke when the esophagus becomes obstructed with food and saliva causing the gas to not be able to escape

freemartin adult cow that is sterile and not able to reproduce

freshening the labor process of dairy animals

freshwater water with little to no salt content

freshwater fish species of fish that live in freshwater with little salt content

frog V-shaped soft pad in the bottom of the hoof

frothy bloat caused by gas being trapped within small bubbles within the rumen causing the abdomen to become swollen and painful

frozen semen method of shipping and storing sperm in liquid nitrogen and may be frozen and preserved for up to 40 years

fry young newborn fish

fryer another term for broiler since they have tender meat

Full Arch giant breeds of rabbits that are over 14 pounds in size

fundus opening of the stomach

fungal class of disease spread by simple-celled organisms or spores that grow on the external body and other areas of the environment, mostly in moist and humid conditions

fungus living organism that invades the external area of the body through direct or indirect contact

FVRCP feline vaccine combination series also known as the feline distemper series; abbreviation for Feline Viral Rhinotracheitis, Calicivirus, and Panleukopenia

gait movement or way a horse moves

gaited horse horse that moves with a specific rhythm and motion

gall bladder organ that stores bile

gallop fast four-beat running gait where each foot hits the ground at a different time

game species of wildlife that are hunted for food or other products

game animals wildlife species that are used for food and other by-products produced by wild animals

game birds wild bird species hunted for food and sport

game commission government agency that monitors the seasons by providing rules, regulations, and licenses for hunting and serves the public in education and conservation purposes

game fish fish species that are caught for food or sport purposes

gander adult male goose

gastric dilation veterinary term for the condition known as bloat in which air or gas fills the stomach, causing the abdomen to become swollen and painful

gastric dilation volvulus (GDV) condition where the stomach and intestinal tract rotate after becoming swollen due to air or gas in the GI tract, causing the intestinal tract's circulation to be cut off

gastrointestinal system (GI) the digestive system that contains the stomach and intestines

gauge the size of a needle measured by the diameter of the needle

geese large water birds used for meat, eggs, feathers, and down

gelding castrated male horse

Gelpi retractor self-retracting instrument with single points on the end that are sharp

genetic flaws undesired traits or characteristics passed from one or both parents to the offspring

genitalia reproductive organs

germinal disc white spot in the egg yolk where the sperm enters the egg cell to allow fertilization

gestation length of pregnancy

gestation cycle length of pregnancy

giant breeds of rabbits that are over 14 pounds in size

giardiasis protozoan disease cause by water contamination

gills organs that act as lungs and allow fish to breathe by filtering oxygen in the water

gilt young female pig that has not yet been bred

gingiva soft gum tissue

gingival hyperplasia excessive amounts of gingival tissue growth

gingivitis inflammation of the gums

gizzard digestive organ of birds that serves as a filter system that breaks down hard foods, such as seed shells or bones; located below the crop

globules particles in milk that reduce the fat content

globulin provide antibodies to help prevent disease

glomerulus capillaries located in a bundle at the renal pelvis

glottis opening into the larynx

glove pack a prepackaged disposable sterile glove pack that is used one time and then discarded

glucose veterinary terminology for blood sugar

goat first type of livestock domesticated that are used for meat, dairy, and fiber products

goat pox viral disease of goats that affects the immune system and causes flu-like symptoms

gonadotropin (GnRH) hormone produced and released to maintain a normal estrus cycle

goose adult female goose

gosling young goose

gown pack holds the sterile gown that is worn by the surgeon and veterinary staff

gram negative bacteria that stains red

gram positive bacteria that stains purple

gram stain tests are evaluated to determine presence of bacteria and type of bacteria in a sample

grass tetany condition in cattle due to eating rich pasture high in nitrogen gases; causes abdominal pain

gravid term for reptiles meaning pregnant

green foods foods such as fruits and vegetables that are green in color and have additional calcium

gregarious instinct or behavior to join a herd or group of like animals

grid technique the plate that holds the film below the X-ray table

grief the emotion of sadness that people feel after the loss of a pet or loved one

grit substance provided to birds to help break down hard substances, such as seed shells

grooming the care of an animal's external body including hair coat, ears, nails, and anal glands

gross examination visible observation of the feces

gross pay total amount of wages earned by an employee before tax deductions

growth diet specialized food formulated to increase the size of muscles, bones, organs, and body weight of young offspring

growth plate area of the bones in young animals that allow the bones to grow and mature as the animal ages and the cartilage joint turns to bone

guilt stage of grief when a person feels they are to blame for the pet's death and that they should have been able to save the pet's life

guinea fowl small wild birds domesticated for meat, eggs, and hunting

habitat area where an animal lives

hairballs large amounts of hair that collects in the digestive system of cats due to excessive grooming

half hitch tie that makes a loop around a stationary location, such as a post or fence

halter equipment applied to the head of the horse for control and handling

hame strap leather parts that attach the harness and collar

hamstring muscle located on the hind limb at the mid- to distal part of the thigh

hand the measurement method of determining the height of a horse equaling 4 inches; hand may be blunt or sharp in shape

hand tanks large containers that hold chemicals and water from developing film

hardware the physical parts of the computer and its equipment

hatchery place where fish are fertilized, bred, hatched, and raised

head part of the body containing the eyes, mouth, and gills

head exam includes examining the eyes, ears, and nose

health certificate documentation that an animal is in good health and can be transported out of state or out of the country

heart organ with four chambers located in the chest between the lungs

heart murmur an abnormal valve that produces an abnormal flow of blood that creates a "swishing" noise on auscultation

heart rate (HR) pumping action of the blood through the heart creating a beating action

heartworm disease disease spread by the mosquito and is of most concern in dogs and cats and occasionally ferrets; causes long, white worms to build up in the heart

heat period of time when a female dog is receptive to a male dog for mating; also known as the estrus cycle

heat stroke serious condition where the body overheats at temperatures of 105 degrees or more

heaves respiratory condition in horses due to an allergic reaction usually to dust

heifer young female cow that has not yet been bred

hematocrit measurement of the percentage of red blood cells in whole, or unclotted blood; also called a PCV

hematology the study of blood

hematoma the rupture of a blood vessel causing a fluid-filled pocket of blood

hematuria blood in the urine

hemipenis the male reproductive organ in reptiles and amphibians

hemoglobin main component of red blood cells that allows oxygen transport and produces iron to allow cells to multiply

hemolysis the rupture of red blood cells causing a pink coloration to develop in the plasma or serum

hemostatic forceps hinged locking instruments that are designed to clamp and hold off blood vessels; also called hemostats

hen adult female chicken

heparinized saline drug that prevents blood from clotting and is mixed with saline to flush into the catheter site to prevent a blood clot from occluding the catheter

hepatic lipidosis condition of birds in which large amounts of fat are deposited in the liver

hepatitis inflammation of the liver

herbivores animals that eat plant-based food sources

herd a group of goats

herd health manager person that maintains the records and health care of a herd of cattle

herder person in charge of a group of sheep

herpetology the study of reptiles and amphibians

hibernate the process of an animal sleeping when its body temperature decreases

hierarchy the place a horse has in the herd

hinge joint that opens and closes

hinny cross of a male horse and female donkey

hip dysplasia common genetic condition of large breed dogs where the ball and socket joint of the femur and pelvis become diseased

histamine chemical released during an allergic reaction

histology the study of tissues

hobbles leg restraints to prevent a horse from kicking

hock common term for the point of the rear leg where the tibia and fibula meet

hog cholera name of the virus that causes brucellosis

hog snare restraint tool with loop on the end of a metal pole used to restrain the pig by pressure over the snout

Holman retractor levers that are flat with a pointed appearance that look similar to an arrow with holes in the handle and are commonly used during orthopedic procedures

homeostasis a balance within the body

hooded variety of rats that are white in the body color with a black or brown coloring that is located on the head and shoulders, giving it a hooded appearance

hoof pick equipment used to remove rocks and debris from horse hooves

hoof wall outer covering of the hoof

hookworms common intestinal parasite of dogs and occasionally cats that have either teeth-like structures or cutting plates with which they attach themselves to the wall of the intestine and feed on the animal's blood

hormone chemicals that are produced by the body that allow for sexual reproduction to occur in both males and females

horn bud area of horn growth on top of the head

horse pulling contests of power to see which horse can pull the heaviest load

horsepower unit of power on an engine or motor that equals 746 watts of electricity or the amount of work done by lifting or moving 550 pounds at a distance of one mile per second

hospital treatment board area in the facility where hospitalized patients are listed along with the skills that are to be completed on them

host the animal that is infected by a parasite and that the parasite feeds off of

house calls appointment scheduled for the veterinarian to examine a patient in the owner's home

housebreaking training a dog to eliminate outside

humane what is considered acceptable by people in regards to an animal's physical, mental, and emotional well-being

humerus large upper bone of the forelimb

hunter horse horse ridden over small fences

hunting the tracking and killing of animals for sport, food, or other resources

hurdle wood, plastic, or metal board used to direct and move pigs

hutches special cage designed for a rabbit that has a covered area to allow the rabbit to escape the weather and an open area that allows proper air and ventilation

hybrid the cross breeding of two breeds to form a new breed not yet recognized by the AKC

hydration status area evaluated over the neck and shoulders, eye sockets, and gums to determine how much water loss has occurred

hydrolysis chemical process of breaking cells down into smaller particles

hyperadrenocorticism increased production problem within the adrenal gland that results in a pituitary gland tumor when the ACTH hormone is increased or an adrenal gland tumor when cortisol hormone production is increased; also called Cushing's disease

hyperglycemia high blood glucose level

hyperkalemia increased potassium in the blood

hypersensitivity increased reaction to allergens causing an allergic reaction

hyperthermia condition where body temperature is above normal and causes the animal to have a fever

hyperthyroidism condition that increases the thyroxine hormone production from the thyroid gland and common in cats

hyphema blood noted inside the eye

hypnotize to place in a trance-like state

hypoadrenocorticism decreased production of steroids within the adrenal glands if the glands fail to function; also called addison's disease

hypocalcemia decrease in blood calcium

hypoglycemia low blood glucose level

hyponatremia decrease of sodium and chlorine in the blood

hypothalamus develops from brain tissue while an animal is in the embryo stage and the gland serves as a reservoir for hormones and allows for the release and regulation of hormones

hypothermia condition where body temperature is below normal, causing the animal to become cold

hypothyroidism condition that decreases the thyroxine hormone production from the thyroid gland and common in dogs

iatrogenic condition caused by humans

ich common aquatic parasite caused by a protozoan that produced white spots on the skin; commonly called whitespot

ideal weight breed standard based on an animal's breed, age, species, purpose, or use and health status

ileum third or last section of the small intestine

ilium first section of the pelvis

IM pin intramedullary pin

immune system responsible for keeping the body healthy and protecting an animal from disease

immunity term used to mean protection; begins at birth during the nursing process

impacted difficult or unable to express

imprinting process of learning through an attachment to an object that will emit adult behaviors and can be generalized to all examples of the object

inappropriate elimination commonly called house soiling when the animal urinates or defecates in the house

incineration the burning of infectious materials or animal carcasses

incinerator a device used to burn the remains of items that have the potential to spread disease

incisors the front teeth located in the upper and lower jaws

incontinence uncontrolled bladder leakage

incubate to keep at a constant warm temperature

incubated the process of keeping eggs warm for hatching

incubation (1) the process of increased temperature to allow an embryo to grow; (2) time an animal becomes infected with disease

incubation period length of time until eggs hatch

incubator equipment used to house eggs at a controlled temperature and humidity

indirect contact spread through ways other than touching an infected animal, such as airborne or through bedding sources

indirect PLR abnormal pupil light response assessment

induced ovulation occurrence of the release of the egg at the time sperm is introduced when fertilization occurs at mating

induced ovulators female cats that become pregnant after breeding with a male as the release of an egg occurs at the time of mating

induction phase the time when the patient is being given anesthesia to make them sleep and remain unconscious and free of sensation during the surgery

infection invasion of the body by foreign matter causing illness and disease

Infectious Bovine Rhinotracheitis (IBR) respiratory virus affecting cattle; commonly called "red nose"

infectious bronchitis respiratory virus that affects only chickens and affects egg production

infertile not able to reproduce; also known as sterile

infertility not capable of reproducing

inflammation causes white blood cells to build up in the site that is affected and may cause pus cells to form, redness around the area, warm to hot skin temperatures, increased body core temperatures, edema, and pain

influenza equine flu virus

infundibulum funnel within the uterus where an egg develops in an egg-laying animal

infusion pump equipment is used to provide a constant flow of fluid at a specific rate throughout the day

inhalation act of inhaling or taking in air to the lungs

in-hand breeding method of holding the mare and stallion for breeding purposes

in-house testing laboratory samples analyzed within the veterinary facility

injection port area on an IV bag used to remove and insert additional substances using a needle and syringe

inner shell membrane thin area located outside of the albumin

insecticides sprays or pour-on chemicals used to control flies and other insects

inspiration to breathe in air and allow the chest to depress

instinctive behaviors behaviors that occur naturally to an animal in a reaction to a stimulus

Institutional Animal Care and Use Committee (IACUC) agency that monitors and approves studies for research prior to being conducted

instrument milk solution that acts as a lubricant to keep the instruments protected from rusting

insulin chemical produced by the liver that is released into the bloodstream and regulates the body's blood sugar

integumentary system includes the skin, hair, pads, horns, beak, nails, hooves, and feathers

intensifying screen located within the film cassette to help produce a better exposure to the film

intensity biomass the number of species in a volume of water

intensive care unit (ICU) area of the facility where critical patients needing immediate and constant care are treated and housed

intermale aggression behavior tendency that is common in adult male cats due to sexual dominance or social ranking

interneurons cells that deliver signals from one neuron to another from the brain to the spinal cord

interpersonal communication allows people to discuss and understand information with other people

intervals a set amount of time that is required to see a patient, such as 15 minutes

intervertebral disc disease (IVDD) common injury of the back in long-backed breeds where inflammation and swelling occurs when one vertebra puts pressure on another

intradermal injection given into the layers of skin

intramedullary pin stainless steel pin or rod placed into the center of a broken bone to keep the bone in place as it heals

intramuscular (IM) injection given into the muscle

intranasal (IN) medication administration into the nasal passage

intraocular into the eyes

intravenous (IV) given within the vein

intravertebral disc disease (IVDD) injury to the spine that causes pressure on the discs and/or spinal cord and may cause partial or complete paralysis

intubation the process of placing a tube into the trachea to establish an airway and allow the patient to continue to inhale gases that keep the patient under general anesthesia as long as necessary

intussusception condition when the stomach or intestine telescopes upon itself cutting off circulation of the digestive tract

inventory physical count of every medicine and supply item within the facility to record accurate records

inventory management knowing the price, manufacturer, pharmaceutical company, amount, and expiration date of each item in the facility

invoice bill given to a client that lists the costs of services and procedures

involuntary reflex occurs without the need for thinking and serves as a necessary body function, such as the heart beating

involution the process of the uterus shrinking after labor

iodine chemical disinfectant used to destroy bacteria

Iospora canis most common coccidia species affecting animals

iris middle layer of the eye that gives an eye its color and holds the pupil

ischium second section of the pelvis

isolation ward area that contains cages and equipment to care for patients that are contagious and may spread disease

isotonic crystalloid fluids given to an animal that is dehydrated and create a balance of elements within the body

isthmus shell membrane of an animal that lays eggs

IV catheter a small plastic piece of equipment that is placed within a vein to administer fluids and medications directly into a patient's bloodstream

jack male donkey

jaundice yellow coloration to the skin or mucous membranes

jejunum second or middle section of the small intestine

jenny female donkey

jird burrowing rodent

jog slow two-beat gait of western horses

Johne's disease digestive and intestinal condition that causes the intestinal wall to thicken

joint ill bacterial infection caused by contamination of the umbilical cord that causes pain and inflammation within a joint

joints where two bones meet and allow a bending motion

jugular vein blood vessel located in the neck or throat area

jumper horse horse ridden over large fences on a set course

juvenile young stage of growth

Kelly hemostats larger in size and are partially serrated to better hold and grasp blood vessels

kennel attendant person working in the kennel area and animals wards, proving food, exercise, and clean bedding

kennel ward area where animals are housed in cages for surgery, boarding, and care

keratin a protein that allows the nail to grow and strengthen

ketosis condition in dairy cattle that causes low blood sugar

keyboard buttons that have letters that allow typing and word formation on a computer

keyboarding act of typing and development of skills used in computer functions and keys

kid young newborn goat

kidding labor process of goats

kidney two organs located in the dorsal abdomen located on either side of the spine that produce urine, are reddish brown in color, and are bean shaped and smooth

killed vaccine manufactured from dead pathogens of a disease and placed into the animal's body in an inactive form

kilocarories amount of energy to raise 1 gram of water by 1 degree and is written as a kcal

kilovoltage peak (kVp) the strength of the X-ray beam

kindness characteristic that means you are helpful, understanding, and work in a friendly manner

kitty taco restraint procedure of wrapping a cat in a towel to restrain it

knee jerk reaction the reaction to a slight tap over the knee joint that causes the knee to jerk in response to assess the nervous system

knee jerk reflex reflex used to assess the spinal nerves

knemidokoptes mite external parasites that occur on the skin, beak, and feathers of birds and are commonly called scaly leg mites

knob projection on the top of the beak

knot tying two pieces of rope to make them not slip and to contain an animal

lab animals animals bred, raised, and cared for specifically in controlled environments to assure accurate results are obtained and genetically suitable for work

labial surface of the teeth located on the front area that is covered by the lips

laboratory location in the facility where tests and samples are prepared or completed for analysis

laboratory animal technician specially trained professionals that care for animals used in research by feeding, cleaning, monitoring, and caring for the overall health

laboratory veterinarian doctors who provide medical care to animals involved in research and perform surgical procedures used in improving techniques for humans and animals

lactated ringer's solution (LRS) fluid of lactic acid that is commonly used to replace fluids, as in dehydration

lactation the process of milk production

lactation diet specialized food provided to breeding females that have completed the gestation phase and are currently producing milk for their offspring

lamb a young sheep under a year of age; also the meat from a young lamb under a year of age

lamb dysentery bacteria diarrhea condition in newborn lambs

lamb feeding specialized production system of raising lambs to weaning and then selling to feed lots

lambing the labor process of sheep

laminitis inflammation of the lamina within the hoof bone and surrounding tissue

laparotomy towel sterile towel used to control bleeding during surgery and as a hand wipe

large size herd a group of more than 100 head of cattle

larvae immature stage of a parasite

laryngoscope a tool made of heavy metal that has a light source on the end to help light the airway view

larynx cartilage that opens into the airway and lies within the throat area

lateral (LAT) an animal positioned on its side

lateral line organ located just under the scales that picks up vibrations in the water

lateral recumbency restraint position of the animal lying on its side

laxative stool softener

layer adult female hen that lays eggs

lead rope applied to a halter to handle, control, and move a horse

leading the act or motion of moving or walking a horse

learned behavior the response to a stimulus that creates a manner by means of watching another animal or person

learning the modification of behavior in response to specific experiences

left atrium thin walled chamber of the heart that passes from the pulmonary artery into the left side of the heart

left ventricle thickest walled chamber of the heart

lens objective microscope viewer that offers a variety of different powers

leptospirosis bacterial disease transmitted through contact with contaminated urine

lethargy inactive; tired

leukocytes veterinary term for white blood cells

liability a legal responsibility

lice tiny insects that live on hair, are species specific, and are contagious

licensed veterinary technician (LVT) a veterinary technician that has passed a state licensing exam

life stage particular periods in one's life characterized by specific needs during that time

lifespan length of time an animal will live; based on body size and health

ligament fibrous strands of tissue that attach bone to bone

ligated to tie off

light horse riding horses between 14.3 and 17 hands in height that are used for pleasure riding

limiting factor missing or lacking one of the essential components to support the life of a wild animal

lipase enzyme produced by the pancreas, which breaks down fats

lipids concentrated source of energy that contains the highest amount of energy of all nutrients

liquid manure system costly waste removal system used on large farms and requires above or below-ground storage tanks with water added for ease of pumping

listening skills involves hearing what is being said by someone and understanding what he or she means

litter wood shavings used as bedding to absorb droppings

litter box equipment filled with pebble-like material and used by a cat to eliminate in

litter trained process of teaching cats to eliminate in a specific area of the house

live cover natural breeding method of the mare and stallion

livebearing fish fish that bear or give birth to live young

liver organ behind the stomach that makes bile and produces glucose

living animals any species of animal that is alive and in some way responds to a stimulus

llamas ruminant animals native to South America and raised for their hair, meat, and used as pack animals

log pulling draft animal contest where logs are pulled to see which team can pull the heaviest amount

loin the lower back or lumbar area

longe line long lead line used to exercise a horse in circles

longhair hair coat that is long and requires regular brushing and grooming

lope three-beat running gait of western horses

lordosis exhibiting a prayer position where the front end of the body is lowered to the ground with the back end held high in the air

lovebirds small songbird of bright colors that originated in Ethiopia; popular cage birds

lower GI a contrast film of the lower digestive tract organs; often given via an enema

lumbar third section of the vertebrae that lie over the lower back

lungs organs that surround the heart and inflate with air for breathing

luteinizing hormone (LH) estrus hormone that allows the production of testosterone, allows for ovulation to occur, and forms the corpus luteum (CL) during the reproduction cycle

luxation a bone that come completely out of place when broken

Lyme disease bacterial disease transmitted through bite of tick

lymphocytes white blood cells that make up the largest part of the bloodstream and aid in immune functions that protect the body from disease and have one large single nucleus that makes up most of the cell

lymphocytosis viral disease that causes white to gray growths to develop over the skin; commonly called cauliflower disease

lysosomes the structures within the cell that digest food and proteins

macaws large bird breed that is brightly colored and originated in South and Central America; playful and easily trained

macrodrip line a regular IV drip that delivers 15 drops of fluid per milliliter or 15 drops/ml

macrominerals minerals needed in large amounts such as calcium

macrophage cells that eat and destroy organisms throughout the body

maiden female horse that had not been bred

maintenance diet specialized diet fed to animals that may be working or competing; the goal is to keep the animal at a specific and constant ideal weight

malignant cancerous tumor

malpractice the act of working below the standards of practice

mammalogy the study of animals

mandible lower jaw

mange dermatitis condition caused by mange mites that cause itching and hair loss over the head and neck and can lead to severe skin infection

manual developing developing X-rays by hand using chemical tanks

manure waste material high in nitrogen, phosphorus, and potassium

marbling appearance of intramuscular fat within the meat

mare the adult intact female horse capable of reproduction

Marek's disease poultry disease caused by a herpes virus also known as range paralysis due to causing wing or leg paralysis

marketability how well a product will sell in an area

market weight adult weight of cattle when they are sold for beef

mastectomy surgical removal of the mammary glands

mastitis inflammation of the mammary glands

mat areas of hair that are interwoven together and form a large clump

mat splitter tool used to break down and separate mats into smaller ones

maxilla upper jaw

Mayo scissors instrument used to cut heavy tissues and sutures that are proportionate in size and when lying on a flat surface curve upward without contact at the end of the blades

Mayo stand the instrument tray that is elevated to hold the surgeon's surgical instruments and supplies

meal fine ground food for young fish and aquatics

meat goat group of goats that produce meat

meat type hog hogs raised for the large amount of meat, known as ham, on their body

mechanical filtration various types of equipment that filters the water to remove harmful particles and keep the water clear

medical record written and recorded information on the care and treatment of the animal

medical waste storage and trash for items contaminated by body fluids and living tissues that may be contaminated with infectious disease, such as surgical drapes or bandage material

Mediterranean class chicken class developed in the Mediterranean for large quantities of egg producers

medium gel where organism growth may occur

medulla center section of a kidney

medulla oblongata part of the brain that controls the body's involuntary functions

medullary cavity hollow center within bone where blood cells are produced

meiosis cell division for breeding and reproductive processes

menance response test a reflex test of the eye where a hand is moved quickly toward the eye without touching the eye or any hair around the eye

meniscus ligament that lies directly over the patella and forms an X shape over the cruciate ligament

mesentery connective tissue from the peritoneum and carries blood vessels and nerves to the small intestine

message an idea being passed along through a route of communication

metabolic diseases conditions that cause a chemical change within the body, usually from stress

metabolism reaction within the body that allows chemicals within the body to break down and be used by the body

metacarpal bones that form the long bones of the feet in the forelimbs

metatarsal bones that form long bones of the feet in the rear limbs

metestrus stage of the heat cycle following estrus when ovulation occurs

Methimazole medicine for a cat to treat hyperthyroidism that blocks the synthesis of thyroxine

Metzenbaum scissors instrument with thin blades and a long handle used in surgery to dissect and remove delicate tissues

microbiology the study of microscopic organisms

microchips electronic identification in a computer chip placed under the skin

microdrip line a small line that delivers 60 drops/ml

microfilaria tiny immature heartworm larvae that live in the bloodstream

microhematocrit measurement of the percentage of red blood cells in whole or unclotted blood; also called a PCV

microhematocrit tube thin, small glass tubes that hold blood within a centrifuge

microminerals minerals needed in small amounts such as iron

microscopic very small; unable to view with the naked eye

midbrain part of the brain that controls the senses

midstream urine collected shortly after urination begins and just prior to the process ending

migratory birds that move from location to location throughout the year

milk fever low blood calcium condition called hypocalcemia

milking herd group of cattle used to produce high-quality amounts of milk

milliamperage (mA) represents the number of X-ray beams based on time

milt fish sperm

mineral block minerals supplied in a block for horses to lick

minerals nutrients needed in every area of the body but are found mostly in the bones and teeth; used by the body based on the animal's needs and mineral availability

miniature small rabbit breed under 2 pounds in size as an adult

misrepresentation working or acting as someone that you are not qualified to be

mite insects that live on skin, hair, or in ears

mitochondria "powerhouse of the cell" and the part of the cell that makes energy

mitosis cell division that divides one cell into two cells and aids in animal growth and tissue repair

mixed breed animal of unknown parentage that is a combination of two or more breeds

mobile services a vehicle equipped with veterinary equipment, tools, and supplies that travels to a client's home to examine a patient

modeling when an animal learns a behavior through watching other animals conduct the behavior

modem box-like equipment that allows access to the Internet

modification techniques training used to retrain the animal and get rid of inappropriate behaviors

modified live vaccine made from altered antigens created from disease pathogens that place small amounts of a disease into the animal's body

mohair long silky locks of hair from the Angora goat used to make clothing and bedding products

molars last set of teeth that are large and located in the back of the mouth

mollusk aquatic species that have only a thick, hard shell

molting process of shedding feathers to allow new feather growth

monitor the computer screen that looks similar to a TV screen

monocyte largest white blood cell; helps neutrophils rid the body of wastes and cell debris

monogastric one simple stomach to digest food

monosaccharide a simple sugar

moral what a person believes is right and wrong

morphology the cell structure, shape, color, and appearance in numbers

mosquito hemostats small forceps with a fine tip that is used to hold delicate tissue

motility movement

motor neuron cells that deliver signals from the central nervous system (CNS) to the muscles to give a response

mouse movable instrument attached to a computer that allows the user to move the cursor

mouth brooder fish that carry their eggs in their mouth for incubation to protect and allow hatching

MSDS safety sheets Material Safety Data Sheets; information published by manufacturers of various products that have the potential to harm humans within the facility

mucosa thin connective tissue that lines the intestinal tract

mucous membrane (mm) the gums of an animal and their color

mule hybrid cross of a female horse and a male donkey

murine veterinary term for mice or rats

murmur abnormal heart sound that produces a swishing sound

Mus musculus genus and species of the mouse

muscle structures located throughout the animal's body that attach to different locations and serve to protect, allow bending and movement, and aid in the strength of the animal

muscular tissue allows movement of body parts

musculoskeletal system body system consisting of bone, muscles, tendons, and ligaments

mutton sheep meat from an animal over a year of age

muzzle equipment placed over the nose to prevent biting

myocardium thick muscle that forms the wall of the heart

nail bed area where the nail growth occurs and is located at the end of the digit near the hair growth of the foot

nail trim cutting the length of the nail to prevent injury and promote comfort

nanny common name for the adult female goat

nares the nostrils or the nose

nasal septum cartilage between the nostrils

nasogastric tube tube passed through nostrils into the stomach to force-feed a patient

National Association of Veterinary Techncians in America (NAVTA) sets the standard of veterinary technology and the care of animals

National Institutes for Health (NIH) agency that specifies animal space requirements of a species housed for research purposes

natural incubation the process of a female or male bird nesting on eggs to hatch them

near side left side of the horse

neck band paper or plastic slip placed around the neck that identifies the animal

neck tag identification tag with numbers worn around the neck of cows to identify them

necropsy procedure done by examining the body of a deceased animal to determine the cause of death

necrotic dead area of tissue

needle holder instrument used to hold a needle while placing sutures into a patient

needle teeth eight long, sharp teeth that are like needles located on each side of the upper and lower jaw

negligence failure to do what is necessary or proper

neoplasia veterinary term for cancer

nephron unit of measurement of the kidney that produces urine

nervous system assessment of the brain and spinal cord through the response of pupil reflexes, the gait, as well as other body reflexes, such as limbs, paws, and digits

nervous tissue contains particles that respond to a stimulus and cause a reaction in the body

net pay amount of money received after tax deductions

neurological pertaining to the brain and spinal cord

neurons specialized cells within the nervous system that control impulses

neuter surgically removal of the reproductive organs or either gender

neuter certificate documentation the animal has been spayed or castrated and is no longer able to reproduce

neutrophils most common white blood cell that destroys microorganisms within tissues that has a nucleus with segments

New World monkeys monkey species that originated in South America

Newcastle disease respiratory virus that causes poultry to twist their necks

night feces slightly softer feces that occur at night and provide nutrition for the digestive system by allowing for an ease of breaking down plant materials

nit larvae of lice

nitrogen cycle a process that converts wastes into ammonia and then into nitrites and then into nitrates

nocturnal term for animals that sleep during the day and are active at night

non-absorbable suture material not broken down within the body and remains intact for long periods

nonessential fatty acids nutrients that are not necessary within the body and are used as additions to a diet when necessary

non-game animals animals that are not hunted for food or other products

non-game birds species of birds and songbirds that are not hunted for food

non-game fish fish species that are either too small or too large to catch and eat as a food source

nonhuman primates animals that resemble humans and have movable thumbs

nonliving systems models that are mechanical in nature and mimic the species of animal that is being studied

non-rebreathing system anesthesia system used for small patients such as rodents, birds and cats that allows them to recirculate oxygen

nonruminant specialized digestive system of the horse that requires small amounts of food to be digested throughout the day

nonruminant system digestive system similar to monogastric animals with a larger, well-developed cecum for breaking down plant fibers

nonruminants animals with a specialized digestive system with a simple stomach built for slow digestion

nonvenomous not poisonous

nonverbal communication behaviors recognized and passed along without the need for speaking

norepinephrine long-acting chemical hormone that increases the heart rate, blood pressure, blood flow, blood glucose, and metabolism

normal saline solution with the same concentration level as salt

nose printing methods similar to finger printing in which the lines of the nose are imprinted for identification

nose ring metal ring placed through the septum of the nose to allow a cow to be lead

nosocomial infection when a human causes the spread of disease and contamination of an animal

NPO nothing by mouth, as in food or water

nucleolus located within the nucleus and forms genetic material

nucleus brain of the cell that is located in the center of a cell

numerical filing system organization of medical records by client or patient last name

nutrient any single class of food or group of like foods that aids in the support of life and makes it possible for animals to grow or provides energy for physiological processes in life

nystagmus condition where the eyes jump back and forth in rhythmic jerks due to damage in the inner ear, brain stem, or cranial nerves

obedience training a dog to obey commands

objective information measured facts; for example, vital signs

oblique view used on areas that need to be placed at an angle to prevent double exposure from other body organs

observation an inspection of an animal at a distance to see what it looks like, acts like, or how it moves naturally

observation skills used to pay attention to a person's body language and speech

occlusal the top surface area of the teeth

Occupational Safety and Health Administration (OSHA) agency that governs laws for safety in the workplace

offspring young newborn animals

oil immersion the use of a specialized oil substance that is placed over a sample to view the contents

Old World monkeys monkeys that originated in Africa and Asia

olecranon the elbow

oliguria decreased amounts of urine production

omasum third section of the ruminant stomach that absorbs nutrients and water

omentum thin lining that surrounds the organs within the abdomen

omnivores animals that eat a meat- and plant-based diet source

onychectomy veterinary term for the declaw surgical procedure of a cat

oocysts single-celled eggs

open gloving process done by removing the left glove using the right hand by grasping it and the edge of the cuff

operculum gill covering that divides the head from the trunk

oral pertaining to the mouth

oral barium study barium solution given by mouth to pass through the digestive system to allow X-rays to be taken over time to view internal structures of the GI tract

oral cavity includes the gums, mouth, teeth, tongue, and throat

orchidectomy castration; surgical removal of the testes

organ a group of tissues

organelle structures that are found within a cell

ornamental fish popular pet kept in aquariums that are small in size and vary in color and appearance

orphaned young animal that is rejected by or loses its mother

orphaned lamb young sheep that has lost its mother or has been rejected by the mother and must be bottle fed

Oryctolagus cuniculus the genus and species of the rabbit

os penis bone within the urethra of male dogs

OSHA Occupational Safety and Health Administration; agency that provides for safe and healthy work environments and conditions for all employees

osmosis water absorption by the body through membranes

ossification the formation of bone

ostrich large flightless bird raised for meat, feathers, oils, and other by-products

outer shell membrane thin area just inside of the shell

oval window membrane that separates the middle and inner ear

ovarian cyst small masses that develop on the ovary of an intact female animal; causes eggs not to develop (sterility)

ovariohysterectomy (OHE) term for a spay; surgical removal of the ovaries and uterus

ovary the part of the uterine horn that allows egg development and is the primary female reproductive organ in egg growth

over-the-counter drugs (OTC) medications that do not require a prescription

oviduct small tube on the end of the uterine horn that is tapered and holds the ovary

ovine the veterinary term for sheep

ovulation release of the egg during estrus in the female to allow reproduction to occur when united with sperm

ovum egg yolk in an animal that lays eggs

oxen members of the bovine family that resemble cattle with long vertical horns on the head

oxygen tank located under the machine or in separate cylinders next to the machine

oxygenation process of keeping air in the water by using dissolved oxygen

oxytocin hormone that releases and causes the muscles of the uterine wall to contract and milk production to begin in the mammary glands

pacemaker system of the heart that controls heart sounds and rhythm

packed cell volume (PCV) measurement of the percentage of red blood cells in whole or unclotted blood; also called a hematocrit

packing slip paper enclosed in a package that details the items in the shipment

paddle long-handled board similar to a hurdle that acts as an extension of the hand to move pigs

palatability how well food tastes and is eaten by an animal

palpation to feel

palpebral reflex evaluated by tapping the fingertip lightly over the surface of the eyelid

pancreas organ that lies next to the stomach and secretes enzymes that aid in digestion; produces and regulates insulin production

panel individual blood test

panleukopenia commonly called feline distemper; systemic viral disease that causes decreased white blood cells in cats

papillae hair on the tongue that acts as taste buds

parakeet small songbird from Australia also known as a Budgie that is brightly colored with bars located over the wings

paralysis partial or complete loss of motor skills

Parascaris equorum roundworm species in large animals are swallowed in contaminated hay or water, hatch in the intestinal tract

parasite living organism that may occur on the internal or external area of the body and feed off the animal causing disease

parasitology the study of organisms that live off of another organism to survive

parasympathetic system spinal cord system responsible for opposite reactions, such as decreased vital signs, regulating the body system back to normal, and controlling peristalsis

parathyroid gland gland located below the thyroid gland

parked standing with the front and rear legs wide apart

parrots variety of medium-sized bird breeds that come from Central America; bright in color, intelligent, playful, and easy to train

parturition the labor process

parvovirus viral infection causing severe and bloody diarrhea; can be fatal

passive immunity developed through antibodies acquired from one animal or source to another, as in colostrum, formula, or plasma

Pasteurella a naturally occurring bacteria in the respiratory tract of rabbits

pasture breeding method of breeding when the mare and stallion are turned out in pasture

patella knee cap that forms joint of the upper and lower rear leg

patency refers to the proper flow and purpose of the item and that it remains intact and usable

pathogen a disease or disease source

pathology the study of cells and diseases

patience trait that means you behave calmly in all situations without any negative complaints

patient the animal being treated

patient history background information of past medical, surgical, nutritional, and behavioral factors that have occurred over the life of the animal

payroll salary paid to each employee by writing checks on a regular basis

payroll deductions money withheld from a paycheck for tax purposes or payment of medical insurance

peacock adult male pea fowl

peafowl poultry raised for colorful, large feathers

peahen adult female pea fowl

pecking order the dominance order of a group of birds

pedal reflex used on the digit or the skin between digits, which is pinched to determine a reflex response

pedigree the parents and breeding lines of animals' offspring

pelleted diet food fed to a bird in a size-appropriate pellet that is composed of the necessary ingredients

pellets fish meal bound into larger particles

pelvic limb rear leg of an animal

pelvis hip bone

pen structure that connects land to water

penile area area in males where urine is passed through to the outside of the body

penis male sex organ and organ that allows urine to be excreted externally

peptide largest hormone that controls proteins in the body

Percheron French horse breed that has a refined build and is a popular carriage horse; black to grey in color

pericardium thin membrane that covers, protects, and maintains the beating action of the heart

perineal urethrostomy (PU) surgical creation of a new opening at the urethra and penis to eliminate the occurrence of urinary blockages

periosteum thin connective tissue covering outer bone

peripheral nervous system (PNS) controls the nerves, detects a stimulus, sends signals, informs the CNS, and causes a response

peristalsis wave-like motion of the stomach that moves food through the intestine in contractions

peritoneum clear, thin lining of the abdomen

permanent adult teeth that are formed after the deciduous teeth are shed

permanent pasture grass source that regrows each year without the need of planting or seeding

permit a form that states a person may legally house and care for a wild animal

personal protective equipment (PPE) equipment used for safety purposes and protection

personnel manual book that describes the staff job descriptions, attendance policy, dress code,, and employer expectations

pesticide chemical used to treat external parasites

pet piller small device, usually made of plastic, that has a long thin handle with a plunger on the end to administer medications to small animals

pH level the quality of water on a scale on 0 to 14, with 70 being neutral

phalanges toes or digits

pharmacy the area where medications are stored and prepared for veterinary patients

pharynx throat area for the passage of food or air

pheromones hormone odors present in the urine giving all who encounter the mark left by others a variety of messages including sexual receptiveness, eating habits, age, and overall health

phlebitis area around the IV catheter site causing redness, swelling, pain, and inflammation

photosynthesis the process of sunlight allowing plants to grow in a water source

physical cleaning the most common method of sanitary control within the veterinary facility

physical examination (PE) an observation of each body system to determine abnormal problems that can cause a health problem

physical hazard safety concerns that may cause physical injuries such as bites, kicks, scratches, or lifting heavy objects

physical restraint use of a person's body to control the position of an animal

physiology the study of the functions of body structures and how they work

pigeons small wild birds raised in captivity for meat and competition purposes

piglets young newborn pigs

pigment the color of the skin

pill counting tray tool used to count out tablets or capsules to send home with a client

pincers pair of structures on the head used for smelling, feeling, and protection

pinna external portion of the ear

pitting edema swelling around the catheter site causing the fluid to be outside of the vein

pituitary gland lies at the base of the brain and controls hormone release in the endocrine glands

pivot joint that rotates around a fixed point

placenta tissue and membrane collection that connect from the fetus to the mother to allow nutrients to be absorbed during the gestation time

plan treatments or procedures provided

plane determines how awake or asleep a patient is under anesthesia

plankton natural plant food source that grows in water

plaque disease affects animals and humans and causes high fever, dehydration, and enlarged lymph nodes that can eventually lead to death

plaque soft buildup of material found over the surfaces of the teeth that is composed of bacteria, food materials, and saliva

plasma formed of various proteins in the body and consists of 60% of the blood system; forms a solid portion of blood

plasma protein measures the ratio of protein within the blood and helps veterinarians determine the hydration status and inflammation occurrence of patients; also called total protein (TP)

platelet blood cell used to clot blood by constricting the vessels to stop bleeding

pleura double membrane covering that lines the lungs

pleural cavity double membrane covering that holds the lungs

pleural friction rub abnormal lung sound that develops with pneumonia and causes a crackling noise

plucked to pull out hair by hand

pneumonia condition of the lungs causing inflammation and may be of viral or bacterial sources

PO by mouth

points sharp edges or corners of teeth

policy manual information provided to all staff members as to what the staff expectations are, such as the dress code, policies on vacation time, personal time, and sick leave

polled no horn growth

polyarthritis condition of young lambs that causes multiple joints to become inflamed and painful

polydactyl having many or multiple toes or digits

polydipsia (PD) increased thirst

polyestrus having multiple heat or estrus cycles during a season

polysaccharide long chain of simple sugars

polyuria (PU) increased urine production

pond contained water area surrounded by human-built dams and levees

pony small equine under 14.2 hands in height, commonly used by children for riding purposes

pool structure that holds a large amount of water and is made of plastic, fiberglass, or concrete and holds large amounts of fish

pop-off valve used to release excess amounts of gas when the pressure becomes too high within the machine system

porcine veterinary term for pigs, hogs, and swine

porcine somatotropin (pST) hormone used to increase protein synthesis; increases food use and produces weight gain

porcine stress syndrome (PSS) condition of heavily muscled pigs that can die from stress, such as overcrowding

posterior chamber area behind the eye that holds the nerves

posterior lobe back lobe of the pituitary gland that controls peptide hormones

poult young turkey

poultry domesticated birds with feather, two legs, two wings, and used as a source of eggs, meat, feathers, and by-products

poultry science study of poultry breeding, incubation, production, and management to learn to produce a quality product at a reasonable price

ppt parts per thousands; determines the amount of salt in water

precocial growth that occurs quickly in bunnies so that they are able to move away from danger in a short amount of time

preconditioning building up an animal's body to prevent stress from occurring

predicted transmission ability estimation record to determine the possible traits that a bull or cow will pass along to offspring

preen to clean the feathers

preexposure vaccine injection given as a preventative to keep an animal from developing a disease before it comes in contact with it

prefix word part at the beginning of a term

premolars wider teeth at the back of the mouth used to grind and tear food

prepuce the skin covering the male anatomy of the penis where urine is excreted

prescription (Rx) type of medication, amount of medication, and directions for use of the medication prepared by the veterinarian

prescription drug medication prescribed by the veterinarian

prescription label label to indicate to the owner how to properly give the medication and identify the medication within the container

pressure valve indicates how much oxygen is left in the tank

primary bandage layer first bandage layer that acts as padding

primary response provides antigens that need 3 to 14 days to build an immune response that begins producing antibodies

primates animals, such as monkeys and apes, that have movable thumbs and in many ways resemble humans

printer allows printing forms and invoices from the computer onto paper

probe method process of passing an instrument into the cloaca to determine the sex of a reptile or amphibian

procedures manual book that describes the rules and regulations of the facility, such as visiting hours, opening, and closing procedures

processing process of developing film that involves being in a darkroom that has no light source or light leaks from the outside area

production cycle the time for young aquatic species to reach full market size

production intensity assurance that the size of the body of water will hold the appropriate amount of food sources and be large enough for a high production level of products

proestrus stage of the heat cycle before estrus that releases the progesterone hormone prior to mating

profitability making money in a business after all other costs have been paid for in purchasing and caring for the product

progeny offspring

progesterone hormone produced to maintain pregnancy

proglotids segments that shed off a tapeworm and appear like small white pieces of flat rice

progress notes documentation chronologically detailing the patient's history, treatments, and outcomes

prolactin hormone that controls milk production

prolapsed uterus condition when the uterus and vaginal tissues separate from the body and allow the uterus body to be passed outside the body due to the stretching of the vagina and vulva

prophylaxis professional cleaning of a patient's teeth

proprioception relates to the body's ability to recognize body position for posture or movement

proteins nutrients that are essential in growth and tissue repair

protozoan class of disease; the simplest forms of life caused by single-celled organisms that are known as parasites that may live inside or outside the body

proventriculus acts as a monogastric stomach and begins the digestion process in birds by releasing excretions to soften food

pseudo rabies condition caused by a virus that shows similar signs as rabies and is spread through oral and nasal secretions

psittacine beak and feather disease (PBFD) contagious and fatal viral disease that affects the beak, feathers, and immune system of birds

psittacosis commonly called "parrot fever" or chlamydia; disease that has zoonotic potential especially in young children and anyone with a poor immune system and causes severe flu-like symptoms

puberty sexual maturity

pubis third section of the pelvis

pullet young chicken raised to lay eggs

pullet production system where hens are raised for fertile egg production

pulmonary artery carry oxygen-poor blood and prevent blood flow

pulp cavity center of the tooth that holds the nerves, veins, and arteries

pulse the heart rate of an animal or each beat created from the heart

pulse oximeter a device used to measure a patient's vital signs by indirectly measuring the oxygen saturation within the blood and any changes in the blood volume

pupae immature larvae stage; go through several molting stages during growth by spinning a cocoon

pupil area in the center of the eye that opens and closes with light responses

pupillary light response test light source shown into the animal's eyes to note constriction or shrinking of the pupil when light is shown into the eye and dilation or expanding when the light source is removed from the eye

purebred animal with known parentage of one breed

purebred business registered breed of cattle that has pedigrees

purebred flock method small production system that raises purebred sheep for reproductive purposes

pushing moving cattle by walking toward them quietly and calmly so they move away into another area

P-wave first peak on an EKG that forms at the SA node (pacemaker) and reflects the flow of electricity and blood through the atrium and a contraction has occurred

Pygmy goat group of small-sized goats used as pets

pylorus exit passageway of the stomach

pyometra severe infection of the uterus in female intact animal that causes the uterus to fill with pus

QRS complex series of peaks that show that the electrical current and blood have flowed through the AV node and have caused a contraction in the ventricles

quadriceps muscle located on the hind limb at the cranial part of the thigh

quail small wild bird that makes the whistle sound "bob-white" and is raised for eggs

quarantined to isolate

quarters sections of the mammary glands that store milk

quick blood supply within the nail bed

quills spines that project and protect from the body of a hedgehog

rabies a deadly virus that can affect any mammal or human; usually transmitted by the saliva of an infected animal transferred into the bloodstream

rabies certificate documentation the animal has received proper vaccination for rabies

rabies laws legal standards in each state to prevent the spread of rabies infections

rabies pole tool used to restrain and control an animal; also called a snare pole

racehorse horse bred for speed and racing purposes

raceways long and narrow water structures with flowing water

radiation using ultraviolet or gamma rays that radiate and kill living organisms; safety hazard that allows X-rays to be produced to take a radiograph

radioactive iodine radiocat; treatment involving iodine being injected

radiology area of the facility where X-rays are taken and developed; this area is regulated due to chemical and radiation exposure

radiolucent appearing black or dark gray in color and may indicate soft tissue or air and may be an exposure factor relating to increased exposure settings or improper patient measurement

radiopaque appearing white to light gray in color and may indicate hard tissue such as bone or may be an exposure factor in that the machine settings were too low or the patient was not properly measured

radius larger bone on the front of the lower forelimb

rain water source of water from the sky that can only be used if not contaminated or polluted

rales abnormal sounds within the chest that sound like musical notes

ram the adult male sheep capable of reproduction

range band method raising large sheep herds on large amounts of land

range paralysis common name for Marek's disease caused by a herpes virus

raptors large birds of prey

rare breeds animals that are bred and raised for a variety of purposes but not found locally

rasp instrument used to file down horse teeth; also called a float

rat tooth forceps instrument with large teeth, similar to rat's, that hold large thick tissues and skin

ration total amount of food an animal needs within a 24-hour time frame

ratite large flightless birds raised for poultry production purposes

Rattus norvegicus genus and species of the rat

reagent chemicals are used to run each individual test in the entire kit

reagent strip chemical test strip used to analyze urine or blood on a long, thin, plastic strip separated by individual square pads containing a chemically treated paper

rearing lifting action of the front limbs off the ground and raising the body into the air

rebreathing system bag that connects to the machine and inflates and deflates as the patient breaths and allows the patient to rebreathe anesthetic gas after it has passed through the CO_2 absorbing canister

receiver person who gets a message in the route of communication

reception area area where clients arrive and check in with the receptionist

receptionist personnel greeting and communicating with patients as they come in and leave the veterinary facility

receptive accepting of

receptor found within the nerves located throughout the body

recumbency term that means to lie as in a restraint position

red blood cells (RBC) most abundant cells within the body that are produced in the bone marrow and transport oxygen throughout the body

redirected aggression refers to an animal's predator instinct where the animal turns its aggressive behaviors on the owner

reduced calorie diet specialized food given to animals that are overweight or less active due to their health status

reefers knot single bow knot that ties in a nonslip but quick-release tie

reefing restraint procedure using rope to place a cow on the ground for restraint

reference lab laboratory samples analyzed outside of the facility in a commercial lab

reflex units of function that produce a movement created by volts

reflex hammer tool used to tap the area of the knees and joints to elicit a response

refractometer tool used to measure the weight of a liquid and determine a liquid's pH level

registered veterinary technician (RVT) a veterinary technician that has passed a state licensing exam

regurgitation process of bringing food into the mouth from the stomach to break down food

renal artery located in the center of the kidney and allows blood flow out of the renal system

renal pelvis innermost section of the kidney that filters blood and urine through the veins and arteries

renal system body system that involves the kidneys and urinary production to rid the body of waste products within the bloodstream

renal vein located in the center of the kidney; allows blood flow into the renal system

re-polarize reset or restart the heart and means to get ready for the next electrical impulse to start

reproduction diet specialized food given to breeding animals for additional nutrient needs

reproductive system the organs in the body responsible for both male and female breeding and reproduction

reptile animal with dry, scaly skin that uses the outside temperature and environment to adjust body temperature

research studies used to answer questions about medical conditions that affect humans and animals

research scientists staff responsible for planning the experiments to be done on animals and ensuring the animals are not in pain

respiration the process of breathing through inhalation and exhalation

respiratory rate (RR) measures the breathing of an animal and evaluates how many breaths an animal takes in a minute

respiratory system part of the body that controls oxygen circulation through the lungs and the body

restraint to hold back, check, or suppress an action and to keep something under control using safety and some means of physical, chemical, or psychological action

retained placenta fetal membranes of waste materials not passed from the uterus after the labor process

reticulum second section of the ruminant stomach that acts as a filter for food

retina inner layer of the eye that contains two types of cells that aid in seeing color and determining depth

retractable capable of projecting inward and outward, as in the nail of a cat

retractor handheld or self-retracting; used during surgeries to hold open an animal's body cavity for exploration

rhea moderate-sized flightless bird native to South America

rhinotracheitis virus that affects the upper respiratory system of cats

rib bone that attaches to individual thoracic vertebrae and protects the heart and lungs

ribosomes structures within the cell that make protein

rickettsial disease caused by parasites that invade the body; spread by biting insects

riding horse horse ridden for show or recreational purposes

right atrium thin-walled chamber of the heart where blood enters

right ventricle thicker walled chamber of the heart that pumps blood to the lungs

right-to-know station area in the facility where the OSHA and MSDS binders are kept with safety and product information

ringworm fungal condition that affects the skin of humans and mammals and is highly contagious

rinsing using water to remove any chemicals from an X-ray film

roaring respiratory condition in horses that sounds like a roar due to the larynx only opening a small amount due to a trauma on the nerves of the throat area

roaster chick over 8 weeks of age and 4 pounds used as meat

Rocky Mountain Spotted Fever rickettsial disease transmitted by the American Dog tick and causes signs that include a fever, joint pain (arthrodynia), depression, and anorexia

rodents mammals that have large front teeth designed for chewing or gnawing, such as rats, mice, and hamsters

rods (1) cells that allow the eye to detect light and depth and make up 95% of the cells within the eye; (2) oblong bacteria shape

rooster adult male chicken

root part of the tooth located below the gum line that holds the tooth in place

root word origin or main part of the word that gives the term its essential meaning

rotors wheels in a centrifuge that spin at a variety of speeds depending on the sample type

roughage hay or grass source fed to livestock and ruminants

round window membrane that receives sound waves through fluid passed through the cochlea

roundworms visible long, white worms that are the most common intestinal parasite in small and large animals

routine appointment appointment scheduled in the appointment book as physical exams during the office appointment schedule; examples would be vaccinations, yearly exams, and rechecks

rugae folds within the stomach when it is empty

rumen first section of the ruminant stomach that acts as a storage vat and softens food for fermentation

ruminant animal with a digestive system that has a stomach with four sections or compartments

rump method restraint method of placing a sheep on its hind end for handling

run group of fish

SA node the term for the pacemaker of the heart

sac fry yolk sac attached to a fry for a day after the hatch

sacral the fourth section of the vertebrae that lies over the pelvis area

sacrum area over the pelvis that is fused together and forms the highest point of the hip joint

safe light a red light that is low in intensity and a filter that doesn't damage the film

salary amount an employee earns each year based on the hourly wage

salinity the amount of salt in a water source

saliva fluid that helps soften and break down food for ease of swallowing and digestion

salivary glands area within the mouth that produces saliva

salmonella infection bacterial infection that causes severe diarrhea

salmonellosis bacterial infection that occurs from touching reptiles or amphibians

salt block salt provided in a block for horses to lick

saltwater water with a salt content of at least 165 ppt

saltwater fish species of fish that require salt in their water source

sanitation the process of keeping an area clean and neat

saphenous vein blood vessel located on the lateral thigh of the rear limb just proximal to the hock

sarcoptic mange disease caused by a mite that lives on hair and skin; is contagious; in humans known as scabies

scabies disease caused by a mite that lives on hair and skin; is contagious; in animals known as sarcoptic mange

scales body covering that serves as skin and protects the outside of the fish

scanner equipment that allows documents and images to be copied and stored electronically

scapula shoulder blade

scavenger hose tube attached to the pop-off valve to prevent gas from leaking from the anesthesia machine and causing unnecessary exposure to humans or animals

scent glands area on the head located behind the horns that produces a strong hormonal odor in male goats

scheduled drug medicine with the potential for abuse and categorized from I to V according to strength; I is highest potential and V lowest potential for abuse

scheduled feeding a set amount of food given at specific times during the day

school group of fish

sclera outer white layer of the eye

screen area that lines the film cassette

scrotum skin over the two testicles that protects and controls the temperature for sperm production

scrub pack sterile supply that contains the necessary items to properly prepare the surgeon, technician, or assistant for surgery for washing hands

scruff the loose skin over the base of the neck of a cat that is sometimes used to restrain a cat

scruff technique used on the base of the neck over the area where the skin is elastic-like and able to be grasped with a fist technique

secondary bandage layer second bandage layer that holds the first layer in place

secondary response provides a quick repeated exposure to an antigen that creates immunity to prevent disease development

sedative medication used to calm an animal

sediment sand-like buildup in the urine composed of crystals and casts

seed diet food composed of natural seeds eaten by birds

seizure loss of voluntary control of the body with an amount of unconsciousness causing uncontrolled violent body activity

selection guidelines set of rules that state the standards of the type of cow and how it is chosen for a production program

semen fluid that transports sperm during reproduction

semen collection procedure using specialized veterinary equipment to collect sperm from male animals to be used in breeding practices

semen morphology process of determining if sperm is adequate for transport and breeding by analyzing the quality and quantity

Semi Arch large-breed rabbits that are between 12 and 14 pounds

semi-aquatic living part of the time in the water and part of the time on land

semicircular canals sensory cells that generate nerve impulses to regulate position

seminiferous vesicle tubes located within the testicle that transport sperm to the testicles

sender person trying to relay an idea or message in the route of communication

senior diet specialized food given to geriatric or older animals over a specific age for their species that require certain nutrient levels due to age

Senn retractor retractor blades located at each end that look like fingers on the end of the instrument and allow body parts to be opened and viewed during surgical procedures

sensory neuron cells that deliver signals from the central nervous system (CNS)

septic buildup of bacteria in the system causing a toxic affect when it gets into the bloodstream

serpents the veterinary term for snakes

serration spaces on the top of a comb

serum liquid portion of blood

sharps sharp instruments and equipment that can injure a human or animal and cause a wound or cut that transmits a contagious disease due to contamination

sharps container area that holds sharp items that may spread disease, such as needles, surgical blades, and glass

shaving the process of taking down an amount of hair close to the skin

shearing method of shaving wool from sheep

shipping fever a condition that occurs in livestock, such as horses and cattle, that are transported

Shire English horse breed that is black or bay in color with feathers on the legs

shoal fish usually live in groups of five or more breeders

shock (1) condition that occurs when an animal does not have enough blood and oxygen reaching the tissues; (2) emotion of a lack of feelings caused by the sudden death of an animal or loved one

shorthair hair coat that is smooth in cats

signalment the patient's name, age or date of birth, gender, color, and breed or species

silos tall storage buildings that keep food free of moisture

silver nitrate stick slim, long wooden applicators with silver nitrate located on the end that is used to clot a bleeding nail by burning the quick

simple fracture a break in the bone that may be complete or incomplete, but does not break through the skin; also known as a closed fracture

simple sugar single particle of carbohydrates that form the building blocks of nutrients and are called monosaccharides

simulated systems models that mimic an animal's behavior and internal or external structures to conduct research

sinus rhythm a normal heart rate and rhythm

sinuses small, open spaces of air located within the skull and nasal bone

sire male parent of a horse

sitting restraint restraint used to keep an animal sitting on its rump and prevent it from standing

skin mite insects that live within hair and skin; not visible to the naked eye

skin turgor process of evaluating an animal for dehydration by lifting the skin over the base of the neck or shoulder blades

skull the bone that holds and protects the brain and head area

slander talking negatively about someone in an improper manner

slicker brush tool used to brush the hair coat

slip knot type of tie used as a quick release

slough tissue that dies and falls off

slurry soft soup-like substance fed to fry

smolt immature salmon greater than an inch in size

SNAP test simple blood test completed in the clinic that tests for certain disease antigens in the bloodstream

snare pole long pole with a noose on the end that acts as a leash and collar to capture a difficult dog; also called rabies pole

sneezing act of removing dust or other particles from the respiratory tract through physical force

snubbing rope rope placed over snout and tied to a post for restraint

SOAP subjective information, objective information, assessment, plan; one method used for documentation in the medical record

socialization interaction with other animals and people to become used to them

soda lime granules within the canister that absorb the carbon dioxide

sodium chloride NaCl; saltwater fluid

software a program or set of instructions the computer follows

solid manure system waste material system low in cost where manure is collected and removed on a daily basis and stored on a pile away from cattle

solitary alone and independent behavior

somatic cells cells that divide in mitosis and help in growth and repair of the body

somatotropin growth hormone that increases protein synthesis in the body causing an increase in the animal's size

sonogram the printout of an ultrasound recording

soremouth highly contagious viral disease that is zoonotic and causes lesions over the mouth and lips of sheep

sorrow positive emotion that leads to sadness over the loss of a pet and causes crying and talking about the pet

sound showing no signs of lameness or injury

sow adult female pig of reproductive age

spawn group of fish eggs in a nest

spawning breeding or reproduction in fish

spay hook instrument with a long, thin handle and a hook on the end used to located the uterus and uterine horns in small female animals during a spay procedure

specialized payment plan a plan approved by the vet that allows a down payment with additional regular payments until the balance is paid

specific gravity (sg) the weight of a liquid measurement

specific pathogen free (SPF) disease free at birth

spectrum a wide variety of factors

speech communication talking in front of others to pass along information verbally

spent hen adult female hen that has aged and decreased or stops egg production and used as meat sources in processed foods

sperm male reproductive cell developed in the testicle

sphygmomanometer device used to measure an animal's blood pressure during surgery; will also help determine tissue perfusion as well as depth of anesthesia; also called a blood pressure monitor

spike port the site on an IV bag where the drip set tubing is placed

spinal column extends from the base of the skull to the end of the tail and allows movement; also called the spine

spinal cord strands of fibers that attach the brain stem to the PNS to send and receive signals through the body; begins at the base of the brain and continues to the lumbar vertebrae

spinal nerve reflexes evaluated for proprioceptive reflexes of the muscles and tendons

splint support applied to a broken bone with bandage and padding to allow a bone to heal

splint bone two small bones on the back of a horse's cannon bone that may become inflamed from large amounts of work and stress on the legs

sponge forceps straight or curved and are used to hold gauze sponges or other absorbent material to clear away bleeding and other debris

spooking startling action

spreadsheet program used to track large quantities of items in one file

spring natural opening in the earth that provides a clean, natural water source

springing heifer young female cow that is pregnant with the first calf

square knot knot tie looped into a square and used to tie and untie cattle easily

squeeze cage wire boxes with small slats that allow injections to be given to a cat, such as vaccines or sedatives

squeeze chute wire cage used to control a cat by sliding the sides closed for such procedures as exams and injection

stag a male pig castrated after maturity

stage the flat section under the lens of a microscope

stallion adult intact male horse capable of reproduction

stanchion head gate that holds the head of a cow in place during restraint

starter food easily digestible food for calves as they begin eating

standing heat the female animal will stand for breeding when a male is present

standing restraint restraint used to keep an animal standing on its feet and prevent it from sitting or lying down

staple remover device placed under the staple and squeezed to open up the staple, which is then gently removed from the site

starches plant or grain materials that provide fiber and bulk in an animal's diet and converts to glucose or sugar during digestion

State Board Examination exam that veterinarians and veterinary technicians are required to pass for state licensure in order to practice

State Board of Veterinary Medicine each state's agency of veterinary members and public members that take disciplinary action in the profession as deemed necessary and monitor the rights of the public

static equilibrium maintenance of the position of the head relative to gravity

statutory laws rules and regulations based on written government law

steer castrated male cow

sterile not able to reproduce

sterile techniques include washing the hands frequently, wearing gloves when handling animals or other possible contaminants, and cleaning all surfaces with disinfectants to prevent spread of disease

sterilization killing all living organisms on a surface

sterilizing destroying *all* microorganisms and viruses on an object using chemicals and/or extreme heat or cold under pressure

sternal recumbency restraint position with the animal lying on its chest

sternum breastbone that protects the organs of the chest

steroid occur naturally in the body and regulate chemicals, such as cholesterol, in the body that control essential life functions

stethoscope instrument used to listen to the heart, lungs, and chest

stirrups pieces of tape at the end of a bandage on a limb that hold the bandage in place and prevent slipping

stock horse horse built to work on a ranch that has athletic ability

stool softeners medication given to produce a bowel movement by softening feces

stretch technique restraint procedure by holding the loose skin of the neck with one hand and pulling the rear limbs backward with the other hand

strongyles common intestinal parasite in large animals that are long, thin, and white in appearance

stud horse adult male intact horse of breeding age

styptic powder chemical used to stop bleeding

subcutaneous (SQ) injection given under the skin

subcutaneous fluids fluids placed under the skin and slowly absorbed by the body

subjective information observations of the animal's appearance and behavior made by the veterinary staff

subluxate bone partially out of the joint

substrate the type of material on the bottom of the cage

suffix the word part at the end of a term

Suffolk English horse breed that is chestnut, smaller in size, and an easy keeper

sugar type of carbohydrate that forms the simplest type of a nutrient

supplements additives placed into the diet in solid or liquid form when needed by the animal

supportive care therapy or treatment that includes such treatment as fluid replacement, medications to decrease vomiting and diarrhea, and pain medicine

surface dwelling fish fish that live and feed near the surface or top of a body of water

surface water runoff excess water drainage from rainfall or precipitation

surgery pack sterile instrument prepared for surgical procedure with tools and supplies required for the procedure

surgical appointment appointment scheduled during the veterinarian's set time for doing surgeries and anesthesia, usually separate from routine office hours

surgical glue sterile substance used for closure of small wounds or surgical sites and is commonly used in feline declaws

surgical log a record of details recording all surgeries performed in the facility

surgical margin area of the patient that should extend 2 to 4 inches beyond the anticipated surgical incision borders

surgical scrub sterile solutions placed onto the surgical margin to clean the patient

surgical staples fast and easy way of using a device that closes the skin with metal pieces

surgical suite area where surgery is performed and harmful anesthesia gases are used and the area must be kept sterile to prevent contamination

suture area of a bone where two fibrous joints meet; fine line where little or no movement occurs

suture reel container of suture material that must be cut and thread into a needle

suture removal removal of sutures within 7–10 days after surgery

suture removal scissors scissors that have one blade that has a small hook shape

swallow reflex the patient licking with the tongue or having increased jaw tone during extubation of the endotrachel tube

swamp fever common name for equine infectious anemia

swan large water birds with long, thin necks raised as ornamental poultry

swedged on needle suture material that has a needle already attached

swim bladder structure that acts like a small balloon that inflates and gives the fish buoyancy to float and balance in all levels of water

swim bladder disease bacterial disease causes a fish to have difficulty swimming to the water's surface and in depths of water

swine the term for pigs or hogs

swine dysentery condition in pigs that causes diarrhea and known as bloody scours and spreads rapidly through contamination

sympathetic system spinal cord system that is responsible for the "fight or flight" reaction, increased vital signs, and certain drug reactions

sympathy sharing feelings with another in the time of sadness

synovial joint moveable joint at the area where two bones meet

synthetic seawater water that is manmade

systemic circulation blood system that delivers nutrients and oxygen to all areas of the body

systole contraction phase of blood pressure or the top number that is normally higher

T3 most potent and active thyroid gland hormone measured in the bloodstream to diagnosis thyroid problems

T4 thyroid hormone converted into tissues; breaks down fats and helps control cholesterol

tabletop technique the distance between the X-ray tube and the top of the X-ray table surface; the film cassette is placed on top of the table and in contact with the patient

tachycardia increased or elevated heart rate

tacky slightly dry, as in the gums

tactfulness characteristic that means doing and saying the appropriate things at the correct time

tadpoles young newborn frogs or toads

taenia species of parasite known as a tapeworm

tail part of the body containing the caudal fin that allows movement and guidance in the water

tail docking surgical removal of all or part of the tail

tail flagging holding the tail high in the air during the estrus cycle

tail switch restraint restraint method of twisting the tail at the base to prevent a cow from kicking and moving

tail tying method of tying the tail to keep it out of the way or to move a sedated horse

talons claws on a bird of prey

tank large round areas of water

tap water water that comes from faucet water

tapeworm internal parasite caused by ingesting fleas; long, segmented bodies

tarsal the point of the rear leg where the tibia and fibula meet; ankle area

tarsus joint of the ankle of the rear limb

tattoos number applied to the skin using an electronic needle with ink

taurine nutrient required by cats for allowing proper development

teats sections of the mammary gland used to produce milk

technique chart a listing of settings on the X-ray machine based on the thickness of the area to be radiographed

temporary pasture grass source that does not regrow each year and need to be planted or seeded

tendon fibrous strands of tissue that attach bone to muscle

terrapins veterinary term for turtles

terrariums plant-sourced housing for reptiles and amphibians

terrestrial living on land

territorial aggression refers to an animal's protective nature of its environment, such as an owner, offspring, or food

territory an area of land that has been claimed by an animal

tertiary bandage layer third bandage player that acts as a protective covering

testicle the primary reproductive organ in male that holds sperm

testosterone main male reproductive hormone that produces male characteristics and is essential for sperm production

tetanus bacterial disease caused by a wound; commonly called lockjaw

thalamus organ within the brain located at the top of the brain stem

thermometer used to measure the temperature of the water

thermostat used to control the water temperature when the water goes below a certain temperature

thiamine nutrient compound of vitamin B that promotes a healthy coat

thoracic second section of vertebrae located over the chest area

thoracic cavity the chest

thrombocyte veterinary term for platelets

thymus a gland in young animals that has an immunologic function

thyroid gland "the master gland"; controls, secretes, and regulates hormone production within the body and is located in the neck area

thyroid stimulating hormone (TSH) hormone produced by the thyroid gland that controls the chemical thyroxine

thyroidectomy surgical removal of the thyroid gland

thyroxine hormone that controls the actions of the thyroid gland and cell metabolism

tibia larger bone in the lower rear leg and lies in front

tick wingless insect; feeds off blood of animal

tidal volume amount of oxygen required by the patient

tissue a group of cells similar in structure and function

tissue adhesive glue-like substance that is used to close incisions

tissue forceps instrument used to grasp tissue and have teeth that grip the tissue

toe pinch response digits can be pinched to elicit a response that should be to pull away from the pressure to assess the nervous system

toner black or colored ink that allows printing in a printer, fax machine, or copier

tongue muscle within the mouth used to hold food

tooth abscess infection of the tooth root

topical a drug that is administered to the skin

total herd cows that have replacement heifers when the adult cows become too old to continue milk production and breeding

total mix ration nutrients needed daily within a mixture of high-quality products

total protein (TP) measures the ratio of protein within the blood and helps veterinarians determine the hydration status and inflammation occurrence of patients; also called plasma protein

towel clamp instrument used to hold the surgical drape in place over the patient

Toxascaris leonina roundworm species with the simplest life cycle

toxic poisonous

Toxocara canis roundworm species with a more complicated life cycle and an effective way of making sure its species will be passed from generation to generation

Toxocara cati roundworm species similar to *T. canis* that may infect the lungs or abdominal tissue

toxoplasmosis protozoan parasite passed in cat feces that may cause abortion in pregnant women

Toxoplasmosis gondi simple-celled organisms that can occur in any mammal but is shed through the feces of cats, which are the only animals that pass the *Toxoplasmosis* through the feces

TPR refers to the temperature, pulse, and respiration

trace minerals minerals needed in small amounts; also called microminerals

trachea the windpipe or tube-like airway established during anesthesia

tract nerves that attach to the spinal cord and run through the body providing sensations and feelings

tragus opening of the ear canal structure; includes the haired area

training teaching an animal to understand specific desired habits

tranquilizer medication used to calm an animal

transducer the instrument used with an ultrasound that scans the body and transmits the waves back to the screen

transverse colon second or middle section of the large intestine

trapping the capture and restraint of wild animals

travel sheet paperwork list that codes services and procedures that allows the staff to track fees to charge the client

treatment area area where animals are treated or prepared for surgery

triage a quick general assessment of an animal to determine how quickly the animal needs a veterinarian's attention

trichobezoar a hairball

Trichuris vulpis whipworm species infect animals through ingesting contaminated food or water

trimester the period of three months of pregnancy when additional nutritional needs must be provided and the development of the fetus occurs

trimming the process of removing a specific amount of hair from one or more locations

trocar plastic or metal pointed instrument placed into the rumen of a ruminant animal that has bloated to relive the air and pressure on the animal's stomach

tropical fish small, brightly colored fish that require warm water and are naturally found in warm, tropical locations

trot slow two-beat gait of English horses

trunk part of the body that contains the body, fins, and anus

trypsin enzyme produced by the pancreas that digests proteins

tumor cell division and development in localized areas

T-wave last peak and the most important part of the EKG; shows that an electrical impulse has traveled through the entire heart and has completed the contraction and re-polarize the heart

twitch chain or rope loop placed over the lip of a horse as a restraint

tympanic bulla osseous chamber at the base of the skull

tympanic membrane eardrum; tissue that separates the outer and middle ear

Tyzzer's disease bacterial infection in rabbits that affects the digestive system that spreads through spores in the air

udders mammary glands in dairy animals that produce milk

ulna smaller bone on the back of the lower forelimb

ultrasonic cleaner instruments are placed within a wire basket and the basket is seated within the cleaner, which holds a disinfectant solution and creates vibrations to clean the items

ultrasound diagnostic tool using ultrasonic sound waves to view images of internal organs and structures

unethical the act of doing something wrong or improper and knowingly doing it

unilateral cryptorchid one retained testicle that has not descended into the scrotum

univalve mollusks with one shell or part of a shell

upland game birds wild species of birds that spend most of their time in the woods and do not move out of an area

upper GI a contrast study of the structure of the upper digestive tract in which the barium is given orally

upper respiratory infection (URI) condition causing sneezing, coughing, nasal discharge, ocular discharge, wheezing, and difficulty breathing

urates urine waste materials of reptiles and amphibians

urbanization disappearance of wildlife due to loss of land, food, and other habitat resources as people take over their land

ureter tube made of smooth muscle and pushes and filters urine out of the kidneys through contractions

urethra narrow tube that connects the male and female reproductive organs to the urinary bladder

urinalysis breakdown of urine components to determine a diagnosis

urogenital system includes the kidneys, urinary bladder, external genitals, mammary glands, and rectal area

urolith term for a bladder stone that form within the urinary bladder

uterine horn two parts of the uterus that hold single or multiple embryos that expand as the embryos grow

uterus large, hollow reproductive organ that holds an embryo during the gestation period and is Y-shaped

vaccination placed into the body to build up resistance in the immune system to disease

vaccine program injections given to protect animals from common diseases and build up the immune system

vacutainer tube tube used to place blood samples in for future sampling

vagina internal female reproductive organ that allows mating and labor to occur

vagus nerve main nerve within the spinal cord system

valve structure in mollusks that acts like gills

vaporizer converts the liquid anesthesia into a gaseous form, which the patient inhales

vas deferens small tube that transports sperm from the testicles to the outside of the body

vasoconstriction diameter of a vessel decreases as blood pressure increases

vasodilatation diameter of a vessel increases as blood pressure decreases

vat large structure that is made of metal or concrete that holds large amounts of water and is commonly used for breeding and reproduction of fish

vein vessels that carry blood to the heart

velvet material shed from antlers of deer and used in medications and nutritional supplements

vena cava large vessel that transports blood into the heart from systemic circulation

venipuncture practice of placing a needle into a blood vessel

venison deer meat

venomous poisonous

vent external area of an avian and reptiles that passes waste materials; also called cloaca and similar to the rectum

ventral root nerve branch of the vertebrae that contain motor nerves

ventral-dorsal (V-D) position where the animal is restrained on its back with the X-ray beam traveling through the ventral aspect first and the dorsal aspect second

ventricular fibrillation (V-fib) condition causing the ventricles to fire electrical currents rapidly; the most serious cause of cardiac arrest

verbal communication the use of speech to pass along information in the communication process

verbal restraint voice commands used to control an animal

vertebrae individual bones of the spine that surround and protect the spinal cord

vertebrates animal with a backbone

vestibule portion of inner ear that contains specialized receptors for balance and position

veterinarian the doctor that holds a state license to practice veterinary medicine and holds the ultimate responsibility within the facility

veterinary animal behaviorist specialized veterinarians that have a special interest and experience in animal behavior problems

veterinary assistant staff member trained to assist veterinarians and technicians in various duties

veterinary hospital manager staff member in charge of the business workings of the entire hospital, including staff scheduling, paying bills, and payroll

Veterinary Practice Act legal document that outlines rules and regulations of veterinary professionals

veterinary technician two-year associate degree (AS) in specialized training to assist veterinarians

veterinary technologist four-year bachelor degree (BS) in specialized training to assist veterinarians

Veterinary–client–patient relationship (VCPR) relationship established between patient, client, and veterinary staff based on trust, expertise, and duty to care for animal

viral class of disease caused by particles that are contagious and spread through the environment

virus causes a disease that is not treatable and must run its course

visceral larval migrans roundworm infection in humans

viscous thick liquid substance

vital signs the signs of life that are measurements to assess the basic functions of an animal's body and include the heart rate, respiratory rate, temperature, blood pressure, mucous membrane color, capillary refill time, and weight of the animal

vitamin C essential nutrient and supplement needed by guinea pigs as they don't produce their own in their body

vitamins nutrients needed in small amounts for life and health maintenance

vitreous humor jelly-like fluid material located behind the iris and lens that controls the pressure of the eye

V-notcher tool used to make cuts in a pig's ear

voided urine sample collected as the animal is urinating

volt electrical units of measurement within the body that create a force within cell

voluntary reflex occurs when an animal asks its body to perform a desired function, such as walking or running

vomiting process of bringing up partially or undigested food that has been in the stomach of monogastric animals

vulva external beginning of the reproductive system in the female; where urine is excreted

waiting room area where clients and patients wait for their appointments

walk relaxed four-beat slow gait

walk-in appointment a client that does not have an appointment and arrives at the hospital wishing to see the veterinarian by walking in the door

want list items that are needed or low in stock within the facility and need to be reordered

warm water water between 65–85 degrees

warm-blooded ability for controlling the body temperature internally

warm housing heated building that holds cattle in the winter and is insulated with individual stalls for each cow

water nutrient that makes up 75% of the body and provides several functions within the body, including body temperature control, body shape maintenance, transporting nutrients within the body's cells, and hydration

water facility areas built to house aquatic animals in natural environments

water soluble vitamins that are not stored within the body, are dissolved by water and therefore needed in daily doses by the body to work; these include vitamins C and D

waterfowl birds that swim in water and spend a large amount of time on the water

waterless shampoo product applied to the hair coat that doesn't require rinsing with water that cleanses the coat

wattle flesh-like area located under the chin

wave schedule appointments maintained by a veterinarian seeing a total number of patients within a set amount of time during the day; based on each hour the facility is open

weaned process of young animal to stop nursing and eat solid foods

weight tape the use of a measurement tape around the girth of an animal that gives an estimate of its body mass

Weitlaner retractor self-retracting instrument with rake-like teeth on the ratchet ends; can be blunt or sharp

welding gloves heavy layers of leather placed over the hands to restrain cats and small animals

well water water pumped from aquifers from deep within the earth where water occurs naturally

West Nile virus a mosquito-borne disease that causes inflammation or swelling of the brain and spinal cord

wet tail condition in hamsters caused by a bacterial disease that spreads rapidly through direct contact or bacterial spores

wether male castrated sheep

wheel barrow technique slightly lifting the animal's hind end off the ground and pushing the body forward in a "wheelbarrow" motion to assess the nervous system

wheezing sound produced during asthma that causes difficulty breathing

whipple tree harness-like attachment that connects the yoke to the wagon by hooks

whipworms common intestinal parasite, occurring mostly in dogs, that gets its name from the whip-like shape of the adult parasite

white blood cells (WBC) five types of blood cells that protect the body from infection and aid in immune system protection

whitespot nickname for ich

whole blood blood sample placed in a lavender top tube to prevent clotting of the sample

wildlife any living animal that has not been domesticated

wildlife biologist person who researches and studies data on all types of wild animals to determine how wildlife can be saved and protected

wildlife management the practice of researching the needs of wildlife, providing them the essentials of life and monitoring their survival

wildlife rehabilitation centers areas that house and care for sick or injured wild animals often run by a veterinary staff

wildlife rehabilitators individual people who care for sick, injured, or orphaned wild animals

wing trim the procedure of clipping the wings to prevent a bird from flying

winking movement of the vulva with frequent vaginal discharge and urination

withdrawal time time when no additives are fed prior to slaughter

wolf-teeth small teeth that may develop in front of the first molars

wool soft fiber of sheep that acts as a coat

words per minute the number of words that can be typed within one minute or 60 seconds word-processing program software that allows typing of a document

work diet specialized food given to animals that use a large amount of energy for work or an activity

written communication the use of writing, emailing, or texting information to pass along in the communication process

X-ray common term for radiograph

X-ray log records the patient's name, client's name, date, X-ray number, X-ray position, thickness of body measurement, and area exposed

X-ray tube part of the machine that holds the radiation source

Y connector plastic device that connects the intake and outtake hose from the anesthesia machine to the endotracheal tube of the patient

yearlings fish that are a year of age

yoke wooden bar that fits over the neck and shoulders of a team of oxen

yolk inner yellow cell membrane created by the hen's ovary

zoo zoological garden that houses wild and exotic animals and plants for people to visit and observe their behaviors and beauty

zookeeper person who manages and cares for animals in a zoo

zoology the study of animal life

zoonosis disease passed from animals to humans

zoonotic a disease that is transmitted from animals to humans

zoonotic hazard safety concerns that allow contagious organisms to be spread to humans, causing infections, viruses, bacterial, fungal, and parasitic transmission

Index

A

B

C

D

E

G

H

I

J

K

L

N

O

P

Q

R

S

T

U

V

W

X

Y

Z